Essentials of Marketing

A Marketing Strategy Planning Approach

17 EDITION

William D. Perreault Jr., PhD
UNIVERSITY OF NORTH CAROLINA

Joseph P. Cannon, PhD
COLORADO STATE UNIVERSITY

E. Jerome McCarthy, PhD
MICHIGAN STATE UNIVERSITY

mheonline.com/advancedplacement

Copyright © 2021 McGraw Hill

All rights reserved. No part of this publication may be reproduced or distributed in any form or by any means, or stored in a database or retrieval system, without the prior written consent of McGraw Hill, including, but not limited to, network storage or transmission, or broadcast for distance learning.

Send all inquiries to:
McGraw Hill
8787 Orion Place
Columbus, OH 43240

ISBN: 978-1-26-619898-4
MHID: 1-26-619898-9

Printed in the United States of America.

5 6 7 8 9 10 11 12 LKV 30 29 28 27 26 25 24

Authors of *Essentials of Marketing*, 17/e

William D. Perreault Jr.

William D. Perreault Jr. is Kenan Professor of Business Emeritus at the University of North Carolina. Dr. Perreault is the recipient of the two most prestigious awards in his field: the American Marketing Association Distinguished Educator Award and the Academy of Marketing Science Outstanding Educator Award. He was also selected for the Churchill Award, which honors career impact on marketing research. He was editor of the *Journal of Marketing Research* and has been on the review board of the *Journal of Marketing* and other journals.

The Decision Sciences Institute has recognized Dr. Perreault for innovations in marketing education, and at UNC he has received several awards for teaching excellence. His books include two other widely used texts: *Basic Marketing* and *The Marketing Game!*

Dr. Perreault is a past president of the American Marketing Association Academic Council, served as chair of an advisory committee to the U.S. Census Bureau, and served as a trustee of the Marketing Science Institute. He has also worked as a consultant to organizations that range from GE and IBM to the Federal Trade Commission.

Joseph P. Cannon

Joseph P. Cannon is a Dean's Distinguished Teaching Fellow and professor of marketing at Colorado State University. He has also taught at the University of North Carolina, Emory University, Instituto de Empresa (Madrid, Spain), INSEAD (Fontainebleau, France), and Thammasat University (Bangkok, Thailand). He has received several teaching awards and the N. Preston Davis Award for Instructional Innovation.

Dr. Cannon's research has been published in the *Journal of Marketing, Journal of Marketing Research, Journal of the Academy of Marketing Science, Journal of Operations Management, Journal of Personal Selling and Sales Management, Journal of Public Policy and Marketing, Antitrust Bulletin,* and the *Academy of Management Review,* among others. He is a two-time recipient of the Louis W. and Rhona L. Stern Award for high-impact research on interorganizational issues. He has also written numerous teaching cases. Dr. Cannon has served on the editorial review boards of the *Journal of Marketing, Journal of the Academy of Marketing Science, Journal of Operations Management, Journal of Personal Selling and Sales Management,* and *Journal of Marketing Education.* The *Journal of Marketing* has honored Dr. Cannon with several distinguished reviewer awards. He served as chair of the American Marketing Association's Interorganizational Special Interest Group (IOSIG). Before entering academics, Dr. Cannon worked in sales and marketing for Eastman Kodak Company.

E. Jerome McCarthy

Marketing lost one of its pioneers when E. Jerome "Jerry" McCarthy passed away at his home in East Lansing, Michigan, in 2015.

After earning a PhD at the University of Minnesota, Dr. McCarthy joined the faculty at Notre Dame and became a Fellow in the prestigious Ford Foundation Program at Harvard, an experience that focused on how to make marketing management practice more rigorous and shaped his thoughts on the needs of students and educators. Dr. McCarthy spent most of his career at Michigan State University, gaining a reputation for working with passion and purpose. He received the AMA's Trailblazer Award in 1987 and was voted one of the "top five" leaders in marketing thought by marketing educators.

Dr. McCarthy was well known for his innovative teaching materials and texts, including *Basic Marketing* and *Essentials of Marketing.* These books changed the way marketing was taught by taking a managerial point of view. He also introduced a marketing strategy planning framework, organizing marketing decisions around the Four Ps—Product, Place, Promotion and Price. As these approaches became the standard in other texts, Dr. McCarthy continued to innovate, including new materials in the digital realm. Today's marketing instructors owe a great debt to this innovative pioneer.

Preface

Essentials of Marketing Is Designed to Satisfy Your Needs

This book is about marketing and marketing strategy planning. At its essence, marketing strategy planning is about figuring out how to do a superior job of satisfying customers. We take that point of view seriously and believe in practicing what we preach. So you can trust that this new edition of *Essentials of Marketing*—and all of the other teaching and learning materials that accompany it—will satisfy your needs. We're very excited about this 17th edition of *Essentials of Marketing* and we hope that you will be as well.

In developing this edition, we've made hundreds of big and small additions, changes, and improvements to the text and all the supporting materials that accompany it. We'll highlight some of those changes in this preface, but first we'll provide background on the evolution of *Essentials of Marketing*.

Building on Pioneering Strengths

Basic Marketing (*Essentials of Marketing*'s parent text) pioneered an innovative structure—the "Four Ps" (Product, Place, Promotion, and Price) with a managerial approach—for the introductory marketing course. It quickly became one of the most widely used business textbooks ever published because it organized the best ideas about marketing so that readers could both understand and apply them. The unifying focus of these ideas is: *How does a marketing manager decide which customers to target and the best way to meet their needs?*

With each new edition of *Essentials of Marketing*, we update the content based on changes in marketing management and the market environment. This book reflects marketing's best practices and ideas. *Essentials of Marketing* and the supporting materials that accompany it have been more widely used than any other teaching materials for the introductory marketing class. It is gratifying that the Four Ps organizing structure has worked well for millions of students and teachers.

The success of *Essentials of Marketing* is not the result of a single strength—or one long-lasting innovation. Other textbooks have adopted our Four Ps framework, and we have continuously improved the book. The text's Four Ps framework, managerial orientation, and strategy planning focus have proven to be foundation pillars that are remarkably robust for supporting new developments in the field, resulting in innovations in the text and package.

Thus, with each new edition of *Essentials of Marketing*, we continue to innovate to better meet the needs of students and faculty. In fact, we have made ongoing changes in how we develop the logic of the Four Ps and the marketing strategy planning process. As always, though, our objective is to provide a flexible, high-quality text and choices from comprehensive and reliable support materials so that instructors and students can accomplish their learning objectives.

What's Different about Essentials of Marketing?

The biggest distinguishing factor about *Essentials of Marketing* is the integrative approach to our teaching and learning package for the introductory marketing course. This integration makes it easier to learn about marketing, teach marketing, and apply it in the real world. For many students, the introductory marketing course will be the only marketing class they ever take. They need to come away with a strong understanding of the key concepts in marketing and how marketing operates in practice. *Essentials of Marketing*:

1. Examines *what* marketing is and *how* to do it.
2. Integrates special topics such as services, international marketing, big data, social media, ethics, and more across the text—with coverage in almost every chapter.
3. Delivers a supplements package completely developed or closely managed by the authors—so each part connects with the text.

The supplements package is extensive—designed to let you *teach marketing your way* (see Exhibit P-1). The integration of these three features delivers a product proven to work for instructors and students. Let us show you what we mean—and why and how instructors and students benefit from the *Essentials of Marketing* teaching and learning package.

What and how of marketing. Marketing operates in dynamic markets. Fast-changing global markets, environmental challenges and sustainability, the blurring speed of technological advances—including an explosion in the use of digital tools by consumers and businesses—are just a few of the current trends confronting today's marketing manager. Whereas some marketing texts merely attempt to describe this market environment, *Essentials of Marketing* teaches students *analytical abilities* and *how-to-do-it skills* that prepare them for success.

Exhibit P-1
Essentials of Marketing: An Integrated Approach to Teaching and Learning Marketing

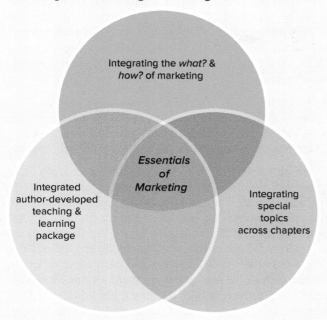

To propel students in this direction, we deliberately include a variety of examples, explanations, frameworks, conceptual organizers, exercises, cases, and how-to-do-it techniques that relate to our overall framework for marketing strategy planning. Taken together, these learning aids speed the development of "marketing sensibility" and enable students to analyze marketing situations and develop marketing plans in a confident and meaningful way. They are practical and they work. And because they are interesting and understandable, they motivate students to see marketing as the challenging and rewarding area it is. In the end, the *Essentials of Marketing* teaching and learning package prepares students to analyze marketing situations and develop exceptional marketing strategies—not just recite endless sets of lists.

Integration of special topics. In contrast to many other marketing textbooks, we emphasize careful integration of special topics. Some textbooks treat "special" topics—such as marketing relationships, international marketing, services marketing, the Internet, digital lifestyles, nonprofit organizations, marketing ethics, marketing analytics, social issues, and business-to-business marketing—in separate chapters (or parts of chapters). We are convinced that treating such topics separately leads to an unfortunate compartmentalization of ideas. For example, to simply tack on a new chapter covering ethics or marketing analytics completely ignores the reality that these are not isolated topics; rather, they must be considered across the rubric of marketing decisions.

Exhibit P-2 shows the coverage of some key topics across specific chapters.

The teaching and learning materials have been designed and developed by the authors—so they seamlessly

Exhibit P-2 Coverage of Special Topics across Chapters*

Special Topic	Chapter																		
	1	2	3	4	5	6	7	8	9	10	11	12	13	14	15	16	17	18	19
Marketing relationships	X	X	X	X	X	X	X	X	X	X	X	X	X	X	X	X	X	X	X
International	X	X	X	X	X	X	X	X	X	X	X	X	X	X	X	X	X		X
Ethics	X	X	X	X	X	X	X	X	X	X	X	X	X	X	X	X	X	X	X
Services	X	X	X	X	X	X	X	X	X	X	X	X	X	X	X	X	X	X	X
B2B	X	X	X	X		X	X	X	X	X	X	X	X	X	X	X	X	X	X
Marketing analytics	X	X	X	X	X	X	X	X	X	X	X	X	X	X	X	X	X	X	X
Technology, Internet, social media, "big data," and digital lifestyle	X	X	X	X	X	X	X	X	X	X	X	X	X	X	X	X	X	X	X
Environment and sustainability	X	X	X	X	X	X	X	X	X	X	X	X	X	X	X	X	X	X	X
Nonprofits		X	X			X	X				X		X		X	X	X		X
Quality		X	X	X	X	X	X	X	X		X		X	X	X	X	X	X	X
Customer value	X	X	X	X	X	X	X	X	X	X	X	X	X	X	X	X	X	X	X
Marketing's link with other functions	X	X	X	X	X	X	X	X	X	X	X	X	X	X	X	X	X	X	X
Marketing for a better world (#M4BW)	X	X	X	X	X	X	X	X	X	X	X	X	X	X	X	X	X	X	X

*X indicates coverage in the form of a section of the chapter, example, illustration, or discussion.

integrate with the textbook. They are integrated to work effectively with *Essentials of Marketing*. We don't tack on extras that have been outsourced and therefore don't mix well with our package. Because of this, you have flexible tools for *teaching and learning marketing your way*. You pick those elements from the package that best fit your students and your teaching approach.

Marketing can be studied in many ways, and the *Essentials of Marketing* text material is only the central component of our *Professional Learning Units System* (*P.L.U.S.*) for students and teachers. Instructors and students can select from our units to develop their own personalized teaching and learning systems. Our objective is to offer you a *P.L.U.S.* "menu" so that you can conveniently select units you want—and disregard what you do not want. Many combinations of units are possible depending on course and learning objectives. Later in this preface we highlight each *P.L.U.S.* element (full details can be found in the Instructor's Manual).

Most business students take only one marketing course in their studies. They deserve the benefits of a highly innovative yet *proven* set of integrated learning materials. Our teaching and learning materials—from the textbook to the online exercises to the test bank to the PowerPoint slides and In-Class Activities—continue to be updated based on what has proven to work for generations of students.

Nineteen Chapters—with an Emphasis on Marketing Strategy Planning

The emphasis of *Essentials of Marketing* is on marketing strategy planning. Nineteen chapters introduce the important concepts of marketing and help students see marketing through the eyes of the manager. The organization of the chapters and topics is carefully planned. We took special care in writing so that:

- It is possible to rearrange and use the chapters in many different sequences to fit different needs.
- All topics and chapters fit together into a clear, overall framework for the marketing strategy planning process.

Broadly speaking, the chapters can be grouped into three sections. The first seven chapters introduce marketing and provide a broad view of the marketing strategy planning process. We introduce the marketing strategy planning process in Chapter 2 and use this framework as a structure for our coverage of marketing throughout the text—see Exhibit P-3. Chapters 3-7 each cover different topics, such as the market environment, competition, segmentation, differentiation, and buyer behavior, as well as how marketing information systems and research provide information about these topics to improve marketing decisions. The second part of the text (Chapters 8-18) goes into the details of planning the Four Ps, with

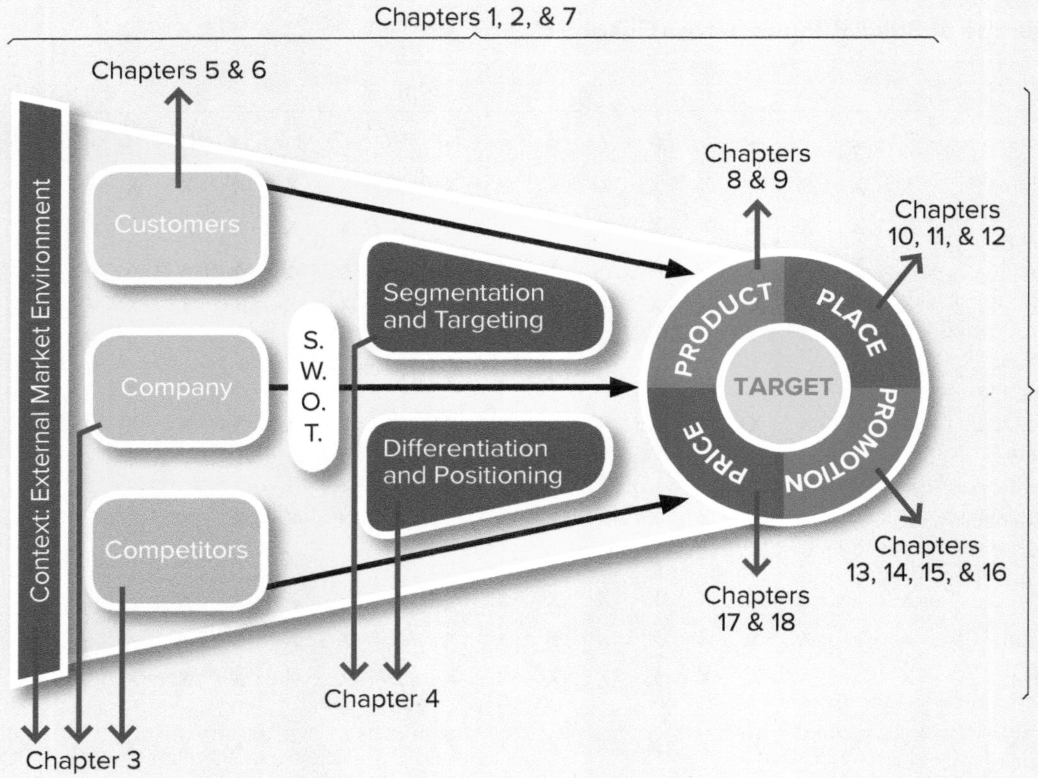

Exhibit P-3 *Essentials of Marketing* and the Marketing Strategy Planning Process

specific attention to the key strategy decisions in each area. Finally, we conclude with an integrative review (Chapter 19) and a critical assessment of marketing's challenges and opportunities.

The first chapter deals with the important role of marketing—focusing not only on how a marketing orientation guides a business or nonprofit organization in the process of providing superior value to customers, but also on the role of macro-marketing and how a market-directed economy shapes choices and quality of life for consumers. Chapter 1 also introduces students to our marketing for a better world (#M4BW) elements—designed to show students examples of companies creatively using marketing to make profits *and* improve the world around them.

Chapter 2 builds on these ideas with a focus on the marketing strategy planning process. It discusses why the marketing strategy planning process involves narrowing down to the selection of a specific target market while blending the Four Ps into a marketing mix to meet the needs of customers. With that foundation in place, Chapter 2 introduces an integrative model of the marketing strategy planning process that serves as an organizing framework for the rest of the text.

Chapter 3 introduces students to the importance of evaluating opportunities in the external environments affecting marketing. This chapter also highlights the critical role of screening criteria to narrow down all possible opportunities to those that the firm will pursue. Chapter 4 shows how market analysis relates to segmentation and differentiation decisions, as well as the criteria used to narrow down to a specific target market and marketing mix. Our approach to segmentation includes defining product-markets—which fits with our strategic planning approach. Analyzing product-markets teaches students how to identify and evaluate opportunities—a key element of marketing strategy.

It is necessary to understand customers in order to segment markets and satisfy target market needs. So the next two chapters take a closer look at customers. Chapter 5 studies the behavioral aspects of the final consumer market. Chapter 6 looks at how business and organizational customers—such as manufacturers, channel members, and government purchasers—are similar to and different from final consumers.

Chapter 7 presents a contemporary view of getting information—from marketing information systems and marketing research—for marketing planning. Chapter 7 includes discussion of how information technology—ranging from intranets to speedy collection of marketing research data—is transforming marketing. This sets the stage for discussions in later chapters about how research and marketing information improve each area of marketing strategy planning.

Chapters 8 through 18 are concerned with developing a marketing mix out of the Four Ps. These chapters are concerned with developing the "right" Product and making it available at the "right" Place with the "right" Promotion at the "right" Price to satisfy target customers while still meeting the objectives of the business. These chapters are presented in an integrated, analytical way—as part of the overall framework for the marketing strategy planning process—so students' thinking about planning marketing strategies develops logically.

Chapters 8 and 9 focus on product planning for goods and services, as well as managing product quality, new-product development, and the different strategy decisions required at different stages of the product life cycle. We emphasize the value of an organized new-product development process for developing truly new products that propel a firm to profitable growth. These chapters also detail how quality management approaches can improve implementation, including implementation of better-quality service.

Chapters 10 through 12 focus on Place. Chapter 10 introduces decisions a manager makes about whether to use direct distribution (for example, selling from the firm's own website) or work with other firms in a channel of distribution. We put special emphasis on the need for channel members to cooperate and coordinate to better meet the needs of customers. Chapter 11 focuses on the fast-changing arena of logistics and the strides that firms are making in using e-commerce to reduce the costs of storing, transporting, and handling products while improving the distribution service they provide customers. Chapter 12 provides a clear picture of retailers, wholesalers, and their strategy planning, including exchanges taking place via the Internet. This chapter helps students see why big changes taking place in retailing are reshaping the channel systems for many consumer products.

Chapters 13 through 16 deal with Promotion. These chapters build on the concepts of integrated marketing communications, direct-response promotion, and customer-initiated digital communication introduced in Chapter 13. Chapter 14 describes the roles of personal selling, customer service, and sales technology in the promotion blend. Chapter 15 covers advertising and sales promotion, including the ways that managers are taking advantage of the Internet and other highly targeted media to communicate more effectively and efficiently. Chapter 16 examines publicity, which we broadly define to include owned, earned, and social media.

Chapters 17 and 18 deal with Price. Chapter 17 focuses on pricing objectives and policies, including the use of information technology to implement flexible pricing; pricing in the channel; and the use of discounts, allowances, and other variations from a list price. Chapter 18 covers cost-oriented and demand-oriented pricing approaches and how they fit in today's competitive environments. The careful coverage of marketing costs helps equip students to deal with the renewed cost-consciousness of the firms they will join.

The final chapter—Chapter 19—considers how efficient the marketing process is. Here we evaluate the effectiveness of both micro- and macro-marketing—and we consider the competitive, technological, ethical, and social challenges facing marketing managers both now and in the future. Chapter 19 also reinforces the integrative nature of marketing management and reviews the marketing strategy planning process that leads to creative marketing plans.

Four appendices can be used to supplement the main text material. Appendix A provides some traditional economic analysis of supply and demand that can be a useful tool in analyzing markets. Appendix B reviews some quantitative tools—or marketing arithmetic—which help marketing managers who want to use accounting data in analyzing marketing problems. Appendix B also reviews forecasting to predict market potential and sales for a company's product. Students especially appreciate Appendix C, which is about career opportunities in marketing. Appendix D provides an example of a marketing plan for Hillside Veterinary Clinic. This example is referenced in Chapter 2 and in end-of-chapter exercises.

Following Appendix D are 39 written cases. The first 5 cases are video cases, available to instructors in video format in the online Instructor Resources. Most of the next 34 short written cases have been updated with new information to make sure they reflect the realities of the current marketplace. The focus of these cases is on problem solving. They encourage students to apply—and engage with—the concepts developed in the text. At the end of each chapter, we recommend the cases that best relate to that chapter's content.

Two bonus chapters (previously available only with our *Basic Marketing* text) are now available online and through custom printing. Bonus Chapter 1 builds on implementation and control, two concepts introduced in Chapter 2. The chapter goes into more detail on these concepts and offers how-to approaches for making implementation and control more effective. The chapter discusses how new information technology tools facilitate these practices and demonstrates how firms use sales analysis, performance analysis, and cost analysis to control marketing strategies and plans. Bonus Chapter 2 includes separate sections that describe how finance, production and operations, accounting, information systems, and human resources interact with marketing to create and implement successful marketing plans.

I love your book—but there are too many chapters for *my* class. *What can I do?*

We are often asked this question by professors who use the quarter system, who prefer to provide students with less breadth of coverage but more depth, or who like the idea of one chapter per week. Three chapters can easily be dropped from the text without harming understanding of other topics:

- Chapter 11 provides coverage of logistics and customer service. We hear that schools with a required course on supply chain management often find this chapter to be redundant. It can also be dropped if there is less desire for this type of coverage. Dropping it does not have an adverse impact on other Place chapters.
- Chapter 18 provides depth in setting prices. Some instructors prefer not to cover this more quantitative treatment and drop the chapter.
- Chapter 19 provides a critical look at marketing and a review of the marketing strategy planning process. If time is an issue, this can be dropped, though we recommend instructors have a concluding lecture that ties together the course.

What's New in This Edition of *Essentials of Marketing?*

Each revision of *Essentials of Marketing* has a few basic themes—areas we try to emphasize across the book. Over the last couple of editions we focused on (1) marketing for a better world, (2) marketing analytics, (3) active learning, and (4) currency. There are several big changes to this edition of *Essentials of Marketing* and hundreds of smaller ones. *Essentials of Marketing* is quick to recognize the many dramatic changes in the market environment and marketing strategy; we are also a leader in pedagogical innovations. Here is a quick overview of what we changed for the 17th edition of *Essentials of Marketing*.

Marketing for a better world. Brand new to this edition—we look at the best of marketing, where marketing practices meet target customer needs and make the world a better place. We have noticed that many of our students come into our class with a negative perception of marketing. We also find that young people today are really interested in the opportunity for business to contribute to making the world a better place. When we look around, we see so many businesses and nonprofit organizations finding ways to make a better world through their marketing efforts. We decided to highlight those efforts, and you will see our "green boxes" and #M4BW across every chapter.

Our perspective on marketing for a better world is not a philanthropic one—we propose that firms do this as an extension of a marketing orientation. A growing number of customers in developed economies want to buy from companies that do good things in the world.

Marketing analytics. The last couple of editions of *Essentials of Marketing* have featured growing coverage of big data and marketing analytics. Following marketing practice and feedback from instructors, this edition builds on that coverage of these topics. Our students—whether

Exhibit P-4 *You Decide How Much Marketing Analytics to Add to Your Class*

I want my students to...	Resources in *Essentials of Marketing*
...understand the importance of analytics and some key terms.	Each chapter features new key terms and/or marketing analytics examples.
...understand how marketing managers use analytics.	Each chapter includes a "Marketing Analytics in Action" activity—a mini case study of analytics in practice.
...calculate some basic marketing analytics.	Each chapter includes an online homework exercise "Marketing Analytics: Data to Knowledge".
...conduct sales and performance analysis.	Assign Bonus Chapter 1, "Implementing and Controlling Marketing Plans: Metrics and Analysis"

marketing majors or in some other area of business—are expected to know what marketing analytics is and how it can be used in marketing strategy planning.

We had discussions with many different instructors about how to add marketing analytics to the introductory marketing course. While everyone agreed that today's student needs to know more about how businesses use marketing analytics, there was disagreement about how much time and focus it should receive. Given that, we make our coverage flexible—allowing *you* the instructor to decide how much attention marketing analytics gets in your class. Most of our extended treatment of marketing analytics can easily be added or left out—as it involves an in-text boxed element, end-of-chapter and online exercises, a marketing simulation, and a bonus chapter. Exhibit P-4 shows how each can contribute to the learning objectives you have for your students.

Currency. Every edition of *Essentials of Marketing* focuses on currency. We find that students like to read current examples. And with technology and customer behavior evolving so quickly, a marketing textbook must keep pace. Instructors and students require up-to-date concepts, content, and examples. In this edition:

- We add hundreds of new examples and images (ads and photos) that engage students. Of course, we continue to include the latest earned, owned, and social media examples—but we also recognize changing consumer behavior, organizational buying behavior, new-product development, the growth and evolution of retail (especially online), advertising, and pricing.
- Current visuals—photos and advertising examples—are particularly important for today's student, so we choose examples that provide a learning experience for the student, not just added color.
- Our *What's Next?* boxes, embedded in each chapter, provide a futuristic perspective on where marketing may be headed.

Active learning exercises. SmartBook offers your students better opportunities to get grounded in the basic concepts of marketing. Many instructors count on this to prepare students for class and utilize more active learning activities inside or outside the classroom. We have designed many active learning exercises so students can learn more critical thinking and application of concepts, which helps them understand how marketing really works. This edition of *Essentials of Marketing* features:

- *Flip Your Marketing Class.* A few years ago, Joe Cannon decided to "flip" his marketing classes. He no longer lectures and now uses about 90 percent of class time on in-class activities. Whether you are ready to completely "flip" your class or are just looking to add more active learning to mix in with your lectures, you will appreciate the materials he has put together. Joe's *Flip Your Marketing Class* e-book can be downloaded from the Instructor's Resource Materials.
- *In-Class Activities.* As part of the *Flip Your Marketing Class* e-book effort, Joe produced a series of exercises that can be used in class. For each chapter there are two to four In-Class Activities. These exercises can be completed in small groups in class. They reinforce and apply concepts learned from *Essentials of Marketing*. Beyond that, many of the following elements of *Essentials of Marketing* could also be used as in-class activities. We have added some brand-new activities, and others were updated.
- *Marketing Analytics in Action.* These in-chapter boxed features place students in the situation many managers face: analyzing and interpreting marketing analytics. The activities show students how marketing analytics are used—but they also ask questions that force students to use critical thinking skills to make marketing decisions. These exercises work particularly well for in-class discussion.
- *Practice Marketing.* The *Practice Marketing* simulation offers a unique way to learn and apply the Four Ps of marketing. Students take on the role of a marketing manager tasked with creating and launching a new product; they do actual marketing strategy planning around a backpack—analyzing customers, competitors, and company—then making target market, product, place, promotion, and price decisions. After

seeing the results of their initial decisions, they can adapt their strategy. The simulation allows students to compete with other students or artificial intelligence characters. Many students have referred to *Practice Marketing* as a virtual internship.
- *What's Next?* Each chapter includes an active learning boxed element. Each *What's Next?* offers an in-depth analysis of some trend or marketing future—and asks students about its implications. These elements have all been updated to make them more forward-looking and active learning focused.
- *Ethical Dilemma.* Several of these critical thinking exercises have been updated from previous editions, with a focus on students making decisions in gray areas, many of which have been introduced by the advancing technologies used in marketing.
- *Questions and Problems.* In past editions, our chapter opening cases have primarily served to motivate a chapter's subject matter. We have added two end-of-chapter "Questions and Problems" (always questions 1 and 2) designed to have students reflect on the opening case studies. Students experience higher-order learning when they are asked to recognize concepts in a case study—so we ask them to do that in question 1. Question 2 turns the chapter opener into a discussion case. Both questions can be used for in-class discussion or homework assignments for instructors looking for higher-order learning objectives for their students.
- *Marketing Analytics*: *Data to Knowledge.* These end-of-chapter exercises—which can be done online—show students how data analysis is used by marketing managers. Using concepts and examples from each chapter, the exercises build higher-order learning skills and demonstrate data-driven marketing decision making. Each practical question walks students through a real-life scenario, shows them how to use a spreadsheet to find answers, and then asks (optional) discussion questions to build critical thinking skills.
- *Cases.* Our video and short cases continue to provide students with opportunities to explore how real companies conduct the marketing strategy planning process.

Chapter-by-chapter. We updated every chapter's opening case scenario. Although we don't have space to list all of our changes, the following are some highlights of the more significant changes we have made to this edition of *Essentials of Marketing:*

Chapter 1. Refreshed with new and updated examples. New discussion and introduction of the idea of marketing for a better world and the associated #M4BW.

Chapter 2. An updated opener and refreshed and more current examples throughout the chapter. We added a new *Ethical Dilemma*. Revised and reorganized coverage of customer lifetime value and customer equity. Added examples of marketing for a better world.

Chapter 3. Updates for currency throughout the chapter. Minor revisions to section on objectives. Major changes throughout the technology section, including the addition of *machine learning* and removal of the ethics coverage from this section. Many changes resulting in more streamlined coverage of the legal environment. Shortened and combined what were three major sections at the end of this chapter. Added examples of marketing for a better world.

Chapter 4. Updates for currency throughout this chapter with new examples. Simplified Exhibit 4-3 and coverage of segmentation. New Exhibit 4-11 provides a detailed explanation of the dynamic behavioral segmentation approach that is being used for online segmentation. Added examples of marketing for a better world.

Chapter 5. Updates for currency across the chapter and new examples. Added examples of marketing for a better world.

Chapter 6. Updates for currency and refreshed examples. New *Ethical Dilemma,* revisions to Exhibits 6-3 and 6-7. New exhibit added to give students more examples of different types of buying processes. Added examples of marketing for a better world.

Chapter 7. Many updates and new examples as marketing research continues to evolve—drawing on new sources of data and new types of marketing analytics. Major revisions to coverage of information systems and two new exhibits designed to reinforce important ideas. Deleted the section on international marketing research, with relevant content integrated elsewhere in the chapter—and dropped the related learning objective. Added examples of marketing for a better world.

Chapter 8. Updates throughout this chapter. New key term *product line length.* New exhibit to better show differences between services and goods. Significantly reduced coverage of warranties—previously had its own learning objective and major section—now briefly covered elsewhere in chapter. Added examples of marketing for a better world.

Chapter 9. Updated for currency. New key terms include *continuous innovation, dynamically continuous innovation,* and *discontinuous innovation*. New *Ethical Dilemma*. Updated coverage of managing service quality with attention to robots replacing service workers. Added examples of marketing for a better world.

Chapter 10. Updated for currency. Major revisions to coverage of selling direct versus indirect—reorganized and new content added. Added two new key terms: *horizontal channel conflict* and *vertical channel conflict*. Moved and updated coverage of *multichannel shopping* and *omnichannel* from Chapter 12 to this chapter as it felt like a better fit. Updated reverse channel coverage. Added examples of marketing for a better world.

Chapter 11. Updated for currency. Added new coverage of food delivery and service level. Major new section at the end of the chapter, "Disaster Relief—Logistics

Saves Lives," provides an example of marketing for a better world but also a nice integration and review of key concepts from this chapter. Added examples of marketing for a better world.

Chapter 12. We always have a lot of revisions in our coverage of retailing—and this edition is no different. To remain current, every edition requires significant rewriting, especially in our section on retailing and the Internet. Other changes include minor updates to Exhibits 12-3, 12-4, and 12-6 and moving coverage of multichannel shopping and omnichannel to Chapter 10. Added examples of marketing for a better world.

Chapter 13. This chapter always has a lot of new examples—students notice dated examples and we aim to eliminate them. Major revision to the section "How Typical Promotion Plans Are Blended and Integrated" that includes a better explanation of this process and new content. Added examples of marketing for a better world.

Chapter 14. Updated for currency. Major rewrite and coverage of customer service. Also major updates and edits to "Information Technology Provides Tools to Do the Job." Added examples of marketing for a better world.

Chapter 15. Because it reflects evolutionary changes in advertising, this chapter always has major changes with each edition. Major updates include (1) updates to the chapter opening case on Domino's; (2) new graphic that really demonstrates the changing advertising media landscape (see Exhibit 15-6); (3) deleted the separate section on advertising agencies, moving some necessary coverage elsewhere in the chapter, but really cutting back this content; (4) many revisions to the "Digital and Mobile Advertising" section; (5) new key term and coverage of *influencers*—including legal issues; (6) revision and updates to sales promotion; and (7) new coverage of legal issues for sales promotion. Added examples of marketing for a better world.

Chapter 16. This chapter is another that requires constant updating with all the changes going on with earned, owned, and social media. There were lots of small changes and updates throughout the chapter, including updates to Exhibits 16-6 and 16-7. Added examples of marketing for a better world.

Chapter 17. Updated for currency as well as major changes that include (1) two new key terms, *benefit corporation* and *B Corporation (B Corp) certification*, (2) new *Ethical Dilemma*, (3) additional coverage on price level policies through the product life cycle, (4) major reorganization including new topics under "Pricing Policies for Price Reductions, Financing, and Transportation," which included (5) streamlined coverage of geographic pricing where we eliminated key terms and coverage of *FOB*, *zone pricing, uniform delivered pricing,* and *freight absorption pricing.* Added examples of marketing for a better world.

Chapter 18. Updated for currency. New exhibit and better explanation of price sensitivity. Added examples of marketing for a better world.

Chapter 19. Modest updates for currency. Added examples of marketing for a better world.

Bonus Chapter 1. The previous edition included (1) adding a new learning objective and section "Marketing Metrics and Analytics Can Guide Marketing Strategy Planning" and (2) *What's Next?* Making better decisions in a world of data analytics. This edition provides updates for currency and new examples of marketing for a better world.

Bonus Chapter 2. Updates for this edition include new examples for currency and examples of marketing for a better world.

Expanded Teaching and Learning Resources for the 17th Edition

The authors of *Essentials of Marketing* and McGraw-Hill Higher Education have put together a variety of resources to supplement your teaching and learning experience. Instructors will find the following resources posted in the Instructor Resources section of the online library for the 17th edition.

- *Active Learning Guide and Exercises*—We hear more and more from instructors seeking to change their marketing course away from being primarily lecture-based to adding more active learning in the classroom. Joe Cannon decided to flip his class and learned a lot in the process. He shares those insights in an e-book that provides some guidance—see *Flip Your Marketing Class*. The ideas work whether you want to completely flip the class or just add more active learning to your current class sessions. One of the most difficult aspects of making this transition is having high-quality activities for students to work on—activities that reinforce important marketing concepts and critical thinking. We have put together many different resources that can help you make this transition—most with teaching notes.
- *Interactive Applications*—An online assignment and assessment solution that connects students with the tools and resources they'll need to achieve success.
- *SmartBook*—Powered by LearnSmart, SmartBook is the adaptive reading experience that helps students learn faster, study more efficiently, and retain more knowledge.
- *Instructor's Manual and Digital Implementation Guide*—Offers a single resource to make it easier for you to decide which resources to use when covering each chapter in the text.
- *PowerPoint Presentation Resources:*
  - *Chapter PowerPoint Slides.* For each chapter there is a set of PowerPoint presentations that includes television commercials and short video clip examples, examples of print advertisements that

demonstrate important concepts, and questions to use with "clickers" or simply to check if students are getting it. These slides cover all the major topics in the chapter.
- *YouTube PowerPoint slides.* We have embedded YouTube videos into a collection with more than 80 slides, which bring virtual guest speakers, viral videos, case studies, and new ads to your classroom presentations.
- *Multimedia Lecture Support Guide and Video Guide.* Now integrated into the Instructor's Manual and Digital Implementation Guide, you will find detailed lecture scripts and chapter outlines for the presentation slides to make getting prepared for class fast and easy.

- *Author Blog*—Connect with the redesigned *Teach the 4 Ps* blog (www.teachthe4ps.com) for links to articles, blog posts, videos, video clips, and commercials—with tips on how to use them with *Essentials of Marketing*. The site is organized by topic—so you can easily find something related to what you are teaching that day—and provides plenty of tips for bringing *active learning* to your classroom.
- *Practice Marketing Simulation*—An online and fully mobile interactive learning environment that simulates the full marketing mix, as well as market segmentation and targeting.
- *Teaching Videos*—The video package includes 17 full-length videos and video cases and 25 shorter animated iSeeIt! videos.
- *Test Bank*—Our test bank includes more than 5,000 objective test questions—every question developed or edited by the authors to ensure it works seamlessly with the text. McGraw-Hill's TestGen program facilitates the creation of tests.

Responsibilities of Leadership

In closing, we return to a point raised at the beginning of this preface. *Essentials of Marketing* has been a leading textbook in marketing since its first edition. We take the responsibilities of that leadership seriously. We know that you want and deserve the very best teaching and learning materials possible. It is our commitment to bring you those materials today with this edition.

We recognize that fulfilling this commitment requires a process of continuous improvement. Because needs change, revisions, updates, and development of new elements must be ongoing. You are an important part of this evolution and of this leadership. We encourage your feedback. The most efficient way to get in touch with us is to send an e-mail message to Joe.Cannon@ColoState.edu. Thoughtful criticisms and suggestions from students and teachers alike have helped make *Essentials of Marketing* what it is today. We hope that you will help us make it what it will be in the future.

William D. Perreault Jr., Joseph P. Cannon, and E. Jerome McCarthy

Acknowledgments

Essentials of Marketing has been influenced and improved by the input of more people than it is possible to list. We want to express our appreciation to those who have played the most significant roles, especially in this edition.

We are especially grateful to our many students who have critiqued and made comments about materials in *Essentials of Marketing.* Indeed, in many ways, our students have been our best teachers.

Many improvements in recent editions were stimulated by feedback from a number of colleagues around the country. Their feedback took many forms. In particular, we would like to recognize the helpful contributions of:

Cliff Ashmead Abdool, *CUNY College of Staten Island*
Roshan (Bob) Ahuja, *Ramapo College of New Jersey*
Thomas Ainscough, *University of South Florida*
Ian Alam, *Ramapo College of New Jersey*
Mary Albrecht, *Maryville University*
David Andrus, *Kansas State University at Manhattan*
Chris Anicich, *Broome Community College*
Maria Aria, *Missouri State University*
April Atwood, *University of Washington*
Ainsworth Bailey, *University of Toledo*
Turina Bakker, *University of Wisconsin*
Jeff Bauer, *University of Cincinnati-Batavia*
Leta Beard, *Washington University*
Amy Beattie, *Nichols College of Champlain*
Cathleen Behan, *Northern VA Community College*
Patty Bellamy, *Black Hills State University*
Suzeanne Benet, *Grand Valley State University*
Shahid Bhuian, *Louisiana Tech University*
John S. Bishop Jr., *Ohio State University*
David Blackmore, *University of Pittsburgh*
Ross Blankenship, *University of California Berkeley*
Maurice Bode, *Delgado Community College*
Jonathan Bohlman, *Purdue School of Management*
William J. Bont, *Grand Valley State University*
Laurie Brachman, *University of Wisconsin-Madison*
Kit Brenan, *Northland Community College*
John Brennan, *Florida State University*
Richard Brien, *De Anza College*
Elten Briggs, *University of Texas-Austin*
Denny Bristow, *St. Cloud State University*
Susan Brudvig, *Ball State University*
Kendrick W. Brunson, *Liberty University*
Gary Brunswick, *Northern Michigan University*
Derrell Bulls, *Texas Women's University*
Michele Bunn, *Collat School of Business, University of Alabama at Birmingham*
Helen Burdenski, *Notre Dame College of Ohio*
Nancy Bush, *Wingate University*
Carmen Calabrese, *University of North Carolina-Pembroke*
Catherine Campbell, *University of Maryland University College*
Amy Caponette, *Pellissippi State Community College*
James Carlson, *Manatee Community College*
Donald Caudill, *Bluefield State College*
Karen Cayo, *Kettering University*
Kenny Chan, *California State University-Chico*
E. Wayne Chandler, *Eastern Illinois University*
Chen Ho Chao, *Baruch College, City University of New York*
Valeri Chukhlomin, *Empire State College*
Margaret Clark, *Cincinnati State Technical and Community College*
Thomas Clark, *University of Houston-Downtown*
Paris Cleanthous, *New York University-Stern School*
Thomas Cline, *St. Vincent College*
Gloria Cockerell, *Collin County Community College*
Linda Jane Coleman, *Salem State College*
Brian Connett, *California State University-Northridge*
Craig Conrad, *Western Illinois University*
Barbara Conte, *Florida Atlantic University*
Sherry Cook, *Southwest Missouri State*
Matt Critcher, *University of Arkansas Community College-Batesville*
Tammy Crutchfield, *Mercer University*
Brent Cunningham, *Jacksonville State University*
Madeline Damkar, *Cabrillo Community College/CSUEB*
Amy Danley, *Wilmington University*
Charles Davies, *Hillsdale College*
J. Charlene Davis, *Trinity University*
Scott Davis, *University of California at Davis*
Dwane Dean, *Manhattan College*
Susan Higgins DeFago, *John Carroll University*
Larry Degaris, *California State University*
Oscar W. DeShields Jr., *California State University-Northridge*
Nicholas Didow, *University of North Carolina-Chapel Hill*
John E. Dillard, *University of Houston-Downtown*
Les Dlabay, *Lake Forest College*
Glenna Dod, *Wesleyan College*
Beibei Dong, *Lehigh University*
Gary Donnelly, *Casper College*
Paul Dowling, *University of Utah*
Laura Downey, *Purdue University*
Phillip Downs, *Florida State University*
Michael Drafke, *College of DuPage*
John Drea, *Western Illinois University*
Colleen Dunn, *Bucks Community College*
Sean Dwyer, *Louisiana Technical University*
Judith Kay Eberhart, *Lindenwood University-Belleville*

Mary Edrington, *Drake University*
Steven Engel, *University of Colorado*
Dr. S. Altan Erdem, *University of Houston-Clear Lake*
Keith Fabes, *Berkeley College*
Peter Fader, *University of Pennsylvania*
Ken Fairweather, *LeTourneau University*
Phyllis Fein, *Westchester Community College*
Lori S. Feldman, *Purdue University*
Mark Fenton, *University of Wisconsin-Stout*
Jodie L. Ferguson, *Virginia Commonwealth University*
Richard Kent Fields, *Carthage College*
Lou Firenze, *Northwood University*
Jon Firooz, *Colorado State University*
Michael Fitzmorris, *Park University*
Richard Fogg, *Kansas State University*
Kim Folkers, *Wartburg College*
Renee Foster, *Delta State University*
Frank Franzak, *Virginia Commonwealth University*
John Gaffney, *Hiram College*
John Gaskins, *Longwood University*
Carol Gaumer, *University of Maryland University College*
Thomas Giese, *University of Richmond*
Karl Giulian, *Fairleigh Dickinson University-Madison*
J. Lee Goen, *Oklahoma Baptist University*
Brent G. Goff, *University of Houston-Downtown*
David Good, *Central Missouri State University*
Pradeep Gopalakrishna, *Pace University*
Keith Gosselin, *California State University-Northridge*
Rahul Govind, *University of Mississippi*
Norman Govoni, *Babson College*
Gary Grandison, *Alabama State University*
Wade Graves, *Grayson County College*
Mitch Griffin, *Bradley University*
Mike Griffith, *Cascade College*
Alice Griswold, *Clarke College*
Barbara Gross, *California State University-Northridge*
Pranjal Gupta, *University of Tampa*
Susan Gupta, *University of Wisconsin at Milwaukee*
John Hadjmarcou, *University of Texas at El Paso*
Khalil Hairston, *Indiana Institute of Technology*
Adam Hall, *Western Kentucky University*
Bobby Hall, *Wayland Baptist University*
Joan Hall, *Macomb Community College*
David Hansen, *Schoolcraft College*
John Hansen, *University of Alabama at Birmingham*
Dorothy Harpool, *Wichita State University*
LeaAnna Harrah, *Marion Technical College*
James Harvey, *George Mason University*
John S. Heise, *California State University-Northridge*
Lewis Hershey, *University of North Carolina-Pembroke*
James Hess, *Ivy Tech Community College*
Wolfgang Hinck, *Louisiana State University-Shreveport*
Pamela Homer, *California State University-Long Beach*
Ronald Hoverstad, *University of the Pacific*
John Howard, *Tulane University*
Doug Hughes, *Michigan State University-East Lansing*
Deborah Baker Hulse, *University of Texas at Tyler*
Janet Hunter, *Northland Pioneer College*
Phil Hupfer, *Elmhurst College*
Hector Iweka, *Lasell College*
Annette Jajko, *Triton College/College of DuPage*

Jean Jaymes, *West California State University-Bakersfield*
Carol Johanek, *Washington University*
Timothy Johnston, *University of Tennessee at Martin*
Keith Jones, *North Carolina A&T State University*
Sungwoo Jung, *Saint Louis University*
Fahri Karakaya, *University of Massachusetts*
Gary Karns, *Seattle Pacific University*
Pat Karush, *Thomas College*
Josette Katz, *Atlantic Cape Community College*
Eileen Kearney, *Montgomery County Community College*
James Kellaris, *University of Cincinnati*
Robin Kelly, *Cuyahoga Community College*
Courtney Kernek, *Texas A&M University-Commerce*
Imram Khan, *University of Nebraska-Lincoln*
Anthony Kim, *California State Polytechnic University, Pomona*
Brian Kinard, *PennState University-University Park*
Rob Kleine, *Ohio Northern University*
Ken Knox, *Ohio State University-Athens*
Kathleen Krentler, *San Diego State University*
Claudia Kubowicz-Malhotra, *University of North Carolina at Chapel Hill*
Dmitri Kuksov, *Washington University*
Jean Laliberte, *Troy State University*
Linda Lamarca, *Tarleton State University*
Kevin Lambert, *Southeast Community College*
Tim Landry, *Kutztown University of Pennsylvania*
Geoffrey Lantos, *Oregon State University*
Richard LaRosa, *Indiana University of Pennsylvania*
Donald Larson, *The Ohio State University*
Dana-Nicoleta Lascu, *Richmond University*
Debra Laverie, *Texas Tech University*
Marilyn Lavin, *University of Wisconsin-Whitewater*
Freddy Lee, *California State University-Los Angeles*
Steven V. LeShay, *Wilmington University*
David Levy, *Bellevue University*
Dr. Jason Little, *Franklin Pierce University*
Doug Livermore, *Morningside College*
Guy Lochiatto, *California State University*
Lori Lohman, *Augsburg College*
Paul James Londrigan, *Mott Community College*
Sylvia Long-Tolbert, *University of Toledo*
Terry Lowe, *Heartland Community College*
Harold Lucius, *Rowan University*
Navneet Luthar, *Madison Area Technical College*
Richard Lutz, *University of Florida*
W. J. Mahony, *Southern Wesleyan University*
Phyllis Mansfield, *Pennsylvania State University-Erie*
Rosalynn Martin, *MidSouth Community College*
James McAloon, *Fitchburg State University*
Lee McCain, *Shaw University*
Christina McCale, *Regis University*
Michele McCarren, *Southern State Community College*
Kevin McEvoy, *University of Connecticut-Stamford*
Rajiv Mehta, *New Jersey Institute of Technology*
Sanjay Mehta, *Sam Houston State University*
Matt Meuter, *California State University-Chico*
Michael Mezja, *University of Las Vegas*
Margaret Klayton Mi, *Mary Washington College*
Herbert A. Miller Jr., *University of Texas-Austin*
Linda Mitchell, *Lyndon State College*
Ted Mitchell, *University of Nevada-Reno*

Robert Montgomery, *University of Evansville*
Todd Mooradian, *College of William and Mary*
Kelvyn A. Moore, *Clark Atlanta University*
Marlene Morris, *Georgetown University*
Brenda Moscool, *California State University-Bakersfield*
Ed Mosher, *Laramie Community College*
Reza Motameni, *California State University-Fresno*
Amit Mukherjee, *Providence College*
Steve Mumford, *Gwynedd-Mercy College*
Clara Munson, *Albertus Magnus*
Thomas Myers, *University of Richmond*
Cynthia Newman, *Rider University*
Philip S. Nitse, *Idaho State University at Pocatello*
J. R. Ogden, *Kutztown University*
David Oh, *California State University-Los Angeles*
Lisa O'Halloran, *Northeast Wisconsin Technical College*
Sam Okoroafo, *University of Toledo*
Jeannie O'Laughlin, *Dakota Wesleyan University*
Okey Peter Onyia, *Lindenwood University*
Louis Osuki, *Chicago State University*
Daniel Padgett, *Auburn University*
Esther S. Page-Wood, *Western Michigan University*
Karen Palumbo, *University of St. Francis*
Terry Paridon, *Cameron University*
Dr. Amy Patrick, *Wilmington University*
Terry Paul, *Ohio State University*
Sheila Petcavage, *Cuyahoga Community College*
Stephen Peters, *Walla Walla Community College*
Man Phan, *Cosumnes River College*
Linda Plank, *Ferris State University*
Lucille Pointer, *University of Houston-Downtown*
Nadia Pomirleanu, *UNLV*
Brenda Ponsford, *Clarion University*
Joel Poor, *University of Missouri*
Teresa Preston, *University of Arkansas at Little Rock*
Tracy Proulx, *Park University*
Brook Quigg, *Peirce College*
Anthony Racka, *Oakland Community College*
Kathleen Radionoff, *Cardinal Stritch University*
Daniel Rajaratnam, *Baylor University*
Catherine Rich-Duval, *Merrimack College*
Brent Richard, *Ramapo College of New Jersey*
Charles W. Richardson Jr., *Clark Atlanta University*
Lee Richardson, *University of Baltimore*
Daniel Ricica, *Sinclair Community College*
Darlene Riedemann, *Eastern Illinois University*
Sandra Robertson, *Thomas Nelson Community College*
Kim Rocha, *Barton College*
Amy Rodie, *University of Nebraska-Omaha*
Carlos Rodriguez, *Governors State University*
Robert Roe, *University of Wyoming*
Ann R. Root, *Florida Atlantic University*
Mark Rosenbaum, *Northern Illinois University*
Donald Roy, *Middle Tennessee State University*
Joel Saegert, *University of Texas at San Antonio*
C. M. Sashi, *Florida Atlantic University*
Tracey King Schaller, *Georgia Gwinnett College*
David Schalow, *University of Wisconsin-Stevens Point*
Nate Scharff, *Grossmont College*
Erika Schlomer-Fischer, *California Lutheran University*
Lewis Schlossinger, *Community College of Aurora*

Henry Schrader, *Ramapo College of New Jersey*
Charles Schwepker, *Central Missouri State University*
Murphy Sewell, *University of Connecticut-Storrs*
Kenneth Shamley, *Sinclair College*
Doris Shaw, *Northern Kentucky University*
Donald Shifter, *Fontbonne College*
Jeremy Sierra, *New Mexico State University*
Toye Simmons, *University of Houston-Downtown*
Lisa Simon, *California Polytech-San Luis Obispo*
Rob Simon, *University of Nebraska*
James Simpson, *University of Alabama in Huntsville*
Aditya Singh, *Pennsylvania State University-McKeesport*
Mandeep Singh, *Western Illinois University*
Jill Slomski, *Mercyhurst College*
James Smith, *University of Houston-Downtown*
Robert Smoot, *Lees College*
Don Soucy, *University of North Carolina-Pembroke*
Roland Sparks, *Johnson C. Smith University*
Melissa St. James, *California State University Dominguez Hills*
Joseph R. Stasio, *Merrimack College*
Gene Steidinger, *Loras College*
Jim Stephens, *Emporia State University*
Tom Stevenson, *University of North Carolina*
Geoffrey Stewart, *University of Louisiana at Lafayette*
Karen Stewart, *The Richard Stockton College of New Jersey*
Stephen Strange, *Henderson Community College*
Randy Stuart, *Kennesaw State University*
Rajneesh Suri, *Drexel University*
John Talbott, *Indiana University*
Uday Tate, *Marshall University*
A. J. Taylor, *Austin Peay State University*
Janice Taylor, *Miami University*
Kimberly Taylor, *Florida International University*
Scott Taylor, *McHenry County College*
Steven Taylor, *Illinois State University*
Jeff Thieme, *Syracuse University*
Scott Thompson, *University of Wisconsin-Oshkosh*
Dennis Tootelian, *California State University-Sacramento*
Gary Tschantz, *Walsh University*
Fran Ucci, *Triton College/College of DuPage*
Sue Umashankar, *University of Arizona*
David Urban, *Virginia Commonwealth University*
Kristin Uttech, *Madison Area Technical College*
Laura Valenti, *Nicholls State University*
Peter Vantine, *Georgia Tech*
Ann Veeck, *Western Michigan University*
Steve Vitucci, *Tarleton State University*
Sharon Wagner, *Missouri Western State College*
Russell Wahlers, *Ball State University*
Suzanne Walchli, *University of the Pacific*
Jane Wayland, *Eastern Illinois University*
Danny "Peter" Weathers, *Louisiana State University*
Alan Weber, *University of Missouri at Kansas City, Bloch School of Management*
Yinghong (Susan) Wei, *Oklahoma State University*
John Weiss, *Colorado State University*
M. G. M. Wetzeis, *Universiteit Maastrict, The Netherlands*
Fred Whitman, *Mary Washington College*
Michelle Wicmandy, *University of Houston-Downtown*
Judy Wilkinson, *Youngstown State University*
Phillip Wilson, *Midwestern State University*

Robert Witherspoon, *Triton College*
John Withey, *Indiana University–South Bend*
Jim Wong, *Shenandoah University*
Joyce H. Wood, *N. Virginia Community College*
Brent Wren, *Manhattanville College*
Doris Wright, *Troy University*
Newell Wright, *James Madison University*
Joseph Yasaian, *McIntosh College*
Gary Young, *Worcester State College*

We've always believed that the best way to build consistency and quality into the text and the other P.L.U.S. units is to do as much as possible ourselves. With the growth of multimedia technologies, it's darn hard to be an expert on them all. But we've had spectacular help in that regard.

The lecture-support PowerPoints have been a tremendous effort over many editions. We appreciate the efforts of Shannon Lemay-Finn, Luis Torres, Jay Carlson, Mandy Noelle Carlson, David Urban, Milt Pressley, and Lewis Hershey for their creative work on the lecture-support PowerPoint presentation slides.

We have had a great team of people work on LearnSmart questions. I would particularly like to thank Leroy Robinson from University of Houston–Clear Lake for his leadership and work on LearnSmart and Jon Firooz of Colorado State University who updated our Computer-Aided Problems that are now titled *Marketing Analytics: Data to Knowledge*.

We consider our "best in the business" video package a true team effort. Although the authors had input, the project has been led in recent years by Nick Childers at Shadows and Light Creative Services. For several editions, Judy Wilkinson has played a big role as producer of the video series for the book. In that capacity, she worked closely with us to come up with ideas, and she provided guidance to the talented group of marketing professors and managers who created or revised videos for this edition.

Of course, like other aspects of *Essentials of Marketing*, the video series has evolved and improved over time, and its current strength is partly due to the insights of Phil Niffenegger, who served as producer for our early video efforts. The video series also continues to benefit from the contributions of colleagues who developed videos in earlier editions. They are

Gary R. Brockway
James Burley
David Burns
Debra Childers
Martha O. Cooper
Carolyn Costley
Angie Fenton
W. Davis Folsom
Pam Girardo
Brenda Green
Douglas Hausknecht
Jean Jaymes
Scott Johnson
Bart Kittle
Claudia Kubowicz
Gene R. Lazniak
Freddy Lee
Bill Levy
Charles S. Madden
W. Glynn Mangold
Becky Manter
Don McBane
Robert Miller
J. R. Montgomery
Linda Mothersbaugh
Michael R. Mullen
Phillip Niffenegger
Okey Peter Onyia
Deborah Owens
Thomas G. Ponzurick
George Prough
Peter Rainsford
Jane Reid
Clinton Schertzer
Roger Schoenfeldt
Thomas Sherer
Jeanne M. Simmons
Walter Strange
Jeff Tanner
Ron Tatham
Rollie O. Tillman
Carla Vallone
Yinghong (Susan) Wei
Robert Welsh
Holt Wilson
Poh-Lin Yeou

Faculty and students at our current and past academic institutions—Michigan State University, University of North Carolina, Colorado State University, Emory University, University of Notre Dame, University of Georgia, Northwestern University, University of Oregon, University of Minnesota, and Stanford University—have significantly shaped the book. Professor Andrew A. Brogowicz of Western Michigan University contributed many fine ideas to early editions of the text and supplements. Neil Morgan, Charlotte Mason, Rich Gooner, Gary Hunter, John Workman, Nicholas Didow, Barry Bayus, Jon Firooz, Ken Manning, L. A. Mitchell, and Ajay Menon have provided a constant flow of helpful suggestions.

We are also grateful to the colleagues with whom we collaborate to produce international adaptations of the text. In particular, Lindsey Meredith, Lynne Ricker, Stan Shapiro, Ken Wong, and Pascale G. Quester have all had a significant impact on *Essentials of Marketing*.

The designers, artists, editors, and production people at McGraw-Hill who worked with us on this edition warrant special recognition. All of them share our commitment to excellence and bring their own individual creativity to the project. First, we should salute Christine Vaughan, who has done a great (and patient) job as content production manager for the project. Without her adaptive problem solving, we could not have succeeded with a (very) rapid-response production schedule—which is exactly what it takes to be certain that teachers and students get the most current information possible.

David Ploskonka worked as product developer on this edition; his insight and project management skills are much appreciated. Our executive brand manager, Meredith Fossel, was new to this edition and brought great enthusiasm, energy, and ideas. We appreciated her valuable perspective on the *Essentials of Marketing* franchise.

The layout and design of the print and online versions of the text included a dedicated team of professionals. Keith McPherson is a long-time creative and valued contributor to *Essentials of Marketing*. He is a great talent and we sincerely appreciate his past efforts that continue to be reflected in the book's design. We sincerely appreciate the talents of Pam Verros who created the interior and Egzon Shaqiri who updated interior design and designed the cover for this edition of *Essentials of Marketing*. We also appreciate David Tietz from Editorial Image who tracked down photos, ads, and permissions for the images we selected to illustrate important ideas.

We owe an ongoing debt of gratitude to Lin Davis. The book probably wouldn't exist without her—without her help, the book would've been just too overwhelming and we'd have quit! Lin was part of this team for more than 25 years. During that time, she made contributions in every aspect of the text and package. Kendra Miller was with the team for the 14th and 15th editions of *Essentials of Marketing,* and her copyediting and insights were immensely valuable. The most recent editions have greatly benefited from technical editing and comments from Jennifer Collins from Molly Words & Widgets who helped copyedit the manuscript—and brought many ideas and insights to this edition.

We are indebted to all the firms that allowed us to reproduce their proprietary materials here. Similarly, we are grateful to associates from our business experiences who have shared their perspectives and feedback and enhanced our sensitivity to the key challenges of marketing management.

Our families have been patient and consistent supporters through all phases in developing *Essentials of Marketing.* The support has been direct and substantive. Pam Perreault and Chris Cannon have provided valuable assistance and more encouragement than you could imagine. Our kids—Suzanne, Will, Kelly, Ally, and Mallory—provided valuable suggestions and ideas as well as encouragement and support while their dads were too often consumed with a never-ending set of deadlines.

Our product must capsulize existing knowledge while bringing new perspectives and organization to enhance it. Our thinking has been shaped by the writings of literally thousands of marketing scholars and practitioners. In some cases, it is impossible to give unique credit for a particular idea or concept because so many people have played important roles in anticipating, suggesting, shaping, and developing it. We gratefully acknowledge these contributors—from the early thought-leaders to contemporary authors and researchers—who have shared their creative ideas. We respect their impact on the development of marketing and more specifically this book.

To all of these persons—and to the many publishers who graciously granted permission to use their materials—we are deeply grateful. Responsibility for any errors or omissions is certainly ours, but the book would not have been possible without the assistance of many others. Our sincere appreciation goes to all who contributed.

William D. Perreault Jr.,
Chapel Hill, North Carolina, U.S.A.

Joseph P. Cannon,
Fort Collins, Colorado, U.S.A.

Brief Contents

1. Marketing's Value to Consumers, Firms, and Society 2
2. Marketing Strategy Planning 32
3. Evaluating Opportunities in the Changing Market Environment 58
4. Focusing Marketing Strategy with Segmentation and Positioning 86
5. Final Consumers and Their Buying Behavior 114
6. Business and Organizational Customers and Their Buying Behavior 144
7. Improving Decisions with Marketing Information 172
8. Elements of Product Planning for Goods and Services 202
9. Product Management and New-Product Development 234
10. Place and Development of Channel Systems 264
11. Distribution Customer Service and Logistics 294
12. Retailers, Wholesalers, and Their Strategy Planning 318
13. Promotion—Introduction to Integrated Marketing Communications 350
14. Personal Selling and Customer Service 378
15. Advertising and Sales Promotion 406
16. Publicity: Promotion Using Earned Media, Owned Media, and Social Media 438
17. Pricing Objectives and Policies 470
18. Price Setting in the Business World 500
19. Ethical Marketing in a Consumer-Oriented World: Appraisal and Challenges 526

Appendix A Economics Fundamentals 552
Appendix B Marketing Arithmetic 564
Appendix C Career Planning in Marketing 579
Appendix D Hillside Veterinary Clinic Marketing Plan 592
BC1 Bonus Chapter 1: Implementing and Controlling Marketing Plans: Metrics and Analysis 621
BC2 Bonus Chapter 2: Managing Marketing's Link with Other Functional Areas 622

Video Cases 623
Cases 635
Glossary 674
Notes 685
Author Index 733
Company Index 745
Subject Index 749

Contents

CHAPTER ONE

Marketing's Value to Consumers, Firms, and Society 2

Marketing—What's It All About? 4
Marketing Is Important to You 5
How Should We Define Marketing? 6
Macro-Marketing 9
The Role of Marketing in Economic Systems 13
Marketing's Role Has Changed a Lot over the Years 15
What Does the Marketing Concept Mean? 17
The Marketing Concept and Customer Value 20
The Marketing Concept Applies in Nonprofit Organizations 23
The Marketing Concept, Social Responsibility, and Marketing Ethics 24

Conclusion 29
Key Terms 29
Questions and Problems 30
Suggested Cases 30
Marketing Analytics: Data to Knowledge 30

CHAPTER TWO
Marketing Strategy Planning 32

The Management Job in Marketing 34
What Is a Marketing Strategy? 35
Selecting a Market-Oriented Strategy Is Target Marketing 36
Developing Marketing Mixes for Target Markets 37
The Marketing Plan Guides Implementation and Control 42
Recognizing Customer Lifetime Value and Customer Equity 45
What Are Attractive Opportunities? 48
Marketing Strategy Planning Process Highlights Opportunities 49
Types of Opportunities to Pursue 52
International Opportunities Should Be Considered 54

Conclusion 55
Key Terms 56
Questions and Problems 56
Marketing Planning for Hillside Veterinary Clinic 57
Suggested Cases 57
Marketing Analytics: Data to Knowledge 57

CHAPTER THREE
Evaluating Opportunities in the Changing Market Environment 58

The Market Environment 60
Objectives Should Set Firm's Course 61
Company Resources May Limit Search for Opportunities 63
Analyzing Competitors and the Competitive Environment 65
The Economic Environment 67
The Technological Environment 68
The Political Environment 71
The Legal Environment 72
The Cultural and Social Environment 74
Screening Criteria Narrow Down Strategies 81

Conclusion 83
Key Terms 84
Questions and Problems 84
Marketing Planning for Hillside Veterinary Clinic 85
Suggested Cases 85
Marketing Analytics: Data to Knowledge 85

CHAPTER FOUR
Focusing Marketing Strategy with Segmentation and Positioning 86

Search for Opportunities Can Begin by Understanding Markets 88
Naming Product-Markets and Generic Markets 91
Market Segmentation Defines Possible Target Markets 93
Target Marketers Aim at Specific Targets 96
What Dimensions Are Used to Segment Markets? 99
More Sophisticated Techniques May Help in Segmenting and Targeting 104
Differentiation and Positioning Take the Customer Point of View 107

Conclusion 111
Key Terms 111
Questions and Problems 111
Marketing Planning for Hillside Veterinary Clinic 112
Suggested Cases 112
Marketing Analytics: Data to Knowledge 112

CHAPTER FIVE
Final Consumers and Their Buying Behavior 114

Consumer Behavior: Why Do They Buy What They Buy? 116
Economic Needs Affect Most Buying Decisions 118
Psychological Influences within an Individual 119
Social Influences Affect Consumer Behavior 128
Culture, Ethnicity, and Consumer Behavior 131
Individuals Are Affected by the Purchase Situation 134
The Consumer Decision Process 135

Conclusion 141
Key Terms 141
Questions and Problems 141
Marketing Planning for Hillside Veterinary Clinic 142
Suggested Cases 142
Marketing Analytics: Data to Knowledge 142

xxi

6

CHAPTER SIX

Business and Organizational Customers and Their Buying Behavior 144

Business and Organizational Customers—A Big Opportunity 146
Organizational Customers Are Different 148
A Model of Business and Organizational Buying 154
Step 1: Defining the Problem 154
Step 2: The Decision-Making Process 155
Step 3: Managing Buyer–Seller Relationships in Business Markets 159
Manufacturers Are Important Customers 163
Producers of Services—Smaller and More Spread Out 165
Retailers and Wholesalers Buy for Their Customers 166
The Government Market 168

Conclusion 170
Key Terms 170
Questions and Problems 170
Marketing Planning for Hillside Veterinary Clinic 171
Suggested Cases 171
Marketing Analytics: Data to Knowledge 171

7

CHAPTER SEVEN

Improving Decisions with Marketing Information 172

Effective Marketing Requires Good Information 174
Changes Are Under Way in Marketing Information Systems 176
The Scientific Method and Marketing Research 184
Five-Step Approach to Marketing Research 185
Step 1: Defining the Problem 186
Step 2: Analyzing the Situation 187
Step 3: Getting Problem-Specific Data 190
Step 4: Interpreting the Data 197
Step 5: Solving the Problem 199

Conclusion 200
Key Terms 200
Questions and Problems 200
Marketing Planning for Hillside Veterinary Clinic 201
Suggested Cases 201
Marketing Analytics: Data to Knowledge 201

CHAPTER EIGHT
Elements of Product Planning for Goods and Services 202

The Product Area Involves Many Strategy Decisions 204
What Is a Product? 205
Differences between Goods and Services 208
Technology and Intelligent Agents Add Value to Products 210
Branding Is a Strategy Decision 212
Achieving Brand Familiarity 215
Branding Decisions: What Kind? Who Brands? 219
Packaging Promotes, Protects, and Enhances 221
Product Classes Help Plan Marketing Strategies 223
Consumer Product Classes 224
Business Products Are Different 227
Business Product Classes—How They Are Defined 228

Conclusion 230
Key Terms 231
Questions and Problems 231
Marketing Planning for Hillside Veterinary Clinic 232
Suggested Cases 232
Marketing Analytics: Data to Knowledge 232

CHAPTER NINE
Product Management and New-Product Development 234

Innovation and Market Changes Create Opportunities 236
Managing Products over Their Life Cycles 238
Product Life Cycles Vary in Length 240
Planning for Different Stages of the Product Life Cycle 242
New-Product Planning 246
An Organized New-Product Development Process Is Critical 248
New-Product Development: A Total Company Effort 256
Need for Product Managers 257
Managing Product Quality 258

Conclusion 261
Key Terms 262
Questions and Problems 262
Marketing Planning for Hillside Veterinary Clinic 262
Suggested Cases 263
Marketing Analytics: Data to Knowledge 263

xxiii

CHAPTER TEN

Place and Development of Channel Systems 264

Marketing Strategy Planning Decisions for Place 266
Place Decisions Are Guided by "Ideal" Place Objectives 267
Channel System May Be Direct or Indirect 269
Channel Specialists May Reduce Discrepancies and Separations 274
Channel Relationships Must Be Managed 276
Vertical Marketing Systems Focus on Final Customers 281
The Best Channel System Should Achieve Ideal Market Exposure 282
Multichannel Distribution and Reverse Channels 285
Entering International Markets 290

Conclusion 292
Key Terms 292
Questions and Problems 292
Marketing Planning for Hillside Veterinary Clinic 293
Suggested Cases 293
Marketing Analytics: Data to Knowledge 293

CHAPTER ELEVEN

Distribution Customer Service and Logistics 294

Physical Distribution Gets It to Customers 296
Physical Distribution Customer Service 297
Physical Distribution Concept Focuses on the Whole Distribution System 300
Coordinating Logistics Activities among Firms 302
The Transporting Function Adds Value to a Marketing Strategy 305
Which Transporting Alternative Is Best? 307
The Storing Function and Marketing Strategy 310
Specialized Storing Facilities May Be Required 312
The Distribution Center—A Different Kind of Warehouse 313
Disaster Relief—Logistics Saves Lives 314

Conclusion 316
Key Terms 316
Questions and Problems 316
Marketing Planning for Hillside Veterinary Clinic 317
Suggested Cases 317
Marketing Analytics: Data to Knowledge 317

CHAPTER TWELVE
Retailers, Wholesalers, and Their Strategy Planning 318

Retailers and Wholesalers Plan Their Own Strategies 320
The Nature of Retailing 322
Planning a Retailer's Strategy 323
Conventional Retailers—Try to Avoid Price Competition 325
Expand Assortment and Service—To Compete at a High Price 326
Evolution of Mass-Merchandising Retailers 327
Some Retailers Focus on Added Convenience 330
Retailing and the Internet 331
Why Retailers Evolve and Change 335
Differences in Retailing in Different Nations 338
What Is a Wholesaler? 339
Wholesaling Is Changing with the Times 340
Wholesalers Add Value in Different Ways 341
Merchant Wholesalers Are the Most Numerous 342
Agents Are Strong on Selling 345

Conclusion 346
Key Terms 347
Questions and Problems 347
Marketing Planning for Hillside Veterinary Clinic 348
Suggested Cases 348
Marketing Analytics: Data to Knowledge 348

CHAPTER THIRTEEN
Promotion—Introduction to Integrated Marketing Communications 350

Promotion Communicates to Target Markets 352
Several Promotion Methods Are Available 353
Someone Must Plan, Integrate, and Manage the Promotion Blend 355
Which Methods to Use Depends on Promotion Objectives 356
Promotion Requires Effective Communication 360
When Customers Initiate the Communication Process 363
How Typical Promotion Plans Are Blended and Integrated 366
Adoption Processes Can Guide Promotion Planning 371
Promotion Blends Vary over the Product Life Cycle 372
Setting the Promotion Budget 374

Conclusion 375
Key Terms 376
Questions and Problems 376
Marketing Planning for Hillside Veterinary Clinic 377
Suggested Cases 377
Marketing Analytics: Data to Knowledge 377

14

CHAPTER FOURTEEN
Personal Selling and Customer Service 378

The Importance and Role of Personal Selling 380
What Kinds of Personal Selling Are Needed? 383
Order Getters Develop New Business Relationships 383
Order Takers Nurture Relationships to Keep the Business Coming 384
Supporting Sales Force Informs and Promotes in the Channel 385
Customer Service Promotes the Next Purchase 387
The Right Structure Helps Assign Responsibility 389
Information Technology Provides Tools to Do the Job 392
Sound Selection and Training to Build a Sales Force 396
Compensating and Motivating Salespeople 397
Personal Selling Techniques—Prospecting and Presenting 399

Conclusion 403
Key Terms 404
Questions and Problems 404
Marketing Planning for Hillside Veterinary Clinic 405
Suggested Cases 405
Marketing Analytics: Data to Knowledge 405

15

CHAPTER FIFTEEN
Advertising and Sales Promotion 406

Advertising, Sales Promotion, and Marketing Strategy Planning 408
Advertising Is Big Business 410
Advertising Objectives Are a Strategy Decision 411
Objectives Determine the Kinds of Advertising Needed 413
Choosing the "Best" Medium—How to Deliver the Message 416
Digital and Mobile Advertising 419
Planning the "Best" Message—What to Communicate 424
Measuring Advertising Effectiveness Is Not Easy 426
Avoid Unfair Advertising 428
Sales Promotion—Do Something Different to Stimulate Change 430
Managing Sales Promotion 434

Conclusion 435
Key Terms 435
Questions and Problems 436
Marketing Planning for Hillside Veterinary Clinic 436
Suggested Cases 436
Marketing Analytics: Data to Knowledge 437

CHAPTER SIXTEEN

Publicity: Promotion Using Earned Media, Owned Media, and Social Media 438

Publicity, the Promotion Blend, and Marketing Strategy Planning 440
Paid, Earned, and Owned Media 442
Customers Obtain Information from Search, Pass-Along, and Experience 446
Create Owned Media Content Your Customers Can Use 448
Earned Media from Public Relations and the Press 452
Earned Media from Customer Advocacy 454
Social Media Differ from Traditional Media 457
Major Social Media Platforms 459
Software Can Manage, Measure, and Automate Online Media 466

Conclusion 468
Key Terms 468
Questions and Problems 468
Marketing Planning for Hillside Veterinary Clinic 469
Suggested Cases 469
Marketing Analytics: Data to Knowledge 469

CHAPTER SEVENTEEN

Pricing Objectives and Policies 470

Price Has Many Strategy Dimensions 472
Objectives Should Guide Strategy Planning for Price 475
Profit-Oriented Objectives 475
Sales-Oriented Objectives 477
Status Quo-Oriented Objective 478
Most Firms Set Specific Pricing Policies—To Reach Objectives 478
Price Flexibility Policies 479
Price-Level Policies and the Product Life Cycle 482
Discount Policies—Reductions from List Prices 484
Allowance Policies—Off List Prices 486
Pricing Policies for Price Reductions, Financing, and Transportation 487
Pricing Policies Combine to Impact Customer Value 491
Legality of Pricing Policies 494

Conclusion 497
Key Terms 498
Questions and Problems 498
Marketing Planning for Hillside Veterinary Clinic 499
Suggested Cases 499
Marketing Analytics: Data to Knowledge 499

xxvii

18

CHAPTER EIGHTEEN
Price Setting in the Business World 500

Price Setting Is a Key Strategy Decision 502
Some Firms Just Use Markups 503
Average-Cost Pricing Is Common and Can Be Dangerous 506
Marketing Managers Must Consider Various Kinds of Costs 507
Break-Even Analysis Can Evaluate Possible Prices 510
Marginal Analysis Considers Both Costs and Demand 512
Additional Demand-Oriented Approaches for Setting Prices 514
Pricing a Full Line 522

Conclusion 523
Key Terms 524
Questions and Problems 524
Marketing Planning for Hillside Veterinary Clinic 525
Suggested Cases 525
Marketing Analytics: Data to Knowledge 525

19

CHAPTER NINETEEN
Ethical Marketing in a Consumer-Oriented World: Appraisal and Challenges 526

How Should Marketing Be Evaluated? 528
Can Consumer Satisfaction Be Measured? 529
Micro-Marketing Often Does Cost Too Much 531
Macro-Marketing Does Not Cost Too Much 533
Marketing Strategy Planning Process Requires Logic and Creativity 536
The Marketing Plan Brings All the Details Together 538
Today's Marketers Face Challenges and Opportunities 541
How Far Should the Marketing Concept Go? 549

Conclusion 550
key term 550
Questions and Problems 550
Marketing Planning for Hillside Veterinary Clinic 551
Suggested Cases 551

APPENDIX A
Economics Fundamentals 552

APPENDIX B
Marketing Arithmetic 564

APPENDIX C
Career Planning in Marketing 579

APPENDIX D
Hillside Veterinary Clinic Marketing Plan 592

BONUS CHAPTER 1

Bonus Chapter 1: Implementing and Controlling Marketing Plans: Metrics and Analysis (full chapter content accessible online through SmartBook) 621

BONUS CHAPTER 2

Bonus Chapter 2: Managing Marketing's Link with Other Functional Areas (full chapter content accessible online through SmartBook) 622

Video Cases

1. Potbelly Sandwich Works Grows through "Quirky" Marketing 624
2. Suburban Regional Shopping Malls: Can the Magic Be Restored? 625
3. Strategic Marketing Planning in Big Brothers Big Sisters of America 627
4. Invacare Says "Yes, You Can!" to Customers Worldwide 630
5. Segway Finds Niche Markets for Its Human Transporter Technology 632

Cases

1. McDonald's "Seniors" Restaurant 636
2. Nature's Own Foods, Inc. 636
3. NOCO United Soccer Academy 637
4. Petoskey Tech Support 638
5. Resin Dynamics 639
6. Dynamic Steel 640
7. Lake Pukati Lodge 641
8. Carmine's Italian Restaurant 642
9. Quiet Night Motel 643
10. Cousin's Ice Center 644
11. Running On 645
12. DrV.com—Custom Vitamins 646
13. Paper Products, Inc. (PPI) 647
14. Schrock & Oh Design 648
15. The Scioto Group 649
16. Hanratty Company 650
17. Wise Water, Inc. 651
18. West Tarrytown Volunteer Fire Department (WTVFD) 653
19. MyOwnWedding.com 654
20. Lake Russell Marine & Camp 656
21. GeoTron International (GTI) 657
22. Bright Light Innovations: The Starlight Stove 658
23. Wire Solutions 659
24. Fresh Harvest 659
25. QXR Tools (QXR) 660
26. AAA Custom Castings, Inc. 661
27. Canadian Mills, Ltd. 662
28. Kingston Home Health Services (KHHS) 663
29. Kennedy & Gaffney (K&G) 665
30. Paglozzi's Pizza Pies 666
31. Silverglade Homes 668
32. Mallory's Lemonade Stand (A) 670
33. Mallory's Lemonade Stand (B) 670
34. Working Girl Workout 671

Glossary 674
Notes 685
Author Index 733
Company Index 745
Subject Index 749

TOC images: p. xxi: Pixtal/AGE Fotostock; p. xxii (left): Lionel Bonaventure/AFP/Getty Images; (right): Rodrigo Reyes Marin/AFLO/Newscom; p. xxiii (left): William Howard/Shutterstock; (right): Romsvetnik/Shutterstock; p. xxiv (left): Design Pics Inc/Alamy Stock Photo; (right): Bill Greene/The Boston Globe/Getty Images; p. xxv (left): McGraw-Hill Education; (right): eans/Shutterstock; p. xxvi (left): Imaginechina/AP Images; (right): Marco Di Lauro/Getty Images; p. xxvii (left): Scott Olson/Getty Images; (right): Anton Novoderezhkin/TASS/Getty Images; p. xxviii (left): William D. Perreault Jr.; (right): Sam Mellish/Getty Images; p. xxix (left): Courtesy of The Chopping Block; (right): Sheila Fitzgerald/Shutterstock; p. xxx (left): MikeDotta/Shutterstock; (right): maggiegowan.co.uk/Alamy Stock Photo; p. xxxi (left): Helen H. Richardson/The Denver Post/Getty Images; (right): YOSHIKAZU TSUNO/AFP/Getty Images

Essentials of Marketing

A Marketing Strategy Planning Approach

CHAPTER ONE

Believe in something.
Even if it means sacrificing everything.

Just do it.

Source: Nike, Inc.

Marketing's Value to Consumers, Firms, and Society

When it's time to roll out of bed in the morning, does the Spotify app on your Samsung Galaxy phone play your "Waking Up Happy" playlist, or is it your roommate blasting The Chainsmokers? Do you throw on your Nike Flyknit shoes and get in a short run before breakfast? Maybe not this morning. You slept in and now you barely have time to throw on your Levi's jeans, shirt from Zara, and Chaco sandals as you race off to class. You are hungry. Will you open your GE refrigerator and choose a Chobani pineapple-flavored Greek yogurt and an Einstein's bagel with Philadelphia Cream Cheese? Or maybe you grab a Chicken Apple Sausage breakfast sandwich at Caribou Coffee in the student union. If you hurry, your roommate can give you a ride to school in her new Ford Fiesta, although you could ride your Big Shot Fixie bike or take the bus that the city bought from Mercedes-Benz. So many choices.

When you think about it, you can't get very far into a day without bumping into marketing—and what the whole marketing system does for you. It affects every aspect of our lives—often in ways we don't even consider.

In other parts of the world, people wake up each day to different kinds of experiences. A family in a rural African village may have little choice about what food they will eat or where their clothing will come from. In the world's more economically developed countries, consumers find plenty of choices on store shelves. And if no one buys a particular color, size, or style, then companies stop producing it. So, you may have trouble finding a Chicken Apple Sausage sandwich in Hangzhou, China, where the locals are more likely lined up waiting for GanQiShi's steamed buns.

One brand found around the world is Nike. How has Nike become the choice for so many professional and casual athletes around the world? Is it the more than $1 billion Nike spends each year for endorsements from star athletes like LeBron James and Cristiano Ronaldo? Maybe it's the innovations, like Nike's self-lacing sneakers, lightweight Flyknit shoes, and its Nike+ software apps. What part do the 24,000 retailers that carry Nike products play? Do Nike's connections with tens of millions of customers on Twitter, Instagram, and Facebook build customer relationships? And just how much (if at all) do these marketing strategy decisions affect Nike's sales and profits?

More than 50 years ago, Phil Knight and his college track coach, Bill Bowerman, founded Blue Ribbon Sports (later renamed Nike) to distribute Japanese running shoes. A few years later they were designing, producing, and selling athletic shoes.

Nike really took off after signing basketball star Michael Jordan to endorse its basketball shoes. The Air Jordan line took the market by storm. Nike raced further ahead when its advertising agency came up with the "Just Do It!" slogan and an ad campaign that covered television, magazines, and billboards around the world. "Just Do It" helped carry Nike through the 1990s while profits soared on rising sales aided by low-cost foreign production.

Things haven't always gone smoothly for Nike. In the late 1990s, the company came under attack when it was reported that some of its suppliers used child labor. At first Nike denied responsibility, claiming it couldn't control how its suppliers operated. But public protest showed that society expected more from a large, successful corporation, and Nike began to closely monitor its suppliers' labor practices. Since then, Nike's social responsibility efforts have turned around its reputation. For example, Nike recently set an ambitious sustainability goal: to double its sales while halving its environmental impact. Already, three-quarters of its shoes and apparel contain some recycled material. Strategy decisions like these don't immediately increase Nike's profits, but Nike makes a better world when it recognizes "the future of sport is interlocked with the future of our planet" (Noel Kinder, Nike's chief sustainability officer).

Another Nike marketing decision was quite a bit more controversial. One of Nike's sponsored athletes, NFL quarterback Colin Kaepernick, became a divisive figure when he kneeled during the pregame playing of the national anthem. That choice divided a nation—with some supporting his stand for social justice and others claiming he was unpatriotic. It eventually cost Kaepernick his job when no team wanted to sign the controversial player.

Then Nike jumped into the fray with an advertising campaign featuring Kaepernick and the tagline "Believe in something. Even if it means sacrificing everything." Almost instantly, athletes, consumers, celebrities, and even the president of the

United States reacted—jumping in with full-throated support or harsh criticism of the ad. Nike thought many of its target customers—urban Millennials—would rally behind the cause and the Nike brand. Nike was right: sales jumped 10 percent and net income leapt 15 percent. The campaign brought attention to Nike and racial injustice, forcing more debate on just what makes the world a better place.

Innovation remains important to Nike's culture. For example, its Nike+ apps for smartphones include workouts hosted by athletes like tennis star Serena Williams and Chinese track and field star Su Bingtian, who act as virtual trainers. The apps encourage working out and build long-term customer relationships. When Nike released a new, limited edition Air Jordan basketball shoe, it experimented with direct-to-consumer distribution. The shoes could only be purchased through Snapchat—and sold out in just 23 minutes.

Nike's marketing adds value—it consistently ranks among the 20 most valuable brands in the world and annual sales exceed $36 billion. But when it comes to athletic clothing and shoes, customers have plenty of choices. If it wants to stay ahead of a strong field of competitors that includes Adidas, Under Armour, Skechers (now the number two seller of athletic footwear in the United States), and Chinese upstart Li Ning, then Nike needs to continue to innovate and stay ahead of its customers' needs and wants.[1]

LEARNING OBJECTIVES

In this chapter, you'll learn what marketing is all about and why it's important to you as a consumer. We'll also explore why it is so crucial to the success of individual firms and nonprofit organizations and the impact that it has on the quality of life in different societies.

When you finish this chapter, you should be able to

1. know what marketing is and why you should learn about it.
2. understand the difference between marketing and macro-marketing.
3. know the marketing functions and why marketing specialists—including intermediaries and collaborators—develop to perform them.
4. understand what a market-driven economy is and how it adjusts the macro-marketing system.
5. know what the marketing concept is—and how it should guide a firm or nonprofit organization.
6. understand what customer value is and why it is important to customer satisfaction.
7. know how social responsibility and marketing ethics relate to the marketing concept.
8. understand the important new terms (shown in red).

Marketing—What's It All About?

LO 1.1

Marketing is more than selling or advertising

How did all those bicycles get here?

Many people think that marketing means "selling" or "advertising." It's true that these are parts of marketing. But *marketing is much more than selling and advertising.*

To illustrate some of the other important things that are included in marketing, think about all the bicycles being pedaled with varying degrees of energy by bike riders around the world. Most of us don't make our own bicycles. Instead, they are made by firms such as Trek, Specialized, Canyon, and Electra.

Most bikes do the same thing—get the rider from one place to another. But a bike rider can choose from a wide assortment of models. They are designed in different sizes and with or without gears. Off-road bikes have large knobby tires. Kids and older people may want more wheels—to make balancing easier. Some bikes need baskets or even trailers for cargo. You can buy a basic bike for less than $100. Or you can spend more than $5,000 for a custom frame.

Marketing helps make sure that each customer gets the bicycle that best meets his or her needs.
(left): Jonathan Gelber/FStop Images GmbH/Alamy Stock Photo; (middle): Monkey Business Images/Shutterstock; (right): Pixtal/AGE Fotostock

This variety of styles and features complicates the production and sale of bicycles. The following list shows some of the things a manager should do before and after deciding to produce and sell a bike.

1. Analyze the needs of people who might buy a bike and decide if they want more or different models.
2. Determine how many of these people will want to buy bicycles, where in the world they live, and when they will want to buy.
3. Identify competing companies that also produce bikes, what kind they sell, and at what prices.
4. Predict the designs of bikes—frame and handlebar styles, derailleurs, types of wheels, brakes, and other accessories—different customers will want, and decide which of these people the firm will try to satisfy.
5. Determine whether to sell bikes directly to consumers or through retailers—and if retailers, which ones should be used.
6. Decide how to tell potential customers about the firm's bikes.
7. Estimate the prices potential customers are willing to pay for their bikes and if the firm can make a profit selling at those prices.
8. Figure out how to provide customer service if a customer has a problem after buying a bike.

The above activities are not part of **production**—actually *making* goods or *performing* services. Rather, they are part of a larger process—called *marketing*—that provides needed direction for production and helps make sure that the right goods and services are produced and find their way to consumers.

You'll learn much more about marketing activities in Chapter 2. For now, it's enough to see that marketing plays an essential role in providing consumers with need-satisfying goods and services and, more generally, in creating customer satisfaction. Simply put, **customer satisfaction** is the extent to which a firm fulfills a customer's needs, desires, and expectations.

Marketing Is Important to You

Marketing is important to every consumer

Marketing affects almost every aspect of your daily life. The choices you have among the goods and services you buy, the stores where you shop, and the radio and TV programs you tune in to are all possible because of marketing. In the process of providing all these choices, marketing drives organizations to focus on what it takes to satisfy you, the customer. Most of the things you want or need are available conveniently *when* and *where* you want or need them.

Some courses are interesting when you take them but not directly relevant to your life once they're over. That's not so with marketing—you'll be a consumer dealing with marketing for the rest of your life regardless of what career you pursue. Moreover, as a consumer, you pay for the cost of marketing activities. In advanced economies, marketing costs about 50 cents of every consumer dollar. For some goods and services, the percentage is much higher. It makes sense to be an educated consumer and to understand what you get and don't get from all that spending.

Marketing will be important to your job

Another reason for studying marketing is that it offers many exciting and rewarding career opportunities. Throughout this book, you will find information about opportunities in different areas of marketing (see especially Appendix C).

If you're aiming for a nonmarketing job, knowing about marketing will help you do your job better. Throughout the book, we'll discuss ways that marketing interacts with other parts of the company—including finance, accounting, human resources, computer information systems, research and development, and more. Further, marketing is important to the success of every organization.

Beyond that, the same basic principles used to sell soap or breakfast cereal are used to "sell" ideas, politicians, health care services, environmental sustainability, museums, and even colleges. No matter what job you end up doing, you are very likely to have to understand others' needs, and perhaps persuade people to behave differently or change their minds about something. Doctors and nurses often need to persuade patients to take their medicine and change their eating habits. Managers have to understand people that work with them and convince them to change behaviors to increase their job performance. Marketing principles will help you achieve those goals and maybe help you get your next job.

A marketing approach can help you get your next job

You will probably be seeking a job sometime soon, offering your services—as an accountant, a salesperson, a computer programmer, a financial analyst, or perhaps a store manager. Or maybe you will be looking for an opportunity with more responsibility or higher pay where you currently work. You will have more success getting what you want when you take a marketing approach and try to figure out how to best satisfy the needs, interests, and desires of a current or prospective employer the same way a business looks at its customers. Much of what you learn about how businesses market their products and services to customers can be applied in the job market. Even your résumé is part of a marketing campaign to sell yourself to an employer. See Appendix C for more details on how to write your personal marketing plan.[2]

Marketing affects innovation and standard of living

An even more basic reason for studying marketing is that marketing plays a big part in economic growth and development. One key reason is that marketing encourages research and **innovation**—the development and spread of new ideas, goods, and services. As firms offer new and better ways of satisfying consumer needs, customers have more choices among products, which fosters competition for consumers' money. This competition drives down prices. Moreover, when firms develop products that really satisfy customers, fuller employment and higher incomes can result. The combination of these forces means that marketing has a big impact on consumers' standard of living—and it is important to the future of all nations.[3]

How Should We Define Marketing?

There are micro and macro views of marketing

In our bicycle example, we saw that a producer of bicycles has to perform many customer-related activities besides just making bikes. The same is true for an insurance company or an art museum. This supports the idea of marketing as a set of activities done by an individual organization to satisfy its customers.

On the other hand, people can't survive on bicycles and art museums alone! In advanced economies, it takes goods and services from thousands of organizations to satisfy the many needs of society. Further, a society needs some sort of marketing system to organize the efforts of all the producers, wholesalers, and retailers required to satisfy the varied needs of all its citizens. So marketing is also an important social process.

We can view marketing in two ways: *from a micro view as a set of activities performed by organizations* and also *from a macro view as a social process.* Yet, in everyday use when people talk about marketing, they have the micro view in mind. So that is the way we will define marketing here. However, the broader macro view that looks at the whole production-distribution system is also important, so later we will provide a separate definition and discussion of macro-marketing.

Marketing defined

Marketing is the performance of activities that seek to accomplish an organization's objectives by anticipating customer or client needs and directing a flow of need-satisfying goods and services from producer to customer or client. Let's look at this definition.[4]

Applies to profit and nonprofit organizations

Marketing applies to both profit and nonprofit organizations. Profit is the objective for most business firms. But other types of organizations may seek more members or acceptance of an idea. Customers or clients may be individual consumers, business firms, nonprofit organizations, government agencies, or even foreign nations. Although most customers and clients pay for the goods and services they receive, others may receive them free of charge or at a reduced cost through private or government support.

More than just persuading customers

Marketing isn't just selling and advertising. Unfortunately, some executives still think of it that way. They feel that the job of marketing is to "get rid of" whatever the company happens to produce. In fact, the aim of marketing is to identify customers' needs and meet those needs so well that the product almost "sells itself." This is true whether the product is a physical good, a service, or even an idea. If the whole marketing job has been done well, customers don't need much persuading. They should be ready to buy. And after they buy, they'll be satisfied and ready to buy the same way the next time.

Begins with customer needs

Marketing should begin with potential customer needs—not with the production process. Marketing should try to anticipate needs. And then marketing, rather than production, should determine what goods and services are to be developed—including decisions about product features; design and packaging; prices or fees; transporting and storing policies; advertising and sales tactics; and, after the sale, installation, customer service, warranty, and perhaps even disposal and recycling policies.

Consider France's Sodebo, maker of fresh packaged meals (sandwiches and pasta) sold in grocery stores. Sodebo's consumer research found some customers had an

Marketing aims to identify customer needs and then meet those needs. NoDoz knows that at times, some people need to make sure they don't fall asleep.
Source: Lil' Drug Store Products, Inc.

unmet need: they wanted to satisfy midday hunger in a healthy way. The research showed that whereas busy consumers were willing to buy a high-quality to-go salad, they found current choices skimpy and unsatisfying. Sodebo developed new salads that were convenient, delicious, and filling. On the way to market, Sodebo tested different recipes and packages to find what consumers and retailers liked best. By starting with customer needs, Sodebo's Salade & Compagnie line of 10 different boxed salads became a best seller.[5]

Does not do it alone

This does not mean that marketing should try to take over production, accounting, and financial activities. Rather, it means that marketing—by interpreting customers' needs—should provide direction for these activities and try to coordinate them.

Marketing involves exchanges

The idea that marketing involves a flow of need-satisfying offerings from a producer to a customer implies that there is an exchange of the need-satisfying offering for something else, such as the customer's money. Marketing focuses on facilitating exchanges. In fact, *marketing doesn't occur unless two or more parties are willing to exchange something for something else.* For example, in a **pure subsistence economy**—where each family unit produces everything it consumes—there is no need to exchange goods and services and no marketing is involved. (Although each producer–consumer unit is totally self-sufficient in such a situation, the standard of living is typically relatively low.)

Builds a relationship with the customer

Keep in mind that a marketing exchange is usually part of an ongoing relationship, not just a single transaction. Rather, the goal is continuing sales and an ongoing *relationship* with the customer. In the future, when the customer has the same need again—or some other need that the firm can meet—other sales will follow. A consumer does not visit her local Shell station once, but perhaps every week or two—as long as Shell's gas, coffee, or service satisfies her, she is likely to keep going to the same gas station. This *flow* of need-satisfying goods and services builds a long-lasting relationship that benefits both the firm and the customer.

The focus of this text—management-oriented micro-marketing

Because you are probably preparing for a career in management, the main focus of this text will be on managerial marketing, or the micro view of marketing. We will see marketing through the eyes of the marketing manager.

The marketing ideas we will be discussing throughout this text apply to a wide variety of situations. They are important for new ventures started by one person as well as big corporations, in domestic and international markets, and regardless of whether the focus is on marketing physical goods, services, or an idea or cause. They are equally critical whether the relevant customers or clients are individual consumers, businesses, or some other type of organization. For editorial convenience, we will sometimes use the term *firm* as a shorthand way of referring to any type of organization, whether it is a business, political party, a religious organization, a government agency, or the like. However, to reinforce the point that the ideas apply to all types of organizations, throughout the book we will illustrate marketing concepts in a wide variety of situations.

One of the challenges for many of today's consumer products companies involves addressing the needs of a growing market in developing countries. Makers of many health and beauty products, for example, have found a potential market with the rural poor. For an example of how effective micro-marketing can be used to appeal to this growing market, read *What's Next?* Marketing to developing countries' rural poor.

Although marketing within individual firms (micro-marketing) is the primary focus of the text, marketing managers must remember that their organizations are just small parts of a larger macro-marketing system. Therefore, next we will briefly look at the macro view of marketing. Then we will develop the managerial view more fully in later chapters.

What's Next? Marketing to developing countries' rural poor

In recent decades India has experienced rapid economic growth. Many of its citizens have more income and enjoy a higher quality of life. That helps explain why Unilever's Indian subsidiary, Hindustan Unilever Limited (HUL), has worked hard to build a 40 percent share of the Indian market with its product lines that include soaps, toothpaste, and packaged foods.

Previously, HUL focused on India's cities, where customers with money were concentrated. Yet, almost three-fourths of India's 1 *billion* plus people still live in rural areas. About a third of these rural villagers still lack access to electricity—and less than half have basic sanitation. Many of them have an income of less than $2 a day. Conventional wisdom suggests that these poor rural villagers have too little money to be an attractive market. And it's expensive to distribute products to far-flung villages.

But now that is changing. HUL's marketing managers decided that Indian villagers represent an opportunity for growth—and that villagers might benefit if they could purchase the soaps, toothpaste, and packaged food products that HUL is successfully selling in urban areas of India.

HUL tailored a new marketing strategy to this target market. Many products are repackaged in "sachets"—small bags that contain a one- or two-day supply. HUL prices the small sachets so that villagers can afford them—and that in turn gives customers a chance to try quality products that were previously priced out of their reach.

HUL created its "Shakti Ammas" (women entrepreneurs) program to communicate the benefits of its products and distribute them in remote rural areas. The program sets up rural women as home-based distributors and sales agents. These women stock HUL products at their homes and go door-to-door to sell them. They also organize meetings in local schools and at village fairs to educate fellow villagers on health and hygiene issues.

This program continues to evolve. To provide wider distribution, male entrepreneurs (Shaktimaan) were recruited and given bicycles. The bikes allow each Shaktimaan to cover five or six nearby villages—far more than the Shakti Ammas previously covered on foot. A partnership with a leading Indian telecom provider gives Shakti entrepreneurs an additional product to sell while on their rounds.

Today, more than 80,000 micro-entrepreneurs have their own businesses—operating in 162,000 villages and reaching over 4 million rural households across India. HUL's success in India spurred Unilever to adapt the model to developing countries around the globe. The Shakti have a new source of income and are learning about business—while they bring the health benefits of improved hygiene to rural villages. And, of course, HUL hopes to clean up with a new source of growth. To see and hear more about Project Shakti, check out this video: http://youtu.be/E7Hvp_CCtYY.

Through its marketing, HUL helps deliver a better quality of life to millions of India's poor. After seeing Unilever's success, many of its competitors developed similar programs around the world. Together, such changes create a better world for hundreds of millions of people.[6]

How do we see parts of the definition of marketing represented in this case study? What customer needs does the Shakti Ammas program address? How does it build a relationship between HUL and the customer?

Macro-Marketing

LO 1.2

Macro-marketing is a social process that directs an economy's flow of goods and services from producers to consumers in a way that effectively matches supply and demand and accomplishes the objectives of society.[7]

Emphasis is on whole system

With macro-marketing we are still concerned with the flow of need-satisfying goods and services from producer to consumer. However, the emphasis with macro-marketing is not on the activities of individual organizations. Instead, the emphasis is on *how the whole marketing system works*. This includes looking at how marketing affects society and vice versa.

Every society needs a macro-marketing system to help match supply and demand. Different producers in a society have different objectives, resources, and skills. Likewise, not all consumers share the same needs, preferences, and wealth. In other words, within every society there are both heterogeneous (highly varied) supply capabilities and heterogeneous demands for goods and services. The role of a macro-marketing system is to effectively match this heterogeneous supply and demand *and* at the same time accomplish society's objectives.

An effective macro-marketing system delivers the goods and services that consumers want and need. It gets products to them at the right time, in the right place, and at a price they're willing to pay. It keeps consumers satisfied after the sale and brings them back to purchase again when they are ready. That's not an easy job—especially if you think about the variety of goods and services a highly developed economy can produce and the many kinds of goods and services consumers want.

Separation between producers and consumers

Effective marketing in an advanced economy is difficult because producers and consumers are often separated in several ways. As Exhibit 1-1 shows, exchange between producers and consumers is hampered by *spatial separation; separation in time; separation of information; separation in values;* and *separation of ownership.* You may love your cell phone, but you probably don't know when or where it was produced or how it got to you. The people in the factory that produced it don't know about you or how you live. The producer knows it wants to make that phone at a low cost and isn't sure what features and benefits you are seeking.

In addition, most firms specialize in producing and selling large amounts of a narrow assortment of goods and services. This allows them to take advantage of mass production with its **economies of scale**—which means that as a company produces larger numbers of a particular product, the cost of each unit of the product goes down. Yet most

Exhibit 1-1 Marketing Facilitates Production and Consumption

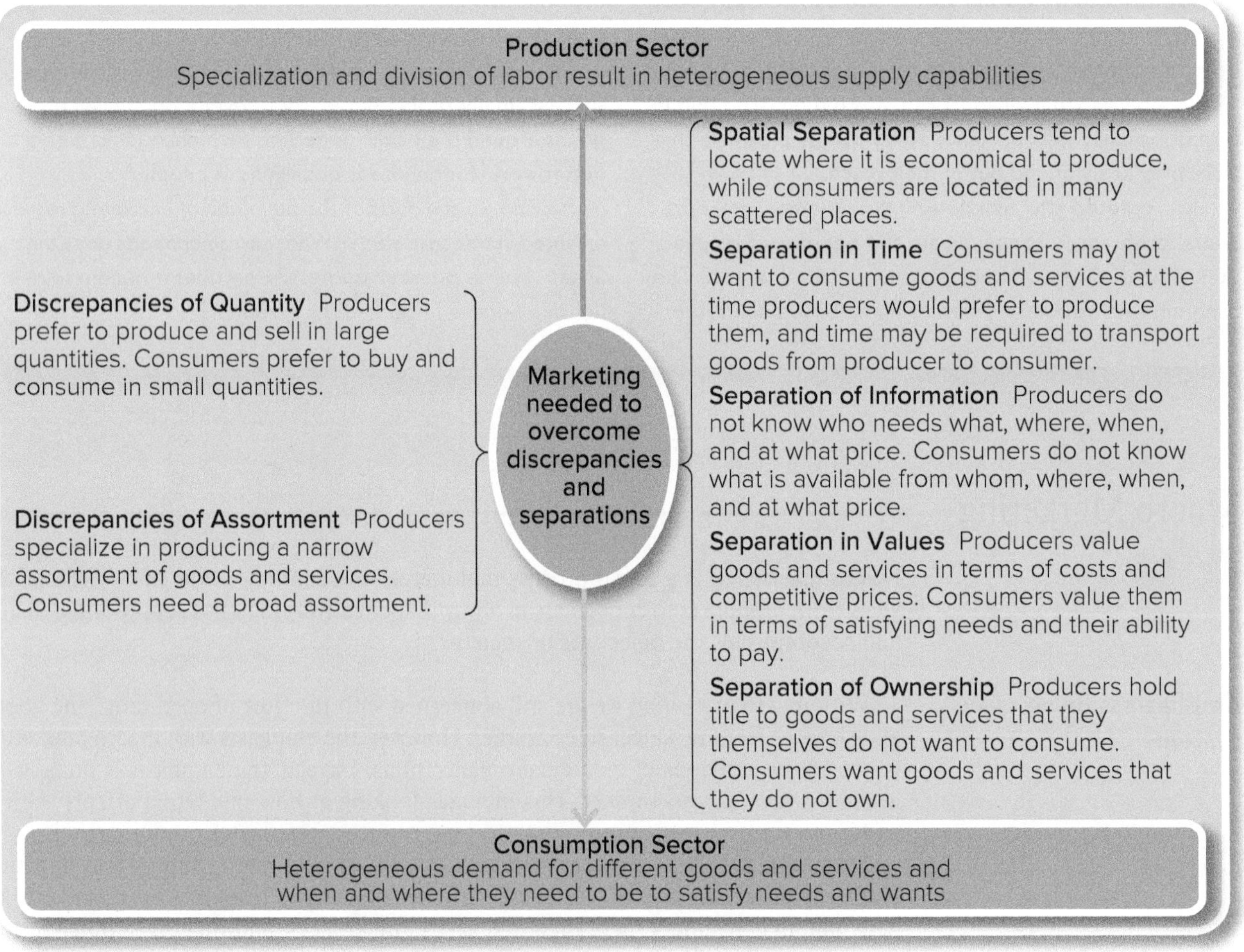

Production Sector
Specialization and division of labor result in heterogeneous supply capabilities

Discrepancies of Quantity Producers prefer to produce and sell in large quantities. Consumers prefer to buy and consume in small quantities.

Discrepancies of Assortment Producers specialize in producing a narrow assortment of goods and services. Consumers need a broad assortment.

Marketing needed to overcome discrepancies and separations

Spatial Separation Producers tend to locate where it is economical to produce, while consumers are located in many scattered places.

Separation in Time Consumers may not want to consume goods and services at the time producers would prefer to produce them, and time may be required to transport goods from producer to consumer.

Separation of Information Producers do not know who needs what, where, when, and at what price. Consumers do not know what is available from whom, where, when, and at what price.

Separation in Values Producers value goods and services in terms of costs and competitive prices. Consumers value them in terms of satisfying needs and their ability to pay.

Separation of Ownership Producers hold title to goods and services that they themselves do not want to consume. Consumers want goods and services that they do not own.

Consumption Sector
Heterogeneous demand for different goods and services and when and where they need to be to satisfy needs and wants

consumers want to buy only a small quantity; they also want a wide assortment of different goods and services. Apple makes millions of iPhones in a few factories in China, but most customers want to buy just one. And most smartphone buyers like to go to a store with several different phones from different companies. These "discrepancies of quantity" and "discrepancies of assortment" further complicate exchange between producers and consumers (Exhibit 1-1). That is, each producer specializes in producing and selling large amounts of a narrow assortment of goods and services, but each consumer wants only small quantities of a wide assortment of goods and services.[8]

The purpose of a macro-marketing system is to overcome these separations and discrepancies. The "universal functions of marketing" help solve these problems.

Marketing functions help narrow the gap

LO 1.3

The **universal functions of marketing** are buying, selling, transporting, storing, standardization and grading, financing, risk taking, and market information. They must be performed in all macro-marketing systems. *How* these functions are performed—and *by whom*—may differ among nations and economic systems. But they are needed in any macro-marketing system. Let's take a closer look at them now.

Any kind of exchange usually involves buying and selling. The **buying function** means looking for and evaluating goods and services. The **selling function** involves promoting the product. It includes the use of personal selling, advertising, customer service, and other direct and mass-selling methods to tell customers about the product. This is probably the most visible function of marketing.

The **transporting function** means the movement of goods from one place to another. The **storing function** involves holding goods until customers need them.

Standardization and grading involve sorting products according to size and quality. This makes buying and selling easier because it reduces the need for inspection and sampling. **Financing** provides the necessary cash and credit to produce, transport, store, promote, sell, and buy products. **Risk taking** involves bearing the uncertainties that are part of the marketing process. A firm can never be sure that customers will want to buy its products. Products can also be damaged, stolen, or outdated. The **market information function** involves the collection, analysis, and distribution of all the information needed to plan, carry out, and control marketing activities, whether in the firm's own neighborhood or in a market overseas. Together these universal functions of marketing address the discrepancies and separations in Exhibit 1-1.

Acai berries are popular in the United States as a supplement and as an ingredient in juices and smoothies. A range of marketing functions are needed to overcome the spatial separation between the Central and South American farms, where the fruit is harvested from acai palms, and U.S. consumers. Consider the process: standardizing and grading the berries, transporting and storing the fruit and its juice, financing production of the acai end products, as well as the buying and selling functions—all are necessary to bring a tasty fruit smoothie to a thirsty lady. *(top-left): Brasil2/iStock/Getty Images; (bottom-left): Andre Penner/AP Images; (right): zjuzjaka/Shutterstock*

Producers, consumers, and marketing specialists perform functions

Producers and consumers sometimes handle some of the marketing functions themselves. However, exchanges are often easier or less expensive when a marketing specialist performs some of the marketing functions. For example, both producers and consumers may benefit when an **intermediary**—someone who specializes in trade rather than production—plays a role in the exchange process. In Chapters 10, 11, and 12 we'll cover the variety of marketing functions performed by the two basic types of intermediaries: retailers and wholesalers. Imagine what it would be like to shop at many different factories and farms for the wide variety of brands of packaged foods that you like rather than at a well-stocked local grocery store. Although wholesalers and retailers must charge for services they provide, this charge is usually offset by the savings of time, effort, and expense that would be involved without them. So these intermediaries can help make the whole macro-marketing system more efficient and effective.

A wide variety of other marketing specialists may also help smooth exchanges among producers, consumers, or intermediaries. These specialists are **collaborators**—firms that facilitate or provide one or more of the marketing functions other than buying. These collaborators include advertising agencies, marketing research firms, independent product-testing laboratories, Internet service providers, public warehouses, transporting firms, communications companies, and financial institutions (including banks). Walmart and Google recently partnered to allow customers to shop by voice with Google Assistant. The collaboration helps both parties advance their marketing goals.

Functions can be shifted and shared

From a macro-marketing viewpoint, all of the marketing functions must be performed by someone—an individual producer or consumer, an intermediary, a marketing collaborator, or, in some cases, even a nation's government. No function can be completely eliminated. However, from a micro viewpoint, not every firm must perform all of the functions. Rather, responsibility for performing the marketing functions can be shifted and shared in a variety of ways. Further, not all goods and services require all the functions at every level of their production. "Pure services"—like a plane ride—don't need storing, for example. But storing is required in the production of the plane and while the plane is not in service.

Intermediaries and collaborators develop and offer specialized services that facilitate exchange between producers and consumers. Instagram offers services that help a firm create advertising campaigns. Instagram is a collaborator; it facilitates its clients' selling function. Companies that produce building supplies work with an intermediary such as Home Depot to better serve the electricians, plumbers, and carpenters that buy and use its products.
Sources: (left) Instagram: The Rocket Science Group; (right) Home Depot: Homer TLC, Inc.

Regardless of who performs the marketing functions, in general they must be performed effectively and efficiently or the performance of the whole macro-marketing system will suffer. With many different possible ways for marketing functions to be performed in a macro-marketing system, how can a society hope to arrive at a combination that best serves the needs of its citizens? To answer this question, we can look at the role of marketing in different types of economic systems.

The Role of Marketing in Economic Systems

LO 1.4

All societies must provide for the needs of their members. Therefore, every society needs some sort of **economic system**—the way an economy organizes to use scarce resources to produce goods and services and distribute them for consumption by various people and groups in the society.

How an economic system operates depends on a society's objectives and the nature of its political institutions.[9] But regardless of what form these take, all economic systems must develop some method—along with appropriate economic institutions—to decide what and how much is to be produced and distributed by whom, when, to whom, and why.

Government officials may make the decisions

Forty or fifty years ago, many countries' economies—including China's and Russia's—were directed by government officials. In a **command economy**, government officials decide what and how much is to be produced and distributed by whom, when, to whom, and why. These decisions are usually part of an overall government plan, so command economies are also called "planned" economies. It sounds good for a government to have a plan, but as a practical matter, attempts by a government to dictate an economic plan often don't work out as intended. Countries with command economies often had little variety, so consumers had few choices. And because the government struggled to forecast what consumers would want, there were often shortages of goods and services. As a result, consumers were often dissatisfied with the systems. Over time, most planned economies transitioned to a system that left decision making to customers and producers.

A market-directed economy adjusts itself

In a **market-directed economy**, the individual decisions of the many producers and consumers make the macro-level decisions for the whole economy. In a pure

Artists and craftworkers in developing economies do not often have a large local market for their products. Ten Thousand Villages operates retail stores and an e-commerce website that help artisans reach more customers and promote economic development in their communities.
Source: Ten Thousand Villages

market-directed economy, consumers make a society's production decisions when they make their choices in the marketplace. They decide what is to be produced and by whom—through their dollar "votes."

In a market-directed economy, consumers choose from a range of options that may meet their needs. They are not forced to buy any goods or services, except those judged as good for society—things such as national defense, police and fire protection, roads, public schools, and public health services. These are provided by the community—and citizens' taxes pay for them.

Similarly, producers are free to do whatever they wish—provided that they stay within the rules of the game set by government *and* receive enough dollar "votes" from consumers. If they do their job well, they earn a profit and stay in business. But profit, survival, and growth are not guaranteed.

In a market-directed economy, the system is governed by the words and actions of consumers and citizens, with the government setting the boundaries.

Prices customers pay determine a measure of value

Prices in the marketplace are a rough measure of how society values particular goods and services. If consumers are willing to pay the market prices, then they are getting at least their money's worth. If prices and profits are high, then consumers either stop buying or competitors enter the market. Similarly, the cost of labor and materials is a rough measure of the value of the resources used in the production of goods and services to meet these needs. New consumer needs that can be served profitably—not just the needs of the majority—will probably be met by some profit-minded businesses.

Customers use "voices" to gain power in the market

The rise of digital and social media gives customers more power in the market. These days if customers feel they have been wronged by a company, they can share their dissatisfaction with more than just their immediate friends and family. A review on a website or a social media post may be read by hundreds or thousands of other people.

When musician Dave Carroll's guitar was damaged by United Airlines—and baggage handlers and the airline refused to compensate him—Carroll's band made a YouTube video. The funny video, "United Breaks Guitars," went viral, millions saw it, and lo and behold, United offered Carroll a new guitar.

Not every customer will make a video of a bad experience, but many will make social media posts—or "like" or pass along someone else's story. Prospective customers often search the web to learn about companies before buying. They can easily find positive and negative information on the Internet, giving firms an extra incentive to play by the socially accepted rules of the game.

Public interest groups

In many Western economies, public interest groups and consumers provide an additional check on a market-directed economy. For example, the Center for Science in the Public Interest (CSPI) is a consumer watchdog group that pressures food companies to make healthier products. CSPI regularly issues reports on unhealthy foods like the high fat content in movie theater popcorn and Chinese food. The consulting firm UL TerraChoice regularly exposes firms that make false or exaggerated environmental claims about their products.

Government enforces the "rules of the game"

The American economy and most other Western economies are mainly market-directed—but not completely. Society assigns supervision of the system to the government. For example, besides setting and enforcing the "rules of the game," government agencies control interest rates and the supply of money. They also set import and export rules that affect international competition, regulate radio and TV broadcasting, sometimes control wages and prices, and so on. Government also tries to be sure that property is protected, contracts are enforced, individuals are not exploited, no group unfairly monopolizes markets, advertising is honest, and producers deliver the kinds and quality of goods and services they claim to be offering.

Exhibit 1-2 Model of a Market-Directed Macro-Marketing System

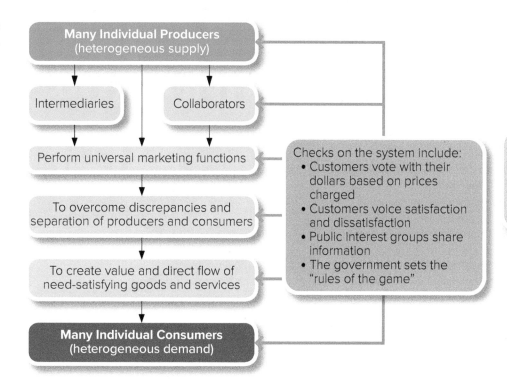

Macro-marketing systems adjust over time

So far, we have described how a market-directed macro-marketing system adjusts to become more effective and efficient by responding to customers and public interest groups and staying within government regulations (see Exhibit 1-2). As you read this book, you'll learn more about how marketing affects society and vice versa. You'll also learn more about specific marketing activities and be better informed when drawing conclusions about how fair and effective the macro-marketing system is. For now, however, we'll return to our general emphasis on a managerial view of the role of marketing in individual organizations.

Marketing's Role Has Changed a Lot over the Years

It's clear that marketing decisions are very important to a firm's success. But marketing hasn't always been so complicated. In fact, understanding how marketing thinking has evolved makes the modern view clearer, so we will discuss five stages in marketing evolution: (1) the simple trade era, (2) the production era, (3) the sales era, (4) the marketing department era, and (5) the marketing company era. We'll talk about these eras as if they applied generally to all firms—but keep in mind that *some managers still have not made it to the final stages*. They are stuck in the past with old ways of thinking.

Simple trade era to production era

When societies first moved toward some specialization of production and away from a subsistence economy where each family raised and consumed everything it produced, traders played an important role. Early "producers for the market" made products that were needed by themselves and their neighbors. As bartering became more difficult, societies moved into the **simple trade era**—a time when families traded or sold their "surplus" output to local distributors. These specialists resold the goods to other consumers or other distributors. This was the early role of marketing—and it is still the focus of marketing in many less-developed areas of the world. In fact, even in the United States, the United Kingdom, and other more advanced economies, marketing didn't change much until the Industrial Revolution brought larger factories more than 150 years ago.

In the past, many banks created products but left it to the marketing department to sell them. In today's more competitive environment, banks conduct research to discover customer needs, then develop products that meet those needs. This leads to more satisfied customers. When Bank of the Wichitas added an online branch, it decided to use an untraditional approach and called it Redneck Bank. Redneck Bank offers customers "flat out free checking" and convenience served with a healthy dollop of humor. The approach has been successful, and the small bank from Oklahoma quickly added customers from all over the United States.
Source: Bank of the Wichitas®

From the production to the sales era

From the Industrial Revolution until the 1920s, many companies were in the production era—some still are today. The **production era** is a time when a company focuses on production of a few specific products—perhaps because few of these products are available in the market. "If we can make it, it will sell" is management thinking characteristic of the production era. Because of product shortages, many nations—including some of the postcommunist republics of eastern Europe—continue to operate with production era approaches.

By about 1930, most companies in the industrialized Western nations had more production capability than ever before. Now the problem wasn't just to produce—but to beat the competition and win customers. This led many firms to enter the sales era. The **sales era** is a time when a company emphasizes selling because of increased competition.

To the marketing department era

For most firms in advanced economies, the sales era continued until at least 1950. By then, sales were growing rapidly in most areas of the economy. The problem was deciding where to put the company's effort. Someone was needed to tie together the efforts of research, purchasing, production, shipping, and sales. As this situation became more common, the sales era was replaced by the marketing department era. The **marketing department era** is a time when all marketing activities are brought under the control of one department to improve short-run policy planning and to try to integrate the firm's activities.

To the marketing company era

Since 1960, most firms have developed at least some managers with a marketing management outlook. Many of these firms have even graduated from the marketing department era into the marketing company era. The **marketing company era** is a time when, in addition to short-run marketing planning, marketing people develop long-range plans—sometimes five or more years ahead—and the whole-company effort is guided by the marketing concept.

What Does the Marketing Concept Mean?

LO 1.5

The **marketing concept** means that an organization aims *all* of its efforts at satisfying its *customers*—at a *profit*. The marketing concept is a simple but very important idea.

The marketing concept is not a new idea—it has been around for a long time. But some managers show little interest in customers' needs. These managers still have a **production orientation**—making whatever products are easy to produce and *then* trying to sell them. They think of customers existing to buy the firm's output rather than of firms existing to serve customers and—more broadly—the needs of society.[10]

Well-managed firms have replaced this production orientation with a marketing orientation. A **marketing orientation** means trying to carry out the marketing concept. Instead of just trying to get customers to buy what the firm has produced, a marketing-oriented firm tries to offer customers what they need.

Three basic ideas are included in the definition of the marketing concept: (1) customer satisfaction, (2) a total company effort, and (3) profit (and/or other measures of long-term success)—not just sales—as an objective. See Exhibit 1-3. These ideas warrant further discussion.

Customer satisfaction guides the whole system

"Give the customers what they need" seems so obvious that it may be hard for you to see why the marketing concept requires special attention. However, people don't always do the logical thing—especially when it means changing what they've done in the past. When customers have choices, they choose companies that best meet their needs, desires, and expectations. They purchase again from these companies, and they tell their friends about the great experience. This all starts with customer satisfaction.

The whole company works together to satisfy customers

Ideally, all managers should work together as a team. Every department may directly or indirectly impact customer satisfaction. But some managers tend to build "fences" around their own departments. There may be meetings to try to get them to work together—but they come and go from the meetings worried only about protecting their own turf.

Exhibit 1-3
Organizations with a Marketing Orientation Carry Out the Marketing Concept

We use the term *production orientation* as a shorthand way to refer to this kind of narrow thinking—and lack of a central focus—in a business firm. But keep in mind that this problem may be seen in sales-oriented sales representatives, advertising-oriented agency people, finance-oriented finance people, directors of nonprofit organizations, and so on. It is not a criticism of people who manage production. They aren't necessarily any more guilty of narrow thinking than anyone else.

The fences come down in an organization that has accepted the marketing concept. There may still be departments because specialization often makes sense. But the total system's effort is guided by satisfying customers' needs and wants—instead of what each department would like to do. The marketing concept provides a guiding focus that *all* departments adopt. It must be a philosophy of the whole organization, not just an idea that applies to the marketing department.

Profit, revenues, and costs are examples of **marketing metrics** which refer to numeric data that allow marketing managers to evaluate performance, often against a set target or goal.[11] At a basic level, marketing managers use these metrics to calculate how well a strategy performs. For example, consider a student who wants to sell T-shirts on campus to support a student organization. The student could estimate profits in advance and then evaluate the profits later. The Marketing Analytics in Action: Revenue, Cost, and Profit exercise describes this in more detail.

Survival and success require a profit

Firms must satisfy customers. But keep in mind that it may cost more to satisfy some needs than any customers are willing to pay. Or it may be much more costly to try to attract new customers than it is to build a strong relationship with—and encourage repeat purchases from—existing customers. So profit—the difference between a firm's revenue and its total costs—is the bottom-line measure of the firm's success and ability to survive. It is the balancing point that helps the firm determine what needs it will try to satisfy with its total (sometimes costly!) effort.

Some organizations have other measures of long-term success. Most nonprofit organizations need revenues to exceed costs to survive long term—but they may have other goals valued by employees and supporters. For example, a food bank may measure success by the number of meals it serves to the homeless. Many for-profit organizations consider more than profit; for example, many are concerned with the environment or their community.

Marketing Analytics in Action: Revenue, Cost, and Profit

Marketing managers pay close attention to profit—that also means monitoring revenues and costs. And whereas marketing managers want to know a company's overall profits, they are also likely to examine revenues, costs, and profits for individual products, perhaps in different geographic markets. This suggests some simple metrics that can be calculated with the following formulas:

- Revenue = Price × Quantity sold
- Profit = Revenue − Cost

To better understand how a marketing manager might use these numbers, let's look at a simple example. Julie Tyler is a college student in charge of fundraising for a student organization. Last year the group successfully sold T-shirts to raise money and awareness for the group. They made 250 T-shirts for $12 each and sold them at a price of $20 per T-shirt. This year Julie wants to buy a higher-quality shirt that will cost $15, and she believes she can sell them for $30. However, at the higher price she expects to sell only 200 shirts.

1. *Calculate the profits the group earned last year.*
2. *Calculate the profits Julie would earn this year using these prices and assuming her sales projections are accurate.*
3. *What else could Julie do to increase profits?*

Many organizations go beyond profit

Even many for-profit businesses have measures that go beyond profit(s) alone. Many organizations explicitly consider a **triple bottom line**—which measures an organization's economic, social, and environmental outcomes—as a measure of long-term success. *Profit* is the economic outcome. *Social* refers to how the company's business activities affect its employees and other people in the communities where it operates. The third bottom line takes into account *environmental responsibility,* usually seeking to at least not harm the natural environment. Together, these are sometimes referred to as measures of people, planet, and profit.[12]

Consider Namaste Solar, which actively seeks economic profits while also measuring the impact on its local community and the planet. For example, the company gives 20 percent of its after-tax earnings to support projects such as a community bicycle-recycling program or the local children's museum. Namaste Solar's core products help the planet, but the company also offers grants of up to $30,000 to help schools and nonprofits install their own solar systems. The company also self-monitors and reduces its waste. It recycles or reuses 90 percent of its office supplies and employs a zero waste company kitchen. Many of Namaste Solar's customers and employees value its triple bottom line orientation.[13]

Adoption of the marketing concept is not universal

The marketing concept may seem obvious, but it is easy to maintain a production-oriented way of thinking. Producers of industrial commodities such as steel, coal, and chemicals have tended to remain production-oriented in part because buyers see little difference among competitors. Some industries with limited competition, including electric utilities and cable television providers, have also been slow to adopt the marketing concept. When an industry gets competitive, consumers have choices and flock to those that deliver customer satisfaction. This provides an incentive for more firms to practice the marketing concept.[14]

Take a look at Exhibit 1-4. It shows some differences in outlook between adopters of the marketing concept and typical production-oriented managers. As the exhibit suggests, the marketing concept forces the company to think through what it is doing—and why.

Exhibit 1-4 Some Differences in Outlook between Adopters of the Marketing Concept and Typical Production-Oriented Managers

Topic	Marketing Orientation	Production Orientation
Attitudes toward customers	Customer needs determine company plans	Customers should be glad we exist, trying to cut costs and bringing out better products
Product offering	Company makes what it can sell	Company sells what it can make
Role of marketing research	To determine customer needs and how well company is satisfying them	To determine customer reaction, if used at all
Interest in innovation	Focus is on locating new opportunities	Focus is on technology and cost cutting
Customer service	Satisfy customers after the sale and they'll come back again	An activity required to reduce consumer complaints
Focus of advertising	Need-satisfying benefits of goods and services	Product features and how products are made
Relationship with customer	Customer satisfaction before and after sale leads to a profitable long-run relationship	Relationship ends when a sale is made
Costs	Eliminate costs that do not give value to customer	Keep costs as low as possible

The Marketing Concept and Customer Value

LO 1.6

Take the customer's point of view

A manager who adopts the marketing concept sees customer satisfaction as the path to profits. To better understand what it takes to satisfy a customer, it is useful to take the customer's point of view.

A customer may look at a market offering from two perspectives. One deals with the potential benefits of that offering; the other concerns what the customer has to give up to get those benefits. Consider a student who has just finished an exam and is thinking about getting an iced coffee from Starbucks. Our coffee lover might see this as a great-tasting treat, a quick pick-me-up, a quiet place to relax and meet friends, or as a way to get to know an attractive classmate. Clearly, different needs are associated with these different benefits. The cost of getting these benefits would include the price of the coffee and any tip to the server, but there might be other nonmonetary costs. For example, how difficult it is to find parking is a convenience cost. Slow service could be an aggravation.

Customer value reflects benefits and costs

As this example suggests, both benefits and costs can take many different forms. They also may vary depending on the situation. However, it is the *customer's view* of the various benefits and costs that is important. We want to repeat this important distinction: it is the *customer's view* of costs and benefits that is important. A company may want to offer a more fun atmosphere at its coffee shop—but whether it succeeds depends on the customer's perception.

This leads us to the concept of **customer value**—the difference between the benefits a customer sees from a market offering and the costs of obtaining those benefits. Exhibit 1-5 shows the wide range of different types of benefits; for example *functional benefits* could save a customer time or effort, *emotional benefits* might deliver fun and entertainment or design and aesthetics, and *life-changing benefits* could increase motivation or feelings of belonging.[15]

Exhibit 1-5 also recognizes costs. Some people think that higher customer value comes from a low price. But that may not be the case at all. A good or service that doesn't meet a customer's needs results in low customer value, even if the price is very low. A high price may be more than acceptable when it obtains the desired benefits. Think again about our Starbucks example. You can get a cup of coffee for a much lower price, but Starbucks offers more than just a cup of coffee.

Exhibit 1-5 Customer Value Equals Benefits Minus Costs

Customer Value = Benefits − Costs

Examples of different types of benefits:

Functional
- Save time
- Simplify
- Provide information
- Reduce cost

Emotional
- Provide fun/entertainment
- Lower anxiety
- Offer superior design/aesthetics
- Provide rewards in some form

Life-changing
- Give hope
- Offer motivation
- Provide sense of affiliation/belonging
- Support an organization that makes the world a better place

Examples of different types of costs:

Monetary
- Money
- Interest rate
- Fees

Inconvenience
- Time delay to receive the benefit
- Effort required to receive benefit

Our definition of customer value might help you start thinking about how Starbucks could offer more customer value. Starbucks probably needs to start by providing a great-tasting cup of coffee—but what else can increase customer value? Although Starbucks could easily increase value by lowering prices, it prefers to seek out ways to increase benefits. Do some Starbucks customers experience benefits when a drive-through window saves them time in the morning? Does the Starbucks app, with the option of preordering your coffee and skipping the line, save a customer time and reduce effort? Does the Starbucks Rewards program, which leads to free cups of coffee, deliver benefits? Does Starbucks' commitment to ethically sourced coffee make a customer *feel* better? Any of these potential value-enhancing activities depends on Starbucks' customers' interpretation of their value.

Customers may not think about it very much

It is useful for a manager to evaluate ways to improve the benefits, or reduce the costs, of what the firm offers customers. However, this doesn't mean that customers stop and compute some sort of customer value score before making each purchase. If they did, there wouldn't be much time in life for anything else. So a manager's objective and thorough analysis may not accurately reflect the customer's impressions. Yet it remains the customer's view that matters—even when the customer has not thought about it.

Where does competition fit?

You can't afford to ignore competition. Consumers usually have choices about how they will meet their needs. If a firm hopes to win and keep customers over the long haul, that firm must offer a level of customer value greater than that offered by competitors. Consumers have choices—and they tend to choose options that deliver the most value.

Companies recognize that an offering with high value can change if the competition offers something better. Competition drives innovation and continuous improvements in most markets. When Uber entered the "taxi" market with a more convenient way to hail a ride, many taxi companies developed their own apps and improved service. After Tesla introduced its $50,000 Model S60 electric car with a range of 208 miles per charge, General Motors (GM) upped its game to bring out the $30,000 Chevrolet Bolt with a 238-mile range. Of course, there are other benefits to car ownership, and Tesla hopes its newest model delivers what some customers see as a better value than the Bolt.[16]

Build relationships with customer value

Firms that embrace the marketing concept seek ways to build a profitable long-term relationship with each customer. Even the most innovative firm faces competition sooner or later. Trying to get new customers by taking them away from a competitor is usually more costly than retaining current customers by really satisfying their needs. Satisfied customers buy again and again, whereas dissatisfied customers often tell others not to buy. With a long-term relationship, the customer's buying job is easier, and it also increases the selling firm's profits.

Building relationships with customers requires that everyone in a firm work together to provide customer value before *and after* each purchase. If there is a problem with a customer's bill, the accounting people can't just leave it to the salesperson to straighten it out or, even worse, act like it's "the customer's problem." These hassles raise customers' costs of doing business. The long-term relationship with the customer—and the lifetime value of the customer's future purchases—is threatened unless everyone works together to make things right for the customer. Similarly, the firm's advertising might encourage a customer to buy once, but if the firm doesn't deliver on the benefits promised in its ads, the customer is likely to go elsewhere the next time the need arises. In other words, anytime the customer value is reduced—because the benefits to the customer decrease or the costs increase—the relationship is weakened.[17]

Exhibit 1-6 Satisfying Customers with Superior Customer Value to Build Profitable Relationships

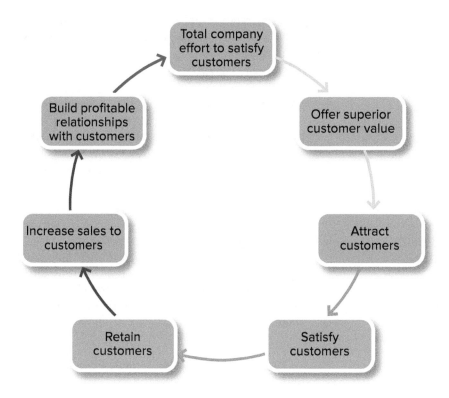

Exhibit 1-6 summarizes these ideas. In a firm that has adopted the marketing concept, everyone focuses on customer satisfaction. They look for ways to offer superior customer value. That helps attract customers in the first place—and keeps them satisfied after they buy. So when they are ready to make repeat purchases, the firm is able to keep them as customers. Sales may increase further because satisfied customers are likely to buy other products offered by the firm. In this way, the firm builds profitable relationships with its customers. In other words, when a firm adopts the marketing concept, it wins and so do its customers.

REI's value delivers satisfied customers—again and again

One company that illustrates these ideas is the retail store Recreational Equipment, Inc., better known as REI. REI specializes in outdoor recreation, including camping supplies, sporting goods, travel gear, and clothing. The company has built enduring relationships with customers who shop at its more than 140 stores or its website. REI customers see superior value in the company's knowledgeable store staff and large selection of high-quality outdoor gear. The company attracts customers who learn about it from already satisfied customers, who see its stores in their town, or who hear about its clever promotions—like when it tells its customers it is closing on Black Friday (one of the biggest shopping days of the year) because everyone should go outside and enjoy nature.

Customers typically enjoy their shopping experiences at REI—and they become satisfied and repeat customers after using the outdoor equipment they purchase. REI's outdoorsy customers also appreciate that REI leads hundreds of outdoor companies in defending public lands. REI is operated as a co-op and is owned by its six million active members, who become lifetime members by paying $20. This sense of community also helps REI to retain customers. Members get special members-only coupons and an annual 10 percent rebate on all purchases. These extras enhance customer value and keep customers coming back. And when these loyal customers buy more next year, REI becomes even more profitable. Everyone at REI—from buyers who choose the merchandise it sells to

Source: Recreational Equipment, Inc.

associates offering advice on the sales floor—focuses on increasing customer satisfaction. This virtuous cycle (see Exhibit 1-6) has helped REI become one of the hottest brands in retail.[18]

The Marketing Concept Applies in Nonprofit Organizations

Newcomers to marketing thinking

The marketing concept is as important for nonprofit organizations as it is for business firms. In fact, marketing applies to all sorts of public and private nonprofit organizations—ranging from government agencies, health care organizations, educational institutions, and religious groups to charities, political parties, and fine arts organizations.

Support may not come from satisfied "customers"

As with any business firm, a nonprofit organization needs resources and support to survive and achieve its objectives. Yet support often does not come directly from those who receive the benefits the organization produces. For example, the World Wildlife Fund protects animals. If supporters of the World Wildlife Fund are not satisfied with its efforts—if they don't think the benefits are worth what it costs to provide them—they will put their time and money elsewhere.

Just as most firms face competition for customers, most nonprofits face competition for the resources and support they need. The Air Force faces a big problem if it can't attract new recruits. A shelter for homeless individuals may fail if supporters decide to focus on some other cause, such as AIDS education.

What is the "bottom line"?

As with a business, a nonprofit must take in as much money as it spends or it won't survive. However, a nonprofit organization does not measure "profit" in the same way as a firm. And its key measures of long-term success are also different. The YMCA, colleges, symphony orchestras, and the United Way, for example, all seek to achieve different objectives and need different measures of success. When everyone in an organization agrees to *some* measure(s) of long-run success, it helps the organization focus its efforts. Most nonprofit organizations have defined some way they hope to make the world a better place.

May not be organized for marketing

Some nonprofits face other challenges in organizing to adopt the marketing concept. Often no one has overall responsibility for marketing activities. Even when some leaders do the marketing thinking, they may have trouble getting unpaid volunteers with many different interests to all agree with the marketing strategy. Volunteers tend to do what they feel like doing![19]

#M4BW Although nonprofits are newcomers to marketing, some are finding creative approaches to raise money. The United Nations' World Food Programme (WFP) aims to fight hunger. WFP used marketing to reduce the number of hungry children. It developed a smartphone app that allows people to "share the meal" with hungry children. Each time they "tap" the gold medallion, they donate $0.50, which feeds one child for a day. The app has led to sharing of more than 10 million meals, with many more shared each day.
Source: World Food Program; (smartphone frame): Oleg GawriloFF/Shutterstock

The Marketing Concept, Social Responsibility, and Marketing Ethics

LO 1.7

Society's needs must be considered

The marketing concept is so logical that it's hard to find fault with it. Yet when a firm focuses its efforts on satisfying some consumers—to achieve its objectives—there may be negative effects on society. For example, producers and consumers making free choices can cause conflicts and difficulties. This is called the **micro-macro dilemma**. What is "good" for some firms and consumers may not be good for society as a whole.

For instance, many people in New York City buy bottled water because they like the convenience of easy-to-carry disposable bottles with spill-proof caps. On the other hand, the city already provides citizens with good-tasting, safe tap water at a fraction of the cost. Is this just a matter of free choice by consumers? It's certainly a popular choice! Yet, critics point out that it is an inefficient use of resources to waste oil making and transporting millions of plastic bottles that end up in landfills where they leach chemicals into the soil. That kind of thinking, about the good of society as a whole, explains why New York City has run ads that encourage consumers to "get your fill" of free city water. What do you think? Should future generations pay the environmental price for today's consumer conveniences?[20]

Questions like these are not easy to answer. The basic reason is that many different people may have a stake in the outcomes—and social consequences—of the choices made by individual managers *and* consumers in a market-directed system. This means that marketing managers should be concerned with **social responsibility**—a firm's obligation to improve its positive effects on society and reduce its negative effects. As you read this book and learn more about marketing, you will also learn more about social responsibility in marketing—and why it must be taken seriously. You'll also see that being socially responsible sometimes requires difficult trade-offs.

Social responsibility can be difficult to assess when the effects on society are mixed. Take hydraulic fracturing (fracking), a process that sends pressurized liquid deep into the ground to remove oil and natural gas. Some argue that fracking helps America become more energy independent, produces cleaner fuel compared to alternatives, and lowers gas prices, thus stimulating the economy. On the other hand, the chemicals and procedures used in fracking may harm the environment and cause earthquakes. The government and many oil and gas producers are working to minimize the negative effects, but as you can imagine there are no easy answers for these conflicts.[21]

The issue of social responsibility in marketing also raises other important questions—for which there are no easy answers.

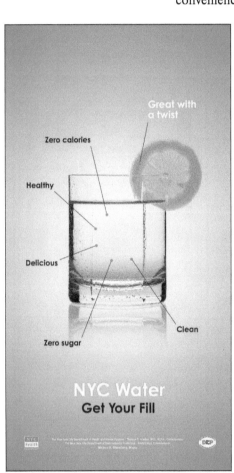

Source: New York City Department of Health/DCF Advertising

Should all consumer needs be satisfied?

Some consumers want products that may not be safe or good for them in the long run. Some critics argue that businesses should not offer high-heeled shoes, alcoholic beverages, marijuana, gambling, or sugar-coated cereals because they aren't "good" for some consumers in the long run. Similarly, bicycles and roller blades are among the most dangerous products identified by the Consumer Product Safety Commission. Who should decide if these products will be offered to consumers?

What if it cuts into profits?

There are times when being socially responsible can increase not only a firm's profits, but also its costs. Even though some consumers will pay premium prices to buy "fair trade" coffee, napkins made from recycled paper, hybrid cars that pollute less, or products made in overseas factories that pay a fair wage, these items may cost more to produce. At the same time, other consumers attach little or no value to these social measures and refuse to pay a higher price for such products. Consequently, many firms selling to this second group are reluctant to be more socially responsible.

When a society feels that the social benefits are important, it may add regulations to create a level playing field for all firms and to ensure that these benefits are provided. For example, there are laws that protect rivers from water pollution and that restrict the use of child labor. Still, it is difficult for governments to impose regulations that govern all circumstances where such trade-offs occur. So is the marketing concept really desirable?

Socially conscious marketing managers are trying to answer these questions. Their definition of customer satisfaction includes long-range effects—as well as immediate customer satisfaction. They try to balance consumer, company, and social interests.

The marketing concept guides marketing ethics

A manager cannot be truly consumer-oriented and at the same time intentionally unethical. However, at times, problems and criticism may arise because a manager did not fully consider the ethical implications of a decision. In either case, there is no excuse for sloppiness when it comes to **marketing ethics**—the moral standards that guide marketing decisions and actions. Each individual develops moral

#M4BW Ben & Jerry's runs its business with a very clear set of values. These values guide the ice cream maker's employees, including its marketing managers' ethics—the moral standards that guide marketing decisions and actions.
Source: Ben & Jerry's United States, Inc.

standards based on his or her own values. That helps explain why opinions about what is right or wrong often vary from one person to another, from one society to another, and among different groups within a society. It is sometimes difficult to say whose opinions are "correct." Even so, such opinions may have a very real influence on whether an individual's (or a firm's) marketing decisions and actions are accepted or rejected. So marketing ethics are not only a philosophical issue; they are also a pragmatic concern.

Problems may arise when one or a few individual managers don't share the same marketing ethics as others in the organization. A small number of people—or even one person operating alone—can damage a firm's reputation and even survival. Consider the scandal faced by Volkswagen when it came to light that the company had found a way to "cheat" on emissions tests. When the company could not find a way to meet emissions reduction targets in its diesel vehicles, a few supervisors approved development of software designed to "cheat" emission testing systems. For years, about 11 million Volkswagen cars spewed as much as 40 times the allowable amount of pollutants. After this was discovered, Volkswagen paid out more than $23 billion in fines and fixes.[22]

To be certain that standards for marketing ethics are as clear as possible, many organizations have developed their own written codes of ethics. These codes usually state—at least at a general level—the ethical standards that everyone in the firm should follow in dealing with customers and other people. Many professional societies also have such codes. For example, the American Marketing Association's code of ethics—see Exhibit 1-7—sets specific ethical standards for many aspects of marketing.[23]

Exhibit 1-7 Summary of American Marketing Association Statement of Ethics

Preamble

The American Marketing Association commits itself to promoting the highest standard of professional ethical norms and values for its members (practitioners, academics, and students). Norms are established standards of conduct that are expected and maintained by society and/or professional organizations. Values represent the collective conception of what communities find desirable, important, and morally proper. Values also serve as the criteria for evaluating our own personal actions and the actions of others. As marketers, we recognize that we not only serve our organizations but also act as stewards of society in creating, facilitating, and executing the transactions that are part of the greater economy. In this role, marketers are expected to embrace the highest professional ethical norms and the ethical values implied by our responsibility toward multiple stakeholders (e.g., customers, employees, investors, peers, channel members, regulators, and the host community).

Ethical Norms

As Marketers, we must:
- Do no harm.
- Foster trust in the marketing system.
- Embrace ethical values.

Ethical Values

- Honesty—to be forthright in dealings with customers and stakeholders.
- Responsibility—to accept the consequences of our marketing decisions and strategies.
- Fairness—to balance justly the needs of the buyer with the interests of the seller.
- Respect—to acknowledge the basic human dignity of all stakeholders.
- Transparency—to create a spirit of openness in marketing operations.
- Citizenship—to fulfill the economic, legal, philanthropic, and societal responsibilities that serve stakeholders.

Source: American Marketing Association, "Codes of Conduct | AMA Statement of Ethics," 2019. https://www.ama.org/.

Understand your own ethics

Throughout the text, we will be discussing the types of ethical issues individual marketing managers face. But we won't be moralizing and trying to tell you how you should think on any given issue. Rather, by the end of the course we hope that *you* will have some firm personal opinions about what is and is not ethical in micro-marketing activities. To help you think through your opinions, you will find *Ethical Dilemmas* in each chapter. As in the one below, the answer is not always obvious.[24]

> **Ethical Dilemma**
>
> *What would you do?* A customer purchases a GE Profile refrigerator that comes with a one-year warranty on parts and labor. The salesperson suggests that the customer consider the store's three-year extended service plan to cover any problems with the refrigerator. The customer replies, "I'm getting a GE Profile because it's a reputable brand—and I have read these extended warranties aren't necessary." Fourteen months later, the customer returns to the store and complains that the ice maker on the refrigerator doesn't work right—and argues "the store needs to make it right."
>
> *If you were the store manager, what would you say? Would your response be any different if you knew that the customer was going to post a negative review on the company website?*

Marketing has its critics

We must admit that marketing—as it exists in the United States and other developed societies—has many critics. Marketing activity is especially open to criticism because it is the part of business most visible to the public.

A number of typical complaints about marketing are summarized in Exhibit 1-8. Think about these criticisms and whether you agree with them or not. What complaints do you have that are not covered by one of the categories in Exhibit 1-8?

Such complaints should not be taken lightly. They show that many people are unhappy with some parts of the marketing system. Certainly, the strong public support for consumer protection laws proves that not all consumers feel they are being treated like royalty.

Exhibit 1-8 Sample Criticisms of Marketing

- Advertising is everywhere, and it's often annoying, misleading, or wasteful.
- The quality of products is poor, and often they are not even safe.
- There are too many unnecessary products.
- Retailers add too much to the cost of distribution and just raise prices without providing anything in return.
- Marketing serves the rich and exploits the poor.
- Marketers overpromise service, and when a consumer has a problem, nobody cares.
- Marketing creates interest in products that pollute the environment.
- Private information about consumers is collected and used to sell them things they don't need.
- Marketing makes people too materialistic and motivates them toward "things" instead of social needs.
- Easy consumer credit makes people buy things they don't need and can't afford.

#M4BW

Can marketing help create a better world?

As the criticisms above suggest, some people believe that marketing's primary activity is convincing people to buy stuff they don't need and can't afford. We are not so naïve that we don't recognize that at least some marketing managers and companies share this philosophy. That said, *we believe that the vast majority of businesspeople and companies believe that good marketing contributes to a better world.*

What do we mean by a "better world"? Our definition follows our discussion of the marketing concept, social responsibility, and marketing ethics. We believe a better world is one where: (1) buyers and sellers make better decisions, are healthier, and through their consumption choices (or decision not to consume) experience a better quality of life; (2) buyers and sellers make decisions that have less adverse impacts on others; and (3) marketing strategy decisions address some of the world's most challenging problems—including hunger, poverty, and climate change.[25]

Exhibit 1-9 Examples of Companies Using Marketing for a Better World (#M4BW)

Company	How It Uses Marketing for a Better World (#M4BW)[26]
Vodafone and Safaricom	Created the mobile phone–based money platform M-Pesa, which allows people who lack access to bank accounts to join the larger economy. Kenya alone boasts more than 25 million users.
Nestlé	Recognizing the harms caused by sugar, Nestlé has prioritized reducing sugar in its products. Its R&D department has created a prototype hollowed-out sugar crystal that may reduce the amount of sugar in some of its products (including chocolate) by 40 percent.
Novartis	Brought essential medicines and education on hygiene to India's rural poor.
Volvo	Acknowledging the automobile's contribution to environmental problems, Volvo announced that starting in 2019 every Volvo would run partially or completely on an electric engine.
Cemex	Overhauled its business model and figured out how to help a half-million Latin American families build their own home for one-third the usual time and cost—and still make a profit.

Many companies are stepping up to these challenges—and often find that such endeavors are profitable. We are not saying this is a charitable endeavor. Exhibit 1-9 shows a few examples of marketing strategies that have made for a better world. Throughout this text, we will point out people, brands, and organizations that use marketing for a better world. We highlight these examples with a green box around the text and our #M4BW (Marketing for a Better World) icon. We want to remind (and hopefully inspire) you to practice marketing for a better world.

#M4BW Some businesses look to the United Nations Sustainable Development Goals for guidance on how to align making profits with helping create a better world.
Source: United Nations

CONCLUSION

The basic purpose of this chapter is to introduce you to marketing and highlight its value for consumers, firms, and society. In Chapter 2, we introduce a marketing strategy planning process that is the framework for ideas developed throughout the rest of the text—and that will guide your marketing thinking in the future. This chapter sets the stage for that by introducing basic principles that guide marketing thinking.

You've learned about two views of marketing, both of which are important. One takes a micro view and focuses on marketing activities by an individual business (or other type of organization). This is what most people (including most business managers) have in mind when they talk about marketing. But it's important to understand that marketing also plays a more macro role. Macro-marketing is concerned with the way the whole marketing system works in a society or economy. It operates to make exchanges and relationships between producers and their customers more effective.

We discussed the functions of marketing and who performs them. This includes not only producers and their customers but also marketing specialists who serve as intermediaries between producers and consumers and other specialists (such as product-testing labs and advertising agencies) who are collaborators and facilitate marketing functions.

We explained how a market-directed economy works, through the macro-marketing system, to provide consumers with choices. We introduced macro-marketing in this chapter, and we'll consider macro-marketing issues throughout the text. But the major focus of this book is on marketing by individual organizations. Someone in an organization must plan and manage its activities to make certain that customer needs are satisfied.

That's why understanding the marketing concept is another objective. The marketing concept is the basic philosophy that provides direction to a marketing-oriented firm. It stresses that the company's efforts should focus on satisfying some target customers—at a profit. Production-oriented firms tend to forget this. The various departments within a production-oriented firm let their natural conflicts of interest get in the way of customer satisfaction.

We also introduced the customer value concept. It is marketing's responsibility to make certain that what the firm offers customers really provides them with value that is greater than they can obtain somewhere else. In today's competitive markets, a firm must offer superior customer value if it wants to attract customers, satisfy them, and build beneficial long-term relationships with them.

A final objective was for you to see how social responsibility and marketing ethics relate to the marketing concept. The chapter ends by considering criticisms of marketing—both of the way individual firms work and of the whole macro system. When you have finished reading this book, you will be better able to evaluate these criticisms.

By learning more about marketing-oriented decision making, you will be able to make more efficient and socially responsible decisions. This will help improve the performance of individual firms and organizations (your employers). Eventually it will help our macro-marketing system work better.

KEY TERMS

production, 5
customer satisfaction, 5
innovation, 6
marketing, 7
pure subsistence economy, 8
macro-marketing, 9
economies of scale, 10
universal functions of marketing, 11
buying function, 11
selling function, 11
transporting function, 11
storing function, 11

standardization and grading, 11
financing, 11
risk taking, 11
market information function, 11
intermediary, 12
collaborators, 12
economic system, 13
command economy, 13
market-directed economy, 13
simple trade era, 15
production era, 16
sales era, 16

marketing department era, 16
marketing company era, 16
marketing concept, 17
production orientation, 17
marketing orientation, 17
marketing metrics, 18
triple bottom line, 19
customer value, 20
micro-macro dilemma, 24
social responsibility, 24
marketing ethics, 25

QUESTIONS AND PROBLEMS

1. The case that opens this chapter features Nike. Review this case and pull out as many examples as you can of different key terms and concepts. For example, Nike demonstrated social responsibility (key term) when it began to monitor its suppliers' labor practices.

2. Review the Nike case study that opens this chapter. In what ways does Nike appear to follow the marketing concept? Suggest three other activities Nike could do to follow the marketing concept.

3. If a producer creates a revolutionary new product and consumers can learn about it and purchase it on a website, is any additional marketing effort really necessary? Explain your thinking.

4. Distinguish between the micro and macro views of marketing. Then explain how they are interrelated, if they are.

5. Refer to Exhibit 1-1 and give an example of a purchase you made recently that involved separation of information and separation in time between you and the producer. Briefly explain how these separations were overcome.

6. Define the functions of marketing in your own words. Using an example, explain how they can be shifted and shared.

7. Distinguish between how economic decisions are made in a command economy and how they are made in a market-directed economy.

8. Explain why a market-directed macro-marketing system encourages innovation. Give an example.

9. Define the marketing concept in your own words, and then explain why the notion of profit is usually included in this definition.

10. Define the marketing concept in your own words, and then suggest how acceptance of this concept might affect the organization and operation of your college.

11. Give examples of some of the benefits and costs that might contribute to the customer value of each of the following products: (a) a wristwatch, (b) a weight-loss diet supplement, (c) a cruise on a luxury liner, and (d) a checking account from a bank.

12. What are examples of the benefits that you can provide to a prospective employer in your field? What are examples of the costs that employer would incur if it hired you? How do you think you can increase the value you offer a prospective employer?

13. Give an example of a recent purchase you made where the purchase wasn't just a single transaction but rather part of an ongoing relationship with the seller. Discuss what the seller has done (or could do better) to strengthen the relationship and increase the odds of you being a loyal customer in the future.

14. Discuss how the micro-macro dilemma relates to each of the following products: high-powered engines in cars, nuclear power, bank credit cards, and pesticides that improve farm production.

15. Think of three companies that you feel make the world a better place. Explain what they do to make the world a better place. Then explain whether you think this makes them more profitable or not.

SUGGESTED CASES

1. McDonald's "Seniors" Restaurant
2. Nature's Own Foods, Inc.
16. Hanratty Company

MARKETING ANALYTICS: DATA TO KNOWLEDGE

CHAPTER 1: REVENUE, COST, AND PROFIT RELATIONSHIPS

This problem introduces you to the Marketing Analytics exercises, which are available online for this text—and gets you started with the use of spreadsheet analysis for marketing decision making. Connect creates a spreadsheet to help you focus on the use of marketing analytics and not spreadsheet creation. In this case the problem is relatively simple. In fact, you could work it without the software. But by starting with a simple problem, you will learn how to use the program more quickly and see how it will help you with more complicated problems.

Sue Cline, the business manager at Magna University Student Bookstore, is developing plans for the next academic year. The bookstore is one of the university's nonprofit activities, but any "surplus" (profit) it earns is used to support the student activities center.

Two popular products at the bookstore are the student academic calendar and notebooks with the school name. Sue thinks that she can sell calendars to 90 percent of Magna's

3,000 students, so she has had 2,700 printed. The total cost, including artwork and printing, is $11,500. Last year the calendar sold for $5, but she is considering changing the price this year.

Sue thinks that the bookstore will be able to sell 6,000 notebooks if they are priced right. But she knows that many students will buy similar notebooks (without the school name) from stores in town if the bookstore price is too high.

Sue has entered the information about selling price, quantity, and costs for calendars and notebooks in the spreadsheet program so that it is easy to evaluate the effect of different decisions. The spreadsheet is also set up to calculate revenue and profit, based on

Revenue = (Selling price) × (Quantity sold)

Profit = (Revenue) − (Total cost)

Design element: #M4BW box globe icon: ©Vectoryzen/Shutterstock

CHAPTER TWO

Lionel Bonaventure/AFP/Getty Images

Marketing Strategy Planning

There was a time when it didn't seem an exaggeration for Barnum & Bailey's ads to tout the circus as "The Greatest Show on Earth." For 100 years, circuses brought excitement and family entertainment to towns all over the country. Parents hardly noticed the hard benches they sat on as they watched their kids cheer for the acrobats, clowns, and animal acts. But by the 1980s, the popularity of traditional circuses was in decline; many went out of business.

You can imagine why this sad state of affairs would be a concern for Guy Laliberté—a stilt walker, accordion player, and fire-eater—and others in his band of performers. But instead of bemoaning the demise of the circus, they saw an opportunity for a new kind of entertainment—and their idea gave birth to Cirque du Soleil.

Their new style of circus still traveled to audiences and set up a "big top" tent, but costly and controversial animal acts were eliminated. Instead, the entertainment focused on an innovative combination of acrobatics, music, and theater. This more sophisticated offering appealed to adults. Importantly, adults were willing to pay more for tickets when the show targeted them and not just kids—especially when the traditional circus benches were replaced with more comfortable seats.

Cirque du Soleil quickly struck a chord with audiences, and soon the producers were developing new shows and expanding tours to reach new markets. Laliberté recognized and built on this breakthrough opportunity. At any one time, about a dozen different Cirque du Soleil shows travel across Europe, Asia, Australia, and North and South America. Each show performs in a host city from two weeks to three months. Ten other Cirque shows have permanent homes and target tourists visiting New York City; Las Vegas, Nevada; Orlando, Florida; and Riviera Maya, Mexico. Each of the shows has a unique theme. For example, *OVO* looks at the world of insects, *Messi10* is loosely based on the life of Argentinian soccer star Lionel Messi, and *Bazzar* presents an "eclectic lab of infinite creativity."

When considering new shows, Cirque evaluates each idea on its creativity, uniqueness, and likelihood of becoming a blockbuster. New shows can take more than five years and $100 million to develop. Cirque anticipates recouping these development costs over each show's planned 10-year run. Some shows pay back even quicker. *Michael Jackson THE IMMORTAL World Tour* was a big hit—selling more than $140 million in tickets in its first full year on the road. That show's successes led to the development of *Michael Jackson ONE*, now based in Las Vegas.

As all of this suggests, Cirque du Soleil's marketing managers constantly evaluate new opportunities. Some opportunities, like its theme park in Nuevo Vallarta, Mexico, diversify Cirque's product line. A new product designed to attract customers who probably wouldn't go to a traditional performance, the theme park offers an immersive experience where visitors are literally part of the show. Not every opportunity makes it to market, however. Cirque thought its own women's clothing line could appeal to its biggest fans, but this idea was later screened out.

Cirque also reaches new customers through television specials, DVDs, and online streaming. These small-screen shows generate additional revenue while giving customers a taste of Cirque du Soleil—whetting their appetite for a live show.

Once customers see a live Cirque du Soleil show, they want to see more. So, Cirque advertises to encourage customers to see that first show. Ads in airline magazines target travelers heading to cities with permanent shows; traveling shows are heavily advertised in local media. Publicity and word-of-mouth are also important.

Cirque du Soleil knows customers who love the shows will come back again and again. Cirque's marketing managers often use digital tools—both because they are effective with their target market and because they can more easily track and analyze the results. For example, Cirque knows that its biggest fans will download an app to stay informed; the app also tells Cirque when those customers are in Las Vegas. Then Cirque knows to remind those customers to attend a show. Cirque collects data from its social media sites, including Instagram, YouTube, and Facebook. By analyzing these data, Cirque learned that its biggest fans wanted more backstage stories about performers and makeup specialists. So, it created video stories and posted them on social media where fans liked and shared them with friends.

Cirque offers special packages to different groups of customers. For example, it offers lower-price "Family Packs" and premium packages for customers wanting a post-show meet and greet with the performers.

Cirque du Soleil's managers take social responsibility seriously. Within just a few years of its founding, the company adopted policies for community social and cultural action. In its 2017–2020 Corporate Social Strategy, Cirque's CEO expresses pride that the company "uses its creative powers to make the world better." One part of that strategy focuses on making a difference in the lives of young people and encouraging them to "change the world." To help achieve that goal, Cirque du Soleil donates 500 tickets to at-risk youth in each city it visits.

Although Barnum & Bailey's "Greatest Show on Earth" has folded its tent, Cirque du Soleil's success inspired new competitors with similar entertainment fare. The lumberjack-themed Cirque Alfonse and its chainsaw jugglers tour across multiple continents. A former Cirque director started the aquatic-themed Le Rêve in Las Vegas. The well-known Cirque du Soleil brand name still gives the troupe a competitive advantage when it introduces new shows. This carefully crafted marketing mix generates ticket sales totaling about a billion dollars a year.[1]

LEARNING OBJECTIVES

Marketing managers at Cirque du Soleil make many decisions as they develop marketing strategies. Making good marketing strategy decisions is never easy, yet knowing what basic decision areas to consider helps you plan a more successful strategy. This chapter will get you started by giving you a framework for thinking about marketing strategy planning—which is what the rest of this book is all about.

When you finish this chapter, you should be able to

1. understand what a marketing manager does.
2. know what marketing strategy planning is—and why it is the focus of this book.
3. understand target marketing.
4. be familiar with the Four Ps in a marketing mix.
5. know the difference between a marketing strategy, a marketing plan, and a marketing program.
6. understand what customer lifetime value and customer equity are and why marketing strategy planners seek to increase them.
7. be familiar with the text's framework for marketing strategy planning.
8. know four broad types of marketing opportunities that help in identifying new strategies.
9. understand why strategies for opportunities in international markets should be considered.
10. understand the important new terms (shown in red).

The Management Job in Marketing

LO 2.1

In Chapter 1 you learned about the marketing concept—a philosophy to guide the whole firm toward satisfying customers at a profit. From the Cirque du Soleil case, it's clear that marketing decisions are very important to a firm's success. Let's look more closely at the marketing management process.

The **marketing management process** is the process of (1) *planning* marketing activities, (2) directing the *implementation* of the plans, and (3) *controlling* the plans. Planning, implementation, and control are basic jobs of all managers—but here we will emphasize what they mean to marketing managers.

Exhibit 2-1 shows the relationships among the three jobs in the marketing management process. The jobs are all connected to show that the marketing management process is continuous. In the planning job, managers set guidelines for the implementing job and specify expected results. They use these expected results in the control job to determine if everything has worked out as planned. The link from the control job to the

Exhibit 2-1 The Marketing Management Process

Strategic management planning concerns the whole firm

planning job is especially important. This feedback often leads to changes in the plans or to new plans.

The job of planning strategies to guide a whole company is called **strategic (management) planning**—the managerial process of developing and maintaining a match between an organization's resources and its market opportunities. This is a top-management job. It includes planning not only for marketing but also for production, finance, human resources, and other areas.

Although marketing strategies are not whole-company plans, company plans should be market-oriented. And the marketing plan often sets the tone and direction for the whole company. Thus, we will use *strategy planning* and *marketing strategy planning* to mean the same thing.[2]

What Is a Marketing Strategy?

LO 2.2

Marketing strategy planning means finding attractive opportunities and developing profitable marketing strategies. But what is a "marketing strategy"? We have used these words rather casually so far. Now let's see what they really mean.

A **marketing strategy** specifies a target market and a related marketing mix. It is a big picture of what a firm will do in some market. Two interrelated parts are needed:

Exhibit 2-2 A Marketing Strategy

1. A **target market**—a fairly homogeneous (similar) group of customers to whom a company wishes to appeal.
2. A **marketing mix**—the controllable variables the company puts together to satisfy this target group.

The importance of target customers in this process can be seen in Exhibit 2-2, where the target market is at the center of the diagram. The target market is surrounded by the controllable variables that we call the "marketing mix." A typical

urbanbuzz/Shutterstock

marketing mix includes some product, offered at a price, with some promotion to tell potential customers about the product, and a way to reach the customer's place.

The marketing strategy for Herbal Essences hair care products aims at a specific group of target customers: young women in their teens and early 20s. The products include various shampoos, conditioners, gels, and hairspray for different types of hair. The product names and brightly colored packaging grab customers' attention. For example, Body Envy (in orange bottles) adds body to flat hair, and Hello Hydration (in blue bottles) features a coconut fragrance and extra moisture. The curvy, nested shampoo and conditioner bottles subtly encourage customers to buy the products together. By seeking eye-level placement at Target and Walmart, Herbal Essences locates its products where most of its target customers shop for hair care essentials. The brand's print, television, and online advertising incorporate a mythical quality that supports the products' organic origins. Herbal Essences' home page (www.herbalessences.com) includes hundreds of customer ratings and reviews, links to a Facebook Fan page (with more than 1 million fans), a Twitter feed, and a YouTube channel. The shampoo and conditioner retail for $3 to $8 a bottle, with occasional dollar-off coupons to encourage new customer trial. Fast-growing sales suggest this marketing mix hits the bulls-eye for this target market.[3]

Selecting a Market-Oriented Strategy Is Target Marketing

LO 2.3

Target marketing is not mass marketing

A marketing strategy specifies some *particular* target customers. This approach is called "target marketing" to distinguish it from "mass marketing." **Target marketing** says that a marketing mix is tailored to fit some specific target customers. In contrast, **mass marketing**—the typical production-oriented approach—vaguely aims at "everyone" with the same marketing mix. Mass marketing assumes that everyone is the same—and it considers everyone to be a potential customer. It may help to think of target marketing as the "rifle approach" and mass marketing as the "shotgun approach." See Exhibit 2-3.

Exhibit 2-3 Production-Oriented and Marketing-Oriented Managers Have Different Views of the Market

Production-oriented manager sees everyone as basically similar and practices "mass marketing"

Marketing-oriented manager sees everyone as different and practices "target marketing"

A marketing mix does not need to appeal to all customers. Target marketers recognize that customer needs vary. They identify a target market to serve, and then create a marketing mix to match that market's needs. Whereas a $400 cooler is not for everyone, YETI identified customers (including people dedicated to fishing) that value a high-quality, durable, "indestructible" cooler "built for the wild."
Source: YETI

Mass marketers may do target marketing

Commonly used terms can be confusing here. The terms *mass marketing* and *mass marketers* do not mean the same thing. Far from it! *Mass marketing* means trying to sell to "everyone," as we explained earlier. *Mass marketers* such as Kraft Foods and Walmart aim at large but clearly defined target markets. The confusion with mass marketing occurs because their target markets usually are large and spread out.

Target marketing can mean big markets and profits

Target marketing is not limited to small market segments—only to fairly homogeneous ones. A very large market—even what is sometimes called the "mass market"—may be fairly homogeneous, and a target marketer will deliberately aim at it. For example, parents of young children are a large group that is homogeneous on many dimensions, including their attitudes about changing baby diapers. In the United States alone, this group spends more than $6 billion a year on disposable diapers—so it should be no surprise that it is a major target market for companies such as Kimberly-Clark (Huggies) and Procter & Gamble (Pampers).

The basic reason to focus on some specific target customers is so that you can develop a marketing mix that satisfies those customers' *specific* needs better than they are satisfied by some other firm. For example, E*TRADE uses a website (www.etrade.com) to target knowledgeable investors who want a convenient, low-cost way to buy and sell stocks online without a lot of advice (or pressure) from a salesperson.

Developing Marketing Mixes for Target Markets

LO 2.4

There are many marketing mix decisions

There are many possible ways to satisfy the needs of target customers. A product might have many different features. It could be sold directly to customers via the Internet, offered only in stores, or both. Customer service levels before or after the sale can be adjusted. The package, brand name, and warranty can be changed. Various advertising media—newspapers, magazines, cable, the Internet—may be used. The company can develop social media sites on Facebook, Snapchat, or Instagram. A company's own sales force or other sales specialists can be used. The price can be changed, discounts can be given, and so on. With so many possible variables, is there any way to help organize all these decisions and simplify the selection of marketing mixes? The answer is yes.

The "Four Ps" make up a marketing mix

It is useful to reduce all the variables in the marketing mix to four basic ones:

Product
Place
Promotion
Price

It helps to think of the four major parts of a marketing mix as the "*Four Ps.*"

Exhibit 2-4
A Marketing Strategy—Showing the Four Ps of a Marketing Mix

Customer is not part of the marketing mix

The customer—the target market—is shown surrounded by the Four Ps in Exhibit 2-4. Some students assume that the customer is part of the marketing mix—but this is not so. The customer should be the *target* of all marketing efforts. For example, many soft drink customers are concerned about how much sugar they consume—so Coke Zero Sugar targets customers who love the taste of Coke but don't want sugar.

Exhibit 2-5 shows some of the strategy decision variables organized by the Four Ps. These will be discussed in later chapters. For now, let's just describe each P briefly.

Product—the good or service for the target's needs

The Product area is concerned with developing the right "product" for the target market. This offering may involve a physical good, a service, or a blend of both. Whereas Coke Zero Sugar, Jeep Wrangler, and the Samsung Galaxy phone are physical goods, the product for T-Mobile is the communication *service* it provides—sending texts, completing phone calls, and connecting customers to the Internet. The Product of a political party is the policies it works to achieve. The important thing to remember is that your good or service should satisfy some customers' needs.

(left): Twin Design/Shutterstock; *(right):* McGraw-Hill Education

Exhibit 2-5 Strategy Decision Areas Organized by the Four Ps

Product	Place	Promotion	Price

• Physical good
• Service
• Features
• Benefits
• Quality level
• Accessories
• Installation
• Instructions
• Warranty
• Product lines
• Packaging
• Branding
	• Objectives
	• Channel type
	• Market exposure
	• Kinds of intermediaries
	• Kinds and locations of stores
	• How to handle transporting and storing
	• Service levels
	• Recruiting intermediaries
	• Managing channels
		• Objectives
		• Promotion blend
		• Salespeople
		  Kind
		  Number
		  Selection
		  Training
		  Motivation
		• Advertising
		  Targets
		  Kinds of ads
		  Media type
		  Copy thrust
		  Prepared by whom
		• Publicity
		• Sales promotion
			• Objectives
			• Flexibility
			• Level over product life cycle
			• Geographic terms
			• Discounts
			• Allowances

Coca-Cola introduced Coke Zero Sugar to replace Coke Zero after sales of Zero didn't meet expectations and research showed customers didn't understand the benefits. Some of the Product decisions the marketing manager for Coke Zero Sugar made include the brand name (to make very clear that sugar was not an ingredient), product features (flavor similar to original Coke), and the packaging colors (changed to look more similar to Coke to attract traditional Coke customers).

Place—reaching the target

Place is concerned with all the decisions involved in getting the right product to the target market's Place. A product isn't much good to a customer if it isn't available when and where it's wanted.

A product reaches customers through a channel of distribution. A **channel of distribution** is any series of firms (or individuals) that participate in the flow of products from producer to final user or consumer.

Sometimes a channel of distribution is short and runs directly from a producer to a final user or consumer. This is common in business markets and in the marketing of services. For example, T-Mobile sells its phone services directly to final consumers. However, as shown in Exhibit 2-6, channels are often more complex—as when Coke Zero Sugar goes from Coca-Cola to wholesaler/bottlers and then to wholesalers and retailers before reaching consumers. When serving different target markets, a company may use more than one channel of distribution. For example, Coke Zero Sugar is distributed through a different channel of distribution (not shown) when customers pour a fountain drink in a McDonald's or Chick-fil-A restaurant.

We will also see how physical distribution service levels and decisions concerning logistics (transporting, storing, and handling products) relate to the other Place decisions and the rest of the marketing mix.

Promotion—telling and selling the customer

The third P—Promotion—is concerned with telling the target market or others in the channel of distribution about the right product. Sometimes promotion is focused on acquiring new customers and sometimes it's focused on retaining current customers. Promotion includes personal selling, mass selling, and sales promotion. It is the marketing manager's job to blend these methods of communication.

Personal selling involves direct spoken communication between sellers and potential customers. Personal selling may happen face-to-face, over the telephone, or even via an online conference. Sometimes personal attention is required *after the sale*. **Customer service**—a personal communication between a seller and a customer who wants the

Exhibit 2-6 Four Examples of Basic Channels of Distribution for Consumer Products

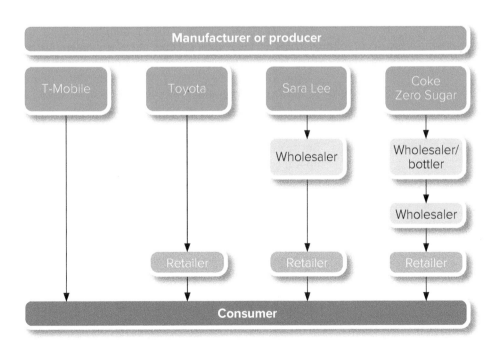

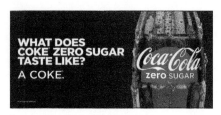

Source: The Coca-Cola Company

seller to resolve a problem with a purchase—is often a key to building repeat business. Individual attention comes at a price; personal selling and customer service can be very expensive. This personal effort is often blended with mass selling and sales promotion.

Mass selling is communicating with large numbers of customers at the same time. The main form of mass selling is **advertising**—any *paid* form of nonpersonal presentation of ideas, goods, or services by an identified sponsor. Here the advertiser *pays* a media source (newspapers, television network, or website) for permission to place an advertisement. **Publicity**—any *unpaid* form of nonpersonal presentation of ideas, goods, or services—includes earning favorable coverage in newspaper stories and creating web pages that provide product information for interested customers. A brand page on social media—perhaps Facebook, Twitter, or Pinterest—might also tell customers about the right product.

Sales promotion refers to those promotion activities—other than advertising, publicity, and personal selling—that stimulate interest, trial, or purchase by final customers or others in the channel. This can involve the use of coupons, point-of-purchase materials, samples, signs, contests, events, catalogs, novelties, and circulars.

When Coke Zero became Coke Zero Sugar, TV, print, digital, and social media advertising as well as sales promotion let customers know what was new. Salespeople talked to buyers at fast-food, grocery, and convenience stores and often set up in-store sales promotions.

Price—making it right

In addition to developing the right Product, Place, and Promotion, marketing managers must also set the right Price. Price setting must consider the kind of competition in the target market and the cost of the whole marketing mix. A manager must also try to estimate customer reaction to possible prices. Besides this, the manager must know current practices as to markups, discounts, and other terms of sale. And if customers won't accept the Price, all of the planning effort is wasted.

The price of Coke Zero Sugar varies depending on the Place it is sold and the amount of customer convenience delivered. For example, the cost might be more than $0.25 per ounce ($4.00 for a 16-ounce cup) at a sports stadium or as low as $0.02 per ounce for warm cans at the grocery store ($2.99 per 12-pack).

The Four Ps deliver value

In Chapter 1 we introduced customer value. We described how customer value increases when customers perceive more benefits or lower costs. Marketing managers use the Four Ps to deliver customer value to a target market. For example, customers may perceive more value if the Product is of higher quality or includes more features; Place decisions make a product more convenient for customers to purchase; Promotion effectively communicates value; or Price is lowered. A firm must understand its target market to know how to best adjust the Four Ps to deliver value at a price that also delivers profits to the company.

Each of the Four Ps contributes to the whole

All Four Ps are needed in a marketing mix. In fact, they should all be tied together. But is any one more important than the others? Generally speaking, the answer is no—all contribute to one whole. When a marketing mix is being developed, all (final) decisions about the Four Ps should be made at the same time. That's why the Four Ps are arranged around the target market in a circle—to show that they all are equally important.

Let's sum up our discussion of marketing mix planning thus far. We develop a *Product* to satisfy the target customers. We find a way to reach our target customers' *Place*. We use *Promotion* to tell the target customers (and others in the channel) about the product that has been designed for them. And we set a *Price* after estimating expected customer reaction to the total offering and the costs of getting it to them. Together, a marketing mix should deliver customer value for the target market.

Strategy jobs must be done together

It is important to stress—and cannot be overemphasized—that selecting a target market *and* developing a marketing mix are interrelated. Both parts of a marketing strategy must be decided together. It is *strategies* that must be evaluated against the company's objectives—not alternative target markets or alternative marketing mixes.

Ethical Dilemma

What do you think? You are the marketing manager for Jones Department Stores, a 125-store chain in the northwestern United States. Your company used an outside developer to write the store's smartphone app. The developer just added some new location-based features. For example, customers can enter their "favorite Jones store" location into the app. When customers shop at the store, they can open the app to find a map of the store's layout. The store can also deliver coupons to customers while they shop in the store (the app knows they are there).

The developer also points out another feature. He tells you that another major retailer shows customers different prices on the app—depending on whether the customer is in the store or not. So, for example, the app might show the price of a particular sweater as $79 while in the store (matching the price tag in the store). But, when a customer who is not in the store looks at the same sweater on the app, the price would be lower—say $59. This other retailer found that in-store customers are much less price sensitive than online shoppers. Online shoppers can click somewhere else and buy—so Jones could offer more discounts to online shoppers compared to those in the store. Because the app knows when a customer is in the store, higher prices can be shown there. When the customer is using the app away from the store, lower prices could be shown. Jones can decide which products have lower online prices and how much to discount the price. This helps the store maximize profits. After all, why give customers a discount if they are willing to pay full price?

Would you recommend including this feature on the new version of the app? How would customers paying the higher price feel if they found out other customers paid lower prices? Explain.

Understanding target markets leads to good strategies

The needs of a target market often virtually determine the nature of an appropriate marketing mix, so marketers must analyze their potential target markets with great care. This book will explore ways to identify attractive market opportunities and develop appropriate strategies.

Let's look at the strategy planning process more closely in the classic case of Jeff Silverman and Toddler University (TU), Inc., a shoe company he started. During high school and college, Silverman worked as a salesperson at local shoe stores. He also gained valuable experience during a year working for Nike. From these jobs he learned a lot about customers' needs and interests. He also realized that some parents were not satisfied when it came to finding shoes for their preschool children.

Silverman saw there were many different types of customers for baby shoes, each group with a different set of needs. From his observations, there was one market that was underserved. These *Attentive Parents* wanted shoes that met a variety of needs. They wanted shoes to be fun and fashionable and functional. They didn't want just a good fit but also design and materials that were really right for baby play and learning to walk. These well-informed, upscale shoppers were likely to buy from a store that specialized in baby items. They were also willing to pay a premium price if they found the right product. Silverman saw an opportunity to serve the Attentive Parents target market with a marketing mix that combined, in his words, "fit and function with fun and fashion." He developed a detailed marketing plan that attracted financial backers, and his company came to life.

Toddler University's marketing strategy was successful because it developed a distinctive marketing mix that was precisely relevant to the needs of its target market.
Source: Toddler University

Silverman contracted with a producer in Taiwan to make shoes to his specs with his Toddler University brand name. TU's specs were different—they improved the product for this target market. Unlike most rigid high-topped infant shoes, TU's shoes were softer with more comfortable rubber soles. The shoes were stitched rather than glued so they lasted longer and included an extra-wide opening so the shoes slipped more easily onto squirming feet. A patented special insert allowed parents to adjust the shoes' width. The insert also helped win support from retailers. With 11 sizes of children's shoes—and five widths—retailers usually stocked 55 pairs of each style. TU's adjustable width reduced the stocking requirements, making the line more profitable than competing shoes. TU's Product and Place decisions worked together to provide customer value and also to give TU a competitive advantage.

For promotion, TU's print ads featured close-up photos of babies wearing the shoes and informative details about their special benefits. Creative packaging promoted the shoe and attracted customers in the store. For example, TU put one athletic-style shoe in a box that looked like a gray gym locker. TU also provided stores with "shoe rides"—electric-powered rocking replicas of its shoes. The rides attracted kids to the shoe department, and because they were coin-operated, they paid for themselves in a year.

TU priced most of its shoes at $35 to $40 a pair. This is a premium price, but the Attentive Parents typically have smaller families and are willing to spend more on each child.

In just four years, TU's sales jumped from $100,000 to more than $40 million. To keep growth going, Silverman expanded distribution to reach new markets in Europe. To take advantage of TU's relationship with its satisfied target customers, TU expanded its product line to offer shoes for older kids. Then Silverman made his biggest sale of all: He sold his company to Genesco, one of the biggest firms in the footwear business.[4]

The Marketing Plan Guides Implementation and Control

LO 2.5

Marketing plan fills out marketing strategy

As the Toddler University case illustrates, a marketing strategy sets a target market and a marketing mix. It is a big picture of what a firm will do in some market. A marketing plan goes farther. A **marketing plan** is a written statement of a marketing strategy *and* the time-related details for carrying out the strategy. It should spell out the following in detail: (1) what marketing mix will be offered, to whom (that is, the target market), and for how long; (2) what company resources (shown as costs) will be needed at what rate (month by month perhaps); and (3) what results are expected (sales and profits perhaps monthly or quarterly, customer satisfaction levels, and the like). The plan should also include some control procedures so that whoever is to carry out the plan will know if things are going wrong. This might be something as simple as comparing actual sales against expected sales with a warning flag to be raised whenever total sales fall below a certain level.

Appendix D provides a sample marketing plan for a veterinary clinic. At the end of each chapter, there is an exercise titled *Marketing Planning for Hillside Veterinary Clinic* that introduces you to aspects of a marketing plan as related to the topics in that chapter. This gives you a step-by-step way to learn how chapter concepts apply to marketing planning and to develop your plan-building skills as you progress through the text. In Chapter 19 we review all of the elements in a marketing plan. At that point, you will have learned about all of the major strategy decision areas (Exhibit 2-5) and how to blend them into an innovative strategy.

Campbell Soup Company develops different soups (and related marketing mixes) for the specific needs of different target markets. Customers looking for a soup that fills them up choose from the Campbell's Chunky line; Campbell's Soup on the Go offers convenience, and #M4BW Campbell's Well Yes! soups feature clean, simple, and nutritious ingredients.
Source: CSC Brands, L.P.

Implementation puts plans into operation

After a marketing plan is developed, a marketing manager knows *what* needs to be done. Then the manager is concerned with **implementation**—putting marketing plans into operation. Strategies work out as planned only when they are effectively implemented. Many **operational decisions**—short-run decisions to help implement strategies—may be needed.

Managers should make operational decisions within the guidelines set down during strategy planning. They develop product policies, place policies, and so on as part of strategy planning. Then operational decisions within these policies probably will be necessary—while carrying out the basic strategy. Note, however, that as long as these operational decisions stay within the policy guidelines, managers are making no change in the basic strategy. If the controls show that operational decisions are not producing the desired results, however, the managers may have to reevaluate the whole strategy—rather than just working harder at implementing it.

It's easier to see the difference between strategy decisions and operational decisions if we illustrate these ideas using our Toddler University example. Possible Four Ps or basic strategy policies are shown in the left-hand column in Exhibit 2-7, and examples of operational decisions are shown in the right-hand column.

It should be clear that some operational decisions are made regularly—even daily—and such decisions should not be confused with planning strategy. Certainly, a great deal of effort can be involved in these operational decisions. They might take a good

Exhibit 2-7 Relation of Strategy Policies to Operational Decisions for a Baby Shoe Company

Marketing Mix Decision Area	Strategy Policies	Likely Operational Decisions
Product	Carry as limited a line of colors, styles, and sizes as will satisfy the target market.	Add, change, or drop colors, styles, and/or sizes as customer tastes dictate.
Place	Distribute through selected "baby products" retailers that will carry the full line and provide good in-store sales support and promotion.	In market areas where sales potential is not achieved, add new retail outlets and/or drop retailers whose performance is poor.
Promotion	Promote the benefits and value of the special design and how it meets customer needs.	When a retailer hires a new salesperson, send current training package with details on product line; increase use of local newspaper print ads during peak demand periods (e.g., before holidays).
Price	Maintain a "premium" price, but encourage retailers to make large-volume orders by offering discounts on quantity purchases.	Offer short-term introductory price "deals" to retailers when a new style is first introduced.

part of the sales or advertising manager's time. But they are not the strategy decisions that will be our primary concern.

Analytical tools provide control

Our focus in this text is on developing marketing strategies. But eventually marketing managers must control the marketing plans that they develop and implement. The control job provides feedback so managers can change marketing strategies to better meet customer needs. To maintain control, a marketing manager uses a number of analytical tools to learn more about customers and their buying habits—and how they respond to changes to a firm's marketing mix.[5]

Marketing analytics—measure, manage, and analyze marketing performance

In Chapter 19 we examine control of whole marketing plans in a bit more detail. Yet marketing managers are often looking to measure specific customer activities along the customer's path toward making a purchase. This can help managers to make better segmenting, positioning, or marketing mix decisions. In Chapter 1 we introduced the term *marketing metrics*. Marketing metrics are used to perform **marketing analytics**—the practice of measuring, managing, and analyzing marketing performance to maximize its efficiency and effectiveness.[6]

Marketing managers have access to more and more data—especially when customers use computers and mobile devices to research and make purchases. This behavior can be tracked, compiled, and used for marketing analytics. Throughout the text we will discuss how these new technologies are used and also explore the ethical and legal questions they raise. To better understand how marketing managers use metrics and conduct marketing analytics, each chapter includes one *Marketing Analytics in Action* activity. Each of these activities describes a marketing metric and explains how it can be analyzed. Some of these activities include a link to an online calculator or spreadsheet to demonstrate the math.

Several plans make a whole marketing program

Most companies implement more than one marketing strategy—and related marketing plan—at the same time. Procter & Gamble (P&G) targets users of laundry detergent with at least three different strategies. Some consumers want Tide's superior cleaning capabilities; others prefer the color protection of Cheer or the pleasant scents of Gain. Each detergent has a different formulation and a different approach for letting its target market know about its benefits. Yet P&G must implement each of these marketing strategies at the same time—along with strategies for Bounty, Olay, Charmin, and many other brands.

Exhibit 2-8 Elements of a Firm's Marketing Program

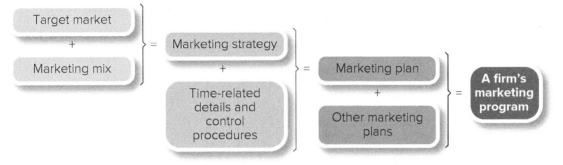

A **marketing program** blends all of the firm's marketing plans into one "big" plan (see Exhibit 2-8). But the success of the marketing program depends on the care that goes into planning the individual strategies and how well they work together to create value for customers and the firm.

Recognizing Customer Lifetime Value and Customer Equity

LO 2.6

In Chapter 1 we introduced the idea that building customer value attracts customers, and satisfying those customers builds profitable long-term relationships. Let's take that a step further and understand how marketing strategies and marketing programs that build relationships with customers create ongoing financial value for a marketer.

Relationships increase customer lifetime value

Loyal customers continue to buy brands that satisfy them, often seeking out other products from that same company. Many firms recognize this and measure the **customer lifetime value (CLV)**, which is the total profits a single customer contributes to a firm over the length of the relationship. For example, Mallory, a 22-year-old college graduate, might purchase a new Honda Fit for $16,000. A few years later, Mallory could be in the market for a small sport-utility vehicle; if she was satisfied with the Fit and the service she received, Mallory might buy a Honda CR-V for $25,000. After a couple of kids, a Honda Odyssey minivan starts to look appealing, potentially followed by more Honda cars and trucks over the next few decades. A few Honda lawn mowers might also be purchased along the way. If Honda continues to provide good value to this customer, the lifetime total of her purchases of Honda products could exceed $400,000. If Honda makes a 10 percent profit (on average), the customer lifetime value for Mallory could approach $40,000.

This insight—the lifetime value of a customer—changes how many firms approach marketing strategy planning. This marketing analytic demonstrates how important it is for a company to retain customers over time. If Mallory was not satisfied with her initial Honda Fit, she may not have purchased more Honda products and may have had much lower customer lifetime value. A good marketing program satisfies customers and develops strategies to make it easy for satisfied customers to buy more from the company. Companies seek to build *long-term relationships* with customers. For example, Honda has vehicles to satisfy customers across different stages of life.

How can we estimate customer lifetime value?

Many firms collect data and measure customer lifetime value. There are different ways to calculate CLV. Which method works best depends on customer buying behavior, marketing practices, and available data. No matter the approach, three marketing metrics are needed to estimate customer lifetime value: (1) average profit margin, (2) retention rate, and (3) acquisition cost. Let's talk about these metrics next.

In Chapter 1 we talked about profit margin—and we will talk more about this when we get into the details of pricing in Chapter 17. For now, we will assume each customer has an *average profit margin* that can be estimated based on his or her purchases over some time period.

Of course, Mallory could keep buying Honda products or she might stop and buy a Toyota, Chevrolet, or BMW instead. The **retention rate** refers to the percentage of customers retained as compared to the total number of customers. So, for example, if this year Local Joe's coffee shop has 1,000 customers and next year, 800 of *those* customers are still patronizing Local Joe's for coffee (some might find a new coffee shop, move to a new city, or just stop drinking coffee), then the coffee shop has an 80 percent (800/1,000) retention rate.

The last marketing metric used to calculate CLV is **acquisition cost**—the expense required to acquire each new customer. For example, our coffee shop could spend $10,000 this year on advertising designed to attract new customers—and attract 100 new customers. The acquisition cost for each customer is the total advertising dollars spent divided by the number of new customers ($10,000/100 = $100) and results in an acquisition cost of $100 per customer.

Just to see how it works, we provide a simple example in the Marketing Analytics in Action: Customer Lifetime Value activity next.[7]

Marketing Analytics in Action: Customer Lifetime Value

Many firms use marketing metrics to calculate customer lifetime value (CLV). Recall that CLV is the total stream of purchases that a customer could contribute to the company over the length of the relationship.

To better understand how CLV is calculated, let's walk through an example. Consider Local Joe's—a small coffee shop on the lake in a midsized town in Minnesota. The owner, Joe, wonders how valuable one of his "regular" customers is—and how important it is for customer retention. He defines regular customers as those coming in at least once a week for more than a few months. Reading about CLV, Joe learns that he needs a few numbers to get started. He estimates these numbers from data and his knowledge of his customers:

1. Average profit margin per customer per year (M), estimated at $250 per year
2. Retention rate (R), estimated at 60 percent
3. Acquisition cost (AC), estimated at $125 per customer

These numbers are entered into a CLV equation to estimate CLV:

$$CLV = M*[R/(1 - R)] - AC$$

If we plug the numbers from 1 to 3 above into this equation, we estimate customer lifetime value as $250 per customer.

$$CLV = \$250 * [0.6/(1 - 0.6)] - \$125 = \$250$$

1. *Calculate the customer lifetime value if Local Joe's increases its retention rate to 70 percent. To do this, use the equation above and change 0.6 to 0.7.*
2. *If Local Joe's charged higher prices, it could possibly earn a higher average profit margin per customer. Calculate customer lifetime value if average profit margin per customer rose to $300 per year (note, assume the 60 percent retention rate).*
3. *Describe two marketing tactics that Local Joe's could use to increase customer lifetime value. Tie your answers to how each might change the numbers in the customer lifetime value equation.*

Customer equity considers lifetime value of all current and future customers

Customer lifetime value is an estimate of a *single* customer's value. We can carry this idea a step further by considering *all of a firm's current and future customers* and the revenue and costs associated with each. **Customer equity** is the expected earnings stream (profitability) of a firm's current and prospective customers over some period of time.

While it is possible to estimate customer equity, the math is beyond the scope of this book. Still, the idea of customer equity can be a useful guide for developing marketing strategy. Following a customer equity approach guides the marketing manager to make marketing decisions that enhance the firm's long-term profits—not just for the next quarter or year. By estimating the impacts that different marketing strategies and marketing programs have on customer equity, a firm can make marketing decisions with long-run financial implications in mind.

Customer equity focuses on the revenues and costs of acquiring, retaining, and enhancing customers

The customer equity approach also suggests three potential paths for growth: (1) *acquiring* new customers, (2) *retaining* current customers, and (3) *enhancing* the customer value by increasing their purchases. Each of these efforts (acquiring, retaining, and enhancing) has its own costs and benefits as well.

Some firms conducting this analysis find that it costs less to retain a customer than to acquire a new one. This insight can lead a firm to increase investments to retain customers—for example, spending more on customer service. A marketing manager should evaluate the effectiveness of a marketing mix in achieving each of these objectives, considering long-term revenues and costs. Typically, a marketing plan includes different marketing strategies to address each of these goals. So, for example, a plan should recognize that as it acquires more customers, it will soon need to increase investments in customer retention because there will be more customers to retain.

Ally Bank wants to grow its revenue and profitability from each customer—leading to greater customer equity. Ally Bank *acquires* many new customers who seek higher interest rates on their savings accounts—see the first advertisement. Those interest rates and Ally's 24 hours a day, 7 days a week telephone customer service help the bank *retain* customers. As an online bank, customers regularly visit the bank's website. Customer equity also grows when current customers buy other products from a company; Ally *enhances* customer equity when customers purchase additional financial services from Ally including CDs, IRAs, and auto loans.
(top-left and bottom): Source: Ally Financial Inc.; (top-right): Blue jean images RF/Getty Images

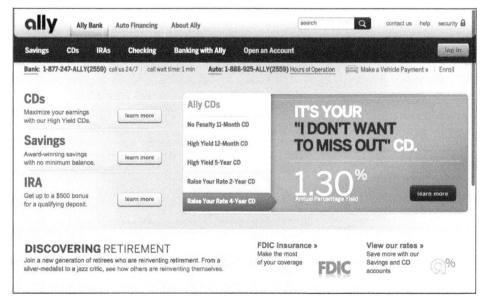

Different marketing strategies for acquiring, retaining, and enhancing

Acquiring customers requires a marketing strategy that targets customers who are not currently using a firm's products. Retaining and enhancing customers are different efforts that target current customers and attempt to keep or grow their purchases. Consider Harry's, an online seller of shaving supplies:

Source: Harry's, Inc.

- *Customer acquisition:* Harry's targets new customers with a special "Starter Set" that offers a razor handle, a few blades, and small bottle of shaving gel at a special price.
- *Customer retention:* Customers are automatically signed up for refills every few months—which serves as a strategy for retaining current customers. High-quality razors and blades at a low price also keep customers coming back.
- *Enhanced customer value:* Harry's added more products, including bar soap, body wash, and post-shave balm—and uses e-mails to encourage customers who buy its razors and blades to buy even more. This further enhances an individual's lifetime value and the portfolio of customers' equity.

What Are Attractive Opportunities?

LO 2.7

Effective marketing strategy planning matches opportunities to the firm's resources (what it can do) and its objectives (what top management wants to do). Successful strategies get their start when a creative manager spots an attractive market opportunity. Yet an opportunity that is attractive for one firm may not be attractive for another. Attractive opportunities for a particular firm are those that the firm has some chance of doing something about—given its resources and objectives.

Breakthrough opportunities are best

Throughout this book, we will emphasize finding **breakthrough opportunities**—opportunities that help innovators develop hard-to-copy marketing strategies that will be very profitable for a long time. That's important because there are always imitators who want to "share" the innovator's profits—if they can. It's hard to continuously provide *superior* value to target customers if competitors can easily copy your marketing mix.

Competitive advantage is needed—at least

Even if a manager can't find a breakthrough opportunity, the firm should try to obtain a competitive advantage to increase its chances for profit or survival. **Competitive advantage** means that a firm has a marketing mix that the target market sees as better than a competitor's mix. A competitive advantage may result from

Uber discovered a breakthrough opportunity and developed a hard-to-copy marketing strategy. Uber's smartphone app connects people looking for rides with drivers who will pick them up and take them to their desired location. Uber is taking that strategy global—this ad targets new drivers in India. Translation: "I earn even more than 90,000 rupees. I am an Uber driver partner."
Source: Uber

Exhibit 2-9 Overview of Marketing Strategy Planning Process

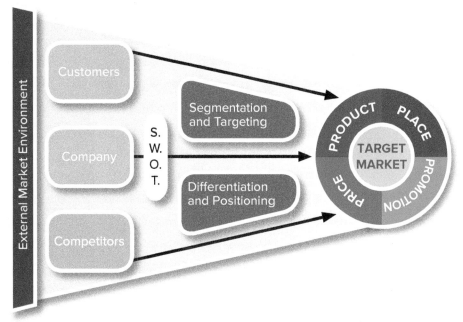

Note: The marketing manager narrows down from screening broad market opportunities to develop a focused marketing strategy. S.W.O.T. denotes Strengths, Weaknesses, Opportunities, and Threats.

efforts in different areas of the firm—cost cutting in production, innovative R&D, more effective purchasing of needed components, or financing for a new distribution facility. Similarly, a strong sales force, a well-known brand name, or good dealers may give it a competitive advantage in pursuing an opportunity. Whatever the source, an advantage succeeds only if it allows the firm to provide superior value and satisfy customers better than some competitor.

Walgreens' source of competitive advantage emerges from being the largest drug retailing chain in the United States. The retailer operates 8,300 stores spread across all 50 states. Almost anywhere in the United States, consumers will find a Walgreens nearby. Many of Walgreens' target customers place a high value on convenience, making this source of competitive advantage particularly relevant. Because it would be costly to build as many stores, this strategy is hard for competitors to copy.

Avoid hit-or-miss marketing with a logical process

You can see why a manager *should* seek attractive opportunities. But that doesn't mean that everyone does—or that everyone can turn an opportunity into a successful strategy. It's all too easy for a well-intentioned manager to react in a piecemeal way to what appears to be an opportunity. Then by the time the problems are obvious, it's too late.

Developing a successful marketing strategy doesn't need to be a hit-or-miss proposition. And it won't be if you learn the marketing strategy planning process developed in this text. Exhibit 2-9 summarizes the marketing strategy planning process we'll be developing throughout the rest of the chapters.

Marketing Strategy Planning Process Highlights Opportunities

We've emphasized that a marketing strategy requires decisions about the specific customers the firm will target and the marketing mix the firm will develop to appeal to that target market. We can organize the many marketing mix decisions (review Exhibit 2-5) in terms of the Four Ps—Product, Place, Promotion, and Price. Thus, the "final" strategy decisions are represented by the target market surrounded by the Four Ps. However, the idea isn't just to come up with *some* strategy. After all, there are hundreds or even thousands of combinations of marketing mix decisions and target markets (i.e., strategies) that a firm might try. Rather, the challenge is to zero in on the *best* strategy.

Process narrows down from broad opportunities to specific strategy

As Exhibit 2-9 suggests, it is useful to think of the marketing strategy planning process as a narrowing-down process. Later in this chapter and in Chapters 3 and 4 we will go into more detail about strategy decisions relevant to each of the terms in this figure. Then, throughout the rest of the book, we will present a variety of concepts and "how to" frameworks that will help you improve the way you make these strategy decisions. As a preview of what's coming, let's briefly overview the general logic of the process depicted in Exhibit 2-9.

The process starts with a broad look at a market—paying special attention to customer needs, the firm's objectives and resources, and competitors. This helps identify new and unique opportunities that might be overlooked if the focus is narrowed too quickly.

Customers want to buy a product that is different from what competitors offer. Bounty wants its target customers to know that its more absorbent paper towels help customers save by using less than a "leading ordinary brand."
Source: Procter & Gamble

Screening criteria make it clear why you select a strategy

There are usually more opportunities—and strategy possibilities—than a firm can pursue. Each one has its own advantages and disadvantages. Trends in the external market environment may make a potential opportunity more or less attractive. These complications can make it difficult to zero in on the best target market and marketing mix. However, developing a set of specific qualitative and quantitative screening criteria can help a manager define what business and markets the firm wants to compete in. We will cover screening criteria in more detail in Chapter 3. For now, you should realize that the criteria you select in a specific situation grow out of an analysis of the company's objectives and resources.

S.W.O.T. analysis highlights advantages and disadvantages

A useful aid for organizing information from the broader market and developing relevant screening criteria is the **S.W.O.T. analysis**—which identifies and lists the firm's strengths, weaknesses, opportunities, and threats. The name *S.W.O.T.* is simply an abbreviation for the first letters in the words *s*trengths, *w*eaknesses, *o*pportunities, and *t*hreats. Strengths and weaknesses come from assessing the company's resources and capabilities. For example, a local farmer's market might have a great reputation in its community (strength) but have limited financial resources (weakness).

Opportunities and threats emerge from an examination of customers, competition, and the external market environment. The farmer's market might see an opportunity when a growing number of customers in its community show an interest in eating locally grown fruits and vegetables, whereas a threat could be a drought that limits local farmers' production. With a S.W.O.T. analysis, a marketing manager can begin to identify strategies that take advantage of the firm's strengths and opportunities while avoiding weaknesses and threats.

Segmentation helps pinpoint the target

In the early stages of a search for opportunities, we're looking for customers with needs that are not being satisfied as well as they might be. Of course, potential customers are not all alike. They don't all have the same needs—nor do they always want to meet needs in the same way. Part of the reason is that there are different possible types of customers with many different characteristics. In spite of the many possible differences, there often are subgroups (segments) of consumers who are similar and could be satisfied with the same marketing mix. Thus, we try to identify and understand these different subgroups with market segmentation. We will explain approaches for segmenting markets later in Chapter 4. Then, in Chapters 5 and 6, we delve into the

What's Next? Offer more by offering less

In increasingly competitive markets, many companies struggle to find a marketing mix that genuinely differs from the competition. For example, just how differently do customers perceive Shell and Exxon gasoline? Levi's and Lee jeans? Visa and MasterCard? American Airlines, Delta Air Lines, and United Airlines? Whereas some customers see big differences and have a favorite, many see little difference and buy the lowest-priced option.

Sometimes *What's Next?* involves figuring out how to go in a different direction from the competition. Some clever brands differentiate by offering *less* than the competition in most areas—and then add a little something *extra* and *unexpected* for the category. Take a look at JetBlue. When JetBlue entered the air travel market in 2000, all of the major airlines offered free meals on every flight; the choice of flying in first class, business class, or coach; and a wide range of fares. JetBlue offered none of these benefits. But JetBlue wasn't a budget carrier, either. Every plane featured something extra—plush leather seats from the front to the back of the plane and satellite television in every seat. JetBlue also promised to never bump anyone from a flight. Over time, competitors followed JetBlue's lead, so its initial differentiation declined—yet many customers still see differences and remain loyal.

If you live in the western United States, you might find an In-N-Out Burger nearby. Unlike other fast-food joints, In-N-Out doesn't have kids' meals, salads, or desserts. In fact, the menu includes just six items! Yet there is more here than meets the eye. In-N-Out makes everything on the menu from scratch using fresh (not frozen) ingredients. Plus, it has a "secret menu"; only insiders know what it means to ask for an order "Protein Style" or "Animal Style." This combination differentiates it from McDonald's—and keeps loyal customers coming back for more.

When IKEA stores began selling furniture, appliances, and home accessories in the United States, there was nothing like it. American furniture shoppers were accustomed to high levels of customer service and a wide selection of styles. IKEA arrived with only Danish-style furnishings, little sales help on the showroom floor, and no delivery service—and by the way, you had to put the furniture together yourself when you got it home. The furniture was cheap, just like the prices. IKEA told customers they should plan to replace it in a few years. Yet IKEA stores are not bare bones—each features a restaurant with smoked salmon and Swedish meatballs and free day care for the kids of shopping parents.

Some businesses differentiate by offering more than the competition—test-drive a 740-horsepower Ferrari or stay at a luxurious Mandarin Oriental Hotel. On the other hand, JetBlue, In-N-Out, and IKEA went the other direction. When competitors added more, these companies offered less—except for something special—and their most loyal customers appreciate the difference.[8]

What other brands can you think of that have differentiated by offering less?

---

many interesting aspects of customer behavior. For now, however, you should know that understanding customers is at the heart of using market segmentation to narrow down to a specific target market. In other words, segmentation helps a manager decide to serve some segment(s)—subgroup(s) of customers—and not others.

Customers want something different

A marketing mix won't get a competitive advantage if it *just* meets needs in the same way as some other firm. Marketing managers want to identify customer needs that are not being addressed or might be met better than the competition. Combining analyses of customers, competitors, and the company helps the marketing manager identify possible strategies that differentiate a marketing mix from the competition. **Differentiation** means that the marketing mix is distinct from what is available from a competitor. For some examples of differentiation, read *What's Next?* Offer more by offering less. In Chapter 4 we will discuss differentiation and how it is used to help position a brand in a customer's mind.

Narrowing down to a superior marketing mix

Sometimes difference is based mainly on one important element of the marketing mix—say, an improved product or faster delivery. However, differentiation often requires that the firm fine-tune all of the elements of its marketing mix to the specific needs of a distinctive target market. Target customers are more likely to recognize differentiation when there is a consistent theme integrated across the Four Ps decision areas. The theme should emphasize the differences so target customers will think of the firm as being in a unique position to meet their needs.

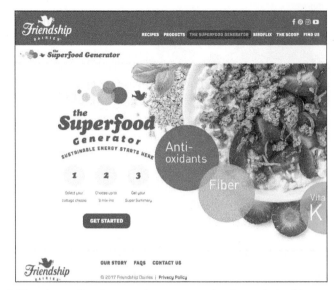

Friendship Dairies (FD) saw its mainstay cottage cheese product fall in sales as Greek yogurt became popular. An analysis of its traditional customers revealed that many now purchased "superfoods" like nuts and berries. So Friendship Dairies repositioned its cottage cheeses as "the original superfood," a perfect "mixer" with other superfoods. This positioning was reinforced with FD promotions and a website.
Source: Friendship Dairies

Let's take a look at how Friendship Dairies' marketing strategy planning and differentiation revived sales for its brand of cottage cheese. Friendship Dairies (FD) is a strong regional brand of cottage cheese found in the northeastern United States and Florida. When Greek yogurt burst on the scene around 2006, it took a lot of sales from cottage cheese—FD's sales fell by more than half in the next five years. When FD's marketing managers analyzed customers, they found that many of its target customers were looking for "superfoods"—like nuts, berries, and vegetables. As a company, Friendship Dairies knew its cottage cheese had been a healthy food since FD started making it in 1917. So it repositioned the brand as the "original superfood." FD wanted customers to see that cottage cheese mixed well with other superfoods for customers seeking a healthy lifestyle.

Friendship Dairies' marketing strategy sought to demonstrate to its target customers how it differed from the competition and created customer value. Among its operational decisions, FD posted cottage cheese-centric recipes featuring other superfood mix-ins on its website and made them easy to share on social media. FD's online Superfood Generator encouraged customers to find their own mix-ins, and demonstrated the health benefits of each combination. It created joint promotions with "mix-in" brands like Dole Pineapple. New bright and colorful packaging attracted attention to FD on grocery store shelves. Coupons provided an incentive for customers to put FD cottage cheese in their shopping carts.

Starting with a broad look at the market (customers, competitors, company, and context), FD's marketing managers narrowed down to a specific strategy. And it worked. Customers liked this superfood—data showed many new customers bought cottage cheese for the first time, visits to FD's website jumped 600 percent, and sales increased 9 percent in just one year.[9]

Exhibit 2-9 focuses on planning each strategy by carefully narrowing down to a specific marketing strategy. Of course, this same approach works well when several strategies are to be planned. This may require different managers working together to develop a marketing program that increases customer equity.[10]

Types of Opportunities to Pursue

LO 2.8

Many opportunities seem obvious only after someone else identifies them. So, early in the marketing strategy planning process it's useful for marketers to have a framework for thinking about the broad kinds of opportunities they may find. Exhibit 2-10 shows four broad possibilities: market penetration, market development, product development, and diversification. We will look at these separately to clarify the ideas. These opportunities differ depending on whether a firm targets its current customers or new ones—and whether it uses existing or new products. However, some firms pursue more than one type of opportunity at the same time.

Exhibit 2-10 Four Basic Types of Opportunities

	Present products	New products
Present markets	Market penetration	Product development
New markets	Market development	Diversification

Market penetration

Market penetration means trying to increase sales of a firm's present products in its present markets—probably through a more aggressive marketing mix. The firm may try to strengthen its relationship with customers to increase their rate of use or repeat purchases or try to attract competitors' customers or current nonusers.

The ridesharing service Uber knew a prime opportunity was to get its current customers to use the Uber service more often, so it introduced Uber Rewards. Customers earn points each time they use an Uber service. Points can earn customers cash back, priority pickups at airports, and complimentary upgrades. Customers often spend more to earn points.

Market development

Market development means trying to increase sales by selling present products in new markets. This may involve searching for new uses for a product. For example, Vievu makes body cameras for law enforcement officers, but it found a new market among plumbers, electricians, and home appliance repair people. These professionals use the body cameras to show customers the work they did on their homes.

Once it had a solid footing in the U.S. market, Uber moved into many international markets—developing these markets by replicating the app and model it used in the United States in other countries.

Product development

Product development means offering new or improved products for present markets. By knowing the present market's needs, a firm may see new ways to satisfy customers. A few years after it started as a ridesharing service, Uber developed a food ordering and delivery service called Uber Eats. Uber was able to build on its well-known brand and keep its drivers happy with a new source of income.

Innocent Drinks started producing smoothies in the United Kingdom. After its success there, Innocent moved the same products into other European countries including Germany. This is an example of market development. Arm & Hammer already made kitty litter, but the company's research and development produced a completely new product, Clump & Seal, for its present customers. This is an example of product development, a new product for present markets.
(left): Source: Innocent Germany; (right): Source: Arm&Hammer

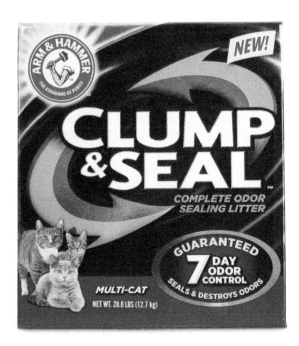

Diversification

Diversification means moving into totally different lines of business—perhaps entirely unfamiliar products, markets, or even levels in the production-marketing system. Products and customers that are very different from a firm's current base may look attractive to the optimists—but these opportunities are usually hard to evaluate. That's why diversification usually involves the biggest risk. McDonald's, for example, opened two hotels in Switzerland. The plan was to serve families on the weekend, but business travelers were the target during the week. Business travelers are not the group that McDonald's usually serves, and an upscale hotel is very different from a fast-food restaurant. This helps explain why operation of the Golden Arch hotels was taken over by a hospitality management company after two years.

Uber has also sought new opportunities for growth through diversification. Uber added a tractor-trailer rental business and now leases hundreds of tractor-trailers, connecting this new product line to trucking firms in need of a trailer. The technology underlying the Uber app makes the hookup.[11]

Which opportunities come first?

Usually firms find attractive opportunities fairly close to markets they already know. Most firms think first of greater market penetration. They want to increase profits and grow customer equity where they already have experience and strengths. On the other hand, many firms find that market development—including the move into new international markets—profitably takes advantage of current strengths.

International Opportunities Should Be Considered

LO 2.9

It's easy for a marketing manager to fall into the trap of ignoring international markets, especially when the firm's domestic market is prosperous. Yet there are good reasons to go to the trouble of looking elsewhere for opportunities.

The world is getting smaller

Many countries have reduced barriers, such as taxes on imports, which in the past made international trade more costly. These moves have increased international trade all over the world. In addition, advances in e-commerce, transportation, and communications are making it easier and cheaper to reach international customers. With a website and e-mail, even the smallest firm can provide international customers with a great deal of information—and easy ways to order—at very little expense.

Develop a competitive advantage at home and abroad

If customers in other countries are interested in the products a firm offers—or could offer—serving them may improve economies of scale. Lower costs (and prices) may give a firm a competitive advantage both in its home markets *and* abroad. Black & Decker, for example, uses electric motors in many of its tools and appliances. By selling overseas as well as in the United States, it achieves economies of scale and the cost per motor is very low.

Get an early start in a new market

A company facing tough competition, thin profit margins, and slow sales growth at home may get a fresh start in another country where demand for its product is just beginning to grow. A marketing manager may be able to transfer marketing know-how—or some other competitive advantage—the firm has already developed. Consider JLG, a Pennsylvania-based producer of equipment used to lift workers and tools at construction sites. Faced with tough competition, JLG's profits all but evaporated. By cutting costs, the company improved its domestic sales. But it got an even bigger boost from expanding overseas. In the first five years, its international sales were greater than what its total sales had been before. Then, when JLG added distribution in China, international sales grew to be half of its business. JLG sales in Europe benefit from new safety rules that require workers to be on an aerial platform if they're working up high. JLG continues to enjoy global growth.[12]

Find better trends in variables

Unfavorable trends in the market environment at home—or favorable trends in other countries—may make international marketing particularly attractive. For example, population growth in the United States has slowed and income is leveling off. In other places in the world, population and income are increasing rapidly. Many U.S. firms can no longer rely on the constant market growth that once drove increased domestic sales. Growth—and perhaps even survival—will come only by aiming at more distant customers. It doesn't make sense to casually assume that all of the best opportunities exist at home.[13]

A growing middle class and an interest in Western fashion are making China a more attractive opportunity for Levi's. Its "Live in Levi's" campaign, here showing a Chinese New Year celebration, is designed to reinforce the fun and rebellious image the brand wants to represent.
Source: Levi Strauss & Co.

Weigh the risks of going abroad

Marketing managers should consider international opportunities, but risks are often higher in foreign markets. Many firms fail because they don't know the foreign country's culture. Learning foreign regulations can be difficult and costly. Political or social unrest makes it difficult to operate in some countries. Venezuela is a striking example. Current Venezuelan leaders have threatened to nationalize some international businesses that have located there. Careful planning can help reduce some of these risks, but ultimately managers must assess both the risks and opportunities that exist in each international market.

 #M4BW

International markets can be opportunities for making the world better

There are plenty of opportunities for companies to improve the quality of life in developing countries—and still make a profit. For example, the French food company Danone has long had an interest in developing markets. In Brazil, it reformulated its best-selling cheese (reducing sugar and adding vitamins) to improve nutrition for Brazilian youth. In Bangladesh, Danone developed a nutrient-rich Danone yogurt for children.[14]

Norwegian fertilizer maker Yara helped launch a green revolution in the East African country of Tanzania. Yara led a public–private partnership to revitalize an area of Tanzania the size of Italy—making it more farming friendly and boosting the incomes of small local farmers. Yara sells fertilizer and teaches better farming techniques. Many farmers have seen a 10-fold increase in crop yields since the program began, and Yara how holds a 50 percent market share in the country. These kinds of win–win (good for a business and good for society) opportunities are at the heart of the marketing concept and marketing for a better world.[15]

CONCLUSION

This chapter introduced you to the basic decision areas involved in marketing strategy planning and explained the logic for the marketing strategy planning process summarized in Exhibit 2-9. In the remainder of this book we'll rely on this exhibit as a way to highlight the organization of the topics we are discussing.

In this chapter, you learned that the marketing manager must constantly study the market environment—seeking attractive opportunities and planning new strategies. You learned about the difference between target marketing and a target marketer. A marketing strategy specifies a target market and the marketing mix the firm will offer to provide that target market with superior customer value. A marketing mix has four major decision areas: the Four Ps—Product, Place, Promotion, and Price.

Controls are needed to ensure that the marketing plans are carried out successfully. If anything goes wrong along the way, continual feedback should cause the process to be started over again—with the marketing manager planning more attractive marketing strategies. Thus, the job of marketing management is one of continuous planning, implementation, and control. Strategies are not permanent; changes should be expected as market conditions change.

The chapter also introduced the ideas of customer lifetime value and customer equity. These concepts reinforce the importance of retaining customers by delivering superior customer value. The importance of building customer equity suggests different marketing strategies might aim at acquiring, retaining, and enhancing customers. Another framework suggested that opportunities might be based on growing current customers and on developing new markets, new products, or both.

There are usually more potential opportunities than a firm can pursue, so possible target markets must be matched with marketing mixes the firm can offer. This is a narrowing-down process. The most attractive strategies—marketing plans and whole marketing programs—are chosen for implementation.

Firms need effective strategy planning to survive in our increasingly competitive markets. The challenge isn't just to come up with some strategy, but to zero in on the strategy that is best for the firm given its objectives and resources—and taking into consideration its strengths and weaknesses and the opportunities and threats that it faces. To improve your ability in this area, this chapter introduces a framework for marketing strategy planning. The rest of this text is organized to deepen your understanding of this framework and how to use it to develop profitable marketing mixes for clearly defined target markets. After several chapters on analyzing target markets, we will discuss each of the Four Ps in greater detail.

KEY TERMS

marketing management process, 34
strategic (management) planning, 35
marketing strategy, 35
target market, 35
marketing mix, 35
target marketing, 36
mass marketing, 36
channel of distribution, 39
personal selling, 39
customer service, 39
mass selling, 40

advertising, 40
publicity, 40
sales promotion, 40
marketing plan, 42
implementation, 43
operational decisions, 43
marketing analytics, 44
marketing program, 45
customer lifetime value (CLV), 45
retention rate, 46
acquisition cost, 46

customer equity, 45
breakthrough opportunities, 48
competitive advantage, 48
S.W.O.T. analysis, 50
differentiation, 51
market penetration, 53
market development, 53
product development, 53
diversification, 54

QUESTIONS AND PROBLEMS

1. Review the Cirque du Soleil case study that opens the chapter. From this case, identify examples of different key terms and concepts covered in the chapter. For example, both the shows and DVDs are examples of Product.

2. Review the Cirque du Soleil case study that opens the chapter. Offer an example of each of the four basic types of opportunities (see Exhibit 2-10 and related discussion) that Cirque du Soleil could pursue. These do not have to be examples that Cirque du Soleil is currently pursuing.

3. Distinguish clearly between a marketing strategy and a marketing mix. Use an example.

4. Distinguish clearly between mass marketing and target marketing. Use an example.

5. Why is the target market placed in the center of the Four Ps in the text diagram of a marketing strategy (Exhibit 2-4)? Explain using a specific example from your own experience.

6. If a company sells its products only from a website, which is accessible over the Internet to customers from all over the world, does it still need to worry about having a specific target market? Explain your thinking.

7. Explain, in your own words, what each of the Four Ps involves.

8. Evaluate the statement "A marketing strategy sets the details of implementation."

9. Distinguish between strategy decisions and operational decisions, illustrating for a local retailer.

10. In your own words, explain what customer equity means and why it is important.

11. Consider two vastly different companies—one sells oral health care products (toothbrushes, toothpaste, and mouthwash), the other is a ridesharing service (like Lyft or Uber). For each of these companies, describe the firm's tactics (consider the Four Ps) in (a) acquiring customers, (b) retaining customers, and (c) enhancing sales from customers.

12. Distinguish between a strategy, a marketing plan, and a marketing program, illustrating for a local retailer.

13. Outline a marketing strategy for each of the following new products: (a) a radically new design for a toothbrush, (b) a new fishing reel, (c) a new resort hotel on a beach in Mexico, and (d) a new industrial stapling machine.

14. Provide a specific illustration of why marketing strategy planning is important for all businesspeople, not just for those in the marketing department.

15. Research has shown that only about three out of every four customers are, on average, satisfied by a firm's marketing programs. Give an example of a purchase you made where you were not satisfied and what the firm could have changed to satisfy you. If customer satisfaction is so important to firms, why don't they score better in this area?
16. Distinguish between an attractive opportunity and a breakthrough opportunity. Give an example.
17. Explain how new opportunities may be discovered by defining a firm's markets more precisely. Illustrate with a situation where you feel there is an opportunity—namely, an unsatisfied market segment—even if it is not very large.
18. In your own words, explain why the book suggests that you should think of marketing strategy planning as a narrowing-down process.
19. Explain the major differences among the four basic types of growth opportunities discussed in the text and cite examples for two of these types of opportunities.
20. Explain why a firm may want to pursue a market penetration opportunity before pursuing one involving product development or diversification.
21. In your own words, explain several reasons why a marketing manager should consider international markets when evaluating possible opportunities.
22. Give an example of a foreign-made product (other than an automobile) that you personally have purchased. Give some reasons why you purchased that product. Do you think that there was a good opportunity for a domestic firm to get your business? Explain why or why not.

MARKETING PLANNING FOR HILLSIDE VETERINARY CLINIC

Appendix D (the Appendices follow Chapter 19) includes a sample marketing plan for Hillside Veterinary Clinic. Skim through the different sections of the marketing plan. Look closely at the "Marketing Strategy" section.

1. What is the target market for this marketing plan?
2. What is the strategy Hillside Veterinary Clinic intends to use?
3. What are your initial reactions to this strategy? Do you think it will be successful? Why or why not?

SUGGESTED CASES

3. NOCO United Soccer Academy
4. Petoskey Tech Support
5. Resin Dynamics

12. DrV.com—Custom Vitamins
17. Wise Water, Inc.

Video Case 1. Potbelly Sandwich
Video Case 3. Big Brothers Big Sisters of America

MARKETING ANALYTICS: DATA TO KNOWLEDGE

CHAPTER 2: TARGET MARKETING

Marko, Inc.'s managers are comparing the profitability of a target marketing strategy with a mass marketing "strategy." The spreadsheet gives information about both approaches.

The mass marketing strategy is aiming at a much bigger market. But a smaller percentage of the consumers in the market will actually buy this product—because not everyone needs or can afford it. Moreover, because this marketing mix is not tailored to specific needs, Marko will get a smaller share of the business from those who do buy than it would with a more targeted marketing mix.

Just trying to reach the mass market will take more promotion and require more retail outlets in more locations—so promotion costs and distribution costs are higher than with the target marketing strategy. On the other hand, the cost of producing each unit is higher with the target marketing strategy—to build in a more satisfying set of features. But because the more targeted marketing mix is trying to satisfy the needs of a specific target market, those customers will be willing to pay a higher price.

In the spreadsheet, "quantity sold" (by the firm) is equal to the number of people in the market who will actually buy one each of the product—multiplied by the share of those purchases won by the firm's marketing mix. Thus, a change in the size of the market, the percentage of people who purchase, or the share captured by the firm will affect quantity sold. And a change in quantity sold will affect total revenue, total cost, and profit.

Design element: #M4BW box globe icon: ©Vectoryzen/Shutterstock

CHAPTER THREE

Rodrigo Reyes Marin/AFLO/Newscom

Evaluating Opportunities in the Changing Market Environment

Back in 1995, when Amazon.com first went online, its founder Jeff Bezos saw a breakthrough opportunity in changing the retail shopping experience. Amazon's strategy took advantage of trends Bezos observed, especially a growing number of consumers shopping on the Internet. Laws back then didn't require the collection of sales taxes on online sales, which offset shipping and handling charges. Early investors were patient as Amazon lost $3 billion from 1995 to 2003. They shared Bezos' belief that if Amazon consistently delivered value and satisfaction, customers would come back, buy more, and tell their friends. Bezos' and his investors' predictions were on target.

Amazon wasn't just "lucky" to be in the right place and time. Amazon understands its mission and its strengths and weaknesses. Amazon's mission statement helps the company keep its attention on what matters most: "We strive to offer our customers the lowest possible prices, the best available selection, and the utmost convenience." With this focus, Amazon pioneered community features such as customer reviews and customer questions and answers, which help customers discover new products and make more informed decisions. Amazon Marketplace gives many sellers the opportunity to compete directly with Amazon. Amazon also expanded into new product markets, over time adding electronics, home and garden, health and beauty aids, industrial tools, clothing and shoes, cloud computing services, and more. Amazon now has more than 50 percent market share of online sales in the United States.

One of Amazon's most important resources is its access to and use of data about customers—especially how customers behave on the Amazon site. For example, when customers visit its site, Amazon knows what they search for, which pages they visit, what they buy and don't buy, and whether they read reviews or shop around before purchasing. These data and Amazon's artificial intelligence systems help it deliver shopping experiences unique to every customer. Amazon combines these data with knowledge gained from observing tens of millions of other customers to anticipate its customers' wants and needs. These insights help Amazon sell highly targeted and personalized advertising directly to many companies. For example, when a Seattle-based physical therapy chain was looking to expand, Amazon showed its ads only to local customers who had recently bought knee braces.

Amazon recognized opportunities in other countries, initially launching new sites in the United Kingdom and Germany. Like the United States, these markets offered a large number of customers with access to the Internet and high incomes. Amazon is now in 17 countries, including Australia, Canada, France, India, and China. Not all of these efforts have gone well. Whereas many Chinese consumers have access to the Internet and growing incomes, Amazon's share of China's e-commerce sales are stuck at less than 2 percent. In China, Amazon faces strong domestic competition from Alibaba and JD.com, which control 43 percent and 20 percent of China's online market share, respectively. In India, Amazon faces headwinds from regulations crafted to protect local retailers. A new Indian law targets Amazon and prohibits foreign companies from holding online inventory and selling it directly to customers.

Amazon continually strives to make shopping convenient. For a $119 annual fee, Amazon Prime customers receive free two-day shipping on select products and free access to select videos, music, and more. Amazon is also setting up warehouses near major metro areas and testing different delivery approaches. For example, local inventory enables Amazon's Prime Now program to offer one-hour delivery in some markets and same-day delivery in others. Amazon hopes to someday use drone aircraft to deliver small packages in 30 minutes or less. Right now, a patchwork of local laws—rather than technology—is holding up this plan.

Striving for "utmost convenience," Amazon knows it needs physical stores, too. Amazon hopes to leverage two of its greatest resources—its technological expertise and financial resources—to find ways to improve the in-store shopping experience. Amazon is testing a new grocery store concept called Amazon Go, which lets customers walk in, grab what they want, and walk out—without having to swipe a card or stand in a checkout line. These customers are not getting everything for free—after all, Amazon is in business to make a profit. Artificial intelligence and sensors placed throughout the store track items customers place in their physical baskets, and the Amazon Go app places those same items in their virtual shopping carts. When finished shopping, a customer simply walks out, and everything is charged to his or her Amazon account. Is this what's next for retail? And when will Amazon bring this to Whole Foods, which it recently purchased?

Another innovation, the Amazon Echo, was created to make life (and shopping) better for Amazon customers. Echo is a nine-inch-tall smart speaker. When a customer calls out "Alexa," Echo wakes up and connects to a personal digital assistant. From there, a user can ask Alexa to play music, turn on lights, deliver news and sports scores, report the weather—and of course place an order on Amazon. It's so easy a six-year-old can do it—and one did. When a little girl from Dallas, Texas, asked Alexa to "send me a KidKraft Sparkle Mansion dollhouse and some sugar cookies," they arrived at her home two days later. After the surprise delivery, Mom and Dad activated parental controls to add another step to the ordering process.

Amazon's diverse product line means it needs to monitor a wide range of strong competitors. Amazon engages in price wars with Walmart over books, toys, and DVDs; fights Netflix and Hulu in video streaming; and battles Facebook and Google for online advertising. Amazon has never been afraid to tackle markets with strong competitors. On the contrary, it seeks large and growing markets where it can leverage its strengths—company size, technology skills, efficient operations, and creative personnel. To grow in these competitive and fast-changing market environments, Amazon must continue to anticipate market trends and adjust its marketing strategies.[1]

LEARNING OBJECTIVES

The Amazon case shows that a marketing manager must understand customer needs and choose marketing strategy variables within the framework of a changing market environment. Opportunities need to fit with a firm's objectives and resources, and managers should screen for opportunities where there is a chance for competitive advantage and, if possible, favorable trends in the external environment.

When you finish this chapter, you should be able to

1. know the variables that shape the environment of marketing strategy planning.
2. understand why company objectives are important in guiding marketing strategy planning.
3. see how the resources of a firm affect the search for opportunities.
4. know how to conduct a competitor analysis and how different types of competition affect strategy planning.
5. understand how the economic and technological environments can affect strategy planning.
6. know how elements of the political and legal environment affect marketing strategy planning.
7. understand the cultural and social environment and how demographic trends affect strategy planning.
8. understand how to screen and evaluate marketing strategy opportunities.
9. understand the important new terms (shown in **red**).

The Market Environment

LO 3.1

The marketing strategy planning process (see Exhibit 2-9) requires narrowing down to the best opportunities and developing a strategy that gives the firm a competitive advantage and provides target customers with superior customer value. This narrowing-down process should consider the important elements of the market environment and how they are shifting.

A large number of forces shape the market environment. The direct market environment includes customers, the company, and competitors. The external market environment is broader and includes four major areas:

1. Economic environment
2. Technological environment
3. Political and legal environment
4. Cultural and social environment

Exhibit 3-1 Marketing Strategy Planning, Competitors, Company, and External Market Environment

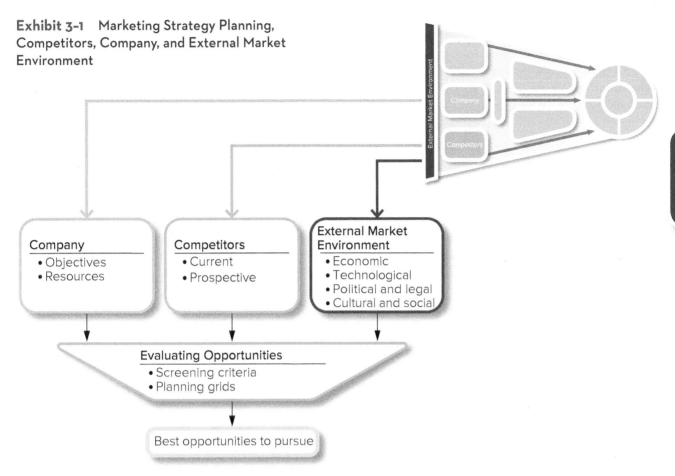

In this chapter we'll look at the key market environment variables shown in Exhibit 3-1 in more detail. We'll see how they shape opportunities—limiting some possibilities but making others more attractive. Managers can't alter the variables of the market environment. That's why it's useful to think of them as uncontrollable variables. On the other hand, a manager should analyze the environment when making decisions that can be controlled. For example, a manager can select a strategy that leads the firm into a market where competition is not yet strong or where trends in the external market are likely to support market growth.

Objectives Should Set Firm's Course

LO 3.2

Company objectives should shape the direction and operation of the whole business. It is difficult to set objectives that really guide the present and future development of a company. While marketing managers should be heard, the responsibility for setting whole-company objectives resides with top management. Top management must look at the whole business, relate its present objectives and resources to the external environment, and then decide what the firm wants to accomplish in the future.

A mission statement helps set the course

Each firm needs to develop its own objectives based on its own situation. This is important, but top executives often don't state their objectives clearly. If objectives aren't clear from the start, different managers may hold unspoken and conflicting objectives.

Many firms try to avoid this problem by developing a **mission statement**, which sets out the organization's basic purpose for being. A good mission statement focuses on a

#M4BW Warby Parker was founded with a rebellious spirit and a lofty mission statement: "to offer designer eyewear at a revolutionary price, while leading the way for socially conscious businesses."
Source: *Warby Parker*

Company objectives give direction to marketing objectives

few key goals rather than embracing everything. It should guide managers in determining which opportunities to pursue.

The mission statement for the American Red Cross provides this direction: "The American Red Cross prevents and alleviates human suffering in the face of emergencies by mobilizing the power of volunteers and the generosity of donors." This tells a Red Cross employee she should not offer services supporting children's literacy—even if it's a good cause. On the other hand, the opportunity to develop a smartphone app that accepts donations for aid to Nepal following an earthquake fits right in with this charity's mission. A mission statement may need to be revised as new market needs arise or as the market environment changes. This would be a fundamental change and not one that is made casually.[2]

A mission statement is important, but it is not a substitute for more specific objectives that provide guidance in screening possible opportunities. Company objectives guide managers as they search for and evaluate opportunities—and later plan marketing strategies.

Particular *marketing* objectives should be set within the framework of larger company objectives. As shown in Exhibit 3-2, firms need a hierarchy of objectives—moving from company objectives to marketing department objectives. For each marketing strategy, firms also need objectives for each of the Four Ps—as well as more detailed objectives. For example, in the Promotion area, we may need objectives for personal selling, mass selling, publicity, *and* sales promotion.

Insurance company USAA provides an example. One of USAA's top company objectives is to deliver high levels of customer satisfaction. Because customer service is a key driver of satisfaction, it is critical to have highly qualified and well-trained customer

Exhibit 3-2 A Hierarchy of Objectives

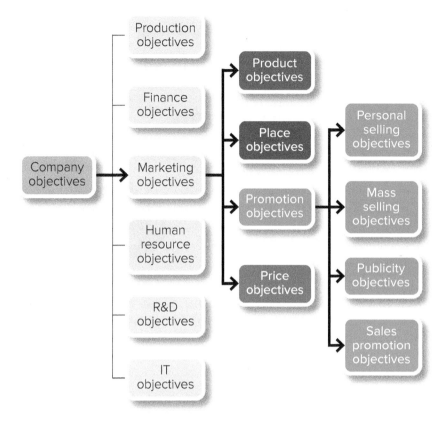

service representatives (CSRs) responding to customers' questions or problems. This, in part, drives human resource objectives that guide the hiring and training of CSRs. It also suggests USAA's information technology (IT) group could develop an application that increases the effectiveness of CSRs in their dealings with customers. In this case, USAA developed new software that allows CSRs to view the same online screens that customers see on their own devices. Customer service objectives (part of marketing and personal selling objectives) call for improved ability to solve any problems on a customer's first call—leading to higher levels of customer satisfaction. Achieving these objectives helps USAA deliver even higher levels of customer service, satisfaction, and customer retention in a competitive insurance market.[3]

When setting objectives, managers should keep in mind what is realistic and achievable. Overly ambitious objectives are useless if the firm lacks the resources to achieve them.

Company Resources May Limit Search for Opportunities

LO 3.3

Every firm has some resources—hopefully some unique ones—that set it apart. Breakthrough opportunities—or at least some competitive advantage—come from making use of these strengths while avoiding direct competition with firms having similar strengths.

To find its strengths or recognize weaknesses, a firm must evaluate its functional areas (production, research and engineering, marketing, general management, and finance) as well as its present products and markets. The knowledge of people at the firm can also be a unique resource. By analyzing successes or failures in relation to the firm's resources, management can discover why the firm was successful—or why it failed—in the past. Let's take a closer look at three specific types of resources that guide marketing strategy planning—financial, production, and marketing.

The opportunities a firm decides to pursue may depend on its resources and capabilities. For example, General Mills was already producing a key ingredient for its new Chocolate Chex Mix, so it went ahead and launched Chex Mix Turtle. When 3-D printing technology became increasingly practical, Proto Labs added this production capability and faster services for its customers.
(left): Source: General Mills, Inc.; (right): Used with permission of Proto Labs, Inc.

Financial resources determine opportunities

Some opportunities require large amounts of capital just to get started. Money may be required for R&D, production facilities, marketing research, or advertising before a firm makes its first sale. And even a really good opportunity may not be profitable for years—so cash flow matters, too. If financial resources are limited, a firm may have to develop lower-cost and potentially riskier marketing plans. Lack of financial strength is often a barrier to entry into an otherwise attractive market.

What and how many can be produced? At what cost?

Production resources also affect opportunities. In many businesses, the cost of producing and selling each unit decreases as the quantity produced increases. This can place smaller firms at a production cost disadvantage when seeking to win business from larger competitors. On the other hand, new—or smaller—firms sometimes have the advantage of flexibility. They are not handicapped with large, special-purpose facilities that are obsolete or poorly located.

Some firms gain flexibility by not doing their own manufacturing. Hanes used to have its own factories produce its underwear and T-shirts. Now it has other companies make these products. And when Hanes needs to make a change, it has the flexibility to work with whatever suppliers around the world are best able to meet its specifications.

Marketing strategies often build on existing marketing resources

Marketing resources can create opportunities for a firm. In the product area, for example, a familiar brand can be a big strength. Starbucks is famous for its coffee beverages. When Starbucks introduced its coffee ice cream, many people quickly tried it because they knew what Starbucks flavor meant.[4]

IBM drew on its strong customer relationships when it introduced its Watson artificial intelligence software. IBM had a decades-long track record of helping businesses solve problems through the application of technology. This helps IBM get the attention of companies looking for an expert to help them add artificial intelligence to their goods and services.

Firms can and often do develop new resources and capabilities when entering new markets. The lack of such resources should not rule out an opportunity, but an organization should recognize the costs required to develop the new capabilities.

Analyzing Competitors and the Competitive Environment

LO 3.4

Avoid head-on competition

The **competitive environment** affects the number and types of competitors the marketing manager faces and how they may behave. Although marketing managers usually can't control these factors, they can choose strategies that avoid head-on competition. And where competition is inevitable, they can plan for it.

Kinds of competitive situations

Economists describe four basic kinds of market (competitive) situations: monopoly, oligopoly, pure competition, and monopolistic competition. Understanding the differences among these market situations is helpful in analyzing the competitive environment, and our discussion assumes some familiarity with these concepts. (For a review, see Exhibit A–11 and the related discussion in Appendix A, following Chapter 19.)

Competitor-free environments are rare

Monopoly situations occur when one firm completely controls a broad product-market. These are relatively rare in market-directed economies. When monopolies are necessary, monopolists often face a great deal of government regulation. For example, in many parts of the world prices set by utility companies (electricity and water) must be approved by a government agency. Monopolists can be tempted to ignore customer needs, but a customer orientation can fend off increased government regulation and help the firm if competitors later enter the market.

Pure competition and oligopoly when there is little or no differentiation

Many product-markets head toward *pure competition*—or *oligopoly*—over the long run. In these situations, competitors offer very similar marketing mixes, and customers see the alternatives as close substitutes. In other words, competitors have failed to differentiate their offerings. In this situation, managers usually compete on low prices, and profit margins shrink. In oligopoly there are a few large firms competing, whereas in pure competition there are often many firms.

Avoiding these competitive situations is sensible and fits with our emphasis on finding a competitive advantage. Marketing managers can't just adopt the same "good" marketing strategy being used by other firms. That leads to head-on competition and a downward spiral in prices and profits. So target marketers try to offer a marketing mix better suited to target customers' needs than competitors' offerings.

Monopolistic competition is typical and a challenge

In *monopolistic competition,* different firms offer marketing mixes that at least some customers see as different. Each competitor tries to get control (a monopoly) in its "own" target market. But competition still exists because some customers see the various alternatives as substitutes. Most marketing managers in developed economies face monopolistic competition.

In monopolistic competition, marketing managers sometimes try to differentiate very similar products by focusing on other elements of the marketing mix. For example, many consumers believe that most brands of gasoline are similar. This makes it difficult for a Texaco station to attract customers by claiming it offers better gasoline. So a Texaco station might compete on other parts of its offering, perhaps by serving Seattle's Best Coffee, giving discounted car washes, or maintaining longer hours of operation.

Such approaches may not work for long if they are easily copied by competitors. Thus, marketing managers should actively seek a **sustainable competitive advantage**, a marketing mix that customers see as better than a competitor's mix and cannot be quickly or easily copied. For example, it is difficult for a competitor to imitate the computer operating system used by Apple or to build the brand awareness of Taco Bell. But if Taco Bell introduces a new burrito, another Mexican fast-food restaurant might be able to produce a similar-tasting dish.

Analyze competitors to find competitive advantage

The best way for a marketing manager to avoid head-on competition is to find new or better ways to satisfy customers' needs and provide value. The search for a breakthrough opportunity—or some sort of competitive advantage—requires an understanding not only of company and customers but also of competitors. That's why marketing managers turn to **competitor analysis**—an organized approach for evaluating the strengths and weaknesses of current or potential competitors' marketing strategies.

The basic approach to competitor analysis is simple. You compare the strengths and weaknesses of your current (or planned) target market and marketing mix with what competitors are currently doing or are likely to do in response to your strategy.

The initial step in competitor analysis is to identify potential competitors. It's useful to start broadly and from the viewpoint of target customers. Companies may offer quite different products to meet the same needs, but they are competitors if customers see them as offering close substitutes. For example, Avis car rental might recognize that customers arriving for business in a new city might consider renting a car from Enterprise or hailing a ride from Lyft.

Keep a closer eye on competitive rivals

Marketing managers usually narrow the focus of their analysis to the set of **competitive rivals** that are their closest competitors. Rivals offering similar products are usually easy to identify. However, with a really new and different product concept, the closest competitor may be a firm that is currently serving similar needs with a different type of product. Netflix realizes that video delivery platforms like HBO and YouTube are not its only competition. It also recognizes that customers could choose to spend "entertainment time" playing an online video game like *Fortnite*.

Marketing managers will often assess their progress against competitive rivals by monitoring market share. **Market share** is the portion of total sales in a product category accounted for by a particular brand. So the market share of the Apple iPhone in the United States is the portion of iPhones sold in the United States in a given period of time as compared to all smartphones sold in the United States in the same time period. It is usually expressed as a percentage. Market share may be based on units or dollars sold. Marketing managers often track market share over time to monitor how well a product is doing in a particular market. The Marketing Analytics in Action: Market Share activity provides an example of this.

Actively seek information about competitors

A marketing manager should actively seek information about current and potential competitors. Although most firms try to keep the specifics of their plans secret, much public information may be available. Sources of competitor information include the competitor's own website, trade publications, sales reps, suppliers, and

---

Marketing Analytics in Action: Market Share

Marketing managers often keep a close eye on market share, comparing how market share changes over time and typically sorting out market share in different geographic locations. Consider the data below, which show the Apple iPhone's market share in four countries in the three months ending December 2017 and 2018.[5]

Country	iPhone Market Share		
	December 2017	December 2018	Market Share Change
United States	43.9%	43.7%	−0.2%
Mexico	4.4	5.5	+1.1
China	28.6	23.3	−5.3
Japan	55.2	46.4	−8.8

1. *What insights would this provide for a marketing manager with responsibility for global sales of the iPhone?*
2. *Speculate on why the iPhone's market share declined so much over one year in China and Japan.*
3. *Why do you think market share for the iPhone is much higher in the United States and Japan as compared to Mexico and China? Look at data in Exhibit 3-3 for ideas.*

other industry experts. In business markets, customers may be quick to explain what competing suppliers are offering.

Ethical issues may arise

The search for information about competitors can raise ethical issues. For example, people who change jobs and move to competing firms may have a great deal of information, but is it ethical for them to use it? Similarly, some firms have been criticized for going too far—like waiting at a landfill for competitors' trash to find copies of confidential company reports or "hacking" a competitor's computer network.

Beyond the moral issues, spying on competitors to obtain trade secrets is illegal, and damage awards can be huge. In one example, the courts ordered competing firms to pay Procter & Gamble about $125 million in damages for stealing secrets about its Duncan Hines soft cookies.[6]

> **Ethical Dilemma**
>
> *What would you do?* You are a salesperson for a company that manufactures industrial lighting used in factories and warehouses. During a recent sales call on one of your customers, ABC Supply, an ABC engineer tells you that ABC is testing a competitor's top secret new product. The competitor is making light bulbs with artificial intelligence and motion detectors to track employees throughout a large facility. The engineer shows you a box full of the light bulbs. When the engineer leaves the room, you have the opportunity, without anyone seeing or knowing, to put one of the bulbs in your briefcase. Your company promotes getting competitive information any way possible and gives bonuses for competitive information like this. Based on what you have heard, this "sample" might be worth $1,000.[7]
>
> *Would you take one of the light bulbs? What do you think of this company's "bonus" policy and attitude?*

The Economic Environment

LO 3.5

The **economic environment** refers to macroeconomic factors, including national income, economic growth, and inflation, that affect patterns of consumer and business spending. The rise and fall of the economy in general, within certain industries, or in specific parts of the world can have a big impact on what customers buy.

Economic conditions change rapidly

The economic environment can, and does, change quite rapidly. The effects can be far-reaching and require changes in marketing strategy.

Even a well-planned marketing strategy may fail if a country or region goes through a rapid business decline. You can see how quickly this can occur by considering what happened in the U.S. housing market during the economic recession that began near the end of 2007. Earlier in the decade the economy was growing, household incomes were increasing, and interest rates were low. As a result, the housing market was hot. Manufacturers of building materials, home builders, real estate firms, and mortgage companies all enjoyed strong profits as they scrambled to keep up with demand. In 2008 the housing market abruptly collapsed. Firms that had done so well a year earlier were suffering huge losses—and many went bankrupt. Worse, millions of people lost their homes when they could not afford rising payments for variable-rate mortgages.

Interest rates affect buying for big-ticket items

Changes in the economy are often accompanied by changes in the interest rate—the charge for borrowing money. Many consumers do not have the money to pay the full price of an expensive purchase, like a new truck or house. They may choose to borrow money and pay back the loan with interest. Interest rates directly affect the total price borrowers must pay for products. So the interest rate affects when, and if, they will buy.

Consider the situation of a home buyer who might borrow $150,000 on a 30-year mortgage at 4 percent interest. Monthly payments would be $716. If the interest rate on that same 30-year loan jumped 7 percent, monthly payments would be $998—almost $300 more per month. You can see how high interest rates can cause many consumers to not buy a home, buy a smaller home, or buy the home but not new furniture or landscaping.

Inflation makes customers behave differently

Inflation refers to the rate at which the prices of goods and services rise. If the rate of inflation was 3 percent last year, this means that, on average prices rose 3 percent in the prior year. In the last couple of decades, the United States and most developed countries have experienced inflation rates well below 5 percent. These rates are considered "normal" and generally do not affect customer behavior.

In some Latin American countries, inflation has exceeded 400 percent a year in recent times. This would mean that a customer might expect the price of bread or a car to *double* in just three months. When consumers anticipate fast-rising prices, they behave differently, usually spending money as soon as they get it for fear it will buy less tomorrow. When price increases outpace income gains, consumers have no choice but to buy less.

Marketing managers must watch the economic environment carefully. In contrast to the cultural and social environment, economic conditions can move rapidly and require immediate strategy changes.[8]

The Technological Environment

Technology affects opportunities

Technology is the application of science to convert an economy's resources to output. Technology affects marketing in two basic ways: it creates opportunities for new products, and it drives the development of new processes (ways of doing things). Anticipating technologies helps firms spot threats and opportunities.

Anticipate technologies and plan for the future

Although most technological developments don't come out of nowhere, it is not always clear how a technology just over the horizon might change a business. For example, producers of cameras and GPS might not have seen how quickly smartphones would cut into their business. Marketing managers should monitor technologies that might impact their industry—and engage in contingency planning for potential opportunities and threats.

Consider the pending impact of driverless cars. Many expect that by the year 2030, most of us will be transported in robot-driven cars. Automakers need to prepare new vehicles with this technology. Cars may be redesigned around getting tasks done—perhaps with pop-up workstations that convert into baby-changing tables. Robots are expected to be much better drivers than humans, so auto insurance companies need to prepare for lower premiums. In the United States, auto accidents account for about 2 million emergency room visits each year. Because driverless cars are expected to have substantially fewer accidents, hospitals should anticipate significant declines in demand for emergency services. Bars and restaurants may see more business—and sell more alcohol—when patrons don't need to worry about driving home under the influence. Marketing managers who anticipate the impact of new technology can plan and adapt marketing strategies for the future.[9]

Artificial intelligence changes marketing processes

Many of the fastest-growing new technologies use **artificial intelligence (AI)**, which refers to having machines operate like humans with respect to learning and decision making. Artificial intelligence is implemented by an **intelligent agent**, a device that observes an environment and acts to achieve a goal. Many companies today utilize AI to deliver better service to customers; for example, AI is behind the driverless car we just discussed. The car operates as an intelligent agent that makes decisions about when to stop, accelerate, brake, and turn. Over time it can also learn passenger preferences for a slower "driver" or more scenic routes. The Amazon Echo includes an intelligent agent that learns where a customer lives, so it chooses the right city when it is asked about "today's weather" or "local news." In the *What's Next?* Intelligent agents learn to create and deliver customer value box, we take a closer look at AI and intelligent agents.

What's Next? Intelligent agents learn to create and deliver customer value

Faster computer processing and more data allow intelligent agents—powered by artificial intelligence—to deliver more value to customers. Intelligent agents are not new. For example, a home thermostat is an intelligent agent. Since 1927, thermostats could sense a room's temperature and activate heating or cooling to maintain a constant temperature. These days, thermostats are smarter and deliver more value. The Nest thermostat automatically learns a customer's schedule and preferences—its sensors observe when you go to sleep, wake up, and go to work. It also learns your preferred temperature for each of these times of day. Customers save money (and the environment) by using heat and air-conditioning only when they need it.

Let's look at a few more examples of intelligent agents delivering customer value.

Many people enjoy Instagram as a place to see photos and video stories from friends and family. While some users "follow" hundreds or thousands of "friends," most just want to see a small subset of those friends' posts. Instagram's intelligent agent learns a user's preferences and places those posts at the top of the feed.

Intelligent agents also help shoppers save time and money. For example, the Honey app automates the process of searching for an online coupon. When customers check out of an online store, they simply click on the "Apply Coupons" button and the Honey app searches its database for the best coupon for that purchase. It applies the coupon and customers save money and the time looking for a coupon. These examples are here now, so *What's Next?*

Soon intelligent agents will deliver value by taking on marketing activities—perhaps doing a better job at a lower cost than humans. For example, intelligent agents will answer every customer service phone call on the first ring—computers can handle unlimited customers at the same time. The agents will be smart enough to answer most questions in any language. Other intelligent agents will operate like retail salespeople. Someday soon, a customer's smartphone might also be a salesperson. Imagine Juli telling her phone, "I need red sneakers." The phone responds with follow-up questions to better narrow down her needs: "What will you use them for?" "How important is comfort?" "What athletic shoes have you liked in the past?" The intelligent agent "remembers" Juli wears a size 8, is price sensitive, and places a high value on free shipping and easy returns. The agent searches online for coupons and compares retailer offers before presenting the customer with the best "red sneaker" options given her wants and needs. Juli clicks "buy this" next to the Adidas Gazelle from Zappos, and the agent places her order.

The use of artificial intelligence and intelligent agents in marketing is just beginning. As intelligent agents learn more about customers, their recommendations and advice will improve. Buyers will save time and money when intelligent agents offer personalized recommendations and sage buying advice.[10]

 #M4BW

Machine learning teaches intelligent agents dermatology

Machine learning refers to a type of computer algorithm where a software application becomes more accurate in predicting outcomes without actual programming. Through machine learning, a program makes predictions, receives feedback on whether it was correct or not, and then updates the program. With access to a lot of data, machine learning can help a software program learn a specific skill rather quickly. For example, differentiating between a benign (not dangerous) skin lesion and a cancerous melanoma (which can be very dangerous) can be challenging for patients and doctors. SkinVision trained its software by showing it hundreds of thousands of pictures of cancerous and noncancerous skin lesions. Machine learning "taught" the software, and it became more accurate than most dermatologists in identifying cancer. Now SkinVision's smartphone app lets a user snap a photo of the area of concern and instantly receive an initial diagnosis. SkinVision could help many people get a fast diagnosis and earlier treatment at a lower cost than visiting a dermatologist.[11]

#M4BW

Technology cares for aging Japanese population

Japan is concerned about how it will care for its elderly—especially those with dementia. The Japanese population is aging quicker than most of the world. Already, a quarter of the Japanese population is 65 or older (compared to just 15 percent of Americans). By 2025, Japan predicts a shortfall of almost 400,000 senior caregivers. Some companies are developing different intelligent agents to help fill the void.

Robots are particularly popular. Sony's robotic pet dog Aibo, and SoftBank's Paro, a robotic baby seal, have been shown to elicit emotional responses and yield benefits similar to live animal therapy (without all the owner responsibility). Soft-Bank Robotics' humanoid companion "Pepper" recognizes faces and human emotions and asks and answers questions; seniors enjoy chatting with Pepper, which also monitors patients' mental health for doctors and family members. Advances like these can provide health care at lower costs and make for better experiences for an aging world.[12]

Technology comes with challenges

Although technological change opens up many new opportunities, it also poses challenges for marketers. For some marketing managers new technology is scary, and they avoid what they do not understand. For others it is easy to fall in love with the latest thing—whether from the firm's R&D lab or a consultant selling social media tools—and blindly add it to the firm's marketing strategy. Both approaches can lead to a production-oriented way of thinking. That makes it more important than ever for a marketing orientation to guide the process. Finding the best applications of technology still requires that marketing managers begin with customer needs.

#M4BW Researchers at Coca-Cola developed a new technology for the plastic bottles used for many of its products. PlantBottle packaging looks and functions just like traditional PET plastic bottles—and it is recyclable. But because these bottles are partially made from plant material, they have a much smaller carbon footprint, making them better for the environment and an example of marketing for a better world. To accelerate future development of plant-based plastics and to promote marketing for a better world, Coke shared this technology with Ford, Heinz, Nike, and Procter & Gamble (but not Pepsi).
Source: The Coca-Cola Company

Changes in the technological environment can be a threat or an opportunity. Forward-looking companies see opportunity. Deluxe Corporation was founded more than 100 years ago and soon became a leader in the printing of paper checks and business forms used by consumers and businesses. Nowadays, financial transactions are increasingly handled by electronic means; demand for paper checks has fallen off a cliff. Deluxe saw this coming and adapted its marketing strategy. Today, Deluxe serves small businesses with a wide range of custom products and marketing services.
Source: Deluxe Enterprise Operations

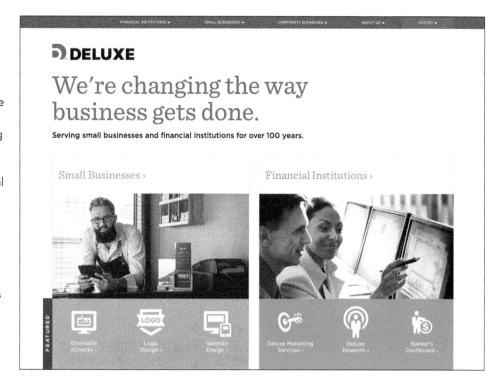

The Political Environment

LO 3.6

The attitudes and reactions of people, social critics, and governments all affect the political environment. Consumers in the same country usually share a common political environment, but the political environment can also affect opportunities at a local or international level.

Nationalism can be limiting in international markets

Strong sentiments of **nationalism**—an emphasis on a country's interests before everything else—affect how macro-marketing systems work. They can affect how marketing managers work as well. Nationalistic feelings can reduce sales—or even block all marketing activity—in some international markets. For many years, China has made it difficult for outside firms to do business there—in spite of the fact that the Chinese economy has experienced explosive growth as its factories have turned out larger and larger portions of the goods sold in the United States, Europe, and other parts of the world.

The "Buy American" policy in many government contracts and business purchases reflects this same attitude in the United States. There is broad consumer support for protecting U.S. producers—and jobs—from foreign competition. Automaker Chrysler drew on this sentiment in its "Imported from Detroit" campaign, which included Detroit rap music artist Eminem and a two-minute Super Bowl commercial "Halftime in America" featuring Clint Eastwood. Both promoted Chrysler's ties to Detroit, Michigan, Chrysler's American home.[13]

Trade agreements free the flow of goods between countries

Important dimensions of the political environment are likely to be similar among nations that have banded together to have common regional economic boundaries. **Free trade** refers to agreements between countries to not restrict imports and exports. Some countries have restrictions that raise the costs of importing, but free trade agreements drop those barriers. Since its inception in 1994, the North American Free Trade Agreement (NAFTA), and later its replacement United States-Mexico-Canada Agreement (USMCA), reshaped the rules of trade among the United States, Canada, and Mexico.

A similar agreement exists across Europe. Twenty years ago, each country in Europe had its own unique trade rules and regulations. These differences made it difficult and expensive to move products from one country to another. Now, the almost 30 countries

Adero wants marketers to keep in mind that a website that targets prospects from all over the world won't be successful in turning them into customers if it ignores nationalism and cultural differences.
Source: Adero, Inc.

of the European Union (EU) are reducing conflicting laws, taxes, and other obstacles to trade within Europe. This, in turn, is reducing costs and prices, and it is creating new jobs. Many of the member countries use the same currency (the euro), simplifying inter-European commerce and trade yet more.

Free trade backlash

Times are changing. After a few decades of movement toward more free trade, the United States and some other countries have recently begun to impose trade restrictions. After free trade moved many jobs out of the country, American workers sought greater protection for their jobs. Emerging economies often have fewer laws that protect workers and the environment; opponents of free trade argue this subjects women and children to grueling working conditions and degrades natural resources. This debate continues to play out, and time will tell whether the world moves toward more open or closed trade policies. Marketing managers may need to be prepared for either outcome as changes in the political environment often lead to changes in the legal environment.[14]

The Legal Environment

Our discussion of the legal environment focuses on important laws in the United States (other countries often have similar laws) that regulate marketing practices. Most of these laws are ultimately designed to protect consumers from less scrupulous marketing practices by promoting competition or prohibiting deceptive marketing practices. Many laws are designed to specifically limit what a marketing manager can do with respect to one or more of the 4 Ps. The chapters on the 4 Ps will address specific laws in more detail.

Trying to encourage competition

American economic and legislative thinking is based on the idea that competition among many firms helps the economy. For example, if only one company sold all the gasoline in the country, it might be tempted to significantly raise prices to maximize profits—consumers would have little choice but to accept the higher prices. Instead, the competition between sellers of gasoline keeps prices in check.

Efforts by businesses to limit competition are considered contrary to the public interest. More than a century ago, Congress passed a series of antimonopoly laws that were designed to keep one company from dominating a market. The Sherman Act (1890) outlawed anticompetitive activities (for example, when two or more companies conspire to set prices) and actions that create or attempt to monopolize a market. The Clayton Act (1914) focused on anticompetitive practices that did not fall under the Sherman Act.

Other laws protect consumers

Other laws protect consumers from less scrupulous marketing practices. Some consumer protections are built into the English and U.S. common law systems. A seller has to tell the truth (if asked a direct question), abide by contracts, and stand behind the firm's product (to some reasonable extent). Beyond this, it is expected that vigorous competition in the marketplace will protect consumers—*so long as they are careful.*

Laws are also created in situations where the average consumer may not have the ability or information to make informed judgments. For example, various laws regulate packaging and labels, telemarketing, credit practices, prices, environmental claims, dangerous products (tobacco and alcohol), and consumer privacy. Some of these laws will be discussed in more detail in subsequent chapters.

Laws govern product safety

For some products, consumers may not be able to accurately assess the safety of a product before they buy—automobiles, for example. To protect consumers, Congress passed the Consumer Product Safety Act (of 1972), which set up the Consumer Product Safety Commission (CPSC). The CPSC has broad power to set safety standards and can impose penalties for failure to meet these standards. The commission has the power to *force* a product off the market—or require expensive recalls to correct problems.

This puts pressure on marketing managers to consider safety in product design. And safety must be treated seriously by marketing managers. There is no more tragic example of this than the recalls of Firestone tires originally used on Ford's Explorer SUV. Hundreds of consumers were killed or seriously injured in accidents.[15]

Prosecution is serious— you can go to jail

Businesses and *individual managers* are subject to both criminal and civil laws. Penalties for breaking civil laws are limited to blocking or forcing certain actions—along with fines. Where criminal law applies, jail sentences can be imposed. For example, several managers at Beech-Nut Nutrition Company were fined $100,000 each and sent to jail. In spite of ads claiming that Beech-Nut's apple juice was 100 percent natural, they tried to bolster profits by secretly using low-cost artificial ingredients.[16]

State and local laws vary

Besides federal legislation—which affects interstate commerce—marketers must be aware of state and local laws. There are state and city laws regulating minimum prices and the setting of prices, regulations for starting up a business (licenses, examinations,

After many consumers became more interested in buying organic products, the United States Department of Agriculture (USDA) created specific standards companies must adhere to in order to use the "USDA Organic" label. Capri Sun Organic juice meets those standards and displays the label on its package.
Keith Homan/Shutterstock

and even tax payments), and in some communities, regulations prohibiting certain activities—such as telephone selling or selling on Sundays or during evenings.

This often creates local opportunities or threats. For example, some states have legalized marijuana for medical or recreational use, while others continue to prohibit the drug. When electric carmaker Tesla Motors tried to bypass auto dealers and sell its vehicles directly to consumers, many states enacted laws prohibiting direct sales of motor vehicles. Some cities opposed companies like Lime and Bird bringing scooter-sharing services to their city. The city of Milwaukee banned Bird scooters after they were placed around the city without warning. For larger firms, a tapestry of state and local laws may necessitate adapting marketing strategies to local requirements.[17]

The Cultural and Social Environment

LO 3.7

The **cultural and social environment** affects how and why people live and behave as they do—which affects customer buying behavior and eventually the economic, political, and legal environments. Many variables make up the cultural and social environment. Some examples are the languages people speak; the type of education they have; their religious beliefs; what type of food they eat; the style of clothing and housing they have; and how they view work, marriage, and family. Because the cultural and social environment has such broad effects, most people don't stop to think about it—how it may be changing or how it may differ for other people.

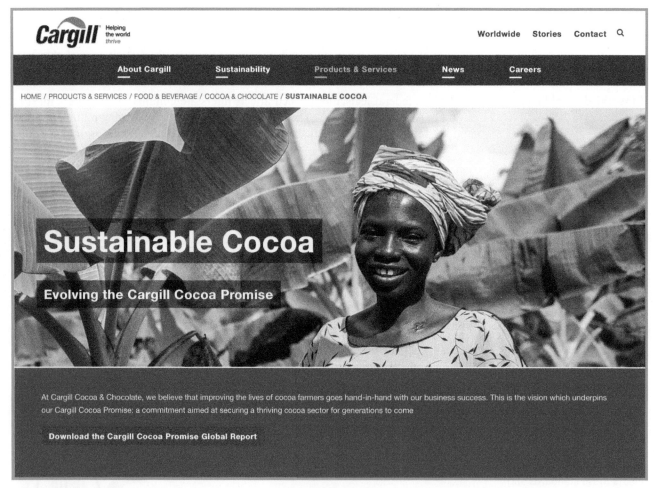

#M4BW Cargill sells ingredients to food producers. If you eat a can of soup, ice cream, yogurt, chewing gum, or beer, odds are it includes an ingredient from Cargill. And these days, Cargill often ensures that ingredient is sustainable—so the food producer can make the same claim.
Source: Cargill, Incorporated

#M4BW

Sustainability matters

One growing social and cultural trend follows from rising concerns about climate change. When shopping these days, more and more consumers consider **sustainability**—the idea that it's important to meet present needs without compromising the ability of future generations to meet their own needs. Two recent studies found that about two out of three consumers are willing to pay more for a product that is deemed more sustainable. Consumers care about the environment and expect companies to do something about it; in another global study, 81 percent of respondents believed that "companies should help improve the environment." These attitudes put more pressure on companies to make profits while also creating a better world.[18]

#M4BW

Companies respond to calls for sustainability

Businesses are getting the message. McDonald's has announced that it will soon buy all of its coffee from sources that meet international sustainability standards certified by Fair Trade and Rainforest Alliance. Consumers are concerned about plastic containers—we throw away lots of plastic. Recently, Unilever, Procter & Gamble (P&G), Nestlé, and PepsiCo began testing reusable containers for their products. PepsiCo will use refillable glass bottles for its Tropicana orange juice, and P&G will use aluminum bottles for its Pantene shampoo and stainless steel containers for Tide detergent. Products will be delivered to customers' homes with empty packaging returned, cleaned, and refilled.[19]

Where people are around the world

Other social and cultural changes relate to demographic trends. We examine evolving trends related to population, age, income, literacy, and technology adoption across the globe and within the United States. These types of demographic data are also important to marketing strategy planning.

Exhibit 3-3 summarizes data for a number of representative countries from different regions around the world. Even with a population of more than 316 million, the United States makes up only 4.9 percent of the world's over 7 billion people. Marketing managers in any country may seek growth opportunities in other countries.

Although the size of a market is important, the population trend also matters. The world's population is growing fast, but that population growth varies dramatically from country to country. For example, between 2013 and 2025, the population is projected to grow 35 percent or more in Ethiopia and Nigeria and about 25 percent in Egypt, Singapore, and Bangladesh. During this same period, growth will be about 9 percent in the United States and 3 percent in China, whereas the population is predicted to decline in Japan, Russia, Germany, and Finland. These trends have many marketing managers paying increased attention to fast-growing developing countries in Africa and Southeast Asia.[20]

Marketing managers for eBay have found many opportunities for growth in fast-growing markets like Vietnam.
NetPhotos/Alamy Stock Photo

Exhibit 3-3 Demographic Dimensions and Characteristics for Selected Countries

Country	2013 Population (000s)	2025 Projected Population (000s)*	2013–2025 Projected Population Change (%)	Population Density per Square Km	Percent of Population in Urban Areas	GNI per Capita (U.S. $)	GDP (billions of $U.S.)	Estimated Literacy % of Population (age 15+)	Mobile Phones/100 Inhabitants, 2017	% Mobile Phone/100 Growth, 2010-2017	% Population Using Internet, 2015	% Internet User Growth, 2010-2015
Algeria	39,210	45,841	17	17	71	4,870	167	80	121	33	38	206
Argentina	41,450	47,165	14	16	92	12,460	583	98	140	1	69	54
Australia	23,130	25,054	8	3	89	60,070	1,339	99	113	11	85	11
Bangladesh	156,600	197,674	26	1,237	34	1,190	195	61	88	97	14	289
Brazil	200,400	218,259	9	25	86	9,850	1,775	93	113	13	59	45
Canada	35,160	37,559	7	4	82	47,540	1,551	99	86	14	88	10
China	1,357,000	1,394,639	3	146	56	7,930	11,008	96	105	66	50	47
Egypt	82,060	103,742	26	92	43	3,340	331	76	106	26	36	66
Ethiopia	94,100	131,261	39	99	19	590	62	49	60	663	12	1,447
Finland	5,439	5,251	-4	18	84	46,550	232	99	132	-15	93	7
France	6,031	68,860	4	122	80	40,540	2,419	99	106	16	85	10
Germany	80,620	79,226	-2	234	75	45,940	3,363	99	129	18	88	7
Haiti	10,320	11,252	9	389	59	810	9	61	59	48	12	46
India	1,252,000	1,396,046	12	441	33	1,600	2,095	72	87	43	26	247
Iran	77,450	90,481	17	49	73	n/a	n/a	87	107	48	44	177
Israel	8,059	8,984	12	387	92	35,770	299	97	127	3	79	17
Italy	59,830	62,591	5	206	69	32,810	1,821	99	141	-10	66	22
Japan	127,300	123,386	-3	348	93	38,840	4,383	99	133	39	93	19
Kenya	4,350	53,196	20	81	26	1,340	63	78	86	43	46	226
Mexico	122,300	134,829	9	65	79	9,710	1,144	95	89	14	57	85
Nigeria	173,600	234,363	35	200	48	2,820	481	90	76	38	47	98
Norway	5,084	5,682	11	14	80	93,740	387	100	108	-6	97	4
Pakistan	182,100	228,385	20	245	39	1,440	271	56	73	26	18	125
Romania	19,960	20,872	5	86	55	9,500	178	99	115	-4	56	40
Russia	143,500	128,180	-11	9	74	11,450	1,331	100	158	-5	73	71
Saudi Arabia	28,830	31,877	11	15	83	23,550	646	95	122	-35	70	70
Singapore	5,399	6,733	25	7,807	100	52,090	293	97	148	2	82	16
Spain	46,650	51,415	10	93	80	28,530	1,199	98	113	3	79	20
Turkey	74,930	90,498	21	102	73	9,950	718	96	96	13	54	35
United Kingdom	64,100	67,244	5	269	83	43,390	2,858	99	120	-1	92	8
United States	316,100	346,407	9	35	82	55,980	18,037	99	122	32	75	4
Vietnam	89,710	102,459	14	296	34	1,990	194	95	126	0	53	72

Note: Data for projected population is from U.S. Census Bureau International Data Base. Urbanization data and literacy data from the CIA *World Factbook* and World Bank. All other data from the World Bank. Data for 2015 unless otherwise noted. Literacy data from 2015, except for Australia (2003), Canada (2003), Finland (2000), France (2003), Germany (2003), Israel (2004), Japan (2002), Somalia (2001), Switzerland (2003), United Kingdom (2003), and United States (2003). Internet users are individuals who have used the Internet (from any location) in the last 12 months. Internet can be used via a computer, mobile phone, personal digital assistant, games machine, digital TV, etc.

*Data from 2013.

Sources: U.S. Census Bureau, CIA *World Factbook*, and World Bank.

A shift from rural to urban areas

Just 50 years ago, about two-thirds of the world's population lived in rural areas. Today about half live in urban areas, as more people move to cities for better job opportunities. The extent of urbanization varies widely across countries. While about 82 percent of U.S. residents live in urban areas, more than 90 percent do in Japan, Singapore, Israel, and Argentina (see Exhibit 3-3). By contrast, in Ethiopia and Kenya, 26 percent or less of the population lives in urban areas. The concentration of people in major cities often simplifies Place and Promotion decisions.

There's no market when there's no income

Profitable markets require income as well as people. The amount of money people can spend affects the products they are likely to buy. When considering international markets, income is often one of the most important demographic dimensions. There are a variety of different measures of national income. One widely used measure is **gross domestic product (GDP)**—the total market value of all goods and services provided in a country's economy in a year by both residents and nonresidents of that country. **Gross national income (GNI)** is a measure that is similar to GDP, but GNI does not include income earned by foreigners who own resources in that nation. By contrast, GDP does include foreign income.

When you compare countries with different patterns of international investment, the income measure you use can make a difference. For example, Ford has a factory in Thailand. The GDP measure for Thailand would include the profits from that factory because they were earned in that country. However, Ford is not a Thai firm, and most of its profit will ultimately flow out of Thailand. The Thai GNI would not include those profits. You should see that using GDP income measures can give the impression that people in less-developed countries have more income than they really do. In addition, in a country with a large population, the income of the whole nation must be spread over more people. So *GNI per capita* (per person) is a useful figure because it gives some idea of the income level of people in the country.

Exhibit 3-3 gives an estimate of GNI per capita and GDP for each country listed. You can see that the larger and more developed industrial nations—including the United States, China, Japan, and Germany—account for the biggest share of the world's GDP. In these countries, except for China, the GNI per capita is also quite high. This explains why so much trade takes place among these countries—and why many firms see them as the more important markets. In general, markets like these offer the best potential for products that are targeted at consumers with higher income levels. The GNI per capita in the United States is $55,980.[21]

Many managers, however, see great potential—and less competition—where GNI per capita is low. For example, Coca-Cola has made a push in Africa, where it hopes to establish a relationship with consumers now and turn that brand loyalty into profitable growth as consumer incomes rise.

Reading, writing, and marketing problems

The ability of a country's people to read and write has a direct influence on the development of its economy—and on marketing strategy planning. The degree of literacy affects the way information is delivered, which in marketing means promotion. The United Nations estimates that 14 percent of adults (aged 15 or older) in the world cannot read and write. About two-thirds of them are women. But that is changing; among youth aged 15 to 24, only 9 percent cannot read and write. You may be surprised by the low literacy rates for some of the countries in Exhibit 3-3. Illiteracy creates challenges for product labels, instructions, and print advertising.[22]

Technology adoption races across continents

Cell phone and Internet usage have increased rapidly around the world. These technologies may have their greatest impact in developing countries where, for example, they've completely skipped the adoption of landline phones and instead moved directly to reliance on mobile phones.

Adoption of these technologies varies across the globe (see Exhibit 3-3). Take cell phones, for example. In many countries, people have separate phones for work and personal use, so the number of cell phones actually exceeds the population. While penetration is lower in many of the poorest nations, those countries are experiencing very fast adoption rates. In Ethiopia, for instance, the 2017 ownership rate of 60 phones per 100 people represents a more than 663 percent increase from just seven years earlier. Similar differences can be observed in Internet access. Marketing managers need to recognize how target markets utilize this technology to determine its role in marketing strategy. For example, setting up a website in another language may be more useful in some countries than others.[23]

#M4BW

Technology and marketing raise living standards in India

One country seeing rapid adoption of mobile phones is India, where cell phone adoption results in a better quality of life for many of India's poorest citizens. For example, after fishermen in India began using cell phones, they called ashore to find out which ports had the most demand for their catch. This helped match supply and demand while lowering waste and stabilizing prices.[24]

India wants to reduce the mortality rate for mothers and young children; each day about 150 mothers die of complications from pregnancy and 3,500 children under age five die from a range of maladies. In India, even the lowest-income families have one mobile device they can share. Johnson & Johnson (J&J) has leveraged mobile phones to expand prenatal and early childhood care across India. Working with the Mobile Alliance for Maternal Action (MAMA), J&J launched a free program that leaves voice messages twice a week for pregnant women and new mothers. Since its 2014 launch, more than 700,000 women have learned more about how to make healthy choices that help get their children off to a good start in life.[25]

Population trends in the U.S. consumer market

While the U.S. population is not growing as quickly as in some other countries, Exhibit 3-4 shows that current population and population growth vary a lot in different regions of the country. The states shaded blue and green are growing at the fastest rate.

Exhibit 3-4   2010 Population (in millions) and Projected Percentage Change by State 2010–2020

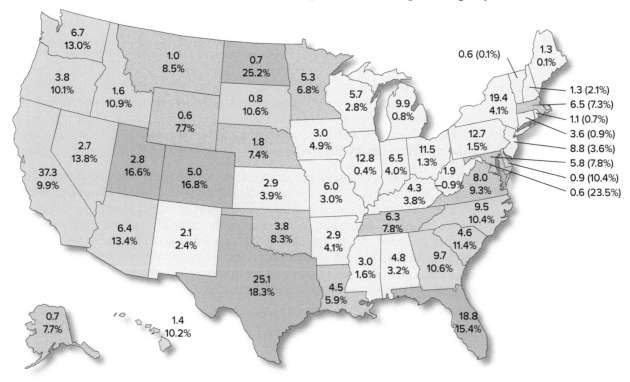

Percentage change in population:
☐ Decreasing–5.0%   ☐ 5.1%–9.5%   ☐ 9.5%–15.0%   ☐ 15.1% or higher

Source: Data from Weldon Cooper Center for Public Service, University of Virginia, Demographics & Workforce Group, May 2016 update.

Exhibit 3-5 Population Distribution by Age Group in 2005, 2015, and 2025

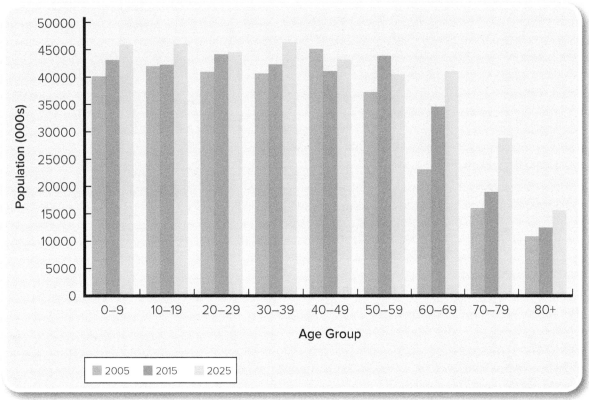

Source: U.S. Census Bureau.

The greatest growth is in the western and southeastern United States. These different rates of growth are important to marketers. Sudden growth in one area may create a demand for new shopping centers and furniture stores—whereas retailers in declining areas face tougher competition for a smaller number of customers.[26]

Boomers drive the graying of America

Another important dimension of U.S. society is its age distribution. In 1980, the median age of the U.S. population was 30—but by 2018 the median age rose to 38. The median age is growing because the percentage of the population in older age groups has increased. Exhibit 3-5 shows population trends by age groups. The graphic shows the number of people in the United States at various age groups in 2005, 2015, and 2025.

In Exhibit 3-5, these changes can be seen most dramatically when looking at the population of those older than age 60. The three age bands (60-69, 70-79, and 80+) each show a large increase by 2025. This will significantly increase the number of **senior citizens** (people older than 65). This increase is partly due to better health care and Americans living longer lives.

It is also because **Baby Boomers**, those born between 1946 and 1964, began to reach age 65 in 2011. Baby Boomers are a powerful demographic force, as there are large numbers of people in this group. This group has a lot of income and cannot be ignored by firms that sell more expensive products. Baby Boomers spend 80 percent of all leisure travel dollars and buy two-thirds of all new cars, half of all new computers, and one-third of movie tickets. They also still like to shop in a physical store—much more so than other generations.[27]

Generation X—fewer in number

Generation X (Gen X) refers to the generation born immediately following the Baby Boom—from 1965 to 1977. This group is much smaller in number than the Baby Boomers it follows—notice the decline in 40- to 49-year-olds from 2005 to 2015 and the decline in 50- to 59-year-olds from 2015 to 2025. This group tends to be raising kids and often

Many products target people of specific generations. Many Millennials are starting to have children and have new needs; Commerce Bank wants them to know it can help. Some Baby Boomers have a different set of needs. Perhaps their hearing is not what it used to be but they don't like others to see their hearing aid. Recognizing these needs, Lyric developed "the world's only 100% invisible hearing aid."
(left): Source: Commerce Bancshares, Inc.; *(right)*: Source: Sonova USA Inc.

spends their money on their kids—though they also like to fix up their homes, which works out well for stores like Lowe's and Pottery Barn.[28]

Generation Y—experiences over possessions

Generation Y (Gen Y), sometimes called Millennials, refers to those born from 1978 to 1994. This group emerged from the echo boom—when Baby Boomers started having kids. In Exhibit 3-5, this can be seen in the rise of those in their 20s from 2005 to 2015 and in their 30s from 2015 to 2025. Millennials are an especially important group for many types of products. Millennials are relatively comfortable with technology. Millennials like convenience and have been credited with sparking a resurgence in frozen food. They are health conscious—which is good for health clubs. They also prefer to spend more on experiences than possessions—bad for pricey cars and good for travel destinations. They also don't mind "sharing," which bodes well for services like Uber and Airbnb. On the other hand, most members of this generation don't like bar soap or chewing gum.[29]

Generation Z—more cautious

Generation Z (Gen Z) refers to those born since 1995. This group is still young, so ideas about its emerging values are more speculative. These "digital natives" were born into a world that already used text messaging, cell phones, and the Internet. The Zs are also part of a more ethnically diverse United States than their parents knew, and they appear to be more accepting of different cultures, races, and religions. Growing up in the shadow of the 9/11 terrorist attacks and during the Great Recession, this group tends to have realistic (as opposed to optimistic) views of the world. The Zs may prefer brands that suggest long-term value, safety, and security. Gen Z is comfortable with social media and often uses different platforms for different reasons—Snapchat for real-life moments, Twitter for the news, and Instagram to share whom they aspire to be. And more than half of Gen Z consider themselves socially conscious and take that into consideration when making purchase decisions. Marketing communications should recognize why Gen Z is on social media and craft messages appropriately.[30]

Don't generalize too much about generations

Marketing managers must be cautious in using these generational generalizations as anything more than a starting point. They should ultimately focus on a more specific target market when developing a strategy. For example, whereas about half of all Millennials are comfortable with technology, more than half do not feel the same. If most marketing managers follow the "general wisdom" about Millennials, there will be subgroups that are not well served.[31]

Social and cultural changes come slowly

The trends and demographic data we've been reviewing show that changes in cultural values and social attitudes evolve slowly. They also reflect large groups of customers—not necessarily a firm's target market. Opportunities and threats can be identified early when marketing managers monitor and anticipate these changes.

Screening Criteria Narrow Down Strategies

LO 3.8

Developing and applying screening criteria

After analyzing a firm's resources (for strengths and weaknesses), the environmental trends the firm faces, and the objectives of top management, marketing managers may compile that information into a set of product-market screening criteria. These criteria should include both quantitative and qualitative components. The quantitative components summarize the firm's objectives: sales, profit, and return on investment (ROI) targets. (*Note:* ROI analysis is discussed briefly in Chapter 9 and Appendix B.) The qualitative components summarize what kinds of businesses the firm wants to be in, what businesses it wants to exclude, what weaknesses it should avoid, and what resources (strengths) and trends it should build on.[32]

Developing screening criteria is difficult but worth the effort. They summarize in one place what the firm wants to accomplish. When a manager can explain the specific criteria that are relevant to selecting (or screening out) an opportunity, others can understand the manager's logic. Thus, marketing decisions are not just made or accepted based on intuition and gut feel. The criteria should be realistic—that is, they should be achievable. Opportunities that pass the screen should be viable strategies that the firm can implement with the resources it has available.

Whole plans should be evaluated

When forecasting the probable results of implementing a marketing strategy, a marketing manager may consider the quantitative part of the screening criteria because only implemented plans generate sales, profits, and return on investment. For a rough screening, the likely results of implementing each opportunity over a logical planning period may be estimated. If a product's life is likely to be three years, for example, a good strategy may not produce profitable results for 6 to 12 months. But evaluated over the projected three-year life, the product may look like a winner. When evaluating the potential of possible opportunities (product-market strategies), it is important to apply similar criteria—that is, whole plans over longer time periods.

General Electric strategic planning grid identifies attractive opportunities

When a firm has many potential strategies to evaluate, comparisons can be made with graphical approaches—such as the nine-box strategic planning grid developed by General Electric (GE) and used by many other companies. Such grids can help evaluate a firm's whole portfolio of strategic plans or businesses.

GE's strategic planning grid—see Exhibit 3-6—forces company managers to make three-level judgments (high, medium, and low) about the business strengths and industry attractiveness of all proposed or existing product-market plans. As you can see from Exhibit 3-6, this approach helps a manager organize information about the company's marketing environments (discussed earlier in this chapter), along with information about its strategy, and translate it into relevant screening criteria.

Exhibit 3-6 General Electric's Strategic Planning Grid

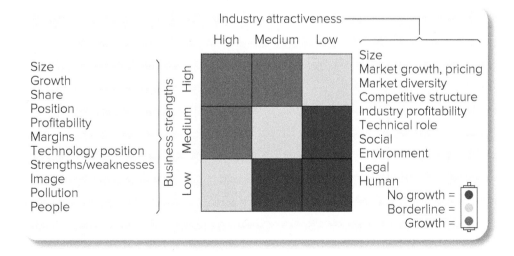

The industry attractiveness dimension helps managers understand: Does this product-market plan look like a good idea? Will it achieve our objectives? To answer these questions, managers have to judge such factors (screening criteria) as the size of the market and its growth rate, the nature of competition, the plan's potential environmental or social impact, and how laws might affect it. Note that an opportunity may be attractive for *some* companies—but not well suited to the strengths (and weaknesses) of a particular firm. That is why the GE grid also considers the business strengths dimension.

The business strengths dimension focuses on the ability of the company to pursue a product-market plan effectively. To make judgments along this dimension, a manager evaluates whether the firm has people with the right talents and skills to implement the plan, whether the plan is consistent with the firm's image and profit objectives, and whether the firm could establish a profitable market share given its technical capability, costs, and size. Here again, these factors suggest screening criteria specific to this firm and market situation.

The GE grid identifies opportunities as favorable for growth when they fall into the green boxes in the upper left-hand corner of the grid. Managers give these opportunities high marks on both industry attractiveness and business strengths—the market is favorable and the company has relevant strengths. The red boxes in the lower right-hand corner of the grid, on the other hand, suggest a no-growth policy. Existing red businesses may continue to generate earnings, but they no longer merit much investment. Yellow businesses are borderline cases—they can go either way. A company may continue to support an existing yellow business but will probably reject a proposal for a new one. It simply wouldn't look good enough on the relevant screening criteria.

This "stoplight" evaluation method is a subjective, multiple-factor approach. It avoids the traps and possible errors of trying to use oversimplified, single-number criteria—such as ROI or market share. Instead, top managers review detailed written summaries of many different screening criteria that help them make summary judgments. This approach helps everyone understand why the company supports some new opportunities and not others.[33]

The factors listed here reflect GE's objectives. Another firm might modify the evaluation to emphasize other screening criteria—depending on its objectives and the type of product-market plans it is considering. In addition, the screening criteria may have different weight or levels of importance.

Screen for unique risks in international markets

The approaches we've discussed so far apply to international markets as well as domestic ones. But in international markets it is often harder to fully understand the market environment variables. This may make it more difficult to see the risks involved in particular opportunities. Some countries are politically unstable; their governments and constitutions come and go. An investment safe under one government might become a takeover target under another.

To reduce the risk of missing some basic variable that may help screen out a risky opportunity, marketing managers sometimes need a detailed analysis of the market environment they are considering entering. Such an analysis can reveal facts about an unfamiliar market that a manager in a distant country might otherwise overlook. Further, a local citizen who knows the market environment may be able to identify an "obvious" problem ignored even in a careful analysis. Thus, it is very useful for the analysis to include inputs from locals—perhaps cooperative distributors.[34]

International risks depend on environmental sensitivity

The farther you go from familiar territory, the greater the risk of making big mistakes. But not all products, or marketing mixes, involve the same risk. Think of the risks as running along a "continuum of environmental sensitivity." See Exhibit 3-7.

Exhibit 3-7 Continuum of Environmental Sensitivity

Insensitive ←——————————————————————→ Sensitive

| Industrial products | Basic commodity-type consumer products | Consumer products that are linked to cultural variables |

Industrial products, like the bulldozers that John Deere sells, are relatively insensitive to different economic or cultural environments. So, the marketing strategy is similar across international markets. On the other hand, food products are often quite sensitive to cultural variables. McDonald's often sells different foods around the world. For example, in the Philippines, you can find the McRice Burger on the menu.
(left): Source: Deere & Company; (right): Source: McDonald's

Some products are relatively insensitive to the economic and cultural environment into which they're placed. These products may be accepted as is—or they may require just a little adaptation to make them suitable for local use. Most industrial products are near the insensitive end of this continuum.

At the other end of the continuum, we find highly sensitive products that may be difficult or impossible to adapt to all international situations. Consumer products that are closely linked to other social or cultural variables are at this end. For example, some cultures view dieting as unhealthy; that explains why products such as Diet Pepsi, although popular in the United States, have sometimes done poorly in other countries. Many American quick-serve restaurants succeed in international markets only after adapting their menus to foreign tastes. In Japan, KFC serves much more of the preferred dark meat chicken and offers meals in rice bowls and bento boxes, which are traditional in the culture. In China, KFC restaurants are much larger to accommodate families and groups that often dine together; kitchens are also larger, to accommodate an expanded menu that often includes many local items.[35]

This continuum helps explain why many of the early successes in international marketing were basic commodities such as gasoline, soap, vehicles, mining equipment, and agricultural machinery. It also helps explain why some consumer products firms have been successful with basically the same promotion and products in different parts of the globe.[36]

CONCLUSION

Businesses need innovative strategy planning to survive in our increasingly competitive markets. In this chapter, we discussed the variables that shape the broad environment of marketing strategy planning and how they may affect opportunities. First we looked at how the firm's own resources and objectives may help guide or limit the search for opportunities. Then we went on to look at the need to understand competition and how to do a competitive analysis. Next we shifted our focus to the external market environments. They are important because changes in these environments present new opportunities, as well as problems, that a marketing manager must deal with in marketing strategy planning.

The economic environment—including chances of recession or inflation—also affects the choice of strategies. And the marketer must try to anticipate, understand, and deal with these changes—as well as changes in the technology underlying the economic environment. We saw how emerging technology and artificial intelligence might impact how marketing managers deliver value now and in the future. The marketing manager must also be aware of the political environment and legal restrictions.

The cultural and social environment affects how people behave and what marketing strategies will be successful. We consider the rise of sustainability and how many firms have adapted or developed new marketing strategies to address customers' willingness to pay more for goods and services that help make the world a better place. Demographic data—including trends in population, income, and technology usage—provide an indicator of social trends, so we looked at these more closely.

Once many potential options are identified, managers use procedures for screening and evaluating opportunities. We explained an approach for developing qualitative and quantitative screening criteria—from an analysis of the strengths and weaknesses of the company's resources, the environmental trends it faces, and top management's objectives. We also discussed ways for evaluating and managing quite different opportunities using the GE strategic planning grid. Finally, we considered how and when international opportunities create challenges.

Now we can go on in the rest of the book to discuss how to turn opportunities into profitable marketing plans and programs.

KEY TERMS

LO 3.9

mission statement, 61
competitive environment, 65
sustainable competitive advantage, 65
competitor analysis, 66
competitive rivals, 66
market share, 66
economic environment, 67

technology, 68
artificial intelligence (AI), 68
intelligent agent, 68
machine learning, 69
nationalism, 71
free trade, 71
cultural and social environment, 74

sustainability, 75
gross domestic product (GDP), 77
gross national income (GNI), 77
senior citizens, 79
Baby Boomers, 79
Generation X (Gen X), 79
Generation Y (Gen Y), 80
Generation Z (Gen Z), 80

QUESTIONS AND PROBLEMS

1. Review the Amazon case study that opens the chapter. From this case, identify examples of different key terms and concepts covered in the chapter. For example, Amazon's mission statement is noted and guides other decisions in the company.

2. Review the Amazon case study that opens this chapter. Create a table like the one below—and fill in the empty cells. If necessary look back at Chapter 2 to remind yourself of the definitions of threat and opportunity. Then describe how each of the three trends could be seen as a threat or an opportunity for Amazon.

Trends	Threat	Opportunity
eBay becomes a larger competitive rival		
Artificial intelligence (see What's Next?)		
Faster Internet connections emerge across Africa		

3. Do you think it makes sense for a firm to base its mission statement on the type of product it produces? For example, would it be good for a division that produces electric motors to have as its mission: "We want to make the best (from our customers' point of view) electric motors available anywhere in the world"?

4. Explain how a firm's resources may limit its search for opportunities. Cite a specific example for a specific resource.

5. In your own words, explain how a marketing manager might use a competitor analysis to avoid situations that involve head-on competition.

6. Discuss the probable impact on your hometown if a major breakthrough in air transportation allowed foreign producers to ship into any U.S. market for about the same transportation cost that domestic producers incur.

7. Identify three products without artificial intelligence (AI), that might add value by adding AI to their marketing mix. For each product, explain how AI could be utilized in the marketing mix and the value it could provide to customers.

8. Will the elimination of trade barriers among countries in Europe eliminate the need to consider submarkets of European consumers? Why or why not?

9. What and whom is the U.S. government attempting to protect in its effort to preserve and regulate competition?

10. Name three specific examples of firms that developed a marketing mix to appeal to customers seeking to purchase more sustainable products.

11. Drawing on data in Exhibit 3-3, do you think that Romania would be an attractive market for a firm that produces home appliances? What about Finland? Discuss your reasons.

12. Discuss the value of gross domestic product and gross national income per capita as measures of market potential in international consumer markets. Refer to specific data in your answer.
13. Discuss how the worldwide trend toward urbanization is affecting opportunities for international marketing.
14. Explain the product-market screening criteria that can be used to evaluate opportunities.
15. Explain General Electric's strategic planning grid approach to evaluating opportunities.

MARKETING PLANNING FOR HILLSIDE VETERINARY CLINIC

Appendix D (the Appendices follow Chapter 19) includes a sample marketing plan for Hillside Veterinary Clinic. The "Situation Analysis" section of the marketing plan includes sections labeled "Competitor Analysis" and "Analysis of the Market Context—External Market Environment." Review those sections and answer the following questions.

1. In the "Competitor Analysis" section, what dimensions were used to analyze competitors? What other dimensions might have been examined?
2. How was competitor information gathered? How else could Hillside have gathered information about its competitors?
3. What aspects of the external market environment are included in the marketing plan? What do you think is the most important information in this section?

SUGGESTED CASES

2. Nature's Own Foods, Inc.
6. Dynamic Steel
22. Bright Light Innovations: The Starlight Stove
29. Kennedy & Gaffney (K&G)

Video Case 1. Potbelly Sandwich
Video Case 2. Suburban Regional Shopping Malls

MARKETING ANALYTICS: DATA TO KNOWLEDGE

CHAPTER 3: COMPETITOR ANALYSIS

Mediquip, Inc. produces medical equipment and uses its own sales force to sell the equipment to hospitals. Recently, several hospitals have asked Mediquip to develop a laser-beam "scalpel" for eye surgery. Mediquip has the needed resources, and 200 hospitals will probably buy the equipment. But Mediquip managers have heard that Laser Technologies—another quality producer—is thinking of competing for the same business. Mediquip has other good opportunities it could pursue—so it wants to see if it would have a competitive advantage over Laser Tech.

Mediquip and Laser Tech are similar in most ways, but there are a few important differences. Laser Technologies already produces key parts that are needed for the new laser product—so its production costs would be lower. It would cost Mediquip more to design the product—and getting parts from outside suppliers would result in higher production costs.

On the other hand, Mediquip has marketing strengths. It already has a good reputation with hospitals—and its sales force calls on only hospitals. Mediquip thinks that each of its current sales reps could spend some time selling the new product and that it could adjust sales territories so only four more sales reps would be needed for good coverage in the market. In contrast, Laser Tech's sales reps call on only industrial customers, so it would have to add 14 reps to cover the hospitals.

Hospitals have budget pressures—so the supplier with the lowest price is likely to get a larger share of the business. But Mediquip knows that both suppliers' prices will be set high enough to cover the added costs of designing, producing, and selling the new product—and leave something for profit.

Mediquip gathers information about its own likely costs and can estimate Laser Tech's costs from industry studies and Laser Tech's annual report. Mediquip has set up a spreadsheet to evaluate the proposed new product.

Design element: #M4BW box globe icon: ©Vectoryzen/Shutterstock

CHAPTER FOUR

Source: The LEGO Group

Focusing Marketing Strategy with Segmentation and Positioning

In the early 1930s in Billund, Denmark, carpenter Ole Kirk Christiansen started a company that built wooden toys. He named his company LEGO, combining the first two letters from each word in the Danish phrase *leg godt*, which means "play well."

During the 1950s, LEGO articulated its plastic bricks as part of a "system of play" based on learning through imagination, creativity, and problem solving. By 1958 LEGO had refined the design of the classic LEGO brick; those 1958 bricks still click and lock with any of the billions of bricks LEGO has made since. Buoyed by the tail end of the Baby Boom generation, LEGO grew rapidly in the 1960s and 1970s. That growth slowed during the 1980s when the "construction toy market" lost favor to electronic toys and software.

LEGO reacted by broadening its market. In the 1990s, LEGO opened three LEGOLAND theme parks, developed video games and electronic toys, produced books and TV shows, and licensed watches and clothing. Although the new product lines increased sales, costs rose even more quickly and in 2004, LEGO lost more than $300 million. LEGO needed a strategy to turn things around—and fast!

To address the immediate financial weakness, LEGO sold off its LEGOLAND theme parks, outsourced production, and consolidated its product lines. In assessing its strengths, LEGO recognized (1) its brand name was widely trusted and (2) its enduring system of play, building with bricks, provided a platform for innovation. To focus its new strategy, LEGO redefined its market around the iconic brick with a vision to equip children for the future through creative, playful learning. LEGO thought it could profitably grow in the "active play market."

Seeking new opportunities, LEGO sought to get to know its customers better. LEGO researchers embedded themselves with families to understand how seven- to nine-year-old boys in Germany and the United States lived and played. One finding showed LEGO that it had taken the wrong lessons from the growth of video games. Assuming kids wanted immediate gratification in play, LEGO dumbed down many of its toys. But the new research found that kids wanted opportunities to demonstrate mastery—evidenced by the scoring, ranking, and sharing common in many computer games. LEGO also discovered that while kids lived very scheduled lives, they appreciated time to themselves.

Guided by these insights, LEGO introduced new kits. For example, new fire and police station kits and *Star Wars* movie-themed kits targeted the seven- to nine-year-old boy segment. For older boys (and their parents) seeking toys that bridge the physical and digital worlds, LEGO introduced Boost. LEGO Boost kits include simple motors and sensors and introduce kids to basic computer coding with a smartphone or tablet app. With Boost, kids can make Vernie—a robot with a bow tie and moving eyebrows—Frankie the Cat, and other models. At $160, Boost is not cheap, but LEGO knows the kit delivers the value many parents want in their kids' toys. LEGO's research found parents in Germany, the United States, and China were especially interested in toys that combine play *and* learning.

Over the years, LEGO has made several failed attempts to crack the "girls active play market." LEGO research showed that after age 5, most girls lost interest in LEGO products. After watching girls play, LEGO researchers created the LEGO Friends line. Research found that whereas boys like to complete a kit before they start playing make-believe, girls prefer to play and pretend as they build. So, the Friends kits come with bagged parts, allowing the girls to start storytelling and rearranging sooner. Girls also had opinions about the LEGO figurines (LEGO calls them "mini-figs"). Whereas boys play with the mini-figs in the third person, girls project themselves onto each figure. Girls also found the traditional stubby mini-fig ugly, so Friends mini-figs look more like real people. In LEGO Friends kits, many mini-fig characters come with names and backstories described in accompanying books. LEGO Friends work across cultures—showing strong sales in the

United States, Germany, and the fast-growing Chinese market. Thanks to Friends, sales to girls jumped from just 9 percent to 27 percent of LEGO's total sales.

Girls *and* boys age seven and up were targeted with LEGO's Hidden Side sets. These kits combine traditional bricks and augmented reality (AR) technology into a single play experience. Kids use an app to chase digital AR ghosts through eight "haunted" building sets, including a haunted mansion and school bus. With Hidden Side kits, boys and girls solve paranormal mysteries and capture the ghosts that "appear" in their LEGO town.

International markets are important for LEGO. While China presents a big opportunity, it requires a different marketing mix. For example, many American and European parents played with LEGO as kids, whereas Chinese parents were not so fortunate. In China, LEGO has its own stores where everyone gets a hands-on introduction. For promotion, LEGO reinforces the learning value from its toys; for example, it shared and promoted a post on the Chinese social media site WeChat where a father describes how LEGO bricks help him teach his child math. A partnership with Chinese Internet giant Tencent allows LEGO to deliver online videos and games to Chinese children.

Achieving LEGO's learning through play mission can be challenging in places affected by the Rohingya crisis in Myanmar and war-torn Syria. So, LEGO gave $100 million to the Sesame Workshop, funding opportunities for learning and play to millions of affected children. Fostering play and learning, the LEGO Foundation (which owns 25 percent of LEGO) helps make the world a better place.

With sales of over $5.5 billion and profits greater than $1 billion, LEGO is the world's largest toy maker. By "playing well" with different groups of customers around the world, LEGO constructed the most valuable company in the toy business.[1]

LEARNING OBJECTIVES

As the LEGO case illustrates, a manager who develops an understanding of the needs and characteristics of specific groups of target customers within the broader market may see new, breakthrough opportunities. But it's not always obvious how to identify the real needs of a target market—or the marketing mix that those customers will see as different from, and better than, what is available from a competitor. This chapter covers concepts and approaches that will help you succeed in the search for those opportunities.

When you finish this chapter, you should be able to

1. define and describe generic markets and product-markets.
2. know what market segmentation is and how to segment product-markets into submarkets.
3. know three approaches to market-oriented strategy planning.
4. know dimensions that may be useful for segmenting markets.
5. recognize how some computer-aided methods are used in segmenting.
6. know what positioning is and why it is useful.
7. understand the important new terms (shown in **red**).

Search for Opportunities Can Begin by Understanding Markets

LO 4.1

Strategy planning is a narrowing-down process

This text takes a marketing strategy planning approach—with the idea that you will learn both what marketing is and how to do it. In Chapter 2 we provided a framework for a logical marketing strategy planning process. It involves careful evaluation of the market opportunities available before narrowing down to focus on the most attractive target market and marketing mix. In Chapter 3 we focused on approaches for analyzing how competitors and the external market environment shape the evaluation of opportunities.

In this chapter, we discuss concepts that guide the selection of specific target customers. See Exhibit 4-1 for an overview. We start by showing how defining markets suggests opportunities to marketing managers. Next, we suggest some practical approaches marketing managers use to segment and target customers. Finally, we take

Exhibit 4-1 Focusing Marketing Strategy with Segmentation and Positioning

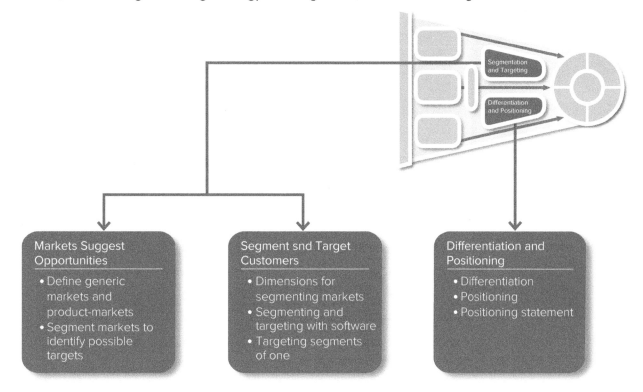

Markets Suggest Opportunities	**Segment snd Target Customers**	**Differentiation and Positioning**
• Define generic markets and product-markets • Segment markets to identify possible targets	• Dimensions for segmenting markets • Segmenting and targeting with software • Targeting segments of one	• Differentiation • Positioning • Positioning statement

a closer look at differentiation and positioning, which were briefly introduced in Chapter 2.

In a broad sense, this chapter is about understanding and analyzing customers in a market. In Chapters 5 and 6 we will look more closely at specific influences on the behavior of both final consumers and organizational customers. However, this chapter sets the stage for that by explaining how marketing managers combine different types of information about customers to guide targeting decisions. A good place to start is by discussing what we really mean when we use the term *market*.

What is a company's market?

Identifying a company's market is an important but sticky issue. In general, a **market** is a group of potential customers with similar needs who are willing to exchange something of value with sellers offering various goods or services—that is, ways of satisfying those needs. However, within a general market, marketing-oriented managers develop marketing mixes for *specific* target markets. Getting the firm to focus on specific target markets is vital.

Don't just focus on the product

Some production-oriented managers don't understand this narrowing-down process. They get into trouble because they ignore the tough part of defining markets. To make the narrowing-down process easier, they just describe their markets in terms of *products* they sell. For example, producers and retailers of greeting cards might define their market as the "greeting card" market. But this production-oriented approach ignores customers—and customers make a market! This leads to missed opportunities.

Hallmark isn't missing these opportunities. Instead, Hallmark aims at the "personal expression" market. The Hallmark.com site, its 2000 Gold Crown stores, and other dealers offer all kinds of products that allow people to express their feelings by capturing and saving memories. As opportunities related to these needs change, Hallmark changes, too. For example, at a Hallmark store you can find both cards and picture frames. At Hallmark.com, subscribers find tips about what to write in a graduation or birthday card, the etiquette to follow in writing a birth announcement, and a

The generic market of "late-night snackers" might be satisfied by very different product types: a trip to Taco Bell, cooking up a frozen DiGiorno pizza, or eating an apple from the refrigerator.
(left): William Howard/Shutterstock; (middle): VStock LLC /Age fotostock; (right): © Kontrec/Getty Images

blog post on how to "Banish boredom with a summer fun list." They can also create custom eCards with video greetings from their favorite characters from *Modern Family* or *Star Wars*. They can send the card by e-mail or post the wishes on Facebook. Hallmark uses the hashtag #CareEnough in its advertisements and social media campaigns. Hallmark's partnership with the Sporting KC soccer club puts "more caring into the world" by helping kids with cancer (#M4BW).[2]

From generic markets to product-markets

To understand the narrowing-down process, it's useful to think of two basic types of markets. A **generic market** is a market with *broadly* similar needs and sellers offering various, *often diverse,* ways of satisfying those needs. In contrast, a **product-market** is a market with *very* similar needs and sellers offering various *close substitute* ways of satisfying those needs.[3]

A generic market description looks at markets broadly and from a customer's viewpoint. Entertainment-seekers, for example, have many very different ways to satisfy their needs. An entertainment-seeker might buy a new 4K TV and subscribe to Hulu, book a vacation on Carnival Cruise Line, or reserve season tickets for the symphony. Any one of these *very different* products may satisfy this entertainment need. Sellers in this generic entertainment-seeker market have to focus on the need(s) customers want satisfied—not on how one seller's product (4K TV, vacation, or live music) is better

Unilever targets customers all over the world. Some of its products have an appeal only in certain countries. Marmite is a British food spread made from yeast extract, a nonalcoholic by-product of beer brewing. The product has a very distinct flavor that even people in Britain love or hate. Similar products are sold in Australia, New Zealand, Switzerland, and Germany, but the products' acquired taste has not caught on in other parts of the world.
Source: Unilever

than that of another producer. It is sometimes hard to understand and define generic markets because *product types that are quite different may compete with one another.*

Broaden market definitions to find opportunities

Broader market definitions—including both generic market definitions and product-market definitions—can help firms find opportunities. But deciding *how* broad to go isn't easy. Too narrow a definition limits a firm's opportunities—but too broad a definition makes the company's efforts and resources seem insignificant. Consider, for example, the mighty Coca-Cola Company. It has great success and a huge market share in the U.S. soft drink market. On the other hand, its share of all beverage drinking worldwide is very small.

Here we try to match opportunities to a firm's resources and objectives. So the *relevant market for finding opportunities* should be bigger than the firm's present product-market—but not so big that the firm couldn't expand and be an important competitor. A small manufacturer of screwdrivers in Mexico, for example, shouldn't define its market as broadly as "the worldwide tool users market" or as narrowly as "our present screwdriver customers." But it may have the capabilities to consider "the repairperson hand-tool market in North America." Carefully naming your product-market can help you see possible opportunities.

Naming Product-Markets and Generic Markets

A product-market should define an opportunity. As this suggests, when evaluating opportunities, product-related terms do not—by themselves—adequately describe a market. A complete product-market definition includes a four-part description:

What:	**1. Product type (type of good and type of service)**
To meet what:	**2. Customer (user) needs**
For whom:	**3. Customer types**
Where:	**4. Geographic area**

We refer to these four-part descriptions as product-market "names" because most managers label their markets when they think, write, or talk about them. Such a four-part definition can be clumsy, however, so we often use a nickname. And the nickname should refer to people—not products—because, as we emphasize, people make markets! For example, LEGO's broadly defined "active play market" describes kids who enjoy toys that foster creative, playful learning.

Product-market definition

Product type describes the goods and/or services that customers want. Sometimes the product type is strictly a physical good or strictly a service. But marketing managers who ignore the possibility that *both* are important can miss opportunities.

Customer (user) needs refer to the needs the product type satisfies for the customer. At a very basic level, product types usually provide functional benefits such as nourishing, protecting, warming, cooling, transporting, cleaning, holding, and saving time. Although we need to identify such "basic" needs first, in advanced economies, we usually go on to emotional needs—such as needs for fun, excitement, pleasing appearance, or status. Correctly defining the need(s) relevant to a market is crucial and requires a good understanding of customers. We discuss these topics more fully in Chapters 5 and 6.

Customer type refers to the final consumer or user of a product type. Here we want to choose a name that describes all present (possible) types of customers. To define customer type, marketers should identify the final consumer or user of the product type, rather than the buyer—if they are different. For instance, producers should avoid treating intermediaries as a customer type—unless intermediaries actually use the product in their own business.

The *geographic area* is where a firm competes, or plans to compete, for customers. Naming the geographic area may seem trivial, but understanding the geographic boundaries of a market can suggest new opportunities. A firm aiming only at the domestic market, for example, may want to expand to other countries.

Product-market boundaries provide focus

This idea of making a decision about the boundaries of a market applies not just to geographic areas served but also to decisions about customer needs and product and customer types. Thus, naming the market is not simply an exercise in assigning labels. Rather, the manager's market definition sets the limits of the market(s) in which the firm will compete. For example, Canon targets a wide range of photographic needs and might approach a product-market named the "digital photographers market," defined as digital cameras (product type) that are easy to use and take high-quality digital photographs (customer needs) for amateur photographers (customer types) in developing countries (geographic area). Canon's product line should appeal to this broad product-market. On the other hand, camera maker GoPro focuses on a more specific set of customer needs and has tighter boundaries on its market definition. GoPro's video cameras attach to surfboards, handlebars, or helmets so that adventure sport enthusiasts can capture all the action. GoPro might define the "adventure video market" as digital video cameras (product type) that are rugged, waterproof, and don't need to be operated by hand (customer needs) for adventure sports enthusiasts (customer types) who live anywhere in the world (geographic area). These product-market definitions give each company a focus that sharpens marketing strategy.

No product type in generic market names

A generic market description *doesn't include any product-type terms.* It consists of only three parts of the product-market definition. This emphasizes that any product type that satisfies the customer's needs can compete in a generic market. Exhibit 4-2 shows the relationship between generic market and product-market definitions.

Generic market names open up new opportunities

When marketing managers define a product-market, they already suggest a particular product type. The advantage of the generic market is that it opens up new opportunities. For example, marketing managers at Eastman Kodak paid too much attention to competition in the market for photographic film (its product) and failed to see how quickly digital imaging was replacing it. The music industry, long defined by sales of physical records, cassettes, or CDs, failed to anticipate how digital downloads and streaming would change how consumers buy music.

The generic market definition, by leaving out the product type, opens the idea of using other products to address customer needs. If the music industry had not been focused on its product—records, cassettes, and CDs—it might have developed its own music-streaming service.

You should now see that defining markets only in terms of current products is not the best way to find new opportunities. Focusing customer needs and customer types

Exhibit 4-2
Relationship between Generic and Product-Market Definitions

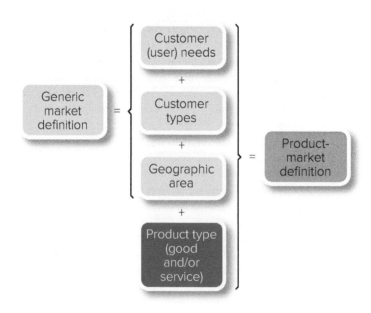

within a geographic area suggests potential opportunities. The most effective way to find these opportunities is through market segmentation.

Market Segmentation Defines Possible Target Markets

LO 4.2

Market segmentation is a two-step process

Naming broad product-markets is disaggregating

Market segmentation is a two-step process of (1) *naming* broad product-markets and (2) *segmenting* these broad product-markets in order to select target markets and develop suitable marketing mixes. Let's dig into a process used to achieve market segmentation.

The first step in effective market segmentation involves naming a broad product-market of interest to the firm. Marketers must break apart—disaggregate—all possible needs into some generic markets and broad product-markets in which the firm may be able to operate profitably (see Exhibit 4–3).

No one firm can satisfy everyone's needs. So, the naming—disaggregating—step involves brainstorming about very different solutions to various generic needs and selecting some broad areas—broad product-markets—where the firm has some resources and experience. Let's see how a bicycle maker might approach market segmentation following the process in Exhibit 4–3.

1. *All customer needs*—narrow down to the need to transport a person from point A to point B.
2. *Some generic market*—the "people transportation market."
3. *Broad product-market*—the "bicycle-riders market."
4. *Similar, narrow product-markets*—these might include:
   - *Submarket 1:* Exercisers
   - *Submarket 2:* Off-road adventurers
   - *Submarket 3:* Commuters
   - *Submarket 4:* Socializers
   - *Submarket 5:* Environmentalists

Exhibit 4-3 Narrowing Down to Target Markets

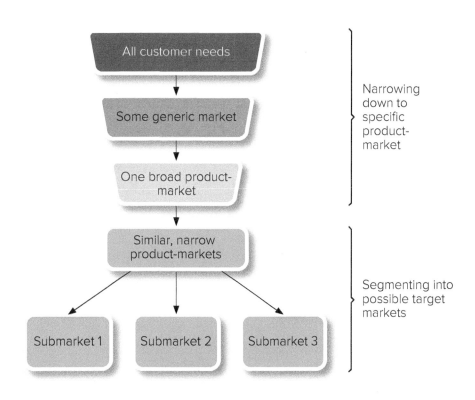

Exhibit 4-4 A Market Grid Diagram with Submarkets

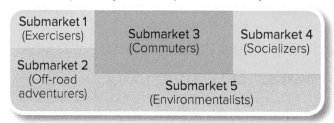

Market grid is a visual aid to market segmentation

Assuming that any broad product-market (or generic market) may consist of submarkets, picture a market as a rectangle with boxes that represent the smaller, more homogeneous product-markets.

Exhibit 4-4, for example, represents the broad product-market of bicycle riders. The boxes show different submarkets. One submarket might focus on people who want to commute to work or school, another on people who want exercise, and so on. Alternatively, in the generic "people transportation market," we might see different product-markets of customers for bicycles, motorcycles, cars, airplanes, ships, buses, and "others."

Segmenting is an aggregating process

Marketing-oriented managers think of **segmenting** as an aggregating process—clustering people with similar needs into a "market segment." A **market segment** is a (relatively) homogeneous group of customers who will respond to a marketing mix in a similar way.

This part of the market segmentation process takes a different approach from the naming part. Here we look for similarities rather than basic differences in needs. Segmenters start with the idea that each person is one of a kind but that it may be possible to aggregate some similar people into a product-market.

Segmenters see each of these one-of-a-kind people as having a unique set of dimensions. Consider a product-market in which customers' needs differ on two important segmenting dimensions: need for status and need for dependability. In Exhibit 4-5A, each dot shows a person's position on the two dimensions. Although each person's position is unique, many of these people are similar in terms of how much status and dependability they want. So a segmenter may aggregate them into three (an arbitrary number) relatively homogeneous submarkets—A, B, and C. Group A might be called

Most consumers associate the Clorox brand with strong cleaning power and bleach. Clorox uses the Green Works brand to target customers who want natural cleaning products that don't have harsh chemical fumes or residue.
Source: The Clorox Company

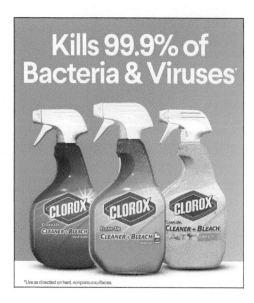

Exhibit 4-5 Every Individual Has His or Her Own Unique Position in a Market—Those with Similar Positions Can Be Aggregated into Potential Target Markets

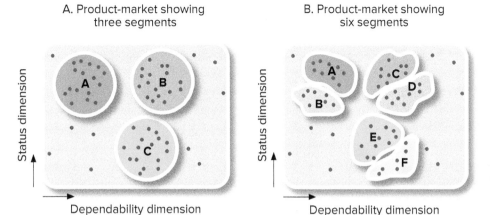

"status-oriented" and Group C "dependability-oriented." Members of Group B want both and might be called the "demanders."

How far should the aggregating go?

The segmenter wants to aggregate individual customers into some workable number of relatively homogeneous target markets and then treat each target market differently.

Look again at Exhibit 4-5A. Remember we talked about three segments. But this was an arbitrary number. As Exhibit 4-5B shows, there may really be six segments. What do you think—does this broad product-market consist of three segments or six?

Another difficulty with segmenting is that some potential customers just don't fit neatly into market segments. For example, not everyone in Exhibit 4-5B was put into one of the groups. Forcing these people into one of the groups would have made these segments more heterogeneous and harder to please. Further, forming additional segments for them probably wouldn't be profitable. They are too few and not very similar in terms of the two dimensions. These people are simply too unique to be catered to and may have to be ignored—unless they are willing to pay a high price for special treatment.

The number of segments that should be formed depends more on judgment than on some scientific rule. But the following guidelines can help.

Criteria for segmenting a broad product-market

Ideally, "good" market segments meet the following criteria:

1. *Homogeneous (similar) within*—the customers in a market segment should be as similar as possible with respect to their likely responses to marketing mix variables *and* their segmenting dimensions.
2. *Heterogeneous (different) between*—the customers in different segments should be as different as possible with respect to their likely responses to marketing mix variables *and* their segmenting dimensions.
3. *Substantial*—the segment should be big enough to be profitable.
4. *Operational*—the segmenting dimensions should be useful for identifying customers and deciding on marketing mix variables.

It is especially important that segments be *operational*. This leads marketers to include demographic dimensions such as age, gender, income, location, and family size. In fact, it is difficult to make some Place and Promotion decisions without such information. For example, magazines, websites, and television shows have data on the demographic characteristics of their readers, visitors, and viewers. This makes it easier for a marketing manager to match these qualities with the firm's target market.

Avoid segmenting dimensions that have no practical operational use. For instance, you may find a personality trait such as moodiness among the traits of heavy buyers of a product, but how could you use this fact? Salespeople can't give a personality test to each buyer. Similarly, advertising couldn't make much use of this information. So

 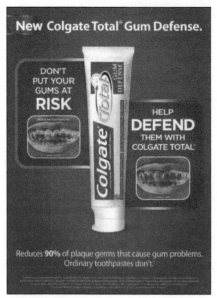 

Firms that compete in the oral health care market have developed a variety of products that appeal to the needs of different customer segments. Sensodyne targets consumers with sensitive teeth, Colgate Total Gum Defense aims at those concerned about the health of their gums, and Tom's Children's toothpaste emphasizes its use of natural ingredients.
(left): Source: GlaxoSmithKline plc.; (middle): Source: Colgate-Palmolive Company; (right): Source: Tom's of Maine

although moodiness might be related in some way to previous purchases, it would not be a useful dimension for segmenting.

Target Marketers Aim at Specific Targets

LO 4.3

Once you accept the idea that broad product-markets may have submarkets, you can see that target marketers usually have a choice among many possible target markets. Marketing managers may choose to aim at one or more more targets. Let's look more closely at the options and how marketing managers make this decision.

Three approaches to market-oriented strategic planning

There are three basic ways to develop market-oriented strategies in a broad product-market.

1. The **single target market approach**—segmenting the market and picking one of the homogeneous segments as the firm's target market.
2. The **multiple target market approach**—segmenting the market and choosing two or more segments, then treating each as a separate target market needing a different marketing mix.
3. The **combined target market approach**—combining two or more submarkets into one larger target market as a basis for one strategy.

Note that all three approaches involve target marketing. They all aim at specific, clearly defined target markets (see Exhibit 4-6). For convenience, we call people who follow the first two approaches the "segmenters" and people who use the third approach the "combiners."

Combiners try to satisfy "pretty well"

Combiners try to increase the size of their target markets by combining two or more segments. Combiners look at various submarkets for similarities rather than differences. Then they try to extend or modify their basic offering to appeal to these "combined" customers with just one marketing mix. For example, a combiner who faces the broad bicycle-riders product-market shown in Exhibit 4-4 might try to develop a marketing mix that would do a pretty good job of appealing to both the *Exercisers* and the *Off-road adventurers*.

**Exhibit 4-6
Segmenters and
Combiners Aim at
Specific Target Markets**

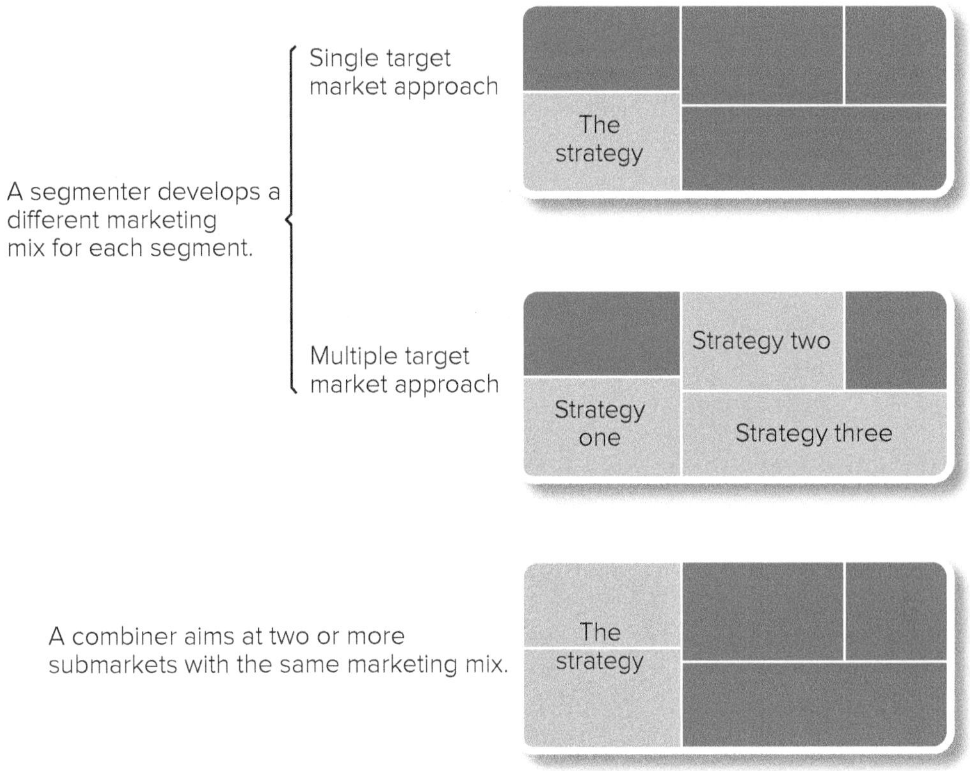

A segmenter develops a different marketing mix for each segment.
- Single target market approach
- Multiple target market approach

A combiner aims at two or more submarkets with the same marketing mix.

The combined market would be bigger than either segment by itself. And although both submarkets might like fatter tires and multiple gears, the *Off-road adventurer* prefers a wider range of gears, a suspension system to handle bumps, and a more durable frame than the *Exerciser* might want or need. On the other hand, comfort is more important to the *Exerciser* who favors a more upright seating position. A single model of bicycle designed for both segments might not be the *best bike* for either segment.

Thus, a combiner makes compromises in developing the marketing mix. The combiner doesn't try to fine-tune each element of the marketing mix to appeal to each of the smaller submarkets. Rather, the marketing mix is selected to work fairly well with each segment.

A combined target market approach may help achieve some economies of scale. It may also require less investment than developing different marketing mixes for different segments—making it especially attractive for firms with limited resources.

Segmenters try to satisfy "very well"

Segmenters aim at one or more homogeneous segments and try to develop a different marketing mix for each segment. Segmenters usually fine-tune their marketing mixes for each target market—perhaps making basic changes in the product itself—because they want to satisfy each segment very well.

Instead of assuming that the whole market consists of a fairly similar set of customers (like the mass marketer does) or merging various submarkets together (like the combiner), segmenters believe that aiming at one, or some, of these smaller markets makes it possible to provide superior value and satisfy them better. This then provides greater profit potential for the firm.

Segmenting may produce bigger sales

Note that segmenters are not settling for a smaller sales potential or lower profits. Instead, they hope to increase sales by getting a much larger share of the business in the market(s) they target. A segmenter who really satisfies the target market can often build such a close relationship with customers that he or she faces no real competition. A segmenter who offers a marketing mix precisely matched to the needs of the target market can often charge a higher price, thus producing higher profits. Customers are willing to pay a higher price because the whole marketing mix provides better customer value.

Consider the recent success of HP's computer business. HP had traditionally tried to serve lots of different customers with a range of personal computers—often hoping that different market segments might like the same machines. HP decided to focus on customers seeking sleeker designs and gamers who sought high-performance computers. For example, its new brand, the OMEN line, includes high-end graphics and high-resolution screens—and is priced at up to $6,000. This focus on key segments helped HP grow its share of the PC market from 18 to 21 percent.[4]

Which approach works best?

Which approach should a firm use? This depends on the firm's resources, the nature of competition, and—most important—the similarity of customer needs, attitudes, and buying behavior. It's usually safer to be a segmenter—that is, to try to satisfy a smaller number of customers *very* well instead of many just *fairly* well.

Resources may determine a single target market versus multiple target markets

A larger firm may have the resources to develop different marketing mixes to appeal to multiple target markets. For example, see the different versions of Dawn dishwashing soap in the photo.

On the other hand, a smaller firm competing against larger competitors may have little choice but to use the single target market approach, focusing all its efforts at a smaller segment where it sees the best opportunity. For example, sneaker maker Kaepa's all-purpose sneaker sales plummeted as Nike and Reebok stole customers with a multiple target market approach. Nike and Reebok developed innovative products and aimed their promotion at specific needs, such as jogging, aerobics, cross-training, and walking. Kaepa turned things around by catering to the needs of cheerleaders. Cheerleading squads can order Kaepa shoes with custom team logos and colors. The soles of the shoes feature finger grooves that make it easier for cheerleaders to build human pyramids. Kaepa also carefully targets its marketing research and promotion. Kaepa salespeople attend cheerleading camps each summer that draw 40,000 enthusiasts. Kaepa even arranges for the cheering teams it sponsors to do demos at retail stores. This generates publicity and pulls in buyers, so retailers put more emphasis on the Kaepa line.[5]

How many segments depends on costs and benefits

If a firm wants to use a multiple-segment approach, it must consider the cost to offer different marketing mixes to different segments. Offering multiple marketing mixes can raise a seller's costs, and the amount can vary across the Four Ps (Product, Place, Promotion, and

Procter & Gamble (P&G) uses the multiple target market approach with its dishwashing liquids. For example, P&G makes several versions of Dawn liquid dishwashing soap. The various versions include different scents (for example, "New Zealand Springs Scent"); cleaning powers (2× or 4× more "Grease Cleaning Power" or "50% Less Scrubbing"); and formulas (including "Antibacterial," "Gentle Clean," and "Botanicals"). Product, Place, Promotion, and Price can also vary across Dawn detergents.
McGraw-Hill Education

Price). Some changes are easier and less costly to make than others. In addition, marketing managers need to know whether the added cost of one more marketing mix delivers value to particular market segments. Let's consider this with examples across the Four Ps.

Product. First, consider the costs of producing many versions of automobiles. It can be expensive to make different automobile models. There are significant economies of scale in producing automobiles, and customers are not usually willing to pay the higher price for a unique version. Consequently, most automakers create a few models, each with a few different packages of options to give customers some choice—but not too much choice. On the other hand, the Product most barbers deliver, a haircut, can be customized to each customer—a good barber offers a blowout, crew cut, faux hawk, high fade, pompadour, or many other styles upon request. Providing a wide choice of cuts may require a more experienced and talented barber, but most customers are happy to pay the modest premium for that choice.

Place. Place may be the most difficult strategy element to vary across multiple target markets. It often adds to costs to sell a product through more retail outlets. That said, a brand like Pepsi wants its drinks available whenever a customer is thirsty. Pepsi can be found at fast-food restaurants, stadiums, vending machines in colleges and universities, grocery stores, and many more outlets.

Promotion. Many companies gather information about individual customers, making it easier to change this element of the marketing mix for smaller groups of customers. For example, online retailer Amazon knows which products each customer views on each visit, and highlights "recently viewed" items the next time that customer visits Amazon.com. This literally suggests that each customer sees a different set of advertisements. On the other hand, it would be costly for a brand of cereal (for example, Cheerios) to deliver different promotional messages to different customer segments. It might choose to advertise "low-sugar" products to all parents or a "heart healthy" message to senior citizens, but many more segments may not be practical.

Price. Business customers often negotiate Price. In these circumstances, customers may pay different prices depending on their perceived level of price sensitivity. The agreed-upon price may be necessary to close a sale. On the other hand, charging different prices in a grocery store can be costly for the store, except where shopper loyalty programs offer discounts to more price-sensitive customers who make sure they have the right card. Even then, all loyalty card holders pay the same price—simply adding one more segment.

Profit is the balancing point

In practice, cost considerations often encourage more aggregating and favor combining submarkets; costs often drop due to economies of scale. On the other hand, some customers prefer to have their needs satisfied more exactly—and will be more satisfied by a segmenter that develops a marketing mix that more closely matches their needs. When the marketing mix better matches their needs, customers are often willing to pay a higher price. Profit is the balancing point. It determines how much segmenting will be done and how unique a marketing mix the firm can afford to offer to each market segment.

What Dimensions Are Used to Segment Markets?

LO 4.4

Segmenting dimensions guide marketing mix planning

Market segmentation forces a marketing manager to decide which product-market dimensions might be useful for planning marketing strategies. The dimensions should help guide marketing mix planning. Exhibit 4-7 shows the basic kinds of dimensions we'll be talking about in Chapters 5 and 6—and their probable effect on the Four Ps. Ideally, we want to describe any potential product-market in terms of all three types of customer-related dimensions—plus a product-type description—because these dimensions help us develop better marketing mixes.

Segmenting draws on why, what, and who

As a marketing manager narrows down to particular segmenting dimensions, some basic questions can help guide the process. First, a marketing manager should understand *why* customers are making a purchase decision. This might take into account customers'

Exhibit 4-7 Relation of Potential Target Market Dimensions to Marketing Strategy Decision Areas

Potential Target Market Dimensions	Effects on Strategy Decision Areas
1. Behavioral needs, attitudes, and how present and potential goods and services fit into customers' consumption patterns.	Affect *Product* (features, packaging, product line assortment, branding) and *Promotion* (what potential customers need and want to know about the firm's offering, and what appeals should be used).
2. Urgency to get need satisfied and desire and willingness to seek information, compare, and shop.	Affect *Place* (how directly products are distributed from producer to customer, how extensively they are made available, and the level of service needed) and *Price* (how much potential customers are willing to pay).
3. Geographic location and other demographic characteristics of potential customers.	Affect size of *target markets* (economic potential), *Place* (where products should be made available), and *Promotion* (where and whom to target with advertising, publicity, sales promotion, and personal selling).

needs, preferences, and consumer decision process. Second, a manager wants to know *what* customers have done before making a decision. For example, are customers buying this product for the first time or are they repeat customers? How frequently do they make this purchase? Finally, *who* are the customers? Where do they live and shop, what is their age or gender, and so on? Answering these why, what, and who questions allows a marketing manager to better identify the dimensions on which customers are similar and different.

Many segmenting dimensions may be considered

Customers can be described by many specific dimensions. Exhibit 4-8 shows some dimensions useful for segmenting consumer markets. A few are behavioral dimensions; others are geographic and demographic. We discuss these final consumer segmenting dimensions in Chapters 3 and 5. Exhibit 4-9 shows some additional dimensions for segmenting markets when the customers are businesses, government agencies, or other types of organizations. These dimensions for segmenting organizational customers are covered in Chapter 6. Regardless of whether customers are final consumers or organizations, segmenting a broad product-market *usually* requires using several different dimensions at the same time.[6]

Exhibit 4-8 Possible Segmenting Dimensions and Typical Breakdowns for Consumer Markets

Behavioral	Needs, benefits sought, thoughts (attitudes), rate of use, purchase relationship, brand familiarity, kind of shopping, type of problem solving, information required, purchase situation, psychographics or lifestyle (see Chapter 5)
Geographic	Country, region, city, size of city, urban/rural (see Chapter 3)
Demographic	Income, gender, age, marital status, family size, family life cycle (see Chapter 5), occupation, education, ethnicity, social class (see Chapter 3)

Exhibit 4-9 Possible Segmenting Dimensions for Business/Organizational Markets

Kind of relationship	Weak/strong loyalty to vendor, single source/multiple vendors, "arm's length"/close partnership (see Chapter 6)
Type of customer	Manufacturer, service producer, government agency, military, nonprofit, wholesaler, or retailer (see Chapter 6)
Geographic	Region of world, country, region within country, urban/rural
Demographic	Size (number of employees or sales volume), primary business or industry (NAICS—see Chapter 6), number of facilities
How product will be used	Installations, components, accessories, raw materials, supplies, professional services (see business product classes in Chapter 8)
Type of buying situation	Decentralized/centralized, buyer/multiple buying influence, straight rebuy/modified rebuy/new-task buying (see Chapter 6)
Purchasing methods	Vendor analysis, purchasing specifications, Internet bids, negotiated contracts (see Chapter 6)

What are the qualifying and determining dimensions?

To select the important segmenting dimensions, think about two different types of dimensions. **Qualifying dimensions** are those relevant to including a customer type in a product-market. **Determining dimensions** are those that actually affect the customer's purchase of a *specific* product or brand in a product-market.

A prospective car buyer, for example, has to have enough money—or credit—to buy a car and insure it. Our buyer also needs a driver's license. This still doesn't guarantee a purchase. He or she must have a real need—such as a job that requires "wheels" or kids who have to be carpooled. This need may motivate the purchase of *some* car. But these qualifying dimensions don't determine what specific brand or model car the person might buy. That depends on more specific interests—such as the kind of safety features, performance, or appearance the customer wants. Determining dimensions related to these needs affect the specific car the customer purchases. If safety is a determining dimension for a customer, a Volvo wagon that offers side impact protection, air bags, and all-wheel drive might be the customer's first choice.

Consider Ford's efforts to sell cars in Vietnam. When Ford arrived in the country a few years ago, many consumers didn't even know how to drive a car—an important *qualifying* dimension when you are selling cars. To increase the size of its market, Ford created a free driving school for Vietnamese wanting to learn how to drive. Ford hoped that training drivers using Ford cars would build brand preference that would become a *determining* dimension for future Vietnamese drivers.

Determining dimensions may be very specific

How specific the determining dimensions are depends on whether you are concerned with a general product type or a specific brand (see Exhibit 4-10). The more specific you want to be, the more particular the determining dimensions may be. In a particular case, the determining dimensions may seem minor. But they are important because they *are* the determining dimensions.

For example, marketing managers at Chicago South Loop Hotel know that some customers drive into downtown Chicago and worry about the cost of parking. For these customers, manner of arrival might be a determining dimension in choosing a particular hotel for their next trip to the Windy City. So Chicago South Loop Hotel offers free parking to encourage this segment to choose its hotel.

Qualifying dimensions are important too

The qualifying dimensions help identify the "core benefits" that must be offered to everyone in a product-market. For example, people won't choose to stay at the Chicago South Loop Hotel unless they have plans to travel to the city of Chicago. Qualifying and determining dimensions work together in marketing strategy planning.

Different dimensions needed for different submarkets

Note that each different submarket within a broad product-market may be motivated by a different set of dimensions. In the snack food market, for example, health food enthusiasts are interested in nutrition, dieters worry about calories, and economical shoppers with lots of kids may want volume to "fill them up."

Exhibit 4-10 Finding the Relevant Segmenting Dimensions

Segmenting dimensions become more specific when the target segment seeks to purchase a particular brand of the product →

All potential dimensions	Qualifying dimensions	Determining dimensions (product type)	Determining dimensions (brand specific)
Dimensions generally relevant to purchasing behavior	Dimensions relevant to including a customer type in the product-market	Dimensions that affect the customer's purchase of a specific type of product	Dimensions that affect the customer's choice of a specific brand

People who are out in the sun should protect their skin. This safety need is a qualifying dimension of consumers in the market for sunscreen products. However, parents may especially seek Banana Boat Baby and Kids Sunblock Lotion and Baby Sprays because they are easy on kids' eyes and don't cause tearing. For these parents, this is a determining need. There are a variety of ways that Coffee-Mate is different from cream, but this ad focuses on the idea that it does not need to be refrigerated. For some consumers, that determines what they will buy.
(left): Source: Edgewell; *(right)*: Source: Nestlé

Sometimes a marketing manager must decide whether a firm should serve customers it really doesn't want to serve. For example, banks sometimes offer marketing mixes that are attractive to wealthy customers but drive off low-income consumers.

Personas give segments life

After creating segments, some firms create *buyer personas*—fictional depictions of customers illustrative of each target segment. These depictions often turn into elaborate background "stories" that describe a typical segment member's demographic characteristics and behavior patterns—related (and unrelated) to buying, motivations, and goals. A buyer persona emerges from customer research, usually personal interviews with several members of the target market segment.

For example, a bicycle maker might develop buyer personas for different target segments. If it targets commuters, the following might reflect the beginnings of a buyer persona:

> Sam is a single 29-year-old male who lives in Seattle, Washington. He rides 10 miles each way to his job where he earns $85,000 a year as a computer programmer. On weekends he likes to hang out with his girlfriend and race mountain bikes. Sam is price sensitive with his commuter bike (he spends more lavishly on his mountain bikes). The commuter bike must be safe and reliable; it should have fenders and a chain guard to deal with the regular rains in Seattle.

This description of Sam might continue with details about Sam's favorite websites, social media behavior, reading selections, long- and short-term goals, and his history of riding. Personas often include a photo. Because people relate best to people (rather than abstract "segments"), personas help everyone across the organization (research and development, customer service, advertising, website development, and more) to better empathize with target customers and develop a marketing orientation.[7]

International marketing requires even more segmenting

Success in international marketing requires even more attention to segmenting. There are more than 192 nations with their own unique cultures! And they differ greatly in language, customs (including business ethics), beliefs, religions, race, and income distribution patterns. (We discuss some of these differences in Chapters 3, 5, and 6.) These additional differences can complicate the segmenting process. Even worse, critical data are often less available—and less dependable—as firms move into international markets. This is one reason why some firms insist that local operations and decisions be handled by natives. They, at least, have a feel for their markets.

Segmenting international markets may require more dimensions. But one practical method adds just one step to the approach discussed earlier. First, marketers segment by country or region—looking at demographic, cultural, and other characteristics, including stage of economic development. This may help them find regional or national submarkets that are fairly similar. Then—depending on whether the firm is aiming at final consumers or business markets—they apply the same basic approaches presented before.

Smart segmenters win in international markets

Many firms make only modest changes to marketing mixes when they enter developing countries, which means that sellers can really stand out when they show a respect for and focus on local customers. A few years ago Korea's LG Electronics started tailoring products to the Indian market. As one example, LG's Charcoal Lighting Heater Microwave includes an autocook menu with settings for more than 130 Indian dishes. The move paid off when LG was chosen as the most trusted consumer durable brand in India.

#M4BW

Targeting the underserved makes a better world

For a long time, the developing world received little attention from many large businesses. These days, more companies are finding that targeting underserved populations with goods and services designed to meet their unique needs generates profits, creates value for customers, and makes a better world.

For example, workers from poor countries may work in wealthier countries where wages are higher and where there may not be enough domestic workers to fill all positions. These workers often wire some of what they earn back to family in their home country. This fuels a $600 billion remittance business. Traditionally, remittance transfers were inconvenient, expensive, and it took days for money to transfer. While much of the banking world used electronic transfers, this service was not readily available for the poor. PayPal's Xoom app targets the remittance business with convenient, fast (within minutes), and low-cost remittance transfers. Xoom brings PayPal profits and saves poor migrant workers millions of dollars in fees.[8]

Ethical issues in selecting segmenting dimensions

Marketing managers sometimes face ethical decisions when selecting segmenting dimensions. Problems may arise if a firm targets customers who are somehow at a disadvantage in dealing with the firm or who are unlikely to see the negative effects of their own choices. For example, some people criticize athletic shoe companies for targeting poor, inner-city kids who see expensive athletic shoes as an important status symbol. Many firms, including producers of infant formula, have been criticized for targeting consumers in less-developed nations. Some nutritionists criticize firms that market soft drinks, candy, and snack foods to children.

#M4BW Schneider Electric believes that access to energy is a basic human right. That said, almost 1.2 billion people in the world don't have access to electricity. Schneider Electric is trying to change that. An off-grid solar solution brought electricity to 128 schools in Kenya. It has also provided energy management training (on topics ranging from "basic electrician" to "advanced industrial automation") to more than 100,000 people in Asia, South America, and Africa, giving them better jobs and accelerating their goal of energy for all. By targeting underserved market segments in Africa, Schneider Electric generates profits and makes for a better world.
Source: Schneider Electric

> **Ethical Dilemma**
>
> *What would you do?* You have just started working as assistant brand manager for Silky Smooth skin cream, and you are responsible for one Southeast Asian country. Silky Smooth is a skin moisturizer that also lightens skin tone. However, the lightening feature has not been noted on Silky Smooth product packaging or in advertising. In this country there is discrimination against people with darker skin tones, so many consumers actively seek to lighten their skin. Two major skin cream competitors promote "skin lightening" by showing ads featuring an unhappy woman who starts using the product and becomes very happy and more successful with her lighter skin. Activists have protested against skin-lightening products because they reinforce a negative stereotype. Silky Smooth's consumer research predicts that promoting skin-lightening benefits to customers with darker skin tones would increase Silky Smooth sales by 20 percent in less than a year. You are wondering whether to target the market segment of women seeking lighter skin with advertising and packaging that emphasizes these benefits. Your boss says it is your call, while also reminding you of the bonus you receive for growing sales by 10 percent or more.
>
> *Do you have any concerns about this issue? Would you target this darker-skinned segment of the population? Why or why not?*

More Sophisticated Techniques May Help in Segmenting and Targeting

LO 4.5

Marketing managers and marketing researchers often turn to computerized methods for help with the segmenting job. A detailed review of the possibilities is beyond the scope of this book. But a brief discussion will give you a flavor of some different computer applications that support segmentation.

Clustering usually requires a computer

One approach involves **clustering techniques**, which try to find similar patterns within sets of data. Clustering groups customers who are similar on their segmenting dimensions into homogeneous segments. Clustering approaches use computers to do what previously was done mainly with intuition and judgment.

The data to be clustered might include such dimensions as demographic characteristics, the importance of different needs, attitudes toward the product, and past buying behavior. The computer searches all the data for homogeneous groups of people. When it finds them, marketers study characteristics of the people in the groups to see why the computer clustered them together. The results sometimes suggest new, or at least better, marketing strategies. Check out Marketing Analytics in Action: Cluster Analysis to see the results of a cluster analysis and segmentation of holiday gift shoppers.[9]

Marketing Analytics in Action: Cluster Analysis

A few years ago, Forbes Consulting Group surveyed 612 Americans aged 18 to 64 who were personally responsible for holiday shopping. The survey was conducted in mid-November, just before typical holiday shopping began. The research used cluster analysis to identify three groups of shoppers:

- "Giving Is Love": These people give gifts as a way of expressing love and affection and want to be admired for picking great gifts. Eighty-three percent pride themselves on giving unique gifts, and 52 percent splurge on luxury items.
- "Obligated Posers": These people mostly just want to avoid being scorned by peers. They are most likely to view shopping as an obligation (53 percent). They don't want to spend a lot of money but they want to look like they did.
- "Utilitarian Grinches": These people feel a great deal of anxiety about gift giving, prefer to give practical gifts that someone "really needs," and prefer to buy online.[10]

1. Name a retail store where each segment might be more likely to shop.
2. Give an example of a product that might be bought by each shopper type.
3. Describe a buyer persona (covered earlier in the chapter) for each of these groups.

What's Next? Target reads its customers' minds

A few years ago, a man walked into a Target store outside Minneapolis demanding to see a manager. The man was upset that Target was sending his teenage daughter coupons for baby clothes and cribs. "Are you trying to encourage her to get pregnant?" he asked the manager. The store manager promised to look into it, but a few days later the father called back and sheepishly admitted that there was more going on in his house than he knew about. His daughter was due in August. How did Target know? Why would Target care?

Consumers' shopping habits are hard to break. Once customers start shopping at a particular store for groceries, personal care products, or greeting cards, they get into a routine and don't even consider shopping at other stores. It turns out that a few major life events—new job, new home, marriage, and yes, new baby—significantly weaken those habits, at least for a short time. At those vulnerable moments, a competing store has an opportunity to steal a customer away—perhaps encouraging a switch with some well-timed promotions—and maybe help the customer develop a new shopping habit.

So how does a retail store such as Target segment its market and target pregnant shoppers? Target's marketing managers wondered if their customer relationship management (CRM) database might hold an answer. The database includes information on the shopping behavior of customers who sign up for the store's loyalty card or use a credit card. Along with additional data Target purchases from third parties, the database offers a detailed profile of millions of Target customers.

A small number of these customers told Target they were pregnant by signing up for the store's baby registry, where they also recorded the date the baby was due to be born. Target's statisticians used data from the customers they knew were pregnant and "mined" their CRM data to try to determine how their shopping behavior changed during pregnancy.

The statisticians discovered 25 products that were purchased more frequently during pregnancy. For example, although many people buy lotions, about six months before their due date purchases of unscented lotions bump up. Around that same time pregnant women stock up on supplements such as calcium, magnesium, and zinc. A few months later there is an uptick in purchases of scent-free soap, extra-large bags of cotton balls, hand sanitizers, and washcloths.

The statisticians use this information to develop a prediction score for each customer in Target's CRM database—predicting whether each is pregnant and when she will deliver. Customers with high scores receive customized direct mailings with coupons for a new crib or disposable diapers. Target wants these customers to start buying groceries and other products at Target, too. So it makes sure the mailings also include other products these customers haven't yet bought at Target. Many of these customers soon have a new habit and make Target their primary shopping location. So the next time you open a promotion for something you are thinking about buying, the store may be reading your mind—via its CRM database.[11]

When this story became public, some customers questioned whether Target had crossed an ethical line. They found Target's behavior creepy. *What do you think? Are Target's practices ethical? Should a retailer use data to predict customer behavior? Why or why not?*

A customer database can focus the effort

A variation of the clustering approach relies on **customer relationship management (CRM)**, where the seller fine-tunes the marketing effort with information from a detailed customer database. The database stores information that is useful for segmentation. Analytic software aids in identifying customer segments, so that each segment can be delivered to a different marketing mix.

Netflix's CRM database includes the specific shows and movies each customer has actually streamed to his or her TV, computer, or mobile device. Netflix has found that these behavioral data predict future viewing behavior much better than demographic information such as age, gender, or geographic location. So Netflix uses analytics software and artificial intelligence to decide what TV shows and movies appear as suggestions on each customer's Netflix home page.[12]

Target's CRM combines previous purchases with demographic data to predict customers' future needs. Read more in *What's Next?* Target reads its customers' minds.[13]

Dynamic behavioral segmentation creates segments based on customers' behavior

Many firms utilize **dynamic behavioral segmentation**, which refers to the use of real-time data to continuously update a customer's placement in a market segment. This is often used by e-commerce companies that collect and store data on a customer—and then update those data each time the customer visits the website. Websites have the power to track and store what a customer does on a website—as well as how the customer came to the website (for example, what search term or advertisement brought them), where in the world they are located, if they have visited before, and much more.

For example, look at the three customers in Exhibit 4-11. Each engages in different behaviors when visiting an e-commerce website that sells home supplies, including vacuum cleaners. When a customer clicks to its website, the site's CRM begins to collect information on each customer—recording and analyzing the customer's behavior (clicks) on this visit and possibly previous visits to the site.

In the first case, "Roseanne's" behaviors indicate she is very concerned about what a new vacuum cleaner would cost her. Her actions are quickly analyzed and she is placed in the "price-conscious shopper" segment. Based on this, she receives a message reinforcing the great deal she received and points out a 30-day price match guarantee. The other customers in Exhibit 4-11 behave differently at the site—and receive different

Exhibit 4-11 Dynamic Behavioral Segmentation: E-commerce Example

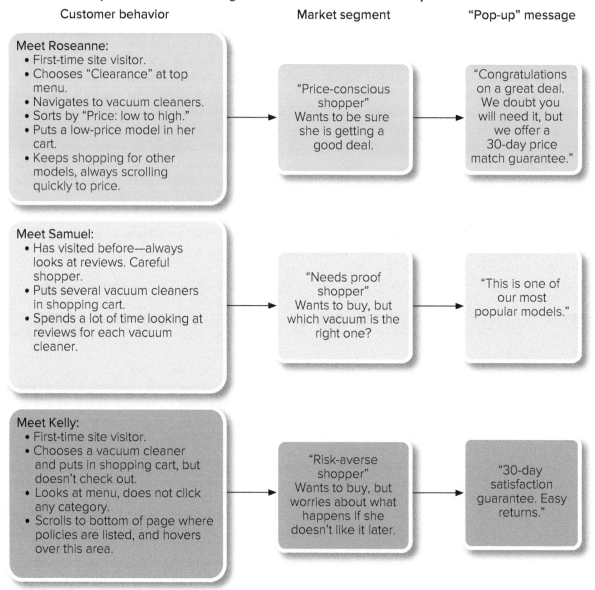

promotional messages. Review Exhibit 4-11 and note how Samuel and Kelly behave and how that determines the pop-up message each receives. An e-commerce site might define a few or many different segments. By giving customers the right message at the right time, a seller is more likely to make a sale.[14]

Targeting a segment of one

The ultimate in personalization occurs when firms consider each customer as its own segment. Business-to-business markets, with very large customers, often treat customer firms individually as "segments of one." Many CRM systems use artificial intelligence to do the same for large numbers of consumers. For example, Macy's mines customer data to adapt marketing communications to *individual* customers. Macy's CRM database includes a great deal of insight about each customer: previous purchases, responses to e-mails, online shopping behavior, and other information collected over time. Macy's uses these data to customize promotions in e-mails and direct mail. For example, Macy's sent out 30,000 different versions of a catalog mailing; the pieces varied from 32 to 76 pages long and featured merchandise designed to appeal to each customer. As a result, a footwear fanatic saw more shoes and moms saw more kids' clothes.[15]

Airlines have begun to compile data, but many are still deciding how and when to use those data to personalize service for individual customers. Many flight attendants carry a tablet with individual information about each passenger, including a customer's frequent flier level, connecting flight information, favorite red wine, and birthday. Airlines are still trying to figure out whether wishing a customer "Happy Birthday!" or offering "Your usual red wine, sir?" is good service—or maybe just a bit too creepy. As airlines start offering more personalized service, they will closely watch how customers react.[16]

Differentiation and Positioning Take the Customer Point of View

LO 4.6

Differentiate the marketing mix—to serve customers better

As we've emphasized throughout, the reason for focusing on a specific target market—by using marketing segmentation approaches or tools such as cluster analysis or CRM—is so you can fine-tune the whole marketing mix to provide some group of potential customers with superior value. By *differentiating* the marketing mix to do a better job meeting customers' needs, the firm builds competitive advantage. When this happens, target customers view the firm's marketing mix as uniquely suited to their preferences and needs.

Although the marketing manager may want customers to see the firm's offering as unique, that is not always possible. Me-too imitators may come along and copy the firm's strategy. Further, even if a firm's marketing mix is different, consumers may not know or care. They're busy and, simply put, the firm's product may not be that important in their lives. This is where another important concept, *positioning*, comes in.

Positioning is based on customers' views

Positioning refers to how customers think about proposed or present brands in a market. Without a realistic view of how customers think about offerings in the market, it's hard for the marketing manager to differentiate. At the same time, the manager should know how he or she *wants* target customers to think about the firm's marketing mix. Positioning issues are especially important when competitors in a market appear to be very similar. For example, many people think that there isn't much difference between one provider of home owner's insurance and another. But State Farm Insurance uses advertising to emphasize the value of the service and personal attention from its agents, who live right in the customer's neighborhood. Low-price insurers who sell from websites or toll-free numbers can't make that claim.

Figuring out what customers really think about competing products isn't easy, but there are approaches that help. Most of them require some formal marketing research. The results are usually plotted on graphs to help show how consumers view the competing products. Usually, the products' positions are related to two or three product features that are important to the target customers.

Sanuk wants customers to think about its shoes as "fun." The brand takes its name from the Thai word for "fun," its products are known for a funky sense of style, and its promotions are creative and humorous. These contribute to the brand's positioning, and Sanuk's target customers perceive the brand as fun and irreverent. Stove Top stuffing wants to expand its sales by repositioning its product from a holiday specialty to a food that can be eaten year-round. *(left): Source: Deckers Brands; (right): Source: The Kraft Heinz Company*

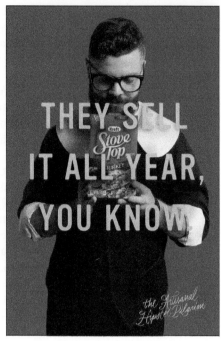

Managers make the graphs for positioning decisions by asking consumers to make judgments about different brands—including their "ideal" brand—and then use computer programs to summarize the ratings and plot the results. The details of positioning techniques—sometimes called *perceptual mapping*—are beyond the scope of this text. But Exhibit 4-12 shows the possibilities.[17]

Exhibit 4-12 shows the "product space" for different brands of bar soap using two dimensions—the extent to which consumers think the soaps moisturize and deodorize their skin. For example, consumers see Dove as quite high on moisturizing but low on deodorizing. Dove and Tone are close together—implying that consumers think of them as similar on these characteristics. Dial is viewed as different from Dove and Tone and

Exhibit 4-12 "Product Space" Representing Consumers' Perceptions for Different Brands of Bar Soap

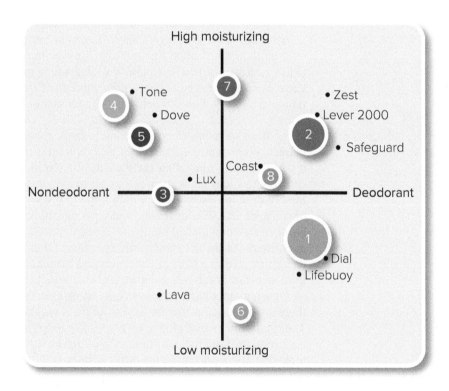

at a distance from them on the graph. Remember that positioning maps are based on *customers' perceptions*—the actual characteristics of the products (as determined by a chemical test) might be different!

Each segment may have its own preferences

The circles in Exhibit 4-12 show different sets (submarkets) of consumers clustered near their ideal soap preferences. Groups of respondents with a similar ideal product are circled to show apparent customer concentrations. In this graph, the size of the circles suggests the size of the segments for the different ideals.

Ideal clusters 1 and 2 are the largest and are close to two popular brands—Dial and Lever 2000. It appears that customers in cluster 2 want more moisturizing than they see in Dial. However, exactly what this brand should do about this isn't clear. Perhaps Dial should leave its physical product alone—but emphasize moisturizing more in its promotion to make a stronger appeal to those who want moisturizers. A marketing manager talking about this approach might simply refer to it as "positioning the brand as a good moisturizer." Of course, whether the effort is successful depends on whether the whole marketing mix delivers on the promise of the positioning communication.

Note that ideal cluster 7 is not near any of the present brands. This may suggest an opportunity for introducing a new product—a strong moisturizer with some deodorizers. A firm that chooses to follow this approach would be making a segmenting effort.

Combining versus segmenting

Positioning analysis may lead a firm to combining—rather than segmenting—if managers think they can make several general appeals to different parts of a "combined" market. For example, by varying its promotion, Coast might try to appeal to segments 1, 2, and 8 with one product. These segments are all quite similar (close together) in what they want in an ideal brand. On the other hand, there may be clearly defined submarkets—and some parts of the market may be "owned" by one product or brand. In this case, segmenting efforts may be practical—moving the firm's own product into another segment of the general market area where competition is weaker.

Positioning as part of broader analysis

A positioning analysis helps managers understand how customers see their market. It is a visual aid to understanding a product-market. The first time such an analysis is done, managers may be shocked to see how much customers' perceptions of a market differ from their own. For this reason alone, positioning analysis may be crucial. But a positioning analysis usually focuses on specific product features and brands that are close competitors in the product-market. Thus, it is a product-oriented approach. Important *customer*-related dimensions—including needs and attitudes—may be overlooked.

Premature emphasis on product features is dangerous in other ways as well. As our bar soap example shows, starting with a product-oriented definition of a market and how bar soaps compete against other bar soaps can make a firm miss more basic shifts in markets. For example, bars have lost popularity to liquid soaps. Other products, such as bath oils or body washes for use in the shower, are now part of the relevant competition also. Managers wouldn't see these shifts if they looked only at alternative bar soap brands—the focus is just too narrow.

It's also important to realize that the way consumers look at a product isn't just a matter of chance. Let's return to our bar soap example. Although many consumers do think about soap in terms of moisturizing and deodorizing, other needs shouldn't be overlooked. For example, some consumers are especially concerned about wiping out germs. Marketers for Dial soap recognized this need and developed ads that positioned Dial as "the choice" for these target customers. This helped Dial win new customers, including those who switched from Lifebuoy—which was otherwise similar to Dial (see Exhibit 4-12). In fact, what happened to Lifebuoy highlights what happens if managers don't update their marketing strategy as customer needs and competition change. Lifebuoy was the first deodorant soap on the market; it was a leading brand for more than 100 years. But it gradually lost sales to competitors with stronger marketing mixes (clearer differentiation, better positioning, and superior customer value) until sales declined and Unilever stopped selling it in the United States.

Sometimes repositioning is needed

Sometimes research shows the marketing manager that target customers are not viewing the brand in the desired way—it needs to be *repositioned*. Changing customers' perceptions of a brand is not easy—and requires changes to the marketing mix. In the mid-2000s Korean carmaker Hyundai found that its positioning was not helping it sell more vehicles. When it conducted research that showed pictures of the Hyundai Veracruz, with the Hyundai name hidden, to a group of potential car buyers, 71 percent said they would buy one. Another group of buyers was shown the same pictures *with* the Hyundai name—and only 52 percent were interested this time. Hyundai learned that it had a reputation as a "cheap, low-quality" car, even though data showed its vehicles were among the best made in the world. To reposition the brand, Hyundai gave buyers a 10-year, 100,000-mile warranty on every car and ran ads that emphasized its high safety and quality ratings. Soon Hyundai wasn't seen as cheap; instead, it was "more value for your money." With the new positioning, Hyundai's market share jumped 70 percent in three years.[18]

Positioning statement provides direction for marketing strategy

Sometimes marketing managers use a positioning statement to provide focus for a marketing mix. A **positioning statement** concisely identifies the firm's desired target market, product type, primary benefit or point of differentiation, and the main reasons a buyer should believe the firm's claims. The one or two benefits highlighted in the statement should be those most important to the target market—and unique to the brand. It's important that everyone involved in planning the marketing strategy agree with the positioning statement because it helps narrow options and guide the selection of a marketing mix.

Some firms use a template like the following to aid in preparation of a positioning statement:

> For (*our target market*), (*our brand*) of all (*product type*) delivers (*key benefit or point of differentiation*) because (*our brand*) is (*reasons to believe*).

A few years ago, marketing managers for Mountain Dew used this template to develop the following positioning statement:

> For 16- to 24-year-old males who embrace excitement, adventure, and fun, Mountain Dew, of all carbonated soft drinks, delivers great taste that exhilarates like no other because Mountain Dew is energizing, thirst-quenching, and has a one-of-a-kind citrus flavor.

The positioning statement provided Mountain Dew's advertising agency with a direction to follow that led to a series of television and print advertisements that reinforced this positioning. The statement guided decisions about packaging, point-of-purchase promotion, sponsorships, the look and feel of the website, and choices for new flavors. The marketing strategy based on this positioning helped Mountain Dew gain market share with this target market.[19]

As we emphasize throughout the text, you must understand potential needs and attitudes when planning marketing strategies. If customers treat different products as substitutes, then a firm has to position itself against those products, too. Customers won't always be conscious of all the detailed ways that a firm's marketing mix might be different, but careful positioning can help highlight a unifying theme or benefits that relate to the determining dimensions of the target market. Thus, it's useful to think of positioning as part of the broader strategy planning process—because the purpose is to ensure that the whole marketing mix is positioned for competitive advantage.

Source: PepsiCo, Inc.

CONCLUSION

Chapters 2 and 3 introduced a framework for strategy planning that starts with analysis of the broad market and then narrows down to a specific target market and marketing mix. The basic purpose of this chapter is to show how marketing managers use market segmentation and positioning to guide that narrowing-down process.

Now that you've read this chapter you should understand how to carefully define generic markets and product-markets and how that can help in identifying and evaluating opportunities. We stressed the shortcomings of a too-narrow, product-oriented view of markets and explained why it's better to take a broader view that also includes consideration of customer needs, the product type, the customer type, and the geographic area.

We also discussed approaches for market segmentation—the process of naming and then segmenting broad product-markets to find potentially attractive target markets. Some people try to segment markets by starting with the mass market and then dividing it into smaller submarkets based on a few demographic characteristics. But this can lead to poor results. Instead, market segmentation should first focus on a broad product-market and then group similar customers into homogeneous submarkets. The more similar the potential customers are, the larger the submarkets can be. Four criteria for evaluating possible product-market segments were presented.

Once a broad product-market is segmented, marketing managers can use one of three approaches to market-oriented strategy planning: (1) the single target market approach, (2) the multiple target market approach, or (3) the combined target market approach. In general, we encourage marketers to be segmenters rather than combiners.

We also introduced different dimensions that are used to segment markets. This included behavioral, geographic, and demographic dimensions for consumers. For business and organizational markets, the kind of relationship, type of customer, how a product will be used, and type of buying situation were just some of the categories of dimensions. This section also explained how target market dimensions are tied to marketing strategy decision areas.

We also covered computer-aided approaches such as clustering techniques, CRM, and dynamic behavioral segmentation. We emphasized the role of positioning in providing a focus or theme to the various elements of a differentiated marketing mix that fits the preferences of target customers. We also described how online customers can be grouped in real time using dynamic behavioral segmentation.

In summary, good marketers should be experts on markets and the dimensions that segment them. By thoughtfully segmenting markets, they may spot opportunities—even breakthrough opportunities—and help their firms succeed against aggressive competitors offering similar products.

In Chapters 5 and 6 you'll learn more about the buying behavior of final consumers and organizational customers. As you enrich your understanding of customers and how they behave, you will gain command of a broader set of dimensions that are important for segmentation and positioning.

KEY TERMS

LO 4.7

market, 89
generic market, 90
product-market, 90
market segmentation, 93
segmenting, 94
market segment, 94
single target market approach, 96

multiple target market approach, 96
combined target market approach, 96
combiners, 96
segmenters, 97
qualifying dimensions, 101
determining dimensions, 101
clustering techniques, 104

customer relationship management (CRM), 105
dynamic behavioral segmentation, 106
positioning, 107
positioning statement, 110

QUESTIONS AND PROBLEMS

1. Review the LEGO case study that opens the chapter. From this case, identify examples of different key terms and concepts covered in the chapter. For example, LEGO operates in the "active play market."

2. Review the LEGO case study that opens this chapter. Applying concepts from the chapter, how else could LEGO segment the market? Use at least two segmenting dimensions not discussed in the current case and describe each market segment. Describe three marketing tactics that could be used in developing a strategy for each target market.

3. Distinguish between a generic market and a product-market. Illustrate your answer with an example.

4. Explain what market segmentation is.

5. List the types of potential segmenting dimensions, and explain which you would try to apply first, second, and third in a particular situation. If the nature of the situation would affect your answer, explain how.

6. Explain why segmentation efforts based on attempts to divide the mass market using a few demographic dimensions may be very disappointing.

7. Illustrate the concept that segmenting is an aggregating process by referring to the admissions policies of your own college and a nearby college or university.

8. Review the types of segmenting dimensions listed in Exhibits 4-8 and 4-9, and select the ones you think should be combined to fully explain the market segment you personally would be in if you were planning to buy a new watch today. List several dimensions and try to develop a shorthand name, like "fashion-oriented," to describe your own personal market segment. Then try to estimate what proportion of the total watch market would be in your market segment. Next, explain if there are any offerings that come close to meeting the needs of your market. If not, what sort of a marketing mix is needed? Would it be economically attractive for anyone to try to satisfy your market segment? Why or why not?

9. Identify the determining dimension or dimensions that explain why you bought the specific brand you did in your most recent purchase of (a) a soft drink, (b) shampoo, (c) a shirt or blouse, and (d) a larger, more expensive item, such as a bicycle, camera, or boat. Try to express the determining dimension(s) in terms of your own personal characteristics rather than the product's characteristics. Estimate what share of the market would probably be motivated by the same determining dimension(s).

10. Marketing for a better world (#M4BW) may include efforts by some firms to offer value to underserved market segments. Give examples of two companies that are demonstrating this.

11. Consider the market for off-campus apartments in your city. Identify some submarkets that have different needs and determining dimensions. Then evaluate how well the needs in these market segments are being met in your geographic area. Is there an obvious breakthrough opportunity waiting for someone?

12. Explain how positioning analysis can help a marketing manager identify target market opportunities.

13. Write a *personal* positioning statement using the format described for a positioning statement. In this case, replace "our target market" with an industry or type of job that you want to do; replace "our brand" with your name; replace the "key benefit or point of differentiation" with something you can bring to an employer in this situation, and replace "reasons to believe" with proof that you can deliver your key benefit or point of differentiation.

MARKETING PLANNING FOR HILLSIDE VETERINARY CLINIC

Appendix D (the Appendices follow Chapter 19) includes a sample marketing plan for Hillside Veterinary Clinic. Look through the "Customer Analysis" section.

1. How does the marketing plan segment the market?
2. Can you think of other segmentation dimensions that could be used?
3. What do you think of the approach Hillside used to determine target markets? Is it using a single target market, multiple target market, or combined target market approach?
4. How does Hillside plan to differentiate and position its offering?

SUGGESTED CASES

2. Nature's Own Foods, Inc.
3. NOCO United Soccer Academy
7. Lake Pukati Lodge
10. Cousin's Ice Center
14. Schrock & Oh Design
27. Canadian Mills, Ltd.

Video Case 1. Potbelly Sandwich
Video Case 4. Invacare
Video Case 5. Segway

MARKETING ANALYTICS: DATA TO KNOWLEDGE

CHAPTER 4: SEGMENTING CUSTOMERS

The marketing manager for Audiotronics Software Company is seeking new market opportunities. He is focusing on the voice recognition market and has narrowed down to three segments: the Fearful Typists, the Power Users, and the Professional Specialists. The Fearful Typists don't know much about computers—they just want a fast way to create e-mail messages, letters, and simple reports without errors. They don't need a lot of special features. They want simple instructions and a program that's easy to learn. The Power Users know a lot about computers, use them often, and want a voice recognition program with many special features. All computer programs seem easy to them, so they aren't worried about learning to use the various features. The Professional Specialists have jobs that require a lot of writing. They don't know much about computers but are willing to learn. They want special features needed for their work—but only if they aren't too hard to learn and use.

The marketing manager prepared a table summarizing the importance of each of three key needs in the three segments (see table that follows).

Market Segment	Importance of Need (1 = not important; 10 = very important)		
	Features	Easy to Use	Easy to Learn
Fearful Typists	3	8	9
Power Users	9	2	2
Professional Specialists	7	5	6

Audiotronics' sales staff conducted interviews with seven potential customers who were asked to rate how important each of these three needs were in their work. The manager prepared a spreadsheet to help him cluster (aggregate) each person into one of the segments—along with other similar people. Each person's ratings are entered in the spreadsheet, and the clustering procedure computes a similarity score that indicates how similar (a low score) or dissimilar (a high score) the person is to the typical person in each of the segments. The manager can then "aggregate" potential customers into the segment that is most similar (that is, the one with the *lowest* similarity score).

Potential Customer	Importance of Need (1 = not important; 10 = very important)			Type of Computer
	Features	Easy to Use	Easy to Learn	
A.	8	1	2	Dell laptop
B.	6	6	5	HP desktop
C.	4	9	8	Apple
D.	2	6	7	Apple
E.	5	6	5	HP desktop
F.	8	3	1	Dell laptop
G.	4	6	8	Apple

Design element: #M4BW box globe icon: ©Vectoryzen/Shutterstock

CHAPTER FIVE

Romsvetnik/Shutterstock

Final Consumers and Their Buying Behavior

When Sony introduced its Walkman in the late 1970s, it quickly became a popular way for music lovers to play cassette tapes with their favorite songs—anywhere they went. Competing players quickly emerged, but Sony kept its lead by improving its Walkman and then offering models for CDs when that medium came on the market.

In the late 1990s, the new MP3 format offered quality music from a digital file that played on a computer or portable player. Diamond Multimedia's Rio was the first MP3 player. The Rio was innovative, but users had to download music from virus-ridden websites—or use special software to "rip" CDs to MP3 format. Many music buffs liked the idea of having songs at their fingertips, but getting the digital files was just too complicated. Further, music companies filed lawsuits charging some with illegally downloading MP3 music files. All of this slowed the initial adoption of MP3 players.

Attitudes quickly changed when Apple offered an innovative marketing mix that addressed these customer needs. iTunes offered legal downloads of songs at a reasonable price without the risk of viruses and made it easy to organize digital music on a computer and transfer it to Apple's iPod. The iPod was stylish and easy to use. Apple's advertising generated awareness of and interest in the new concept among customers with the promise of "a thousand songs in your pocket."

Apple is known for getting ahead of trends. When Apple reinvented the cell phone, it didn't do marketing research to discover consumers' needs and wants. Apple figured that consumers would be unable to foresee something so radically new. Apple's iPhone was more than just a cell phone that also played music—its colorful touchscreen display, web browsing, and music player wowed consumers. Now the iPhone lets them play games like *Fortnite* and *Donut County*, check social media like Snapchat or TikTok, watch videos on YouTube, or improve health and wellness by clicking on the Calm or 10% Happier: Meditation apps.

The iPhone's success created attention and opportunities for other Apple products. Once someone owns an iPhone, he or she pays more attention to Apple—lingering at a store display, watching an Apple ad, or asking friends about their AirPods. A positive iPhone experience raises a customer's expectations and trust in the entire Apple product line; many iPhone buyers go on to buy a MacBook, iPad, or Apple Watch, and maybe subscribe to Apple TV and Apple Music. Once customers join the Apple ecosystem, they often don't think much about which brand to buy when it's time to upgrade their phones. They trust that Apple products are of high quality and meet their needs.

While Apple's iPod and iTunes were innovative, consumer beliefs and attitudes on how to buy and listen to music changed. Apple thought consumers expected to "own" their music—and would continue paying for downloads. But streaming music services like Spotify and Pandora changed consumers' attitudes. They found consumers were happy just to "rent" music rather than buy and "own" a digital file.

By the time the Apple Music streaming service launched, consumers had a lot of choice among streaming services. Customers with economic concerns often chose the free, advertising-supported option from Spotify. Others found that Spotify didn't carry Taylor Swift's music catalog, and that Tidal had exclusive access to songs from Kanye West and Beyoncé. Decisions, decisions. More chose Spotify, which had a head start, but Apple Music soon became the number two player globally.

When Apple introduced a new iPhone without a headphone jack, many long-time customers were upset; using wired headphones required a special dongle, and the phone couldn't be charged while listening. At $159, Apple's pricey AirPods wireless headphones with a fancy charger-case were its answer. Unfortunately for Apple, AirPods became a punchline in viral posts on Twitter, like this one: "Top 10 richest people: (1) AirPod users, (2) Amazon CEO, (3) Bill Gates . . . ," and sales fell short of targets. A couple years later, the Christmas season sparked a turnaround in customers' attitudes toward AirPods. Gift givers often spend more on a gift than they might for themselves. AirPods became the trendy gift and went from joke to the must-have fashion accessory. Soon Apple had another hit product.

As growth in the United States slowed, Apple turned more attention to international markets, with particular attention on China, the largest market in the world. When Apple first entered China, it quickly became the "it" brand; the iPhone was the ultimate status symbol. Apple fueled that belief by getting iPhones in the hands of many opinion leaders and celebrities.

More recently, the iPhone has struggled in China. While competitors like Huawei and Xiaomi adapted phones for the Chinese market, Apple stuck with its one-size-fits-all model. For example, Chinese consumers like making mobile payments (using the phone like a credit or debit card), but this requires special software the iPhone didn't initially support. Also, many Chinese want two phone numbers—one for business and one for personal use—and one phone. While competitors' phones featured two SIM card slots, making it easy to use one phone with two numbers, the iPhone didn't have this feature. While iPhone fixed those shortcomings, Huawei phones with features similar to the iPhone sold for almost 50 percent less. Chinese consumers turned away from the iPhone.

Strong competitors like Huawei, Samsung, and Spotify are all seeking a piece of Apple's pie. Continued success will depend on Apple's ability to understand consumer behavior, anticipate customer needs, and exceed customer expectations all around the world.[1]

LEARNING OBJECTIVES

Many variables influence consumer buying behavior. As the Apple case highlights, successful marketing strategy planning requires a clear understanding of how target consumers buy—and what factors affect their decisions. The learning objectives for this chapter will help you develop that understanding.

When you finish this chapter, you should be able to

1. describe how economic needs influence the buyer decision process.
2. understand how psychological variables affect an individual's buying behavior.
3. understand how social influences affect an individual's buying behavior.
4. describe how culture and ethnicity influence consumer buying behavior.
5. explain how characteristics of the purchase situation influence consumer behavior.
6. explain the process by which consumers make buying decisions.
7. understand important new terms (shown in **red**).

Consumer Behavior: Why Do They Buy What They Buy?

The Apple case shows the in-depth customer knowledge needed for effective marketing strategy planning. Without that understanding, it would be difficult to zero in on the right target market, or to develop and adapt a marketing mix that will be the best value for customers.

Understanding consumer behavior can be a challenge. Specific behaviors vary a great deal for different people, products, and purchase situations. In today's global markets the variations are countless. That makes it impractical to catalog all the possibilities. But there are general behavioral principles—frameworks—that marketing managers can apply to better understand their specific target markets. In this chapter, our approach focuses on developing your skill in working with these frameworks by exploring thinking from economics, psychology, sociology, and other behavioral disciplines. We'll take a look at the topics shown in Exhibit 5-1, which includes a simplified model of consumer behavior.

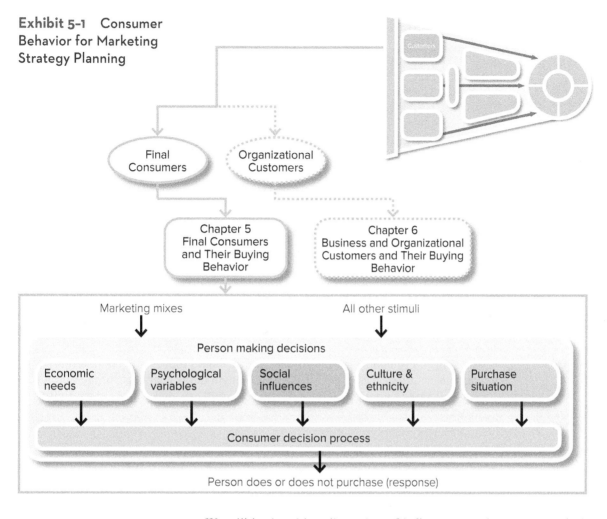

Exhibit 5-1 Consumer Behavior for Marketing Strategy Planning

We will begin with a discussion of influences on the consumer decision process: economic needs, psychological variables, social influences, culture/ethnicity, and the purchase situation. Exhibit 5-2 provides an expanded look at these influences. Following our discussion of these different categories of influences, we will look more closely at the consumer decision process.

Exhibit 5-2 A Model of Influences on Consumer Behavior

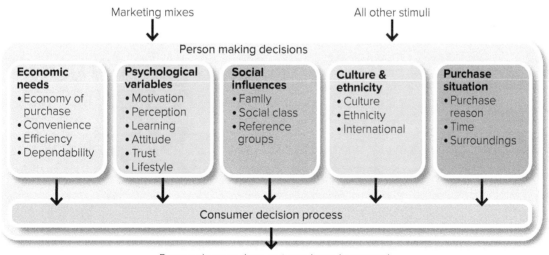

Economic Needs Affect Most Buying Decisions

LO 5.1

Economic buyers seek the best uses of time and money

Some economists assume that consumers are primarily **economic buyers**—people who know all the facts and logically compare choices to get the greatest satisfaction from spending their time and money. The economic-buyer theory says that consumers decide what to buy based on **economic needs**, which are concerned with making the best use of a consumer's time and money—as the consumer judges it. Some consumers look for the lowest price. Others will pay extra for convenience. And others may weigh price and quality for the best value. Some economic needs include:

1. Economy of purchase or use (buying a less expensive store brand of frozen pizza).
2. Efficiency in operation or use (a Dyson vacuum cleaner's power makes for faster cleanup).
3. Dependability in use (Toyotas have lower repair costs over the years).
4. Improvement of earnings (Ally Bank offers a better interest rate).
5. Convenience (a drive-through lane at a Starbucks makes it convenient to pick up coffee on the way to work in the morning).

Clearly, marketing managers must be alert to new ways to appeal to economic needs. Most consumers appreciate firms that offer them improved economic value for the money they spend. But improved value does not just mean offering lower and lower prices. For example, Products can be designed to work better, require less service, or last longer. Promotion can inform consumers about product benefits in terms of measurable factors such as operating costs, the length of the guarantee, or the time a product will save. Carefully planned Place decisions can make it easier and faster for customers who face a lack of time to make a purchase.

Many firms adjust their marketing mixes for target markets that place a high value on convenience. Whole Foods Market sells more takeout food than most restaurants. Tide to Go is an instant stain remover that fits easily in a purse or briefcase and requires no water. Many consumers like the convenience of online shopping at Zappos.

Income affects needs

The ability to satisfy economic needs largely depends on how much money a consumer has available—which in turn depends a great deal on household income. In the United States, income distribution varies widely. In 2017, 25 percent of households reported an income of less than $30,001, and the median income—where half the households earned

Economic needs are concerned with making the best use of a consumer's time and money. For some consumers, that means getting an inexpensive sandwich. For other consumers, it means buying a more reliable car, like the Honda Brio.
(left): Source: McDonald's; (right): Source: Honda Motor Co., Inc.

more and half earned less—was $61,822. After taking account of inflation, the median income has been flat over the last two decades. During this same period, America's middle-income consumers have been hit hard by the rising costs of necessities.[2]

In most households, people don't have enough income to buy everything they want. For many products, these people can't be customers even if they want to be. For example, most families spend a good portion of their income on such necessities as food, rent or house payments, home furnishings, transportation, and insurance. A family's purchase of "luxuries" comes from **discretionary income**—what is left of income after paying taxes and paying for necessities.

Discretionary income is an elusive concept because the definition of necessities varies from family to family and over time. It depends on what they think is necessary for their lifestyle. High-speed Internet service purchased out of discretionary income by a lower-income family may be considered a necessity by a higher-income family.[3]

Consumer confidence affects spending

Economic conditions affect consumer confidence and spending. In a strong economy, consumers feel confident and secure in their jobs, so they are more likely to borrow money to buy a larger house or a new car or to vacation in an exotic locale. On the other hand, when consumers worry about job prospects or their retirement savings decline, they are more cautious spenders.

Economic value and income are important factors in many purchase decisions. But most marketing managers think that buyer behavior is not as simple as the economic-buyer model suggests. Consumer confidence shows how economic factors are related to psychological influences, so let's look more closely at the psychological variables that influence buying behavior (see Exhibit 5-2).

Psychological Influences within an Individual

LO 5.2

Needs motivate consumers

Everybody is motivated by needs and wants. **Needs** are the basic forces that motivate a person to do something. Some needs involve a person's physical well-being; others the individual's self-view and relationship with others. Needs are more basic than wants. **Wants** are "needs" that are learned during a person's life. For example, everyone needs water or some kind of liquid, but some people also have learned to want LaCroix Cerise Limón (cherry lime) sparkling water.

When a need is not satisfied, it may lead to a drive. The need for liquid, for example, leads to a thirst drive. A **drive** is a strong stimulus that encourages action to reduce a need. Drives are internal—they are the reasons behind certain behavior patterns. In marketing, a product purchase results from a drive to satisfy some need.

Some critics imply that marketers can somehow manipulate consumers to buy products against their will. But trying to get consumers to act against their will is a waste of time. Instead, a good marketing manager studies what consumer drives, needs, and wants already exist and how they can be satisfied better.

Consumers seek benefits to meet needs

We're all a bundle of needs and wants. Exhibit 5-3 lists some important needs

Whether a product is a want or a need depends on what a consumer has learned during his or her life. Someone may *need* a car to drive to work or school. If he or she has learned to love sports cars, the consumer may *want* a Nissan Z.
Source: Nissan Motor Company

Exhibit 5-3 Possible Needs Motivating a Person to Some Action

Types of Needs	Specific Examples			
Physiological needs	Food Sex	Rest Liquid	Activity Self-preservation	Sleep Warmth/coolness
Psychological needs	Nurturing Playing/relaxing Self-identification	Curiosity Order Power	Independence Personal fulfillment Pride	Love Playing/competition Self-expression
Desire for . . .	A better world Acceptance Affiliation Esteem Knowledge	Respect Status Achievement Appreciation Sympathy	Beauty Happiness Self-satisfaction Variety Affection	Companionship Distinctiveness Recognition Sociability Fun
Freedom from . . .	Fear Pain	Depression Stress	Loss Sadness	Anxiety Illness

that might motivate a person to some action. This list, of course, is not complete. But thinking about such needs can help you see what *benefits* consumers might seek from a marketing mix.

When a marketing manager defines a product-market, the needs may be quite specific. For example, the food need might be as specific as wanting a Domino's thick-crust pepperoni pizza delivered to your door hot and ready to eat.

Marketing managers appeal to customer needs

Consumer psychologists often argue that a person may have several reasons for buying—sometimes at the same time. Maslow is well known for his five-level hierarchy of needs. We will discuss a similar four-level hierarchy that is easier to apply to consumer behavior. Exhibit 5-4 illustrates the four levels along with an example

Exhibit 5-4 The PSSP Hierarchy of Needs

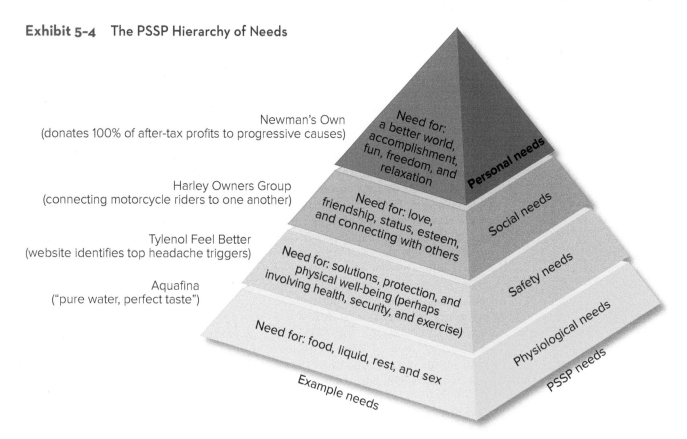

120

showing how a company has tried to appeal to each need. Generally, lower-level needs must be addressed before consumers seek to satisfy the higher-level needs. The lowest-level needs are physiological, followed by safety, social, and personal needs (PSSP).[4]

#M4BW

Marketing managers appeal to customer needs and create a better world

Many marketing managers are finding ways to appeal to customer needs at all levels—while simultaneously creating value for customers and society. **Physiological needs** are concerned with biological needs—food, liquid, rest, and sex. Marketers that offer *solutions* to consumer problems build brand loyalty. Charmin addressed such a need for holiday shoppers in New York City. Because it can be difficult to find a public restroom—and coffee shops don't like it when noncustomers use their facilities—Charmin's "holiday gift to the City" was luxury public restrooms located in Times Square. Charmin made the world a better place, and users will also remember Charmin's gift the next time they buy toilet tissue.

Safety needs are concerned with protection and physical well-being (perhaps involving health, financial security, medicine, and exercise). Under Armour's wide range of fitness products promote health and exercise. Northwestern Mutual Life offers life insurance and financial planning for those seeking financial security.

Social needs are concerned with love, friendship, status, and esteem—things that involve a person's interaction with others. Marketers that help people *connect* with others inspire positive feelings about their own brands. Coca-Cola addressed this need with a series of viral videos around the theme of delivering happiness. One featured a bright red Coca-Cola truck that stopped on streets in Brazil. A sign on the back of the truck told consumers to "Push" a large button. When they did, the truck dispensed happiness in the form of Cokes and fun items such as T-shirts, soccer balls, and even a surfboard. The truck created on-the-street social connections. The fun video helped associate Coke with fun and social connections when viewers forwarded a link to friends and family. Coke built its brand and fostered a better world.

Personal needs, on the other hand, are concerned with an individual's need for personal satisfaction—unrelated to what others think or do. Examples include accomplishment, fun, freedom, and relaxation—as well as a desire to make the world a better place.

Many people want to make food choices that enable them to achieve a healthier balance in their lives. ConAgra Foods (maker of healthy foods such as SmartPop! popcorn and Egg Beaters) developed an online tool to help. Members complete a survey at "Start Making Choices" and then receive feedback and advice on nutrition, activity, and well-being based on a calculated "Balanced Life Index." Weekly contact between the company and its customers keeps the relationship alive. ConAgra builds customer relationships by making their lives healthier.

Often marketing managers try to address multiple needs. Consider the California Milk Processor Board (creators of the infamous "Got Milk?" campaign) and its online game *Get the Glass*. The game appealed to kids—the group that needs milk the most. Besides being fun (personal needs), it educated players on why milk is important (safety needs). It could be played with family—or forwarded to friends (social needs). Of course, drinking milk addresses thirst (physiological needs) and promotes strong bones (safety needs). In the online game's first two months, more than 6 million people visited the site, with 650,000 reaching the Glass at Fort Fridge. The game was credited with helping boost sales of milk by more than 10 million gallons compared to the same period a year earlier.[5]

Perception determines what consumers see and feel

Consumers select varying ways to meet their needs, sometimes because of differences in **perception**—how we gather and interpret information from the world around us.

We are constantly bombarded by marketing stimuli—ads, products, signs, stores—yet we notice only a small number of these. This is because we apply the following selective processes:

1. **Selective exposure**—our eyes and minds seek out and notice only information that interests us. This often relates to current needs. When we are hungry, we notice signs for restaurants; research has shown that advertising is much more effective when customers are about to make a purchase.[6]
2. **Selective perception**—we screen out or modify ideas, messages, and information that conflict with previously learned attitudes and beliefs. So if we believe that Apple products are cool, we ignore ads by Samsung that suggest Apple is uncool.
3. **Selective retention**—we remember only what we want to remember. Even when we notice stimuli in the market, we may not place it in memory if we don't believe it or don't feel that information will be useful to us. If we read an article about the restaurant Panera Bread removing artificial ingredients from its food, but we don't eat at Panera, we won't commit this new information to memory.

Recognizing that consumers use these selective processes, marketing managers try to engage with customers when they are likely to be open to their messages. For example, Mazda advertises its sports car on websites such as Edmunds.com where consumers research cars.

Learning determines what response is likely

Learning is a change in a person's thought processes caused by prior experience. Learning is often based on direct experience: A little girl tastes her first cone of Ben & Jerry's Cherry Garcia flavor ice cream, and learning occurs! Learning may also be based on indirect experience or associations. If you watch an ad that shows other people enjoying Ben & Jerry's Chocolate Fudge Brownie low-fat frozen yogurt, you might conclude that you'd like it too.

Consumer learning may result from things that marketers do, or it may result from stimuli that have nothing to do with marketing. Either way, almost all consumer behavior is learned.[7]

#M4BW The Consumer Safety Institute in the Netherlands wants to remind parents that children see those cleaning supplies under the sink differently. As consumers approached this outdoor ad, the image changed from the one on the left to the one on the right and back.
Source: eiligheidNL; Agency: Lemz/ Netherlands

The ad shows two different images when viewed from different angles (lenticular effect).

Exhibit 5-5 The Learning Process

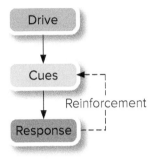

Sales take off after Febreze learns a lesson

Source: Procter & Gamble

Ethics question: Do consumers learn needs from marketing?

Attitudes relate to buying

Experts describe a number of steps in the learning process. We've already discussed the idea of a drive as a strong stimulus that encourages action. Depending on the **cues**—products, signs, ads, and other stimuli in the environment—an individual chooses some specific response. A **response** is an effort to satisfy a drive. The specific response chosen depends on the cues and the person's past experience.

Reinforcement of the learning process occurs when the response is followed by satisfaction—that is, reduction in the drive. Reinforcement strengthens the relationship between the cue and the response. And it may lead to a similar response the next time the drive occurs. Repeated reinforcement leads to development of a habit—making the individual's decision process routine. Exhibit 5-5 shows the relationships of the important variables in the learning process.

Marketing managers for Febreze used the learning process to develop a best-selling product, but first they had to get the drive, cue, response, and reinforcement just right. Febreze contains a chemical very effective at eliminating odors—spray it on clothing, furniture, or carpet and smells magically disappear. Febreze's marketing managers figured that most consumers desired a home free of offensive odors (drive), so the first television ads for Febreze focused on smells around the house. For example, one showed a dog on a couch and a woman saying, "Sophie will always smell like Sophie, but now my couch doesn't have to" (cue). In the ad, the woman sprays Febreze on the couch (response) and the dog smell disappears (reinforcement).

In spite of an aggressive advertising campaign, Febreze sold poorly when it was initially launched. To learn why, marketing researchers interviewed customers in their homes. They discovered two problems. First, many people living in the smelliest homes simply didn't realize their homes smelled at all. They had become desensitized to the smells of their pets, cigarette smoke, or other odors, so for them, the desired cue didn't exist. Second, what Febreze provided—lack of a smell—didn't provide positive reinforcement for the customer.

So Febreze's marketing managers went back to the drawing board. They interviewed more customers and reviewed video of people cleaning their homes. They identified another market, people who wanted a clean house (drive). When they saw a messy home (cue), they cleaned the house (response) and finished by rewarding themselves with something relaxing or happy—maybe just a smile (reinforcement). Here marketing managers realized that a scented version of Febreze might provide positive reinforcement—something that could signal the cleaning job was done. So Febreze added scents and new advertising that showed homemakers spraying a room with Febreze after cleaning (reinforcement). Sales took off; now Febreze sells more than $1 billion a year. And Febreze's marketing managers learned a valuable lesson.[8]

Some critics argue that marketing efforts encourage people to spend money on learned wants totally unrelated to any basic need. Consumer needs are internal, but advertising can make consumers aware of their needs—and suggest a way to satisfy these needs. For example, a customer may have thirst as a basic need; advertising helps him become aware of Simply Lemonade brand as a tasty way to satisfy his need. The next time he is thirsty, he may want Simply Lemonade. So advertising does not create needs—but it can make consumers more aware of an existing need and create a want for a specific product.[9]

An **attitude** is a person's point of view toward something. The "something" may be a product, an advertisement, a salesperson, a firm, or an idea. Attitudes are an important topic because they affect the selective processes, learning, and the buying decisions.

#M4BW Consumer attitudes can affect their purchasing (and donating) behavior. The World Wildlife Fund knows consumers have more favorable attitudes toward panda bears as compared to bluefin tuna—but both species are endangered and need support.
Source: World Wildlife Fund

Because attitudes are usually thought of as involving liking or disliking, they have some action implications. Beliefs are not so action-oriented. A **belief** is a person's opinion about something. Beliefs may help shape a consumer's attitudes but don't necessarily involve any liking or disliking. It is possible to have a belief—say, that Listerine PocketPak strips have a medicinal taste—without really caring what they taste like. On the other hand, beliefs about a product may have a positive or negative effect in shaping consumers' attitudes. For example, promotion for Splenda, a no-calorie sweetener in a yellow packet, informs consumers that it's "made from sugar so it tastes like sugar." A dieter who believes that Splenda will taste better because it is made from sugar might try it instead of just routinely rebuying another brand, like Equal. On the other hand, a person with diabetes might believe that he should avoid Splenda—like he avoids other products made from sugar—even though Splenda is actually suitable for people with diabetes.[10]

In an attempt to relate attitude more closely to purchase behavior, some marketers stretch the attitude concept to include consumer "preferences" or "intention to buy." Managers who must forecast how much of their brand customers will buy are particularly interested in the intention to buy. Forecasts would be easier if attitudes were good predictors of intentions to buy. Unfortunately, the relationships usually are not that simple. A person may have a positive attitude toward hot tubs, but have no intention to buy one.

 #M4BW

"Green" attitudes change consumer behavior—the "5 Rs"

Many consumers are showing increasingly positive attitudes toward sustainability—and these attitudes are changing some people's behavior. While some are choosing brands they view as more sustainable, others are changing their behavior more radically. Some advocates of sustainability encourage people to practice the "5 Rs of sustainability": (1) *refuse*—stop buying stuff; (2) *reduce* overall consumption; (3) *reuse*—choose reusable, not disposable, products; (4) *repurpose*—use product packaging, for example, for some other purpose; and (5) *recycle*. While most of us are familiar with the last R, the other 4 Rs are moving from fringe behaviors to being increasingly common. This may have implications for marketing managers: greater practicing of the 5 Rs may lead consumers to buy less stuff and to be more selective in what they buy.

Another way to reduce consumption is to *share* instead of *own*. Partially driven by environmental concerns, car sharing is growing. Read more on this trend in *What's Next?* Will sharing replace owning?

What's Next? Will sharing replace owning? (#M4BW)

In much of the United States and Europe consumers long viewed car *ownership* as a necessity. That is starting to change with evolving needs, attitudes, and lifestyles. Whether seeking a simpler life, a desire to save the planet, or just plain economics, many consumers are becoming more comfortable with car *sharing*.

There is an economic case for turning in the keys. Monthly payments, insurance, parking, gas, depreciation, and maintenance add up to an average of $8,000 per year for car owners—while sharing costs far less. There is also a sustainability case; producing fewer cars and driving fewer miles would both help the environment. Let's look at a few new ways customers are getting around in cars without owning one.

Car-sharing services like Zipcar are growing fast. Signing up for Zipcar is simple: customers hop online, fill out an application, and pay a $25 one-time application fee. Applicants with a good driving record receive a Zipcard in the mail a few days later. Renting a car is even easier. A member (called a Zipster) logs on to Zipcar's website, searches for cars in his or her area, and makes a reservation. Cars are easily located in a parking lot or on a street. At the reserved time, a Zipster unlocks the car with a wave of a smartphone (equipped of course with the Zipcar app) and drives off, later returning the car to the same general area when done.

Another approach lets individual car owners share their vehicles (and earn a few bucks) through a *peer-to-peer* network. Car (or truck) owners can list a vehicle at Turo.com by posting photos and a brief description. They also set daily and weekly rental rates. Prescreened renters search the site, click on a car, and wait to hear if it's available. Turo takes a cut and provides insurance coverage. The company facilitates hundreds of thousands of transactions a year. Car owners can earn $4,000 or more a year.

Although carpooling has been around a long time, *ridesharing* uses the Internet to connect riders with drivers going to the same destination at the same time. Frédéric Mazzella, founder of BlaBlaCar (more on the name in a moment), came up with the idea on a holiday trip to see his family. Mazzella noticed that most of the cars on the road contained only the driver. He thought there had to be a way to fill some of those empty passenger seats.

BlaBlaCar members create a brief profile with a photo and an indication of how much they like to talk—a single "Bla" means not very chatty, "BlaBla" people are talkative, and "BlaBlaBlas" can't shut up. Drivers post where they are going, when, and how much a customer needs to "chip in" on expenses (usually a fraction of the cost of a corresponding rail or bus ticket). Passengers search for rides, and drivers have a choice about whom they carry. After the ride, drivers and passengers rate and review each other to help future riders make better choices.

Consumers' attitudes on ownership and sharing are changing. Recorded music has largely moved from owning physical goods—records, tapes, and CDs—to renting music from streaming services like Spotify and Apple Music. And some people share a bedroom or their entire home on services such as Airbnb and VRBO.[11]

Identify other product-markets where consumers are sharing, borrowing, or renting instead of owning. What product-markets do you think might see a shift from ownership to sharing? Why?

#M4BW

"Green" beliefs change marketing mixes

Consumers increasingly believe that their consumption decisions can help or harm the planet. With more consumers considering sustainability in their purchase decisions, marketing managers are responding with marketing mixes designed to address these ecological needs. For example, Subway sandwich shops added recycling bins. Almost all major carmakers sell hybrid or all-electric cars. Malt-O-Meal breakfast cereals, a low-cost brand that has always been packaged in bags, now touts the environmental benefits of this packaging. Its "Bag the Box" campaign garnered publicity on a small budget. Many consumers have positive attitudes toward brands that try to make a difference in this area.[12]

It isn't easy to change negative attitudes

As companies offering new services, like car sharing, quickly discover, it's not easy to change customer attitudes. That can be especially difficult when those attitudes are negative. A couple of decades ago, frozen foods were common in American households. Then many consumers developed negative attitudes toward frozen (as compared to fresh) vegetables. They thought fresh food tasted better. In spite of the value consumers placed on convenience, sales of frozen foods were in decline for many years.

Recently, some new marketing mixes and promotional efforts from several frozen food companies have begun to change those negative attitudes. High-end brands like Evol and Lyfe Kitchen showed consumers that frozen food could be tasty. Nestlé introduced Wildscape, a product line that features better ingredients, like honey-bourbon brisket, cauliflower in gochujang sauce, and more. They also feature semitransparent plastic jars that look like ice cream containers. Responding to consumer concerns about microwaving in plastic containers, Conagra's Healthy Choice Power Bowls now feature plant-based compostable packaging. And after more than a decade of declining sales, more positive attitudes are changing that trend.[13]

Ethical issues may arise

Part of the marketing job is to inform and persuade consumers about a firm's offering. An ethical issue sometimes arises, however, if consumers have *inaccurate* beliefs. For example, promotion of a "children's cold formula" may play off parents' fears that adult medicines are too strong—even though the basic ingredients in the children's formula are the same and only the dosage is different.

Marketers must also be careful about promotion that might encourage false beliefs, even if the advertising is not explicitly misleading. For example, ads for Ultra SlimFast low-fat beverage don't claim that anyone who buys the product will lose weight and look like the slim models who appear in the ads—but some critics argue that the advertising gives that impression.

Meeting expectations is important

Attitudes and beliefs sometimes combine to form an **expectation**—an outcome or event that a person anticipates or looks forward to. Consumer expectations often focus on the benefits or value that the consumer expects from a firm's marketing mix. This is an important issue for marketers because a consumer is likely to be dissatisfied if his or her expectations are not met. Promotion that overpromises can create this problem. Finding the right balance, however, can be difficult. A few years ago Van Heusen came up with a new way to treat its wash-and-wear shirts so that they look better when they come out of the wash. Van Heusen promoted these shirts as "wrinkle-free." The new shirt was an improvement, but consumers who expected it to look as if it had been ironed were disappointed.[14]

Ethical Dilemma

What would you do? You are a marketing assistant for Auntie Em's Cookie Company, which makes and distributes packaged cookies through grocery stores. Your company recently ran a test market for a new brand of low-fat cookies called Tastee DeeLites. The new brand meets government standards to be labeled and advertised as "low fat," so the ads and package used in the test market highlighted that benefit. Test-market sales were very promising. However, now a consumer activist group has created a website (www.TasteeDeeLIES.com) that claims the Tastee DeeLites package and ads are misleading because the product's high calories make it even more fattening than most other cookies. Your boss has asked you to recommend how Auntie Em's should handle this situation.

Drawing on what you've learned about consumer behavior, do you think consumers are being misled? Does your company have any responsibility to respond to these charges? Should changes be made to the product, package, or promotion?

Building trust builds sales

Trust is the confidence a person has in the promises or actions of another person, brand, or company. Trust drives expectations because when people trust, they *expect* the other party to fulfill promises or perform capably.

Customers may come to trust through experience with a company or person. For example, if your dog always happily chows down Purina dog food—and remains healthy and active—you might come to trust the Purina brand. A customer might trust a travel agent who previously planned enjoyable vacations for her family. Trust can also relate to recommenders—for example, many consumers trust *Consumer Reports* magazine and put a lot of stock in its product reviews and ratings.

Highly trusted people, brands, and companies have many advantages in the marketplace. Consumers prefer to buy from and are more loyal to brands they trust. Trusted brands are less susceptible to price-based competition. For example, few people would switch from a trusted hair stylist just to save a few dollars.

Morton knows that consumers have positive attitudes toward trusted brands. Because the Morton brand is trusted, customers are more likely to try its new Safe-T-Plus and other products to melt icy sidewalks and driveways. This ad is directed at retailers, encouraging them to order the new product.
Source: Morton Salt, Inc.

Trust takes time to build, but it can be destroyed quickly. Consider what happened to Samsung a few years ago when some of its Galaxy Note 7 smartphones started catching on fire. Customers were justifiably worried. While Samsung quickly recalled the bad phones, many consumers lost faith in a brand they previously trusted. It took time for Samsung to rebuild trust in its Galaxy phones. Trust can be fragile, and companies must work to maintain customer trust that has been built—sometimes over many decades.[15]

Psychographics focus on activities, interests, and opinions

Psychographics or **lifestyle analysis** is the analysis of a person's day-to-day pattern of living as expressed in that person's *A*ctivities, *I*nterests, and *O*pinions—sometimes referred to as *AIOs*. Exhibit 5-6 shows a number of variables for each of the AIO

Exhibit 5-6 Lifestyle Dimensions (and Some Related Demographic Dimensions)

Dimension	Examples		
Activities	Work	Vacation	Surfing web
	Hobbies	Entertainment	Shopping
	Social events	Club membership	Sports
Interests	Family	Community	Food
	Home	Recreation	Media
	Job	Fashion	Achievements
Opinions	Themselves	Business	Products
	Social issues	Economics	Future
	Politics	Education	Culture
Demographics	Income	Geographic area	Occupation
	Age	Ethnicity	Family size
	Family life cycle	Dwelling	Education

dimensions—along with some demographics used to add detail to the lifestyle profile of a target market.

Understanding the lifestyle of target customers has been especially helpful in providing ideas for advertising themes. Let's see how it adds to a typical demographic description. It may not help Dodge marketing managers much to know that an average member of the target market for a Dodge Durango is 34.8 years old, married, lives in a three-bedroom home, and has 2.3 children. Lifestyles help marketers paint a more human portrait of the target market. For example, lifestyle analysis might show that the 34.8-year-old is also a community-oriented consumer with traditional values who especially enjoys spectator sports and spends a lot of time in other family activities. An ad might show a happy family unloading from a Durango at a softball game so the target market could really identify with the ad. And the ad might be placed during an ESPN show whose viewers match the target lifestyle profile.[16]

Social Influences Affect Consumer Behavior

LO 5.3

We've been discussing some of the ways needs, attitudes, and other psychological variables influence the buying process. Now we'll look at how a consumer's family, social class, and reference groups influence the consumer decision process.[17]

Family life cycle influences needs

Relationships with other family members influence many aspects of consumer behavior. Family members may share many attitudes and values, consider one another's opinions, and divide various buying tasks. Marital status, age, and the age of any children in the family have an especially important effect on how people spend their income. Put together, these dimensions tell us about the life-cycle stage of a family. Exhibit 5-7 shows a summary of stages in the family life cycle. Let's take a closer look at how a few of these stages influence buying behavior.

Exhibit 5-7 Stages in Modern Family Life Cycles

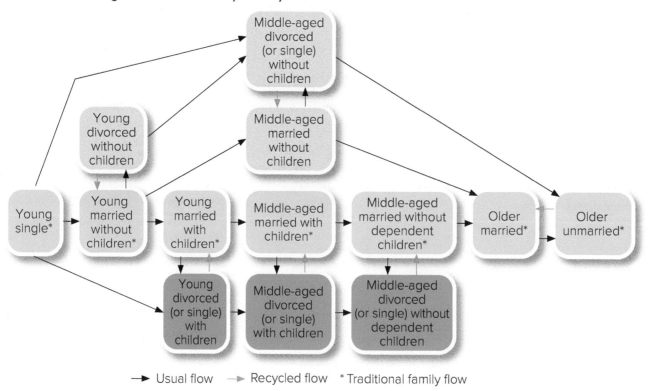

Many customer needs are closely related to the family life cycle. Pampers knows that "Young marrieds with children" need diapers. Edward Jones knows that "empty nesters" start thinking more about investing and retirement.
(left): Source: Procter & Gamble; (right): Source: Edward Jones Investments

Young people and families accept new ideas

Singles and young couples seem to be more willing to try new products and brands—and they are careful, price-conscious shoppers. Although many young people are waiting longer to marry, most tie the knot eventually. These younger families—especially those with no children—are still accumulating durable goods, such as automobiles and home furnishings. Only as children arrive and grow does family spending shift to soft goods and services, such as education, medical, and personal care. To meet expenses, people in this age group often make more purchases on credit, and they save less of their income.

Divorce disrupts the traditional family life-cycle pattern. The mother usually has custody of the children, and the father may pay child support. The mother and children typically have much less income than two-parent families. Such families spend a larger portion of their income on housing, child care, and other necessities, leaving little for discretionary purchases. If a single parent remarries, the family life cycle may start over again.[18]

Reallocation for teenagers

Once children become teenagers, further shifts in spending occur. Teenagers eat more, want to wear expensive clothes, like music, and develop recreation and education needs that are hard on the family budget. American teens currently spend more than $200 billion a year. Teens have a lot of influence on household spending for clothes, shoes, food, and beverages. They are more likely to be spending their own money on entertainment categories like apps, toys, games, books, and music.[19]

Selling to the empty nesters

Another important category is the **empty nesters**—people whose children are grown and who are now able to spend their money in other ways. This tends to be a high-income period, especially for white-collar workers. Empty nesters are an attractive market for many items. Often they spend more on travel and other things they couldn't afford earlier in life. Much depends on their income, of course.[20]

Who is the real decision maker in family purchases?

In years past, most marketers in the United States targeted the wife as the family purchasing agent. Now, with sex-role stereotypes changed and with night and weekend shopping more popular, men and older children take more responsibility for shopping and decision making. Family roles vary from one culture to another.

Buying responsibility and influence vary greatly depending on the product and the family. Although often only one family member goes to the store for a specific

Social class affects attitudes, values, and buying

purchase, other family members may influence the decision or really decide what to buy. Still others may use the product.²¹

Up to now, we've been concerned with individuals and their family relationships. Now let's consider how society looks at an individual and perhaps the family—in terms of social class. A **social class** is a group of people who have approximately equal social position as viewed by others in the society.

Almost every society has some social class structure. In most countries, social class is closely related to a person's occupation, but it may also be influenced by education, community participation, where a person lives, income, possessions, social skills, and other factors—including what family a person is born into.

In most countries—including the United States—there is *some* general relationship between income level and social class. But people with the same income level may be in different social classes. So income by itself is usually not a good measure of social class. And people in different social classes may spend, save, and borrow money in very different ways.

Marketers want to know what buyers in various social classes are like. In the United States, simple approaches for measuring social class groupings are based on a person's *occupation, education,* and *type and location of housing.* By using marketing research surveys or available census data, marketers can get a feel for the social class of a target market.

It is estimated that teenagers make or directly influence more than $200 billion in annual spending in the United States. Tyson recognizes this fact and appeals to parents who have heard their teens cry, "I'm still hungry."
Source: Tyson Foods, Inc.

What do these classes mean?

Although many people think of America as a middle-class society, studies suggest that in many marketing situations distinctive social class groups do exist. Various classes shop at different stores. They prefer different treatment from salespeople. They buy different brands of products—even though prices are about the same. And they have different attitudes toward spending and saving—even when they have the same income level.

Reference groups help people form attitudes

A **reference group** refers to the people to whom an individual looks when forming attitudes about a particular topic. Such people may influence a purchase decision. People normally have several reference groups for different topics.

Three different types of reference groups might influence purchase decisions. An *aspirational* reference group includes people an individual desires to be like—for example, performers or athletes. Many brands use famous personalities to endorse their products, as when Gatorade encouraged individuals to "Be Like Mike" (basketball star Michael Jordan). An *associative* reference group includes people who more realistically reflect an individual—perhaps friends, family members, or coworkers. Teens, for example, often check out what friends wear when deciding what clothes to purchase. Finally, a *dissociative* reference group includes people an individual does not want to be like. An ad for the Ford S-Max implored, "Don't be the minivan guy." Reference groups most often influence purchases that others can observe such as automobiles, clothing, or places to live.²²

Opinion leaders influence buyers

An **opinion leader** is a person who influences others. Opinion leaders aren't necessarily wealthier or better educated. And opinion leaders on one subject aren't

Marketing managers often target opinion leaders because they influence others. The Syfy television network wants advertisers to know that many of its viewers are opinion leaders. Some people use social media, like Instagram, to follow opinion leaders on subjects that interest them. More than 100,000 people follow "detoxinista" on Instagram, where they find recipes that "naturally support the body's detoxification system." They can also discover she is a big fan of her Instant Pot.
(Syfy): Source: Syfy, a division of NBCUniversal; (Instagram post): Source: Megan Gilmore/Instagram

necessarily opinion leaders on another. For example, you may have a friend who is ahead of the curve in knowing about video games, but you might not want that friend's opinion about new clothing styles and cosmetics. Opinion leaders can influence both what product categories a consumer considers and what brands he or she purchases.

Social media boosts social influence

Social media such as Facebook, Twitter, Instagram, and Snapchat make it easier for consumers to influence one another. On social media it is easy to follow people from various reference groups, whether that means Justin Bieber's Twitter feed or an old high school friend's Facebook posts. These give consumers access to new reference groups that may influence their buying decision process.

Social media can also amplify the voices of opinion leaders. A traveler can have a lot of influence by posting a long, detailed review of a sight that "can't be missed" in Thailand. A friend can rave (or rage) about the new Marvel movie—and influence who goes (and who stays home). Marketing managers are trying to figure out how to manage this type of influence, which we will discuss more in Chapter 16.[23]

Culture, Ethnicity, and Consumer Behavior

LO 5.4

Cultural variation across ethnic groups or countries can result in differences in how consumers make purchase choices. As we discussed in Chapter 3, these differences create opportunities for resourceful firms. In this section we discuss how culture, ethnicity, and international boundaries influence consumer behavior.

Culture surrounds the other influences

Culture is the whole set of beliefs, attitudes, and ways of doing things of a reasonably homogeneous set of people. In this chapter, as well as Chapter 3, we look at the broad impact of culture.

We can think of the American culture, the French culture, or the Chinese culture. People within these cultural groupings tend to be more similar in outlook and behavior. But often it is useful to think of subcultures within such groupings. For example, within the American culture, there are various religious, ethnic, and regional subcultures.

Failure to consider cultural differences, even subtle ones, can result in problems. To promote their product and get people to try it, marketers for the antacid medication Pepto-Bismol often provide free samples at festivals and street fairs. Their idea is that people tend to overindulge at such events. However, when they distributed sample packets at a festival in San Francisco's Chinatown, they insulted many of the people they wanted to influence. Booths with Chinese delicacies lined the streets, and many of the participants interpreted the sample packets (which featured the word *Nauseous* in large letters) as suggesting that Chinese delicacies were nauseating. The possibility of this misinterpretation may seem obvious in hindsight, but if it had been that obvious in advance, the whole promotion would have been handled differently.[24]

Do ethnic groups buy differently?

America may be called a melting pot, but ethnic groups deserve special attention when analyzing markets. One basic reason is that people from different ethnic groups may be influenced by very different cultural variables. They may have quite varied needs and their own ways of thinking. Moreover, many Americans value diversity, and the United States is becoming a multicultural market. As a result, rather than disappearing in a melting pot, some important cultural and ethnic dimensions are being preserved and highlighted. This creates both opportunities and challenges for marketers.

Ethnic markets are increasingly important

Marketers are paying more attention to ethnic groups because the number of ethnic consumers is growing at a much faster rate than the rest of the United States. Although some of this growth results from immigration, a much larger share follows from higher birthrates among ethnic minorities. Asian Americans, African Americans, and Hispanics are also seeing their wealth grow. Together these trends point to growing buying power for these ethnic groups. Ethnic and cultural groups often cluster in certain areas and share buying patterns that can be important to some businesses. Let's take a closer look at the three largest ethnic groups in the United States.[25]

Hispanic market is large and growing

In 2017, Hispanics, the largest ethnic group in the United States, totaled about 18 percent of the population. In 2017, the median household income of Hispanics grew to $50,486. Hispanics are also much younger, with a median age of 29 compared to 41 for non-Hispanics. It was estimated that in 2019, Hispanic buying power in the United States totaled $1.7 trillion. More than half of all Hispanics in the United States live in three states—California, Texas, and Florida. That concentration makes promotion easier. Hispanics are known for embracing technology, and they often buy brands and products that reflect their family values. Data show that Hispanics are disproportionately likely to buy vegetables and dried grains, hot sauce, and women's fragrances. About half of Hispanics in the United States speak Spanish more than English.

Many companies are responding to this trend. For example, Wells Fargo branches in the Southwest have added Latin-style decor, bicultural tellers, and Spanish-language promotion. Efforts such as these have significantly increased Wells Fargo's business. P&G found Hispanic consumers like using fragrances in their homes and developed Febreze Destinations Collections, a line of air fresheners targeting this market.

African Americans appreciate cultural traditions

Forty-four million African Americans comprise almost 14 percent of the U.S. population—but in some southern states that percentage is about 30 percent. African Americans have a median household income of $40,258; by 2019 they were estimated to have $1.4 trillion in buying power. African American consumers often buy products tied to cultural traditions and family gatherings. African Americans are more likely to buy unprepared meat and juice drinks, among other products.

P&G launched a campaign specifically targeting African American women. "My Black Is Beautiful" includes a cable TV show, website, and a Facebook page with more than 2 million members. These efforts have helped P&G brands reach out to African Americans.

Asian Americans are smaller in number, higher in income

There are about 19 million Asian Americans in the United States, about one-third of whom live in California. Although just 6 percent of the population, this group boasts the highest average household income at $81,331, which projected to over $1 trillion in buying power in 2018. Asian Americans disproportionately purchase more fresh produce, organic foods, and electronic goods, and are also more likely to shop online as compared to other groups. Many Asian Americans are first or second generation in the United States, and three-quarters speak a language other than English at home.

Online news and entertainment site BuzzFeed has targeted the fast-growing Asian American segment with content appealing to this group. BuzzFeed, known for its "listicles" (articles composed of lists), has recently published "22 Signs You Grew Up with Immigrant Chinese Parents," "21 Annoying Comments Filipinos Are Tired of Hearing," and more to get Asian Americans to click through.[26]

Stereotypes are common and misleading

A marketer needs to carefully study ethnic dimensions, because they can be subtle and change quickly. For example, second-generation immigrants—those born in the United States—often differ significantly from their parents who immigrated to the country. And marketing managers must be careful not to stereotype ethnic consumers. Hispanics include people with roots in the Caribbean Islands and Mexico as well as Central and South America. The "black" market is not a single group; marketing managers should use other segmenting dimensions to more accurately reflect the great variability in African American households. The 19 million Asian Americans include people with Chinese, Filipino, Indian, and other backgrounds—yet no single group makes up more than 20 percent of this total. And, of course, a growing number of Americans are mixed race. Marketing managers might start with these general categories, but other segmenting variables should be used to better understand and target customers.

Toyota used similar but different advertisements for its Camry. All of the ads featured the same car, with different advertising designed to appeal to different ethnic groups. The version targeting African Americans was based on the theme "Strut." It includes an image of a peacock and the music that wrestler John Cena uses when he enters the ring. In the Hispanic version, the driver enjoys his ride so much that he declines to take a call from his mother. Given the reverence this ethnic group has for moms, it was seen as rebellious. Finally, a version targeting Asian Americans showed a father picking up his daughter from baseball practice. The goal here was to show how the Camry brought out a more affectionate side of the Asian American dad.
Source: Toyota Motor Sales, U.S.A., Inc.

Customer behavior differs across international markets

Planning strategies that consider cultural and behavioral differences across international markets can be a challenge. Marketing managers need to understand the differences. Each foreign market may need to be treated as a separate target market with its own submarkets. Ignoring cultural differences—or assuming that they are not important—almost guarantees failure in international markets.

Consider the situation faced by marketers as they introduced Swiffer, the fast-selling wet mop, in Italy. Research showed that Italian women wash their floors four times more often than Americans. Based on that, you might predict a big success for Swiffer in Italy. Yet many new cleaning products flop there. Fortunately, the research suggested a reason. Many Italians have negative attitudes about ad claims that a product makes cleaning *easier*. Although this benefit has wide appeal in the United States, many Italian women doubt that something that works easily will meet their standards for cleanliness. So for the Italian market, Swiffer was modified and beeswax was added to polish floors after they have been mopped. Now Swiffer is a top seller in Italy.[27]

Sometimes an understanding of local culture points out new ways to blend the Four Ps. For example, Nestlé knew that free samples would kickstart the adoption process of its new line of food flavorings in Brazil. In the United States, it's common to distribute free samples at stores, but that would seem strange to Brazilians. Local Nestlé managers in Brazil proposed a more culturally acceptable approach to distribute the samples. In Brazil, people cook on gas stoves and delivery people regularly bring canisters of gas to their homes, so Nestlé paid these local workers to offer their customers samples of the flavorings and explain how to use them. Consumers showed more interest in the samples when the conversation took place with someone they trusted right by the stove where the flavorings would be used.[28]

Individuals Are Affected by the Purchase Situation

LO 5.5

The **purchase situation** takes into account the purpose, time available, and location where a purchase is made.

Consumers consider the purpose for the purchase

Consumers can buy the same type of product, yet they may have completely different *purposes* for the purchase. This influences buying behavior. For example, how a consumer buys a bunch of flowers depends on whether they are for a mother who lives far away, the host of a dinner party, or a girlfriend following a little disagreement. The reason for the purchase influences how the buyer evaluates options and, ultimately, the choice made.

Time affects what happens

Time influences a purchase situation. When consumers make a purchase—and the time they have available for shopping—will influence their behavior. Socializing with friends at a Starbucks induces different behavior than grabbing a quick cup of 7-Eleven coffee on the way to work. The urgency of the need is another time-related factor. Shopping for tires is much more urgent when one has a flat as compared to when one notices that the tread is just starting to wear thin. When the need is urgent, customers will be less inclined to evaluate a range of options—and place more emphasis on speed and convenience.

Some retail stores have discovered that customers spend more when they shop slower. Origins stores found that customers who sat spent 40 percent more than those who did not, and the beauty care retailer took notice. Its stores were redesigned to slow down the shopping experience—adding a wall specially lit for selfies, a large sink to facilitate sampling soaps, and plenty of seating.[29]

Surroundings affect buying too

Where a purchase is made and the surrounding location can affect buying behavior. The excitement at an on-site auction may stimulate impulse buying. Checking out an

Consumer behavior varies with situational factors. Consumers shop differently when buying gifts—and BoConcept tries to make gift buying easier by presenting a range of ideas. Another situational factor is the surroundings. The more resortlike experience offered by the Park Hyatt may influence a consumer's buying decisions.
(left): Source: BoConcept; (right): Source: Hyatt Corporation

auction online might lead to a different response. A restaurant could have a loud and fun atmosphere or a quiet, romantic environment—each affects how customers behave. Consider how the warehouse-like setting in a Sam's Club store signals low-cost operations and price reductions.[30]

Needs, benefits sought, attitudes, motivation, and even how a consumer selects certain products all vary depending on the purchase situation. So different purchase situations may require different marketing mixes—even when the same target market is involved.

The Consumer Decision Process

LO 5.6

The model in Exhibit 5-2 organizes the many different influences on consumer behavior. It helps explain *why* consumers make the decisions they make. Now, we'll expand that model with a closer look at the steps in the consumer decision process and a focus on *how* consumers make decisions.[31] See Exhibit 5-8.

Recognizing a need creates a problem for the consumer

The consumer decision process begins when a consumer becomes aware of an unmet need. The consumer's problem-solving process then focuses on how best to meet that need. Problem recognition often happens quickly. A student on the way to class, for example, may realize that she's thirsty and wants something to drink. Or

Exhibit 5-8 An Expanded Model of Consumer Behavior

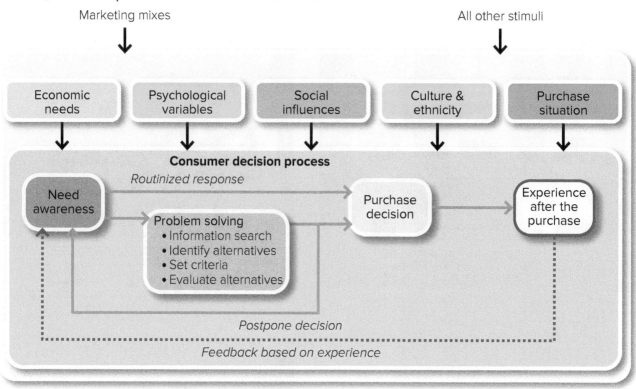

problem recognition may take shape over time. For example, a recent grad with a new apartment might want a comfortable place to sit while watching TV in the evening. These situations present problems that may be solved with a purchase. But what purchase would it be?

Steps in consumer problem solving

How a consumer solves the problem depends on the situation. Exhibit 5-9 highlights the four basic problem-solving steps a consumer may go through to satisfy a need. As the exhibit indicates, consumers do not always move forward in this process—sometimes they will postpone a decision or start over later. Marketing managers want their product to be in consumers' minds when they engage in these steps:

Exhibit 5-9 Consumer Problem Solving

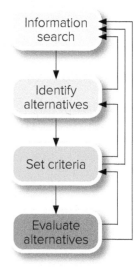

1. *Information search:* During this step, customers seek information about a solution for their problem or about specific brands. Customers may find information from other customers, experts, or a company's promotion (advertising, salespeople, or website, for example).
2. *Identify alternatives:* During this step, customers recognize available options for meeting a need.
3. *Set criteria:* During this step, customers identify criteria and the relative importance of each in preparation for the final step.
4. *Evaluate alternatives:* During this step, customers consider the costs and benefits of various alternatives.

Exhibit 5-10 Problem-Solving Continuum

Low involvement
Frequently purchased
Inexpensive
Little risk
Little information needed

→ Routinized response behavior | Limited problem solving | **Extensive problem solving**

High involvement
Infrequently purchased
Expensive
High risk
Much information desired

Three levels of problem solving are useful

These activities may require a lot of time and effort on the part of a consumer, very little time, or they may be skipped altogether. How much effort is put into a buying decision depends in part on the economic needs, psychological variables, social influences, and purchase situation factors discussed earlier in this chapter. It's also tied to the amount of risk a buyer sees in making a wrong choice. For instance, when a buyer has little involvement in the purchase or the expense is small, a poor buying decision will result in little financial or social risk. Under these circumstances the buyer usually puts little effort into the buying decision. On the other hand, when a buyer is highly involved and cares deeply about the choice, or when the price is high, there may be more risk in making a bad decision. So the buyer often extends more effort in order to be more confident. Exhibit 5-10 suggests three levels of problem solving that relate to the amount of effort the buyer puts into the decision.

Consumers use **extensive problem solving** when they put *much* effort into deciding how to satisfy a need—as is likely for a completely new purchase or to satisfy an important need. At this level of problem solving, consumers conduct a more comprehensive information search, identify more alternatives, have more criteria for evaluation, and spend more time evaluating options (see Exhibit 5-9). For example, an avid computer gamer may put a great deal of effort into buying a new gaming computer. Our gamer might solicit friends' opinions about the graphics speed and audio quality for different models before going online to compare options and prices and read technical reviews. Then the gamer might visit a local store for a hands-on demo of a favorite game on a few computers. To narrow down to a final choice, the gamer could evaluate customer service support and warranties. This is certainly not an impulse purchase! The decision to buy—and what to buy—comes only after an extensive effort.

Pe3k/Shutterstock

Limited problem solving is used by consumers when *some* effort is required in deciding the best way to satisfy a need. This is typical when the consumer has some previous experience with a product but isn't quite sure which choice to make at the moment. A seasoned computer gamer, for instance, may already know that he likes sports games and what store has the newest releases. At the store, he might get the salesperson's advice and check out the video quality on a few games before deciding which to buy. This is a deliberate purchase, but only a limited amount of effort is expended before making the decision.

A consumer uses **routinized response behavior** when he or she regularly selects a particular way of satisfying a need when it occurs. Routinized response is typical when a consumer has considerable experience in how to meet a specific need and requires no new information. For example, our gamer might automatically buy the latest version of *Madden NFL* as soon as EA Sports makes it available. Routinized response also may occur when a buyer trusts a company, brand, or a friend's recommendation. Because trust lowers the risk of making the wrong choice, less effort is required from the buyer.

Routinized response behavior is also typical for **low-involvement purchases**—purchases that have little importance or relevance for the customer. Let's face it, for most of us, buying a box of salt is probably not one of the burning issues in our life.[32]

Buying isn't always rational

The idea of a decision process does *not* imply that consumers always apply *rational* processes in their buying decisions. To the contrary, consumers don't always seek accurate information or make smart choices that provide the best economic value. This is often because of the influences on consumer behavior that we discussed earlier in the chapter. For example, most sport-utility vehicles never leave the paved road, but buyers like the image of driving an SUV and *knowing* they can get off the paved road if they want to. When a tourist spends 1,000 euros on a Loewe leather purse, it may simply be that she loves the style, can afford it, and "has to have it." Needs are operating in such a purchase, but they are higher-level needs and not some sort of functional "requirement."

To buy or not to buy

A customer who is ready to buy may have to decide which brand to purchase and where to make the purchase (that is, which store or location). There may be related decisions about colors, features, or other options as well. Buying can be complicated—and sometimes the complexity of all the choices causes customers to delay buying.

When a consumer decides to delay a purchase, often the sale is lost. He or she may never return. For that reason, some marketers make it easy for customers to come back later and purchase. For example, Walmart recently instituted a Christmas layaway program. When using layaway, a customer makes incremental payments on a purchase—and does not take it home until it is paid in full. Similarly, online retailers often allow customers to place items on a "Wish List" or "Save to Favorites" so they are easy to find when they return to the site.

When a customer gets close to making a purchase, marketing managers want to help "tip the scales." At this point, offers of special financing or free delivery might be helpful in overcoming a delayed purchase.

Consumers can have second thoughts after a purchase

After making a purchase, buyers often have second thoughts and wonder if they made the right choice. The resulting tension is called **dissonance**—a feeling of uncertainty about whether the correct decision was made. (In lay terms, this is known as "buyer's remorse.") This may lead a customer to seek additional information to confirm the wisdom of the purchase.[33]

Post-purchase regret is a bigger problem

Sometimes uncertainty isn't the issue. Rather, the consumer is certain about being unhappy with a purchase. When a post-purchase experience fails to live up to expectations, a customer will be disappointed. For example, going out for an expensive dinner—and then getting slow service and cold food—can result in considerable dissatisfaction and regret over the choice. The diner is much less likely to visit that restaurant again. A consumer may regret making a purchase for a variety of reasons that the consumer didn't anticipate when making the purchase. Whatever the reason, regret is not likely to lead to the same decision in the future.

Some consumers spread the word after they buy

Many consumers talk about their purchases and share opinions about their good and bad experiences. Recommendations from friends can have a big influence on whether we try a new restaurant, buy a hybrid car, or choose a veterinarian. Consumers are even more likely to share stories about being dissatisfied than satisfied. This is important for a marketer to remember. At times, his or her career might call for countering negative publicity—doing damage control—to save a product's image.[34]

Predictive analytics and the consumer decision process

Some firms already utilize marketing analytics and intelligent agents during different stages of the consumer decision-making process. Many firms compile data useful for **predictive analytics**—a process to analyze data to make predictions about unknown future events. For a marketing manager, unknown future events include customers' future needs, the information useful to problem solving, and whether a customer is considering a switch to a competitor. Let's look at a few examples of how such predictive analytics could operate in consumer decision making:

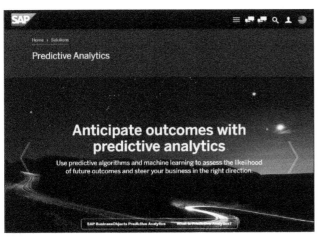

SAP is one of many software companies that use data to make predictions about future events—including consumers' purchasing plans.
Source: SAP SE

- *Need awareness.* A sporting goods store might gather mileage data from a customer's fitness app to predict when the customer will be ready for a replacement set of running shoes. Under Armour and Nike have already invested in fitness tracking apps.
- *Information search.* Based on a customer's past car shopping behavior, a website for an automaker could present features the customer is most concerned with. For example, a customer whose previous behavior indicates a concern for gas mileage and rear seat legroom might have this information presented first.
- *Alternatives identification.* A customer's Netflix home page offers viewing recommendations based on his or her past viewing behavior.
- *Post-purchase.* Marketing managers often try to predict if customers will stop using a product. For example, a bank may notice changes in a customer's credit card usage—possibly indicating he or she has a new card. The bank may intervene by offering the customer extra "points" or a temporary low interest rate.[35]

New concepts require an adoption process

When consumers face a really new concept, their previous experience may not be relevant. These situations involve the **adoption process**—the steps individuals go through on the way to accepting or rejecting a new idea. Although the adoption process is similar to the decision-making process, learning plays a clearer role, and promotion's contribution to a marketing mix is more visible.

In the adoption process, an individual moves through some fairly definite steps:

1. *Awareness*—the potential customer comes to know about the product but lacks details. The consumer may not even know how it works or what it will do.
2. *Interest*—if the consumer becomes interested, he or she will gather general information and facts about the product.
3. *Evaluation*—a consumer begins to give the product a mental trial, applying it to his or her personal situation.
4. *Trial*—the consumer may buy the product to experiment with it in use. A product that is either too expensive to try or isn't available for trial may never be adopted.
5. *Decision*—the consumer decides on either adoption or rejection. A satisfactory evaluation and trial may lead to adoption of the product and regular use. According to psychological learning theory, reinforcement leads to adoption.
6. *Confirmation*—the adopter continues to rethink the decision and searches for support for the decision—that is, further reinforcement.[36]

Dropbox had to work with the adoption process when it introduced cloud (online) data storage—an idea that was not familiar to many computer users. Dropbox lets users create a file folder on their devices—say their laptop, tablet, or smartphone—and share it with others. For example, if a consumer places a photo, song, or document in the Dropbox folder on their laptop, it would automatically be replicated on his or her phone and tablet computer. The file is also backed up on the Internet. Furthermore, folders can also be shared with friends or work colleagues.

Dropbox first targeted techies, people who are comfortable with technology. The techies quickly downloaded Dropbox because it filled a need—these users tend to have

Lynne Cameron/PA Images/Alamy Stock Photo

many digital devices. Dropbox came with 2 GB of free online storage. To encourage early customers to spread *awareness* and generate *interest* in Dropbox, they were given additional storage space for each person they referred to Dropbox. Many techies found good value in Dropbox and were happy to spread the word. A video on its website demonstrated how Dropbox works, making it easy for customers to *evaluate* its usefulness and show how easy (low cost) it was to install. This and the free 2 GB of storage made *trial* easy, too. As others used Dropbox, many began to share files with friends, family, and coworkers. Some found Dropbox so useful they made the *decision* to buy more storage space on a subscription basis (paying each year). As others adopted Dropbox it became easier to share large files, which provided users with *confirmation* of Dropbox's value, leading them to renew their subscription.

Marketing analytics can monitor adoption

Marketing managers may use analytics to monitor how a new idea is being adopted by the market. Marketing metrics at each step can be used to evaluate this progress. Marketing Analytics in Action: Adoption Process examines how this might be useful in the case of Dropbox.

Marketing Analytics in Action: Adoption Process

The adoption process might be modeled using marketing analytics. Let's see how this might work for Dropbox, which could use this information to forecast sales or to fine-tune a marketing mix. Let's examine possible metrics in four steps along the adoption process. (Note that to simplify, we have combined interest and evaluation and left off the decision step.)

- *Awareness:* An advertising campaign might initially be used to make customers aware. Assume an advertising campaign makes 1,000,000 people in the target market ("techies") aware of Dropbox.
- *Interest and evaluation:* In this case, interest and evaluation could be inferred from the number of customers who visit the website and view the video describing the product. Assume 10 percent of the 1,000,000 (in Step 1) who were made aware from the advertising subsequently visit the website and watch the video.
- *Trial:* In this case, trial could be measured as the percentage of people who watch the video and subsequently download and install the Dropbox free trial. Assume 10 percent of people who watch the video go on to install the free software.
- *Confirmation:* For the Dropbox scenario, that is the percentage of trial customers who make a purchase—we assume 5 percent of those who complete the trial will make a purchase.

Using the numbers above, Dropbox could estimate the number of users and the number of initial sales it would make. The equation using the assumptions in Steps 1–4 is as follows:

Number of customers buying Dropbox = 1,000,000 × 10% (who watch the video) × 10% (who download the software) × 5% (who purchase the software)

Use this equation and the insights just discussed to answer the following questions:

1. *Using the numbers presented above, how many of the initial 1,000,000 aware customers eventually make a purchase?*
2. *Assume Dropbox makes a new video that customers find more compelling. This increases the 10 percent in Step 3 to 15 percent. What is the number of buying customers with this new assumption?*
3. *Besides a more compelling video, identify two other changes a marketing manager might make to change the percentages shown in Steps 2 to 4. Describe how your proposed change will affect the percentage.*

CONCLUSION

In this chapter, we analyzed the individual consumer as a problem solver who is influenced by economic needs, psychological variables, social influences, cultural factors, and the purchase situation. We showed how these variables influence the consumer decision process, what the steps in that process are, and why it is important to consider these steps when planning a marketing strategy. For example, we discussed three levels of problem solving that you can use to evaluate how consumers approach different types of purchase decisions—from the simple, routinized response behavior to the complex, extensive problem-solving behavior. We also discussed how the consumer's experience after the purchase impacts what that consumer will do in the future.

From a broader perspective, this chapter makes it clear that each consumer and purchase decision is somewhat unique. It isn't possible to catalog all of the individual possibilities. Rather, the overall focus of this chapter is to provide general frameworks you can use to analyze consumers regardless of the particular product or decision. This also helps you identify the dimensions of consumer behavior most important for segmenting the market and developing a targeted marketing mix.

By now it should be clear that expensive marketing errors can be made when you assume that other consumers will behave in the same manner as you or your family and friends. That's why we rely on the social and behavioral sciences for insight about consumer behavior and why marketing research is so important to marketing managers when they are developing a marketing strategy for a particular target market. When managers understand how and why consumers behave the way they do, they are better able to develop effective marketing mixes that really meet the needs of their target market.

KEY TERMS

LO 5.7

economic buyers, 118
economic needs, 118
discretionary income, 119
needs, 119
wants, 119
drive, 119
physiological needs, 121
safety needs, 121
social needs, 121
personal needs, 121
perception, 121
selective exposure, 122
selective perception, 122

selective retention, 122
learning, 122
cues, 123
response, 123
reinforcement, 123
attitude, 123
belief, 124
expectation, 126
trust, 127
psychographics, 127
lifestyle analysis, 127
empty nesters, 129
social class, 130

reference group, 130
opinion leader, 130
culture, 132
purchase situation, 134
extensive problem solving, 137
limited problem solving, 137
routinized response behavior, 137
low-involvement purchases, 138
dissonance, 138
predictive analytics, 138
adoption process, 138

QUESTIONS AND PROBLEMS

1. Review the Apple case study that opens the chapter. From this case, identify examples of different key terms and concepts covered in the chapter. For example, on-the-spot demos of the iPod can be conducted by an "opinion leader."

2. Review the Apple case study that opens this chapter. Applying concepts from the chapter, what else could Apple do to enhance the launch and subsequent sales of Apple Music? Offer some specific strategy suggestions and justify each by referring to concepts from this chapter.

3. In your own words, explain economic needs and how they relate to the economic-buyer model of consumer behavior. Give an example of a purchase you recently made that is consistent with the economic-buyer model. Give another that is not explained by the economic-buyer model. Explain your thinking.

4. Explain what is meant by a hierarchy of needs and provide examples of one or more products that enable you to satisfy each of the four levels of need.

5. Cut out or photocopy two recent advertisements: one full-page color ad from a magazine and one online ad from a website. In each case, indicate to which needs the ads appeal.

6. Explain how an understanding of consumers' learning processes might affect marketing strategy planning. Give an example.
7. Briefly describe your own *beliefs* about the potential value of an all-electric car, your *attitude* toward them, and your *intention* about buying one the next time you need to buy a car.
8. Give an example of a recent purchase experience in which you were dissatisfied because a firm's marketing mix did not meet your expectations. Indicate how the purchase fell short of your expectations—and also explain whether your expectations were formed based on the firm's promotion or on something else. Will it affect how much you trust that firm or brand in the future?
9. Explain psychographics and lifestyle analysis. Explain how they might be useful for planning marketing strategies to reach college students, as opposed to average consumers.
10. Illustrate how the reference group concept may apply in practice by explaining how you personally are influenced by some reference group for some product. What are the implications of such behavior for marketing managers?
11. Give two examples of recent purchases where the specific purchase situation influenced your purchase decision. Briefly explain how your decision was affected.
12. Give an example of a recent purchase in which you used extensive problem solving. What sources of information did you use in making the decision?
13. On the basis of the data and analysis presented in Chapter 5, what kind of buying behavior would you expect to find for the following products: (a) a haircut, (b) a shampoo, (c) a digital camera, (d) a tennis racket, (e) a dress belt, (f) a cell phone, (g) life insurance, (h) an ice cream cone, and (i) a new checking account? Set up a chart for your answer with products along the left-hand margin as the row headings and the following factors as headings for the columns: (a) how consumers would shop for these products, (b) how far they would travel to buy the product, (c) whether they would buy by brand, (d) whether they would compare with other products, and (e) any other factors they should consider. Insert short answers—words or phrases are satisfactory—in the various boxes. Be prepared to discuss how the answers you put in the chart would affect each product's marketing mix.
14. Describe a purchase situation where you, a friend, or a family member made a purchase choice that considered the implications of your choice on others (society, the environment, or a friend or family member). How did those influences affect steps in the purchase decision process shown in Exhibit 5-8?
15. Interview a friend or family member about two recent purchase decisions. One decision should be an important purchase, perhaps the choice of an automobile, a place to live, or a college. The second purchase should be more routine, such as a meal from a fast-food restaurant or a regularly purchased grocery item. For each purchase, ask your friend questions that will help you understand how the decision was made. Use the model in Exhibit 5-8 to guide your questions. Describe the similarities and differences between the two purchase decisions.

MARKETING PLANNING FOR HILLSIDE VETERINARY CLINIC

Appendix D (the Appendices follow Chapter 19) includes a sample marketing plan for Hillside Veterinary Clinic. Look through the "Customer Analysis" section and consider the following questions.

1. Based on the marketing plan, what do we know about the consumer behavior of the target market?
2. What additional information do you think would be helpful before developing a marketing strategy for Hillside?

SUGGESTED CASES

1. McDonald's "Seniors" Restaurant
3. NOCO United Soccer Academy
8. Carmine's Italian Restaurant
9. Quiet Night Motel
10. Cousin's Ice Center
11. Running On
12. DrV.com—Custom Vitamins
27. Canadian Mills, Ltd.

MARKETING ANALYTICS: DATA TO KNOWLEDGE

CHAPTER 5: SELECTIVE PROCESSES

Submag, Inc. uses direct-mail promotion to sell magazine subscriptions. Magazine publishers pay Submag $3.12 for each new subscription. Submag's costs include the expenses of printing, addressing, and mailing each direct-mail advertisement plus the cost of using a mailing list. There are many suppliers of mailing lists, and the cost and quality of different lists vary.

Submag's marketing manager, Shandra Debose, is trying to choose between two possible mailing lists. One list has been generated from phone directories. It is less expensive than the other list, but the supplier acknowledges that about 10 percent of the names are out-of-date (addresses where people have moved away). A competing supplier offers a list of active members of professional associations. This list costs 4 cents per name more than the phone list, but only 8 percent of the addresses are out-of-date.

In addition to concerns about out-of-date names, not every consumer who receives a mailing buys a subscription. For example, *selective exposure* is a problem. Some target customers never see the offer—they just toss out junk mail without even opening the envelope. Industry studies show that this wastes about 10 percent of each mailing—although the precise percentage varies from one mailing list to another.

Selective perception influences some consumers who do open the mailing. Some are simply not interested. Others don't want to deal with a subscription service. Although the price is good, these consumers worry that they'll never get the magazines. Submag's previous experience is that selective perception causes more than half of those who read the offer to reject it.

Of those who perceive the message as intended, many are interested. But *selective retention* can be a problem. Some people set aside the information and then forget to send in the subscription order.

Submag can mail about 25,000 pieces per week. Shandra Debose has set up a spreadsheet to help her study the effects of the various relationships discussed earlier and to choose between the two mailing lists.

Design element: #M4BW box globe icon: ©Vectoryzen/Shutterstock

CHAPTER SIX

Design Pics Inc/Alamy Stock Photo

Business and Organizational Customers and Their Buying Behavior

MetoKote Corporation specializes in protective coating applications, such as powder-coat and liquid paint, that other manufacturers need for the parts and equipment they make. For example, when you see John Deere agricultural, construction, or lawn care equipment, many of the components have likely been coated (painted) in a MetoKote facility. In fact, Deere and MetoKote have a close buyer-seller relationship. Although Deere uses a variety of methods to identify suppliers and get competitive bids for many items it needs, it's different with MetoKote. Deere isn't going to switch to some other supplier just because other options provide cheaper coatings. MetoKote not only provides protective coatings for many Deere products, it has also built facilities right next to some Deere plants. When it's time for a component to be coated, a conveyer belt moves the part out of the Deere plant and into the MetoKote facility. A short time later it's back—and it's green or yellow.

Many people were involved in the decision to purchase coating services in this way. The responsibility for choosing a painting vendor didn't rest only with Deere's purchasing department; it involved input from people in finance, quality control, and production. They worked together—often with MetoKote personnel—to figure out exactly what was needed to provide the high-quality protective finish Deere's customers expect on each piece of Deere equipment. A decision like this was not obvious, and before selecting MetoKote, Deere's buying team gathered information and reviewed different suppliers. Once the team decided to work with MetoKote, it was a long-term decision.

Deere does all this because it knows how demanding the customers in its agricultural, commercial, and construction equipment markets can be. Deere's organizational customers differ from the consumers that buy its riding lawn mowers. While there are fewer business customers, many of them spend more than $100,000 a year on Deere equipment. And they often conduct a financial analysis to ensure those investments pay off in the long run. They want good value—and Deere is known for delivering that. These business buyers are experts, highly trained in what they buy.

Consider the market for Deere agricultural equipment. The highly competitive agricultural market puts pressure on farmers to produce more, lower their costs of operation, and operate sustainably. Farmers can buy lower-*priced* farm equipment from Deere's competitors—including AGCO and Kubota. Yet many choose John Deere products because they create value by saving farmers money in other ways.

One way they save money is by using data, analytics, and artificial intelligence to help farmers run more efficient and effective operations. John Deere is at the forefront of this high-tech trend. For example, Deere's automated driving technology allows a farmer riding in the cab to focus on the crops instead of driving. Deere's See & Spray system uses smart cameras that can distinguish weeds from crops. It then targets chemicals at the weeds only. This reduces herbicide usage by 80 to 90 percent, lowers farmers' costs, and reduces chemicals on the food we eat.

Data from different pieces of John Deere equipment flow seamlessly into the farmer's MyJohnDeere.com site, creating an "operations center" for the farm. A farm's JDLink (another John Deere tool) manages all of a farmer's equipment (even non-Deere equipment). For example, JDLink tracks usage and tells a farmer when a piece of equipment needs maintenance. Optimizing equipment reduces downtime and lowers service costs—delivering tangible economic value to farmers.

John Deere's dealer network is also critical to its success. Although some customers buy directly from John Deere, most purchase from one of the 2,000-plus John Deere dealers located in more than 100 countries. These dealers buy and stock what their customers want. And the dealers know that John Deere's reputation for quality service is just as important as the company's reputation for quality products—Deere drops dealers who don't measure up to its goals for customer satisfaction. Deere also encourages dealers to consolidate so they can lower costs with greater economies of scale in purchasing.

Dealers appreciate that Deere set up the website www.machinefinder.com and a Facebook page, which connect customers with dealers selling used Deere equipment. The sale of a used piece of equipment usually means a customer has already bought a new piece of Deere equipment to replace it. Deere even finances used purchases through John Deere Credit.

Deere designs marketing strategies that solve particular customer problems. For example, golf course managers buy Deere equipment to maintain their fairways, greens, and sand traps because they value the reliability and durability of the brand's products. However, they also need a variety of operating supplies—ranging from grass seed and irrigation equipment to ball washers and chemicals. Golf courses don't usually have one person dedicated to purchasing; it's handled by course managers who have many responsibilities. They appreciate Deere's One Source service, which provides golf course managers with a "one-stop shop" for the things they need—all backed by the trusted Deere name. One Source saves time and strengthens Deere's relationships with both its dealers and their golf course customers. Of course, it creates new challenges for Deere's purchasing department, which must identify, evaluate, monitor, and recommend the suppliers for the products that golf courses need. Deere's dealers and golf course customers rely on Deere's purchasing specialists to make the right decisions, but that's part of the value that Deere provides.

While John Deere delivers customer value to many different customers, farmers remain its bread and butter. Helping farmers become more efficient and effective will be needed as they figure out how to feed the 10 billion people expected to be living on Earth in 30 years (up from about 8 billion today). Enhanced productivity delivers value to farmers and the world.[1]

LEARNING OBJECTIVES

As the John Deere case illustrates, the buying behavior of organizational customers can be very different from the buying behavior of final consumers. Marketing strategy planning requires a solid understanding of who these organizational customers are and how they buy.

When you finish this chapter, you should be able to

1. name and give examples of the different types of business and organizational buyers.
2. describe how organizational and business markets differ from consumer markets.
3. describe each step in the model of organizational/business buying.
4. explain the different types of buying processes.
5. understand the different types of buyer-seller relationships and their benefits and limitations.
6. know about the number and distribution of manufacturers and why they are an important customer group.
7. know how buying by service firms, retailers, wholesalers, and governments is similar to—and different from—buying by manufacturers.
8. understand important new terms (shown in **red**).

Business and Organizational Customers—A Big Opportunity

LO 6.1

Most people think about an individual final consumer when they hear the term *customer*. But many marketing managers aim at customers who are not final consumers. In fact, the dollar value of purchases made by businesses and other organizations in the United States is more than double that made by final consumers.[2]

Exhibit 6-1 Understanding Business and Organizational Customers for Marketing Strategy Planning

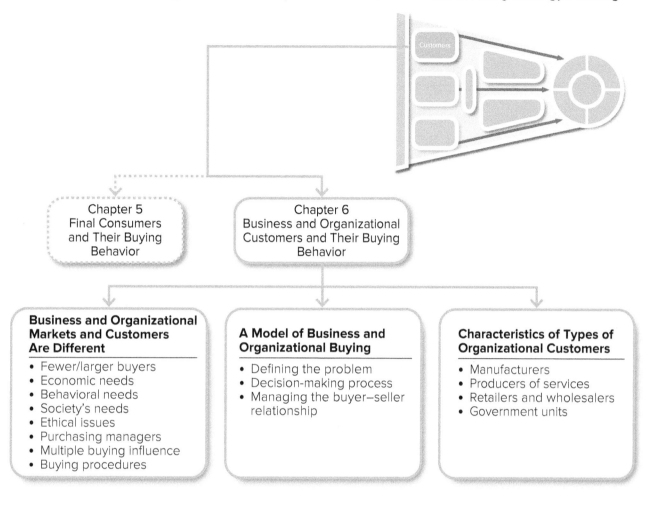

Business and organizational customers are any buyers who buy for resale or to produce other goods and services. There are many different types of organizational customers, including the following:

- *Producers of goods and services*—including manufacturers, farmers, real estate developers, hotels, banks, and even doctors and lawyers.
- *Intermediaries*—wholesalers and retailers.
- *Government units*—federal agencies in the United States and other countries as well as state and local governments.
- *Nonprofit organizations*—national organizations such as the American Red Cross and Girl Scouts as well as local organizations such as museums and churches.

As this suggests, not all organizational customers are business firms. Even so, they are sometimes loosely referred to as *business buyers, intermediate buyers,* or *industrial buyers*—and marketing managers often refer to organizational customers collectively as the "business-to-business" market, or simply, the *B2B market.* We will also use the terms *purchasing managers, organizational buyers,* or just *buyers* to refer to various people in the organization responsible for doing the buying.

In this chapter, we'll focus on organizational customers and their buying behavior (see Exhibit 6-1). In Chapter 5 we focused on buying by final consumers, so here we'll start by covering important ways that organizational buying differs from buying by final consumers. Next, we will look at a model of business and organizational buying. Finally, we'll focus on some key differences among the specific types of organizational customers.

Keep in mind that, for many firms, marketing strategy planning is about meeting the needs of organizational customers, not final consumers. A firm can target both final consumers and organizations, but different marketing mixes may be needed. As you learn about the buying behavior of organizations, think about how a firm's marketing mix may need to be different and how it may be adjusted.

Organizational Customers Are Different

LO 6.2

In some ways, organizational buying is similar to the model of consumer behavior introduced in Chapter 5. Businesses and organizations make purchases to solve problems, they seek information, and the decision-making process varies depending on the nature of the purchase. Yet organizational customers and their buying behaviors differ from final consumers and their buying behaviors. These differences have important implications for the marketing strategy planning process (see Exhibit 6-2).

Customers are fewer in number, but each spends more dollars

As compared to the consumer market, the organizational market includes far fewer buyers—yet the dollars spent by each are, on average, much greater. Consider that Procter & Gamble's sales to Walmart are more than $11 billion. Although that case reflects an extreme, most B2B sellers have many customers that provide each of them with thousands—if not hundreds of thousands—of dollars in revenue. This justifies business marketers investing more marketing resources into satisfying each customer.

Economic needs are primary

Organizations intently focus on economic factors when they make purchase decisions. Buyers try to account for the total cost of selecting a supplier and its particular marketing mix, not just the initial price of the product. For example, a hospital that needs a new type of digital X-ray equipment might look at both the original cost and

Exhibit 6-2 Differences between Organizational Customers and Final Consumers

ongoing costs, how it would affect doctor productivity, and, of course, the quality of the images it produces. The hospital might also consider the seller's reliability and its ability to provide speedy maintenance and repair.

The matter of reliability deserves further emphasis. Companies count on reliable suppliers—suppliers that deliver consistent quality on time. An organization may not be able to function if purchases don't arrive when they're expected. For example, imagine what would happen on a Buick assembly line if Nexteer Automotive was a few days late delivering its adaptive steering wheels. Buick might have to shut down the assembly line. This would affect other suppliers—raising their costs—and Buick's dealers might not be happy if they lost car sales.

The cost of poor quality can be significant. Consider the costs to automakers when air bags made by Takata were found to be defective. In the United States alone more than 30 million cars were already on the road with the faulty safety device. The total cost of replacing all those air bags was more than half a billion dollars, not counting the harm to the reputations of the automakers involved and, most importantly, lives lost.[3]

Behavioral needs still matter

Although economic needs are usually primary, B2B sellers know that individual buying firms and individual buyers are also influenced by noneconomic factors. Some of the behavioral dimensions mentioned in Chapter 5, such as security, status, and a sense of accomplishment, are relevant here, too. Purchasing managers and others involved in buying decisions are also human—they worry about their jobs, want respect from peers, and usually seek friendly relationships with suppliers.

For example, some purchasing people want to be seen—inside and outside their company—as more forward thinking. These buyers are often eager to imitate progressive competitors or even to be the first to try new products. Such innovators deserve special attention when new products are being introduced. Organizational buyers want to do their jobs well, and good suppliers try to help them accomplish that. Along those lines, a recent survey found that buyers are more motivated to receive and open e-mails from marketers who offer fresh insight and ideas.[4]

For many purchases, organizational buyers know that choosing a supplier that provides inconsistent quality or late deliveries can be very costly—but it also means the buyer may be blamed. Buyers want to avoid this risk—and a marketer who develops a marketing mix that lowers the buyer's personal risk puts himself or herself in a stronger position.

A seller's marketing mix may need to consider both the needs of the company and the individual. Buyers want to avoid taking risks that reflect badly on their decisions. Honda reminds customers that its reputation for reliability is something they can count on—which lowers the risk a buyer might feel buying an unknown brand. A. O. Smith knows that organizational buyers want to save their company money, but they also want to feel good about their choices.
(left): Source: American Honda Motor Co., Inc.; (right): Source: A. O. Smith

#M4BW

Sometimes society's needs matter

Some businesses take societal needs into consideration when making purchasing decisions. For example, New Belgium Brewing Company put sustainable values in its mission statement. That means its purchasing people select more energy-efficient (but higher-priced) brew kettles, buy wind-powered electricity, and build facilities that are more costly but use the latest green ideas—even when they might cost more.

Other firms find they can save money and help the environment. For example, when Falconbridge Limited's aluminum smelter changed to more efficient (but initially more expensive) light bulbs, it saved almost $100,000 per year in energy bills.

It isn't just environmental sustainability that buyers can consider. Other buying organizations think about the quality of life for their suppliers—economic viability, their employees and their suppliers' communities. Buying organizations that bring these concerns into their purchasing decisions and behaviors can make a better world. For more on this topic, see *What's Next?* Buying for a better world.

A buying firm *may* take into account three areas of need—company, individual, and society. See Exhibit 6-3 for a summary of this idea. Note the dashed line around "society's needs" suggests that societal needs may not be relevant for all buying firms or buying situations.

Ethical conflicts may arise

While consumers have ethical issues to consider when making purchases, organizational buyers must be careful to avoid a conflict between their own self-interest and company outcomes. This can be difficult; sometimes the ethical line can be blurry. Consequently, some firms set strict ethical guidelines. Marketers must be careful here. A salesperson offering a company pen to a prospect may view the giveaway as part of the promotion effort—but the customer firm may have a policy against employees accepting any gift from a supplier.

Exhibit 6-3 Overlapping Needs of Individual Influencers and the Customer Organization

150

What's Next? Buying for a better world  (#M4BW)

While B2B buyers care about doing what's best for their company, many organizations want to do more. Creative buyers are finding ways to help their companies *and* make the world a better place.

For example, South African grocery chain Woolworths (no relationship to the U.S. drugstore chain) follows its "Global Business Journey." This strategic initiative identifies key areas where Woolworths can make sub-Saharan Africa (where it operates) a better place. For example, to achieve objectives in three of those key areas—sustainable farming, ethical sourcing, and water—Woolworths began to work closely with the farmers who grow the fruits and vegetables on its store shelves. Many of these suppliers are small businesses with limited financial resources. They regularly contend with South African droughts, so Woolworths developed its "Farming for the Future" program to train farmers on water and fertilizer management. Those who attended reduced water usage by 16 percent and cut pesticide and herbicide use in half. To give its black- and women-owned suppliers a leg up, its "Supplier and Enterprise Development Program" offers them low-cost loans. Programs like these bring Woolworths a more stable supply chain, better-quality produce, *and* make a better South Africa.

Levi Strauss, known best for its blue jeans, uses its influence to help suppliers become a force for good. Most of Levi's suppliers are in developing countries and employ large numbers of young women. Levi's wanted to help improve the quality of life for hundreds of thousands of garment workers across its supply chain. It designed "Improving Worker Well-Being," a 10-week course with topics like health, hygiene, communication, and critical thinking. Initially reluctant, factory managers were pleasantly surprised to find that graduates of the program were happier, healthier, *and* more productive and loyal.

A few large and influential aluminum buyers put pressure on the notoriously *un*sustainable aluminum mining industry. Nestlé decided it wanted all the aluminum in its Nespresso coffee capsules to be sustainably sourced. German automaker Audi introduced sustainability ratings for all of its suppliers. Apple sought a greener iPhone. These demands put mining giant Rio Tinto and aluminum producer Alcoa on notice—and both found more sustainable mining approaches. *And* all these companies can tell customers their products are now more environmentally friendly.

One final example comes from Brazilian beauty products maker Natura, which has a mission focused on sustainability. Natura is a founding member of the Union for Ethical BioTrade, which advocates "sourcing with respect for a world in which people and biodiversity thrive." Toward that end, Natura carefully sources ingredients from the Amazon River area, helping many indigenous people set up new businesses that foster economic development in a poor part of the country. These efforts to promote sustainability for people and the planet appeal to Natura's target markets, and the fast-growing company is now the largest Brazilian cosmetics company *and* making for a better Brazil.[5]

What should a company do when a purchasing decision benefits society but comes at an economic cost to the buying firm? In what situations (if any) should economics take precedence in such a decision?

Although most organizational buyers act in an ethical manner, there have been highly publicized abuses. For example, Omnicare, Inc., which provides pharmaceutical drugs to customers that include nursing homes, was charged with soliciting and receiving kickback payments from its nursing home customers and drugmaker suppliers, including Johnson & Johnson (J&J). J&J and Omnicare apparently sold the drug Risperdal to nursing homes as a treatment for dementia and anxiety, conditions for which it was not approved. The two companies also agreed to downplay risks of diabetes and weight gain when talking to patients and doctors. The case may cost the two firms more than $2 billion in fines and other charges—plus untold damage to their reputations.

Marketers need to take concerns about conflict of interest very seriously. Part of the promotion job is to persuade different individuals who may influence an organization's purchase. Yet the whole marketing effort may be tainted if it even *appears* that a marketer has encouraged a person who influences a decision to put personal gain ahead of company interest. The pressure marketing managers feel to make sales goals must be tempered by ethical choices.[6]

Ethical Dilemma

What would you do? Assume that you work for a small manufacturing firm in the purchasing department. Your manager tells you that your supplier of cleaning supplies and paper products (paper towels, toilet paper) is going out of business and your job is to choose a new supplier. She tells you that these supplies are not central to operations, so you should "just find the cheapest source." In evaluating three options, you see that one supplier (GreenCleanNow) sells environmentally friendly cleaning products and paper products made with recycled products. The salesperson's demonstration clearly shows these products are functionally equivalent to the other products and clearly better for the environment. But the prices they charge are 2 percent higher than the best price from an alternative supplier offering products that are not so environmentally friendly. If you ask your boss, you are pretty sure she will say, "I told you to get the cheapest products."

What would you do? Explain your choice. What are the pros and cons of your decision?

Purchasing managers are specialists

Many organizations rely on specialists to ensure that purchases are handled sensibly. These specialists have different titles in different firms (such as procurement officer, supply manager, purchasing agent, or buyer), but basically they are all **purchasing managers**—buying specialists for their employers. In large organizations, they usually specialize by product area and are real experts.

Some people think purchasing is handled by clerks who sit in cubicles and do the paperwork to place orders. That view is out-of-date. Today, most firms look to their procurement departments to help cut costs and provide competitive advantage. In this environment, purchasing people have a lot of clout. And there are good job opportunities in purchasing for capable business graduates.

Salespeople often have to see a purchasing manager first—before they contact any other employee at the buying firm. These buyers hold important positions and take a dim view of sales reps that try to go around them. Rather than being "sold," these buyers want salespeople

BASF plastics can be used in automobile parts—replacing heavier metal parts. The lower weight can increase fuel efficiency. BASF personnel work with purchasing managers and engineers at automakers to determine where plastic parts offer the greatest benefit to the end user (car buyer).
Source: BASF Corporation 2015

to provide accurate information that will help them solve problems and buy wisely. They like information on new goods and services, as well as tips on potential price changes, supply shortages, and other changes in market conditions. Sometimes all it takes for a sales rep to keep a buyer up-to-date is to send an occasional e-mail. An intuitive buyer can tell when a sales rep has the customer firm's best interests at heart.

Although purchasing managers usually coordinate relationships with suppliers, other people may also play important roles in influencing the purchase decision.[7]

Multiple buying influence in a buying center

Multiple buying influence means that several people—perhaps even top management—play a part in making a purchase decision. Possible buying influences include the following:

1. *Users*—perhaps production line workers or their supervisors.
2. *Influencers*—perhaps engineering or R&D people who help write specifications or supply information for evaluating alternatives.
3. *Buyers*—purchasing managers who have the responsibility for working with suppliers and arranging the terms of the sale.
4. *Deciders*—people in the organization who have the power to select or approve the supplier—often a purchasing manager but perhaps top management for larger purchases.
5. *Gatekeepers*—people who control the flow of information within the organization. Gatekeepers might include purchasing managers, receptionists, secretaries, research assistants, bookkeepers, and many more.

The following example portrays how different buying influences work in the purchase of a corporate jet. A company that already owns a corporate jet may decide it is time to lease a new one. The current lease is expiring, and a newer jet would have better fuel efficiency and be more reliable. Middle and upper management (users) would be likely to provide comments on features they would like to see in the new model. Thus, they play the roles of influencers. The expert knowledge of the chief pilot and the airplane mechanic would make them significant influencers on the purchase. The chief pilot might also serve as gatekeeper by deciding which information and how much of his expertise he shares with others. A finance executive would also be likely to weigh in (another influencer) with thoughts on costs and financing the purchase. An assistant to the purchasing manager (gatekeeper) does online searches for information and decides what information to put into a report. After recommendations have been made, a large multimillion-dollar purchase like a corporate jet might go before the board of directors (deciders) for final approval. After all these buying influences are considered and a particular jet is selected, one of the purchasing agents for the firm (the buyer) works with the chosen supplier to arrange the terms of the sale, including delivery dates, payment, installation, and after-sale service.

It is helpful to think of a **buying center** as all the people who participate in (or influence) a purchase. Because different people may make up a buying center from one decision to the next, the salesperson must study each case carefully. Just learning whom to talk with may be hard, but thinking about the various roles in the buying center can help.

The salesperson may have to talk to every member of the buying center—stressing different topics for each. This not only complicates the promotion job but also lengthens it. Approval of a routine order may take anywhere from a day to several months. On very important purchases—a new building, major equipment, or a new information system—the selling period may take a year or more.[8]

Buying procedures are standardized

Buying organizations often establish formal procedures governing how purchases are made. For example, if a contract exists with an office supply vendor, everyone in the organization may be required to make their purchases through a special website. A

A company's purchase of a corporate jet is likely to involve multiple buying influence—meaning several people from the buying firm will play a part in making a purchase decision.
Source: Cessna Aircraft Company

person who needs to purchase something may have to complete a **requisition**—a request to buy something. For larger purchases, there may be rules requiring that multiple suppliers are contacted so that prices can be compared.

If a large organization has facilities at dispersed locations, much of the purchasing work may be done at a central location. With centralized buying, a sales rep may be able to sell to facilities all over a country—or even across several countries—without leaving the base city. Walmart handles many of the purchase decisions for stores in its retail chain from its headquarters in Arkansas. Many purchasing decisions for agencies of the U.S. government are handled in Washington, DC.

The general procedures for organizational buying are outlined in the next section.

A Model of Business and Organizational Buying

LO 6.3

Business and organizational buying generally follows the three-step approach shown in Exhibit 6-4. Depending on the nature of the purchase, how each step plays out can vary—and the third step is not always part of the process. The next few sections of this chapter will examine these three steps more closely. Let's begin by looking at how organizations define the problem by narrowing down from problem recognition to specifying a particular solution.[9]

Step 1: Defining the Problem

Organizations make purchases to satisfy needs. Generally speaking, most organizations make purchases for the same basic reason: they buy goods and services that will help them meet the demand for the goods and services that they in turn supply to their markets. In other words, their basic need is to satisfy their own customers and clients. A producer buys because it wants to earn a profit by making and selling goods or services. A wholesaler or retailer buys products it can profitably resell to its customers. A town government wants to meet legal and social obligations to its citizens.

From recognizing problems to describing needs

At their earliest stages, problems may not be well understood. A manufacturer might be seeking a way to speed up production. A dentist might wonder if there is a better way to bill her customers. A clothing retailer may be looking for spring fashions to appeal to

Exhibit 6-4 A Model of Organizational Buying

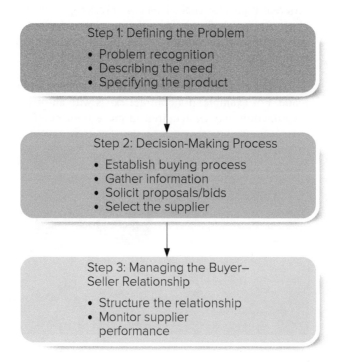

Eastman Chemical Co. developed Eastman Tritan, an innovative copolyester (plasticlike) material. Tritan is durable, heat resistant, and offers design flexibility and ease of processing. It is ideal for the molded parts needed by kitchen appliance manufacturers. Promotion from Eastman Chemical, possibly through ads like this one or a salesperson calling on a buyer, will help the buying firm recognize a need and see that Tritan is the solution.
Source: Eastman Chemical Company

its target market. A school district may be seeking a way to provide more students with access to computers. Each of these realizations can trigger the organization to seek more details about the problem at hand.

Ultimately, the problem should be narrowed down to a more specific need. What product or service can solve the client's problem? Selling organizations often try to get in at this stage. Some sellers even help buyers realize they have a problem—for example, a seller of medical office management software might send direct mail to dentists describing the benefits other dental offices have received after using its software.

Specifications describe the product

In some situations, an organization will formalize the problem with a detailed set of purchase requirements. In this case, organizational buyers may buy on the basis of a set of **purchasing specifications**—a written description of what the firm wants to buy. When quality is highly standardized, as is often the case with manufactured items, the specification may simply consist of a brand name or part number.

Often, however, when the purchase requirements are more complicated, the specifications may set out detailed information about the performance standards the product must meet. Purchase specifications for services tend to be detailed because services are less standardized and usually are not performed until after they're purchased.

Step 2: The Decision-Making Process

LO 6.4

After the buying organization recognizes the problem, describes the need, and specifies the product, the next step involves the decision-making process—how a firm decides whether to buy, what to buy, and what criteria to consider when evaluating suppliers. Buyers then gather information, solicit proposals from suppliers, and finally choose a supplier. The decision-making process can vary depending on the nature of the purchase—so we start by looking at different buying processes (see Exhibit 6-5).[10]

Three kinds of buying processes are useful

It is useful to think about three types of buying processes. **New-task buying** occurs when a customer organization has a new need and wants a great deal of information. The company may never have purchased this type of product before. A **straight rebuy** is a routine repurchase that may have been made many times before. Buyers probably don't bother looking for new information or new sources of supply. Most of a company's small or recurring purchases are of this type—but they take only a small part of an organized buyer's time. The **modified rebuy** is the in-between process where some

Exhibit 6-5 Organizational Buying Processes

Characteristics	Type of Process		
	New-Task Buy	Modified Rebuy	Straight Rebuy
Time Required	Much	Medium	Little
Multiple Influence	Much	Some	Little
Review of Suppliers	Much	Some	None
Information Needed	Much	Some	Little

review of the buying situation is done—though not as much as in new-task buying. This might be a buying decision that has not been done before, or the criteria used to make a purchase decision may have changed.

It helps to compare and contrast these buying processes with some examples. Exhibit 6-6 shows examples of the different situations and associated buying processes for companies that sell software, beer, or batteries.

Straight rebuys often use e-commerce order systems

E-commerce computer systems *automatically* handle a large portion of straight rebuys. Buyers program decision rules that tell the computer how to order and leave the details of following through to the computer. For example, when a machine tool maker receives an order that requires certain materials or parts, the computer information system automatically orders them from the appropriate suppliers, sets the delivery date, and schedules production.

If conditions change, buyers modify the computer instructions. When nothing unusual happens, however, the computer system continues to routinely rebuy as needs develop—electronically sending purchase orders to the regular supplier.

New-task buying requires the most information

Whereas buyers want to stay current on all facets of purchasing, new-task buying situations motivate them to seek specific information. Often a new-task buy starts with a *user* who becomes aware of a need and begins researching solutions. Even though a wide variety of information sources are available (see Exhibit 6-7), business buyers will use the sources they trust. To build trust, a marketer must make sure its information is reliable and useful to the buyer. Sellers should present information objectively; if it appears too self-serving and biased toward the suppliers' offering, it may not be trusted or used by the buyer.

How much information a customer collects depends on the importance of the purchase and the level of uncertainty about what choice might be best. The time and

Exhibit 6-6 Examples of Different Products with Different Buying Processes

	Product Being Sold		
	Office Automation Software	Beer to Grocery Store	Batteries to a Car Manufacturer
New-task buy	A dentist's office has never used any type of office management software.	A grocery store chain begins selling beer for the first time.	A carmaker is designing its first electric car and needs to buy batteries for it.
Modified rebuy	A dentist's office adds a new module to existing software.	A grocery store adds a new brand of beer.	A carmaker has made electric cars but seeks out a more powerful battery for a new model.
Straight rebuy	A dentist's office renews an annual subscription for software it has used for many years.	A grocery store reorders a brand and size of beer it has purchased before.	A carmaker is ordering the same batteries every week for a production model vehicle.

Exhibit 6-7 Major Sources of Information Used by Organizational Buyers

	Marketing sources	Nonmarketing sources
Personal sources	• Salespeople • Others from supplier firms • Trade shows	• Buying center members • Outside business associates • Consultants and outside experts
Impersonal sources	• Online events and virtual trade shows • Sales literature and catalogs • E-mails and newsletters • Website content including blogs, video, case studies, and white papers • Social media (a company LinkedIn page)	• Online searches • Rating services • Trade associations • News publications • Product directories • Online communities • Online review sites • Social media (comments on LinkedIn)

expense of searching for information may not be justified for a minor purchase. But a major purchase often involves real detective work by the buyer.

New-task buying situations provide a good opportunity for a new supplier to make inroads with a customer. With a buyer actively searching for information, the seller's promotion has a much greater chance of being noticed and having an impact. Advertising, trade show exhibits, e-mails, and salespeople can all help build the buyer's attention, but an informative website may be essential for getting attention in the first place.[11]

Younger buyers gather information differently

In Chapter 3 we described how the large number of Millennials—born between 1978 and 1994—and their particular worldview affects *consumer* buying behavior. Millennials are increasingly responsible for *organizational* buying—and they bring new attitudes and approaches. It's no surprise that Millennials are more likely to jump online when researching a potential supplier. Research also shows they are more comfortable turning to social media sites—and they love video content. Marketing managers who adapt marketing strategies to better fit the needs of the new Millennial buyers will see long-term benefits.[12]

Search engines—a first step to gathering information

Most purchasing managers start with an Internet search when they need to identify new suppliers, better ways to meet needs, or information to improve decisions. Buyers often rely on highly specialized search engines—like one that finds all types of steel that meet certain technical specifications and then compares prices. But buyers also use general-purpose search engines such as Google. A search across the whole web can often locate off-the-shelf products that eliminate the need to buy expensive, custom-made items. For example, a firm in Saudi Arabia ordered $1,000 worth of tiny rubber grommets from Allstates Rubber & Tool, a small firm in the suburbs of Chicago. If the buyer's search hadn't located Allstates' website, the only alternative would have been to pay much more for custom-made grommets—and Allstates wouldn't have picked up a new customer.[13]

Marketing managers know that it is critical to have a website that buyers can find. That's why suppliers often pay for a sponsored link (an ad) that appears when certain keywords are included in a search. A supplier might also change its website so that it is more likely to appear high on a list of searches.

Buyers want sites with useful content

Having useful content on a website not only moves it higher on the buyer's search results, but it also gives buyers a reason to fully explore the seller's website. Instead of trying to sell the customer, a supplier needs to educate buyers about their needs and present the advantages and disadvantages of the seller's products. Buyers like to read targeted **white papers** (an authoritative report or guide that addresses important issues in an industry and offers solutions). White papers may advocate a seller's solution, but if they appear objective, they can help establish a firm as a thought leader in a particular area. Buyers also like to read case studies to learn about how other companies have addressed similar needs. Video content and blogs can also make a seller's website more useful.

Buyers share experiences in online communities

Buyers especially value recommendations from others who have already dealt with a similar need. Sometimes they turn to their LinkedIn network (an online social media site for businesspeople) to solicit advice from colleagues. Others turn to online communities. For example, at Spiceworks Community (www.spiceworks.com), IT professionals ask and answer technical questions, research best practices, and learn what others think about a product or service they are considering. Buyers trust this source of information, and similar online communities have developed around other industries. As buyers rely more on social networks, communications from marketers may have less influence on buyers' attitudes and choices.[14]

Online reviews growing in B2B

Buying organizations used to be reluctant to share their experiences in a public forum. Now some firms are gathering reviews, opinions, and experiences that help organizational buyers make better decisions. Capterra has more than a million reviews of a wide range of software products—categories include job applicant tracking, construction management, church management, and more. VendOp is an online tool where buying organizations can search for suppliers, read reviews, and view ratings on speed, price, and quality.

Solicit proposals from suppliers

When buyers in B2B markets identify potential suppliers, they contact one or more suppliers for proposals. A supplier may simply send a brochure or product catalog, or it may have a salesperson contact the buyer over the phone. In more complex buying situations, there may be formal presentations from different suppliers that also submit detailed proposals.

Buyers ask for competitive bids to compare offerings

Sometimes buyers will ask suppliers to submit a **competitive bid**—the terms of sale offered by the supplier in response to the purchase specifications posted by a buyer. If different suppliers' quality, dependability, and delivery schedules all meet the specs, the buyer will select the low-price bid. But a creative marketer needs to look carefully at the buyer's specs—and the need—to see if changing other elements of the marketing mix could provide a competitive advantage.

Source: State of California

Rather than search for suppliers, buyers sometimes post their requirements and invite qualified suppliers to submit a bid. Some firms set up or participate in a procurement website that directs suppliers to companies (or divisions of a company) that need to make purchases. These sites make it easy for suppliers to find out about the purchase needs of the organizations that sponsor the sites. This helps increase the number of suppliers competing for the business, which can drive down prices or provide more beneficial terms of sale. For example, when the California Department of Transportation was planning $4 billion in new construction projects, it established a procurement site so that potential suppliers knew each project's requirements for submitting a competitive bid.

Evaluating and selecting suppliers

Considering all of the factors relevant to a purchase decision can be very complex. A supplier or product that is best in one way may not be best in others. To try to deal with these situations, many firms use **vendor analysis**—a formal rating of suppliers on all relevant areas of performance. The purpose isn't just to get a low price from the supplier on a given part or service. Rather, the goal is to lower the *total costs* associated with purchases. Analysis might show that the best vendor is the one that helps the customer reduce costs of excess inventory, retooling of equipment, or defective parts.[15]

Buying procedures can vary

From our earlier discussion, you can see that buyers make important decisions about how to deal with one or more suppliers. At one extreme, a buying firm might want to rely on competition among all available vendors to get the best price on every order it places. The buyer cares little about which supplier fills the order; only the price and delivery time matter, and maybe a different vendor is chosen to fill each order. On the other hand, there are situations where it makes sense to routinely buy everything from a single vendor with which a buyer already has a good relationship. In practice, there are many important and common variations between these extremes. To better understand the variations, let's take a closer look at the benefits and limitations of different types of buyer–seller relationships.

Step 3: Managing Buyer–Seller Relationships in Business Markets

LO 6.5

After selecting a supplier, the buyer and seller must figure out how to structure and manage the relationship. Many organizational purchases are ongoing—a buyer will make regular purchases. Other purchases might involve a single transaction—buying a replacement part, for example. Buyers and suppliers should work together to figure out the best way to structure the relationship. In this section we look at the many different considerations for managing buyer–seller relationships.

Close relationships solve problems

There are often significant benefits of a close working relationship between a supplier and a customer firm. And such relationships are becoming common. Many firms are reducing the number of suppliers with whom they work—expecting more in return from the suppliers that remain. The best relationships involve real partnerships where there's mutual trust and a long-term outlook. Closely tied firms often share tasks at lower total cost than would be possible working at arm's length.

The partnership between AlliedSignal and Betz Laboratories, for example, shows the benefits of a good relationship. At one time, Betz was just one of several suppliers that sold Allied chemicals to keep the water in its plants from gunking up pipes and rusting machinery. But Betz didn't stop at selling commodity powders. Teams of Betz experts and Allied engineers studied each plant to find places where water was being wasted. In less than a year, a team in one plant found $2.5 million in potential cost savings. For example, by adding a few valves to recycle the water in a cooling tower, Betz was able to save 300 gallons of water per minute, resulting in savings of more than $100,000 a year and reduced environmental impact. Because of ideas like this, Allied's overall use of water treatment chemicals decreased. However, Betz's sales to Allied doubled because it became Allied's only supplier of water chemicals.[16]

In today's business markets, many suppliers of goods and services seek to build long-term relationships that provide value for their customers. Kaleidoscope provides services to help its B2B customers with brand strategy and design. In this ad, Kaleidescope says "'Bite-sized' relationships are a dime a dozen. But there is nothing sweeter than working with people you like who deliver consistent results year after year."
Courtesy of Kaleidoscope

Relationships can involve many from both sides

As the Allied–Betz Laboratories collaboration shows, some buyer–seller relationships can involve multiple buying influence—more than just a purchasing agent and a salesperson. To develop effective solutions, those closest to the problems should be directly involved. This may mean bringing people together from accounting, finance, production, information systems, and/or other functional areas of both the buyer and seller firms.

Relationships may not make sense

Although close relationships can produce benefits, they are not always best. For buyers, long-term commitments can also reduce flexibility. When competition drives down prices and spurs innovation, customers may be better off letting suppliers compete for their business. It may not be worth the customer's investment to build a relationship for purchases that are not particularly important or made that frequently. Besides that, close relationships take time and attention to build and manage.

Sellers don't usually want closer relationships with all of their customers, either. Some customers may place orders that are too small or require so much special attention that the relationship would never be profitable for the seller. Also, in situations where a customer doesn't want a relationship, trying to build one may cost more than it's worth. Buyers and sellers should choose closer relationships where the benefits outweigh the costs.[17]

Relationships have many dimensions

Relationships are not "all or nothing" arrangements. Many firms may have a close relationship in some ways and not in others. Thus, it's useful to think about five key dimensions that help characterize most buyer–seller relationships: cooperation, information sharing, operational linkages, legal bonds, and relationship-specific adaptations. Purchasing managers for the buying firm and salespeople for the supplier usually coordinate the different dimensions of a relationship. However, as shown in Exhibit 6-8, close relationships often involve direct contacts between a number of people from other areas in both firms.[18]

Cooperation treats problems as joint responsibilities

In cooperative relationships, the buyer and seller work together to achieve both mutual and individual objectives. The two firms treat problems that arise as a joint responsibility. National Semiconductor (NS) and Siltec, a supplier of silicon wafers, found clever ways to cooperate and cut costs. Workers at the NS plant used to throw away the expensive plastic cassettes that Siltec uses to ship the silicon wafers. Now Siltec and NS cooperate to recycle the cassettes. This helps the environment and also saves more than $300,000 a year. While Siltec passes along most of that to the larger NS in the form of lower prices, it also lowers its own costs.[19]

Shared information is useful but may be risky

Some relationships involve open sharing of information. This might include the exchange of proprietary cost data or demand forecasts or working jointly on new product designs. Information might be shared in discussions between personnel, or through information systems connected via the Internet, a key facet of B2B e-commerce. The electronic approach has a big advantage in that it is fast and easy to update the information. It also saves time. A customer can check detailed product specs or the status of a job on the production line without having to wait for someone to respond.

Exhibit 6-8
Key Dimensions of Relationships in Business Markets

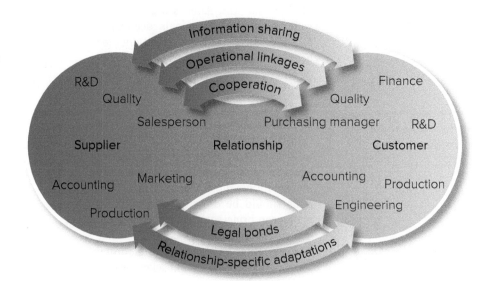

Information sharing can lead to better decisions and better planning. However, firms don't want to share information if there's a risk that a partner might misuse it. For example, one of General Motors' suppliers shared some of its planned technology with the automaker. It later found out that General Motors' purchasing chief showed blueprints of that secret technology to competing suppliers. Violations of trust in a relationship are an ethical matter and should be taken seriously.

Suppliers can be wary of sharing bad news with an important customer. For example, a supplier may wait to tell a customer of financial problems—and that can cause trouble. When Edscha, a German supplier of sunroofs to BMW, unexpectedly filed for insolvency, BMW had a crisis in the making. BMW was about to launch its new Z4 convertible—and Edscha supplied all the tops. Getting a new supplier up to speed would have taken six months, so BMW scrambled to support Edscha and launch the Z4 on schedule. A more trusting and open relationship might have provided more lead time to work out problems.[20]

Operational linkages share functions between firms

Operational linkages are direct ties between the internal operations of the buyer and seller firms. These linkages usually involve ongoing coordination of activities between the firms. Shared activities are especially important when neither firm, working on its own, can perform a function as well as the two firms can working together.

Business customers often require operational linkages to reduce total inventory costs, maintain adequate inventory levels, and keep production lines moving. On the other hand, keeping too much inventory is expensive. Providing a customer with inventory when it's needed may require that a supplier be able to provide **just-in-time delivery**—reliably getting products there *just* before the customer needs them. We'll discuss just-in-time systems in more detail in Chapter 11. For now, it's enough to know that closer relationships between buyers and sellers involve operational linkages that lower costs and increase efficiency.

Contracts spell out obligations

Many purchases in business markets are simple transactions. The seller's responsibility is to transfer title to goods or perform services, and the buyer's responsibility is to pay the agreed price. However, more complex relationships may be spelled out in detailed legal contracts. An agreement may apply only for a short period, but long-term contracts are also common.

For example, a customer might ask a supplier to guarantee a 6 percent price reduction for a particular part for each of the next three years and pledge to virtually eliminate defects. In return, the customer might offer to double its orders and help the supplier boost productivity.

Sometimes the buyer and seller know roughly what is needed but can't fix all the details in advance. For example, specifications or total requirements may change over

Spain's EmpordAigua offers consulting services to help firms manage water quality and usage. EmpordAigua believes that no two drops of water are alike—nor are any two customers' water needs. So EmpordAigua develops close buyer-seller relationships based on cooperation, information sharing, and relationship-specific adaptations to develop custom water solutions for each of its client's needs.
Source: EmpordAigUA® Water Design & Technology

time. Then the relationship may involve **negotiated contract buying**, which means agreeing to contracts that allow for changes in the purchase arrangements. In such cases, the general project and basic price are described but with provision for changes and price adjustments up or down.

Some managers figure that even a detailed contract isn't a good substitute for regular good-faith reviews to make sure that neither party gets hurt by changing business conditions. Harley-Davidson used this approach when it moved toward closer relationships with a smaller number of suppliers. Purchasing executives tossed out detailed contracts and replaced them with a short statement of principles to guide relationships between Harley and its suppliers.

Specific adaptations invest in the relationship

Relationship-specific adaptations involve changes in a firm's product or procedures that are unique to the needs or capabilities of a relationship partner. Industrial suppliers often custom design a new product for just one customer; this may require investments in R&D or new manufacturing technologies. MetoKote, in its relationship with John Deere described in the case at the beginning of this chapter, made a specific adaptation by building its coating plant right next door to Deere's factory.

Buying firms may also adapt to a particular supplier. When Apple designed MacBook computers with an Intel computer chip, it made it difficult to change to a different chipmaker later. However, buyers are often hesitant to make big investments that increase dependence on a specific supplier. Typically, they do it only when there isn't a good alternative—perhaps because only one or a few suppliers are available to meet a need—or if the benefits of the investment are clear before it's made.

Specific adaptations are usually made when the buying organization chooses to **outsource**—contract with an outside firm to produce goods or services rather than producing them internally. Many firms have turned to outsourcing to cut costs, which is why much outsourcing is handled by suppliers in countries where labor costs are lower. For example, many American companies are outsourcing production to firms in China and customer service to India.[21]

A powerful customer may control the relationship

Although a marketing manager may want to work in a cooperative partnership, that may be impossible with large customers who have the power to dictate how the relationship will work. Often a powerful customer negotiates lower prices from suppliers. For example, when Hewlett-Packard (HP) started selling personal computers through Walmart stores, it knew a rock-bottom price would be required. So it used its large sales volume to demand lower prices from its suppliers and the contract manufacturers that assembled the PCs. This helped HP offer lower-priced computers while maintaining higher profit margins than its competition.[22]

Buyers usually use several sources to spread their risk

Even if a marketing manager develops the best possible marketing mix and cultivates a close relationship with the customer, the customer may not give *all* of its business to one supplier. Buyers often look for several dependable sources of supply to protect themselves from unpredictable events such as strikes, fires, or floods in one of their suppliers' plants. For example, Western Digital, the world's largest provider of hard-disk drives, produces most of its drives in Thailand. When flooding in Thailand closed down many of Western Digital's manufacturing plants, it significantly tightened the world's supply of personal computers.[23]

Even when buyers don't want a single source of supply, a good marketing mix is still likely to win a larger share of the total business—which can prove to be very important. From a buyer's point of view, it may not seem like a big deal to give a particular supplier a 40 percent share of the orders rather than a 20 percent share. But for the seller that doubles its sales with that one customer, that represents a 100 percent increase.

Buyers monitor supplier performance

Organizations routinely monitor supplier performance. Suppliers that fail to meet a buyer's performance expectations may be dropped. The process can be good for both buyer and supplier. Suppliers that perform well may find their share of a buyer's purchases go up. Suppliers that listen to feedback from customers can learn how to improve and increase the value of the relationship. Good buying organizations provide regular

Marketing Analytics in Action: Supplier Scorecards

HighFly Drones manufactures drone aircraft for the hobbyist market. A particular printed circuit board (part number PX5534) is used in three of HighFly's most popular drone models. Because this product is so important, HighFly regularly purchases it from two different suppliers: Ace Electronics and Charter Components. The price for each PX5534 is $9.62, and each supplier is paid that same price.

Last year, each supplier received half of HighFly's orders. For the coming year, HighFly wants to work more closely with one PX5534 supplier, giving it 75 percent of the orders, while the other supplier would receive 25 percent. To help decide which supplier will get the greater part of the orders, the purchasing manager reviews the scorecards shown below for the two suppliers over the last three months. The suppliers are rated on three performance dimensions using a 0 to 100 scale, with 100 being the highest level of performance.

	Ace Electronics			Charter Components		
	Sept.	Oct.	Nov.	Sept.	Oct.	Nov.
Quality	85	88	92	98	86	78
Communication	81	72	84	92	94	92
Delivery	71	94	82	92	96	94
Average	**79**	**85**	**86**	**97**	**92**	**88**

Review the performance of Ace Electronics and Charter Components over the last three months, then answer the following questions:

1. *What kind of feedback do you think is warranted for Ace Electronics? What concerns do you have with its performance?*
2. *What kind of feedback do you think is warranted for Charter Components? What concerns do you have with its performance?*
3. *Which supplier should receive the larger share of HighFly's business, and why? Under what circumstances would you be inclined to choose the other supplier to receive the larger share?*

---

feedback to their suppliers without being asked—and smart suppliers listen closely and respond to them. Honda, for example, provides a monthly *scorecard* that details the supplier's performance in five areas: quality, delivery, quantity delivered, performance history, and any special incidents. The report fosters an ongoing dialogue between Honda suppliers about ways that both Honda and the supplier can work together to improve supplier performance.[24]

Let's take a closer look at how a supplier scorecard works in practice in the Marketing Analytics in Action: Supplier Scorecards activity.

Variations in buying by customer type

We've been discussing aspects of relationships and buying approaches that generally apply to different types of customer organizations—in both the United States and internationally. However, it's also useful to have more detail about specific types of customers. Knowing the size, number, geographic location, and buying procedures of manufacturers, service firms, retailers, wholesalers, and governments helps marketing managers segment markets, identify targets, and create more effective marketing mixes.

Manufacturers Are Important Customers

LO 6.6

There are not many big ones

One of the most striking facts about manufacturers is how few there are compared to final consumers. This is true in every country. In the United States, for example, there are about 250,000 factories. The majority of these are quite small—less than 2 percent have more than 500 employees and about three-quarters have fewer than 20 employees. Smaller manufacturing firms have a much less formal buying process than large firms.[25]

Customers cluster in geographic areas

In addition to concentration by company size, industrial markets are concentrated in certain geographic areas. Internationally, industrial customers are concentrated in countries that are at the more advanced stages of economic development. From all the talk in the news about the United States shifting from an industrial economy to a service and information economy, you might conclude that the United States is an exception—that the industrial market in this country is shrinking. It is true that the number of people employed in manufacturing has been shrinking, but U.S. manufacturing output is higher than at any other time in the nation's history. The rate of growth, however, is fastest in countries where labor is cheapest.[26]

Within a country, there is often further concentration of manufacturing in specific areas. In the United States, many factories are concentrated in big metropolitan areas—especially in New York, Pennsylvania, Ohio, Illinois, Texas, and California. There is also concentration by industry. In Germany, for example, the steel industry is concentrated in the Ruhr Valley. Most automobile manufacturing occurs in Michigan, Indiana, and Ohio.

Business data often classify industries

The products an industrial customer needs to buy depend on the business it is in. Because of this, sales of a product are often concentrated among customers in similar businesses. This fact helps business marketing managers to segment their markets. For example, apparel manufacturers are the main customers for zippers. Marketing managers must focus their marketing mixes on prospective customers who exhibit characteristics similar to their current customers.

Detailed information is often available to help a marketing manager learn more about customers in different lines of business. The U.S. government collects and publishes data by the **North American Industry Classification System (NAICS) codes**—groups of firms in similar lines of business. (NAICS is pronounced like "nakes.") The number of establishments, sales volumes, and number of employees—broken down by geographic areas—are given for each NAICS code. A number of other countries collect similar data, and some of them try to coordinate their efforts with an international variation of the NAICS system. However, in many countries data on business customers are incomplete or inaccurate.

A firm like Alcoa is likely to find that the majority of customers that use aluminum are concentrated within a few industries that can be identified by their North American Industry Classification System (NAICS) code number.
(left): Fotosearch/Getty Images; (top right): Alejandro da Silva Farias/ E+/Getty Images; (bottom right): Thomas Trutschel/Photothek/ Getty Images

Exhibit 6-9 Illustrative NAICS Code Breakdown for Apparel Manufacturers

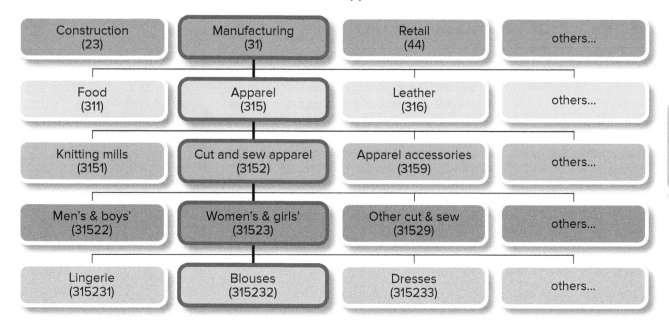

Let's take a closer look at how the NAICS codes work. The NAICS code breakdowns start with broad industry categories such as construction (23), manufacturing (31), retail trade (44), and so on (see Exhibit 6-9). Within each two-digit industry breakdown, much more detailed data may be available for three-digit industries (that is, subindustries of the two-digit industries). For example, within the two-digit manufacturing industry (code 31) there are manufacturers of food (311), leather (316), and others, including apparel manufacturers (315). Then each three-digit group of firms is further subdivided into more detailed four-, five-, and six-digit classifications. For instance, within the three-digit apparel manufacturers (315) there are four-digit subgroups for knitting mills (3151), cut and sew apparel (3152), and producers of apparel accessories (3159). Exhibit 6-9 illustrates that breakdowns become more detailed as you move to codes with more digits. However, detailed data (say, broken down at the four-digit level) aren't available for all industries in every geographic area. The government does not provide detail when only one or two plants are located in an area.

Many firms find their *current* customers' NAICS codes and then look at NAICS-coded lists for similar companies that may need the same goods and services. Other companies look at which NAICS categories are growing or declining to discover new opportunities.[27]

Producers of Services—Smaller and More Spread Out

LO 6.7

The service side of the U.S. economy is large and has been growing fast. Service operations are also growing in some other countries. There are many good opportunities to provide these service companies with the products they need to support their operations. But there are also challenges.

The United States has almost 6 million service firms—more than 17 times as many as it has manufacturers. Some of these are big companies with international operations. Examples include AT&T, Hilton Hotels, Prudential Insurance, PwC, Wells Fargo, and Accenture. These firms have purchasing departments that are like those in large manufacturing organizations. But as you might guess given the large number of service firms, most of them are small—Foothills Family Dental, North Coast Maid Service, and Computer Doctors. They're also more spread out around the country than manufacturing concerns. Factories often locate where transportation facilities are good, raw materials are available, and it is less costly to produce goods in quantity. Service operations, in contrast, often have to be close to their customers.

Buying may not be as formal

Purchases by small service firms are often handled by whoever is in charge or their administrative assistant. This may be a doctor, lawyer, owner of a local insurance agency, hotel manager, or office manager. Suppliers who usually deal with purchasing specialists in large organizations may have trouble adjusting to this market. Personal selling is still an important part of promotion, but reaching these customers in the first place often requires more advertising. And small service firms may need much more help in buying than a large corporation.

Small service customers like Internet buying

Small service companies that don't attract much personal attention from salespeople often rely on e-commerce for many of their purchases. Purchases by small customers can add up—so for many suppliers these customers are an important target market. Increasingly suppliers cater to the needs of these customers with specially designed websites. A well-designed website can be efficient for both customers and suppliers. Customers can get information, place orders, or follow up with a call or e-mail for personal attention from a salesperson or customer service rep when it's needed.

In a smaller service organization, purchases may be made by the person who is in charge rather than a person with full-time responsibility for purchasing. This ad recognizes this reality and tries to make renting from Hertz easy for this target market.
Source: The Hertz Corporation

Retailers and Wholesalers Buy for Their Customers

Most retail and wholesale buyers see themselves as purchasing agents for their target customers—remember the old saying that "Goods well bought are half sold." They are experts at what their customers want—and won't be persuaded by a sales rep for a manufacturer that can't provide it.

#M4BW

Millennials want to buy sustainable products

Walmart understands consumers' desire for sustainable products and knows that one of its target markets includes the environmentally conscious Millennials. Walmart doubled down on its long-standing sustainability goals and looked to its suppliers for help. Walmart's high-volume buying gives it a lot of power with suppliers—when it asks suppliers to change, they do. When Walmart told its suppliers that it would not buy products containing potentially dangerous chemicals (like triclosan and formaldehyde), it forced many suppliers to change ingredients. This followed a campaign—that also included suppliers—that successfully diverted 82 percent of waste materials away from landfills. These moves will help Walmart grow its business with a key target market *and* make a better world.[28]

Committee buying is impersonal

Space in retail stores is limited, and buyers for retail chains simply are not interested in carrying every product that some salesperson wants them to sell. In an average week, 150 to 250 new items are offered to the buying offices of a large chain such as Safeway. If the chain accepted all of them, it would add 10,000 new items during a single year! Obviously, these firms need a way to deal with this overload. Most retailers carefully evaluate products. A product has to prove itself before gaining widespread adoption.

Decisions to add or drop lines or change buying policies may be handled by a *buying committee*. The seller still calls on and gives a pitch to a buyer—but the buyer does not have final responsibility. Instead, the buyer prepares forms summarizing proposals for new products and passes them on to the committee for evaluation. The seller may not get to present her story to the buying committee in person. This rational, almost cold-blooded, approach certainly reduces the impact of a persuasive salesperson. On the other hand, it may favor a firm that has hard data on how its whole marketing mix will help the retailer attract and keep customers.

The entrepreneurs who started PenAgain (a uniquely shaped writing tool), for example, had to have more than a distinctive product to get shelf space at Walmart. Their presentation to Walmart had to include hard data showing their marketing mix already worked well in other retail stores and evidence of their ability to supply the large quantities a retailer the size of Walmart would need. After that, Walmart might sell PenAgain in a limited number of stores, seeking evidence the product met sales targets, before selling it in all of its stores.[29]

Buyers watch computer output closely

Most larger firms use sophisticated computerized inventory replenishment systems. Scanners at retail checkout counters keep track of what goes out the door—and computers use these data to update the records. Even small retailers and wholesalers use automated control systems that create daily reports showing sales of every product. Buyers with this kind of information know, in detail, the profitability of competing products. If a product isn't moving, the retailer isn't likely to be impressed by a salesperson's request for more in-store attention or added shelf space.

Reorders are straight rebuys

Retailers and wholesalers usually carry a large number of products. A drug wholesaler, for example, may carry up to 125,000 products. Because they deal with so many products, most intermediaries buy their products on a routine, automatic reorder

Retailers see themselves as buying agents for their target customers—but they also have to determine what they can profitably sell. Retailers are attracted to well-known brands such as Dove because they know their consumers are more likely to buy brands they know. Chiquita wants to remind convenience stores that selling bananas can bring profits.
(left): Source: Mars, Incorporated; (right): Source: Chiquita Brands L.L.C.

basis—straight rebuys—once they make the initial decision to stock specific items. Automatic computer ordering is a natural outgrowth of computerized checkout systems. Sellers to these markets must understand the size of the buyer's job and have something useful to say and do when they call.

The Government Market

Size and diversity

Some marketers ignore the government market because they think that government red tape is more trouble than it's worth. They probably don't realize how big the government market really is. Government is the largest customer group in many countries—including the United States. About 30 percent of the U.S. gross domestic product is spent by various government units. Different government units buy almost every kind of product. They run not only schools, police departments, and military organizations, but also supermarkets, public utilities, research laboratories, offices, hospitals, and even liquor stores. These huge government expenditures cannot be ignored by an aggressive marketing manager.

Competitive bids may be required

Government buyers in the United States are expected to spend money wisely—in the public interest—so their purchases are usually subject to much public review. To avoid charges of favoritism, most government customers buy by specification using a competitive bidding procedure. Often the government buyer must accept the lowest bid that meets the specifications. You can see how important it is for the buyer to write precise and complete specifications. Otherwise, sellers may submit a bid that fits the specs but doesn't really match what is needed. By law, a government unit might have to accept the lowest bid—even for an unwanted product.

The approved supplier list

Specification and bidding difficulties aren't problems in all government orders. Items that are bought frequently—or for which there are widely accepted standards—are purchased routinely. The government unit simply places an order at a previously approved price. To share in this business, a supplier must be on the list of approved suppliers and agree on a price that will stay the same for a specific period—perhaps a year.[30]

The City of San Jose worked with a number of suppliers when it built the Lake Cunningham Regional Skate Park. Among the city's vendors for this project:
1. California Skateparks of Upland, California, used 800 cubic yards of a special spray concrete mixture to create the park's bowls, halfpipes, and walls.
2. Robert A. Bothman, Inc., from San Jose, California, created planters that were filled with wildflowers, native grasses, and palms to give the park more color and shade.
3. Musco Lighting from Oskaloosa, Iowa, lit the park with 50-foot poles, each bearing three 1,500-watt fixtures.

Courtesy of Dan Sparagna

Learning what government wants

In the United States, there are about 90,000 local government units (school districts, cities, counties, and states) as well as many federal agencies that make purchases. Some firms sell directly to government agencies; others work with prime contractors that resell to the government. Both are good opportunities for many businesses. Small and medium-sized businesses often find it easier to sell to prime contractors rather than win a contract with a large government agency. Potential suppliers should focus on the government units or prime contractors they want to cater to and learn the bidding methods of those units. Then they should monitor websites where government contracts are advertised. Target marketing can make a big contribution here—making sure the marketing mixes are well matched with the different bid procedures.

Because government agencies want to promote competition for their business, marketers are provided with a lot of information—in print form and online. The U.S. government has a central source, and USA.gov is a great place to start. There are also resources at FedBizOpps.gov (www.fbo.gov), including videos, publications, and a search tool to find opportunities to fit a firm's strategy. The General Services Administration handles vendor contracts for off-the-shelf goods and services; information for vendors is available at www.gsa.gov. Every federal government agency has a separate office devoted to small businesses with direct links to prime contractors. Various state and local governments also offer guidance, as do government units in other countries.[31]

Source: USA.gov

Dealing with foreign governments

Foreign governments offer a good opportunity for some organizations. But sellers should recognize that selling to government units in foreign countries can be a real challenge, one that should be approached with caution. In many cases, a firm must get permission from the government in its own country to sell to a foreign government. Moreover, most government contracts favor domestic suppliers if they are available. Even if such favoritism is not explicit, public sentiment may make it very difficult for a foreign competitor to get a contract. Or the government bureaucracy may simply bury a foreign supplier in so much red tape that there's no way to win.

Is it unethical to "buy help"?

In some countries, government officials expect small payments (grease money) just to speed up processing of routine paperwork, inspections, or decisions from the local bureaucracy. Outright influence peddling—where government officials or their friends request bribe money to sway a purchase decision—is common in some markets. In the past, marketers from some countries have looked at such bribes as a cost of doing business. However, the **Foreign Corrupt Practices Act**, passed by the U.S. Congress in 1977, prohibits U.S. firms from paying bribes to foreign officials. A person who pays bribes, or authorizes an agent to pay them, can face stiff penalties. However, the law was amended in 1988 to allow small grease money payments if they are customary in a local culture. Since 1998, the law applies to foreign firms or foreign individuals who accept payments while in the United States. In recent years, prosecutions under this statute have increased thanks to more rigorous enforcement and greater cooperation with foreign governments. For example, in 2012 more than 80 firms were under investigation for violating the Foreign Corrupt Practices Act. This included Wal-Mart Stores, Inc., which was alleged to have bribed officials in Mexico to quickly obtain permits to open stores. The Sarbanes-Oxley Act of 2002 makes individual executives responsible for their company's financial disclosures; and a bribe mischaracterized as a legitimate expense may violate the law. Managers need to be careful and up front about such payments.[32]

CONCLUSION

In this chapter we examined organizational buying and how it differs from final consumer buying. We saw that organizational buyers rely heavily on economic factors and cost-benefit analysis to make purchase decisions. Organizational buyers generally prefer to avoid risk. This chapter showed how multiple influences are important in buying decisions—and how marketing managers must recognize and attend to the needs of all members of the buying center.

The chapter introduced a three-step model of business and organizational buying. The first step—defining the problem—described how buyers recognize problems and then find solutions by determining buying needs. The second step—the decision-making process—examined three organizational buying processes: new-task buying, modified rebuy, and the straight rebuy. Each process involves different levels of information gathering, and this section reviewed how buyers gather information for buying. The third step—managing the buyer-seller relationship—looked at different aspects of these relationships in practice. Buying behavior and marketing opportunities may change when there's a close relationship between a supplier and a customer. There are different ways that a supplier can build close relationships with its customers. We identified key dimensions of relationships and their benefits and limitations.

We saw that organizational buyers buy for resale or to produce other goods and services—this group includes manufacturers, farmers, distributors, retailers, government agencies, and nonprofit organizations. The chapter concluded by providing insights about buying practices particular to manufacturers, service firms, intermediaries, and governments.

Understanding how organizations buy can help marketing managers identify logical dimensions for segmenting markets and developing marketing mixes. The unique nature of each product will likely require further adjustments in the mix. Different product classes are discussed in Chapter 8. Variations by product may provide additional segmenting dimensions to help a marketing manager fine-tune a marketing strategy.

KEY TERMS

LO 6.8

business and organizational customers, 147
purchasing managers, 152
multiple buying influence, 153
buying center, 153
requisition, 154
purchasing specifications, 155
new-task buying, 155
straight rebuy, 155
modified rebuy, 155
white paper, 157
competitive bid, 158
vendor analysis, 158
just-in-time delivery, 161
negotiated contract buying, 162
outsource, 162
North American Industry Classification System (NAICS) codes, 164
Foreign Corrupt Practices Act, 169

QUESTIONS AND PROBLEMS

1. Review the John Deere case study that opens the chapter. From this case, identify examples of different key terms and concepts covered in the chapter. For example, when MetoKote built a paint plant right next door to John Deere's manufacturing plant, this was an example of a relationship-specific adaptation.

2. Review the John Deere case study that opens this chapter. Applying concepts from the chapter, what else could be done to build a more effective buyer-supplier relationship with MetoKote? How should John Deere manage a relationship with a supplier of cleaning supplies for its manufacturing plant? Offer some specific recommendations based on what you learned in this chapter.

3. Compare and contrast the buying behavior of final consumers and organizational buyers. In what ways are they most similar and in what ways are they most different?

4. Briefly discuss why a marketing manager should think about who is likely to be involved in the buying center for a particular purchase. Is the buying center idea useful in consumer buying? Explain your answer.

5. If a nonprofit hospital were planning to buy expensive MRI scanning equipment (to detect tumors), who might be involved in the buying center? Explain your answer and describe the types of influence that different people might have.

6. Describe the situations that would lead to the use of the three different buying processes for a particular product—a manufacturer of bicycles that will be purchasing handlebars for a mountain bike.

7. How likely is it that each of the following would use competitive bids: (a) a small town that needs a road resurfaced, (b) a scouting organization that needs a printer to print its scouting handbook, (c) a hardware retailer that wants to add a new lawn mower line, (d) a grocery store chain that wants to install new checkout scanners, and (e) a sorority that wants to buy a computer to keep track of member dues? Explain your answers.

8. Discuss the advantages and disadvantages of just-in-time supply relationships from an organizational buyer's point of view. Are the advantages and disadvantages merely reversed from the seller's point of view?

9. Explain why a customer might be willing to work more cooperatively with a small number of suppliers rather than pitting suppliers in a competition against one another. Give an example that illustrates your points.

10. Would a tool manufacturer need a different marketing strategy for a big retail chain such as Home Depot than for a single hardware store run by its owner? Discuss your answer.

11. Discuss the importance of target marketing when analyzing organizational markets. How easy is it to isolate homogeneous market segments in these markets?

12. Explain how NAICS codes might be helpful in evaluating and understanding business markets. Give an example.

13. Considering the nature of retail buying, outline the basic ingredients of promotion to retail buyers. Does it make any difference what kinds of products are involved? Are any other factors relevant?

14. The government market is extremely large, yet it is often slighted or even ignored by many firms. Red tape is certainly one reason, but there are others. Discuss the challenges and opportunities in selling to the government market and be sure to include the possibility of segmenting in your analysis.

MARKETING PLANNING FOR HILLSIDE VETERINARY CLINIC

Appendix D (the Appendices follow Chapter 19) includes a sample marketing plan for Hillside Veterinary Clinic. Hillside decided to focus on final consumers and their pets rather than include organizational customers who might need veterinary care for animals. Such customers might range from dog breeders and farmers to animal protection shelters and law enforcement agencies who work with dogs. Would it be easy or hard for Hillside to expand its focus to serve customers who are not final consumers? Explain your thinking.

SUGGESTED CASES

5. Resin Dynamics
6. Dynamic Steel
14. Schrock & Oh Design
21. GeoTron International (GTI)

MARKETING ANALYTICS: DATA TO KNOWLEDGE

CHAPTER 6: VENDOR ANALYSIS

CompuTech Inc. makes circuit boards for personal computers. It is evaluating two possible suppliers of electronic memory chips.

The chips do the same job. Although manufacturing quality has been improving, some chips are always defective. Both suppliers will replace defective chips, but the only practical way to test for a defective chip is to assemble a circuit board and "burn it in" (run it and see if it works). When one chip on a board is defective at that point, it costs $2 for the extra labor time to replace it. Supplier 1 guarantees a chip failure rate of not more than 1 per 100 (that is, a defect rate of 1 percent). The second supplier's 2 percent defective rate is higher, but its price is lower.

Supplier 1 has been able to improve its quality because it uses a heavier plastic case to hold the chip. The only disadvantage of the heavier case is that it requires CompuTech to use a connector that is somewhat more expensive.

Transportation costs are added to the price quoted by either supplier, but Supplier 2 is farther away so transportation costs are higher. And because of the distance, delays in supplies reaching CompuTech are sometimes a problem. To ensure that a sufficient supply is on hand to keep production going, CompuTech must maintain a backup inventory—and this increases inventory costs. CompuTech figures inventory costs—the expenses of finance and storage—as a percentage of the total order cost.

To make its vendor analysis easier, CompuTech's purchasing agent has entered data about the two suppliers on a spreadsheet. He based his estimates on the quantity he thinks he will need over a full year.

Design element: #M4BW box globe icon: ©Vectoryzen/Shutterstock

CHAPTER SEVEN

Bill Greene/The Boston Globe/Getty Images

Improving Decisions with Marketing Information

In 1946, Bill Rosenberg launched Industrial Luncheon Services, delivering meals, snacks, and coffee to factory workers in the Boston area. From his daily sales totals, Rosenberg knew his working-class customers loved his coffee and doughnuts. Rosenberg thought a sit-down restaurant with these two menu items could be a success, and a few years later, Dunkin' Donuts was born. Back then Dunkin' charged just 5 cents for one of its 52 varieties of doughnuts and 10 cents for a cup of coffee.

Over time, Dunkin' expanded its menu. Adding flavored coffees, lattes, Chai, iced coffees, and other beverages put it in direct competition with Starbucks. At the time, most of Dunkin's stores were in the northeastern United States, and it saw a big opportunity for growth across the United States and internationally. Dunkin's marketing managers wondered if they should change the company's marketing strategy to better fit evolving customer behavior, changing competition, and new target markets.

Dunkin' managers knew they had many options: for example, adding new sandwiches, offering catering and delivery services, and providing cozier seating. They hired Copernicus Marketing to conduct research and help design Dunkin's "store of the future." Product design software evaluated more than 2 billion combinations by varying portion sizes, exterior store design, interior music selection, and more. With data from a nationally representative sample of more than 1,000 customers and prospects, sales and costs were forecast for each combination, and the most profitable options were identified. The decision support software suggested a marketing mix that emphasized Dunkin's quality coffee, speedy service at the counter, and drive-through windows would best deliver sales—and a new strategy was born.

As Dunkin' Donuts grew, it sought to better understand its loyal coffee drinkers—and those of its chief rival, Starbucks. Dunkin' paid dozens of its most dedicated customers to buy coffee at Starbucks, while simultaneously paying a similar number of Starbucks loyalists to come to Dunkin' Donuts. After debriefing interviews, the two groups were found to be so different that Dunkin' researchers dubbed them "tribes."

What each tribe detested about its rival's store was exactly what made it love its usual outlet. For example, Starbucks' regulars found Dunkin' outlets boring, austere, and unoriginal. They didn't like that workers dumped standard amounts of milk and sugar in their drinks; they didn't feel special at Dunkin' Donuts. Although Dunkin' tribe members wanted newer-looking stores, the Starbucks experience turned them off. All those laptop users made it hard to find a seat—and they wondered why coffee shops needed couches. They complained about Starbucks' higher prices and slower service. They didn't like Starbucks' "tall," "grande," and "venti" lingo; just give us "small," "medium," and "large" please! This exercise convinced Dunkin' that there were customers out there who wanted an alternative to Starbucks—and Dunkin' could fine-tune a marketing strategy to provide it.

A psychographic survey offered further insight into the attitudes, values, and interests of Dunkin' tribe members. They're busy, love routine, prefer no frills, and see themselves as down-to-earth folks. While one-third of Americans fit this profile, these people are more common in the Midwest and southern United States. So Dunkin' focused its new store openings in these areas. The research also guided the "America Runs on Dunkin'" advertising campaign featuring office and construction workers getting through their days with the chain's help.

Before Dunkin' Donuts adds new items to its menu, it tests the ideas with customers. Dunkin' chefs used focus groups to test a new line of hearty snacks for drive-through customers looking for an on-the-go snack. They found these customers liked the smoothies and hot flatbreads, but the tiny stuffed pinwheels did not satisfy their hunger. When Dunkin' came back with larger-sized "bites" filled with pork and other ingredients, customers approved, and new items appeared on Dunkin's menus.

Dunkin' encourages customer feedback and listens closely for ideas to fine-tune its marketing strategy. A group of its best customers serve on the Dunkin' Advisory Panel, where they regularly complete surveys and participate in online focus groups. The company also monitors its website and social media efforts, reading what Facebook fans and Twitter followers write and using analytical software to monitor Dunkin's buzz on the Internet.

Data even drive the precise location of each Dunkin' shop; a software program analyzes demographics, competition, and traffic patterns at the neighborhood level. For one store, this program predicted sales would increase by moving the store just 100 yards—from one end of a strip mall to the other—and adding a drive-through lane. After the move, sales jumped more than 50 percent.

When Dunkin' wondered if the costs and discounts of a loyalty program would be offset by higher sales and profits, it conducted an experiment. A group of customers' purchases were monitored before and after they joined a prototype Dunkin' Donuts loyalty program. Customers in the program spent 40 percent more. Decision made. Now millions use the DD Perks app, which also gives Dunkin' data on individual customers.

It shouldn't be a surprise that research guides Dunkin's sustainability and corporate social responsibility (CSR) efforts. Recent CSR strategies were developed after reviews of internal reports, interviews with senior managers, and surveys of employees and customers. The research identified what mattered most to internal and external stakeholders and how Dunkin' could have a positive impact. High-priority topics focused Dunkin's efforts on sustainable and ethical sourcing, security of customer data, and packaging. It led to decisions to trade out polystyrene for eco-friendly double-wall paper cups and to collaborate to support coffee sustainability.

Research was also needed when Dunkin' Donuts considered changing its name to Dunkin'. Marketing managers wanted to simplify branding, signal a focus on beverages, and broaden its appeal to Millennials. Dunkin' tested the new name in a few locations and monitored customer reactions. When Dunkin' finally took the plunge, research by YouGov suggested it might be working. YouGov regularly surveys consumers about 1,600 brands, asking if they have "heard something positive about a brand and talked about the brand with friends and family in the past two weeks." As Dunkin's new branding was launched, YouGov reported that Dunkin' posted the biggest increase in positive word-of-mouth among Millennials.

Dunkin' now has more than 11,500 stores in 41 countries. To keep growing, Dunkin' knows it has to keep collecting, analyzing, and acting on data for its marketing strategy planning, implementation, and control.[1]

LEARNING OBJECTIVES

The Dunkin' case shows that successful marketing strategies require information about potential target markets and their likely responses to marketing mixes—as well as information about competitors and other market environment variables. Managers also need information for implementation and control. Without good information, managers are left to guess—and in today's fast-changing markets, that invites failure. In this chapter, we discuss how marketing managers get the information that they need.

When you finish this chapter, you should be able to

1. discuss marketing information systems.
2. understand the scientific approach to marketing research.
3. cite methods for collecting secondary and primary data.
4. understand the role of observing, questioning, and using experimental methods in marketing research.
5. understand the challenges to interpreting marketing research data.
6. understand important new terms (shown in **red**).

Effective Marketing Requires Good Information

LO 7.1

Marketing managers have questions

To make good marketing decisions, managers need accurate information about what is happening in the market. They usually can't get all of the information they'd like, but part of their job is to find cost-effective ways to get the information that is imperative.

Marketing managers look to marketing research to help answer at least four types of questions. First, they are often interested in knowing *what value* a marketing mix can deliver and *to whom?* The answers to these questions inform decisions about market segmentation and the financial return to the company. For example, smartphone maker OnePlus' customer research identified battery life as an important attribute among its customers in India. Many people in India have long commutes and work jobs without nearby electrical outlets. Offering longer battery life than the iPhone helped OnePlus sell more phones than its more well-known competitor.[2]

Second, they need to know *which measures matter?* Marketing managers can track a wide range of data, so they often want to find out which measures are most closely tied to desired outcomes, like higher awareness or increased sales. For example, should a marketing manager for Taco Bell be more concerned with Instagram likes or brand awareness? A third category of questions follows from the question—*what works?* This category of questions answers which marketing strategy decisions are working and which are not. This allows a marketing manager for Frito-Lay to evaluate the effectiveness of a recent advertising campaign or whether in-store promotions for the new Cheetos Flamin' Hot Asteroids were worth the cost.

Finally, marketing managers often want to know *what if?* For example, what could happen if prices are raised (or lowered), if a new store is opened, or if a new product line is launched. These questions help a marketing manager make informed choices about potential strategies.[3]

Information is a bridge to the market

In this chapter, we'll focus on the two resources that marketing managers utilize for information to help them answer these questions and make better decisions. One source is **marketing research**—procedures that develop and analyze new information about a market. Marketing research involves a wide range of techniques, including Internet searches, customer surveys, experiments, direct observation of customers, and many more. Marketing managers often find that one-at-a-time marketing research projects can be too costly or take too long to get the desired information. So, in many companies marketing managers also routinely get help from a **marketing information system (MIS)**—which is an organized way of continually gathering, accessing, and analyzing information that marketing managers need to make ongoing decisions.

Marketing managers may need marketing research, an MIS, or a combination of both to get the information they need to make decisions during any step in the marketing strategy planning process—or to improve implementation and control (see Exhibit 7-1). In this chapter, we'll discuss ways to make marketing research and an MIS more useful, and the key issues that marketing managers face in using them.

Big companies turn to marketing research specialists

Most large companies have a separate marketing research department to plan and manage research projects. People in these departments usually rely on outside specialists to carry out the work on particular projects. Further, they may call in specialized marketing consultants and marketing research organizations to take charge of a whole project.

Smaller companies usually don't have separate marketing research departments. They depend on their salespeople or managers to conduct what research they do. In the past, nonprofit organizations rarely did marketing research, but more have begun to see its value and often use outside specialists.

IT and marketing cooperate to create an MIS

When it comes to setting up an MIS, a company usually turns to information technology (IT) experts. Sometimes this expertise resides in an organization's IT department. Even small firms may have a person who handles most of the technical work on its computer systems. Many small and large firms look to outside consultants and service providers for help with increasingly complex information systems.

Whether these activities are conducted inside the company or are outsourced, it is very important that marketing managers be closely involved in the design of marketing research or an MIS. Specialists in research and IT can make sure that the technical aspects are handled correctly, but marketing managers need to make sure they can access information to guide marketing strategy planning or implementation and control.

Exhibit 7-1 Marketing Information Inputs to Marketing Strategy Planning Decisions

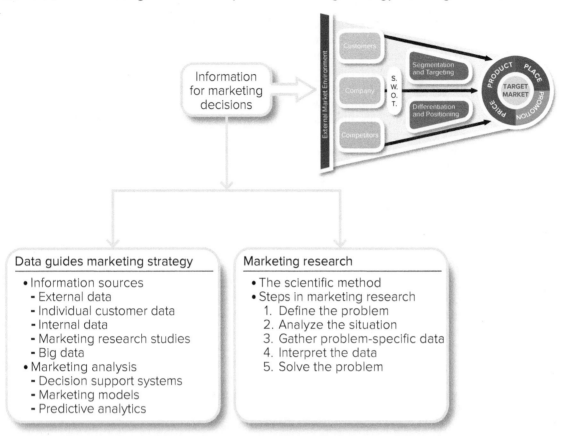

Data guides marketing strategy
- Information sources
  - External data
  - Individual customer data
  - Internal data
  - Marketing research studies
  - Big data
- Marketing analysis
  - Decision support systems
  - Marketing models
  - Predictive analytics

Marketing research
- The scientific method
- Steps in marketing research
  1. Define the problem
  2. Analyze the situation
  3. Gather problem-specific data
  4. Interpret the data
  5. Solve the problem

Changes Are Under Way in Marketing Information Systems

Marketing managers for some companies make decisions based heavily on their own judgment—with very little hard data. When it's time to make a decision, they may wish they had more information. But by then it's too late, so they do without. Marketing managers with ready access to appropriate and timely information make better marketing strategy decisions. Modern MIS practices, especially collecting data on individual customers, raise important legal and ethical questions about customers' privacy, which we discuss in detail at the end of this section.

An MIS makes information available and accessible

Many firms realize that it doesn't pay to wait until you have important questions you can't answer. They anticipate the information they will need. They work to develop a continual flow of information that is available and quickly accessible from an MIS when it's needed.

We won't cover all of the technical details of planning for an MIS, but you should understand what an MIS is so you know some of the possibilities. Exhibit 7-2 shows that a good MIS stores data collected from inside and outside the company and from various marketing research studies. The data can be analyzed to guide the marketing strategy planning process.

Internal data are already collected

An MIS should have a wide range of internal data readily accessible to marketing managers. For accounting, tax, billing, and production planning, data on sales or cost information for different products, in specific geographic areas, by customers and more are already in a company's computer system. This kind of information is required and

Exhibit 7-2 Elements of a Marketing Information System

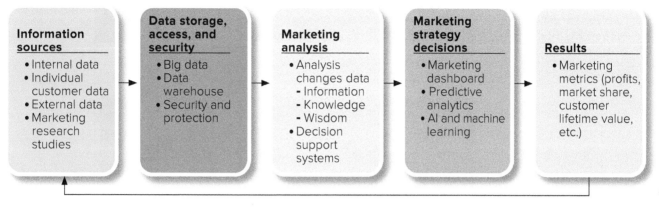

common. A company can also collect data on visits to its website, from order placement to delivery, prices, and more. While these data are internal, they must be tied to the MIS to give the marketing managers access.

A customer relationship management (CRM) system (see Chapter 4) includes customer information tied to specific customers. This can help place customers into segments or even allow for customized marketing mixes.

Gather external data on a routine basis

An MIS should also collect external information from the marketplace. This information can include competitors' new product announcements or pricing changes and articles about customers. There may be information on industry growth trends and relevant legal and regulatory issues. New types of data are becoming available with advances in technology. For example, many firms collect and store large amounts of data generated from inexpensive digital sensors placed on industrial equipment, automobiles, and shipping crates. Other firms store and track all kinds of activity from each visit to a firm's website. Marketing managers should also have a system in place to routinely scan the external market environment, customers, and competitors to ensure that important information is available when needed for marketing strategy planning.

Marketing research studies focus on problems

The third source of data comes from a company's previous marketing research studies. In the second half of this chapter, we dig deeper into various kinds of marketing research and the application of each. For now, it helps to know that marketing research studies usually start with a particular problem a manager is trying to solve. Past studies should be "on file" so that a manager can find out if there is information to help make a decision. For example, if a study was conducted two years ago to gauge customer responses to a price increase, that information should be available to a marketing manager considering a price change today. Although not exactly the same situation, the information might be useful enough to guide a strategy decision.

The big data explosion

Advances in computing power and the Internet have created an explosion of data for today's businesses. Estimates suggest that organizations process more than 1,000 times as much data today as they did in 2000. The explosion of data comes as organizations collect and store more information from internal sources, external sources, and marketing research studies. At the same time that the amount of data has grown, the cost of storing data has plummeted, making it easier for companies to save almost all data for possible later use. This explosion is often referred to as **big data**—data sets too large and complex to work with typical database management tools.[4]

Big data's complexity starts with the Four Vs

Big data typically are collected and stored with the idea of learning more about customers and better serving up ads, offers, or products that capture their interest. For

Many firms look to outside experts, like Adobe, to help them analyze and draw insights from big data.
Source: Adobe Systems Incorporated

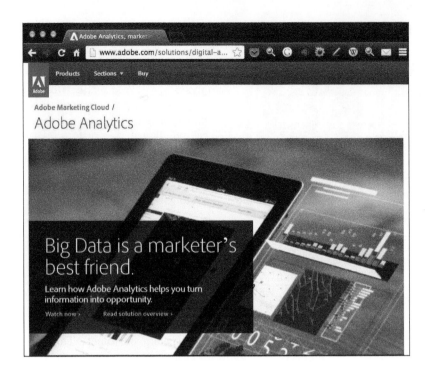

marketing managers to use the data, they need to recognize their opportunities and complexity. Thus, it helps to understand big data's four dimensions—the Four Vs (see Exhibit 7-3).

The Four Vs reflect the challenges for information systems managers (in charge of storing and facilitating access to data) and opportunities for marketing managers who must decide what information can be stored, how quickly it can be analyzed, and how it can be utilized in marketing strategy planning. We have already referred to the first V, *volume*, which refers to the quantity of data generated and stored.[5]

Big data's variety creates opportunity and challenge

Marketing managers can also collect a wide *variety* of data. These days data are more than just sales numbers. For example, Coca-Cola takes data from webcams that monitor consumers' facial expressions while watching its ads. Other companies have machines that analyze photos on sites such as Instagram. From photos, software identifies social media users' emotions, the activity they are engaged in, logos on their clothing, and products they might be carrying. New wearable technology (such as the Fitbit or Apple Watch) collects health data. It is estimated that 30 billion pieces of content (such as articles, photos, songs, and emotions) are shared on Facebook each month. More than 400 million tweets are sent each day on Twitter. This wide variety makes it difficult to analyze but also offers a new opportunity for marketing managers to learn more about customers, competitors, and the external market environment.[6]

Exhibit 7-3 The Four Vs Dimensions of Big Data

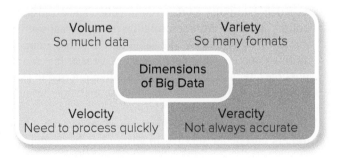

Velocity creates opportunity if firms can act fast

Big data *velocity* refers to the speed at which data are generated and can be processed. Velocity is important for marketing managers seeking to do predictive analytics (see Chapter 5) and trying to anticipate what customers might be interested in buying when using real-time data. So, for example, if a customer conducts a Google search for "Wilson tennis racket" and then looks at a particular model and reads reviews from other customers, artificial intelligence could instantaneously infer the customer is interested in this model of tennis racket. Based on the shopper's previous shopping behavior, Wilson might want to offer the customer a coupon to "buy now." The ability to make this offer when a customer is actually shopping depends on velocity. Can the company gather and analyze these data in real time, so that they can make the customer a relevant offer? The data are there if a company can act quickly enough.

Veracity—can we trust big data?

Finally, many managers question the *veracity* or quality of the data. How accurate can so much data be? Which data can be relied on, and which are more suspect? According to one study, about one in three business leaders don't trust making decisions with this information. If they do not trust this information, will they use it to make marketing strategy decisions?

Some marketing managers will blindly follow the numbers from big data; they forget their intuition. This may be a useful time to employ a caution from a wise sociologist: *"Not everything that counts can be counted, and not everything that can be counted counts."*[7] While written more than 50 years ago, it seems particularly relevant with respect to big data. We have seen that the volume, variety, and velocity of big data create a lot of "countable" data. Marketing managers should remember that their own intuition and wisdom are important, too—and that at least some of the data they collected may not be. A good marketing manager should sort through all those data and determine what matters most.

Store data so they're easy to access

An MIS organizes all of the incoming data and information in a **data warehouse**—a place where databases are stored so that they are available when needed. You can think of a data warehouse as an electronic library, where all of the information can be searched with some sort of search engine. With big data, firms have had to build larger data warehouses and find new ways to store, index, and access a wide variety of data.[8]

Data security matters

Storing data—especially data on individual customers—is a responsibility that a company must take seriously. While many of the technical aspects of protecting data fall on the information technology staff, marketing managers need to recognize the harm that may come from data breeches. A firm's reputation can be severely damaged when a hacker accesses its customers' personal data. High-profile data breeches at Uber, Target, Under Armour, Panera Bread, and Saks Fifth Avenue did not engender customer trust in those firms. Companies should also be wary of internal threats to data security. Amazon investigated employees who were leaking data in exchange for bribes.[9]

New software, like that sold by SAS, helps marketing managers analyze complex "big data."
Source: SAS Institute Inc.

Analysis changes data to information, knowledge, and wisdom

All those *data* in raw form are not useful to marketing managers. An MIS therefore includes software programs that analyze raw data and convert them to something managers can use to make marketing strategy decisions. Data become *information* when they provide answers to questions of "who," "what," "where," "how much," and "when." So, for example, sales data might be presented in a table that shows the sales of different products over time and in different geographic locations, answering questions such as: *Who is buying our product? Where do they live? How much do they buy? What products are they buying?*[10]

Information becomes *knowledge* when it helps marketing managers answer "how" and "why" questions. Marketing managers combine experience with information to generate knowledge. So, a marketing manager may observe growth from a particular target market (information from looking at sales data) but combine this with experience (or a marketing research study) to find out why the target customers purchased more. Other questions that require knowledge include: *How did our largest competitor gain access to an international market? Why did that Facebook post generate so many comments? Why did a competitor sell its product through grocery stores?*

Wisdom involves an ability to accurately predict the future. An experienced marketing manager may gain wisdom over time. For example, a marketing manager at Frito-Lay might have *information* indicating that customers in the northeast United States are not satisfied with a recently launched snack food. This might cause the marketing manager to invest in marketing research to find out the reasons (why) for the dissatisfaction—generating *knowledge*. As marketing managers gain experience and understanding of why things happened in the past, they can put that together to build *wisdom* and better predict changes in a marketing strategy to turn things around. Along the same lines, artificial intelligence and software can "learn" how customers respond to different Facebook ads and automatically show future customers an ad that has proven more effective. The software uses past consumer behavior to predict future behavior. Exhibit 7-4 summarizes this move from data to wisdom.

Decision support systems can convert information to knowledge

With so many data available, marketing managers often look to software programs for aid in interpreting what they find in the MIS. It can be helpful to have a **decision support system (DSS)**—a computer program that makes it easy for a marketing manager to obtain and use information. A DSS helps marketing managers convert data into information and knowledge that allow them to make informed choices about marketing strategy. Many decision support systems blur the line between marketing analysis and marketing strategy decisions—as they can be set up to automate and implement strategy decisions based on analysis. Let's look at a few types of decision support systems.

Exhibit 7-4 Analysis Changes Data to Information, Knowledge, and Wisdom

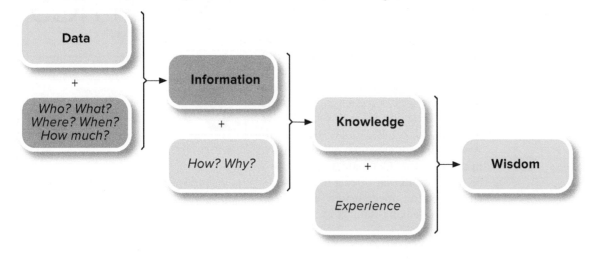

Marketing dashboards display up-to-the-minute marketing information in an easy-to-read format. The marketing manager chooses the information most helpful to decision making. In this dashboard from TechnologyAdvice, the marketing manager can track visits to a website, view how customer engagement is influencing sales, and see how digital channels contribute to web traffic. This information can help the marketing manager fine-tune a marketing strategy.
Source: TechnologyAdvice

Marketing dashboards monitor the market

One way to interface with a data warehouse and a DSS is through a **marketing dashboard**, which displays up-to-the-minute marketing information in an easy-to-read format—much like a car's dashboard shows the speedometer and fuel gauge. A marketing dashboard can be constructed to show data and provide analysis. Marketing dashboards are usually customized to a manager's areas of responsibility. For example, a dashboard for a customer service manager at Verizon Wireless might show the percentage of customer calls dropped by cell towers, the location of repair trucks in the field, and the number of callers "on hold" waiting for customer service help. With early warning about potential problems, the manager can quickly make corrections. For example, a manager might call in extra customer service help if too many customers are on hold.[11]

Marketing models help with marketing strategy planning

Some decision support systems go even further, helping marketing managers improve all aspects of their planning—blending individual Ps, combining the Four Ps into mixes, developing and selecting plans. Further, they can monitor the implementation of current plans, comparing results against plans and making necessary changes more quickly. Drawing from data on the demographics in a geographic area, customers' online behavior, or how a marketing strategy worked in the past, a DSS can make estimates using a marketing model. A **marketing model** is a statement of relationships among marketing variables. It enables a manager to look at the sales (and costs) expected with different types of promotion and select the marketing mix that is best for a particular target market.[12]

AI and machine learning can generate information and implement strategies

As we learned in Chapters 3 and 5, data, artificial intelligence, and software can be used for predictive analytics—potentially indicating when a particular customer is interested in making a purchase. So, for example, Subaru's data warehouse contains information about a particular customer who leases a Forrester with a lease that will expire in three months. It may also gather data on the customer's online search behavior and have data showing searches for a "Subaru Outback" and "Toyota Highlander." Subaru's marketing analytics software predicts the customer is in the market to buy a new car and may be considering a Subaru and a competitor's vehicle. Marketing strategies can target this particular customer—for example, sending an e-mail with a discount on a new Outback to spur action. This type of precise targeting can be automated with artificial intelligence software that recognizes specific types of opportunities based on data analysis. Machine learning may occur if the software tries different tactics—in this case it might "learn" that e-mailing a promotional offer to a customer is most effective 40 days before the lease expires. This learning occurs after trying to mail offers earlier and later than this timing and discovering the most appropriate lead time. Similar learning may occur around the type of offer or wording of an offer. In this case, the software does not know why the time frame works—no knowledge is generated—but it does know what works best and creates a marketing strategy based on that.

Predictive analytics and marketing strategy

Marketing models can provide predictive analytics by forecasting customer behavior. A model can predict what will happen if a new product is introduced, an advertisement is run, or a price is lowered. For example, auto parts retailer AutoZone uses marketing models to optimize the product mix at *each* of its 5,000 stores. Although AutoZone offers more than a million items, each store can hold only about 40,000, so it helps to anticipate what items customers will be looking for when they come in. AutoZone supplements sales histories at each outlet with external data. For example, it gets vehicle registration data by zip code to determine the year and model of cars being driven in the local areas around each outlet. When all these data are plugged into a marketing model, AutoZone has the wisdom to predict customer demand at each store, and it adjusts its product mix to better meet local customer needs.[13]

Although data drive many marketing strategy decisions, one might not expect to be able to predict a song's success. But that is just what some music industry people are doing. To learn more, check out *What's Next?* Big data and predictive analytics "hear" the future.

Knowledge and wisdom guide planning, implementation, and control

We do not want to overstate the use of marketing technology. While these practices are becoming more common, most marketing managers do not have this luxury. These marketing managers analyze data and information using spreadsheet software. This gives managers knowledge that helps them make marketing strategy planning decisions. The concepts driving marketing strategy planning do not change, and, in many situations, marketing managers can provide insights superior to various models and predictive analytics.

Analysis guides planning and drives results

The purpose of collecting, storing, and analyzing data is to guide the strategic planning process. After implementing a marketing strategy, marketing managers should make sure the strategy and the results are also fed back into the data warehouse. This will help future marketing managers facing a similar problem to learn from the company's previous experience.

An MIS raises questions about customers' privacy

So far, we have seen that with an MIS, a firm collects, stores, and analyzes data to assist marketing strategy planning. We also noted that many firms collect data on individual customers so they can do more precise targeting, focus on a segment of one, and use dynamic behavioral targeting (see more in Chapter 4). We briefly offered examples of how firms utilize data and information on individual customers to deliver customized products and services, promotional messages, web pages, prices, and more.

The use of customers' **personal data**, information that can be used by itself or in combination with other information to identify someone, raises questions about privacy. One definition of *privacy* is simply the right to be left alone. With respect to the data and information that may be gathered and stored in an MIS, we more specifically consider **informational privacy**, anything that limits others' access to personal data

What's Next? Big data and predictive analytics "hear" the future

Forecasting the success of a new product can be a dicey proposition, but it's particularly challenging to anticipate the fickle tastes of music junkies. Traditionally, the music industry relied on the intuition of record label executives. These music veterans traveled the country listening to acts in bars and clubs trying to find the next Beyoncé, Taylor Swift, or Ed Sheeran. Record labels are motivated to find that next superstar—the top 1 percent of musicians earn about three-quarters of all recorded music revenue. With so much of music being listened to online, data can be easily compiled and analyzed. Big data and predictive analytics can offer the music industry a peek at *What's Next?*

One company bringing data to this quest is the smartphone app Shazam. Shazam listens to a brief sample of an unfamiliar song, compares the clip to a large database, and tells the user the song's title and artist. While users learn something new, so do music executives. These data might predict the next big hit. With more than 20 million searches each day, Shazam gives a pretty good picture of what people don't yet recognize but know they like.

Shazam can also help a record label decide if it really has a hit on its hands. For example, Republic Records thought it might have a future star in SoMo, and his first hit could be "Ride." So it tested the song in the small (population 63,000) town of Victoria, Texas, where just one radio station played "Ride." Soon it was the most tagged song on Shazam in Victoria, and the company knew it had a hit. With added promotion, "Ride" reached number 20 on U.S. R&B charts. With Shazam data, record labels have a better idea about how to improve returns on their promotion investments.

Another music-related data collection venture, Next Big Sound (NBS), evolved from a student project in a Northwestern University entrepreneurship class. NBS scours the web for all kinds of music data—the number of plays a song has on Pandora, its digital downloads, and the number of new Facebook likes or Twitter mentions of an artist. NBS compiles these data and creates a digital dashboard for its clients. Years before they became household names, NBS data predicted stardom for Iggy Azalea, A$AP Rocky, and Macklemore & Ryan Lewis. NBS' Find mines data to identify obscure bands with breakout potential. Its patented "likelihood of success" algorithm predicts dollar sales better than traditional methods. NBS' tracking identifies the best promotional opportunities for new music. For example, NBS analysis showed that getting on Conan O'Brien's show offers a bigger boost than other late-night shows.

Big data and predictive analytics are being used across the music industry, from personalized radio to songwriting. Music-streaming services such as Pandora and Spotify use proprietary algorithms to determine which songs a listener hears next. Hit Song Science tries to predict hits. It analyzes a song for patterns in rhythm, chord progression, length, lyrics, and more, yielding a "hit score" from 1 to 10. Norah Jones is suspected of having used this service for her debut album *Come Away With Me*, which sold over 27 million copies. Some songwriters revise songs to get higher scores. Therein lies a concern of some in the music industry. Listeners already favor the familiar—"Today's Top Hits" and "Today's Hits" are the top stations on Spotify and Pandora, respectively.[14]

If songwriters and streaming services are copying what worked in the past, will unusual songs never get airplay? Are we likely to hear minor variations on what we already know and like? What would that mean for the future of creativity around music?

What are legal rights to privacy?

that people consider sensitive or confidential. As marketing managers gather personal data, they must consider the legal and ethical issues around informational privacy.[15]

As marketing managers seek to collect and utilize personal data for marketing strategy planning, they must be careful of overstepping and breaking the law. In the United States, as of this writing, the laws governing this area remain a patchwork of state and federal laws that cover specific data categories (financial and medical data, for example) or marketing practices (texting, telemarketing, personalized advertising). That said, some general themes emerge, in that firms should give *notice* and offer *choice*. Notice means that—even if it is given in the fine print on an agreement that almost no one reads—customers must be informed about how their personal data may be used. Given that notice, customers must be offered a choice about whether the firm adequately protects those data. The choice may be that they can choose not to use an app or visit a website.

These standards are lower and less clear than informational privacy laws in the European Union (EU), Canada, and Japan, which are each covered by their own comprehensive privacy law. Let's take a closer look at one of these laws, the European Union's **General Data Protection Regulation (GDPR)**, which is a set of laws on data protection and privacy for all individuals within the European Union. Because this law also protects EU citizens who may visit a website located in countries outside the EU, the law has implications for many firms located outside the EU.

While a complete review of the GDPR is beyond the scope of this book, a few cautions are warranted. Like U.S. laws, notice and choice are required. Beyond that, the GDPR expects firms gathering personal data to be user friendly so that it is clear what data are being collected and why. There are other restrictions, and marketing managers developing websites that may be used by EU citizens should consult an expert.[16]

Is it ethical for apps to share user data?

Ethical questions by their nature fall into a gray area. The vast majority of websites offer notice and choice and stay within the law. Questions are raised when an app or website goes beyond what a user expects. For example, some mobile phone apps were found to share customer data with Facebook—which then tied the information to a Facebook profile, allowing Facebook to do more precise targeting. For example, a real estate app shared which houses a user "saved." Another app shared the menstrual cycle information women entered into a health tracking app. This type of private data might be very useful for targeted promotion efforts. Advertising that targets the right customers at the right time will be more effective. Based on the examples above, a real estate agent could promote his services to people actively looking for homes in a certain area or a maker of tampons could promote feminine care products to women in a timely manner.[17]

How should we feel about these examples? Users might appreciate advertising that specifically targets their wants and needs at exactly the right time; they learn about goods and services that might help them solve problems. On the other hand, a user might not want a private company to have that specific, personal information. The Ethical Dilemma that follows gives you an opportunity to think about these types of situations.

Ethical Dilemma

What would you do? Imagine you are the founder of GamifyNow, a startup that makes simple games for playing on smartphones. GamifyNow is struggling financially and may not make it even two more months. You are approached by one of your programmers who tells you that she has identified a way for GamifyNow to make some extra money. She tells you a social media site approached her and offered to pay GamifyNow cash up front to add some programming code to each of its games. The extra cash would help the company get through the next six months, giving the company enough time to gain a foothold in the highly competitive casual gaming market. The new code would automatically send data from each user's phone back to the social media site. The data would include the names and phone numbers of all of a user's contacts as well as other data the user enters into other smartphone apps. Your company would only have to update its user agreement and it would all be legal. While you know that few users read your user agreement, you know that most probably wouldn't care anyway. They might even expect this because your games are free to download.[18]

Would you install this code on GamifyNow's games? Why or why not? Explain your thinking.

The Scientific Method and Marketing Research

LO 7.2

Marketing research—combined with the strategy planning framework we discussed in Chapter 2—can also help marketing managers make better decisions.

Marketing research is guided by the **scientific method**, a decision-making approach that focuses on being objective and orderly in *testing* ideas before accepting them. With the scientific method, managers don't just *assume* that their intuition is correct. Instead,

Nielsen is a global marketing research firm. Nielsen's clients want to know that a scientific approach to marketing research will give them a better idea of "What's Next."
Source: The Nielsen Company

they use their intuition and observations to develop **hypotheses**—educated guesses about the relationships between things or about what will happen in the future. Then they test their hypotheses before making final decisions.

A manager who relies only on intuition might introduce a new product without testing consumer response. But a manager who uses the scientific method might say, "I think (hypothesize) that consumers currently using the most popular brand will prefer our new product. Let's run some consumer tests. If at least 60 percent of the consumers prefer our product, we can introduce it in a regional test market. If it doesn't pass the consumer test there, we can make some changes and try again." With this approach, decisions are based on evidence, not just hunches.

The scientific method forces an orderly research process. Some managers don't carefully specify what information they need. They blindly move ahead—hoping that research will provide "the answer." Other managers may have a clearly defined problem or question but lose their way after that. These hit-or-miss approaches waste both time and money.

Five-Step Approach to Marketing Research

The **marketing research process** is a five-step application of the scientific method that includes:

1. Defining the problem
2. Analyzing the situation
3. Getting problem-specific data
4. Interpreting the data
5. Solving the problem

Exhibit 7-5 shows the five steps in the process. Note that the process may lead to a solution before all of the steps are completed. Or, as the feedback arrows show, researchers may return to an earlier step if needed. For example, the interpreting step may point to a new question—or reveal the need for additional information—before a final decision can be made.

Effective research usually requires cooperation

Good marketing research requires cooperation between researchers and marketing managers. Researchers must be sure their research focuses on real problems.

Marketing managers must be able to explain what their problems are

Exhibit 7-5 Five-Step Scientific Approach to the Marketing Research Process

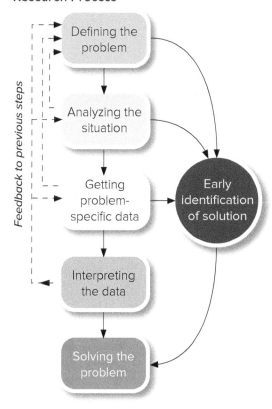

and what kinds of information they need. They should be able to communicate with specialists in the specialists' language. Marketing managers may only be "consumers" of research, but they should be informed consumers—able to explain exactly what they want from the research. They should also know about some of the basic decisions made during the research process so they know the limitations of the findings.

For this reason, our discussion of marketing research won't emphasize mechanics but rather how to plan and evaluate the work of marketing researchers.[19]

Step 1: Defining the Problem

The first step in the research process is defining the problem or reason for the research. Defining the problem is often the most difficult step in the marketing research process, but it's important for the objectives of the research to be clearly defined. The best research job on the wrong problem is wasted effort.

Pinpointing and identifying the problem is half the battle

Our strategy planning framework is useful for guiding the problem definition step. It can help the researcher identify the real problem area and what information is needed. Do we really know enough about our target markets to work out all of the Four Ps? Before deciding how to position our product, do we understand our competitors' strengths and weaknesses? Do we know enough to decide what celebrity to use in an ad or how to handle the price war in New York City or Tokyo? If not, we may want to do research rather than rely on intuition.

Don't confuse problems with symptoms

The problem definition step sounds simple—and that's the danger. It's easy to confuse problems with symptoms.

Marketers for Kiwi shoe polish encountered this situation. Marketing managers blamed sluggish sales on a lack of innovation in shoe polish. So Kiwi managers conducted a number of consumer studies on the problem of how to improve its polish. Yet none of the new product ideas clicked with consumers. Things changed when a new CEO realized that the slow sales and lack of interest in new polish products was just a symptom of the fact that today's footwear is more frequently made from synthetic materials than from leather. Today's shoes need less polish. So the CEO changed the focus of research to another problem—figure out what foot- and footwear-related needs were not currently being met. That research showed that having comfortable, fresh-smelling shoes was a top priority for women. Further research focused on how to solve that problem led to the development of several new products, including Kiwi Fresh'ins, which are lightly scented, disposable, and ultrathin shoe inserts that keep feet feeling fresh and comfortable all day.[20]

Source: S. C. Johnson & Son, Inc.

Setting research objectives may require more understanding

Sometimes the research objectives are very clear. A manager wants to know if the targeted households have tried a new product and what percent of them bought it a second time. But research objectives aren't always so simple. The manager might also want to know *why* some didn't buy or whether they had even heard of the product. Companies rarely have enough time and money to study everything. A manager must narrow the research objectives. One good way is to develop a list of research questions that includes all possible problem areas. Then the manager can consider the items on the list more completely—in the situation analysis step—before narrowing down to final research objectives.

Step 2: Analyzing the Situation

LO 7.3

What information do we already have?

When the marketing manager thinks the real problem has begun to surface, a situation analysis is useful. A **situation analysis** is an informal study of what information is already available in the problem area. It can help define the problem and specify what additional information, if any, is needed.

The situation analysis may begin with quick research—perhaps an Internet search; a closer look at information in an MIS; and phone calls or informal talks with people familiar with the industry, problem, or situation.

Let's consider a situation facing the marketing manager for the Hershey's candy company. She reviews last month's sales and finds that sales in the Chicago region are down by 5 percent, which concerns her. She could begin a situation analysis by digging deeper into the MIS and discover that sales of Hershey chocolate bars at one large grocery store chain are down by 25 percent. Otherwise sales in this region are up. Her next step might be an e-mail or phone call to the salesperson who calls on this grocery store chain. A phone call might reveal that last month the store had a big promotion going with a competitor's candy bars. Through this type of situation analysis, you can see that the marketing manager turned data into information and knowledge—she now knows why sales are down. This type of analysis might lead to an early solution—perhaps talking with the salesperson about how to set up a future promotion to try to gain back lost sales.

Situation analysis helps educate a researcher

The situation analysis is especially important if the marketing manager is dealing with unfamiliar areas or if the researcher is a specialist who doesn't know much about the management decisions to be made. They *both* must be sure they understand the problem area—including the nature of the target market, the marketing mix, competition, and other external factors. Otherwise, the researcher may rush ahead and make costly mistakes or simply discover facts that management already knows. The following case illustrates this danger.

A marketing manager at the home office of a large retail chain hired a research firm to do in-store interviews to learn what customers liked most and least about some of its stores in other cities. Interviewers diligently filled out their questionnaires. When the results came in, it was apparent that neither the marketing manager nor the researcher had done their homework. No one had even talked with the local store managers! Several of the stores were in the middle of some messy remodeling—so all the customers' responses concerned the noise and dust from the construction. The research was a waste of money.

Secondary data may provide the answers—or some background

The situation analysis should also find relevant **secondary data**—information that has been collected or published already. Later, in Step 3, we will cover **primary data**—information specifically collected to solve a current problem. Too often researchers rush to gather primary data when much relevant secondary information is already available—at little or no cost (see Exhibit 7-6)!

Secondary data are often readily available

Secondary data are often readily available from the firm's MIS. Secondary data in the MIS may also include routinely collected external data or the big data discussed earlier in this chapter. So, for example, a company's routine monitoring of the number of "hits" a web page receives and what search terms were used to find the web page should already be in the MIS. Data that have not been organized in an MIS may be available from the company's files and reports.

Secondary data are also available from libraries, trade associations, government agencies, and private research organizations. Increasingly, these organizations put their information online, so one of the first places a researcher should look for secondary data is on the Internet.

Monitor the web and anticipate trends

While a business can look at recent sales data to assess market trends, more important may be to get ahead of the next trend. Online tools can be useful in this regard. For example, jewelry retailer Moriarty's Gem Art examines free Google Trends reports to identify which styles of rings or necklaces or types of metal are increasingly searched.

Exhibit 7-6 Examples of Sources, Techniques, and Tools for Gathering Secondary and Primary Data

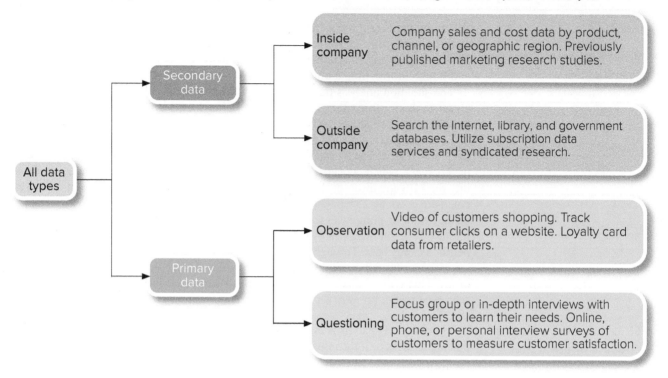

Marketing managers may also watch social media to spot the next hot trend. Clothing retailers can monitor Instagram and Twitter posts from fashion trendsetters. One retail consultant suggests following well-known Hollywood stylists to see what clothes they are putting on celebrities. Online used bookseller SilverFire Books monitors book trends by following #bookstagram on Instagram—the hashtag recently had more than 30 million posts. Getting ahead of the next trend helps marketing managers identify opportunities in new target markets, new products, and more.[21]

Search the web better

Marketing managers can find a treasure trove of useful information available on the Internet. But "available" is not the same as "reliable." Anyone can post anything on the Internet. So, as with any other research source, managers must carefully evaluate the accuracy of online information.

The trick to the Internet is finding what's needed. When managers have a pretty clear idea about what they want, search engines such as Google are a good place to start. A major problem is that searches often identify too many irrelevant sources. Everyone searches the web—and most of us think we're good at it. But it makes sense to learn a few tips for conducting better online searches. For some advice, check out "How to search on Google" (https://goo.gl/CQ2Mhz).

Gemma Talbot is a fashion blogger who lives in London. Her blog and Instagram page could be followed by fashion designers or retailers to monitor trends.
Source: Gemma Talbot/Instagram

Marketing Analytics in Action: Sentiment Analysis

This activity utilizes a free online sentiment analysis tool called Social Mention.
First follow these steps:

- Go the Social Mention website (www.socialmention.com).
- From the home page, click on the FAQ link and read about the measures that Social Mention reports—strength, sentiment, passion, and reach.
- Go back to the main website and conduct searches for three different fitness activity trackers (Fitbit, Garmin Vivosmart, and Acer Liquid Leap). For each, search in "All." (Note that this analysis might take up to a minute to complete.)
- Record the measures of strength, sentiment, passion, and reach for each activity tracker.
- On the left, under Sentiment, click on "negative" for each to learn the reasons for negative comments.

Next answer the questions that follow:

1. *Look at the date and timing of the data for each activity tracker. How are they similar or different? What might be reasons for any differences?*
2. *Which of the three activity trackers has the best score on each measure?*
3. *How could this information be used by a marketing manager for Fitbit?*

Monitor chatter on the web with sentiment analysis	Sometimes it can be useful to monitor and review what others are saying about a company or its products. One way to do this is to conduct a **sentiment analysis**—an automated process of analyzing and categorizing social media to determine the amount of positive, negative, and neutral online comments a brand receives. To better understand how to use sentiment analysis, see the Marketing Analytics in Action: Sentiment Analysis activity next.[22]
Be cautious when interpreting sentiment analysis	Tools that measure sentiment analysis are alluring; they can quickly generate a lot of information, and many are free. Yet this information is selective—it includes Internet users who choose to be vocal. Measures may not accurately represent the sentiment of the whole target market. For example, when Skittles launched a new website, its marketing managers were alarmed by early criticism of the site in social media outlets including blogs, Facebook, and Twitter. But when Skittles managers tapped a broader set of customers, they found little need for concern.
Government data are inexpensive	Federal and state governments publish data on many subjects. Government data are often useful in estimating the size of markets. In Chapters 3 and 5, we gave a number of examples of the different types of data available. Almost all government data are immediately available in inexpensive print and digital publications, on websites, and in downloads ready for further analysis. Sometimes it's more practical to use summary publications with links or references that lead to more detailed reports. For the U.S. market, the United States Census Bureau Topics page (www.census.gov/topics.html) and its Data and Statistics page (www.usa.gov/statistics) are good starting places. Another useful summary reference is the *Statistical Abstract of the United States,* which provides a summary of statistics on the social, political, and economic organization of the United States—though it must be purchased or accessed through a library. Some city and state governments have similar agencies for local data. Internationally, similar reports can be found for advanced economies, although less-developed countries have less information.
Subscribe to data services	Marketing research companies, including the Nielsen Company and Information Resources, Inc. (IRI), collect, compile, and sell subscription services with retail sales data. Retail checkout scanners help researchers collect very specific, and useful, information. Often this type of data feeds directly into a firm's MIS. Managers of a large chain of stores can see exactly what products have sold each day and how much money each department

in each store has earned. Companies that sell products through these retailers can also subscribe to the data. But the scanner also has wider applications for marketing research.

Nielsen (www.nielsen.com) and other research companies build **consumer panels**—a group of consumers who provide information on a continuing basis. Whenever a panel member shops for groceries, he or she gives an ID card to the clerk, who scans the number. Then the scanner records every purchase—including brands, sizes, prices, and any coupons used. In a variation of this approach, consumers use a handheld scanner to record purchases once they get home. Sometimes members of a panel answer questions and the answers are merged with the scanner data.

Data captured by electronic scanners are equally important to e-commerce in business-to-business markets. Increasingly, firms mark their shipping cartons and packages with computer-readable bar codes that make it fast and easy to track inventory, shipments, orders, and the like. As information about product sales or shipments becomes available, it is instantly included in the MIS and accessible over the Internet.

Syndicated research shares data collection costs

Some private research firms specialize in collecting data and then sell them to managers in many different client firms. Often the marketing manager subscribes to the research service and gets regular updates. This information may be added to a firm's MIS. About 40 percent of marketing research spending is for syndicated research, which helps explain why it can be an economical approach when marketing managers from many different firms need the same type of data. For example, many different auto producers use J.D. Power's (www.jdpower.com) surveys of customer satisfaction—often as the basis for advertising claims. Subscription data services are available for many different industries—ranging from food services to prescription drugs to microelectronic devices.

Situation analysis yields a lot—for very little

The virtue of a good situation analysis is that it can be very informative and takes little time. It's also inexpensive compared to more formal research efforts—like a large-scale survey. A phone, access to the Internet, and time might be all a marketing manager needs to gather a lot of insight. Situation analysis can help focus further research or even eliminate the need for it entirely. The situation analyst is really trying to determine the exact nature of the situation and the problem. This can lead the researcher back to Step 1, defining the problem; to an early identification of the solution; or to identifying problem-specific data needs.

Determine what else is needed

At the end of the situation analysis, you can see which research questions—from the list developed during the problem definition step—remain unanswered. Then you have to decide exactly what information you need to answer those questions and how to get it. This may require discussion between technical experts and the marketing manager. Often companies use a written **research proposal**—a plan that specifies what information will be obtained and how—to be sure no misunderstandings occur later. The research plan may include information about costs, what data will be collected, how they will be collected, who will analyze them and how, and how long the process will take.

Step 3: Getting Problem-Specific Data

LO 7.4

Gathering primary data

There are different methods for collecting primary data. Which approach to use depends on the nature of the problem and how much time and money are available.

In most primary data collection, the researcher tries to learn what customers think about some topic or how they behave under some conditions. There are two basic methods for obtaining information about customers: *questioning* and *observing*. Questioning can be qualitative or quantitative; observing can take many forms.

Qualitative questioning—open-ended to get more depth

Qualitative research seeks in-depth, open-ended responses, not yes or no answers. The researcher wants to get people to share their thoughts on a topic, without giving them many directions or guidelines about what to say. The real advantage of this approach is *depth*. Each person can be asked follow-up questions so the researcher really understands what *that* respondent is thinking. The qualitative approach gets at the details, even if the researcher needs a lot of judgment to summarize it all.

Journey maps follow the whole buying decision process

A researcher can use a qualitative, in-depth interview to learn more about a buyer's decision process. In Chapters 5 and 6, we learned about a general buying process for consumers and organizations. Often a company wants to learn more about a customer's experience throughout the entire the purchase process. A series of open-ended questions can lead to a **customer journey map**, the story and graphic diagram of a customer's experience in the buying process from need awareness through the purchase process and post-purchase relationship. For example, a cable TV provider might develop several journey maps based on interviews that start with why they needed cable TV (for instance, moving to a new city, dissatisfaction with satellite), the information sources consulted (friends, family, social media, online reviews), and so on, with questions delving into the installation and early usage experiences. The interviewer probes to get each customer's thoughts, emotions, and behaviors throughout the process. A journey map identifies opportunities to improve the entire purchase experience. For another example, see the nearby customer journey map.[23]

Shondra Portis and her family of three are moving to a new state and need to purchase broadband Internet service.

Phases:	Need Awareness	Problem Solving	Decision	Post-Purchase
Doing	Figuring out new services needed	Going online to read reviews	I follow Tom's advice	Arrange appointment for installation
	Surfing online for options	Asking Facebook friends for advice	Call to place order – wait on hold for 5 minutes	Installer arrives two hours late
	Busy with lots of moving activities			
Thinking	We need Internet as soon as we arrive	How do I make this decision?	Is long wait a sign of future problems?	I cannot believe I have to wait two weeks for installation.
	One more thing to add to my to-do list	How fast of service do we need?	Tom is a technology expert. I can trust him	Did I make the right choice? Poor customer service.
		Should I do what my friend Tom recommends?	Why do they need so much information?	
Feeling	I feel overwhelmed with this whole move—too much to do in too little time.	Confused at times	I felt good about my decision at first.	Dissonance. Uncertainty about choice
		Anxious about making right choice	Annoyed at long phone wait. Worried?	Happy! Finally have service
		Excited about faster options	Relieved–one more thing done	

This example of a journey map shows a customer moving through a decision to purchase broadband Internet service as she moves to a new home in another city. The interviewer here uses Post-it Notes as the interviewee shares what she was doing, thinking, and feeling as she moved through the purchase process.

Frank Merfort/Alamy Stock Photo

Focus groups stimulate discussion

One widely used form of qualitative questioning in marketing research is the **focus group interview**, which involves simultaneously interviewing 6 to 10 people in an informal group setting. A focus group also uses open-ended questions, but here an interviewer seeks group interaction—to stimulate thinking and get immediate reactions. Focus groups can be conducted in person or online. Focus groups can be conducted quickly and at relatively low cost—on average about $4,000 each.

MOD-PAC, a large printing company, faced a problem: business from its traditional large customers was drying up. MOD-PAC managers wanted to know if there were niche markets willing to buy printing services over the Internet. So they conducted online focus groups, each with prospects from a different target market. The focus groups helped MOD-PAC managers see that each group had different needs and used different terms to discuss its problems. Each focus group also indicated that there was interest in buying printing services online. In response, MOD-PAC developed its "Print Lizard" website; the home page uses tabs to route customers from different segments to the part of the site that caters to their specific needs.[24]

Online focus groups help offset some of the limitations of traditional face-to-face focus groups. Participants who meet online usually feel freer to express their honest thoughts—and an aggressive individual is less likely to dominate the group. Regardless of how a focus group is conducted, conclusions reached from a session usually vary depending on who watches it. A typical problem—and serious limitation—with qualitative research is that it's hard to measure the results objectively. The results seem to depend largely on the viewpoint of the researcher. In addition, people willing to participate in a focus group—especially those who talk the most—may not be representative of the broader target market.

Focus groups are probably being overused. It's easy to fall into the trap of treating an idea arising from a focus group as a "fact" that applies to a broad target market. To avoid this trap, many researchers use qualitative research to prepare for quantitative research. Qualitative research can provide good ideas—hypotheses. But we often need other approaches—based on more representative samples and objective measures—to test the hypotheses.

FocusVision Research is a marketing research provider that offers both qualitative and quantitative (Q+Q) research. Qualitative research offers more depth, while quantitative research provides more representative samples and objective measures to test hypotheses.
Source: FGI Research Inc.

Structured questioning gives more objective results

When researchers use identical questions and response alternatives, they can summarize the information quantitatively. Samples can be larger and more representative, and various statistics can be used to draw conclusions. For these reasons, most survey research is **quantitative research**, which seeks structured responses that can be summarized in numbers, such as percentages, averages, or other statistics. For example, a marketing researcher might calculate what percentage of respondents have tried a new product and then figure an average score for how satisfied they were.

Fixed responses speed answering and analysis

Survey questionnaires usually provide fixed responses to questions to simplify analysis of the replies. This multiple-choice approach also makes it easier and faster for respondents to reply. Simple fill-in-a-number questions are also widely used in quantitative research. Fixed responses are also more convenient for computer analysis, which is how most surveys are analyzed.

Surveys come in many forms

Decisions about what specific questions to ask and how to ask them usually depend on how respondents will be contacted—by mail, e-mail, on the phone, or in person. What question and response approach is used may also affect the survey. There are many possibilities. For example, whether the survey is self-administered or handled by an interviewer, the questionnaire may be on paper or online. The online survey can be programmed to skip certain questions, depending on answers given. Online surveys also allow the researcher to show pictures or play audio/video clips (for example, to get reactions to a television ad).

Mail and online surveys are common and convenient

A questionnaire distributed by mail or online is useful when extensive questioning is necessary. Respondents can complete the questions at their convenience. They may be more willing to provide personal information—because a questionnaire can be completed anonymously. But the questions must be simple and easy to follow because no interviewer is there to help. If the respondent is likely to be a computer user, a questionnaire on a website can include a help feature with additional directions for people who need them.

A big problem with questionnaires is that many people don't complete them. The widespread use of surveys means that many are now ignored. The **response rate**—the percentage of people contacted who complete the questionnaire—is often low and respondents may not be representative. Low response rates can generate misleading information if the respondents are not representative. For example, people who complete online questionnaires tend to be younger and better educated than the population as a whole. Sometimes companies will pay respondents or offer an opportunity to win a gift card to encourage responses.

A well-designed survey with a representative response can provide strategic insight. Consider Schoology, a learning management system (LMS) seeking to grow sales in public K-12 schools. Schoology designed an online survey that was completed by 369 K-12 public school teachers and administrators. Schoology found teachers' biggest classroom challenge was managing diverse student needs and varied learning styles. These insights led to an overhaul of promotional materials that emphasized Schoology's strengths in lesson planning and student assessment. The more relevant message increased sales.[25]

Surveys have limitations. Sometimes it takes awhile to compile data—though online surveys are faster. It can be difficult to get respondents to expand on particular points in mail or online surveys. In markets where illiteracy is a problem, it may not be possible to get any response. In spite of these limitations, the convenience and economy of self-administered surveys makes them popular.

Telephone surveys—fast but maybe not trustworthy

Telephone interviews can be effective for getting quick answers to simple questions or when there is a need to probe and really learn what the respondent is thinking. With computer-aided telephone interviewing, answers are immediately recorded on a computer, resulting in fast data analysis.

Unfortunately, many respondents simply don't answer the phone when they don't recognize the number—or refuse to participate in a phone survey. A recent study found that only 9 percent of households initially contacted actually completed a survey—down from 36 percent in the late 1990s. Because of these concerns, telephone survey use is in decline.

Marketing managers often turn to specialists to help with marketing research. FocusVision promises to help learn more about "what customers do, how they think, feel, act." Doyle Research wants marketing managers to know that it is a "qualitative research strategist."
(left): Source: FocusVision; (right): Source: 20/20 Research

Personal interview surveys—can be in-depth

A personal interview survey is usually much more expensive per interview than online, mail, or telephone surveys. But it's easier to get and keep the respondent's attention when the interviewer is right there. The interviewer can also help explain complicated directions and perhaps get better responses. For these reasons, personal interviews are commonly used for research on business customers.

Researchers have to be careful that having an interviewer involved doesn't affect the respondent's answers. Sometimes people won't give an answer they consider embarrassing. Or they may try to impress or please the interviewer. Further, in some cultures people don't want to give any information. For example, many people in Africa, Latin America, and eastern Europe are reluctant to be interviewed.

Personal interviews may also include open-ended questions that allow a researcher to get richer insights. A researcher for Pinterest went to France to learn more about how customers there used the social media site. Pinterest had translated the site into French and was feeling good about the user experience. After interviewing users of the site, the researcher discovered that translation alone was not enough. Some respondents mentioned they had done a search for "gâteau"—which is literally translated as "cake." While users saw a bunch of cakes, none were what French users were looking for. In France, a "gâteau" refers to a very particular style of cake that is single-layered, ugly, and brown, though reportedly quite tasty.[26]

Questioning has limitations

Questioning—whether qualitative or quantitative—has its limitations. Respondents sometimes give answers that they think the questioner wants to hear, or they may not accurately recall past events. When that is the case, observing may be more accurate or economical.

Observing—what you see is what you get

Observing—as a method of collecting data—focuses on a well-defined problem. Here we are not talking about the casual observations that may stimulate ideas in the early steps of a research project. With the observation method, researchers try to see or record what the subject does naturally. They don't want the observing to *influence* the subject's behavior.

For example, marketing managers for Heinz Ketchup knew they had an opportunity if they could make it easier to eat French fries while driving. To learn more, marketing managers sat behind a one-way mirror and watched consumers sitting in minivans put ketchup on fries, burgers, and chicken nuggets during a simulated driving experience. They saw some people tear the corners off packets with their teeth and others squirt ketchup in their mouth before adding fries. Insights from this research fueled development of Heinz Dip & Squeeze ketchup packets, which can be squeezed from one end like the traditional foil pouch or opened by peeling back the lid, for those who prefer dipping. Consumers like the new packaging, and fast-food restaurants are selling more French fries out of the drive-through window.[27]

McGraw-Hill Education

Ethnographic research looks at customers in their own environment

Ethnographic research has roots in anthropology, which studies different cultures by observing participants in their natural habitat. In marketing research, ethnography involves studying customers in their homes or at work.

Video cameras can be useful for ethnography because subjects often don't know their behavior is being recorded. Some researchers use this technique to study the routes consumers follow through a store or how they select products. A dog food manufacturer put

Remesh.ai offers a unique platform for marketing research. Its artificial intelligence is guided by real people. It allows a marketing manager to engage and understand a live audience. Remesh claims to combine the depth of qualitative research with the larger sample size typical of quantitative research.
Source: Remesh

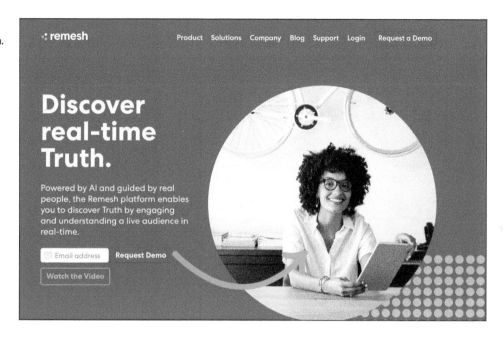

video cameras in the pet food aisle in supermarkets to learn more about how people choose dog food and treats. The videos showed that kids often picked the treats, but that the kids' parents chose the food. The videos also revealed that kids couldn't reach treats when they were on higher shelves. Sales increased when the treats were moved to lower shelves.[28]

Walk a mile in the customer's shoes

Sometimes the best way to observe a customer is for a marketing manager to become a customer. By calling his firm's own customer service number and pretending to be a customer, a marketing manager can discover how long it takes someone to answer the phone—or the ability of a customer service rep to politely and efficiently solve a problem.

Marketing managers at Walgreens drugstores wanted to better understand the challenges facing older adults—an important and growing target market. So they donned glasses that blurred vision and wore large rubber gloves that simulated arthritic hands. After experiencing firsthand the difficulties facing seniors, Walgreens managers began to use larger print in the weekly circular and installed call buttons near heavy items such as bottled water and laundry detergent.[29]

Experimental method controls conditions

A marketing manager can get a different kind of information—with either questioning or observing—using the experimental method. With the **experimental method**, researchers compare the responses of two (or more) groups that are similar except on the characteristic being tested. The experimental method is sometimes called *A/B testing,* because it compares options labeled A and B. Researchers want to learn if the specific characteristic—which varies among groups—*causes* differences in some response among the groups.

For example, online retailer Zappos wanted to find out if adding short product demonstration videos to each product page would increase sales (a hypothesis). Zappos knew that there would be some cost to creating so many videos, so it needed to know whether the videos would increase sales. Customers who clicked on the tested products were randomly shown a product page with a video or an identical page without a video. Zappos found that purchases increased about 10 percent when pages included the video description. That experiment justified the production of more than 50,000 videos—and higher sales.[30]

For another example, see the nearby photos of two home pages used in an experiment run by by California Closets. Which one do you think generated more leads?

A combination of research methods may be needed

Using one research method to solve an initial problem may identify new questions that are best answered with different research methods. Consider WD-40, a popular all-purpose lubricant sold in a blue-and-yellow spray can. To find potential new uses of WD-40, researchers visited mechanics and watched them as they worked. These

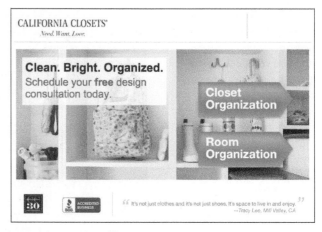

California Closets builds custom closets. Its marketing manager wanted to know which of two possible pages would generate more leads. Version A on the left provided copy that matched the copy of a click-through advertisement. Version B used a different headline. Which of the two pages do you think generated more customer interest? Why? To find the answer, see the bottom of the page.*

Source: California Closet Company, Inc.

observers realized that even small cans of WD-40 were difficult for mechanics to handle in tight spaces. In addition, the spray created drips and messes because it was difficult to control the amount being applied.

To address these problems, the new-product team developed a prototype for the No Mess Pen, a small marker that delivers a precise amount of the lubricant. Then, researchers held focus groups to get reactions from mechanics. They weren't encouraging. Mechanics didn't think that the small unit would handle their large application needs. Yet many thought their spouses might like the pen for small household lubrication jobs. To follow up on this idea, WD-40 conducted online surveys. More than two-thirds of the women respondents said they would buy the product. To fine-tune targeting and promotion, WD-40 then conducted more than 40 in-home studies to learn how families actually used the No Mess Pen. This research confirmed that women were the primary target market, but that men used the pens as well. Moreover, the pen didn't replace the can of WD-40 already found in most households; rather, pens were stored in desk drawers, cars, and toolboxes so they'd be handy. WD-40 used different research methods to address different problems, but in combination they contributed to making the No Mess Pen a great success.[31]

Evaluate costs and benefits of research

Whether collecting secondary data for a situation analysis or primary data from a focus group or survey, marketing research takes time and money. A good marketing manager knows that the value of additional information lies in the ability to design more effective marketing strategies. Similarly, different research methods provide different insights—and come at different costs. Marketing managers do not need to spend $100,000 to determine whether to spend $50,000 on advertising.

There are also benefits to getting information quickly—particularly in some markets. For example, competition and customer behavior can change quickly in high-technology markets. If a study takes six months to complete, the marketing manager may be developing a marketing strategy for a market that no longer exists—or, at best, looks quite different.

Small companies with limited budgets must be especially creative in identifying low-cost ways to get the information they need. Marketing managers can learn a lot by simply making sure they listen to what their firm's customers have to say. Even when doing

*California Closets found that Version A generated 115 percent more leads than Version B. The lesson it learned was that customers came to the page motivated by a particular message (from the ad) and the page needed to reinforce that message to get customers to show more interest by filling out a form.
Source: "12 Surprising A/B Test Results to Stop You Making Assumptions," Unbounce, September 19, 2012.

informal research, marketers should have objectives for their questions. They can also learn a lot by studying competitors' websites and promotional materials or by shopping in their stores. When conducting formal research, they should be careful not to be too frugal. Skipping the pretesting of a questionnaire or using employees instead of real customers in a focus group may give managers a false sense of confidence—and worse, the wrong solution to a problem.

Step 4: Interpreting the Data

LO 7.5

What do they really mean?

After someone collects the data, they have to be analyzed to decide what they all mean. In quantitative research, this step usually involves statistics. **Statistical packages**—easy-to-use computer programs that analyze data—have made this step easier. As we noted earlier, some firms provide *decision support systems* so managers can use a statistical package to interpret data themselves. More often, however, technical specialists are involved at the interpretation step.

Cross-tabulation is one of the most frequently used approaches for analyzing and interpreting marketing research data. It shows the relationship of answers to two different questions. Exhibit 7-7 is an example. The cross-tab analysis shows that smartphone and cell phone usage is similar by gender. However, when you look at the analysis by age, younger people are more likely to have a cell phone or a smartphone.

There are many other approaches for statistical analysis—the best one depends on the situation. The details of statistical analysis are beyond the scope of this book. But a good manager should know enough to understand what a research project can and can't do.[32]

Is your sample representative?

It's usually impossible for marketing managers to collect all the information they want about everyone in a **population**—the total group they are interested in. Marketing researchers typically study only a **sample**, a part of the relevant population. How well a sample *represents* the total population affects the results. Results from a sample that is not representative may not give a true picture.

The manager of a retail store might want a phone survey to learn what consumers think about the store's hours. If interviewers make all of the calls during the day, consumers who work outside the home during the day won't be represented. Those interviewed might say the limited store hours are "satisfactory." Yet it would be a mistake to assume that *all* consumers are satisfied.

Research results are not exact

An estimate from a sample, even a representative one, usually varies somewhat from the true value for a total population. Managers sometimes forget this. They assume that survey results are exact. Instead, when interpreting sample estimates, managers should think of them as *suggesting* the approximate value.

Exhibit 7-7 Cross-Tabulation Breakdown of Responses on a Survey of Smartphone Usage

Demographic Group	% of U.S. Adults Who Own the Following Devices	
	Any Cell Phone	Smartphone
Total	95%	77%
Men	95%	80%
Women	94%	75%
Ages 18–29	100%	94%
Ages 30–49	98%	89%
Ages 50–64	94%	73%
Ages 65+	85%	46%

Source: "Mobile Fact Sheet," Pew Research Center, February 6, 2018, www.pewinternet.org/fact-sheet/mobile/.

If random selection is used to develop the sample, researchers can use statistical methods to help determine the likely accuracy of the sample value. This is done in terms of **confidence intervals**—the range on either side of an estimate that is likely to contain the true value for the whole population. Some managers are surprised to learn how wide that range can be.

Consider a wholesaler that has 2,000 retail customers and wants to learn how many of these retailers carry a product from a competing supplier. If the wholesaler randomly samples 100 retailers and 20 say yes, then the sample estimate is 20 percent. But with that information the wholesaler can be only 95 percent confident that the percentage of all retailers is in the confidence interval between 12 and 28 percent. The larger the sample size, the greater the accuracy of estimates from a random sample. With a larger sample, a few unusual responses are less likely to make a big difference.[33]

Validity problems can destroy research

Even if the sampling is carefully planned, it is also important to evaluate the quality of the research data. Managers and researchers should be sure that research data really measure what they are supposed to measure. Many of the variables marketing managers are interested in are difficult to measure accurately.

Validity concerns the extent to which data measure what they are intended to measure. Validity problems are important in marketing research because many people will try to answer even when they don't know what they're talking about. Further, a poorly worded question can mean different things to different people and invalidate the results. Often, pretests of a research project are required to evaluate the quality of the questions and measures and to ensure that potential problems have been identified.

Consider the validity of some customer satisfaction surveys. For example, Allstate insurance company uses customer satisfaction surveys to monitor how well its agents take care of customers—and gives bonuses to agencies with high scores. But some agencies try to game the system. At least one agency regularly e-mails customers to request a perfect score on the survey: "Each time we obtain a 'Perfect 10' on the survey, our agency is given additional resources from Allstate that will allow us to provide our clients with an even higher level of service." With that "nudge," Allstate may not be actually measuring what it thinks it is measuring.[34]

Ethics involved in interpreting and presenting results

Marketing managers want information they can trust when they make marketing decisions. But research often involves many hidden details. A person who wants to misuse research to pursue a personal agenda can often do so.

Surveys may not be valid when the sample does not accurately represent the target population. Survey Sampling and i.think_inc. help marketing researchers develop samples that are highly representative of the target market.
(left): Source: Survey Sampling International; (right): Source: i.think.com 2015

Perhaps the most common ethical issues concern decisions to withhold certain information about the research. For example, a manager might selectively share only those results that support his or her viewpoint. Others involved in a decision might never know that they are getting only partial truths.[35]

Step 5: Solving the Problem

The last step is solving the problem

In the problem solution step, managers use the research results to make marketing decisions.[36]

Some researchers, and some managers, are fascinated by the interesting tidbits of information that come from the research process. They are excited if the research reveals something they didn't know before. But if research doesn't have action implications, it has little value and suggests poor planning by the researcher and the manager.

When the research process is finished, the marketing manager should be able to apply the findings in marketing strategy planning—the choice of a target market or the mix of the Four Ps. If the research doesn't provide information to help guide these decisions, the company has wasted research time and money.

We emphasize this step because it is the reason for and logical conclusion to the whole research process. This final step must be anticipated at each of the earlier steps. To conclude this chapter, let's briefly examine a case study of research—and a solution—that show how marketing can make for a better world (#M4BW).

#M4BW

Solving a problem creates a better world

Millions of people in Ghana live without the sanitary benefits of in-home toilets. The "Clean Team," a collaboration between Unilever, Water and Sanitation for the Urban Poor (WSUP), and IDEO.org set out to solve this problem. This problem was a systems-level challenge; not only were toilets required, but a solution would have to include a way to safely transport and dispose of the waste (adding indoor plumbing and sewage lines was cost prohibitive). The Clean Team conducted marketing research to understand the challenges and build a viable solution.

The research began with interviews of dozens of poor, urban Ghanaians. The team wanted to know more about what a toilet should look like and how waste should be collected. One key piece of learning was of historical note. For many years Ghana relied on "night soil collectors" who would go house-to-house every night emptying bucket latrines. But after many of the night soil collectors ended up dumping the waste in city streets, the practice was banned because of the threat to public health.

The Clean Team quickly developed prototype toilets, showed them to people, received feedback, and made further changes. They also asked customers if they were comfortable with servicemen coming into their homes and asked where a toilet would go in their home. The back-and-forth with real customers identified how the service would work and generated knowledge about how to promote and price the offering. After putting all this learning together, the Clean Team launched its toilet subscription service, which includes regular pickups and disposal. The Clean Team solves an important problem for tens of thousands of Ghanaians.[37]

Source: Industrial Designers Society of America

CONCLUSION

Marketing managers face difficult decisions in selecting target markets and managing marketing mixes, but they rarely have all the information they would like to have before making those decisions. This doesn't mean that managers have to rely solely on intuition; they can usually obtain some good information that will improve the quality of their decisions. Both large and small firms can take advantage of Internet and intranet capabilities to develop marketing information systems (MIS) that help ensure routinely needed data are available and quickly accessible.

Some questions can be answered only with marketing research. Marketing research should be guided by the scientific method. This approach to solving marketing problems involves five steps: (1) defining the problem, (2) analyzing the situation, (3) getting problem-specific data, (4) interpreting the data, and (5) solving the problem. This objective and organized approach helps keep problem solving on task. It reduces the risk of doing costly and unnecessary research that doesn't achieve the desired end—solving the marketing problem.

Our strategy planning framework can be very helpful in evaluating marketing research. By finding and focusing on real problems, researchers and marketing managers may be able to move more quickly to a useful solution during the situation analysis stage—without the costs and risks of gathering primary data. With imagination, they may even be able to find answers in their MIS or in readily available secondary data. However, primary data from questioning, observing, or conducting experiments may be needed. Qualitative data often provide initial insights or hypotheses—which might be tested with more representative samples and quantitative approaches.

KEY TERMS

LO 7.6

marketing research, 175
marketing information system (MIS), 175
big data, 177
data warehouse, 179
decision support system (DSS), 180
marketing dashboard, 181
marketing model, 181
personal data, 182
informational privacy, 182
General Data Protection Regulation (GDPR), 184

scientific method, 184
hypotheses, 185
marketing research process, 185
situation analysis, 187
secondary data, 187
primary data, 187
sentiment analysis, 189
consumer panel, 190
research proposal, 190
qualitative research, 190
customer journey map, 191

focus group interview, 192
quantitative research, 192
response rate, 193
experimental method, 195
statistical packages, 197
population, 197
sample, 197
confidence interval, 198
validity, 198

QUESTIONS AND PROBLEMS

1. Review the Dunkin' case study that opens the chapter. From this case, identify examples of different key terms and concepts covered in the chapter. For example, the first paragraph describes "daily sales totals," which are examples of secondary data.

2. Review the Dunkin' case study that opens this chapter. Imagine you are in charge of Dunkin's expansion into Russia. Identify five different types of research you would conduct before opening your first store in Moscow.

3. Discuss the concept of a marketing information system and why it is important for marketing managers to be involved in planning the system.

4. In your own words, explain why a decision support system (DSS) can add to the value of a marketing information system. Give an example of how a decision support system might help.

5. Discuss how output from a marketing information system (MIS) might differ from the output of a typical marketing research department.

6. Discuss some of the likely problems facing the marketing manager in a small firm who plans to search the Internet for information on competitors' marketing plans.

7. Explain the key characteristics of the scientific method and show why these are important to managers concerned with research.

8. Distinguish between primary data and secondary data and illustrate your answer.

9. With so much secondary information now available free or at low cost over the Internet, why would a firm ever want to spend the money to do primary research?

10. Explain why a company might want to do focus group interviews rather than individual interviews with the same people.
11. Distinguish between qualitative and quantitative approaches to research—and give some of the key advantages and limitations of each approach.
12. Define response rate and discuss why a marketing manager might be concerned about the response rate achieved in a particular survey. Give an example.
13. Would a firm want to subscribe to a subscription data service if the same data were going to be available to competitors? Discuss your reasoning.
14. Explain how you might use different types of research (focus groups, observation, survey, and experiment) to forecast market reaction to a new kind of disposable baby diaper, which is to receive no promotion other than what the retailer will give it. Further, assume that the new diaper's name will not be associated with other known products. The product will be offered at competitive prices.
15. Discuss the concept that some information may be too expensive to obtain in relation to its value. Illustrate.

MARKETING PLANNING FOR HILLSIDE VETERINARY CLINIC

Appendix D (the Appendices follow Chapter 19) includes a sample marketing plan for Hillside Veterinary Clinic. Look through the "Customer Analysis" and "Competitor Analysis" sections in the "Situation Analysis" section and consider the following questions.

1. What different types of marketing research were conducted to fill out these sections of the marketing plan?
2. What are the strengths of the research conducted? What are the weaknesses?
3. Keeping in mind probable cost and time to complete, what additional research would you recommend?

SUGGESTED CASES

3. NOCO United Soccer Academy
9. Quiet Night Motel

MARKETING ANALYTICS: DATA TO KNOWLEDGE

CHAPTER 7: MARKETING RESEARCH

Texmac, Inc. has an idea for a new type of weaving machine that could replace the machines now used by many textile manufacturers. Texmac has done a telephone survey to estimate how many of the old-style machines are now in use. Respondents using the present machines were also asked if they would buy the improved machine at a price of $10,000.

Texmac researchers identified a population of about 5,000 textile factories as potential customers. A sample of these were surveyed, and Texmac received 500 responses. Researchers think the total potential market is about 10 times larger than the sample of respondents. Two hundred twenty of the respondents indicated that their firms used old machines like the one the new machine was intended to replace. Forty percent of those firms said that they would be interested in buying the new Texmac machine.

Texmac thinks the sample respondents are representative of the total population, but the marketing manager realizes that estimates based on a sample may not be exact when applied to the whole population. He wants to see how sampling error would affect profit estimates. Data for this problem appear in the spreadsheet. Quantity estimates for the whole market are computed from the sample estimates. These quantity estimates are used in computing likely sales, costs, and profit contribution.

Design element: #M4BW box globe icon: ©Vectoryzen/Shutterstock

CHAPTER EIGHT

Source: Under Armour, Inc.

Elements of Product Planning for Goods and Services

Kevin Plank was a business major and football player at the University of Maryland when he spotted an opportunity. The cotton T-shirts he and his teammates wore under their football pads quickly became sweat-soaked, heavy, and uncomfortable during practices and games. Plank looked for a product that performed better than a T-shirt; he discovered new types of fabrics and performance clothing worn by bicyclists and hikers.

In New York City's garment district, Plank learned about a Polyester-Lycra blend fabric that didn't trap moisture. He developed several prototype shirts and asked friends who were players in the NFL to try them. The players really liked the skintight compression shirts. They were comfortable under football gear and wicked away sweat—keeping the players cooler, drier, and lighter. Plank knew he was on to something when his friends clamored for more shirts. At the time, he couldn't afford a big ad campaign to tout the benefits of his product, and he didn't have relationships with retailers who could help build demand with final consumers. So Plank focused on the target market he knew best: college football teams.

Plank went back to New York and ordered 500 shirts. These were the first products with the Under Armour brand name and the start of what became the HeatGear warm weather product line. He loaded his shirts into his SUV and traveled to colleges across the Southeast. Plank tried to persuade coaches, players, and equipment managers about the benefits of his unique shirts. Many were not initially convinced of their value—especially because the price was three to five times that of a T-shirt. However, it took players only one practice to see the shirt's advantages, and praise for the product quickly spread.

Under Armour's success soon attracted Nike and Adidas to the performance clothing market. Later, dealer brands such as Kohl's Tek Gear and JCPenney's Simply for Sports moved in with lower prices that appealed to customers who saw little differentiation among the brands. To combat this, Under Armour put more emphasis on innovation and creative promotion to build customer preference for its brand. For example, Under Armour ran ads featuring professional athletes such as Dallas Cowboy (and Plank's former teammate) Eric Ogbogu. When the muscular Ogbogu barked the firm's tagline "Protect This House," it instantly became part of popular sports culture and a rallying cry of players and fans across the country.

Under Armour's success led to an expansion of its product assortment—adding shoes, bottoms, backpacks, and accessories and moving into new sports including baseball, golf, soccer, and volleyball. It even went right at one of Nike's strengths—the basketball shoe market. Under Armour's signing of NBA star Stephen Curry to an endorsement deal turned into a coup when Curry later won the league's Most Valuable Player award and his team won the NBA championship.

Under Armour also stepped up investment in offerings for the women's market. It hired a new product manager, increased the marketing team from 1 person to 10, and created a 20-person design team. These employees focused solely on women's apparel, accessories, and footwear. New products were fashion forward and designed to fit the female body. To appeal to women more likely to work out at the gym or do yoga, its "What's Beautiful" advertising campaign encouraged women with the "No Matter What, Sweat Every Day" tagline. Social media promoted an online community where women created profiles, set goals, and provided encouragement. Subsequently, its "I Will What I Want" campaign featured American Ballet Theater's Misty Copeland and Olympic skier Lindsey Vonn. The campaign highlighted the physical and mental fortitude driving their success. The strategy worked, and Under Armour's share of the female athlete market grew quickly; it soon accounted for 30 percent of Under Armour's sales.

The market for many of Under Armour's core products has matured; sales growth has slowed, and profit margins have declined. It faces many strong competitors. To spur growth, Under Armour invested in new products operating in fast-growing markets. It purchased apps like MapMyFitness and UA Record to support its customers' fitness goals. Under Armour hopes to deliver more value to its customers by integrating this technology into its core products—entering into the "wearable technology" market.

For example, besides a water-repellent and insulated upper for running in rain and snow, the HOVR ColdGear

Reactor running shoes include Under Armour's built-in Record Sensor technology. Record Sensor tracks a runner's cadence, distance, pace, stride, and steps—the data hard-core runners use to get an extra edge. A runner reads results on the MapMyRun app and checks advice offered by an AI-powered running "coach." Under Armour's research and development team uses the data it collects to design its next generation of technology-enhanced products. Adding services like this to its core products gives customers more value and differentiates Under Armour from its competitors.

To stay ahead of those competitors, Under Armour's marketing managers monitor various marketing metrics to ensure the brand remains on track. For example, they track each ad's YouTube views and the market share for everything from HeatGear shirts to ColdGear shoes. These numbers contribute to the overall value of the Under Armour brand, recently valued at more than $3 billion. Under Armour is also aggressive in protecting its brand's "UA" trademark; it recently won a lawsuit against a Chinese upstart with a "copycat" logo.

While Nike, Adidas, and dealer brands put pressure on Under Armour's prices and margins, Kevin Plank enjoys taking on big competitors. During his college football days, the 5-foot 11-inch, 229-pound Plank once had to block his 6-foot 4-inch, 269-pound buddy Eric Ogbogu, and the bigger man ended up on his back. Watch out, Nike![1]

LEARNING OBJECTIVES

Developing a marketing mix that provides superior customer value for target customers is ultimately the focus of the marketing strategy planning process. To develop a successful marketing mix requires that each of the Four Ps is carefully planned and works well with all the others. Many important strategy decisions are required. This chapter focuses on key ideas related to strategy planning for Product.

When you finish this chapter, you should be able to

1. understand what "Product" really means.
2. cite the key differences between goods and services.
3. describe how technology augments a product to add value for customers.
4. understand what branding is and how to use it in strategy planning.
5. understand the importance of packaging in strategy planning.
6. know the differences among various consumer and business product classes and how product classes can help a marketing manager plan marketing strategy.
7. understand important new terms (shown in **red**).

The Product Area Involves Many Strategy Decisions

The Under Armour case highlights some of the important topics and strategy decision areas that we'll discuss in this chapter and in Chapter 9. As shown in Exhibit 8-1, there are many strategy decisions in the Product area. They're the focus of this chapter. Then, in Chapter 9 we'll take a "how to" look at developing new products and also explain how strategy usually changes as products move through their life in the market.

We'll start here by looking at Product through the eyes of the customer. This focuses attention on the customers' total experience with the Product—regardless of whether it is a physical good, a service, or both. We then examine different ways that technology is being integrated into many products. Next, we examine one of the most important Product strategy areas, branding, which includes many different decisions. Then we look at packaging. Finally, we'll consider product classes, which show how strategy decisions for Product relate to decisions for Place, Promotion, and Price.

Exhibit 8-1 Product Decisions for Marketing Strategy Planning

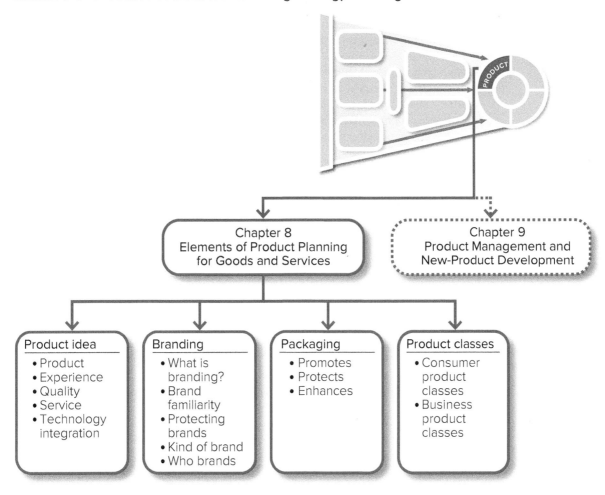

What Is a Product?

LO 8.1

Customers buy satisfaction

When Jif sells its peanut butter, is it just selling roasted peanuts, sugar, salt, and oil? When Air Jamaica sells a ticket for a flight to the Caribbean, is it just selling wear and tear on an airplane and pilot fatigue? The answer to these questions is no. Instead, what these companies are really selling is the satisfaction, use, or benefit the customer wants.

Consumers care that their Jif peanut butter tastes great, easily spreads on bread for a fast sandwich, and offers a source of protein in their diet. If they worry about too much fat or sugar in their diet, they can buy Jif Reduced Fat or Simply Jif. Of course, retailers who sell peanut butter are satisfied because Jif tells customers about the brand in advertising and packages the peanut butter in bright, eye-catching colors that help the brand sell.

Similarly, Air Jamaica's customers want a safe, comfortable, on-time flight—but they also want easy online reservations, low prices, smooth check-in at the airport, and luggage that arrives undamaged and on time. In other words, purchases deliver the highest level of satisfaction when the customer's entire *experience* with the product meets or exceeds the customer's needs.

Product means the need-satisfying offering of a firm. The idea of "Product" as potential customer satisfaction or benefits is very important. Many business managers get wrapped up in the technical details involved in producing a product. But most customers think about a product in terms of the total satisfaction it provides. That satisfaction may require a "total" product offering that is really a combination of a physical good with the right features, helpful instructions, a convenient package, a trustworthy warranty, excellent service, and perhaps even a familiar name that has satisfied the customer in the past.

Product quality and customer needs

Product quality should also be determined by how customers view the product. From a marketing perspective, **quality** means a product's ability to satisfy a customer's needs or requirements. This definition focuses on the customer—and how the customer thinks a product will fit some purpose. For example, the "best" satellite TV service may not be the one with the highest number of channels but the one that includes a local channel that a consumer wants to watch. Similarly, the best-quality clothing for casual wear on campus may be a pair of jeans, not a pair of dress slacks made of a higher-grade fabric.

Among different types of jeans, the one with the most durable fabric might be thought of as having the highest grade or *relative quality* for its product type. Marketing managers often focus on relative quality when comparing their products to competitors' offerings. However, a product with better features is not a higher-quality product if the features aren't what the target market wants.

In Chapter 9, we'll look at ways to manage product quality. For now, however, it is important to see that quality and satisfaction depend on the total product offering. If potato chips get stale on the shelf because of poor packaging, the consumer will be dissatisfied. A broken button on a shirt will disappoint the customer—even if the laundry did a nice job cleaning and pressing the collar.[2]

The Product includes the entire experience. Oracle's database software can help companies deliver better customer experiences. So it reminds its target customers of the benefits of a great experience.
Source: Oracle Corporation

Ethical Dilemma

What would you do? Your construction firm was the low-price bidder on a plan to build three new runways at an airport. After winning the contract, you assured the airport commissioner that your work would far exceed the minimum quality specs in the contract. However, a test of the batch of concrete for the second runway shows that it's not as strong as the concrete you've been using. It does exceed the specs in the contract, but barely. Throwing away the concrete would eat up most of the profit expected from the job and also delay the airport in using the runway. There are various options. You could proceed with the project and be quiet about it, later admitting what happened. Alternatively, you could call the commissioner, reveal everything, and then ask for approval to proceed (chancing denial—an outcome that would seriously hurt your company). With or without approval, you could offer a special warranty.[3]

Explain what you would do. What, if anything, would you say to your employees about your decision?

Warranty puts quality promise in writing

A **warranty** explains what the seller promises about its product. A marketing manager should decide whether to offer a specific warranty and, if so, what the warranty will cover and how it will be communicated to target customers. Some companies will use a long-term warranty to signal quality. For example, Yardbird makes and sells high-quality outdoor furniture; its warranty promises to replace any defective products within three years of purchase—with longer coverage for its Sunbrella fabric and aluminum frames.

 #M4BW

Quality evolves toward better world

As consumers' needs and requirements change, their definition of quality changes, too. For example, for a long time, antibiotics were liberally used in treating the animals that provide our meat, eggs, and milk. Keeping animals healthy lowers costs. These days, many more consumers are concerned that the overuse of antibiotics promotes "superbugs." These consumers want foods raised without antibiotics. Recognizing this change, some fast-food companies have adjusted to the new quality requirements. For example, McDonald's reduced antibiotics in its beef supply and Pizza Hut did the same with the chicken it uses in its wings and pizza.[4]

With growing interest in sustainability, some products redefine quality and add more sustainable ingredients. This is especially important when the target market includes Millennials (see Chapter 3), who place a high value on sustainability. For example, IKEA wants a sustainable positioning, so it uses only recycled or FSC certified (from responsibly managed forests) wood in its products. Allbirds developed a sustainable version of EVA foam for its flip-flops. While the EVA foam used in other flip-flops is made using fossil fuels, Allbirds' SweetFoam is made from sugar cane. To promote a better world, Allbirds allows any company, even its competitors, to use SweetFoam.[5]

Sorting out product lines, product assortments, and individual products

An **individual product** is a particular product within a product line. A **product line** is a set of individual products that are closely related. It usually is differentiated by brand, level of service offered, price, or some other characteristic. For example, each size and scent of a brand of soap is an individual product. Intermediaries usually think of each separate product as a stock-keeping unit (SKU) and assign it a unique SKU number.

A **product assortment** is the set of all product lines and individual products that a firm sells. The seller may see the products in a line as related because they're produced or operate in a similar way, sold to the same target market, sold through the same types of outlets, or priced at about the same level. Sara Lee, for example, has many product lines in its product assortment—including beverages, lunch meats, desserts, insecticides, body care, air care, and shoe care. But Crocs makes only footwear and Maclaren focuses on baby strollers. See Exhibit 8-2 for examples of these three terms for the Coca-Cola Company.

Product lines require strategy decisions

Marketing managers have several decisions to make about product lines. One strategy decision involves **product line length**—the number of individual products in a product line. Sometimes extending the length of the product line offers marketing managers new opportunities. Firms can add more colors, flavors, styles, and sizes to appeal to a wider range of customers. For example, Taco Bell offers breakfast meals, which appeal to a new market segment. Coca-Cola in recent years has added Coca-Cola Orange Vanilla, Twisted Mango Diet Coke, and Coca-Cola Zero Sugar to appeal to a wider range of customers.

Each individual product and target market may require a separate strategy. For example, Clorox's strategy for selling Clorox Scented Bleach will differ from its strategy for selling Clorox Regular Bleach. In this book and the first marketing course, we'll focus mainly on developing one marketing strategy at a time. But remember that a marketing manager may have to plan *several* strategies to develop an effective marketing program for a product line or a whole company.[6]

Exhibit 8-2 Examples of Product Assortment, Individual Product, and Product Line

This photo shows part of the *product assortment* offered by Coca-Cola. Each individual size and brand represents an *individual product*. All of these carbonated soft drinks—Coca-Cola, Coke Zero Sugar, Orange Vanilla Coke, Coca-Cola Life, and Cherry Coke—are part of the same *product line*.
Photos: McGraw-Hill Education

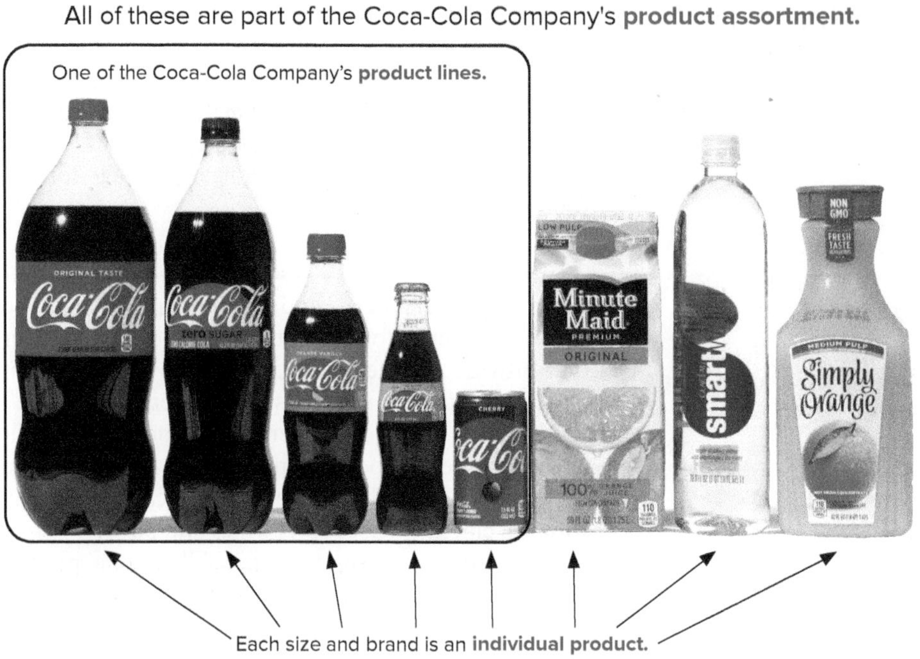

All of these are part of the Coca-Cola Company's **product assortment.**

One of the Coca-Cola Company's **product lines.**

Each size and brand is an **individual product.**

Differences between Goods and Services

LO 8.2

Goods and/or services are the product

A product may be a tangible, physical *good*. Or a product may be a **service**—an intangible offering involving a deed, performance, or effort. A service is not physical or tangible. Many products are a *blend of both*—part good and part service. Exhibit 8-3 shows that a product can range from a 100 percent emphasis on a physical good (for a commodity like steel pipe) to a 100 percent emphasis on service (for a product like satellite radio from SiriusXM). Many products include a combination of goods and services. When you eat out, you are buying food (a physical good) that is prepared and served by a restaurant's staff (a service).

Regardless of the blend of goods and services involved, the marketing manager must consider most of the same elements in planning products and marketing mixes. Given this, we usually won't make a distinction between goods and services but will call all of them Products. However, understanding key differences in goods and services can help fine-tune marketing strategy planning. Exhibit 8-4 summarizes some of the main differences between goods and services and the implications for marketing strategy planning. Review the exhibit and then read ahead for more details.

Exhibit 8-3 Examples of Possible Blends of Physical Goods and Services in a Product

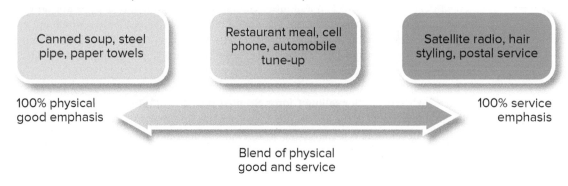

Canned soup, steel pipe, paper towels

Restaurant meal, cell phone, automobile tune-up

Satellite radio, hair styling, postal service

100% physical good emphasis

100% service emphasis

Blend of physical good and service

Exhibit 8-4 Major Differences between Goods and Services and Implications for Marketing Strategy Planning

Most goods are...	Most services are...	Implications for Marketing Strategy Planning
Tangible	Intangible	Services cannot be touched, so customers look for recommendations from trusted sources or tangible features of the service provider.
Produced in a factory far from consumers	Produced locally in front of customers	Service providers should be well trained and, in more complex service systems, should be in place to facilitate coordination. Locate service near customers.
Nonperishable	Perishable	Services use price to manage supply and demand. Services manage wait experience.

How tangible is the product?

Because a good is a physical thing, it can be seen and touched. You can try on a pair of Vans shoes, smell Starbucks beans as they roast, and page through the latest issue of *People* magazine. A good is a *tangible* item. It's usually easy to know exactly what you will get before you decide to buy it. And once you've bought it, you own it.

In contrast, services are not physical—they are *intangible*. When you provide a customer with a service, the customer can't keep it. Rather, a service is experienced, used, or consumed. You go to a Marvel movie, but afterward all you have is a memory. You can buy a pass to ski at Vail and enjoy the experience, but you don't own the ski lift. Sometimes it's a challenge that customers can't see, feel, or smell a service before they buy it. For example, a person who wants advice from an accountant doesn't know in advance how good the advice will be.

To reduce this uncertainty, service customers often seek referrals from friends or advice from online reviews. They may also look for cues to help them judge the quality of a service before they buy. That's why some service providers emphasize physical evidence of quality. A lawyer is likely to have diplomas on the wall, shelves loaded with books, and furnishings that suggest success.

A consumer can't hold a service and look at it before purchasing, so service firms often use messages and images in their promotion that help make the benefits of the service experience more tangible. Allstate's "Mayhem" ads are a vivid reminder of the value of a good insurance policy.
Source: Allstate Insurance Company

Where the product is produced

Goods are typically mass-produced in a factory far away from the customer. A service is usually produced in person—where the customer is located—*after* the customer has committed to making the purchase. It is often difficult to achieve economies of scale with personal services. One reason is that service suppliers often need duplicate equipment and staff at places where the service is actually provided. Charles Schwab sells investment advice along with financial products worldwide. That advice could, perhaps, be produced more economically in a single building in New York City and made available only on its website. But Charles Schwab has offices all over the world because many customers want a personal touch from their stockbroker.

Local production affects quality

A worker in a factory that makes Whirlpool appliances can be in a bad mood and customers will never know. Even if there are production problems, quality controls are likely to catch defective goods before they leave the factory. Service quality often isn't that consistent; one reason is that it's hard to separate the service experience from the person who provides it. A rude teller in a bank can drive away customers. Service providers also vary in their ability, and problems with the service they deliver are usually obvious to customers. In addition, when many people must all work well together—as in a hospital or on a cruise ship—it's even more of a challenge to deliver consistent service quality. In these more complex service delivery situations, systems should be set up to facilitate coordination.

Services cannot be stored

Services are perishable. They can't be produced and then stored to sell at some future time when more customers want to buy. This makes it difficult to balance supply and demand, especially if demand varies a lot. At Thanksgiving, Southwest Airlines has to turn away customers because most of its flights are fully booked; potential passengers may use a different mode of transportation or not travel at all. Perhaps Southwest could buy more planes and hire more pilots, but most of the time that would result in costly excess capacity—planes flying with empty seats. Southwest conducts financial analysis to optimize the cost from lost customers at peak times against the costs from empty seats at other times.

Because of problems like this, airlines, doctors, hotels, and other service firms sometimes charge fees to clients who don't show up when they say they will. Service organizations also use a variety of approaches to shift customer demand to less busy times. Movie tickets are cheaper for afternoon shows, restaurants offer early-bird specials, and hotels that cater to business travelers promote weekend getaways. Firms also try to reduce the dissatisfaction that customers may feel if they must wait for service. Golf courses provide practice greens, and some doctors' offices provide comfortable seating and magazines.

Adding services to goods to differentiate

When competitors focus only on physical goods, a firm may differentiate its offering by adding a service valued by the target market. Many companies make high-quality televisions, but Panasonic's research revealed that some consumers worried about how to set up a fancy new TV when they got it home. So Panasonic added Plasma Concierge service to support its customers with well-trained advisors and priority in-home service. Along the same lines, some customers liked IKEA furniture (the "goods") but really didn't like that they were supposed to assemble it on their own. So, for an additional fee, IKEA now offers assembly services.[7]

Technology and Intelligent Agents Add Value to Products

LO 8.3

Firms often seek to add customer value to a core product offering. These days, many find that technology—often a website or software application—provides additional value and improves a customer's experience and satisfaction. The addition of technology can also help differentiate a product and offer competitive advantage in markets where competing products are seen as undifferentiated. Accordingly, many firms look at integrating technology with a core product as a major product decision area. Let's take a closer look at how some firms use technology to supplement a core offering.

Technology integrated into a product

Many products that were previously not thought of as "tech" products now include technology. Cars have been called "computers on wheels." Computers now track maintenance and deliver entertainment. Some cars even parallel park on their own. Another example is in photography. Taking good pictures used to require knowledge of f-stops, shutter speed, and film ISO, but now all that "expertise" is in the camera's software: it evaluates the scene and decides the best settings for taking a picture. Even home appliances are getting computerized. LG recently introduced a refrigerator with a 29-inch touchscreen that stores recipes and sends text messages alerting users of needed ingredients.[8]

Apps add services to a core product

Another opportunity to provide value follows from software applications that offer value-adding services to a core product. For example, the app for Subway can help customers find a nearby restaurant, place an order, and even pay. When customers told GoPro they loved the video camera but hated editing their recordings, GoPro bought the software app Splice to make editing easier. These services supplement the core products at Subway and GoPro and add value for customers.

Augmented reality adds to an experience

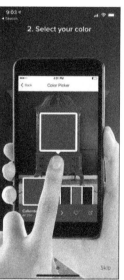

Augmented reality (AR) overlays a computer-generated image, sound, text, or video on a user's view of the physical world. Popularized in the game *Pokémon Go,* augmented reality can occur when users view the "real world" through their smartphone while an AR app overlays that picture with useful information. For example, with the Discover Moscow photo app on their smartphones, travelers in Moscow can virtually "find" famous Russians from history and take selfies with them. The app offers tips on where to look as travelers experience an interactive historical tour of the Russian capital.

AI anticipates customer needs

In earlier chapters we introduced three emerging technology advances that have begun to change many aspects of marketing—artificial intelligence, intelligent agents, and analytics. *Artificial intelligence (AI)* refers to having machines operate like humans with respect to learning and decision making; *intelligent agents* are devices that observe an environment and act to achieve a

Paint company Benjamin Moore's Color Capture app helps customers find the paint that matches a color they find anywhere. Take a photo of the color you want to match (say a dresser) and the app finds the perfect match with a Benjamin Moore paint. Yelp's augmented reality feature allows customers to hold their smartphone up to a street and view restaurant reviews for all restaurants on the street. *(top-left and top-right):* Source: Benjamin Moore & Co.; *(bottom):* Source: Yelp Inc.

goal. *Analytics* are used to help AI and intelligent agents anticipate customer needs. Today many products try to forecast what a customer will want next, perhaps even before the customer knows it.

Google Now is an intelligent agent that operates on Android smartphones. The "digital assistant" turns information from a user's online search activity, location, e-mail, and calendar into wisdom to predict what the user needs next. So when Google Now pulls together information that a business traveler is flying home today and is currently driving a rental car to the airport, it can suggest gas stations near the airport where the tank can be refilled—before the traveler even thinks of this idea. This ability to predict customer needs brings wisdom to products.[9]

Google is a leader in offering AI services to make people's lives easier. For example, AI can save Gmail users time. After opening an e-mail, "Smart Reply" suggests three possible e-mail responses. A user simply clicks on whichever one is preferred. Over time, Smart Reply learns a user's habits and preferences, and future suggestions become more customized and reflect the user's natural communication style.[10]

#M4BW

AI helps individuals with disabilities

AI services can make the world a better place for people with disabilities. For example, to help blind customers shop, Chinese retailing giant Alibaba developed an intelligent, interactive overlay for smartphones. The clear silicone sheet gives users easy access to three buttons that provide shortcuts for getting around on an app. The smart screen also "reads" a page's content to the user, who can then click buttons in response. Better yet, it costs less than $0.05 to make one.[11]

AI often requires integrating information about customers from different sources. The Internet offers a way to connect those information sources and may offer even more ways for technology to offer value. Read more in *What's Next?* The Internet of Things connects products.

Sellers get data too

While customers see value from the integration of technology into products, it also offers value to firms that collect and analyze the data. Tracking customers' everyday behavior through these apps might provide ideas about underlying customer needs and potential new product features. It might also help a company like Google deliver more targeted promotions. Think back to the example we offered of the business traveler and Google Now. Google knows what airline that customer flies, how he or she purchases flights (price sensitive or time sensitive), the customer's rental car choice, and even his or her driving habits. Of course that information helps the AI to offer better service, but it can also be sold to advertisers who want to deliver targeted messages to the traveler. Companies must manage the legal and ethical responsibilities that come with collecting and storing these data (see Chapter 7).

Branding Is a Strategy Decision

LO 8.4

There are so many brands—and we're so used to seeing them—that we take them for granted. But branding is an important decision area, so we will treat it in some detail.

What is branding?

Branding means the use of a name, term, symbol, or design—or a combination of these—to identify a product. It includes the use of brand names, trademarks, and practically all other means of product identification.

Brand name has a narrower meaning. A **brand name** is a word, letter, or a group of words or letters. Examples include Verizon Wireless, WD-40, 3M Post-its, and GoPro Hero.

What's Next? The Internet of Things connects products

Think about some of the physical objects you see and use every day: a car tire, T-shirt, stop sign, thermostat, coffeemaker, smartphone, garage door, traffic light, or a lawn sprinkler system. Inside each of these objects, a sensor, actuator, or data communications technology can be embedded to connect the object with the Internet or some other data network. Once connected, the objects can be tracked, coordinated, and controlled. This connection allows these goods to deliver new services. These interconnected objects are called the *Internet of Things (IoT)*. Just now getting started, many believe the Internet of Things is *What's Next?* in delivering customer value. Let's take a look at some of the features and benefits the IoT could soon deliver.

Imagine your alarm clock rings and tells your coffeemaker to start brewing your morning cup so it is ready by the time you get to the kitchen. In fact, with its artificial intelligence, the clock learns (by simply observing when it rings and when you grab the coffee) that you take 20 minutes to shower and brush your teeth before you get to the coffee—so it delays the brewing to be fresh and hot at that time. The sprinkler system is smart, too. It knows how long and when to run—its sensors know about recent rainfall *and* the weather forecast for the next few days. As you leave home to go shopping, the front door locks, garage door closes, and lights turn off—all automatically. The opposite happens when you return.

As you enter the grocery store parking lot, cameras recognize your license plate and your phone lights up with coupons based on your past purchases. In the store, your smartphone loads a shopping list prepared by your refrigerator and pantry—where sensors know what you like to keep on hand and can tell what's there and what you need. Arriving at the cash register, you find that your cartload has already been rung up and charged to your bank account—just bag it all up and be on your way. That evening, as you approach your favorite local watering hole for a drink, your phone warns you that your "ex" is inside. Uh oh, the next move is yours.

Businesses and governments already leverage IoT technology for sustainability and efficiency. Cameras at businesses know when everyone has gone home; at that point, the building automatically turns out the lights and adjusts the heating or cooling to save energy. Truck delivery routes are programmed to minimize delay, adjusting on the fly as traffic patterns change throughout a day. Visual sensors on railroad cars inspect tracks in real time, providing feedback on sections of track most in need of repair. Sensors in roads adjust the timing of traffic lights to optimize traffic flow (#M4BW).

With sensor prices plummeting, analysts predict that by the early 2020s, more than 50 billion "things" will contain this technology. Opportunities for enhancing products with technology services has the potential to offer individuals, businesses, and governments more convenience and greater efficiency. The challenge for marketing managers will be to recognize real customer needs—and figure out if and how the IoT can deliver them.[12]

How could a connection to the Internet of Things add value for owners of an automobile? A refrigerator? A lawn mower?

Trademark is a legal term. A **trademark** includes only those words, symbols, or marks that are legally registered for use by a single company. For example, PowerPoint, Bubble Wrap, and Q-Tips are all trademarks that refer to products from Microsoft, Sealed Air Corporation, and Unilever, respectively. A **service mark** is the same as a trademark except that it refers to a service offering.[13]

The word *FedEx* can be used to explain these differences. The FedEx overnight delivery service is branded under the brand name FedEx (whether it's spoken or printed in any manner). When "FedEx" is printed in a certain typeface, however, it becomes a trademark. A trademark need not be attached to the product. It need not even be a word—it can be a symbol. Exhibit 8-5 shows some common trademarks.

Brands meet needs

Well-recognized brands make shopping easier. Think of trying to buy groceries, for example, if you had to evaluate each of the 25,000 items every time you went to a supermarket. Many customers are willing to buy new things—but having gambled and lost, they like to buy a sure thing the next time. Brand names connect a product with the

Exhibit 8-5 Recognized Trademarks and Symbols Help in Promotion

A trademark includes only those words, symbols, or marks that are legally registered for use by a single company. Companies carefully design logos. Once they are established and well recognized, they usually change very little over the years. In 2014, the Pepsi globe logo was updated to give it a more contemporary look, though it is similar to previous logos. The Red Bull logo has changed little from its origins as a drink to help Thai workers work longer. While Adobe's original logo featured the company's full name, "Adobe Systems Incorporated," on a gray background, its more familiar white "A" on a red background has been in use since 1990. Goodyear has used its winged foot logo since 1900. Audi's four rings represented four different automakers that merged—only Audi survives, but it retains the historic logo. Marvel started as a comic book brand, and for many years the logo included the word *Comics*. The brand is now more well known for its movie franchise and uses a simpler logo.
Bashigo/Shutterstock Images, LLC

benefits a customer can expect. The connection may be learned from past consumer experience, from the firm's promotion, or in other ways. Customers *trust* the brand name if they consistently have a positive experience with the brand or they hear good things from the firm's promotion or other customers. If their marketing mixes work, customers trust that every time they eat *Certified Angus Beef* it will be "tender, high-quality meat" and that their oil changes at *Jiffy Lube* will always be "fast and convenient." If a brand consistently delivers on a promise that target customers consider important, those customers will pay a premium price for the certainty that comes with that brand name.

Brand promotion has other advantages for branders as well as customers. A good brand reduces the marketer's selling time and effort. Good brands can also improve the company's image—speeding acceptance of new products marketed under the same name. For example, many consumers quickly tried Listerine PocketPaks breath fresheners when they appeared because they already knew they trusted Listerine mouthwash.[14]

What conditions favor branding?

Although branding helps customers make buying decisions, branding is not always easy to do. Building a well-respected brand name can be costly. It doesn't always make sense as part of a marketing strategy. In some industries, well-known brands are just not that common. Think about it. Can you recall a brand name for file folders, bed frames, electric extension cords, or nails? The following are conditions that can make a market more favorable to successful branding:

1. The product is easy to label and identify by brand or trademark.
2. The product quality is easy to maintain and the best value for the price.
3. Dependable and widespread availability is possible. When customers start using a brand, they want to be able to continue using it.
4. Demand is strong enough that the market price can be high enough to make the branding effort profitable.
5. There are economies of scale. If the branding is really successful, costs should drop and profits should increase.
6. Favorable shelf locations or display space in stores help. This is something retailers can control when they brand their own products.

Branding may be more or less valuable in different countries

In general, these conditions are less common in developing economies, and that may explain why efforts to build brands in emerging markets often fail. For example, one study found Chinese consumers willing to pay a premium of only 2 percent for branded products they regularly purchase—as compared to premiums of 20 percent or more in developed countries.[15]

Achieving Brand Familiarity

Today, familiar brands exist for most product categories, ranging from crayons (Crayola) to real estate services (RE/MAX). Nevertheless, brand acceptance must be earned with a good product and regular promotion. **Brand familiarity** means how well customers recognize and accept a company's brand. The degree of brand familiarity affects the planning for the rest of the marketing mix—especially where the product should be offered and what promotion is needed.

Five levels of brand familiarity

Understanding the five levels of brand familiarity can be useful for strategy planning: (1) rejection, (2) nonrecognition, (3) recognition, (4) preference, and (5) insistence.

Some brands have been tried and found wanting. **Brand rejection** means that potential customers won't buy a brand unless its image is changed—or if the customers have no other choice. Rejection may suggest a change in the product or perhaps only a shift to target customers who have a better image of the brand. Overcoming a negative image is difficult and can be very expensive. Sometimes customers have no other choice. Many music and sports fans would like to reject Ticketmaster, which sells tickets to sports and music events. But customers often have little choice when it comes time to buy tickets to see their favorite band or team perform live. Those issues have helped jump-start competitors for Ticketmaster, though the company is working hard to reverse that negative customer attitude.[16]

Brand rejection is a big concern for service-oriented businesses because it's hard to control the quality of service. A business traveler who gets a dirty room in a Hilton Hotel in Rio de Janeiro, Brazil, might not return to a Hilton anywhere. Yet it's difficult for Hilton to ensure that every housekeeper does a good job every time.

Thinking its image might be getting a bit stale, Tropicana updated the packaging for its orange juice. It ditched the longtime Tropicana brand symbol of a colorful orange pierced by a straw for the package on the right showing a muted glass of orange juice. Sales fell as some loyal Tropicana customers failed to recognize their regular brand on store shelves. Soon the original package design was back in stores. Customers appreciate a familiar trademark.
Courtesy of Joseph P. Cannon, Ph.D

Brand nonrecognition means customers are not aware

Some products are seen as basically the same. **Brand nonrecognition** means final consumers don't recognize a brand at all—even though intermediaries may use the brand name for identification and inventory control. Examples of categories where brand nonrecognition is common include school supplies, inexpensive dinnerware, many of the items that you'd find in a hardware store, and thousands of dot-coms on the Internet.

Brand recognition means that customers remember the brand. This may not seem like much, but it can be a big advantage if there are many "nothing" brands on the market. Even if consumers can't recall the brand without help, they may be reminded when they see it in a store among other less familiar brands.

Branders seek brand preference or brand insistence

Most branders would like to win **brand preference**—which means that target customers usually choose the brand over other brands, perhaps because of habit or favorable past experience. **Brand insistence** means customers insist on a firm's branded product and are willing to search for it. Loyalty is one benefit; branders also know that customers who insist on a brand are most likely to let others know of their "love" and spread positive word-of-mouth about the product.[17]

A brand is likely to have target customers at each level of brand familiarity. Ideally, customers move to higher levels of brand familiarity (brand preference and brand insistence) over time. Marketing strategies often aim to encourage this movement. Consider the café-bakery chain Panera Bread. It's possible that some target customers may reject the brand because of a bad previous experience. However, if Panera provides consistently excellent service and tries to recognize and compensate for (hopefully) rare instances of poor service, those who reject the brand should be a very small group. Other customers in the target market may not recognize the brand and need promotion that is directly aimed at them, possibly in the form of advertising. With successful promotion, these customers will likely move to the brand recognition level. Because higher levels of brand familiarity will probably require firsthand experience, coupons or additional promotion efforts may be needed to encourage an actual visit to a café. If Panera delivers a good value (great-tasting food and good service at a fair price), then more customers are likely to develop brand preference. Customers who are truly delighted may move to brand insistence and recommend Panera to their friends and family.[18]

Marketing managers measure brand awareness

While marketing managers wish that all target customers insisted on their brand, they know that process starts with brand recognition. To assess that, firms often use a measure of *brand awareness*. To learn how marketing managers measure brand awareness and how they use it to aid in strategy development, see Marketing Analytics in Action: Brand Awareness.

For most consumers, roofing shingles fall in the category of brand nonrecognition. Consumers do not buy roofing shingles often, and few know the names of any brand. Apple not only has very high brand recognition, but a high level of brand insistence. These customers are lined up outside an Apple Store to be among the first to get their hands on the latest iPhone.
(left): Nycshooter/iStock/Getty Images; (right): Jessicakirsh/Shutterstock

Marketing Analytics in Action: Brand Awareness

A marketing manager for Rio Centro, a Tennessee-based chain of Mexican restaurants, wanted to measure brand awareness of Rio Centro restaurants in three Tennessee cities: Memphis, Nashville, and Knoxville. A telephone survey was conducted. To focus on a target market of people who eat at Mexican restaurants, the survey did not include those who answered no to the question, "In the past year, have you eaten at a Mexican restaurant?" Those answering yes to the question were then asked two additional questions:

- "Would you please name all the Mexican restaurants in your city that come to mind?" (The percentage of people answering "Rio Centro" to this question is the level of "*unaided* brand awareness" in the following table.)
- If participants in the survey did not mention Rio Centro in question 1, they were then asked, "Have you ever heard of Rio Centro Mexican restaurant?" (Column 2 "Aided Brand Awareness" includes the percentage of people who answered yes to this question, *plus*, those listing Rio Centro in question 1.)

	Unaided Brand Awareness	Aided Brand Awareness
Knoxville	64%	81%
Nashville	18%	41%
Memphis	31%	42%

1. *What are some reasons why you see differences across the three cities with respect to unaided brand awareness?*
2. *What are some reasons why you see differences across the three cities with respect to aided brand awareness?*
3. *Why do you think Memphis and Nashville differ by so much with respect to unaided awareness, but not in aided brand awareness?*

The right brand name can help

A good brand name can help build brand familiarity. It can help tell something important about the company or its product. For more than 50 years, Kraft Foods used the name A.1. Steak Sauce. As steak declined in popularity, Kraft dropped "Steak" from the name to give the product broader appeal. A name change to an iconic 50-year-old brand is not taken lightly.

Naming a brand can be art and science. Exhibit 8-6 lists some characteristics of a good brand name. Lululemon, Outback, Jelly Belly, Bonobos, and DieHard are examples of successful brand names that fit some of these criteria. Can you think of others?

Companies that compete in international markets face a special problem in selecting brand names. A name that conveys a positive image in one language may be meaningless in another. Or, worse, it may have unintended meanings. British food company Sharwood discovered this after spending millions of dollars launching a curry sauce called *Bundh*. Unfortunately one of the target markets were Punjabi speakers, who thought the word sounded like the Punjabi word for "backside." IKEA uses Swedish and Norwegian words as brand names, bringing a unique character to the brand. But when IKEA opened a store in Thailand, it discovered the *Redalen* bed and *Jättebra* plant pot sounded a lot like crude Thai terms related to sex.[19]

Exhibit 8-6 Characteristics of a Good Brand Name

- Short and simple
- Easy to spell and read
- Easy to recognize and remember
- Easy to pronounce
- Can be pronounced in only one way
- Can be pronounced in all languages (for international markets)

- Suggestive of product benefits
- Adaptable to packaging/labeling needs
- No undesirable imagery
- Always timely (does not go out-of-date)
- Adaptable to any advertising medium
- Legally available for use (not in use by another firm)

A respected name builds brand equity

Because it's costly to build brand recognition, some firms prefer to acquire established brands rather than try to build their own. The value of a brand to its current owner or to a firm that wants to buy it is sometimes called **brand equity**—the value of a brand's overall strength in the market. For example, brand equity is likely to be higher if many satisfied customers insist on buying the brand and if retailers are eager to stock it. Customers are more likely to buy and pay higher prices. That almost guarantees ongoing profits.

The most valuable brands are familiar and include Apple, Google, McDonald's, AT&T, Coca-Cola, Disney, and SAP. While consulting firms that estimate brand value use different methodologies, they agree that brand value is significant. One study of brands recently assigned the following values for the top five brands: Google's brand at $302 billion, Apple's at $301 billion, Amazon's at $208 billion, Microsoft's at $201 billion, and China's Tencent at $179 billion.[20]

Laws protect brand names and trademarks

With all that value wrapped up in a brand, firms must protect brand names and trademarks. Fortunately for those investing in brands, U.S. common law and civil law protect the rights of trademark and brand name owners. The **Lanham Act** (of 1946) spells out what kinds of marks (including brand names) can be protected and the exact method of protecting them. The law applies to goods shipped in interstate or foreign commerce.

The Lanham Act does not force registration, but registering under the Lanham Act is often a first step toward protecting a trademark to be used in international markets. That's because some nations require that a trademark be registered in its home country before they will register or protect it.

Brands need to be actively protected

While there are laws in place, brand owners must carefully monitor their brands. For example, a brand name or trademark can become public property if it falls into common usage for a product category. This happened with (formerly) brand names like cellophane, aspirin, shredded wheat, and kerosene.

Burger chain In-N-Out's menu and the colors in its restaurant are part of its trademark. Before In-N-Out had restaurants in Utah, a restaurant called Chadder's opened using In-N-Out's same menu and color scheme. If In-N-Out were to let this stand, these features might become public property. In-N-Out sued; a judge ruled against Chadder's, and it soon went out of business.[21]

Counterfeiting is accepted in some cultures

Even when products are properly registered, counterfeiters may make unauthorized copies. Counterfeit products cause a brand to lose sales and jeopardize its reputation. Many well-known brands—ranging from Levi's jeans to Rolex watches to Zantac medicine—face this problem. International trade in counterfeit and pirated goods may exceed $500 billion annually. Counterfeiting continues to grow and is especially common in developing countries where regulation is weak or cultural values differ. In Azerbaijan and Bulgaria, BP discovered counterfeit BP service stations—with low-quality fuel. And at a knockoff Apple Store in China, many employees think they are working at the real thing.[22]

Counterfeit products are not unusual in China. This counterfeit LEGO toy was being sold in Beijing, China. LEGO makes strong efforts to curb sales of these products.
Tatchaphol/Shutterstock

Branding Decisions: What Kind? Who Brands?

Marketing managers have some different options when it comes to branding. Two questions involve the kind of brand to use and who should do the branding (see Exhibit 8-7). Let's explore these two questions to understand the choices.

Keep it in the family

Branders of more than one product must decide whether they are going to use a **family brand**—the same brand name for several products—or individual brands for each product. Examples of family brands are Keebler snack food products and Whirlpool appliances.

The use of the same brand for many products makes sense if all are similar in type and quality. The main benefit is that the goodwill attached to one or two products may help the others. Money spent to promote the brand name benefits more than one product, which cuts promotion costs for each product.

A special kind of family brand is a **licensed brand**—a well-known brand that sellers pay a fee to use. For example, the familiar Sunkist brand name has been licensed to many companies for use on more than 400 products in 30 countries. Betty Crocker sells Sunkist Lemon Bar mix, Jelly Belly makes Sunkist Fruit Gems candy, and Dr Pepper produces Sunkist orange soda. Sunkist earns extra revenue and its

Pillsbury licensed the Girl Scouts brand to produce Thin Mints Brownie Mix. Thin Mints is a trademark and the name of one of the Girl Scout Cookies.
Keith Homan/Shutterstock

Exhibit 8-7 Different Types of Branding Questions

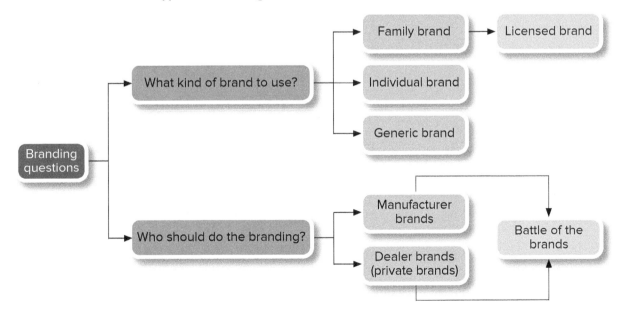

partners get instant brand recognition. That said, Sunkist must carefully choose partners that sell quality products and promote the brand in ways that won't harm its carefully curated image.²³

Individual brands for outside and inside competition

A company uses **individual brands**—separate brand names for each product—when it's important for the products to each have a separate identity, as when products vary in quality or type.

If the products are really different, such as Tide (laundry detergent) and Pampers (baby diapers), then the one company that owns them (in this case P&G) is likely to choose individual brands and avoid confusion. Some firms use individual brands with similar products to make segmentation and positioning efforts easier. For example, when General Mills introduced a line of organic cereals, it used the Cascadian Farm name and the Big G logo was not on the box. The rationale was that consumers who try to avoid additives might not trust a big corporate brand.²⁴

Generic "brands"

Products that some consumers see as commodities may be difficult or expensive to brand. Some manufacturers and intermediaries have responded to this problem with **generic products**—products that have no brand at all other than identification of their contents and the manufacturer or intermediary. Generic products are usually offered in plain packages at lower prices. They are quite common in less-developed nations.²⁵

The online retailer Brandless sells only generic products. None of the products Brandless carries have a brand name.
Source: Brandless, Inc.

Manufacturer brands versus dealer brands

Manufacturer brands are brands created by producers. These are sometimes called *national brands* because the brand is promoted all across the country or in large regions. Note, however, that many manufacturer brands are now distributed globally. Such brands include Nabisco, Colgate, Northwestern Mutual Life, Marriott, Mastercard, and McDonald's.

Dealer brands, also called **private brands** (or private label), are brands created by intermediaries. Examples of dealer brands include Primo Taglio and Priority Pet (Safeway), Up & Up (Target), and Sam's Choice and Equate (Walmart).

Who's winning the battle of the brands?

Manufacturer and dealer brands often sell in the same retail stores. The **battle of the brands**—the competition between dealer brands and manufacturer brands—is just a question of which brands will be more popular and who will be in control. At one time, manufacturer brands were much more popular than dealer brands. In some categories, such as soft drinks, they still are. But in categories such as milk and cheese, dealer brands are very strong contenders. In the United States, almost 30 percent of grocery store purchases are now dealer brands, with percentages much higher in many European countries and very low in most Asian and South American countries. Over the past 30 years, dealer brands have been slowly gaining the upper hand. During the recent economic downturn, more consumers were motivated to try dealer brands—and many were pleasantly surprised by the quality and the variety they found.

Brand names tend to do better in some categories. For example, in the United States, private label accounts for less than 2 percent of hair care sales. Hair care brands tend to be innovative and have a high level of advertising, which results in higher perceived product differentiation and more loyal customers. On the other hand, in developed

When customers come to a store, they often have a choice between similar products where a dealer brand is offered alongside manufacturer brands. While Post's Raisin Bran and Cascadian Farm are manufacturer brands, in Walmart they compete with the Great Value dealer brand.
Editorial Image, LLC

countries about 40 percent of all milk sales are from dealer brands. Milk buyers tend to purchase frequently and see little differentiation among brands.[26]

Retailers motivated to develop dealer brands

There is plenty of motivation for intermediaries to develop dealer brands. The intermediary usually earns a better margin on the sale of a dealer brand. Strong dealer brands also give an intermediary leverage in negotiations with the manufacturer brand.

In the grocery trade, retailers are putting more attention behind new dealer brand products. Kroger and Safeway have hired marketing managers with experience at major national brands. Recently, almost a third of all new-product introductions were for dealer brands, up from less than 10 percent just a few years ago. Traditionally, dealer brands were knockoffs of manufacturer brands—often of lower quality with lower prices. Now some dealer brands sell premium products. Kroger's popular "Private Selection" line of frozen pizzas is priced similar to the manufacturer brand DiGiorno. Kroger conducts its own marketing research, which recently led to changes in the packaging of its Pet Pride pet food brand. It learned that dog owners like to see packages with dogs and owners playing together, while cat owners just want to see a happy cat. New packages reflect those changes.

Intermediaries have some advantages in this battle. With the number of large retail chains growing, they are better able to arrange reliable sources of supply at low cost. They can also give dealer brands special shelf position and promotion. Because manufacturer brands typically need to build strong brand preference or brand insistence through heavy promotion and quality ingredients—both of which raise costs and lead to higher prices—national brands' prices are about 30 percent higher on average. Still, national brands are not going to disappear. Intermediaries know that many customers are still attracted to Mountain Dew, Colgate, Tide, and Cheerios.[27]

Packaging Promotes, Protects, and Enhances

LO 8.5

Packaging involves promoting, protecting, and enhancing the product. Packaging can be important to both sellers and customers (see Exhibit 8-8). It can make a product more convenient to use or store. It can prevent spoiling or damage. Good packaging makes products easier to identify and promotes the brand at the point of purchase and even in use.

Exhibit 8-8 Some Ways Packaging Benefits Consumers and Marketers

Opportunity to Add Value	Some Decision Factors	Examples
Promoting	Link product to promotion	The bunny on the Energizer battery package is a reminder that it "keeps going and going."
	Branding at point of purchase or consumption	Coke's logo greets almost everyone each time the refrigerator is opened.
	Product information	Nabisco's nutrition label helps consumers decide which cookie to buy, and a UPC code reduces checkout time and errors.
Protecting	For shipping and storing	Cardboard inserts and a box protect the Roku streaming player.
	From tampering	Tylenol's safety seal prevents tampering.
	From shoplifting	Cardboard hang-tag on Gillette razor blades is too large to hide in hand.
	From spoiling	Kraft's shredded cheese has a resealable zipper package to keep it fresh.
Enhancing product	The environment	Starbucks develops compostable coffee cups and eliminates straws from cold drink cups.
	Convenience in use	The package for Better Oats brand oatmeal includes a built-in measuring cup pouch so consumers don't need a measuring cup.
	Added product functions	Plastic tub is useful for refrigerator leftovers after the Cool Whip is gone.

Packaging can lower distribution costs

Better protective packaging is very important to manufacturers and wholesalers. They sometimes have to pay the cost of goods damaged in shipment. Retailers need protective packaging too. It can reduce storing costs by cutting down on breakage, spoilage, and theft. Good packages also save space and weight so they are easier to transport, handle, and display—and better for the environment.[28]

#M4BW

Greener packaging creates value for buyers and sellers

U.S. shoppers generate much more trash per person than anywhere else on earth. Much of what is tossed out is packaging. It overloads landfills, litters streets, and pollutes the environment. For years plastic seemed to be the perfect packaging material because it is clean, light, and durable; unfortunately, this means it's everywhere and lasts forever. Even colorful package graphics are troublesome. The ink to print them often has toxins that later creep into the soil and water. Firms should try to give consumers what they want, but in applying that logic to packaging, many people have been shortsighted. Businesses and consumers alike have acted as if there was nothing an *individual* could do to reduce environmental problems. Now that attitude is changing.

A growing number of consumers are interested in making greener choices. Firms are making packaging choices that are better for the environment. For example, Timberland makes its shoe boxes with 100 percent recycled materials, soy-based inks, and water-based glues. Whole Foods Market uses salad bar containers made from sugarcane waste that safely turn into compost within about 90 days. It makes sense for firms to publicize such efforts to attract like-minded consumers. Publicity also calls attention to the idea that even small changes can add up to big improvements.

Laws reduce confusion

The **Federal Fair Packaging and Labeling Act** (of 1966) requires that consumer goods be clearly labeled in easy-to-understand terms to give consumers more information. The law also calls on industry to try to reduce the confusing number of package sizes and make labels more useful. Since then, there have been further guidelines. The most far-reaching are based on the Nutrition Labeling and Education Act of 1990, which requires food manufacturers to use a uniform format that allows consumers to compare the nutritional value of different products. Recently there have been more changes, including requirements to clearly show the fat content of food and ingredients that trigger common food allergies.

The courts tend to rule against misleading labels. For example, the Supreme Court ruled that Coca-Cola (which makes many different lines of juices) could be sued by POM Wonderful for calling one of its juices "Pomegranate-Blueberry" even though it contained less than 1 percent pomegranate juice.[29]

Ethical decisions remain

Although various laws provide guidance on many packaging issues, many areas still require marketing managers to make ethical choices. For example, some firms have been criticized for designing packages that conceal a downsized product, giving consumers less for their money. Similarly, some retailers design packages and labels for their private-label products that look just like, and are easily confused with, manufacturer brands. Are efforts such as these unethical, or are they simply an attempt to make packaging a more effective part of a marketing mix? Different people will answer differently.

Laws provide details about how nutrition facts must be reported on food packages.
Editorial Image, LLC

Many critics think that labeling information is too often incomplete or misleading. For example, what does it really mean if a label says a food product is "organic" or "low fat"? How far should a marketing manager go in putting potentially negative information on a package? Should Häagen-Dazs affix a label that says "This product will clog your arteries"? That sounds extreme, but what type of information *is* appropriate?

Some products require a great deal of packaging—which may not be good for the environment. Meal delivery services like HelloFresh and Blue Apron have to package their products in heavy-duty cardboard boxes, with insulated liners, and dry ice. That is a lot of "recycling" for one dinner. Many consumers like the convenience that accompanies the myriad product and packaging choices available. Is it unethical for a marketing manager to give consumers with different preferences a choice? Some critics argue that it is. Others praise firms that give consumers choices.[30]

Product Classes Help Plan Marketing Strategies

LO 8.6

So far in this chapter, we've focused on key strategy decisions for Product (see Exhibit 8-1). Managers usually try to blend those decisions in a unique way to differentiate the firm's offering and create superior customer value. However, you don't have to treat *every* product as unique when planning strategies—some classes of products benefit from similar marketing mixes. So now we'll introduce these product classes and show why they are a useful starting point for developing marketing mixes for new products and for evaluating present mixes.

Product classes start with type of customer

All products fit into one of two broad groups—based on the type of customer that will use them. **Consumer products** are products meant for the final consumer. **Business products** are products meant for use in producing other products. The same product *might* be both a consumer product and a business product. See the pair of ads for Bush's Baked Beans and read the caption for an example.

There are product classes within each group. Consumer product classes are based on how consumers think about and shop for products. Business product classes are based on how buyers think about products and how they'll be used.

Many items in Bush's Best line of food products sell as both consumer and business products. Consumers can purchase Bush's Baked Beans in a 28-ounce can at their local grocery store and serve them for dinner. On the other hand, a restaurant buys a case of six 117-ounce cans of Bush's Baked Beans from a wholesaler. A restaurant uses them as a side dish and sells them as part of a meal. Bush Brothers & Company uses different marketing mixes to reach the different target markets.
Source: Bush Brothers & Company

Consumer Product Classes

Consumer product classes divide into four groups: (1) convenience, (2) shopping, (3) specialty, and (4) unsought. *Each class is based on the way people think about and shop for products.* See Exhibit 8-9 for a summary of how these product classes relate to marketing mixes.[31]

Convenience products—purchased quickly with little effort

Convenience products are products a consumer needs but isn't willing to spend much time or effort shopping for. These products are bought often, require little service or selling, don't cost much, and may even be bought by habit. A convenience product may be a staple, impulse product, or emergency product.

Staples are products that are bought often, routinely, and without much thought—such as breakfast cereal, canned soup, and most other packaged foods used almost every day in almost every household.

Impulse products are products that are bought quickly—as *unplanned* purchases—because of a strongly felt need. True impulse products are items that the customer hadn't planned to buy, decides to buy on sight, may have bought the same way many times before, and wants right now. If the buyer doesn't see an impulse product at the right time, the sale may be lost.[32]

Emergency products are products that are purchased immediately when the need is great. The customer doesn't have time to shop around when a traffic accident occurs, a thunderstorm begins, or an impromptu party starts. The price of the ambulance service, raincoat, or ice cubes won't be important.

Shopping products are compared

Shopping products are products that a customer feels are worth the time and effort to compare with competing products. Shopping products can be divided into two types, homogeneous or heterogeneous, depending on what customers are comparing. The exact same products can be homogeneous or heterogeneous. It just depends on how a particular target market thinks about and shops for the product.

Homogeneous shopping products are items the customer sees as basically the same and wants at the lowest price. So, for example, a target market that believes "all gasoline is the same" seeks out and buys gas at the station with the lowest prices. Another target

Exhibit 8-9 Consumer Product Classes and Marketing Mix Planning

Consumer Product Class	Marketing Mix Considerations	Consumer Behavior
Convenience Products		
Staples	Maximum exposure with widespread, low-cost distribution; mass selling by producer; usually low price; branding is important.	Routinized (habitual); low effort; frequent purchases; low involvement.
Impulse	Widespread distribution with display at point of purchase.	Unplanned purchases bought quickly.
Emergency	Need widespread distribution near probable point of need; price sensitivity low.	Purchase made with time pressure when a need is great.
Shopping Products		
Homogeneous	Need enough exposure to facilitate price comparison; price sensitivity high.	Customers see little difference among alternatives and seek lowest price.
Heterogeneous	Need distribution near similar products; promotion (including personal selling) to highlight product advantages; less price sensitivity.	Extensive problem solving; consumer may need help in making a decision (salesperson, website, etc.).
Specialty Products		
	Price sensitivity is likely to be low; limited distribution may be acceptable, but should be treated as a convenience or shopping product (in whichever category product would typically be included) to reach persons not yet sold on its specialty product status.	Willing to expend effort to get specific product, even if not necessary; strong preferences make it an important purchase; Internet becoming important information source.
Unsought Products		
New unsought	Must be available in places where similar (or related) products are sought; needs attention-getting promotion.	Need for product not strongly felt; unaware of benefits or not yet gone through adoption process.
Regularly unsought	Requires very aggressive promotion, usually personal selling.	Aware of product but not interested; attitude toward product may even be negative.

market could believe that all 32-inch Ultra HD televisions are basically the same—so they shop the Internet for the lowest price. A low-cost producer might try to promote that its products are "just as good as" higher-priced alternatives. Vizio has used this strategy with considerable success in the television market.

Heterogeneous shopping products are items the customer sees as different and wants to inspect for quality and suitability—for most market segments, furniture, clothing, and membership in a spa are good examples. In this situation, target consumers seek information from a knowledgeable salesperson or a reputable website. Case in point: *Some* customers who *do* see 32-inch Ultra HD televisions as different will visit local stores to see various models and talk to a salesperson, then go to websites like CNET.com to read professional and user reviews.

Branding may be less important for heterogeneous shopping products. The more carefully consumers compare price and quality, the less they rely on brand names or labels. Some retailers carry competing brands so consumers won't go to a competitor to compare items.

Specialty products—no substitutes please!

Specialty products are consumer products that the customer really wants and makes a special effort to find. Shopping for a specialty product doesn't mean comparing—the buyer wants that special product and is willing to search for it. It's the customer's *willingness to search*—not the extent of searching—that makes it a specialty product.

Any branded product that consumers insist on by name is a specialty product. Marketing managers want customers to see their products as specialty products and ask for

Impulse products are bought quickly as unplanned purchases. They are more likely to be found if they are distributed at point of purchase. For its target customers, Christian Louboutin shoes are a specialty product. Price sensitivity is usually low, and customers will make a special effort to find them.
(left): Alamy, Inc.; (right): smartphone frame: Shutterstock Images, LLC; shoe: Neiman Marcus

them over and over again. Building that kind of relationship isn't easy. It means satisfying the customer every time. However, that's easier and a lot less costly than trying to win back dissatisfied customers or attract new customers who are not seeking the product at all.

Unsought products need promotion

Unsought products are products that potential customers don't yet want or know they can buy, so they don't search for them at all. In fact, consumers probably won't buy these products if they see them—unless promotion can show their value.

There are two types of unsought products. **New unsought products** are products offering really new ideas that potential customers don't know about yet. Informative promotion can help convince customers to accept the product, ending its unsought status. Dannon's yogurt and Litton's microwave ovens are popular items now, but initially they were new unsought products.

Regularly unsought products are products—such as gravestones, life insurance, and nursing homes—that stay unsought but not unbought forever. There may be a need, but potential customers aren't motivated to satisfy it. For this kind of product, personal selling is *very* important.

Many nonprofit organizations try to "sell" their unsought products. For example, the American Red Cross regularly holds blood drives to remind prospective donors of how important it is to give blood.

Brand managers for pain reliever Panadol Ultra faced two challenges in rural Guatemala: (1) a high level of illiteracy and (2) an unsought product because many in the target market didn't know about using pills for pain relief. Panadol created a sponge in the shape of the pill that local women could use as a cushion when carrying heavy items on their heads—a common practice. Soon women became a walking ad for the brand and sales increased by 45 percent among the target market.
Source: Panadol Ultra; Agency: Saatchi & Saatchi/Guatemala

One product may be seen in several ways

The same product might be seen in different ways by different target markets at the same time. For example, a product viewed as a staple by most consumers in the United States, Canada, or some similar affluent country might be seen as a heterogeneous shopping product by consumers in another country. The price might be much higher when considered as a proportion of the consumer's budget, and the available choices might be very different. Similarly, for some people salsa is seen as a staple; for others—who, for example, think it is worth tracking down W. B. Williams Georgia Style Peach Salsa on the Internet—it is a specialty product.

Business Products Are Different

Business product classes are different from consumer product classes because they relate to how and why business firms make purchases. Thus, knowing the specific classes of business products helps in strategy planning. First, however, it's useful to note some important ways that the market for business products is different from the market for consumer products.

One demand derived from another

The big difference between the consumer products market and the business products market is **derived demand**—the demand for business products derives from the demand for final consumer products. For example, car manufacturers buy about one-fifth of all steel products. But if demand for cars drops, they'll buy less steel. Then even the steel supplier with the best marketing mix is likely to lose sales.[33]

Price increases might not reduce quantity purchased

Total *industry* demand for business products is fairly inelastic. Business firms must buy what they need to produce their own products. Even if the cost of basic silicon doubles, for example, Intel needs it to make computer chips. However, sharp business

Business product classes are different from consumer product classes because they relate to how and why business firms make purchases. The demand for many business products is derived and depends on the demand for products where it is used. When Philips sells its television sets to hotels, demand depends on how many new hotels are being built—or how many rooms may need upgrading. For a hotel, the television would be a capital item that can be used and depreciated for many years. Boston Scientific's "Target Detachable Coils with Tenzing Technology" are often used in surgery, so the demand for the product depends on consumer demand for surgery.
(left): Source: Koninklijke Philips N.V.; (right): Source: Boston Scientific Corporation

buyers try to buy as economically as possible. So the demand facing *individual sellers* may be extremely elastic—if similar products are available at a lower price.

Tax treatment affects buying too

How a firm's accountants—and the tax laws—treat a purchase is also important to business customers. An **expense item** is a product whose total cost is treated as a business expense in the year it's purchased. A **capital item** is a long-lasting product that can be used and depreciated for many years. Often it's very expensive. Customers pay for the capital item when they buy it, but for tax purposes the cost is spread over a number of years. This may reduce the cash available for other purchases.

Business Product Classes—How They Are Defined

Business product classes are based on how buyers think about products and how the products will be used. The classes of business products are (1) installations, (2) accessories, (3) raw materials, (4) components, (5) supplies, and (6) professional services. Exhibit 8-10 relates these product classes to marketing mix planning.

Installations—a boom-or-bust business

Installations—such as buildings, land rights, and major equipment—are important capital items. One-of-a-kind installations—such as office buildings and custom-made machines—generally require special negotiations for each sale. Negotiations often involve top management and can stretch over months or even years. Standardized major equipment is treated more routinely.

Exhibit 8-10 Business Product Classes and Marketing Mix Planning

Business Product Classes	Marketing Mix Considerations	Buying Behavior
Installations	Usually requires skillful personal selling by producer, including technical contacts, or understanding of applications; leasing and specialized support services may be required.	Multiple buying influence (including top management) and new-task buying are common; infrequent purchase, long decision period, and boom-or-bust demand are typical.
Accessory equipment	Need fairly widespread distribution and numerous contacts by experienced and sometimes technically trained personnel; price competition is often intense, but quality is important.	Purchasing and operating personnel typically make decisions; shorter decision period than for installations; Internet sourcing.
Raw materials	Grading is important, and transportation and storing can be crucial because of seasonal production and/or perishable products; markets tend to be very competitive.	Long-term contract may be required to ensure supply; online auctions.
Component parts and materials	Product quality and delivery reliability are usually extremely important; negotiation and technical selling typical on less-standardized items; replacement aftermarket may require different strategies.	Multiple buying influence is common; online competitive bids used to encourage competitive pricing.
Maintenance, repair, and operating (MRO) supplies	Typically require widespread distribution or fast delivery (repair items); arrangements with appropriate intermediaries may be crucial.	Often handled as straight rebuys, except important operating supplies may be treated much more seriously and involve multiple buying influence.
Professional services	Services customized to buyer's need; personal selling very important; inelastic demand often supports high prices.	Customer may compare outside service with what internal people could provide; needs may be very specialized.

Installations are a boom-or-bust business. During growth periods, firms may buy installations to increase capacity. But during a downswing, sales fall off sharply.[34]

Specialized services are needed as part of the product

Suppliers sometimes include special services with an installation at no extra cost. A firm that sells (or leases) equipment to dentists, for example, may install it and help the dentist learn to use it.

Accessories—important but short-lived capital items

Accessories are short-lived capital items—tools and equipment used in production or office activities—such as Canon's small copy machines, Rockwell's portable drills, and Steelcase's filing cabinets. Accessories are more standardized than installations and they're usually needed by more customers.

Because these products cost less and last a shorter time than installations, multiple buying influence is less important. Operating people and purchasing agents, rather than top managers, may make the purchase decision. As with installations, some customers may wish to lease or rent—to expense the cost.

Raw materials become part of a physical good

Raw materials are unprocessed expense items—such as logs, iron ore, and wheat—that are moved to the next production process with little handling. Unlike installations and accessories, *raw materials become part of a physical good and are expense items*.

There are two types of raw materials: (1) farm products and (2) natural products. **Farm products** are grown by farmers—examples are oranges, sugarcane, and cattle. **Natural products** are products that occur in nature—such as timber, iron ore, oil, and coal.

The need for grading is one of the important differences between raw materials and other business products. Nature produces what it will—and someone must sort and grade raw materials to satisfy various market segments.

Most buyers of raw materials want ample supplies in the right grades for specific uses—fresh vegetables for Green Giant's production lines or logs for Weyerhaeuser's paper mills. To ensure steady quantities, raw materials customers often sign long-term contracts, sometimes at guaranteed prices.

Component parts and materials must meet specifications

Components are processed expense items that become part of a finished product. Component *parts* are finished (or nearly finished) items that are ready for assembly into the final product. Intel's microprocessors included in personal computers and TRW's air bags in cars are examples. Component *materials* are items such as wire, plastic, or textiles. They have already been processed but must be processed further before becoming part of the final product. Quality is important with components because they become part of the firm's own product.

Some components are custom-made. In this case, teamwork between the buyer and seller may be needed to arrive at the right specifications, and a buyer may develop a close partnership with a dependable supplier. In contrast, standardized component materials are more likely to be purchased online using a competitive bidding system.

Because component parts go into finished products, a replacement market often develops. Car tires are components originally sold in the *OEM (original equipment market)* that become consumer products in the *aftermarket*.[35]

Supplies for maintenance, repair, and operations

Supplies are expense items that do not become part of a finished product. Supplies can be divided into three types: (1) maintenance, (2) repair, and (3) operating supplies—giving them their common name: MRO supplies.

Maintenance and small operating supplies are like convenience products. The item will be ordered because it is needed—but buyers won't spend much time on it. For such "nuisance" purchases branding is important, and so are breadth of assortment and the seller's dependability. Intermediaries usually handle the many supply items. They are often purchased via online catalog sites.[36]

Important operating supplies, such as coal and fuel oil, receive special treatment. Usually there are several sources for such commodity products—and large volumes may be purchased at global exchanges on the Internet.

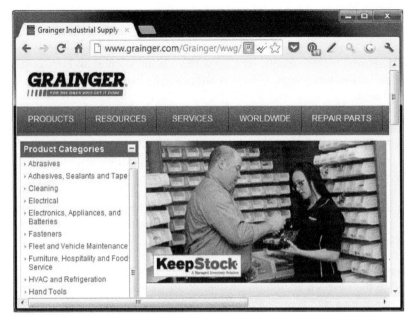 

Business product classes are based on what buyers think about products and how the product will be used. Grainger is a wholesaler that offers a wide range of maintenance, repair, and operating supplies (often called MRO supplies). EY is a consulting company that offers professional services.
(left): Source: W.W. Grainger, Inc.; (right): Source: Ernst & Young LLP

Professional services—
pay to get it done

Professional services are specialized services that support a firm's operations. They are usually expense items. Management consulting services can improve the company's efficiency. Information technology services can maintain a company's networks and websites. Advertising agencies can help promote the firm's products. And food services can improve morale.

Managers compare the cost of buying professional services outside the firm (*outsourcing*) to the cost of having company people do them. Work that was previously done by an employee is now often purchased from an independent specialist. Clearly, the number of service specialists is growing in our complex economy.

CONCLUSION

In this chapter, we looked at Product broadly—which is the right vantage point for marketing strategy planning. We saw that a product may be a good or a service, or some combination of both. And we saw that a firm's Product is what it offers to *satisfy the needs of its target market*—which may include the customer's experience both before and after the purchase. We also described some key marketing differences between goods and services.

We reviewed the Product area strategy decisions required for branding and packaging—and saw how the right decisions can add value for customers and give a product a competitive edge. Customers view a brand as a guarantee of quality, which leads to repeat purchases, lower promotion costs, higher sales figures, and greater customer equity. Packaging offers promotional opportunities and can inform customers. Variations in packaging can also help a product appeal to different segments of the market. And packaging can help protect the product anywhere in the channel of distribution.

The brand familiarity a product earns is a measure of the marketing manager's ability to carve out a separate market. Therefore, ultimately, brand familiarity affects Place, Price, and Promotion decisions. Strategy planning for the marketing mix will vary across product classes. We introduced both consumer product classes (based on *how consumers think about and shop for products*) and business product classes (based on *how buyers think about products and how they'll be used*). In addition, we showed how the product classes affect planning marketing mixes.

230

KEY TERMS

Product, 206
quality, 206
warranty, 207
individual product, 207
product line, 207
product assortment, 207
product line length, 207
service, 208
augmented reality (AR), 211
branding, 212
brand name, 212
trademark, 213
service mark, 213
brand familiarity, 215
brand rejection, 215
brand nonrecognition, 215
brand recognition, 216
brand preference, 216
brand insistence, 216

brand equity, 218
Lanham Act, 218
family brand, 219
licensed brand, 219
individual brands, 220
generic products, 220
manufacturer brands, 220
dealer brands, 220
private brands, 220
battle of the brands, 220
packaging, 221
Federal Fair Packaging and Labeling Act, 222
consumer products, 223
business products, 223
convenience products, 224
staples, 224
impulse products, 224
emergency products, 224

shopping products, 224
homogeneous shopping products, 224
heterogeneous shopping products, 225
specialty products, 225
unsought products, 226
new unsought products, 226
regularly unsought products, 226
derived demand, 227
expense item, 228
capital item, 228
installations, 228
accessories, 229
raw materials, 229
farm products, 229
natural products, 229
components, 229
supplies, 229
professional services, 230

QUESTIONS AND PROBLEMS

1. Review the Under Armour case study that opens this chapter. From this case, identify examples of different key terms and concepts covered in the chapter. Kohl's Tek Gear is an example of a dealer brand.

2. Review the Under Armour case study that opens this chapter. Using three different consumer product classes as an example, describe how a customer would think about and buy Under Armour products. Then examine the marketing mix considerations in Exhibit 8-8 for each of the product classes you chose. What would Under Armour need to do to support the marketing strategy for each?

3. Discuss several ways in which physical goods are different from pure services. Give an example of a good and then an example of a service that illustrates each of the differences.

4. What products are being offered by a shop that specializes in bicycles? By a travel agent? By a supermarket? By a new car dealer?

5. Consumer services tend to be intangible, and goods tend to be tangible. Use an example to explain how the lack of a physical good in a pure service might affect efforts to promote the service.

6. Explain some of the different aspects of the customer experience that could be managed to improve customer satisfaction if you were the marketing manager for (a) an airport branch of a rental car agency, (b) a fast-food restaurant, (c) an online firm selling software directly to consumers from a website, and (d) a hardware store selling lawn mowers.

7. Explain how technology could be used to create value for the following: (a) hair salon, (b) quick oil change shop, (c) passenger car tires, and (d) toaster.

8. Is there any difference between a brand name and a trademark? If so, why is this difference important?

9. List five brand names and indicate what product is associated with the brand name. Evaluate the strengths and weaknesses of the brand name.

10. Explain family brands. Should Best Buy carry its own dealer brands to compete with some of the popular manufacturer brands it carries? Explain your reasons.

11. What does the degree of brand familiarity imply about previous and future promotion efforts? How does the degree of brand familiarity affect the Place and Price variables?

12. Give an example where packaging costs probably (a) lower total distribution costs and (b) raise total distribution costs.

13. How would the marketing mix for a staple convenience product differ from the mix for a homogeneous shopping product? How would the mix for a specialty product differ from the mix for a heterogeneous shopping product? Use examples.

14. In what types of stores would you expect to find (a) convenience products, (b) shopping products, (c) specialty products, and (d) unsought products?
15. For the most relevant target market, what kinds of consumer products are the following: (a) smartwatches, (b) automobiles, and (c) toothpastes? Explain your reasoning.
16. What kinds of business products are the following: (a) lubricating oil, (b) electric motors, and (c) a firm that provides landscaping and grass mowing for an apartment complex? Explain your reasoning.
17. For the kinds of business products described in this chapter, complete the following table (be brief, using one or a few well-chosen words).
    A. *Kind of distribution facility (or facilities) needed and functions it (they) will provide.*
    B. *Caliber of salespeople required.*
    C. *Kind of advertising required.*

Products	A	B	C
Installations			
Buildings and land rights			
Major equipment			
Standard			
Custom-made			
Accessories			
Raw materials			
Farm products			
Natural products			
Components			
Supplies			
Maintenance and small operating supplies			
Important operating supplies			
Professional services			

MARKETING PLANNING FOR HILLSIDE VETERINARY CLINIC

Appendix D (the Appendices follow Chapter 19) includes a sample marketing plan for Hillside Veterinary Clinic. Look through the "Marketing Strategy" section.

1. What goods does Hillside Veterinary Clinic sell?
2. What services does Hillside Veterinary Clinic sell?
3. What consumer product classes are offered by Hillside Veterinary Clinic?
4. The discussion of product classes in this chapter indicates what marketing mix is typical for different classes of products. Does the marketing strategy recommended in Hillside's marketing plan fit with those considerations? Why or why not?

SUGGESTED CASES

1. McDonald's "Seniors" Restaurant
3. NOCO United Soccer Academy
7. Lake Pukati Lodge
13. Paper Products, Inc. (PPI)
28. Kingston Home Health Services (KHHS)

Video Case 4. Invacare

MARKETING ANALYTICS: DATA TO KNOWLEDGE

CHAPTER 8: BRANDING DECISION

Wholesteen Dairy, Inc. produces and sells Wholesteen brand condensed milk to grocery retailers. The overall market for condensed milk is fairly mature, and there's sharp competition among dairies for retailers' business. Wholesteen's regular price to retailers is $8.88 a case (24 cans). FoodWorld—a fast-growing supermarket chain and Wholesteen's largest customer—buys 20,000 cases of Wholesteen's condensed milk a year. That's 20 percent of Wholesteen's total sales volume of 100,000 cases per year.

FoodWorld is proposing that Wholesteen produce private-label condensed milk to be sold with the FoodWorld brand name. FoodWorld proposes to buy the same total quantity as it does now, but it wants half (10,000 cases) with the Wholesteen brand and half with the FoodWorld brand. FoodWorld wants Wholesteen to reduce costs by using a lower-quality can for the FoodWorld brand. That change will cost Wholesteen $0.01 less per can than it costs for the cans that Wholesteen uses for its own brand. FoodWorld will also provide preprinted labels with its brand name—which will save Wholesteen an additional $0.02 a can.

Wholesteen spends $70,000 a year on promotion to increase familiarity with the Wholesteen brand. In addition, Wholesteen gives retailers an allowance of $0.25 per case for their local advertising, which features the Wholesteen brand. FoodWorld has agreed to give up the advertising allowance for its own brand, but it is willing to pay only $7.40 a case for the milk that will be sold with the FoodWorld brand name. It will continue under the old terms for the rest of its purchases.

Sue Glick, Wholesteen's marketing manager, is considering the FoodWorld proposal. She has entered cost and revenue data on a spreadsheet so she can see more clearly how the proposal might affect revenue and profits.

Design element: #M4BW box globe icon: ©Vectoryzen/Shutterstock

CHAPTER NINE

Source: iRobot

Product Management and New-Product Development

The founders of iRobot didn't know that their company would become the world leader in home robots. However, from the start they didn't intend to just imitate other firms' products. Instead, they wanted to create totally new product concepts that would change society. That is where the company is headed, with products that make life easier for its customers. Let's take a closer look at how iRobot got to where it is today.

One of iRobot's early projects involved developing concepts for robotic toys for Hasbro. During this time, it learned the importance of cost control; for two years, most of the toy ideas iRobot pitched to Hasbro were rejected. While technically elegant, they were too expensive. These experiences helped iRobot focus on an idea for a low-priced robotic vacuum cleaner—later named the Roomba®.

Roomba's new-product developers understood the importance of taking the customer's perspective. Early Roomba prototypes went home with iRobot employees, where their spouses, friends, and neighbors could test them. The developers gained valuable feedback and quickly learned that not everyone was technically savvy. When design engineers talked about how best to train customers to use the Roomba, the team realized that such thinking was backward. Instead of training customers, they designed Roomba's software to figure out what to do when a customer pushed the start button. Roomba is so simple to use that most people don't even need the few-pages-long owner's manual.

The original idea was for Roomba to just have sweeper brushes. However, in focus groups, consumers said that Roomba also needed a vacuum. Responding to this consumer input added extra expense late in the project, but iRobot was ready because of the cost-control lessons learned from working with Hasbro. With Roomba, developers had controlled every penny from the outset and could afford to add the vacuum to provide consumers the value they wanted.

Before iRobot introduced Roomba, the vacuum cleaner market was mature, with few breakthrough product ideas. Then iRobot opened up the "robot vacuum market" with Roomba. It's a slick-looking, 15-inch disk that is less than 4 inches tall. It's a robot, but it requires no programming. You simply place it on the floor, press a button to turn it on, and Roomba scurries around the room doing its job. It can scoot under the sofa, avoid the furniture, and return to its battery charger when the job is done.

Marketing managers wanted to sell Roomba to a large market segment—people who hate to vacuum. But they were concerned Roomba's appeal might be limited to a small market of gadget-loving techies. To offset concerns about the complexity of the technology, introductory promotion described Roomba as an intelligent vacuum cleaner. For the first few years, the word *robot* didn't even appear on the package (except in the iRobot company name).

For its launch, iRobot worked with specialty retailers such as Sharper Image and Brookstone. These retailers were willing to show videos of Roomba in action and train their salespeople to explain and demonstrate Roomba in their stores. The extra promotion push was important because initially customers didn't know the brand name and were not looking for this sort of product. However, iRobot quickly generated publicity for Roomba—with everything from appearances on the *Today* show and videos on YouTube, to reviews in magazines, newspapers, and at CNET.com. All of this media attention—and some traditional advertising—propelled sales and quickly made Roomba a familiar brand. Soon iRobot expanded distribution to Bed Bath & Beyond, Target, and Amazon.

By controlling costs, iRobot initially priced Roomba at $200. A higher price might have been acceptable to some customers, but that low price helped fend off potential competitors, at least for a while. Soon the market started to grow and other competitors, including Electrolux, P3, Neato, Eufy, and Dyson, jumped into the market. While competition forced selective price discounting, iRobot's patent-protected technology and new models with new features keep it ahead of the pack; it still commands more than 60 percent of the robotic vacuum cleaner market.

For example, its Internet-connected Roomba 980 includes a smartphone app so it can be controlled from anywhere; just open the app, push the "Clean" button, and voilà! Roomba starts to work. Or, even easier, just ask your Google Home or Amazon Echo to "vacuum the living room." The new Roomba 980 also has more cleaning power than previous models; the

AeroForce 3-stage cleaning system uses agitation, extraction, and suction. Continuous improvement and new features like these differentiate Roomba and give it a competitive advantage. This helps iRobot generate higher margins and a better return on investment on the whole Roomba product line, which ranges in price from $300 to $1,100.

While 90 percent of iRobot's sales come from Roomba, iRobot is bringing robot assistance to other cleaning needs: Looj clears gutters, Mirra scrubs pools, and Braava cleans hardwood and tile floors. But iRobot sees the most potential for its new lawn mower, Terra. Terra was in the new-product development process for more than a decade. This new-product project was challenging; iRobot actually gave up on it twice before overcoming technological barriers. While Terra isn't the first mower in this small but growing product-market, iRobot thinks it might have an advantage being a follower in this market. Terra overcame many of the limitations of the pioneering robot lawn mowers. With the iRobot app, GPS, and its proprietary Imprint Smart Mapping system, Terra is easy to set up and control. iRobot anticipates rapid growth and "Roomba-sized" potential for Terra.

iRobot sees a future that includes all of us getting more help from robots, so iRobot supports STEM (Science, Technology, Education, and Math) education with programs that get kids excited about robotics. For example, the iRobot Create 2 Programmable Robot gives educators and students a prototype robot they can program.

iRobot's creative product development and marketing strategies have fueled remarkable growth. In the last 25 years, iRobot has sold more than 25 million robots worldwide, has sales that exceed $1 billion, and has profits of more than $100 million. To keep that going, iRobot must continue to fill its pipeline with new products to meet tomorrow's customers' needs.[1]

LEARNING OBJECTIVES

Developing new products and managing them for profitable growth are keys to success for most firms. Yet many new products fail. Even products that succeed face new challenges as competition becomes more intense. So the marketing strategy that supported the product's initial success usually needs to change as the market evolves. This chapter will help you understand this evolution and how it relates to effective new-product development and creative strategy changes for existing products—both of which are crucial to attracting and retaining target customers.

When you finish this chapter, you should be able to

1. understand how product life cycles affect strategy planning.
2. describe what is involved in designing new products and what "new products" really are.
3. understand the new-product development process.
4. appreciate the team effort that goes into new-product development.
5. understand the need for product or brand managers.
6. understand how total quality management can improve goods and services.
7. understand important new terms (shown in **red**).

Innovation and Market Changes Create Opportunities

Successful new products, like those in the iRobot case, are critical in driving profitable growth for both new and established companies. iRobot pioneered a fast-growing new product-market—and "computer-controlled cleaning tools" are meeting customer needs in new ways. Similarly, in Chapter 5 we looked at how the iPod, iPhone, and other innovations in digital media have changed personal entertainment. In fact, all around us there is a constant life-and-death struggle where old products are replaced by new products. Before Crest Whitening Strips, the only path to whiter teeth involved undergoing expensive procedures in a dentist's office. And it remains to be seen whether new STōK Cold Brew Coffee will change how we imbibe a morning pick-me-up or whether Halo Top ice cream will lead to better frozen treats. These recent new-product launches have attracted initial interest from consumers.[2]

#M4BW

Innovation is needed to solve global challenges

Think about some of the grand challenges facing our planet. Problems like climate change and hunger will require innovative solutions. Today, many businesses are stepping up with new products that help address these challenges. For example, recent advances have cut the cost of solar power by more than 60 percent. And advances in battery technology will allow us to store electricity produced in the light of day for use at night.

Other innovations may help deal with both hunger and climate change. While the developing world's demand for meat continues to grow, more meat also means more greenhouse gases. Emerging innovations in plant-based "meat" like the Beyond Burger and the Impossible Burger provide eco-friendly alternatives for meat lovers. While there remains a long way to go before these grand challenges are solved, innovative companies and new products offer hope for a better world.[3]

Managing products and new-product development

These innovations show that products, customer behavior, and competition change over time. These changes create opportunities for marketing managers and pose challenges as well. Developing new products and managing existing products to meet changing conditions are important to the success of every firm. In Chapter 8 we looked at important strategy planning decisions that need to be made for new products and sometimes changed for existing products.

In this chapter, we'll look at how successful new products are developed in the first place—and what marketing managers need to know and do to manage their growth. We'll start by explaining the cycle of growth and decline that new-product innovations go through and how marketing strategy typically evolves over that cycle. When you understand the stages in this cycle, you can see *why* it is so critical for a firm to have an effective new-product development process—and why the challenges of managing a product change as it matures (see Exhibit 9-1).

Exhibit 9-1 The Role of Product Management and New-Product Development in Marketing Strategy

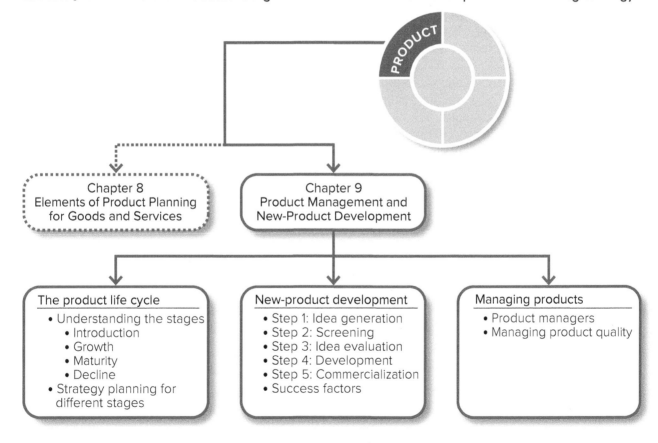

Managing Products over Their Life Cycles

LO 9.1

Revolutionary products create new product-markets. But competitors are always developing and copying new ideas and products—making existing products out-of-date more quickly than ever. Products, like consumers, go through life cycles.

Product life cycle has four major stages

The **product life cycle** describes the stages a new-product idea goes through from beginning to end. The product life cycle is divided into four major stages: (1) market introduction, (2) market growth, (3) market maturity, and (4) sales decline. The product life cycle is concerned with new types (or categories) of products in the market, not just what happens to an individual brand.

A particular firm's marketing mix usually must change during the product life cycle. There are several reasons why customers' attitudes and needs may change over the product life cycle. The product may be aimed at entirely different target markets at different stages. And the nature of competition moves toward pure competition or oligopoly.

Further, total sales of the product—by all competitors in the industry—vary in each of its four stages. They move from very low in the market introduction stage to high at market maturity and then back to low in the sales decline stage. More important, the profit picture changes too. These general relationships can be seen in Exhibit 9-2. Note that sales and profits do not move together over time. *Industry profits decline while industry sales are still rising.*[4]

Market introduction— investing in the future

In the **market introduction** stage, sales are low as a new idea is first introduced to a market. Customers aren't looking for the product. Even if the product offers superior value, customers don't even know about it. Informative promotion is needed to tell potential customers about the advantages and uses of the new-product concept.

Even though a firm promotes its new product, it takes time for customers to learn that the product is available. Most companies experience losses during the introduction stage because they spend so much money for Product, Place, and Promotion development. Of course, they invest the money in the hope of future profits.

Market growth—profits go up and down

In the **market growth** stage, industry sales grow fast—but industry profits rise and then start falling. The innovator begins to make big profits as more and more customers buy. But competitors see the opportunity and enter the market. After East African Breweries created a sensation in Nigeria with its nonalcoholic malt beverage Alvaro, Coca-Cola followed eight months later with its own malt drink Novida.[5] Some just copy the most successful product or try to improve it to compete better. Others try to refine their offerings to do a better job of appealing to some target markets.

Exhibit 9-2 Typical Life Cycle of a New-Product Concept

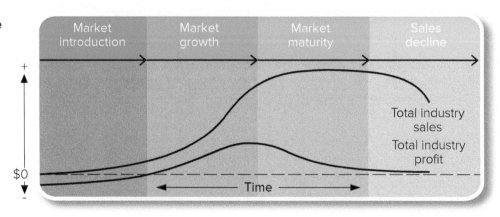

The new entries result in much product variety. So monopolistic competition—with down-sloping demand curves—is typical of the market growth stage.

This is the time of biggest profits *for the industry.* It is also a time of rapid sales and earnings growth for companies with effective strategies. *But it is toward the end of this stage when industry profits begin to decline* as competition and consumer price sensitivity increase (see Exhibit 9-2).

Market maturity—sales level off, profits decline

The **market maturity** stage occurs when industry sales level off and competition gets tougher. Many aggressive competitors have entered the race for profits—except in oligopoly situations. Industry profits go down throughout the market maturity stage because promotion costs rise and some competitors cut prices to attract business. Less efficient firms can't compete with this pressure—and they drop out of the market. There is a long-run downward pressure on prices.

New firms may still enter the market at this stage—increasing competition even more. Note that late entries skip the early life-cycle stages, including the profitable market growth stage. And they must try to take a share of the saturated market from established firms, which is difficult and expensive. The market leaders have a lot at stake, so they fight hard to defend their share. Customers who are happy with their current relationship won't switch to a new brand, so late entrants usually have a tough battle.

Persuasive promotion becomes even more important during the market maturity stage. Products may differ only slightly. Most competitors have discovered effective appeals or just copied the leaders. As the various products become almost the same in the minds of potential consumers, price sensitivity is a real factor.[6]

In the United States, the markets for most cars, most household appliances, and many consumer packaged goods such as breakfast cereal, carbonated soft drinks, and laundry detergent are in market maturity. This stage may continue for many years—until a basically new product idea comes along—even though individual brands or models come and go.

In the market maturity stage, sales level off and more brands may be competing. Brands need to use persuasive advertising and show how their marketing mix is differentiated from those of their many competitors. In this case, Pedigree has made a dog food designed especially for strong dogs; Jardiland has a formula for dogs that may need to lose a few pounds.
(left): Source: Mars, Inc.; (right): Source: Jardiland

Sales decline—a time of replacement

During the **sales decline** stage, new products replace the old. Price competition from dying products becomes more vigorous—but firms with strong brands may make profits until the end because they have successfully differentiated their products.

As the new products go through their introduction stage, the old ones may keep some sales by appealing to their most loyal customers or those who are slow to try new ideas. These conservative buyers might switch later—smoothing the sales decline.

Product life cycles don't relate to individual products

Product life cycles describe *industry* sales and profits for a *product idea* within a particular product-market. The sales and profits of an individual brand may not, and often do not, follow the life-cycle pattern. They may vary up and down throughout the life cycle—sometimes moving in the opposite direction of industry sales and profits.

A firm may introduce or drop a specific product during *any* stage of the product life cycle. For example, Apple launched Apple Music well after Spotify debuted. Its strong brand name and the ecosystem created by its smartphones and computers have helped it gain share and get established. On the other hand, the social media site Google+ arrived after Facebook, failed to offer an appealing marketing mix, and eventually shut down.

Each market should be carefully defined

A product idea can also be in a different life-cycle stage in different markets. For example, in the United States, milk is in the market maturity stage. U.S. consumers drink 18 times more milk than Asian consumers do—where milk is in the market introduction stage. Some firms in the dairy business are trying to grow the Asian market. To appeal to the Asian palate, they are selling milk with added flavors such as ginger and honey.[7]

Strategy planners who naively expect sales of an individual product to follow the general product life-cycle pattern are likely to be surprised. In fact, it might be more sensible to think in terms of "product-market life cycles" rather than product life cycles—but we will use the term *product life cycle* because it is commonly accepted and widely used.

Product Life Cycles Vary in Length

How long a whole product life cycle takes—and the length of each stage—varies widely across products. The cycle may vary from a few years—for example, pocket-size video cameras were quickly replaced by smartphones—to more than 100 years for gas-powered cars.

The product life-cycle concept does not tell a manager precisely *how long* the cycle will last. But a manager can often make a good guess based on the life cycles for similar products. Sometimes marketing research can help too.

Product life cycles are getting shorter

Although the life cycles for different products vary, in general, product life cycles are getting shorter. This is partly due to rapidly changing technology. One new invention may make possible many new products that replace old ones. Tiny electronic microchips led to thousands of new products—from Texas Instruments calculators in the early days to microchip-controlled heart valves now.

Think about how many years it took for some of the world's major innovations to reach 25 percent of the U.S. population: it took electricity 46 years; the telephone, 35 years; the personal computer, 16 years; and the Internet, 7 years. How soon will it be before 25 percent of the U.S. population has a drone or a driverless car?

Why some products grow more quickly

A new-product idea will move quickly through the early stages of the product life cycle when the innovation has certain characteristics. For example, the greater the *comparative advantage* of a new product over those already on the market, the more rapidly its sales will grow. Sales growth is also faster when the product is *easy to use* and if its advantages are *easy to communicate.* If the product *can be tried* on a limited basis—without a lot of risk to the customer—it can usually be introduced more quickly. Finally, if the product is *compatible* with the values and experiences of target customers, they are likely to buy it more quickly.[8]

The fast adoption of smartphones is a good example. When smartphones came onto the scene, they were *compatible* with consumers who were already using mobile phones and computers. Early smartphones were *easy to use,* and customer familiarity made the benefits *easy to communicate* as well. It helped that consumers often saw friends using one—making it easy to *try* one without much risk. Finally, smartphones were compatible with the values and experiences of target customers—though many people still don't appreciate it when friends text or check their Instagram in the middle of dinner.

New products often create new product-markets and new product life cycles but also bring a premature end to another product life cycle. Over time, generations of mobile phones evolved into today's smartphone, which provides the same features of several different products from the 1990s: flashlight, alarm clock, television, calculator, audio recorder, video camera, notepad, music player, camera, and phone. The combination may cost about the same (it depends), but most customers value the convenience of having one device with all those products in their pocket. Looking to the future, GE's tweet asks, "What do you think smartphones will be like in 2050, according to 2015 tech?"
Source: General Electric

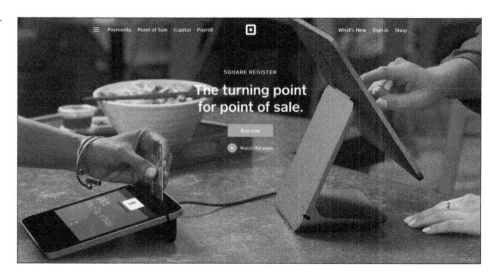

New products that do a better job of meeting the needs of specific target customers are more likely to move quickly and successfully through the introductory stage of the product life cycle. Square Register allows individuals and retailers to accept credit cards on a tablet computer. Such convenience, coupled with low cost, helped the product move swiftly into the market growth stage of the product life cycle.
Source: Square, Inc.

Pioneer or follower—which strategy works best?

The product life cycle means that firms must be developing new products all the time. Further, they must try to use marketing mixes that will make the most of the market growth stage—when profits are highest. The question becomes, is it better to be the *pioneer*—the first to market with a new-product idea—or a *follower*?

On average, pioneers tend to be less profitable over the long run—in part because many do not survive. That said, there are some real success stories among pioneers. For example, FedEx invented overnight delivery and remains the market leader. More often, though, there is an advantage to being the *second-mover*—one that quickly follows the pioneer. Second-movers that have a strong customer focus and respond quickly with a superior marketing mix can build market share during the market growth stage.

There are many examples of successful second-movers. Although many consider Apple to be an innovator, the iPod, iPhone, and iPad were not the first of their kind to market. Yet Apple responded quickly with a better marketing mix in each of these product categories. Likewise, Amazon wasn't the first online bookstore. RC Cola had the first diet cola on the market, but Coke and Pepsi quickly copied the idea and took over the market. A nimble second-mover can learn from the pioneer's mistakes and quickly come to market with a marketing mix that provides superior value.[9]

The short happy life of fashions and fads

The sales of some products are influenced by **fashion**—the currently accepted or popular style. Fashion-related products tend to have short life cycles. What is currently popular can shift rapidly. A certain color or style of clothing—baggy jeans, miniskirts, wayfarer sunglasses, or four-inch-wide ties—may be in fashion one season and outdated the next. Marketing managers who work with fashions often have to make very fast product changes.

A **fad** is an idea that is fashionable only to certain groups who are enthusiastic about it. But these groups are so fickle that a fad is even more short-lived than a regular fashion. Many toys—whether digital pets like Tamagotchi or shaped rubber bands like Silly Bandz—are fads, but they do well during a short-lived cycle.[10]

Planning for Different Stages of the Product Life Cycle

Length of cycle affects strategy planning

Where a product is in its life cycle—and how fast it's moving to the next stage—should affect marketing strategy planning. Marketing managers must make realistic plans for the later stages. Exhibit 9-3 shows the relationship of the product life cycle to the marketing mix variables. The technical terms in this figure are discussed later in the book.

Introducing new products

Exhibit 9-3 shows that a marketing manager has to do a lot of strategy planning to introduce a really new product. Money must be spent developing the new product. Even if the product is unique, this doesn't mean that everyone will immediately come running to the producer's door. The firm will have to build channels of distribution—perhaps offering special incentives to win cooperation. Promotion is needed to build demand *for the whole idea,* not just to sell a specific brand. Because all of this is expensive, it may lead the marketing manager to try to "skim" the market—charging a relatively high price to help pay for the introductory costs.

The correct strategy, however, depends on how quickly the new idea will be accepted by customers—and how quickly competitors will follow with their own versions of the product. When the early stages of the cycle will be fast, a low initial (penetration) price may make sense to help develop loyal customers early and keep competitors out.

Be prepared to pivot to a new marketing mix

It can be difficult to predict customer response to a really new product. So marketing managers need to carefully monitor initial customer reactions and be prepared to

Exhibit 9-3 Typical Changes in the Marketing Mix over the Product Life Cycle

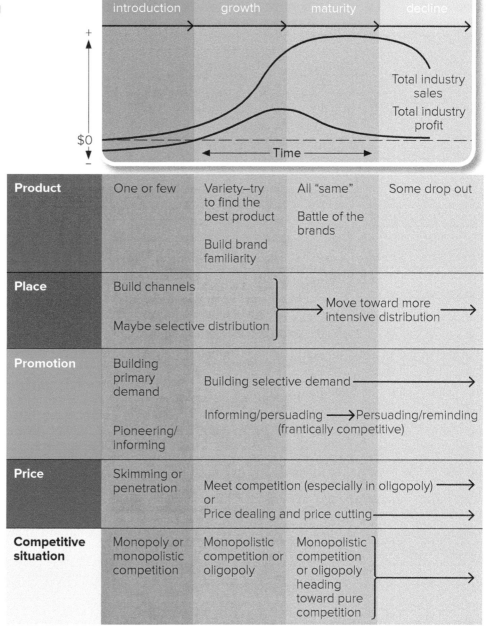

pivot—or move to a new marketing mix. After all the effort of planning a new product, this can be difficult for some marketing managers. A good pivot follows from responding to customer needs. Back in 2010, an app called Burbn allowed users to check in with friends, post plans, and share photos. The founders realized that the most interesting part of the app was photo sharing. So they pivoted the strategy, relaunched the product as Instagram, and the rest, as they say, is history.[11]

Strategy changes in market growth

In the market growth stage, new customers are usually aware of the product, and many are looking for it. During this phase, marketing managers want to build brand familiarity. Promotion often shifts from informing customers to persuading customers to buy their brand. As new competitors enter the market, brands are also forced to price their product competitively.

Managing maturing products

It's important for a firm to have some competitive advantage as it moves into market maturity. Even a small advantage can make a big difference—and some firms do very well by carefully managing their maturing products. They are able to capitalize on a slightly better product or perhaps lower production or marketing costs. Or they are simply more successful at promotion—allowing them to differentiate their more or less homogeneous product from competitors. For example, graham crackers were competing in a mature market and sales were flat. Nabisco used the same ingredients to create bite-size Teddy Grahams and then promoted them heavily. These changes captured new sales and profits for Nabisco.[12]

Product life cycles keep moving. But that doesn't mean a firm should just sit by as its sales decline. There are other choices. A firm can improve its product or develop an innovative new product for the same market. Or it can develop a new strategy targeted at a new mar-

Airbnb is a community marketplace to discover, list, and book unique accommodations in more than 25,000 cities around the world. Airbnb hosts list their extra space—be it a private island, a treehouse, an igloo, or a castle—and guests discover and book these spaces for short- or long-term stays. Because this was a new idea to many of Airbnb's target customers, Airbnb had to build primary demand in the market introduction stage; it used promotion to tell its target market how the service works. E-mail campaigns, listings on Craigslist, publicity, and positive word-of-mouth have helped it rapidly grow in room nights booked.
Source: Airbnb

ket where the life cycle is not so far along. That approach is working for InSinkErator. It has an 80 percent share of all garbage disposals, but in the mature U.S. market, disposals are already in more than half of all homes. In contrast, garbage disposals are

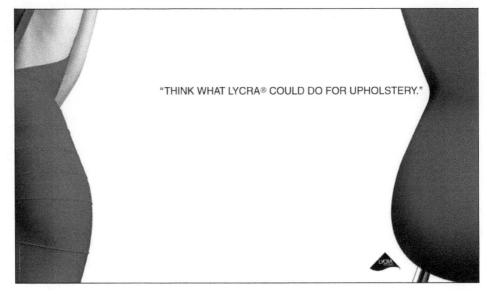

Some companies continue to do well in market maturity by improving their products or by finding new uses and applications. DuPont International's Lycra has expanded from personal apparel to furniture upholstery.
Source: DuPont Textiles & Interiors' LYCRA®

in only about 10 percent of homes in many areas of Europe. Although many households there can afford a disposal, until recently most cities prohibited them on environmental grounds. When research showed that garbage disposals actually provide environmental benefits, InSinkErator adapted its strategy and sales in Europe have grown quickly.[13]

Add value in mature product-markets

In the highly competitive market maturity stage of the product life cycle, many firms slash prices to maintain market share. Creative marketers will find other ways to create value and maintain their margins. That is what Elkay Manufacturing Co. did with the basic water fountain. You know, the one where you lean over, push a button, and slurp a drink. After making those water fountains for decades, Elkay noticed changes in consumer behavior. More consumers were filling water bottles at the fountain, so Elkay designed new fountains that included an additional top-filling nozzle. The new style has become a big seller.[14]

Stimulate growth with new uses

In mature or possibly declining markets, marketing managers can try to find new ways to use a product. Following years of slow growth in Europe, brand managers at Philadelphia Cream Cheese (Philly) looked more closely at the habits of the brand's most frequent buyers. They discovered these customers were using the product as an ingredient—not just as something to smear on bagels. European marketing managers launched an advertising campaign that promoted adding Philly to everything from spaghetti to Spanish tapas. Hoping to inspire creative home cooks, they persuaded Tesco, the UK's leading grocer, to place Philly next to main-dish staples like salmon. Now there are twice as many people in the UK using Philly as an ingredient and sales are again on the rise.[15]

Phase out dying products

Not all strategies are exciting growth strategies. If prospects are poor in a product-market, a phase-out strategy may be needed. The need for phasing out becomes more obvious as the sales decline stage arrives. But even in market maturity, it may be clear that a particular product is not going to be profitable enough to reach the company's objectives. In any case, it is wise to remember that marketing plans are implemented as ongoing strategies. Salespeople make calls, inventory moves in the channel, advertising is scheduled for several months into the future, and so on.

The market for traditional shoe polish has been in decline as leather shoes become less common and people have less time to polish their own shoes. Even in the sales decline stage, market leader Kiwi develops new products like Kiwi Express Shine.
Source: S.C. Johnson & Son, Inc.

Because the firm usually experiences losses if managers end a plan too abruptly, it's sometimes better to phase out the product gradually.

Phasing out a product may involve some difficult implementation problems. But phase-out is also a *strategy*—and it must be market-oriented to cut losses. In fact, it is possible to milk a dying product for some time if competitors move out more quickly and there is ongoing (though declining) demand. Some customers are willing to pay attractive prices to get their old favorite.

New-Product Planning

LO 9.2

In most markets, progress marches on. Firms must develop new products or modify their current products to meet changing customer needs and competitors' actions. Not having an active new-product development process means that, consciously or subconsciously, the firm decides to milk its current products and eventually go out of business. New-product planning is not optional; it is essential for survival in today's dynamic markets.[16]

What is a new product?

In discussing the introductory stage of product life cycles, we focused on the types of really new product innovations that tend to disrupt old ways of doing things. However, each year firms introduce many products that are basically refinements of existing products. So a **new product** is one that is new *in any way* for the company concerned. But customers don't see all new products in the same way.

New products, or innovations, can be loosely grouped into three categories based on the extent to which customers have to change their behavior to adopt the new product.

Customers readily adopt continuous innovations

Continuous innovations are new products that don't require customers to learn new behaviors. Such products usually entail minor variations on existing products. A new toothpaste flavor, a low-calorie iced tea, or a new style of earring would fit in this category. Customers can readily understand and use these new products. Promotion for this type of innovation emphasizes awareness—often with new packaging or advertising that touts an added feature.

#M4BW These new products are all examples of different types of innovation that also make the world a better place. The Xbox Adaptive Controller helps people with limited hand and arm mobility play video games. Its target market will significantly change behavior, making it a discontinuous innovation. Six-pack rings can be dangerous to marine wildlife—potentially trapping and strangling them or being mistaken for food. So a startup company, E6PR (short for "Eco Six Pack Ring"), developed an edible and biodegradable six-pack ring that uses waste from the beer brewing process. This dynamically continuous innovation allows consumers to modestly change their behavior—dispose of six-pack rings differently—while also helping fish and turtles. Rebbl's line of herbal drinks is an example of a continuous innovation—customers do not need to change their behavior to consume them. Rebbl makes the world better by donating 2.5 percent of net sales to efforts to end human trafficking.
(left): Source: Microsoft; (middle): Source: E6PR; (right): Source: Rebbl, Inc.

Dynamically continuous innovation requires some change

Dynamically continuous innovations are new products that require minor changes in customer behavior. For example, 3M developed a Bluetooth-enabled stethoscope, which works like a traditional stethoscope, except that it wirelessly records and stores the sounds of the heart, lung, and other body parts in digital computer files. The innovation requires new behaviors by removing the need to key in information. Doctors can attach the sound files to computer records, as well as send them to other doctors or specialists for help diagnosing and treating patients. Promotion for dynamically continuous innovations needs to clearly communicate the benefits of the innovation.

 #M4BW

Customers must learn new behaviors for discontinuous innovations

Discontinuous innovations are new products that require that customers adopting the innovation significantly change their behavior. This type of innovation often results in a completely new product-market and new-product life cycle. Appliance maker Godrej needed to think differently about refrigeration in rural India. In this part of the world, where most families make less than $5 a day and electricity is unreliable or unavailable, there is little demand for the major appliances Godrej produces. Customers generally don't purchase food that requires staying cool for more than a couple of hours. These consumers have basic needs—they would like to keep milk, vegetables, and fruit cool for a day or two. Then they could shop less frequently and prepare a broader range of meals. For this market, Godrej engineers created ChotuKool ("little cool" in Hindi), a portable cooler that runs on direct current (DC) or an external battery. The design minimizes heat loss and power usage. The innovation is life-altering for people in this part of the world and changes how they shop and cook. Promotion for a discontinuous innovation like ChotuKool usually requires personal selling and product demonstrations to educate customers about new behaviors. The result is a product that makes for a better world.[17]

FTC says product is "new" for only six months

A firm can call its product new for only a limited time. Six months is the limit according to the **Federal Trade Commission (FTC)**—the federal government agency that polices antimonopoly laws. To be called new, according to the FTC, a product must be entirely new or changed in a "functionally significant or substantial respect."[18]

Patent and copyright laws protect inventors

When developing a new product, a firm can file with a government for a **patent**, which grants the inventor the ability to "exclude others from making, using, offering for sale, or selling the invention." Patent law provides an incentive for inventors to share their new technology with the world in exchange for a short-term monopoly on its use. Many firms with a competitive advantage for innovative products file patents and vigorously protect them to keep competitors from stealing their technology. Apple and Samsung have filed various suits against each other regarding technology and innovation in smartphones and tablet computers.[19]

In the United States and many other countries, *copyright* law protects the producer of a creative work (literary, musical, dramatics, or artistic), giving them exclusive rights to reproduce the work. There are some limitations to copyright. These laws are designed to protect inventors and provide an incentive for innovation.[20]

Ethical issues in new-product planning

New-product decisions—and decisions to abandon old products—often involve ethical considerations. For example, some firms (including firms that develop pharmaceutical drugs) have been criticized for holding back important new-product innovations until patents run out, or sales slow down, on their existing products.

At the same time, others have been criticized for *planned obsolescence*—releasing new products that the company plans to soon replace with improved new versions. Similarly, wholesalers and retailers complain that producers too often keep their new-product introduction plans a secret and leave intermediaries with dated inventory that they can sell only at a loss.

Different marketing managers might have very different reactions to such criticisms. However, product management decisions often have a significant effect on customers and intermediaries. A too-casual decision may lead to a negative backlash that affects the firm's strategy or reputation.[21]

> **Ethical Dilemma**
>
> *What would you do?* You have been working for a few months at a fast fashion clothing designer. Your company tries to spot trends in casual fashion and quickly bring new clothes to market. You recently met with a supplier that offers a great cotton fabric that will last twice as long and costs the same price as other cotton materials. The supplier has a patented new weaving process. When you bring up this innovation at a meeting, your manager shoots it down, saying "Look, we don't want our products to last very long. Kids today don't care how long something lasts, they just want the latest thing. And if we make our products last longer, customers will end up buying less." You are somewhat uncomfortable with this argument.
>
> *Should you bring up this subject with your manager again? How would you address it with your coworkers? Your manager?*

An Organized New-Product Development Process Is Critical

LO 9.3

Identifying and developing new-product ideas—and effective strategies to go with them—is often the key to a firm's success and survival. But the costs of new-product development and the risks of failure are high. Experts estimate that consumer packaged-goods companies spend more than $20 million to introduce a new brand—and 80 to 95 percent of those new brands flop. That's a big expense—and a waste. In the service sector, the front-end cost of a failed effort may not be as high, but it can have a devastating long-term effect if dissatisfied consumers turn elsewhere for help.[22]

Why most new products fail

A new product may fail for many reasons. Most often, companies fail to offer a unique benefit or they underestimate the competition. Sometimes the idea is good but the company has design problems—or the product costs much more to produce than was expected. Some companies rush to get a product on the market without developing a complete marketing plan.[23]

But moving too slowly can be a problem too. The longer new-product development takes, the more likely it is that customer needs will be different when the product is actually introduced. This can be problematic for the automobile industry, where new-product development takes years. A few years ago gas prices were much higher, and automakers invested in smaller, more fuel-efficient and electric vehicles. Then gas prices fell and the new vehicles were no longer popular.

A formal new-product development process

To move quickly and also avoid expensive new-product failures, companies should follow an organized new-product development process. The following pages describe such a process, which moves logically through five steps: (1) idea generation, (2) screening, (3) idea evaluation, (4) development (of product and marketing mix), and (5) commercialization. See Exhibit 9-4.

The general process is similar for both consumer and business markets—and for both goods and services. There are some significant differences, but we will emphasize the similarities in the following discussion.

Exhibit 9-4 New-Product Development Process

Idea generation	Screening	Idea evaluation	Development	Commercialization
Ideas from: Customers and users Marketing research Competitors Other markets Company people Intermediaries, etc	Strengths and weaknesses Fit with objectives Market trends Rough ROI estimate	Concept testing Reactions from customers Rough estimates of costs, sales, and profits	R&D Develop model or service prototype Test marketing mix Revise plans as needed ROI estimate	Finalize product and marketing plan Start production and marketing "Roll out" in select markets Final ROI estimate

Process tries to kill new ideas—economically

An important element in the new-product development process is continued evaluation of a new idea's likely profitability and return on investment. The hypothesis tested is that the new idea will *not* be profitable. This puts the burden on the new idea to either prove itself or be rejected. As shown in Exhibit 9-5, as the process moves along, the number of ideas retained declines over time. Such a process may seem harsh, but experience shows that most new ideas have some flaw. Marketers try to discover those flaws early and either find a remedy or reject the idea completely.

Applying this process requires much analysis of the idea *before* the company spends money to develop and market a product. Over time, those ideas that survive receive increasingly large investment (see Exhibit 9-5). The costs in the latter stages of the new-product development can be high, so the process ensures that the best ideas receive the most investment. This is a major departure from the usual production-oriented approach—in which a company develops a product first and then asks sales to "get rid of it."

Step 1: Idea generation

Finding new-product ideas can't be left to chance. Instead, firms need a formal procedure to generate a continuous flow of ideas. For some companies, a constant investment in research and development results in a flow of new technology with new-product

Exhibit 9-5 New-Product Development Process: Number of Ideas and Investment in Each Idea

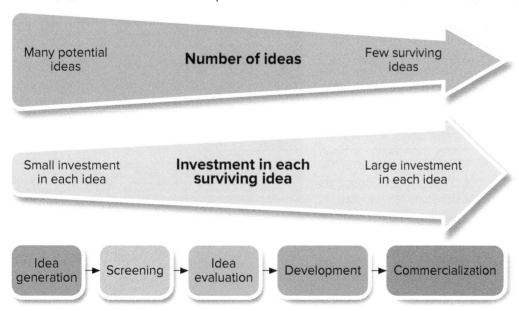

applications. But new ideas can come from a company's own sales or production staff, wholesalers or retailers, competitors, consumer surveys, or other sources such as trade associations, advertising agencies, or government agencies. By analyzing new and different views of the company's markets and studying present consumer behavior, a marketing manager can spot opportunities that have not yet occurred to competitors or even to potential customers. Let's look at some examples of how this works.[24]

Ideas from customers

Customers can be a great source of new ideas—and if they suggest the idea, they are more likely to buy it. Companies such as Rubbermaid and eBags comb through customer reviews for new-product ideas. T-shirt maker Threadless takes this approach to another level. Threadless community members submit T-shirt designs as part of a contest. Winning designs are chosen based on ratings and comments from other members—a process called *crowdsourcing*. Threadless then produces the winning designs and sells them on its website. Because community members tell them which shirts to make, almost every product sells out.

Ideas from customer data

Data that track customer behavior suggest what new products they might be interested in buying. For example, online retailer CafePress sells themed gifts and user-customized products. Because its products tie to popular culture and current events, CafePress examines what customers enter into its search bar. Recently "Class of 2021" and "Zombies" were popular searches on the site, so CafePress quickly added products to address these customer needs. As an offline example, Coca-Cola gathers data from its Freestyle drink machines. The fountain machines let consumers choose from 125 different flavors and drinks. The data on what they buy are sent back to Coke. When Coke sees a trend or regional variation, it can bottle a new drink for wider distribution.[25]

Ideas from other companies and markets

Because no one firm can always be first with the best new ideas, companies should pay attention to what competitors are doing. Some firms use what's called *reverse engineering*. For example, new-product specialists at Ford Motor Company buy other firms' cars as soon as they're available. Then they take the cars apart to look for new ideas or improvements. British Airways talks to travel agents to learn about new services offered by competitors.

Sometimes ideas from one product-market can be adapted for another—where it will be a *new* product. To get such ideas, some firms "shop" in international markets. For instance, food companies in Europe are experimenting with an innovation from Japan—a clear, odorless, natural film for wrapping food. Consumers don't have to unwrap it; when they put the product in boiling water or a microwave, the wrapper vanishes. For more sources of ideas, see *What's Next?* New-product ideas are natural.[26]

A computer system breach by a hacker or a computer virus can be very costly for a business. As these hazards increased, customers asked insurance companies for help in reducing that risk. Insurers developed new products to meet those needs.

Courtesy of Poindexter Surety Group

What's Next? New-product ideas are natural  (#M4BW)

Over the last 3 billion or so years, nature has found ways to solve a wide range of problems. When humans have a problem, biomimicry asks how nature solved a similar problem—and often discovers an idea for a *What's Next?* product.

How does nature create a flexible fastener? Perhaps the best-known example of biomimicry is Velcro, which was invented in 1941 by a Swiss engineer who was curious why burrs stubbornly stuck to his dog's fur following a hike in the woods. Placing the burrs under a microscope, he saw the miniature hooks that grabbed tiny loops found on clothes, socks, and animal fur. This inspired the two-part Velcro fastener: one side with tiny hooks and the other with matching loops.

How does nature create strong bonds? A professor at Oregon State University wondered how mussels keep such a tight grip on underwater rocks with waves persistently pounding down on them. Subsequent research found the mussels secrete proteins known as byssal threads. The threads create an adhesive with superior strength and flexibility. This finding inspired development of a soy-based adhesive Columbia Forest products used to create its PureBond plywood. The adhesive replaces formaldehyde, a toxic chemical used in most plywood.

How does nature hear specifics in a noisy space? Voice-controlled devices like Google Home and Amazon Echo need microphones that identify a voice command while filtering out other sounds. When startup Soundskrit asked the question, it discovered that mosquitoes and crickets use tiny hairs to detect the direction of a sound, and then focus on that sound while filtering out others. Soundskrit developed hardware that mimics this process and built better directional microphones.

How is nature quiet while also aerodynamic? Engineers for Japan's 200-mph Shinkansen commuter train found the train was loud—especially when traveling through tunnels. The project's engineer, an avid birdwatcher, noticed how the Kingfisher's long, narrow beak allowed it to dive into the water without making a splash. This insight inspired a redesign of the train. Besides reducing noise, the new trains are more aerodynamic, traveling 10 percent faster and using 15 percent less electricity.

How does nature manage extreme temperature variation? On the African savannas, temperatures can range from 35 to 110 degrees Fahrenheit in a single day. Termites there build and live in mounds that can reach more than 15 feet in height and maintain virtually constant temperatures inside. Looking closer at the termite mounds, scientists discovered the insects' secrets; they constantly open and close vents to draw in cool air at the base and ventilate hot air through chimneys. Drawing on these insights, architects designed Eastgate Center, a high-rise building in Zimbabwe that consumes less than 10 percent of the energy used in similarly sized buildings.

Nature's problem-solving experience draws on millions of years of adaptation. Its answers are usually efficient *and* sustainable. *What's Next?* Maybe a better world.[27]

Ask your own questions of nature at www.asknature.org. You may be surprised by the answers.

 #M4BW

Technology speeds global solutions to market

Searches for new ideas in some industries may be aided by artificial intelligence and machine learning (see Chapter 3). AI can lead to discoveries that would take humans alone too long to identify. In areas like pharmaceutical drugs and clean energy, hard-working scientists can only examine a small set of opportunities. Machine learning thrives in these situations, as the technology can examine many possibilities in a short period of time. Machine learning also "thinks" differently and may suggest paths that scientists wouldn't have considered.

For example, producing new medicines is a very complicated process. Many parts of this research and development process can be sped up by applying machine learning technology. For example, machine learning can identify and evaluate many different molecules in a fraction of the time it would take a team of scientists. One of the challenges facing renewable energy is the more efficient solar cells and new battery technology for more effective energy storage. Breakthrough opportunities may emerge from the identification of new raw materials—something machine learning excels at. Breakthroughs in either of these areas may lead to new medicines and lower-cost and cleaner energy.[28]

Step 2: Screening

Screening involves evaluating the new ideas with the type of S.W.O.T. analysis described in Chapter 2 and the product-market screening criteria described in Chapter 3. Recall that these criteria include the combined output of a resources (strengths and weaknesses) analysis, a long-run trends analysis, and a thorough understanding of the company's objectives. See Exhibit 2-9 and Exhibit 3-7. A "good" new idea should eventually lead to a product (and marketing mix) that will give the firm a competitive advantage—hopefully, a lasting one.

The life-cycle stage at which a firm's new product enters the market has a direct bearing on its prospects for growth. So screening should consider how the strategy for a new product will hold up over the whole product life cycle.

Safety must be considered

Real acceptance of the marketing concept prompts managers to screen new products on the basis of how safe they are. Safety is not a casual matter. The U.S. Consumer Product Safety Act (of 1972) set up the Consumer Product Safety Commission to encourage safety in product design and better quality control. The commission has a great deal of power. It can set safety standards for products and order costly repairs or the return of unsafe products. And it can back up its orders with fines and jail sentences. The Food and Drug Administration has similar powers for foods and drugs.

Product safety complicates strategy planning because not all customers—even those who want better safety features—are willing to pay more for safer products. Some features cost a lot to add and increase prices considerably. These safety concerns must be considered at the screening step because a firm can later be held liable for unsafe products.

Products can turn to liabilities

Product liability means the legal obligation of sellers to pay damages to individuals who are injured by defective or unsafe products. Product liability is a serious matter. Liability settlements may exceed not only a company's insurance coverage but its total assets!

Relative to most other countries, U.S. courts enforce a strict product liability standard. Sellers may be held responsible for injuries related to their products no matter how the items are used or how well they're designed. In one widely publicized judgment, McDonald's paid a huge settlement to a woman who suffered third-degree burns when her coffee spilled on her. The court concluded that there was not enough warning about how hot the coffee was.

Product liability is a serious ethical and legal matter. Many countries are attempting to change their laws so that they will be fair to both firms and consumers. But until product liability questions are resolved, marketing managers must be even more sensitive when screening new-product ideas.[29]

ROI is a crucial screening criterion

Getting by the initial screening criteria doesn't guarantee success for the new idea. But it does show that at least the new idea is in the right ballpark *for this firm*. If many ideas pass the screening criteria, a firm must set priorities to determine which ones go on to the next step in the process. This can be done by comparing the ROI (return on investment) for each idea—assuming the firm is ROI-oriented. The most attractive alternatives are pursued first.

At this stage, ROI involves making rough estimates of profits and development cost. To better understand how ROI is used in new-product screening, review the Marketing Analytics in Action: Return on Investment (ROI) next.

Step 3: Idea evaluation

When an idea moves past the screening step, it is evaluated more carefully. This stage involves getting more reactions from customers, even though at this stage an actual product has yet to be developed. Although this can make getting customer input more difficult, firms need extensive feedback before adding the expense of producing the product.

Marketing Analytics in Action: Return on Investment (ROI)

Return on investment (ROI) is a way for managers to evaluate an investment. *Investment* refers to the dollar resources the firm invests in a project or business. For example, with respect to new-product development, a new product may require $4 million for research and development, inventory, promotion to launch the product, and so on. That investment is expected to generate a net profit of $1 million in the product's first year. In this example, the return on investment (ROI) is 25 percent or ($1 million ÷ $4 million).[30]

ROI can be increased in one of three ways:

- Increase the profit margin.
- Increase the sales revenue.
- Decrease the investment in the new product.

Let's see how this concept can help us evaluate two new-product ideas at the screening step of new-product development. The new-product development manager at National Grains is evaluating two new breakfast cereals. The two products are (1) an artificially sweetened chocolate cereal (internal code name = Choco) and (2) a gluten-free cereal (GluFree). As part of the screening step, the new-product development manager estimates ROI for each new-product idea.

Drawing on experience and marketing research, the new-product manager estimates the first-year sales and profit margin (the return) and new-product development and launch costs (the investment). Choco is expected to have greater sales volume than GluFree, but at a lower profit margin because it will face more competition. GluFree is expected to target a focused segment of gluten-free shoppers, which lowers the costs of launching the product and therefore overall product development costs. The estimated ROI is shown in the following spreadsheet:

	Choco	GluFree
First-year sales revenue*	$20.0	$12.0
Profit margin (%)	15%	20%
Net profit (Revenue × Profit margin)	$ 3.0	$ 2.4
Investment in product development	$10.0	$ 6.0
ROI (Net profit ÷ Investment)	30%	40%

*All dollar figures in millions.

1. What are two specific ways the new-product development manager could increase the ROI for Choco?
2. What are two specific ways the new-product development manager could increase the ROI for GluFree?
3. What other financial information should the new-product development manager consider in evaluating these options?

At this stage, marketing managers should describe and relate the assumptions they are making about each new idea. The product idea represents a hypothesis, or educated guess, about how to meet customer needs. For example, a law firm may assume its corporate clients care enough about convenience to pay a premium price for a new web-based document sharing service. By making assumptions—like the value customers place on convenience—explicit, marketing managers can conduct research to find out as quickly as possible if assumptions are true.

Initial evaluation may come from informal focus groups. Their reactions can be helpful—especially if they show that potential users are not excited about the new idea. A more formal method uses **concept testing**—getting reactions from customers about how well a new-product idea fits their needs. Concept testing uses marketing research—ranging from focus groups to surveys of potential customers. Some firms run concept tests online, which can lower costs and speed feedback. It can be fast and easy to show some photos or a video of a concept, along with possible prices, to a sample of target customers. Follow-up survey questions gauge consumer reactions.

Ideas that survive to the development stage receive more attention. An automaker might develop a scale model of a car or a full-fledged, operational prototype. Honda's Micro Commuter and BMW's i8 Spyder may be on the market if they survive continued scrutiny from customers and show signs of a positive ROI.
(left): Andrew Blyth/Asia Photo Connection/Alamy Stock Photo; (right): eans/Shutterstock

Writers for Disney Channel's TV show *Sofia the First* have found concept testing helps. The writers share scripts and plotlines with kids at a nearby preschool to learn what kids like and understand. For example, an episode intended to be called "Sofia's First Slumber Party" didn't resonate; kids were unfamiliar with the phrase *slumber party*. So the episode was changed to "The Big Sleepover."[31]

Marketing research can also help identify the size of potential markets, which helps companies estimate likely costs, revenue, and profitability. Together, all this information helps marketing managers decide whether there is an opportunity, whether it fits with the firm's resources, *and* whether there is a basis for developing a competitive advantage. With such information, the firm can develop a more reliable estimate of ROI in various market segments and decide whether to continue the new-product development process.[32]

Step 4: Development

Product ideas that survive the screening and idea evaluation steps get further investment of time and money. Usually, this involves more research and development (R&D) to design and develop the physical part of the product. Or, in the case of a new service offering, the firm works out the details of what training, equipment, staff, and so on will be needed to deliver on the idea. Input from a firm's earlier efforts helps guide this technical work.

Customers react to prototypes
Passing the idea evaluation phase often leads next to the creation of a **prototype**—an early sample or model built to test a concept. A service firm may try to train a small group of service providers and test the service on real customers. 3-D printing technology allows a computer-designed drawing to quickly be converted into a three-dimensional replica, often at a relatively low cost. For example, automakers can "print" a life-size car fender, door, or even an auto body. This allows for checking fit, finish, and styling.

Customers may even be involved in a *co-creation process*—where customers react to prototypes and suggest improvements. This process uses *rapid prototyping*, where customer input is received and quickly designed into a revision of the product—and then fed back to customers for further input. The repetitive process in rapid prototyping encourages innovations to "fail early and fail often" so the best ideas get to market more quickly. Google uses this approach to test new features it wants to add to its family of online services (Google, Gmail, Chrome, and others)—often incorporating feedback in less than 24 hours and then seeking additional feedback from a test group.[33]

With actual goods and services, potential customers can more realistically react to how well a product meets their needs. Focus groups, panels, and surveys provide feedback on features and the whole product idea. Sometimes that reaction kills the idea. For example, Coca-Cola Foods believed it had a great idea with Minute Maid Squeeze-Fresh—frozen orange juice concentrate in a squeeze bottle. In tests, however, Squeeze-Fresh bombed. Consumers loved the idea but hated the product. In real life, it was messy to use, and no one knew how much concentrate to squeeze in the glass.[34]

Market testing uses real market conditions

Firms often use full-scale *market testing* to get customer reactions under real market conditions or to test variations in the marketing mix. For example, a firm may test alternative brand names, prices, or advertising copy in different test cities. Note that the firm tests the whole marketing mix, not just the product. For example, a hotel chain might test a new service offering at one location to see how it goes over. Running market tests is costly, but *not* testing is risky. Frito-Lay was so sure it understood consumers' snack preferences that it introduced a three-item cracker line without market testing. Even with network TV ad support, MaxSnax suffered overwhelming consumer indifference. By the time Frito-Lay pulled the product from store shelves, it had lost $52 million.[35]

Step 5: Commercialization

A product idea that survives this far can finally be placed on the market. Putting a product on the market is expensive, and success usually requires the cooperation of the whole company. Manufacturing or service facilities have to be set up. Goods have to be produced to fill the channels of distribution, or people must be hired and trained to provide services. Further, introductory promotion is costly—especially if the company is entering a very competitive market.

Because of the size of the job, some firms introduce their products city by city or region by region—in a gradual "rollout"—until they have complete market coverage. Rollouts also permit more market testing, but the main purpose is to do a good job implementing the marketing plan. Marketing managers also need to pay close attention to control—to ensure that the implementation effort is working and that the strategy is on target.

Steps should not be skipped

Because speed can be important, it's always tempting to skip needed steps when some part of the process seems to indicate that the company has a "really good idea." But the process moves in steps—gathering different kinds of information along the way. By skipping steps, a firm may miss an important aspect that could make a whole strategy less profitable or actually cause it to fail.

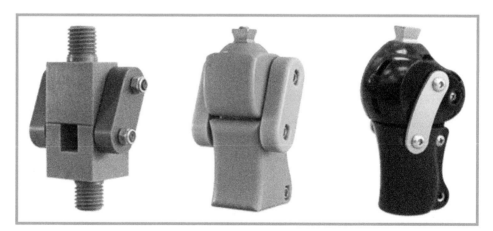

#M4BW D-Rev is a nonprofit that focuses on the health of the world's poor by developing affordable health technologies. It recently addressed the challenge of creating an affordable, high-quality replacement knee for amputees. After a six-year development process that began with prototypes (left and center), D-Rev introduced the Re-Motion Knee (right). The product is available in many developing countries for $80, about a fifth of the price of comparable prosthetic devices.
Source: D-Rev

New-Product Development: A Total Company Effort

LO 9.4

We've been discussing the steps in a logical, new-product development process. However, as shown in Exhibit 9-6, many factors can impact the success of the effort.[36]

Top-management support is vital

Companies that are particularly successful at developing new goods and services seem to have one key trait in common: enthusiastic top-management support for new-product development. New products tend to upset old routines that managers of established products often try in subtle but effective ways to maintain. So someone with top-level support, and authority to get things done, needs to be responsible for new-product development.

A culture of innovation

Top management also drives the organization's culture. A culture of innovation consists of people open to exchanging ideas, willing to listen to others, and pushing good ideas forward. Some companies formally encourage employees to pursue innovation—by giving them time. For example, Google allows all employees the freedom to spend 20 percent of their time working on new ideas, even if the ideas are unrelated to their job description. A culture that supports innovation generates more ideas.[37]

Put someone in charge

Rather than leaving new-product development to anyone in engineering, R&D, or sales who happens to be interested in taking the initiative, successful companies *put* someone in charge. It may be a person, department, or team. But it's not a casual thing. It's a major responsibility of the job.

Balance market needs and company resources

From idea generation to commercialization, a company's R&D specialists, operations, and marketing personnel must work together to evaluate the feasibility of new ideas. Everyone should be guided by a clear understanding of customer needs. It doesn't make sense for R&D people to develop a technology or product that doesn't have potential for the firm and its markets.

Marketing managers must recognize that new-product projects need to meet return-on-investment goals. Team members from R&D and production should provide insights about the company's capabilities and costs to produce. It isn't sensible for a marketing manager to develop elaborate marketing plans for goods or services that the firm simply can't produce—or produce profitably.

Exhibit 9-6 New-Product Development Success Factors

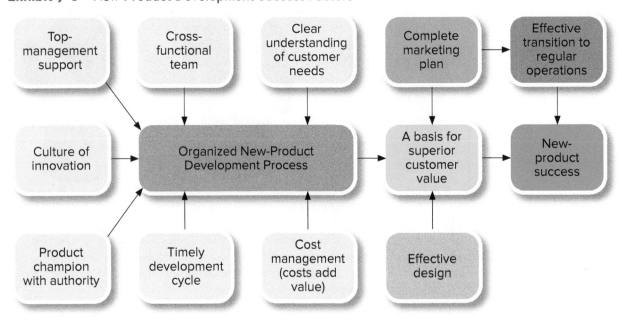

Clearly, a balancing act is involved here. But the critical point is one we've been emphasizing throughout the whole book: *Marketing-oriented firms seek to satisfy customer needs at a profit with an integrated, whole-company effort.*

New-product success

Savvy marketing managers generate products the market values. They must put together comprehensive marketing plans, communicating the features and benefits of a new product, identifying the right price and the best distribution channels, pinpointing competition, and establishing a budget. The same marketing manager who shepherded the product through new-product development may manage it in regular operations. Or there may be a handoff to a product manager. In either case, after a new product launches, activities that ensure long-term success must follow.

Need for Product Managers

LO 9.5

Product variety leads to product managers

When a firm has only one or a few related products, everyone is interested in them. But when a firm has products in several different product categories, management may decide to put someone in charge of each category, or each brand, to be sure that attention to these products is not lost in the rush of everyday business. **Product managers** or **brand managers** manage specific products—often taking over the jobs formerly handled by an advertising manager. That gives a clue to what is often their major responsibility—*Promotion*—because the products have already been developed by the new-product people. However, some brand managers start at the new-product development stage and carry on from there.

Product managers are especially common in large companies that produce many kinds of products. Several product managers may serve under a marketing manager. Sometimes these product managers are responsible for the profitable operation of a particular product's whole marketing effort. Then they have to coordinate their efforts with others, including the sales manager, advertising agencies, production and research people, and even channel members. This is likely to lead to difficulties if product managers have no control over the marketing strategy for other related brands or authority over other functional areas whose efforts they are expected to direct and coordinate.

To avoid these problems, in some companies the product manager serves mainly as a "product champion"—concerned with planning and getting the promotion effort implemented. A higher-level marketing manager with more authority coordinates the efforts and integrates the marketing strategies for different products into an overall plan.[38]

Adapting products for international markets

The activities of product managers vary a lot depending on their experience and aggressiveness and the company's organizational philosophy. Today, companies are emphasizing marketing experience—because this important job takes more than

P&G has some of the world's best-known brands, including Crest toothpaste; Gillette razors; Head & Shoulders shampoo; Downy, Gain, Tide, and Ariel laundry detergents; and many more. Each product or brand has a manager who primarily manages Promotion for that brand.
Mark Dierker/McGraw-Hill Education

academic training and enthusiasm. But it is clear that someone must be responsible for developing and implementing product-related plans, especially when a company has many products.

Product managers may work with managers in other countries to decide whether and how to adapt products for different markets. Some adaptations may be required and fairly simple—for example, many countries have different electrical voltages, necessitating different plugs for electrical products. Other adaptations demand more creativity. Nestlé adapted its KitKat chocolate candy bars for the Japanese palate with 19 unique flavors including soy sauce, miso, sweet potato, and blueberry. In Germany, KitKat lemon yogurt bars are popular; in the UK, it's peanut butter. Nestlé adapts many of its other products to international preferences too. Whereas in the United States fruit-flavored ice "popsicles" are a summertime treat all kids crave, in China Nestlé created flavored milk ice on a stick to fit local preferences. The most popular flavor? Banana.[39]

Managing Product Quality

LO 9.6

Total quality management meets customer requirements

In Chapter 8 we explained that product quality means the ability of a product to satisfy a customer's needs or requirements. Now we'll expand that idea and discuss some ways a manager can improve the quality of a firm's goods and services. We'll develop these ideas from the perspective of **total quality management (TQM)**, the philosophy that everyone in the organization is concerned about quality, throughout all of the firm's activities, to better serve customer needs.

The cost of poor quality is lost customers

Most of the early attention in quality management focused on reducing defects in goods produced in factories. At one time most firms assumed defects were an inevitable part of mass production. They saw the cost of replacing defective parts or goods as just a cost of doing business—an insignificant one compared to the advantages of mass production. However, many firms were forced to rethink this assumption when Japanese producers of cars, electronics, and cameras showed that defects weren't inevitable. Much to the surprise of some production-oriented managers, the Japanese experience showed that it is less expensive to do something right the first time than it is to pay to do it poorly and then pay again to fix problems. And their success in taking customers away from established competitors made it clear that the cost of defects wasn't just the cost of replacement!

From the customer's point of view, getting a defective product and having to complain about it is a big headache. The customer can't use the defective product and suffers the inconvenience of waiting for someone to fix the problem—*if* someone gets around to it. It certainly doesn't deliver superior value. Rather, it erodes goodwill and leaves customers dissatisfied, less trusting of the brand, and possibly spreading their dismay to others via word-of-mouth or the Internet. The big cost of poor quality is the cost of lost customers.

Firms that adopted TQM methods to reduce manufacturing defects soon used the same approaches to overcome many other problems. Their success brought attention to what is possible with TQM—whether the problem concerns poor customer service, flimsy packaging, or salespeople who can't answer customers' questions.

Getting a handle on doing things right the first time

The idea of identifying customer needs and doing things right the first time seems obvious, but it's easier said than done. Problems always come up, and it's not always clear what isn't being done as well as it could be. People tend to ignore problems that don't pose an immediate crisis. But firms that adopt TQM always look for ways to improve implementation with **continuous improvement**—a commitment to constantly make things better one step at a time. Once you accept the idea that there *may* be a better way to do something and you look for it, you may just find it!

Technology can have a big impact on service delivery. Savioke's Relay robot delivers towels, coffee, and late-night bottles of wine in some Aloft hotels. The Woebot app uses artificial intelligence to provide mental health services. The app includes natural language processing (it can understand what you tell it), psychological expertise, and a sense of humor to help people concerned about their mental health.
(left): Courtesy of Savioke; (right): Source: Woebot

Data and artificial intelligence aid quality efforts

Some firms use data and artificial intelligence (AI) to support continuous improvement efforts. For example, industrial equipment makers have added smart meters and digital sensors to products; they are constantly monitored and provide early warnings of failure. This means parts can be replaced before they fail—and the data generate knowledge about improvements for future models. Internet-connected sensors have even been placed in cows' stomachs to monitor their health. The system sends a farmer an e-mail if the cow falls ill. The sensors often identify health issues days before they would have been recognized by a farmer.[40]

Health care generates many data, and some startups are figuring out how to analyze those data to improve the quality of health care. For example, Enlitic uses large medical databases and artificial intelligence to help radiologists vastly improve how they read and interpret medical images. Deep Genomics leverages huge genetic databases and artificial intelligence to identify patterns that foreshadow health problems. These insights allow doctors to offer interventions with their patients before a disease or condition is otherwise visible.[41]

Building quality into services

Services create particular quality management challenges. Most products involve some service component whether it is primarily a service, primarily a physical good, or a blend of both. Even a manufacturer of ball bearings isn't just providing wholesalers or producers with round pieces of steel. Customers need information about deliveries, they need orders filled properly, and they may have questions to ask the firm's accountant or engineers. Because almost every firm must manage the service it provides customers, let's focus on some of the special concerns of managing service quality.

Deliver quality at each touchpoint

Marketing managers must deliver quality throughout the entire customer experience. If they do not, even one problem can affect how the customer perceives the firm's quality. Think about how an otherwise great restaurant experience could be ruined by a rude waiter. The purchase and consumption process for many products involves multiple *touchpoints,* or points where there is contact between the customer and the company. There are usually more touchpoints for services than for goods.

Consider everything that needs to go right for a customer getting an oil change at a Jiffy Lube. The experience might begin at the Jiffy Lube website, where a customer seeks information about the services offered, prices, hours of operation, and the phone number and address of a local Jiffy Lube. Is this information

David McNew/Getty Images

easy to find? A phone call to a local Jiffy Lube outlet should be promptly answered by a friendly, knowledgeable person. Upon arrival, is the driveway well marked? Is there a long queue for service? The attendant may have entered the license plate number and could already have the customer's details on his computer screen—or not. Is the check-in process quick? Is the attendant courteous? The waiting area should be clean and perhaps offer Wi-Fi. What happens if another customer has noisy young children—does that affect the quality of the oil change experience? The payment process should be fast—and offer the customer information about his or her car. The car should be clean when it is picked up. Finally, the car should not experience any service problems immediately following the oil change. Whew! That is a lot of things to get right, every time, if Jiffy Lube is to deliver high quality to its customers. Other types of products can have greater or fewer touchpoints, but a marketing manager should recognize that all of them contribute to the customer's perception of quality.

Train people and empower them to serve

A service provider usually deals directly with the customer, making it difficult to provide consistent service quality. People are just not as consistent in their actions as machines or computers are. In addition, service quality often depends on a service provider interpreting each customer's needs. For example, a hair stylist has to ask good questions and successfully interpret the customer's responses before it's possible to provide the right cut. As a result, a person doing a specific service job may perform one specific task correctly but fail the customer in a host of other ways. Two keys to improving service quality are (1) training and (2) empowerment.

All employees in contact with customers need training—many firms see 40 hours a year of training as a minimum. Good training usually includes role-playing on handling different types of customer requests and problems. A rental car attendant who fails to resolve or ignores a customer complaint may leave the customer dissatisfied—even if the rental car was perfect.

Companies can't afford an army of managers to inspect how each employee implements a strategy—and such a system usually doesn't work anyway. Quality cannot be "inspected in." It must come from the people who do the service jobs. Firms that commit to service quality empower employees to figure out how to best satisfy customers' needs. **Empowerment** means giving employees the authority to correct a problem without first checking with management. At a hotel, for instance, an empowered room-service employee knows it's okay to run across the street to buy the specific brand of bottled water a guest requests.[42]

Specify jobs and measure performance

Managers who develop successful quality programs clearly specify and write out exactly what tasks need to be done, how, and by whom. This may seem unnecessary. After all, most people know, in general, what they're supposed to do. However, if the tasks are clearly specified, it's easier to see what criteria should be used to measure performance. Once criteria are established, there needs to be some basis on which to evaluate the job being done.

Will robots replace service workers?

Many service tasks can be specified to the point that they can be automated using artificial intelligence (AI). The most basic level of AI involves routine and repetitive tasks. For example, phone calls that used to be answered by human customer service agents may now be handled by intelligent agents that respond to natural language requests. Just ask "How do I return these pajamas? They are the wrong size," and a computer interprets the question and gives the necessary details.

As AI capabilities improve, even service jobs requiring analytical skills may be replaced. Many people talk to a (human) financial advisor for advice on how to invest their money. Now more customers have "robo-advisor" software managing their portfolio. An investor simply visits the Wealthfront website, answers a series of questions, and the software suggests a portfolio tailored to the customer's goals and risk tolerance. As these machines learn, they will continuously deliver higher-quality service.[43]

Getting a return on quality is important

While the cost of poor quality is lost customers, the type of quality efforts we've been discussing can also increase costs. It's easy to fall into the trap of running up unnecessary costs trying to improve some facet of quality that really isn't that important to customer satisfaction or customer retention. When that happens, customers may still be satisfied, but the firm can't make a profit because of the higher costs. In other words, there isn't a financial return on the money spent to improve quality. A manager should focus on quality efforts that really provide the customer with superior value—quality that costs no more to provide than customers will ultimately be willing to pay.[44]

CONCLUSION

This chapter introduced the product life-cycle concept and showed how life cycles affect marketing strategy planning. The product life-cycle concept shows why new products are so important to growth in markets and also helps explain why different strategies—including strategies for new, improved products—need to be developed over time. Innovators—or fast copiers—who successfully bring new products to market are usually the ones who achieve the greatest growth in customer equity.

In today's highly competitive marketplace it is no longer profitable to simply sell "me too" products. Markets, competition, and product life cycles are changing at a fast pace. New products help a company appeal to new target markets by appealing to unmet needs. New products can also encourage current customers to purchase more. In addition, they can help retain customers by adapting to changing customer needs.

Just because a product is new to a company doesn't mean that it is a really new innovation and starts a new-product life cycle. However, from a marketing manager's perspective, a product is new to the firm if it is new in any way or to any target market. Firms don't just develop and introduce new products; they do so within the context of the whole marketing strategy.

Many new products fail, but we presented an organized new-product development process that helps prevent that fate. The process makes it clear that new-product success isn't just the responsibility of people from R&D or marketing, but rather requires a whole-company effort.

We also described product and brand management. To help a product or brand grow, managers in these positions usually recommend ways to adjust all of the elements of the marketing mix, but the emphasis is often on Promotion.

Poor product quality results in dissatisfied customers. Alert marketers look for ways to design better quality into new products and to improve the quality of ones they already have. Approaches developed in the total quality management (TQM) movement can be a big help in this regard. Ultimately, the challenge is for the manager to focus on aspects of quality that really matter to the target customer. Otherwise, the cost of the quality offered may be higher than what target customers are willing to pay.

In combination, this chapter and Chapter 8 introduce strategy decision areas for Product and important frameworks that help you see how Product fits within an overall strategy. These chapters also start you down the path to a deeper understanding of the Four Ps. In Chapter 10, we expand on that base by focusing on the role of Place in the marketing mix.

KEY TERMS

product life cycle, 238
market introduction, 238
market growth, 238
market maturity, 239
sales decline, 240
fashion, 242
fad, 242
new product, 246
continuous innovations, 246
dynamically continuous innovations, 247
discontinuous innovations, 247
Federal Trade Commission (FTC), 247
patent, 247
Consumer Product Safety Act, 252
product liability, 252
concept testing, 253
prototype, 254
product managers, 257
brand managers, 257
total quality management (TQM), 258
continuous improvement, 258
empowerment, 260

QUESTIONS AND PROBLEMS

1. Review the iRobot case study that opens the chapter. From this case, identify examples of different key terms and concepts covered in the chapter. For example, the Roomba 980 robot appears to be a discontinuous innovation.

2. Review the iRobot case study that opens this chapter. Assume that iRobot is trying to develop a home health care robot. This robot would help older adults around the house with tasks such as reminders to take medicines and basic cleaning. Describe how this might look at each step in the new-product development process. Provide as much detail—even if you are speculating—as possible.

3. Explain how industry sales and industry profits behave over the product life cycle.

4. Cite two examples of products that you think are currently in each of the product life-cycle stages. Consider services as well as physical goods.

5. Explain how you might reach different conclusions about the correct product life-cycle stage(s) in the worldwide automobile market.

6. Explain why individual brands may not follow the product life-cycle pattern. Give an example of a new brand that is not entering the life cycle at the market introduction stage.

7. Discuss the life cycle of a product in terms of its probable impact on a manufacturer's marketing mix. Illustrate your answer using personal computers.

8. What characteristics of a new product will help it move through the early stages of the product life cycle more quickly? Briefly discuss each characteristic—illustrating with a product of your choice. Indicate how each characteristic might be viewed in some other country.

9. Starbucks asks customers for new-product ideas at https://ideas.starbucks.com. Go to this site and click on "View Ideas." Next, click on "Popular Ideas." Choose two popular ideas and read the ideas and comments. How do you think Starbucks benefits from this site? What could be done to improve the site?

10. What is a new product? Illustrate your answer.

11. Explain the importance of an organized new-product development process and illustrate how it might be used for (a) a new hair care product, (b) a new children's toy, and (c) a new fast-food restaurant.

12. Discuss how you might use the new-product development process if you were thinking about offering some kind of summer service to residents in a beach resort town.

13. Explain the role of product or brand managers. When would it make sense for one of a company's current brand managers to be in charge of the new-product development process? Explain your thinking.

14. If a firm offers one of its brands in a number of different countries, would it make sense for one brand manager to be in charge, or would each country require its own brand manager? Explain your thinking.

15. Discuss the social value of new-product development activities that seem to encourage people to discard products that are not all worn out. Is this an economic waste? How worn out is "all worn out"? For example, must a shirt have holes in it? How big?

16. What are the major advantages of total quality management as an approach for improving the quality of goods and services? What limitations can you think of?

MARKETING PLANNING FOR HILLSIDE VETERINARY CLINIC

Appendix D (the Appendices follow Chapter 19) includes a sample marketing plan for Hillside Veterinary Clinic. Look through the "Marketing Strategy" section.

1. Hillside offers many different products. Identify several of these products and indicate where you think each one is in its product life cycle.

2. Exhibit 9-3 summarizes some marketing mix characteristics based on where a product fits in the product life cycle. Is Hillside's marketing plan consistent with what this exhibit suggests? Why or why not?

SUGGESTED CASES

6. Dynamic Steel
20. Lake Russell Marine & Camp
22. Bright Light Innovations: The Starlight Stove

MARKETING ANALYTICS: DATA TO KNOWLEDGE

CHAPTER 9: GROWTH STAGE COMPETITION

AgriChem, Inc. has introduced an innovative new product—a combination fertilizer, weed killer, and insecticide that makes it much easier for soybean farmers to produce a profitable crop. The product introduction was quite successful, with 1 million units sold in the year of introduction. And AgriChem's profits are increasing. Total market demand is expected to grow at a rate of 200,000 units a year for the next five years. Even so, AgriChem's marketing managers are concerned about what will happen to sales and profits during this period.

Based on past experience with similar situations, they expect one new competitor to enter the market during each of the next five years. They think this competitive pressure will drive prices down about 6 percent a year. Further, although the total market is growing, they know that new competitors will chip away at AgriChem's market share—even with the 10 percent a year increase planned for the promotion budget. In spite of the competitive pressure, the marketing managers are sure that familiarity with AgriChem's brand will help it hold a large share of the total market and give AgriChem greater economies of scale than competitors. In fact, they expect that the ratio of profit to dollar sales for AgriChem should be about 10 percent higher than for competitors.

AgriChem's marketing managers have decided the best way to get a handle on the situation is to organize the data in a spreadsheet. They have set up the spreadsheet so they can change the "years in the future" value and see what is likely to happen to AgriChem and the rest of the industry. The starting spreadsheet shows the current situation with data from the first full year of production.

Design element: #M4BW box globe icon: ©Vectoryzen/Shutterstock

CHAPTER TEN

Imaginechina/AP Images

Place and Development of Channel Systems

In the 1970s, the early "microcomputers" were hard to set up and difficult to use, so few people wanted them. That also explains why there were no computer stores. Altair, one of the first brands, initially sold mainly at "electronics fairs." Most of these gatherings were in California; often buyers and sellers just met in an open market on Saturday mornings. Then Heath introduced its more powerful H89 computer through mail-order catalogs as a build-it-yourself kit. Heath added value with good telephone technical support.

Soon after that, Xerox used its competitive advantage in distribution to introduce its 820 model. Business customers liked buying computers from the same wholesalers that regularly handled their Xerox copiers. By 1980, Radio Shack's large retail store network helped Xerox's easy-to-use TRS-80 become the best-selling computer. Customers appreciated the accessibility of Radio Shack's in-store staff and tech support specialists.

As you read this, it probably occurs to you that most of the firms mentioned no longer sell computers. These early firms couldn't adjust quickly enough when IBM introduced its first PC. The pull of IBM's familiar brand gave more customers the confidence to buy. Sales of PCs surged as IBM established a chain of its own retail stores and worked closely with select dealers who promised to pay special attention to the IBM brand. Big-business customers bought in quantity directly from IBM's aggressive sales force. After IBM's design became an industry standard, firms such as Compaq, Hewlett-Packard (HP), and Toshiba quickly jumped in with PC models of their own—often selling through independent computer dealers.

Soon after, Michael Dell, then just a first-year college student, started buying and reselling computers from his dorm room. Dell figured a target market of price-conscious customers would respond to a different marketing mix. He used direct-response advertising in computer magazines; customers called a toll-free number to order a computer with the exact features they wanted. Then Dell used UPS to ship directly to the customer. Prices were kept low because the direct channel eliminated retailer markup, and the build-to-order approach reduced inventory costs. Dell also built reliable machines and delivered superior customer service. It would have been tough to centralize all of this if he had been working with thousands of retailers.

Although Dell's direct model worked well with tech-savvy consumers and small businesses, it struggled to make inroads with big government and corporate buyers. Over time Dell adapted its strategy, using a sales force and building relationships, to make progress with these buyers.

At the same time, HP and other firms tried to imitate Dell's successful direct-order approach. However, this move created conflict with the retailers already selling most of HP's PCs; retailers were not happy with competition from their own supplier! When these retailers retaliated by pushing other brands, HP limited the models sold online and added programs to support its dealers; for example, it provided quick turnaround for retailers placing orders for custom-built computers.

Looking for new ways to grow, Dell saw prospects in international markets. After success in Europe, it went to Asia. China was a huge market with a small but growing middle class and many small and medium-sized businesses. But Dell ran into an entrenched Chinese competitor; state-supported PC maker Legend (later renamed Lenovo) already owned the small but growing market. Dell struggled to gain distribution outside of big Chinese cities where Lenovo had a strong distribution network.

This is where our story of the PC market shifts; Lenovo built a strong foundation in China to eventually become a global PC powerhouse. Back in the 1990s, it was a distributor that adapted foreign PCs with local software and helped customers with Internet connectivity. Later, Lenovo moved up the value chain by manufacturing its own computers. In China at that time, personal computers were at the introductory stage of the product life cycle. Lenovo's presence in thousands of retail shops across the country reassured first-time computer buyers who wanted advice and also needed to feel and touch a computer before purchasing.

Over time some Chinese customers became more comfortable with PC buying and sought the lower-cost option of buying direct from the manufacturer. So Lenovo added a direct selling channel. A more agile supply chain and just-in-time delivery system also kept Lenovo's costs low, an important strategy for the price-sensitive Chinese consumer market. At the time, it took several months' salary for a middle-class Chinese consumer to buy a PC. Yet many Chinese consumers found a way, because education is so highly valued in Chinese culture.

Back in 2005, Lenovo was little known outside of China. Then it made a bold move: Lenovo bought IBM's then struggling PC business. IBM's move out of the highly competitive

PC market was just what Lenovo needed to jump-start a worldwide distribution network. IBM's ThinkPad was popular with business customers around the world, and the acquisition gave Lenovo access to IBM's network of value-added resellers—firms that purchase PCs, tailor software and accessories, and finalize pricing to fit business customer needs. Lenovo's supply chain lowered the ThinkPad's production costs without sacrificing its reputation for power and durability.

After discovering that value-added resellers in the United States made more sales at lower selling costs as compared to Lenovo's own sales force, Lenovo reinvested in these channel partners. It grew the number to more than 35,000, and now Lenovo trains these dealers to sell its computer servers and services.

Lenovo recently led the battle in the now mature (and maybe declining) global PC market with about 24 percent market share, just ahead of HP, while Dell remained in third place with 17 percent. Lenovo's strategy of different products and distribution models in different parts of the world continues to help it grow. Lenovo's supply chain and distribution network continue to deliver the goods.[1]

LEARNING OBJECTIVES

This case shows that offering customers a good product at a reasonable price is not the whole story. Marketing managers must make decisions about how they will make goods and services available to a target customer's Place when the customer wants them. This chapter's learning objectives will help you understand the role of Place in marketing strategy.

When you finish this chapter, you should be able to

1. understand what product classes suggest about Place objectives.
2. understand why some firms use direct channel systems whereas others work with intermediaries and indirect systems.
3. understand how and why marketing specialists develop strategies to make channel systems more effective.
4. understand how to develop cooperative relationships and avoid conflict in channel systems.
5. know how channel members in vertical marketing systems shift and share functions to meet customer needs.
6. understand the differences between intensive, selective, and exclusive distribution.
7. know how multichannel distribution and reverse channels operate.
8. know the main approaches firms use to reach customers in international markets.
9. understand important new terms (shown in **red**).

Marketing Strategy Planning Decisions for Place

Managers must think about **Place**—making goods and services available in the right quantities and locations, when customers want them. And when different target markets have different needs, a number of Place variations may be required. Our opening case makes it clear that new Place arrangements can dramatically change the competition in a product-market. This is especially important in business today because information technology, including websites and e-commerce, makes it easier for firms to work together more efficiently and also to reach customers directly.

In this chapter and the two that follow, we'll deal with the many important marketing strategy decisions that a marketing manager must make concerning Place. Exhibit 10-1 gives an overview. We'll start here with a discussion of Place objectives and how they relate to product classes and the product life cycle—ideas introduced in the Product chapters (8 and 9). We'll then discuss the type of channel that most aptly meets

Exhibit 10-1 Marketing Strategy Planning Decisions for Place

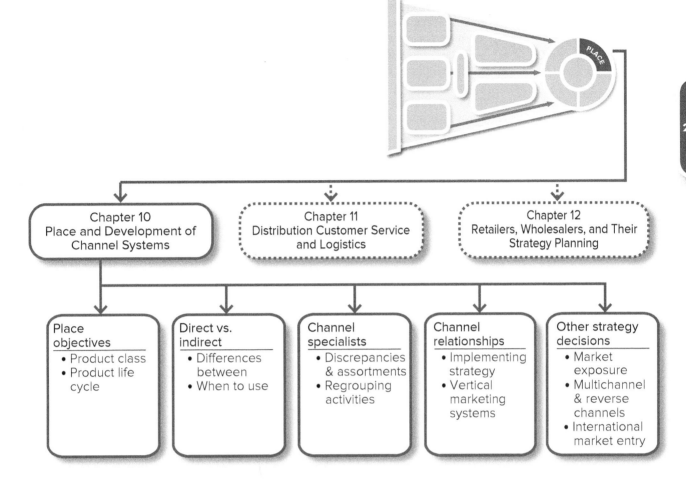

customers' needs. We'll show why specialists are often involved and how they come together to form a **channel of distribution**—any series of firms or individuals who participate in the flow of products from producer to final user or consumer. We'll also consider how to manage relations among channel members to reduce conflict and improve cooperation. This chapter concludes by considering the desired level of market exposure (and how many channel outlets are needed) as well as approaches for reaching customers in international markets.

In Chapter 11, we'll expand the Place discussion to decisions that a marketing manager makes about physical distribution, including customer service level, transporting, and storing. Then, in Chapter 12, we'll take a closer look at the many different types of retailing and wholesaling firms. We'll consider their role in channels as well as the strategy decisions they make to satisfy their own customers.

Place Decisions Are Guided by "Ideal" Place Objectives

LO 10.1

All marketing managers want to be sure that their goods and services are available in the right quantities and locations—when customers want them. But customers may have different needs in these areas as they make different purchases.[2]

Product classes suggest Place objectives

In Chapter 8 we introduced the product classes, which summarize consumers' urgency to have needs satisfied and their willingness to seek information, shop, and compare. Now you should be able to use the product classes to handle Place decisions.

Exhibit 8-9 shows the relationship between consumer product classes and ideal Place objectives. Similarly, Exhibit 8-10 shows the business product classes and how they relate to customer needs. Study these exhibits carefully. They set the framework for making Place decisions. In particular, the product classes help us decide how much market exposure we'll need in each geographic area.

Recall from Chapter 8 that customers won't spend much time shopping for convenience products; thus, widespread distribution is needed to make products available when the need strikes. If customers want to compare shopping products, some distribution channels can make that process easier. For example, online sellers often carry a broad assortment of products to make this easier, and some physical stores offer similar advantages. *Heterogeneous* shopping goods should have outlets that provide information—knowledgeable salespeople who can provide insights about different brands and models and online stores with videos or information. *Homogeneous* shopping goods favor low-cost retailers as customers focus mostly on low prices. Customers will search for specialty goods, so the higher cost of widespread distribution may not be needed. Unsought products should be sold at locations where other, related products are available.

Business product classes reflect what buyers think about the products and how they are used. Accessory equipment and maintenance, repair, and operating (MRO) supplies are used frequently, requiring widespread distribution. Raw materials often have special transportation needs that must be considered. Reliable delivery is critical for component parts and materials. Professional services are usually delivered in person, so location matters.

Place system is not automatic

Several different product classes may be involved if different market segments view a product in different ways. Thus, marketing managers may need to develop several strategies, each with its own Place arrangements. There may not be one Place arrangement that is best.

For example, many consumers view Tide laundry detergent as a staple for doing laundry at home. To meet their needs, it is widely distributed in grocery and discount stores. Customers who run out of detergent while doing wash at a Laundromat see Tide as an emergency product and want small boxes available in a vending machine. And

 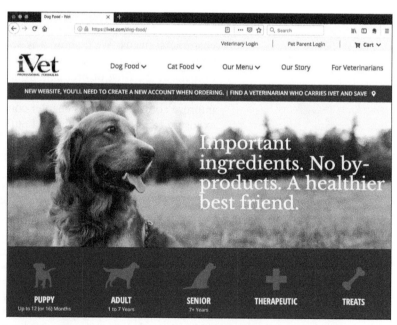

For most target customers, the Iams brand of premium dog food is a staple or heterogeneous shopping product. So Iams distributes its dog food through supermarkets, pet superstores, and mass merchandisers, displaying it near competing products. On the other hand, most customers view iVet's super-premium dog food as a specialty product. Consequently, it is sold only through veterinarians, who give the product an extra push with their recommendation.
(left): Source: The Procter & Gamble Company; (right): Source: iVet Professional Formulas

some hotels look at Tide as an operating supply that the housekeeping department needs to provide guests with clean sheets and towels. For these hotels, Tide comes in large drums that are sold and delivered by wholesalers of housekeeping supplies.

Place decisions have long-run effects

The marketing manager must also consider Place objectives in relation to the product life cycle; see again Exhibit 9-3. Place decisions often have long-run effects. They're usually harder to change than Product, Promotion, and Price decisions. Many firms that thought they could quickly establish effective websites for direct online sales, for example, found that it took several years and millions of dollars to work out the kinks. It can take even longer and cost more to develop effective working relationships with others in the channel. Legal contracts with channel partners may limit changes. And it's hard to move retail stores and wholesale facilities once they are set up. Yet, as products mature, they typically need broader distribution to reach different target customers. Marketing managers need to keep the future in mind when they develop Place objectives.

Channel System May Be Direct or Indirect

LO 10.2

One of the most basic Place decisions producers must make is whether to handle the whole distribution job themselves—perhaps by relying on direct-to-customer e-commerce selling or opening their own stores—or use wholesalers, retailers, and other specialists.

Many firms prefer to distribute directly to the final customer or consumer because they want to control the whole marketing job. They may think that they can serve target customers at a lower cost or do the work more effectively than intermediaries. Although this can be true, marketing managers should carefully assess the value intermediaries add for end customers. In this section, we will first look at why some businesses sell direct and then explore the advantages of using intermediaries.

Girl Scouts sell their cookies direct to customers. But customers need to know when and where a Girl Scout will be selling the cookies in order to buy, so Girl Scouts of America created the Cookie Finder app for the iPhone. Customers just enter their zip code, and they are directed to the nearest source of the tasty treats.
Source: Girl Scouts Cookie Finder App; (smartphone frame): Oleg GawriloFF/ Shutterstock

Direct distribution maintains control

One reason a producer chooses direct distribution is because it wants to maintain control of the marketing mix. Wholesalers and retailers often carry competing brands and make decisions that are in their own interests. This may not always be aligned with the interests of an individual producer. For example, Goodyear wants its target customers to know that its Assurance TripleTred All-Season tires get 21 percent better wet traction than a leading competitor. But if Roadmasters Auto and Tire Centers make a bigger profit margin on that "leading competitor," the salesperson at Roadmasters may not deliver that "better wet traction" message. If Goodyear has a promotion to sell this tire at a lower price, Roadmasters may decide to keep the extra margin and not pass the discount along to its customers. At its own online store, Goodyear can be more certain that customers receive the message and price Goodyear prefers.

Ela Family Farms in Hotchkiss, Colorado, sells its organic tree fruits—along with jams, sauces, and nectars—direct to consumers. Consumers can find Ela's tasty product line at their farmer's market stand or by ordering online with direct home delivery. Business products are often sold direct. GE Aviation has only a few very large customers that buy its commercial jet engines direct from the producer.
(left): Source: Ela Family Farms; *(right):* Source: General Electric

Direct customer contact generates data and knowledge

Another benefit to direct distribution is that it puts a firm in direct contact with its customers. By working directly with customers, the company generates data, information, and knowledge about its market. Sometimes this happens because a company's salespeople and customer service workers talk regularly to customers. What they hear quickly gets back to the company's marketing managers. Companies with a direct-to-customer e-commerce channel have ready access to how customers behave on their websites. For example, what products do they buy? Which pages or messages attract the most attention? How do customers respond to price changes? All of these data can be turned into information that can be used to change, adapt, or fine-tune a marketing mix. An intermediary may choose to keep this information to itself, perhaps feeling it gives them some power and control. For the company to obtain similar information may require costly marketing research, which may also take longer to acquire.

Common with business customers

Many business products are sold direct to customer. Alcan sells aluminum directly to General Motors. Woodward produces products such as pumps and fuel nozzles that it sells directly to aircraft makers such as Boeing, Airbus, and Gulfstream. This is understandable because in business markets there are fewer transactions, orders are larger, and customers may be concentrated in one geographic area. Marketing mixes may be customized to each customer. Further, once relationships are established, e-commerce systems can efficiently handle orders.

Service firms often sell direct

Service firms often use direct channels. If the service must be produced in the presence of customers, there may be little need for intermediaries. For example, accounting and consulting services are typically sold directly to customers. On the consumer side, small-business service providers like hairstylists, landscapers, and cleaning services usually don't utilize intermediaries.[3]

E-commerce sparks direct distribution of consumer products

While most consumer products are still sold through intermediaries, a growing number of consumer products are being sold direct to consumer. This reflects two trends. First, the growth of website-based e-commerce systems and delivery services such as UPS and FedEx give many firms direct access to customers whom it would have been impossible to reach in the past. This new option lowers the cost for direct distribution.

In addition, consumers have grown increasingly comfortable purchasing products online. Many customers no longer feel the need to see, touch, and feel a product before a purchase. Trust in online reviews and guarantees of free returns have lowered the risk of buying before trying.

 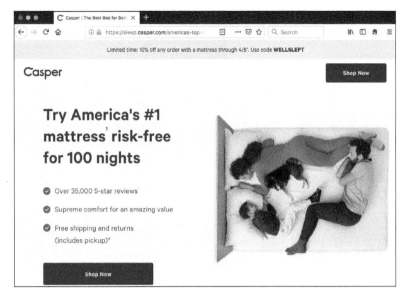

A few years ago, many customers would have felt they needed to try on a pair of shoes or lie down on a mattress before making a purchase. Now some companies sell these shopping products online without those opportunities to try before buying. To make customers more comfortable with the purchase, Allbirds offers a 30-day money-back guarantee and Casper lets customers try a new mattress "risk-free for 100 nights."
(left): Source: Allbirds, Inc.; (right): Source: Casper

In recent years, some startup businesses have used a direct-to-customer business model to disrupt markets. Warby Parker produces and sells eyeglasses online for about $100 each. The company lets customers try them on at home before making a final decision. Dollar Shave Club and Harry's both sell razors directly to customers on a subscription model, where customers receive a new set of blades every few months.

When suitable intermediaries are not available

A firm may have to go direct if suitable intermediaries are not available or will not cooperate. Sometimes intermediaries that have the best contacts with the target market are hesitant to add unproven vendors or new products. There is only so much space in a physical store, so adding one product usually requires dropping another. Intermediaries stock what they believe their customers want to buy. Adding unproven goods can be risky. Even major consumer product companies like Chobani, General Mills, and Nestlé cannot necessarily expect Walmart or Kroger to stock their new products. The challenge can be even greater for less well-known brands. A firm must demonstrate demand for its products. Sometimes that can be achieved by using direct distribution.

For example, when Glacéau began selling its now popular Vitaminwater, wholesale distributors had no interest in carrying it. So the owner of the company delivered the bottled water directly to small retailers around New York City. Soon he had data that proved his product would sell, and distributor interest grew. Glacéau became a success by slowly building support from retailers and wholesalers, but many new products fail because the producer can't find willing channel partners and doesn't have the resources to handle direct distribution.[4]

When indirect channels are best

While some direct-to-customer businesses claim to lower prices by eliminating the intermediary, that is often not possible. Sometimes customers prefer to buy from intermediaries, or the intermediary performs necessary activities more effectively or for a lower cost. Let's look at some situations where intermediaries are a better choice.

Customers want to buy some products at the same time

Customers often have established buying patterns. They prefer to buy products from different producers at the same time. For example, most consumers prefer the convenience of picking up milk, bread, meat, fruit, and vegetables in a single shopping trip.

Keurig and Nespresso are two companies that sell single-serve coffee and the equipment for brewing it. Keurig sells most of its products through retailers, with some direct sales via its website. Nespresso uses a direct model, selling through its website and a few Nespresso stores. Both companies have been successful with these business models.
(left): Source: Keurig, Inc.; (right): Source: Nestlé Nespresso SA

Similarly, Square D, a producer of electrical supplies, might want to sell directly to electrical contractors. It can certainly set up a website for online orders or even open sales offices in key markets. But if contractors like to make all of their purchases in one convenient stop—at a local electrical wholesaler—the only practical way to reach them is through a wholesaler.

Intermediaries have knowledge

Intermediaries can be specialists who provide information to bring buyers and sellers together. For example, most consumers don't know much about the wide variety of home and auto insurance policies available. A local independent insurance agent may help them decide which policy, and which insurance company, best fits their needs.

Intermediaries who are close to their customers are

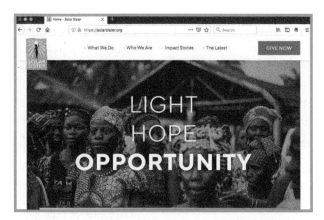

#M4BW It was difficult for Solar Sister to find distribution channel partners when it tried to sell its solar lamps, solar phone chargers, and solar panels in Uganda, Tanzania, and Nigeria. Because Solar Sister's target market needed to be informed about how the solar products could save them time and money, they wanted salespeople on the ground. Solar Sister decided that women in local villages, who were already trusted by their neighbors, could be trained to sell its products. These women entrepreneurs purchase Solar Sister's products and resell them to others in their villages. The women are not employed by Solar Sister, so this is not direct distribution. By empowering these women as distributors, Solar Sister makes the world a better place for its customers and the women entrepreneurs.
Source: Solar Sister

often able to anticipate customer needs and forecast demand more accurately. This information can reduce inventory costs in the whole channel—and helps the producer smooth out production.

Most producers seek help from specialists when they first enter international markets. Specialists can provide crucial information about customer needs and insights into differences in the market environment.

Intermediaries provide working capital and credit

Direct distribution often requires a significant investment in facilities, people, and information technology. A company that has limited financial resources or wants to retain flexibility may want to avoid that investment by working with established intermediaries.

Intermediaries may further reduce a producer's need for working capital by buying the producer's output and carrying it in inventory until it's sold. If customers want a good "right now," there must be inventory available to make the sale. And if customers are spread over a large area, it will probably be necessary to have widespread distribution.

Some intermediaries play a critical role by providing credit to customers at the end of the channel. A wholesaler who knows local customers can help reduce credit risks. It's an unhappy day when the marketing manager learns that a customer who was shipped goods based on an online order can't pay the invoice.

Many firms use direct and indirect channels

Selling direct or indirect is not necessarily an either/or decision. When a company serves multiple target markets, it may choose to sell direct and through intermediaries. For example, a company like Autolite uses direct channels to sell its spark plugs to automakers like Ford. Yet it also sells to consumers through intermediaries like Pep Boys. Deciding whether to sell directly, indirectly, or both requires an understanding of discrepancies and separations (introduced briefly in Chapter 1), which will be discussed later in this chapter.

Cost and effectiveness should determine direct/indirect decision

The most important reason for using an indirect channel of distribution is when an intermediary helps producers serve customer needs better and at lower cost. Marketing managers should carefully evaluate each target market's needs and determine its

For a long time, HP sold ink for HP InkJet computer printers only through intermediaries like Amazon and Office Depot. While it still uses those channels, customers can now buy ink directly from HP. In fact, HP printers can now order on their own. An HP printer can automatically place an order for a replacement ink cartridge when it recognizes the ink supply is low. A customer receives a new cartridge in the mail before the old one goes dry. Many home and small-business customers appreciate the added convenience.
Source: HP Development Company, L.P.

Don't get confused...

customers' willingness to pay. A firm should also understand its capabilities and those of willing intermediaries. This will help the marketing manager determine whether to use intermediaries and, if so, which intermediaries would work best.

Before we leave the topic of direct and indirect distribution, let's clarify two sometimes confusing topics. First, sometimes products are sold directly to customers at home or work. Rodan + Fields skincare products, Scentsy scented products, or LuLaRoe women's clothing all rely on direct selling. This selling involves a salesperson talking directly to a customer. However, most of these "salespeople" are *not* company employees. Rather, they work independently, and the companies they sell for refer to them as *dealers, distributors, agents,* or some similar term. So in a strict technical sense, this is not really direct producer-to-consumer distribution.

Another potentially confusing topic is the term **direct marketing**—direct communication between a seller and an individual customer using a promotion method other than face-to-face personal selling. Sometimes direct marketing promotion is coupled with direct distribution from a producer to consumers. However, many firms that use direct marketing promotion distribute their products through intermediaries. For example, Logitech, which produces personal computer peripherals, uses e-mails to communicate directly with customers who purchase Logitech products through retailers. So the term *direct marketing* is primarily concerned with the Promotion area, not Place decisions. We'll talk about direct marketing promotion in more detail in Chapter 13.[5]

Deciding whether to sell directly, indirectly, or both requires an understanding of the discrepancies and separations (introduced briefly in Chapter 1). Now we'll go into more detail.

Channel Specialists May Reduce Discrepancies and Separations

LO 10.3

The assortment and quantity of products customers want may be different from the assortment and quantity of products companies produce. Producers are often located far from their customers and may not know how best to reach them. Customers in turn may not know about their choices. Specialists develop strategies to adjust for these discrepancies and separations.[6]

Because FedEx offers specialized services that facilitate exchange between producers and consumers, it helps to reduce discrepancies and separations.
Reed/Bloomberg/Getty Images

Discrepancies of quantity and assortment

Discrepancy of quantity means the difference between the quantity of products it is economical for a producer to make and the quantity final users or consumers normally want. For example, most manufacturers of golf balls produce large quantities—perhaps 200,000 to 500,000 in a given time period. The average golfer, however, wants only a few balls at a time. Adjusting for this discrepancy usually requires intermediaries—wholesalers and retailers.

Producers typically specialize by product—and therefore another discrepancy develops. **Discrepancy of assortment** means the difference between the lines a typical producer makes and the assortment final consumers or users want. Most golfers, for example, need more than golf balls. They want golf shoes, gloves, clubs, a bag, and, of course, a golf course to play on. And they usually don't want to shop for each item separately. So, again, there is a need for wholesalers and retailers to adjust these discrepancies.

Channel specialists adjust discrepancies with regrouping activities

Regrouping activities adjust the quantities or assortments of products handled at each level in a channel of distribution. There are four regrouping activities: accumulating, bulk-breaking, sorting, and assorting. When one or more of these activities is needed, a marketing specialist may develop them to fill this need.

Adjusting quantity discrepancies by accumulating and bulk-breaking

Accumulating involves collecting products from many small producers. Much of the coffee that comes from Colombia is grown on small farms in the mountains. Accumulating the small crops into larger quantities is a way of getting the lowest transporting rate and making it more convenient for distant food-processing companies to buy and handle them. Accumulating is especially important in less-developed countries and in other situations, like agricultural markets, where there are many small producers.

Accumulating is also important with professional services because they often involve the combined work of a number of individuals, each of whom is a specialized producer. A hospital makes it easier for patients by accumulating the services of a number of health care specialists, many of whom may not actually work for the hospital.

Many wholesalers and retailers who operate from Internet websites focus on accumulating. Specialized sites for everything from Chinese art to Dutch flower bulbs bring together the output of many producers.

Bulk-breaking involves dividing larger quantities into smaller quantities as products get closer to the final market. The bulk-breaking may involve several levels in the channel. Wholesalers may sell smaller quantities to other wholesalers or directly to retailers. Retailers continue breaking bulk as they sell individual items to their customers.

Adjusting assortment discrepancies by sorting and assorting

Different types of specialists adjust assortment discrepancies. They perform two types of regrouping activities: sorting and assorting.

Sorting means separating products into grades and qualities desired by different target markets. For example, an investment firm might offer its customers shares in a mutual fund made up only of stocks for companies that pay regular dividends. Similarly, a wholesaler that specializes in serving convenience stores may focus on smaller packages of frequently used products.

Assorting means putting together a variety of products to give a target market what it wants. This usually is done by those closest to the final consumer or user—retailers or wholesalers who try to supply a wide assortment of products for the convenience of their customers. Thus, a wholesaler selling Yazoo tractors and mowers to golf courses might also carry Pennington grass seed and Scotts fertilizer.

Digital products need distribution too

While most digital products are intangible, they also need distribution. Music, television programs, movies, books, video games, and software all exist in an intangible, digital form—though some are also sold as physical goods. Many service products—airline tickets and banking, for example—are often shopped for and delivered online.

Distribution costs can be very low for digital products because they can travel over the Internet at little or no cost. Yet even without a physical form, distribution remains an important marketing strategy decision for firms with digital products.

Virtual products often face the same discrepancies and separations as tangible goods. Consequently, channel specialists add value to the distribution of digital products by performing regrouping activities. Learn more about how this works for distributing video entertainment in *What's Next?* Bits and bytes need distribution too.

Adding or subtracting channels can add value and differentiate

Specialists develop strategies to adjust separations and discrepancies if they need to be adjusted. Sometimes spotting such a need creates an opportunity. For example, many office workers in big cities don't have time to run out to a restaurant for lunch every day. The growth of food trucks in urban areas emerged as a new channel for some restaurants.

On the other hand, there is no point in having intermediaries just because that's the way it has always been done. Eliminating intermediaries might create other opportunities. Some manufacturers of business products now reach more customers in distant markets with a website than was previously possible for them to reach with independent manufacturers' agents who sold on commission (but otherwise left distribution to the firm). The website cost advantage can translate to lower prices and a marketing mix that is a better value for some target segments.[7]

Channel Relationships Must Be Managed

LO 10.4

Marketing manager must choose type of channel relationship

Intermediary specialists can help make a channel more efficient. But there may be problems getting the different firms in a channel to work together well. How well they work together depends on the type of relationship they have. This should be carefully considered because marketing managers usually have choices about what type of channel system to join or develop.

The whole channel should have a product-market commitment

Ideally, all of the members of a channel system should have a shared *product-market commitment*—with all members focusing on the same target market at the end of the channel and sharing the various marketing functions in appropriate ways. When members of a channel do this, they are better able to compete effectively for the customer's business. Unfortunately, many marketing managers overlook this idea because it's not the way their firms have traditionally handled channel relationships.

Traditional channel systems involve weak relationships

In **traditional channel systems**, the various channel members make little or no effort to cooperate with one another. They buy and sell from one another—and that's the extent of their relationship. Each channel member does only what it considers to be in its own best interest. It doesn't worry about other members of the channel. This is shortsighted, but it's easy to see how it can happen. The objectives of the various channel members may be different. For example, Cooper Industries wants a wholesaler of electrical building supplies to sell Cooper products. But a wholesaler who works with different producers may not care whose products get sold. The wholesaler just wants happy customers and a good profit margin.[8]

Conflict gets in the way

Specialization can make a channel more efficient—but not if the specialists are so independent that the channel doesn't work smoothly. Because members of traditional channel systems often have different objectives—and different ideas about how things should be done—conflict is common.

What's Next? Bits and bytes need distribution too

These days, many people consume video entertainment (television programs or movies) in a digital format—via their satellite or cable television provider or in a streaming format via services like Netflix or Hulu. The video marketplace is characterized by a vast amount of programming on hundreds of networks. It is almost impossible to keep up with all the choices. Even the variety of programming shown on a single network such as ESPN or HBO is a lot for most viewers to evaluate. As a result, channel specialists such as Hulu are emerging. As with other intermediaries, Hulu collects programming from television networks and movie studios and streams it over the Internet to smartphones, computers, and TV sets. Are more channel specialists *What's Next?* for digital distribution?

As a channel specialist, Hulu reduces discrepancies and separations between consumers (viewers of video content) and producers (primarily creators of television programming and movies). Viewers face *discrepancies of quantity*—there are millions of hours of television programming produced, but most viewers watch only a couple hours or less at a time. Viewers also face *discrepancies of assortment;* for instance, a viewer may enjoy watching nature programs, but that type of program is shown on Discovery Channel, National Geographic Channel, the Public Broadcasting Service, and MSNBC.

Hulu performs *regrouping* activities to better satisfy viewer needs. For example, by bringing together programming from multiple networks, Hulu *accumulates* content, making it easier for consumers to visit one place to seek and find the desired programming. Hulu does *bulk-breaking;* it receives programming from networks in bulk—all the episodes of *Family Guy*, for example—and then offers them for viewing one episode at a time. By allowing viewers to identify networks, programming, or sports teams to follow, Hulu also performs a *sorting* activity—where consumers identify what they see as higher quality. This system combines with other categories developed by Hulu to perform *assorting* that helps consumers view collections of programs that match their interests.

As you can see, producers of digital products face many of the same challenges facing producers of physical goods.[9]

Now see *if you can conduct the same type of analysis (identifying discrepancies and regrouping activities) on the airline travel and e-book markets. Go to the websites* **Expedia.com** *(for airline tickets) and* **Amazon.com** *and its Kindle product (for e-books) to learn how these sellers operate. What do these sites do to reduce discrepancies of quantity and assortment?*

There are two basic types of conflict in channels of distribution (see Exhibit 10-2). **Vertical channel conflict** occurs between firms at different levels in the channel of distribution. A producer and a retailer may have different goals—for example, a producer might want to maximize the number of units sold while a retailer could seek to maximize profits. A producer could offer a discount to a retailer hoping that the retailer will pass the price cut on to customers, leading to increased sales. However, the retailer may decide to buy more but keep prices at the same level. Another example occurs when a producer and a retailer disagree about how much promotion effort the retailer should give the producer's product. The rise of Sephora as a major retailer of luxury beauty products has created conflict with producers. Sephora stores have added private branded products, and its traditional cosmetic brands (Estée Lauder and Lancôme) lost shelf space, salesperson effort, and sales to the Sephora Collection brand.[10]

Horizontal channel conflict occurs between firms at the same level in a distribution channel. This might include conflict between two competing retailers. For example, a bicycle store that keeps a complete line of bikes on display, has a knowledgeable sales staff, and lets customers take test rides isn't happy to find out that an online store with little inventory and no salespeople offers customers lower prices on the same models. The online retailer gets a free ride on the competing store's investments in inventory and sales staff.

Managing channel conflict

Some level of conflict may be inevitable—or even useful if that is what it takes for customers at the end of the channel to receive better value. However, most marketing managers try to avoid conflicts that harm relationships with channel partners. Often different channels serve different target markets—and this alone minimizes conflict. For example, convenience stores like 7-Eleven and membership stores like Costco both sell Coke—but one offers it cold in individual servings at many convenient locations, whereas the other offers it warm and 24 cans at a time at a warehouse store. Each offers a different marketing mix to a different target market.

Another strategy manages conflict by offering different products through each channel. Gibson produces guitars. It was selling its guitars and a range of accessories (strings, picks, pick holders, etc.) on its Gibson website, but this created conflict with the music stores that carried Gibson products. The stores did not appreciate the added competition. Gibson also wanted all of its dealers to carry all the various accessories, but dealers found that carrying so many different individual products was costly in inventory carrying costs. So Gibson stopped selling guitars on its website in response to complaints from its traditional retailers and distributors. On the other hand, Gibson's website continued to sell a wide variety of accessories that most retailers did not want to stock. Gibson's competitor, Fender, also decided to sell direct to consumers and created conflict with its dealers. Many Fender dealers think customers will test out a new guitar in their stores and buy on the Fender website. While Fender promises not to undercut its dealers' prices, some stores may find it safer to nudge customers toward Gibson guitars instead of Fender.[11]

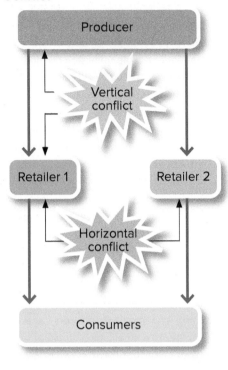

Exhibit 10-2 Types of Channel Conflict

STIHL sells its high-quality chain saws through 8,000 independent STIHL dealers nationwide. These dealers provide product demonstrations, good advice, and expert on-site service. STIHL limits horizontal conflict by not selling through home improvement warehouses or online stores that put more emphasis on price competition and less on service.
Source: Andreas Stihl AG & Company KG

Sometimes firms will eliminate the conflict by dropping competing product lines. After Walmart purchased Moosejaw, a small outdoor retailer, it began selling some of its premium products, like $250 Deuter hiking bags and $100 Leki hiking poles, on the Walmart.com website. Deuter and Leki's other retail outlets were concerned Walmart would start cutting prices and take sales from them. After hearing complaints, Leki and Deuter asked Walmart not to sell the brands—online or in store—and Walmart complied.[12]

In general, treating channel partners fairly—even when one partner is more powerful—tends to build trust and reduce conflict. Trusting relationships lead to greater cooperation in the channel. As a consequence, channel members are better able to satisfy the needs of target customers at the end of the channel.[13]

Channel captain can guide channel relationships

Although each channel system should act as a unit, some firms are in a better position to take the lead in the relationship and in coordinating the whole channel effort. This situation calls for a **channel captain**—a manager who helps direct the activities of a whole channel and tries to avoid or solve channel conflicts.

For example, when Harley-Davidson wanted to expand sales of fashion accessories, it was difficult for motorcycle dealers to devote enough space to all of the different styles. Harley considered selling the items directly from its own website, but that would take sales away from dealers who were working hard to help Harley sell both cycles and fashions. Harley's president asked a group of dealers and Harley managers to work together to come up with a plan they all liked. The result was a website that sells Harley products through the dealer that is closest to the customer.[14]

Acting as channel captains in their respective channels, both Peterson and Electrolux are able to get cooperation from many independent wholesalers and retailers as well as big chains like Lowe's because they develop marketing strategies that help the whole channel compete more effectively. This also helps everyone in the channel do a better job of meeting the needs of target customers at the end of the channel.
(left): Source: Peterson Manufacturing Company; (right): Source: AB Electrolux

The concept of a single channel captain is logical, but most traditional channels don't have a recognized captain. The various firms don't act as a coordinated system. Yet firms are interrelated, even if poorly, by their policies. So it makes sense to try to avoid channel conflicts by planning for channel relations. The channel captain arranges for the necessary functions to be performed in the most effective way.[15]

Some producers lead their channels

Sometimes producers take the lead in channel relations. Typically this occurs when the producer is large, dominates a category, or wields a powerful brand name. In this case, intermediaries usually take a back seat and support the producer. They often negotiate a situation that is profitable for both parties, but the producer usually has a stronger hand.

Exhibit 10-3A shows this type of producer-led channel system. Here the producer has selected the target market and developed the Product, set the Price structure, invested in consumer and channel Promotion, and developed the Place setup. Intermediaries are then expected to finish the Promotion job in their respective places. Of course, in a retailer-dominated channel system, the marketing jobs would be handled in a different way.

Boar's Head, maker of premium meats, cheeses, and condiments sold at deli counters across the United States, offers an example of a producer-led channel system. Boar's Head gained this position by outspending its rivals on advertising and earning brand preference and brand insistence from consumers. Boar's Head negotiates with retailers so that its signage dominates the deli department, and it is often the only premium brand in the store. The brand commands a premium price. Prudent marketing decisions help Boar's Head maintain good margins.

Powerful retailers can lead channels

Sometimes wholesalers or retailers take the lead. They are closer to the final user or consumer and are in an ideal position to assume the channel captain role. Intermediaries find that the ability to gather, analyze, and interpret data generates insights about customer needs. They can then seek out producers who can meet these needs with products at reasonable prices. Powerful retail chains such as Walmart, Lowe's, Kroger, and Tesco now dominate the channel systems for many products in the United States, Asia, and Europe. In these situations, the retailer

Exhibit 10-3 How Channel Functions May Be Shifted and Shared in Different Channel Systems

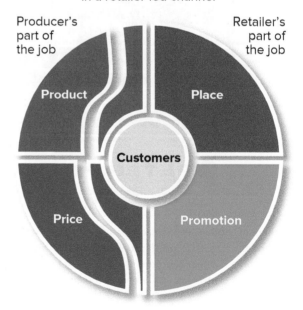

A. How strategy decisions are handled in a producer-led channel

B. How strategy decisions are handled in a retailer-led channel

280

often suggests new products suppliers should make. They also have a strong influence on prices. Exhibit 10-3B shows how marketing strategy might be handled in this sort of retailer-led channel system.[16]

Vertical Marketing Systems Focus on Final Customers

LO 10.5

Many marketing managers accept the view that a coordinated channel system can help everyone in the channel. These managers are moving their firms away from traditional channel systems and instead are developing or joining vertical marketing systems. In a traditional system, each member of the channel focuses on its own interests—typically maximizing its own profits without paying much attention to others in the channel. This can result in greater conflict and less total profit for the whole channel. **Vertical marketing systems** are channel systems in which the whole channel focuses on the same target market at the end of the channel. To achieve these goals, there must be higher levels of coordination and cooperation among the members of the channel.

Such systems make sense, because if the final customer doesn't buy the product, the whole channel suffers. There are three types of vertical marketing systems—corporate, administered, and contractual. Exhibit 10-4 summarizes some characteristics of these systems and compares them with traditional systems.

Corporate channel systems are highly integrated

Let's start on the right side of Exhibit 10-4. This represents the most tightly integrated type of vertical marketing system. Some firms develop their own vertical marketing systems by internal expansion or by buying other firms, or both. One company owns the whole channel system. With **corporate channel systems**—corporate ownership all along the channel—we might say the firm is going "direct." But actually, the firm may be handling manufacturing, wholesaling, *and* retailing—so it's more accurate to think of the firm as a vertical marketing system.

For example, Sherwin-Williams produces paint and owns the distribution centers that hold inventory and the retail stores where they are sold. In these channels, there is a high level of coordination and information sharing. Sherwin-Williams Paint Stores' computer systems share sales data with the company's home office and its paint production facilities. Promotional events, such as sales of certain styles, are closely coordinated because the whole channel is owned by the same company. So when Duration Exterior Acrylic Latex goes on sale, ads are run locally, signage is visible in all the stores, and the distribution centers and stores are well stocked for the increased demand.

Exhibit 10-4 Characteristics of Traditional and Vertical Marketing Systems

	Type of Channel			
		Vertical Marketing Systems		
Characteristics	Traditional	Administered	Contractual	Corporate
Amount of cooperation	Little or none	Some to good	Fairly good to good	Complete
Data and information sharing	Little or none	Some	Moderate	Extensive
Control maintained by	None	Economic power and leadership	Contracts	Ownership by one company
Examples	Typical channel of "independents"	General Electric, MillerCoors, Scotts Miracle-Gro	McDonald's, Holiday Inn, Ace Hardware, SuperValu, Coca-Cola, Chevrolet	Florsheim Shoes, Sherwin-Williams

Sherwin-Williams is an example of a corporate channel system. The paint manufacturer owns the whole channel system—from its product to its retail stores.
Eric Glenn/Shutterstock

Sometimes a corporate channel system is set up that way from the start. Sometimes it happens through **vertical integration**—acquiring firms at different levels of channel activity. For example, IKEA purchased forests in Romania to have greater control over an important raw material.

Vertical integration has potential advantages—stable sources of supplies, better control of distribution and quality, greater buying power, faster and more reliable flow of information and data, and lower executive overhead. Provided that the discrepancies of quantity and assortment are not too great at each level in a channel, vertical integration can be profitable. However, many managers have found that it's hard to be really good at running manufacturing, wholesaling, and retailing businesses that are very different from one another. Instead, they try to be more efficient at what they do best and focus on ways to get cooperation in the channel for the other activities.[17]

Administered and contractual systems may work well

Firms can often gain the advantages of vertical integration without building a costly corporate channel. A manager can develop administered or contractual channel systems instead. In **administered channel systems**, the channel members informally agree to cooperate with one another. They can agree to routinize ordering, share inventory and sales information over computer networks, standardize accounting, and coordinate promotion efforts. This usually occurs when one member of the channel system, perhaps a channel captain, has enough power to influence others toward greater cooperation. Retailers like Target or large manufacturers like P&G often control this type of vertical marketing system.

In **contractual channel systems**, the channel members agree by contract to cooperate with one another. For example, Ford Motor Company has detailed contracts with its dealers that outline Ford's responsibilities and promises as well as the dealer's requirements. McDonald's restaurants are owned locally, but all have a contract with the company to ensure they operate in a similar way. With both of these systems, the members retain some of the flexibility of a traditional channel system.

Vertical marketing systems dominate the marketplace

Vertical systems in the consumer products area have a healthy majority of retail sales and should continue to increase their share in the future. Vertical marketing systems are becoming the major competitive units in the U.S. distribution system and are growing rapidly in other parts of the world as well.[18]

The Best Channel System Should Achieve Ideal Market Exposure

LO 10.6

You may think that all marketing managers want their products to have maximum exposure to potential customers. This isn't true. Some product classes require much less market exposure than others. **Ideal market exposure** makes a product available widely enough to satisfy target customers' needs but not exceed them. Too much exposure only increases the total cost of marketing.

Exhibit 10-5 Comparing Levels of Market Exposure

	Level of Market Exposure		
	Intensive	Selective	Exclusive
Number of outlets	Many	Some—a few in each geographic area	Few—one per geographic area
Used when…	Convenience products, business supplies, and higher sales volume offset higher distribution costs	Widespread coverage is needed, but also special attention from intermediaries	Fewer customers so support needed from intermediaries or when franchisees demand protected markets
Examples	Coca-Cola, Wrigley's chewing gum, Butterfinger candy bars	Oakley sunglasses, Vera Bradley handbags, Goodyear tires	Audi cars, Sub-Zero appliances, Piaget watches, Five Guys restaurants

Ideal exposure may be intensive, selective, or exclusive

Intensive distribution is selling a product through all responsible and suitable wholesalers or retailers who will stock or sell the product. **Selective distribution** is selling through only those intermediaries who will give the product special attention. **Exclusive distribution** is selling through only one intermediary in a particular geographic area. For an overview, see Exhibit 10-5. As we move from intensive to exclusive distribution, we give up exposure in return for some other advantage—including, but not limited to, lower cost.

Intensive distribution—sell it where they buy it

Intensive distribution is commonly needed for convenience products and business supplies—such as laser printer cartridges, three-ring binders, and copier paper—used by all offices. Customers want such products nearby. For example, Rayovac batteries were not selling well even though their performance was very similar to other batteries. Part of that was due to heavier advertising for Duracell and Energizer. But consumers usually don't go shopping for batteries. They're purchased on impulse 83 percent of the time. To get a larger share of purchases, Rayovac had to be in more stores. It offered retailers a marketing mix with less advertising and a lower price. In three years, the brand moved from being available in 36,000 stores to 82,000 stores—and that increase gave sales a big charge.[19]

Products like breakfast cereal have selective distribution. They sell where customers expect to buy these products, in grocery stores and convenience stores. On the other hand, a car like the Bugatti uses exclusive distribution. There may be only one Bugatti dealer for hundreds of miles. The luxury car is likely to be sought out by those interested in buying one.
(left): quiggyt4/Shutterstock; (right): Gisela Schober/Getty Images

Selective distribution—sell it where it sells best

Selective distribution covers the broad area of market exposure between intensive and exclusive distribution. It may be suitable for all categories of products. Only the better intermediaries are used here. Companies commonly use selective distribution to gain some of the advantages of exclusive distribution—while still achieving fairly widespread market coverage.

Reduce costs and get better partners

A selective policy might be used to avoid selling to wholesalers or retailers that (1) place orders that are too small to justify making calls, (2) make too many returns or request too much service, (3) have a poor credit rating, or (4) are not in a position to do a satisfactory job.

Selective distribution is becoming more popular than intensive distribution as firms see that they don't need 100 percent coverage of a market to support national advertising. Interested customers without a local store can often buy online. Often the majority of sales come from relatively few customers—and the others buy too little compared to the cost of working with them. This is called the 80/20 rule—80 percent of a company's sales often come from only 20 percent of its customers.

Get special effort from channel members

Selective distribution can produce greater profits not only for the producer but for all channel members. Wholesalers and retailers are more willing to promote products aggressively if they know they're going to obtain the majority of sales through their own efforts. They may carry wider lines, do more promotion, and provide more service—all of which lead to more sales. Viper makes systems that remotely start a car—sit in your house on a cold day, push a button, and your car starts. Its selective distribution (just a few outlets per city) gets it stronger support from its dealers.

Selective often moves to intensive as market grows

In the early part of the life cycle of a new unsought good, a producer may have to use selective distribution. Well-known wholesalers and retailers may have the power to get such a product introduced, but that often means limiting the number of competing wholesalers and retailers. The producer may be happy with such an arrangement at first but dislike it later when more retailers want to carry the product.

Exclusive distribution sometimes makes sense

Exclusive distribution is just an extreme case of selective distribution—the firm selects only one wholesaler or retailer in each geographic area. Besides the various advantages of selective distribution, producers may want to use exclusive distribution to help control prices and the service offered in a channel. Franchisors such as McDonald's and 1-800-GOT-JUNK? offer franchisees exclusive territories.

Is limiting market exposure legal? It depends

Exclusive distribution is an area considered under U.S. antimonopoly laws. Courts currently focus on whether an exclusive distribution arrangement hurts competition.

Horizontal arrangements—among *competing* retailers, wholesalers, or producers—to limit sales by customer or territory have consistently been ruled illegal by the U.S. Supreme Court. Courts consider such arrangements obvious collusion that reduces competition and harms customers.

The legality of vertical arrangements—between producers and intermediaries—is not as clear-cut. A 1977 Supreme Court decision (involving Sylvania and the distribution of TV sets) reversed an earlier ruling that it was always illegal to set up vertical relationships limiting territories or customers. Now courts can weigh the possible good effects against the possible restrictions on competition. They look at competition between whole channels rather than just focusing on competition at one level of distribution. The Sylvania decision does not mean that all vertical arrangements are legal. Rather, it says that a firm has to be able to legally justify any exclusive arrangements.

Thus, firms should be cautious about entering into *any* exclusive distribution arrangement. The courts can force a change in relationships that were expensive to develop. And even worse, the courts can award triple damages if they rule that competition has been hurt.

The same cautions apply to selective distribution. Here, however, less formal arrangements are typical—and the possible impact on competition is more remote. It is now more acceptable to carefully select channel members when building a channel system. Refusing to sell to some intermediaries, however, should be part of a logical plan with long-term benefits to consumers.[20]

Multichannel Distribution and Reverse Channels

LO 10.7

Trying to achieve the desired degree of market exposure can lead to complex channels of distribution. Firms may need different channels to reach different segments of a broad product-market or to be sure they reach each segment at different stages of the purchase process. Sometimes this results in competition and conflict among different channels. Sometimes additional channels must be developed to return or recycle products. Let's look closer at multichannel distribution and then reverse channels.

Many firms use more than one channel of distribution

Consider the different channels used by a company that publishes computer books. See Exhibit 10-6. This publisher sells through a general book wholesaler that in turn sells to Internet and independent book retailers. The publisher might also sell through a computer supplies wholesaler that serves electronics superstores like Best Buy. It may also sell some of its best sellers through a large chain or even to consumers who order directly from its website. An Internet retailer might offer books in print and as e-book downloads. Of course, all this exposure helps the publisher to reach more customers. On the other hand, it can cause problems and potential conflict because different wholesalers and retailers want different markups. It also increases competition, including price competition. And the competition among different intermediaries may lead to conflicts between the intermediaries and the publisher. Managers must consider the impact of conflict if they choose to increase market exposure by using more channels of distribution.

Multichannel distribution is becoming more common

Multichannel distribution occurs when a producer uses several competing channels to reach the same target market—perhaps using several intermediaries in addition to selling directly. Multichannel distribution is becoming more common. A single target market can buy a new Apple iPad at Apple's website, an Apple store, an online retailer, or a physical store such as Target or a college bookstore. Dr Pepper also uses multichannel distribution; you can buy a Dr Pepper at a convenience store, grocery store, vending machine, or restaurant.

Sometimes producers use multichannel distribution because their present channels do a poor job or fail to reach some potential customers. For example, Reebok International had been relying on local sporting goods stores to sell its shoes to high school and college athletic teams. But Reebok wasn't getting much of the business. Sales jumped when it set

Exhibit 10-6 An Example of Different Channels of Distribution Used by a Publisher of Computer Books

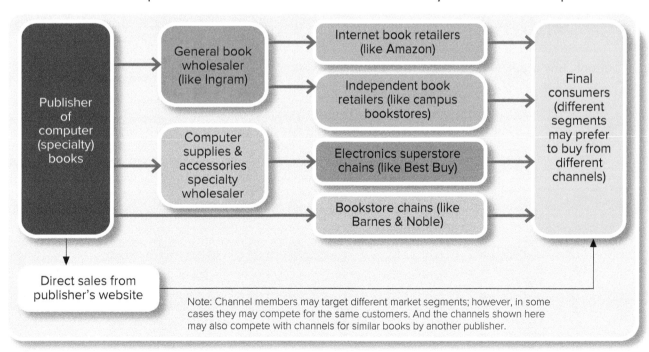

Note: Channel members may target different market segments; however, in some cases they may compete for the same customers. And the channels shown here may also compete with channels for similar books by another publisher.

At one time, *The Wall Street Journal* was distributed through just a few channels, including on the street at newsstands. Now customers can find the "newspaper" for business on its website or Facebook. These channels respond to consumers who are spending more time in front of computer screens and on mobile devices.
(upper left): Michael Brown/Getty Images; (lower left): Source: Wall Street Journal via wsj.com; (right): Source: Wall Street Journal via Facebook

up its own sales department team to sell directly to the schools. The schools could still buy from a local sporting goods store, but now they had a second option.[21]

Multichannel shoppers shop across channels...

One major reason for the growth of multichannel distribution is that it reflects how customers want to shop. For some purchases, customers find that different channels are more effective at different stages in the purchase process. When customers search for information, evaluate products, and make purchases across different channels, they are multichannel shopping. **Multichannel shoppers** use different channels as they move through a purchase process.

For example, a multichannel shopper seeking a new tennis racquet might read customer reviews of the Wilson Blade tennis racquet at Amazon.com, go to a pro shop at a local tennis club to hit some balls with the Blade, and then purchase the tennis racquet at a discount from TennisExpress.com.

...but only buy at one

Multichannel shopping causes problems for those channel members that feel they are helping move a sale along but do not make any profits from their efforts. In the example cited in the purchase of the tennis racquet, Amazon and the tennis club helped the shopper, but only TennisExpress.com made a profit.

When consumers go to a brick-and-mortar store to inspect a product and then purchase from an online retailer with a lower price, this practice is called *showrooming*. Typically, a physical retailer acts as a showroom (providing a customer with a real product to see, try, and hold) for another (often lower-priced and online) competitor. Smartphone apps make the practice easy; with Amazon's app, customers scan product bar codes in a retail store and see the Amazon price. With one click they can make the purchase online.

The practice also works in reverse. *Webrooming* occurs when a customer gathers information at an online store (perhaps by reading reviews) and then purchases at a brick-and-mortar store—maybe for the convenience of getting the product sooner. Of course, the only retailer making money on any shopping trip is the one the customer chooses to purchase from, not the others from whom the customer gleaned information along the purchase journey.[22]

Omnichannel: the seamless multichannel shopping experience

Some wholesalers and retailers are working to keep customers within their ecosystem for the entire buying process. Some online retailers have added physical stores, and many brick-and-mortar stores have online sites with useful buying information if not actual e-commerce stores. Retailers that use this strategy to connect their physical store with online channels can create a competitive advantage. **Omnichannel** is a multichannel selling approach in which a retailer provides a seamless customer shopping experience from computer, mobile device, or brick-and-mortar store. Omnichannel is an example of multichannel distribution designed to appeal to a single consumer who prefers to shop across channels. For example, Bank of America allows customers to apply for loans, check account balances, transfer money, and more from a website, mobile app, or, for old-fashioned folks, at a bank branch. Starting on one channel will carry to another.[23]

Marketing managers monitor all channels

When a firm utilizes multichannel distribution, the marketing manager routinely monitors sales in each channel. One tool is a *sales by channel report,* which lists unit or dollar volume sales for each channel. This report gives marketing managers insight into the performance of each channel. And if each channel targets different segments, it can offer insights into customer behavior. To better understand how this type of report works, see Marketing Analytics in Action: Sales by Channel Report.

Ethical decisions may be required

If competition changes or customers' Place requirements shift, the current channel system may not be effective. The changes required to serve customer needs may hurt one or more members of the channel. Ethical dilemmas in the channel area arise in situations like this—because not everyone in the channel can win.

For example, wholesalers and the independent retailers they serve in a channel of distribution may trust a producer channel captain to develop marketing strategies that will work for the whole channel. However, the producer may decide that consumers, and its own business, are best served by a change (say, dropping current wholesalers and selling directly to big retail chains). A move of this sort, if implemented

---

Marketing Analytics in Action: Sales by Channel Report

Kelly Rizzo is the product manager for BookPacks, which produces small and medium-sized backpacks mostly used by students of all ages as school bookbags. In reviewing her plans for next year, Kelly pulled a sales by channel report showing the number of backpacks sold in each of BookPacks' channels over the last five months. Review the following report and then answer the questions.

	Number of Backpacks Sold by Channel					Total by Channel
	Aug	Sep	Oct	Nov	Dec	
Department stores (e.g., Macy's)	2,644	1,658	1,421	1,854	1,721	9,298
Discount stores (e.g., Walmart)	14,844	7,854	3,956	4,112	4,725	35,531
Online stores (e.g., eBags)	1,354	1,892	2,786	4,025	5,688	15,745
College bookstores	8,632	2,864	1,432	1,874	9,435	24,237
Total by month	**27,514**	**14,268**	**9,595**	**11,865**	**21,569**	**84,811**

1. Which channels of distribution sell the most BookPacks backpacks?
2. Which two months produce the most sales for BookPacks? Speculate on the reasons why each of the two largest months have higher sales.
3. Speculate on reasons for the high month-to-month variation observed in the discount store, online store, and college bookstore channels.
4. Should Kelly consider investing more in one of the channels? If so, which one and why?

immediately, may not give current wholesaler-partners a chance to make adjustments of their own. The more dependent they are on the producer, the more severe the impact is likely to be. It's not easy to determine the best or most ethical solution in these situations. However, marketing managers must think carefully about the consequences of Place strategy changes for other channel members. In channels, as in any business dealing, relationships of trust must be treated with care.[24]

> **Ethical Dilemma**
>
> *What would you do?* Assume you are the sales manager for a small but growing company that makes sustainable fashion. As you arrive at a trade show to promote your firm's new line, your assistant pulls you aside to warn you that some of the owners of shops that usually carry your line are looking for you—and several are hopping mad. They've heard rumors that a big retail chain will be carrying your new line in all of its stores. In the past, your firm has distributed its fashions only through these small, "independent" retail shops, and they see the big chains as threats. Although you have had some negotiations with a buyer for the big retail chain, no deal has been reached yet.
>
> *What will you say to the owners of the small shops, many of whom helped your firm get started and have always supported your marketing plans? If you are certain that most of these small retailers will not place any orders if they think the big chains will be carrying the same line, will you respond differently? Explain your thinking.*

Reverse channels are important too

Most firms focus on getting products to their customers. But some marketing managers must also plan for **reverse channels**—channels used to retrieve products that customers no longer want. The need for reverse channels may arise in a variety of different situations. Toy companies, automobile firms, drug companies, and others sometimes have to recall products because of safety problems. A firm that makes an error in completing an order may have to take returns. If a ViewSonic computer monitor breaks while it's still under warranty, someone needs to get it to the repair center. Soft drink companies may need to recycle empty bottles. Some countries have regulations that require taking back products at the end of their life. And, of course, customers sometimes just want to return something they bought, perhaps because it doesn't fit right or they changed their mind.[25]

Because disposing of batteries in the trash can be harmful to the environment, the Swiss government created INOBAT, an organization that promotes recycling with creative promotion and an easy-return bag.
Source: INOBAT

288

#M4BW

New laws require reverse channels in some industries

Many firms reluctantly add reverse channels—doing so because of laws designed to help the environment. For example, some "take back" laws require manufacturers to recycle or reuse hazardous materials or products at the end of their useful life—at no additional cost to the customer. In Europe, automakers must take back and recycle or reuse 85 percent of any vehicle made after 2004. One of the most well-established laws is the European Community's **Waste Electrical and Electronic Equipment (WEEE) Directive**, which requires producers to take back waste electrical and electronic equipment. This includes products like computers and televisions.

Similar laws are cropping up in the United States. By making producers responsible for recycling, governments create an incentive for manufacturers to design sustainability into new products. It is resulting in more eco-friendly products and contributes to a better world.

#M4BW

Reverse channels—sustainable and profitable

While reverse channels help the environment, many firms find them profitable too. In rural China, for example, it's cheaper for Coke to reuse glass bottles than to rely on plastic packages or cans. Recycling has been even more important to Xerox. Customers responded well to Xerox's offer to dispose of their old copy machines when they bought a new model. Better yet for Xerox, in the first year of the program, it saved $50 million by refurbishing the equipment and then reselling parts from the recycled models.

Some customers see value in purchasing more sustainable products. Shaw Floors targets these customers. Its EcoWorx carpet tiles are manufactured from recycled carpets. Shaw also promises that when the time comes, it will pick up and recycle the old tiles at no cost. To make it easy for the customer to follow up, Shaw's telephone number is printed on the back of each tile.[26]

#M4BW Brazil-based food bank Banco de Alimentos' Reverse Delivery is another example of capacity capture. Before this program, food delivery drivers across São Paulo returned from deliveries empty-handed (a wasted resource). Restaurants participating in the Reverse Delivery program still deliver the food, but they also ask the customer if they have any food to donate to those in need. Drivers bring the food back to the restaurant, where it is later collected by the food bank. Programs like this don't cost these companies much, if anything, yet they make the world a better place and maybe help a restaurant's brand image when it is associated with a good cause.
Source: Banco de Alimentos

#M4BW

Capacity capture—reverse channels capturing value

Reverse channels create new opportunities for forward-thinking companies. *Capacity capture* eliminates waste and utilizes new sources of value in a supply chain. The idea is to find an additional use for an existing resource or activity. For example, in the United Kingdom, Nissan partnered with the power company Enel to create a vehicle-to-grid (V2G) system that will create more electric power across the country. The V2G allows owners of Nissan's electric LEAF and e-NV200 vans to sell unused electric power stored in the vehicles' batteries back to the power company. The vehicles can also share power with a home battery system developed by Nissan.[27]

Plan for reverse channels

When marketing managers don't plan for reverse channels, the firm's customers may be left to solve "their" problem. That usually doesn't make sense. So a complete plan for Place may need to consider an efficient way to return products—with policies that different channel members agree on. It may also require specialists who were not involved in getting the product to the consumer. But if that's what it takes to satisfy customers, it should be part of marketing strategy planning.[28]

Entering International Markets

LO 10.8

All of the strategy decisions for Place (see again Exhibit 10-1) apply whether a firm is just focused on its domestic market or is also trying to reach target customers in international markets. However, marketing managers typically face differences in international markets that require additional choices. In the external market environment, culture and laws are almost always different from those with which the marketing manager is familiar. Developing countries may have less stable economies, and political environments involve more risk. Financial reporting requirements may or may not be as rigorous—so evaluating a customer's creditworthiness can be a challenge. Still, many small and medium-sized businesses see international sales as a growth opportunity that is worth the risk.

There are five basic ways to enter international markets (see Exhibit 10-7). As a rule, the approaches with greater risk and required investment offer the benefit of greater control over the marketing mix used.

Exporting often comes first

Some companies get into international marketing just by **exporting**—selling some of what the firm produces to foreign markets. Some firms start exporting just to take advantage of excess capacity—or even to get rid of surplus inventory. Some firms decide to change little if anything about the product, the label, or even the instructions. This explains why some early efforts at exporting are not very satisfactory. Other firms work closely with intermediaries who develop appropriate marketing mix changes and handle problems such as customs, import and export taxes, shipping, exchange rates, and recruiting or working with wholesalers and retailers in the foreign country.

American firm Sono-Tek Corporation produces ultrasonic spray coating technology used in many electronic manufacturing processes. The company gets 60 percent of its

Exhibit 10-7 Basic Approaches for Entering International Markets

Exporting → Licensing → Management contracting → Joint venture → Direct investment

Generally increasing investment, risk, and control of marketing

revenue from exporting, mostly to Europe and many parts of Asia. Even exporting has higher costs, though, as Sono-Tek estimates that it spends 80 percent of its sales and marketing budget overseas.[29]

Licensing is often an easy way

Licensing means selling the right to use some process, trademark, patent, or other right for a fee or royalty. The licensee in the foreign market takes most of the risk because it must make some initial investment to get started. The licensee also does most of the marketing strategy planning for the markets it is licensed to serve. If good partners are available, this can be an effective way to enter a market. Gerber entered the Japanese baby food market this way, but it exports to other countries.

Management contracting sells know-how

Management contracting means that the seller provides only management and marketing skills—others own the production and distribution facilities. Some mines and oil refineries are operated this way—and Hilton operates hotels all over the world for local owners using this method. This is another relatively low-risk approach to international marketing. The low level of commitment to fixed facilities makes the approach attractive in developing nations or ones where the government is less stable.

Joint venturing increases involvement

In a **joint venture**, a domestic firm enters into a partnership with a foreign firm. As with any partnership, there can be honest disagreements over objectives—for example, how much profit is desired and how fast it should be paid out—as well as operating policies. Where a close working relationship can be developed—perhaps based on one firm's technical and marketing know-how and the foreign partner's knowledge of the market and political connections—this approach can be very attractive to both parties. Typically the two partners must make significant investments and agree on the marketing strategy. Once a joint venture is formed, it can be difficult to end if things aren't working out. American investment bank JPMorgan Chase used this approach to enter China, where stiff regulations prohibited a foreign firm from owning a controlling share of a Chinese investment bank or money management firm.

Direct investment involves ownership

When a foreign market looks really promising, a firm may want to take a bigger step with a direct investment. **Direct investment** means that a parent firm has a division (or owns a separate subsidiary firm) in a foreign market. This gives the parent firm complete control of marketing strategy planning. Direct investment is a big commitment and usually entails greater risks. If a local market has economic or political problems, the firm cannot easily leave. On the other hand, by providing local jobs, a company builds a strong presence in a new market. This helps the firm develop a good reputation with the government and customers in the host country. And the firm does not have to share profits with a partner.

Direct investment also helps a firm learn more about a new market. For example, U.S.-based quick-lube chain Grease Monkey entered the Chinese market by opening company-owned locations. It quickly discovered that Chinese customers wanted a much broader range of services than was customary in the United States. Grease Monkey added the services—and greater profits followed.[30]

Some firms tailor their websites to sell directly to export customers in foreign countries. Nike used this approach in Argentina.
Source: Nike, Inc.

CONCLUSION

In this chapter, we discussed the role of Place in marketing strategy. Place decisions are especially important because they may be difficult and expensive to change. So marketing managers must make Place decisions very carefully.

We discussed how product classes and the product life cycle are related to Place objectives. This helps us determine how much a firm should rely on indirect channel systems with intermediaries or direct systems.

Marketing specialists and channel systems develop to adjust discrepancies of quantity and assortment. Their regrouping activities are basic in any economic system. Adjusting discrepancies provides opportunities for creative marketers.

Channels of distribution tend to work best when there is cooperation among the members of a channel—and conflict is avoided—so we discussed the importance of planning channel systems and the role of a channel captain. We stressed that channel systems compete with one another and that vertical marketing systems seem to be winning.

Channel planning also requires firms to decide on the degree of market exposure they want. The ideal level of exposure may be intensive, selective, or exclusive. We discussed how more producers are using multichannel strategies and more customers are becoming multichannel shoppers. We also described how reverse channels operate and why they are needed. Finally, we examined different approaches for entering international markets.

KEY TERMS

Place, 266
channel of distribution, 267
direct marketing, 274
discrepancy of quantity, 275
discrepancy of assortment, 275
regrouping activities, 275
accumulating, 275
bulk-breaking, 275
sorting, 275
assorting, 275
traditional channel systems, 276
vertical channel conflict, 277

horizontal channel conflict, 277
channel captain, 279
vertical marketing systems, 281
corporate channel systems, 281
vertical integration, 282
administered channel systems, 282
contractual channel systems, 282
ideal market exposure, 282
intensive distribution, 283
selective distribution, 283
exclusive distribution, 283
multichannel distribution, 285

multichannel shoppers, 286
omnichannel, 287
reverse channels, 288
Waste Electrical and Electronic Equipment (WEEE) Directive, 289
exporting, 290
licensing, 291
management contracting, 291
joint venture, 291
direct investment, 291

QUESTIONS AND PROBLEMS

1. Review the PC industry case study that opens the chapter. From this case, identify examples of different key terms and concepts covered in the chapter. For example, channel conflict is shown when HP added direct distribution in response to Dell.

2. Review the PC industry case study that opens this chapter. Applying Place concepts from the chapter, what should Lenovo do to gain a stronger position in the smartphone market?

3. Give two examples of service firms that work with other channel specialists to sell their products to final consumers. What marketing functions can the specialist provide in each case?

4. Discuss some reasons why a firm that produces installations might use direct distribution in its domestic market but use intermediaries to reach overseas customers.

5. Explain discrepancies of quantity and assortment using the clothing business as an example. How does the application of these concepts change when selling steel to the automobile industry? What impact does this have on the number and kinds of marketing specialists required?

6. Explain the four regrouping activities with an example from the building supply industry (nails, paint, flooring, plumbing fixtures, etc.). Do you think that many specialists develop in this industry, or do producers handle the job themselves? What kinds of marketing channels would you expect to find in this industry, and what functions would various channel members provide?

7. Insurance agents are intermediaries who help other members of the channel by providing information and handling the selling function. Does it make sense for an insurance agent to specialize and work exclusively with one insurance provider? Why or why not?

8. Discuss the Place objectives and distribution arrangements that are appropriate for the following products (indicate any special assumptions you have to make to obtain an answer):

a. A postal scale for products weighing up to 2 pounds.
b. Children's toys: (1) radio-controlled model airplanes costing $100 or more and (2) small rubber balls.
c. Heavy-duty, rechargeable, battery-powered robotic nut tighteners for factory production lines.
d. Fiberglass fabric used in making roofing shingles.

9. Give an example of a producer that uses two or more different channels of distribution. Briefly discuss what problems this might cause.
10. Explain how a channel captain can help traditional independent firms compete with a corporate (integrated) channel system.
11. Find an example of vertical integration within your city. Are there any particular advantages to this vertical integration? If so, what are they? If there are no advantages, how do you explain the integration?
12. How does the nature of the product relate to the degree of market exposure desired?
13. Why would intermediaries want to be exclusive distributors for a product? Why would producers want exclusive distribution? Would intermediaries be equally eager to get exclusive distribution for any type of product? Why or why not? Explain with reference to the following products: candy bars, batteries, golf clubs, golf balls, steak knives, televisions, and industrial woodworking machinery.
14. Discuss the promotion a new grocery products producer would need in order to develop appropriate channels and move products through those channels. Would the nature of this job change for a new producer of dresses? How about for a new, small producer of installations?
15. Describe the advantages and disadvantages of the approaches to international market entry discussed in this chapter.

MARKETING PLANNING FOR HILLSIDE VETERINARY CLINIC

Appendix D (the Appendices follow Chapter 19) includes a sample marketing plan for Hillside Veterinary Clinic. Look through the "Marketing Strategy" section.

1. Why does Hillside sell its product directly instead of indirectly?
2. Hillside has a small selection of pet supplies that it sells to people who bring in their pets. What products does it resell at retail? What channel functions does it provide, and what channel functions are performed by its suppliers?

SUGGESTED CASES

13. Paper Products, Inc. (PPI)   15. The Scioto Group   16. Hanratty Company

MARKETING ANALYTICS: DATA TO KNOWLEDGE

CHAPTER 10: INTENSIVE VERSUS SELECTIVE DISTRIBUTION

Hydropump, Inc. produces high-quality pumps and sells them to business customers. Its marketing research shows a growing market for a similar type of pump aimed at final consumers—for use with Jacuzzi-style tubs in home remodeling jobs. Hydropump will have to develop new channels of distribution to reach this target market because most consumers rely on a retailer for advice about the combination of tub, pump, heater, and related plumbing fixtures they need. Hydropump's marketing manager, Robert Black, is trying to decide between intensive and selective distribution. With intensive distribution, he would try to sell through all the plumbing supply, bathroom fixture, and hot tub retailers who will carry the pump. He estimates that about 5,600 suitable retailers would be willing to carry the new pump. With selective distribution, he would focus on about 280 of the best hot tub dealers (2 or 3 in the 100 largest metropolitan areas).

Intensive distribution would require Hydropump to do more mass selling—primarily advertising in home renovation magazines—to help stimulate consumer familiarity with the brand and convince retailers that Hydropump equipment will sell. The price to the retailer might have to be lower too (to permit a bigger markup) so they will be motivated to sell Hydropump rather than some other brand offering a smaller markup.

With intensive distribution, each Hydropump sales rep could probably handle about 300 retailers effectively. With selective distribution, each sales rep could handle only about 70 retailers because more merchandising help would be necessary. Managing the smaller sales force and fewer retailers with the selective approach would require less manager overhead cost.

Going to all suitable and available retailers would make the pump available through about 20 times as many retailers and have the potential of reaching more customers. However, many customers shop at more than one retailer before making a final choice—so selective distribution would reach almost as many potential customers. Further, if Hydropump is using selective distribution, it would get more in-store sales attention for its pump and a larger share of pump purchases at each retailer.

Black has decided to use a spreadsheet to analyze the benefits and costs of intensive versus selective distribution.

Design element: #M4BW box globe icon: ©Vectoryzen/Shutterstock

CHAPTER ELEVEN

Marco Di Lauro/Getty Images

Distribution Customer Service and Logistics

If you want a Coca-Cola, there's usually one close by—no matter where you might be in the world. And that's no accident. An executive for the best-known brand name in the world stated the goal simply: "Make Coca-Cola available within an arm's reach of desire." To achieve that level of customer service, Coke works with many different channels of distribution. But that's just the start. Think about what it takes for a bottle, can, or cup of Coke to be there whenever you're thirsty. In warehouses and distribution centers, on trucks, in restaurants and sports arenas, and in thousands of other retail outlets, Coke handles, stores, and transports more than 600 billion soft drink servings per year. Getting all of its product to consumers is a huge undertaking, but Coke does it effectively and at a low cost.

Readily accessible information about what the market needs helps keep Coke's distribution on target. Coke uses an Internet-based data system that links about 1 million retailers and other sellers to Coke and its bottlers. The system lets Coke bottlers and retailers exchange orders, invoices, and pricing information online. Coke sells more than 500 brands and 3,500 products in 206 countries, making for what could be a logistical nightmare. But Coke's marketing and logistics managers coordinate with its computer information systems group to integrate this system worldwide.

It works. Computer systems show Coke managers exactly what's selling in each market, and sophisticated software programs accurately predict future sales. Analytics help factor in the effects of seasonal changes, weather forecasts, and planned promotions to fine-tune and plan inventories and deliveries. Orders are processed instantly, so sales to consumers at the end of the channel aren't lost because of stock-outs. Coke moves products efficiently through the channel. In Cincinnati, for example, Coke built the beverage industry's first fully automated distribution center. And when Coke's truck drivers get to a retail store, they knowingly stock the shelves with the correct mix of products.

Coke's strategies in international markets have similar objectives. But because cultural differences and the stage of market development vary across countries, Coke's emphasis varies as well. To increase sales in France, for example, Coke installed thousands of soft drink coolers in French supermarkets. In Great Britain, Coke emphasizes multipacks because it wants to have more inventory at the point of consumption—in consumers' homes. In Japan, by contrast, Coke relies heavily on an army of truck drivers to constantly restock 1 million Coke vending machines, more per capita than anywhere else in the world. Japanese customers can even use their cell phones to buy a Coke from one of these vending machines. In Thailand, Coke's bottlers are buying more trucks and increasing the frequency of delivery to keep up with fast growth.

In less-developed areas, the focus may be on different challenges—especially if the limitations of the Place system can make Coke products hard to find or costly. In Africa, where per capita consumption of Coke products is one-tenth of that in the United States, Coca-Cola wants its products to be cold and available everywhere—even in the lowest-income areas. For example, in Kabira, a Kenyan slum, a local distributor uses three shipping containers as a portable warehouse. Dirt and gravel roads make it easier to deliver by hand, so delivery personnel stack a couple dozen crates onto two-wheeled carts and fan out to 345 small shops delivering Coke, Fanta, and Stoney Ginger Beer. Deliveries might occur once or twice a day because the small shops hold so little inventory. Coke provides shops with plenty of "red"—Coke signage, tablecloths, and refrigerated coolers. Returnable glass bottles help keep down costs, and at prices of less than 25 cents (U.S.) Coke becomes a treat on hot days.

While technology facilitates Coke's complex distribution system, retailers and wholesalers in developing countries don't always have scanner checkout systems and electronic data interchange. When this is the case, Coke utilizes a technology readily accessible to most channel partners: cell phones. For example, in Vietnam, Coke uses an app to help retailers and wholesalers manage inventory and ordering and speed up order fulfillment. When a retailer uses the app to order a case of Coke, the message immediately goes to multiple wholesalers. The wholesaler that can fill the order most quickly responds; deliveries—often by motorcycle—now happen in hours instead of days. This means local retailers don't run out and more consumers have a Coke within that "arm's reach of desire."

Fortune magazine recently put the spotlight on Coca-Cola's commitment to sustainability. Its decisions in the logistics area have had big environmental effects. For example, Coca-Cola helped develop vending machines that are HFC-free and

40 to 50 percent more energy efficient than conventional beverage equipment. Similarly, in the United States and Canada, Coca-Cola's gas-electric hybrid delivery trucks cut emissions and fuel consumption by a third.

In spite of positive steps like these, Coca-Cola still faces some challenges concerning sustainability and logistics. Critics argue that in a society where there is already a safe supply of tap water, it doesn't make sense to bottle water (Coke's brand is Dasani), transport it in trucks that consume fuel and contribute to pollution, and then add worry about how best to dispose of the empty bottles and bottle tops.

Coca-Cola's worldwide distribution network delivers a sustainable competitive advantage. Of course, Pepsi, Dr Pepper, and Hansen's Natural Sodas are competing for distributors' attention and retail shelf space. Coke is pushing new products as well. The battle to quench customers' thirsts depends on each brand's marketing programs—but clearly Place has an important role to play.[1]

LEARNING OBJECTIVES

Choosing the right distribution channels is crucial in getting products to the target market's Place. But, as the Coca-Cola case shows, that alone doesn't ensure that products are placed "within an arm's reach of desire"—when, where, in the quantities that customers want them, and at a price they're willing to pay. In this chapter, we discuss how marketing managers ensure that they also have physical distribution systems that meet their customers' needs—at both an acceptable service level and an affordable cost.

When you finish this chapter, you should be able to

1. understand why logistics (physical distribution) is such an important part of Place and marketing strategy planning.
2. understand why the physical distribution customer service level is a key marketing strategy variable.
3. understand the physical distribution concept and why the coordination of storing, transporting, and related activities is so important.
4. see how firms can cooperate and share logistics activities that will provide added value to their customers.
5. know about the advantages and disadvantages of various transportation methods.
6. know how inventory and storage decisions affect marketing strategy.
7. understand the distribution center concept.
8. understand important new terms (shown in **red**).

Physical Distribution Gets It to Customers

LO 11.1

Whenever Product includes a physical good, Place requires logistics decisions. **Logistics** is the transporting, storing, and handling of goods in ways that match target customers' needs with a firm's marketing mix—both within individual firms and along a channel of distribution. **Physical distribution (PD)** is another common name for logistics.

There are many different combinations of logistics decisions. Each combination can result in a different level of distribution service and different costs, so companies must determine the best way to provide the level of distribution service that customers want and are willing to pay for. We start this chapter by considering these critical logistics decisions (see Exhibit 11-1). Next, we describe the choice among different modes of transportation. Each has its own costs and benefits. We conclude with decisions about inventory and the use of distribution centers.

Exhibit 11-1 The Role of Logistics and Physical Distribution Customer Service in Marketing Strategy

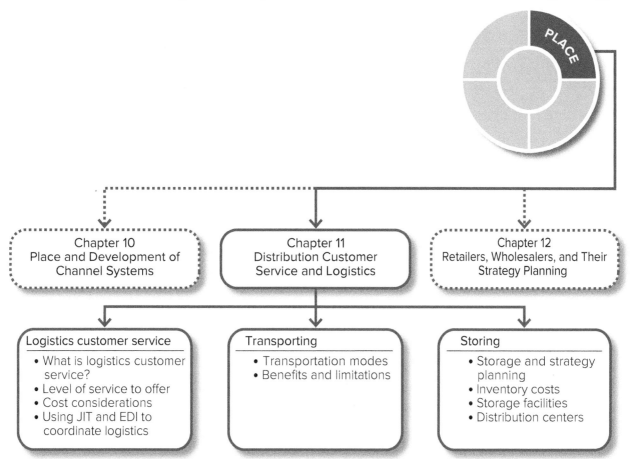

Logistics costs are very important to both companies and consumers. These costs vary from firm to firm and, from a macro-marketing perspective, from country to country. For some products, a company may spend half or more of its total marketing dollars on physical distribution activities. As lawmakers debate the high cost of health care, logistics costs represent more than one-third of the expense attached to hospital supplies. Although this is down from about 50 percent a few decades ago, there's still room for improvement.

Differences in logistics costs around the world can be substantial. Developed economies in the United States and Europe spend about 9 to 15 percent of GDP on logistics-related costs—whereas in the developing economies of Latin America and Africa, costs are 30 percent or more. Marketing managers must carefully consider these costs when entering new markets.[2]

Physical Distribution Customer Service

LO 11.2

From the beginning, we've emphasized that marketing strategy planning is based on meeting customers' needs. Planning for logistics and Place is no exception. So let's start by looking at logistics through a customer's eyes.

Customers want products, not excuses

Customers don't care how a product was moved or stored or what some channel member had to do to provide it. Rather, customers think in terms of the physical distribution **customer service level**—how rapidly and dependably a firm can deliver what they, the customers, want.

The best shipping method for any given product is usually a balancing act between service and cost. Trains provide a slower, less expensive shipping approach as compared to airfreight.
(left): Scanrail/IStockphoto/Getty Images; (right): grifco/123RF

What does this really mean? It means that Toyota wants to have enough windshields delivered to make cars *that* day—not later so production stops *or* earlier so there are a lot of extras to move around or store. It means that business executives who rent cars from Hertz want them to be ready when they get off their planes. It means that when you order a blue shirt at the Lands' End website you receive blue, not pink. It means you want your Tostitos to be whole when you buy a bag at the snack bar—not crushed into crumbs from rough handling in a warehouse.

Physical distribution is invisible to most consumers

Physical distribution is, and should be, a part of marketing that is "invisible" to most consumers. It gets their attention only when something goes wrong. At that point, it may be too late to do anything that will keep them happy.

In countries where physical distribution systems are inefficient, consumers face shortages of the products they need. By contrast, most consumers in the United States and Canada don't think much about physical distribution. This probably means that these market-directed macro-marketing systems work pretty well—that a lot of individual marketing managers have made good decisions in this area. But it doesn't mean that the decisions are always clear-cut or simple. In fact, many trade-offs may be required.

Trade-offs of costs, service, and sales

Most customers would prefer very good service at a very low price. But that combination is hard to provide because it usually costs more to provide higher levels of service. So most physical distribution decisions involve trade-offs among costs, customer service level, and sales.

If you want a new HP computer and the Best Buy store where you would like to buy it doesn't have it on hand, you're likely to buy it elsewhere; or if that model HP is hard to get, you might just switch to some other brand. Assuming your Best Buy store can locate your desired computer at another Best Buy or in stock at HP's factory, perhaps it could keep your business by guaranteeing two-day delivery to your home. This will cost the store manager some airfreight charges. In this case, the manager trades the savings of storing inventory for the extra cost of speedy delivery. Missing one sale may not seem that important, but it all adds up; consider that, a few years ago a computer company lost more than $500 million in sales because its computers weren't available when and where customers were ready to buy them.

Exhibit 11-2 illustrates trade-off relationships like the one highlighted in the HP example just mentioned. The exhibit shows costs on the vertical axis and customer

Exhibit 11-2 Trade-offs among Physical Distribution Costs, Customer Service Level, and Sales

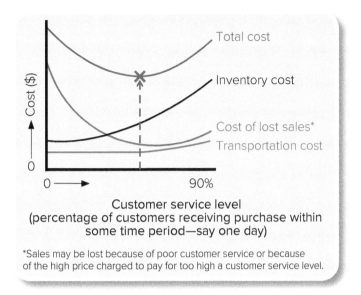

*Sales may be lost because of poor customer service or because of the high price charged to pay for too high a customer service level.

service level (measured as the percentage of customers receiving delivery within one day) on the horizontal axis. Note the following:

- As service level increases (moving left to right), transportation cost (in blue) is flat for a bit before increasing as shipments become more frequent to ensure there are no stock-outs.
- Inventory cost (black line) rises as the retailer carries more computers to meet any level of demand.
- Cost of lost sales (green line) initially falls because customers are not likely to buy a different product when the model they want is in stock. However, as customer service climbs to higher levels (mostly a result of rising inventory carrying costs), the seller must raise prices to cover the higher service levels. This leads to some customers choosing alternatives—and raises the cost of lost sales.
- When all of these costs are added together, the total cost (red line) falls as service level increases—but then rises. The optimum level of customer service (based on cost) occurs where the total cost is minimized (X on the graph).

Exhibit 11-2 provides a starting point to discuss the trade-offs marketing managers must make. Often transportation and inventory carrying costs can be reliably determined, but the cost of lost sales is an estimate. Marketing managers should work closely with distribution managers to discuss and monitor the trade-offs. To better understand how this analysis occurs in practice, complete the Marketing Analytics in Action: Trade-offs and Customer Service Levels activity.

Trade-offs more complicated with more products

Trade-offs that must be made in physical distribution can be even more complicated. The approach demonstrated in Exhibit 11-2 focuses on the sale of a single item. But a customer dissatisfied by not finding the item she wanted may soon shop elsewhere—reducing the lifetime customer value. For example, a few years ago cost-cutting efforts at Walmart resulted in reduced inventory—and some empty shelves. Some of those disappointed Walmart customers decided to shop elsewhere, and losses were far greater than the failure to sell particular items on a particular day or two. Similar problems occurred when Target entered Canada a few years ago. Target could not deliver quickly enough, and empty shelves made a bad first impression. Target never recovered and closed all 133 stores it operated there.

Information technology, data, and analytics can improve service levels and cut costs at the same time. Better information flows make it easier to coordinate activities, improve efficiency, and add value for the customer.[3]

Marketing Analytics in Action: Trade-offs and Customer Service Levels

Customer service levels are a major challenge for stores that sell shoes. Consider that for each style, the store may stock different colors and many sizes. In this case, we define customer service level as the percentage of customers who find the shoe style, color, and size they wish to purchase in stock on the day they shop. All the possible combinations can make 100 percent customer service costly. How might a store selling shoes decide on the best level of customer service?

Ronnie Chin is a marketing manager for Wanderlust Outdoor, a small chain of outdoor shops. She manages the availability of outdoor shoes and boots for all stores. Ronnie generated the following table from internal data. The left side of the table shows the current costs for transportation, inventory, and lost sales, as well as the total cost.

Ronnie is considering adding in-store kiosks that would make ordering out-of-stock shoes easier and faster—plus Wanderlust would include free two-day shipping to a customer's home on all kiosk orders. This would change the customer service cost structure. The right side of the table lists costs at different customer service levels using the kiosks. Note the following:

- With kiosks, transportation costs are higher at lower levels of customer service. Because Wanderlust pays for the two-day shipping to a customer's home, this raises its transportation costs. This happens more often at low customer service levels.
- The kiosks would not change inventory cost.
- With kiosks, costs for lost sales would fall, especially at lower levels of customer service. Because more customers make a kiosk purchase instead of leaving the store empty-handed, fewer sales are lost at lower levels of customer service.

Customer Service Level	Current Costs* at Different Customer Service Levels				Costs* at Different Customer Service Levels with Kiosks			
	Transport	Inventory	Lost Sales	Total Cost	Transport	Inventory	Lost Sales	Total Cost
50%	$12	$22	$60	$ 94	$14	$22	$38	$ 74
60%	$12	$24	$50	$ 86	$14	$24	$30	$ 68
70%	$12	$27	$43	$ 82	$14	$27	$30	$ 71
80%	$13	$31	$37	$ 81	$14	$31	$31	$ 76
90%	$14	$37	$44	$ 95	$14	$37	$42	$ 93
100%	$15	$45	$50	$110	$15	$45	$50	$110

*All costs in 000s

1. What customer service level do you recommend? Why?
2. What other factors might Ronnie want to consider before making her decision regarding desired customer service level?
3. Do you recommend buying the kiosks? Why or why not?

Physical Distribution Concept Focuses on the Whole Distribution System

LO 11.3

The physical distribution concept

The **physical distribution (PD) concept** says that all transporting, storing, and product-handling activities of a business and a whole channel system should be coordinated as one system that seeks to minimize the cost of distribution for a given customer service level. Both lower costs and better service help increase customer value. This seems like common sense, but until recently most companies treated physical distribution functions as separate and unrelated activities.

Exhibit 11-3 Examples of Factors That Affect PD Service Levels

- Advance information on product availability
- Time to enter and process orders
- Back-order procedures
- Where inventory is stored
- Accuracy in filling orders
- Damage in shipping, storing, and handling
- Online status information
- Advance information on delays
- Time needed to deliver an order
- Reliability in meeting delivery date
- Compliance with customer's instructions
- Defect-free deliveries
- How needed adjustments are handled
- Procedures for handling returns

Traditionally, responsibility for different logistics activities was spread among various departments in a firm—production, shipping, sales, warehousing, purchasing, and others. No one person was responsible for coordinating storing and shipping decisions or customer service levels. It was even rarer for different firms in the channel to collaborate. Each just did its own thing. Unfortunately, in too many firms these old-fashioned ways persist—with a focus on individual functional activities rather than the whole physical distribution system.[4]

Decide what service level to offer

With broader adoption of the physical distribution concept, this is changing. Firms work together to decide what aspects of service are most important to customers at the end of the channel. Then they focus on finding the least expensive way to achieve the target level of service.

Exhibit 11-3 shows a variety of factors that may influence the customer service level (at each level in the channel). The most important aspects of customer service depend on target market needs. Xerox might focus on how long it takes to deliver copy machine repair parts once it receives an order. When a copier breaks down, customers want the repair "yesterday." The service level might be stated as "we will deliver 90 percent of all emergency repair parts within 24 hours." This might require that commonly needed parts be available on the service truck, that order processing be very fast, and that parts not available locally be sent by airfreight. Obviously, supplying this service level will affect the total cost of the PD system. But it may also beat competitors.

Fast PD service can be critical for retailers that appeal to consumers who are eager to get a new product that is in hot demand—the latest movie release, best-selling book, or popular toy or video game.

Food delivery trying to match needs with service level

There is a big market in food delivery. Every American eats. Every day. And now many firms are trying to figure out how to offer greater convenience to those customers. PD can be a source of competitive advantage if firms match needs and service level.

Sauder tries to help customer firms do a better job of tracking the status of orders and making certain that products are where they are needed at the right time.
Source: Sauder Woodworking Company

Grocery stores are one place consumers get their food. These days, many customers expect more convenience and speed as part of their PD service—and at a low cost. Walmart customers can fill their shopping cart online. Prices are the same as those found in the store. Their order is picked out by a personal shopper at the store of their choice. Customers can pick up their order within a few hours. Or, for a fee, they can have it delivered to their home. Walmart contracts with delivery services to make that happen.[5]

Some customers want food to arrive already prepared. For ready-made food, delivery used to be limited mostly to pizza and Chinese food. Now, for about $5, delivery services like Grubhub, Uber Eats, and DoorDash deliver prepared meals from local restaurants to hungry folks who don't want to cook (or leave home). Customers who are willing to plan a little more in advance and do a little more of the work themselves are turning to meal delivery services like Blue Apron and HelloFresh. These services ship out a meal prep kit that includes a recipe and all the ingredients (right down to herbs and spices) in preportioned packages. The meal kits need to be ordered at least a few days in advance, and customers have to prepare the meal on their own.

Most of these services are new. Each is trying to find a target market and service level that appeals to customers. And, of course, customers have to be willing to pay for the additional convenience.

Find the lowest total cost for the right service level

In selecting a PD system, the **total cost approach** involves evaluating each possible PD system and identifying *all* of the costs of each alternative. This approach uses the tools of cost accounting and economics. Costs that otherwise might be ignored—like inventory carrying costs—are considered. The possible costs of lost sales due to a lower customer service level may also be considered.

For example, Vegpro Kenya compared different PD systems for shipping ready-to-eat fresh produce from fields in Kenya to grocery stores in major European cities. The analysis showed that the costs of airfreight transportation were significantly higher than using trucks and ships. But the firm also found that costs of spoilage and inventory could be much lower when airfreight is used. The faster airfreight-based PD system brought customers fresher produce at about the same total cost. So Vegpro cleans, chops, and packages vegetables in its 27,000-square-foot, air-conditioned facility at the Nairobi airport—using low-cost African labor. And the next day, fresh beans, baby carrots, and other vegetables are on store shelves in Madrid, London, and Paris.[6]

Coordinating Logistics Activities among Firms

LO 11.4

Functions can be shifted and shared in the channel

As a marketing manager develops the Place part of a strategy, it is important to decide how physical distribution functions can and should be divided within the channel. Who will store, handle, and transport the goods—and who will pay for these services? Who will coordinate all of the PD activities?

There is no right sharing arrangement. Physical distribution can be varied endlessly in a marketing mix and in a channel system. And competitors may share these functions in different ways—with different costs and results.

How PD is shared affects the rest of a strategy

How the PD functions are shared affects the other three Ps—especially Price. The sharing arrangement can also make (or break) a strategy. Consider Channel Master, a firm that wanted to take advantage of the growing market for the dishes used to receive TV signals from satellites. The product looked like it could be a big success, but the small company didn't have the money to invest in a large inventory. So Channel Master decided to work only with wholesalers who were willing to buy (and pay for) several units—to be used for demonstrations and to ensure that buyers got immediate delivery.

In the first few months Channel Master earned $2 million in revenues—just by providing inventory for the channel. And the wholesalers paid the interest cost of carrying inventory—more than $300,000 the first year. Here the wholesalers helped share the risk of the new venture, but it was a good decision for them, too. They won many sales from a competing channel whose customers had to wait several months for delivery. And by getting off to a strong start, Channel Master became a market leader.

JIT requires a coordinated effort

If firms in the channel do not plan and coordinate how they will share PD activities, PD is likely to be a source of conflict rather than a basis for competitive advantage. Let's consider this point by taking a closer look at *just-in-time (JIT) delivery systems*, which we introduced in Chapter 6 and defined as reliably getting products there just before the customer needs them.

A key advantage of JIT for business customers is that it reduces their PD costs—especially storing and handling costs. However, if the customer doesn't have any backup inventory, there's no security blanket if a supplier's delivery truck gets stuck in traffic, there's an error in what's shipped, or there are any quality problems. Thus, a JIT system requires that a supplier have extremely high quality control in every PD activity.

A JIT system usually requires that a supplier respond to very short order lead times and the customer's production schedule. Thus, e-commerce order systems and information sharing over computer networks are often required. JIT suppliers often locate their facilities close to important customers. Trucks may make smaller and more frequent deliveries—perhaps even several times a day.

A JIT system shifts greater responsibility for PD activities backward in (up) the channel. If the supplier can be more efficient than the customer can be in controlling PD costs—and still provide the customer with the service level required—this approach can work well for everyone in the channel. However, JIT is not always the best approach. It may be better for a supplier to produce and ship in larger, more economical quantities—if the savings offset the distribution system's total inventory and handling costs.[7]

Supply chain may involve even more firms

In our discussion, we have taken the point of view of a marketing manager. This focuses on how logistics should be coordinated to meet the needs of customers at the end of the channel of distribution. Now, however, we should broaden the picture somewhat because the relationships within the distribution channel are sometimes part of a broader network of relationships in the **supply chain**—the complete set of firms and facilities and logistics activities that are involved in procuring materials, transforming them into intermediate or finished products, and distributing them to customers.

For example, Manitowoc is one of the world's largest manufacturers of cranes. Its huge mobile cranes are used at construction sites all around the world. Robert Ward, who is in charge of purchasing for Manitowoc, must ensure an unbroken flow of parts and materials so that Manitowoc can keep its promises to customers regarding crane delivery dates. This is difficult because each crane has component parts from many suppliers around the globe. Further, any supplier may be held up by problems with its own suppliers. In one case, Manitowoc's German factory was having trouble getting key chassis parts from two suppliers in Poland. Ward traced the problem back to a Scandinavian steel mill that was behind on shipments to the Polish firms. Manitowoc buys a lot of steel and has a lot of leverage with steel distributors, so Ward scoured Europe for distributors who had extra inventory of steel

Source: Manitowoc Company, Inc.

Companies such as Marsh and IBM provide consulting services and analytical tools to help firms manage complex supply chains.
(left): Source: Marsh LLC.; (right): Source: IBM

plate. The steel he found was more expensive than buying the steel directly from the mill, but the mill couldn't keep supplies flowing and the distributors could. By helping coordinate the whole supply chain, Manitowoc was able to keep its promises and deliver cranes to its customers on schedule.[8]

Work to meet the needs of the customer at the end of the supply chain

Ideally, all of the firms in the supply chain should work together to meet the needs of the customer at the very end of the chain. That way, at each link along the chain, the shifting and sharing of logistics functions and costs are handled to result in the most value for the final customer. This makes a supply chain more efficient and effective—giving it a competitive advantage against other supply chains competing to serve the same target market.[9]

Better information helps coordinate PD

Coordinating all of the elements of PD has always been a challenge—even in a single firm. Trying to coordinate orders, inventory, and transportation throughout the whole supply chain is even tougher. But information shared over the Internet and at websites has been important in finding solutions to these challenges. Physical distribution decisions will continue to improve as more firms are able to have their computers "talk" to each other directly and as websites help managers access up-to-date information whenever they need it.

Electronic data interchange sets a standard

Until recently, differences in computer systems from one firm to another hampered the flow of information. Many firms attacked this problem by adopting **electronic data interchange (EDI)**—an approach that puts information in a standardized format easily shared between different computer systems. In many firms, purchase orders, shipping reports, and other paper documents were replaced with computerized EDI. With EDI, a customer transmits his or her order information directly to the supplier's computer. The supplier's computer immediately processes the order and schedules production, order assembly, and transportation. Inventory information is automatically updated, and status reports are available instantly. The supplier might then use EDI to send the updated information to the transportation provider's computer. In fact, most international transportation firms rely on EDI links with their customers.[10]

Improved information flow and better coordination of PD activities are key reasons for the success of Pepperidge Farm's line of premium cookies. Most of the company's delivery truck drivers use handheld computers to record the inventory at each stop along their routes. They use a wireless Internet connection to instantly transmit the information to a computer at the bakeries, and cookies in short supply are produced. The right assortment of fresh cookies is quickly shipped to local markets, and delivery trucks are loaded with what retailers need that day. Pepperidge Farm moves cookies from its bakeries to store shelves in about three days; most cookie producers take about 10 days. That means fresher cookies for consumers and helps support Pepperidge Farm's high-quality positioning and premium price.[11]

Ethical issues may arise

Some ethical issues that arise in the PD area concern communications about product availability. For example, some critics say that Internet sellers too often take orders for products that are not available or that they cannot deliver as quickly as customers expect. Yet a marketing manager can't always know precisely how long it will take before a product will be available. It doesn't make sense for the marketer to lose a customer if it appears that he or she can satisfy the customer's needs. But the customer may be inconvenienced or face added cost if the marketer's best guess isn't accurate. Similarly, some critics say that stores too often run out of products that they promote to attract consumers to the store. Yet it may not be possible for the marketer to predict demand or to know when placing an ad that deliveries won't arrive. Different people have different views about how a firm should handle such situations. Some retailers just offer rainchecks.

> **Ethical Dilemma**
>
> *What would you do?* Many major firms, ranging from Nike and Starbucks to Walmart and IKEA, have been criticized for selling products from overseas suppliers whose workers toil in bad conditions for long hours and at low pay. Defenders of the companies point out that overseas sourcing provides jobs that are better than what workers would have without it. Critics think that companies that sell products in wealthy countries have a social responsibility to see that suppliers in less-developed nations pay a fair wage and provide healthy working conditions.[12]
>
> *What do you think? Should U.S. firms be required to monitor the employment practices of suppliers in their supply chains? Should all suppliers be held to Western legal or moral standards? What solutions or compromises might be offered?*

The Transporting Function Adds Value to a Marketing Strategy

LO 11.5

Transporting aids economic development and exchange

Transporting is the marketing function of moving goods. Transportation makes products available when and where they need to be—at a cost. But the cost is less than the value added to products by moving them, or there is little reason to ship in the first place.

Transporting can help achieve economies of scale in production. If production costs can be reduced by producing larger quantities in one location, these savings may more than offset the added cost of transporting the finished products to customers. Without low-cost transportation, both within countries and internationally, there would be no mass distribution as we know it today.

Transporting can be costly

Transporting costs limit the target markets a marketing manager can serve. Shipping costs increase delivered cost—and that's what really interests customers. Transport costs add little to the cost of products that are already valuable relative to their

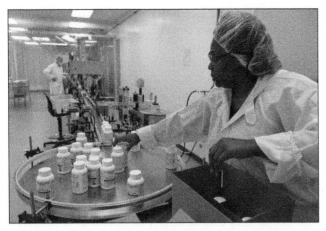 

The cost of transportation adds little to the total cost of products like pharmaceutical drugs or supplements, which are already expensive relative to their size and weight. On the other hand, transporting costs can be a large part of the total cost for heavy products that are low in value, like sheet aluminum.
(left): Mark Elias/Bloomberg/Getty Images; (right): Steve Smith

size and weight. A case of medicine, for example, might be shipped to a drugstore at low cost. But transporting costs can be a large part of the total cost for heavy products of low value—such as many minerals and raw materials. You can imagine that shipping a massive roll of aluminum to a producer of soft drink cans is an expensive proposition. Exhibit 11-4 shows transporting costs as a percentage of total sales dollars for several products.[13]

Governments may influence transportation

Government often plays an important role in the development of a country's transportation system, including its roads, harbors, railroads, and airports. And different countries regulate transportation differently, although regulation has in general been decreasing.

As regulations decreased in the United States, competition in the transportation industry increased. As a result, a marketing manager generally has many carriers in one or more modes competing for the firm's transporting business. Or a firm can do its own transporting. Knowing about the different modes is important.[14]

Exhibit 11-4
Transporting Costs as a Percentage of Selling Price for Different Products

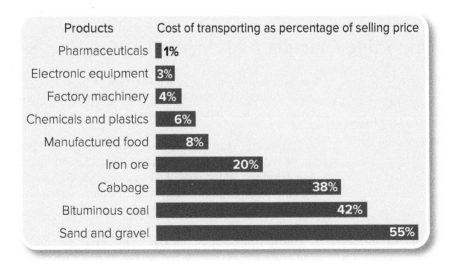

Products	Cost of transporting as percentage of selling price
Pharmaceuticals	1%
Electronic equipment	3%
Factory machinery	4%
Chemicals and plastics	6%
Manufactured food	8%
Iron ore	20%
Cabbage	38%
Bituminous coal	42%
Sand and gravel	55%

Which Transporting Alternative Is Best?

Transporting function must fit the whole strategy

The transporting function should fit into the whole marketing strategy. But picking the best transporting alternative depends on the product, other physical distribution decisions, and what service level the company wants to offer. The best alternative should provide the level of service (for example, speed and dependability) required at as low a cost as possible. Exhibit 11-5 shows that different modes of transportation have different strengths and weaknesses. Low transporting cost is *not* the only criterion for selecting the best mode.[15]

Railroads—large loads moved at low cost

Railroads are still the workhorse of the U.S. transportation system. They carry more freight over more miles than any other mode. However, they account for less than 10 percent of transport revenues. They carry heavy and bulky goods—such as coal, steel, and chemicals—over long distances at relatively low cost. Because railroad freight usually moves more slowly than truck shipments, it is not as well suited for perishable items or those in urgent demand. Railroads are most efficient at handling full carloads of goods. Less-than-carload (LCL) shipments take a lot of handling, which means they usually move more slowly and at a higher price per pound than carload shipments.[16]

Trucks are more expensive, but flexible and essential

The flexibility and speed of trucks make them better at moving small quantities of goods for shorter distances. They can travel on almost any road. They go where the rails can't. They are also reliable in meeting delivery schedules, which is an essential requirement for logistics systems that provide rapid replenishment of inventory after a sale. In combination these factors explain why at least 75 percent of U.S. consumer products travel at least part of the way from producer to consumer by truck. And in countries with good highway systems, trucks can give extremely fast service. Trucks compete for high-value items. Trucks are essential for firms that deliver to people's homes—the so-called "last

Low labor costs in developing countries can make transportation costs relatively low—especially for short distances. In Egypt, one-third of McDonald's sales involve delivery.
Peter Macdiarmid/Getty Images

Exhibit 11-5 Benefits and Limitations of Different Transport Modes

		Transporting Features				
Mode	Cost	Delivery Speed	Number of Locations Served	Ability to Handle a Variety of Goods	Frequency of Scheduled Shipments	Dependability in Meeting Schedules
Rail	Medium	Average	Extensive	High	Low	Medium
Truck	High	Fast	Very extensive	High	High	High
Water	Very low	Very slow	Limited	Very high	Very low	Medium
Pipeline	Low	Slow	Very limited	Very limited	Medium	High
Air	Very high	Very fast	Extensive	Limited	High	High

mile" for e-commerce. They will become even more critical as online sales and same-day delivery become important parts of some retailers' marketing strategies.[17]

Ship it overseas, but slowly

Water transportation is the slowest shipping mode, but it is usually the lowest-cost way to ship heavy freight. Water transportation is very important for international shipments and often the only practical approach. This explains why port cities such as Boston, New York City, Rotterdam, Osaka, and Singapore are important centers for international trade.

Inland waterways are important too

Inland waterways (such as the Mississippi River and Great Lakes in the United States and the Rhine and Danube in Europe) are also important, especially for bulky, nonperishable products such as iron ore, grain, and gravel. However, when winter ice closes freshwater harbors, alternate transportation must be used.

Pipelines move oil and gas

Pipelines are used primarily to move oil and natural gas, which means they are important both in oil-producing and oil-consuming countries. Only a few major cities in the United States, Canada, Mexico, and Latin America are more than 200 miles from a major pipeline system. However, the majority of the pipelines in the United States are located in the Southwest, connecting oil fields and refineries.

Airfreight is expensive, but fast and growing

The most expensive cargo transporting mode is the airplane—but it is fast! Airfreight rates are on average three times higher than trucking rates—but the greater speed may offset the added cost.

High-value, low-weight goods—such as high-fashion clothing and parts for the electronics industry—are often shipped by air. Perishable products that previously could not be shipped are now being flown across continents and oceans. Flowers and bulbs from Holland, for example, now are jet-flown to points all over the world. And airfreight has become very important for small emergency deliveries, like repair parts.

But airplanes may cut the total cost of distribution

Using planes may reduce the cost of packing, unpacking, and preparing goods for sale and may help a firm reduce inventory costs by eliminating outlying warehouses. Valuable benefits of airfreight's speed are less spoilage, theft, and damage. Although the *transporting* cost of air shipments may be higher, the *total* cost of distribution may be lower. As more firms realize this, airfreight firms—such as DHL Worldwide Express, FedEx, and Emery Air Freight—have enjoyed rapid growth.[18]

Will drones deliver soon?

Small unmanned aircraft, called drones, have been used in the military and by hobbyists. They are now being used for delivery services all over the world. In Germany, DHL uses drones to deliver medicine to a small island off the German coast. Swiss Post has been testing drone delivery of mail. In China, online retailer JD.com has developed heavy-duty drones to help it deliver to rural China. And Amazon is testing drones for delivery in Canada (the Federal Aviation Administration has slowed testing in the United States).[19]

Put it in a container—and move between modes easily

Products often are moved by several different modes and carriers during their journey. This is especially common for international shipments. Japanese firms, like Sony, ship stereos to the United States, Canada, and Europe by boat. When they arrive at the dock, they are loaded on trains and sent across the country. Then the units are delivered to a wholesaler by truck or rail.

To better coordinate the flow of products between modes, transportation companies like

Opla/E+/Getty Images

CSX offer customers a complete choice of different transportation modes. Then CSX, not the customer, figures out the best and lowest-cost way to shift and share transporting functions among the modes.[20]

Loading and unloading goods several times used to be a real problem. Parts of a shipment would become separated, damaged, or even stolen. It would take a week and many hours of high-priced labor to unload a cargo ship. All this handling of the goods, perhaps many times, raised costs and slowed delivery. Many of these problems are reduced with **containerization**—grouping individual items into an economical shipping quantity and sealing them in protective containers for transit to the final destination. This protects the products and simplifies handling during shipping. Shipping containers also facilitate intermodal transport. Some containers are as large as truck bodies.

#M4BW

Advances lower transportation costs in developing countries

Transportation choices are usually not so good in developing countries. Roads are often poor, rail systems may be limited, and ports may be undeveloped. Local firms that specialize in logistics services may not exist at all. Even so, firms that are willing to invest the effort can reap benefits and help their customers overcome the effects of these problems.

Metro AG, a firm based in Germany that has opened wholesale facilities in Bangalore and several other major cities in India, illustrates this point. Metro focuses on selling food products and other supplies to the thousands of restaurants, hotels, and other small businesses in the markets it serves. When Metro started in India, 40 percent of the fruits and vegetables it purchased from farmers were spoiled, damaged, or lost by the time they got to Metro. These problems piled up because the produce traveled from the fields over rough roads and was handled by as many as seven intermediaries along the way. To overcome these problems, Metro gave farmers crates to protect freshly picked crops from damage and to keep them away from dirt and bacteria that would shorten their shelf life. Further, crates were loaded and unloaded only once because Metro bought its own refrigerated trucks to pick up produce and bring it directly to its outlets. Metro used the same ideas to speed fresh seafood from fishermen's boats. Many of Metro's restaurant customers previously bought what they needed from a variety of small suppliers, many of whom would run out of stock. Now the restaurant owners save time and money with one-stop shopping at Metro. Metro is growing fast in India because it has quality products that are in stock when they're needed, and it's bringing down food costs for its customers. Sounds like marketing for a better world.[21]

#M4BW

Transportation choices have environmental costs too

Marketing managers should be sensitive to the environmental effects of transportation decisions. Trucks, trains, airplanes, and ships contribute to air pollution and global warming; estimates suggest that on average more than half of most firms' total carbon emissions come from transportation. Other environmental issues may spring from transportation as well. For example, a damaged pipeline or oil tanker can spew thousands of gallons of oil before it can be repaired.

Many firms are taking steps to reduce such problems. FedEx and UPS are revamping their fleets to use more electric and alternative-fuel vehicles. Whereas rail is usually the cleanest way to move land freight a long distance, General Electric's recently introduced Evolution locomotives have 5 percent better fuel economy and 40 percent lower emissions compared to previous models. Truck manufacturers are also working to improve fuel efficiency and environmental impact. Peterbilt and International are among firms working to build diesel-hybrid 18-wheelers. The U.S. government supports these initiatives with the Environmental Protection Agency's SmartWay program (www.epa.gov/smartway/learn-about-smartway). It helps freight carriers, shippers, and logistics companies improve fuel efficiency and reduce environmental impact. Today the public *expects* companies to manufacture, transport, sell, and dispose of products in an environmentally sound manner.[22]

#M4BW Menasha, a supplier of innovative packaging, wants its client firms to realize that its package designs can help reduce both logistics costs and environmental impacts.
Source: Menasha Corporation

 #M4BW

Transportation analytics aid the environment and lower costs

Transportation is made more efficient with analytics. Transportation companies routinely place sensors on trucks and train cars to monitor their movement. Managers use the data generated by these sensors to identify opportunities to more effectively manage the economic and environmental costs of transportation.

For example, UPS uses global positioning data to learn and recommend routes for its drivers—routes that maximize service and minimize fuel consumption. The system even adapts on the fly to changes in traffic or customer requests.[23] Soon this type of analysis may not be relayed to a *human* driver. To learn more, see *What's Next? Who's—or rather what's—driving that truck?*

The Storing Function and Marketing Strategy

LO 11.6

Storing can smooth out sales and increase profits and consumer satisfaction

Storing is the marketing function of holding goods so they're available when they're needed. **Inventory** is the amount of goods being stored.

Storing is necessary when production of goods doesn't match consumption. This is common with mass production. Nippon Steel, for example, might produce thousands of steel bars of one size before changing the machines to produce another size. It's often cheaper to produce large quantities of one size, and store the unsold quantity, than to have shorter production runs. Thus, storing goods allows the producer to achieve economies of scale in production.

Storing varies in the channel system

Storing allows producers and intermediaries to keep stock at convenient locations, ready to meet customers' needs. In fact, storing is one of the major activities of some intermediaries.

Most channel members provide the storing function for some length of time. Even final consumers store some things for their future needs. Which channel members store the product, and for how long, affects the behavior of all channel members. For example, the producer of Snapper lawn mowers tries to get wholesalers to inventory a wide

What's Next? Who's—or rather what's—driving that truck?

Someday not too far in the future, you are going to be driving down the highway in your car. You'll look to your right as you pass a truck, but there will be nobody sitting in the driver's seat. Who will be driving this truck? As it turns out, *what's* driving that truck will be a better question because artificial intelligence (AI) may be doing the driving—and it will be safer than most truck drivers. That might sound a bit far-fetched, but tests are already under way. Experts in the field have little doubt that driverless trucks are *What's Next?* in transportation. Let's take a closer look at what is happening with this technology.

There are many benefits to removing drivers from trucks. The primary motivation, as with many technologies, is economic. About 75 percent of the cost of shipping a truckload of goods across the United States goes to paying the driver. Most truck drivers have a built-in incentive to drive faster because they are paid by the mile; but by driving slower, driverless trucks can optimize fuel economy. Federal regulations also limit human truckers to 11 hours of driving per day before a required 8-hour break. A driverless truck can operate practically 24/7, making much better utilization of an expensive capital asset. Another benefit is safety. Each year about 350,000 accidents and 4,000 fatalities on U.S. roads are attributed to large trucks. About 90 percent of those are caused by human error; AI will be able to drive much more safely.

The move to driverless trucks is under way. In fact, mining companies Rio Tinto and BHP already use autonomous vehicles on their private land. Controlled road tests have been conducted on highways in Europe and the United States. Right now, truck producer Freightliner's demonstrations include a "highway pilot" or driver on board—though drivers will disappear as the artificial intelligence learns from early experience and gets smarter.

Big truck makers like Daimler and Volvo, together with tech startups, are helping to quickly make this a reality. For example, Otto, a firm founded by former Google executives, developed a $30,000 retrofit kit that adds driverless functionality to any truck built since 2013. That beats buying a new truck for $200,000. Another startup, Peloton Technology, produces software that helps self-driving vehicles facilitate multiple self-driving trucks driving closely behind one another in a convoy, allowing multiple trucks to drive as a "platoon." Sensors and direct vehicle-to-vehicle communications allow trucks to drive closer together, operating as a single unit. The "drafting" effect of planned "tailgating" lowers wind resistance and saves 10 to 15 percent on fuel.

Experts estimate that by 2030, autonomous trucks will be widespread and you won't even look twice when passing one on that cross-country drive. *What's next* after that? Safer roads, cleaner air, and cheaper deliveries if this AI technology delivers on its promises.[24]

---

selection of its machines. That way, retailers can carry smaller inventories because they can be sure of dependable local supplies from wholesalers. And the retailers might decide to sell Snapper—rather than Toro or some other brand that they would have to store at their own expense.

If consumers "store" the product, more of it may be used or consumed. That's why Breyer's likes customers to buy its half-gallon packages of ice cream. The "inventory" is right there in the freezer—and ready to be eaten—whenever the impulse hits.

Goods are stored at a cost

Storing can increase the value of goods, but *storing always involves costs* too. Different kinds of cost are involved (see Exhibit 11-6). Car dealers, for example, must store cars on their lots—waiting for the right customer. The interest expense of money tied up in inventory is a major cost. In addition, if a new car on the lot is dented or scratched, there is a repair cost. If a car isn't sold before the new models come out, its value drops. There is also a risk of fire or theft—so the retailer must carry insurance. And, of course, dealers incur the cost of leasing or owning the display lot where they store the cars.

In today's competitive markets, most firms watch their inventories closely. They try to cut unnecessary inventory because it can make the difference between a profitable strategy and a losing one. On the other hand, a marketing manager must be very careful in making the distinction between unnecessary inventory and inventory needed to provide the distribution service level customers expect. A few years ago, shipments of

Exhibit 11-6 Many Expenses Contribute to Total Inventory Cost

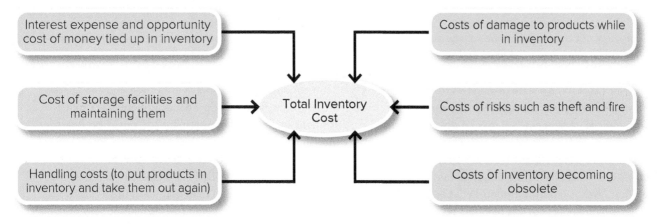

Boeing's 787 Dreamliner aircraft were significantly delayed by a shortage of bolts that hold the jet airplanes together. The lack of availability of this relatively small part cost Boeing millions of dollars.[25]

Rapid response cuts inventory costs

Many firms are finding that they can cut inventory costs and still provide the desired customer service level—if they can reduce the time it takes to replace items that are sold. This is one important reason that JIT has been widely adopted. The firms involved use EDI, the Internet, and similar computerized approaches to share information and speed up the order cycle and delivery process.

Specialized Storing Facilities May Be Required

New cars can be stored outside on the dealer's lot. Fuel oil can be stored in a specially designed tank. Coal and other raw materials can be stored in open pits. But most products must be stored inside protective buildings. Often, firms can choose among different types of specialized storing facilities. The right choice may reduce costs and serve customers better.

Private warehouses are common

Private warehouses are storing facilities owned or leased by companies for their own use. Most manufacturers, wholesalers, and retailers have some storing facilities either in their main buildings or in a separate location. A sales manager often is responsible for managing a manufacturer's finished-goods warehouse, especially if regional sales branches aren't near the factory.

Firms use private warehouses when a large volume of goods must be stored regularly. Yet private warehouses can be expensive. If the need changes, the extra space may be hard, or impossible, to rent to others.

Public warehouses fill special needs

Public warehouses are independent storing facilities. They can provide all the services that a company's own warehouse can provide. A company might choose a public warehouse if it doesn't have a regular need for space. For example, Tonka Toys uses public warehouses because its business is seasonal. Tonka pays for the space only when it is used. Public warehouses are also useful for manufacturers that must maintain stock in many locations, including foreign countries. See Exhibit 11-7 for a comparison of private and public warehouses.[26]

Warehousing facilities cut handling costs too

The cost of physical handling is a major storing cost. Goods must be handled once when put into storage and again when removed to be sold. To reduce these costs,

Exhibit 11-7
A Comparison of Private Warehouses and Public Warehouses

Characteristics	Type of Warehouse	
	Private	Public
Fixed investment	Very high	No fixed investment
Unit cost	High if volume is low; very low if volume is very high	Low: charges are made only for space needed
Control	High	Low managerial control
Adequacy for product line	Highly adequate	May lack convenience
Flexibility	Low: fixed costs have already been committed	High: easy to end arrangement

modern one-story buildings away from downtown traffic have replaced most old multi-story warehouses. They eliminate the need for elevators and permit the use of power-operated lift trucks, battery-operated motor scooters, roller-skating order pickers, electric hoists for heavy items, and hydraulic ramps to speed loading and unloading.

Bar codes, Universal Product Code (UPC) numbers, and electronic radio-frequency identification (RFID) tags make it easy for computers to monitor inventory, order needed stock, and track storing and shipping costs. For example, clothing retailer Zara puts RFID tags on every piece of merchandise. The tags allow Zara to know where every piece of merchandise is at any given time—greatly simplifying the process of taking inventory. Each time Zara sells a garment, the tag immediately notifies the stockroom to bring up another from the stockroom. Chips in the tags make it easier to find similar garments—in that store or at a nearby store.

The Distribution Center—A Different Kind of Warehouse

LO 11.7

Is storing really needed?

Discrepancies of assortment or quantity between one channel level and another are often adjusted at the place where goods are stored. It reduces handling costs to regroup and store at the same place—*if both functions are required.* But sometimes regrouping is required when storing isn't.

Don't store it; distribute it

A **distribution center** is a special kind of warehouse designed to speed the flow of goods and avoid unnecessary storing costs. Today, the distribution center concept is widely used by firms at all channel levels. Many products buzz through a distribution center without ever tarrying on a shelf; workers and equipment immediately sort the products as they come in and then move them to an outgoing loading dock and the vehicle that will take them to their next stop. Technology is key to making distribution centers efficient.

Consider how food distributor Sysco addresses the challenge of shipping more than 21.5 million tons of fruit, vegetables, meats, and other food-related products each year.

Robotic order fulfillment systems are common in many large warehouses.
Baloncici/iStock/Getty Images

Sysco's customers include one in every three restaurants, school cafeterias, and other food service outlets in the United States and Canada. Because Sysco delivers so many products in so many different forms—from boxes of frozen French fries, crates of Granny Smith apples, and 80-pound tubs of flour to pots, pans, and utensils—its handling needs vary widely.

To move all these goods efficiently and quickly, Sysco relies on technology. It uses supply chain management software and an EDI system to automatically direct supplies from vendors such as Kraft and Kellogg to one of two Sysco "redistribution centers." There, orders are quickly consolidated and moved to one of Sysco's 177 distribution centers. The 400,000-plus-square-foot distribution centers are organized by weight and temperature; heavier items—such as cans weighing 40 to 50 pounds—are housed on one side, and lighter items—such as boxes of potato chips—on the other. To fill orders, custom software supplies forklift drivers with printouts telling them which items to pick first and how to stack them on a pallet based on the weight of the items, their location in the distribution center, and their ultimate destination. After calculating delivery routes that optimize time and fuel use, the software also provides instructions for loading the trucks.

Sysco delivers more than food to its customers. On the "Customer Solutions" section of Sysco's website, restaurateurs are connected with business resources (such as payroll and marketing tools) and managers of school lunch programs or nursing home meals can learn about Sysco's profitability consultation, menu analysis, and safety training. Together, Sysco's combination of efficient operations, service and support, and quality products provide excellent value for its target market.[27]

Direct store delivery skips the distribution center

Some firms prefer to skip the distribution center altogether and ship products directly from where they are manufactured to retail stores. This may move products more quickly, but usually at a higher cost. Frito-Lay uses this approach. It handles more than 10,000 direct delivery routes to more than 200,000 small-store customers. The close relationships route drivers build with many small retailers help Frito-Lay better understand end consumers and adapt product mixes to particular stores. These extra services result in more shelf space and higher prices at the small stores.[28]

Managers must be innovative to provide customers with superior value

More competitive markets, improved technology, coordination among firms, and efficient new distribution centers are bringing big improvements to the PD area. Yet the biggest challenges may be more basic. As we've emphasized here, physical distribution activities transcend departmental, corporate, and even national boundaries. So taking advantage of ways to improve often requires cooperation all along the channel system. Too often, such cooperation doesn't exist—and changing ingrained ways of doing things is hard. But marketing managers who push for innovations in these areas are likely to win customers away from firms and whole channel systems that are stuck doing things the old way.[29]

Disaster Relief—Logistics Saves Lives

To conclude this chapter, we bring together a physical distribution context you may not have thought about. In fact, you may not think about it unless you, a friend, or a family member finds themselves in a disaster area. In disaster situations, the customer service level required is very high, and performing well requires implementing the physical distribution concept at the highest levels. Yet all this happens in very trying conditions. Many of the ideas we have talked about in this chapter are relevant to these situations—and advances in logistics applied to a disaster save lives and make for a better world.

Disasters create logistics challenges

Hurricanes, tsunamis, earthquakes, and other natural disasters create immediate needs for emergency relief supplies. And the logistics involved in delivering them includes many of the same activities found in the physical delivery of consumer goods. However, in a disaster situation, life and death often hinge upon the speed with which food, water, and medical supplies can be delivered. Yet, when bridges, roads, and airports are destroyed, local transportation can be complicated, if not impossible. And, even worse, there is no advance warning when or where aid will be needed. Imagine what it would be like for one business to be instantly ready to distribute millions of products to a target market that usually doesn't exist, moves around the world, and then, without notice, pops up somewhere else with insatiable needs.

Source: Brian P. Seymour/U.S. Navy

Partners are needed

People in advanced societies have high expectations that help will be immediate when disaster strikes. Yet, it's nearly impossible for relief agencies to meet those expectations. Still, improved performance is on the way from both disaster relief agencies and private businesses, which have learned from recent efforts. For example, instead of stockpiling drugs, tents, and blankets, agencies are learning to rely on outsourcing. Agencies arrange open orders with suppliers who must be prepared to instantly ship supplies whenever and wherever they are needed.

Organizations with logistics expertise also lend a helping hand. Immediately following disasters in all parts of the world, transportation giants like FedEx, DHL, and China Southern Airlines have responded quickly with planes and trucks that facilitate delivery of needed supplies. Walmart and its Japanese Seiyu stores have responded more quickly than governmental organizations to disasters like Hurricane Katrina and Japan's Fukushima earthquake.

Information key to getting the right stuff to the right place

When chaos hits, coordination of relief efforts is possible only if there is good information. Agencies need to know what supplies are available, where they're located, what needs are greatest, and where and how quickly deliveries can be made. Having one central communication hub—to collect and share this type of information—and IT systems specifically dedicated to the task are key. A new system called SUMA allows relief workers to manage incoming donations, put them in the right storage places, and establish shipping priorities. SUMA also helps relief agencies receive communications from victims, and solar phone chargers and lanterns are often dropped from planes along with tents and water.

Other physical distribution solutions are decidedly low-tech, but equally important. For example, boxes need to be color coded so it's obvious which ones contain critical medical supplies and perishable food. And donated goods must be packed in cartons light enough to be carried manually in locations that have damaged roads and no power or equipment.

Physical distribution, done right, saves lives.[30]

CONCLUSION

This chapter explained the major logistics activities and how they contribute to the value of products by getting them to the place that customers want or need them. If the distribution customer service level meets their needs and can be provided at a reasonable cost, customers may not even think about the logistics activities that occur behind the scenes. But if products are not available when and where they need to be, a strategy will fail. So decisions in these areas are an important part of Place and marketing strategy planning.

We emphasized the relationship among customer service level, transporting, and storing. The physical distribution concept focuses on coordinating all the storing, transporting, and product-handling activities into a smoothly working system—to deliver the desired service level and customer value at the lowest total cost.

Marketing managers often want to improve service and may select a higher-cost alternative to improve their marketing mix. The total cost approach might reveal that it is possible both to reduce costs and to improve customer service—perhaps by working closely with other members of the supply chain.

We discussed various modes of transporting and their advantages and disadvantages. We also discussed ways to reduce inventory costs. For example, distribution centers are an important way to cut storing and handling costs, and computerized information links—within firms and among firms in the channel—are increasingly important in blending all of the logistics activities into a smooth-running system.

Effective marketing managers make important strategy decisions about physical distribution. Creative strategy decisions may result in lower PD costs while maintaining or improving the customer service level. And production-oriented competitors may not even understand what is happening.

KEY TERMS

LO 11.8

logistics, 296
physical distribution (PD), 296
customer service level, 297
physical distribution (PD) concept, 300
total cost approach, 302

supply chain, 303
electronic data interchange (EDI), 304
transporting, 305
containerization, 309
storing, 310

inventory, 310
private warehouses, 312
public warehouses, 312
distribution center, 313

QUESTIONS AND PROBLEMS

1. Review the Coca-Cola case study that opens the chapter. From this case, identify examples of different key terms and concepts covered in the chapter. For example, a distribution center is discussed in the second paragraph.

2. Review the Coca-Cola case study that opens this chapter. Assume that Coca-Cola is adding packaged fresh fruit snacks (they have a two-week shelf life) to its product line. Would the current physical distribution system for its drinks work for snacks? Why or why not? What changes might be needed?

3. Explain how adjusting the customer service level could improve a marketing mix. Illustrate with your own example.

4. Briefly explain which aspects of customer service you think would be most important for a producer that sells fabric to a firm that manufactures furniture.

5. Briefly describe a purchase you made where the customer service level had an effect on the product you selected or where you purchased it.

6. Discuss the types of trade-offs involved in PD costs, service levels, and sales.

7. Give an example of why it is important for different firms in the supply chain to coordinate logistics activities.

8. Discuss some of the ways computers are being used to improve PD decisions.

9. Explain why a just-in-time delivery system would require a supplier to pay attention to quality control. Give an example to illustrate your points.

10. Discuss the problems a supplier might encounter in using a just-in-time delivery system with a customer in a foreign country.

11. Review the list of factors that affect PD service levels in Exhibit 11-3. Indicate which factors are most likely to be improved by EDI links between a supplier and its customers.

12. Explain the total cost approach and why it may cause conflicts in some firms. Give examples of how conflicts might occur between different departments.

13. Discuss the relative advantages and disadvantages of railroads, trucks, and airlines as transporting methods.

14. Discuss why economies of scale in transportation might encourage a producer to include a regional merchant wholesaler in the channel of distribution for its consumer product.

15. Discuss some of the ways that air transportation can change other aspects of a Place system.

16. Explain which transportation mode would probably be most suitable for shipping the following goods to a large Los Angeles department store:
    a. 300 pounds of Maine lobster
    b. 15 pounds of screwdrivers from Ohio
    c. Three dining room tables from High Point, North Carolina
    d. 500 high-fashion dresses from the fashion district in Paris
    e. A 10,000-pound shipment of exercise equipment from Germany
    f. 600,000 pounds of various appliances from Evansville, Indiana
17. Indicate the nearest location where you would expect to find large storage facilities. What kinds of products would be stored there? Why are they stored there instead of some other place?
18. When would a producer or intermediary find it desirable to use a public warehouse rather than a private warehouse? Illustrate, using a specific product or situation.
19. Discuss the distribution center concept. Is this likely to eliminate the storing function of conventional wholesalers? Is it applicable to all products? If not, cite several examples.
20. Clearly differentiate between a warehouse and a distribution center. Explain how a specific product would be handled differently by each.
21. If a retailer operates only from a website and ships all orders by UPS, is it freed from the logistics issues that face traditional retailers? Explain your thinking.

MARKETING PLANNING FOR HILLSIDE VETERINARY CLINIC

Appendix D (the Appendices follow Chapter 19) includes a sample marketing plan for Hillside Veterinary Clinic. Look through the "Marketing Strategy" section. To provide veterinary care to pets, Hillside needs to have a variety of medical supplies on hand. To handle that, it relies on deliveries from suppliers and its own inventory decisions. It also sells some retail pet products to customers, and that requires a separate set of decisions about how it will handle inventory.

1. What logistics issues related to medical supplies should Hillside consider? Can you think of ways in which delivery from its suppliers or its own inventory decisions will be important in its ability to help its patients?
2. With respect to the retail pet products that Hillside sells, what level of customer service should customers expect?
3. What issues are involved in the storage of pet supplies?

SUGGESTED CASES

16. Hanratty Company

24. Fresh Harvest

MARKETING ANALYTICS: DATA TO KNOWLEDGE

CHAPTER 11: TOTAL DISTRIBUTION COST

Proto Company has been producing various items made of plastic. It recently added a line of plain plastic cards that other firms (such as banks and retail stores) will imprint to produce credit cards. Proto offers its customers the plastic cards in different colors, but they all sell for $40 per box of 1,000. Tom Phillips, Proto's product manager for this line, is considering two possible physical distribution systems. He estimates that if Proto uses airfreight, transportation costs will be $7.50 a box, and its cost of carrying inventory will be 5 percent of total annual sales dollars. Alternatively, Proto could ship by rail for $2 a box. But rail transport will require renting space at four regional warehouses—at $26,000 a year each. Inventory carrying cost with this system will be 10 percent of total annual sales dollars. Phillips prepared a spreadsheet to compare the cost of the two alternative physical distribution systems.

Design element: #M4BW box globe icon: ©Vectoryzen/Shutterstock

CHAPTER TWELVE

Scott Olson/Getty Images

Retailers, Wholesalers, and Their Strategy Planning

Back in the 1970s, Bernie Marcus and Arthur Blank ran a chain of Southern California home improvement centers called Handy Dan. While running some pricing experiments, they found that lower prices raised sales revenue and lowered costs as a percentage of sales. They thought the lower prices could increase Handy Dan's profits compared to the "buy low and sell high" pricing commonly practiced in their industry. However, before they could expand the experiment, new ownership at Handy Dan came in and swept out most of top management—including Marcus and Blank.

From a Los Angeles coffee shop, the two drafted a business plan for a new kind of home improvement center. A New York City investment firm funded the idea, and the first two Home Depot stores opened outside of Atlanta, Georgia. Marcus and Blank helped usher in a new "mass-merchandising" approach to retailing, demonstrating that low prices could generate high sales volume and profitability.

Even in the company's early days, Home Depot's 100,000-square-foot warehouse stores featured a wide variety of merchandise—from paint to plumbing to electrical and lumber. Experienced contractors and serious do-it-yourselfers loved the wide selection and low prices. To appeal to a less experienced and less confident target market, Home Depot hired and trained a knowledgeable staff to teach customers how to lay a tile floor, handle a power tool, or build a fence. As this target market gained confidence, they tackled more household jobs—after going to Home Depot, of course.

After Marcus and Blank retired in the early 2000s, Home Depot sought to grow profits by opening stores in international markets like China and cutting costs at home. These strategies were not successful. After six years of disappointing sales in China, Home Depot pulled out, citing cultural differences. Chinese consumers still preferred to shop at smaller, specialty shops. Back in the United States, cost control included cutting the number and training of its sales associates. While this increased profits in the short term, it alienated employees and customers. When customers couldn't find knowledgeable help at Home Depot, many headed to rival Lowe's, which offered better service. Sales lagged and then dropped when the economic recession and housing crash came in 2008.

At this point, Home Depot returned to a familiar marketing strategy: low prices and high customer service. More sales associates and spruced-up stores created a better shopping experience. Store managers' incentives emphasized customer satisfaction. Soon employee morale improved; customers noticed the changes and returned to Home Depot.

Home Depot faces a wide range of competitors. Fellow home improvement center Lowe's goes after a similar target market. Other competitors seek a slice of Home Depot's business. For example, Sherwin-Williams Paint Stores offer a limited product line of paint, stain, painting supplies, and wallpaper. Many formerly independent hardware stores joined the Ace Hardware cooperative, where pooling purchases and marketing expenses lowers costs. When seeking just a few items for a small project, some customers like the convenience of Ace's smaller stores. Other customers pick up those supplies at their local Walmart or Target supercenter.

For many retailers, the most feared competitor is Amazon. The store's name has become a verb; getting "Amazoned" refers to losing significant market share to the online retail giant. Home Depot's marketing strategy tries to avoid direct competition with Amazon. This strategy starts with experts in orange aprons. It also builds brand preference for exclusive store brands like Ryobi power tools, Hampton Bay fans, and Behr paints.

Home Depot also understands the advantages and disadvantages of online shopping. It recognizes online shopping can offer customers access to a rich variety of information, customer reviews, and 24/7 shopping. On the other hand, some products need to be seen and touched before making a purchase, and others don't ship well via UPS. Some customers just need to talk to a knowledgeable salesperson before buying. Home Depot knows many of its target customers want the best of both worlds. These multichannel shoppers want to shift back and forth between a phone, computer, and local store before making a purchase.

"One Home Depot" appeals to multichannel shoppers by combining the best of online and in-store for a better shopping experience. For example, a redesign of Home Depot's website offers improved search capabilities, facilitates comparison shopping, and speeds up online checkout. A new feature offers customers fast, accurate delivery dates and times, and Home Depot's 2,000+ stores have become "distribution centers." Looking ahead, it invested in Roadie, which is kind of an Uber for delivery. A customer can go online and search for a refrigerator, compare models and prices, then visit her local store and type the item into the Home Depot app, which offers a store map that guides her to the refrigerator. From there she can open the refrigerator's French doors and see how the freezer drawers actually work. The app also lets her arrange a two-hour window for next-day delivery. For other purchases, customers buy online and pick up their order in a locker at their local store, bypassing the checkout lines. Home Depot management finds that improving customers' online experience offers a better return on investment than opening new stores. Now 8 percent of sales and most of its growth comes from online shopping.

Professional contractors remain an important growth opportunity for Home Depot. Although they make up only 3 percent of Home Depot's customers, these B2B buyers account for 40 percent of its sales. Over the years, Home Depot has used different strategies for this target market. At one time, a separate wholesale channel (HD Supply) targeted the housing supply market. Eventually, it sold off HD Supply, which now targets different customers than Home Depot. Today, Home Depot uses a different marketing mix to sell to its contractor target market through its retail stores. For example, its Pro Xtra loyalty program offers contractors volume pricing discounts and the option of free direct-to-job delivery services or in-store pickup.

Home Depot seeks opportunities to help its business while making a better world. For example, to help address a shortage of construction workers, Home Depot donated $50 million to train 20,000 workers—with most of that going to veterans returning to civilian life.

Home Depot credits its success to an almost "paranoid" focus on target customers and eliminating friction in the shopping process. Measures of retail success suggest the company's tactics are working. Recently, comparable store sales—which measure sales growth at stores open a year or more—were more than 5 percent, more than triple that of Lowe's.[1]

LEARNING OBJECTIVES

Retail and wholesale organizations exist as members of marketing channel systems. But they also do their own strategy planning as they compete for target customers. As the Home Depot case shows, these firms make decisions about Product, Place, Promotion, and Price. This chapter overviews the strategy planning decisions of different types of retailers and wholesalers. The chapter shows how retailing and wholesaling are evolving—to give you a sense of how things may change in the future.

When you finish this chapter, you should be able to

1. understand the nature and basic structure of retailing.
2. understand how retailers plan their marketing strategies.
3. know about the many kinds of retailers that work with producers and wholesalers as members of channel systems.
4. understand what is different about retailing on the Internet.
5. understand how and why retailers evolve, including the roles of technology, scrambled merchandising, and the "wheel of retailing."
6. understand some of the differences in retailing in different nations.
7. know what progressive wholesalers are doing to modernize their operations and marketing strategies.
8. know the various kinds of merchant and agent wholesalers and the strategies they use.
9. understand important new terms (shown in **red**).

Retailers and Wholesalers Plan Their Own Strategies

As we saw in Chapter 10, retailers and wholesalers perform a vital role in channel systems. Now we'll look at the decisions that retailers and wholesalers make in developing their own strategies. But first, study Exhibit 12-1 for a visual of how everything fits together.

Exhibit 12-1 Marketing Strategy Planning for Retailers and Wholesalers

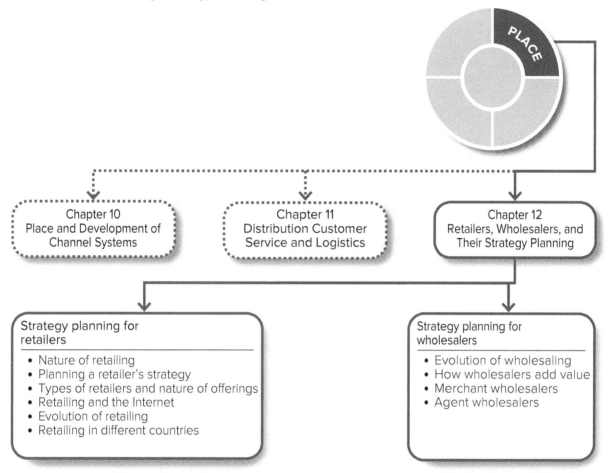

Producers can choose a variety of different paths to make goods and services available to customers when, where, and in the right quantities (see Exhibit 12-2). In Chapter 10 we discussed how some producers deliver goods or services directly to customers. We also saw how in some industries, retailers and/or wholesalers add value so it makes sense to use indirect channels. And of course, many producers use more than one of these paths to get products to customers.

When marketing managers decide to use intermediaries in their marketing strategy, they can choose among many different types of retailers and wholesalers. Retailers and wholesalers have different target markets and different marketing mixes, and a producer should select intermediaries that best facilitate its strategy.

To better understand these options, it helps to understand strategy planning for retailers (see Exhibit 12-1). Retailers create a marketing mix that provides value for a target market. We also discuss how retailing has evolved and where it is heading, including some important international differences. Chapter 12 concludes by considering strategy planning and the services offered by different types of wholesalers.

Exhibit 12-2 Wholesalers and Retailers in a Typical Channel of Distribution

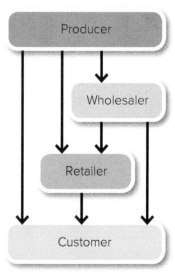

The Nature of Retailing

LO 12.1

Retailing covers all of the activities involved in the sale of products to final consumers. Retailers range from large chains of specialized stores, such as Bath & Body Works, to individual merchants like the woman who sells baskets from an open stall in the central market in Ibadan, Nigeria. Some retailers operate from stores and others operate without a store—by selling online, on TV, with a printed catalog, from vending machines, or even in consumers' homes. Most retailers sell physical goods produced by someone else. But in the case of service retailing—such as dry cleaning, fast food, tourist attractions, online bank accounts, or hair salons—the retailer is often the producer. Because they serve individual consumers, even the largest retailers face the challenge of handling many small transactions.

The nature of retailing and its rate of change are generally related to the stage and speed of a country's economic development. In the United States, retailing is more varied and dynamic than in most other countries. By studying the U.S. system, you will better understand where retailing is headed in other parts of the world.

Retailing is big business

Retailing is crucial to consumers in every macro-marketing system. For example, consumers spend about $5.5 *trillion* (that's $5,500,000,000,000!) a year buying goods and services from U.S. retailers.

A few big retailers do most of the business

There are more than a million retailers in the United States, but most of these retailers are small; more than half report annual sales of less than $1 million. The larger retail stores—those selling more than $5 million annually—do most of the business. Less than 15 percent of the retail stores are this big, yet they account for almost 75 percent of all retail sales.[2]

Big chains have market clout

One reason for the dominance of large retailers is that many achieve economies of scale from a corporate chain. A **corporate chain** is a firm that owns and manages more than one store—and often it's many. Chains—including Nordstrom, Walmart, Kroger, 7-Eleven, and Chipotle—have grown rapidly and now account for about half of all retail sales. Chains can buy in larger quantities and earn lower prices, they get operational efficiencies from running many stores, and they can apply advertising and other promotion costs over many stores. You can expect chains to continue to grow and take business from independent stores.

Some businesspeople like to open retail franchises with brand names consumers already know—brands with track records of success, like Dunkin' and Baskin-Robbins.
Source: DD IP Holder LLC AND BR IP Holder LLC

Retail through cooperatives and franchisors

Because size offers such an advantage, many retailers not connected to a chain join cooperatives or franchises. Retailers' cooperatives pool purchasing and marketing expenses to increase efficiency and effectiveness. Piggly Wiggly grocery, Ace Hardware, NAPA Auto Parts stores, and Carpet One are examples of cooperatives.

In a **franchise operation**, the franchisor develops a good marketing strategy, and the retail franchise holders carry out the strategy in their own units. Each franchise holder benefits from its relationship with the larger company and its experience, buying power, promotion, and image. In return, the franchise holder signs a contract agreeing to pay fees and commissions and strictly follow franchise rules designed to continue the successful strategy. Franchise holders' sales account for about a third of all retail sales. Hampton Hotels, Anytime Fitness, Subway, and Supercuts are examples of franchises.[3]

Small retailers specialize

While big chains and franchises dominate the sales volume, 85 percent of all retailers are small. But it is not easy for small retailers; the failure rate of small retailers is high. A recent study found less than half were operating four years after starting in business. Without the same economies of scale, small retailers that survive usually develop a specialty and become highly knowledgeable about a market niche. For example, online retailer Wicker Central specializes in outdoor wicker furniture. With four toy stores in Charleston, South Carolina, Wonder Works survives because it offers visitors a unique experience on each visit. These stores compete with larger competitors based on intimate knowledge of their markets and high levels of customer service.[4]

Planning a Retailer's Strategy

LO 12.2

Retailers interact directly with final consumers, so strategy planning is critical to their survival. If a retailer loses a customer to a competitor, the retailer is the one that suffers. Producers and wholesalers still make *their* sale regardless of which retailer sells their product. Let's take a closer look at strategy planning for retailers.

Consumers have reasons for buying from particular retailers

Different consumers favor different kinds of retailers. But many retailers either don't know or don't care why. Successful retailers identify possible target markets and try to understand why these people buy where they do. They also understand their own company's strengths and weaknesses as well as those of its competitors. These insights guide the retailer's efforts to fine-tune its marketing mix to the needs of specific target markets.[5]

Marketing mix decisions for retailers

In developing a strategy, a retailer should identify its target market and consciously make decisions on each of the Four Ps; Exhibit 12-3 outlines some of the options

Exhibit 12-3 Examples of Marketing Strategy Decisions for Retailers

Product
- Product selection (width and depth of assortment, brands, quality)
- After-sale service (set-up)
- Special services (special orders, entertainment, gift wrap)
- Packaging/packaging-free options

Place
- Physical stores and/or sales over the Internet
- Number and location of stores
- Shopping atmosphere (comfort, safety)
- Store size, layout, and design
- Home delivery options

Price
- Credit cards—whether to offer a store card
- Discount policies
- Frequency and level of sales prices
- Charge (or not) for delivery or other services

Promotion
- Advertising
- Publicity (Facebook, Pinterest)
- Salespeople (number, training)
- In-store/online–displays, online videos, reviews
- Sales promotion–buy one donate one

available. The chosen combination determines how target customers perceive the retailer's offering (its positioning) and how it is differentiated from other retailers. The marketing mix must provide superior value to some target market, or the retailer will fail.[6]

Strategy planning for two successful shoe retailers

Consider the marketing strategies for two very different, yet successful, shoe retailers. Usually located in malls, Foot Locker is a specialty shop that sells athletic footwear and some athletic apparel. Its primary target market is young males aged 12 to 20, though it also sells to other active individuals. Other segments are targeted more directly through its Lady Foot Locker and Kids Foot Locker stores. Although its shoes can be ordered online, most sales occur in one of the more than 3,000 stores located around the world. Each store typically carries less than 300 styles of athletic shoes with a focus on those designed for running, basketball, and casual wear. Whereas Foot Locker carries several brands, about 70 percent of its sales are of Nike shoes. Its promotion includes a sales staff at each store and advertising featuring famous athletes like basketball player Russell Westbrook and football player Tom Brady. Foot Locker's biggest fans—almost 6 million of them—follow the store on Facebook. Foot Locker's prices are competitive, and regular sales offer larger discounts. The formula seems to work; Foot Locker's same store sales keep growing.[7]

Zappos.com offers a different marketing mix to a different target market. Zappos operates several online storefronts in addition to its core website—including Zappos Couture, selling high-fashion shoes, and Zappos Running, selling athletic footwear. Although Zappos recently added clothing, handbags, housewares, and beauty supplies to its product assortment, its core product line remains shoes. Zappos' wide product line includes more than 150,000 styles and 1,400 brands—including well-known brands such as Nike, Timberland, and Bandolino. Because many shoppers are reluctant to buy shoes without first trying them on, Zappos devotes a whole page to each shoe it sells. Each page features eight or more photos from different angles, a video, customer reviews, detailed descriptions, and suggestions for other shoes "You may also like." Zappos also offers free shipping and free returns—for up to 365 days after purchase! The retailer wins praise for its customer service, which is available 24/7 on the phone or the web. That service helps build positive word-of-mouth for Zappos, and more than 75 percent of its customers are repeat buyers. It has a relatively small budget for its advertising, seen on television, print, online, and even on the bottom of the plastic bins in airport security lines. Zappos and Foot Locker developed very different marketing mixes, but each keeps target customers coming back.[8]

Zappos creatively utilizes its small advertising budget—in this case, placing ads at the bottom of airport security bins. Foot Locker advertises on television using well-known athletes like New England Patriots quarterback Tom Brady. *(top): Source: Zappos LLC; (bottom): Source: Foot Locker Retail, Inc.*

#M4BW

Stores reduce or eliminate packaging

As more customers seek to make sustainable purchase choices, some stores are finding ways to reduce waste. Many supermarkets (and some communities) discourage the use of disposable plastic bags. Amazon asks all its suppliers to reduce packaging materials and pressured laundry detergents to create lighter-weight, more sustainable packaging. Tide, for example, made a more concentrated formula and changed packages from plastic jugs to a cardboard box that is four pounds lighter. That reduces waste, lowers shipping costs, and helps the environment.

Some stores take it one step further. At Nada, a grocery store in Vancouver, British Columbia, customers won't find anything in packages. Herbs don't come in packages—a customer takes a sprig (or two) of basil, toothpaste is in glass jars (bring your own package), and customers are encouraged to bring their own cup for coffee (though you can borrow one of Nada's mugs if you forget—please return it when you are done). In Singapore, UnPackt offers a similar model. Customers bring their own containers, and prices are lower than conventional supermarkets. These examples suggest the global appeal of sustainability and opportunities for marketing for a better world.[9]

The ethics of "better world" claims

We have noted throughout this text that many customers are willing to pay a premium to purchase products that meet their needs *and* help create a better world. This provides an incentive for many firms to claim they engage in "better world" practices—even if their efforts are largely superficial. Consumers have to decide how much they trust a company's claims. Some claims are bolstered with certifications like "fair trade" or "organic." Other claims are not regulated. How much should a company need to contribute to a "better world" before claiming a product it sells is "fair trade" or "organic" or "environmentally friendly"? These questions don't have easy answers; wrestle with them as you read the Ethical Dilemma that follows.

Ethical Dilemma

What would you do? Farmers in poor countries get very little money for crops—such as coffee, cocoa, and bananas—that they grow for export. Some consumers in prosperous nations are willing to pay retailers higher prices for "fair trade" goods so the farmers will receive greater compensation. But critics question whether fair trade works as it should. For example, Sainsbury's is a popular British food retailer. It was charging $2.74 per pound for "fair trade" bananas versus only $0.69 per pound for regular bananas. Farmers, however, got only $0.16 extra from that $2.05 price premium. Critics charge that Sainsbury's makes more from the "fair trade" promotion than the farmers it is supposed to help. Many retailers have similar programs.[10]

Do you think that Sainsbury's is acting ethically? What do you think Sainsbury's and other similar retailers should do? Why?

Different types of retailers emphasize different strategies

Retailers have an almost unlimited number of ways to alter their offerings—their marketing mixes—to appeal to a target market. Because of all the variations, it's oversimplified to classify retailers and their strategies based on a single characteristic—such as merchandise, services, sales volume, or even whether they operate in cyberspace. But a good place to start is by considering basic types of retailers and some differences in their strategies.

Conventional Retailers—Try to Avoid Price Competition

LO 12.3

Single-line, limited-line retailers specialize by product

About 150 years ago, **general stores**—which carried anything they could sell in reasonable volume—were the main retailers in the United States. But with the growing number of consumer products after the Civil War, general stores couldn't offer enough variety in all their traditional lines. So some stores began specializing in dry goods, apparel, furniture, or groceries.

Exhibit 12-4 Types of Retailers and the Nature of Their Offerings (*with examples*)

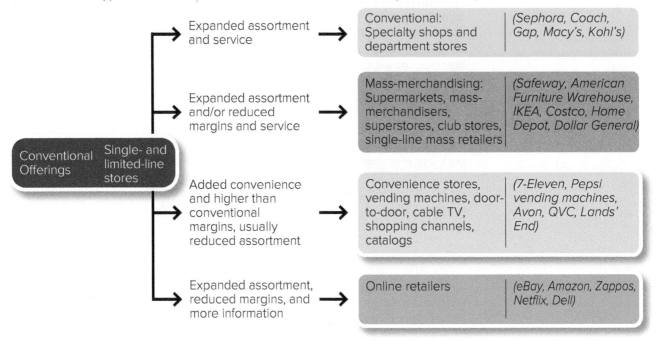

Now most *conventional* retailers are **single-line** or **limited-line stores**—stores that specialize in certain lines of related products rather than a wide assortment. Many specialize not only in a single line, such as clothing, but also in a *limited line* within the broader line. Within the clothing line, a retailer might carry *only* shoes, formal wear, or even neckties but offer depth in that limited line.

Single-line, limited-line stores are being squeezed

The main advantage of limited-line retailers is that they can satisfy some target markets better. Perhaps some are just more conveniently located, but most adjust to suit specific customers. They build a relationship with their customers and earn a position as *the* place to shop for a certain type of product. But these retailers face the costly problem of having to stock some slow-moving items in order to satisfy their target markets. Many of these stores are small—with high expenses relative to sales. So they try to keep their prices up by avoiding competition on identical products.

Conventional retailers like this have been around for a long time and are still found in every community. They are a durable lot and clearly satisfy some people's needs. In fact, in most countries conventional retailers still handle the vast majority of all retailing sales. However, this situation is changing fast. Nowhere is the change clearer than in the United States. Conventional retailers are being squeezed by retailers who modify their mixes in the various ways suggested in Exhibit 12-4. Let's look closer at some of these other types of retailers.

Expand Assortment and Service—To Compete at a High Price

Specialty shops usually sell shopping products

A **specialty shop**—a type of conventional limited-line store—is usually small and has a distinct "personality." Specialty shops sell special types of shopping products, such as high-quality sporting goods, exclusive clothing, baked goods, or even antiques. They aim at a carefully defined target market by offering a unique product assortment, knowledgeable salesclerks, and better service.

Department stores combine many limited-line stores and specialty shops

Catering to certain types of customers whom the management and salespeople know well simplifies buying, speeds turnover, and cuts costs due to obsolescence and style changes. Specialty shops probably will continue to be part of the retailing scene as long as customers have varied tastes and the money to satisfy them.[11]

Department stores are larger stores that are organized into many separate departments and offer many product lines. Each department is like a separate limited-line store and handles a wide variety of shopping products, such as menswear or housewares. They are usually strong in customer services, including credit, merchandise return, delivery, and sales help. Leading department store chains include Macy's, Dillard's, and Nordstrom.

Department stores are still a major force in big cities. However, in the United States, the number of department stores as well as the average sales per store and their share of retail business have declined

Specialty shops and department stores often rely on well-trained salespeople who deliver expertise and customer service.
John Henley/Getty Images

significantly since the 1970s. Well-run limited-line stores compete with good service and often carry the same brands. In the United States and many other countries, mass-merchandising and online retailers have posed an even bigger threat.[12]

Evolution of Mass-Merchandising Retailers

Mass-merchandising is different from conventional retailing

The conventional retailers just discussed think that demand in their area is fixed—and they have a "buy low and sell high" philosophy. Many modern retailers reject these ideas. Instead, they accept the **mass-merchandising concept**, which says that retailers should offer low prices to get faster turnover and greater sales volumes by appealing to larger markets. The mass-merchandising concept applies to many types of retailers, including both those that operate stores and those that sell online. But to understand mass-merchandising better, let's look at its evolution from the development of supermarkets to modern mass-merchandisers such as Walmart in the United States, Tesco in the UK, and Amazon.com on the Internet.

Supermarkets started the move to mass-merchandising

The basic idea for **supermarkets**, large stores specializing in groceries with self-service and wide assortments, developed in the United States during the 1930s Depression. In earlier days, customers entered a store, and a clerk behind a counter fetched requested items. Then some innovators introduced self-service as a way to cut costs while also providing a broad assortment in large bare-bones stores.[13]

Today's supermarkets carry 20,000 to 40,000 items. Per-store sales are about $20 million a year, with about 75 percent of that in food. The average size of a supermarket is 40,000 square feet. In the United States, there are about 35,000 supermarkets, and competition in most areas is intense. More recently, supermarket operators have opened superstores of 50,000 to 100,000 square feet with extensive selection.[14]

#M4BW Unfortunately, in some parts of the United States, poverty is a way of life. Many people live in what is termed a "food desert," a part of the country where it is difficult to purchase high-quality fresh food. Food deserts exist when supermarkets fail to adequately meet the needs of low-income, urban target markets—often finding this market unwilling to pay higher prices for fresh foods. The California-based 99 Cents Only chain of grocery stores bucks that trend. Located mostly in low-income, inner-city neighborhoods, the chain sells healthy food at reasonable prices—giving this underserved market a healthy alternative. It does this, in part, by selling the "ugly" produce that is often rejected by higher-end grocers; while these fruits and vegetables don't always look "pretty," they are still nutritious and tasty. 99 Cents Only makes for a better world with a strategy that reduces food waste and meets the needs of an underserved target market.[15]
Ringo Chiu/ZUMAPRESS/Newscom

Supermarkets are planned for maximum efficiency. Checkout scanners make it possible to carefully analyze the sales of each item and allocate more shelf space to faster-moving and higher-profit items. Survival depends on efficiency and high sales volume. Net profits in supermarkets usually run a thin 1 to 2 percent of sales or less!

Mass-merchandisers add more assortment

Mass-merchandisers are large self-service stores with many departments that emphasize "soft goods" (housewares, clothing, and fabrics) and staples (like health and beauty aids) and offer lower prices to get faster turnover. Mass-merchandisers, such as Walmart and Target, have checkout counters in the front of the store and little sales help on the floor. Today, the average mass-merchandiser has nearly 60,000 square feet of floor space, but many new stores are 100,000 square feet or more. Mass-merchandisers grew rapidly—and they've become the primary place to shop for many frequently purchased consumer products. To move into new markets—big cities or small towns—some of these retailers are opening stores with smaller footprints.

Supercenters meet all routine needs

Some supermarkets and mass-merchandisers have moved toward becoming **supercenters (hypermarkets)**—very large stores that try to carry not only food and drug items but all goods and services that the consumer purchases *routinely*. These superstores look a lot like a combination of the supermarkets, drugstores, and

Supermarkets offer wide assortments and self-service, while warehouse club stores like Costco provide their target customers a limited assortment, warehouse ambience, and low prices.
(left): Jeff Greenough/Blend Images/Getty Images; (right): Rick Bowmer/AP Images

mass-merchandisers from which they have evolved, but the concept is different. A supercenter is trying to meet *all* the customer's routine needs at a low price. Supercenter operators include Meijer, Fred Meyer, Target, and Walmart. In fact, Walmart's supercenters have turned it into the largest food retailer in the United States.

Supercenters average more than 150,000 square feet and carry about 50,000 items. Their assortment in one place is convenient, but many time-pressured consumers think that the crowds, lines, and "wandering around" time in the store are not. Some supercenters have responded by reducing product line depth. For example, Walmart recently decided that it didn't need to carry 24 tape measures and now carries just 4.[16]

New mass-merchandising formats keep coming

The warehouse club is another retailing format that quickly gained popularity. Sam's Club and Costco are two of the largest. Consumers usually pay an annual membership fee to shop in these large, no-frills facilities. Each warehouse club carries about 3,500 items, including food, appliances, yard tools, tires, and other items that many consumers see as homogeneous shopping items and want at the lowest possible price. The growth of these clubs has also been fueled by sales to small-business customers. That's why some people refer to these outlets as wholesale clubs. However, when half or more of a firm's sales are to final consumers, it is classified as a retailer, not a wholesaler.[17]

Single-line mass-merchandisers rise and fall

Since 1980, many retailers focusing on single product lines have adopted the mass-merchandisers' approach with great success. Now out of business, Toys "R" Us pioneered this trend. Similarly, IKEA (furniture), Lowe's (home improvement), Best Buy (electronics), and Staples (office supplies) attract large numbers of customers with their large assortment and low prices in specific product categories. These stores are called *category killers* because it's so hard for less specialized retailers to compete. Now online retailers offering greater assortment and even lower prices threaten the category killers.[18]

Analytics for mass-merchandising

We have seen the growth and evolution of mass-merchandising, with its focus on low prices and fast turnover. To monitor that turnover, retailing managers utilize a measure of **stockturn rate** (also called *inventory turnover*)—the number of times the average inventory is sold in a year. Calculation of this measure is discussed in more detail in Appendix B. For now, it helps to understand that stockturn rate is one indication of the health of a retailer as it measures how quickly its inventory sells. Rising inventory (and

Marketing Analytics in Action: Stockturn Rate

Retail managers keep a close eye on the stockturn rate because a falling rate typically foreshadows a decline in other financial measures.[19] Stockturn rate can be measured at the level of a product category, for example, women's shoes in a department store. In this case, comparisons are best made over time within the category. The marketing manager in charge of women's shoes would look at the stockturn rate over time; if this month's number is higher than last month's, this provides some evidence the marketing mix is performing better than before.

Stockturn rate can also be measured at the level of the retail chain. In this case, performance is best evaluated over time (comparing current month to the previous month) or by comparing across similar retailers. The following table shows actual stockturn rates for different retail chains organized by type of retailer. Review the table and answer the questions that follow.

	Stockturn Rate[20]
Supermarkets	
Kroger	17.28
SuperValu	12.57
Mass-merchandisers	
Walmart	8.08
Target	5.38
Single-line mass-merchandisers	
Home Depot	4.97
Lowe's	3.87
Specialty shops	
Foot Locker	4.30
Finish Line	3.83
O'Reilly's Automotive	1.44
Advance Auto Parts	1.08
Department stores	
Macy's	2.69
Dillard's	2.47

1. *Why might the two supermarkets have a considerably higher stockturn rate than the other retailers listed in this table?*
2. *Why might auto parts dealers have a lower stockturn rate as compared to athletic shoe stores?*
3. *Does Walmart's higher stockturn rate necessarily mean that it is more profitable than Target? Why or why not?*

a lower stockturn rate) means that product is not selling quickly enough. Retailing managers often monitor stockturn rate from month to month or in comparison to similar retailers. To see how this operates in practice, see Marketing Analytics in Action: Stockturn Rate.

Some Retailers Focus on Added Convenience

Convenience (food) stores must have the right assortment

Convenience (food) stores are a convenience-oriented variation of the conventional limited-line food stores. Instead of expanding their assortment, however, convenience stores limit their stock to pickup or fill-in items such as bread, milk, beer, and eat-on-the-go snacks. Many also sell gas. Stores such as 7-Eleven and Stop-N-Go aim to fill consumers' needs between trips to a supermarket, and many of them are competing with fast-food outlets. They offer convenience, not assortment, and often charge prices 10 to 20 percent higher than nearby supermarkets. However, as many other retailers have expanded their hours, intense competition is driving down convenience store prices and profits.[21]

Vending machines are convenient

Automatic vending is selling and delivering products through vending machines. Vending machine sales account for only about 1.5 percent of total U.S. retail sales. Yet for some target markets this retailing method can't be ignored.

Some target markets value the convenience of automatic vending or convenience stores.
(left): Keith Beaty/Toronto Star/Getty Images; (right): Miguel Candela/SOPA Images/LightRocket/Getty Images

Although vending machines can be costly to operate, consumers like their convenience. Many vending machines are becoming more convenient by accepting credit cards and mobile payment. Traditionally, soft drinks, candy bars, and snack foods have been sold by vending machines. Now some higher-margin products are beginning to use this channel. For example, Standard Hotels uses poolside vending machines to sell bathing suits.[22]

Stores come to the home—in person, on television, and in catalogs

In-home shopping in the United States started in the pioneer days with **door-to-door selling**—a salesperson going directly to the consumer's home. Variations on this approach are still important for firms such as Amway Global and Mary Kay. It meets some consumers' need for convenient personal attention. Although gaining popularity in some international markets such as China and parts of Africa, it now accounts for less than 1 percent of retail sales in the United States.

Customers can also shop at home by watching cable television channels dedicated to shopping and then calling in their orders by phone. QVC and Home Shopping Network operate in the United States, Japan, and some European countries. Similarly, catalogs allow customers to page through merchandise and place orders over the phone or online. Deliveries will usually occur a few days later. These home shopping methods often use a multichannel approach by adding a website.[23]

Let's take a closer look at online retailing, which has quickly become the most popular way to shop at home (and sometimes at work).

Retailing and the Internet

LO 12.4

Internet retailing is growing fast, but varies by category

Internet retailing is growing fast; online sales are growing about 15 percent per year, whereas total retail sales are growing around 4 percent. More and more people are comfortable shopping online—with more than a third buying something online once a week. That said, online sales still make up only about 10 percent of all retail sales.[24]

Online shopping still is not very common for many of the most frequent purchases people make—groceries and health and beauty, for example. It is also not common for many of the biggest purchases people make—like cars and trucks. In some categories, the share of online sales is much greater. For example, more than half of books, movies, and music are sold online; about 40 percent of computer and consumer electronics and a third of office equipment and supplies and toys and hobby are bought online. The impact of the Internet on retail sales is even greater; many shoppers gather information on the Internet—often at online retailers—before purchasing in a physical store.[25]

Exhibit 12-5 Comparing the Advantages of Shopping Online with Shopping at a Physical Store

Advantages of Shopping Online	Advantages of Shopping at a Physical Store
• Rich variety of product information from many sources	• Ability to hold, try on, or test products before purchase
• Reviews and tips from customers who've already used the product	• Personal help and interaction
• Wide assortment	• Edited assortment—easier to decide
• Fast and convenient purchase/checkout	• Convenient product returns
• Simplified price and product feature comparison	• Help with setup and ongoing service needs
• Anytime and anywhere access	• Instant access to products and instant gratification
	• Shopping with friends or family as a social activity

Online retailers versus "brick-and-mortar" stores

These days there are online retailers, physical stores (sometimes called *brick-and-mortar retailers*), and many stores that use both approaches. **Online retailers** are stores that sell exclusively or almost exclusively online. These stores usually don't have a physical store that consumers can visit.

Consumers see advantages and disadvantages to shopping online as compared to shopping in a physical store. Exhibit 12-5 offers a comparison of these two shopping methods.[26] These days innovative online retailers are finding ways to offset many of the advantages of physical stores—at the same time physical stores are utilizing the Internet to add value to their offerings. This battle for consumer dollars is fueling a period of innovation and evolution unlike anything retailing has ever seen. Let's take a closer look at how online and brick-and-mortar retailers use the Internet to improve customers' shopping experiences.[27]

Online retail offers low prices and wide assortment

With no storefront and limited sales help, online retailers often have lower operating costs than brick-and-mortar stores. These cost advantages lead many online stores to use low prices to attract customers. But because the Internet makes it easy to compare products and prices from different online sellers, price-sensitive shoppers often choose the store with the lowest price. That puts constant price pressure on Internet sellers to figure out how to differentiate their offerings.

Online stores can also offer a very wide assortment of products—often from different sellers. It is easy to click from site to site or simply conduct a search for a specific brand/model and find multiple sites offering the same thing. Amazon and eBay bring different sellers together to offer customers more choices on a single retail platform. To counter that online advantage, many brick-and-mortar stores are adding in-store computer terminals (kiosks) that connect to the Internet so customers can access the store's website and buy what they can't find in the store.

Lowering the risk of shopping online

Some customers still see risk in online shopping. They worry about unexpected shipping costs, whether their purchase will look the same when it arrives as it did online, how easy it will be to return, or whether an online retailer will mishandle personal information such as credit card numbers.

Successful online retailers are overcoming these customer concerns. Most sell well-known brands and provide free shipping. Some now include bar-coded return labels with the original shipment to simplify returns—which may be even more convenient than returning them to a local store. Many online stores provide excellent customer service and build a good reputation—because word-of-mouth matters online. And as customers do more online shopping, they gain confidence. Together these efforts are eliminating the worries and hassles of online shopping.

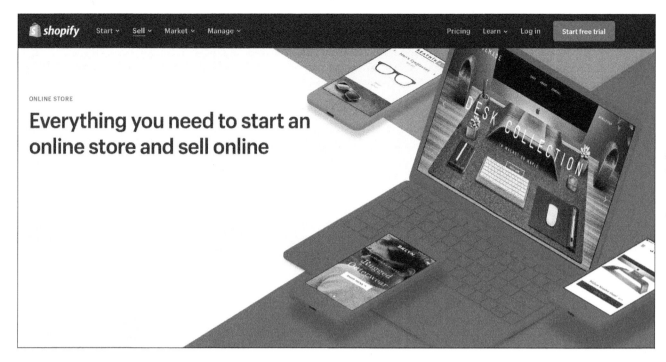

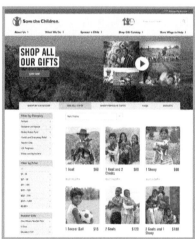

The Internet presents many new opportunities for online shopping. Costs can be relatively low to start an online store, as there is no need to commit to a long-term lease on a building. Some small businesses—more than 800,000—use the Shopify platform to quickly set up an online store. Shopify offers "Everything you need to start an online store and sell online," including tools to help with payments, marketing, shipping, and customer engagement. #M4BW As customers have gotten used to online retail, a natural extension was for online donations. At the Save the Children Federation's online store, shoppers/donors can "buy" aid for children around the world—$180 pays for "Two goats and a sheep [to] help families feed and care for their children."

(top): Source: Shopify; (bottom left): Source: Save the Children Federation, Inc.

Obtaining purchase information

For many purchases, consumers seek information to help them make a purchase decision. This is especially important when a customer buys in a product category for the first time. It also matters more with clothing or fashion items, where fit matters. Physical stores offer the advantage of being able to touch, feel, and perhaps try on something before purchasing. And physical stores have a salesperson there to answer customer questions. Yet, online retailers are addressing many of these customer concerns as well. Online reviews, virtual salesclerks, and "fit guides" provide information to make buying easier. Many sites post photos and video demonstrations to give customers more information. At online eyeglass retailer EyeBuyDirect.com, customers can upload a photo of themselves and then see what they look like when they virtually "try on" different frames. Another eyeglass retailer, Warby Parker, encourages customers to choose five frames and try them on at home before placing an order.[28]

What is convenience?

Traditional thinking about retail stores considered shopping convenience from the perspective of product assortment, store location, and store hours. By contrast, when shopping online, a customer can access a wide assortment on a single retailer's site or by clicking from one website to another. In addition, online stores are always open—and a customer does not even need to leave home to shop.

However, the Internet makes shopping inconvenient in other ways. You have to plan ahead. When you buy something, you've actually just ordered it and you have to wait for delivery. Online retailers are trying to overcome these disadvantages with faster shipping. It started with low-cost two-day shipping; then one-day shipping became more reasonable. Now some online retailers offer same-day delivery in a few large cities.

Big data and analytics personalize shopping

Online retailers have lots of information on their customers' shopping behavior. For example, an online retailer can track (and store) data showing how a customer moves through its website, which product pages the customer views, whether she reads the reviews or watches a video, what competitive products she considers, and what (if anything) she ultimately purchases. Cookies (small data files placed on a customer's computer) can even tell a retailer what a shopper does at a competitor's website, pull information from a customer's Facebook page, and even track his movements around town via his cell phone. These data potentially offer a retailer a great deal of information on each customer.

The ability to collect and analyze big data and adapt a marketing mix in real time (before that customer's next "click") gives online retailers the opportunity to deliver a personalized shopping experience. Using predictive analytics, retailers anticipate their customers' needs. For example, a customer known to have shopped at a competitor's site might be seen as a "price-oriented shopper" and could receive a targeted (only for that customer) discount. Knowing that a customer previously purchased a sweater from the store, an online retailer might show more sweaters of a similar style on the front page of its site when that customer returns. When a customer signs in to Netflix to watch a movie, the site already makes recommendations for other videos based on that customer's previous viewing history and the viewing habits of other Netflix customers. These stored customer data make recommendations more on-target and relevant—and encourage additional purchases. All of these activities deliver customers a shopping experience tailored to their needs and interests.

Physical retailers add online feature

In Chapter 10 we introduced the idea that many consumers are *multichannel shoppers*—using different channels as they move through the purchase process. To address this trend, many physical retailers add new features to offset the advantages of online stores. These days most brick-and-mortar retailers have an online presence. Their websites may simply provide product information, or they may be full-blown e-commerce websites. Earlier in this chapter, we described how Foot Locker and Home Depot use their online stores to offer a wider product assortment than can be carried in a single store. Target, Walmart, Macy's, and Best Buy all do substantial sales volume through their online stores. An important marketing strategy planning decision for a retailer with physical stores involves deciding what online capabilities (if any) to build.

Smartphones are the new online store

Consumers—especially Millennials—are increasingly comfortable shopping on their smartphones. The behavior may be as simple as texting a photo of a pair of shoes to a friend for feedback or researching prices across stores. Some brick-and-mortar retailers see an opportunity to leverage this trend with websites and apps that deliver extra services to customers in physical stores. For example, with Sam's Club's Scan & Go app customers skip the checkout line. They scan products as they shop and click a button to pay directly from the app. Shopping on mobile devices is a new consumer behavior, and leading retailers are watching closely for opportunities to develop services that enhance target customers' shopping experience.

Omnichannel brings it all together

The concept of *omnichannel,* a multichannel selling approach in which a retailer provides a seamless customer shopping experience from computer, mobile device, or brick-and-mortar store, was also introduced in Chapter 10. Many brick-and-mortar retailers implement an omnichannel strategy to offer customers a better shopping experience. Companies implement omnichannel in different ways. For example, some customers use a site to check what is in stock at their local store. The websites for Office Depot and Best Buy show inventory at a local store and allow for online purchase and a choice of immediate in-store pickup or home delivery. For its prescription drug customers, the Walgreens app integrates the user experience. Customers can check on or

Omnichannel connects the shopping experiences of a store's online and brick-and-mortar stores, delivering a better customer experience. For example, the Chick-fil-A app remembers a customer's "favorite order." Upon entering a Chick-fil-A store, a customer skips the line, clicks a button on the app to place his order (payment is automatic), grabs a seat, and waits for his meal to arrive. King Soopers (a grocery store) allows customers to do all their shopping online and then choose one-hour home delivery or curbside pickup at their local store. *(left): Source: Chick-fil-A; (right): Source: The Kroger Co.*

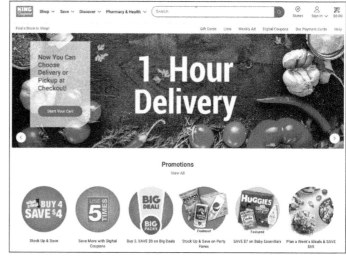

renew prescriptions as well as set reminders to take a drug or reorder. Omnichannel strategies that meet customers' needs win over multichannel shoppers.[29]

Online retailers open physical stores

A few formerly online-only retailers are investing in brick-and-mortar stores to deliver a better shopping experience to their customers. For example, after struggling to build its online grocery business, Amazon decided that having physical stores might be helpful and purchased Whole Foods Market. Amazon has also opened physical bookstores in malls and on some college campuses. Eyeglass seller Warby Parker and home furnishings retailer Wayfair started online and later opened brick-and-mortar stores. The future may lead more retailers to try to combine the best in online and physical stores to better serve customers.[30]

Online retailers partner with physical stores to access new target markets

In a strategy that aims to quickly gain physical store locations, some online retailers expand their reach by partnering with smaller retailers that cannot develop their own online presence. This practice is growing quickly in Asia and offers online retailers access to target customers (often in rural areas) that don't have easy access to the Internet while offering retailers a much larger product assortment. Chinese retailing giant JD.com is placing kiosks (small booths with a computer and Internet access) in mostly rural store locations across the country—they plan to open 1,000 locations *per day*. Similarly, Indonesia's Kioson partnered with post offices across the country, quickly gaining more than 30,000 kiosk partners. Creative options like these expand online retailing to new target markets.[31]

Why Retailers Evolve and Change

LO 12.5

The Internet and online shopping—and the competitive threat they pose to traditional retailing—have been a big motivator of evolution and change in retailing. In this section, we discuss some broader retailing trends and consider how in-store technology influences shopping.

The wheel of retailing keeps rolling

The **wheel of retailing theory** says that new types of retailers enter the market as low-status, low-margin, low-price operators and then, if successful, evolve into more conventional retailers offering more services with higher operating costs and higher prices. Then they're threatened by new low-status, low-margin, low-price retailers—and the wheel turns again. Department stores, supermarkets, and mass-merchandisers went through this cycle. Some online retailers are on this path.

The wheel of retailing theory doesn't explain all major retailing developments. Vending machines enter as high-cost, high-margin operations. Convenience food stores are high-priced. Suburban shopping centers don't emphasize low price.

Scrambled merchandising—mixing product lines for higher profits

Conventional retailers tend to specialize by product line. But many modern retailers are moving toward **scrambled merchandising**—carrying any product lines they think they can sell profitably. Supermarkets and drugstores sell anything they can move in volume—pantyhose, phone cards, motor oil, and potted plants. Hardware stores sell pretzels and coffee. Mass-merchandisers don't just sell everyday items but also cell phones, computer printers, and jewelry.[32]

Product life-cycle concept applies to retailer types too

A retailer with a new idea may have big profits—for a while. But if it's a really good idea, the retailer can count on speedy imitation and a squeeze on profits. Other retailers will copy the new format or scramble their product mix to sell products that offer them higher margins or faster turnover. That puts pressure on the original firm to change or lose its market.

Some conventional retailers are in decline as these life and death cycles continue. More recent innovators, like the online retailers, are still in the market growth stage (see Exhibit 12-6). Many retailing formats that are mature in the United States are only now beginning to grow in other countries.

Technology drives ongoing retail evolution

While many of the changes in retailing have been driven by the Internet, some creative retailers utilize other technology to enhance customers' shopping experiences. For example, Kroger is a leader in the use of technology to enhance customers' in-store shopping experience. A few years ago, the average wait in a checkout line at its supermarkets was four minutes. Then Kroger installed special infrared cameras at each store entrance and above cash registers. A computer algorithm interprets the camera data, and software automatically signals a manager when more than two customers are waiting in line. The store manager can then open additional checkout lanes. The home office monitors service levels at every Kroger store. Using this technology, Kroger has reduced the average wait time to less than 30 seconds.[33]

Some stores are investing in artificial intelligence technology to improve the shopping experience. The Caper smart shopping cart includes a built-in scanner and card swipe, so customers can avoid the traditional checkout process. Based on the items a customer has already purchased, the cart can make suggestions for additional items (a handle-mounted touchscreen conveys messages).[34]

Brick-and-mortar retailers saw the advantage online retailers gained by analyzing data showing how customers click through the online store. Now physical stores are adding in-store tracking and using that knowledge to guide store layout, merchandising, and sales promotion. Some stores utilize video cameras; others use sensors or track cell phones to monitor customer movement around stores. Jeweler Alex and Ani learned where customers linger in the store and which items were picked up most often—leading to changes in merchandising or where products were placed in the store—and increased sales.

Exhibit 12-6 Retailer Life Cycles—Timing and Years to Market Maturity

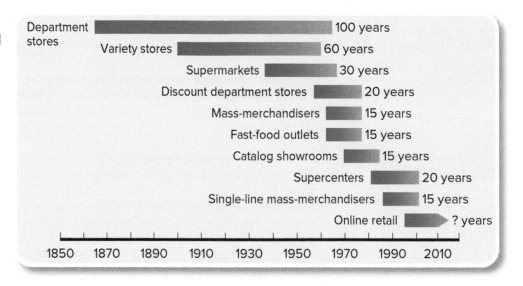

What's Next? Psychic retailers

Brick-and-mortar retailers are afraid of online retailers. Online retailers can track and monitor customers' movement and behavior across the web and integrate that data with social media and other purchase information. As online retailers figure out how to apply predictive analytics to these data, it could put brick-and-mortar retailers at a significant disadvantage. Online competition is inspiring some brick-and-mortar retailers to collect data of their own. And this might be *What's Next?* in retailing—a future where brick-and-mortar retailers combine individual customer data and customers' smartphones, apps, and in-store technology to deliver customized marketing mixes for each customer.

The first step for brick-and-mortar retailers is to collect data to learn more about customer needs and shopping behaviors. This data collection often begins with loyalty cards. Many retailers give discounts to customers who use loyalty cards as an enticement to sign up. Add to that a retailer's shopping app, which can track customers' movements through the store. The app knows which aisles a customer visits and where they linger. Some retailers are experimenting with video data that will let them know what products customers pick up, how long they look at them—even whether they are smiling or frowning. Someday soon, you may find yourself stopping at the lawn mower display in Home Depot, smiling as you look at an electric mower. You don't buy, but the next day Home Depot's app sends you a message suggesting "it's time to buy a new electric Greenworks lawn mower." Or perhaps Home Depot might combine your purchase history with your store wanderings and determine you are thinking about a major kitchen remodel. The store might send you detailed kitchen makeover guides to tempt you to start that project.

To learn even more about consumers, retailers can add data from other sources to create a "big data" file. They can purchase credit card data to add in a customer's purchase history from any store. A mapping app shares how many times a customer has visited the store's main competitor. And mining a customer's social media gives additional insights. When all those data are run through predictive analytics, retailers can decide what products to show each customer and what discounts will encourage a customer to buy.

For example, the store downloads the purchase history from your credit card company and information from social media. These are combined with in-store tracking and loyalty card data, creating a "big data" salad. From these data, a retailer might know you "Liked" an Instagram photo of a friend's new shirt. Previous history suggests your favorite colors are red and blue—and that 15 percent discounts are usually enough to motivate you to buy. The next time you visit the mall, a clothing store there could send you a text message with a photo of a blue and red sweater stating "15% off today only."

That's not all. Someday retailers may know what you want—before you even want it. Amazon has a patent for "anticipatory shipping" where it *ships* customers' orders *before* they click "buy." Perhaps someday soon, you will walk into your favorite Nordstrom store just to look around. A salesperson calls you by name and takes you to a changing room where you'll find three outfits in your size and style (how did they know you needed a new suit?). You love the outfits and purchase two on the spot! Psychic retailing. Maybe that's *What's Next?*[35]

Consider a coffee shop and a college bookstore—how could some of the ideas discussed in this section be used to improve customer experiences at each of these stores? What other technologies could they employ?

Retailers are only scratching the surface of possibilities with technologies like these; to get a peek into the future, check out *What's Next?* Psychic retailers.

Ethical issues for retailers

Let's not forget that these new technologies may raise ethical questions for retailers and consumers. As you read in our discussion of online retailing and in the *What's Next?* box, retailers gather and use information to offer customers more personalized shopping experiences. This appears to be consistent with the marketing concept. But is it ethical to monitor customers' shopping behavior without their knowledge? Should retailers need customers' permission? Does it depend on how this information is used? And is it fair to offer discounts to some customers but not others? Many retailers will also track individual customers' returns and calculate a "risk score." Several retailers deny returns for customers who have a high score. Should customers who return purchases more often be treated differently?

Retailers can use technology to help customers through the buying process. The IKEA Place app uses augmented reality so customers "see" how a piece of furniture would look in their own home. McDonald's purchased a machine learning and predictive analytics company to help it deliver unique customer messages via its stores' drive-through menu boards. The menu can be changed with each customer, even while a customer is ordering, offering a unique marketing mix depending on what the customer orders, the time of day, weather, nearby events, and historical sales data. The menu offers suggestions. For example, a customer ordering two Happy Meals at 5 p.m. is probably a parent ordering dinner for her kids. If the parent is not ordering her own meal, she may not be interested in McDonald's for dinner, but she might be tempted by an offer of a coffee or snack as a pick-me-up. The menu can be programmed to make this suggestion. Over time, the machine learning identifies which offers generate the most incremental sales.[36]
(left): Source: Inter IKEA Systems B.V.; (right): jax10289/Shutterstock

The marketing concept should guide firms away from unethical treatment of customers. However, a retailer on the edge of going out of business may lose perspective on the need to satisfy customers in both the short and the long term.[37]

Differences in Retailing in Different Nations

LO 12.6

New ideas spread across countries

New retailing approaches that succeed in one part of the world are often quickly adapted to other countries. Self-service approaches that started with supermarkets in the United States are now found in retail operations worldwide. The supercenter concept, on the other hand, initially developed in Europe.

Mass-merchandising requires mass markets

The low prices, selections, and efficient operations offered by mass-merchandisers might be attractive to consumers everywhere, but consumers in less-developed nations often don't have the income to support mass distribution. The small shops that survive in these economies sell in very small quantities, often to a small number of consumers.

Retailers moving to international markets must adapt marketing strategies

Slow growth at home has prompted some large retail chains to move into international markets. They think that the competitive advantages that worked well in one market can provide a similar advantage in another country. But legal and cultural differences in international markets can make success difficult. Despite success in Latin America and Canada, Walmart has struggled in Germany and Japan. Similarly, French mass-merchandiser Carrefour expanded in Europe and South America, but its U.S. stores failed and it experienced legal problems in Indonesia.

Other retailers, such as California-based My Dollarstore, have successfully adapted for quick international growth. My Dollarstore franchises the "dollar store" concept worldwide and adapts its marketing strategy to local markets. In India, the price of each

Coca-Cola's desire to put its products "within an arm's reach of desire" anywhere in the world requires emphasizing different retail channels in different places.
(left): chanonnat srisura/Shutterstock; (middle): Dominika zarzycka/Shutterstock; (right): Peter Morgan/AP Images

product is 99 rupees or about two dollars. Dollar stores in the United States target lower-income consumers, but in India the "Made in America" label attracts many higher-income consumers. Initially the merchandise in the Indian stores was the same as in U.S. stores. However, My Dollarstore quickly discovered what sold (Hershey's syrup is a hit) and what didn't (papaya and carrot juice). It also offered money-back guarantees, an unusual practice in India. Adaptations like these helped entice consumers into My Dollarstore's Indian franchises.[38]

Online retailing varies across nations

Online shopping behavior, and therefore online retailing, varies considerably across countries. For example, in the United Kingdom about half of all consumers regularly purchase online, whereas in emerging markets online sales are almost nonexistent. The most obvious prerequisite for online retailing is access to the Internet—in particular, broadband (fast) connections. Chapter 3 discussed technology and the uneven adoption of the Internet and mobile phones across the globe. A reliable national postal system (lacking in many countries) is also needed for delivering online purchases.

Even when infrastructure is in place, cultural factors influence preferences for online shopping. For example, a European study of 20,000 clothing shoppers identified seven segments based on their shopping needs. One segment, nicknamed "time-pressed optimizers," was particularly interested in online shopping. With very busy lives, this group had less time for in-store shopping so they sought the best possible product by doing research online. Yet only 3 percent of Italian shoppers and 6 percent of French shoppers fell into this segment—as compared to 16 percent of Brits and 18 percent of Germans. Another segment, "price-oriented bargain hunters," enjoyed going to many stores and rummaging to find bargain merchandise. This group was most common in Italy (31 percent of shoppers) and helps explain why only about 10 percent of Italians regularly purchased online.[39]

What Is a Wholesaler?

LO 12.7

It's hard to define what a wholesaler is because there are so many different wholesalers doing different jobs. Some of their activities may even seem like manufacturing. As a result, some wholesalers describe themselves as "manufacturer and dealer." Some like to identify themselves with such general terms as *merchant, agent, dealer,* or *distributor.* And others just take the name commonly used in their trade—without really thinking about what it means.

To avoid a long technical discussion on the nature of wholesaling, we'll use the U.S. Census Bureau definition: **Wholesaling** is concerned with the *activities* of those persons or establishments that sell to retailers and other merchants, or to industrial, institutional, and commercial users, but that do not sell in large amounts to final consumers. So **wholesalers** are firms whose main function is providing wholesaling activities. Wholesalers sell to all of the different types of organizational customers described in Chapter 6.

Wholesaling activities are just variations of the basic marketing functions—gathering and providing information, buying and selling, grading, storing, transporting, financing, and risk taking—we discussed in Chapter 1. You can understand wholesalers' strategies better if you look at them as members of channels. They add value by doing jobs for their customers and their suppliers.

Wholesaling Is Changing with the Times

A hundred years ago wholesalers dominated distribution channels in the United States and most other countries. The many small producers and small retailers needed their services. This situation still exists in less-developed economies.

Wholesaling is in decline

However, in developed nations, as producers became larger many bypassed the wholesalers. Cost-conscious buyers at some large retail chains, including Walmart and Lowe's, even refuse to deal with some wholesalers who represent small producers. Efficient delivery services from UPS and FedEx make it easier for many producers to ship directly to their customers, even those in foreign markets. The Internet also puts pressure on wholesalers whose primary role is providing information to bring buyers and sellers together. These factors have combined to decrease the number of wholesalers in the United States; there are about 400,000 wholesalers today—following a steady decline over the last 20 years.

Opportunities remain for progressive wholesalers

Those that have survived are adapting their marketing strategies and finding new ways to add value in the channel. Progressive wholesalers are becoming more concerned with their customers and with channel systems. Many of the same strategies used by retailers are also being utilized by wholesalers; for example, some wholesalers are using omnichannel because their customers are multichannel shoppers. Some of the biggest B2B e-commerce sites are wholesaler operations, and many wholesalers are enjoying significant growth. Others develop voluntary chains that bind them more closely to their customers.

Frieda's innovates in wholesale produce

Frieda's Inc. is a good example; it is a wholesaler that supplies supermarkets and food service distributors with $30 million worth of exotic fruits and vegetables every year. It was started by Frieda Caplan in 1962; today her daughters Karen and Jackie run the company. It is a sign of the marketing savvy of these women that artichokes, Chinese donut peaches, alfalfa sprouts, and spaghetti squash no longer seem very exotic. All of these crops were once viewed as unusual. Few farmers grew them, supermarkets didn't handle them, and consumers didn't know about them. Caplan helped change all of that. She realized that some supermarkets wanted to attract less price-sensitive consumers who preferred more interesting choices in the

Frieda Caplan and her daughters Karen and Jackie continue to bring innovative fruits and vegetables to grocery stores across the United States.
Courtesy of Fear No Fruit Productions

produce department, so she looked for products that would help her retailer-customers meet this need. For example, the funny-looking kiwifruit with its fuzzy brown skin was popular in New Zealand but virtually unknown to U.S. consumers. Caplan worked with small farmer-producers to ensure that she could provide her retailer-customers with a steady supply. She packaged kiwifruit with interesting recipes and promoted it to consumers. Because of her efforts, demand has grown, and most supermarkets now carry kiwifruit. That has attracted competition from larger wholesalers. But that doesn't bother the Caplans. When one of their specialty items becomes a commodity with low profit margins, another novel item replaces it. In a typical year, Frieda's introduces about 40 new products—some of the latest include Kiwano melons and Stokes Purple sweet potatoes.

Frieda's also has an advantage because of the special services it provides. It was the first wholesaler to routinely use airfreight for orders and send produce managers a weekly "hot sheet" about its best sellers. Today the Caplans use online videos and a blog to let customers know what's new. Frieda's website attracts final consumers with helpful tips and recipes. And now that more consumers are eating out, Frieda's has established a separate division to serve the special needs of food service distributors.[40]

Wholesalers need to add value

Progressive wholesalers need to be efficient, but that doesn't mean they all have low costs. Some wholesalers' higher operating expenses result from the strategies they select, including the special services they offer to some customers. Let's look more closely at different types of wholesalers and the different ways they add value to channels of distribution.

Wholesalers Add Value in Different Ways

LO 12.8

Exhibit 12-7 compares the number, sales volume, and costs of some major types of wholesalers. The differences in operating costs suggest that each of these types performs, or does not perform, certain wholesaling functions. But which ones and why? And why do manufacturers use merchant wholesalers—costing 13.1 percent of sales—when agent wholesalers cost only 3.7 percent?

To answer these questions, we must understand what these wholesalers do and don't do. Exhibit 12-8 gives a big-picture view of the major types of wholesalers we'll be discussing. There are many more specialized types, but our discussion will give you

Exhibit 12-7 U.S. Wholesale Trade by Type of Wholesale Operation

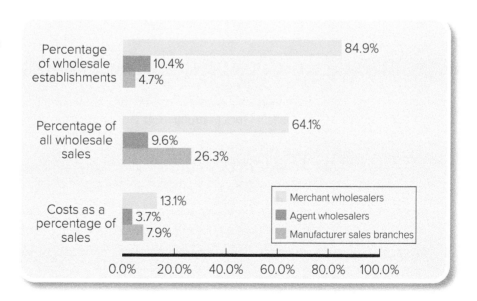

Exhibit 12-8 Types of Wholesalers

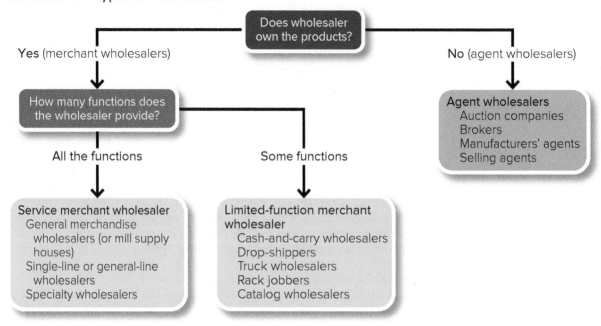

a sense of the diversity. Note that a major difference between merchant and agent wholesalers is whether they *own* the products they sell. Before discussing these wholesalers, we'll briefly consider producers who handle their own wholesaling activities.

Manufacturers' sales branches are considered wholesalers

Manufacturers who just take over some wholesaling activities are not considered wholesalers. However, when they have **manufacturers' sales branches**—warehouses that producers set up at separate locations away from their factories—they're classified as wholesalers by the U.S. Census Bureau and by government agencies in many other countries.

In the United States, these manufacturer-owned branch operations account for about 4.7 percent of wholesale facilities—but they handle 26.3 percent of total wholesale sales. One reason sales per branch are so high is that the branches are usually placed in the best market areas. This also helps explain why their operating costs, as a percentage of sales, are often lower. It's also easier for a manufacturer to coordinate information and logistics functions with its own branch operations than with independent wholesalers.[41]

Merchant Wholesalers Are the Most Numerous

Merchant wholesalers own (take title to) the products they sell. They often specialize by certain types of products or customers. For example, Fastenal is a wholesaler that specializes in distributing threaded fasteners used by a variety of manufacturers. It owns (takes title to) the fasteners for some period before selling to its customers. If you think all merchant wholesalers are fading away, Fastenal is proof that they can serve a needed role. In the last decade Fastenal's profits have grown at about the same pace as Microsoft's.[42]

Exhibit 12-7 shows that almost 85 percent of the wholesaling establishments in the United States are merchant wholesalers—and they handle more than 64 percent of wholesale sales. Merchant wholesalers are even more common in other countries. Japan is an extreme example. Products are often bought and sold by a series of merchant wholesalers on their way to the business user or retailer.[43]

Service wholesalers provide all the functions

Service wholesalers are merchant wholesalers that provide all the wholesaling functions. Within this basic group are three types: (1) general merchandise, (2) single-line, and (3) specialty.

General merchandise wholesalers are service wholesalers that carry a wide variety of nonperishable items such as hardware, electrical supplies, furniture, drugs, cosmetics, and automobile equipment. With their broad line of convenience and shopping products, they serve hardware stores, drugstores, and small department stores. *Mill supply houses* operate in a similar way, but they carry a broad variety of accessories and supplies to serve the needs of manufacturers.

Single-line (or general-line) wholesalers are service wholesalers that carry a narrower line of merchandise than general merchandise wholesalers. For example, they might carry only food, apparel, or certain types of industrial tools or supplies. In consumer products, they serve the single- and limited-line stores. In business products, they cover a wider geographic area and offer more specialized service.

Specialty wholesalers are service wholesalers that carry a very narrow range of products and offer more information and service than other service wholesalers. For example, a firm that produces specialized lights for vehicles might rely on specialty wholesalers to help reach automakers in different countries. A consumer products specialty wholesaler might carry only health foods instead of a full line of groceries. Some limited-line and specialty wholesalers are growing by helping independent retailer-customers compete with mass-merchandisers. But in general, many consumer-products wholesalers have been hit hard by the growth of retail chains that set up their own distribution centers and deal directly with producers.

A specialty wholesaler of business products might limit itself to fields requiring special technical knowledge or service. Richardson Electronics is an interesting example. One of its specialties is in distributing replacement parts, such as electron tubes, for old equipment that many manufacturers still use on the factory floor. Richardson describes itself as "on the trailing edge of technology," but many of its customers operate in countries where new technologies are not yet common. Richardson gives them easy access to information from its website (www.rell.com) and makes its products available quickly by stocking them in locations around the world.[44]

Henry Schein is the largest distributor of health care products and services to office-based practitioners in North America. It also serves many international countries. It operates as a full-service wholesaler for medical, dental, and veterinary professionals.
Source: Henry Schein, Inc.

Limited-function wholesalers provide some functions

Limited-function wholesalers provide only *some* wholesaling functions. In the following paragraphs, we briefly discuss the main features of these wholesalers. Although less numerous in some countries, these wholesalers are very important for some products.

Cash-and-carry wholesalers want cash

Cash-and-carry wholesalers operate like service wholesalers—except that the customer must pay cash. In the United States, big warehouse clubs have taken much of this business. But cash-and-carry operators are common in less-developed nations where very small retailers handle the bulk of retail transactions. Full-service wholesalers often refuse to grant credit to small businesses that may have trouble paying their bills.

Drop-shippers do not handle the products

Drop-shippers own (take title to) the products they sell—but they do *not* actually handle, stock, or deliver them. These wholesalers are mainly involved in selling. They get orders and pass them on to producers. Then the producer ships the order directly to the customer. Drop-shippers commonly sell bulky products (like lumber) for which additional handling would be expensive and possibly damaging. Drop-shippers in the United States are already feeling the squeeze from buyers and sellers connecting directly via the Internet. But the progressive ones are fighting back by setting up their own websites and getting fees for referrals.

Truck wholesalers deliver—at a cost

Truck wholesalers specialize in delivering products that they stock in their own trucks. Their big advantage is that they promptly deliver perishable products that regular wholesalers prefer not to carry. A 7-Eleven store that runs out of potato chips on a busy Friday night doesn't want to be out of stock all weekend! Truck wholesalers help retailers keep a tight rein on inventory, and they seem to meet a need.

Rack jobbers sell hard-to-handle assortments

Rack jobbers specialize in hard-to-handle assortments of products that a retailer doesn't want to manage—and rack jobbers usually display the products on their own wire racks. For example, a grocery store or mass-merchandiser might rely on a rack jobber to decide which paperback books or magazines it sells. The wholesaler knows which titles sell in the local area and applies that knowledge in many stores.

Catalog wholesalers reach outlying areas

Catalog wholesalers sell through catalogs that may be distributed widely to smaller industrial customers or to retailers that might not be called on by other wholesalers. Customers place orders at a website or by mail, e-mail, fax, or telephone. These wholesalers sell lines such as hardware, jewelry, sporting goods, and computers. For example, Inmac uses a catalog that is printed in six languages and a website (www.inmac.com) to sell a complete line of computer accessories. Many of its customers don't have a local wholesaler, but they can place orders from anywhere in the world. Most catalog wholesalers quickly adapted to the Internet. It fits what they were already doing and makes it easier. But they're facing more competition too; the Internet allows customers to compare prices from more sources of supply.[45]

 #M4BW

Wholesalers find ways to add "better world" value

Merchant wholesaler Henry Schein sells dental supplies (along with products for other health care professionals). It firmly believes that access to quality health care makes a difference in people's lives, so Henry Schein supports Give Kids A Smile, which provides free oral health services to needy children around the world. Henry Schein donates supplies and supports volunteer dentists. Since it started in 2003, the program has served more than 5.5 million kids around the world. Henry Schein can feel good about both how it made the world better *and* about its financial returns; sales to dentists who participate in the program jumped 10 percent.[46]

Agents Are Strong on Selling

They don't own the products

Agent wholesalers are wholesalers that do *not* own the products they sell. Their main purpose is to help in buying and selling. Agent wholesalers normally specialize by customer type and by product or product line. But they usually provide even fewer functions than the limited-function wholesalers. They operate at relatively low cost—sometimes 2 to 6 percent of their selling price—or less in the case of website-based agents who simply bring together buyers and sellers.

They are important in international trade

Agents are common in international trade. Many markets have only a few well-financed merchant wholesalers. The best many producers can do is get local representation through agents and then arrange financing through banks that specialize in international trade.

Agent wholesalers are usually experts on local business customs and regulations in their own countries. Sometimes a marketing manager can't work through a foreign government's red tape without the help of a local agent.

Manufacturers' agents provide selling expertise

A **manufacturers' agent** sells similar products for several noncompeting producers—for a commission on what is actually sold. Such agents work almost as members of each company's sales force, but they're really independent wholesalers. More than half of all agent wholesalers are manufacturers' agents. Their big plus is that they already call on some customers and can add another product line at relatively low cost—and at no cost to the producer until something sells! If an area's sales potential is low, a company may use a manufacturers' agent because the agent can do the job at low cost. Small producers often use agents everywhere because their sales volume is too small to justify their own sales force.

Agents can be especially useful for introducing new products. For this service, they may earn 10 to 15 percent commission. (In contrast, their commission on large-volume established products may be quite low—perhaps only 2 percent.) A 10 to 15 percent commission rate may seem small for a new product with low sales. Once a product sells well, however, a producer may think the rate is high and begin using its own sales reps.

Manufacturers' agents sell similar products for several noncompeting producers. They are usually specialists in particular product-markets. By representing multiple producers, manufacturers' agents provide a more effective and efficient selling process for many firms. The MRA is a nonprofit organization that promotes the use of manufacturers' agents in the paper/plastic disposables, packaging, and sanitary maintenance industries.
Source: Manufacturers Representatives of America

Export and import agents are experts in international trade

Export agents and **import agents** are basically manufacturers' agents who specialize in international trade. These agent wholesalers operate in every country and help international firms adjust to unfamiliar market conditions in foreign countries.

Manufacturers' agents will continue to play an important role in businesses that need an agent to perform order-getting tasks. But manufacturers' agents everywhere are feeling pressure when it comes to routine business contacts. More producers are turning to telephone selling, websites, e-mail, teleconferencing, and faxes to contact customers directly.[47]

IronPlanet uses auctions to connect buyers and sellers of used heavy equipment such as cranes, tractors, and trucks.
Source: IronPlanet, Inc.

Brokers offer market information and knowledge

Brokers bring buyers and sellers together. Brokers usually have a *temporary* relationship with the buyer and seller while a particular deal is negotiated. They are especially useful when buyers and sellers don't come into the market very often. The broker's product is information about what buyers need and what supplies are available. If the transaction is completed, they earn a commission from whichever party hired them. **Export brokers** and **import brokers** operate like other brokers, but they specialize in bringing together buyers and sellers from different countries. Smart brokers quickly saw new opportunities to expand their reach by using the Internet. As the Internet causes consolidation, it will also provide more value. A smaller number of cyberbrokers will cut costs and dominate the business with larger databases of buyers and sellers.

Selling agents—almost marketing managers

Selling agents take over the whole marketing job of producers—not just the selling function. A selling agent may handle the entire output of one or more producers, even competing producers, with almost complete control of pricing, selling, and advertising. In effect, the agent becomes each producer's marketing manager.

Financial trouble is one of the main reasons a producer calls in a selling agent. The selling agent may provide working capital and may also take over the affairs of the business. But selling agents also work internationally. A **combination export manager** is a blend of manufacturers' agent and selling agent—handling the entire export function for several producers of similar but noncompeting lines.

Auction companies speed up the sale

Auction companies provide a place where buyers and sellers can come together and bid to complete a transaction. Traditionally they were important in certain lines—such as livestock, fur, tobacco, and used cars—where demand and supply conditions change rapidly.

CONCLUSION

In this chapter, we explored evolving approaches to retailing and wholesaling. We also examined how marketers are finding an efficient and effective balance of technology and the personal touch that works for their target markets.

There are many different types of retailers, each offering different marketing mixes to appeal to different target customers. Lower margins and faster turnover are the modern philosophy for mass-merchandisers, but this is no guarantee of success as retailers' life cycles mature. Online retailing is growing quickly, and many brick-and-mortar retailers are taking advantage of the Internet to create more appealing marketing mixes.

Retailing tends to evolve in predictable patterns—and we discussed the wheel of retailing theory to help understand this. But the growth of chains and scrambled merchandising continues as retailing evolves to meet changing consumer demands.

Wholesalers can provide functions for those both above and below them in a channel of distribution. These services are closely related to the basic marketing functions. Different types of wholesalers perform different marketing functions, with some providing all the functions and others providing few. Eliminating wholesalers does not eliminate the need for the functions they provide. The chapter also discussed how many wholesalers are using technology to perform these functions in more efficient ways.

One thing is certain: The evolving Internet and yet-to-emerge technologies won't allow marketers to rest on their laurels. Successes will be short-lived as retail and wholesale competitors keep looking for new ways to satisfy customer needs. The most agile retailers and wholesalers are the most likely to survive.

KEY TERMS

retailing, 322
corporate chain, 322
franchise operation, 322
general stores, 325
single-line stores, 326
limited-line stores, 326
specialty shop, 326
department stores, 327
mass-merchandising concept, 327
supermarkets, 327
mass-merchandisers, 328
supercenters (hypermarkets), 328
stockturn rate, 329
convenience (food) stores, 330
automatic vending, 330

door-to-door selling, 331
online retailers, 332
wheel of retailing theory, 335
scrambled merchandising, 336
wholesaling, 339
wholesalers, 339
manufacturers' sales branches, 342
merchant wholesalers, 342
service wholesalers, 343
general merchandise wholesalers, 343
single-line (or general-line) wholesalers, 343
specialty wholesalers, 343
limited-function wholesalers, 344
cash-and-carry wholesalers, 344

drop-shippers, 344
truck wholesalers, 344
rack jobbers, 344
catalog wholesalers, 344
agent wholesalers, 345
manufacturers' agent, 345
export agents, 345
import agents, 345
brokers, 346
export brokers, 346
import brokers, 346
selling agents, 346
combination export manager, 346
auction companies, 346

QUESTIONS AND PROBLEMS

1. Review the Home Depot case study that opens the chapter. From this case, identify examples of different key terms and concepts covered in the chapter. For example, Home Depot is an example of a category killer.

2. Review the Home Depot case study that opens this chapter. Go to the HomeDepot.com website and consider the discussion in this chapter of online retailing and omnichannel. What else could Home Depot do to improve its online and offline shopping experience for customers?

3. Compare and contrast the marketing mix and target market for a bike shop in your community with the online bicycle retailer Performance Bicycle (www.performancebike.com). Use your best judgment to identify each retailer's primary target market and some of its Product, Place, Promotion, and Price decisions.

4. What sort of a "product" are specialty shops offering? What are the prospects for organizing a chain of specialty shops?

5. Discuss a few changes in the market environment that you think help explain why online retailing has been growing so rapidly.

6. Using your own shopping experience (or that of friends or family), use examples to explain advantages to online shopping and advantages to shopping in a brick-and-mortar store.

7. For each of the following products and target markets, explain which type of retailer or wholesaler you believe would be most appropriate to sell directly to: (a) shoes for college and professional basketball teams, (b) milk to families, (c) personal computers to first-time computer buyers, (d) high-end gaming computers to people who enjoy online gaming, (e) nuts and bolts to manufacturing firms.

8. What advantages does a retail chain have over a retailer who operates with a single store? Does a small retailer have any advantages in competing against a chain? Explain your answer.

9. Consider the evolution of wholesaling in relation to the evolution of retailing. List several changes that are similar and several that are fundamentally different.

10. Do wholesalers and retailers need to worry about new-product planning just as a producer needs to have an organized new-product development process? Explain your answer.

11. What risks do merchant wholesalers assume by taking title to goods? Is the size of this risk about constant for all merchant wholesalers?

12. Why would a manufacturer set up its own sales branches if established wholesalers were already available?

13. What is an agent wholesaler's marketing mix?

14. Why do you think many merchant wholesalers handle competing products from different producers, whereas manufacturers' agents usually handle only noncompeting products from different producers?

15. What alternatives does a producer have if it is trying to expand distribution in a foreign market and finds that the best existing merchant wholesalers won't handle imported products?

MARKETING PLANNING FOR HILLSIDE VETERINARY CLINIC

Appendix D (the Appendices follow Chapter 19) includes a sample marketing plan for Hillside Veterinary Clinic. Look through the "Marketing Strategy" section.

1. What kind of retail operation is the vet clinic? Does it fit any of the types described in this chapter?
2. How could Hillside make use of a website?
3. The marketing plan notes future plans to offer kennel (boarding) services and pet supplies. How will this change Hillside's current strategy? Does the marketing plan provide a good sense of what needs to be done? Do you have other recommendations for Hillside?

SUGGESTED CASES

11. Running On
12. DrV.com—Custom Vitamins
14. Schrock and Oh Design
15. The Scioto Group
16. Hanratty Company

Video Case 2. Suburban Regional Shopping Malls

MARKETING ANALYTICS: DATA TO KNOWLEDGE

CHAPTER 12: SELECTING CHANNEL INTERMEDIARIES

Art Glass Productions, a producer of decorative glass gift items, wants to expand into a new territory. Managers at Art Glass know that unit sales in the new territory will be affected by consumer response to the products. But sales will also be affected by which combination of wholesalers and retailers Art Glass selects. There is a choice between two wholesalers. One wholesaler, Giftware Distributing, is a merchant wholesaler that specializes in gift items; it sells to gift shops, department stores, and some mass-merchandisers. The other wholesaler, Margaret Degan & Associates, is a manufacturers' agent who calls on many of the gift shops in the territory.

Art Glass makes a variety of glass items, but the cost of making an item is usually about the same—$5.20 per unit. The items would sell to Giftware Distributing at $12.00 each—and in turn the merchant wholesaler's price to retailers would be $14.00—leaving Giftware with a $2.00 markup to cover costs and profit. Giftware Distributing is the only reputable merchant wholesaler in the territory, and it has agreed to carry the line only if Art Glass is willing to advertise in a trade magazine aimed at retail buyers for gift items. These ads will cost $8,000 a year.

As a manufacturers' agent, Margaret Degan would cover all of her own expenses and would earn 8 percent of the $14.00 price per unit charged the gift shops. Individual orders would be shipped directly to the retail gift shops by Art Glass using United Parcel Service (UPS). Art Glass would pay the UPS charges at an average cost of $2.00 per item. In contrast, Giftware Distributing would anticipate demand and place larger orders in advance. This would reduce the shipping costs, which Art Glass would pay, to about $0.60 per unit.

Art Glass' marketing manager thinks that Degan would be able to sell only about 75 percent as many items as Giftware Distributing—because she doesn't have time to call on all of the smaller shops and doesn't call on any department stores. On the other hand, the merchant wholesaler's demand for $8,000 worth of supporting advertising requires a significant outlay.

The marketing manager at Art Glass decided to use a spreadsheet to determine how large sales would have to be to make it more profitable to work with Giftware and to see how the different channel arrangements would contribute to profits at different sales levels.

CHAPTER THIRTEEN

Source: Geico

Promotion—Introduction to Integrated Marketing Communications

Back in the 1930s, during the depths of the Great Depression, Leo and Lillian Goodwin started the Government Employees Insurance Company—GEICO. GEICO kept operating costs low by selling auto insurance to only two low-risk target markets: federal employees and military personnel. GEICO passed on the savings in the form of lower premiums—and sales steadily grew for decades.

After becoming a wholly owned subsidiary of Berkshire Hathaway in 1996, GEICO's management sought to accelerate earnings growth by targeting new markets. However, achieving growth in the mature auto insurance market meant that GEICO would need to take customers away from better-known competitors such as Allstate and State Farm. As if that wasn't difficult enough, many prospects didn't even know about GEICO. Ted Ward, GEICO's vice president of marketing, discussed this situation with the firm's ad agency, the Martin Agency of Richmond, Virginia. Together they decided that an aggressive advertising campaign could achieve the objectives of increasing awareness of GEICO and bringing in new customers.

The GEICO campaign that emerged used an animated, talking lizard to help get attention and communicate the firm's message. In the first commercial, the charming reptile with the British accent stated, "I am a gecko, not to be confused with GEICO, which could save you hundreds on car insurance. So stop calling me!" The humorous ads quickly achieved GEICO's objectives: generating awareness and interest among target customers. The original plan was for the Gecko campaign to run for a short time, but research showed that customers loved the Gecko, and the company saw a bump in its sales. The Gecko continues to be an important part of GEICO's image and promotions.

GEICO wants customers to know it offers good value—great car insurance at low prices. But to prove this, customers must reach out for a price quote. It's quick and easy to get a quote at GEICO's website, so to spur the target audience to action, GEICO ads remind them that "15 minutes could save you 15 percent or more on car insurance." GEICO likes to have a variety of ads—and has had all kinds of animals and inanimate objects talking in its ads; pigs, camels, bumper stickers, and dinner plates star in its humorous television spots.

In addition, GEICO's promotion objectives include everyone driving safer. So its Smartdogs campaign featured dogs that were trained to stop people from using their smartphones while driving (no selfies behind the wheel!). A microsite (www.geico.com/distracted/) shows funny videos of the Smartdogs and demonstrates how to enable the "Do Not Disturb While Driving" mode on a smartphone. This campaign raises awareness for GEICO *and* makes the world a better place by discouraging distracted driving (#M4BW).

Although GEICO aims to reach its target market with advertising, it knows that many customers interested in insurance start the information gathering process on their own. GEICO knows when people type "auto insurance" into Google, they are probably looking to buy car insurance. GEICO uses various techniques to make sure GEICO appears near the top of the organic search results. Just in case customers don't see that, it also pays for a sponsored link (advertisement) ensuring that GEICO will join other ads at the top of the search results. GEICO can pay more than $75 each time a customer clicks on one of these ads, so keeping customers interested is important; the page customers land on after clicking one of these ads asks for their ZIP Code, the first step toward getting a price quote from GEICO.

All this promotion helped make GEICO a familiar name, but many insurance buyers still want to talk to a real person before deciding what to do. These customers visit with a GEICO salesperson at one of its local offices, or they call GEICO's national call center and talk with an inside sales rep. GEICO selects capable salespeople who are licensed insurance agents and trains them to develop an understanding of each customer's needs and concerns so that the agents can then persuasively explain GEICO's benefits to the customer.

Of course, GEICO seeks to build ongoing customer relationships after they sign up for a policy. Regular contacts and updates are handled with promotional e-mails. Later, if a customer who purchases a policy has a problem, GEICO's highly rated customer service team quickly works to resolve it. Customers can also turn to their GEICO Mobile App for roadside assistance or to talk to its artificial intelligence–powered virtual assistant "Kate." Kate answers questions about policy coverage, billing information, and more.

GEICO's website also fosters that relationship with current customers. The website is a type of "owned media"—GEICO owns and manages it—unlike television and radio *advertising* that GEICO pays to access. Many current customers use GEICO's website to manage and update their policy or make an insurance claim. They can also learn more about GEICO's other insurance products, various discounts, local gas prices, and more. The site also links to GEICO's social media pages.

GEICO's social media, including a blog, Facebook and Instagram pages, a YouTube channel, and a Twitter feed, also build relationships with its current customers and connect it with potential new customers. By following GEICO's Facebook page (or the Gecko's own page), customers can see the latest television ads and photos of the Gecko. Customers can also "Like" or "Share" an ad or other post with their own Facebook friends. Potential customers are more likely to consider GEICO when they have seen a friend "Like" the brand.

GEICO's promotion brings together advertising, personal selling, and online media; in combination, they help GEICO acquire and retain customers. That recipe—along with a great product—has moved GEICO's market share from 3 percent to 13 percent in just over 20 years. GEICO's successful advertising campaigns have inspired similar efforts from competitors, including Progressive ("Hi Flo!") and Allstate ("Yikes! It's Mayhem.").[1]

LEARNING OBJECTIVES

As the GEICO example shows, there are many decisions that a marketing manager must make concerning Promotion, and it is an important part of marketing strategy planning. Marketing managers usually blend a variety of different promotion methods to achieve promotion objectives because each method has its own strengths and limitations. In this chapter we introduce the major promotion options and how to integrate them into an effective whole.

When you finish this chapter, you should be able to

1. know the advantages and disadvantages of the promotion methods a marketing manager can use in strategy planning.
2. understand the integrated marketing communications concept and why firms use a blend of different promotion methods.
3. understand the importance of promotion objectives.
4. know how the traditional communication process affects promotion planning.
5. understand how customer-initiated interactive communication affects promotion planning.
6. know how typical promotion plans are blended to get an extra push from wholesalers and retailers, as well as help from customers in pulling products through the channel.
7. understand how promotion blends typically vary over the adoption curve and product life cycle.
8. understand how to determine how much to spend on promotion efforts.
9. understand important new terms (shown in **red**).

Promotion Communicates to Target Markets

LO 13.1

Promotion is communicating information between the seller and potential buyer or others in the channel to influence attitudes and behavior. The promotion part of the marketing mix involves telling target customers that the right Product is available at the right Place at the right Price. Promotion should be fine-tuned for specific target markets, fit with the other variables of the marketing mix, and reinforce the strategy's differentiation and positioning.

This is the first of four chapters that discuss issues pertinent to Promotion. We begin with an overview of the major promotion methods. Marketing managers can choose

Exhibit 13-1 Promotion and Marketing Strategy Planning

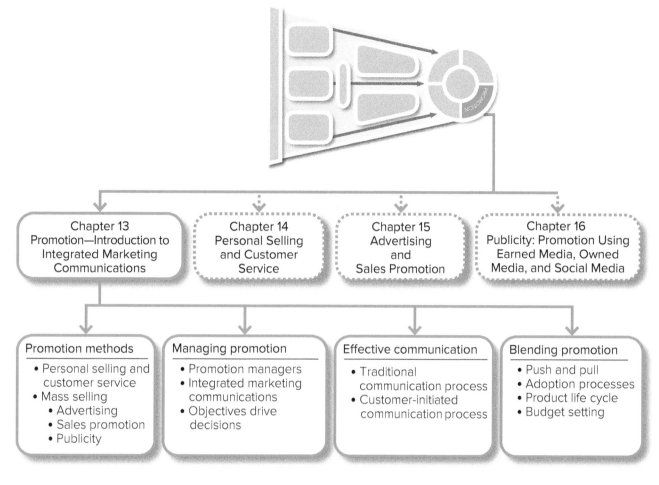

from several basic types of promotion: personal selling, mass selling, sales promotion, and publicity (see Exhibit 13-1). Because these methods have different strengths and limitations, a marketing manager typically uses them in combination to achieve specific objectives. We also discuss the specialists who are involved in managing different types of promotion and why it is important for them to work together as a team. This chapter also provides models that will help you understand how communication works—including when companies broadcast information to target markets and when target customers seek information on their own. Finally, we look at some concepts that help marketing managers develop the best promotion blend.

In Chapter 14, we take a closer look at the important promotion strategy decisions marketing and sales managers make in personal selling and customer service. Chapter 15 provides a closer look at advertising and sales promotion, and Chapter 16 examines publicity, which may occur through a range of nontraditional media including a company's website, social media, and more.

In this chapter we'll go into some detail about the different promotion methods—a key challenge for marketing managers is how best to blend them. It's helpful to begin with a brief overview of the promotion methods available.

Several Promotion Methods Are Available

Personal selling—flexibility is its strength

Personal selling involves direct spoken communication between sellers and potential customers. Customer service is a form of personal communication between a customer and seller to resolve a problem with a purchase. Salespeople get immediate feedback, which helps them adapt. Although some personal selling is included in most marketing

 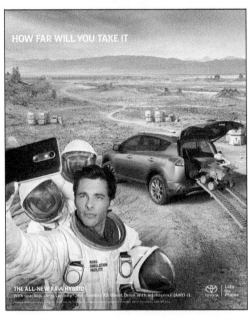 

Promotion is communicating information between the seller and potential buyer or others in the channel to influence attitudes and behavior. There are several promotion methods available. Personal selling involves direct spoken communication between the seller and potential customers. Salespeople can directly engage with individual customers and answer their specific questions. Advertising is any paid form of nonpersonal presentation of ideas, goods, or services by an identified sponsor. It is a form of mass selling that involves sending the same message out to a large group of potential customers at the same time. Toyota wants to let target market customers know that its new RAV4 Hybrid can go anywhere a customer wants to take it. Sales promotion includes activities other than advertising, publicity, and personal selling that seek to stimulate interest, trial, or purchase. HelloFresh wants customers to try its meal delivery service. It offers $30 off a first and second box, then $10 off a third and fourth box. This promotion encourages customers to give the service a long enough trial to make using it a habit.
(left): Tyler Olson/Shutterstock; (middle): Source: Toyota Motor Sales, U.S.A., Inc.; (right): Source: HelloFresh

mixes, it can be very expensive. So it's often desirable to combine personal selling with advertising, sales promotion, and/or publicity.

Mass selling involves advertising and publicity

Mass selling is communicating with large numbers of potential customers at the same time. It's less flexible than personal selling, but when the target market is large and scattered, mass selling can be less expensive.

Advertising is a primary form of mass selling. **Advertising** is any *paid* form of nonpersonal presentation of ideas, goods, or services by an identified sponsor. It includes the use of traditional media such as magazines, radio and TV, signs, and direct mail as well as new media such as podcasts, Facebook, and Snapchat. Marketing managers pay for advertising to be placed on specific media.

Sales promotion tries to spark immediate interest

Sales promotion refers to promotion activities—other than advertising, publicity, and personal selling—that stimulate interest, trial, or purchase by final customers or others in the channel. Sales promotion may be aimed at consumers, at intermediaries, or at a firm's own employees. Examples include contests and coupons aimed at consumers, trade shows or calendars for wholesalers or retailers, or sales contests and meetings aimed at a company's own sales force. Relative to other promotion methods, sales promotion can usually be implemented quickly and get results sooner. In fact, most sales promotion efforts are designed to produce immediate results.

Publicity avoids media costs

Publicity is any *unpaid* form of nonpersonal presentation of ideas, goods, or services. Of course, publicity people are paid. But they try to attract attention to the firm and its offerings *without having to pay media costs*. Publicity includes a wide range of different types of media: a company's website and the material it posts on the website, viral videos, word-of-mouth communication, a company's Facebook page, and its "tweets" on

Twitter. It can also include coverage it receives in the press—for example, when a movie or restaurant is reviewed in the newspaper.

Less is spent on advertising than personal selling or sales promotion

Many people incorrectly believe that promotion money gets spent primarily on advertising—because advertising is all around them. But all the special sales promotions—coupons, sweepstakes, trade shows, and the like—add up to even more money. Similarly, much personal selling happens in the channels and in other business markets. In total, most firms spend less money on advertising than on personal selling or sales promotion.

Someone Must Plan, Integrate, and Manage the Promotion Blend

LO 13.2

Each promotion method has its own strengths and weaknesses. In combination, they complement one another. Each method also involves its own distinct activities and requires different types of expertise. As a result, it's usually the responsibility of specialists—such as sales managers, advertising managers, sales promotion managers, public relations managers, and social media managers—to develop and implement the detailed plans for the various parts of the overall promotion blend.

Sales managers manage salespeople

Sales managers are concerned with managing personal selling. Often the sales manager is responsible for building good distribution channels and implementing Place policies. In smaller companies, the sales manager may also act as the marketing manager and be responsible for advertising and sales promotion.

Advertising managers work with ads and agencies

Advertising managers manage their company's mass-selling effort—in television, newspapers, magazines, online, and advertising on social media sites. Their job is choosing the right media and messaging and to design those into advertising. Advertising may be handled in-house, by departments existing within the firms, or it may be contracted to outside advertising agencies.

Sales promotion managers need many talents

Sales promotion managers manage their company's sales promotion effort. In some companies, a sales promotion manager has independent status and reports directly to the marketing manager. If a firm's sales promotion spending is substantial, it probably *should* have a specific sales promotion manager. Sometimes, however, the sales or advertising departments handle sales promotion efforts—or sales promotion is left as a responsibility of individual brand managers. Regardless of who the manager is, sales promotion activities vary so much that many firms use both inside and outside specialists.

Publicity may be managed by many

An advertising manager may handle publicity, but in larger firms there may be someone who manages **public relations**—communication with noncustomers, including the press, labor, public interest groups, stockholders, and the government. A *social media manager* may be in charge of a company's social media (Facebook page, Instagram, LinkedIn page and posts, etc.) and possibly its website. Any of these jobs may be outsourced—though a specialist that manages public relations probably differs from one that manages social media or a firm's website.

Marketing managers talk to all, blend all

Although many specialists may be involved in planning for and implementing specific promotion methods, determining the blend of promotion methods is a strategy decision—and it is the responsibility of the *marketing manager*. With all of the promotion options, determining the best blend is a challenging responsibility.

The marketing manager weighs the pros and cons of the various promotion methods and then devises an effective promotion blend—fitting in the various departments and personalities and coordinating their efforts. Then the advertising, sales, and sales promotion managers should develop the details consistent with what the marketing manager wants to accomplish.

Starbucks intentionally coordinates a consistent message that it communicates to its target customers. Starbucks' advertising, its app, and its baristas convey a message that the brand represents a high-quality food and beverage experience.
(top left and right): Source: Starbucks Coffee Company; (bottom left): Anton Novoderezhkin/TASS/Getty Images

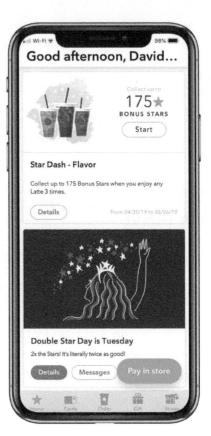

Send a consistent and complete message with integrated marketing communications

Effective blending of all the firm's promotion efforts should produce **integrated marketing communications**—the intentional coordination of every communication from a firm to a target customer to convey a consistent and complete message.

The GEICO case at the start of this chapter is a good example of integrated marketing communications. Different promotion methods handle different parts of the job. Yet the methods are coordinated so that the sum is greater than the parts. The separate messages are complementary, but also consistent.

It seems obvious that a firm's different communications to a target market should be consistent. However, when a number of different people are working on different promotion elements, they are likely to see the same big picture only if a marketing manager ensures that it happens. Achieving consistency is harder when different firms handle different aspects of the promotion effort. For example, different firms in the channel may have conflicting objectives.

To get effective coordination, everyone involved with the promotion effort must clearly understand the plan for the overall marketing strategy. They all need to understand how each promotion method will contribute to achieve specific promotion objectives.[2]

Which Methods to Use Depends on Promotion Objectives

LO 13.3

Overall objective is to affect behavior

A marketing manager usually has to set priorities for the promotion objectives. The ultimate objective is to encourage customers to choose a *specific* product. However, which promotion objectives are of highest priority will depend on the market situation and target market. For example, as we saw in Chapters 5 and 6, customers often move along a step-by-step buying path—and the path may differ for different types of purchases. Sometimes customers are familiar with the product, and sometimes it is completely new to them. Sometimes customers are buying a product for the very first time, and sometimes they have bought it many times before. Objectives should be guided by what we know about target customers. In this section, we discuss different types of promotion objectives and tie them to frameworks describing the purchase process.

Informing, persuading, and reminding are basic promotion objectives

Promotion objectives must be clearly defined—because the right promotion blend depends on what the firm wants to accomplish. It's helpful to think of three basic promotion objectives: *informing, persuading,* and *reminding* target customers about the company and its marketing mix. All try to affect buyer behavior by providing customers with information.

It's also useful to set more specific promotion objectives that state *exactly whom* you want to inform, persuade, or remind, and *why.* This is unique to each company's strategy—and specific objectives vary by promotion method. We'll talk about more specific promotion objectives in Chapters 14, 15, and 16. Here we'll focus on the three basic promotion objectives and then link them to the adoption process and a new model.

Informing is educating

Potential customers must know something about a product if they are to buy it. A firm with a really new product may not have to do anything but inform consumers of the product's features and benefits. An *informing* objective can show that it meets customer needs better than other products can. A small business might want to have a service regularly clean its offices. The owner might begin by searching on the Internet. There she might find a local cleaning company's website with information about various services.

Some companies use informing to indirectly support their brand. Educational content is useful to customers, so they are likely to search for it, engage with it, and maybe share it with friends. For example, Fidelity Investments has many articles and videos that offer advice on investing more generally—not just in Fidelity's products. As another example, Australian online fashion retailer Showpo's YouTube channel regularly posts fun tutorial videos on topics from "How to Dress Like Ariana Grande" to "10 Ways to Nail the Bike Shorts Trend." They also slip in videos featuring Showpo products, like "We're taking fashion [to the] next level . . . a dress that changes colour!" The videos all have a common thread—they appeal to Showpo's target market and keep them coming back for more.

Persuading usually becomes necessary

When competitors offer similar products, the firm must not only inform customers that its product is available but also persuade them to buy it. A *persuading* objective means the firm will try to develop a favorable set of attitudes so customers will buy, and keep buying, its product. A persuading objective often tries to demonstrate how one brand is better than others. To convince consumers to buy Brawny paper towels, ads show Brawny as the towel that's best for tough cleanup jobs. A salesperson for Andersen Windows tries to convince home builders about the quality and affordability of Andersen Windows as compared to those of a competitor, so the builder will choose Andersen for future housing projects.

Reminding may be enough

If target customers already have positive attitudes about a firm's marketing mix—or a good relationship with a firm—a *reminding* objective might be suitable. Customers who have been attracted and sold to once are still targets for competitors' appeals. Reminding them of their past satisfaction may keep them from shifting to a competitor. An accountant working for a small local firm might phone her customers once every few months to "check in" and see if they have any questions. This serves as a reminder that the accountant is there and available.

Promotion objectives relate to adoption processes

In Chapter 5 we looked at consumer buying as a step-by-step problem-solving process through which buyers go on the way to adopting (or rejecting) an idea or product. The three basic promotion objectives relate to these steps (see the first two columns in

The objective of the advertisement on the left is to educate customers by telling them chocolate milk comes "With natural protein, carbs, and electrolytes. That's real recovery backed by real science." Fans of the Milwaukee Bucks professional basketball team follow the team on Instagram. The team can use its Instagram to remind customers about the team's activities.
(left): Source: America's Milk Companies; (right): Source: Milwaukee Bucks/Instagram

Exhibit 13-2). *Informing* and *persuading* may be needed to affect the potential customer's knowledge and attitudes about a product and then bring about its adoption. Later, promotion can simply *remind* the customer about that favorable experience and confirm the adoption decision.

The AIDA model is a practical approach

The basic promotion objectives and adoption process fit very neatly with another action-oriented model—called AIDA—that we will use in this chapter and in Chapters 14, 15, and 16 to guide some of our discussion. The **AIDA model** consists of four promotion jobs: (1) to get *Attention,* (2) to hold *Interest,* (3) to arouse *Desire,* and (4) to obtain *Action.*

The first and third columns in Exhibit 13-2 show the relationship of the promotion objectives and adoption process to the AIDA jobs. Getting attention is necessary to make consumers aware of the company's offering. Holding interest gives the communication a chance to build the consumer's interest in the product. Arousing desire affects the evaluation process, perhaps building preference. And obtaining action includes gaining trial, which may lead to a purchase decision. Continuing promotion is needed to confirm the decision and encourage an ongoing relationship and additional purchases.

British marketers for Pampers disposable diapers generated attention and interest with TV ads that showed the world from a baby's perspective. To encourage desire and action, they used creative in-store and point-of-purchase advertising. For example, on the doors of restrooms with baby-changing facilities, fake doorknobs were placed unreachably high, with the message: "Babies have to stretch for things. That's why they like the extra

Exhibit 13-2 Relation of Promotion Objectives, Adoption Process, AIDA Model, and Organizational Buying Model

Promotion Objectives	Adoption Process	AIDA Model	Organizational Buying Model
Informing	Awareness Interest	Attention Interest	Defining the problem
Persuading	Evaluation Trial	Desire	Making the decision
Reminding	Decision Confirmation	Action	Managing the buyer-seller relationship

What's Next? Once upon a time . . .

Does telling a story get customers' attention and interest? Research finds that people remember stories more than statistics. Another study discovered that the most popular Super Bowl ads follow a classic story pattern: exposition, rising action, climax, falling action, and dénouement (resolution). Budweiser's "Lost Dog" Super Bowl ad follows this arc. The ad begins on a ranch, showing the connection between a puppy and one of the Budweiser Clydesdale horses (exposition). When the puppy gets lost in the city, the rancher posts "lost dog" posters (rising action). The puppy journeys toward home, but with the ranch in the distance, we see a wolf stalking the pup (more rising action). Back at the ranch, the Clydesdales hear the pup, so they escape the barn, coming to the puppy's rescue (climax). The Clydesdales and puppy run home together (falling action) and the muddy puppy gets a bath (dénouement). That one-minute story won the *USA Today* Super Bowl Ad Meter—a survey of viewers—and generated lots of attention for Budweiser.

Other stories get customers' interest because they explain how a company solves problems. This approach is behind many B2B firms' use of case studies that describe how they solved a specific customer's problems. Networking hardware giant Cisco Systems' "Never Better" campaign creates digital stories that combine video, photography, and text to tell powerful stories that provide emotional resonance. The documentary style demonstrates how Cisco's technology helps save rhinos (https://goo.gl/cDGNIz), makes cities smarter (https://goo.gl/Vadxsp), and more.

Founding stories can help a brand communicate its values and positioning. One construction company shares the story of how its founder would pick up two employees in his battered pickup truck, after getting up before his two little girls woke up, drinking a steaming cup of coffee. Rain or snow, he met workers at the gate each morning. The details—steaming hot coffee, little girls, battered pickup truck, waving to employees at that gate—created memorable images and let customers (and employees) know what this company stood for.

What's Next? A full-length feature movie? No, wait, LEGO has already done that—five times now! Maybe we can expect a mystery novel from construction machinery company Caterpillar. Probably not soon—but you should see its Jenga ad (https://youtu.be/DWc8dUl7Xfo). Stories that resonate with target markets grab attention and interest; many are passed on to friends and family. But it isn't easy coming up with a good story. Do you remember the 2013 flop *Movie 43*? Almost no one does. And marketing stories that fail to resonate will not help a brand either.[3]

Describe a story your college or university could tell that might resonate with prospective students. How could that story be turned into a short television ad or YouTube video?

---

stretchiness of Pampers Active Fit." And on store shelves, in a play on babies' disobedient nature, pull-out Pampers information cards were marked "Do Not Pull."[4]

Promotion objectives and B2B

In Chapter 6 we showed how organizational buyers go through a similar step-by-step process on the way to purchase. The three steps in our organizational buying model also correspond to the three basic promotion objectives; see Exhibit 13-2.

Stories grab attention and hold interest

While there is growing interest in data and analytics among marketing managers, they have not forgotten the power of a good story in getting customers' attention, holding their interest, and arousing desire. Many managers believe that *What's Next?* is actually a return to good storytelling—see *What's Next?* Once upon a time. . . .

#M4BW

Story can draw attention to global problems (and a brand)

Lifebuoy soap used storytelling to focus attention all over the world on an important problem. Each year diarrhea and pneumonia claim the lives of two million children before they reach age five. Research shows that regular handwashing with soap significantly cuts the risk of diarrhea and infection. Lifebuoy soap wanted to get this message to people in third-world countries, so it created its "Help a Child Reach 5" campaign. As a demonstration, Lifebuoy adopted Thesgora, a village in rural India, and showed how handwashing lowered diarrhea incidence from 36 percent to 5 percent. Lifebuoy created a three-minute video story (see https://youtu.be/UF7oU_YSbBQ) showing a father walking on his hands through town to a religious shrine to offer thanks. We then see that this is his son's fifth birthday—a milestone his previous children never reached. This is also the Lifebuoy story—the essence of the brand—that handwashing can save lives.[5]

These two images are part of an advertisement in the social media app Snapchat. They demonstrate elements of the AIDA model. The image on the left shows a camel chewing his cud. It is funny and grabs attention. When the viewer clicks "READ" at the bottom of this screen, the image on the right appears. The headline, photo, and details below hold interest and arouse desire for a trip to Dubai—hopefully on an Emirates flight.
Source: Emirates/Snapchat

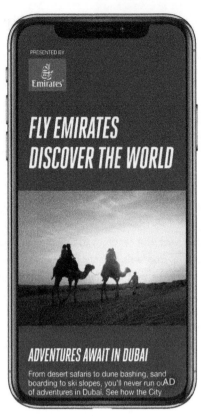

Promotion Requires Effective Communication

LO 13.4

Communication starts with a trusted source and a receiver

Promotion is wasted when it doesn't communicate effectively. There are many reasons why a promotion message can be misunderstood or not heard at all. To understand this, it's useful to think about a whole **communication process**—which means a source trying to reach a receiver with a message. Exhibit 13-3 shows the elements of the communication process. Here we see that a **source**—the sender of a message—is trying to deliver a message to a **receiver**—a potential customer. These elements can be used as a diagnostic tool, where a marketing manager can think about the various factors affecting a marketing communication, and serves as a reminder of everything that has to go right to get a message across.

Customers evaluate the source of the message in terms of trustworthiness and credibility. For example, American Dental Association (ADA) studies show that Listerine mouthwash helps reduce plaque buildup on teeth. Listerine mentions the ADA endorsement in its promotion to help make the promotion message credible. If a salesperson (source) is not trusted, then the message being delivered will not be used by the receiver.

Exhibit 13-3 The Traditional Communication Process

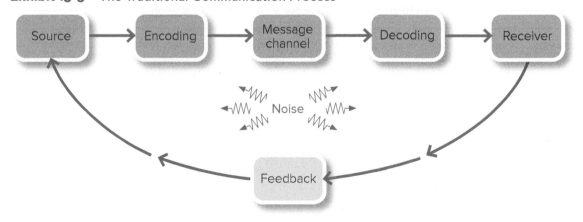

A major advantage of personal selling is that the source—the seller—can get immediate feedback from the receiver. It's easier to judge how the message is being received and to change it if necessary. Mass sellers usually must depend on marketing research or total sales figures for feedback—and that can take too long. Many marketers include toll-free telephone numbers and website addresses as ways of building direct-response feedback from consumers into their mass-selling efforts.

Exhibit 13-4 This Same Message May Be Interpreted Differently

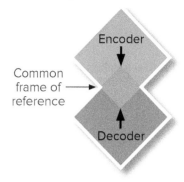

Noise distracts the message

Noise—shown in Exhibit 13-3—is any distraction that reduces the effectiveness of the communication process. Conversations and snack-getting during TV ads are noise. An industrial buyer reading a text message during a salesperson's presentation is noise. Advertisers who plan messages must recognize that many possible distractions—noise—can interfere with communications.

Encoding and decoding depend on a common frame of reference

The basic difficulty in the communication process occurs during encoding and decoding. **Encoding** is the source deciding what it wants to say and translating it into words or symbols that will have the same meaning to the receiver. **Decoding** is the receiver translating the message. This process can be very tricky. The meanings of various words and symbols may differ depending on the attitudes and experiences of the two groups. People need a common frame of reference to communicate effectively (see Exhibit 13-4).

Pepsi encountered this problem with an advertisement featuring fashion model Kendall Jenner. In the ad, Jenner is involved in a photo shoot but becomes distracted by a passing group of protesters. She joins the protest and later brings a smile to the face of a stoic police officer by giving him a can of Pepsi. While Pepsi tried to send a message of unity, peace, and understanding, it missed the mark in the eyes of many viewers and was roundly criticized in the media. Many thought that it downplayed and commercialized protests on important social issues. The message Pepsi thought it encoded was not the one decoded by many in its target market.[6]

The same message may be interpreted differently

Different audiences can interpret a message differently. Such differences are common in international marketing where cultural differences and translation are problems. KFC's "finger-lickin' good"[7] chicken turned into "eat your fingers off" when first translated to Chinese. The Swedish brand Samarin thought it would avoid these problems with ads showing three simple drawings—and no words. The first picture shows a man holding his stomach in obvious pain, in the second he is drinking Samarin, and in the last he is smiling and happy. The ads worked great—except in the Middle East where people read from right to left. Uh oh! Many firms run into problems like this.[8]

Message channel is important too

The communication process is complicated even more because the message is coming from a source through some **message channel**—the carrier

The message of "unity, peace, and understanding" that Pepsi tried to encode in this advertisement was decoded differently by many viewers.
Source: PepsiCo

of the message. A source can use many message channels to deliver a message. The salesperson does it in person with voice and action. Advertising must do it with media such as magazines, TV, e-mail, or Internet websites. A particular message channel may enhance or detract from a message. A TV ad, for example, can *show* that Dawn dishwashing detergent "takes the grease away"; the same claim might not be as convincing (or even opened) if it arrives in a consumer's e-mail.

Feedback takes many forms

The last element in the traditional communication process is *feedback*—communication from the receiver back to the source. Feedback may take many different forms: a customer may simply have a different attitude, seek more information, visit a store, or purchase the product. For this reason, objectives should state the desired feedback, and marketing managers should measure whether the communication is having the anticipated response.

Integrated direct response seeks immediate feedback

Sometimes marketing managers want to get immediate feedback from *specific* customers. This prompts firms to turn to direct marketing—direct communication between a seller and an individual customer using a promotion method other than face-to-face personal selling. Most direct marketing communications are designed to prompt immediate feedback—a direct response—by customers. That's why this type of communication is often called *direct-response promotion*.

Direct mail and e-mail are two tools commonly used for direct-response promotion. A carefully selected mailing list—perhaps from the firm's customer relationship management (CRM) database—allows advertisers to reach customers with specific interests. For the best response, marketing managers carefully segment their customers and use mail or e-mail to deliver targeted messages.

E-mail can be low cost and targeted

Most customers use e-mail these days, and communicating via e-mail offers benefits to buyers and sellers. Sellers like the relatively low cost compared to direct mail. E-mails give customers a chance to click through to photos and videos. Media tools such as these help them decide whether to make a donation or place an order.

Of course, getting to the message in the e-mail necessitates a customer opening the e-mail. Personalized and creative messages can get customers' attention in a crowded e-mail inbox. To get opened, an e-mail subject line should grab target customers' attention. For example, when a customer buys eyeglasses at Warby Parker, the retailer asks when the prescription expires. Two weeks before it expires, you get an e-mail with the subject line: "Uh-oh, your prescription is expiring." This catches the reader's attention—the subject line is true—and increases the odds of the e-mail being opened.[9]

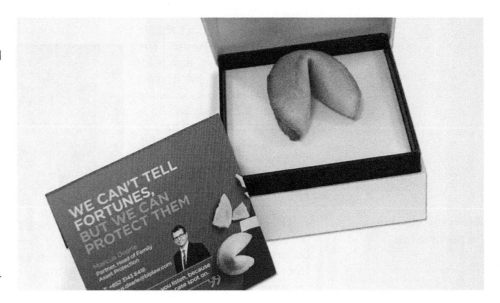

The law firm Bryan Cave Leighton Paisner (BCLP) wanted to launch a new financial services product for a target market of advisers to high-net-worth individuals in Asia (so this was a B2B, not a B2C, campaign). The target market was limited in size and easily identified—so BCLP used direct marketing. It sent each adviser the simple message and a fortune cookie to drive that message home. The campaign generated a 1,400 percent return on investment.
Source: Bryan Cave Leighton Paisner

When Customers Initiate the Communication Process

LO 13.5

The traditional process in Exhibit 13-3 assumes that it's the seller ("source") who initiates communication. Let's look closer at another communication model, shown in Exhibit 13-5, where a customer ("receiver") initiates the communication process. For example, a consumer enters "pizza delivery" into Google or asks a retail salesperson for help, or organizational buyers phone salespeople to ask questions or request bids. The growth of the Internet makes it more common for buyers to start the information gathering process. Although this "customer-initiated" process has many of the same components as the traditional process we considered earlier, the differences and implications for Promotion are significant.

Customer initiates communication with a search process

In the process in Exhibit 13-5, a customer (*receiver*) initiates communication with a decision to *search* for information in a particular message channel. The most common and far-reaching *message channel* used for information searching is the Internet—usually queried through a search engine such as Google or Bing and accessed by a personal computer, tablet, or cell phone. Sometimes a buyer immediately links to a particular seller's web page to seek out information. The message channel is still the carrier of the message, as was the case before, but searchable message channels usually feature an archive of existing messages on a number of topics. There may be many available topics—even millions.

Customer decides how much information to get

The receiver then reviews and screens the various options and decides which messages to pursue. A search engine returns a list of results that includes paid advertisements and free listings. A search can also occur at the website of a producer, retailer, or wholesaler. For example, a business customer may visit a distributor of janitorial supplies and search for industrial cleaning solutions—and then click on pricing information or perhaps the safety of the ingredients. Similarly, a consumer might visit Netflix and search for horror movies and then look for the director, actors, or the comments of people who have already seen the movie. The customer chooses the information of interest to her—not the marketing manager.

Marketers must grab attention to be selected

Because a customer now has so many options, it's important for a marketer to be among the first to grab that customer's attention. That means that an online retailer, like Backcountry, which sells outdoor clothing, wants to be near the top of the search results when someone searches for "Patagonia flannel shirt." In this case,

Exhibit 13-5 A Model of Customer-Initiated Interactive Communication

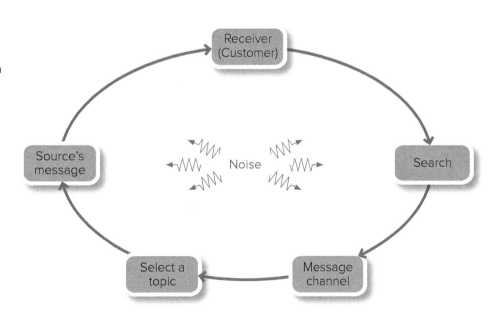

Backcountry pays a search engine company to put a sponsored (advertising) link near the top of the list or make changes to its website so it appears at the top of search results.

When the receiver looks more closely at the information on-screen, he decides whether to stay or leave the site. Usually more information is just a click away—and that click can be to dig deeper into the marketer's site or to click away to another website. Marketing managers need to make sure that the page holds the receiver's interest or begins to arouse desire. Noise can still be a problem as well—and lead the searcher in a different direction. For example, a confusing website may make it difficult to figure out a retailer's return policy and discourage a customer from buying.

As this short scenario depicts, consumers can collect information from a wide range of sources along the journey to making a purchase decision. Some of those information sources are outside the control of a marketing manager. That doesn't mean those sources should be ignored. Marketing managers can still make sure that they have plenty of product pictures—and that they are easy to post on Pinterest and Instagram. They can also monitor what is being said about the brand online. Whereas online compliments let the firm know what it is doing well, complaints can highlight unmet customer needs—and give the company a chance to turn things around.

Promotion timing and relevance have an impact

Communication that customers receive during the customer-initiated communication process is often timely—it frequently occurs when a customer is actively seeking information related to a buying decision. Consequently, customers more readily pay attention and are more likely to be interested in promotional information.

When customers surf the Internet, they often signal they are ready to make a purchase. For example, reading articles about different kinds of bicycles, visiting a couple of websites, and "Liking" the local bike shop on Facebook might all be indications of an impending bicycle purchase. Those actions tell Google and Facebook this customer may be buying a bike soon. A bike maker or local bike shop would like to target communication at this customer now—before she makes a purchase. Customers often leave

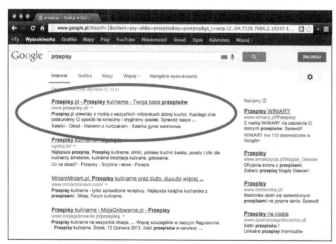 

Understanding customers' online search behavior helps brands develop websites that attract target customers. Polish women love to search for and share recipes—the best-selling magazine in Poland is all about recipe sharing. International food brand Knorr wanted to appear more local and trustworthy to Polish consumers, so it decided to become the country's best source of recipes and cooking ideas. Knorr found that the Polish word for recipe, *przepisy*, was available as a URL. Knorr then built a branded website, a Facebook page, and mobile apps around recipe sharing. Whenever someone searches Google.pl for "przepisy" (recipe), Knorr's site comes up on top—helping drive more traffic to the branded site. As online recipe sharing grows, so do visitors to the site. The average visitor spends almost eight minutes on the site, the Facebook page has more than a half million "Likes," and the mobile app has been downloaded more than 300,000 times. Most important, the Knorr brand has closed the gap on its local competition with respect to trustworthiness and brand image.
(left): Source: Google; (right): Source: Unilever

telltale signs of buying interest. Marketing managers who can identify target customers already interested in buying their products can then deliver timely and relevant online (and offline) advertising to these customers.

Prepare for customers who initiate communication

When customers have the power to decide where to click next, marketing managers need to create a different kind of promotion. The first challenge is for marketing managers to figure out how to appear high on target customers' searches. Second, the marketing managers must have truly useful information—or customers will quickly click to something that will help them make a better purchase decision. And finally, promotion needs to keep them there to learn more about the seller's offering.

A customer searching for tips on packing for a safari in South Africa's popular Kruger National Park is likely to find this blog post by the Discover Africa travel agency.
Source: Discover Africa Blog

We will discuss specific tools that can be used when we cover publicity in Chapter 16. Until then, an example will demonstrate what we mean by capturing and retaining the inbound customer. The travel agency Discover Africa creates content that attracts customers to its website and keeps them coming back. Discover Africa started by identifying nine different target markets that it thought it could serve (for example, "Mr. Business," "The Campers," and "The Student/Backpacker"). Then Discover Africa created articles, blog posts, videos, and travel guides for its website—each designed to appeal to one or more target groups. With a little advertising of the discoverafrica.com website and its content, others started to notice and link to the site. This helps Discover Africa appear near the top of online searches on travel in Africa. Discover Africa was soon viewed as an authority on African travel, more customers sought out the website's information, and sales climbed.[10]

Ethical issues in marketing communications

Promotion is one of the most frequently criticized areas of marketing. Many criticisms focus on whether communications are honest and fair. Marketers must sometimes make ethical judgments in considering these charges and in planning their promotion.

For example, we often look to the source of a communication to decide whether we believe it is credible. When a TV news program broadcasts a video publicity release, consumers don't know it was prepared to achieve marketing objectives. They think the news staff is the source. That may make the message more credible, but is it fair? Many say yes—as long as the publicity information is truthful. Similar concerns are raised about the use of celebrities in advertisements. A person who plays the role of an honest and trustworthy person on a popular TV series may be a credible message source in an ad, but is using such a person misleading to consumers? Some critics believe it is. Others argue that consumers recognize advertising when they see it and know celebrities are paid for their endorsements.

The most common criticisms of promotion relate to exaggerated claims. If an advertisement or a salesperson claims that a product is the "best available," is that just a personal opinion or should every statement be backed up by proof? What type of proof should be required? Some promotions do misrepresent the benefits of a product. Customers look to online reviews for objective opinions. What about posting fake favorable reviews for your company or its products—or negative reviews about a competitor? Real reviews offer a way for consumers to see beyond potential exaggerated claims. But if consumers feel that at least some reviews are phony, they may distrust all reviews. Some people believe that the Internet will eventually expose dishonest players in the market. This question is not always easy to answer; see the following Ethical Dilemma.

> **Ethical Dilemma**
>
> *What would you do?* A friend of your family owns an upscale Italian restaurant called Giupetto's. Because you are a student of marketing, he asks for your help. Business has suddenly fallen off—their "regulars" are still coming in, but they don't see as many new customers as before. His cook has heard that the owner of a competing restaurant has been pressuring his employees to post negative reviews of Giupetto's food, service, and prices. When you check the website, there are a number of unfavorable, anonymous reviews. You also notice that there are about 30 very upbeat reviews of the competing restaurant. Your friend wants you to help give the competitor "what he deserves" and write some negative reviews about that restaurant.
>
> *Would you do what he asks? Why or why not? What else could you do?*

Most marketing managers realize that the ultimate proof comes when the customer makes the purchase. Customers won't spread positive word-of-mouth or come back if the marketing mix doesn't deliver what the promotion promises. As a result, most marketing managers work to make promotion claims specific and believable.[11]

How Typical Promotion Plans Are Blended and Integrated

LO 13.6

There is no one right blend

There is no one *right* promotion blend for all situations. Each one must be developed as part of a marketing mix and should be designed to achieve the firm's promotion objectives in each marketing strategy. So let's take a closer look at typical promotion blends for different situations.

Get a push in the channel with promotion to intermediaries

When a channel of distribution involves intermediaries, their cooperation can be crucial to the success of the overall marketing strategy. **Pushing** (a product through a channel) means using normal promotion effort—personal selling, advertising, and sales promotion—to help sell the whole marketing mix to possible channel members. See Exhibit 13-6. This approach emphasizes the importance of securing the wholehearted cooperation of channel members to promote the product in the channel and to the final user.

Producers usually take on much of the responsibility for the pushing effort in the channel. However, wholesalers often handle at least some of the promotion to retailers. Similarly, retailers often handle promotion in their local markets. The overall effort is most likely to be effective when all of the individual messages are carefully integrated.

Promotion to intermediaries emphasizes personal selling and sales promotion

Salespeople usually handle most of the important communication with wholesalers and retailers. These clients don't want empty promises. They want to know what they can expect in return for their cooperation and help. A salesperson can answer questions about what promotion will be directed toward the final consumer, each channel member's part in marketing the product, and important details on pricing, markups, promotion assistance, and allowances. A salesperson can also help the firm determine when it should adjust its marketing mix from one intermediary to another.

When suppliers offer similar products and compete for attention and shelf space, intermediaries usually pay attention to the one with the most profit potential. Sales promotions targeted at intermediaries often focus on short-term arrangements that will improve the intermediary's profits. For example, a soft drink bottler might offer a convenience store a free case of drinks with every two cases it buys. The free case improves the store's profit margin on the whole purchase. And the increased inventory encourages the convenience store to get behind selling the soft drinks.

Exhibit 13-6 Promotion May Encourage Pushing in the Channel, Pulling by Customers, or Both

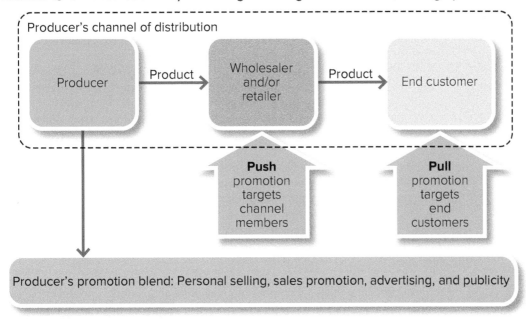

Producers will also run ads in trade magazines to recruit new intermediaries or to inform channel members about a new offering. Trade ads usually encourage intermediaries to contact the supplier for more information or visit its website, where they can gather information and perhaps fill out a contact form. A salesperson would usually take over from there.

Push within a firm—with promotion to employees

Some firms emphasize promotion to their own employees—especially salespeople or others in contact with customers. This type of *internal marketing* effort is basically a variation on the pushing approach. One objective of an annual sales meeting is to inform reps about important elements of the marketing strategy—so they'll work together as a team to implement it. Some firms use promotion to motivate employees to provide better customer service or achieve higher sales.

#M4BW Promotion to employees is helpful in a service company, where the quality of employees' efforts is a big part of the product. When the accounting firm KPMG wanted its employees to feel a greater sense of purpose in their work, it instituted an internal marketing campaign. On this poster, the smaller text reads "When family farms and ranches need loans, KPMG works with the credit system to help secure them, furthering America's proud tradition of family farming." Other posters noted how KPMG was helping combat terrorism and championing democracy. Feeling a greater sense of purpose, employees were rededicated to solving customers' problems.[12]
Source: KPMG International Cooperative

Pulling policy—customer demand pulls the product through the channel

Most producers also focus a significant amount of promotion on customers at the end of the channel. This helps stimulate demand and pulls the product through the channel of distribution. Intermediaries want to carry products that customers are ready to buy. **Pulling** means getting customers to ask intermediaries for the product. Pulling is promotion targeted at end customers (see Exhibit 13-6).

Pulling and pushing are usually used in combination. However, if intermediaries won't work with a producer—perhaps because they're already carrying a competing brand—a producer may try to use a pulling approach by itself. This involves a more aggressive promotion effort to final consumers or users—perhaps using coupons or samples—temporarily bypassing intermediaries. If the promotion works, the intermediaries are forced to carry the product to satisfy customer requests. However, this approach is risky. Customers may lose interest before reluctant intermediaries make the product available. At a minimum, intermediaries should be told about the planned pulling effort—so they can be ready if the promotion succeeds.

Who handles promotion to final customers at the end of the channel varies in different channel systems depending on the mix of pushing and pulling. Further, the promotion blend typically varies depending on whether customers are final consumers or business users.[13]

Promotion to final consumers

For most products, the large number of consumers almost forces producers of consumer products and retailers to emphasize advertising, sales promotion, and publicity. Sales promotion—such as coupons, contests, or free samples—builds consumer interest and short-term sales of a product. An informative website that includes customer reviews and video demonstrations helps customers build favorable impressions of a company and its products. Effective mass selling may build enough brand familiarity so that little personal selling is needed, as in self-service and discount operations.[14]

Personal selling can be effective too. Often mass selling brings customers to a store where personal selling may be used. Personal selling to final consumers is often found in channel systems for more expensive products, such as those for financial services, furniture, fine jewelry, consumer electronics, designer clothing, and automobiles.

Green Mountain Coffee uses both push and pull to support its coffees. Green Mountain's website and membership program promote a "Deal of the Week" directly to consumers. The ad on the right targets purchasing agents at grocery stores, letting them know that Green Mountain's family of brands will bring volume and profits to their stores.
Source: Keurig Green Mountain, Inc.

Promotion to business customers

Producers and wholesalers that target business customers often emphasize personal selling. This is practical because there are fewer of these customers and their purchases are typically larger. Sales reps can be more flexible in adjusting their companies' appeals to suit each customer—and personal contact is usually required to close a sale. A salesperson is also able to call back later to follow up, resolve any problems, and nurture the relationship with the customer.

Although personal selling dominates in business markets, mass selling is necessary too. A typical sales call on a business customer costs about $500. That's because salespeople spend less than half their time actually selling. The rest is consumed by such tasks as traveling, paperwork, sales meetings, and strictly service calls. So it's seldom practical for salespeople to carry the whole promotion load.

Business buyers often engage in Internet searches to identify solutions for current needs, so it is important for a seller's website to appear near the top of the search results. Otherwise, it has no opportunity to draw buyers' attention and build interest for the firm's products. Online case studies and informative web pages can often attract their attention and interest. Ads in trade magazines might also be used to inform potential customers about a solution. Domestic and international trade shows also help identify prospects. Many of these forms of promotion funnel a customer to a salesperson who may be needed to answer specific questions.

Promotion starts by getting attention and interest

While promotion methods change as customers move through the buying process, one way to get started is with an e-mail marketing campaign. Interested customers may click on an e-mail and follow a link to a website where they can gather information and fill out an online form if they want even more information. E-mail campaigns like these are closely

Online file sharing and collaboration service Dropbox Business was looking to develop a new target market—marketers. While many marketing professionals were aware of and often used the free Dropbox product, few were aware of the unique benefits of Dropbox Business in helping teams collaborate. Research into marketing professionals found that about 75 percent indicated that personality clashes or communication problems had a detrimental effect on team performance. To help those teams function better, Dropbox started by developing an online personality quiz—Marketing Dynamix. After members of a team took the quiz, each received a 20-page "Self-Portrait" e-book with personality insights—and the team better understood individual differences and how members might collaborate more effectively. The program was kicked off with e-mails to marketing managers and advertising in marketing trade magazines and on Snapchat. They also engaged 50 "influencers," who were asked to share the quiz across their social media channels. Soon marketers were taking the quiz and sharing their results on social media sites like Facebook, Twitter, and LinkedIn, where others saw it and soon took the quiz. Upon visiting the Marketing Dynamix site, many wanted to "find out more about Dropbox Business" and clicked on the "Contact Sales" button. From there they filled out a form and were contacted by a Dropbox salesperson. The creative campaign generated more than twice the target number of leads (people who filled out the forms).
Source: Dropbox

Marketing Analytics in Action: Measures of E-mail Performance

Let's examine how health care insurance provider UltraCare uses integrated marketing communications to sell its insurance plans to small and medium-sized businesses. UltraCare uses e-mail, a website, and its sales force to perform the four promotion jobs in the AIDA model. To monitor the performance of the e-mails it sends out and its website, UltraCare tracks the following measures:

- *E-mail open rate:* the percentage of sent e-mails that were opened by a recipient. Because customers see only the e-mail's "subject line" before opening, this measures how well the subject line gets the customer's *attention*.
- *E-mail click-through rate:* the percentage of customers who open the e-mail and subsequently click on a link to UltraCare's website. Each e-mail describes the benefits of UltraCare and asks recipients to click for more information at the UltraCare website. E-mail click-through rate measures whether the e-mail's content captures the customer's *interest*.
- *Form completion rate:* the percentage of customers who click through to the website and then fill out a request for more information. Once at the UltraCare website, customers can watch a short video with more information and are then asked to leave their name, e-mail address, and phone number for a follow-up call from an UltraCare salesperson. Filling out the form indicates a *desire* for more information.
- The sales force will have the job of obtaining *action* with a follow-up sales call to interested customers.

The marketing manager obtains a mailing list of e-mail addresses and names of human resource (HR) managers. HR managers are usually important first contacts when selling health insurance plans. The marketing manager decides to conduct a test of two different e-mail approaches that focus on two different features of UltraCare's health plans: low cost or quality of care. E-mails (and web pages for those who clicked through) with each appeal were created and sent to 2,000 HR managers. Review the results in the table below and answer the questions that follow.

		Appeal Used	
		Low Cost	Quality of Care
a.	E-mails sent out	2,000	2,000
b.	E-mails opened	1,232	864
c.	E-mail open rate (=a/b)*	61.6%	43.2%
d.	E-mails clicked through	151	147
e.	E-mail clicked-through rate (=d/b)**	12.3%	17.0%
f.	Completed forms	36	71
g.	Form completion rate (=f/d)**	23.8%	48.3%

*Calculated by dividing a/b where a = e-mails sent out and b = e-mails opened.

**Note that sometimes e-mail click-through rate and form completion rate are calculated as a percentage of all e-mails sent out. Our calculations are the percentage from the previous step.

1. Which of the two e-mails' subject lines (low cost or quality of care) does a better job of grabbing HR managers' attention?
2. Which of the two e-mails does a better job of generating interest?
3. Which of the two web pages and videos does a better job of arousing desire?

---

monitored. To see how this works in action and how marketing managers use analytics to track the process, see Marketing Analytics in Action: Measures of E-mail Performance.

Each market segment may need a unique blend

Knowing what type of promotion is typically emphasized with different targets is useful in planning the promotion blend. But each unique market segment may need a separate marketing mix and a different promotion blend. You should be careful not to slip into a shotgun approach (using a single promotion across market segments) when what you really need is a rifle approach (a specific mix for a particular target market). This can result from too much combining, as discussed in Chapter 4.

Marketing managers blend push and pull activities because of the specific needs of those target markets. Another way to segment a market depends on where customers are in the adoption process, and we discuss these needs next.

Adoption Processes Can Guide Promotion Planning

LO 13.7

The AIDA and adoption processes (for review, see Exhibit 13-2) look at individuals. This emphasis on individuals helps us understand how promotion affects the way that people behave. But it's also useful to look at markets as a whole. Different segments of customers within a market may behave differently—with some taking the lead in trying new products and, in turn, influencing others.

Promotion must vary for different adopter groups

Research on how markets accept new ideas has led to the adoption curve model. The **adoption curve** shows when different groups accept ideas. It emphasizes the relations among groups and shows that individuals in some groups act as leaders in accepting a new idea. Promotion efforts usually need to change over time to adjust to differences among the adopter groups.

Exhibit 13-7 shows the adoption curve for a typical successful product. Some of the important characteristics of each of these customer groups are discussed next. Which one are you? Does your group change for different products?

Innovators don't mind taking some risks

The **innovators** are the first to adopt. They are eager to try a new idea and willing to take risks. Innovators tend to be young and well educated. They are likely to be mobile and have many contacts outside their local social group and community. Business firms in the innovator group are often specialized and willing to take the risk of doing something new.

Innovators tend to rely on impersonal and scientific information sources, or other innovators, rather than salespeople. They often search for information on the Internet, read technical websites and publications, or look for informative ads in special-interest magazines.

Early adopters are often opinion leaders

Early adopters are well respected by their peers and often are opinion leaders. They tend to be younger, more mobile, and more creative than later adopters. But unlike innovators, they have fewer contacts outside their own social group or community. Business firms in this category also tend to be specialized.

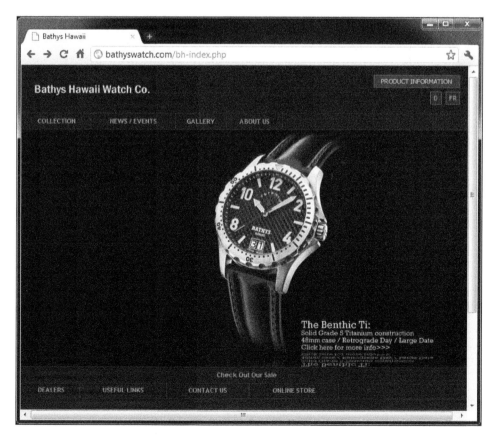

The website for Bathys Hawaii Watch Co. was getting about 60 hits a day until Gizmodo, a blog on new consumer technology that appeals to technology opinion leaders, wrote about the Bathys watch designed especially for surfers. After that, website hits jumped to 1,800 per day, and sales increased by 300 percent.
Source: Bathys Hawaii

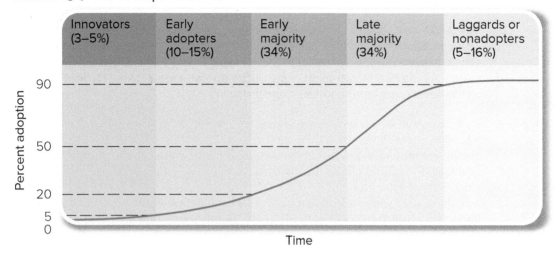

Exhibit 13-7 The Adoption Curve

Along with the innovators, this group is often most interested in the technical performance of any new idea or product. Of all the groups, this one tends to have the greatest contact with salespeople. Mass media are important information sources too. Marketers should be very concerned with attracting and selling to the early adopter group. Their acceptance is crucial. The next group, the early majority, look to the early adopters for guidance. The early adopters can help the promotion effort by spreading *word-of-mouth* information and advice among other consumers.

Early majority group is deliberate

The **early majority** avoids risk and waits to consider a new idea after many early adopters have tried it—and liked it. Average-sized business firms that are less specialized often fit in this category. If successful companies in their industry adopt the new idea, they will too.

The early majority is often most interested in knowing how adopting a new idea will help them solve a problem or offer greater convenience. The early majority has a great deal of contact with mass media, salespeople, and early adopter opinion leaders. Members usually aren't opinion leaders themselves.

Late majority is cautious

The **late majority** is cautious about new ideas. Often they are older and more set in their ways, so they are less likely to follow early adopters. In fact, strong social pressure from their own peer group may be needed before they adopt a new product. Business firms in this group tend to be conservative, smaller-sized firms with little specialization.

The late majority makes little use of marketing sources of information—mass media and salespeople. They tend to be oriented more toward other late adopters rather than outside sources they don't trust.

Laggards or nonadopters hang on to tradition

Laggards or **nonadopters** prefer to do things the way they've been done in the past and are very suspicious of new ideas. They tend to be older and less well educated. The smallest businesses with the least specialization often fit this category. They cling to the status quo and think it's the safe way.

The main source of information for laggards is other laggards. This certainly is bad news for marketers. In fact, it may not pay to bother with this group.[15]

Promotion Blends Vary over the Product Life Cycle

Stage of product in its life cycle

The adoption curve helps explain why a new product goes through the product life-cycle stages described in Chapter 9. Promotion blends usually have to change to achieve different promotion objectives at different life-cycle stages.

Market introduction stage—"this new idea is good"

During market introduction, the basic promotion objective is informing. If the product is a really new idea, the promotion must build **primary demand**—demand for the general product idea—not just for the company's own brand. Rooftop solar panels, "smart" appliances (that connect to the Internet), and self-driving long-haul trucks are good examples of product concepts where primary demand is just beginning to grow. There may be few potential innovators during the introduction stage, and personal selling can help find them. Firms also need salespeople to find good channel members and persuade them to carry the new product. Sales promotion may be targeted at salespeople or channel members to get them interested in selling the new product. Customers often view new-product ideas as risky, so sales promotion may be helpful to stimulate interest and trial.

Market growth stage—"our brand is best"

In the market growth stage, more competitors enter the market, and promotion emphasis shifts from building primary demand to stimulating **selective demand**—demand for a company's own brand. The main job is to persuade customers to buy, and keep buying, the company's product.

Now that there are more potential customers, mass selling becomes more economical. But salespeople and personal selling must still work in the channels, expanding the number of outlets and cementing relationships with channel members.

The evolution of promotion for Banquet Homestyle Bakes illustrates these first two stages. When ConAgra Foods introduced Homestyle Bakes, it was the first shelf-stable meal kit with the meat already in the package. ConAgra, also the producer of Armour processed meats, had the expertise to create a tasty product that a consumer could prepare in a few minutes and then just stick in the oven. When Homestyle Bakes came out, there was no direct competition. The sales force used marketing research data to convince retailers to give the product shelf space, and ads used humor to highlight that the package was unusually heavy because it already included meat. However, over time new competition entered the market with similar offerings. So promotion shifted to emphasize why Homestyle Bakes was better, with a variety of new flavors and 10 percent more meat. Similarly, to keep customers interested in the Homestyle brand, the sales force shifted its efforts to get retailers to participate in Homestyle Bakes' "Super Meals/Super Moms" contests, which offered harried moms prizes such as a visit to a spa.[16]

Market maturity stage—"our brand is better, really"

In the market maturity stage, mass selling and sales promotion may dominate the promotion blends of consumer products firms. Business products may require higher levels of personal selling—perhaps supplemented by more advertising. The total dollars allocated to promotion may rise as competition increases.

If a firm already has high sales—relative to competitors—it may have a real advantage in promotion at this stage. For example, sales of Tylenol tablets are about four times the sales of Motrin competing tablets. If both Tylenol and Motrin spend the same percentage of sales (say 35 percent) on promotion, Tylenol will spend four times as much as its smaller competitor and will probably communicate to more people.

Firms that have differentiated their marketing mixes may favor mass selling because they have something to talk about. For instance, a firm with a strong brand may use reminder-type advertising or target frequent-buyer promotions at current customers to strengthen the relationship and keep customers loyal. This may be more effective than costly efforts to win customers away from competitors.

However, as a market drifts toward pure competition, some companies resort to price-cutting. This may temporarily increase the number of units sold, but it is also likely to reduce total revenue and the money available for promotion. The temporary sales gains disappear and prices are dragged down even lower when competitors retaliate with their own short-term sales promotions, like price-off coupons. As cash flowing into the business declines, spending may have to be cut back.[17]

Sales decline stage—"let's tell those who still want our product"

During the sales decline stage, the total amount spent on promotion usually decreases as firms try to cut costs to remain profitable. Because some people may still want the product, firms need more targeted promotion to reach these customers.

On the other hand, some firms may increase promotion to try to slow the cycle, at least temporarily. Crayola had almost all of the market for children's crayons, but sales were slowly declining as new kinds of markers came along. Crayola increased ad spending to urge parents to buy their kids a "fresh box."

When introducing a new product, in-store sampling can be used to introduce the product to customers and build primary demand. Getting a "taste" of a new product reduces the risk inherent in buying something a customer has never tried. In a mature product category—like cheese—a brand like Tillamook wants customers to know why its brand is better than the many other brands of cheddar cheese.
(left): Vicki Beaver/Alamy Stock Photo; (right): Source: Tillamook County Creamery Association

Setting the Promotion Budget

LO 13.8

Size of budget affects promotion efficiency and blend

There are some economies of scale in promotion. An ad on national TV might cost less *per person* reached than an ad on local TV. Similarly, citywide radio, TV, and newspapers may be cheaper than neighborhood newspapers or direct personal contact. But the *total cost* for some mass media may force small firms, or those with small promotion budgets, to use promotion alternatives that are more expensive per contact. More precisely targeted promotion may be more effective anyway—as the "wide net" of some advertising may result in many people who were not targeted, and may not be potential customers, seeing an ad. Marketing managers need to consider both their budget and their target market.

Find the task, budget for it

The most common method of budgeting for promotion expenditures is to compute a percentage of either past sales or sales expected in the future. The virtue of this method is its simplicity. However, just because this mechanical approach is common doesn't mean that it's smart. It leads to expanding marketing expenditures when business is good and cutting back when business is poor. When business is poor, this approach may just make the problem worse—if weak promotion is the reason for declining sales.

In light of our continuing focus on planning marketing strategies to reach objectives, the most sensible approach to budgeting promotion expenditures is the **task method**—basing the budget on the job to be done. It helps a marketing manager to set priorities so that the money spent on promotion produces specific and desired results. In fact, this approach makes sense for *any* marketing expenditure, but here we'll focus on promotion.

A practical approach is to determine which promotion methods are most economical and effective for the tasks that need to be completed to achieve communication objectives. The costs of these tasks are then totaled to determine how much should be budgeted for promotion (just as money is allocated for other marketing activities required by the strategy). In other words, the firm can assemble its total promotion budget directly from detailed plans rather than by simply relying on historical patterns or ratios.

Hostess Snacks uses social media, including an Instagram and Twitter, to connect with its customers. Posts have no media costs and can get hundreds or even thousands of likes. While there is no cost to placing images and captions on their social media, marketing managers must remember the costs of maintaining these pages. For every brand with hundreds of thousands (or millions) of followers, there are many more with fewer than a thousand followers.
(left): Source: Hostess Brands, LLC/Instagram, Inc; (right): Source: Hostess Brands, LLC/Twitter

Marketing research clarifies budgeting estimates

Some promotion activities and objectives can be readily measured. Often a firm has measured the results and return on investment from previous promotion efforts. These can provide a good estimate of the cost to achieve a similar objective. For example, if Taco Bell sets a target of adding 10,000 new "Likes" to its Facebook page and has used coupons to grow it in the past, it has some previous results to compare new campaigns to. Or perhaps a firm's advertising agency has experience that can provide better estimates. Taco Bell can measure the cost of the promotion (coupons plus any advertising) and the number of new "Likes" on the Facebook page. Figuring out the value of a "Like" might be more difficult.

Other objectives can be more challenging to predict and measure. What if the objective is to raise the level of brand familiarity of Bonobos pants for fashion-conscious men. Bonobos might think running ads in select men's magazines makes sense. But what if Bonobos has never run magazine ads? Although not impossible, predicting the results of the advertising for the first time can be a challenge. Even when evaluation is difficult, marketing managers should still try to evaluate costs and returns to develop a realistic budget.[18]

CONCLUSION

Promotion is an important part of any marketing mix. Most consumers and intermediate customers can choose from among many products. To be successful, a producer must not only offer a good product at a reasonable price but also inform potential customers about the product and where they can buy it. Further, producers must tell wholesalers and retailers in the channel about their product and marketing mix. These intermediaries, in turn, must use promotion to reach their customers. And the promotion blend must fit with the rest of the marketing mix and the target market.

In this chapter, we introduced different promotion methods and discussed the advantages and disadvantages of each method. We also discussed the integrated marketing communications concept and explained why most firms use a blend of different promotion methods. Although the overall promotion objective is to affect buying behavior, the basic promotion objectives are informing, persuading, and reminding. These objectives help guide the marketing manager's decisions about the promotion blend.

Models from the behavioral sciences help us understand the communication process and how it can break down.

These models recognize different ways to communicate. We discussed direct-response promotion for developing more targeted promotion blends. And we described an approach where customers initiate and interact with the marketer's communications and how this provides new and different challenges for marketing managers.

This chapter also recognized other factors that influence decisions about promotion blends. Marketing managers must make decisions about how to split promotion that is directed at final consumers or business customers—and at channel members. Promotion blends are also influenced by the adoption curve and the product life-cycle stages. Finally, we described how promotion budgets are set and influence promotion decisions.

In this chapter, we considered some basic concepts that apply to all areas of promotion. In Chapters 14, 15, and 16 we'll discuss personal selling, customer service, advertising, publicity, and sales promotion in more detail.

KEY TERMS

LO 13.9

Promotion, 352
personal selling, 353
mass selling, 354
advertising, 354
sales promotion, 354
publicity, 354
sales managers, 355
advertising managers, 355
sales promotion managers, 355
public relations, 355
integrated marketing communications, 356

AIDA model, 358
communication process, 360
source, 360
receiver, 360
noise, 361
encoding, 361
decoding, 361
message channel, 361
pushing, 366
pulling, 368
adoption curve, 371

innovators, 371
early adopters, 371
early majority, 372
late majority, 372
laggards, 372
nonadopters, 372
primary demand, 372
selective demand, 373
task method, 374

QUESTIONS AND PROBLEMS

1. Review the GEICO case study that opens the chapter. From this case, identify examples of different key terms and concepts covered in the chapter. For example, advertising is represented by the Gecko ads.

2. Review the GEICO case study that opens this chapter. Then go to GEICO's Facebook page and Twitter feed. Suggest three different posts for each site—one each aiming primarily to (a) inform, (b) persuade, and (c) remind.

3. Briefly explain the nature of the three basic promotion methods available to a marketing manager. What are the main strengths and limitations of each?

4. In your own words, discuss the integrated marketing communications concept. Explain what its emphasis on "consistent" and "complete" messages implies with respect to promotion blends.

5. Relate the three basic promotion objectives to the four jobs (AIDA) of promotion using a specific example.

6. If a company wants its promotion to appeal to a new group of target customers in a foreign country, how can it protect against its communications being misinterpreted?

7. Promotion has been the target of considerable criticism. What specific types of promotion are probably the object of this criticism? Give a particular example that illustrates your thinking.

8. With direct-response promotion, customers provide feedback to marketing communications. How can a marketing manager use this feedback to improve the effectiveness of the overall promotion blend?

9. How can a promotion manager aim a message at a certain target market using social media (like Facebook or Instagram) or with a search engine (like Google)? Give an example.

10. What promotion blend would be most appropriate for producers of the following established products? Assume average- to large-sized firms in each case and support your answer: (a) chocolate candy bars, (b) car batteries, (c) castings for truck engines, (d) a special computer used by manufacturers for control of production equipment, (e) inexpensive plastic rain hats, and (f) a phone case that has achieved specialty product status.

11. A small company has developed an innovative new spray-on glass cleaner that prevents the buildup of electrostatic dust on computer screens and TVs. Give examples of some low-cost ways the firm might effectively promote its product. Be certain to consider both push and pull approaches.

12. Would promotion be successful in expanding the general demand for (a) almonds, (b) air travel, (c) golf clubs, (d) walking shoes, (e) high-octane unleaded gasoline, (f) single-serving, frozen gourmet dinners, and (g) bricks? Explain why or why not in each case.

13. Explain how an understanding of the adoption process would help you develop a promotion blend for digital tape recorders, a new consumer electronics product that

produces high-quality recordings. Explain why you might change the promotion blend during the course of the adoption process.

14. Discuss how the adoption curve should be used to plan the promotion blend(s) for a new automobile accessory—an electronic radar system that alerts a driver if he or she is about to change lanes into the path of a car that is passing through a blind spot in the driver's mirrors.

15. If a marketing manager uses the task method to budget for marketing promotions, are competitors' promotion spending levels ignored? Explain your thinking and give an example that supports your point of view.

MARKETING PLANNING FOR HILLSIDE VETERINARY CLINIC

Appendix D (the Appendices follow Chapter 19) includes a sample marketing plan for Hillside Veterinary Clinic. Look through the "Marketing Strategy" section.

1. What are Hillside's promotion objectives? How do they differ for the various goods and services the company offers?
2. Do the promotion activities recommended in the plan fit with the promotion objectives? Create a table to compare them. Label the columns: good/service, promotion objective, and promotion activities.
3. Based on the situation analysis, target market, and intended positioning, recommend other (low-cost) promotion activities for Hillside.

SUGGESTED CASES

14. Schrock & Oh Design
18. West Tarrytown Volunteer Fire Department (WTVFD)
19. MyOwnWedding.com

MARKETING ANALYTICS: DATA TO KNOWLEDGE

CHAPTER 13: SELECTING A COMMUNICATIONS CHANNEL

Helen Troy, owner of three Sound Haus stereo equipment stores, is deciding what message channel (advertising medium) to use to promote her newest store. Her current promotion blend includes direct-mail ads that are effective for reaching her current customers. She also has knowledgeable salespeople who work well with consumers once they're in the store. However, a key objective in opening a new store is to attract new customers. Her best prospects are professionals in the 25-44 age range with incomes over $38,000 a year. But only some of the people in this group are audiophiles who want the top-of-the-line brands she carries. Troy has decided to use local advertising to reach new customers.

Troy narrowed her choice to two advertising media: Pandora, a music streaming service that offers localized advertising services, and a biweekly magazine that focuses on entertainment in her city. Many of the magazine's readers are out-of-town visitors interested in concerts, plays, and restaurants. They usually buy stereo equipment at home. But the magazine's audience research shows that many local professionals do subscribe to the magazine. Troy doesn't think that the objective can be achieved with a single ad. However, she believes that ads in six issues will generate good local awareness with her target market. In addition, the magazine's color format will let her present the prestige image she wants to convey in an ad. She thinks that will help convert aware prospects to buyers. Specialists at a local advertising agency will prepare a high-impact ad for $2,000, and then Troy will pay for the magazine space.

Pandora can target an audience similar to Troy's own target market. She knows repeated ads will be needed to be sure that most of her target audience is exposed to her ads. Troy thinks it will take daily ads for several months to create adequate awareness among her target market. Pandora will provide an announcer and prepare a recording of Troy's ad for a one-time fee of $200. All she has to do is tell them what the message content for the ad should say.

Both Pandora and the magazine gave Troy reports summarizing recent audience research. She decides that comparing the two media in a spreadsheet will help her make a better decision.

Design element: #M4BW box globe icon: ©Vectoryzen/Shutterstock

CHAPTER FOURTEEN

William D. Perreault Jr.

Personal Selling and Customer Service

As a student in the College of Business at the University of Illinois, Pooja Gupta wanted a job that would offer interesting challenges, give opportunities for professional growth, and value her enthusiasm. She found what she wanted with Ferguson. Ferguson was actively recruiting on college campuses to find the brightest and best candidates for its sales jobs—so, in a way, the job found her.

Gupta knew that motivated young people often find the best opportunities in fast-growing companies. She didn't expect, however, that her fast-growing company would be a wholesaler of plumbing supplies, pipes, valves, and fittings. To the contrary, she'd heard that many wholesalers were declining. But that didn't apply to Ferguson. For decades it has doubled in size about every five years—and now it's the largest U.S. distributor of plumbing products. And in a business that serves such a wide variety of customer types—large industrial firms, city waterworks, commercial builders and subcontractors, kitchen and bath dealers, and final consumers—you don't get that kind of growth without an effective sales force.

It's Ferguson's sales force that gets the initial orders with new customers, builds the relationships that instill customer loyalty, and provides the customer service support that Ferguson emphasizes. Its sales reps understand their customers' business problems and how Ferguson's products, e-commerce, and state-of-the-art logistics systems can help solve them. This expertise and focus on customer needs make Ferguson salespeople trusted partners.

An effective sales force like the one at Ferguson doesn't just happen. Someone needs to figure out the promotion jobs that require personal selling and then get the right people on the job. That is why Ferguson's marketing managers work closely with sales managers.

Ferguson carries more than a million products, provides service centers at 1,400 locations, and has divisions that specialize by different customer segments. It would be futile for sales reps to try to be experts in everything. Instead, sales managers carefully match each salesperson to particular territories, customers, and product lines. Gupta, for example, helps contractors in the Virginia market figure out how to satisfy the needs of final consumers for whom they are building or remodeling homes. She knows the current fashions for kitchen and bath renovations, how to reduce "behind-the-wall" plumbing installation costs for a big new apartment building, and the advantages and limitations of hundreds of brands from companies such as Kohler, Elkay, Moen, and Jacuzzi.

Ferguson's salespeople have very different days depending on the types of customers they serve. Some Ferguson salespeople work with cities and huge waterworks contractors on infrastructure projects such as updating water purification facilities. And salespeople for Ferguson's Integrated Systems Division (ISD) are really selling a big-business idea rather than "pipe." They show top executives at customer firms why they should invest millions of dollars in a full-service supply relationship where Ferguson does all the purchasing and warehousing for entire manufacturing facilities. Other salespeople work as inside sales reps. They don't meet with customers face-to-face, but instead communicate with customers on the phone, by e-mail, and via social media. These technical experts require a different skill set than the salesclerks who work in one of Ferguson's self-service Xpress outlets, where they primarily take customer orders, find the product in the warehouse, and ring up sales.

To recruit talented people for these varied jobs, Ferguson's sales managers use a wide variety of methods. For example, the "Careers" section of Ferguson's website collects job applicant profiles on an ongoing basis. When a position opens up, qualified candidates are notified. And Ferguson actively recruits on college campuses, typically hiring hundreds of graduates every year. After a pre-interview on campus, select candidates go to a regional office and meet a number of managers from that area.

After the best people are selected, Ferguson provides the sales training to make them even better. Of course, the training is different for different people. For example, most new college recruits go through a 10- to 12-month training program that rotates them through a range of jobs, from working in the warehouse learning shipping and logistics, to counter sales that helps them learn about products and Ferguson systems, and a stint in inside sales. The training utilizes a range of training methods, from self-study online modules, to role-playing, to working in the field with experienced managers who help them build professional problem-solving skills as well as technical knowledge.

Even experienced sales reps need ongoing training on new strategies or policies. When Ferguson's management saw an opportunity to enhance customer loyalty, a training program was developed and implemented with Ferguson's

23,000 associates. They participated in face-to-face and online courses covering topics such as "the difference between customer satisfaction and loyalty" and "how to earn loyal customers." Ferguson estimated the return on investment for the training program at more than 400 percent.

To be sure that each salesperson is highly motivated, Ferguson's sales managers make certain that sales compensation arrangements and benefits reward salespeople for producing needed results. For example, the evaluation considers how well individuals work within a team—because in the customer service culture at Ferguson, great teamwork is critical. Ferguson's experienced reps continue to help contractors, builders, and home buyers choose the right supplies and fuel Ferguson's growth.[1]

LEARNING OBJECTIVES

Promotion is communicating with potential customers and others in the channel. As the Ferguson case suggests, personal selling is often the best way to do it. While face-to-face with prospects, salespeople can adjust what actions they take in response to the prospect's interests, needs, questions, and feedback. If, and when, the prospect is ready to buy, the salesperson is there to ask for the order. And afterward, the salesperson works to be certain that the customer is satisfied and will buy again in the future. In this chapter, you'll learn about the key strategy decisions related to personal selling that marketing managers and the sales managers who work with them make.

When you finish this chapter, you should be able to

1. understand the importance and nature of personal selling.
2. know the three basic sales tasks—order-getting, order-taking, and supporting—and what various kinds of salespeople can be expected to do.
3. understand why customer service presents different challenges than other personal selling tasks.
4. know the different ways sales managers can organize salespeople so that personal selling jobs are handled effectively.
5. know how sales technology affects the way sales tasks are performed.
6. know what the sales manager must do, including selecting, training, and organizing salespeople to carry out the personal selling job.
7. understand how the right compensation plan can help motivate and control salespeople.
8. understand when and where to use the three types of sales presentations.
9. understand important new terms (shown in **red**).

The Importance and Role of Personal Selling

LO 14.1

Personal selling requires strategy decisions

In this chapter, we'll discuss the importance and nature of personal selling and customer service so you'll understand the strategy decisions in this area (see Exhibit 14–1).

We'll also discuss a number of frameworks and how-to approaches that guide these strategy decisions. Because these approaches apply equally to domestic and international markets, we won't emphasize that distinction in this chapter. This does not mean, however, that personal selling techniques don't vary from one country to another. On the contrary, in dealing with *any* customer, the salesperson must adjust for cultural influences and other factors that might affect communication. For example, a Japanese customer and an Arab customer might respond differently to subtle aspects of a salesperson's behavior. The Arab customer might expect to be physically very close to a salesperson, perhaps only two feet away, while they talk. The Japanese customer might consider that distance rude. Similarly, what topics of discussion are considered

Exhibit 14-1 Strategy Planning and Personal Selling

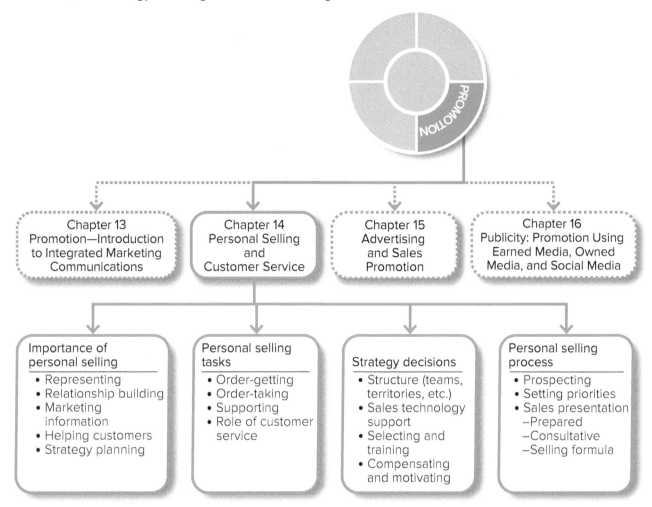

sensitive, how messages are interpreted, and which negotiating styles are used vary from one country to another. A salesperson must know how to communicate effectively with each customer—wherever and whoever that customer is.[2]

Personal selling is important

Personal selling is absolutely essential in the promotion blends of some firms. Consider how you would feel if you regularly had to meet payrolls and somehow, almost miraculously, your salespeople kept coming in with orders just in time to keep the business profitable.

Personal selling is often a company's largest single operating expense. This is another reason why it is important to understand sales management decisions. Bad ones are costly in both lost sales and in actual out-of-pocket expenses.

Every economy needs and uses many salespeople. In the United States, *1 person of every 10 in the total labor force is involved in sales work*. By comparison, that's about 20 times more people than are employed in advertising. Any occupation that employs so many people and is so important to the economy deserves study. Looking at what salespeople do is a good way to start.

Helping to buy is good selling

Good salespeople don't just try to *sell* the customer. Rather, they try to *help the customer buy*—by understanding the customer's needs and presenting the advantages and disadvantages of their products. Such helpfulness results in satisfied customers and long-term relationships. And strong relationships often form the basis for a competitive advantage, especially in business markets.

Good salespeople help customers find solutions to problems. Often that requires a trusting relationship that comes from carefully listening to customers and offering expertise customers count on. Boise wants its customers to know that its salespeople are good listeners, and True Value reminds customers they can count on the know-how of its salesclerks.
(left): Boise Cascade Company; (right): True Value Company

Not only the salespeople sell

Selling may be done by almost anyone in the organization. When you ask the butcher at your local supermarket for advice, he helps you buy the right cut of meat. The butcher is selling. When your auto mechanic identifies the appropriate motor oil for your oil change, the mechanic is selling. When a web page designer helps a marketing manager build a site that meets her needs, the designer is selling. When a doctor talks to a patient about the importance of changing his diet and taking his medication to lower his blood pressure, the doctor is selling. We may not call each of these people a salesperson, but they are all helping customers buy products or ideas. Although this chapter focuses on more traditional salesperson roles, smart companies make sure that everyone in the organization who talks to customers knows at least something about selling.[3]

Salespeople represent the whole company—and customers too

The salesperson is often a representative of the whole company—responsible for explaining its total effort to customers rather than just pushing products. The salesperson may provide information about products, explain company policies, and even negotiate prices or diagnose technical problems.

The sales rep is often the only link between the firm and its customers, especially if customers are far away. When a number of people from the firm work with the customer organization—which is common when suppliers and customers form close relationships—it is usually the sales rep who coordinates the relationship for his or her firm (see Exhibit 6-7).

As evidence of these changing responsibilities, some companies give their salespeople such titles as account representative, field manager, sales consultant, market specialist, or sales engineer.

Sales force aids in marketing information function as well

The salesperson also represents the customers' interests inside the firm. The sales force can aid in the marketing information function too. The sales rep may be the first to hear about a new competitor or a competitor's new strategy. And sales reps who are well attuned to customers' needs can be a key source of ideas for new products or new uses for existing products.

It was a salesperson for Material Sciences Corporation (MSC) who spotted an opportunity for the firm in the automotive market. The sales rep worked closely with the Ford team developing the new F-150 truck. When he asked for more information about how Ford was trying to reduce road noise, he in effect opened the door for a potential application of MSC's Quiet Steel product. Quiet Steel had not previously been used in the automotive market. It was a perfect fit. These insights helped MSC gain a foothold with other automakers. A couple of questions from a salesperson helped consumers get quieter cars and trucks.[4]

Salespeople can be strategy planners too

Some salespeople are expected to be marketing managers in their own territories. And some become marketing managers by default because top management hasn't provided detailed strategy guidelines. Either way, the salesperson may have choices about (1) which customers to target, (2) which particular products to emphasize, (3) which intermediaries to rely on for help, (4) what message to communicate and how to use promotion money, and (5) how to adjust prices. A salesperson who can put together profitable strategies and implement them well can rise very rapidly. The opportunity is there for those prepared and willing to work.[5]

What Kinds of Personal Selling Are Needed?

LO 14.2

If a firm has too few salespeople, or the wrong kind, some important personal selling tasks may not be completed. And having too many salespeople wastes money. In addition, the right balance may change over time with other changes in strategy or the market environment. That's why many firms have to restructure their sales forces.

One of the difficulties of determining the right number and kind of salespeople is that every sales job is different. Generally, though, each sales job is some mix of three basic types of sales tasks (see Exhibit 14-2). This gives us a starting point for understanding what sales tasks need to be done and how many people are needed to do them.

Exhibit 14-2 Basic Sales Tasks

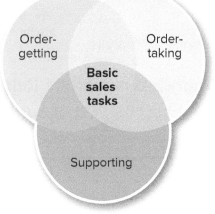

Personal selling is divided into three tasks

The three **basic sales tasks** are order-getting, order-taking, and supporting. For convenience, we'll describe salespeople by these terms—referring to their primary task—*although one person may do two or all three tasks in some situations.*

Order Getters Develop New Business Relationships

Order getters are concerned with establishing relationships with new customers and developing new business. **Order-getting** means seeking possible buyers with a well-organized sales presentation designed to sell a good, service, or idea.

Order getters must know what they're talking about, not just be personal contacts. Order-getting salespeople normally are well paid—many earn more than $100,000 a year!

Producers' order getters—find new opportunities

Producers of all kinds of products, especially business products, have a great need for order getters. They use order getters to locate new prospects, open new accounts, see new opportunities, and help establish and build channel relationships.

Most customers are more interested in ways to save or make more money than in technical details. Good order getters cater to this interest. They help the customer identify ways to solve problems; then they sell concepts and ideas, not just physical products. The goods and services they supply are merely the means of achieving the customer's end.

To be effective at this sort of "solutions selling," an order getter often needs to understand a customer's whole business as well as technical details about the product and its applications. For example, a salesperson for automated manufacturing equipment must understand a prospect's production process as well as the technical details of converting to computer-controlled equipment.

Order getters for professional services—and other products where service is a crucial element of the marketing mix—face a special challenge. The customer usually can't inspect a service before deciding to buy. The order getter's communication and relationship with the customer may be the only basis for evaluating the quality of the supplier.

Wholesalers' order getters—almost hand it to the customer

Salespeople for agent wholesalers are often order getters—particularly the more aggressive manufacturers' agents and brokers. They face the same tasks as producers' order getters. But, unfortunately for them, once the order-getting is done and the customers become established and loyal, producers may try to eliminate the agents and save money with their own order takers.

Retail order getters—influence consumer behavior

Convincing consumers about the value of products they haven't seriously considered takes a high level of personal selling ability. Order getters for unsought consumer products must help customers see how a new product can satisfy needs now being filled by something else. Without order getters, many common products—ranging from mutual funds to air conditioners—might have died in the market introduction stage. The order getter helps bring products out of the introduction stage into the market growth stage.

Order getters are also helpful for selling *heterogeneous* shopping products. Consumers shop for many of these items on the basis of suitability and value. They welcome useful information.

Order Takers Nurture Relationships to Keep the Business Coming

Order takers sell to regular or established customers, complete most sales transactions, and maintain relationships with their customers. After a customer becomes interested in a firm's products through an order getter, supporting salesperson, or through advertising or sales promotion, an order taker usually answers any final questions and completes the sale. **Order-taking** is the routine completion of sales made regularly to target customers. It usually requires ongoing follow-up to make certain that the customer is totally satisfied.

Producers' order takers—train, explain, and collaborate

Order takers work on improving the whole relationship with their accounts, not just on completing a single sale. Even in e-commerce, where customers place routine orders with computerized order systems, order takers do a variety of important jobs that are essential to the business relationship. Someone has to explain details, make adjustments, handle complaints, explain new prices or terms, place sales promotion materials, and keep customers informed of new developments. An order taker who fails to meet a customer's expectations on any of these activities might jeopardize the relationship and future sales.

Firms sometimes use order-taking jobs to train potential order getters and managers. Such jobs give them an opportunity to meet customers and better understand their needs. And, frequently, they run into some order-getting opportunities.

Order takers who are alert to order-getting possibilities can make a big difference in generating new sales. Some firms lose sales just because no one ever asks for the order. Banks try to avoid this problem. For example, when a customer walks into a First Bank branch to make a deposit, the teller's computer screen shows information about the customer's accounts. If the balance in a checking account is high and the customer does not use any of the bank's other investment services, the teller is trained to ask if the customer would be interested in learning about the bank's certificates of deposit. Some firms use more sophisticated customer relationship management (CRM) database systems that figure out which specific financial service would be best for the teller to recommend.[6]

Friendly and capable retail order takers can play an important role in building good relations with customers.
Blue Jean Images/Alamy Stock Photo

Wholesalers' order takers—not getting orders but keeping them

Whereas producers' order takers usually handle relatively few items, wholesalers' order takers often sell thousands of items. Sales reps who handle that many items may single out a few of the newer or more profitable items for special attention, but it's not possible to give extensive sales effort to many. So the main job of wholesalers' order takers is to maintain close contact with customers, place orders, and check to be sure the company fills orders promptly. Order takers also handle any adjustments or complaints and generally act as liaisons between the company and its customers.

In retail, salesclerks are usually order takers

Order-taking may be almost mechanical at the retail level—for example, at the supermarket checkout counter. Some retail clerks perform poorly because they receive little training. Turnover can be high, too, as many receive little more than the minimum wage. Even so, retail order takers play a vital role in a retailer's marketing mix. Customers expect prompt and friendly service. They will find a new place to shop, do their banking, or have their car serviced rather than deal with a salesclerk who is inept or rude or acts annoyed by having to complete a sale.

Supporting Sales Force Informs and Promotes in the Channel

Supporting salespeople help the order-oriented salespeople, but they don't try to get orders themselves. Their activities are aimed at enhancing the relationship with the customer and getting sales in the long run. For the short run, however, they are ambassadors of goodwill who may provide specialized services and information. There are three types of supporting salespeople: missionary salespeople, technical specialists, and customer service reps.

When a customer firm, like a supermarket chain, buys Hobart equipment for a new store, Hobart has supporting salespeople every step of the way to be certain that all customer needs are met. The ad tagline: "Another successful store opening. See you bright and early."
Hobart Corporation

Missionary salespeople can increase sales

Missionary salespeople are supporting salespeople who work for producers—calling on intermediaries and their customers. They try to develop goodwill and stimulate demand, help intermediaries train their salespeople, and often take orders for delivery by intermediaries. Missionary salespeople are sometimes called *merchandisers* or *detailers*.

Producers who rely on merchant wholesalers or e-commerce to obtain widespread distribution often use missionary salespeople. The sales rep can give a promotion boost to a product that otherwise wouldn't get much attention because it's just one of many. A missionary salesperson for Vicks' cold remedy products, for example, might visit pharmacists during the cold season and encourage them to use a special end-of-aisle display for Vicks' cough syrup—and then help set it up. The wholesaler that supplies the drugstore would benefit from any increased sales but might not take the time to urge use of the special display.

An imaginative missionary salesperson can double or triple sales for a company. Naturally, this doesn't go unnoticed. In some companies, a missionary sales job is a first step on a career path to order-oriented jobs.

Technical specialists are experts who know product applications

Technical specialists are supporting salespeople who provide technical assistance to order-oriented salespeople. Technical specialists are often science or engineering graduates with the know-how to understand the customer's applications and explain the advantages of the company's product. They are usually more skilled in showing the technical details of their product than in trying to persuade customers to buy it. Before the specialist's visit, an order getter probably has stimulated interest. The technical specialist provides the details.

Customer service reps solve problems after a purchase

Customer service reps work with customers to resolve problems that arise with a purchase, usually after the purchase has been made. Unlike other supporting sales activities, which are needed only in certain selling situations, *every* marketing-oriented company needs good people to handle customer service. Customer service is important to both business customers and final consumers. There are times when a customer's problem simply can't be resolved without a personal touch.

In general, all types of personal selling help win customers, but effective customer service is especially critical in keeping them. It's useful to think of customer service reps as *the salespeople who promote a customer's next purchase by being sure that the customer is satisfied with a previous purchase.* In this chapter, you'll see that the strategy decisions for customer service reps are the same as for others involved in personal selling. In spite of this, some firms don't view customer service as a personal selling activity—or as part of the firm's integrated marketing communications. They manage it as a production operation where output consists of responses to questions from "problem customers." That approach is one reason that customer service is often a problem area for firms, which is why it is useful to take a closer look at why customer service activities are so important and why firms should manage them as part of the personal selling effort.

Customer Service Promotes the Next Purchase

LO 14.3

Customer service is not the product

People sometimes use the term *customer service* as a catch-all expression for anything that helps customers. Our focus here is on the service that is required *to solve a problem that a customer encounters with a purchase.* This highlights an important distinction in how customers look at their purchase experience. In that regard, it is useful to think about the difference between customer service and the service (or support) that is part of the product that a customer buys.

In Chapter 8, we discussed the idea that a firm's product is its need-satisfying offering, and that it may be a physical good, a service, or a combination of the two (see Exhibit 8-3). Chase Bank offers consumers credit card services for a fee. Pizza Hut delivers a (hopefully) hot pizza to your door. Dell sells computer hardware and software that is supported with telephone or website technical support for some period of time after the purchase. In all of these situations, customers see service as an important aspect of what they are purchasing.

However, from a customer's perspective, that kind of service is different from the customer service that is required to fix a problem when something doesn't work as the customer hopes or expects. For example, our customer doesn't expect the Chase Bank ATM to eat her credit card when she's on a trip, doesn't want the Pizza Hut pizza to arrive late with the wrong toppings, and isn't planning on Dell sending the wrong computer. These problems are breakdowns in the firms' marketing mixes. What the customer expected from the seller is not what the customer got.

When a customer service rep works to solve a customer's problem, it often involves taking steps to remedy what went wrong. But repairing a negative experience is fundamentally different from providing a positive experience in the first place. No matter how effective the customer service solution, the problem is an inconvenience or involves other types of costs to the customer. Thus, the customer value from the firm's marketing mix is lower than what the customer bargained for. Often it's also less than the value the firm *intended* to provide.

In Spain, cell phone operator Vodafone has set up customer service hotlines with representatives who can speak 11 languages from Arabic to Romanian. That's because Spain has one of Europe's fastest-growing immigrant populations, with more than 600,000 foreigners arriving annually. Vodafone's customer service efforts help the company acquire and retain these new customers.
Vodafone Group Plc

Promoting the next sale builds customer equity

Recall that in Chapter 2 we introduced the idea that marketing managers seek to increase customer equity, or the expected earnings stream (profitability), of a firm's current and prospective customers over some period of time. The customer equity approach suggests three paths for a business to grow: (1) acquiring new customers, (2) retaining current customers, and (3) enhancing the purchases from current customers. Good customer service helps marketing managers with all three!

Great customer service and positive word-of-mouth adds new customers

Think of the power of positive word-of-mouth. We will talk about this more in Chapter 16, but we are all more likely to go to a new dentist, try a restaurant, or buy a new jacket after a friend gives an endorsement. And we are all likely to avoid one of these when a friend tells us about a bad experience. Customer service is an opportunity for a company to turn a customer problem into positive word-of-mouth—which leads to acquiring new customers.

Customer service retains and enhances

Customers who have a problem that is not resolved to their satisfaction are much more likely to not buy again. Great customer service fixes those problems and gives the customer a reason to give a company another chance.

A well-trained customer service rep not only solves the customer's immediate problem but could ask enough questions to find out if there are other products or services the company sells that the customer might purchase. A happy customer may be open to these cross-selling opportunities.

#M4BW

Zappos delivers happiness *and* customer equity

Let's see how online retailer Zappos' outstanding customer service creates customer equity. Zappos empowers its customer service representatives to do whatever it takes to make customers happy. One customer service rep sent flowers to a customer who had recently lost her mother. Another ordered (and paid for) pizza for a whole neighborhood following Hurricane Sandy. While on the phone with a Zappos customer service rep, a customer described how he wrote a blog where he shared the challenge he was facing in losing weight. He also mentioned to the Zappos rep that Internet trolls often made nasty comments on his blog. For weeks after that, the Zappos rep went to his blog and posted encouraging comments. The blogger/customer later wrote a whole blog post about this Zappos employee. What did any of those activities have to do with selling shoes? Or solving problems with an order for shoes? Nothing! But they brought customers happiness at a troubled time. Selling shoes is not life-changing, but the *way* Zappos sells them impacts people's lives. Those experiencing the kindness of a Zappos rep who really cares, or simply hearing the story, end up buying more often or telling their friends—all of which also helps Zappos' bottom line. This kind of customer service delivers happiness *and* customer equity.[7]

Big data and social media help customer service get proactive

We have portrayed customer service as a reactive form of communication—companies waiting for customers to call or write in with a complaint. Many customers don't bother contacting a company—they just stop doing business with the offending firm or voice their complaints online. Then there are companies that try to solve customers' problems even when the customers don't ask for help. They are monitoring the web and using big data and predictive analytics to provide *proactive* customer service. These companies reach out to unhappy customers before their frustration boils over.

Consider Comcast, which developed a well-earned reputation for poor customer service over the last couple of decades. Now the company has begun to turn around that reputation—though turning around a reputation takes time. Whereas Comcast continues to use inbound telemarketing to answer and respond to customer questions, another group inside the company trolls the Internet and accesses big data from Facebook, Twitter, blogs, and discussion forums to find anyone complaining about Comcast.

Analytics help group members turn these data into information and knowledge that helps them anticipate customer problems. Then Comcast reaches out to those customers to address their complaints. This proactive approach is beginning to turn around Comcast's reputation.[8]

Customer service reps are customer advocates

A breakdown in any element of the marketing mix can result in a requirement for customer service. Ideally, a firm should deliver what it promises, but marketing is a human process and mistakes do happen. Consider, for example, a customer who decides to use Verizon cell phone service because its ad—or salesperson—said that the first month of service would be free. If Verizon bills the customer for the first month, is it a pricing problem, a promotion problem, or a lack of coordination in the channel? From the customer's perspective, it really doesn't matter. What does matter is that expectations have been dashed. The customer doesn't need explanations or excuses but instead needs an advocate to make things right.

Sometimes the marketing mix is fine, but the customer makes a purchase that is a mistake. Or customers may simply change their minds. Either way, customers usually expect sellers to help fix purchasing errors. Firms need policies about how customer service reps should deal with customer errors. But most firms simply can't afford to alienate customers, even ones who have made an error, if they expect them to come back in the future. Sometimes the toughest sales job is figuring out how to keep a customer who is unhappy.

Regardless of whether the firm or customer causes the problem, customer service reps need to be effective communicators, have good judgment, and realize that they are advocates not only for their firm but also for its customers. As that implies, the rest of the company needs to be organized to provide the support reps need to fix problems. When customer service is done right, it promotes the customer's next purchase.

The Right Structure Helps Assign Responsibility

LO 14.4

We have described three sales tasks—order-getting, order-taking, and supporting. A sales manager must organize the sales force so that all the necessary tasks are done well. In many situations, a particular salesperson might be given two, or all three, of these tasks. For example, 10 percent of a particular job may be order-getting, 80 percent order-taking, and the additional 10 percent customer service. On the other hand, organizations are often structured to have different salespeople specialize in different sales tasks and by the target markets they serve.[9]

Sales tasks may be handled by a team

If different people handle different sales tasks, firms often rely on **team selling**—when different people work together on a specific account. Sometimes members of a sales team are not from the sales department at all. If improving the relationship with the customer calls for input from the quality control manager, then that person becomes a part of the team, at least temporarily. Producers of big-ticket items often use team selling. IBM uses team selling to sell information technology solutions for a whole business. Different specialists handle different parts of the job—but the whole team coordinates its efforts to achieve the desired result.

Big accounts get special treatment

Very large customers often require special sales efforts—and relationships with them are treated differently. Moen, a maker of plumbing fixtures, has a regular sales force to call on building material wholesalers and an elite **major accounts sales force** that sells directly to large accounts—such as Lowe's, Home Depot, Ferguson, or other major retail chains that carry plumbing fixtures.

You can see why this sort of special attention is justified when you consider Procter & Gamble's relationship with Walmart. Walmart is a $10 billion customer for P&G. Walmart accounts for a third or more of the total national sales in many of the

Some sellers use sales teams to sell to their very large accounts. These teams collaborate to address the customers' needs.
Morsa Images/Getty Images

product categories in which P&G competes. For instance, Walmart sells about one-third of the laundry detergent in the United States—including lots of P&G's brand Tide. Recently Walmart added a new brand of laundry detergent, Persil, and placed it on the shelf right next to Tide. P&G's major account sales team assessed the situation and increased Tide's promotion budget 30 percent to maintain its market share at Walmart.[10]

Some salespeople specialize in telephone selling

Some firms have a group of salespeople who specialize in **telemarketing**—using the telephone to call on customers or prospects. Fortunately, for those of us not wanting our dinner to be interrupted, the National Do Not Call Registry in the United States and similar laws in other countries have largely eliminated telemarketing to consumers. Registered users cannot be called except by nonprofits and a few other select groups. However, the reception to telephone selling in business markets is often quite different.

Inside sales forces are efficient and effective

Telephone selling, along with e-mail, video presentations, text messaging, and social media, is often used in B2B markets by an **inside sales force**—a sales force that meets with customers in a manner that is not face-to-face. Inside salespeople are often used with small or hard-to-reach customers a firm might otherwise promote to with mass selling or just ignore. But now many firms find an inside sales force to be more efficient and effective than a traditional face-to-face sales force. The big advantage of telephone selling by an inside sales group is that it saves time and money for the seller and it gives customers a fast and easy way to solve a purchasing problem. For example, many customers just call in to the inside sales force for assistance or to place an order. Telephone contact may supplement a good website; the website provides standard information, and an inside salesperson answers specific questions on the phone.

Companies that produce goods and services for final consumers also rely heavily on toll-free telephone lines to give final consumers easy access to customer service reps. In most cases, there is no other practical way for the producer to be sure that retailers are taking care of customers or their problems. A customer service call center provides a way for the producer to get direct feedback from customers—and perhaps find solutions to potential problems.[11]

Sales tasks are done in sales territories

Often companies organize selling tasks on the basis of a **sales territory**—a geographic area that is the responsibility of one salesperson or several working together. A territory might be a region of a country, a state, or part of a city, depending on the market potential. An airplane manufacturer like Boeing might consider a whole country as *part* of a sales territory for one salesperson.

Carefully set territories can reduce travel time and the cost of sales calls. Assigning territories can also help reduce confusion about who has responsibility for a set of sales tasks. Consider the Hyatt hotel chain. At one time, each hotel had its own salespeople to get bookings for big conferences and business meetings. That meant that people who had responsibility for selecting meeting locations might be called on by sales reps from 20 or 30 different Hyatt hotels. Now, the Hyatt central office divides up responsibility for working with specific accounts; one rep calls on an account and then tries to sell space in the Hyatt facility that best meets the customer's needs.

Specializing by product lines brings more expertise to customers

Sometimes simple geographic division doesn't serve customers' best interests. When companies sell different products that require very different knowledge or selling skills, it can make sense to have salespeople specialized by product line. For example, DuPont makes special films for hospital X-ray departments as well as chemicals used in laboratory blood tests. DuPont uses different salespeople (each an expert in his or her product line) who call on different people in the same hospitals. It might be a little less efficient for DuPont, but the salespeople are more effective and customers are happier to have experts in each product area.

Size of sales force depends on workload

Once the important sales tasks are specified and the responsibilities divided, the sales manager decides how many salespeople are needed. Sales managers engage in a systematic process to determine the optimal number. The first step is estimating how much work can be done by one person in a given time period. Then the sales manager figures out how many customers need attention. This information helps determine how many salespeople are required in total. To see how this analysis works in practice, see Marketing Analytics in Action: Workload and Sales Force Size.

Monitor sales tasks to keep the right number of salespeople

Some managers forget that over time the right number of salespeople may change as sales tasks change. Then when a problem becomes obvious, they try to change everything in a hurry—a big mistake. Consideration should be ongoing as to what type and how many salespeople are needed. If the sales force needs to be reduced, it doesn't make sense to let a lot of people go all at once, especially when that could be avoided with some planning.

Marketing Analytics in Action: Workload and Sales Force Size

For many years, the Parker Jewelry Company was very successful selling its silver jewelry to department and jewelry stores in the southwestern United States. But top managers wanted to expand into the big urban markets in the northeastern United States. They realized that most of the work for the first few years would require order getters. They felt that a salesperson would need to call on each account at least once a month to get a share of this competitive business. Based on their experience in the Southwest, they estimated that a salesperson could make only five calls a day on prospective buyers and still allow time for travel, waiting, and follow-up on orders that came in. This meant that a sales rep who made calls 20 days per month could handle about 100 stores (5/day × 20 days). The sales manager purchased a database that showed there were 1,182 jewelry stores in the Northeast that would be good targets for their jewelry.

1. *How many salespeople would Parker Jewelry need to hire to adequately serve its new northeastern United States target market?*
2. *How many salespeople should Parker Jewelry hire to pursue a more focused target market of the 309 jewelry stores in New York State?*
3. *The manager thought that an inside sales force could use telemarketing and handle 12 calls per day (for 20 days per month). The salespeople would still try to make contact once per month. If Parker Jewelry wanted to target those original 1,182 jewelry stores, how many inside salespeople would it need to hire?*

Information Technology Provides Tools to Do the Job

LO 14.5

Sometimes technology substitutes for or complements personal selling

Some sales tasks that have traditionally been handled by a human can now be handled effectively (often more effectively) and at lower cost by using artificial intelligence, machine learning, an e-commerce system, or other technology. Whether technology can replace a human in doing the personal selling can be explained in large part by two factors reflected in the two-by-two diagram in Exhibit 14-3. Organizations might examine traditional selling tasks based on the amount of standardized information (orders, invoices, delivery status, product information, and prices) regularly exchanged and the level of relationship building (problem solving, coordination, support, and cooperation).

High relationship + low information sharing = emphasis on personal selling

In many selling situations, there is a great deal of problem solving and coordination needed. These more complicated sales require a salesperson to create and build relationships. When information is not standardized—in other words, the information swapped back and forth is changing all the time—we need to focus on personal selling alone (see lower right box in Exhibit 14-3). A salesperson selling trucking services may find that there are relatively fewer transactions and each transaction and price is unique—so technology cannot easily substitute for personal selling.

Increased information sharing adds e-commerce

If relationships are needed, but there is a great deal of standard information—for example, routine ordering, pricing, and delivery status—then the salesperson should be supplemented by a customized e-commerce system or electronic data interchange (see Chapter 11) that allows the buyer continuous access to needed information (see upper right box in Exhibit 14-3). Walmart may have this type of relationship with a large supplier like Procter & Gamble. There is plenty of planning and coordination that requires a salesperson (or sales team), but with thousands of products and hundreds of stores, regular orders and invoicing are pretty standardized. Computers are the best choice to manage that flow of information.

When relationship building is not required

There are a growing number of situations where little or no salesperson relationship building is needed. If there is a lot of standardized information flowing, but no relationship building is needed (upper left box in Exhibit 14-3), then setting up a standard e-commerce or electronic data interchange where computers exchange information about stock levels and orders may be sufficient. Customer service can be a phone call or e-mail away and might be contacted for exceptions. Many supplier–wholesaler relationships find this type of exchange to be efficient and effective.

Exhibit 14-3 Examples of Possible Personal Selling Emphasis

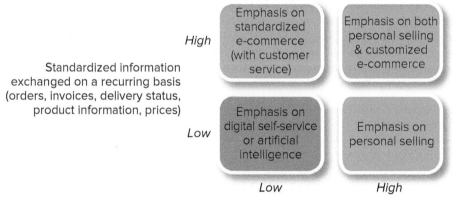

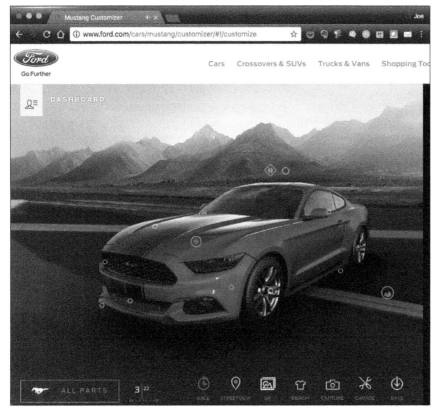 

Ford Motor Company's Mustang Customizer offers customers self-service technology that allows them to build a custom Ford Mustang. After building their car, they can create a gif (image of their car) they can display on social media. They can also "drive" their virtual car in an online racing game. Most important for Ford, the customer can find a local dealer and purchase the real car. Drift is a company that makes chatbots—artificial intelligence that answers customer questions. Drift's own chatbot, named Driftbot, initiates a text message conversation when a customer visits Drift's website. Once Driftbot learns why the customer has visited the site, it transitions the customer to a salesperson to continue the chat.
(left): Ford Motor Company; (right): Drift.com, Inc.

Digital self-service works when relationship building and standardized information needs are low

Other situations have less frequent transactions and no need for a relationship at all (purple box in Exhibit 14-3). This is a situation where digital self-service makes sense. This is the role of ATMs for banks. If the customer needs money at an airport in the middle of the night, the ATM provides better support than the customer could get with a real person at the bank. Many firms provide self-service at their websites. For example, at HPShopping.com customers configure computers to their own specs. When looking at options such as "How much memory to buy," customers simply click on the "Help me decide" button for more information. Some firms tap into artificial intelligence and chatbots to deliver anytime–anywhere communication with customers.

Intelligent agents make recommendations and . . .

Often salespeople are placed in a role of recommending a particular product to a customer. A salesperson asks questions, carefully listens to a customer's answers, and then makes a recommendation about which product might best serve a customer's unique needs. In Chapter 3 we introduced the concept of an intelligent agent, a device that observes an environment and achieves a goal. In the sales context, an intelligent agent can either observe a customer's behavior or ask questions and then offer recommendations.

At fine hotels, guests may talk to a concierge (a kind of "salesperson" with a desk in the lobby) who offers advice about dining or entertainment options. The Cosmopolitan Hotel created an artificial intelligence chatbot with a sassy personality named Rose. Hotel guests text Rose with questions. Rose helps guests with restaurant reservations, spa treatments, and event tickets—even offering inside information like secret menu items. Rose is a good salesperson; guests who engage with Rose spend almost 40 percent more than those who don't engage.[12]

... deliver on-demand customer service

Intelligent agents are particularly adept at delivering high-quality customer service—at a low cost. These chatbots answer customer questions quickly and accurately—which is why most customers come to customer service in the first place. For example, when customers call Autodesk for customer assistance, an intelligent agent understands and answers most verbal questions. While questions can be elevated to a real human customer service agent, the vast majority are handled quickly by an intelligent agent.

Artificial intelligence and sales managers' judgments

Some activities traditionally performed by sales managers may be replaced by intelligent agents. For example, sales managers sometimes assign inside salespeople to specific customers—seeking to match a salesperson's personality and skill set with customers' needs. For outside salespeople, a sales manager may try to match a salesperson with a territory. And data and predictive analytics might be able to match customers with salespeople. For more on this idea, see *What's Next?* Analytics puts the right sales rep with the right customer.

Changes in how salespeople work

New sales technology tools also change how sales tasks and responsibilities are planned and handled. It is usually the sales manager's job—perhaps with help from specialists in technology—to decide what types of tools are needed and how they will be used.

To get a sense of what is involved, consider a day in the life of a sales rep for a large consumer packaged goods firm. Over a hasty breakfast, she plans the day's sales calls on her laptop's organizer, checks a LinkedIn contact to learn more about a buyer she will meet for the first time that afternoon, watches a 15-minute video presentation for a new product, and sorts through a dozen e-mail messages. One e-mail is from a buyer for a supermarket chain. Sales in the chain's paper towel category are off 10 percent, and he wants to know if the rep can help. The rep downloads sales trend data for the chain and its competitors from her firm's intranet. A spreadsheet analysis of the data reveals that the sales decline is due to new competition from warehouse clubs.

After a videoconference with a brand manager and a company sales promotion specialist to seek advice, she prepares a PowerPoint presentation, complete with a proposed shelf-space plan that recommends the buyer promote larger-sized packages of both her company's and competitors' brands. Before leaving home, the rep e-mails an advance copy of the report to the buyer and her manager. In her car, she calls the buyer to schedule an appointment. After her appointment, she opens Salesforce.com and enters information about this customer. Technology is thoroughly integrated into the day of the modern salesperson.[13]

Salespeople use CRM software, like that provided by Salesforce.com, to track and manage customer relationships.
Salesforce.com, Inc.

Software and hardware provide competitive advantage

The sales rep in this example relies on support from an array of software and hardware, some of which wasn't even available a decade ago. Software for customer relationship management, spreadsheet sales analysis, digital presentations, time management, sales forecasting, customer contact, and shelf-space management is at the salesperson's fingertips—most of it available right online. Commonplace hardware includes everything from smartphones and tablet computers to personal videoconferencing systems. In

What's Next? Analytics puts the right sales rep with the right customer

If matching the right salesperson with the right customer is *What's Next?*, Assurant Solutions might be leading the way to this future. Assurant Solutions sells credit insurance. Credit insurance pays insured people's mortgage, car, or credit card payments if they lose their job or experience major medical problems. Customers often buy credit insurance as an add-on when making large purchases, such as a home or car. When customers make payments, they see an additional monthly fee for the credit insurance. Sometimes that monthly fee gets some customers second-guessing their need for the insurance; they decide they might be better off taking their chances. When that happens, they call Assurant's customer service team to cancel the policy. With customer retention important to profitability, this team's job is to prevent customers from dropping the coverage.

Assurant's customer service group used to be operationally optimized—staffed to ensure 80 percent of customers waited less than 20 seconds to speak to a customer service representative (CSR). Inbound phone calls were routed to CSRs with expertise in the product the customer owned, and computer screens displayed scripts that CSRs followed and offers they could make to try to keep a customer from following through on canceling a policy. With all those efforts, Assurant's CSRs retained just 16 percent of those calling to cancel—above average for the industry, but not good enough. Assurant figured there was an opportunity for improvement. So it turned to analytics.

Assurant's customer service managers have access to lots of data: on customers, products, CSRs, and of course the outcome from each transaction. Assurant also has plenty of statisticians who know how to predict outcomes by mining data for information. After analyzing past transactions, they identified patterns that helped predict success (retaining a customer) and failure when customers call in to cancel a policy. Assurant discovered that different CSRs were more successful dealing with certain kinds of customers and problems.

Assurant's analysts built models that predict which CSRs will be most effective with each type of customer. Inbound calls are analyzed in real time, so each call is instantly routed to the CSR expected to have the best chance of keeping the customer's business. If the best CSR for a specific customer is busy on another call, then it rolls to the next-best CSR, and so on down the line. By better matching customers and CSRs, Assurant *doubled* its retention rate. The new system also prioritizes customers with the biggest premiums, and this *tripled* the dollars retained.

Although the analytics don't tell *why* some reps outperform others, matching customers and CSRs works. Assurant's customer service managers hypothesize the success comes from CSRs building better rapport with customers. The analytics know enough about each customer and CSR to create matches where the possibility of connecting is greatest. Creating the right match makes it easier to build a foundation of trust, and persuasion and retention follow.[14]

Is this *What's Next?* in other sales situations? Discuss how analytics like those described here could be used at a company like Ferguson Enterprises (the chapter opening case). How could it be done? Would it be practical?

many situations, these technologies give sales reps new ways to meet customers' needs while achieving the objectives of their jobs.

These tools change how well the job is done. Yet this is not simply a matter that is best left to individual sales reps. Use of these tools may be necessary just to compete effectively. For example, if a customer expects a sales rep to access data on past sales and provide an updated sales forecast, a sales organization that doesn't have this capability will be at a real disadvantage in keeping that customer's business.

On the other hand, these tools have costs. There is an obvious expense of buying the technology. But there is also the training cost of keeping everyone up-to-date. Often that is not an easy matter. Some salespeople who have done the sales job well for a long time "the old-fashioned way" resent being told that they have to change what they are doing, even if it is what customers expect. So if a firm expects salespeople to be able to use these technologies, that requirement needs to be included in selecting and training people for the job.[15]

Sound Selection and Training to Build a Sales Force

LO 14.6

Selecting good salespeople takes judgment

It is important to hire *well-qualified* salespeople who will do a good job. But selection in many companies is done without serious thought about exactly what kind of person the firm needs. Managers may hire friends and relations, or whoever is available, because they feel that the only qualification for a sales job is a friendly personality. This approach leads to poor sales, lost customers, and costly sales force turnover.

Progressive companies are more careful. They constantly update a list of possible job candidates. They invite applications at the company's website. They schedule candidates for multiple interviews with various executives and do thorough background checks. Unfortunately, such techniques don't guarantee success. But a systematic approach based on several different inputs results in a better sales force.

One problem in selecting salespeople is that two different sales jobs with identical titles may involve very different selling or supporting tasks and require different skills. A carefully prepared job description helps avoid this problem.

Job descriptions should be in writing and specific

A **job description** is a written statement of what a salesperson is expected to do. It might list 10 to 20 specific tasks—as well as routine prospecting and sales report writing. Each company must write its own job specifications. And it should provide clear guidelines about what selling tasks the job involves. This is critical to determine the kind of salespeople who should be selected—and later it provides a basis for seeing how they should be trained, how well they are performing, and how they should be paid.

Good salespeople are trained, not born

The idea that good salespeople are born that way may have some truth—but it isn't the whole story. A salesperson needs to be taught about the company and its products, giving effective sales presentations, using appropriate sales technology, and building relationships with customers. But this isn't always done. Many salespeople do a poor job because they haven't had good training. Firms often hire new salespeople and immediately send them out on the road, or the retail selling floor, with no grounding in the basic selling steps and no information about the product or the customer. They just get a price list and a pat on the back. This isn't enough!

All salespeople need some training

It's up to sales and marketing management to be sure that salespeople know what they're supposed to do and how to do it. HP faced this problem. For years the company was organized into divisions based on different product lines—printers, network servers,

Customers who rent heavy construction equipment want to deal with a knowledgeable salesperson. CAT selects salespeople who have experience with the applications for which the equipment will be used and gives them training on CAT products and new developments in the market.
Caterpillar

and the like. However, sales reps who specialized in the products of one division often couldn't compete well against firms that could offer customers total solutions to computing problems. When a new top executive came in and reorganized the company, all sales reps needed training in their new responsibilities, how they would be organized, and what they should say to their customers about the benefits of the reorganization.

Sales training should be modified based on the experience and skills of the group involved. But the company's sales training program should cover at least the following areas: (1) company policies and practices, (2) product information, (3) how to build relationships with customer firms, and (4) professional selling skills.

Selling skills can be learned

Many companies spend the bulk of their training time on product information and company policy. They neglect training in selling techniques because they think selling is something anyone can do. But training in selling skills can pay off. Estée Lauder, for example, has selling skills for the "beauty advisors" who sell its cosmetics down to a fine art—and its training manual and seminars cover every detail. Estée Lauder advisors who take the training seriously immediately double their sales.[16] Training can also help salespeople learn how to be more effective in cold calls on new prospects, in listening carefully to identify a customer's real objections, in closing the sale, and in working with customers in difficult customer service situations.

Training often starts in the classroom with lectures, case studies, and streamed video of trial presentations and demonstrations. But a complete training program adds on-the-job observation of effective salespeople and coaching from sales supervisors. Many companies also use web-based training, weekly sales meetings or work sessions, annual conventions, and regular e-mail messages and newsletters, as well as ongoing training sessions, to keep salespeople up-to-date.[17]

Compensating and Motivating Salespeople

LO 14.7

To recruit, motivate, and keep good salespeople, a firm has to develop an effective compensation plan. Ideally, sales reps should be paid in such a way that what they want to do—for personal interest and gain—is in the company's interest too. Most companies focus on financial motivation—but public recognition, sales contests, and simple personal recognition for a job well done can be highly effective in encouraging greater sales effort.[18] Our main emphasis here, however, will be on financial motivation.[19]

Two basic decisions must be made in developing a compensation plan: (1) the level of compensation and (2) the method of payment.

Compensation varies with job and needed skills

To build a competitive sales force, a company must pay at least the going market wage for different kinds of salespeople. To be sure it can afford a specific type of salesperson, the company should estimate—when the job description is written—how valuable such a salesperson will be. A good order getter may be worth more than $100,000 or more to one company but only $25,000 to $35,000 to another—just because the second firm doesn't have enough to sell! In such a case, the second company should rethink its job specifications, or completely change its promotion plans, because the going rate for order getters is much higher than $35,000 a year.

If a job requires extensive travel, aggressive order-getting, or customer service contacts with troublesome customers, the pay may have to be higher. But the salesperson's compensation level should compare, at least roughly, with the pay scale of the rest of the firm. Normally, salespeople earn more than the office or production force but less than top management.

Payment methods vary

Given some competitive level of compensation, there are three basic methods of payment: (1) straight salary, (2) straight commission (incentive), or (3) a combination plan. A straight salary offers the most security for the salesperson. Commission pay, in contrast, offers the most incentive and is tied to results actually achieved. A commission is often based on a percentage of dollar sales, but it may be a financial incentive based on other outcomes—such as the number of new accounts, customer satisfaction ratings, or customer service problems resolved in some time period. Most salespeople want some security, and most companies want salespeople to have some incentive to work smarter and harder, so the most popular method is a combination plan that includes some salary and some commission. Bonuses, profit sharing, pensions, stock plans, insurance, and other fringe benefits may be included too.

Salary gives control— if there is close supervision

A salesperson on straight salary earns the same amount regardless of how he or she spends time. So the salaried salesperson is expected to do what the sales manager asks—whether it is order-taking, supporting sales activities, solving customer service problems, or completing sales call reports. However, the sales manager maintains control *only* by close supervision. As a result, straight salary or a large salary element in the compensation plan increases the amount of sales supervision needed.

Commissions can both motivate and direct

If personal supervision would be difficult, a firm may get better control with a compensation plan that includes some commission, or even a straight commission plan, with built-in direction. One trucking company knows its most profitable product is to fill trucks that would otherwise return empty. So its sales incentive plan pays higher commissions on business needed to balance freight shipments—depending on how heavily traffic has been moving in one direction or another. Another company that wants to motivate its salespeople to devote more time to developing new accounts could pay higher commissions on shipments to new customers. However, a salesperson on a straight commission tends to be his or her own boss. The sales manager is less likely to get help on sales activities that won't increase the salesperson's earnings.

Incentives should link efforts to results

The incentive portion of a sales rep's compensation should be large only if there is a direct relationship between the salesperson's efforts and results. Otherwise, a salesperson in a growing territory might have rapidly increasing earnings, while the sales rep in a poor area will have little to show for the same amount of work. Such a situation isn't fair, and it can lead to high turnover and much dissatisfaction. A sales manager can take such differences into consideration when setting a salesperson's **sales quota**—the specific sales or profit objective a salesperson is expected to achieve. Often a salesperson receives a bonus for meeting the sales quota.

Commissions reduce need for working capital

Small companies that have limited working capital or uncertain markets often prefer straight commission, or combination plans with a large commission element. When sales drop off, costs do too. Such flexibility is similar to using manufacturers' agents who get paid only if they deliver sales. This advantage often dominates in selecting a sales compensation method. Exhibit 14-4 shows the general relation between personal selling expense and sales volume for each of the basic compensation alternatives.

In order to motivate salespeople, some companies reward those who meet sales targets. Rewards might include travel—like a Royal Caribbean cruise.
Royal Caribbean International

Compensation plans should be clear

Salespeople are likely to be dissatisfied if they can't see the relationship between the results they produce and their pay.

Exhibit 14-4 Relation between Total Selling Expense and Sales Volume

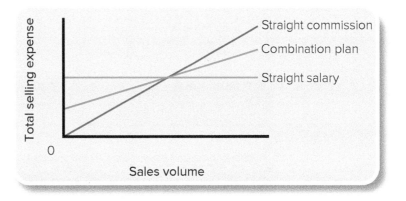

A compensation plan that includes different commissions for different products or types of customers can become quite complicated. Simplicity is best achieved with straight salary. But in practice, it's usually better to sacrifice some simplicity to gain some incentive, flexibility, and control. The best combination of these factors depends on the job description and the company's objectives.

To make it easier for a sales rep to see the relationship between effort and compensation, some firms provide the rep with that information online. For example, sales reps at Oracle, a company that sells database systems, can check a website and see how they are doing. As new sales results come in, the report at the website is updated. Sales managers can also make changes quickly—for example, by putting a higher commission on a product or more weight on customer satisfaction scores.[20]

Sales managers must plan, implement, and control

Managers must regularly evaluate each salesperson's performance and be certain that all the needed tasks are being done well. The compensation plan may have to be changed if the pay and work are out of line. And by evaluating performance, firms can also identify areas that need more attention—by the salesperson or management.[21]

Personal Selling Techniques—Prospecting and Presenting

LO 14.8

We've stressed more detail so you understand the basic steps each salesperson should follow—including prospecting and selecting target customers, planning sales presentations, making sales presentations, and following up after the sale. Exhibit 14-5 shows the steps we'll consider. You can see that the salesperson is just carrying out a planned communication process, as we discussed in Chapter 13.[22]

Prospecting—narrowing down to the right target

Narrowing down the personal selling effort to the right target requires constant, detailed analysis of markets and much prospecting. Basically, **prospecting** involves following all the leads in the target market to identify potential customers.

Finding live prospects who will help make the buying decision isn't as easy as it sounds. In business markets, for example, the salesperson may need to do some hard detective work to find the real purchase decision makers.

Some companies provide prospect lists or a customer relationship management (CRM) database to make this part of the selling job easier. The CRM database may be integrated with other marketing communication tools to help salespeople spend more time working on the best prospects. ThoughtLava, a website design firm, uses its CRM database to initially contact prospects by e-mail. It uses software that tracks which prospects open the e-mail, which click through to the firm's website, and even which pages they visit. Given this information, ThoughtLava's salespeople know in advance which of the firm's services interest each prospect, and that helps them decide which prospects to focus on.[23]

Exhibit 14-5 Key Steps in the Personal Selling Process

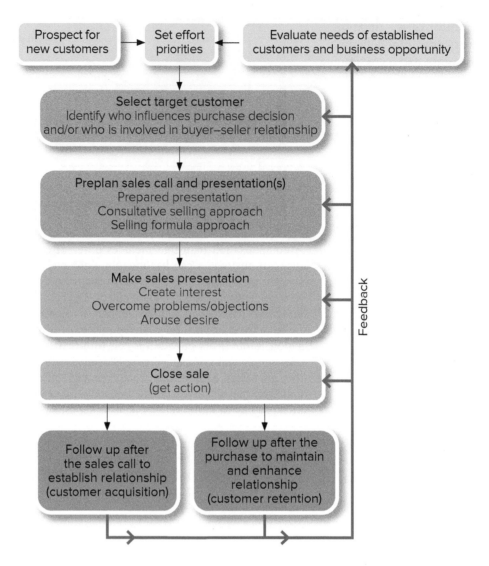

All customers are not equal

Although prospecting focuses on identifying new customers, established customers require attention too. It's often time-consuming and expensive to establish a relationship with a customer, so once established it makes sense to keep the relationship healthy. That requires the rep to routinely review active accounts, rethink customers' needs, and reevaluate each customer's long-term business potential. Some small accounts may have the potential to become big accounts, and some accounts that previously required a lot of costly attention may no longer warrant it. So a sales rep may need to set priorities both for new prospects and existing customers.

How long to spend with whom?

Once a set of prospects and customers who need attention has been identified, the salesperson must decide how much time to spend with each one. A sales rep must qualify customers—to see if they deserve more effort. The salesperson usually makes these decisions by weighing the potential sales volume as well as the likelihood of a sale. This requires judgment. But well-organized salespeople usually develop some system because they have too many demands on their time.[24]

Many firms provide their reps with CRM systems to help with this process. Most of them use some grading scheme. A sales rep might estimate how much each prospect is likely to purchase and the probability of getting and keeping the business, given the competition. The computer then combines this information and grades each prospect. Attractive accounts may be labeled A—and the salesperson may plan to call on them weekly until the sale is made, the relationship is in good shape, or the customer is moved into a lower category. B customers might offer somewhat lower potential and be called on monthly. C accounts might be called on only once a year—unless they happen to contact the salesperson. And D accounts might be transferred to a telemarketing group.[25]

Three kinds of sales presentations may be useful

Once the salesperson selects a target customer, it's necessary to make a **sales presentation**—a salesperson's effort to make a sale or address a customer's problem. But someone has to plan what kind of sales presentation to make. This is a strategy decision. The kind of presentation should be set before the sales rep goes calling. And in situations where the customer comes to the salesperson—in a retail store, for instance—planners have to make sure that prospects are brought together with salespeople.

Salespeople are constantly looking for ways to be more efficient in identifying sales leads and prospects. Some companies look to an outside company like Capterra to help generate leads and prospects.
Capterra Inc.

A marketing manager can choose two basically different approaches to making sales presentations: the prepared approach or the consultative selling approach. Another approach, the selling formula approach, is a combination of the two. Each of these has its place.

The prepared sales presentation

The **prepared sales presentation** approach uses a memorized presentation that is not adapted to each individual customer. This approach says that a customer faced with a particular stimulus will give the desired response—in this case, a yes answer to the salesperson's prepared statement, which includes a **close**, the salesperson's request for an order.

If one trial close doesn't work, the sales rep tries another prepared presentation and attempts another closing. This can go on for some time—until the salesperson runs out of material or the customer either buys or decides to leave. Exhibit 14-6 shows the relative participation of the salesperson and customer in the prepared approach. Note that the salesperson does most of the talking.

Firms may rely on this canned approach when only a short presentation is practical. It's also sensible when salespeople aren't very skilled. The company can control what they say and in what order. The approach is often used by telemarketers who may have only a few minutes to get a customer's attention and generate interest. For example, Novartis uses missionary salespeople to tell doctors about new drugs when they're introduced. Doctors are busy,

Exhibit 14-6 Prepared Approach to Sales Presentation

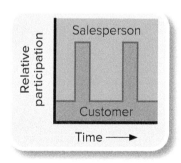

so they give the rep only a minute or two. That's just enough time to give a short, prepared pitch and leave some samples. To get the most out of the presentation, Novartis refines it based on feedback from doctors whom it pays to participate in focus groups.[26]

But a canned approach has a weakness. It treats all potential customers alike. It may work for some and not for others. A prepared approach may be suitable for simple order-taking—but it is no longer considered good selling for complicated situations.

Consultative selling builds on the marketing concept

The **consultative selling approach** involves developing a good understanding of the individual customer's needs before trying to close the sale. This name is used because the salesperson is almost acting as a consultant to help identify and solve the customer's problem. With this approach, the sales rep makes some general benefit statements to get the customer's attention and interest. Then the salesperson asks questions and *listens carefully* to understand the customer's needs. Once they agree on needs, the seller tries to show the customer how the product fills those needs and to close the sale. This is a problem-solving approach—in which the customer and salesperson work together to satisfy the customer's needs. That's why it's sometimes called the *need-satisfaction approach*. Exhibit 14-7 shows the participation of the customer and the salesperson during such a sales presentation.

Exhibit 14-7 Consultative Selling Approach to Sales Presentation

The consultative selling approach takes skill and time. The salesperson must be able to analyze what motivates a particular customer and show how the company's offering would help the customer satisfy those needs. The sales rep may even conclude that the customer's problem is really better solved with someone else's product. That might result in one lost sale, but it also is likely to build real trust and more sales opportunities over the life of the relationship with the customer.

Although consultative selling is commonly used in B2B markets, retailers can build relationships with customers by using this approach. Salespeople at Nordstrom department stores are well known for their attention to customers. They often know their regular customers and will ask a range of questions to make sure they get exactly what they need. A customer looking for a pair of dress pants may be asked a half-dozen questions by a salesperson—"What will you be using them for?" "Do you expect to travel with them?" "Do you prefer something lightweight for summer, or do you want to wear them year-round?" And more. These questions help the salesperson recommend pants that fit the customer's precise needs. This retail salesperson is the customer's consultant on pants. Customers are rewarded with the "perfect fit," and the salesperson and Nordstrom earn the customer's loyalty.[27]

Selling formula approach—some of both

The **selling formula approach** starts with a prepared presentation outline—much like the prepared approach—and leads the customer through some logical steps to a final close. The prepared steps are logical because we assume that we know something about the target customer's needs and attitudes.

Exhibit 14-8 shows the selling formula approach. The salesperson does most of the talking at the beginning of the presentation—to communicate key points early. This part of the presentation may even have been prepared as part of the marketing strategy. As the sales presentation moves along, however, the salesperson brings the customer into the discussion to help clarify just what needs this

Exhibit 14-8 Selling Formula Approach to Sales Presentation

customer has. The salesperson's job is to discover the needs of a particular customer to know how to proceed. Once it is clear what kind of customer this is, the salesperson comes back to show how the product satisfies this specific customer's needs and to close the sale.

AIDA helps plan sales presentations

Most sales presentations follow the AIDA sequence: Attention, Interest, Desire, Action. The time a sales rep spends on each of the steps varies depending on the situation and the selling approach being used. But it is still necessary to begin a presentation by getting the prospect's *attention* and, hopefully, to move the customer to *action*.[28]

Ethical issues may arise

As in every other area of marketing communications, ethical issues arise in the personal selling area. The most basic issue, plain and simple, is whether a salesperson's presentation is honest and truthful. But addressing that issue is a no-brainer. No company is served well by a salesperson who lies or manipulates customers to get their business.

> **Ethical Dilemma**
>
> *What would you do?* Assume that you are a sales rep and sell costly electronic systems for use in automated factories. You made a sales presentation to a customer, but he didn't place an order—and then wouldn't take your calls when you tried to inform him that your company was coming out with a more reliable model at the same price. Months later, he e-mails a purchase order for immediate delivery on the model you originally discussed. You have the old model in stock, and it will be difficult to sell once the new model arrives in two weeks. In fact, your company has doubled the usual commission rate to clear out the old model.
>
> *Do you try to contact the customer again to tell him about the new model, or do you do what he has requested and immediately fill the order with the old model? Either way, if you make the sale, the commission will pay for your upcoming vacation to the Caribbean. Explain what you would do and why.*

Most sales reps sooner or later face a sales situation in which they must make more difficult ethical decisions about how to balance company interests, customer interests, and personal interests. Conflicts are less likely to arise if the firm's marketing mix really meets the needs of its target market. Similarly, they are less likely to occur when the firm sees the value of developing a long-term relationship with the customer. Then the salesperson is arranging a happy marriage. By contrast, ethical conflicts are more likely when the sales rep's personal outcomes (such as commission income) or the selling firm's profits hinge on making sales to customers whose needs are only partially met by the firm's offering. A number of financial services firms, for example, have garnered bad publicity—and even legal problems—from situations like this.

Ideally, companies can avoid the whole problem by supporting their salespeople with a marketing mix that really offers target customers unique benefits. Moreover, top executives, marketing managers, and sales managers set the tone for the ethical climate in which a salesperson operates. If they set impossible goals or project a "do-what-you-need-to-do" attitude, a desperate salesperson may yield to the pressure of the moment. When a firm clearly advocates ethical selling behavior and makes it clear that manipulative selling techniques are not acceptable, the salesperson is not left trying to swim against the flow.[29]

CONCLUSION

In this chapter, we discussed the importance and nature of personal selling. Selling is much more than just getting rid of the product. In fact, a salesperson who is not given strategy guidelines may have to become the strategy planner for the market he or she serves. Ideally, however, the sales manager and marketing manager work together to set some strategy guidelines: the kind and number of salespeople needed; what sales technology support will be provided; the kind of sales presentation desired; and selection, training, and motivation approaches.

We discussed the three basic sales tasks: (1) order-getting, (2) order-taking, and (3) supporting. Most sales jobs combine at least two of these three tasks. We also considered the role of customer service and why it is so important to a firm and its customers. Once a firm specifies the important tasks, it can decide on the structure of its sales organization and the number of salespeople it needs. The nature of the job and the level and method of compensation also depend on the blend of these tasks. Firms should develop a job description for each sales job. This, in turn, provides guidelines for selecting, training, and compensating salespeople.

Once the marketing manager agrees to the basic plan and sets the budget, the sales manager must implement the plan, including directing and controlling the sales force. This includes assigning sales territories and controlling performance. A sales manager is deeply involved with the basic management tasks of planning and control—as well as ongoing implementation of the personal selling effort. We also explored the role of artificial intelligence, intelligent agents, and other types technology in making personal selling and sales management more effective and efficient.

We also reviewed some basic selling techniques and identified three kinds of sales presentations. Each has its place—but the consultative selling approach seems best for higher-level sales jobs. In these kinds of jobs, personal selling is achieving a new, professional status because of the competence and level of personal responsibility required of the salesperson. The day of the old-time glad-hander is passing in favor of the specialist who is creative, industrious, persuasive, knowledgeable, highly trained, and therefore able to help the buyer. This type of salesperson always has been, and probably always will be, in short supply. And the demand for high-level salespeople is growing.

KEY TERMS

LO 14.9

basic sales tasks, 383
order getters, 383
order-getting, 383
order takers, 384
order-taking, 384
supporting salespeople, 385
missionary salespeople, 386
technical specialists, 386

customer service reps, 386
team selling, 389
major accounts sales force, 389
telemarketing, 390
inside sales force, 390
sales territory, 390
job description, 396
sales quota, 398

prospecting, 399
sales presentation, 401
prepared sales presentation, 401
close, 401
consultative selling approach, 402
selling formula approach, 402

QUESTIONS AND PROBLEMS

1. Review the Ferguson case study that opens the chapter. From this case, identify examples of different key terms and concepts covered in the chapter. For example, it mentions that salespeople specialize by customer segments—part of getting the right structure.

2. Review the Ferguson case study that opens the chapter. The case does not mention how the sales force is compensated—salary, commission, or some combination. Assume they are compensated by straight (100 percent) salary and the company changes to a straight (100 percent) commission compensation. How might this affect how the salespeople behave? What behaviors might become more common? What behaviors might become less common?

3. What strategy decisions are needed in the personal selling area? Why should the marketing manager make these strategy decisions?

4. What kind of salesperson (or what blend of the basic sales tasks) is required to sell the following products? If there are several selling jobs in the channel for each product, indicate the kinds of salespeople required. Specify any assumptions necessary to give definite answers. (a) laundry detergent, (b) corporate jets, (c) office furniture, (d) men's underwear, (e) mattresses, (f) a new software that delivers artificial intelligence in customer service, and (g) life insurance.

5. Distinguish among the jobs of producers', wholesalers', and retailers' order-getting salespeople. If one order getter is needed, must all the salespeople in a channel be order getters? Illustrate.

6. Discuss the role of the manufacturers' agent in a marketing manager's promotion plans. What kind of salesperson is a manufacturers' agent? What type of compensation plan is used for a manufacturers' agent?

7. Discuss the future of the specialty shop if producers place greater emphasis on mass selling because of the inadequacy of retail order-taking.

8. Compare and contrast missionary salespeople and technical specialists.

9. Think about a situation when you or a friend or family member encountered a problem with a purchase and tried to get help from a firm's customer service representative. Briefly describe the problem, how the firm handled it, and what you think about the firm's response. How could it have been improved?

10. Would it make sense for your school to have a person or group whose main job is to handle "customer service" problems? Explain your thinking.
11. A firm that produces mixes for cakes, cookies, and other baked items has an incoming toll-free line for customer service calls. The manager of the customer service reps has decided to base about a third of their pay on the number of calls they handle per month and on the average amount of time on the phone with each customer. What do you think are the benefits and limitations of this incentive pay system? What would you recommend to improve it?
12. Explain how a compensation plan could be developed to provide incentives for experienced salespeople and yet make some provision for trainees who have not yet learned the job.
13. Cite an actual local example of each of the three kinds of sales presentations discussed in the chapter. Explain for each situation whether a different type of presentation would have been better.
14. Describe a consultative selling sales presentation that you experienced recently. How could it have been improved by fuller use of the AIDA framework?
15. How would our economy operate if personal salespeople were outlawed? Could the economy work? If so, how? If not, what is the minimum personal selling effort necessary? Could this minimum personal selling effort be controlled by law?

MARKETING PLANNING FOR HILLSIDE VETERINARY CLINIC

Appendix D (the Appendices follow Chapter 19) includes a sample marketing plan for Hillside Veterinary Clinic. Look through the "Marketing Strategy" section.

1. What personal selling tasks are performed at Hillside Veterinary Clinic and who does them?
2. If Hillside wanted to put more emphasis on "order-getting" to promote growth, what ideas do you have for how to do it?
3. Based on the situation analysis, target market, and intended positioning, recommend some ways that Hillside could actively work to improve its reputation for customer service.

SUGGESTED CASES

12. DrV.com—Custom Vitamins
16. Hanratty Company
17. Wise Water, Inc.
21. GeoTron International (GTI)
25. QXR Tools (QXR)

MARKETING ANALYTICS: DATA TO KNOWLEDGE

CHAPTER 14: SALES COMPENSATION

Franco Welles, sales manager for Nanek, Inc., is trying to decide whether to pay a sales rep for a new territory with straight commission or a combination plan. He wants to evaluate possible plans—to compare the compensation costs and profitability of each. Welles knows that sales reps in similar jobs at other firms make about $36,000 a year.

The sales rep will sell two products. Welles is planning a higher commission for Product B—because he wants it to get extra effort. From experience with similar products, he has some rough estimates of expected sales volume under the different plans and various ideas about commission rates. The details are found in the spreadsheet. The program computes compensation and how much the sales rep will contribute to profit. "Profit contribution" is equal to the total revenue generated by the sales rep minus sales compensation costs and the costs of producing the units.

Design element: #M4BW box globe icon: ©Vectoryzen/Shutterstock

CHAPTER FIFTEEN

Source: Domino's IP Holder LLC

Advertising and Sales Promotion

Looking for a small-business venture to help pay for college, brothers Tom and James Monaghan borrowed $900 and bought Domi-Nick's Pizza in Ypsilanti, Michigan. A year later, Tom bought out his brother, swapping him a Volkswagen Beetle for his share of the business. He renamed the business Domino's and focused on delivering hot pizza. Pizza delivery was still new back then, so Domino's used pioneering advertising to build customer awareness of this service. Over time, Domino's developed a delivery and marketing "system" to sell pizzas to the price-sensitive college student target market. Domino's signed up franchisees in college towns and the business took off.

Once customers got used to delivery, Domino's figured out how to make, bake, and deliver pizzas fast. Domino's advertised a "30 minutes or it's free" delivery guarantee. That promise was a major part of Domino's advertising until customers got the impression its delivery drivers valued speed over safety. To ensure pizzas still arrived hot, Domino's introduced an innovative packaging concept called HeatWave. Domino's also rolled out flavored crusts as limited-time promotions to stimulate customer interest. Supported by advertising that built demand for the Domino's brand, the company kept growing.

With all that growth, it was an exciting time to open a Domino's franchise. Franchisees especially valued Domino's well-known brand name and ongoing investments in advertising. Domino's regularly ran national advertising and subsidized franchisees' local promotional efforts through a cooperative advertising program. The package was appealing, and franchisees kept signing up.

But by 2008, even with advertising expenses at an all-time high, Domino's growth stalled in the face of competition from Papa John's, Pizza Hut, and other chains. It was more than competition that slowed Domino's. While focusing on Promotion, Place, and Price, the company had not paid enough attention to its Product. Pizza marketing research showed Domino's ranked number one in price and convenience but last in taste. Focus groups told Domino's the crust tasted "like cardboard" and the sauce "like ketchup."

Changing the recipe that had grown the brand for almost 50 years wasn't an easy decision. But Domino's chefs listened to customers and reinvented Domino's pizza from the crust up. Testing with consumers and franchisees found the new product tasted much better. But would customers try the new pizza?

Domino's was eager to tell customers about its improved product, so advertising was needed. Domino's advertising agency Crispin Porter Bogusky proposed an honest campaign: admit the old pizza was awful, tell them you listened and have something new for them to try. A $50 million advertising campaign—mostly on highly rated sports and entertainment TV programs—spread word of the new pizza far and wide. In one series of ads, Domino's head chef shows up at the doors of the harshest critics in those original focus groups and asks them to try the new pizza. They loved it!

Although advertising generated awareness and interest in its new recipe, Domino's used sales promotion to motivate customer trial. It introduced a money-back guarantee, franchisees offered free samples, and advertising touted a special: two medium, two-topping pizzas for just $5.99. All the promotion (and the new recipe) worked. Same-store sales soared 14 percent in the first three months after the new pizza was introduced, and Domino's marketing managers knew they achieved their objectives. More importantly, people who tried the new pizza liked it; they kept ordering, and sales stayed high even after the advertising blitz and sales promotion tailed off.

Next, Domino's set out to improve the *experience* of ordering and delivery. With the goal of making ordering and receiving a hot pizza as fast and easy as possible, Domino's AnyWare was born. Customers can order a Domino's pizza in more than 15 ways that don't involve human contact: on Facebook, Twitter with emojis, text, Apple Watch, a voice-activated app, Amazon Echo, and more. Time is the enemy of cooked food, so fast delivery makes for better pizza. To make improvements, franchisees in Australia are targeting 10-minute (out of oven to customer door) delivery times. Robots deliver in Australia and Europe, and self-driving cars are used in Miami. Domino's wants to satisfy customers anywhere they want a pizza, so they now deliver to more than 200,000 "hotspots" (locations without usual street addresses), including parks and beaches.

AnyWare collects customer data that Domino's uses in targeted promotions. Domino's combines that information with third-party data from other sources, including the U.S. Post Office, demographic and competitor intelligence, and more. Analysts mine the data and turn them into knowledge about where to open new stores, how to segment customer markets, and where and when to deliver targeted advertising and sales promotion efforts. For example, Domino's data can tell that Cheryl on First Avenue is a vegetarian and responds to dollar-off discounts on veggie pizzas, while John on Main Street buys a meat eater's pizza on Saturday afternoons during college football season. Domino's Piece of the Pie Rewards loyalty program encourages loyalists to eat more pizza with members-only discounts—while collecting additional customer data.

Domino's seeks to be a valued member in its communities. For example, many franchisees participate in Domino's Delivering the Dough, which offers fundraising opportunities for community groups or nonprofits. A national campaign offers "Paving for Pizza" grants to fix potholes: "Bad roads shouldn't happen to good pizza." Cities and towns get road repair and Domino's gains publicity, goodwill, and advertising exposure.

Domino's has all Four Ps working together for its target market. It boasts the highest customer loyalty of all major pizza brands. In the last decade, its stock price increased 60-fold. Now lots of people want a piece of the pie; there are more than 16,000 stores in 85 other countries. Domino's growth continues thanks to its pizza, technology, advertising, and sales promotion.[1]

LEARNING OBJECTIVES

The Domino's case shows that advertising and sales promotion are often critical elements in the success or failure of a strategy. But many firms do a poor job with both, so just copying how other firms handle these important strategy decisions is not "good enough." There is no sense in imitating bad practices. This chapter helps you understand important decisions that make effective use of advertising and sales promotion.

When you finish this chapter, you should be able to

1 recognize the importance of advertising in the economy.
2 understand why a marketing manager sets specific objectives to guide the advertising effort.
3 understand when the various kinds of advertising are needed.
4 understand how to choose the "best" medium.
5 understand the main ways that digital advertising differs from advertising in other media.
6 understand how to plan the "best" message—that is, the copy thrust.
7 describe the challenges and methods used to measure advertising effectiveness.
8 understand how to advertise legally.
9 understand the importance and nature of sales promotion.
10 understand the important new terms (shown in **red**).

Advertising, Sales Promotion, and Marketing Strategy Planning

Advertising and sales promotion can play a central role in the Promotion blend because they reach many customers at the same time. On a per-contact basis, these promotion methods provide a relatively low-cost way to inform, persuade, and activate customers. Advertising and sales promotion can position a firm's marketing mix as the one that meets customer needs. They can help motivate channel members or a firm's own employees, as well as final customers.

Unfortunately the results that marketers *actually achieve* with advertising and sales promotion are very uneven. It has been said that half of the money spent on these activities is wasted—but most managers don't know which half. Mass selling can be exciting and involving, or it can be downright obnoxious. Sometimes it's based on careful research, yet much of it is based on someone's pet idea. A creative idea may produce great results or be a colossal waste of money. Ads can stir emotions or go unnoticed.[2]

This chapter explains approaches to help you understand how successful advertising and sales promotion work (see Exhibit 15-1). After a brief overview of the

Exhibit 15-1 Marketing Strategy Planning, Advertising, and Sales Promotion

business of advertising, we look at the different decisions marketing managers have to make, including (1) desired objectives, (2) kind of advertising, (3) media to reach target customers, and (4) what to say (message). We take a deeper dive into digital and mobile advertising because of their rapid growth and unique capabilities. Then we examine measuring advertising effectiveness and legal issues. Finally, the last section examines sales promotion.

International dimensions are important

The basic strategy planning decisions for advertising and sales promotion are the same regardless of where in the world the target market is located. However, the look and feel can vary a lot in different countries. The choices available to a marketing manager within each of the decision areas may also vary dramatically from one country to another.

Commercial television may not be available. If it is, government rules may limit the type of advertising permitted or when ads can be shown. For example, privacy laws are stricter in Europe as compared to the United States, making it more difficult to track consumers' online behavior and deliver highly targeted

In some parts of the world, there has not been a long tradition of toothbrush and toothpaste use. Signal's advertising educates consumers on the importance of regular toothbrushing.
Source: Unilever Group

advertising. Radio broadcasts in a market area may not be in the target market's language. The target audience may not be able to read. Access to interactive media such as the Internet may be nonexistent. Cultural influences may limit ad messages. Ad agencies that already know a nation's unique advertising environment may not be available.

Sales promotion can also differ in some international markets. For example, a typical Japanese grocery retailer with only 250 square feet of space doesn't have room for *any* special end-of-aisle displays. Consumer promotions may be affected too. In some developing nations, samples can't be distributed through the mail because they're routinely stolen before they get to target customers. And some countries ban consumer sweepstakes because they see them as a form of gambling.

In this chapter we'll consider a number of these international issues, but we'll focus on the array of choices available in the United States and other advanced economies.

Advertising Is Big Business

LO 15.1

Total spending is big—and growing internationally

As an economy grows, advertising becomes more important—because more consumers have income and advertising can get results. But good advertising results cost money. And spending on advertising is significant. In 1946, U.S. advertising spending on media (the largest and most easily tracked expense) alone was slightly more than $3 billion. It is now over $200 billion—and projected to continue growing about 3 percent per year.[3]

Over the last decade, the rate of advertising spending has increased even more rapidly in other countries. Total ad media spending worldwide is more than than $500 billion. China, the world's second-largest advertising market, is growing fast and now accounts for more than $100 billion. Most advertising money is spent in North America, Asia, and Europe.[4]

Most advertisers aren't really spending that much

Although total spending on advertising seems high, U.S. corporations spend an average of only about 2.5 percent of their sales dollars on advertising. Worldwide, the percentage is even smaller. Exhibit 15-2 shows, however, that advertising spending as a percentage of sales dollars varies significantly across product categories. Producers of consumer products generally spend a larger percentage than firms that produce business products. For example, U.S. companies that make perfume and cosmetics spend about 21.2 percent of their sales dollars on advertising whereas household furniture producers spend about 8.4 percent. However, companies that sell construction machinery and equipment, plastics and resins, and computer and office equipment spend less than 1 percent of their sales on advertising.

Retailers also vary significantly in how much they spend on advertising. In general, the percentage is smaller for retailers and wholesalers than for producers. Although some large chains such as Kohl's, Macy's, and JCPenney spend more than 5 percent on advertising, other retailers and wholesalers spend 2 percent or less. Individual firms may spend more or less than others in the industry, depending on the role of advertising in their promotion blend. Although Exhibit 15-2 does not break out online stores, they typically spend much less than physical retail stores.[5]

Advertising agencies often do the work

An advertising manager manages a company's advertising effort. Many advertising managers, especially those working for large retailers, have their own advertising departments that plan specific advertising campaigns and carry out the details. Others turn over much of the advertising work to specialists—the advertising agencies.

Advertising agencies are often used to develop and implement advertising. **Advertising agencies** are specialists in planning and handling mass-selling details for advertisers. Agencies play a useful role. They are independent of the advertiser and have an outside viewpoint. They bring experience to an individual client's problems because they work for many other clients. They can often do the job more economically than a company's own department. And if an agency isn't doing a good job, the client can select another.

Exhibit 15-2 Advertising Spending as Percentage of Sales for Illustrative Product Categories

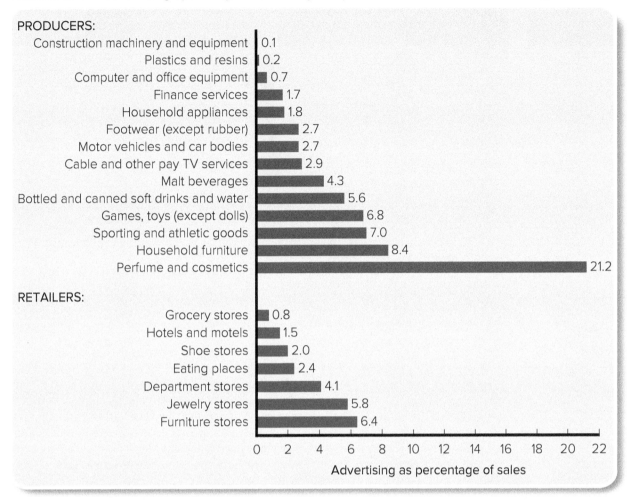

Some full-service agencies handle any activity related to advertising, publicity, or sales promotion. Other agencies are more specialized. For example, in recent years there has been rapid growth of digital agencies that specialize in developing Internet ads and online publicity and sales promotion.

Advertising doesn't employ that many people

Although total advertising expenditures are large, the advertising industry itself employs relatively few people. The major expense is for media time and space. Many students hope for a glamorous job in advertising, but there are fewer jobs in advertising than you might think. In the United States, only about 500,000 people work directly in the advertising industry. Advertising agencies employ less than half of all these people.[6]

Advertising Objectives Are a Strategy Decision

LO 15.2

Advertising objectives must be specific

Every ad and every advertising campaign should have clearly defined objectives. These should grow out of the firm's overall marketing strategy and the promotion jobs assigned to advertising. It isn't enough for the marketing manager to say "Promote the product." The marketing manager must decide exactly what advertising should do.

The marketing manager sets the overall direction

Advertising objectives usually start with the marketing manager, who works with the advertising manager to develop the objectives and an appropriate budget to accomplish

Exhibit 15-3 Examples of Different Types of Advertising over Adoption Process Stages

Awareness	Interest	Evaluation and trial	Decision	Confirmation
Teaser campaigns Pioneering ads Jingles/slogans Viral advertising Announcements	Informative or descriptive ads Image/celebrity ads E-mail ads Demonstration of benefits	Competitive ads Persuasive copy Comparative ads Testimonials Search ads	Direct-action retail ads Point-of-purchase ads Price deal offers	Reminder ads Informative "why" ads

them. Good advertising objectives are specific, measurable, and include a timeframe. The following list contains some potential advertising objectives:

1. For our target market, position the new product as the most technologically advanced in the industry by January 1.
2. Increase unaided brand awareness to 50 percent in our target market by July 1.
3. Obtain trial of our product from 5 percent in the state of Oregon by the end of the year.
4. Get 100,000 new visitors to our website by December 1.

Objectives guide implementation too

The specific objectives obviously affect what type of advertising is best. Exhibit 15-3 shows that the type of advertising that achieves objectives for one stage of the adoption process may be off target for another. For example, Taco Bell used informative television ads and in-store point-of-purchase materials to encourage consumers to try its new Doritos Locos Tacos. The aggressive campaign helped Taco Bell sell 100 million of the tacos in their first 10 weeks on the market. On the other hand, Dassault Aviation ran ads in *Bloomberg Businessweek* magazine targeting business leaders with its Falcon line of jets. The manufacturer of corporate jets hoped to raise awareness and drive customers to its website for more information.

Coordinating advertising across the channel to achieve objectives

Sometimes advertising objectives can be accomplished more effectively or more economically by someone else in the channel. Firms should work closely with other channel members to coordinate advertising efforts to get the best results. For example, a producer

Mini Cooper's Dutch advertising agency had multiple objectives when it left large empty boxes with torn ribbons and wrapping paper around Amsterdam shortly after Christmas. The boxes created awareness for the Mini Cooper and reinforced its positioning as small, different, and "fun"—like a toy.
Source: Cooper Car Company, Ubachswisbrun/JWT Amsterdam

of office supplies like Avery Dennison may find that the most economical use of its advertising dollars is the weekly flyer of a retail chain like Office Depot. So Avery Dennison offers **advertising allowances**—price reductions to firms further along in the channel—to encourage them to advertise or otherwise promote the firm's products locally.

Cooperative advertising involves producers sharing in the cost of ads with wholesalers or retailers. Franchisors might also use cooperative advertising to help local franchisees build their business. This helps the intermediaries compete in their local markets. It also helps the producer get more promotion for its advertising dollars because media usually give local advertisers lower rates than national or international firms. In addition, a retailer or wholesaler who is paying a share of the cost is more likely to follow through.[7]

Objectives Determine the Kinds of Advertising Needed

LO 15.3

The chosen advertising objectives largely determine which of two basic types of advertising to use—product or institutional (see Exhibit 15-4). **Product advertising** tries to sell a product. We will discuss three categories of product advertising—pioneering, competitive, and reminder—which focus on getting consumers to know, like, and remember something. Then we will discuss **institutional advertising**, which promotes an organization's image, reputation, or ideas rather than a specific product.

Pioneering advertising builds primary demand

Pioneering advertising tries to develop primary demand for a product category rather than demand for a specific brand. Pioneering advertising is usually done in the early stages of the product life cycle; it informs potential customers about the new product and helps turn them into adopters. For example, when Netflix launched one of the first major video streaming services, it needed to educate customers how a streaming (not cable or satellite) service worked.

Competitive advertising emphasizes selective demand

Competitive advertising tries to develop selective demand for a specific brand. A firm is forced into competitive advertising as the product life cycle moves along—to hold its own against competitors. For example, as smartphones moved to the growth stage of the product life cycle, advertising emphasized features and benefits to persuade target customers they needed more megapixels for the camera, more space, or lower prices.

Competitive advertising may be either direct or indirect. **Direct competitive advertising** aims for immediate buying action. **Indirect competitive advertising** points out product advantages to affect future buying decisions.

#M4BW While plant-based "burgers" have been available for many years, they targeted vegetarians and did not try to imitate meat. The Impossible Burger is a plant-based meat that mimics the flavor and experience of eating meat. The target market is meat eaters who are mostly unaware of plant-based meat. Impossible Burger's pioneering advertising educates meat eaters about the product with informative advertising copy: "Impossible Foods examined every part of the burger experience, making sure the sights, sounds, smells, textures, and most importantly flavors, make meat lovers rejoice. Even better, because we use 0% animals, we can make it using 95% less land, 74% less water, and with 87% less greenhouse gas emissions."
Source: Impossible Foods Inc.

Exhibit 15-4 Types of Advertising

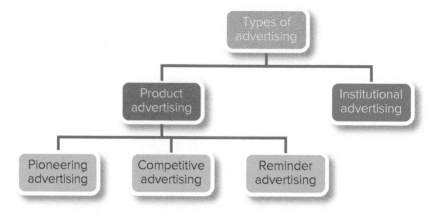

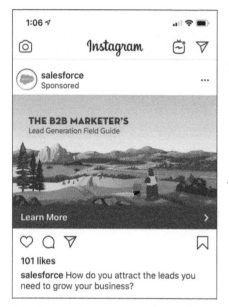

Salesforce sells customer relationship management software. Its Instagram ad is an example of a direct competitive ad that aims for immediate action. Clicking on "Learn More" takes customers to a page where they fill out an online form with contact information so they can be contacted by a Salesforce salesperson. The indirect competitive ad for Swedish Rail points out advantages of taking the train as compared to driving.
(left): Source: Salesforce.com, Inc./Instagram; (right): Source: Swedish Transport Administration

Most of Delta Air Lines' advertising is of the competitive variety. Much of it tries for immediate sales—so the ads are the direct type with prices, timetables, and website addresses or phone numbers to call for reservations. Some of its ads are the indirect type. They focus on the quality of service and suggest you check Delta's website the next time you travel.

Comparative advertising means making specific brand comparisons—using actual product names. Verizon touted its superior coverage with ads showing maps that highlighted its service as compared to AT&T. An Audi television ad shows a guy driving down the highway in his BMW. At full speed, he jumps out of his car and climbs onto a passing truck that is carrying a delivery of new Audi Q5s.

Some countries forbid comparative advertising. But in the United States, the Federal Trade Commission decided to encourage comparative ads because it thought they would increase competition and provide consumers with more useful information. Superiority claims are supposed to be supported by research evidence—but the guidelines aren't clear. Comparative ads can also backfire by calling attention to competing products that consumers had not previously considered.

Burger King uses a comparative ad to show that its Whopper is bigger than McDonald's Big Mac.
Source: Burger King

Reminder advertising reinforces a favorable relationship

Reminder advertising tries to keep the product's name before the public. It may be useful when the product has achieved brand preference or insistence, perhaps in the market maturity or sales decline stages. It is used primarily to reinforce previous promotion. Here the advertiser may use soft-sell ads that just mention or show the name—as a reminder. Hallmark, for example, often relies on reminder ads because most consumers already know the brand name and, after years of promotion, associate it with high-quality cards and gifts.

LEGO's reminder ad keeps the brand's name in front of customers.
Source: The Lego Group

Institutional advertising—remember our name

Institutional advertising usually focuses on the name and prestige of an organization or industry. It may seek to inform, persuade, or remind. Its basic objective is to develop goodwill or improve an organization's relations with various groups—not only customers but also current and prospective channel members, suppliers, shareholders, employees, and the general public. The British government, for instance, uses institutional advertising to promote England as a place to do business. Many Japanese firms, like Hitachi, emphasize institutional advertising, in part because they often use the company name as a brand name.

Companies sometimes rely on institutional advertising to present the company in a favorable light, perhaps to overcome image problems. Other organizations use institutional advertising to advocate a specific cause or idea. Insurance companies and organizations such as Mothers Against Drunk Driving, for example, use these advocacy ads to encourage people not to drink and drive.[8]

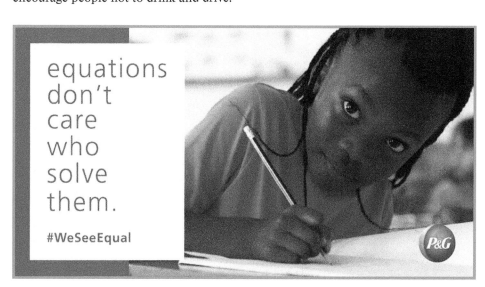

#M4BW P&G's #WeSeeEqual campaign supported gender equality and ran around International Women's Day. This institutional advertising campaign promotes P&G's image by aligning it with a cause; it does not promote a specific product.
Source: Procter & Gamble

Choosing the "Best" Medium—How to Deliver the Message

Advertising media are the various means by which the message is communicated to the target market. Exhibit 15-5 lists the most common kinds of media—digital (mobile phones and computers), television, radio, magazines, and out-of-home (including billboards, cinema, bus stops, etc.). It also provides a brief overview of each medium, listing advantages and disadvantages of each. Marketing managers seek the best medium, which varies with the situation. Effectiveness depends on how well the medium fits with the rest of a marketing strategy—that is, it depends on (1) your promotion objectives, (2) what target markets you want to reach, (3) the funds available for advertising, and (4) the nature of the media, including whom they *reach,* with what *frequency,* with what *impact,* and at what *cost.*[9]

Advertising media's relative importance is changing over time

Advertising dollars follow target customers' eyeballs and attention. Customers are spending much more time on the Internet—and more of that time has moved to mobile devices (smartphones). In addition, many traditionally print media now publish online.

Exhibit 15-6 shows estimates of the relative share of ad spending for each type of media from 2010 to 2022. As the exhibit shows, the media favored by advertising have changed a lot in the last decade. For example, in 2010, digital advertising on mobile devices was less than 1 percent of all advertising spending; however, it is expected to reach almost 50 percent by 2022. About 39 percent of advertising dollars in 2010 were spent on television, but its share will fall below 25 percent by 2022. Advertising in newspaper, radio, magazines, and directories is also in decline. While the data in Exhibit 15-6 are for the United States, these trends are similar worldwide.

Still, creative advertising managers have many media choices, and all of these options (and more) merit some consideration. Let's consider some of the factors that influence media choice.

Medium should fit promotion objectives

The advertising medium should support the promotion objectives. If the objective requires demonstrating product benefits, TV or online ads with video may be the best option. When "action" and a sale are important, online search advertising—which targets customers as they are looking to buy—often works best. Newspapers, radio, and local online advertising work for businesses operating in local markets, and online ads can also target by geography.

Exhibit 15-5 Advantages and Disadvantages of Major Advertising Media

Kind of Media	Advantages	Disadvantages
Digital: Mobile	Ads link to more detailed website, some "pay for results," easier to track results, can be highly targeted, precise-location based	Hard to compare costs with other media, more focused on later stages of purchase process
Digital: Desktop/laptop	Same as above (except not as precisely location based)	Same as above
TV	Demonstrations, image building, good attention, wide reach, cable can be targeted	"Clutter"—ads compete for attention, expensive, limited time (usually 30 seconds or less), often skipped by viewers
Radio	Wide reach, segmented audience, inexpensive, use of sound and voices can help create certain image of business	Weak attention, many different rates, short exposure, declining audience, listeners cannot review the ad, ads interrupt entertainment
Magazines	High reader involvement, very targeted, good detail, some "pass along," image quality high—better for photos	Inflexible, long lead times, cost can be high, limited flexibility on location of ad within magazine
Out-of-home (including billboards, cinema, bus stops, etc.)	Captive audience, can be geographic and local	Outdoor: "glance" medium Cinema: primarily a younger audience

Exhibit 15-6 Share of Ad Spending in the United State by Medium

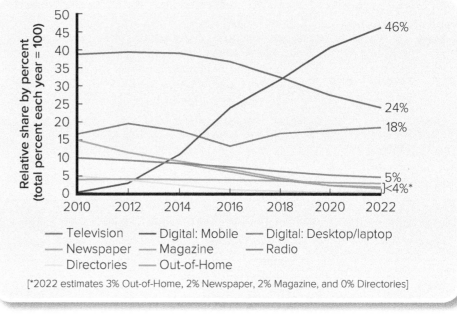

[*2022 estimates 3% Out-of-Home, 2% Newspaper, 2% Magazine, and 0% Directories]

Source: eMarketer Inc.

Advertisers pay for the whole audience

Advertisers pay for the whole audience a chosen media delivers, including those who aren't potential customers. When carmaker Audi runs a Super Bowl ad, it pays for everyone watching the game, not just those with the income and interest in a luxury car. When an ad appears in someone's Facebook feed, the advertiser pays whether the customer ignores it or not.[10]

Match your market with the media

To ensure good media selection, the advertiser first must *clearly* specify its target market. Then the advertiser can choose media that reach those target customers. Most media firms use marketing research to develop profiles of their audiences. Generally, this research focuses on demographic characteristics rather than the segmenting dimensions specific to the planning needs of *each* different advertiser.

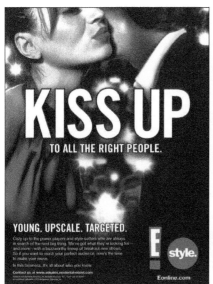

Cable and streaming television channels offer messages targeting specific demographic, cultural, and special interest groups. This allows advertising to be aimed at people who are "Young" and "Upscale" on the E! Network, Hispanics in the United States with Univision, and tennis fans through the Tennis Channel.
(left): Source: E! Entertainment Television, LLC; (middle): Source: Univision Communications Inc.; (right): Source: Tennis Channel

Select media to zero in on specific target markets

Today, advertisers direct more attention to reaching smaller, more defined target markets. Even traditional media are becoming more targeted. TV is a good example. Cable TV channels—such as CNN, Nickelodeon, HGTV, and the Golf Channel—are taking advertisers away from the networks because they target specific audiences. The Golf Channel, for example, averages around 100,000 viewers at a time, but these upscale consumers are *the* target market for makers of golf clubs, balls, and accessories. So advertising for golf products on a major network isn't nearly as efficient as advertising on a single cable channel.

Using big data, TV tailors ads to individual households

As we discussed in Chapter 7, research firms collect big data from consumers' credit and loyalty cards, online activities, and other sources to build profiles of individual households. *Addressable TV* is a technology that allows cable television to deliver specific ads to each household instead of advertising to everyone watching a particular show.

The U.S. Army has used this technology to aim recruitment ads at specific customers. For example, its "family influencers" target market received a TV ad showing a daughter discussing her enlistment decision with her parents. At the same time, households fitting the "youth ethnic 1" segment saw an ad showing African American men testing and repairing military equipment. Other groups saw ads customized to their demographic background and role in the enlistment decision.[11]

Magazines sort readers by special interests

Many magazines serve only special interest groups, such as cooks, new parents, or personal computer users. Trade magazines target people working in particular fields such as furniture retailing and electrical wholesaling. The most profitable magazines seem to be the ones aimed at clearly defined target markets. With the right campaign, magazines deliver results.

For example, LEGO had a promotion objective of getting parents to use its Duplo line of building blocks as a way for them to play *with* their kids—instead of sending kids off to play on their own. So they turned to *Parents* magazine. LEGO managers created a pull-out calendar that featured 31 activities for parents to do with their kids. LEGO asked parents to post pictures of themselves doing the Duplo activities with their kids to the Duplo Facebook page. Following the ad, LEGO research saw a jump in parents' purchase intention and belief that Duplo made their kids smarter.[12]

Advertising managers often look to grab attention with creative alternative media. This billboard ad offers "free screen replacement when you buy the latest phones" from O_2.
Sam Mellish/Getty Images

The advertising media listed in Exhibit 15-5 attract the vast majority of advertisers' budgets. But advertising specialists always look for cost-effective new media that will help advertisers reach their target markets. For example, one company successfully sells space for signs on bike racks that it places in front of 7-Eleven stores. A new generation of ATMs shows video ads while customers wait to get their money.

Specialized media are small, but gaining

Digital and Mobile Advertising

LO 15.5

Customers spend more time online

As Exhibit 15-6 shows, the Internet (particularly via mobile devices) is where advertisers spend an increasing share of their media budgets. This change in media spending follows Americans' screen-time behavior. On average, Americans spend more time online and on mobile devices than they do watching television—and much more than reading newspapers and magazines. That varies by age range, with younger generations (Millennials and Generation Z) being more active in digital media as compared to Baby Boomers and senior citizens. So marketing managers need to know their target market as well.[13]

The Internet allows consumers to socialize, review information, enjoy entertainment, and shop. This change in behavior creates challenges and opportunities for marketing managers targeting specific customers or segments through digital media on computers and mobile devices. Chapter 16 discusses how marketing managers use *unpaid* media—often on the Internet—to communicate with customers. In this section, we look at how *paid digital media*—online and mobile *advertising*—are used in marketing strategy planning. We begin by highlighting some of digital advertising's unique qualities and then describe some of the main types of digital and online advertising.

Most advertisers pay only if ads deliver

Advertisers often pay for online ads differently than other media. Most media charge customers based on the number of people who see an ad—the number of viewers of a television show or readers of a magazine, for example. Most websites, on the other hand, use a **pay-per-click** advertising model, where advertisers pay media costs only when a customer clicks on the ad that links to the advertiser's website. Many firms like this ability to directly track the cost of advertising and resulting sales—allowing a marketing manager to set a budget and track return on investment.

Online ads know where customers live and work

Computers and smartphones give out their geographic location—and that helps advertisers. Many advertisers are interested in targeting customers who live or work in a particular geographic location, and often at a particular time of day. That information can help in targeting the right customers with the right message. Consider Miller's Bakery in New Jersey, which sells assorted pastries and custom cakes. Using the Google AdWords advertising service, Miller's controls when and where the bakery's ads show up in search advertising. For example, Miller's wanted to target customers who live or work near one of its locations. So it showed ads only to customers who use Google and were located within a two-mile radius of one of the bakery's two stores. The targeted ads helped increase website visits and—more importantly for Miller's—call-in orders doubled.[14]

Online ads know where customers have been on the web

When customers shop online, their clickstream (what sites and pages they visit) can be tracked. Many websites place a small file called a *cookie* on the computers of people who visit their sites. The customer's online behavior is then tracked and monitored. **Retargeting** (or *behavioral retargeting*) displays ads to a web user based on sites he or she previously visited. Retargeting delivers ads to customers who may have browsed on a website but not purchased. A customer can receive ads at other (unrelated) websites reminding him or her of previous website visits. So, for example, when a customer views desk chairs at online retailer Overstock.com—but decides not to purchase—the customer may soon see ads for those same chairs on other websites. Perhaps you have experienced advertising "stalking" you online.

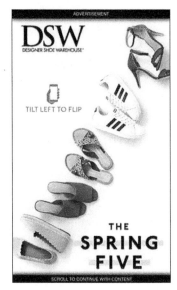

Digital and mobile offer unique opportunities for advertisers. The ad on the left appeared on the website NBCNews.com after one of the authors made a purchase at the online clothing retailer Territory Ahead. The shirt featured in the ad was one the author considered but did not purchase. DSW took advantage of mobile phone technology to make the ads in the middle and on the right. When viewers tilt their phone to the left or right, the images are flipped through like a catalog (no swiping or scrolling required). An interested shoe shopper can see many different shoes, hopefully holding interest and arousing desire.
(left): Source: Territory Ahead; (middle and right): Source: Designer Brands

"Programmatic advertising": artificial intelligence meets media planning

Retargeting can occur because artificial intelligence guides online advertising placement. Traditionally, media planners made decisions about where advertising would be placed. Media planners decided which magazines, television programs, or alternative media delivered the target customers an advertiser wanted to reach. Artificial intelligence is changing this. *Programmatic delivery* refers to the use of software and artificial intelligence to automate placing online advertising on websites or in social media to target users. The process increases the efficiency and effectiveness of media planning. The low cost also makes it easier for advertisers to target customers more precisely.[15]

Different types of digital advertising

Most online advertisers seek a direct response—they entice a customer to click on a link, gather information, and engage. Advertising on the Internet continues to evolve quickly, with marketing managers receiving rapid feedback in the form of clicks and other engagement measures. This information allows an advertiser to put more efforts behind effective advertising and withdraw from those that are not working as well. Further, advances in technology continually open doors for new types of advertising. To get the attention of web surfers, Internet advertisers have created different types of ads that work to achieve different objectives. Most of these ads also utilize pay-per-click and engage in retargeting as described earlier. Exhibit 15-7 lists examples of the main types of digital advertising. Each offers some advantages and disadvantages.[16]

Banner and pop-up ads build brands

Banner (or *display*) *ads* are a type of online advertising that places an ad on a web page, often across the top or to the side of the page's primary content. *Pop-up ads* are similar to banner ads but usually cover an entire browser window—"popping up" in front of (or behind) a user's web page. For these ads, the **click-through rate**—the number of people who click on the ad divided by the number of people the ad is presented to—can be low (less than 1 in 1,000).

Although click-through rates are low, banner ads often have the goal of generating awareness and building a brand, so click-through may not be the best measure of effectiveness. Evidence suggests that these ads can be combined with offline advertising to increase purchase intent. Click-through rates are also higher when video content is used in an ad.[17]

Exhibit 15-7 Types of Digital Advertising, Examples, and Advantages and Disadvantages

Type of Digital Ad	Advantages	Disadvantages
Banner ads (including pop-up ads)	Relatively low cost; banner ads work best for building a brand and image; good tracking tools can measure effectiveness; targeted ads can bring in customers	Some types (especially pop-up) can be seen as intrusive; browser ad blockers may limit reach; banner ads so prevalent that people tend to ignore them
Directories and classifieds (e.g., Craigslist)	Low cost; highly targeted; best for customers ready to buy	Does not work for brand building; does not fit for large companies
Search advertising (e.g., Google Ads and Amazon)	Easy to set budget and control costs; targets users in information-gathering stage of the buying process; easy to measure return on investment; highly targeted; high credibility	Bidding process for keywords can become costly; information overload for customers; limited space for copy thrust; limited/no image capability
Social media advertising (on Facebook, Snapchat, and other social media sites)	High knowledge of individuals; potentially wide reach; can stimulate viral; can be narrowly targeted; relatively low cost; endorsements from friends' "Likes" can be powerful; easy to set a budget	Users of social media actively avoid advertising; advertiser generally has fewer data than from other sources
Mobile advertising (often includes search and social networks advertising on smartphones)	A large and growing audience; works best for high-involvement/utilitarian products; potential for location-based targeting; works well for immediate purchase	Technical limitations (small screen, bandwidth, data transfer); not user friendly, not very interactive; measurement tools still emerging

Directory and classified ads

When customers know what they want, they may turn to a directory or classified ad. People looking for a used car might look on Craigslist or a directory service such as Cars.com. These sites can be low cost and work well for local businesses, but they don't generally work as well when the objective is generating awareness or building a brand.

Search engine ads know *what* you want

Google, Amazon, and Facebook are battling for the future of advertising. The advantage Google (and other search engines such as Bing and Yahoo!) has is that customers often turn to the site to help with a purchase decision. An online search may occur when a customer has a problem—but no solution—or it may occur when a customer knows exactly what he or she wants to buy. More than 80 percent of consumers use search engines for at least some of their shopping. When customers are shopping, the sponsored search ads that appear along with the organic search results are usually relevant and helpful. Retailers like REI capitalize on this knowledge by paying a search engine firm for a sponsored search ad. For example, REI will pay for one of its ads to appear above or next to the results of a search on "hiking boots."

The click-through rate on Google search ads averages 1 to 3 percent. Advertisers bid on search "keywords," paying more for those search terms that may be associated with leads for expensive purchases. For example, while the average cost-per-click-through on Google is around $2, advertisers can pay more than $45 for each person clicking through on an ad placed next to searches for "insurance," "loans," or "mortgage."[18]

Customers can also conduct product searches within a retailer—especially at Amazon, which has grown to become the second-largest search advertising site (after Google). While searching for "hiking boots" at Amazon, a few "sponsored" options show up before other listings. After clicking a particular boot, a customer is again presented with "Sponsored products related to this item." Advertisers have the opportunity to pay for one or both of these slots.[19]

Social media sites know *who* you are

Facebook's advantage follows from the detailed knowledge it has of each of its users. When Facebook users fill out a profile, they give Facebook demographic information. And with each new status update, photo, friend, "Like," and clicked link, users give Facebook more insights about themselves, their preferences, and their values. Facebook knows most of its users' ages, ethnicities, jobs, relationship statuses, restaurant preferences, favorite music groups, birthdays, and more. Facebook analyzes this information to offer advertisers specific profiles that help them target ads. The challenge for Facebook is that users don't go to the social network with buying in mind—they go to see what their friends are doing. Response to Facebook ads has been increasing as advertisers and Facebook learn more about how to use them; click-through rates average between 0.5 and 1.5 percent.[20]

"Influencers" leverage social media to become advertisers

Marketing has long relied on endorsements or testimonials from athletes, celebrities, or subject matter experts within advertising—"Hey Aaron Rodgers, tell me more about State Farm Insurance." Social media have created a whole new opportunity for endorsers of all kinds. People from a wide range of backgrounds have built up large followings on social media. Perhaps because they are celebrities or experts on a particular topic, some people have hundreds of thousands or even millions of followers. Many use this platform to influence others—perhaps by letting others know their favorite brand of makeup or chewing gum. Some brands have begun to pay **influencers**, trusted or well-known figures who can sway attitudes or purchase decisions among a particular target market—to promote a brand.

These influencers are opinion leaders who can pass along a brand's message via their own social media channels. Some brands seek (and usually pay for) a social media endorsement and access to an influencer's followers. Some influencers have very large followings. For example, Hollywood-trained makeup artist Huda Kattan uses her blog and Instagram account (@hudabeauty) to share product reviews and tutorials. She has more than 35 million followers. She now sells her own brand of cosmetics, and it is one of Sephora's top sellers. Sometimes a brand will choose to work with "micro-influencers" who do not have a large following.

Smartphones know *where* you are right now

Smartphones are increasingly ubiquitous in developed countries. Almost four out of every five American adults own a smartphone, and they spend more than four hours a day on them. This creates a new opportunity for advertisers. The biggest players in mobile advertising are Google and Facebook, which offer search and social networking services through mobile devices—and deliver advertising at the same time.[21]

Because mobile devices track customers' whereabouts through GPS, the segmenting dimension of physical location at a specific time can be added to target marketing. Many businesses—bars, restaurants, and retail stores—find it useful to target customers based on this information. For example, a downtown restaurant can advertise only to customers walking within a few blocks of the restaurant around lunchtime. At the very least, because customers conduct searches and access social media from mobile devices, advertisers must have mobile-friendly versions of their ads.[22]

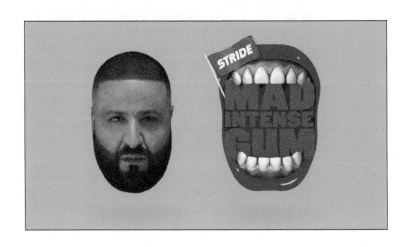

Hip-hop artist DJ Khaled is an influencer. He has many followers on Snapchat, where his "snaps" average 3 million views. Stride gum used that influence in a campaign promoting its "mad intense gum."
Source: DJ Khaled

Big data serve timely mobile ads

A smartphone can tell Walmart when a customer is physically present in one of its stores. It also knows customers are open to receiving and using coupons while they are in a Walmart store, so Walmart encourages its customers to download its smartphone app—offering coupons and e-receipts as an added incentive. When a customer is near a Walmart store, advertising (often with a coupon attached) is delivered and often nudges them to visit the store and spend more.

Companies are also experimenting with integrating big data and mobile advertising. Google was once again involved as it discovered that flu-related searches suggested where flu outbreaks were occurring two weeks before published reports from the Centers for Disease Control and Prevention. Advertising managers for Vicks' Behind Ear Thermometer saw an opportunity to leverage Google's big data on flu trends. Vicks wanted to target mothers—and the music streaming service Pandora had data that allowed them to identify listeners with kids. Vicks further segmented moms by delivering ads only in areas of the country experiencing flu outbreaks with the message: "Flu levels in your area are high. Be prepared with Vicks' revolutionary Behind Ear Thermometer." Tapping the ad would direct the mom to the nearest store that sold the thermometer: "Buy at Walgreens 0.5 mile away." Big data and mobile advertising allowed Vicks to narrowly segment the market and deliver timely, relevant messages to mothers.[23]

Specialized media offer opportunities for highly targeted messages. Ford Motorcraft services Ford cars needing repair or maintenance. The need for such services increases in bad weather, so Ford Motorcraft worked with the Weather Channel UK to place ads on the backdrops for the Weather Channel app. The ads would appear only during specific weather conditions—including snow, rain, and fog.
Source: Ford Motor Company/The Weather Channel LLC

Ethics and digital advertising

Digital and mobile advertising raise many ethical questions. For example, the U.S. government has been critical of Google, Yahoo!, and Bing, claiming these search engines do little to distinguish the organic search results offered by the search engine from the paid advertising. Because customers tend to favor organic results, this can be misleading and potentially deceptive. In 2002, the Federal Trade Commission (FTC) asked search engines to use shading, borders, text differences, or labels to clearly differentiate the search results from paid advertising. Over time, search engines have made such differences less apparent.[24]

Cookies and mobile advertising also raise questions about privacy. Cookies track a person's whereabouts on the Internet. Mobile advertising is based on an advertiser knowing where a customer is physically located at any given time. Some question whether it is appropriate for advertisers to track people—what they do online and even where they go—especially because most have not knowingly given the advertiser permission to track their whereabouts. Research has suggested that these are serious concerns for many consumers; marketing managers need to tread carefully or they may invite more scrutiny from regulators and consumer advocacy groups.[25]

Ethical questions also exist for the media companies in this space. Advertisers are often paying for fake customers. They don't mind paying when interested customers click on their ads. However, *click fraud* occurs when a person or software program automatically clicks on an ad without having any interest in the ad's subject. The intent is to defraud the advertiser and make money for an unscrupulous website.

Influencers are also paid based on their number of followers. Some influencers have been caught buying more followers. Others have used "bots" (computer programs) that engage with their posts—to make it look like they have a more engaged following. Social media platforms have been making efforts to clean up these practices.[26]

Do consumers have any ethical responsibilities? Technology allows consumers to block advertising. Some mobile phone companies in Europe have started to block all mobile advertising. They argue the ads slow customers' online experience and increase their data usage. Most web browsers allow customers to block advertising. People access free content on many ad-supported websites. Is it ethical for *consumers* to use ad-blocking software on their computers and smartphones?[27]

Ad-Blocker Pro is one of several products that people can use to block advertising from websites they visit online.
Source: Ad-Blocker.org; smartphone frame: SPF/Shutterstock

Planning the "Best" Message—What to Communicate

LO 15.6

Specifying the copy thrust

Once you decide *how* the messages will reach the target audience, you have to decide on the **copy thrust**—what the words and illustrations should communicate. Carrying out the copy thrust is the job of advertising specialists. But the advertising manager and the marketing manager need to understand the process to be sure that the job is done well.[28]

Let AIDA help guide message planning

Basically, the overall marketing strategy should determine *what* the message should say. Then management judgment, perhaps aided by marketing research, can help decide how to encode this content so it will be decoded as intended. As a guide to message planning, we can use the AIDA concept: getting Attention, holding Interest, arousing Desire, and obtaining Action.

Getting attention

Getting attention is an ad's first job. Many readers leaf through magazines without paying attention to any of the ads, and viewers get snacks during TV commercials. When watching a program on a DVR, they may zip past the commercial with the flick of a button. Online they may use a pop-up blocker or click on the next website before the ad message finishes loading onto the screen. Advertisers must get their attention.

Many attention-getting devices are available. A large headline, computer animations, shocking statements, attractive models, animals, online games, special effects—anything different or eye-catching—may do the trick. However, the attention-getting device can't detract from, and hopefully should lead to, the next step: holding interest.

424

The copy thrust—both the text and the image—in this ad for Lurpak butter grabs attention and holds interest.
Source: Lurpak

Holding interest

Holding interest is more difficult. A humorous ad, an unusual video effect, or a clever photo may get your attention—but once you've seen it, then what? If there is no relation between what got your attention and the marketing mix or the ad does not address your needs, you'll move on. To hold interest, the tone and language of the ad must fit with the experiences and attitudes of the target customers and their reference groups. As a result, many advertisers develop ads that relate to specific emotions. They hope that the good feeling about the ad will stick, even if its details are forgotten.

Arousing desire

Arousing desire to buy a particular product is one of an ad's most difficult jobs. The ad must convince customers that the product can meet their needs. Testimonials may persuade a consumer that other people with similar needs like the product. Product comparisons may highlight the advantages of a particular brand.

To arouse desire, an ad should usually focus on a *unique selling proposition,* the main point of differentiation from competitors. It should aim at an important unsatisfied need. This can help differentiate the firm's marketing mix and position its brand as offering superior value to the target market. Too many advertisers ignore the idea of a unique selling proposition. Rather than using an integrated blend of communications to tell the whole story, they cram too much into each ad—and then none of it has any impact.

Obtaining action

Getting action is the final requirement—and not an easy one. From communication research, we now know that prospective customers must be led beyond considering how the product *might* fit into their lives to actually trying it.

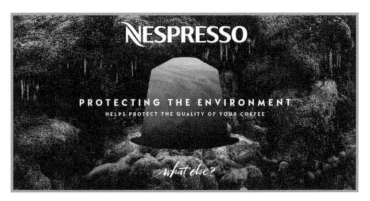

#M4BW Advertising can promote more than one message—possibly supporting a product and a cause. For example, Ben & Jerry's ran advertising supporting climate change, "If it's melted, it's ruined!," with images of melting ice cream cones. Nespresso's "The Choices We Make" campaign encouraged customers to choose Nespresso, in part because of its efforts to protect the environment. These campaigns communicate to target customers that the brands care about profits and a better world.
(left): Source: Ben & Jerry's Homemade Inc.; (right): Nestlé Nespresso S.A.

Direct-response ads can sometimes help promote action by encouraging interested consumers to do *something* even if they are not ready to make a purchase. For example, Fidelity Investments has run TV ads featuring colorful graphs, a sign with "Wow!" and the company's phone number and website address. And just in case viewers don't "get it," Blondie's song "Call Me" plays in the background. Fidelity wants to encourage interested consumers to take the first step in building a relationship.[29]

Is native advertising ethical?

Sometimes the best messages are disguised—meaning they are not obviously advertising. Consumers tend to have their guard up when they know something is advertising. To sneak past that "guard," advertisers might make advertising look like media content. *Native advertising* is advertising designed to not look like ads; it mimics the look and feel of the platform on which the ad appears. For example, a newspaper ad may look a lot like a newspaper article.

As another example, automaker Mini USA published "25 Places That Look Not Normal," in collaboration with online news site *BuzzFeed*. The post said, "Prepare to have your mind blown. We tip our hats to those who see things differently" (http://goo.gl/wMEBgA). Below the title, you see Mini USA listed unobtrusively in text. Following 25 pretty awesome-looking places, you see Mini's slogan "We tip our hats to those who see things differently. MINI. NOT NORMAL," followed by a video advertisement for Mini.

Native advertising gets attention with catchy headlines and holds interest with engaging content. Consumers find these ads more believable than traditional advertising. However, critics charge that native advertising deceives viewers and is therefore unethical. The FTC has issued guidelines, which are often ignored by advertisers.[30]

Ethical Dilemma

What would you do? You are the director of marketing for an online dating company. Your company is an upstart in the industry—still trying hard to get attention in a world where Match.com, OkCupid, Tinder, and others are well known. *NewsSpace*, an online site known for its cleverness in attracting readers with its entertaining and provocative headlines, has approached your marketing team with an offer to advertise an article it will post on the site. Your company slogan is "Mating Habits for Humans." *NewsSpace* wants to post an article "9 Awesome Animal Mating Habits"—which would provide a photo of how different animals show care and concern for others of their kind. *NewsSpace* proposes that number 9 on the list would be something for humans and feature your company's website. A video ad for your company would be in the number 10 slot. The company has ideas for ranking 1 through 8, but you can influence those in any way you like. The ad would appear to be just another article at the site—using the same format and function as other articles—but in small type it would be noted "Sponsored by...." A couple of people on your team think this is a great idea. A few others wonder if the practice might be deceptive or possibly even illegal. The representative from *NewsSpace* says that it is legal and it gets few complaints.[31]

Would you run this type of ad? If yes, explain why and what you would tell those who oppose the ad. If no, explain why and what you would tell those who are in favor.

Measuring Advertising Effectiveness Is Not Easy

LO 15.7

Success depends on the total marketing mix

It would be convenient if we could measure the results of advertising by looking at sales. Some breakthrough ads do have a very direct effect on a company's sales—and the advertising literature is filled with success stories that "prove" advertising increases sales. Similarly, marketing research firms such as Information Resources can sometimes compare sales levels before and after the period of an ad campaign. Yet we usually can't measure advertising success just by looking at sales. Advertising managers should measure effectiveness against the advertising objectives. The total marketing mix—not just advertising—is responsible for the sales result. Sales results are also affected by what competitors do and by other changes in the external market environment.

Research and pretesting improve the odds

Ideally, advertisers should pretest advertising before it runs rather than relying solely on their own guesses about how good an ad will be. The judgment of creative people or advertising experts may not help much. They often judge only on the basis of originality or cleverness of the copy and illustrations.

Many progressive advertisers now demand laboratory or market tests to evaluate an ad's effectiveness. A concept test of an ad might ask for feedback on a rough mockup of a print ad or a storyboard (think comic book version) of a television ad. With a laboratory test, people are brought into a place where they are shown ads. Some-

Some companies use big data and artificial intelligence to learn how individual customers shop, what brands they prefer, and when they are in the market to buy particular products. Customers respond to advertising messages relevant to their current needs. Rocket Fuel is an ad technology company that helps advertisers use artificial intelligence to make their advertising campaigns more efficient and effective.
Source: RocketFuel Inc.

times, physiological measures (like pupil dilation or eye tracking) are done while people watch or read an advertisement. While this gives a researcher a lot of control, it isn't very representative of real-world viewing. So field tests, perhaps in a sample of the target market's homes with the natural distractions that occur, offer other insights.

Hindsight may lead to foresight

After ads run, researchers may try to measure how much consumers recall about specific products or ads. The response to radio or television commercials or magazine readership can be estimated using various survey methods to check the size and composition of audiences. Similarly, most Internet advertisers keep track of how many "hits" on the firm's website come from ads placed at other websites.[32]

Ads must be seen to have any chance

Marketing managers choose media outlets to ensure that advertising gets in front of target customers. In making these decisions, the first decision is whether target customers see the media. A second consideration is how well the media fit with their desired promotion objective. For example, a firm with an "action" objective runs advertising that targets customers further along in the purchase cycle. Online search ads often work here, whereas billboard advertising may not. Other ads designed to build a brand's image often work better in media like TV, magazines, and online where photos or videos can be used. Finally, an advertiser must consider the costs to get to customers. Some of these types of decisions are demonstrated in Marketing Analytics in Action: Advertising Impressions and Media Cost.

Marketing Analytics in Action: Advertising Impressions and Media Cost

With traditional media, marketing managers typically pay for *impressions*, or the number of people an ad reaches. For example, impressions refer to the number of people who view a television commercial, view a newspaper ad, hear a radio ad, or see a billboard. For some media, impressions are inferred by using data from a ratings agency that has information on the number of viewers of a TV show or drivers that pass a billboard on a highway.

That information is then used to calculate *cost per one thousand impressions (CPM)*, which measures the cost to place an ad where it receives 1,000 impressions. For example, a newspaper advertisement that cost $300 to run and reaches 10,000 readers would have a CPM of $30 ($300/10). CPM allows an advertiser to compare the costs of different media.

Although CPM provides a starting point for comparing advertising media, the CPM can vary across media. Some media are more valuable—perhaps because they reach a very specific target market or very desirable set of customers. And some media can be valuable because they catch customers just as they are about to make a large purchase—so there is a lot of competition for those media. For example, we earlier noted how search advertising can be very expensive for search terms like "insurance." Consider these issues in the following brief scenario.

Giles Lu is the marketing manager for a chain of four men's clothing stores in a medium-sized U.S. city. He is trying to decide which media option to use in advertising the stores. He has previously used local newspapers and radio to run ads. A friend who works in marketing at a restaurant suggested he check out billboards and Yelp—an app with reviews of all kinds, but mostly of restaurants. When Lu contacts the media, he receives the CPM costs shown below. Review the numbers and consider the questions that follow.[33]

Media	CPM
Local newspaper	$ 12
Local radio	$ 10
Billboard	$ 3
Yelp	$450

1. Why is the CPM for Yelp so much higher than the other three options?
2. Why is the CPM for billboard advertising so low?
3. What else should Giles Lu consider before deciding which media to use?

Avoid Unfair Advertising

LO 15.8

Government agencies may say what is fair

In most countries, the government takes an active role in deciding what kinds of advertising are allowable, fair, and appropriate. For example, France and Japan limit the use of cartoon characters in advertising to children, and Canada bans *any* advertising targeted directly at children. In Switzerland, an advertiser cannot use an actor to represent a consumer. New Zealand limits political ads on TV. In the United States, print ads must be identified so they aren't confused with editorial matter; in other countries ads and editorial copy can be intermixed. Most countries limit the number and length of commercials on broadcast media.

What is seen as positioning in one country may be viewed as unfair or deceptive in another. For example, when Pepsi was advertising its cola as "the choice of the new generation" in most countries, Japan's Fair Trade Committee didn't allow it—because in Japan Pepsi was not "the choice." China's relatively unrestricted advertising climate has recently become more restrictive. P&G was fined almost $1 million for promising its Crest whitening toothpaste would bring whiter teeth in "just one day."[34]

Differences in rules mean that a marketing manager may face very specific limits in different countries, and local experts may be required to ensure that a firm doesn't waste money developing ads that will never be shown or that consumers will think are deceptive.

FTC controls unfair practices in the United States

In the United States, the Federal Trade Commission (FTC) has the power to control unfair or deceptive business practices, including deceptive advertising. The FTC has been policing deceptive advertising for many years. It is getting results now that advertising agencies, as well as advertisers, share equal responsibility for false, misleading, or unfair ads.

This is a serious matter. If the FTC decides that a particular practice is unfair or deceptive, it has the power to require affirmative disclosures—such as the health warnings on cigarettes—or **corrective advertising**—ads to correct deceptive advertising. The FTC forced Bayer to run corrective advertising after Bayer made unproven claims about its drug Yaz's efficacy for treating acne and premenstrual syndrome. Other advertisers learn from these cases. The possibility of large financial penalties or the need to pay for corrective ads causes most advertisers to stay well within the law.[35]

However, sometimes ad claims seem to get out of hand anyway. The FTC has started to crack down on claims related to weight loss and health. For example, KFC quickly stopped running several of its TV ads after the FTC objected to the ads and opened an investigation. KFC's ads positioned fried chicken as a healthy choice in fast food, but there was also lots of small print at the bottom of the screen to qualify the claims. Skechers and Reebok claimed that their shoes would help customers lose weight and strengthen muscles. The FTC took issue with the unfounded claims and the two firms had to return millions of dollars to consumers.[36]

Regulations added for influencers

As discussed earlier in this chapter, many firms pay for endorsements from social media "influencers" who have a large number of "followers." While it is usually clear that a television or magazine advertisement *is* an advertisement, it is not always obvious if a social media post has been paid for. Maybe someone posts about a new pair of jeans because they really like them—and were not paid to say that. FTC guidelines now make celebrity endorsers liable for false statements about a product. They shouldn't talk or post about a product unless they have experience with it. The FTC guidelines also require disclosure when there is a "material connection" between the influencer and the brand. This includes their receiving free products as well as any financial payment.

betina_goldstein has more than 70,000 Instagram followers. Some of her Instagram posts are advertisements. By rule, the FTC requires her to disclose when she is compensated for promoting a brand (in this case Nail inc), which she does with the hashtag #sponsored.
Source: Betina Goldstein/Instagram

What is unfair or deceptive is changing

What constitutes unfair and deceptive advertising is a difficult question. The law provides some guidelines, but the marketing manager must make personal judgments as well. The social and political environment is changing worldwide. Practices considered acceptable some years ago are now questioned or considered deceptive. Saying or even implying that your product is *best* may be viewed as deceptive. And a 1988 revision of the Lanham Act protects firms whose brand names are unfairly tarnished in comparative ads.

Supporting ad claims is a fuzzy area

It's really not hard to figure out how to avoid criticisms of being unfair and deceptive. Making a vague claim that cannot be proved or disproved (sometimes called "puffery") is acceptable—the FTC won't bring charges for ads that claim "the best hamburger in the world." But marketing managers need to put a stop to the typical production-oriented approach of trying to use advertising to differentiate me-too products that are not different and don't offer customers better value.[37]

Sales Promotion—Do Something Different to Stimulate Change

LO 15.9

The nature of sales promotion

Sales promotion refers to Promotion activities—other than advertising, publicity, and personal selling—that stimulate interest, trial, or purchase by final customers or others in the channel. Sales promotion generally complements other promotion methods. Although advertising campaigns and sales force strategy decisions tend to have longer-term effects, a sales promotion activity usually lasts for a limited time. Sales promotion can often be implemented quickly and get sales results sooner than advertising. Further, sales promotion objectives usually focus on prompting some short-term action. For an intermediary, such action might be a decision to stock a product, provide a special display space, or give the product extra sales emphasis. For a consumer, the desired action might be to try a new product, switch from another brand, or buy more of a product. The desired action by an employee might be a special effort to satisfy customers. Exhibit 15-8 shows examples of typical sales promotions targeted at final customers, channel members, or a firm's own employees. Let's walk through some examples of how sales promotion works for each type of customer.[38]

Tie consumer sales promotions to objectives

Much of the sales promotion aimed at final consumers tries to increase demand—perhaps temporarily—or to speed up the time of purchase. Such promotion might involve developing materials to be displayed in retailers' stores, including point-of-purchase materials, rewards programs, or aisle displays. It might include sweepstakes contests as well as coupons designed to get customers to buy a product by a certain date.

Exhibit 15-8 Examples of Sales Promotion Activities

Aimed at final consumers or users	Aimed at wholesalers or retailers	Aimed at company's own sales force
Contests	Price deals	Contests
Coupons	Promotion allowances	Bonuses
Aisle displays	Sales contests	Meetings
Samples	Calendars	Portfolios
Trade shows	Gifts	Displays
Point-of-purchase materials	Trade shows	Sales aids
Frequent buyer programs	Meetings	Training materials
Sponsored events	Catalogs	
Partnering with causes (#M4BW)	Merchandising aids	
Limited availability products	Videos	

What's Next? Brand-nonprofit partnerships raise awareness and money (#M4BW)

Driving home from a charity auction, Ryan Cummins and Matt Pohlson were disappointed they didn't win the chance to play basketball with former NBA star Magic Johnson; they couldn't come close to competing against the winning $15,000 bid. They thought about the auction and wondered if there was a better way raise awareness and money for good causes. Questioning *What's Next?* got their creative juices flowing and they soon launched Omaze. The fundraising platform designs win-win-win-win opportunities by connecting charities, brands, celebrities, and donors.

What Omaze does is pretty straightforward. It raises money for charities by offering once-in-a-lifetime experiences that anyone who buys a ticket can win—ten dollars buys 100 entries and the good feeling of helping a good cause. For example, Disney and director J.J. Abrams offered a chance to win a walk-on part in *Star Wars: The Force Awakens*. That kind of offer generates lots of online buzz for the new movie—and plenty of donations. In just 10 weeks the promotion raised $4.2 million for UNICEF Innovation Labs—and got someone in the movie.

Coca-Cola was looking for a creative way to promote the brand's social conscience and raise money to help end mother-to-child transmission of the HIV virus. It teamed up with Omaze and (Red), an AIDS-focused nonprofit started by musician Bono. Coca-Cola sponsored "Share the Sounds of an AIDS Free Generation," which shared a previously unreleased Queen song, "Let Me in Your Heart Again." Omaze's platform offered chances to win products and experiences, including Bono's guitar, backstage passes at a U2 concert, and an opportunity to spend a day with the pop rock band OneRepublic. The event raised money and awareness for HIV prevention and tied Coca-Cola to a good cause.

Another project offered a chance to win a Tesla Model 3 donated by Kimbal Musk (brother of Tesla founder Elon Musk). Kimbal wanted to raise money for his Big Green charity. Big Green builds Learning Gardens in low-income schools across the country. The gardens teach kids about real food and community engagement. Lots of folks entered to win Kimbal's Model 3, raising attention and $2.1 million for Big Green and awareness for Tesla.

Is Omaze *What's Next?* for fundraising, brand building, and sales promotion? Omaze carefully curates the once-in-a-lifetime experiences it offers—making sure each aligns with the values of the brand, celebrity, cause, and donors. It helps when something is "cool," as this generates more online buzz. In the end, we all win when these promotions make the world a better place.[39]

Think of some brand-celebrity-cause partnerships that you think make sense. Explain why you think these partnerships would work well for each party.

Often, sales promotion has the goal of encouraging consumers to try a new product. Consumers may see less risk in trying something new if it comes with a price reduction through a coupon or free sample. The marketing manager hopes that buying something new changes a buying habit. Another objective of sales promotion can be to encourage a brand's current customers to purchase more frequently. Loyalty cards or offering a free sandwich after you purchase five are ways to encourage customers to increase purchases.

#M4BW

Brands ally with causes and advance both

Some brands build alliances with nonprofit organizations that advance the brand's objectives (perhaps gaining awareness or supporting positioning) and those of a cause. For example, the nonprofit Service Year Alliance works to make a year of national service easily accessible and more common for young Americans. Going on a service year often involves moving to a new city. So Airbnb partnered with Service Year Alliance, offering two weeks of free housing to members while they looked for permanent housing. Alliances between nonprofits and brands are gaining traction and making for a better world. Read more in *What's Next?* Brand-nonprofit partnerships raise awareness and money.

Consumer sales promotions come in many forms. One type of consumer promotion is a rewards program. Citi offers its customers "double cash" back to reward their loyalty and encourage them to use the credit card more often. Another way to encourage customers to purchase is to make products available for a limited time. McDonald's Shamrock Shakes are only available for about a month each year (around St. Patrick's Day), so fans are encouraged to get one while they can.
(left): Source: Citigroup Inc.; (right): Source: McDonald's

B2B promotions move products or generate leads

Sales promotion directed at industrial customers might use the same kinds of ideas—for example, temporarily lowering a price to encourage trial of a new product. In B2B, the sales promotion people might also set up and staff trade show exhibits to generate attention and interest for a firm and its products. Trade shows usually occur in one city over a period of three to seven days. Customers attend to learn from industry experts. For example, the Consumer Electronics Show in Las Vegas each January lasts four days and attracts more than 4,000 electronics exhibitors and 180,000 attendees. After the show, many of the attendees become leads for the company's salespeople, who try to convert customer interest to actual sales. Trade shows are big events that require significant planning.

Some sellers give promotion items—pens, watches, or clothing (usually with the firm's brand name on them)—to remind business customers of their products. This is common, but it can be a problem because some companies do not allow buyers to accept any gifts.[40]

Trade promotion aims at intermediaries

Trade promotion refers to sales promotion aimed at intermediaries. A range of tools may be used depending on the promotion objective. Sometimes a producer uses trade promotion to get the intermediary's sales force to pay more attention to the producer's products. Sales contests can help achieve that goal, encouraging intermediary salespeople to earn prizes by exceeding sales targets. Another approach is to train intermediary salespeople on the product, perhaps at a producer-sponsored sales meeting in a desirable location (so that attendance has its own rewards, in addition to professional growth).

About half of sales promotion spending targeting intermediaries has the effect of reducing the price in some manner or another. Discounting products can give a wholesaler or retailer an incentive to sell the product more quickly—with either extra effort or a lower price. We will discuss various types of price-related promotions in Chapter 17.

Sales promotion for own employees

Sales promotion aimed at the company's own sales force might try to encourage sales reps to provide better service, get new customers, sell a new product, or sell the company's whole line. Depending on the promotion objectives, the tools might be contests, bonuses on sales or number of new accounts, or holding sales meetings at fancy resorts.

Objectives and situation influence sales promotion decisions

As we have seen, there are many different types of sales promotion, but what type is appropriate depends on the situation and objectives. For example, Exhibit 15-9 shows some possible ways that a short-term promotion might affect sales. The sales pattern in the graph on the left might occur if Hellmann's issues coupons to help clear its excess mayonnaise inventory. Some consumers might buy earlier to take advantage of the coupon, but unless they use extra mayonnaise, their next purchase will be delayed and sales will fall after the promotion period. In the center graph, kids might convince parents to eat more Happy Meals while McDonald's has a Monopoly game promotion, but when it ends, sales go back to normal. The graph on the right shows a Burger King marketer's dream come true: free samples of a new style of french fries quickly pull in new customers who like what they try and keep coming back after the promotion ends. This is the best-case scenario—a sales promotion encourages trial, and customers change their habits. From these examples, you can see that the situation and the objective of the sales promotion should determine what type is best. Marketing managers should monitor and analyze the short- and long-term financial costs and benefits of sales promotion to properly evaluate its role in the promotion blend.

Sales promotion spending grows in mature markets

Sales promotion involves so many different types of activities that it is difficult to estimate accurately how much is spent in total. There is consensus, however, that the total spending on sales promotion exceeds spending on advertising.[41]

Exhibit 15-9 Some Possible Effects of a Sales Promotion on Sales

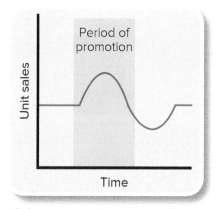

Sales temporarily increase, then decrease, then return to regular level

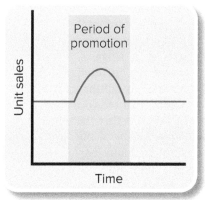

Sales temporarily increase and then return to regular level

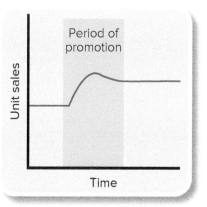

Sales increase and then remain at higher level

One reason for increased use of sales promotion by many consumer product firms is that they are generally competing in mature markets. There's only so much soap that consumers want to buy, regardless of how many brands there are vying for their dollars. There's also only so much shelf space that retailers will allocate to a product category.

The competitive situation is intensified by the growth of large, powerful retail chains. They have put more emphasis on their own dealer brands and also demand more sales promotion support for the manufacturer brands they do carry.

It is perhaps because of this competition that many consumers have become more price sensitive. Many sales promotions, like coupons, have the effect of lowering the prices consumers pay. Sales promotion has been used as a tool to overcome consumer price resistance, creating a downward cycle of pricing and promotion.

Managing Sales Promotion

Does sales promotion erode brand loyalty?

Some experts think that marketing managers—especially those who deal with consumer packaged goods—put too much emphasis on sales promotion. They argue that the effect of most sales promotion is temporary and that money spent on advertising and personal selling helps the firm more over the long term. When the market is not growing, sales promotion may just encourage "deal-prone" customers (and intermediaries) to switch back and forth among brands. Here, all the expense of the sales promotion simply contributes to lower profits. It also increases the prices that consumers pay because it increases selling costs.

Sales promotion is hard to manage

Another problem in the sales promotion area is that it is easy to make big, costly mistakes. Because sales promotion includes such a wide variety of activities, it's difficult for the typical company to develop skill in managing all of them. Even large firms and agencies that specialize in sales promotion run into difficulties because each promotion is typically custom-designed and then used only once. Mistakes caused by lack of experience can be costly or hurt relationships with customers.

In a promotion for Pampers diapers designed to reward loyal buyers and steal customers away from competing Huggies, marketing managers offered parents Fisher-Price toys if they collected points printed on Pampers' packages. At first the promotion seemed to be a big success because so many parents were collecting points. But that turned into a problem when Fisher-Price couldn't produce enough toys to redeem all the points. Pampers had to add 50 toll-free phone lines to

Auntie Anne's is a chain of pretzel shops. When it wanted to promote awareness and trial of its new Honey Whole Grain Pretzel, it turned to HelloWorld, an agency that specializes in sales promotion. The promotion generated 1.6 million sweepstakes entries and more than 200,000 registrations.
Source: Auntie Anne's

Legal concerns add to the complexity

handle all the complaints, and a lot of angry parents stopped buying Pampers for good. Problems like this are common.[42]

Consumer and trade promotions are both subject to regulation by the Federal Trade Commission and often state and local laws. For example, in most states, contests and sweepstakes must be carefully designed so they are not considered a form of gambling. There are also rules covering premiums or giveaways, and companies need to be careful they are not misrepresenting their true value. Trade promotions often involve direct or indirect adjustments to prices that could be seen as discriminatory pricing, which is covered by the Robinson-Patman Act.[43]

Not a sideline for amateurs

Sales promotion mistakes are likely to be worse when a company has no sales promotion manager. If the personal selling or advertising managers are responsible for sales promotion, they often give it less attention. They allocate money to sales promotion if there is any "left over" or if a crisis develops. This approach misuses a valuable element of the promotion blend.

Making sales promotion work is a learned skill, not a sideline for amateurs. That's why specialists in sales promotion have developed, both inside larger firms and as outside consultants. Some of these people are real experts. But it's the marketing manager's responsibility to set sales promotion objectives and policies that will fit in with the rest of the overall marketing strategy.[44]

CONCLUSION

It may seem simple to develop an advertising campaign. Just pick the medium and develop a message. But it's not that easy.

This chapter discussed why marketing managers should set specific objectives to guide the entire advertising effort. Knowing what they want to achieve, marketing managers can determine what kind of advertising—product or institutional—to use. We also discussed three basic types of product advertising: pioneering, competitive (direct and indirect), and reminder.

Marketing managers must also choose from various media—so we discussed their advantages and disadvantages. Because the Internet and mobile devices offer new advertising opportunities and challenges, we discussed how they are similar to and different from other media. And, of course, marketing managers must determine the message—or copy thrust—that will appear in ads. This chapter also discussed how to make sure that advertising is done legally.

Sales promotion spending is big and growing. This approach is especially important in prompting action—by customers, intermediaries, or salespeople. There are many different types of sales promotion, and it is a problem area in many firms because it is difficult for a firm to develop expertise with all of the possibilities.

Advertising and sales promotion are often important parts of a promotion blend, but as many customers tune out advertising, other forms of media are gaining a greater share of the promotion budget. In Chapter 16 we will explore different forms of publicity: earned media, owned media, and social media.

KEY TERMS

advertising agencies, 410
advertising allowances, 413
cooperative advertising, 413
product advertising, 413
institutional advertising, 413
pioneering advertising, 413
competitive advertising, 413

direct competitive advertising, 413
indirect competitive advertising, 413
comparative advertising, 414
reminder advertising, 415
advertising media, 416
pay-per-click, 419
retargeting, 419

click-through rate, 420
influencers, 422
copy thrust, 424
corrective advertising, 429
trade promotion, 432

QUESTIONS AND PROBLEMS

1. Review the Domino's case study that opens the chapter. From this case, identify examples of different key terms and concepts covered in the chapter. For example, it is mentioned that Domino's used sales promotion, including offering free samples, to motivate customer trial.

2. Review the Domino's case study that opens the chapter. Consider the "Digital and Mobile Advertising" section of the chapter. Develop a plan with three specific ideas about how Domino's could develop digital and mobile programs consistent with its target markets and positioning. Explain why your ideas would work.

3. Identify the strategy decisions a marketing manager must make in the advertising arena.

4. Discuss the relation of advertising objectives to marketing strategy planning and the kinds of advertising actually needed. Illustrate with examples.

5. List several media that might be effective for reaching consumers in a developing nation with low per capita income and a high level of illiteracy. Briefly discuss the limitations and advantages of each medium you suggest.

6. Give three examples where advertising to intermediaries might be necessary. What is (are) the objective(s) of such advertising?

7. What does it mean to say that "money is invested in advertising"? Is all advertising an investment? Illustrate with examples.

8. Find advertisements to final consumers that illustrate the following types of advertising: (a) institutional, (b) pioneering, (c) competitive, and (d) reminder. What objective(s) does each of these ads have? List the needs each ad addresses.

9. Describe the type of media that might be most suitable for promoting (a) tomato soup, (b) greeting cards, (c) a business component material, and (d) playground equipment. Specify any assumptions necessary to obtain a definite answer.

10. Briefly discuss some of the pros and cons an advertising manager for a producer of sports equipment might want to think about in deciding whether to advertise on the Internet.

11. Discuss the use of testimonials in advertising. Which of the four AIDA steps might testimonials accomplish? Are testimonials suitable for all types of products? If not, for which types are they most suitable?

12. Does advertising cost too much? How can this be measured?

13. Is it unfair to criticize a competitor's product in an ad? Explain your thinking.

14. Discuss some ways that a firm can link its sales promotion activities to its advertising, personal selling, and publicity efforts—so that all of its promotion efforts result in an integrated effort.

15. Indicate the type of sales promotion that a producer might use in each of the following situations and briefly explain your reasons:

    a. A firm has developed an improved razor blade and obtained distribution, but customers are not motivated to buy it.

    b. A competitor is about to do a test market for a new brand and wants to track sales in test market areas to fine-tune its marketing mix.

    c. A big grocery chain won't stock a firm's new popcorn-based snack product because it doesn't think there will be much consumer demand.

MARKETING PLANNING FOR HILLSIDE VETERINARY CLINIC

Appendix D (the Appendices follow Chapter 19) includes a sample marketing plan for Hillside Veterinary Clinic. Look through the "Marketing Strategy" section.

1. What are Hillside's advertising objectives?
2. What types of advertising and media are being proposed? Why are these types used and not others?
3. What type of copy thrust is recommended? Why?
4. What sales promotion activities are being planned? What are the goals of sales promotion?

SUGGESTED CASES

18. West Tarrytown Volunteer Fire Department (WTVFD)
20. Lake Russell Marine & Camp
31. Silverglade Homes

MARKETING ANALYTICS: DATA TO KNOWLEDGE

CHAPTER 15: SALES PROMOTION

As a community service, disc jockeys from radio station WMKT formed a basketball team to help raise money for local nonprofit organizations. The host organization finds or fields a competing team and charges $5.00 admission to the game. Money from ticket sales goes to nonprofit organization.

Ticket sales were disappointing at recent games, averaging only about 300 people per game. When WMKT's marketing manager, Bruce Miller, heard about the problem, he suggested using sales promotion to improve ticket sales. The PTA for the local high school—the sponsor for the next game—is interested in the idea but is concerned that its budget doesn't include any promotion money. Miller tries to help the PTA by reviewing his idea in more detail.

Specifically, he proposes that the PTA give a free T-shirt (printed with the school name and date of the game) to the first 500 ticket buyers. He thinks the T-shirt giveaway will create a lot of interest. In fact, he says he is almost certain the promotion would help the PTA sell 600 tickets, double the usual number. He speculates that the PTA might even have a sellout of all 900 seats in the school gym. Further, he notes that the T-shirts will more than pay for themselves if the PTA sells 600 tickets.

A local firm that specializes in sales promotion items agrees to supply the shirts and do the printing for $2.40 per shirt if the PTA places an order for at least 400 shirts. The PTA thinks the idea is interesting but wants to look at it more closely to see what will happen if the promotion doesn't increase ticket sales. To help the PTA evaluate the alternatives, Miller sets up a spreadsheet with the relevant information.

CHAPTER SIXTEEN

Courtesy of The Chopping Block

Publicity: Promotion Using Earned Media, Owned Media, and Social Media

Brian Halligan and Dharmesh Shah met as MBA students at MIT in 2004. While helping entrepreneurs with their marketing, Halligan noticed the declining effectiveness of many traditional marketing tactics—sending out e-mails, advertising, cold calling customers, and attending trade shows. Customers avoided many of these communications. Around the same time, Shah attracted thousands of readers to his *OnStartups* blog without any traditional promotion.

Inbound promotion works best when customers are looking for information before making a purchase. Many consumers and business buyers start the purchase process this way. If inbound promotion reflects customer behavior and is so much more effective, why weren't more businesses using it? Halligan and Shah found that for most businesses, starting the process was daunting—build a website, write a blog, design a Facebook page. It seemed easier to keep doing things the old way. So, Halligan and Shah created HubSpot software to simplify inbound promotion.

HubSpot's promotion blend follows the same four steps it recommends to its customers: (1) acquiring—getting awareness and interest; (2) converting—moving customers to desire and action (i.e., purchasing a subscription to HubSpot software); (3) onboarding—helping customers with installation and training on the software; and (4) ongoing—working with customers analyzing results and increasing product usage. Promotion methods change as customers move through the process.

For HubSpot, acquisition begins with making customers aware of its inbound philosophy and the problems it solves, thus fostering interest in the software. The founders got some attention with their book, *Inbound Marketing: Get Found Using Google, Social Media, and Blogs*. Other customers discovered HubSpot through its free online "Marketing Grader." This tool evaluates the effectiveness of a user's website and offers pointers for improving it—tips HubSpot software can implement.

Customers also discover HubSpot when searching for ideas about how to improve their marketing strategy—HubSpot's content appears high in search results. They discover the free content HubSpot regularly publishes online, including blogs, case studies, white papers, podcasts, online seminars, and e-books. Some web pages ask for customer contact information, which HubSpot compiles in a database to learn more about each customer's needs and interests. Some of HubSpot's customers keep up with the company on social media; when followers "Like" a HubSpot Facebook post, "favorite" a tweet, or comment on the Inbound Marketers group on LinkedIn (which HubSpot runs), they endorse HubSpot to their friends and colleagues. HubSpot tracks every customer interaction, from website visits to e-mails and phone calls.

Finding this valuable trove of marketing insight, many customers sign up for HubSpot's e-mail list. HubSpot customizes e-mails, sending each customer information relevant to his or her job, industry, and interests. It monitors customers to see which have the highest levels of engagement. For example, it records which customers open e-mails and follows those who frequently return to the HubSpot website. These customers are usually the most promising leads, and a salesperson contacts these prospects. Using a consultative selling approach, the salesperson learns the customer's needs, often giving an online demonstration of the HubSpot software. The sales process may continue for weeks (or months) as HubSpot's salesperson educates a customer about inbound promotion and the software and recommends the best combination of HubSpot services.

The next two phases—onboarding and ongoing—are central to customer retention. If customers don't see value in HubSpot's software, they cancel their subscription. So HubSpot works closely with customers during installation and training. It monitors customer usage and offers tips and training to customers who don't take full advantage of the software. Other customers find value in HubSpot's Success Community, where fellow customers and HubSpot technical support people answer questions and share tips.

A typical HubSpot customer is Shelley Young, CEO and founder of The Chopping Block, a recreational cooking school in Chicago. When her business first opened more than two decades ago, Shelley was able to connect one-on-one with customers, helping them pick the right class and continuing the relationship after that first class. As the business grew, she found that was no longer the case. The Chopping Block turned to a digital marketing agency, which redesigned its website with HubSpot's Content Management System (CMS).

The Chopping Block website always had plenty of great content for cooks (and wannabe cooks)—but it wasn't always easy to find on the site. The CMS organizes content on the website, and each customer can easily find exactly the right information. The Chopping Block's website caters to several market segments; for example, different content draws in the adult cooking class student, the cooking party planner, and the in-store shopper. Finding useful content, customers often wander the site and discover even more. They may follow The Chopping Block on social media sites like Instagram or subscribe to its YouTube channel for just-in-time video lessons. HubSpot software also helps on the back end where The Chopping Block's marketing team analyzes the behavior of the site's 73,000 monthly visitors. The team sees what keyword searches brought customers to the site, which content is shared most on social media, and what actions lead to class sign-ups. These insights guide the development of new content. The low-pressure, high-information content of the site fosters customer trust and whets their appetites for a class or a party.

The Chopping Block is just one of many satisfied HubSpot customers. HubSpot ranks number one in customer satisfaction among marketing automation software vendors. All the useful content on the website makes HubSpot easy to find—and its great service and proven results keeps customers there.[1]

LEARNING OBJECTIVES

As the HubSpot case points out, many organizations are finding new approaches to telling customers about their products. This chapter provides you with a broad overview of new promotion tools that utilize earned media, owned media, and social media.

When you finish this chapter, you should be able to

1. explain how publicity fits into the promotion blend.
2. understand the differences between paid, earned, and owned media.
3. explain how customers obtain information from search, pass-along, and experience.
4. describe different types of owned media.
5. describe how to use public relations to earn attention from the press.
6. describe how to earn word-of-mouth from customers.
7. explain what social media are and how they differ from traditional media.
8. compare and contrast the major social media platforms.
9. describe how firms use software to measure, manage, and automate online promotion.
10. understand important new terms (shown in **red**).

Publicity, the Promotion Blend, and Marketing Strategy Planning

LO 16.1

This is the fourth and final chapter covering Promotion (see Exhibit 16-1). Previous chapters have introduced integrated marketing communications (Chapter 13), personal selling and customer service (Chapter 14), and advertising and sales promotion (Chapter 15). In this chapter we examine *publicity*, which refers to any *unpaid* form of nonpersonal presentation of ideas, goods, or services and includes what we call *earned*, *owned*, and *social media*.

Publicity requires strategy decisions

This chapter takes a close look at different types of publicity. The chapter begins by comparing advertising (which uses paid media) to different types of publicity (unpaid media). Then the concepts of paid media, earned media, and owned media are introduced—explaining how each has advantages and disadvantages in the promotion blend. After briefly describing how information flows through search, pass-along, and branded experiences, we provide more details on different types of owned and earned media. Next, we explain the different types of social media and describe software used to manage online media.

Exhibit 16-1 Marketing Strategy Planning for Publicity, Earned Media, Owned Media, and Social Media

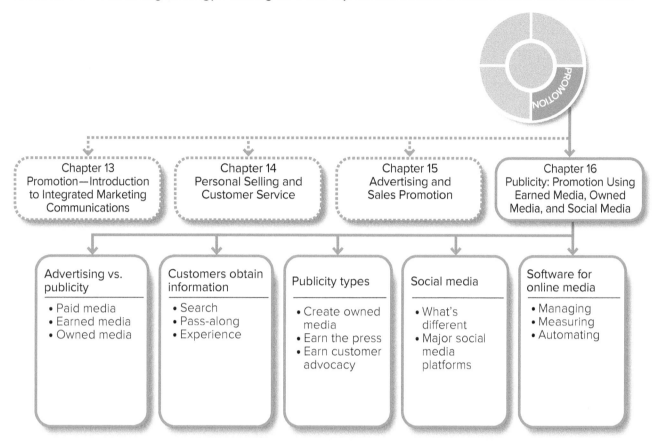

Integrating publicity into the promotion blend

Source: Osprey Packs; (smartphone frame): Oleg GawriloFF/Shutterstock

Publicity works best when it is integrated into the entire marketing mix. A marketing manager must understand how the right promotion moves a customer through the buying process. As a customer moves through the process, different types of information are needed.

Consider the different types of promotion involved in Kat's purchase of a backpack. To highlight the role of publicity, we *italicize* those types of promotion.

Kat likes to hike and was thinking about buying a new backpack. She sees a magazine ad for an Osprey backpack and recalls a *friend's recommendation* of the brand—capturing Kat's attention and interest. So she *visits the Osprey website*, sees and reads about different backpacks, compares features, and watches a short video. While there, Kat finds links to stores that sell the packs. Then she *reads reviews* at eBags.com and learns of some new features she should consider when buying a new backpack. Not quite ready to buy, Kat puts the purchase on hold. A week later, while scrolling through Facebook, a sponsored post (advertisement) by Osprey Packs appears in her newsfeed and rekindles Kat's interest and desire for a new backpack. By the time she arrives at her local REI store, Kat has narrowed her choices to three Osprey models. A helpful salesperson shows her the bags, answers some questions, and tells her how she could get a free Osprey water bottle with a backpack purchase (sales promotion); she buys the Osprey Aura AG 65 EX. After all this, Kat feels like part of the Osprey family—she likes to *show fellow hikers* all the features of her new backpack whenever asked. Kat also likes to stay informed and *follows Osprey Packs on Facebook and Instagram*—often "*Liking*" their posts and photos.

Various elements of Osprey's promotion blend combine to help move Kat through the purchase process for an Osprey backpack. At different stages, advertising, sales promotion, publicity, and personal selling contribute to her decision to purchase. And after her purchase, Kat helped with more publicity as she generated positive word-of-mouth for Osprey. This chapter adds some new media to our discussion of Promotion and shows how they can be integrated to achieve promotion objectives.

Paid, Earned, and Owned Media

LO 16.2

To better understand publicity, it helps to compare it with advertising (covered in Chapter 15). Advertising and publicity are both types of mass selling—in this chapter, we will differentiate them. We have already mentioned that advertising uses *paid media;* advertisers pay for placement on television, radio, online, and so on. Within publicity, we have two types of unpaid media—owned media and earned media. This section defines, describes, and compares paid, owned, and earned media (see Exhibit 16-2).

Advertising uses *paid* media

Media refers to the means used for mass communication. For example, advertising utilizes **paid media** or messages generated by a brand (or company or nonprofit organization) and communicated through a message channel the brand pays to access. The source of the message is the organization doing the advertising.

Publicity relies on *unpaid* media—*owned* or *earned*

Owned media refers to promotional messages generated by a brand (or company or nonprofit organization) communicated through a message channel the brand directly controls. Owned media include the brochures and catalogs a company mails to target customers. A brand's product website, blog, and social media pages—YouTube channel, Facebook page, or Instagram—are also examples of owned media. As with paid media, the selling company is the source of the message.

Earned media refers to promotional messages *not* directly generated by the company or brand, but rather by third parties such as journalists or customers. The company or brand is assumed to *earn* the attention of journalists or customers because there is an interesting story to tell or because others would value this information. Earned media can be positive or negative—for example, poor customer service might earn a restaurant a bad review on Yelp.

Journalists' messages can appear as editorial content in a newspaper, magazine, or on a website that provides news. Many newspapers and magazines (online and offline) publish restaurant, theater, book, and movie reviews—or review new products. Trade magazines write articles about new goods and services of interest to their readers.

Exhibit 16-2 Paid, Owned, and Earned Media

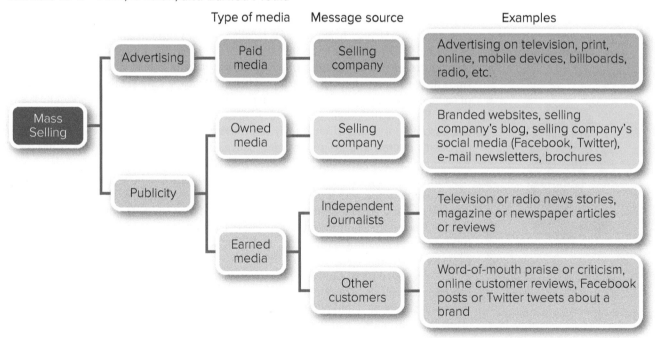

A company or brand can also *earn* word-of-mouth communication from customers. **User-generated content** refers to any type of communication created by customers for other customers. User-generated content can take many forms. Customers can recommend their hairstylist to a friend or demonstrate their new Samsung Galaxy phone to a coworker. A customer can also send a tweet praising a new book—or post a review at Amazon describing what he likes (and doesn't like) about his new nonstick frying pan. All of these types of user-generated content are examples of earned media.[2]

Promotion and customer trust

Chapters 5 and 6 describe how customers gather and use information needed for a buying decision. In Chapter 13 we noted that customers' trust of a source influences whether they believe and act on information. Exhibit 16-2 shows three general types of media. Research finds that customers generally view earned and owned media as more credible and trustworthy than paid media.

A study of 30,000 Internet users from 60 countries found the four most highly trusted sources for buying information are forms of publicity, not advertising (see Exhibit 16-3). Consumers trust information from *earned* media sources most:

Exhibit 16-3 Customer Trust in Various Sources of Information

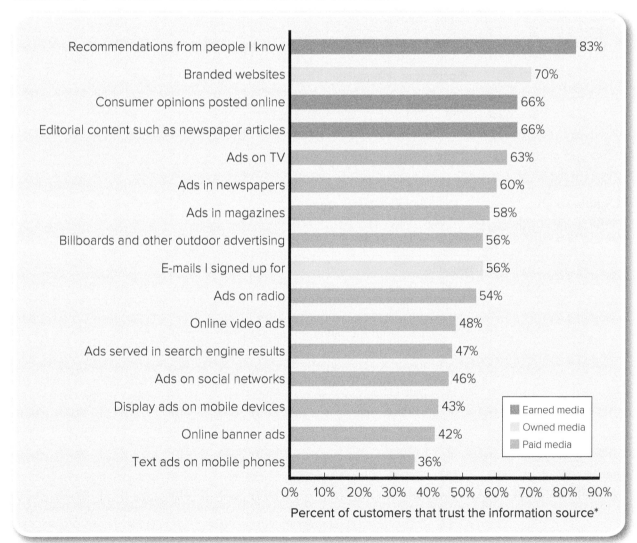

Note: Sources of earned or owned media; others are forms of paid media.
*Percentages reflect those responding that they "completely trust" or "somewhat trust" the information source.

Exhibit 16-4 Benefits and Challenges of Advertising Media, Owned Publicity Media, and Earned Publicity Media

Media Type	Benefits	Challenges
Paid media (advertising)	• High message control • More precise targeting • Potentially large audience	• Not trusted • Customers easily avoid • More costly, declining effectiveness
Owned media	• High message control • Relatively low cost • Niche audiences • Versatile in message content and format	• Need to drive or attract customers to sites and create value to ensure their return • Require resources to manage and maintain
Earned media	• Most trusted information source • Customers most likely to act on this information	• Very little message control • Can be negative toward brand • Difficult to measure • Difficult to create • Difficult to target

83 percent of respondents trust recommendations from people they know, 66 percent trust online consumer opinions (reviews), and 66 percent trust editorial content such as newspaper articles. Branded websites (a form of *owned* media) are trusted by 70 percent of respondents. These levels of trust are significantly higher than various types of advertising (paid) media. Among various advertising (paid) media, trust ranged from 36 percent for text ads on mobile phones to 63 percent for television advertising.[3]

Benefits and challenges for each media type

The general differences among these three types of media extend beyond trust. Let's look at the benefits and challenges facing each (see Exhibit 16-4).

Paid media—high message control yet becoming less effective

Chapter 15 addressed many of the benefits of advertising. Marketing managers (or their advertising agencies) develop the copy thrust and therefore control the words and images in the messages customers receive. Most advertising is designed to target a large number of customers at once, whereas specialized media target fewer customers more precisely.

On the other hand, customers are getting better at avoiding advertising messages. Most customers see advertising as an interruption of something they would rather be doing—watching their favorite TV program, not a television commercial; listening to a popular song on the radio, not an advertising jingle; getting to a favorite website without waiting for an ad to play. New technology helps many customers sidestep advertising messages. Digital video recorders (DVRs) let consumers skip right past television commercials. Online music streaming services such as Spotify allow subscribers to listen ad-free, and pop-up blockers or simply clicking to another page lets Internet surfers bypass ads. Combine these behaviors with customers' lack of trust in advertising and it's easy to understand why advertising effectiveness has been declining.

Owned media—high message control and lower cost

Owned media allow a marketing manager to craft the precise message customers receive. Marketing managers can choose what information is displayed in a brochure or on the company website, YouTube channel, Instagram, Pinterest page, or Twitter feed. A website may include an information page that provides details about a product's features with information to show how a customer can find retailers that carry the product. Owned media also allow a manager a great deal of versatility with the messaging.

For example, a web page might contain a short video overview of a product with a link to a detailed, text-heavy web page for those customers desiring technical specifications. While owned media have low media costs—and usually lower costs overall—marketing managers need to remember that there is a cost to create and maintain owned media channels.

The challenges for marketing managers include maintaining owned media and making sure customers find their information when it is needed in the buying process. A website or social media feed needs regular updating and monitoring to ensure the site contains content customers find helpful. Additionally, the site must be easy to navigate and search. Later in this chapter we discuss approaches for developing owned media that attract customers by creating value for them.[4]

Customers act on earned media, but it has risks

Marketing managers are usually thrilled when positive articles are written about their products or customers post positive reviews on social media sites. These are the most powerful messages a target customer can receive—customers trust and act on this information. Think about your own experience. How likely are you to visit a new restaurant when you see an advertisement in the newspaper? How does that compare to having a friend tell you, "We had a great dinner last night at that new Italian restaurant on Main Street"?

On the other hand, earned media create challenges for a brand. These influential message sources are out of a marketing manager's control. When a restaurant critic writes a negative review in the newspaper or an unhappy customer posts a less-than-flattering review on Yelp, customers are much less likely to visit that business. This chapter also includes discussion about how marketing managers can foster positive earned media while managing negative media.[5]

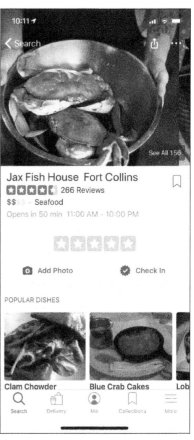

Jax Fish House, a restaurant in Fort Collins, Colorado, maintains an Instagram page (image on left). Jax uses this form of owned media to post pictures of its delicious food and drinks. Jax controls what gets posted on Instagram. On the other hand, Jax does not directly control what is said in the 266 reviews it has on Yelp (image on right). Fortunately, delivering a great dining experience helps the restaurant earn high praise from customer/reviewers.
(left): Source: Big Red F Restaurant Group/Instagram; (right): Source: Big Red F Restaurant Group/Yelp

Customers Obtain Information from Search, Pass-Along, and Experience

LO 16.3

Whereas advertising media are generally broadcast to customers, customers discover owned and earned media through different processes. Before we look at different types of media, it is helpful to understand three other processes by which customers obtain information—search, pass-along, and experience. By understanding how each process works, marketing managers can select the best tools and practices to achieve particular promotion objectives.

Marketers want to be found when customers search

Because customers often search for information during the shopping process (see Exhibit 13-5), marketing managers need to make sure that when target customers search online, material useful to them is "found." Most customers use a search engine such as Google, Bing, or Yahoo!—and most favor the organic search results rather than the paid ads at the side (or top) of the search (see Chapter 15 for more on search advertising). When conducting an online search, most users don't look past the 10 results on a typical first page of search; studies find that half of those who click on an organic search result choose the first or second result and about three-quarters pick one of the top five. So, when customers are looking, it pays to get near the top of that first page of results.

To appear high on organic search results, marketing managers often use **search engine optimization (SEO)**, which is the process of designing a website so that it ranks high in a search engine's unpaid results. SEO considers how target customers search, what keywords they use in a search, and how the search engine prioritizes its results. Although the technical elements of SEO are beyond the scope of this textbook, a range of techniques can be employed to influence standing in search results.[6]

Customers search offline too

Customers search for answers offline as well. This may lead them to a salesperson who can answer their questions. It might also lead them to a neighbor or friend who might help them "find a new dentist" or "recommend a preschool." In these latter two situations, the information often leads to other customers sharing information with one another directly—a form of pass-along.

Pass-along from trusted sources

Pass-along occurs when one customer passes information on to one or more other customers. Customers pass along videos, articles, coupons, or websites to other customers. They also recommend brands when the seller merits an endorsement—or advise against a brand that deserves it. Sellers can earn more recommendations by making it easy for customers to share recommendations or content.[7]

Customers had a positive experience with Samsung when the electronics maker put charging stations (with Samsung signage) in places like airports and convention centers. The charging stations positioned Samsung as a "lifesaver" and built favorable attitudes with lucrative business travelers—who tend to buy a lot of consumer electronics. More recently, Samsung put charging stations on college campuses to target another gadget-loving group—students. These are examples of branded services.[8]
David McNew/Getty Images

What's Next? Take STEPPS to go viral

Customers spreading a positive message far and wide—that's a marketing manager's dream. For every highly contagious message, hundreds don't get much pass-along. Can we increase the odds? According to Jonah Berger, there are characteristics of an idea or message that make it contagious. He suggests six STEPPS to increase pass-along: (1) **S**ocial currency, (2) **T**riggers, (3) **E**motion, (4) **P**ublic, (5) **P**ractical value, and (6) **S**tories.

Social currency occurs when the sharer looks good by sharing. When a new bar called Please Don't Tell opened in New York City, it had a secret. There was no front door to get into the bar; the only entry was through a phone booth in the corner of a neighboring hot dog restaurant. Those who knew the secret gained social currency when they shared this tip with friends. Many did and now the bar is a hit—without advertising, a street sign, or a front door.

Triggers make a topic easy to remember and encourage others to talk about it. When Philadelphia's fine-dining steakhouse, Barclay Prime, added a $100 cheesesteak sandwich to its menu, people started talking. Cheesesteak sandwiches typically sell for $5. The home of the original cheesesteak, Philadelphia has hundreds of cheesesteak shops whose existence naturally triggers conversations about Barclay Prime's budget buster.

Happiness is one *emotion* people like to share—and humor brings that about. Poo-Pourri sells fragrant sprays to clear the air after using the toilet. When the company created the funny video, "Girls Don't Poop," it was widely shared online—getting more than 40 million views, lots of laughs, and plenty of awareness for Poo-Pourri (see https://youtu.be/ZKLnhuzh9uY).

When a product's features are noticeably *public*, they spread more quickly. When Apple introduced the iPod, most headphones were black with black cords, so Apple's white headphones stood out. Designer Christian Louboutin's shoes have red soles that draw attention when wearers walk down a street. These public displays stand out and the idea spreads.

Public service announcements have *practical value* but are often kind of boring. When Metro (commuter) Trains in Melbourne, Australia, wanted to remind customers of safe practices around trains, it decided to grab and hold people's attention. The video "Dumb Ways to Die" features animated animal characters and a catchy tune (see https://youtu.be/IJNR2EpSOjw). The video makes a serious point in a humorous way—and has been viewed more than 175 million times. The practical and funny video conveys an important public message and makes the world a better place (#M4BW).

Finally, people also like to pass along interesting *stories*. Stories capture the imagination and make an idea easy to remember and share. A classic story describes a customer who returned a set of tires to a Nordstrom department store in Alaska. Nordstrom didn't sell tires, but the customer bought them from a now-closed tire store that used to be located in the same location. A Nordstrom employee gave the customer his money back, the story spread, and the store's reputation for taking care of customers grew.[9]

Marketing managers are often looking for What's Next? *Are these STEPPS practical? Think of a video or idea that has recently gone viral. Which of the STEPPS does it reflect? Describe how a nonprofit organization focused on raising money for AIDS prevention in Africa could use each of the STEPPS to increase likelihood of its message and mission going viral.*

The ultimate in pass-along—going viral

When customers spread a message far and wide, it's called *viral promotion*. Getting a video to go viral is not easy. That said, marketing science has identified some tricks that improve the odds. Read more in *What's Next?* Take STEPPS to go viral.

Customers experience branded services

Customers gain experience with a brand in many ways. Most often customers gain experience with a brand by using its Product—its goods and/or services. Some firms use Promotion to enhance customer experiences. **Branded services** are valued services a brand provides that are not directly connected to a core product offering. The services are designed to share a message—typically that the brand cares about its customers by offering them some free service. This helps customers perceive a company or brand as caring about the customers' welfare. Customers *experience* these benefits and develop a positive attitude that fosters trust.[10]

#M4BW

Branded service shows what engineers can do

Branded services are usually free or low cost and offer target customers benefits that leave a positive impression. Designed right, they can also make the world a better place. Consider what the University of Engineering and Technology (UTEC) in Lima, Peru, did to promote an upcoming enrollment period. In Lima, potable water is a precious commodity; the city lies in a coastal desert that gets less than an inch of rain a year—though it also experiences 98 percent humidity. So the engineering school created a billboard with a special technology; it produces drinking water from the humid air. The billboard generates thousands of liters of water per month, enough for hundreds of families. The message it sends reinforces UTEC's tagline, "We will continue changing the world through engineering," though in this case, it also changed the world through marketing.[11]

Create Owned Media Content Your Customers Can Use

LO 16.4

When solving a problem, customers have choices about where to go to find information. When faced with a new problem or need, many customers turn first to the Internet. When a business or organization creates owned media content that helps customers solve their problems, their messages get into buyers' minds. In this section, we will examine different types of owned media.

Exhibit 16-5 provides an overview of the types of owned media discussed in this section. Social media (such as Facebook, Instagram, and Twitter) often operate as owned

Exhibit 16-5 Owned Media—How Customers Find Each Type and Typical Promotion Objectives

Type of Owned Media	How Customers Typically Discover Each Type of Owned Media	Typical Promotion Objectives
Educational web pages	• Customers searching for answers	• Build awareness • Educate and inform customers • Foster customer trust
White papers and case studies	• Customers searching for ideas, information, or solutions to problems	• Build awareness • Educate and inform customers • Build reputation as thought leader • Foster customer trust
Landing pages	• Advertising drives customers to landing pages • Customers searching for ideas, information, or solutions to problems	• Build awareness • Start a relationship • Generate leads
Blogs	• Pass-along from friends' content on social media • Customers searching for ideas, information, or solutions to problems • Return or monitor a blog that user found valuable previously	• Maintain interest • Build reputation as thought leader
Infographics	• Customers searching for solutions • Practical value can motivate pass-along	• Grab attention • Inform customers
Branded apps	• Practical value can motivate pass-along	• Support brand positioning
Brand communities	• Discover through search • Return to communities where value was found previously • Practical value can motivate pass-along	• Provide value for customers • Remind customers
E-mail newsletter	• Subscribe to valuable e-mail content • Pass-along valuable content	• Maintain relationship • Remind customers • Cross-sell to customers

media and are covered later in this chapter. Exhibit 16-5 also describes how customers typically discover each type of media and representative promotion objectives. Understanding how customers discover owned media helps a firm develop other promotion efforts that direct target customers to owned media.

Educate customers and start a conversation

A website, salesperson, or book that helps a customer solve a problem also earns that customer's trust. Customers who find content useful will pass along the site to others. Those types of connections also help a firm with its search engine optimization—creating a virtual circle. Content can fill a customer's need without requiring that the customer make a purchase. When customers see a company providing useful information but not asking for something in return, customers trust the content provider. This eventually leads to more sales.

Online jewelry retailer Blue Nile's website educates customers. Most of its target customers, young men looking to buy an engagement ring, have never purchased a diamond before. A video tutorial and separate sections describing characteristics of diamonds help them feel more comfortable with making such a large purchase. That appreciation often leads to a diamond purchase from Blue Nile.

Business customers seek solutions

Organizational buyers often jump right on the Internet when they discover a new need. Often they seek out "thought leaders" in the industry to learn more about their options. For a company to establish itself as such a leader, it might publish white papers. A **white paper** is an authoritative report or guide that addresses important issues in an industry and offers solutions. Some companies will self-publish an extended version of a white paper as an *e-book*. E-books and white papers are most successful when, in addition to describing a problem, they help the customer solve it without promoting a particular company's products. An objective tone helps the provider build trust and credibility with potential customers. Business customers also like to read **case studies**—success stories about how a company helped another customer.[12]

Some firms ask customers to leave contact information before downloading a white paper, case study, or even an e-book. For example, a seller may ask for a customer's name, e-mail address, and job title. This feeds into a lead management system the seller can use for follow-up with an e-mail or phone call from a salesperson.

Landing pages to engage customers

A **landing page** is a customized web page that logically follows from clicking on an organic search result, online advertisement, or other link. Landing pages may be linked from social media, an e-mail campaign, or search engine optimization around specific key terms. The goal of a landing page is to move customers along in the purchase process. Because customers can easily click away from a web page, a landing page that directly addresses a customer's needs minimizes click-away. Often a landing page will make an offer to the customer—perhaps the opportunity to download a white paper—and ask for his or her contact information in return.[13]

Some customers "land" on the home page

Some organizations and brands are so well known that customers simply jump to a home page. When a company serves multiple target markets, the home page should quickly help each target market get to the information it needs. Consider the New York Public Library, which serves, among others, academic researchers from around the world who seek access to the library's online resources, Spanish-speaking residents of the Bronx who are interested in local classes (in Spanish) on how to use a computer, tourists in New York City who want to visit the architecturally stunning main library building on Fifth Avenue, and individuals and foundations that donate to the library. The library's home page has links to "Research," "Locations," "Classes and Events," and "Support the Library" that take each target market to the information it needs with one click.

Product web pages drive desire and action

By the time a customer comes to a web page developed for a specific product, the customer is probably already aware and interested. The product page should be designed to give customers the kind of information needed to promote desire and

 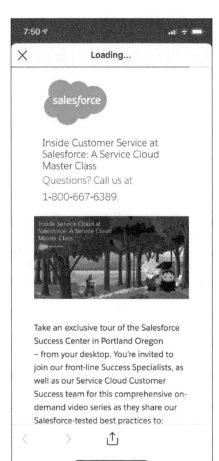 

This series of images provides an example of how a landing page works. The image on the left was a sponsored (paid media) advertisement for CRM software company Salesforce. The image and copy attract the attention of the target market (someone interested in customer service). Clicking on "Learn More" reveals the middle screen shown above. This image stokes interest and desire with details about a customer service class and a video. Scrolling down on this page (image on right) wants to spur a target customer to action—by entering contact information in exchange for the opportunity to watch a video.
(left): Source: Salesforce.com, Inc./Instagram; (middle and right): Source: Salesforce.com, Inc.

action—possibly detailed product information, pictures, videos, and/or links to where a purchase can be made. In other situations, a recent buyer of the product may need a software update or owner's manual.

Blogs informally communicate with customers

A **blog** is a regularly updated website, usually managed by one person or a small group and written in an informal, conversational style. Many companies use blogs as a way to regularly communicate with customers. A blog allows a company or individual to get their ideas out to interested target customers. Some firms use a blog to position a firm or individual as a thought leader on a particular topic or to just demonstrate their helpfulness and concern. Customers like to do business with experts and caring people and companies.

Larry McGlynn, president and CEO of McGlynn, Clinton & Hall Insurance Agencies, uses his *Massachusetts Family Insurance* blog to connect with current and potential customers. Larry is a funny guy who cares about his customers. Some of his blog posts, like the one with the YouTube video from a demolition derby titled, "When does an insurance agent enjoy a car crash?," highlight his humor. Other posts offer serious information—like when he explained law on distracted driving in "no txt'g while drv'g!" The posts get McGlynn new business indirectly—he uses the posts as a way to stay in touch with clients by e-mailing them a link to each new post. Many pass along the e-mail to friends and family—some of whom become customers.[14]

NeoMam created this infographic to explain how and why infographics work.
Source: Neomam Studios Limited

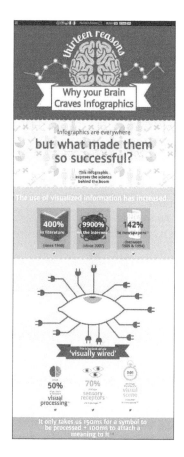

Infographics get attention and pass-along

In a world of text messages and short attention spans, many firms turn to images to capture customers' attention and interest. One type of image is an **infographic**—a visual image such as a chart or diagram used to represent information or data. The visual images, often combined with minimal text, grab attention, simplify complex information, and highlight trends or patterns. A well-designed infographic quickly conveys useful information and gets passed along more readily than a text-heavy website.[15]

Branded apps solve customer problems

Some brands are finding ways to deliver branded services directly to where customers can use them most easily—through apps on their mobile devices. **Branded apps** are sponsored software applications that benefit customers by providing entertainment, solving a problem, and/or saving time. A branded app can be a positive reminder for the brand, especially when its function ties closely to the core product.

Charmin toilet paper takes a decidedly humorous approach to helping on-the-go customers with their personal needs. The SitOrSquat app for smartphones uses GPS to show customers nearby public restrooms that, based on customer reviews, are clean (you can sit) or dirty (you'd better squat).[16]

Sephora's branded app offers many features. One of the most popular helps women choose makeup by virtually trying on shades and styles before buying.
Source: Sephora USA, Inc.

Online communities connect customers

A **brand community** is a group of customers joined around a particular brand or common set of shared interests. A brand community should be designed to serve the interests of its members—and the brand often enjoys increased customer loyalty, larger purchases, or simply being viewed as a trusted partner. One long-running brand community is the Harley Owners Group (HOG). HOGs operate locally to organize motorcycle owners around the lifestyle and attitude represented by Harley-Davidson. The group's social interactions and activities enhance the value customers get from owning a Harley and ultimately increase customer loyalty. Companies as diverse as LEGO (toys), Oracle (database software), and H&R Block (tax preparation services) have also developed successful brand communities.[17]

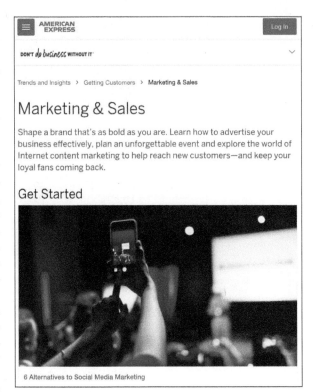

American Express helps its target market of small-business owners thrive with its Open Forum website. The site is filled with helpful content on everything from leadership and compensation to advertising and social media. An open community allows contributions from guest bloggers who offer advice, tips, and tricks these customers appreciate. While visiting the site, small-business owners can easily link to all of American Express' business services. The site helps American Express build a closer relationship with its small-business target market.
Source: American Express Company

Build customer relationships with e-mail

When people think of companies sending out e-mail, many think of *spam*—unwanted, irrelevant messages sent out to a large number of recipients. These days effective spam filters limit unwanted e-mails, and marketers have found that customers like to receive only pertinent and timely messages from companies they want to do business with. However, effective e-mail newsletters help companies build ongoing customer relationships that lead to increased purchases. E-mail newsletters are an important part of many retailers' promotion blends.

Newsletters can tell customers about new products, existing products, or special sale prices. Some companies develop different e-mail newsletters tailored to specific customer interests—and desired frequency of communication. For example, online retailer Lands' End allows customers to share their interests (Men's, Women's, Kids', School Uniforms, or Home) and determine mailing frequency (daily, weekly, monthly, or stop e-mailing me). Ethical companies ask customers for permission before sending e-mails and make it easy to unsubscribe from newsletters.[18]

Earned Media from Public Relations and the Press

LO 16.5

We mentioned that media coverage can be earned from customers—or from the press. Let's look first at how it works with the press and then from the vantage point of customer advocacy.[19]

| Get coverage in the press | Some companies *earn* editorial coverage—with their goods and services or good works featured in articles, stories, and reviews in magazines, newspapers, or on television. If a firm has a really new message, a published article may get more attention than advertising can. Trade magazines carry articles featuring the newsworthy products of interest to people in a particular job or industry.

This kind of coverage doesn't usually happen by luck or accident. A good public relations effort often lays the groundwork for companies that get positive attention in the press. Sometimes a firm's public relations people write the basic copy of an article and then try to convince magazine or newspaper editors to print it. |

| Get found when the press searches for a story | Southwest Airlines uses press releases to generate publicity in the popular press. When it wants to promote its special fares and new routes, its public relations (PR) staff uses a targeted approach to get attention from news reporters. Because many reporters research story ideas on specialized search engines, such as PRWeb and Google News, the PR staff at Southwest writes press releases so they appear at the top of the reporters' search lists. Southwest's PR staff practices SEO and researches the keywords reporters use most frequently on these search engines, later using those words in press releases. For example, Southwest's press releases use the phrase "cheap airfare" because it is in four times as many search requests as "cheap airline tickets." Southwest also puts a hot link to its special low-fare web page at the very start of each press release. The link allows Southwest PR staff to track which press releases work best—then it uses that information to fine-tune other messages. These extra efforts have paid off. In one case Southwest generated $1.5 million in online ticket sales with just four press releases.[20] |

| Online pressrooms help journalists | A public relations person or staff can make it easier for the press. A PR group can develop a press kit (sometimes called a *media kit*), including promotional materials specifically designed for the media. Many businesses and nonprofits have a "Press" section on their website that posts press releases, photos, and sometimes videos geared to the press. For its traveling performances, Cirque du Soleil relies on local newspaper and television to produce stories that stoke interest and demand for tickets. Cirque du Soleil's online pressroom provides the press with hundreds of photos and videos of its performances. |

#M4BW

Do something good

Another way to encourage press coverage is to do something good—and let the press know. The press enjoys covering do-good stories. By making a positive impact on a community, a business can earn positive press coverage. Consider Surgery on Sunday, a nonprofit organization based in Lexington, Kentucky, that offers outpatient surgical procedures at no cost for uninsured or underinsured patients. The Lexington Surgery Center, part of the KentuckyOne Health network, opens its doors one Sunday a month to offer these services. Their generosity generates goodwill, receives positive coverage from local and national press, and has created a better world for thousands of patients and their families.[21]

| Bloggers are the new press | Independent bloggers often "publish" their own stories on topics of interest. Readers often view these citizen "journalists" as objective sources of information. Many bloggers become experts on a particular topic and have readers interested in what they have to say. Identifying influential bloggers who have the attention of target customers can be helpful.

Moms are a large target market for Kimberly-Clark, maker of disposable diapers and other baby products. It recognized a growing cadre of "mommy bloggers," stay-at-home moms who write blogs about child-rearing. With a loyal readership, these bloggers can |

influence many people. So Kimberly-Clark sent samples of its new Huggies Pure & Natural premium diapers to 500 of them, hoping many would like the product and write about their positive experiences.[22]

Ethics require disclosure of conflicts of interest

The power of earned media from the press comes from people assuming that such coverage is objective. Unfortunately, there is evidence that some journalists are paid to provide positive brand mentions. This clearly violates the ethics policies of legitimate news organizations and may violate the law. If such practices continue, they threaten the credibility of particular publications and journalism more generally.[23]

The same advice goes for bloggers. When a blogger (or someone posting on social media) is paid for an endorsement or testimonial for a product, this must be disclosed. Besides being an ethical, and perhaps legal issue, lack of disclosure can cause embarrassment for the blogger and the brand. When two bloggers road-tripped across country and wrote nice stories about Walmart, it gave the impression they were completely objective. It was later discovered that the bloggers were supported by the organization Working Families for Walmart—and paid indirectly by Walmart. Walmart suffered through some negative publicity.[24]

New moms looking for advice might go to the *Rookie Moms* blog. The site offers tips on pregnancy and parenting.
Source: Rookie Moms

Earned Media from Customer Advocacy

LO 16.6

Customers trust and act on recommendations from friends, family, and colleagues. Even recommendations from strangers often have a stronger influence on buying decisions than advertising does. Research has also found that customers acquired from word-of-mouth are more loyal and have greater lifetime value than customers acquired from other means.[25]

That said, although information from other customers is influential, it is much more difficult to control than the messages a marketing manager crafts for an advertisement, sales promotion, or owned media. When customers control the message, the words might be exactly what the marketing manager wants, or they could be incorrect, inappropriate, or harmful to the company. Let's take a closer look at how customer-to-customer communication works and some strategies for fostering the more positive forms of this promotion.

Opinion leaders spread the word—good and bad

Opinion leaders like to share their views—and they also get attention from other customers who respect their views. Marketers value personal recommendations from opinion leaders. For example, some movie fans like to be the first to see new flicks. If they like a movie, they quickly tell their friends and word-of-mouth publicity does the real selling job. When online grocer FreshDirect opened in New York City, positive word-of-mouth was key to its fast growth.

454

However, consumers are even more likely to talk about negative experiences than positive ones. So, if early groups reject the product, it may never get off the ground. In one study, 64 percent of consumers said they would not shop at a store after being told about someone else's negative experience there.[26]

Paying for opinion leaders

To help motivate consumers to spread the word, a company called BzzAgent helps marketing managers start conversations. BzzAgent works with about 800,000 "agents" in the United States, Canada, and the United Kingdom. Agents who sign up to help with a particular campaign receive product samples and information. If they like the product, they are urged to pass the word. Although such behavior could be deceptive, BzzAgent encourages its agents to be ethical and disclose their status as "BzzAgents." Kraft Foods, General Mills, and Dockers are among the companies that have run campaigns through BzzAgent.[27]

> **Ethical Dilemma**
>
> *What would you do?* Your friend Sheila tells you that she has become a "Buzzer." She works for Buzz Generator, a company that helps marketing managers spread the word about new products. Buzz Generator gives Sheila samples of new products and pays her between $25 and $50 per product to use the product and share her "enthusiasm" with her friends. Buzz Generator provides her with ideas about how she can bring up the products in conversations and suggests what she might say about the products. Sheila writes a report highlighting her experience and any feedback she receives. The company asks Sheila to disclose her role as a "Buzzer" to friends when she brings up the new products, but Sheila tells you that this tends to make people not listen to her message. Plus she thinks friends find her more credible and more of an innovator if she leaves out the part about getting the product for free and getting paid. You recall a recent time when she raved about a new frozen pizza she had tried—and now wonder if that rave was real. She asks if you want to become a "Buzzer."
>
> *Would you sign up to do this work for Buzz Generator? Why or why not? Do you think your friend is acting ethically? Why or why not?*

Most customers are reluctant to recommend

Most customers are reluctant to recommend products—even when they do have a great experience. There are a number of reasons why customers hold back. Some worry a recommendation may result in a dissatisfying experience for a friend. This fear of negative consequences often outweighs the benefits of a friend having a good experience. People also worry about the personal costs of recommendations. If a business they recommend gets busier, service quality may decline or prices might go up. Some products can be difficult to explain to others—and people don't want to be put on the spot when trying to explain them. People also hold back if they feel the recommendation doesn't make them look good. Or they worry that others might question their ethics, perhaps believing they receive some type of kickback for making the recommendation.[28]

Encourage word-of-mouth

For customers to share their great experiences, they must see doing so as safe, fun, and worthwhile. First, a company must offer a great product. Consistently high performance gives a customer confidence that it happens all the time—and increases confidence that anyone they recommend it to will have a similar experience. Second, companies must give customers a specific story to share—anything from delighting (far exceeding customer expectations) to creating an unusual experience. Sometimes unusual is simple—imagine if the cable TV repairperson called to say he would be 20 minutes *early* and asked if that was inconvenient. You might be surprised enough to tell some friends. Third, companies shouldn't be afraid to ask for recommendations. It doesn't hurt to come out and ask satisfied customers to leave reviews or tell their

Marketers often seek ways to encourage customers to share positive experiences. Besides having a good product, a business might find other ways to promote sharing. These days, Millennials often share dining experiences with friends and family on social media—usually with a photo. Restaurants are responding by providing food (or other features) worthy of (and fun for) sharing on social media. New York City bistro Jack's Wife Freda has embraced this idea with colorful food beautifully presented and clever sugar packets—both of which contribute to plenty of earned media on Instagram.
Source: *Jack's Wife Freda/Instagram*

friends. A link to a review site following a purchase makes this easier. Fourth, some organizations offer easily shareable content on their social media sites—making it easy for customers to click and share information. Social media sharing "buttons" can be placed next to online content to facilitate pass-along.

Referral programs give customers incentives to recommend

Another way to motivate positive word-of-mouth is through a referral program. A **referral program** offers a current customer an incentive for recommending a new customer to a business. Usually referral programs require the new customer to make a purchase. To give both the person giving and the person receiving the referral an incentive, both are usually offered some sort of discount. So, for example, after a customer gives a positive review at Airbnb (a website where customers can rent lodging), they are asked if they would like to refer a friend. The friend is sent a $25 discount—and after the friend books a place to stay through Airbnb, a similar credit is given to the person who made the referral.[29]

These types of programs are often used by online companies because of the simplicity and low cost of giving and receiving referrals over the Internet. It is also easy for a company to experiment with different levels of compensation and find the offers that work best.

Ratings and reviews let customers do the selling

Reviews allow customers to sell to one another. Reviews can have a strong influence on purchase behavior because readers view reviews as objective reports that share first-hand knowledge. As a result, many companies include rating and review capabilities on their owned media. Most major retailers find reviews increase sales.

Manage negative reviews when they come along

Whereas online compliments let the firm know what it is doing well, complaints can highlight unmet customer needs—and give managers a chance to turn things around. Marketing managers should not ignore negative reviews—they should respond and fix the problems.

That is what the owner of a California spa did. She was horrified to learn her spa had only a two-and-a-half-star rating (out of five) on the popular review site Yelp.com. She set out to fix things. First, she e-mailed the unhappy reviewers in an effort to make things right. Then she encouraged her satisfied customers to post reviews. She also fixed areas of concern. Soon the spa had an acceptable four-star rating on Yelp, and rather than scaring off prospects, the review site was spurring them to action.[30]

The ethics and danger of falsifying earned media

Companies need to be careful they are not too zealous in seeking positive word-of-mouth. Not only is it unethical, firms are often easily exposed when they engage in unethical practices. For example, a Honda manager went on the Facebook page for the Honda Crosstour and raved about the new car's design—when most other comments were negative. Some readers recognized the manager's name from articles in the automobile press and cried foul. Honda ended up embarrassed when the deception was reported on *Autoblog* and other sites. Sometimes TripAdvisor suspects some of the reviews on its site are fake—like when a hotel's employees post reviews that build up their facility and tear down a competitor. To prevent fake reviews from harming the credibility of its own site, TripAdvisor flags suspicious reviews with a red disclaimer.[31]

Social media simplify and amplify pass-along

People are more likely to pass along reviews when it is easy to do. So social media sites such as Facebook and Twitter make it as easy as clicking on a "Like," "Share," or "Retweet" button. Social media also amplify customer word-of-mouth—as one person can have hundreds or even thousands of followers. Let's take a closer look at social media and their role in the promotion blend.

Social Media Differ from Traditional Media

LO 16.7

In Chapter 15 we learned about traditional media—including television, magazines, radio, and so on. Advertisers place their messages on these media to communicate to customers. The growth of the Internet has led to a rise in different types of social media. **Social media** refers to websites or software applications that allow users to create and share ideas, information, photos, and videos and interact in a social network. Examples of social media include Facebook, Twitter, Instagram, Pinterest, Snapchat, YouTube, and LinkedIn. About 7 in 10 Americans are on at least 1 social media platform. Among those who are 18 to 29 years old, the number is 9 in 10. Let's look closely at how social media differ from traditional media before we focus on how social media can work in the promotion blend.[32]

Social media—paid, owned, *and* earned

Paid, owned, and earned media may appear on social media. Chapter 15 describes advertising (paid media) on social media sites. On Facebook, for example, a seller can choose to run a "sponsored message" (paid media) on the site. That same seller might also create a Facebook page (owned media)

Advertising (paid media) for Sonos speakers appears in target customers' Facebook newsfeeds.

Source: Sonos, Inc.

that allows a brand's fans to "Like" the page and keep up with messages it sends out. When a customer clicks "Like" on a brand page or writes something on that brand's wall, the customer's friends see it (earned media). Most of the major social networking sites provide similar paid, earned, and owned media opportunities.

Social media—no media placement cost but not free

Most social media have no *direct* media cost, but that doesn't mean use of social media platforms is free. A company needs to hire staff or an agency to manage its presence on social media. Once a company starts a Facebook page or Twitter stream, it should regularly maintain the page—or customers may wonder if the company has gone out of business.

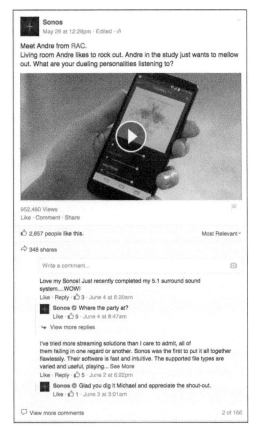

Sonos has its own Facebook brand page (owned media). Interested customers can "Like" the page to get news from Sonos in their Facebook feeds.
Source: Sonos, Inc.

Social media—many-to-many communication

Advertising is a one-to-many model of communication—a brand sends a message to many customers at once. Social media operate in a different way; social media offer a many-to-many model of communication. When a brand posts on Facebook, some customers engage and respond—and others simply follow the back-and-forth interaction. Many brands respond to tweets and Facebook posts in order to engage with customers. Those conversations appear in public—so other customers can read and (if they like) enter into the conversation.

Advertising plans, social media reacts

Advertising through paid media typically involves a relatively long planning cycle. Evaluating concepts, approving ideas, and producing a print, radio, television, or online advertisement can take months. Although marketing managers should have a social media plan, this form of communication offers the ability to react to current or emerging events/comments immediately. Many social media managers monitor news and events that may have a relationship to, or potential interest in, the brand's target market.

It is those in-the-moment reactions that can get some social media posts more attention. A few years ago, the Super Bowl had a blackout in the middle of the football game. Oreo's fast-thinking social media manager tweeted "Power Out? No Problem" and posted a photo of an Oreo cookie with the caption: "You can still dunk in the dark." The clever joke was retweeted more than 15,000 times and generated plenty of conversations around Oreo.

Customers interact with one another while commenting on Facebook posts by or about Sonos (earned media).
Source: Sonos, Inc.

#M4BW

Social media spread positive messages

Social media users like to spread positive messages to their social media followers. Brand managers for Mattel's Barbie leveraged this when it developed its #MoreRoleModels campaign. The promotion celebrated International Women's Day and the 60th anniversary of the Barbie doll. The effort highlights professions where women are underrepresented; for example, only 30 percent of scientists, 7 percent of pilots, and 11 percent of film directors are women. Barbie honored 17 historical and contemporary women role models—including conservationist Bindi Irwin, snowboarding champion Chloe Kim, and aviation pioneer Amelia Earhart. Besides creating a new doll for each woman, the campaign asked fans to tag women who inspired them with #MoreRoleModels. Facebook and Instagram provided a natural platform for sharing their role model choices. More than 8 million social media impressions made lots of people feel good about the Barbie brand and its role in making a better world for women everywhere.[33]

Major Social Media Platforms

LO 16.8

There are many different social media platforms—and there seem to be new ones appearing regularly. Seven social media sources attract considerable attention in the United States: YouTube, Facebook, Instagram, Pinterest, LinkedIn, Snapchat, and Twitter. Each of these sites offers something different to users and to a marketing manager's promotion blend. This section describes how these social networking leaders can be used in a firm's promotion blend. See the summary in Exhibit 16-6.

YouTube—video for all

YouTube is a video-sharing website where users upload, view, rate, share, create playlists, and comment on videos. YouTube is used by more online Americans (73 percent of U.S. adults) than any other social media. That said, among users, it is checked less frequently than most other social media—with 51 percent of users checking it at least daily. The site is used by almost all ages, with a falloff only occurring among those 65 and older.

Many videos feature a 30-second video ad (often skippable after 5 seconds) before playing. Brands can create their own YouTube channels. Anyone can create an account on YouTube and then subscribe to a channel. The most highly subscribed-to YouTube channels include those for entertainers, including musicians (Justin Bieber and Ed Sheeran rank high), record labels (India's T-Series), comedy (PewDiePie), and sports (WWE)—as well as education (Cocomelon—Nursery Rhymes). These channels each have more than 40 million subscribers. As can be seen here, YouTube offers a way for entertainers to deliver their Product and communicate with customers.

Some brands use YouTube to stay engaged with customers. For example, the LEGO YouTube channel has more than 7 million subscribers (see also the Chapter 4 opening case study). LEGO has five different brand channels and other related accounts. The videos it posts highlight new products, promote LEGO movies, announce build challenges, and more. Its subscribers love to watch, keeping LEGO's most ardent fans engaged. Even B2B brands can use YouTube to demonstrate products, tell great customer stories, and keep customers informed. Cisco Systems' channel has just over 200,000 subscribers; the videos it posts give customers an opportunity to learn more about the brand and how other customers use its products.

Facebook—the king of social media

Facebook is an online social networking website that allows registered users to create profiles, upload photos and video, and send messages to connected friends.[34] Facebook dominates social media; the site has more than 2.3 billion users globally; 69 percent of the U.S. adult population has used the platform. Facebook continues to grow its user base, though growth has slowed. Facebook is visited more regularly than any other social media site; 74 percent of Facebook users check the site at least once a day. Facebook is the most popular social media site with older Americans—with 68 percent of those age 50–64 and 46 percent of those 65 and older reporting they have used Facebook. It also appeals more to women (75 percent) than to men (63 percent).

Exhibit 16-6 Big Seven Social Media Platforms—User Demographics

	YouTube	Facebook	Instagram	Pinterest	LinkedIn	Snapchat	Twitter
Global active users (millions)[a]	1,900	2,271	1,000	250	303	287	326
% of U.S. adult population using	73	69	37	28	27	24	22
Frequency of use (% of U.S. users)							
Check in: daily/less often[b]	51/49	74/26	63/37	N/A	N/A	61/39	42/58
Demographic data (U.S. users)							
Income < $30/$30–$75/$75+ (000s)	68/75/83	69/72/74	35/39/42	18/27/41	10/26/49	27/26/22	20/20/31
Male/female usage (%)	78/68	63/75	31/43	15/42	29/24	24/24	24/21
White/Black/Hispanic	71/77/78	70/70/69	33/40/51	33/27/22	28/24/16	22/28/29	21/24/25
Age 18–24 (%)	90	76	75	38	17	73	44
Age 25–29 (%)	93	84	57	28	44	47	31
Age 30–49 (%)	87	79	47	35	37	25	26
Age 50–64 (%)	70	68	23	27	24	9	17
Age 65+ (%)	38	46	8	15	11	3	7
Owned media—how companies, nonprofits, or brands create owned media on platform	Company can host its own page	Company and/or brand pages	Build an Instagram page and add images	Create a Pinterest page and pinboards	Build a company page; post company news	Company can create page, post stories, sponsor filters and lenses	Create one or more Twitter profiles, create hashtags
Earned media—how customers demonstrate support	Users can give videos thumbs-up/down; easy to share with others	Customers "Like," "Share," and comment	"Follows," "Likes," and comments	Customers save ("Pin") favorites to their own pinboards; follow others	News gets "Likes" or comments from other LinkedIn users	Users may share images with branded filters or lenses	Customers can "Follow," "Tweet," "Retweet," and comment

[a] Statista, January 2019, number of active accounts.
[b] These data not reported for Pinterest and LinkedIn.

Sources: Unless otherwise noted, data from survey in the United States from January 8 to February 7, 2019, by Pew Research Center and reported in "Share of U.S. Adults Using Social Media, Including Facebook, Is Mostly Unchanged since 2018," Pew Research Center. Questions asked as percentage of U.S. adults who say they use.

Users typically scroll through their "newsfeed," viewing posts (videos, photos, status updates) from "Friends," "Liked" businesses or organizations, and sponsored posts (paid advertising). Because the average user has hundreds of friends, Facebook has an algorithm that determines the content placed in the user's newsfeed. The algorithm favors posts from friends and brands that the user actively engages with by clicking or commenting on posts.

On Facebook anyone can set up pages to promote a business or nonprofit. Business pages might feature photos, videos, links to a company's home page, reviews, and other information. After a business or nonprofit sets up and publishes a Facebook page, users can choose to click the "Like" button and become a "fan" of that page. Fans receive updates from the brand in their newsfeed. Some large brands, including McDonald's, MTV, Disney, and Red Bull, boast more than 40 million fans.

All of this user engagement gives Facebook a great deal of data on individual customer interests. Those big data allow firms to precisely target customers with advertising. That said, Facebook has received a great deal of negative publicity for the way it has used and taken care of customer data. This may lead to greater regulation of the company.

Facebook's size makes presence a must for many companies

Facebook's sheer size—and its use by many different customer segments—makes it necessary for many organizations to have at least some presence. Because customers have to choose to "Like" a brand page, the target for messages should be current customers. Sales contests and discounts are sometimes used to engage fans. Most brands try to post engaging content to encourage fans to click through or share the content with their friends, which (by coming from a friend) offers additional credibility. Nike maintains a corporate Facebook page, but it also keeps specialty pages dedicated to specific sports including Nike Golf, Nike Basketball, Nike Running, and more.

#M4BW St. Jude Children's Research Hospital uses its Facebook page to foster ongoing relationships with its patients, volunteers, and donors. Posts, photos, and videos on the page often share stories of patients who were helped by the hospital and donations from Facebook fans. There is also a link that makes donating easy.
Source: St. Jude Children's Research Hospital

Instagram—when a picture is worth a thousand words

Images and video have long been a prominent part of the copy thrust in print and television advertising. Social media users often skim through content; image and video sharing capture people's attention. Photographs and videos can tell short stories and connect people to a brand or seller—it can inspire them, solve everyday problems, or simply entertain.

While many sites allow for sharing photos and videos, one social media tool focuses on this form of communication. **Instagram** is a free online photo- and video-sharing service geared to mobile phones. Photos and videos can be taken with the Instagram app and shared on other social networking sites including Facebook, Twitter, and Tumblr. Instagram has more than 1 billion user accounts worldwide, and 37 percent of American adults have used the app. Sixty-three percent of users check the app at least once a day. Instagram skews more female, more Hispanic, and more to younger age groups (see Exhibit 16-6). Brands including *National Geographic* magazine, Nike, Victoria's Secret, and the soccer team Real Madrid all have more than 40 million followers on Instagram.[35]

Instagram works well for many small businesses, where an owner or part-time marketing manager may find it less time-consuming to take pictures as compared to writing stories or posts. Businesses in the restaurant, clothing, fashion, travel, and entertainment

Wayfair is an online retailer that sells home furnishings. Its Instagram has more than a million followers. Many of its posts feature home decorating ideas. Some of them, when tapped, show the names and prices of the products shown in the photo. Customers can read or make comments. Clicking on items gives more product information and the option to click to the product on Wayfair's website.
Source: Wayfair LLC

 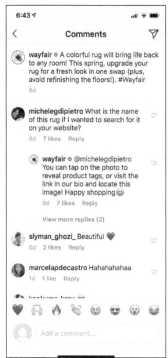 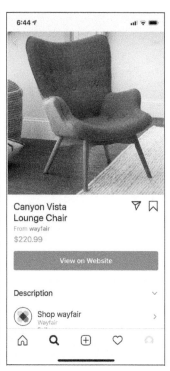

fields often post images on Instagram to maintain or stoke customer interest in their goods and services. Brands including LaCroix water and Glossier beauty products have built businesses largely through Instagram.

Pinterest—bulletin board wish lists

Pinterest is a website that allows registered users to share ideas and images they find online with fellow users. The ideas and images are organized into "pinboards"—a kind of online bulletin board. Pinterest has about 250 million users globally, and about 28 percent of online Americans use the site. Users of the site are mostly female (42 percent of American women as opposed to 15 percent of men), white (as compared to black and Hispanic), and generally have a higher income. Although the site is more frequently used by younger people, it maintains appeal to some in older age groups.[36]

Users create boards around a wide range of topics that work well with images, including art, fashion, travel, and home décor. For many users, it is kind of a "wish list"—so marketing managers want to make sure they post images on their own websites that can easily pin to Pinterest. Pinterest uses artificial intelligence to learn about individual customers and suggest other images to users. This helps keep pinners active on the site longer and lets retailers make recommendations Pinterest users find useful. Retailers can put "Pin It" buttons next to images so that customers can create a wish list that other followers see as well—promoting pass-along and earned media. Many of the biggest brands on Pinterest are retailers; among the largest are L.L.Bean, Nordstrom, and Lowe's, each with more than 4 million followers.[37]

LinkedIn—when business gets social

LinkedIn is a networking website for businesspeople who create personal or company profiles. Globally it boasts more than 300 million registered users and about 27 percent of online American adults. The site has more high-income and professional users. Very young (18-24) and older (over 65) individuals are least likely to have a LinkedIn page. Individuals may have hundreds or thousands of connections; companies such as Microsoft, Apple, and HP have more than 2 million followers. Media companies that target businesspeople, including *The Wall Street Journal* and *Forbes* magazine, have used LinkedIn to distribute articles.[38]

Businesspeople used to think of LinkedIn as a place to network for jobs. Now many companies and salespeople use LinkedIn in their marketing strategy. Companies and

Home improvement retailer Lowe's Pinterest page includes pinboards that might inspire an around-the-house project. The company hopes that inspiration leads customers to their local Lowe's store to pick up supplies.
Source: Lowe's/Pinterest

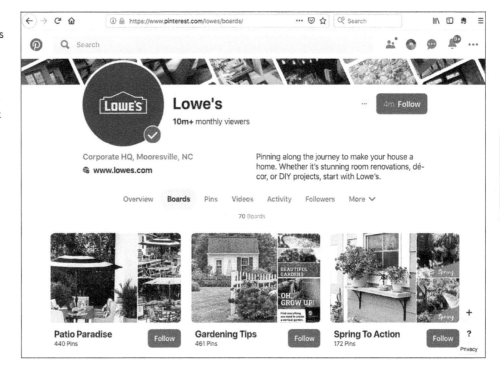

salespeople post news, white papers, and reports to develop a reputation as trusted industry experts. Salespeople can also identify and qualify leads on the platform. LinkedIn has also become a useful place for advertising that targets businesspeople, allowing for precise targeting by job type, industry, or professional interest.

Snapchat—focused on Gen Z

Snapchat is a mobile app and service for sharing photos, videos, and texts with other people. There are almost 300 million Snapchat accounts worldwide, and about 24 percent of online Americans use the service, with 61 percent checking it at least daily. The site clearly skews young with 73 percent of those age 18-24 using Snapchat. Use by those under age 18 is the same or higher. Among other age groups, usage is less than 50 percent; for those over 50, that shrinks to less than 10 percent.[39]

Snapchat is designed for mobile device users. Shared photos can be enhanced with visual effects, stickers, and filters that are sometimes brand-sponsored. While Snapchat started as a photo messaging service, commercial applications in its "Discover" section allow brands to post advertising-supported content. Snapchat continues to work on adding features to make its service more friendly to marketers; most opportunities are forms of paid media. Owned media opportunities are possible but make it more difficult to attract users, and earned media opportunities include developing creative branded lenses or filters that users can add to their photos.

Twitter microblogging service—short "tweets"

Twitter is a social networking microblogging service that allows registered users to send short (280 characters or less) messages called "tweets."[40] Twitter has over 300 million users globally and reaches about 22 percent of the U.S. adult population. However, only about 42 percent of users check the app daily. Twitter users skew younger; 44 percent of Americans age 18-24 and 31 percent of those age 25-29 are registered users, while only 7 percent of people over age 65 have Twitter accounts.

Twitter operates like a public message board, with most tweets available for anyone (even people not signed up for Twitter) able to see them. Most Twitter users follow friends, celebrities, news sites, and brands. CNN, *The New York Times,* and BBC Breaking News, for example, have the largest number of followers. Among active users, tweets tend to be more frequent and less personal as compared to Facebook.

Because of Twitter's character limit, tweets are concise—for example, this tweet from Starbucks: "Buy any fall drink (hot or iced), get one free. 2-6 pm, 9/17-9/21, participating US stores." However, at times brand tweets include a link to more details on a landing page. The payroll management company ADP tweets links to content (produced by ADP or not) relevant to the payroll industry; many of its salespeople retweet (a way of forwarding a tweet) these links to their follower-customers. By frequently passing along non-ADP content, ADP and its sales force earned a reputation for being trusted experts in their field. A simple "hashtag" (#) allows fans to pass along support for well-known brands and helps marketing managers assess engagement and interact with potential leads. Nike fans can post using a Nike hashtag (#Nike) or the Nike tagline (#justdoit).

MoonPie is a sweet snack and not a large company. It has leveraged its humorous social media presence to bring more awareness to the brand.
Source: MoonPie/Twitter

Other social networks

We have highlighted some of the most popular social networks, but depending on the target market and promotion objectives, other social networks should be considered. Tumblr is a microblogging site that makes it easy for users to share all types of media from their browser, phone, desktop, or e-mail. Video- and photo-sharing services are among the fastest-growing social networks and include TikTok, Vimeo, and Flickr. Other large social networks like WhatsApp and Facebook Messenger focus on voice or text communications.

Specialized social networks target very specific audiences. For example, Allrecipes can get a food brand, nutritionist, caterer, or personal chef in front of relevant target markets. Home remodelers, architects, and home furnishings and accessory brands show their wares on Houzz—a site with more than 6 million photos of different home remodeling projects.

Different networks around the world

While all of the social networks discussed here have a significant global presence, marketing managers doing business internationally should be aware of various local social media. Some of these can be quite large. For example, China's WeChat—with more than 1 billion users—includes messaging, social media, and mobile payment. Chinese tech giant Tencent has two social media products: QQ is an instant messaging service with more than 800 million active users, and Qzone has 500 million users who use the site to write blogs, maintain diaries, share photos, watch videos, and listen to music. Although available in several languages, the Facebook-like site VK is especially popular among Russian speakers and has more than 500 million registered users. Each provides a promotion opportunity if target customers use the site.

Deliver value to social media fans and followers

Ultimately, social media need fans and followers—or no one sees the content. Target customers follow companies and brands that deliver some sort of value, and followers will show support for valued posts by "liking," sharing, retweeting, or commenting. There are many ways to deliver value; a brand can offer discounted prices, make

Exhibit 16-7 Different Types of Social Media Content and Examples

Type of Content	Examples
Humor	• Snapchat: Taco Bell created a funny "lens" so users could turn their head into a taco. • YouTube: Sam Adams introduced a new beer using helium as an ingredient—April Fool's! • Twitter: KFC (long advertised a secret recipe with "11 herbs and spices") had a Twitter account that followed only 11 people—5 Spice Girls and 6 guys named Herb. When it was "outed" by a keen-eyed user, her tweet received 700,000 "Likes."
Deals and discounts	• Facebook: Online beauty retailer Sephora offers Facebook-fan-only deals. • Twitter: In the UK, Domino's pizza reduced the price of pizza based on how many people tweeted in time for lunch.
Contests	• Instagram: Habitat for Humanity ran a photo contest to highlight on-site volunteers. • Facebook, Twitter, Pinterest, Instagram, and YouTube: Lay's "Do Us a Flavor" contest asked customers to submit and vote on new flavors for its chips.
Insider knowledge	• Instagram: Fashion designer Oscar de la Renta offers followers a sneak peek of its upcoming fashion lines. • LinkedIn: Microsoft's LinkedIn home page includes behind-the-scenes peeks at the company, blogs, and Q&A.
Educational	• Snapchat: The World Wildlife Fund highlighted endangered species with its #LastSelfie campaign. • YouTube: GE released a video series that highlights the work of its employees. The episode "Fabiana Garcia Brings GE Ultrasound to the Amazon" has more than 700,000 views (#M4BW).
Useful ideas, information, and practices	• LinkedIn: The Four Seasons Hotels company page features travel videos, articles, and tips. • Pinterest: Whole Foods has pinboards on topics ranging from baking to gardening.
Product launches and updates	• Twitter: Retailer H&M shows many of its latest fashions on Twitter. • Snapchat: Madonna launched a new song on Snapchat to target its younger user base.
Customer service	• Twitter: Delta Air Lines provides customer service on Twitter, assisting customers whose flights have been delayed or canceled. • Twitter: Xbox video game's customer support team has more than 600,000 followers and responds quickly to customer questions.
Image and lifestyle	• Pinterest: Retailer Nordstrom posts products and lifestyle inspiration images daily. • Instagram: GoPro posts a customer "Photo of the Day" that highlights the lifestyle of GoPro users.

followers laugh, or provide fast customer service. The TEDx Youth Conference in Sydney, Australia, gave free coffee to anyone who posted a photo from the event or a comment on Instagram, Twitter, or Facebook—that got a lot of "Likes." Some creative examples of value delivery through social media are shown in Exhibit 16-7.

Social media objectives—build relationships, not just fans

Some companies make the mistake of focusing on short-term objectives like generating millions of Facebook fans or thousands of followers on Instagram. Although these objectives are easy to measure, they don't necessarily have any impact on the company's bottom line.

Promotion efforts utilizing social media are best focused on increasing target customers' purchase likelihood. And because customers generally follow brands they already know and like, social media work best when the target market is already a customer. In most situations, social media as earned and owned media are *not* aimed at making target customers aware of the brand or encouraging trial. Instead, promotion objectives for social media usually focus on (a) enhancing brand familiarity by growing the number of customers that prefer or insist on a brand or (b) growing customer equity by encouraging customers to increase their purchases.

Software Can Manage, Measure, and Automate Online Media

LO 16.9

Implementing paid, owned, earned, and social media requires that managers follow many of the principles discussed elsewhere in this textbook. Segmentation remains important—with the lower costs of owned media and social media allowing firms to efficiently target smaller segments. Promotion objectives must be kept in mind. For most organizations, owned, earned, and social media are parts of a larger promotion blend—the marketing manager must carefully pull these pieces together to be efficient and effective. The AIDA model (see Exhibit 13-2) and promotion objectives should guide the copy thrust as well as the amount and type of information provided. For firms that have earned, owned, and social media online, new software helps manage, measure, and automate these media.

Software can manage social media

Because different types of earned, owned, and social media occur online, software can be used to monitor and manage the process. One popular tool is Hootsuite, a dashboard that offers marketing managers the ability to submit messages to Facebook, Twitter, LinkedIn, and other social media on a scheduled basis. The posts can be scheduled weeks or months in advance. Hootsuite and other software can also collect and analyze data from various social media.[41]

Measure results from online media

For many firms, earned, owned, and social media happen online—which can make monitoring and measuring the results faster and easier. So, for example, when Dollar Shave Club posts on Facebook, it sees how many users "Like" the post and what comments they make. When the company posts a video on Twitter, it counts how many followers retweet (forward) the message, how many watch the video, and how many link back to the site and make a purchase.

Software tools allow marketing managers to track different online marketing measures. For example, managers can track the number of visitors to a site, how long those customers stay, and how they found the site—a link from another page? An online search? If so, what keywords were used in the search? Marketing managers can also track what customers purchase and how much they spend. These are just a few of the online measures available for online media. Tracking these data allows marketing managers to see what works—and what doesn't—and rapidly adjust their marketing strategy. Many firms use the free Google Analytics software designed for this analysis; others choose alternative tools.

Bounce rates measure websites' ability to get attention and hold interest

In some marketing strategies, a website is an important communication media. In this case, a business wants to attract customers to its website *and* hold their interest. Attracting customers who simply click away without using the site can be costly if the customer arrived via a pay-per-click advertisement. One measure of a website's success is *bounce rate,* which indicates the percentage of visitors who only visit a single page in a website. A high bounce rate indicates customers are not interested enough to click on other page links. Marketing managers can change how they attract customers to the site or characteristics of the site to lower the bounce rate. To see how this might operate in practice, see Marketing Analytics in Action: Bounce Rate.

Software carries customers through the purchase process

Software can also be programmed to help firms respond to particular customer actions. **Marketing automation software** tracks individual customers' behavior and triggers actions in response to specific customer actions. For organizations with customers who spend a lot of time online, this software provides a low-cost way to learn more about customers and deliver the right messages at the right time. For example, a customer who visits a site three times might automatically be delivered a "pop-up" question "Can I answer any questions for you?" that links to a waiting salesperson.

Another example deals with a troubling issue for online retailers. Studies estimate that about two-thirds of all online shopping carts are abandoned before checkout. This means a customer has come to an online store, placed something in the cart, and then

Marketing Analytics in Action: Bounce Rate

Rosa Winters is the product manager for ProjectPro project management software. Her primary source of leads comes from Facebook advertising and Google search ads. Both media charge on a pay-per-click basis (see Chapter 15). When customers click on a ProjectPro advertisement on one of these sites, they are brought to the ProjectPro home page. Although it is great to get customers to the site, Rosa needs them to click on links at the site to learn more about the software and ultimately purchase it.

ProjectPro's Facebook advertising targets people who "Liked" the Project Management Institute home page, and the advertising then appears in their newsfeeds. The Project Management Institute is an association of professionals dedicated to project management. On Google, ProjectPro's ad is presented above the organic search results when customers search for either (a) "project management software" or (b) "project tracking." The cost-per-click for each of the three ads is different and so is the bounce rate (the percentage of people clicking on the ad who only visited the ProjectPro home page).

The results of the previous month's advertising are shown in the table below. Review the data and answer the questions that follow.[42]

Referral Source	Lure	Cost-per-Click	Bounce Rate
Facebook ad	Project Management Institute fans	$ 1.25	80.0%
Google search ad 1	Keyword: "project management software"	$12.00	21.5%
Google search ad 2	Keyword: "project tracking"	$ 8.00	42.5%

1. *Why do you think the bounce rates differ across the three advertisements? What might be different about the customers' motivation for visiting the site?*
2. *What could be done to lower the bounce rates? Think about how customers were attracted and what is included on the site as well as ideas presented earlier in this chapter.*

left without completing the purchase. Some online retailers use marketing automation to send customers that abandon a cart an e-mail that reminds them of the purchase. The e-mail may even include an offer for free shipping or a discount.[43]

Marketing automation software can be more sophisticated for B2B sellers. Earlier in this chapter, we noted that many companies post white papers and case studies designed to demonstrate solutions to business problems. So imagine a marketing manager seeking

Marketing managers responsible for a firm's social media might use a tool like Falcon. The software helps create content, manages social media, and provides data analytics tools.
Source: Falcon.io ApS

to find a better way to "manage social media." A Google search on the topic might lead to a landing page for Hootsuite (the software mentioned previously) that offers a seven-page white paper titled "8 Tips for Social Business." The paper educates the customer on social media without overly promoting the Hootsuite solution. This paper isn't exactly free—before downloading, an interested customer must surrender some information valuable to Hootsuite: name, e-mail address, company, job title, and phone number. This information is used to start a file on the customer. The white paper also includes a link to other online readings—some of which were not created by Hootsuite—but all of this activity can be tracked. The customer's behavior is used to determine whether a salesperson should phone the prospect—and if the salesperson calls, she can review the file to know what material the customer has found most useful. The HubSpot software described in the chapter-opening case is an example of this type of tool.[44]

CONCLUSION

This is the last chapter on Promotion. The options for promotion have grown in recent years. For decades, most promotion efforts revolved largely around salespeople, advertising, sales promotion, and public relations. This is not to say that earned media (word-of-mouth and professional reviews) and owned media were not important elements of a promotion blend. But the growth of the Internet and social media have made many more promotion tools available to today's marketing managers.

The Internet triggered a great deal of change in promotion and changed customer behavior, with many customers looking online for information to help them make better purchase choices. People's "screen time" moved from television to the computer and now the phone, where they often access the Internet and social media. The Internet also simplifies consumer search and pass-along behaviors that facilitate the use of earned, owned, and social media.

Owned media refers to messages generated by a brand, company, or nonprofit organization communicated through a message channel the brand directly controls. We consider the different types of owned media that can be created on the web and describe how they can be used to create value for customers.

Earned media refers to promotional messages not directly generated by the company or brand, but rather by third parties such as journalists or customers. This includes getting coverage in the press and earning praise from other customers. We discuss the different forms this type of media can take and how marketing managers can encourage more positive earned media.

With *social media* grabbing so much customer attention, the chapter examines seven of the largest platforms—YouTube, Facebook, Instagram, Pinterest, LinkedIn, Snapchat, and Twitter. The chapter concludes with a discussion of software used to manage, monitor, and even automate some of these processes—further lowering their costs and increasing their effectiveness.

KEY TERMS

LO 16.10

paid media, 442
owned media, 442
earned media, 442
user-generated content, 443
search engine optimization (SEO), 446
pass-along, 446
branded services, 447
white paper, 449

case studies, 449
landing page, 449
blog, 450
infographic, 451
branded apps, 451
brand community, 452
referral program, 456
social media, 457

YouTube, 459
Facebook, 459
Instagram, 461
Pinterest, 462
LinkedIn, 462
Snapchat, 463
Twitter, 463
marketing automation software, 466

QUESTIONS AND PROBLEMS

1. Review the HubSpot case study that opens the chapter. From this case, identify examples of different key terms and concepts covered in the chapter. For example, the case mentions customers "Liking" a HubSpot Facebook post—this is an example of pass-along and earned media.

2. Review the HubSpot case study that opens the chapter. Suggest how the company might utilize some other form of earned, owned, or social media in its promotion blend.

3. Describe the general (earned, owned, and paid) and specific (see examples in this chapter) type of media that

might be most suitable for promoting (a) a neighborhood sandwich shop, (b) toothpaste, (c) a hairstylist, and (d) accounting software for small businesses. Specify any assumptions necessary to obtain a definite answer.

4. Which of the following products are good candidates for user-generated content: (a) men's underwear, (b) women's fashion jeans, (c) a brand of salsa, (d) lumber used in home construction.

5. For each of the following products, indicate how search, pass-along, and experience would (or would not) operate for customers seeking information: (a) a brand of shampoo, (b) a pair of jeans, (c) a real estate agent, (d) a company that sells chemicals to a manufacturer, (e) a local shoe store. [Sample Answer for (a): For shampoo, a customer might search online for a shampoo to treat dandruff or oily hair, pass-along might occur if someone asks a friend for a recommendation of a shampoo—but it is unlikely that someone offers an unsolicited recommendation for a shampoo, and branded experience might occur if the shampoo was offered at a hotel or in the showers at a health club.]

6. Briefly discuss some of the pros and cons of earned and owned media for a producer of golf clubs and for a dance studio.

7. Find a web page that you think does a particularly good job of communicating to the target audience. Would it communicate well to an audience in another country (assume it is translated into the foreign language)? Explain your thinking.

8. Find a Facebook page for a brand or company. Evaluate that page. What do you think are the promotion objectives? Does the Facebook page foster engagement (are many followers posting messages directed at the brand)?

9. Name two companies that you think would have success building a Pinterest page. Why do you think they should choose to build a page? What type of content should each company place on its pinboards?

10. What kinds of publicity would work best at a company that markets tractors, combines, and other farm equipment to farmers? Why would the publicity you suggest be effective?

11. Search online and find examples of (a) a case study, (b) an infographic, (c) a blog, and (d) a branded community. Describe the promotion objective you believe each is trying to achieve.

12. Go to the Facebook page for Olive Garden (Italian restaurant chain)—choose the main company page, not an individual location. Scan through the page and open different tabs. Critique the pages. What do you think works well? What could be improved?

MARKETING PLANNING FOR HILLSIDE VETERINARY CLINIC

Appendix D (the Appendices follow Chapter 19) includes a sample marketing plan for Hillside Veterinary Clinic. Look through the "Marketing Strategy" section of the plan.

1. Identify key elements of earned and owned media that Hillside plans to use.
2. Suggest two additional ways that Hillside Veterinary Clinic could promote earned media and two ways it could add owned media.
3. How could Hillside use two additional aspects of social media?

SUGGESTED CASES

3. NOCO United Soccer Academy
14. Schrock & Oh Design
18. West Tarrytown Volunteer Fire Department (WTVFD)
19. MyOwnWedding.com
31. Silverglade Homes
34. Working Girl Workout

MARKETING ANALYTICS: DATA TO KNOWLEDGE

CHAPTER 16: SALES ANALYTICS

Janet runs an online business selling homemade crafts. Looking at the online analytics for her store, she recognized a problem with shopping cart abandonment. Customers would come to her online store, place items in a cart, and then never complete the purchase. Janet began sending reminder e-mails to customers with abandoned carts. Many came back to the store and completed their purchases, but there was still room for significant improvement. Her analysis shows an abandon rate of 65 percent. That is, 65 percent of the people who add items to their online cart didn't actually complete the transaction and place an order. Janet wonders if it would help to offer some sort of discount to these customers to provide an extra incentive to complete the transaction. To evaluate the idea she decides to create a spreadsheet to analyze her options.

Design element: #M4BW box globe icon: ©Vectoryzen/Shutterstock

CHAPTER SEVENTEEN

Sheila Fitzgerald/Shutterstock

Pricing Objectives and Policies

Twenty-somethings Adam Lowry and Eric Ryan shared a San Francisco apartment and an interest in starting a business. Back in 2000, after studying chemistry, Lowry was a climate scientist with a passion for sustainability. Ryan had studied marketing and worked in the advertising business; his interest was design. While hanging out drinking beer and on road trips to the mountains, the two talked about products they thought were cool and uncool—and what markets might be ripe for disruption.

They spotted opportunity in the $17 billion household cleaning product-market. They saw a mature market with little differentiation among brands. Dealer brands were viewed as lower quality at lower prices. National brands gave retailers price deals and advertising allowances in exchange for shelf space. Competitive advertising claimed minor differences in cleaning efficacy, offering coupons and sale prices to woo shoppers whose brand loyalty was more habit than commitment.

The external market environment suggested a path building on the entrepreneurs' strengths. Scientific studies showed that many household cleaning chemicals had unhealthy side effects associated with long-term use. The green movement was picking up momentum. Mass-merchandiser Target was successfully selling everyday items with a designer flair. Lowry and Ryan thought a line of stylish, sustainable products made with safe ingredients that "cleaned like heck and smelled like heaven" could find a niche in this market.

Soon Method Products was born. Lowry and Ryan tested batches of cleaning products they mixed in their bathtub and stored in beer pitchers labeled "Do Not Drink." They filled clear spray bottles with samples of their first four products—cleaners for the kitchen, shower, bath, and glass. To pitch their products to busy independent grocery store managers, they ambushed them in early morning hours. That's how their first customer, the owner of Mollie Stone's Market, discovered Method.

As sales began to take off, Ryan revisited packaging design. Without money to advertise, he thought something unique would garner attention and interest at the point of purchase. Method approached well-known industrial designer Karim Rashid with a pitch to "reinvent soap." Intrigued, Rashid came on board. Rashid's designs got Method noticed. A lot of Method customers discover the brand in a store—they love the cool package and figure at least the soap looks good sitting on the counter. Then customers discover how well the cleaners work and keep buying.

Method's green promise also resonates with customers. Method's "People Against Dirty" campaign tells the brand's story: most cleaners leave behind toxic chemical residues, leaving homes "dirty" after being cleaned. And when these chemicals go down the drain, they contaminate the natural environment. "People Against Dirty" got the message across to Method's target market. Design and green proved to be powerful differentiators and allowed Method to charge a premium price about 20 percent higher than leading brands.

Their first big sale was to Target. Method's stylish packaging was a perfect fit with Target's target market. Following successful test markets in San Francisco and Chicago, Target took Method nationwide. However, Method wasn't a good fit with all retailers; Walmart dropped the line after its customers found Method was too expensive.

Despite high prices and impressive retail contracts, Method struggled to be profitable. By 2006, the brand held only 0.5 percent in market share and less than $100 million in sales. The problem: Method's production and operating costs were higher than giant, megabrand competitors such as Procter & Gamble; brands like Dawn's retail prices were less than Method's production cost. Suppliers didn't give Method the same quantity discounts, and Method didn't have similar manufacturing economies of scale.

Growth had other costs. For example, retailers and wholesalers expected up-front cash payments or a few free cases of product before taking a chance on Method. New products are risky; retailers don't know how much they'll sell.

Fortunately for Method, its financial backers' initial objectives were more concerned with market share and sales volume than profitability. The pressure for profits wasn't as great because as a registered B Corporation, Method made its environmental mission as important as its profits. So Method continued to invest in building support for the brand and developing innovative and environmentally friendly new products.

Method's innovations in the laundry category show how a little company can influence a big—make that huge—market. When Method introduced a laundry soap three times as concentrated as market leader Tide, most consumers saw Method's much smaller bottles for the same price as much more expensive. Method had to teach customers that it took less Method to get the same clean. After learning how to properly "dose" Method, customers came to appreciate the smaller bottles' convenience. Retailers also recognized the value of lower handling costs and a smaller shelf footprint (so they could stock more products) and soon pressured Method's competitors to offer similar concentrates. A few years later, Method launched a new laundry detergent; Method Pump with "smartclean technology" had eight times the concentration of regular detergent.

Method may not look like a bargain to price-sensitive shoppers. At Target.com, a 66-load size of Method Laundry Detergent sells for $12.99 (discounted from its $15.49 list price). That price is higher than the prices for Tide's Clean Breeze 64-load jug ($11.99) or Target's Up&Up 64-load jug ($8.99). Customers wanting a more environmentally friendly option could choose Target's Everspring ($11.99 for 64 loads) and save too. A sale price, a coupon, a package bundled with fabric softener, or a refill with the Method pouch could also add value to the Method purchase.

So why are the higher-priced Method's sales continuing to grow? Because Method's whole marketing mix offers superior value to its target market. Customers might first try Method because of its stylish packaging and green roots, or perhaps it was clever advertising, sales promotion, and publicity efforts that earned the first purchase. But customers keep buying because the products clean well and protect the environment, delivering customer value even at Method's premium price. It also helps that the market segment where Method is cleaning up continues to grow.[1]

LEARNING OBJECTIVES

The Method case demonstrates the importance of Price and how it interacts with the other marketing mix variables to create value and influence customer behavior. This chapter will help you better understand pricing objectives and policies that influence how firms make pricing decisions.

When you finish this chapter, you should be able to

1. explain the dimensions of price and value.
2. understand how pricing objectives should guide strategy planning for pricing decisions.
3. understand choices marketing managers must make about price flexibility.
4. know what a marketing manager should consider when setting the price level for a product in the early stages of the product life cycle.
5. understand the many possible variations of a price structure, including discounts, allowances, and who pays transportation costs.
6. understand the value pricing concept and its role in obtaining a competitive advantage by offering target customers superior value.
7. understand the legality of price-level and price-flexibility policies.
8. understand important new terms (shown in **red**).

Price Has Many Strategy Dimensions

LO 17.1

Price is one of the four major strategy decision variables that a marketing manager controls. Pricing decisions affect both the number of sales a firm makes and how much money it earns. Price is what a customer must give up to get the benefits offered by the rest of a firm's marketing mix, so it plays a direct role in shaping customer value.

Guided by the company's objectives, marketing managers develop specific pricing objectives (see Exhibit 17-1). These objectives drive decisions about key pricing policies: (1) how flexible prices will be, (2) the level of prices over the product life cycle, (3) to whom and when discounts and allowances will be given, and (4) how temporary price reductions, financing, and transportation costs influence customer behavior. After we've looked at these specific areas, we discuss how they combine to impact customer value as well as laws that are relevant. In Chapter 18, we will discuss how prices are set.

It's not easy to define price in real-life situations because price reflects many dimensions. People who don't realize this can make big mistakes.

Exhibit 17-1 Strategy Planning and Pricing Objectives and Policies

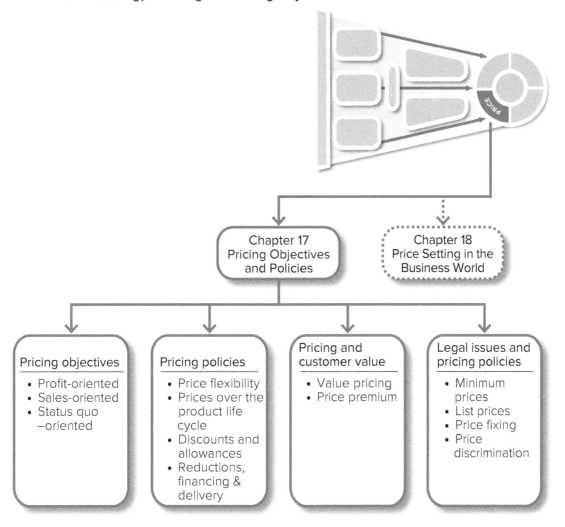

Suppose you've been saving to buy a new car and you see in an ad that, after a $1,000 rebate, the base price for the new-year model is $16,494—5 percent lower than the previous year. At first this might seem like a real bargain. However, your view of this deal might change if you found out you also had to pay a $400 transportation charge and an extra $480 for an extended service warranty. The price might look even less attractive if you discovered that the navigation system, side air bags, and moonroof that were standard the previous year are now options that cost $1,900. The cost of the higher interest rate on the car loan and the sales tax on all of this might come as an unpleasant surprise too. Further, how would you feel if you bought the car anyway and then learned that a friend who just bought the exact same model got a much lower price from the dealer by using a broker he found on the Internet?[2]

IKEA's clever—and carefully located—promotional signs remind customers that its price is a good value compared to what they pay for other everyday goods and services.
Source: Inter IKEA Systems

Exhibit 17-2 Price Exchanged for Something of Value (as seen by consumers or users)

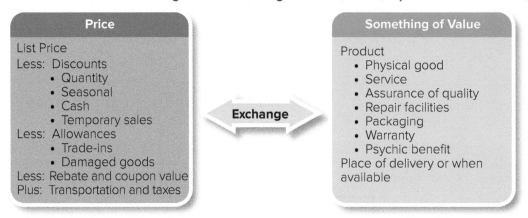

The price equation: Price equals something of value

This example emphasizes that when a seller quotes a price, it is related to *some* assortment of goods and services. So **Price** is the amount of money that is charged for "something" of value. Of course, price may be called different things in different settings. Colleges charge tuition. Landlords collect rent. Motels post a room rate. Magazines have subscriptions. Country clubs get dues. Banks ask for interest when they loan money. Airlines have fares. Doctors set fees. Employees want a wage. People may call it different things, but *almost every business transaction in our modern economy involves an exchange of money—the Price—for something.*

The something can be a physical product in various stages of completion, with or without supporting services, with or without quality guarantees, and so on. Or it could be a pure service—dry cleaning, a lawyer's advice, or insurance on your car.

The nature and extent of this something determines the amount of money exchanged. Some customers pay list price. Others obtain large discounts or allowances because something is *not* provided. Exhibit 17-2 summarizes some possible variations for consumers or users, and Exhibit 17-3 does the same for channel members. These variations are discussed more fully below, and then we'll consider the customer value

Exhibit 17-3 Price Exchanged for Something of Value (as seen by channel members)

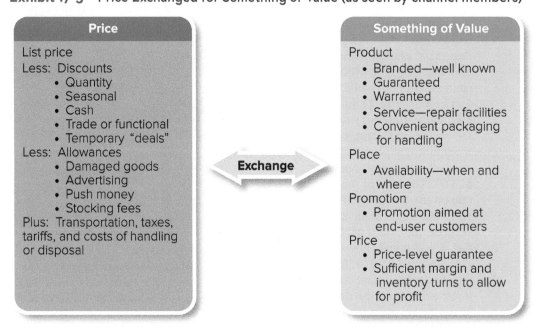

Exhibit 17-4 Possible Pricing Objectives

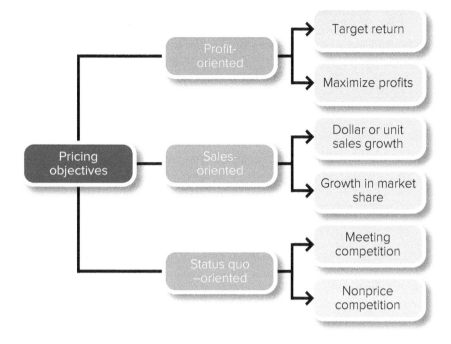

concept more fully—in terms of competitive advantage. But here it should be clear that Price has many dimensions. How each of these dimensions is handled affects customer value. If a customer sees greater value in spending money in some other way, no exchange will occur.[3]

Objectives Should Guide Strategy Planning for Price

Pricing objectives should flow from, and fit within, company-level and marketing objectives. Pricing objectives should be *explicitly stated* because they have a direct effect on pricing policies as well as the methods used to set prices. Exhibit 17-4 shows the various types of pricing objectives we'll discuss.

Profit-Oriented Objectives

Target returns provide specific guidelines

Over the long term, and often over the short term, marketing managers should set objectives oriented toward making a profit. A **target return objective** sets a specific level of profit as an objective. Often this amount is stated as a percentage of sales or of capital investment. A large manufacturer such as Motorola might aim for a 15 percent return on investment. The target for Safeway and other supermarket chains might be a 1 percent return on sales.

A target return objective has administrative advantages in a large company. Performance can be compared against the target. Some companies eliminate divisions, or drop products, that aren't yielding the target rate of return. For example, General Electric sold its small appliance division to Black & Decker because it felt it could earn higher returns in other product-markets.

Some just want satisfactory profits

Some managers aim for only satisfactory returns. They just want returns that ensure the firm's survival and convince stockholders they're doing a good job. Similarly, some small family-run businesses aim for a profit that will provide a comfortable lifestyle.[4]

Many private and public nonprofit organizations set a price level that will just recover costs. In other words, their target return figure is zero. For example, a government agency may charge motorists a toll for using a bridge but then drop the toll when the cost of the bridge is paid.

Similarly, firms that provide critical public services—including many utilities, insurance companies, and defense contractors—sometimes pursue only satisfactory long-run targets. They are well aware that the public expects them to set prices that are in the public interest. They may also have to face public or government agencies that review and approve prices.[5]

Some products have no direct competitors. If customers value those products, then firms will have more flexibility in choosing prices to maximize profits. After arriving at a theater, a moviegoer has no snack options beyond the theater's snack bar.
Source: Just Born, Inc.

Profit maximization can be socially responsible

A **profit maximization objective** seeks to get as much profit as possible. It might be stated as a desire to earn a rapid return on investment—or, more bluntly, to charge all the traffic will bear.

Pricing to achieve profit maximization doesn't always lead to high prices. Low prices may expand the size of the market and result in greater sales and profits. For example, when prices of cell phones were very high, only businesses and wealthy people bought them. When producers lowered prices, nearly everyone bought one.

If a firm is earning a very large profit, other firms will try to copy or improve on what the company offers. Frequently, this leads to lower prices. While many firms follow a profit maximization objective because of legal obligations, others follow another path.

#M4BW

Benefit corporations seek a better world

When a business is legally structured as a corporation, decisions are expected to be made in the shareholders' interests. That has generally been interpreted to mean following the profit maximization objective. In recent years, some corporations have sought an alternative legal structure that allows for explicit consideration of other stakeholders' interests. In most U.S. states, a firm can choose corporate bylaws that refer to the firm as a **benefit corporation**, a legal corporate structure that allows for goals that may include positive impacts on society, employees, the community, and the environment.

Many firms structured as benefit corporations go one step further and seek **B Corporation (B Corp) certification**, a private certification that it meets a high standard for social and environmental performance. An independent company, B Lab, conducts this certification process. Some B Corps are subsidiaries of publicly traded companies. There are more than 2,500 certified B Corps in more than 50 countries, a list that includes small companies and well-known names like Ben & Jerry's, Patagonia, and Method.

Many B Corps use the certification to formally signal to stakeholders, including customers, that the firm is mission driven and makes decisions with the best interests of employees, communities, and the environment—as well as customers—in mind. As we have seen throughout this text, when given a choice some customers want to purchase products from companies that have a "better world" mission. Consequently, many B Corps tell customers of their certification.[6]

Sales-Oriented Objectives

Sales growth doesn't necessarily mean big profits

A **sales-oriented objective** seeks some level of unit sales, dollar sales, or share of market—*without referring to profit.*

Some managers are more concerned about sales growth than profits. They think sales growth always leads to more profits. This sometimes makes sense over the short term. For example, many Procter & Gamble brands in product categories such as shampoo, soap, and diapers lost market share during the recession of 2007–2009. When the economy began to recover, P&G kept prices low and sacrificed profits in order to grow with the recovering economy. It might also work well when products are in the introductory or early growth stages of the product life cycle. However, over the long term this kind of thinking causes problems when a firm's costs are growing faster than sales. Although some firms have periods of declining profits in spite of growth in sales, business managers should usually pay more attention to profits, not just sales.

Some nonprofit organizations set prices to increase market share—precisely because they are *not* trying to earn a profit. For example, many cities set low fares to fill up their buses, reduce traffic, and help the environment. Buses cost the same to run empty or full, and there's more benefit when they're full even if the total revenue is no greater.

Market share objectives are popular

Many firms seek to gain a specified share (percentage) of a market. If a company has a large market share, it may have better economies of scale than its competitors. In addition, it's usually easier to measure a firm's market share than to determine if profits are being maximized.

A company with a longer-run view may aim for increased market share when the market is growing. The hope is that future volume will justify sacrificing some profit in the short run. HP, Dell, and Acer have waged pricing battles in an effort to gain market share in the personal computer market. A similar focus on market share can be seen with Uber and Lyft in ride sharing, and Lime, Bird, and others in dockless scooters. High market share offers economies of scale and negotiating power with suppliers. Companies as diverse as 3M and Coca-Cola look at opportunities in eastern Europe and Southeast Asia this way.

Of course, market share objectives have the same limitations as straight sales growth objectives. A larger market share, if gained at too low a price, may lead to profitless "success."

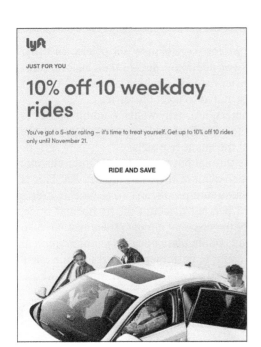

Often new products, like Lyft ride-sharing service and Lime scooter services, want to gain market share quickly. So they offer lower prices to try to get customers into the habit of using their service as opposed to alternatives. The pricing objective for each is to grow sales and/or market share.
(left): Source: Lyft, Inc.; (right): Source: Lime

#M4BW

Sales-oriented objectives a win for schools

Sometimes society benefits when companies seek to grow sales and gain market share. Often these objectives involve low prices to make a product available to more people in a target market. While many K-12 schools in the United States need computers for their students, most also face budgetary constraints. Enter Google's low-priced Chromebook laptop and tablet computers. With Google seeking a foothold among the nation's youngest computer users, the Chromebook's low prices and reliable operating system have attracted a 60 percent share of this target market. Across the United States, school districts save money, students learn to use computers, and Google still makes a profit.

Status Quo-Oriented Objectives

Don't-rock-the-boat objectives

Managers satisfied with their current market share and profits sometimes adopt **status quo-objectives**—don't-rock-the-*pricing*-boat objectives. Managers may say that they want to stabilize prices, or meet competition, or even avoid competition. This don't-rock-the-boat thinking is most common when the total market is not growing.

Sometimes firms in an industry make price changes very carefully—and only if others follow their lead. This tends to prevent price wars, which can drag down all firms' profits. For example, the airline industry typically raises prices collectively. One airline might add a fuel surcharge—if the others do not follow, that airline backs off and all maintain the status quo.

Or stress nonprice competition instead

A status quo-objective may be part of an aggressive overall marketing strategy focusing on **nonprice competition**—aggressive action on one or more of the Ps other than Price. Some companies that sell through the Internet originally thought that they'd compete with low prices and still earn high profits from volume. However, when they didn't get the sales volume they hoped for, they realized that there were also some nonprice ways to compete. For example, Zappos.com offers free shipping and guarantees that it will meet local shoe store prices. But it wins customers with its enormous selection of shoes, a website that makes it easy for customers to find what they want, and excellent customer service before and after the sale.

Most Firms Set Specific Pricing Policies—To Reach Objectives

Administered prices help achieve objectives

Price policies usually lead to **administered prices**—consciously set prices. In other words, instead of letting daily market forces (or auctions) decide their prices, most firms set their own prices. They may hold prices steady for long periods of time or change them more frequently if that's what's required to meet objectives.

If a firm doesn't sell directly to final customers, it usually wants to administer both the price it receives from intermediaries and the price final customers pay. After all, the price final customers pay will ultimately affect the quantity it sells.

Yet it is often difficult to administer prices throughout the channel. Other channel members may also wish to administer prices to achieve their own objectives. This is what happened to Alcoa, one of the largest aluminum producers. To reduce its excess inventory, Alcoa offered its wholesalers a 30 percent discount off its normal price. Alcoa expected the wholesalers to pass most of the discount along to their customers to stimulate sales throughout the channel. Instead, wholesalers bought *their* aluminum at the lower price but passed on only a small discount to customers. As a result, the quantity Alcoa sold didn't increase much, and it still had excess inventory, while the wholesalers made much more profit on the aluminum they did sell.[7]

Some firms don't even try to administer prices. They just meet competition—or worse, mark up their costs with little thought to demand. They act as if they have no choice in selecting a price policy.

Remember that Price has many dimensions. Managers usually *do* have many choices. They *should* administer their prices. And they should do it carefully because, ultimately, customers must be willing to pay these prices before a whole marketing mix succeeds. In the rest of this chapter, we'll talk about policies a marketing manager must set to do an effective job of administering Price.[8]

Price Flexibility Policies

LO 17.3

One-price policy—the same price for everyone

One of the first decisions a marketing manager has to make is whether to use a one-price or a flexible-price policy. A **one-price policy** means offering the same price to all customers who purchase products under essentially the same conditions and in the same quantities. The majority of U.S. firms use a one-price policy—mainly for administrative convenience and to maintain goodwill among customers. But that is changing thanks to technology and the ability to identify different customer segments to which a firm wishes to charge higher or lower prices.

A one-price policy makes pricing easier. But a marketing manager must be careful to avoid a rigid one-price policy. This can amount to broadcasting a price that competitors can undercut, especially if the price is somewhat high. One reason for the growth of mass-merchandisers is that conventional retailers rigidly applied traditional margins and stuck to them. This left an opportunity for mass-merchandisers to undercut conventional retailers on price, allowing them to gain market share.

Flexible-price policy—different prices for different customers

A **flexible-price policy** means offering the same product and quantities to different customers at different prices. When computers are used to implement flexible pricing, the decisions focus more on what type of customer will get a price break. As firms learn how to analyze big data, more organizations are moving from a one-price policy to a flexible-price policy.

Pricing databases make flexible pricing easier

Various forms of flexible pricing are more common now that most prices are maintained in a central computer database. Frequent changes are easier. You see this when supermarket chains give loyalty club cardholders reduced prices on weekly specials. The checkout scanner reads the code on the package, and then the computer looks up the club price or the regular price depending on whether a club card has been scanned.

Dynamic pricing means prices change with demand

Dynamic pricing refers to pricing products at a particular customer's perceived ability to pay. This is the ultimate in price flexibility. The idea is to optimize revenue and profit by charging higher prices to customers willing to pay more and lower prices to those who do not see value at the high price but will buy at lower prices. Some companies use dynamic pricing as a means for changing prices over time. For example, airlines might raise or lower prices over time depending on how many seats are still available on a flight. Tickets purchased months in advance might have a lower price, but as the flight starts filling up,

Slumberland's marketing managers use flexible pricing, including discounts for military veterans and active service members. *Source: Slumberland, Inc.*

prices will rise as supply decreases. It can also work in the other direction; if United Airlines sees that early morning Tuesday flights from St. Louis to Miami are not selling well, the airline can lower the price to stimulate demand—perhaps encouraging some flyers to fly out on Tuesday morning instead of Monday night when there are fewer available seats. It is better to sell some seats at a discount than to have the flight leave with half the seats empty.

Sports teams are increasing ticket revenue by analyzing big data and implementing dynamic pricing. The idea is to charge higher prices for a better product. Initially, these teams simply charged higher prices for more popular weekend games as compared to midweek games. Now teams have adapted and change pricing even days before events—charging more when the weather is better or the matchup more intriguing. For example, the San Francisco Giants baseball team adjusts ticket prices to maximize attendance and revenue for each game. Ticket prices for each seat at each game are adjusted after crunching numbers that examine past ticket sales, the day of the week, time of the game, the opposing team's record, the pitching matchup, the going price at ticket resale sites like StubHub, and even the weather forecast. The Giants were expecting an additional $5 million in revenue the first year using the system. Other teams are following the Giants' lead.[9]

Big data and predictive analytics vary prices by individual

Firms have long adjusted pricing for segments of customers. Now many firms use big data from loyalty cards or online shopping and predictive analytics to deliver customers individual prices. Many sources of big data can be read in real time, so pricing "experiments" can be run. For example, an online retailer can cut prices for a day or even a few hours to see whether it stimulates demand and raises profits. Another example of big data at work in real time is OfficeDepot.com's adjustment of prices depending on customers' browsing history and the physical location of the computer from which they are browsing.

Even brick-and-mortar retailers will use knowledge of individual customers to offer targeted price discounts to certain customers. It is not always a ploy to charge higher prices to customers more likely to pay them. For example, grocery store chain Safeway's Just for U loyalty program mines data to make unique offers to customers based on

While sellers often use artificial intelligence and dynamic pricing to maximize profits, some customers fight back, using apps that predict when lower prices might come about. The Hopper app for Android and iOS uses artificial intelligence and predictive analytics to forecast future airfares. Say a customer wants to fly from Chicago to Milan, Italy, in June (four months from now). She can choose to keep an eye on prices ("Watch This Trip"; see the image on the left) or purchase soon. Hopper predicts the prices will go up and encourages the customer to buy soon (see the image on the right).
Source: Hopper

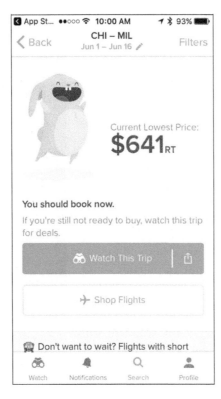
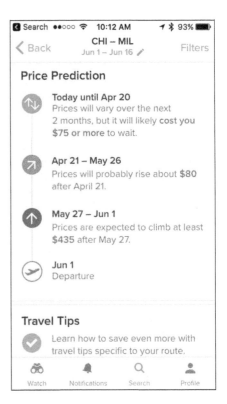

their previous shopping behavior. But they use the offers on products they were likely to buy anyway—with the expectation that customers will buy even more. Customers who buy lots of Cheerios are pleasantly surprised to receive a coupon for one dollar off their favorite breakfast cereal. Tracking data will show where that works and where it doesn't—and may reduce offers to customers that do not increase purchases.[10]

"Surge" prices go up with demand

Car-sharing service Uber employs an analytical model for flexible pricing. Uber's "surge pricing" model adjusts prices to better match supply and demand for its service. When Uber sees demand for rides significantly exceeds supply—for example, on a rainy afternoon in a big city—prices are increased in an effort to raise the supply of drivers and reduce the demand for riders. Drivers, seeing the higher prices they will receive, are encouraged to work later than they might have otherwise wanted; riders, seeing the temporary price increase, may wait until later or seek another transportation option. Prices, which can rise to as much as 10 times the normal fare, fall back to normal when supply and demand stabilize.

> **Ethical Dilemma**
>
> *What would you do?* You are an executive at startup ride-sharing service, JustRide (an upstart Uber competitor). JustRide automatically calculates Uber's price for any ride—and uses that information to set JustRide's price for the same destination. JustRide has gained market share by offering "the lowest-priced ride in town" (a claim made in all of its advertising)—setting prices 10 percent lower than Uber. While JustRide is gaining market share, it is not yet profitable. In a bid to become profitable, management recently changed its pricing policy. The app identifies which of its customers do not check competitor prices (the app knows which other apps were recently used on a smartphone); those customers are then quoted prices 10 percent higher than Uber. Customers who have recently opened Uber (or another ride-sharing app) receive the lower price quote.
>
> Recently, JustRide blew up social media—for the wrong reason. Customers noticed the different price quotes. Some riders would get a quote from JustRide while a companion got one from Uber; they discovered JustRide was priced higher for the same destination. Not exactly keeping their advertising promise. Now #JustLies and #JustHigherPrices are trending on social media—often accompanied by photos showing two different prices.[11]
>
> *Is JustRide's pricing policy unethical? Explain your logic. How is this ethical question similar to or different from Uber's "surge pricing"? How should JustRide respond to this immediately? And in the longer term? In your answer, consider possible changes to advertising and pricing and explain your rationale.*

Salespeople negotiate prices to the situation

Sometimes prices are negotiated and a price is determined by bargaining between the buyer and the seller. This type of flexible pricing is most common in the channels, in direct sales of business products, and at retail for expensive shopping products. Retail shopkeepers in less-developed economies typically use flexible pricing—shopkeepers start with high prices but bargain to try to make a sale at a price the customer will accept while still providing maximum profit to the seller. These situations usually involve personal selling, not mass selling. The advantage of flexible pricing is that the salesperson can adjust price—considering prices charged by competitors, the relationship with the customer, and the customer's bargaining ability. Flexible-price policies often specify a *range* in which the actual price charged must fall.[12]

Too much price-cutting erodes profits

Some sales reps let price-cutting become a habit. This can lead to a lower price level and lower profit. A small price cut may not seem like much, but keep in mind that all of the revenue that is lost would go to profit. If salespeople for a producer that usually earns profits equal to 20 percent of its sales cut prices by an average of about 10 percent, profits would drop by half (see Exhibit 17–5)!

Exhibit 17-5 Impact of Price Cut on Profit

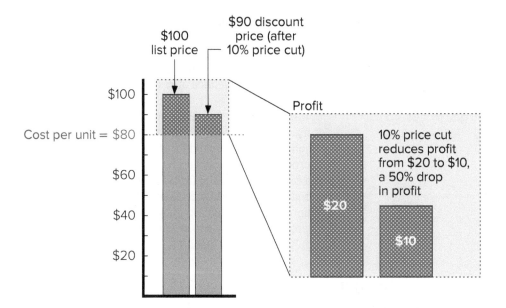

Disadvantages of flexible pricing

Flexible pricing does have disadvantages. A customer who finds that others paid lower prices for the same marketing mix will be unhappy. This can cause real conflict in channels. For example, the Winn-Dixie supermarket chain stopped carrying products of some suppliers who refused to give Winn-Dixie the same prices available to chains in other regions of the country. Similarly, companies that post different prices for different segments on a website that all can see often get complaints.

If buyers learn that negotiating is in their interest, the time needed for bargaining will increase. This can increase selling costs and reduce profits. It can also frustrate customers. For example, most auto dealers use flexible pricing and bargain for what they can get. Inexperienced consumers, reluctant to bargain, often pay hundreds of dollars more than the dealer is willing to accept. By contrast, CarMax has earned high customer satisfaction ratings by offering haggle-weary consumers a one-price policy.[13]

Price-Level Policies and the Product Life Cycle

LO 17.4

Marketing managers who administer prices must consciously set a price-level policy. As they enter the market, they have to set introductory prices that may have long-run effects. They must consider where the product idea is in its life cycle and how fast it's moving. And they must decide if their prices should be above, below, or somewhere in between relative to the market.[14]

Let's first look at a new product in the market introduction stage of its product life cycle. There are few (or no) direct substitute marketing mixes. So the price-level decision should focus first on the nature of market demand. A high price may lead to higher profit from each sale but also to fewer units sold. A lower price might appeal to more potential customers. With this in mind, should the firm set a high or low price?

Skimming pricing—feeling out demand at a high price

A **skimming price policy** tries to sell the top (skim the cream) of a market—the top of the demand curve—at a high price before aiming at more price-sensitive customers (see Exhibit 17-6). Skimming may maximize profits in the market introduction stage for an innovation, especially if there are few substitutes or if some customers are not price sensitive. Skimming is also useful when you don't know very much about the shape of the demand curve. It's sometimes safer to start with a high price that can be reduced if customers balk.

That is what happened to Apple when it introduced the first iPhones. The phone was originally sold for $600, but customers pushed back and it did not meet sales targets. A few months later Apple lowered the price by $200 and sales took off.

Exhibit 17-6 Alternative Introductory Pricing Policies

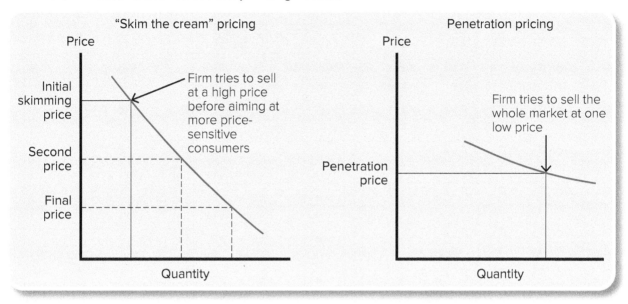

| Price moves down the demand curve | A skimming policy often involves a slow reduction in price over time (see Exhibit 17-6). Note that as price is reduced, new target markets are probably being sought. So as the price level steps down the demand curve, new Place, Product, and Promotion policies may be needed too.

This is what happened when cellular phone service was first introduced. McCaw Cellular was the first to bring this product to market. Wireless phone calls cost $1 per minute and customers paid about $700 for a large clunky phone. At these prices, business customers were about the only ones interested in buying. Over time, more competitors entered the market, prices fell, and more customers found value in the offering. Today, prices are much lower and cellular phones are widely adopted. |
|---|---|
| Penetration pricing—get volume at a low price | A **penetration pricing policy** tries to sell the whole market at one low price. This approach might be wise when the elite market—those willing to pay a high price—is small. This is the case when the whole demand curve is fairly elastic. See Exhibit 17-6. A penetration policy is even more attractive if selling larger quantities results in lower costs because of economies of scale. Penetration pricing may be wise if the firm expects strong competition very soon after introduction. This was the strategy when Amazon first introduced its Echo voice-controlled smart speaker. Anticipating competition from Google, Microsoft, and Apple, Amazon offered the Echo at a low price and grabbed the early market share lead.

Of course, even a low penetration price doesn't keep competitors out of a market permanently. Product life cycles do march on. However, a firm that gets a head start in a new market can often maintain its advantage. |
| Introductory price dealing—temporary price cuts | Low prices tend to attract customers and trial. Therefore, marketers often use **introductory price dealing**—temporary price cuts—to speed new products into a market and get customers to try them. However, don't confuse these *temporary* price cuts with low penetration prices. The plan here is to raise prices as soon as the introductory offer is over. Hopefully by then target customers will have decided it is worth buying again at the regular price. Introductory price dealing should be part of a larger marketing strategy. For example, some developers of software applications (or "apps") for smartphones know that getting onto the store's list of "Top Apps" gets attention, which drives sales. So some app makers price the app low at launch to encourage sales, which moves it up the rankings. Then they raise prices to optimize profits. |

Marketers often use introductory price dealing—in the form of temporary price cuts, introductory coupons, or trade-in allowances—to speed new products into a market. To encourage new customer purchases, Corti-Care offers a 25 percent discount, and the *New York Times* provides a special price to recent college graduates.
(left): Source: Spectrum Brands, Inc.; (right): Source: The New York Times

Established competitors often choose not to meet introductory price dealing—as long as the introductory period is not too long or too successful. However, some competitors match introductory price deals with their own short-term sale prices to discourage customers from shopping around.

Growth and market maturity may require meeting competitors' prices

Competition begins to impact pricing more in the market growth and market maturity stages. Especially in a mature market there is downward pressure on both prices and profit margins. Moreover, differentiating the value a firm offers may not be easy when competitors can quickly copy new ideas. Consider the market for gasoline. Many customers see little difference among the various brands of gasoline. One station choosing to raise prices may lead to a large loss in sales. Cutting prices may lead to similar reductions by competitors, which can cut into profits and lead to a decrease in total revenue for the industry and probably for each firm. In such circumstances there may be no real pricing choice other than to "meet the competition."

Prices fall in sales decline

In the sales decline stage, new products come in to replace the old. As brands compete in a shrinking market, unless a brand has strong differentiation, prices will likely decline. At the same time, costs may decline as well. Many firms back off on Promotion so margins might be maintained.

Discount Policies—Reductions from List Prices

LO 17.5

Prices start with a list price

Most price structures are built around a base price schedule or list price. **Basic list prices** are the prices final customers or users are normally asked to pay for products. In this book, unless noted otherwise, *list price* refers to basic list price.

In Chapter 18, we discuss how firms set these list prices. For now, however, we'll consider variations from list price and why they are made.

Discounts are reductions from list price

Discounts are reductions from list price given by a seller to buyers who either give up some marketing function or provide the function themselves. Discounts can be useful in marketing strategy planning. In the following discussion, think about what function the buyers are giving up, or providing, when they get each of these discounts.[15]

Quantity discounts encourage volume buying

Quantity discounts are discounts offered to encourage customers to buy in larger amounts. This lets a seller get more of a buyer's business, shifts some of the storing function to the buyer, or reduces shipping and selling costs—or all of these. There are two kinds of quantity discounts: cumulative and noncumulative.

Cumulative quantity discounts apply to purchases over a given period—such as a year—and the discount usually increases as the amount purchased increases. Cumulative discounts encourage *repeat* buying by reducing the customer's cost for additional purchases. This is a way to develop loyalty and ongoing relationships with customers. For example, a Lowe's lumberyard might give a cumulative quantity discount to a building contractor who is not able to buy all of the needed materials at once. Lowe's wants to reward the contractor's patronage and discourage shopping around. Along the same lines, Disney World wants to keep you on its properties and not at Universal's Islands of Adventure—so it offers discounts for multi-day passes.

A cumulative quantity discount is often attractive to business customers who don't want to run up their inventory costs. They are rewarded for buying large quantities, even though individual orders may be smaller.

Noncumulative quantity discounts apply only to individual orders. Such discounts encourage larger orders but do not tie a buyer to the seller after that one purchase. Lowe's lumberyard may resell insulation products made by several competing producers. Owens-Corning might try to encourage Lowe's to stock larger quantities of its pink insulation by offering a noncumulative quantity discount.

Seasonal discounts—buy sooner

Seasonal discounts are discounts offered to encourage buyers to buy earlier than present demand requires. If used by a manufacturer, this discount tends to shift the storing function further along in the channel. It also tends to even out sales over the year. For example, Kyota offers wholesalers a lower price on its garden tillers if they buy in the fall, when sales are slow.

Service firms that face irregular demand or excess capacity often use seasonal discounts. For example, some tourist attractions, such as ski resorts, offer lower weekday rates when attendance would otherwise be down.

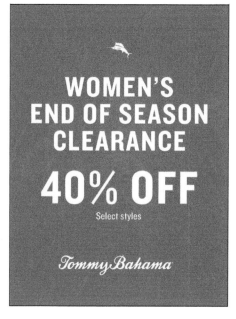

Customers who buy six applications of Frontline Plus flea protection get two more for free—essentially a quantity discount of 25 percent. Clothing brand Tommy Bahama uses seasonal discounts to clear out excess inventory.
(left): Source: Frontline, Merial Inc.;
(right): Source: Tommy Bahama

Payment terms and cash discounts set payment dates

Most sales to businesses are made on credit. The seller sends a bill (invoice) by mail or electronically, and the buyer's accounting department processes it for payment. Some firms depend on their suppliers for temporary working capital (credit). Therefore, it is very important for both sides to clearly state the terms of payment—including the availability of cash discounts—and to understand the commonly used payment terms.

Net means that payment for the face value of the invoice is due immediately. These terms are sometimes changed to net 10 or net 30, which means payment is due within 10 or 30 days of the date on the invoice.

Cash discounts are reductions in price to encourage buyers to pay their bills quickly. The terms for a cash discount usually modify the net terms.

2/10, net 30 means the buyer can take a 2 percent discount off the face value of the invoice if the invoice is paid within 10 days. Otherwise, the full face value is due within 30 days. And it usually is stated or understood that an interest charge will be added after the 30-day free-credit period.

Why cash discounts are given and should be evaluated

Smart buyers carefully evaluate cash discounts. A discount of 2/10, net 30 may not look like much at first. But the buyer earns a 2 percent discount for paying the invoice just 20 days sooner than it should be paid anyway. By not taking the discount, the company in effect is borrowing at an annual rate of 36 percent. That is, assuming a 360-day year and dividing by 20 days, there are 18 periods during which the company could earn 2 percent—and 18 times 2 equals 36 percent a year. A buying company with an average balance of $10,000 per month could save $3,600 by taking the discount.

Trade discounts are often set by tradition

A **trade (functional) discount** is a list price reduction given to channel members for the job they are going to do. A manufacturer, for example, might allow a retailer a 30 percent trade discount from the suggested retail list price to cover the cost of carrying inventory and providing knowledgeable salespeople to demonstrate the manufacturer's products. Similarly, the manufacturer might allow wholesalers a *chain* discount of 30 percent and 10 percent off the suggested retail price. In this case, the wholesalers would be expected to pass the 30 percent discount on to retailers.

Allowance Policies—Off List Prices

Allowances, like discounts, are given to final consumers, business customers, or channel members for doing something or accepting less of something.

Advertising allowances— something for something

Advertising allowances are price reductions given to firms in the channel to encourage them to advertise or otherwise promote the supplier's products locally. For example, Sony might give its retailers an allowance of 3 percent of sales. They, in turn, are expected to spend the allowance on local advertising.

Stocking allowances— get attention and shelf space

Stocking allowances—sometimes called *slotting allowances*—are given to an intermediary to get shelf space for a product. For example, a producer might offer a retailer cash or free merchandise to stock a new item. Stocking allowances are commonly used to get supermarket chains to handle new products. Supermarkets are more willing to give space to a new product if the supplier will offset their handling costs and risks. With a big stocking allowance, the intermediary makes extra profit—even if a new product fails and the producer loses money.[16]

PMs—push for cash

Push money (or prize money) allowances—sometimes called *PMs* or *spiffs*—are given to retailers by manufacturers or wholesalers to pass on to the retailers' salesclerks for aggressively selling certain items. PM allowances are used for new items, slower-moving items, or higher-margin items. They are often used for pushing furniture, clothing, consumer electronics, and cosmetics. A salesclerk, for example, might earn an additional $5 for each new model Panasonic DVD player sold.

Bring in the old, ring up the new—with trade-ins

A **trade-in allowance** is a price reduction given for used products when similar new products are bought. Trade-ins give the marketing manager an easy way to lower the effective price without reducing list price. Sometimes producers want to get older products off the market or move them to a new market. Apple offered trade-in incentives for its iPhones and shipped the older models to developing countries that could not afford the higher price of a new phone.

Pricing Policies for Price Reductions, Financing, and Transportation

Stir customers to action

Marketing managers often use different Price tactics to motivate customers to action. In this section we look at three different ways this can be accomplished: through temporary price reductions, adding convenient ways to finance or pay, and how transportation costs are handled. Let's take a closer look at these pricing policy choices.

Special sales reduce list prices—temporarily

A **sale price** is a temporary discount from the list price. Sale price discounts encourage immediate buying. In other words, to get the sale price, customers give up the convenience of buying when they want to buy and instead buy when the seller wants to sell.

Special sales provide a marketing manager with a quick way to respond to changing market conditions without changing the basic marketing strategy. For example, a retailer might use a sale to help clear extra inventory or to meet a competing store's price.

In recent years, sale prices and deals have become much more common. Some retailers have sales so often that consumers just delay purchase decisions until there's a sale. Others check out a website such as Google Shopping to find out if the product they want is already on sale somewhere else. While producers and retailers may see a bump in sales, prices that change frequently can erode brand loyalty. When customers loyal to one particular brand see a competing brand with a very low price, they may try that brand. Soon they may be "loyal" to several brands. This happened over time in the markets for cold breakfast cereal and other convenience products.

To avoid these problems, some firms selling consumer convenience products offer **everyday low pricing**—setting a low list price rather than relying on frequent sales, discounts, or allowances. Some supermarkets use this approach.

Sale prices should be used carefully and should be consistent with well-thought-out pricing objectives and policies. A marketing manager who constantly uses temporary sales to adjust the price level probably has not done a good job setting the normal price or creating a marketing mix that offers sustainable competitive advantage.[17]

Like most retailers, Harvey Nichols knows that sale prices encourage consumers to purchase products immediately. To get the sale price, customers give up the convenience of buying when they want to buy and instead buy when the seller wants to sell.
Source: Harvey Nichols

Ben & Jerry's knows that coupons lower the risk of trying a new brand or product.
Source: Ben & Jerry's

Coupons or rebates—more for less

Some producers and retailers offer discounts (or free items) through coupons distributed in packages, mailings, print ads, at the store, via e-mail, or online. By presenting a coupon to a retailer, the consumer is given a reduction to the list price. The fastest-growing distribution approach is online—where consumers can search for coupons and print them out, send them to their cell phone, directly add them to their store loyalty card, or just enter the "discount code" to an online order. This is especially common in the consumer packaged goods business—but the use of price-off coupons is also growing in other lines of business.[18]

A variation of targeted price reductions are rebates—refunds paid to consumers after a purchase. Customers often must file some paperwork or apply online to make a claim. The manufacturer later sends the customer a check. Sometimes the rebate is very large. Some automakers offer rebates of $500 to $6,000 to promote sales of slow-moving models. Rebates are also used on lower-priced items, ranging from Duracell batteries to Logitech webcams and Paul Masson wines. Rebates provide a quick way for a manufacturer to offer a short-term price reduction to stimulate sales.

Coupons and rebates give a producer a way to be certain that final consumers actually get the price reduction. If the rebate amount were just taken off the price charged to intermediaries, the intermediaries might not pass the savings along to consumers.

Deals for those who really want them

Coupons and rebates also help marketing managers address another problem. While marketing managers usually prefer to sell at list price where the profit is the highest, they realize that some customers will only purchase at a lower price. In their perfect world, those customers willing to pay a high price pay the high price, and those customers only willing to buy at a lower (but still profitable) discounted price are offered the lower price. The seller earns a high profit from customers paying full price but also earn additional profits from customers only willing to buy at the lower price.

To achieve this goal, a firm might use coupons or rebates to *segment* the market. Bargain hunters, most of whom will only buy the product at a lower price, spend the time and energy necessary to find coupons and fill out rebate forms. Those customers who place a higher value on the whole marketing mix (and perhaps more value on their time) will pay the full price. They don't need the added incentive of a deal and won't actively seek out the discounts. As a result, a marketing manager can increase total profits by segmenting the market and targeting each group separately.

Financing converts "price" to "payment"

Some customers are motivated to action when they can delay paying—or pay small amounts over time. Toward that end, many firms set a pricing policy that includes some form of financing. For buyers who either cannot afford the full outlay or who prefer to pay over time, some form of installment plan can be part of the price. Installment involves making small payments over time—usually with interest payments built in. Many large consumer purchases—like cars, homes, and higher education—are financed. A consumer who cannot afford to spend $20,000 for a new car may be able to afford $300 per month. Business products may also be financed over time, which can help a business with its cash flow. Financing can make a high price appear as a smaller payment.

Consumers say "charge it"

Credit cards can also motivate customers to action. Credit and debit cards offer many benefits to customers. For some customers, the cards offer convenience as they do not have to carry around as much cash. They can also more easily make online purchases. Credit cards offer the additional benefit of allowing customers to finance purchases over time with additional interest charges. Retailers usually accept credit cards from services such as Visa or Mastercard, paying a percentage of the revenue from each credit sale for the service. Some retailers also have aggressive promotions to sign up customers for their own credit cards, because customers who carry a store's credit card usually spend more money at the store. Generous credit terms such as "no interest or payments for one full year" also stimulate sales. Some credit cards offer customers cash back or travel benefits in proportion to their purchases.[19]

Credit raises ethical concerns

There are also ethical concerns about credit card companies and retailers that make it too easy for consumers to buy things they really can't afford. The problem becomes worse when an unpaid balance on a credit card carries a very high interest rate. This can significantly increase the price a consumer pays. Even worse, it leaves many low-income consumers trapped in debt.

Retailers face a challenge when a customer wants to buy something but doesn't quite have all the money at that moment. Sometimes the customer may not have a credit card. When that happens, some retailers nudge customers toward a purchase with Afterpay. Afterpay splits a purchase into four equal installments—paid every two weeks. For some customers, that $100 dress looks a lot more affordable when it is made in four payments of $25 over the next two months. Credit approval is instant and (for better or worse) customers get instant gratification when they can buy instantly.
Source: Afterpay

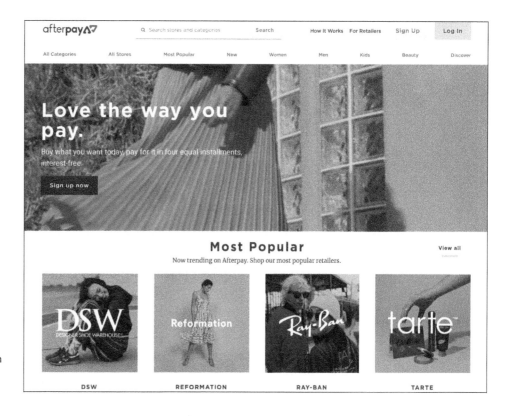

#M4BW

Financing opens opportunities in developing markets

Financing can be a responsible decision when customers use it to save money or invest in their future. For example, in many developing countries, the poor have a hard time saving money. This may be because of having limited resources, a lack of banking options, or perhaps for cultural reasons. Without an ability to save, these customers may be unable to make larger purchases that could improve their lives.

Consider the investment in a child's education or in working longer hours at a business. Many rural villages in Africa and Latin America are located off the electric grid. To work or study in the evening, they may rely on kerosene lamps for light. Kerosene isn't a healthy, environmentally friendly, or even a lower-priced alternative to electricity. But until recently, many of these customers had little choice. While many heard that solar power offered a superior value, most of the target market couldn't afford the $50 to $100 cost to install a small rooftop system. Then many solar suppliers developed installment plans. Customers might pay $10 for installation and then $1.50 a week (half the price of kerosene) for perhaps a year or two. The payment plans opened up new opportunities and made their quality of life better.[20]

Mobile payments are taking off

Mobile payments are payments made at the point of purchase using a mobile device (usually a cell phone or tablet computer). Like credit cards, mobile payments offer customers greater convenience when making purchases. This can increase the value they receive and increase purchases; many retail businesses are offering this service as customers become more accustomed to this form of payment.

In the United States and Europe, where credit cards are more widespread, the adoption of mobile payments has been slow. The Starbucks app, ChasePay, PayPal, Venmo, and Apple Pay are among a growing number of companies offering mobile payment services. With many competing services, the promised convenience for customers may disappear if customers need to use a different service at Starbucks, Walmart, and McDonald's. That said, it is likely that over time a few large services will dominate.

In China and many developing countries, mobile payment use is already common. In China, mobile payments began as a way to help customers purchase online; credit card use was not widespread in China. The most popular mobile payment services are therefore offered by the leading online retailers—Alibaba's Alipay and Tencent's WeChat Pay have more than 90 percent market share. It helped that these payment systems were designed so recipients did not have to purchase hardware. Each retailer has a sign with a unique QR code; customers scan the code on their phones and indicate the payment amount. Now many customers use their phones to handle most transactions—buying online, purchasing groceries, paying for a taxi, getting a bike share, or paying the check at a restaurant. In China, even street vendors and panhandlers accept mobile payments.[21]

Leasing—pay to use

Another way to make a product available to a customer is through a lease agreement. A **lease** gives a customer the right to use something for a specified period of time in exchange for regular payments. In this case, customers do not take ownership of the product, they only pay to use it—in essence renting the product. Leases are common for apartments; many car buyers use them as well. Some businesses will also lease when they do not expect to need something long term or when there might be tax advantages. Regular lease payments are usually lower than a similar financing agreement, and the customer does not have to worry about what to do with the product when they are done with it; they just return it to the owner.

Who pays for transportation?

With many types of goods, a product must be transported from a store or where a product is produced to a customer's home or place of business. Customers typically consider any transportation costs as part of the price—it can be part of the money charged for receiving something of value. Firms have a choice of including these costs as part of the price ("free delivery") or as an additional fee.

In today's competitive markets, a marketing manager must look for ways to enhance the value customers receive from the firm's marketing mix. L.L.Bean uses free shipping; the U.S. Postal Service offers the simplicity and convenience of shipping to any state for one low rate.
(left): Source: L.L.Bean; (right): Source: USPS

The answer to the "who pays" question often ends up following common industry practices. In most business-to-business transactions involving goods, there are industry norms about which party pays for transportation. A seller must understand customer expectations; otherwise, additional charges at the end may lead to dissatisfaction or a last-minute change of heart.

This is what happened when customers first started buying online; many decided not to purchase when they arrived at checkout and discovered an unexpectedly high charge for delivery. Consequently, many online retailers now offer some type of free shipping. In online retail, the norm has evolved that most firms offer free shipping for at least some purchase level ("free shipping on all orders over $50"). This type of offer also encourages customers to buy more. Other firms include paying an annual fee for free delivery (like Amazon Prime). This type of offer encourages customers—who figure they have already paid for delivery—to consolidate more purchases from the same firm. Both of these types of offers are similar to the cumulative and noncumulative quantity discounts described earlier in the chapter.

When buying from brick-and-mortar retailers, customers usually expect to pay for delivery. Customers anticipate paying to have a washing machine transported to their home from Lowe's or a pizza delivery from Pizza Hut. In these situations it probably does not make sense for a firm to offer free delivery, which would cut into margins.

Pricing Policies Combine to Impact Customer Value

LO 17.6

Look at Price from the customer's viewpoint

We've discussed pricing policies separately so far, but from the customer's view they all combine to impact customer value. So when we talk about Price we are really talking about the whole set of price policies that define the real price level. On the other hand, superior value isn't just based on having a lower price than some competitor but rather on the whole marketing mix.

Value pricing leads to superior customer value

Smart marketers look for the combination of Price decisions that results in value pricing. **Value pricing** means setting a fair price level for a marketing mix that really gives the target market superior customer value.

Value pricing doesn't necessarily mean cheap if cheap means bare-bones or low-grade. It doesn't mean high prestige, either, if the prestige is not accompanied by the right quality goods and services. Rather, the focus is on the customer's requirements and how the whole marketing mix meets those needs.

Wendy's $0.99 Everyday Value Menu communicates value to its customers.
Source: The Wendy's Company

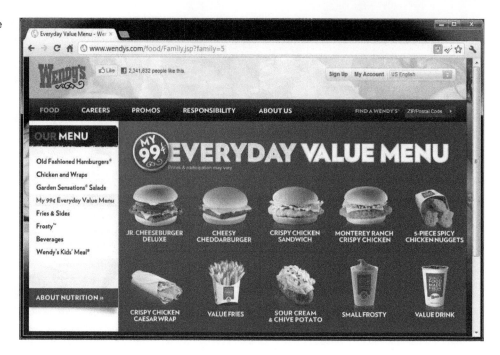

Honda is a firm that has been effective with value pricing. It has different marketing mixes for different target markets. From the $15,000 Fit to the $36,000 S2000 Roadster and $40,000 four-wheel-drive Pilot, Honda offers high quality at reasonable prices. Among fast-food restaurants, Wendy's was one of the first to create a Dollar Menu. This has helped give it a reputation for value pricing.

Companies that use value pricing deliver on their promises. They try to give the consumer pleasant surprises—like an unexpected service—because it increases value and builds customer loyalty. They often give customers their money back if they aren't completely satisfied. They avoid unrealistic price levels—prices that are high only because consumers already know the brand name. They build relationships so customers will come back time and again.

There are Price choices in most markets

Some marketing managers miss the advantages of value pricing. They've heard economists say that in perfect competition it's foolish to offer products above or below the market price. But most firms *don't* operate in perfect competition where what firms offer is exactly the same.

Most operate in monopolistic competition, where products and whole marketing mixes are *not* exactly the same. This means that there are pricing options. At one extreme, some firms are clearly above the market—they may even brag about it. Tiffany's is well known as one of the most expensive jewelry stores in the world. Other firms emphasize below-the-market prices in their marketing mixes. Prices offered by discounters and mass-merchandisers, such as Walmart and Tesco, illustrate this approach. They may even promote their pricing policy with catchy slogans such as "guaranteed lowest prices."

Value pricers define the target market and the competition

In making price decisions and using value pricing, it is important to clearly define the *relevant target market* and *competitors* when making price comparisons.

Consider Walmart prices again from this view. Walmart may have lower prices on electronics products such as flat-panel televisions, but it offers less expertise from the store's sales staff, less selection, and no help installing or setting up a new television. Walmart may appeal to budget-oriented shoppers who compare prices *and* value among different mass-merchandisers. But a specialty electronics store appeals to different customers and may not even be a direct competitor!

A producer of flat-panel televisions with this point of view may offer the specialty electronics store models that are not available to Walmart—to ensure that customers don't view price as the only difference between the two stores. Further, the specialty store needs to clearly communicate to its target market *how* it offers superior value. Walmart is certainly going to communicate that it offers low prices. If that's all customers hear, they will see no differences between retailers except for price. The specialty retailer must emphasize its expertise, selection, or the superior performance of its product line—so that target customers who value these differences know where to find them.

Differentiation earns a price premium

Even though competition can be intense, too many marketers give up too easily. They simply price their product the same as a competitor. There often is a way to differentiate their product, even if it is something that competitors dismiss as less important. Creative brands differentiate and charge a higher price to reflect the additional value customers place on the differentiation. The value customers perceive can be measured as *price premium*; that is, the percentage by which a price exceeds (or falls short of) a benchmark price.[22] The benchmark price may vary, but it usually represents some measure of the average price of all competitors in the market. To see how a brand manager calculates and uses price premium, see Marketing Analytics in Action: Price Premium.

Marketing Analytics in Action: Price Premium

A brand manager may seek to differentiate a product by adding to its perceived value. If that can be accomplished, the brand manager is able to charge a higher price and potentially earn a higher margin.

To demonstrate this example, consider the analysis conducted by Raj Parva, brand manager for Method brand liquid dish detergent in the United Kingdom. Assume the following companies are the only ones in the market:

- Fairy: the market leader; a well-known national brand
- Sunlight: another well-known national brand
- Dealer brands: emphasize low prices
- Method: national brand; known for packaging design and environmentally friendly products

The following table shows Raj's calculations of price premium. He first calculated that the average price per liter in his market was £3.59 and used that as a benchmark. The price premium is based on comparing each brand's average price to the benchmark.

Price premium = (Brand's average price-benchmark price) / (Benchmark price)

Using this figure, he calculated the relative price premium (far right column) for the brands listed above:[23]

Brand	(a) Liters Sold	(b) Average Price per Liter	(c) Annual Sales £ (=a*b)	Price Premium
Fairy	115	£4.01	£461	11.7%
Sunlight	64	£3.83	£245	6.7%
Dealer brands	71	£2.41	£171	−32.9%
Method	24	£4.42	£106	23.1%
Total	274	£3.59	£983	

1. Which brand has the highest price premium? Why do you think this is so?
2. Which brand has the lowest price premium? Why do you think this is so?
3. Raj learned that in Spain and Italy, Method's price premium was 11 percent. What might be two reasons why Method's price premium was lower in these countries?
4. What could a dishwashing liquid do to command a higher price premium?

Price and customer perceptions of quality

Prices also influence how customers view quality. Without other information, most customers associate higher prices with a higher-quality product. Marketing managers should be aware that charging lower prices than competitors might have a detrimental effect on how customers view relative product quality.

Online glasses retailer Warby Parker considered this issue when starting up. The first business plan called for selling eyeglasses that typically cost $500 for less than $50. Warby Parker figured it could be profitable at that price. However, upon reflection, Warby Parker's founders worried that the dramatic price gap might adversely affect customer perceptions of its product quality and optical-related services. Sure enough, research showed Warby Parker that prices below about $100 raised customer suspicions about the quality. So it priced its glasses at $95. It used some of the extra margin to offer better service. For example, it created a "Home Try-On Program" that allows customers to request up to five frames, try them on for five days, and then return the frames, all at no cost to the customer. Even with online rivals such as Zenni Optical and GlassesUSA.com charging half of Warby Parker's prices, Warby Parker continues to capture customers' trust and, consequently, prosper.[24]

Value pricing fits with market-oriented strategy planning

There are times when the marketing manager's hands are tied and there is little that can be done to differentiate the marketing mix. However, most marketing managers do have choices—many choices. They can vary strategy decisions with respect to all of the marketing mix variables, not just Price, to offer target customers superior value. And when a marketer's hands are really tied, it's time to look for new opportunities that offer more promise.

Value can have societal and ethical dimensions

Value is a subjective appraisal. It may differ for each person. This can be more complicated when society has to pay part of the "price." This can be the case for many projects where the government is involved. For example, is it a good investment to build a new interstate highway or build a new school? All taxpayers contribute, but only some benefit.

The issue is the same for health care—especially because Medicare and Medicaid are government programs that pay for the health care of many older U.S. citizens. With an aging population and rising health care costs, value is a growing concern in the health care debate. The breakthrough innovations in health care have costs and benefits that create challenges for a society deciding how they should be valued. Read more about this complicated value issue in *What's Next?* Putting a price (and value) on life.

Legality of Pricing Policies

LO 17.7

This chapter discusses the many pricing decisions that must be made. However, some pricing decisions are limited by government legislation. The first step to understanding pricing legislation is to know the thinking of legislators and the courts. Our focus will be on U.S. legislation, but most countries have similar laws. Still, marketing managers may find that some countries have more restrictive laws. Intel was fined $1.45 billion by European regulators for price discounting practices that were used to keep competitors out of the market.[25]

Minimum prices are sometimes controlled

Unfair trade practice acts put a lower limit on prices, especially at the wholesale and retail levels. They have been passed in more than half the states in the United States. Selling below cost in these states is illegal. Wholesalers and retailers are usually required to take a certain minimum percentage markup over their merchandise-plus-transportation costs. The practical effect of these laws is to protect certain limited-line food retailers—such as dairy stores—from the kind of "ruinous" competition supermarkets might offer if they sold milk as a leader, offering it below cost for a long time.

What's Next? Putting a price (and value) on life

What is the value of one more day of life? One more month? One more year? These are the difficult decisions that doctors, patients, and public policy makers must make. Life-extending medical treatments can be very expensive, and most of the cost falls on insurance companies and governments. As the United States gets older, health care expenses are rising. Decisions about the value of a day, month, or year may be *What's Next?* for lawmakers and voters.

A few years ago, doctors at Memorial Sloan-Kettering Cancer Center made a big decision: they decided to boycott the colorectal cancer drug Zaltrap. The required three-month treatment with Zaltrap cost over $30,000—more than twice as much as an equally effective competitive treatment. Both drugs had similar side effects and offered patients an average of 42 more days of life. Deciding to remove a potentially life-extending medicine from doctors' options because of its cost was unheard of.

Perhaps this choice shouldn't have been a big deal. In most product-markets, an equally effective product costing twice as much has no value, so it would not even be brought to market. But this is the world of medicine. When faced with life-threatening conditions, patients and doctors prefer not to think about cost when deciding on treatment.

The question of "value" gets more complicated when no equivalent treatments exist. Consider Novartis' "miracle pill," Gleevec, that treats a rare cancer—chronic myeloid leukemia. With an 80 percent survival rate 10 years from diagnosis, Gleevec has no equally effective competitors. But an annual supply cost to patients (and their insurer) is $100,000—every year!

Gilead Science's Sovaldi is another breakthrough, boasting a 95 percent rate of *curing* hepatitis C. The cure costs $84,000; a daily dose for 12 weeks breaks down to $1,000 a pill. A less expensive competing treatment has severe side effects and doesn't cure as many patients. Gilead claims that its cure *saves* the health care system money in the long run; patients cured of hepatitis C will never experience—and the health care system won't incur the cost of—the longer-term tolls of this disease, which include liver scarring, cirrhosis, cancer, and organ failure.

Drug companies invest hundreds of millions of dollars in research, development, and in bringing a new product to market. Many new-product efforts fail along the way. High prices cover those costs and ensure future breakthroughs. Critics wonder if these companies are taking advantage of their situation. They wonder if the price justifies the value to both the individual and to society—which ends up paying much of the cost.[26]

This is a What's Next? that no one wants to face. But in the end, someone has to pay for breakthrough drugs. What do you think? Are the prices of these drugs fair? Do they deliver proportional value? How should a company determine prices for breakthrough drugs? Should society—because it pays through insurance and taxes—have a say about whether such drugs are covered by insurance plans?

The United States and most other countries control the minimum price of imported products with antidumping laws. **Dumping** is pricing a product sold in a foreign market below the cost of producing it or at a price lower than in its domestic market. These laws are usually designed to protect the country's domestic producers and jobs. U.S. steelmakers have accused China of dumping and encouraged the U.S. government to take action.

Regulators can set prices

In some industries, often where there is little competition, a state government may control prices. For example, many utilities—such as electric and gas—have little or no competition to keep prices in check, so states will usually have a board that governs the prices they charge. To ensure all citizens have access to affordable auto insurance, most states control auto insurance rates as well. Still, to maintain competition, there are several important regulations in the pricing area.

You can't lie about prices

Phony list prices are prices customers are shown to suggest that the price has been discounted from list. Some customers seem more interested in the supposed discount than in the actual price. Most businesses, trade associations, and government agencies consider the use of phony list prices unethical. In the United States, the FTC tries to stop such pricing—using the **Wheeler-Lea Act**, which bans "unfair or deceptive acts in commerce."[27]

A few years ago, some electronics retailers were criticized on these grounds. They would advertise a $300 discount on a computer when the customer signed up for an Internet service provider, but they did not make clear to the consumer that a three-year commitment—costing more than $700—was required.

Price fixing is illegal—you can go to jail

Difficulties with pricing—and violations of pricing legislation—usually occur when competing marketing mixes are quite similar. When the success of an entire marketing strategy depends on price, there is pressure (and temptation) to make agreements with competitors (conspire). And **price fixing**—competitors getting together to raise, lower, or stabilize prices—is common and relatively easy. *But it is also completely illegal in the United States.* It is considered "conspiracy" under the Sherman Act and the Federal Trade Commission Act. To discourage price fixing, both companies and individual managers are held responsible. In a recent case, an executive at Archer Daniels Midland (ADM) Company was sentenced to three years in jail and the company was fined $100 million.[28]

The Informant is a movie made about the Archer Daniels Midland Company price fixing case.
Groundswell Prods/Kobal/Shutterstock

Different countries have different rules concerning price fixing, which has created problems in international trade. Japan, for example, allows price fixing, especially if it strengthens the position of Japanese producers in world markets.

Producers may set minimum retail prices

Manufacturers usually suggest a retail list price and then leave it up to retailers to decide what to charge in their local markets. In fact, until recently the courts prohibited manufacturers from imposing a minimum price at which their goods could be sold. This was viewed as a form of price fixing and a violation of the Sherman Antitrust Act. However, the ruling in a recent Supreme Court case changes that. The case involved Leegin Creative Leather Products and Kay's Kloset, a retailer that had been discounting Leegin's handbags. To prevent the discounting, Leegin stopped selling to Kay's Kloset. Kay's Kloset brought suit and said that retailers should be free to set their own prices—which, in turn, would keep prices lower for consumers. However, Leegin argued that its strategy focused on building the reputation of its brand with excellent service and advertising. Its retailers couldn't provide that level of service and promotion if they didn't charge a price that offered a sufficient profit margin. Leegin also argued that if one retailer ignored the strategy and cut its price on Leegin products, other retailers would follow suit—and soon the retailers wouldn't be able to provide the backing Leegin bags needed to compete. The court ruling, which supported Leegin, marks an important change because it gives manufacturers more power to control retail pricing.[29]

U.S. antimonopoly laws ban price discrimination unless . . .

Price-level and price-flexibility policies can lead to price discrimination. The **Robinson-Patman Act** (of 1936) makes illegal any **price discrimination**—selling the same products to different buyers at different prices—*if it injures competition.* The law does permit some price differences—but they must be based on (1) cost differences or (2) the need to meet competition. Both buyers and sellers are considered guilty if they know they're entering into discriminatory agreements.

What does "like grade and quality" mean?

Firms in businesses as varied as transportation services, book publishing, and auto parts have been charged with violations of the Robinson-Patman Act in recent, nationally publicized cases. Competitors who have been injured by a violation of the law have incentive to go to court because they can receive a settlement that is three times larger than the damage suffered.

William D. Perreault, Jr., Ph.D.

The Robinson-Patman Act allows a marketing manager to charge different prices for similar products if they are *not* of "like grade and quality." But the FTC says that if the physical characteristics of a product are similar, then they are of like grade and quality. A landmark U.S. Supreme Court ruling against the Borden Company upheld the FTC's view that a well-known label *alone* does not make a product different from one with an unknown label. The company agreed that the canned milk it sold at different prices under different labels was basically the same.

But the FTC's victory in the Borden case was not complete. The U.S. Court of Appeals found no evidence of injury to competition and further noted that there could be no injury unless Borden's price differential exceeded the "recognized consumer appeal of the Borden label." How to measure "consumer appeal" was not spelled out, so producers who want to sell several brands—or dealer brands at lower prices than their main brand—probably should offer physical differences, and differences that are really useful.[30]

Can cost analysis justify price differences?

The Robinson-Patman Act allows price differences if there are cost differences—say, for larger-quantity shipments or because intermediaries take over some of the physical distribution functions. But justifying cost differences is a difficult job. And the justification must be developed *before* different prices are set. The seller can't wait until a competitor, disgruntled customer, or the FTC brings a charge. At that point, it's too late.[31]

Can you legally meet price cuts?

Under the Robinson-Patman Act, meeting a competitor's price is permitted as a defense in price discrimination cases. A major objective of antimonopoly laws is to protect competition, not competitors. And "meeting competition in good faith" still seems to be legal.

Special promotion allowances might not be allowed

Some firms violate the Robinson-Patman Act by providing push money, advertising allowances, and other promotion aids to some customers and not others. The act prohibits such special allowances, *unless they are made available to all customers on "proportionately equal" terms.*[32]

How to avoid discriminating

Because price discrimination laws are complicated and penalties for violations heavy, many business managers follow the safest course by offering few or no quantity discounts and the same cost-based prices to *all* customers. This is *too* conservative a reaction. But when firms consider price differences, they may need a lawyer involved in the discussion!

CONCLUSION

The Price variable offers an alert marketing manager many possibilities for varying marketing mixes. This chapter began by discussing how a firm's pricing objectives may be oriented toward profit, sales, or maintaining the status quo. Clear pricing objectives help in making decisions about the firm's important pricing policies.

This chapter discussed the pros and cons of flexible pricing and some of the approaches that firms use to implement it. It also considered the initial price-level decision—skim the cream or penetration—that the marketing manager must make with new products at the introductory stage of their life cycle, and now pricing typically evolves over the product life cycle. We also discussed a variety of ways that marketing managers adjust the basic list price under different circumstances—by using different types of discounts and allowances. These policies need to be clearly defined by the marketing manager.

The chapter described how the different components of price are traded off against the other marketing mix variables to create something of value for the customer. We also discussed value pricing and how to create a competitive advantage by offering customers superior value—which isn't the same as just offering lower and lower prices.

Pricing comes under greater scrutiny from the law than some other marketing mix variables. So it is important to understand key legal constraints that influence pricing decisions.

This chapter provided a foundation for understanding the objectives and policies that guide pricing decisions. This information provides input into the price-setting process, which we describe in greater detail in the following chapter when we look at both cost- and demand-oriented approaches to pricing.

KEY TERMS

LO 17.8

Price, 474
target return objective, 475
profit maximization objective, 476
benefit corporation, 476
B Corporation (B Corp) certification, 476
sales-oriented objective, 477
status quo–objectives, 478
nonprice competition, 478
administered prices, 478
one-price policy, 479
flexible-price policy, 479
dynamic pricing, 479
skimming price policy, 482
penetration pricing policy, 483
introductory price dealing, 483

basic list prices, 484
discounts, 485
quantity discounts, 485
cumulative quantity discounts, 485
noncumulative quantity discounts, 485
seasonal discounts, 485
net, 486
cash discounts, 486
2/10, net 30, 486
trade (functional) discount, 486
allowances, 486
advertising allowances, 486
stocking allowances, 486
push money (or prize money) allowances, 486

trade-in allowance, 487
sale price, 487
everyday low pricing, 487
rebates, 488
mobile payments, 490
lease, 490
value pricing, 491
unfair trade practice acts, 494
dumping, 495
phony list prices, 495
Wheeler-Lea Act, 495
price fixing, 496
Robinson-Patman Act, 496
price discrimination, 496

QUESTIONS AND PROBLEMS

1. Review the Method case study that opens the chapter. From this case, identify examples of different key terms and concepts covered in the chapter. For example, early on the company had a sales-oriented pricing objective thanks to financial backers concerned with market share and sales volume.

2. Review the Method case study that opens this chapter. Considering what you have learned in the class thus far (other elements of the marketing mix) and, more specifically, pricing lessons in this chapter, offer strategy suggestions for (1) prices and other marketing mix decisions over the product life cycle in the laundry category, and (2) how value pricing would work for Method's laundry products—how does it deliver and communicate value?

3. Identify the strategy decisions a marketing manager must make in the Price area. Illustrate your answer for a local retailer.

4. How should the acceptance of a profit-oriented, a sales-oriented, or a status quo-oriented pricing objective affect the development of a company's marketing strategy? Illustrate for each.

5. Distinguish between one-price and flexible-price policies. Which is most appropriate for a hardware store? Why?

6. What pricing objective(s) is a skimming pricing policy most likely implementing? Is the same true for a penetration pricing policy? Which policy is probably most appropriate for each of the following products: (a) a new type of home lawn-sprinkling system, (b) a skin patch drug to help smokers quit, (c) a DVD of a best-selling movie, and (d) a new children's toy?

7. Are seasonal discounts appropriate in agricultural businesses (which are certainly seasonal)?

8. What are the effective annual interest rates for the following cash discount terms: (a) 1/10, net 20; (b) 1/5, net 10; and (c) net 25?

9. Why would a manufacturer offer a rebate instead of lowering the suggested list price?

10. Give an example of a marketing mix that has a high price level but that you see as a good value. Briefly explain what makes it a good value.

11. Think about a business from which you regularly make purchases even though there are competing firms with similar prices. Explain what the firm offers that improves value and keeps you coming back.
12. Would consumers be better off if all nations dropped their antidumping laws? Explain your thinking.
13. camelcamelcamel (camelcamelcamel.com) is a website that automatically tracks prices at Amazon.com. The site allows users to track prices, showing a product's price history for a year or more. Users can also request e-mail alerts when the price drops to a certain level. Go to Amazon.com and find a popular toy (pick one that has been out at least a year). Copy the URL for that toy and paste it into the bar at the top of the camelcamelcamel site. How much has the price varied in the last year? What was the highest price? Lowest price? Current price?
14. How would our marketing system change if manufacturers were required to set fixed prices on *all* products sold at retail and *all* retailers were required to use these prices? Would a manufacturer's marketing mix be easier to develop? What kind of an operation would retailing be in this situation? Would consumers receive more or less service?
15. Is price discrimination involved if a large oil company sells gasoline to taxicab associations for 2.5 cents less than the price charged to retail service stations? The gas sold to the cab associations will be resold to taxi drivers. What happens if the cab associations resell gasoline not only to taxicab operators but also to the general public?

MARKETING PLANNING FOR HILLSIDE VETERINARY CLINIC

Appendix D (the Appendices follow Chapter 19) includes a sample marketing plan for Hillside Veterinary Clinic. Look through the "Marketing Strategy" section.

1. A veterinary clinic located in another town gives its customers a 10 percent discount on their next vet bill if they refer a new pet owner to the clinic. Do you think that this would be a good idea for Hillside? Does it fit with Hillside's strategy?
2. The same clinic offered customers a sort of cumulative discount—an end-of-year refund if their total spending at the clinic exceeded a certain level. That clinic sees it as a way of being nice to people whose pets have had a lot of problems. Do you think that this is a good idea for Hillside? Why or why not?

SUGGESTED CASES

13. Paper Products, Inc. (PPI) 17. Wise Water, Inc.

MARKETING ANALYTICS: DATA TO KNOWLEDGE

CHAPTER 17: CASH DISCOUNTS

Joe Tulkin owns Tulkin Wholesale Co. He sells paper, tape, file folders, and other office supplies to about 120 retailers in nearby cities. His average retailer-customer spends about $900 a month. When Tulkin started the business in 1991, competing wholesalers were giving retailers invoice terms of 3/10, net 30. Tulkin never gave the issue much thought—he just used the same invoice terms when he billed customers. At that time, about half of his customers took the discount. Recently, he noticed a change in the way his customers were paying their bills. Checking his records, he found that 90 percent of the retailers were taking the cash discount. With so many retailers taking the cash discount, it seems to have become a price reduction.

In addition, Tulkin learned that other wholesalers were changing their invoice terms.

Tulkin decides he should rethink his invoice terms. He knows he could change the percentage rate on the cash discount, the number of days the discount is offered, or the number of days before the face amount is due. Changing any of these, or any combination, will change the interest rate at which a buyer is, in effect, borrowing money if he does not take the discount. Tulkin decides that it will be easier to evaluate the effect of different invoice terms if he sets up a spreadsheet to let him change the terms and quickly see the effective interest rate for each change.

Design element: #M4BW box globe icon: ©Vectoryzen/Shutterstock

CHAPTER EIGHTEEN

MikeDotta/Shutterstock

Price Setting in the Business World

Back in the 1990s, American consumers saw Samsung as a second-tier producer of commodity electronics. In the United States it was known for little more than cheap microwave ovens and color televisions. The company operated on thin margins—and even lost money during the Asian economic crisis in the late 1990s. Samsung figured it could earn higher margins by developing more product lines and clearly differentiating them. Samsung decided to reposition itself as a high-quality producer known for innovation and design.

Samsung set up design studios around the world, developed a more globally oriented corporate culture, and focused on quickly getting new technology to market. This change in strategy worked. Samsung is now the largest consumer electronics company in the world—with the leading market share in many product categories and a reputation for quality and design. The TV market is a microcosm of many changes in Samsung and consumer electronics. Let's look at how the evolution of the TV influenced Samsung's marketing strategy planning, especially pricing.

Samsung was a major player in the LCD panel market in the 1990s—producing the LCD panels for computer monitors and the first flat-panel televisions. In fact, Samsung sold LCD panels to many of its competitors in the TV market. Because it was costly to produce the larger LCD panels needed for televisions, prices were high. By 2001, the average price for flat-panel TVs was about $10,000. At this price, demand was limited. Flat screens were mostly sold by commercial audio-video suppliers to business customers.

Over the next few years prices came down and consumer interest grew. Samsung started distributing its flat-panel TVs through electronics chains such as Best Buy and Circuit City. That helped bring down selling costs, and more importantly, economies of scale in production kicked in to further lower costs and prices.

Anticipating rapid growth in the television market, suppliers of LCD panels invested billions of dollars in new factories to make panels for the ever-expanding range of TV screens. LCD suppliers (including Samsung) wanted to protect those investments, so they secretly met to set prices (and were later found guilty of price fixing). Even those efforts didn't keep prices of LCD panels from dropping as suppliers desperately tried to capture sales with pricing incentives—though they probably fell slower than they would have in a perfectly competitive market.

By 2005, the price of a 30-inch high-definition Samsung LCD television had dropped below $3,000. Still, with a similar-size tube-style television available for less than $800, the size of the consumer market for LCD televisions remained limited. The new technology received strong support from retailers, who enjoyed the 30 percent margins they earned on LCD TVs. Samsung kept pushing technology boundaries and gained a name for itself with its high-quality, ultra-thin LCD TVs. And because Samsung produced the LCD panels for its own TVs, its costs were lower than most of its competitors'. Samsung used this advantage to offer low prices and build market share.

Over the next couple of years, Samsung saw almost 100 new competitors enter the market. Although most were short-lived, Vizio found a niche. Vizio handled design and marketing itself and left production to contract manufacturers in China. As a result, its overhead costs were less than 1 percent of sales; Samsung's overhead costs were 10 to 20 percent of sales. To gain market share, Vizio settled for slim profit margins of just 2 percent, much lower than Samsung's desired returns. Vizio didn't try to offer a TV that had the most advanced video standards or technology. Rather, it just tried to have an inexpensive TV in each of the most popular sizes. This strategy gave Vizio a significant price advantage on the retail floor and helped it gain distribution in discount stores such as Walmart and Costco. Vizio didn't advertise, leaving the promotion job to retailers that found Vizio's low prices attracted customers to their stores.

Retailers pressured other TV makers, like Samsung, to cut prices too. This caused a price war, and prices of LCD flat screens dropped 40 percent in less than a year. Over time, most of the low-priced competitors left the market, and prices stabilized. Samsung and Sony then tried to reduce competition for their retailers by imposing a minimum price strategy—a price below which no retailer could sell. Although it assured retailers better margins, Samsung dropped the plan for fear it would lose market share to Panasonic and Vizio. And soon retailers complained of "selling $2,000 TVs and making $10."

The ongoing focus on price in this market threatened to turn TVs into a commodity—where brand and features didn't matter, only price. To counter that trend, Samsung continues to focus on innovation, offering a full line of TVs. For example, it has 65-inch TVs with suggested retail prices ranging from $829 to $4,999—though street prices (what most retailers actually charge) are from $699 to $4,998. Samsung knows that innovation differentiates its TVs and will move consumers to the higher-priced models in its line, where its margins—and those of its retailers—are higher. Samsung listens to customers and tries to anticipate their needs. For example, its "smart TVs" have built-in Internet connections and apps to allow for easy viewing of services such as Hulu, Vudu, and Netflix. Samsung's QLED 8K TVs command a premium price because aficionados will pay for this viewability.

Communicating the benefits of new features has been a challenge for Samsung and others in this market. Samsung may innovate, but competitors soon follow. Samsung continues to be attacked from the low end by firms like Vizio, while also getting strong competition from Panasonic, Sony, and LG at the high end. Yet Samsung keeps effectively adapting its marketing strategy, including prices, so that customers and retailers get good value and Samsung makes good profits.[1]

LEARNING OBJECTIVES

Ultimately, the price that a firm charges must cover the costs of the whole marketing mix or the firm will not make a profit. So it's important for marketing managers to understand costs and how they relate to pricing. On the other hand, it doesn't make sense to set prices based just on costs, because customers won't buy a product if they don't think that it represents a good value for their money. So setting prices is not an easy task, and marketing managers must carefully balance costs, customer price sensitivity, and other factors. This chapter will help you better understand these issues and how to evaluate them in setting prices.

When you finish this chapter, you should be able to

1. understand how most wholesalers and retailers set their prices by using markups.
2. understand the advantages and disadvantages of average-cost pricing.
3. know how to use break-even analysis to evaluate possible prices.
4. understand the advantages of marginal analysis and how to use it for price setting.
5. understand other demand-oriented factors that influence price setting.
6. explain full-line pricing and factors influencing such decisions.
7. understand important new terms (shown in **red**).

Price Setting Is a Key Strategy Decision

In Chapter 17 we discussed the idea that pricing objectives and policies should guide pricing decisions. We described different dimensions of the pricing decision and how they combine to create value for customers. This chapter builds on those concepts—all of which influence price setting—and gives you additional frameworks that will help you understand how marketing managers set prices. See Exhibit 18-1.

There are many ways to set prices. But, for simplicity, they can be reduced to two basic approaches—*cost-oriented* and *demand-oriented* price setting. We will discuss cost-oriented approaches first because they are most common. Also, understanding the problems of relying on a cost-oriented approach shows why a marketing manager must consider customer demand and price sensitivity to make good Price decisions. The chapter concludes with a discussion of product-line pricing.

Exhibit 18-1 Price Setting and Strategy

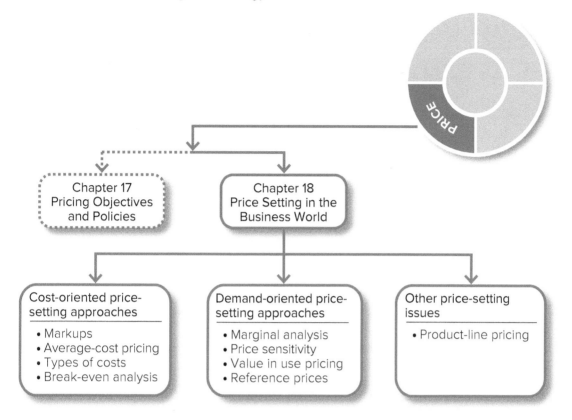

Some Firms Just Use Markups

LO 18.1

Markups guide pricing by intermediaries

Some firms, including most retailers and wholesalers, set prices by using a **markup**—a dollar amount added to the cost of products to get the selling price. For example, suppose that a CVS drugstore buys a bottle of Pantene Pro-V shampoo and conditioner for $3.00. To make a profit, the drugstore obviously must sell Pantene Pro-V for more than $3.00. If it adds $1.50 to cover operating expenses and provide a profit, we say that the store is marking up the item $1.50.

Markups, however, usually are stated as percentages rather than dollar amounts. And this is where confusion sometimes arises. Is a markup of $1.50 on a cost of $3.00 a markup of 50 percent? Or should the markup be figured as a percentage of the selling price—$4.50—and therefore be 33⅓ percent? A clear definition is necessary.

Markup percentage is based on selling price—a convenient rule

Unless otherwise stated, **markup (percent)** means the percentage of selling price that is added to the cost to get the selling price. So the $1.50 markup on the $4.50 selling price is a markup of 33⅓ percent. Markups are related to selling price for convenience.

Note that in some industries, you will find markup calculated on the basis of the purchase price in each level of the channel, but to be consistent we use the definition and calculation it implies. There's nothing wrong with the idea of calculating markup on cost. However, to avoid confusion, it's important to state clearly which markup percent you're using.

A manager may want to change a markup on selling price to one based on cost, or vice versa. The calculations used to do this are simple. (See the section on markup conversion in Appendix B on marketing arithmetic.)[2]

Many use a standard markup percent

Many intermediaries select a standard markup percent and then apply it to all their products. This makes pricing easier. When you think of the large number of items the

To determine selling prices, retail stores often add a standard markup percent to products they resell.
(left): Justin Sullivan/Getty Images; (right): UrbanImages/Alamy Stock Photo

average retailer and wholesaler carry—and the small sales volume of any one item—this approach may make sense. Spending the time to find the best price to charge on every item in stock (day to day or week to week) might not pay.

Moreover, different companies in the same line of business often use the same markup percent. There is a reason for this: Their operating expenses are usually similar. So they see a standard markup as acceptable as long as it's large enough to cover the firm's operating expenses and provide a reasonable profit.

Markup chain may be used in channel pricing

Different firms in a channel often use different markups. A **markup chain**—the sequence of markups firms use at different levels in a channel—determines the price structure in the whole channel. The markup is figured on the *selling price* at each level of the channel.

For example, Black & Decker's selling price for a cordless electric drill becomes the cost the Ace Hardware wholesaler pays. The wholesaler's selling price becomes the hardware retailer's cost. And this cost plus a retail markup becomes the retail selling price. Each markup should cover the costs of running the business and leave a profit.

Exhibit 18-2 illustrates the markup chain for a cordless electric drill at each level of the channel system. The production (factory) cost of the drill is $43.20. In this case, the

Exhibit 18-2 Example of a Markup Chain and Channel Pricing

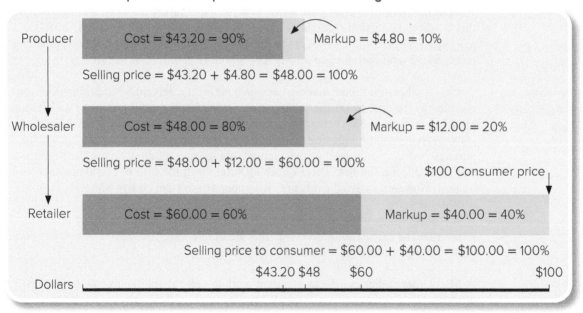

producer takes a 10 percent markup and sells the product for $48. The markup is 10 percent of $48 or $4.80. The producer's selling price now becomes the wholesaler's cost—$48. If the wholesaler is used to taking a 20 percent markup on selling price, the markup is $12—and the wholesaler's selling price becomes $60. The $60 now becomes the cost for the hardware retailer. And a retailer who is used to a 40 percent markup adds $40, and the retail selling price becomes $100.

High markups don't always mean big profits

Some people, including many conventional retailers, think high markups mean big profits. Often this isn't true. A high markup may result in a price that's too high—a price at which few customers will buy. You can't earn much if you don't sell much, no matter how high your markup on a single item. So high markups may lead to low profits.

Markups and stockturn rate important to mass-merchandisers

Some retailers and wholesalers, however, try to speed turnover to increase profit—even if this means reducing their markups. Walmart might mark up the price on a box of Cheerios by $0.50. If that size sells 10,000 boxes in a year, then Walmart earns $5,000 ($0.50 × 10,000). On the other hand, the $24.99 retail price Hamilton Beach SmartToast toaster might include a $5 markup for Walmart. If Walmart sells 200 of these toasters in a year, the retailer earns $1,000 ($5 × 200). The important thing to recognize is that markup combines with stockturn rate to determine what the product actually earns.

In Chapter 12 we noted that this is the practice of mass-merchandisers. They put lower markups on fast-selling items and higher markups on items that sell less frequently. Whether a **stockturn rate** is high or low often depends on the industry and the product involved. For example, an electrical parts wholesaler may expect an annual stockturn rate of 2, whereas a supermarket might expect 8 stockturns on average but 20 stockturns for soaps and 70 stockturns for fresh fruits and vegetables.

Legend Valve's 24-hour delivery and no minimum order policy allow wholesalers to carry less inventory and lower costs. Along with a higher stockturn rate, the wholesalers can make greater profits on Legend's product line.
Source: Legend Valve

Producers consider markups too

Some markups eventually become standard in a trade. Most channel members tend to follow a similar process—adding a certain percentage to the previous price. But who sets price in the first place? The firm that brands a product is usually the one that sets its basic list price. It may be a large retailer, a large wholesaler, or most often, the producer. Producers still face the same questions as retailers–do they charge a higher price, with a higher margin per unit, and lower number of units sold *or* do they charge a lower price, with lower margins, and a higher number of units sold?

 #M4BW

-Low margin, high turnover, better world

While not always the case, there are many situations where low margins and high turnover make the world a better place. Consider the efforts of Vanguard in driving down investing costs. Vanguard is a mutual fund company that manages more than $5 trillion in assets—much of that in retirement accounts for American workers. The founder of Vanguard, John Bogle, is credited with popularizing the index fund and low-cost investing. Vanguard did that by having very low management fees (its margin). It has been estimated that these innovations save investors more than $100 billion per year. The savings are welcomed by many retirees.[3]

Producers have to start with production cost

The challenging question for a producer is figuring out a starting point. Some producers just start with a cost per unit figure and add a markup—perhaps a standard markup—to obtain their selling price. Or they may use some rule-of-thumb formula such as:

Selling price = Average production cost per unit × 3

A producer who uses this approach might develop rules and markups related to its own costs and objectives. Yet even the first step—selecting the appropriate cost per unit to build on—isn't easy. Let's discuss several approaches to see how cost-oriented price setting really works.

Average-Cost Pricing Is Common and Can Be Dangerous

LO 18.2

Although later in this chapter we will discuss better methods for setting prices, average-cost pricing is so commonly employed that it is important for you to understand this method and its weaknesses. And when demand is constant and predictable, average-cost pricing may be appropriate.

Average-cost pricing means adding a reasonable markup to the average cost of a product. A manager usually finds the average cost per unit by studying past records. Dividing the total cost for the last year by all the units produced and sold in that period gives an estimate of the average cost per unit for the next year. If the cost was $32,000 for all labor and materials and $30,000 for fixed overhead expenses—such as selling expenses, rent, and manager salaries—then the total cost is $62,000. See Exhibit 18-3A. If the company produced 40,000 items in that time period, the average cost is $62,000 divided by 40,000 units, or $1.55 per unit. To get the price, the producer decides how much profit per unit to add to the average cost per unit. If the company considers $0.45 a reasonable profit for each unit, it sets the new price at $2.00. Exhibit 18-3A shows that this approach produces the desired profit if the company sells 40,000 units.

It does not make allowances for cost variations as output changes

It's always a useful input to pricing decisions to understand how costs operate at different levels of output. Further, average-cost pricing is simple. But it can also be dangerous. It's easy to lose money with average-cost pricing. To see why, let's follow this example further.

First, remember that the average price of $2.00 per unit was based on output of 40,000 units. But if the firm is able to produce and sell only 20,000 units in the next year, it may

Exhibit 18-3 Results of Average-Cost Pricing

A. Calculation of Planned Profit If 40,000 Items Are Sold		B. Calculation of Actual Profit If Only 20,000 Items Are Sold	
Calculation of Costs:		**Calculation of Costs:**	
Fixed overhead expenses	$30,000	Fixed overhead expenses	$30,000
Labor and materials ($0.80 a unit)	32,000	Labor and materials ($0.80 a unit)	16,000
Total costs	$62,000	Total costs	$46,000
"Planned" profit	18,000		
Total costs and planned profit	$80,000		
Calculation of Profit (or Loss):		**Calculation of Profit (or Loss):**	
Actual unit sales × Price ($2.00*)	$80,000	Actual unit sales × Price ($2.00*)	$40,000
Minus: Total costs	62,000	Minus: Total costs	46,000
Profit (loss)	$18,000	Profit (loss)	($6,000)
Result:		**Result:**	
Planned profit of $18,000 is earned if 40,000 items are sold at $2.00 each.		Planned profit of $18,000 is not earned. Instead, $6,000 loss results if 20,000 items are sold at $2.00 each.	

*Calculation of "reasonable" price: $\dfrac{\text{Expected total costs and planned profit}}{\text{Planned number of items to be sold}} = \dfrac{\$80,000}{40,000} = \$2.00$

be in trouble (see Exhibit 18-3B). Twenty thousand units sold at $2.00 each ($1.55 cost plus $0.45 for expected profit) yield a total revenue of only $40,000. The overhead is still fixed at $30,000, and the variable material and labor cost drops by half to $16,000—for a total cost of $46,000. This means a loss of $6,000, or $0.30 per unit. The method that was supposed to allow a profit of $0.45 per unit actually causes a loss of $0.30 per unit!

The basic problem with the average-cost approach is that it doesn't consider cost variations at different levels of output. In a typical situation, costs are high with low output, and then economies of scale set in—the average cost per unit drops as the quantity produced increases. This is why mass production and mass distribution often make sense. It's also why it is important to develop a better understanding of the different types of costs a marketing manager should consider when setting a price.

Software products, like Adobe's Creative Cloud, have high fixed overhead expenses relative to the variable cost for each additional user. This cost structure can make average-cost pricing especially dangerous.
Source: Adobe Systems, Inc.

Marketing Managers Must Consider Various Kinds of Costs

Average-cost pricing may lead to losses because there are a variety of costs—and each changes in a *different* way as output changes. Any pricing method that uses cost must consider these changes. To understand why, we need to define six types of cost.

There are three kinds of total cost

1. **Total fixed cost** is the sum of those costs that are fixed in total—no matter how much is produced. Among these fixed costs are rent, depreciation, managers' salaries, property taxes, and insurance. Such costs stay the same even if production stops temporarily.
2. **Total variable cost**, on the other hand, is the sum of those changing expenses that are closely related to output—expenses for parts, wages, packaging materials, outgoing freight, and sales commissions. At zero output, total variable cost is zero. As output increases, so do variable costs. If Levi's doubles its output of jeans in a year, its total cost for denim cloth also (roughly) doubles.
3. **Total cost** is the sum of total fixed and total variable costs. Changes in total cost depend on variations in total variable cost, because total fixed cost stays the same.

There are three kinds of average cost

The pricing manager usually is more interested in cost per unit than total cost, because prices are usually quoted per unit.

1. **Average cost (per unit)** is obtained by dividing total cost by the related quantity (that is, the total quantity that causes the total cost).
2. **Average fixed cost (per unit)** is obtained by dividing total fixed cost by the related quantity.
3. **Average variable cost (per unit)** is obtained by dividing total variable cost by the related quantity.

An example shows cost relations

A good way to get a feel for these different types of costs is to extend our average-cost pricing example (Exhibit 18-3A). Exhibit 18-4 shows the six types of cost and how they vary at different levels of output. The line for 40,000 units is highlighted because that was the expected level of sales in our average-cost pricing example. For simplicity, we assume that average variable cost is the same for each unit. Notice, however, that total variable cost increases when quantity increases.

Exhibit 18-5 shows the three average cost curves from Exhibit 18-4. Notice that average fixed cost goes down steadily as the quantity increases. Although the average

Exhibit 18-4   Cost Structure of a Firm

Quantity (Q)	Total Fixed Costs (TFC)	Average Fixed Costs (AFC)	Average Variable Costs (AVC)	Total Variable Costs (TVC)	Total Cost (TC)	Average Cost (AC)
0	$30,000	—	—	—	$ 30,000	—
10,000	30,000	$3.00	$0.80	$ 8,000	38,000	$3.80
20,000	30,000	1.50	0.80	16,000	46,000	2.30
30,000	30,000	1.00	0.80	24,000	54,000	1.80
40,000	**30,000**	**0.75**	**0.80**	**32,000**	**62,000**	**1.55**
50,000	30,000	0.60	0.80	40,000	70,000	1.40
60,000	30,000	0.50	0.80	48,000	78,000	1.30
70,000	30,000	0.43	0.80	56,000	86,000	1.23
80,000	30,000	0.38	0.80	64,000	94,000	1.18
90,000	30,000	0.33	0.80	72,000	102,000	1.13
100,000	30,000	0.30	0.80	80,000	110,000	1.10

$$0.30 \text{ (AFC)} = \frac{30,000 \text{ (TFC)}}{100,000 \text{ (Q)}}$$

$$\begin{matrix} 100,000 \text{ (Q)} \\ \times 0.80 \text{ (AVC)} \\ \hline 80,000 \text{ (TVC)} \end{matrix}$$

$$\begin{matrix} 30,000 \text{ (TFC)} \\ +80,000 \text{ (TVC)} \\ \hline 110,000 \text{ (TC)} \end{matrix}$$

$$1.10 \text{ (AC)} = \frac{110,000 \text{ (TC)}}{100,000 \text{ (Q)}}$$

Exhibit 18-5 Typical Shape of Cost (per unit) Curves When Average Variable Cost per Unit Is Constant

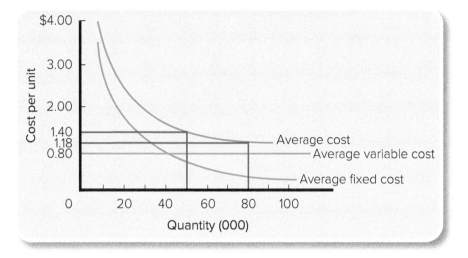

variable cost remains the same, average cost decreases continually too. This is because average fixed cost is decreasing. With these relations in mind, let's reconsider the problem with average-cost pricing.

Ignoring demand is the major weakness of average-cost pricing

Average-cost pricing works well if the firm actually sells the quantity it used to set the average-cost price. Losses may result, however, if actual sales are much lower than expected. On the other hand, if sales are much higher than expected, then profits may be very good. But this will happen only by luck—because the firm's demand is much larger than expected.

To use average-cost pricing, a marketing manager must make *some* estimate of the quantity to be sold in the coming period. Without a quantity estimate, it isn't possible to compute average cost. But unless this quantity is related to price—that is, unless the firm's demand curve is considered—the marketing manager may set a price that doesn't even cover a firm's total cost! You saw this happen in Exhibit 18-3B, when the firm's price of $2.00 resulted in demand for only 20,000 units and a loss of $6,000.

The demand curve is still important even if a manager has not taken time to think about it. For example, Exhibit 18-6 shows the demand curve for the firm we're discussing. This demand curve shows *why* the firm lost money when it tried to use average-cost

Exhibit 18-6 Evaluation of Various Prices along a Firm's Demand Curve

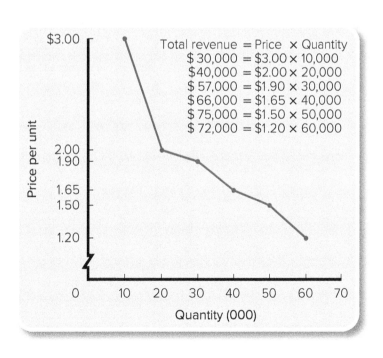

Exhibit 18-7 Summary of Relationships among Quantity, Cost, and Price Using Cost-Oriented Pricing

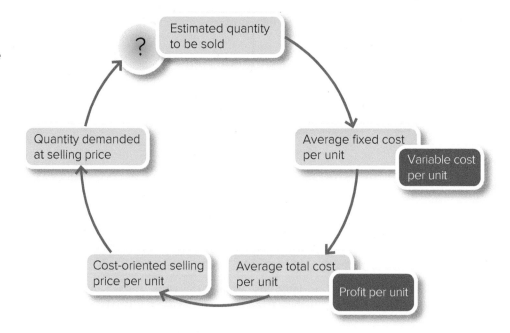

pricing. At the $2.00 price, quantity demanded is only 20,000. With this demand curve and the costs in Exhibit 18-4, the firm will incur a loss whether management sets the price at a high $3.00 or a low $1.20. At $3.00, the firm will sell only 10,000 units for a total revenue of $30,000. But total cost will be $38,000—for a loss of $8,000. At the $1.20 price, it will sell 60,000 units—at a loss of $6,000. However, the curve suggests that at a price of $1.65 consumers will demand about 40,000 units, producing a profit of about $4,000.

In short, average-cost pricing is simple in theory but often fails in practice. In stable situations, prices set by this method may yield profits but not necessarily *maximum* profits. And note that such cost-based prices may be higher than a price that would be more profitable for the firm, as shown in Exhibit 18-6. When demand conditions are changing, average-cost pricing is even more risky.

Exhibit 18-7 summarizes the relationships just discussed. Cost-oriented pricing requires an estimate of the total number of units to be sold. That estimate determines the *average* fixed cost per unit and thus the average total cost. Then the firm adds the desired profit per unit to the average total cost to get the cost-oriented selling price. How customers react to that price determines the actual quantity the firm will be able to sell. But that quantity may not be the quantity used to compute the average cost![4]

Don't ignore competitors' costs

Another danger of average-cost pricing is that it ignores competitors' costs and prices. Just as the price of a firm's own product influences demand, the price of available substitutes may impact demand. By finding ways to cut costs, a firm may be able to offer prices lower than competitors' and still make an attractive profit.

Break-Even Analysis Can Evaluate Possible Prices

LO 18.3

Some price setters use break-even analysis in their pricing. **Break-even analysis** evaluates whether the firm will be able to break even—that is, cover all its costs—with a particular price. This is important because a firm must cover all costs in the long run or there is not much point being in business. This method focuses on the **break-even point (BEP)**—the quantity where the firm's total cost will just equal its total revenue.

Exhibit 18-8 Break-Even Chart for a Particular Situation

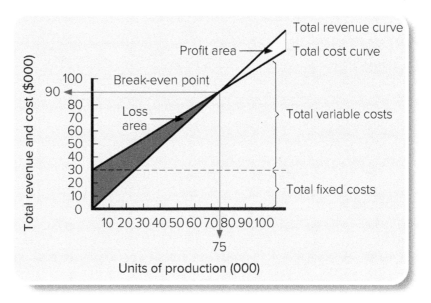

Break-even charts help find the BEP

To help understand how break-even analysis works, look at Exhibit 18-8, an example of the typical break-even chart. *The chart is based on a particular selling price*—in this case $1.20 a unit. The chart has lines that show total costs (total variable plus total fixed costs) and total revenues at different levels of production. The break-even point on the chart is at 75,000 units, where the total cost and total revenue lines intersect. At that production level, total cost and total revenue are the same—$90,000.

The difference between the total revenue and total cost at a given quantity is the profit—or loss! The chart shows that below the break-even point, total cost is higher than total revenue and the firm incurs a loss. The firm would make a profit above the break-even point. However, the firm would reach the break-even point, or get beyond it into the profit area, only *if* it could sell at least 75,000 units at the $1.20 price.

Break-even analysis can be helpful if used properly, so let's look at this approach more closely.

How to compute a break-even point

A break-even chart is an easy-to-understand visual aid, but it's also useful to be able to compute the break-even point.

The BEP, in units, can be found by dividing total fixed costs (TFC) by the **fixed-cost (FC) contribution per unit**—the assumed selling price per unit minus the variable cost per unit. This can be stated as a simple formula:

$$\text{BEP (in units)} = \frac{\text{Total fixed cost}}{\text{Fixed cost contribution per unit}}$$

This formula makes sense when we think about it. To break even, we must cover total fixed costs. Therefore, we must figure the contribution each unit will make to covering the total fixed costs (after paying for the variable costs to produce the item). When we divide this per-unit contribution into the total fixed costs that must be covered, we have the BEP (in units).

To illustrate the formula, let's use the cost and price information in Exhibit 18-8. The price per unit is $1.20. The average variable cost per unit is $0.80. So the FC contribution per unit is $0.40 ($1.20 − $0.80). The total fixed cost is $30,000 (see Exhibit 18-8). Substituting in the formula:

$$\text{BEP} = \frac{\$30,000}{\$0.40} = 75,000 \text{ units}$$

From this you can see that if this firm sells 75,000 units, it will exactly cover all its fixed and variable costs. If it sells even one more unit, it will begin to show a profit—in this case, $0.40 per unit. Note that once the fixed costs are covered, the part of revenue formerly going to cover fixed costs is now *all profit*.[5]

BEP can be stated in dollars too

The BEP can also be figured in dollars. The easiest way is to compute the BEP in units and then multiply by the assumed per-unit price. If you multiply the selling price ($1.20) by the BEP in units (75,000) you get $90,000—the BEP in dollars.

Each possible price has its own break-even point

Often it's useful to compute the break-even point for each of several possible prices and then compare the BEP for each price to likely demand at that price. The marketing manager can quickly reject some price possibilities when the expected quantity demanded at a given price is way below the break-even point for that price.

Break-even analysis is helpful—but not a pricing solution

Break-even analysis is helpful for evaluating alternatives. It is also popular because it's easy to use. Yet break-even analysis is too often misunderstood. Beyond the BEP, profits seem to be growing continually. And the graph—with its straight-line total revenue curve—makes it seem that any quantity can be sold at the assumed price. But this usually isn't true. It is the same as assuming a perfectly horizontal demand curve at that price. In fact, most managers face down-sloping demand situations. And their total revenue curves do *not* keep going straight up.

Break-even analysis is a useful tool for analyzing costs and evaluating what might happen to profits in different market environments. But it is a cost-oriented approach. Like other cost-oriented approaches, it does not consider the effect of price on the quantity that consumers will want—that is, the demand curve. And from earlier in the chapter, we know that average-cost pricing does not accurately take into account cost variations at different levels of output.

To really zero in on the most profitable price, marketers are better off estimating the demand curve by figuring out how much can be sold across a range of feasible prices. Total costs can then be estimated by combining variable and fixed costs across the same range. With better forecasts of sales and cost data, a marketing manager can more accurately estimate profits at different prices. Let's look more closely at this approach, called *marginal analysis,* next.[6]

Marginal Analysis Considers Both Costs and Demand

LO 18.4

Marginal analysis helps find the right price

The best pricing tool marketers have for looking at costs and revenue (demand) at the same time is marginal analysis. **Marginal analysis** focuses on the changes in total revenue and total cost from selling one more unit to find the most profitable price and quantity. Marginal analysis shows how costs, revenue, and profit change at different prices. The price that maximizes profit is the one that results in the greatest difference between total revenue and total cost.[7]

Demand estimates involve "if-then" thinking

Because the price determines what quantity will be sold, a manager needs an estimate of the demand curve to compute total revenue. A practical approach here is for managers to think about a price that appears to be too high and one that is too low. Then, for a number of prices between these two extremes, the manager estimates what quantity it might be possible to sell. You can think of this as a summary of the answers to a series of what-if questions—*What* quantity will be sold *if* a particular price is selected?

Profit is the difference between total revenue and total cost

The first two columns in Exhibit 18-9 give price and quantity combinations (demand) for an example firm. In this case, a manager estimates that sales will be zero at a price of $200; at a price of $175, one unit is predicted to sell; at a price of $160, two units, and so forth. Total revenue in column 3 of Exhibit 18-9 is equal to a price multiplied by its related quantity. Costs at the different quantities are also shown. The total variable costs are $60 per unit, and fixed costs are $200 no matter how many units are sold. Column 6 shows the total cost by adding total variable cost and the fixed cost at

Exhibit 18-9 Revenue, Cost, and Profit at Different Prices for a Firm

(1) Price (P)	(2) Quantity (Q)	(3) Total Revenue (TR = P × Q)	(4) Total Variable Cost (TVC)	(5) Fixed Cost (FC)	(6) Total Cost (TC = TVC + FC)	(7) Profit (Prf = TR − TC)
$200	0	$ 0	$ 0	$200	$200	−$200
175	1	175	60	200	260	−85
160	2	320	120	200	320	0
145	3	435	180	200	380	55
135	4	540	240	200	440	100
125	5	625	300	200	500	125
115	**6**	**690**	**360**	**200**	**560**	**130**
105	7	735	420	200	620	115
95	8	760	480	200	680	80
85	9	765	540	200	740	25
75	10	750	600	200	800	−50
65	11	715	660	200	860	−145
55	12	660	720	200	920	−260

each price and predicted quantity. The profit (column 7) at each quantity and price is the difference between total revenue and total cost. In this example, the best price is $115 (and a quantity of six units are expected to be sold) because that combination results in the highest profit ($130).

Profit maximization with total revenue and total cost curves

Exhibit 18-10 graphs the total revenue, total cost, and total profit relationships for the numbers we've been working with in Exhibit 18-9. The highest point on the total profit curve is at a quantity of six units. This is also the quantity where we find the greatest vertical distance between the total revenue curve and the total cost curve. Exhibit 18-9 shows that it is the $115 price that results in selling six units, so $115 is

Exhibit 18-10 Graphic Determination of the Price Giving the Greatest Total Profit for a Firm

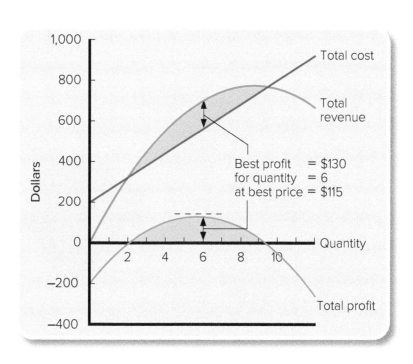

the price that leads to the highest profit. A price lower than $115 results in a higher sales volume, but you can see that the total profit curve declines beyond a quantity of six units. So a profit-maximizing marketing manager would not be interested in setting a lower price.

A profit range is reassuring

Marginal analysis focuses on the price that earns the highest profit. But a slight miss doesn't mean failure, because demand estimates don't have to be exact. There is usually a range of profitable prices. You can see this in Exhibit 18-9 and in the section of Exhibit 18-10 shown in yellow. Although the price that would result in the highest profit is $115, the firm's strategy would be profitable all the way from a price of $85 to $145. So the effort of trying to estimate demand will probably lead to being someplace in the profitable range. In contrast, the mechanical use of average-cost pricing could lead to a price that is much too high—or much too low.

In practice, marginal analysis narrows down options

Some managers don't take advantage of marginal analysis because they think they can't determine the exact shape of the demand curve. But that view misses the point of marginal analysis. Marginal analysis encourages managers to think carefully about what they *do know* about costs and demand. Only rarely is either type of information exact. Yet the practice of marginal analysis helps to narrow down a range of prices for a manager to consider. To see how marginal analysis can operate in practice, see Marketing Analytics in Action: Marginal Analysis.

Marginal analysis identifies a range of potential prices

In practice, the focus of marginal analysis is not on finding the precise price that maximizes profits. It is dependent on sales estimates at different prices. Because these are estimates, they probably get us close to a *range of potential prices*. Further, several practical demand-oriented approaches can help a marketing manager do a better job of understanding the likely shape of the demand curve for a target market. We'll discuss these approaches next.

Additional Demand-Oriented Approaches for Setting Prices

LO 18.5

What makes customers price sensitive?

A marketing manager should know what factors influence target customers' price sensitivity. **Price sensitivity** refers to the degree to which customers' purchase decisions are affected by the price. A manager who understands what influences price sensitivity can better estimate the demand curve the firm faces. Marketing researchers have identified a number of factors that influence price sensitivity across many different market situations. These are shown in Exhibit 18-11 and are discussed next.[8]

Exhibit 18-11 Factors Influencing Customers' Price Sensitivity

Marketing Analytics in Action: Marginal Analysis

Young college student and entrepreneur Stan Slater was trying to determine a price for his gourmet popcorn. After friends and family raved about his special technique and seasoning, Stan decided to try selling Stan's Gourmet Popcorn through convenience stores. Initial meetings with convenience store managers found great interest in his product, though they had questions about the price of the product. Recalling coverage of marginal analysis in his marketing class, Stan wanted to use this technique to find an optimal price.

Stan's first step involved estimating a price that would be too high and one that would be too low. Checking out the competition and his own costs, he figured that $4.00 per bag was probably too high and $1.75 too low. Then he went to several convenience stores to ask the managers their estimates of how many bags would sell at a range of prices from $1.75 to $4.00. Stan added up those estimates and put the information in a spreadsheet (see the first two columns in the spreadsheet below).

Stan then collected estimates for fixed and variable costs to produce, promote, and distribute his popcorn. He estimated his fixed costs at $5,000 and variable costs at $1.00 per bag. From this information, Stan knew he would be able to estimate revenue, costs, and, most importantly, profits at each price level. See the table below where Stan has begun these calculations.

(1) Price (P)	(2) Quantity (Q)	(3) Total Revenue (TR = P × Q)	(4) Total Variable Cost (TVC)	(5) Fixed Cost (FC)	(6) Total Cost (TC = TVC + FC)	(7) Profit (Prf = TR − TC)
$4.00	0	$ 0	$ 0	$5,000	$ 5,000	($5,000)
$3.50	5,000	$17,500	$ 5,000	$5,000	$10,000	$7,500
$3.00	7,500	$22,500	$ 7,500	$5,000	$12,500	$10,000
$2.75	11,000	$30,250	$11,000			
$2.50	15,000					
$2.25	17,500					
$2.00	20,000	$40,000				
$1.75	22,000	$38,500	$22,000			

Fill in the missing cells in the table, then answer the following questions:

1. Which price offers Stan the highest estimated profits?
2. What price range should Stan consider for his product?
3. If you were Stan, what price would you pick? Explain why.

Read on to get ideas about how to further narrow down to a price.

There are many substitutes

The first is the most basic. When customers have *substitute ways* of meeting a need, they are likely to be more price sensitive. When customers have many alternatives, then price matters.

The new CEO of Parker Hannifin (PH), a large industrial parts maker, realized that some of the products PH sold faced many competitors offering substitute parts, whereas others had few or none. Before he arrived, PH managers usually set the price for a part by adding a 35 percent markup. Managers liked this approach; it was easy and also left some room for haggling. The new CEO saw that they could improve on what they were doing. First, he asked them to classify every product—and there were thousands—into categories. At one extreme, a category included PH parts where several competitors offered a similar part. At the other extreme, there was a category of PH parts that were important to customers and were available only from PH. For example, this category included a high-pressure valve that was used on airplane doors; no other supplier had a valve that worked nearly as well for this critical application. In the last category, customers were not price sensitive and prices were increased. On parts with many alternatives (the first category) customers were price sensitive, so prices were lowered and sales increased. Fine-tuning PH prices to consider alternatives was straightforward and increased profits by 25 percent.[9]

Customers can easily compare prices

The impact of substitutes on price sensitivity is greatest when it is easy for customers to *compare prices*. For example, in grocery stores unit prices on the shelves allow customers to see how much a product costs per ounce, pound, or gallon. Many people believe that the ease of comparing prices on the Internet increases price sensitivity and brings down prices. New smartphone technology makes comparisons possible while shopping in brick-and-mortar retail stores. Several smartphone software apps allow consumers to scan a product's bar code and then check prices for the product at other online stores. If nothing else, it may make sellers more aware of competing prices.

The buyer also pays the bill

People tend to be *less* price sensitive when *someone else* pays the bill or shares the cost. Perhaps this is just human nature. Insurance companies think that consumers would reject high medical fees if they were paying all of their own bills. And executives might plan longer in advance to get better discounts on airline flights if their companies weren't footing the bills.

When expenditures are large

Customers tend to be more price sensitive the greater the *total expenditure.* Sometimes a big expenditure can be broken into smaller pieces. Mercedes knows this. When its ads focused on the cost of a monthly lease rather than the total price of the car, more consumers got interested in biting the bullet.

Outcome is less significant

Customers are less price sensitive the greater the *significance of the end benefit* of the purchase. Computer makers will pay more to get Intel processors if they believe that having "Intel inside" sells more machines. Positioning efforts often focus on emotional benefits of a purchase to increase the significance of a benefit. Ads for L'Oréal hair color, for example, show close-ups of beautiful hair while popular celebrities tell women to buy it "because you're worth it." A consumer who cares about the price of a bottle of hair color might still have no question that she's worth the difference in price.

Price sensitivity is higher when it's easier to switch brands

Customers are sometimes less price sensitive if there are *switching costs*—costs that a customer faces when buying a product that is different from what has been purchased or used in the past. For example, if a hair salon raises its prices, many

Most marketing managers want to avoid price competition. To help reduce customers' price sensitivity, this Nescafé ad points out that its coffee is unique and there is "No substitute." Comparison websites like Google Shopping make it easy for customers to compare products and prices, which makes customers more price sensitive.
(left): The Advertising Archives/Alamy Stock Photo; (right): Source: Google Shopping

customers will not switch to a new salon. The hassles of looking for a new stylist—and the risk of ending up with a bad hair day—usually make it easier to simply pay the higher price. The same thing can happen in a business context, where switching costs can be quite high. If Adobe raises the price of its Photoshop software, most managers are reluctant to add the costs of finding an alternative and training employees on a new product. Firms entering new markets like this need to find creative ways to lower switching costs. Incentives such as free trials or introductory price discounts may be used.

These factors apply in many different purchase situations, so it makes sense for a marketing manager to consider each of them in refining estimates of how customers might respond at different prices.[10]

Price sensitivity and ethics

At what point does a marketing manager's consideration of price sensitivity cross an ethical line? Sometimes customers have low price sensitivity; perhaps they must have a product and few or no alternatives exist. Is it ethical for a firm to raise prices? How high can they be raised?

Consider how insensitive many customers are to some pharmaceutical drugs. Drugs that help us live a better life or even save lives (significant outcome) can have few alternatives. The bill is often paid by someone else—an insurance company. Some claimed that Mylan, the distributor of EpiPen, took advantage of this when it raised the price of the device fivefold over seven years. EpiPen, which counteracts allergic reactions for millions of Americans, has few substitutes. The product can be lifesaving in some situations, and most Americans are covered by health insurance. When Mylan pushed the price too far—$600 for a package of two—consumers, insurance companies, and eventually the United States Congress took notice. After some bad publicity, Mylan made efforts to lower prices to end users.[11]

> **Ethical Dilemma**
>
> *What would you do?* You are a pricing specialist for a large grocery store chain that has always charged the same prices in all of its stores. However, average operating costs are higher for its inner-city stores. In addition, having the store nearby is very important to low-income, inner-city consumers who have to rely on public transportation. It's hard for them to shop around, and so they are less price sensitive. Research indicates that these stores can charge prices that are 5 percent higher on average with little effect on sales volume. This would significantly increase profitability.
>
> *Do you think the chain should charge higher prices at its inner-city stores? Why or why not? If prices were increased and antipoverty activists got TV coverage by picketing the chain, how would you respond to a TV reporter covering the story?*

#M4BW

Drive down costs when outcomes matter

Some health care providers can do the right thing—and make more profits. Consider the efforts made by Becton, Dickinson and Company (BD), which manufactures and sells various medical equipment products. Vaccines are effective only if they are properly administered—and for many vaccines that means having a high-quality syringe. In developing countries, the cost of syringes led many health care workers to reuse needles in an effort to vaccinate the largest number of patients. Unfortunately, the practice spreads infectious diseases like hepatitis C and measles—not a good trade-off. To make sure that didn't happen, BD developed injection devices that lock up after a single use and cannot be used on more than one patient. BD drove down the cost of the product with economies of scale—and it also accepts a lower margin. Now it sells these syringes for about five cents each, a price that is not much different from the cost of regular syringes. The single-use format helps BD sell more units at a lower margin, but it is still profitable for BD. More importantly, it ensures better global health practices.[12]

Value in use pricing considers what a customer will save by buying a product. Both ClimateMaster and HYTORC tout the savings realized by switching to their products. *(left): Source: ClimateMaster, Inc.; (right): Source: Hytorc Div Unex Corporation*

Value in use pricing—how much will the customer save?

Organizational buyers think about how a purchase will affect their total costs. Many marketers who aim at business markets keep this in mind when estimating demand and setting prices. They use **value in use pricing**—which means setting prices that will capture some of what customers will save by substituting the firm's product for the one currently being used.

For example, a producer of robots that help assemble cars knows that the robot doesn't just replace a machine. It also reduces labor costs, quality control costs, and—after the car is sold—costs of warranty repairs. The marketer can estimate what each automaker will save by using the machine—and then set a price that makes it less expensive for the automaker to buy the computerized machine than to stick with the old methods. The number of customers who have different levels of potential savings also provides some idea about the shape of the demand curve.

Creating a superior product that could save customers money doesn't guarantee that customers will be willing to pay a higher price. To capture the value created, the seller must convince buyers of the savings—and buyers are likely to be skeptical. A salesperson needs to be able to show proof of the claims.[13]

Auctions reveal what a customer will pay

Auctions have always been a way to determine exactly what some group of potential customers would pay, or not pay, for a product. However, the use of online auctions has dramatically broadened the use of this approach for both consumer and business products. Millions of auctions are on eBay each day. And some firms are setting up their own auctions, especially for products in short supply. The U.S. government is using online auctions as well. For example, the Federal Communications Commission (FCC) auctions rights to use airwaves for cell phones and other wireless devices. Count on more growth in online auctions.[14]

Customers may have reference prices

Some people don't devote much thought to what they pay for the products they buy, including some frequently purchased goods and services. But consumers often have a **reference price**—the price they expect to pay—for many of the products they purchase. And different customers may have different reference prices for the same basic type of purchase. For example, a person who really enjoys reading might have a higher reference price for a popular paperback book than another person who is only an occasional reader.[15]

If a firm's price is lower than a customer's reference price, customers may view the product as a better value and demand may increase. Sometimes a firm will try to position

What's Next? Making money by giving away your product

As competition heats up in many markets, low prices are one way to build market share. *What's Next?* Giving away your product for free? As crazy as that sounds, more firms are using free as a price—and they do it to make money. This not-so-new approach to pricing is gaining popularity. It works best when a product has very low variable costs. For example, many digital products (i.e., software, music, or videos) have near-zero variable costs. It costs almost nothing for Facebook to add another user or for one more customer to watch a video on YouTube.

So how do firms give away something and stay in business? One approach counts on advertisers to provide the revenue. "Free" attracts customer attention, and advertisers pay for access to eyeballs. For example, Japanese photocopy shop Tadacopy offers college students free photocopies—with advertising on the back of each page. Because websites for CNN, Facebook, Google, and ESPN are free, each draws a larger audience than pay-per-view websites. These larger audiences attract advertisers who pay fees to place ads. Facebook and Google provide even more value, offering advertisers detailed information about customers.

Some companies offer a basic product for free while charging more for customers interested in a premium—a strategy called "freemium" (combining the words *free* and *premium*). For example, Dropbox offers customers 2 gigabytes of free storage; customers can pay for more if needed. The revenue from paying customers allows Dropbox to cover the freeloaders. Many smartphone apps are free but include in-app purchases. The "free" app serves as a risk-free customer trial. TurboTax offers free online tax preparation software and uses the data generated from customers' tax returns to target them for premium tax advice.

A "free" strategy must be managed carefully. Once customers get something for free, they can be reluctant to pay for it later. The *New York Times* and other online news sites that once offered all content for free have struggled to get customers to pay a subscription fee. Before a firm offers something for nothing, it's important to think through what the firm gets in exchange—and how that achieves the firm's pricing objectives. "Free" usually works best when customers need to experience the value of the firm's offering to recognize the benefits.[16]

Think about the examples shown here and other "free" products you know about. What is it about these products that allows companies to make money while giving away their products? For what types of products would this business model not be likely to work?

---

the benefits of its product in such a way that consumers compare it with a product that has a higher reference price. Public Broadcasting System TV stations do this when they ask viewers to make donations that match what they pay for "just one month of cable service." Insurance companies frame the price of premiums for homeowners' coverage in terms of the price to repair flood damage—and advertising makes the damage very vivid.

Leader pricing—make it low to attract customers

Leader pricing means setting some very low prices—real bargains—to get customers into retail stores. The idea is not only to sell large quantities of the leader items but also to get customers into the store to buy other products. Certain products are chosen for their promotion value and priced low but above cost. In food stores, the leader prices are the "specials" that are advertised regularly to give an image of low prices. Leader pricing is normally used with products for which consumers do have a specific reference price.

Leader pricing can backfire if customers buy only the low-priced leaders. To avoid hurting profits, managers often select leader items that aren't directly competitive with major lines—as when bargain-priced blank CDs are a leader for an electronics store.[17]

When is "free" a price?

Maybe the ultimate price from a consumer standpoint is "free"—with the expectation that at least some customers will pay later. Some companies use a strategy called **freemium** (a combination of free and premium), which refers to providing a product for no charge, while money is charged for additional features that enhance the product's use. This pricing strategy is gaining popularity; learn more in *What's Next?* Making money by giving away your product.

Bait pricing—offer a steal, but sell under protest

Bait pricing is setting some very low prices to attract customers but trying to sell more expensive models or brands once the customer is in the store. For example, a furniture store may advertise a color TV for $199. But once bargain hunters come to the store, salespeople point out the disadvantages of the low-priced TV and try to convince them to trade up to a better, more expensive set. Bait pricing is something like leader pricing. But here the seller *doesn't* plan to sell many at the low price.

If bait pricing is successful, the demand for higher-quality products expands. This approach may be a sensible part of a strategy to trade up customers. And customers may be well served if—once in the store—they find a higher-priced product offers features better suited to their needs. But bait pricing is also criticized as unethical.

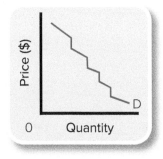

The online video game *Fortnite Battle Royale* uses the freemium pricing approach. The game is free to download and play, but players have the option to pay for items like pickaxes, dance moves, and outfits. About two-thirds of players up-spend—with those who spend averaging $85.
Source: Epic Games, Inc.

Is bait pricing ethical?

Extremely aggressive and sometimes dishonest bait-pricing advertising has given this method a bad reputation. Some stores make it very difficult to buy the bait item. The Federal Trade Commission considers this type of bait pricing a deceptive act and has banned its use in interstate commerce. Even well-known chains like Sears have been criticized for bait pricing.

Psychological pricing—some prices just seem right

Psychological pricing means setting prices that have special appeal to target customers. Some people think there are whole ranges of prices that potential customers see as the same. So price cuts in these ranges do not increase the quantity sold. But just below this range, customers may buy more. Then, at even lower prices, the quantity demanded stays the same again—and so on. Exhibit 18-12 shows the kind of demand curve that leads to psychological pricing. Vertical drops mark the price ranges that customers see as the same. Pricing research shows that there *are* such demand curves.[18]

Odd-even pricing is setting prices that end in certain numbers. For example, products selling below $50 often end in the number 5 or the number 9—such as 49 cents or $24.95. Prices for higher-priced products are often $1 or $2 below the next even dollar figure—such as $99 rather than $100.

Exhibit 18-12 Demand Curve When Psychological Pricing Is Appropriate

Some marketers use odd-even pricing because they think consumers react better to these prices—perhaps seeing them as "substantially" lower than the next highest even price. Marketers using these prices seem to assume that they have a rather jagged demand curve—that slightly higher prices will substantially reduce the quantity demanded.[19]

Subscriptions spread price over time

With **subscription pricing** customers pay on a periodic basis for access to a product. Subscription pricing has long been used by products customers continuously receive on a predictable time frame—for example newspapers, magazines, insurance coverage, and cell phone and pay television service. More recently, other products have used subscription pricing because it gives a seller a predictable stream of revenue and enhances customer lifetime value. With the price broken into smaller pieces, customers may also perceive the price as lower. Now products that previously had a one-time price have turned to subscription models, including software (Adobe, HubSpot, Microsoft); retail (Harry's Razors, Amazon, Blue Apron); and automobiles. GM offers a subscription pricing plan where customers pay $1,500 per month to trade in and out of as many as 18 different Cadillac models.[20]

Price lining—a few prices cover the field

Price lining is setting a few price levels for a product line and then marking all items at these prices. This approach assumes that customers have a certain reference price in mind that they expect to pay for a product. For example, many neckties are priced between $20 and $50. In price lining, there are only a few prices within this range. Ties will not be priced at $20, $21, $22, $23, and so on. They might be priced at four levels—$20, $30, $40, and $50.

Price lining has advantages other than just matching prices to what consumers expect to pay. The main advantage is simplicity—for both salespeople and customers. It is less confusing than having many prices. Some customers may consider items in only one price class. Their big decision, then, is which item(s) to choose at that price.

For retailers, price lining has several advantages. Sales may increase because (1) they can offer a bigger variety in each price class and (2) it's easier to get customers to make decisions within one price class. Stock planning is simpler because demand is larger at the relatively few prices. Price lining can also reduce costs because inventory needs are lower.

Demand-backward pricing and prestige pricing

Demand-backward pricing is setting an acceptable final consumer price and working backward to what a producer can charge. It is commonly used by producers of consumer products, especially shopping products such as women's clothing and

Dunkin' wanted to set a simple, acceptable final consumer price, so it used demand-backward pricing for its breakfast sandwiches. Luxury products, like Cartier watches, often use prestige pricing, where they set a high price to suggest high quality and high status.
(left): Source: DD IP Holder LLC; (right): Source: Cartier

Exhibit 18-13 Demand Curve Showing a Prestige Pricing Situation

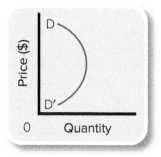

appliances. It is also used with gift items for which customers will spend a specific amount—because they are seeking a $10 or a $15 gift. Many of Mexico's low-income consumers carry only 5 or 10 peso coins, so Ace laundry detergent developed a new version of its product that adjusted features and manufacturing costs to meet a 10-peso price point.[21]

The producer starts with the retail (reference) price for a particular item and then works backward—subtracting the typical margins that channel members expect. This gives the approximate price the producer can charge. Then the average or planned marketing expenses can be subtracted from this price to find how much can be spent producing the item. Candy companies do this. They alter the size of the candy bar to keep the bar at the expected price.

Prestige pricing is setting a rather high price to suggest high quality or high status. Some target customers want the best, so they will buy at a high price. But if the price seems cheap, they worry about quality and don't buy. Prestige pricing is most common for luxury products such as furs, jewelry, and perfume.

It is also common in service industries, where the customer can't see the product in advance and relies on price to judge its quality. Target customers who respond to prestige pricing give the marketing manager an unusual demand curve. Instead of a normal down-sloping curve, the curve goes down for a while and then bends back to the left again (see Exhibit 18-13).[22]

Pricing a Full Line

Our emphasis has been, and will continue to be, on the problem of pricing an individual product mainly because this makes our discussion clearer. But most marketing managers are responsible for more than one product. In fact, their "product" may be the whole company line! So we'll discuss this matter briefly.

Full-line pricing—market- or firm-oriented

Full-line pricing is setting prices for a whole line of products. How to do this depends on which of two basic situations a firm is facing.

In one case, all products in the company's line are aimed at the same general target market, which makes it important for all prices and values to be logically related. This is a common approach with shopping products. A producer of refrigerators might offer several models with different features at different prices to give its target customers some choice. The difference among the prices and benefits should appear reasonable when target customers are evaluating them. Customer perceptions can be important here. A low-priced item, even one that is a good value at that price, may drag down the image of the higher end of the line. Alternatively, one item that consumers do not see as a good value may spill over to how they judge other products in the line. A marketing manager sometimes adds a higher-priced item to an existing product line to influence customer reference prices. The highest-priced product might not get many sales, but it makes the second-highest product in the line appear less expensive by comparison.

In other cases, the different products in the line are aimed at entirely different target markets so there doesn't have to be any relation between the various prices. A chemical producer of a wide variety of products with several target markets, for example, probably should price each product separately.

Costs are complicated in full-line pricing

The marketing manager must try to recover all costs on the whole line—perhaps by pricing quite low on more competitive items and much higher on ones with unique benefits. However, estimating costs for each product is a challenge because there is no single right way to assign a company's fixed costs to each of the products. Regardless of how costs are allocated, any cost-oriented pricing method that doesn't consider demand can lead to very unrealistic prices. To avoid mistakes, the marketing manager should judge demand for the whole line as well as demand for each individual product in each target market.

Complementary product pricing

Complementary product pricing is setting prices on several products as a group. This may lead to one product being priced very low in order to increase profits from another product, thus increasing the product group's total profits. When Gillette introduced the Fusion5 ProShield men's razor, a handle and one blade refill came in at a retail list price of $15.99. However, the blade refill cartridges, which must be replaced frequently, come in at a hefty list price of $32.51 for a package of eight.

The same pricing approach has been used by makers of computer printers, which charge low prices—even taking a loss—on the printer but high prices for ink cartridges. This pricing approach has become increasingly difficult to sustain. Computer printer makers have been undercut by cheap off-brand ink replacement cartridges. Epson's new computer printer models feature refillable ink tanks—saving lots on ink—but the printers cost hundreds of dollars more. Gillette has been forced to lower prices on its razor blades after losing market share to Dollar Shave Club and Harry's, which use a lower-price subscription model.[23]

Product-bundle pricing—one price for several products

A firm that offers its target market several different products may use **product-bundle pricing**—setting one price for a set of products. Firms that use product-bundle pricing usually set the overall price so that it's cheaper for the customer to buy the products at the same time than separately. A bank may offer a product-bundle price for a safe-deposit box, traveler's checks, and a savings account. Bundling encourages customers to spend more and buy products that they might not otherwise buy; because the added cost of the extras is not as high as it normally would be, the value is better.[24]

Most firms that use product-bundle pricing also set individual prices for the unbundled products. This may increase demand by attracting customers who want one item in a product assortment but don't want the extras. Many firms treat services this way. A software company may have a product-bundle price for its software and access to a toll-free telephone assistance service. However, customers who don't need help can pay a lower price and get just the software. Spirit Airlines successfully differentiates its offering from other airlines by *un*bundling the services most airlines offer. Spirit sells deeply discounted airfares and then charges customers for boarding passes, carry-on bags, water, and peanuts—all extras fliers usually receive for "free" on other airlines.[25]

Many fast-food restaurants, like Checkers, use product-bundle pricing by offering one price when customers buy a small fry, small drink, chicken bites, and sandwich.
Source: Checkers Drive-In Restaurants, Inc.

CONCLUSION

In this chapter, we discussed various approaches to price setting. Generally, retailers and wholesalers use markups. Some just use the same markups for all their items because it is simpler, but this is usually not the best approach. It's more effective to consider customer demand, competition, and how markups relate to turnover and profit.

It's important for marketing managers to understand costs; if customers are not willing to pay a price that is at least high enough to cover all of the costs of the marketing mix, the firm won't be profitable. So we describe the different types of cost that a marketing manager needs to understand and how average-cost pricing is used to set prices. But we note that this approach can fail because it ignores demand. We look at break-even analysis, which is a variation of the cost-oriented approach. It is useful for analyzing possible prices. However, managers must estimate demand to evaluate the chance of reaching these possible break-even points.

The major difficulty with demand-oriented pricing involves estimating the demand curve. But experienced managers, perhaps aided by marketing research, can estimate the nature of demand for their products. Even if estimates are not exact, they can help get prices in the right ballpark—and there's usually a profitable range around the most profitable price. So marketers should consider demand when setting prices. We see this with value in use pricing, psychological pricing, odd-even pricing, and full-line pricing. Understanding the factors that influence customer price sensitivity can make these approaches more effective.

Although we do not recommend that cost-oriented approaches be used by themselves, they do help the marketing manager understand the firm's profitability. Pricing decisions should consider the cost of offering the whole marketing mix. But smart marketers do not accept cost as a given—target marketers always look for ways to be more efficient—to reduce costs while improving the value that they offer customers.

KEY TERMS

LO 18.7

markup, 503
markup (percent), 503
markup chain, 504
stockturn rate, 505
average-cost pricing, 506
total fixed cost, 508
total variable cost, 508
total cost, 508
average cost (per unit), 508
average fixed cost (per unit), 508

average variable cost (per unit), 508
break-even analysis, 510
break-even point (BEP), 510
fixed-cost (FC) contribution per unit, 511
marginal analysis, 512
price sensitivity, 514
value in use pricing, 518
reference price, 518
leader pricing, 519
freemium, 519

bait pricing, 520
psychological pricing, 520
odd-even pricing, 520
subscription pricing, 520
price lining, 521
demand-backward pricing, 521
prestige pricing, 522
full-line pricing, 522
complementary product pricing, 523
product-bundle pricing, 523

QUESTIONS AND PROBLEMS

1. Review the Samsung case study that opens the chapter. From this case, identify examples of different key terms and concepts covered in the chapter. For example, "selling $2,000 and making $10" refers to markup.

2. Review the Samsung case study that opens this chapter. Explain how Samsung could make changes to its marketing strategy to address (a) price sensitivity, (b) value in use pricing, and (c) reference prices. How does it use product-line pricing? Go to the section of Samsung's website where its TVs are shown to better understand its television product line.

3. Why do many department stores seek a markup of about 30 percent when some discount houses operate on a 20 percent markup?

4. A producer distributed its riding lawn mowers through wholesalers and retailers. The retail selling price was $800, and the manufacturing cost to the company was $312. The retail markup was 35 percent and the wholesale markup 20 percent. What was the cost to the wholesaler? To the retailer? What percentage markup did the producer take?

5. Relate the concept of stock turnover to the growth of mass-merchandising. Use a simple example in your answer.

6. If total fixed costs are $200,000 and total variable costs are $100,000 at the output of 20,000 units, what are the probable total fixed costs and total variable costs at an output of 10,000 units? What are the average fixed costs, average variable costs, and average costs at these two output levels? Explain what additional information you would want in order to determine what price should be charged.

7. Construct an example showing that mechanical use of a very large or a very small markup might still lead to unprofitable operation whereas some intermediate price would be profitable. Draw a graph and show the break-even point(s).

8. The Davis Company's fixed costs for the year are estimated at $200,000. Its product sells for $250. The variable cost per unit is $200. Sales for the coming year are expected to reach $1,250,000. What is the break-even point? Expected profit? If sales are forecast at only $875,000, should the Davis Company shut down operations? Why or why not?

9. Discuss the idea of drawing separate demand curves for different market segments. It seems logical because each target market should have its own marketing mix. But won't this lead to many demand curves and possible prices? And what will this mean with respect to discounts and varying prices in the marketplace? Will it be legal? Will it be practical?

10. Distinguish between leader pricing and bait pricing. What do they have in common? How can their use affect a marketing mix?

11. Cite a local example of psychological pricing and evaluate whether it makes sense.

12. Cite a local example of odd-even pricing and evaluate whether it makes sense.

13. How does a prestige pricing policy fit into a marketing mix? Would exclusive distribution be necessary?

14. Is a full-line pricing policy available only to producers? Cite local examples of full-line pricing. Why is full-line pricing important?

MARKETING PLANNING FOR HILLSIDE VETERINARY CLINIC

Appendix D (the Appendices follow Chapter 19) includes a sample marketing plan for Hillside Veterinary Clinic. Look through the "Marketing Strategy" section.

1. A veterinary clinic must have some system for dealing with emergencies that occur on weekends and at night when the clinic is closed. Individual vets usually rotate so that someone is always on call to handle emergencies. The price for emergency care is usually 50 percent higher than the price for care during normal hours. Do you think that Hillside should charge higher prices for emergency care? Does it fit with Hillside's strategy?

2. Some customers have expensive pedigree dogs and cats and are less price sensitive than others about fees for veterinary care. Do you think that it would be possible for Hillside to charge higher prices in caring for expensive pets? Why or why not?

SUGGESTED CASES

17. Wise Water, Inc.
23. Wire Solutions
30. Paglozzi's Pizza Pies
32. Mallory's Lemonade Stand (A)
33. Mallory's Lemonade Stand (B)

MARKETING ANALYTICS: DATA TO KNOWLEDGE

CHAPTER 18: BREAK-EVEN/PROFIT ANALYSIS

This problem lets you see the dynamics of break-even analysis. The starting values (costs, revenues, etc.) for this problem are from the break-even analysis example in this chapter (see Exhibit 18-8).

The first column computes a break-even point. You can change costs and prices to figure new break-even points (in units and dollars). The second column goes further. There you can specify target profit level, and the unit and dollar sales needed to achieve your target profit level will be computed. You can also estimate possible sales quantities, and the program will compute costs, sales, and profits. Use a spreadsheet to address these issues.

Design element: #M4BW box globe icon: ©Vectoryzen/Shutterstock

CHAPTER NINETEEN

(left): maggiegowan.co.uk/Alamy Stock Photo; (right): quiggyt4/Shutterstock

Ethical Marketing in a Consumer-Oriented World: Appraisal and Challenges

Although consumers want marketplace choice, we're not there yet globally. Consider people living in a rural village in Mozambique. While inexpensive mobile phone service brings them access to the Internet, access to running water and electricity remains intermittent for many. The only affordable choices for food may be what is grown nearby—a limited selection. The nearest store may be dozens of miles away and contain only a few hundred packaged goods. In Mozambique and many other parts of the world, people are just a natural disaster (perhaps a drought or typhoon) from widespread starvation or malnutrition. While social media and the Internet offer a glimpse of the quality of life in more advanced economies, the vast majority of consumer-citizens in developing countries still wonder if they'll ever have choices among a wide variety of goods and services—and the income to buy them—that most consumers in the United States, Canada, and other advanced economies take for granted.

The challenges faced by consumers and marketing managers in advanced economies seem trivial in comparison. Amazon tries to figure out how to deliver packages in four hours by drone. Walmart must decide which 50,000 items it will stock, including which (or all) of 11 kinds of Cheerios (Original, Frosted, Honey Nut Medley Crunch, Cinnamon Burst . . . you get the picture) and how many styles of frozen French fries (straight cut, crinkle cut, steak cut, waffled, seasoned, and more) to carry. The supercenter uses data to decide which product assortment will deliver the most profits in each store, but what about the consumer trying to decide among all those choices?

Many Americans don't want to make the trip to the grocery store or wait for a delivery—they expect immediate gratification. They expect the corner convenience store to have a selection of frozen gourmet dinners that can be prepared in minutes. Or if that's too much hassle, Grubhub delivers local restaurant fare within the hour. And Starbucks has our caffe latte ready when we pull up at the drive-through at 7:00 in the morning. Few of the world's consumers can expect and get so much of what they want. Do consumers need or want all these options? Are they more satisfied with products that better meet their individual choices—or are they overwhelmed by too much choice?

Much national attention is now directed toward problems of obesity, especially among children. In the United States, more than a third of children and adolescents are overweight. The surge of obesity among children is a global trend. Nutritionists say that in the United States and elsewhere, the key culprits are too much fat, starch, salt, and sugar in diets. World Health Organization experts say that today's levels of childhood obesity will lead to an explosion of illnesses (such as heart disease, type 2 diabetes, and hypertension), drain economies, create enormous suffering, and cause premature deaths. Many nutritionists and public health officials blame "big food"—processed food and fast food.

Whose responsibility is it to fix these global problems? Consider Nestlé, which now defines itself as a "nutrition, health, and wellness" company and cut levels of fat, salt, and sugar from thousands of its products. Is Nestlé making a better world—or is it simply responding to a growing number of health-conscious consumers? Maybe both—as most of its customers don't know Nestlé also works hard to eliminate slavery and child labor from its supply chain. And in Africa, it produces iron-enhanced soup cubes to fight anemia.

What should government's role be? In some places, regulation nudges consumer behavior. For example, France and San Francisco initiated a "soda tax" on sugary drinks. In other places, calories must be prominently displayed on restaurant menus. Should consumers have the right to choose high-calorie or high-fat foods if that's what they want? Should they pay more for their choices? Should government help customers make more informed decisions? And what if their choices raise health care costs for others?

When you think about the contrast between problems of starvation and too much fast food, it's not hard to decide which consumers are better off. But is that just a straw man comparison? Is the situation in less-developed nations one extreme, with the situation in the United States and similar societies just as extreme—only in a different way?

What is the role of marketing? Effective marketing provides customers with choices that meet their wants and needs. But would society be better off if it didn't put quite so much emphasis on marketing? Does all the money we spend on advertising really help consumers? It does keep the cost of television, the Internet, and other media low. Should we expect to order groceries over the Internet and have them delivered to our front door? Conversely, do all these choices just increase the prices consumers pay without adding value? Should companies produce and sell products that make the world a better place? More generally, does marketing serve society well?

We'll discuss these questions in this chapter. Now that you have a better understanding of what marketing is all about, and how the marketing manager contributes to the macro-marketing process, you should be able to decide whether marketing costs too much.[1]

LEARNING OBJECTIVES

Throughout the text, we've discussed ways that marketing can help customers while also meeting a firm's objectives. We've also considered many related ethical and societal questions. In this chapter, we evaluate the overall costs and benefits of marketing to society. This leads into an explanation of how to prepare a marketing plan—because the marketing plan integrates all of the decisions in the marketing strategy planning process and thus guides the firm toward more effective marketing.

When you finish this chapter, you should be able to

1. understand why marketing must be evaluated differently at the micro and macro levels.
2. understand why this text argues that micro-marketing costs too much.
3. understand why this text argues that macro-marketing does not cost too much.
4. understand all of the elements of the marketing strategy planning process and strategy decisions for the Four Ps.
5. know how to prepare a marketing plan and how it relates to the marketing strategy planning process.
6. know some of the challenges marketers face as they work to develop ethical marketing strategies that serve consumers' needs.
7. understand the important new key term (shown in **red**).

How Should Marketing Be Evaluated?

LO 19.1

We must evaluate at two levels

We've stressed the need for marketing to satisfy customers at a cost that customers consider a good value. So in this final chapter we'll focus on both customer satisfaction and the costs of marketing as we evaluate marketing's impact on society (see Exhibit 19-1). As we discussed in Chapter 1, it's useful to distinguish between two levels of marketing. Micro-marketing (managerial) concerns the marketing activities of an individual firm, whereas macro-marketing concerns how the whole marketing system works. Some complaints against marketing are aimed at only one of these levels at a time. In other cases, the criticisms seem to be directed to one level but actually are aimed at the other. Some critics of specific ads, for example, probably would not be satisfied with any advertising. When evaluating marketing, we must treat each of these levels separately.

Nation's objectives affect evaluation

Different nations have different social and economic objectives. Dictatorships, for example, may be mainly concerned with satisfying the needs of society as seen by the political elite. In a socialist state, the objective might be to satisfy society's needs as defined by government planners. In some societies, the objectives are defined by religious leaders. In others, it's whoever controls the military.

Exhibit 19-1 Ethical Marketing in a Consumer-Oriented World

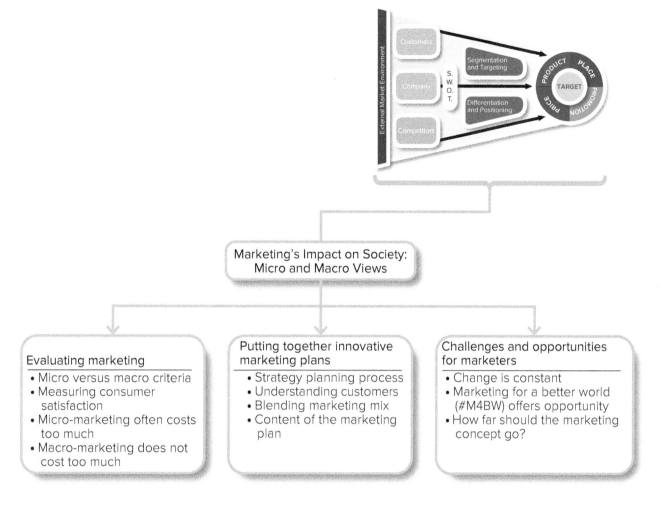

Consumer satisfaction is the objective in the United States

In the United States, *the basic objective of our market-directed economic system has been to satisfy consumer needs as they, the consumers, see them.* This objective implies that political freedom and economic freedom go hand in hand and that citizens in a free society have the right to live as they choose. The majority of American consumers would be unwilling to give up the freedom of choice they now enjoy. The same can be said for Canada, Great Britain, and most other countries in the European Union. To maintain focus, we will concentrate on marketing as it exists in American society.

Therefore, let's try to evaluate the operation of marketing in the American economy—where the present objective is to satisfy consumer needs *as consumers see them*. This is the essence of our system.

Can Consumer Satisfaction Be Measured?

Because consumer satisfaction is our objective, marketing's effectiveness must be measured by *how well* it satisfies consumers. There have been various efforts to measure overall consumer satisfaction—not only in the United States but also in other countries. For example, a team of researchers at the University of Michigan has created the American Customer Satisfaction Index based on regular interviews with tens of thousands of customers of about 230 companies in 43 industries. Similar studies are available for member countries of the European Union.

Satisfaction depends on individual aspirations

This sort of index makes it possible to track changes in consumer satisfaction measures over time and even allows comparison among countries. That's potentially useful. Yet there are limits to interpreting any measure of consumer satisfaction when we try to evaluate macro-marketing effectiveness in any absolute sense. One basic issue is that satisfaction depends on and is *relative to* your level of aspiration or expectation. Less prosperous consumers begin to expect more out of an economy as they see living standards rise around them. Products considered satisfactory one day may not be satisfactory a few years later. Fifty years ago, most people were satisfied with a 21-inch color TV that pulled in three or four channels. But once you become accustomed to a large-screen HD model and enjoy all the options possible with streaming services, that old TV is never the same again.

In addition, consumer satisfaction is a highly personal concept. Thus, looking at the "average" satisfaction of a whole society does not provide a complete picture for evaluating macro-marketing effectiveness. At a minimum, some consumers are more satisfied than others. So although efforts to measure satisfaction are useful, any evaluation of macro-marketing effectiveness has to be in part subjective.

Probably the supreme test is whether the macro-marketing system satisfies enough individual consumer-citizens so that they vote—at the ballot box—to keep it running. So far, we've done so in the United States.[2]

There are many measures of micro-marketing effectiveness

Measuring the marketing effectiveness of an individual firm is also difficult, but it can be done. Expectations may change just as other aspects of the market environment change—so firms have to do a good job of coping with the change. Individual business firms can and should try to measure how well their marketing mixes satisfy their customers (or why they fail). In fact, most large firms now have some type of ongoing effort to determine whether they're satisfying their target markets. For example, the J.D. Power marketing research firm is well known for its studies of consumer satisfaction with different makes of automobiles and computers. And the American Customer Satisfaction Index is also used to rate individual companies.

Many large and small firms measure customer satisfaction with attitude research studies. Other widely used methods include comment cards, e-mail response features on websites, unsolicited consumer responses, opinions of intermediaries and salespeople, market test results, and profits. Of course, customers may be very satisfied with some aspects of what a firm is doing but dissatisfied about other dimensions of performance.[3]

Measures of micro-marketing effectiveness include positive product reviews, like those Tempur Sealy hopes to receive, and awards, like State Farm receiving the J.D. Power Award for its high level of customer satisfaction.
(left): Source: Tempur Sealy, Inc.; (right): Source: State Farm Mutual Automobile Insurance Company

In our market-directed system, it's up to each customer to decide how effectively individual firms satisfy his or her needs. Usually, customers will buy more from firms with an offering that satisfies them—and they'll do it again and again. That's why firms that develop really satisfying marketing mixes are able to develop profitable long-term relationships with the customers they serve. Because efficient marketing plans can increase profits, profits can be used as a rough measure of a firm's efficiency in satisfying customers. Nonprofit organizations have a different bottom line, but they too will fail if they don't satisfy supporters and get the resources they need to continue to operate.

#M4BW

Does marketing have a responsibility to create a better world?

Throughout this book we have highlighted examples of companies that have implemented marketing strategies designed to be profitable *and* make a better world. But do we expect that all firms *should* do this? Do firms have this responsibility?

While some might argue that firms *should* have a "better world" mind-set, we suggest that this isn't necessarily the case. Instead, we provide examples of firms that have developed creative marketing strategies that serve both purposes. We still argue that companies have a responsibility to satisfy the needs of a target market. Marketing has a responsibility to target customers and, as we have said, *to satisfy consumer needs . . . as consumers see them.*

The personal needs of a growing number of consumers in the United States (and many developing countries) include making a better world. These consumers are more likely to support brands and firms that provide a good product *and* act in ways that preserve the environment, support women, advance health care in developing countries, support minorities, or embrace some other cause. These same consumers are much less likely to buy from a firm shown to act unethically or irresponsibly. Many companies engaging in marketing for a better world are therefore responding to customer needs. When those customers advocate and share positive word-of-mouth about firms' good behaviors—and spread negative word-of-mouth about companies' bad behaviors—then businesses that seek marketing strategies that are both profitable and make a better world are acting rationally and in their own best interest.

Evaluating marketing effectiveness is difficult, but not impossible

It's easy to see why opinions differ concerning the effectiveness of micro- and macro-marketing. If the objective of the economy is clearly defined, however—and the argument is stripped of emotion—the big questions about marketing effectiveness probably *can* be answered.

In this chapter we argue that micro-marketing (how individual firms and channels operate) frequently *does* cost too much but that macro-marketing (how the whole marketing system operates) *does not* cost too much, *given the present objective of the American economy—consumer satisfaction.* Don't accept this position as *the* answer but rather as a point of view. In the end, you'll have to make your own judgment.[4]

Micro-Marketing Often Does Cost Too Much

LO 19.2

Throughout the text, we've explored what marketing managers could or should do to help their firms do a better job of satisfying customers—while achieving company objectives. Many firms implement highly successful marketing programs, but others are still too production-oriented and inefficient. For customers of these latter firms, micro-marketing often does cost too much.

Research shows that many consumers are not satisfied; but you know that already. All of us have had experiences when we weren't satisfied—when some firm didn't deliver on its promises. And the problem is much bigger than some marketers want to believe. Research suggests that the majority of consumer complaints are never reported. Worse, many complaints that are reported never get fully resolved.

The failure rate is high

Further evidence that too many firms are too production-oriented—and not nearly as efficient as they could be—is the fact that so many new products fail. New and old businesses—even ones that in the past were leaders in their markets—fail regularly too.

Generally speaking, marketing inefficiencies are due to one or both of these reasons:

1. Lack of interest in or understanding of the sometimes fickle consumer.
2. Improper blending of the Four Ps—a lack of understanding of or adjustment to the market environment, especially what competitors do.

Either of these problems can easily be a fatal flaw—the sort of thing that leads to business failures. A firm can't create value if it doesn't have a clue what customers think or say. Even if a firm listens to the "voice of the customer," there's no incentive for the customer to buy if the benefits of the marketing mix don't exceed the costs. And if the firm succeeds in coming up with a marketing mix with benefits greater than costs, it still won't be a superior value unless it's better than what competitors offer.

Coke spent many millions of dollars to develop and promote Blak, a fusion of cola and coffee that was sold in the energy drink category. After more than a year of investments to build the brand, Coke finally abandoned the effort. Even companies with outstanding marketing talents sometimes make expensive marketing mistakes.
William D. Perreault, Jr., Ph.D.

The high cost of missed opportunities

Another sign of failure is the inability of firms to identify new target markets and new opportunities. A new marketing mix that isn't offered doesn't fail—but the lost opportunity can be significant for both a firm and society. Too many managers seize on whatever strategy seems easiest rather than seeking really new ways to satisfy customers. Too many companies stifle really innovative thinking. Layers of bureaucracy and a "that's not the way we do things" mentality can snuff it out.

On the other hand, not every new idea is a good idea for every company. Many firms have lost millions of dollars with failed efforts to jump on the "what's new" bandwagon—without stopping to figure out how it is going to really satisfy the customer and result in profit for the firm. That is as much a ticket for failure as being too slow or bureaucratic.

Ethical Dilemma

What would you do? Your firm has a new strategy that will make its established product obsolete. However, it will take a year before you are ready to implement the new strategy. If you announce your plan in advance, profits will disappear because many customers and intermediaries will delay purchases until the new product is released. If you don't announce the new plan, customers and intermediaries will continue to buy the established product as their needs dictate, but some will be stuck owning the inferior product and won't do business with you again in the future.

How would you handle this situation? Explain the reasoning for your decisions.

Micro-marketing does cost too much, but things are changing

For reasons like these, marketing does cost too much in many firms. Despite much publicity, the marketing concept is not applied in many places.

But not all firms and marketers deserve criticism. More of them *are* becoming customer-oriented. And many are paying more attention to market-oriented planning to carry out the marketing concept more effectively. Throughout the text, we've highlighted firms and strategies that are making a difference. The successes of innovative firms—such as Nike, Cirque du Soleil, Amazon.com, LEGO, Apple, iRobot, Method,

#M4BW Panera Bread does micro-marketing well. Panera's marketing managers understand that its target customers value 100 percent "clean food." For Panera that means "no artificial preservatives, sweeteners, flavors and no colors from artificial sources." Panera's customers also value its great taste, fast service, and clean restaurants—it shows in Panera's high customer satisfaction scores and profits.
Source: Panera LLC

and GEICO—do not go unnoticed. Yes, they make some mistakes. That's human—and marketing is a human enterprise. But they have also shown the results that market-oriented strategy planning can produce.

Another encouraging sign is that more companies are recognizing that they need a diverse set of backgrounds and talents to meet the increasingly varied needs of their increasingly global customers. They're shedding "not-invented-here" biases and embracing new technologies, comparing what they do with the best practices of firms in totally different industries, and teaming up with outside specialists who can bring a fresh perspective.

Managers who adopt the marketing concept as a way of business life do a better job. They look for target market opportunities and carefully blend the elements of the marketing mix to meet their customers' needs. As more of these managers rise in business, we can look forward to much lower micro-marketing costs and strategies that do a better job of satisfying customer needs.

Macro-Marketing Does Not Cost Too Much

LO 19.3

Some critics of marketing take aim at the macro-marketing system. They typically argue that the macro-marketing system causes an inefficient or ineffective use of resources and leads to an unfair distribution of income. Most of these complaints imply that some marketing activities by individual firms should not be permitted—and because they are, our macro-marketing system does a less-than-satisfactory job. Let's look at some of these positions to help you form your own opinion.

Micro-efforts help the economy grow

Some critics feel that marketing helps create a monopoly, or at least monopolistic competition. Further, they think this leads to higher prices, restricted output, and reduction in national income and employment.

It's true that firms in a market-directed economy try to carve out separate monopolistic markets for themselves with new products. But consumers do have a choice. They don't *have* to buy the new product unless they think it's a better value. Most of the time older products are still available. In fact, to meet the new competition, prices of the older products usually drop. And that makes them even more affordable.

Over several years, an innovator's profits may rise—but rising profits also encourage further innovation by competitors. This leads to new investments, which contribute to economic growth and higher levels of national income and employment. Around the world, many countries failed to achieve their potential for economic growth under command systems because this type of profit incentive didn't exist. The competition between China's two leading online retailers, Alibaba and Tencent, is an example of this. It generated innovations profitable to both firms, while also helping make the world a better place.

 #M4BW

Chinese mobile payment competition brings about a better world

The competition between Chinese mobile payment services Alipay (part of Alibaba's Ant Financial) and Tencent's WeChat Pay is yielding real benefits for low-income Chinese consumers and the world. For one thing, many of China's poorest citizens are "unbanked," meaning they are not served by a bank or other financial institution. For this group, that means greater inconvenience, higher prices, and higher transaction costs. Many of these customers, who often live in rural areas without banks, now have access to low-cost financial services. They are able to save, engage in more efficient buying and selling, and enjoy greater purchasing power, all of which contributes to economic development. Similar services, like Kenya's M-Pesa and Bangladesh's bKash, have had an impact on the poor in other developing countries.

In this case, there is another benefit to the world. Competition also pushed Alipay and Ant Financial to find innovative approaches to keep customers engaged and loyal to the Alipay app. Ant Financial's marketing research showed a growing number of its target consumers were concerned about the environment. To appeal to this market, Ant Financial developed the Ant Forest app. The app rewards Chinese consumers who demonstrate more environmentally friendly habits. The app is tied to Alipay and converts green behaviors—like walking, taking public transportation, and certain online payment activities—into "green energy." When an individual earns enough green energy, Ant Financial plants a real tree. Users also share their "successes" with friends on the network, inspiring others to join China's growing environmental movement. In its first three years of use, Ant Forest use resulted in more than 55 million new trees across China.[5]

Is advertising a waste of resources?

Advertising is the most criticized of all micro-marketing activities. Indeed, many ads *are* annoying, insulting, misleading, and downright ineffective. This is one reason why micro-marketing often costs too much. However, advertising can also make both the micro- and macro-marketing processes work better.

Advertising is an economical way to inform large numbers of potential customers about a firm's products. Provided that a product satisfies customer needs, advertising can increase demand for the product—resulting in economies of scale in manufacturing, distribution, and sales. Because these economies may more than offset advertising costs, advertising can actually *lower* prices to the consumer.[6]

Consumers are not puppets

The idea that firms can manipulate consumers to buy anything the company chooses to produce simply isn't true. A consumer who buys a soft drink that tastes terrible won't buy another can of that brand, regardless of how much it's advertised. In fact, many new products fail the test of the market. Not even large corporations are assured of success every time they launch a new product. Consider, for example, the dismal fates of New Coke, the Amazon Fire Phone, Watermelon Oreos, Burger King Satisfries, and Google Glass. These were well-known companies with good marketing departments that couldn't make customers buy a bad marketing mix.

Needs and wants change

Consumer needs and wants constantly evolve. Few of us would care to live the way our grandparents lived when they were our age. Marketing's job is not just to satisfy consumer wants as they exist at any particular point in time. Rather, marketing must keep looking for new *and* better ways to create value and serve consumers.[7]

Does marketing make people materialistic?

There is no doubt that marketing caters to materialistic values. However, people disagree as to whether marketing creates these values or simply appeals to values already

#M4BW Micro-marketing does not cost too much when firms understand and address customer needs and wants with an effective marketing mix. Because some customers prefer not to use dangerous chemicals on their skin to keep bugs and mosquitoes away, Shieldtox NaturGard uses natural ingredients to offer continuous control of flying insects. Passion Lilie appeals to customers who like its ethical fashion and its values, which empower women.
(left): Source: Reckitt Benckiser Group plc.; (right): Source: Passion Lilie

there. Even in the most primitive societies, people want to accumulate possessions. The tendency for ancient pharaohs to surround themselves with wealth and treasures can hardly be attributed to the persuasive powers of advertising agencies!

Products do improve the quality of life

Clearly, quality of life can't be measured just in terms of quantities of material goods. But when we view products as the means to an end rather than the end itself, they *do* make it possible to satisfy higher-level needs. The Internet, for example, empowers people with information in ways that few of us could have even imagined a few decades ago.

Marketing reflects our own values

For good or bad, marketing reflects our values. Marketing mixes that satisfy customer needs better beat out those that don't. Firms that behave in unethical ways that don't reflect our values tend to lose out in the long run. Firms that seem to provide good products and improve our world more broadly reflect the values we want to see in ourselves and others. Consider how Burger King tied into a set of values that resonate with many of its target customers.

#M4BW

Burger King seeks a world with less bullying

Many people would like to see a world with less bullying. This is a value shared by many parents and families—a target market for Burger King. Each year, 30 percent of schoolkids worldwide are bullied. Burger King decided to take a stand against bullying and created a video to make its point. The video was shot in a real Burger King, with real customers; actors played the bullies, the bullied high school kid, and the Burger King employees. After ordering a burger, customers were given a noticeably beaten up, "bullied" hamburger. While seated in the restaurant, they also witnessed a high school kid being bullied. While 95 percent of the customers went to the counter to report their bullied burger, just 12 percent reported or intervened in the bullying of the high school kid. The video demonstrated how easy it is for most people to look the other way when bullying happens. The video went viral with millions of views—and led to many discussions about how to take a stand against bullying. It also showed that Burger King is a concerned corporate citizen that seeks to make the world a better place.

Not all needs are met

Some critics argue that our macro-marketing system is flawed because it does not provide solutions to important problems, such as questions about how to help people who are homeless or uneducated, dependent children, members of minorities who have suffered discrimination, elderly poor people, and those who are sick. Many of these people do live in dire circumstances. But is that the result of a market-directed system?

There is no doubt that many firms focus their effort on people who can pay for what they have to offer. But, as economies develop and the forces of competition drive down prices, more people are able to afford more of what they want. And the matching of supply and demand stimulates economic growth, creates jobs, and spreads income among more people. Many firms are recognizing that marketing strategies can be profitable and make for a better world—and are developing strategies that do just that (#M4BW).

A market-directed economy makes efficient use of resources. However, it can't guarantee that government aid programs are effective. It doesn't ensure that all voters and politicians agree on which problems should be solved first—or how taxes should be set and allocated. It can't eliminate the possibility of a child being ignored.

These are important societal issues. Consumer-citizens in a democratic society assign some responsibilities to business and some to government. Ultimately, consumer-citizens vote via the ballot box for how they want governments to deal with these concerns—just as they vote with their dollars for which firms to support. As more managers in the public sector understand and apply marketing concepts, we should be able to do a better job meeting the needs of all people.

Marketing Strategy Planning Process Requires Logic and Creativity

LO 19.4

We've said that our macro-marketing system *does not* cost too much, given that customer satisfaction is the present objective of our economy. But we admit that the performance of many business firms leaves a lot to be desired. This presents a challenge to serious-minded students and marketers—and raises the question: What needs to be done?

We hope that this book has convinced you that a large part of the answer to that question is that the *effectiveness and value of marketing efforts in individual firms is improved significantly when managers take the marketing concept seriously—and when they apply the marketing strategy planning process we've presented.* So let's briefly review these ideas and show how they can be integrated into a marketing plan.

Marketing strategy planning process brings focus to efforts

Developing a good marketing strategy and turning the strategy into a marketing plan requires creative blending of the ideas we've discussed throughout this text. Exhibit 19-2 provides a broad overview of the major areas we've been talking about. You first saw this exhibit in Chapter 2—before you learned what's really involved in each idea. Now we must integrate ideas about these different areas to narrow down to a specific target market and marketing mix that represents a real opportunity. This narrowing-down process requires a thorough understanding of the market. That understanding is enhanced by careful analysis of customers' needs, current or prospective competitors, and the firm's own objectives and resources. Similarly, trends in the external market environment may make a potential opportunity more or less attractive.

There are usually more strategy possibilities than a firm can pursue. Each possible strategy usually has a number of different potential advantages and disadvantages. This can make it difficult to zero in on the best target market and marketing mix. However, as we discussed in Chapter 3, developing a set of specific qualitative and quantitative screening criteria—to define what business and market(s) the firm wants to compete in—can help eliminate potential strategies that are not well suited to the firm.

Careful analysis helps the manager focus on a strategy that takes advantage of the firm's strengths and opportunities while avoiding its weaknesses and threats to its

Exhibit 19-2 Overview of Marketing Strategy Planning Process

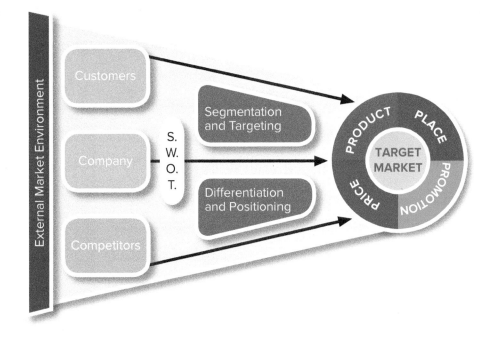

success. These strengths and weaknesses can be compared with the pros and cons of strategies that are considered. For example, if a firm is considering a strategy that focuses on a target market that is already satisfied by a competitor's offering, finding a competitive advantage might require an innovative new product, improved distribution, more effective promotion, or a better price. Just offering a marketing mix similar to the competition's usually doesn't provide any real basis for the firm to position or differentiate its marketing mix as offering superior customer value.

#M4BW Bolthouse Farms saw an opportunity to sell more baby carrots. Because teenagers were the target market, Bolthouse Farms' marketing managers knew that a "healthy snack" positioning was probably not the best approach. So they developed a creative marketing strategy to appeal to this target market: "Eat 'em like junk food." Creative packaging, fun promotion, and distribution in school lunchroom vending machines were just part of the strategy that helped grow a previously stagnant market for baby carrots.
Source: Bolthouse Farms

Exhibit 19-3 Strategy Decision Areas Organized by the Four Ps

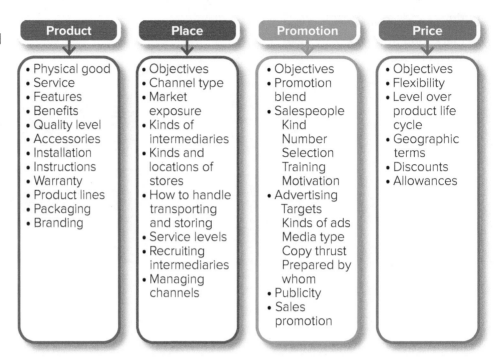

Marketing manager must blend the Four Ps

Exhibit 19-3 reviews the major marketing strategy decision areas organized by the Four Ps. Each of these requires careful decision making. Yet marketing planning involves much more than just independent decisions and assembling the parts into a marketing mix. The Four Ps must be creatively *blended*—so the firm develops the best mix for its target market. In other words, each decision must work well with all of the others to make a logical whole.

In our discussion, we've given the job of integrating the Four Ps strategy decisions to the marketing manager. Now you should see the need for this integrating role. It is easy for specialists to focus on their own areas and expect the rest of the company to work for or around them. This is especially true in larger firms where the size of the whole marketing job is too big for one person. Yet the ideas of the product manager, the advertising manager, the sales manager, the logistics manager, and whoever makes pricing decisions may have to be adjusted to improve the whole mix. It's critical that each marketing mix decision works well with all of the others. A breakdown in any one decision area may doom the whole strategy to failure.

The Marketing Plan Brings All the Details Together

LO 19.5

Marketing plan provides a blueprint for implementation

Once the manager has selected the target market, decided on the (integrated) marketing mix to meet that target market's needs, and developed estimates of the costs and revenues for that strategy, it's time to put it all together in the marketing plan. As we explained in Chapter 2, a marketing plan includes the time-related details—including costs and revenues—for a marketing strategy. Thus, the plan basically serves as a blueprint for what the firm will do.

Exhibit 19-4 provides a summary outline of the different sections of a complete marketing plan. You can see that this outline is basically an abridged overview of the topics we've covered throughout the text. Thus, you can flesh out your thinking for any portion of a marketing plan by reviewing the section of the book where that topic is discussed in more detail. Further, the Hillside Veterinary Clinic case in Appendix D provides a real example of the types of thinking and detail that are included.

Exhibit 19-4 Summary Outline of Different Sections of Marketing Plan

<u>Situation Analysis</u>
Company Analysis
- Company objectives and overall marketing objectives
- Company resources (marketing, production, financial, human, etc.)
- Other marketing plans (marketing program)
- Previous marketing strategy
- Major screening criteria relevant to product-market opportunity selected
  - Quantitative (ROI, profitability, risk level, etc.)
  - Qualitative (nature of business preferred, social responsibility, environment, etc.)
- Major constraints
- Marketing collaborators (current and potential)

Customer Analysis (organizational customers and/or final consumers)
- Product-market definition
- Possible segmenting dimensions (customer needs, other characteristics)
  - Qualifying and determining dimensions
- Identify target market(s) (one or more specific segments)
  - Operational characteristics (demographics, geographic locations, etc.)
- Potential size (number of people, dollar purchase potential, etc.) and likely growth
- Key economic, psychological, social, cultural, and situational influences on buying
- Nature of relationship with customers

Competitor Analysis
- Nature of current/likely competition
- Current and prospective competitors
  - Current strategies and likely responses to plan
- Competitive barriers to overcome and sources of potential competitive advantage

Analysis of the Market Context—External Market Environment
- Economic environment
- Technological environment
- Political and legal environment
- Cultural and social environment

Key Factors from Situation Analysis summarized in S.W.O.T.
- S.W.O.T.: Strengths, weaknesses, opportunities, and threats from situation analysis

<u>Marketing Plan Objectives</u>
- Specific objectives to be achieved with the marketing strategy

<u>Differentiation and Positioning</u>
- How will marketing mix be differentiated from the competition?
- How will the market offering be positioned?
- Positioning statement

<u>Marketing Strategy</u>
Overview of the Marketing Strategy
- General direction for the marketing strategy
- Description of how the Four Ps fit together

Target Market(s)
- Summary of characteristics of the target market(s) to be approached

Product
- Product class (type of consumer or business product)
- Product liability, safety, and social responsibility considerations
- Specification of core physical good or service
  - Features, quality, etc.
- Supporting service(s) needed
- Fit with product line
- Branding (manufacturer versus dealer, family versus individual, etc.)
- Packaging (promotion, labeling, protection, enhancing use)
- Cultural sensitivity of product
- Current product life-cycle stage
- New-product development requirements (people, dollars, time, etc.)

(continued)

Exhibit 19-4 Summary Outline of Different Sections of Marketing Plan—*continued*

Place
- Objectives
  - Degree of market exposure required
  - Distribution customer service level required
- Type of channel (direct, indirect)
  - Other channel members or collaborators required
  - Type/number of wholesalers (agent, merchant, etc.)
  - Type/number of retailers
  - How discrepancies and separations will be handled
  - How marketing functions will be shared
- Coordination needed in company, channel, and supply chain
  - Information requirements (EDI, the Internet, e-mail, etc.)
- Transportation requirements
- Inventory product-handling requirements
- Facilities required (warehousing, distribution centers, etc.)
- Reverse channels (for returns, recalls, etc.)

Promotion
- Objectives
- Major message theme(s) (for integrated marketing communications/positioning)
- Promotion blend
  - Advertising (type, media, copy thrust, etc.)
  - Personal selling (type and number of salespeople, compensation, effort allocation, etc.)
  - Publicity (earned, owned, social media)
  - Sales promotion (for customers, channel members, employees)
- Mix of push and pull required
- Who will do the work?

Price
- Nature of demand (price sensitivity, price of substitutes)
- Demand and cost analyses (marginal analysis)
- Markup chain in channel
- Price flexibility
- Price level(s) (under what conditions) and impact on customer value
- Adjustments to list price (geographic terms, discounts, allowances, etc.)

<u>**Marketing Information Requirements**</u>
- Marketing research needs (with respect to customers, Four Ps, external environment, etc.)
- Secondary and primary data needs
- Marketing information system needs, models to be used, and so on

<u>**Implementation and Control**</u>
Special Implementation Problems to Overcome
- People required
- Manufacturing, financial, and other resources needed

Control
- Marketing information systems and data needed
- Monitoring and analytics needed
- Criterion measures/comparison with objectives (customer satisfaction, sales, cost, performance analysis, etc.)

Budget, Sales Forecasts, and Estimates of Profit
- Costs (all elements in plan, over time)
- Sales (by market, over time, etc.)
- Estimated operating statement (pro forma)

Timing
- Specific sequence of activities and events, and so on
- Likely changes over the product life cycle

Risk Factors and Contingency Plans

Marketing plan spells out the timing of the strategy

Some time schedule is implicit in any strategy. A marketing plan simply spells out this time period and the time-related details. Usually, we think in terms of some reasonable length of time—such as six months, a year, or a few years. But it might be only a month or two in some cases, especially when rapid changes in fashion or technology are important. Or a strategy might be implemented over several years, perhaps the length of the early stages of the product's life.

Although the outline in Exhibit 19-4 does not explicitly show a place for the time frame for the plan or the specific costs for each decision area, these should be included in the plan—along with expected estimates of sales and profit—so that the plan can be compared with *actual performance* in the future. In other words, the plan not only makes it clear to everyone what is to be accomplished and how—it also provides a basis for the control process after the plan is implemented.[8]

A complete plan spells out the reasons for decisions

The plan outline shown in Exhibit 19-4 is quite complete. It doesn't just provide information about marketing mix decisions—it also includes information about customers (including segmenting dimensions), competitors' strategies, other aspects of the market environment, and the company's objectives and resources. This material provides important background information relevant to the "why" of the marketing mix and target market decisions.

Too often, managers do not include this information; their plans just lay out the details of the target market and the marketing mix strategy decisions. This shortcut approach is more common when the plan is really just an update of a strategy that has been in place for some time. However, that approach can be risky.

Managers too often make the mistake of casually updating plans in minor ways—perhaps just changing some costs or sales forecasts—but otherwise sticking with what was done in the past. A big problem with this approach is that it's easy to lose sight of why those strategy decisions were made in the first place. When the market situation changes, the original reasons may no longer apply. Yet if the logic for those strategy decisions is not retained, it's easy to miss changes taking place that should result in a plan being reconsidered. For example, a plan that was established in the growth stage of the product life cycle may have been very successful for a number of years. But a marketing manager can't be complacent and assume that success will continue forever. When market maturity hits, the firm may be in for big trouble—unless the basic strategy and plan are modified. If a plan spells out the details of the market analysis and logic for the marketing mix and target market selected, then it is a simple matter to routinely check and update it. Remember: The idea is for all of the analysis and strategy decisions to fit together as an integrated whole. Thus, as some of the elements of the plan or market environment change, the whole plan may need a fresh approach.

Today's Marketers Face Challenges and Opportunities

LO 19.6

Marketers face the challenge of preparing creative and innovative marketing plans, but that in itself will not improve the value of marketing to society. For marketing managers to be more effective at strategy planning, they must recognize some basic changes and trends affecting marketing strategy planning. These changes and trends range from technology to demographic patterns—all of which require marketing managers to remain flexible and adaptive. We will briefly highlight some of those trends to give you an idea of what marketing managers are facing today and expect to face in the future. We will then dig a little deeper into one of the changes and trends that has been a theme of this book—consumer interest in marketing for a better world.

Change is the only thing that's constant

We need better marketing performance at the firm level. Progressive firms pay attention to changes in the market—including trends in the market environment—and how marketing strategies need to be improved to consider these changes. Exhibit 19-5 lists some of the important trends and changes we've discussed throughout this text.

Exhibit 19-5 Some Important Changes and Trends Affecting Marketing Strategy Planning

Communication Technologies
The Internet and intranets
Satellite communications and Wi-Fi
Videoconferencing and Internet telephony
Text messaging
Growing use of smartphones around the world
Consumer online "search" shopping

Role of Computerization
E-commerce, websites
Wireless networks
Scanners, bar codes, and RFID for tracking
Multimedia integration
"Cloud" computing
Mobile web access
Marketing analytics and predictive analytics
Artificial intelligence and intelligent agents
Augmented reality
Internet of Things

Marketing Research
Growth of marketing information systems
Data warehouses and data mining
Big data
Online research—surveys, focus groups, online communities
Decision support systems
Search engines
Web analytics
Customer relationship management (CRM) systems

Customer Behavior and Demographic Patterns
Growing concern about sustainability and climate change
Consumers expect companies to contribute to a "better world"
Technology usage around the globe
Rapid growth in senior and ethnic submarkets
Aging of the Baby Boomers
Population growth slowdown in United States
Geographic shifts in population
Slower real income growth in United States

Business and Organizational Customers
Greater interest in sustainability and supporting "better world" brands
Closer buyer–seller relationships and single sourcing
Just-in-time inventory systems/EDI
Web portals and Internet sourcing
Interactive bidding and proposal requests
Online as source of information—salespeople involved later in buying process
E-commerce and supply chain management

Product
Role of customer's total experience
More attention to quality
More attention to service technologies
Growth of dealer brands and private label
Category management
More attention to sustainable design
More attention to discontinuous innovation
Faster new-product development
R&D teams with market-driven focus
Rapid prototyping, 3-D printing

Channels and Logistics
Next-day and same-day delivery
Online retailing and wholesaling
Clicks and bricks (multichannel)
Multichannel shopping and omnichannel strategies
Larger, more powerful retail chains
More attention to distribution service
Real-time inventory replenishment
Automated warehousing and handling
Cross-docking at distribution centers
Rise of third-party logistics (outsourcing)

Sales Promotion
Point-of-purchase promotion
Trade promotion increasing
Event sponsorships
Customer loyalty programs
Customer acquisition cost analysis

Personal Selling
Post-sale customer service
Solution-oriented selling
Sales technology
Major accounts specialization
More inside salespeople
Growth in team selling

Mass Selling
More emphasis on earned and owned media
Growth of social media
Role of customer reviews and ratings
Integrated marketing communications
Growth of more targeted media
Consolidation of global advertising agencies
Consolidation of media companies
Shrinking media budgets

Pricing
Value pricing
Overuse of sales and deals
Bigger differences in discounts
Dynamic pricing
Freemium pricing
Comparison price shopping easier with Internet

International Marketing
More international market development
Global competitors—at home and abroad
Global communication over the Internet
New trade rules (WTO, EU, etc.)
Greater use of tariffs
Impact of "pop" culture on traditional cultures
Tensions between "have" and "have-not" cultures
Growing income and population in emerging markets

General
Privacy issues
More attention to profitability, not just sales
Greater attention to delivering superior value
Addressing of environmental concerns
More pressure on marketing to demonstrate financial returns

A constant challenge for food producers is ensuring product safety. Instrumentation, like that sold by Japanese firm Leco, is often used to monitor food quality and safety. One safety lapse can harm a brand's reputation. The headline of the ad reads, "I am inside your taste of goodness."
Source: LECO Japan

Most of the changes and trends summarized in Exhibit 19-5 have a positive effect on how marketers serve society. And this ongoing improvement is self-directing. As consumers shift their support to firms that do meet their needs, laggard businesses are forced to either improve or get out of the way.

One of the needs being expressed by consumers is for businesses to behave ethically and, where possible, to contribute to making the world a better place. As this positive opportunity for marketing is a theme of this textbook, we will look closer at this particular trend.

How can marketing contribute to a better world?

In Chapter 1, we explained that marketing creates a better world when: (1) buyers and sellers make better decisions, are healthier, and through their consumption choices (or decision not to consume) experience a better quality of life; (2) buyers and sellers make decisions that have less adverse impacts on others; and (3) marketing strategy decisions address some of the world's most challenging problems—including hunger, poverty, and climate change. This suggests that buyers need to be well informed when they make buying choices. It also suggests that firms can be strategic by understanding their customers. While firms have no responsibility to create a better world, doing so may be in their best interest as it responds to customer needs.[9]

Throughout this book, we have provided examples of successful marketing strategies that generate profits for firms *and* make for a better world. We hope these examples remind you of some of the good things business does and inspire you to practice marketing for a better world. Data suggest that a growing number of customers want to purchase from businesses making a positive impact on the world. And businesses monitoring these trends are getting ahead of the curve. We also believe that for firms to engage in marketing strategies that achieve both of these objectives, they need to operate in a regulatory environment that allows consumers to better understand if and how firms are really supporting a better world.

Three broad groups' behaviors can make a better world

There are three broad groups relevant to our discussion of effective marketing strategies that contribute to a better world. As shown in Exhibit 19-6, we see opportunities for consumer-citizens, business/marketing, and government/society. When these groups work together, they can address global challenges and contribute to helping marketing for a better world.

Our goal in these last few pages is to suggest that there are opportunities for consumers and businesses to engage in small (or maybe large) changes that can make a better

Exhibit 19-6 Basic Requirements for Marketing for a Better World

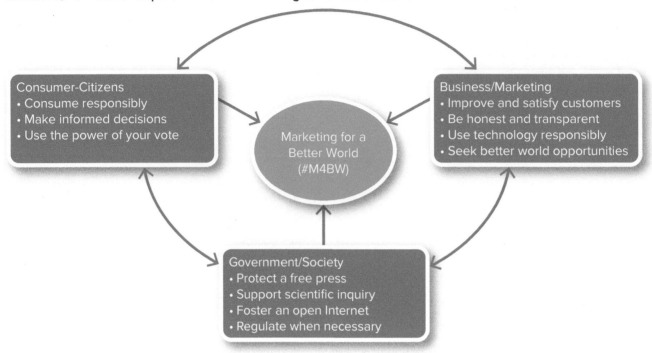

world. If we, as employees or managers of businesses, or as consumers, recognize the challenges, we are more likely to make changes that lead to a movement. We will highlight a few opportunities for business, consumers, and government and society to work together to address these problems. We begin our discussion where we believe marketing should start—with customers.

Use available information

As consumer-citizens, each of us shares the responsibility for preserving an effective macro-marketing system. And we should take this responsibility seriously. That includes the responsibility to make wise choices. While a majority of consumers ignore most of the available information that could help them spend money more wisely (and guide the marketing process), a growing number are paying attention. And there is more information available. Consumerism encouraged nutritional labeling, unit pricing, truth in lending, plain-language contracts and warranties, and so on. Government agencies publish many consumer buying guides on everything from tires to appliances, as do organizations such as Consumers Union. Nonprofit organizations report on the social responsibility of many firms. Organizations like Climate Counts provide advice on making more sustainable purchases. Most of this information is available on the web. It makes sense to use it.

For this movement to take root—and for inefficient and ineffective marketing strategies to disappear and be replaced by more effective strategies—consumers must be more responsible. Americans tend to perform their dual role as consumer-citizens with a split personality. Many behave differently at the cash register than we do on our soapbox. For example, we say that we want to protect the environment, but when it comes to making our own purchase decisions, many still choose the more convenient or lower-priced product rather than the one that is the sustainable choice. We complain when Facebook shares our data without our permission, but we continue to log back onto the site after reading about its questionable ethics.

This is not to suggest that every consumer choice follows the extensive problem-solving, high information-gathering mode described in Chapter 5. There is not enough time in the day to make every choice a well-researched choice. That is why we call for business and government to help consumers make better choices (discussed later in the chapter). Use the power of your vote to ask the government to regulate as needed.

#M4BW Companies that produce products with environmental benefits will find greater success when those products deliver other benefits as well. Owens-Illinois developed a campaign that points out all the benefits of glass. Besides being endlessly recyclable, things taste better in glass. Products packaged in glass are also healthier and higher quality.
Source: Owens-Illinois Inc.

What can marketing do?

We have seen that customers are loyal to brands they believe contribute to a better world—and are more likely to tell friends about such brands. Next, we describe a few other ways that businesses can help address global challenges. This includes having a continuous improvement attitude, being more honest with customers, using technology responsibly, and seeking "better world" opportunities.

Keep getting better

Marketing managers must constantly evaluate their strategies to be sure they're not being left in the dust by competitors who see new and better ways of doing things. It's crazy for a marketing manager to constantly change a strategy that's working well. But too many managers fail to see or plan for needed changes. They're afraid to do anything different and adhere to the idea that "if it ain't broke, don't fix it." But a firm can't always wait until a problem becomes completely obvious to do something about it.

When it makes sense, seek profits *and*...

We have provided many examples of companies with marketing strategies that are both profitable *and* make for a better world. A firm's primary responsibility is to stay in business—and profits are generally necessary to make that happen. We are not suggesting that firms start acting like philanthropic nonprofits. We believe that better world marketing strategies can be profit maximizing *and* a better world strategy can work for businesses targeting customers that care about this. Yet, like other strategy decisions, it should be done to support business objectives.

Promotion should be more honest and transparent

All businesses need to build trust—and that starts by being honest with customers. Promotion provides powerful ways to communicate with customers. Yet too many firms lapse into telling only half the truth. This is most obvious when there is a shift in consumer interest. For example, a firm's advertising may accurately proclaim that its food product has no "trans fat"—but do consumers think that means the product is healthy, low-calorie, or even low in fat?

When target customers factor sustainability into buying decisions, firms must avoid the temptation of **greenwashing**, making false claims that imply a company or its products do more for the environment, or other false claims about making the world a better place. It's good for a firm to create a biodegradable package and promote it, but the cleaning product in the package shouldn't contain chemicals that will be harmful once they're in the sewer system. Firms that are honest, transparent, and authentic with their marketing will be more trusted by customers.

Growing consumer cynicism about promotion is also a problem. As it gets worse, firms as well as consumers suffer. Regulations say that marketing communications shouldn't be false or misleading, but managers need to take seriously the responsibility to be truthful to their customers. Marketing communications should be helpful—not just legal. Managers who don't get this message are likely to learn a hard lesson from activists who spread criticisms of their firms across the web and other media. The potential harm to a brand's reputation from this sort of negative publicity has many firms cautious about overstating their claims, including ones related to sustainability efforts.[10]

Use technology responsibly

We live in a time of dramatic new technologies. Marketing managers need to be careful about how they use this technology. Firms face the challenge of determining which technologies are acceptable and which are not. For example, gene research has opened the door to lifesaving medicines, genetically altered crops that resist drought or disease, and even cloning of human beings. Yet in all of these arenas there is intense conflict among different groups about what is appropriate. How should these decisions be made?

Respect consumer privacy

Marketers also use technology to gather customer information. Marketing managers should be sensitive to consumers' rights and privacy. In this book, we describe how marketing managers collect and analyze data about customers. Many of these data are collected and analyzed without customers' knowledge. Organizations can abuse these rights. For example, credit card records—which reveal much about consumers' purchases and private lives—are routinely sold to anybody who pays for the file. Many people post a great deal of personal information on social networking websites such as Facebook, assuming they have some privacy. This is often not the case. Facebook and most social media platforms share this information with advertisers seeking a specific target market. Consider what is already happening and what might be coming in *What's Next?* Marketers use big data and analytics—cool or creepy?[11]

Is it unethical to make addictive products?

Many firms create technology products with the goal of making their offering as appealing to customers as possible. They want customers to use their product more often. That sounds like a straightforward goal; Facebook, Apple's iPhone, and *Candy Crush* (a game for mobile devices) want to keep customers engaged. These companies, and others, want to keep customers on their site, device, or game longer and more often. It helps Facebook sell more advertising, Apple sell more phones, and *Candy Crush* generate more in-app purchases. Some critics suggest that this is leading to a form of addiction.

Toward that end, programmers at these companies have added features that foster more engagement (addiction?). Whether it is the infinite scroll and "Likes" of Facebook, the bright colors of iPhone's OLED screen, or the "sweet talk" and social design of *Candy Crush,* all of these features make products more engaging and potentially addictive. Is it ethical to make products so appealing that customers just cannot stop using them?[12]

> **Ethical Dilemma**
>
> *What would you do?* You and a friend have developed a word-play game for smartphones. After testing the game on friends and family, you have some data. While they all say the game is fun, 75 percent of them stopped playing it after a week. The average playtime of those still using it is just 12 minutes a day. A top programmer at a leading game developer tells you its games (which are all free to start but feature in-game purchase opportunities) added features that increased average daily gameplay from 22 to 34 minutes. She said that 15 percent of "super-users" are "totally addicted"; they average 3 hours of gameplay per day and generate 90 percent of each game's revenue. She'll agrees to share her tips and tricks in exchange for a 5 percent stake in your game. You wonder if getting people "addicted" is the right thing to do? *Should a company try to make its product more engaging—so customers use it more often? Is this the responsibility of business, the government, or the consumer? Would you "pay" this top programmer for the advice? Why or why not?*

What's Next? Marketers use big data and analytics—cool or creepy?

Throughout this book, we have discussed big data and analytics (see Chapter 7). Marketing managers use big data to build "social profiles" of individual consumers. Right now, a few companies create profiles using customers' clicks, conversations, Facebook "Likes," Twitter tweets, cell phone activity, and credit card and loyalty card purchases. With artificial intelligence and analytics software, they deliver unique promotional messages and offers. In exchange, customers give up some privacy.

This is an early indication of *What's Next?* in advertising and sales promotion. Is this new practice cool? Or is it creepy? To better understand how the big data revolution affects customers, let's take a closer look at how this might play out in real life by looking at how this impacts college students Nick and Shelby.

Nick often eats fast food and regularly receives texts with coupons for fast-food purchases. His family is not wealthy, and saving money is important. Nick's profile shows that he usually spends significantly more than the coupon amount, and computer analysis of Nick's photos and comments on Facebook suggest he is overweight. Soon Nick starts seeing ads for diets; he also receives a special introductory offer from a local health club. Cool or creepy? Nick's Facebook newsfeed and Twitter feed include "articles" from news sites on healthy eating with an emphasis on Kellogg's Special K diet products and weight-loss tips from South Beach Diet products. Cool or creepy? Some days, as Nick walks home from school, his Snapchat flashes a picture of an ice-cream cone (and a $0.50 off deal) just a block before he passes a Baskin-Robbins ice-cream store. Cool or creepy?

Shelby is from a wealthy family. Her friends know she works out regularly and maintains a healthy diet. Last week she was online reading articles about mountain bikes—now when she surfs the web, she sees ads for mountain bikes. Cool or creepy? As Shelby walks into her grocery store, her phone lights up with an offer for $1 off on Ben & Jerry's Greek Frozen Yogurt—she loves that stuff but rarely eats it unless she's feeling down. Did the grocery store know she and her boyfriend just broke up? Yes, it did; she posted a Facebook status update that morning. Cool or creepy? Shelby often receives direct-mail offers for low-interest credit cards; Nick doesn't get those. Does the financial institution figure that healthier behavior is a sign of self-discipline and, therefore, financial responsibility? Or is it just that her family is wealthier? Will insurance companies draw the same conclusions and adjust their rates accordingly? Cool or creepy?

Marketing managers see an opportunity to leverage big data and analytics to customize advertising, sales promotion, and pricing. Tracked by your cell phone's GPS, messages target your location. If a $0.50 coupon motivates Nick to buy ice cream, but it takes $1.00 off to move Shelby, each gets a different coupon. Shelby's wealthier background suggests she has a higher customer lifetime value than Nick, so department stores will offer her better deals, hoping to build brand loyalty. And those "articles" served to your newsfeed might be written by someone paid by a brand prominently featured in the article.

Cool or creepy? It probably depends on your values. As we move into a brave new world where customers surrender privacy (usually without their knowledge), you should have questions.[13]

Do you want to trade your privacy for personalized marketing mixes? What do you have to gain or lose? Who owns your data? Should governments regulate this—or should it be up to you? There are no easy answers. Is this cool? Creepy? A little of both? A lot of both?

The role of the government and social institutions

The role of government in this situation may be to promote the free flow of accurate information. Earlier we suggested that a more effective and efficient micro-marketing system—and for marketing to lead to a better world—we need consumers to make more informed choices. Three institutions in society should be promoted and protected by government to foster these goals: (1) a free press, (2) science, and (3) an open Internet. Each of these seeks to provide more accurate information about what companies do.

A vibrant and free press keeps an eye on business

A vibrant free (and fair) press serves an important role in monitoring and publicizing the actions of both government and business. For example, when members of Uber's top management were reported to be engaged in or ignoring sexual harassment at the company, the press investigated and reported about what happened.

Many consumers, seeking to exercise their concerns, chose to do business with Uber's competitor, Lyft. When Mylan dramatically increased the price of its EpiPen, a crucial antidote for anyone suffering from a life-threatening allergic reaction, its use of monopoly power was called out by the press. The subsequent publicity harmed Mylan and brought its CEO before Congress to testify. The company subsequently lowered prices. The press also publicized Facebook's lack of transparency about how it handles and uses its customers' data. The press provided information that helped consumers make choices to support (or not) businesses that engage in good (or bad) behavior.

Unfortunately, the press may be more valuable to society in general than to individuals. Because subscriptions and advertising in newspapers and magazines and their associated websites has fallen, the press has fewer resources for continuing its watchdog role. As a society, we may have to have a conversation about the role of the press as an institution. If consumers want to buy from "good" companies, then they need accurate information about what companies qualify.

Science can tell us what works

In an increasingly complex world, the best solutions are not always clear. This is why we need to continue to invest in good science to provide objective answers. As consumers seek to make better decisions—and businesses want to do the right thing—they need to know what actually works. For example, science (not industry insiders or interest groups) should tell us if genetically modified foods are the answer (or a threat) to food security. Science can tell us if driving electric cars, managing refrigeration chemicals, switching to LED light bulbs, or restoring tropical forests does more to slow global warming. (The correct answer is all of them help the climate, but managing refrigeration chemicals helps the most. Maybe that information will inspire development and purchase of new products with alternative refrigeration technology.)[14]

Too often in history, business and science are at odds. When evidence pointed to cigarette smoking causing cancer, seat belts saving lives, or gasoline-powered cars causing global warming, these industries fought back for many years. It took years and an increasing amount of scientific evidence for consumers to recognize the potential problems and governments to take action—but eventually scientific evidence won the day. Government and society should support science so that consumers have accurate information about the effects the products they buy have on themselves and the world.

Internet for consumers—sunny or partly cloudy

The Internet promises to create a more transparent business environment for customers by shining "sunlight" on bad practices. The Internet can quickly spread information about a firm that produces a defective product or engages in unethical behavior. Companies that claim to be environmentally friendly without backing up those claims are "outed" by reputable nonprofits and the word spreads. In many ways this has proven to be the case. Early adopter customers review purchases on websites. Reports of a boring book, misleading product descriptions, poorly designed products, or bad customer service help later customers make more informed choices. On the other hand, there are growing reports of companies gaming the review system with fake reviews—leading consumers to question this source of information.

In general, the Internet shines a light that helps consumers make more informed choices. The risk is that increasingly web-savvy brands will find ways to clean up their image without cleaning up their practices.[15]

Regulation necessary at times

One of the advantages of a market-directed economic system is that it operates automatically. But in our version of this system, consumer-citizens, by voting for political leaders, provide certain constraints (laws), which can be modified at any time. Managers who ignore consumer attitudes must realize that their actions may cause new restraints. If consumers want to make more informed choices, there need to be

#M4BW Much of the world's chocolate is produced by slave labor. In Ghana and the Ivory Coast, children and young adults are abducted and taken to cocoa farms where they are forced to work against their will for little or no money. Tony's Chocolonely wants to make "100% slave free the norm in chocolate." The Internet helps Tony's get the word out to customers who care about this issue.
Source: Tony's Chocolonely

penalties for companies that don't tell the truth—a free and open press and Internet are not always enough. Throughout this book we have described many of the laws that keep firms in line. While regulations keep companies honest, they also raise costs. As a society, we have to ask if and when regulations are needed.

How Far Should the Marketing Concept Go?

Should marketing managers limit consumers' freedom of choice?

Achieving a better macro-marketing system is certainly a desirable objective. But what part should a marketer play in deciding what products to offer?

This is extremely important because some marketing managers, especially those in large corporations, can have an impact far larger than they do in their roles as consumer-citizens. For example, should they refuse to produce hazardous products, such as skis or motorcycles, even though such products are in strong demand? Should they install safety devices that increase costs and raise prices that customers don't want to pay?

These are difficult questions to answer. Some things marketing managers do clearly benefit both the firm and consumers because they lower costs or improve consumers' options. But other decisions may actually reduce consumer choice and conflict with a desire to improve the effectiveness of our macro-marketing system.

Consumer-citizens should vote on the changes

It seems fair to suggest, therefore, that marketing managers should be expected to improve and expand the range of goods and services they make available—always trying to add value and better satisfy consumers' needs and preferences. This is the job we've assigned to business.

If pursuing this objective makes excessive demands on scarce resources or has an unacceptable ecological effect, then consumer-citizens have the responsibility to vote for laws restricting individual firms that are trying to satisfy consumers' needs. This is the role that we, as consumers, have assigned to the government—to ensure that the macro-marketing system works effectively.

It is important to recognize that some *seemingly minor* modifications in our present system *might* result in very big, unintended problems. Allowing some government agency to prohibit the sale of products for seemingly good reasons could lead to major changes we never expected and could seriously reduce consumers' present rights to freedom of choice, including "bad" choices.

CONCLUSION

Macro-marketing does *not* cost too much. Consumers have assigned business the role of satisfying their needs. Customers find it satisfactory and even desirable to permit businesses to cater to them and even to stimulate wants. As long as consumers are satisfied, macro-marketing will not cost too much—and business firms will be permitted to continue as profit-making entities. But business exists at the consumer's discretion. It's mainly by satisfying the consumer that a particular firm—and *our* economic system—can justify its existence and hope to keep operating.

In carrying out this role—granted by consumers—business firms are not always as effective as they could be. Many business managers don't understand the marketing concept or the role that marketing plays in our way of life. They seem to feel that business has a basic right to operate as it chooses. And they proceed in their typical production-oriented ways. Further, many managers have had little or no training in business management and are not as competent as they should be. Others fail to adjust to the changes taking place around them. And a few dishonest or unethical managers can do a great deal of damage before consumer-citizens take steps to stop them. As a result, marketing by individual firms often *does* cost too much. But the situation is improving. More business training is now available, and more competent people are being attracted to marketing and business generally. Clearly, *you* have a role to play in improving marketing activities in the future.

The marketing strategy planning process presented in this book provides a framework that will guide you to more effective marketing decisions—and marketing that really does deliver superior value to customers. It also benefits the firm through profits and growth. It's truly a "win-win" situation. And in our competitive, market-driven economy, managers and firms that lead the way in creating these successes will not go unnoticed. As effective marketing management spreads to more companies, the whole macro-marketing system will be more efficient and effective.

Finally, we return to a theme throughout this textbook—that many firms today develop marketing strategies that are profitable and make for a better world (#M4BW). At this point we analyze why this trend has occurred and conclude that it is driven by customers. It is therefore consistent with the marketing orientation and reflects the desire to satisfy customer needs. In most cases, strategies that emphasize making for a better world reflect customers' interest and willingness to purchase from firms that engage in these practices. We then describe how customers need to be responsible in making better purchase choices—and they need to evaluate information in the marketplace. We then outline the responsibilities of marketers to be honest and transparent and of government and society, including the press, to hold companies to higher standards of honesty. In this type of economic system, consumers can make the best choices for them—and that helps make for better marketing and a better world.

KEY TERM

LO 19.7

greenwashing, 545

QUESTIONS AND PROBLEMS

1. Review the case study that opens this chapter. From this case, identify examples of different concepts covered in the chapter. For example, highlight examples of micro- versus macro-marketing criteria.

2. Review the case study that opens this chapter and look at the section "Today's Marketers Face Challenges and Opportunities." If you could advise the U.S. government, would you increase regulation around those challenges? Why or why not? Offer specific examples.

3. Explain why marketing must be evaluated at two levels. What criteria should be used to evaluate each level of marketing? Defend your answer. Explain why your criteria are better than alternative criteria.

4. Discuss the merits of various economic system objectives. Is the objective of the American economic system sensible? Could it achieve more consumer satisfaction if sociologists or public officials determined how to satisfy the needs of lower-income or less educated consumers? If so, what education or income level should be required before an individual is granted free choice?

5. Should the objective of our economy be maximum efficiency? If your answer is yes, efficiency in what? If not, what should the objective be?

6. Why does adoption of the marketing concept encourage a firm to operate more efficiently? Be specific about the impact of the marketing concept on the various departments of a firm.

7. In the short run, competition sometimes leads to inefficiency in the operation of our economic system. Many people argue for monopoly in order to eliminate this inefficiency. Discuss this solution.

8. How would officially granted monopolies affect the operation of our economic system? Consider the effect on allocation of resources, the level of income and employment,

and the distribution of income. Is the effect any different if a firm obtains a monopoly by winning out in a competitive market?

9. Comment on the following statement: "Ultimately, the high cost of marketing is due only to consumers."

10. Distinguish clearly between a marketing strategy and a marketing plan. If a firm has a really good strategy, does it need to worry about developing a written plan?

11. How far should the marketing concept go? How should we decide this issue?

12. Should marketing managers, or business managers in general, refrain from producing profitable products that some target customers want, but that may not be in their long-run best interest? Should firms be expected to produce "good" but less profitable products? What if such products break even? What if they are unprofitable but the company makes other profitable products—so on balance it still makes some profit? What criteria are you using for each of your answers?

13. Should a marketing manager or a business refuse to produce an "energy-gobbling" appliance that some consumers are demanding? Should a firm install an expensive safety device that will increase costs but that customers don't want? Are the same principles involved in both these questions? Explain.

14. Discuss how one or more of the trends or changes shown in Exhibit 19-5 are affecting marketing strategy planning for a specific firm that serves the market where you live.

15. Discuss how slower economic growth or no economic growth would affect your college community—in particular, its marketing institutions.

16. *The Consumerist* (www.consumerist.com) is a widely read blog. It operates as a consumer watchdog and monitors corporate behavior. Go to this blog and read a posting critical of some firm's behavior—click through to the original source if necessary. What do you think of this story? Does it change your attitude toward the firm? How could sites such as these influence consumers? How could they influence firms?

MARKETING PLANNING FOR HILLSIDE VETERINARY CLINIC

Appendix D (the Appendices follow this chapter) includes a sample marketing plan for Hillside Veterinary Clinic. Review the entire marketing plan.

1. How do the pieces fit together?
2. Does the marketing strategy logically follow from the target market dimensions?
3. Does the marketing strategy logically follow from the differentiation and positioning?
4. Does the plan appear reasonable given the stated objectives?

SUGGESTED CASES

17. Wise Water, Inc.
22. Bright Light Innovations: The Starlight Stove
25. QXR Tools (QXR)

26. AAA Custom Castings, Inc.
27. Canadian Mills, Ltd.
28. Kingston Home Health Services (KHHS)

Video Case 1. Potbelly Sandwich
Video Case 3. Big Brothers Big Sisters of America

Design element: #M4BW box globe icon: ©Vectoryzen/Shutterstock

APPENDIX A

Economics Fundamentals

LEARNING OBJECTIVES

The economist's traditional analysis of supply and demand is a useful tool for analyzing markets. In particular, you should master the concepts of a demand curve and demand elasticity. A firm's demand curve shows how the target customers view the firm's Product—really its whole marketing mix. And the interaction of demand and supply curves helps set the size of a market and the market price. The interaction of supply and demand also determines the nature of the competitive environment, which has an important effect on strategy planning. The learning objectives and following sections of this appendix discuss these ideas more fully.

When you finish this appendix, you should be able to

1. understand the "law of diminishing demand."
2. explain elasticity of demand and supply.
3. understand why demand elasticity can be affected by availability of substitutes.
4. understand demand and supply curves and how they set the size of a market and its price level.
5. know the different kinds of competitive situations and understand why they are important to marketing managers.
6. understand important new terms (shown in **red**).

Products and Markets as Seen by Customers and Potential Customers

Economists see individual customers choosing among alternatives

A basic idea from economics is that most customers have a limited income and simply cannot buy everything they want. They must balance their needs and the prices of various products. Economists usually assume that customers have a fairly definite set of preferences and that they evaluate alternatives in terms of whether the alternatives will make them feel better or in some way improve their situation.

But what exactly is the nature of a customer's desire for a particular product?

Exhibit A-1 Demand Schedule for Potatoes (10-pound bags)

Point	(1) Price of Potatoes per Bag (P)	(2) Quantity Demanded (bags per month) (Q)	(3) Total Revenue per Month (P × Q = TR)
A	$1.60	8,000,000	$12,800,000
B	1.30	9,000,000	_____
C	1.00	11,000,000	11,000,000
D	0.70	14,000,000	_____
E	0.40	19,000,000	_____

Usually economists answer this question in terms of the extra utility the customer can obtain by buying more of a particular product—or how much utility would be lost if the customer had less of the product. It is easier to understand the idea of utility if we look at what happens when the price of one of the customer's usual purchases changes.

The law of diminishing demand

LO A.1

Suppose that consumers buy potatoes in 10-pound bags at the same time they buy other foods such as bread and rice. If the consumers are mainly interested in buying a certain amount of food and the price of the potatoes drops, it seems reasonable to expect that they will switch some of their food money to potatoes and away from some other foods. But if the price of potatoes rises, you expect these consumers to buy fewer potatoes and more of other foods.

The general relationship between price and quantity demanded illustrated by this food example is called the **law of diminishing demand**, which says that if the price of a product is raised, a smaller quantity will be demanded, and if the price of a product is lowered, a greater quantity will be demanded. Experience supports this relationship between price and total demand in a market, especially for broad product categories or commodities such as potatoes.

The relationship between price and quantity demanded in a market is what economists call a *demand schedule*. An example is shown in Exhibit A-1. For each row in the table, column 2 shows the quantity consumers will want (demand) if they have to pay the price given in column 1. The third column shows that the total revenue (sales) in the potato market is equal to the quantity demanded at a given price times that price. Note that as prices drop, the total *unit* quantity increases, yet the total *revenue* decreases. Fill in the blank lines in the third column and observe the behavior of total revenue, an important number for the marketing manager. We will explain what you should have noticed, and why, a little later.

The demand curve—usually down-sloping

If your interest is seeing only at which price the company will earn the greatest total revenue, the demand schedule may be adequate. But a demand curve shows more. A **demand curve** is a graph of the relationship between price and quantity demanded in a market, assuming that all other things stay the same. Exhibit A-2 shows the demand

Exhibit A-2 Demand Curve for Potatoes (10-pound bags)

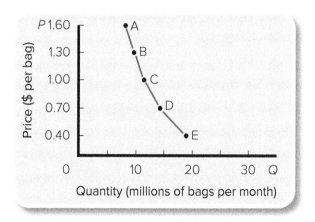

Exhibit A-3 Demand Schedule for 1-Cubic-Foot Microwave Ovens

Point	(1) Price per Microwave Oven (P)	(2) Quantity Demanded per Year (Q)	(3) Total Revenue (TR) per Year (P × Q = TR)
A	$300	20,000	$ 6,000,000
B	250	70,000	17,500,000
C	200	130,000	26,000,000
D	150	210,000	31,500,000
E	100	310,000	31,000,000

curve for potatoes—really just a plotting of the demand schedule in Exhibit A-1. It shows how many potatoes potential customers will demand at various possible prices. This is a "down-sloping demand curve."

Most demand curves are down-sloping. This just means that if prices are decreased, the quantity customers demand will increase.

Demand curves always show the price on the vertical axis and the quantity demanded on the horizontal axis. In Exhibit A-2, we have shown the price in dollars. For consistency, we will use dollars in other examples. However, keep in mind that these same ideas hold regardless of what money unit (dollars, yen, euros, pounds, etc.) is used to represent price. Even at this early point, you should keep in mind that markets are not necessarily limited by national boundaries—or by one type of money.

Note that the demand curve shows only how customers will react to various possible prices. In a market, we see only one price at a time, not all of these prices. The curve, however, shows what quantities will be demanded, depending on what price is set.

Microwave oven demand curve looks different

To get a more complete picture of demand-curve analysis, let's consider another product that has a different demand schedule and curve. A demand schedule for standard 1-cubic-foot microwave ovens is shown in Exhibit A-3. Column 3 shows the total revenue that will be obtained at various possible prices and quantities. Again, as the price goes down, the quantity demanded goes up. But here, unlike the potato example, total revenue increases as prices go down—at least until the price drops to $100.

Every market has a demand curve, for some time period

These general demand relationships are typical for all products. But each product has its own demand schedule and curve in each potential market, no matter how small the market. In other words, a particular demand curve has meaning only for a particular market. We can think of demand curves for individuals, groups of individuals who form a target market, regions, and even countries. And the time period covered really should be specified, although this is often neglected because we usually think of monthly or yearly periods.

The difference between elastic and inelastic

LO A.2

The demand curve for microwave ovens (see Exhibit A-4) is down-sloping—but note that it is flatter than the curve for potatoes. It is important to understand what this flatness means.

We will consider the flatness in terms of total revenue because this is what interests business managers.[1]

When you filled in the total revenue column for potatoes, you should have noticed that total revenue drops continually if the price is reduced. This looks undesirable for sellers and illustrates inelastic demand. **Inelastic demand** means that although the quantity demanded increases if the price is decreased, the quantity demanded will not "stretch" enough—that is, it is not elastic enough—to avoid a decrease in total revenue.

In contrast, **elastic demand** means that if prices are dropped, the quantity demanded will stretch (increase) enough to increase total revenue. The upper part of the microwave oven demand curve is an example of elastic demand.

But note that if the microwave oven price is dropped from $150 to $100, total revenue will decrease. We can say, therefore, that between $150 and $100, demand is inelastic—that is, total revenue will decrease if price is lowered from $150 to $100.

Exhibit A-4 Demand Curve for 1-Cubic-Foot Microwave Ovens

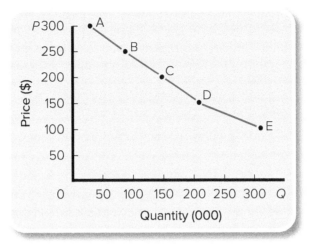

Thus, elasticity can be defined in terms of changes in total revenue. *If total revenue will increase if price is lowered, then demand is elastic. If total revenue will decrease if price is lowered, then demand is inelastic.* (*Note:* A special case known as "unitary elasticity of demand" occurs if total revenue stays the same when prices change.)

Total revenue may increase if price is raised

A point often missed in discussions of demand is what happens when prices are raised instead of lowered. With elastic demand, total revenue will *decrease* if the price is *raised*. With inelastic demand, however, total revenue will *increase* if the price is *raised*.

The possibility of raising price and increasing dollar sales (total revenue) at the same time is attractive to managers. This occurs only if the demand curve is inelastic. Here total revenue will increase if price is raised, but total costs probably will not increase—and may actually go down—with smaller quantities. Keep in mind that profit is equal to total revenue minus total costs. So when demand is inelastic, profit will increase as price is increased!

The ways total revenue changes as prices are raised are shown in Exhibit A-5. Here total revenue is the rectangular area formed by a price and its related quantity. The larger the rectangular area, the greater the total revenue.

Exhibit A-5 Changes in Total Revenue as Prices Increase

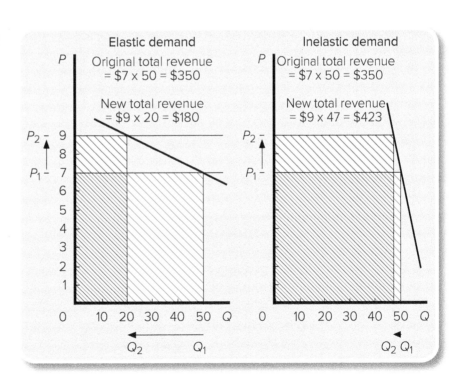

P_1 is the original price here, and the total potential revenue with this original price is shown by the area with blue shading. The area with red shading shows the total revenue with the new price, P_2. There is some overlap in the total revenue areas, so the important areas are those with only one color. Note that in the left-hand figure—where demand is elastic—the revenue added (the red-only area) when the price is increased is less than the revenue lost (the blue-only area). Now let's contrast this to the right-hand figure, when demand is inelastic. Only a small blue revenue area is given up for a much larger (red) one when price is raised.

An entire curve is not elastic or inelastic

It is important to see that it is *wrong to refer to a whole demand curve as elastic or inelastic*. Rather, elasticity for a particular demand curve refers to the change in total revenue between two points on the curve, not along the whole curve. You saw the change from elastic to inelastic in the microwave oven example. Generally, however, nearby points are either elastic or inelastic—so it is common to refer to a whole curve by the degree of elasticity in the price range that normally is of interest—the *relevant range*.

Demand elasticities affected by availability of substitutes and urgency of need

LO A.3

At first, it may be difficult to see why one product has an elastic demand and another an inelastic demand. Many factors affect elasticity, such as the availability of substitutes, the importance of the item in the customer's budget, and the urgency of the customer's need and its relation to other needs. By looking more closely at one of these factors—the availability of substitutes—you will better understand why demand elasticities vary.

Substitutes are products that offer the buyer a choice. For example, many consumers see grapefruit as a substitute for oranges and hot dogs as a substitute for hamburgers. The greater the number of "good" substitutes available, the greater will be the elasticity of demand. From the consumer's perspective, products are "good" substitutes if they are very similar (homogeneous). If consumers see products as extremely different, or heterogeneous, then a particular need cannot easily be satisfied by substitutes. And the demand for the most satisfactory product may be quite inelastic.

As an example, if the price of hamburger is lowered (and other prices stay the same), the quantity demanded will increase a lot, as will total revenue. The reason is that not only will regular hamburger users buy more hamburger, but some consumers who formerly bought hot dogs or steaks probably will buy hamburger too. But if the price of hamburger is raised, the quantity demanded will decrease, perhaps sharply. Still, consumers will buy some hamburger, depending on how much the price has risen, their individual tastes, and what their guests expect (see Exhibit A-6).

In contrast to a product with many "substitutes"—such as hamburger—consider a product with few or no substitutes. Its demand curve will tend to be inelastic. Motor

Exhibit A-6 Demand Curve for Hamburger (a product with many substitutes)

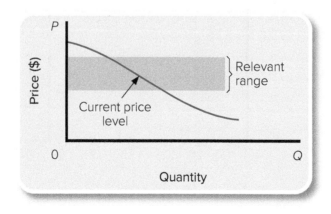

Exhibit A-7 Demand Curve for Motor Oil (a product with few substitutes)

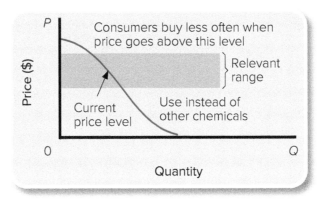

oil is a good example. Motor oil is needed to keep cars running. Yet no one person or family uses great quantities of motor oil. So it is not likely that the quantity of motor oil purchased will change much as long as price changes are *within a reasonable range*. Of course, if the price is raised to a staggering figure, many people will buy less oil (change their oil less frequently). If the price is dropped to an extremely low level, manufacturers may buy more—say, as a lower-cost substitute for other chemicals typically used in making plastic (Exhibit A-7). But these extremes are outside the relevant range.

Demand curves are introduced here because the degree of elasticity of demand shows how potential customers feel about a product—and especially whether they see substitutes for the product. But to get a better understanding of markets, we must extend this economic analysis.

Markets as Seen by Suppliers

Customers may want some product—but if suppliers are not willing to supply it, then there is no market. So we'll study the economist's analysis of supply. And then we'll bring supply and demand together for a more complete understanding of markets.

Economists often use the kind of analysis we are discussing here to explain pricing in the marketplace. But that is not our intention. Here we are interested in how and why markets work and the interaction of customers and potential suppliers. Later in this appendix we will review how competition affects prices, but how individual firms set prices, or should set prices, is discussed fully in Chapters 17 and 18.

Supply curves reflect supplier thinking

Generally speaking, suppliers' costs affect the quantity of products they are willing to offer in a market during any period. In other words, their costs affect their supply schedules and supply curves. Whereas a demand curve shows the quantity of products customers will be willing to buy at various prices, a **supply curve** shows the quantity of products that will be supplied at various possible prices. Eventually, only one quantity will be offered and purchased. So a supply curve is really a hypothetical (what-if) description of what will be offered at various prices. It is, however, a very important curve. Together with a demand curve, it summarizes the attitudes and probable behavior of buyers and sellers about a particular product in a particular market—that is, in a product-market.

Some supply curves are vertical

We usually assume that supply curves tend to slope upward—that is, suppliers will be willing to offer greater quantities at higher prices. If a product's market price is very high, it seems only reasonable that producers will be anxious to produce more of the

Exhibit A-8 Supply Schedule for Potatoes (10-pound bags)

Point	Possible Market Price per 10-lb Bag	Number of Bags Sellers Will Supply per Month at Each Possible Market Price
A	$1.60	17,000,000
B	1.30	14,000,000
C	1.00	11,000,000
D	0.70	8,000,000
E	0.40	3,000,000

Note: These data and the resulting supply curve in Exhibit A-9 are for a single month to emphasize that farmers might have some control over when they deliver their potatoes. There would be different data, and consequently a different curve, for each month.

product and even put workers on overtime or perhaps hire more workers to increase the quantity they can offer. Going further, it seems likely that producers of other products will switch their resources (farms, factories, labor, or retail facilities) to the product that is in great demand.

On the other hand, if consumers are willing to pay a very low price only for a particular product, it's reasonable to expect that producers will switch to other products, thus reducing supply. A supply schedule (Exhibit A-8) and a supply curve (Exhibit A-9) for potatoes illustrate these ideas. This supply curve shows how many potatoes would be produced and offered for sale at each possible market price in a given month.

In the very short run (say, over a few hours, a day, or a week), a supplier may not be able to change the supply at all. In this situation, we would see a vertical supply curve. This situation is often relevant in the market for fresh produce. Fresh strawberries, for example, continue to ripen, and a supplier wants to sell them quickly—preferably at a higher price—but in any case, they must be sold.

If the product is a service, it may not be easy to expand the supply in the short run. Additional barbers or medical doctors are not quickly trained and licensed, and they have only so much time to give each day. Further, the prospect of much higher prices in the near future cannot easily expand the supply of many services. For example, a hit play or an "in" restaurant or nightclub is limited in the amount of product it can offer at a particular time.

Elasticity of supply

The term *elasticity* also is used to describe supply curves. An extremely steep or almost vertical supply curve, often found in the short run, is called **inelastic supply** because the quantity supplied does not stretch much (if at all) if the price is raised.

Exhibit A-9 Supply Curve for Potatoes (10-pound bags)

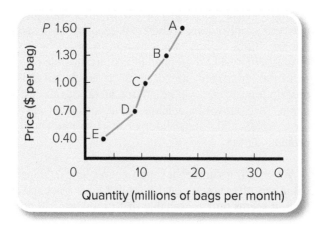

A flatter curve is called **elastic supply** because the quantity supplied does stretch more if the price is raised. A slightly up-sloping supply curve is typical in longer-run market situations. Given more time, suppliers have a chance to adjust their offerings, and competitors may enter or leave the market.

Demand and Supply Interact to Determine the Size of the Market and Price Level

LO A.4

We have treated market demand and supply forces separately. Now we must bring them together to show their interaction. The *intersection* of these two forces determines the size of the market and the market price—at which point (price and quantity) the market is said to be in *equilibrium*.

The intersection of demand and supply is shown for the potato data discussed earlier. In Exhibit A-10, the demand curve for potatoes is now graphed against the supply curve in Exhibit A-9.

In this potato market, demand is inelastic—the total revenue of all the potato producers would be greater at higher prices. But the market price is at the **equilibrium point**—where the quantity and the price sellers are willing to offer are equal to the quantity and price that buyers are willing to accept. The $1.00 equilibrium price for potatoes yields a smaller *total revenue* to potato producers than a higher price would. This lower equilibrium price comes about because the many producers are willing to supply enough potatoes at the lower price. *Demand is not the only determiner of price level. Cost also must be considered—via the supply curve.*

Some consumers get a surplus

Presumably, a sale takes place only if both buyer and seller feel they will be better off after the sale. But sometimes the price a consumer pays in a sales transaction is less than what he or she would be willing to pay.

The reason for this is that demand curves are typically down-sloping, and some of the demand curve is above the equilibrium price. This is simply another way of showing that some customers would have been willing to pay more than the equilibrium price—if they had to. In effect, some of them are getting a bargain by being able to buy at the equilibrium price. Economists have traditionally called these bargains the **consumer surplus**—that is, the difference to consumers between the value of a purchase and the price they pay.

Some business critics assume that consumers do badly in any business transaction. In fact, sales take place only if consumers feel they are at least getting their money's worth. As we can see here, some are willing to pay much more than the market price.

Exhibit A-10
Equilibrium of Supply and Demand for Potatoes (10-pound bags)

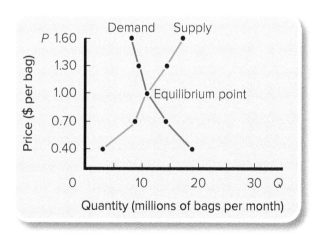

Demand and Supply Help Us Understand the Nature of Competition

LO A.5

The elasticity of demand and supply curves and their interaction help predict the nature of competition a marketing manager is likely to face. For example, an extremely inelastic demand curve means that the manager will have much choice in strategy planning, especially price setting. Apparently customers like the product and see few substitutes. They are willing to pay higher prices before cutting back much on their purchases.

The elasticity of a firm's demand curve is not the only factor that affects the nature of competition. Other factors are the number and size of competitors and the uniqueness of each firm's marketing mix. Understanding these market situations is important because the freedom of a marketing manager, especially control over price, is greatly reduced in some situations.

A marketing manager operates in one of four kinds of market situations. We'll discuss three kinds: pure competition, oligopoly, and monopolistic competition. The fourth kind, monopoly, isn't found very often and is like monopolistic competition. The important dimensions of these situations are shown in Exhibit A-11.

When competition is pure

Many competitors offer about the same thing

Pure competition is a market situation that develops when a market has

1. Homogeneous (similar) products.
2. Many buyers and sellers who have full knowledge of the market.
3. Ease of entry for buyers and sellers; that is, new firms have little difficulty starting in business—and new customers can easily come into the market.

More or less pure competition is found in many agricultural markets. In the potato market, for example, there are thousands of small producers—and they are in pure competition. Let's look more closely at these producers.

Although the potato market as a whole has a down-sloping demand curve, each of the many small producers in the industry is in pure competition, and each of them faces a flat demand curve at the equilibrium price. This is shown in Exhibit A-12.

As shown at the right of Exhibit A-12, an individual producer can sell as many bags of potatoes as he chooses at $1—the market equilibrium price. The equilibrium price is

Exhibit A-11 Some Important Dimensions Regarding Market Situations

Important Dimensions	Types of Situations			
	Pure Competition	Oligopoly	Monopolistic Competition	Monopoly
Uniqueness of each firm's product	None	None	Some	Unique
Number of competitors	Many	Few	Few to many	None
Size of competitors (compared to size of market)	Small	Large	Large to small	None
Elasticity of demand facing firm	Completely elastic	Kinked demand curve (elastic and inelastic)	Either	Either
Elasticity of industry demand	Either	Inelastic	Either	Either
Control of price by firm	None	Some (with care)	Some	Complete

Exhibit A-12 Interaction of Demand and Supply in the Potato Industry and the Resulting Demand Curve Facing Individual Potato Producers

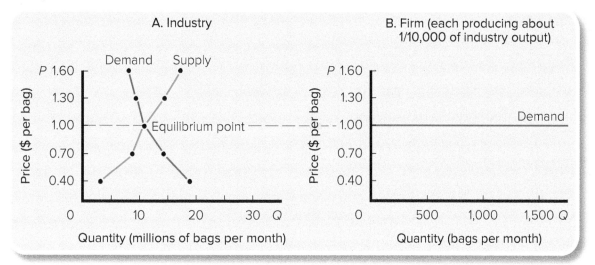

determined by the quantity that all producers choose to sell given the demand curve they face.

But a small producer has little effect on overall supply (or on the equilibrium price). If this individual farmer raises 1/10,000th of the quantity offered in the market, for example, you can see that there will be little effect if the farmer goes out of business—or doubles production.

The reason an individual producer's demand curve is flat is that the farmer probably couldn't sell any potatoes above the market price. And there is no point in selling below the market price! So, in effect, the individual producer has no control over price.

Markets tend to become more competitive

Not many markets are *purely* competitive. But many are close enough so we can talk about "almost" pure competition situations—those in which the marketing manager has to accept the going price.

Such highly competitive situations aren't limited to agriculture. Wherever *many* competitors sell *homogeneous* products—such as textiles, lumber, coal, printing, and laundry services—the demand curve seen by *each producer* tends to be flat.

Markets tend to become more competitive, moving toward pure competition (except in oligopolies—see below). On the way to pure competition, prices and profits are pushed down until some competitors are forced out of business. Eventually, in long-run equilibrium, the price level is high enough only to keep the survivors in business. No one makes any profit—they just cover costs. It's tough to be a marketing manager in this situation!

When competition is oligopolistic

A few competitors offer similar things

Not all markets move toward pure competition. Some become oligopolies. **Oligopoly** situations are special market situations that develop when a market has

1. Essentially homogeneous products—such as basic industrial chemicals or gasoline.
2. Relatively few sellers—or a few large firms and many smaller ones that follow the lead of the larger ones.
3. Fairly inelastic industry demand curves.

The demand curve facing each firm is unusual in an oligopoly situation. Although the industry demand curve is inelastic throughout the relevant range, the demand curve facing each competitor looks "kinked" (see Exhibit A-13). The current market price is at the kink.

Exhibit A-13   Oligopoly–Kinked Demand Curve–Situation

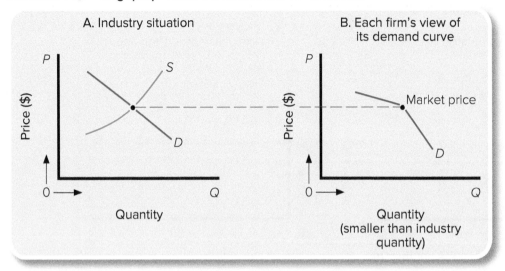

There is a market price because the competing firms watch one another carefully—and they know it's wise to be at the kink. Each firm must expect that raising its own price above the market price will cause a big loss in sales. Few, if any, competitors will follow the price increase. So the firm's demand curve is relatively flat above the market price. If the firm lowers its price, it must expect competitors to follow. Given inelastic industry demand, the firm's own demand curve is inelastic at lower prices, assuming it keeps its share of this market at lower prices. Because lowering prices along such a curve will drop total revenue, the firm should leave its price at the kink—the market price.

Sometimes there are price fluctuations in oligopolistic markets. This can be caused by firms that don't understand the market situation and cut their prices to try to get business. In other cases, big increases in demand or supply change the basic nature of the situation and lead to price-cutting. Price cuts can be drastic, such as DuPont's price cut of 25 percent for Dacron. This happened when DuPont decided that industry production capacity already exceeded demand, and more plants were due to start production.

It's important to keep in mind that oligopoly situations don't just apply to whole industries and national markets. Competitors who are focusing on the same local target market often face oligopoly situations. A suburban community might have several gas stations, all of which provide essentially the same product. In this case, the "industry" consists of the gas stations competing with one another in the local product-market.

As in pure competition, oligopolists face a long-run trend toward an equilibrium level, with profits driven toward zero. This may not happen immediately—and a marketing manager may try to delay price competition by relying more on other elements in the marketing mix.

When competition is monopolistic

A price must be set

You can see why marketing managers want to avoid pure competition or oligopoly situations. They prefer a market in which they have more control. **Monopolistic competition** is a market situation that develops when a market has

1. Different (heterogeneous) products—in the eyes of some customers.
2. Sellers who feel they do have some competition in this market.

The word *monopolistic* means that each firm is trying to get control in its own little market. But the word *competition* means that there are still substitutes. The vigorous competition of a purely competitive market is reduced. Each firm has its own down-sloping demand curve. But the shape of the curve depends on the similarity of competitors' products and marketing mixes. Each monopolistic competitor has freedom—but not complete freedom—in its own market.

Judging elasticity will help set the price

Because a firm in monopolistic competition has its own down-sloping demand curve, it must make a decision about price level as part of its marketing strategy planning. Here, estimating the elasticity of the firm's own demand curve is helpful. If it is highly inelastic, the firm may decide to raise prices to increase total revenue. But if demand is highly elastic, this may mean there are many competitors with acceptable substitutes. Then the price may have to be set near that of the competition. And the marketing manager probably should try to develop a better marketing mix.

CONCLUSION

The economist's traditional demand and supply analysis provides a useful tool for analyzing the nature of demand and competition. It is especially important that you master the concepts of a demand curve and demand elasticity. How demand and supply interact helps determine the size of a market and its price level. The interaction of supply and demand also helps explain the nature of competition in different market situations. We discussed three competitive situations: pure competition, oligopoly, and monopolistic competition. The fourth kind, monopoly, isn't found very often and is like monopolistic competition.

The nature of supply and demand—and competition—is very important in marketing strategy planning. We discuss these topics more fully in Chapters 3 and 4 and then build on them throughout the text. This appendix provides a good foundation on these topics.

KEY TERMS

law of diminishing demand, 553
demand curve, 553
inelastic demand, 554
elastic demand, 554
substitutes, 556

supply curve, 557
inelastic supply, 558
elastic supply, 559
equilibrium point, 559

consumer surplus, 559
pure competition, 560
oligopoly, 561
monopolistic competition, 562

QUESTIONS AND PROBLEMS

1. Explain in your own words how economists look at markets and arrive at the "law of diminishing demand."

2. Explain what a demand curve is and why it is usually down-sloping. Then give an example of a product for which the demand curve might not be down-sloping over some possible price ranges. Explain the reason for your choice.

3. What is the length of life of the typical demand curve? Illustrate your answer.

4. If the general market demand for men's shoes is fairly elastic, how does the demand for men's dress shoes compare to it? How does the demand curve for women's shoes compare to the demand curve for men's shoes?

5. If the demand for perfume is inelastic above and below the present price, should the price be raised? Why or why not?

6. If the demand for shrimp is highly elastic below the present price, should the price be lowered?

7. Discuss what factors lead to inelastic demand and supply curves. Are they likely to be found together in the same situation?

8. Why would a marketing manager prefer to sell a product that has no close substitutes? Are high profits almost guaranteed?

9. If a manufacturer's well-known product is sold at the same price by many retailers in the same community, is this an example of pure competition? When a community has many small grocery stores, are they in pure competition? What characteristics are needed to have a purely competitive market?

10. List three products that are sold in purely competitive markets and three that are sold in monopolistically competitive markets. Do any of these products have anything in common? Can any generalizations be made about competitive situations and marketing mix planning?

11. Cite a local example of an oligopoly, and explain why it is an oligopoly.

APPENDIX B

Marketing Arithmetic

LEARNING OBJECTIVES

Marketing students must become familiar with the essentials of the language of business. Businesspeople commonly use accounting terms when talking about costs, prices, and profit. And using accounting data is a practical tool in analyzing marketing problems. The objectives of this appendix will help you understand these concepts and how they are used by marketing managers.

When you finish this appendix, you should be able to

1. understand the components of an operating statement (profit and loss statement).
2. know how to compute the stockturn rate.
3. understand how operating ratios can be used to analyze a business.
4. understand how to calculate markups and markdowns.
5. understand how to calculate return on investment (ROI) and return on assets (ROA).
6. understand the basic forecasting approaches and why they are used.
7. understand important new terms (shown in **red**).

The Operating Statement

LO B.1

An **operating statement** is a simple summary of the financial results of a company's operations over a specified period of time. Some beginning students may feel that the operating statement is complex, but as we'll soon see, this really isn't true. *The main purpose of the operating statement is determining the net profit figure and presenting data to support that figure.* This is why the operating statement is often referred to as the *profit and loss statement*.

Exhibit B-1 An Operating Statement (Profit and Loss Statement)

Perry Company
Operating Statement
For the Year Ended December 31, 202X

Gross sales			$540,000
Less: Returns and allowances			40,000
Net sales			$500,000
Cost of sales:			
Beginning inventory at cost		$80,000	
Purchases at billed cost	$310,000		
Less: Purchase discounts	40,000		
Purchases at net cost	270,000		
Plus: Freight-in	20,000		
Net cost of delivered purchases		290,000	
Cost of goods available for sale		370,000	
Less: Ending inventory at cost		70,000	
Cost of sales			300,000
Gross margin (gross profit)			200,000
Expenses:			
Selling expenses:			
Sales salaries	60,000		
Advertising expense	20,000		
Website updates	10,000		
Delivery expense	10,000		
Total selling expense		100,000	
Administrative expenses:			
Office salaries	30,000		
Office supplies	10,000		
Miscellaneous administrative expense	5,000		
Total administrative expense		45,000	
General expenses:			
Rent expense	10,000		
Miscellaneous general expenses	5,000		
Total general expense		15,000	
Total expenses			160,000
Net profit from operation			$40,000

Exhibit B-1 shows an operating statement for a wholesale or retail business. The statement is complete and detailed so you will see the framework throughout the discussion, but the amount of detail on an operating statement is *not* standardized. Many companies use financial statements with much less detail than this one. They emphasize clarity and readability rather than detail. To really understand an operating statement, however, you must know about its components.

Only three basic components

The basic components of an operating statement are *sales*—which come from the sale of goods and services; *costs*—which come from the producing and selling process; and the balance (called *profit* or *loss*)—which is just the difference between sales and costs. So there are only three basic components in the statement: sales, costs, and profit (or loss). Other items on an operating statement are there only to provide supporting details.

Time period covered may vary

There is no one time period an operating statement covers. Rather, statements are prepared to satisfy the needs of a particular business. This may be at the end of each day or at the end of each week. Usually, however, an operating statement summarizes results for one month, three months, six months, or a full year. Because the time period does vary, this information is included in the heading of the statement as follows:

> **Perry Company**
> **Operating Statement**
> **For the (Period) Ended (Date)**

Also see Exhibit B-1.

Management's uses of operating statements

Before going on to a more detailed discussion of the components of our operating statement, let's think about some of the uses for such a statement. Exhibit B-1 shows that a lot of information is presented in a clear and concise manner. With this information, a manager can easily find the relation of net sales to the cost of sales, the gross margin, expenses, and net profit. Opening and closing inventory figures are available—as is the amount spent during the period for the purchase of goods for resale. Total expenses are listed to make it easier to compare them with previous statements and to help control these expenses.

All this information is important to a company's managers. Assume that a particular company prepares monthly operating statements. A series of these statements is a valuable tool for directing and controlling the business. By comparing results from one month to the next, managers can uncover unfavorable trends in the sales, costs, or profit areas of the business and take any needed action.

A skeleton statement gets down to essential details

Let's refer to Exhibit B-1 and begin to analyze this seemingly detailed statement to get firsthand knowledge of the components of the operating statement.

As a first step, suppose we take all the items that have dollar amounts extended to the third, or right-hand, column. Using these items only, the operating statement looks like this:

Gross sales	$540,000
Less: Returns and allowances	40,000
Net sales	500,000
Less: Cost of sales	300,000
Gross margin	200,000
Less: Total expenses	160,000
Net profit (loss)	$ 40,000

Is this a complete operating statement? The answer is *yes*. This skeleton statement differs from Exhibit B-1 only in supporting detail. All the basic components are included. In fact, the only items we must list to have a complete operating statement are

Net sales	$500,000
Less: Costs	460,000
Net profit (loss)	$ 40,000

These three items are the essentials of an operating statement. All other subdivisions or details are just useful additions.

Meaning of sales

Now let's define the meaning of the terms in the skeleton statement.

The first item is sales. What do we mean by sales? The term **gross sales** is the total amount charged to all customers during some time period. However, there is always some customer dissatisfaction or just plain errors in ordering and shipping goods. This results in returns and allowances, which reduce gross sales.

A **return** occurs when a customer sends back purchased products. The company either refunds the purchase price or allows the customer dollar credit on other purchases.

An **allowance** usually occurs when a customer is not satisfied with a purchase for some reason. The company gives a price reduction on the original invoice (bill), but the customer keeps the goods and services.

These refunds and price reductions must be considered when the firm computes its net sales figure for the period. Really, we're only interested in the revenue the company manages to keep. This is **net sales**—the actual sales dollars the company receives. Therefore, all reductions, refunds, cancellations, and so forth made because of returns and allowances are deducted from the original total (gross sales) to get net sales. This is shown below.

Gross sales	$540,000
Less: Returns and allowances	40,000
Net sales	$500,000

Meaning of cost of sales

The next item in the operating statement—**cost of sales**—is the total value (at cost) of the sales during the period. We'll discuss this computation later. Meanwhile, note that after we obtain the cost of sales figure, we subtract it from the net sales figure to get the gross margin.

Meaning of gross margin and expenses

Gross margin (gross profit) is the money left to cover the expenses of selling the products and operating the business. Firms hope that a profit will be left after subtracting these expenses.

Selling expense is commonly the major expense below the gross margin. Note that in Exhibit B-1, **expenses** are all the remaining costs subtracted from the gross margin to get the net profit. The expenses in this case are the selling, administrative, and general expenses. (Note that the cost of purchases and cost of sales are not included in this total expense figure—they were subtracted from net sales earlier to get the gross margin. Note, also, that some accountants refer to cost of sales as *cost of goods sold*.)

Net profit—at the bottom of the statement—is what the company earned from its operations during a particular period. It is the amount left after the cost of sales and the expenses are subtracted from net sales. *Net sales and net profit are not the same.* Many firms have large sales and no profits—they may even have losses! That's why understanding costs, and controlling them, is important.

Detailed Analysis of Sections of the Operating Statement

Cost of sales for a wholesale or retail company

The cost of sales section includes details that are used to find the cost of sales ($300,000 in our example).

In Exhibit B-1, you can see that beginning and ending inventory, purchases, purchase discounts, and freight-in are all necessary to calculate cost of sales. If we pull the cost of sales section from the operating statement, it looks like this:

Cost of sales:			
Beginning inventory at cost			$ 80,000
Purchases at billed cost	$310,000		
Less: Purchase discounts	40,000		
Purchases at net cost	270,000		
Plus: Freight-in	20,000		
Net cost of delivered purchases		290,000	
Cost of goods available for sale		370,000	
Less: Ending inventory at cost		70,000	
Cost of sales			$300,000

Cost of sales is the cost value of what is *sold,* not the cost of goods on hand at any given time.

Inventory figures merely show the cost of goods on hand at the beginning and end of the period the statement covers. These figures may be obtained by physically counting goods on hand on these dates or estimated from perpetual inventory records that show the inventory balance at any given time. The methods used to determine the inventory should be as accurate as possible because these figures affect the cost of sales during the period and net profit.

The net cost of delivered purchases must include freight charges and purchase discounts received because these items affect the money actually spent to buy goods and bring them to the place of business. A **purchase discount** is a reduction of the original invoice amount for some business reason. For example, a cash discount may be given for prompt payment of the amount due. We subtract the total of such discounts from the original invoice cost of purchases to get the *net* cost of purchases. To this figure we add the freight charges for bringing the goods to the place of business. This gives the net cost of *delivered* purchases. When we add the net cost of delivered purchases to the beginning inventory at cost, we have the total cost of goods available for sale during the period. If we now subtract the ending inventory at cost from the cost of the goods available for sale, we get the cost of sales.

One important point should be noted about cost of sales. The way the value of inventory is calculated varies from one company to another—and it can cause big differences in the cost of sales and the operating statement. (See any basic accounting textbook for how the various inventory valuation methods work.)

Exhibit B-1 shows the way the manager of a wholesale or retail business arrives at her cost of sales. Such a business *purchases* finished products and resells them. In a manufacturing company, the purchases section of this operating statement is replaced by a section called cost of production. This section includes purchases of raw materials and parts, direct and indirect labor costs, and factory overhead charges (such as heat, light, and power) that are necessary to produce finished products. The cost of production is added to the beginning finished products inventory to arrive at the cost of products available for sale. Often, a separate cost of production statement is prepared, and only the total cost of production is shown in the operating statement. See Exhibit B-2

Exhibit B-2 Cost of Sales Section of an Operating Statement for a Manufacturing Firm

Cost of sales:			
Finished products inventory (beginning)		$ 20,000	
Cost of production (Schedule 1)		100,000	
Total cost of finished products available for sale		120,000	
Less: Finished products inventory (ending)		30,000	
Cost of sales			$ 90,000
Schedule 1, Schedule of cost of production			
Beginning work in process inventory			15,000
Raw materials:			
Beginning raw materials inventory		10,000	
Net cost of delivered purchases		80,000	
Total cost of materials available for use		90,000	
Less: Ending raw materials inventory		15,000	
Cost of materials placed in production		75,000	
Direct labor		20,000	
Manufacturing expenses:			
Indirect labor	$4,000		
Maintenance and repairs	3,000		
Factory supplies	1,000		
Heat, light, and power	2,000		
Total manufacturing expenses		10,000	
Total manufacturing costs			105,000
Total work in process during period			120,000
Less: Ending work in process inventory			20,000
Cost of production			$100,000

for an illustration of the cost of sales section of an operating statement for a manufacturing company.

Expenses

Expenses go below the gross margin. They usually include the costs of selling and the costs of administering the business. They do not include the cost of sales, either purchased or produced.

There is no right method for classifying the expense accounts or arranging them on the operating statement. They can just as easily be arranged alphabetically or according to amount, with the largest placed at the top and so on down the line. In a business of any size, though, it is clearer to group the expenses in some way and use subtotals by groups for analysis and control purposes. This was done in Exhibit B-1.

Summary on operating statements

The statement presented in Exhibit B-1 contains all the major categories in an operating statement—together with a normal amount of supporting detail. Further detail can be added to the statement under any of the major categories without changing the nature of the statement. The amount of detail normally is determined by how the statement will be used. A stockholder may be given an abbreviated operating statement—whereas the one prepared for internal company use may have a lot of detail.

Computing the Stockturn Rate

LO B.2

A detailed operating statement can provide the data needed to compute the **stockturn rate**—a measure of the number of times the average inventory is sold during a year. Note that the stockturn rate is related to the *turnover during a year,* not the length of time covered by a particular operating statement.

The stockturn rate is a very important measure because it shows how rapidly the firm's inventory is moving. Some businesses typically have slower turnover than others. But a drop in turnover in a particular business can be very alarming. It may mean that the firm's assortment of products is no longer as attractive as it was. Also, it may mean that the firm will need more working capital to handle the same volume of sales. Most businesses pay a lot of attention to the stockturn rate, trying to get faster turnover (and lower inventory costs).

Three methods, all basically similar, can be used to compute the stockturn rate. Which method is used depends on the data available. These three methods, which usually give approximately the same results, are shown below.[1]

(1) $$\frac{\text{Cost of sales}}{\text{Average inventory at cost}}$$

(2) $$\frac{\text{Net sales}}{\text{Average inventory at selling price}}$$

(3) $$\frac{\text{Sales in units}}{\text{Average inventory in units}}$$

Computing the stockturn rate will be illustrated only for Formula 1, because all are similar. The only difference is that the cost figures used in Formula 1 are changed to a selling price or numerical count basis in Formulas 2 and 3. (*Note:* Regardless of the method used, you must have both the numerator and denominator of the formula in the same terms.)

If the inventory level varies a lot during the year, you may need detailed information about the inventory level at different times to compute the average inventory. If it stays at about the same level during the year, however, it's easy to get an estimate. For example, using Formula 1, the average inventory at cost is computed by adding the beginning and ending inventories at cost and dividing by 2. This average inventory figure is then divided into the cost of sales (in cost terms) to get the stockturn rate.

For example, suppose that the cost of sales for one year was $1,000,000. Beginning inventory was $250,000 and ending inventory $150,000. Adding the two inventory figures and dividing by 2, we get an average inventory of $200,000. We next divide the cost of sales by the average inventory ($1,000,000 ÷ $200,000) and get a stockturn rate of 5. The stockturn rate is covered further in Chapters 12 and 18.

Operating Ratios Help Analyze the Business

LO B.3

Many businesspeople use the operating statement to calculate **operating ratios**—the ratio of items on the operating statement to net sales—and to compare these ratios from one time period to another. They can also compare their own operating ratios with those of competitors. Such competitive data are often available through trade associations. Each firm may report its results to a trade association, which then distributes summary results to its members. These ratios help managers control their operations. If some expense ratios are rising, for example, those particular costs are singled out for special attention.

Operating ratios are computed by dividing net sales into the various operating statement items that appear below the net sales level in the operating statement. The net sales figure is used as the denominator in the operating ratio because it shows the sales the firm actually won.

We can see the relation of operating ratios to the operating statement if we think of there being another column to the right of the dollar figures in an operating statement. This column contains percentage figures, using net sales as 100 percent. This approach can be seen below.

Gross sales	$540,000	
Less: Returns and allowances	40,000	
Net sales	500,000	100%
Less: Cost of sales	300,000	60
Gross margin	200,000	40
Less: Total expenses	160,000	32
Net profit	$ 40,000	8%

The 40 percent ratio of gross margin to net sales in the preceding example shows that 40 percent of the net sales dollar is available to cover selling and administrative expenses and provide a profit. Note that the ratio of expenses to sales added to the ratio of profit to sales equals the 40 percent gross margin ratio. The net profit ratio of 8 percent shows that 8 percent of the net sales dollar is left for profit.

The value of percentage ratios should be obvious. The percentages are easily figured and much easier to compare than large dollar figures.

Note that because these operating statement categories are interrelated, only a few pieces of information are needed to figure the others. In this case, for example, knowing the gross margin percent and net profit percent makes it possible to figure the expenses and cost of sales percentages. Further, knowing just one dollar amount and the percentages lets you figure all the other dollar amounts.

Markups

LO B.4

A **markup** is the dollar amount added to the cost of sales to get the selling price. The markup usually is similar to the firm's gross margin because the markup amount added onto the unit cost of a product by a retailer or wholesaler is expected to cover the selling and administrative expenses and to provide a profit.

The markup approach to pricing is discussed in Chapter 18, so it will not be discussed at length here. But a simple example illustrates the idea. If a retailer buys an article that costs $1 when delivered to his store, he must sell it for more than this cost if he hopes to make a profit. So he might add 50 cents onto the cost of the article to cover his selling and other costs and, hopefully, to provide a profit. The 50 cents is the markup.

The 50 cents is also the gross margin or gross profit from that item *if* it is sold. But note that it is *not* the net profit. Selling expenses may amount to 35 cents, 45 cents, or even 55 cents. In other words, there is no guarantee the markup will cover costs. Further, there is no guarantee customers will buy at the marked-up price. This may require markdowns, which are discussed later in this appendix.

Markup conversions

Often it is convenient to use markups as percentages rather than focusing on the actual dollar amounts. But markups can be figured as a percent of cost or selling price. To have some agreement, *markup (percent)* will mean percentage of selling price unless stated otherwise. So the 50-cent markup built into the $1.50 *selling price* is a markup of 33⅓ percent. On the other hand, the 50-cent markup is a 50 percent markup on *cost*.

Some retailers and wholesalers use markup conversion tables or spreadsheets to easily convert from cost to selling price, depending on the markup on selling price they want. To see the interrelation, look at the two formulas below. They can be used to convert either type of markup to the other.

(4) $$\text{Percent markup on selling price} = \frac{\text{Percent markup on cost}}{100\% + \text{Percent markup on cost}}$$

(5) $$\text{Percent markup on cost} = \frac{\text{Percent markup on selling price}}{100\% - \text{Percent markup on selling price}}$$

In the previous example, we had a cost of $1, a markup of 50 cents, and a selling price of $1.50. We saw that the markup on selling price was 33⅓ percent—and on cost, it was 50 percent. Let's substitute these percentage figures—in Formulas 4 and 5—to see how to convert from one basis to the other. Assume first of all that we know only the markup on selling price and want to convert to markup on cost. Using Formula 5, we get

$$\text{Percent markup on cost} = \frac{33\frac{1}{3}\%}{100\% - 33\frac{1}{3}\%} = \frac{33\frac{1}{3}\%}{66\frac{2}{3}\%} = 50\%$$

On the other hand, if we know only the percent markup on cost, we can convert to markup on selling price as follows:

$$\text{Percent markup on selling price} = \frac{50\%}{100\% + 50\%} = \frac{50\%}{150\%} = 33\frac{1}{3}\%$$

These results can be proved and summarized as follows:

Markup $0.50 = 50% of cost, or 33⅓% of selling price
+ Cost $1.00 = 100% of cost, or 66⅔% of selling price
Selling price $1.50 = 150% of cost, or 100% of selling price

Note that when the selling price ($1.50) is the base for a markup calculation, the markup percent (33⅓ percent = $0.50/$1.50) must be less than 100 percent. As you can see, that's because the markup percent and the cost percent (66⅔ percent = $1.00/$1.50) sum to exactly 100 percent. So if you see a reference to a markup percent that is greater than 100 percent, it could not be based on the selling price and instead must be based on cost.

Markdown Ratios Help Control Retail Operations

The ratios we discussed previously were concerned with figures on the operating statement. Another important ratio, the **markdown ratio**, is a tool many retailers use to measure the efficiency of various departments and their whole business. But note that it is *not directly related to the operating statement*. It requires special calculations.

A **markdown** is a retail price reduction that is required because customers won't buy some item at the originally marked-up price. This refusal to buy may be due to a variety of reasons—soiling, style changes, fading, damage caused by handling, or an original price that was too high. To get rid of these products, the retailer offers them at a lower price.

Markdowns are generally considered to be due to business errors, perhaps because of poor buying, original markups that are too high, and other reasons. (Note, however, that some retailers use markdowns as a way of doing business rather than a way to correct errors. For example, a store that buys overstocked fashions from other retailers may start by marking each item with a high price and then reduce the price each week until it sells.) Regardless of the reason, markdowns are reductions in the original price—and they are important to managers who want to measure the effectiveness of their operations.

Markdowns are similar to allowances, because price reductions are made. Thus, in computing a markdown ratio, markdowns and allowances are usually added together and then divided by net sales. The markdown ratio is computed as follows:

$$\text{Markdown \%} = \frac{\$ \text{ Markdowns} + \$ \text{ Allowances}}{\$ \text{ Net sales}} \times 100$$

The 100 is multiplied by the fraction to get rid of decimal points.

Returns are *not* included when figuring the markdown ratio. Returns are treated as consumer errors, not business errors, and therefore are not included in this measure of business efficiency.

Retailers who use markdown ratios usually keep a record of the amount of markdowns and allowances in each department and then divide the total by the net sales in each department. Over a period of time, these ratios give management one measure of the efficiency of buyers and salespeople in various departments.

It should be stressed again that the markdown ratio is not calculated directly from data on the operating statement because the markdowns take place before the products are sold. In fact, some products may be marked down and still not sold. Even if the marked-down items are not sold, the markdowns—that is, the reevaluations of their value—are included in the calculations in the time period when they are taken.

The markdown ratio is calculated for a whole department (or profit center), *not* individual items. What we are seeking is a measure of the effectiveness of a whole department, not how well the department did on individual items.

Return on Investment (ROI) Reflects Asset Use

LO B.5

Another off-the-operating-statement ratio is **return on investment (ROI)**—the ratio of net profit (after taxes) to the investment used to make the net profit, multiplied by 100 to get rid of decimals. Investment is not shown on the operating statement. But it is shown on the **balance sheet** (statement of financial condition), which is another accounting statement that shows a company's assets, liabilities, and net worth. It may take some digging or special analysis, however, to find the right investment number.

Investment means the dollar resources the firm has invested in a project or business. For example, a new product may require $4 million in new money—for inventory, accounts receivable, promotion, and so on—and its attractiveness may be judged by its likely ROI (for an extended example of this, see the Marketing Analytics in Action activity in Chapter 9). If the net profit (after taxes) for this new product is expected to be $1 million in the first year, then the ROI is 25 percent—that is, ($1 million ÷ $4 million) × 100.

There are two ways to figure ROI. The *direct* way is

$$\text{ROI (in \%)} = \frac{\text{Net profit (after taxes)}}{\text{Investment}} \times 100$$

The *indirect* way is

$$\text{ROI (in \%)} = \frac{\text{Net profit (after taxes)}}{\text{Sales}} = \frac{\text{Sales}}{\text{Investment}} \times 100$$

This way is concerned with net profit margin and turnover—that is,

$$\text{ROI (in \%)} = \text{Net profit margin} \times \text{Turnover} \times 100$$

This indirect way makes it clearer how to *increase* ROI. There are three ways:

1. Increase profit margin (with lower costs or a higher price)
2. Increase sales
3. Decrease investment

Effective marketing strategy planning and implementation can increase profit margins, sales, or both. And careful asset management can decrease investment.

ROI is a revealing measure of how well managers are doing. Most companies have alternative uses for their funds. If the returns in a business aren't at least as high as outside uses, then the money probably should be shifted to the more profitable uses.

Some firms borrow more than others to make investments. In other words, they invest less of their own money to acquire assets—what we called *investments*. If ROI calculations use only the firm's own investment, this gives higher ROI figures to those who borrow a lot—which is called *leveraging*. To adjust for different borrowing proportions—to make comparisons among projects, departments, divisions, and companies easier—another ratio has come into use. **Return on assets (ROA)** is the ratio of net profit (after taxes) to the assets used to make the net profit—times 100. Both ROI and ROA measures are trying to get at the same thing—how effectively the company is using resources. These measures became increasingly popular as profit rates dropped and it became more obvious that increasing sales volume doesn't necessarily lead to higher profits—or ROI or ROA. Inflation and higher costs for borrowed funds also force more concern for ROI and ROA. Marketers must include these measures in their thinking, or top managers are likely to ignore their plans and requests for financial resources.

Forecasting Target Market Potential and Sales

LO B.6

Effective strategy planning and developing a marketing plan require estimates of future sales, costs, and profits. Without such information, it's hard to know if a strategy is potentially profitable.

The marketing manager's estimates of sales, costs, and profits are usually based on a forecast (estimate) of target **market potential**—what a whole market segment might buy—and a **sales forecast**—an estimate of how much an industry or firm hopes to sell to a market segment. Usually we must first try to judge market potential before we can estimate what share a particular firm may be able to win with its particular marketing mix.

Three levels of forecasts are useful

We're interested in forecasting the potential in specific market segments. To do this, it helps to make three levels of forecasts.

Some economic conditions affect the entire global economy. Others may influence only one country or a particular industry. And some may affect only one company or one product's sales potential. For this reason, a common top-down approach to forecasting is to

1. Develop a *national income forecast* (for each country in which the firm operates) and use this to
2. Develop an *industry sales forecast,* which then is used to
3. Develop forecasts for a *specific company,* its *specific products,* and the *segments* it targets.

Exhibit B-3 Straight-Line Trend Projection—Extends Past Sales into the Future

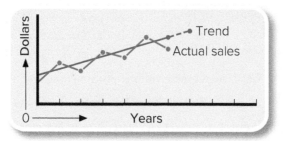

Generally, a marketing manager doesn't have to make forecasts for a national economy or the broad industry. This kind of forecasting—basically trend projecting—is a specialty in itself. Such forecasts are available in business and government publications, and large companies often have their own technical specialists. Managers can use just one source's forecast or combine several. Unfortunately, however, the more targeted the marketing manager's earlier segmenting efforts have been, the less likely that industry forecasts will match the firm's product-markets. So managers have to move directly to estimating potential for their own companies and for their specific products.

Two approaches to forecasting

Many methods are used to forecast market potential and sales, but they can all be grouped into two basic approaches: (1) extending past behavior and (2) predicting future behavior. The large number of methods may seem confusing at first, but this variety has an advantage. Forecasts are so important that managers often develop forecasts in two or three different ways and then compare the differences before preparing a final forecast.

Extending past behavior can miss important turning points

When we forecast for existing products, we usually have some past data to go on. The basic approach, called **trend extension**, extends past experience into the future. With existing products, for example, the past trend of actual sales may be extended into the future (see Exhibit B-3).

Ideally, when extending past sales behavior, we should decide why sales vary. This is the difficult and time-consuming part of sales forecasting. Usually we can gather a lot of data about the product or market or about changes in the market environment. But unless we know the *reason* for past sales variations, it's hard to predict in what direction, and by how much, sales will move. Graphing the data and statistical techniques—including correlation and regression analysis—can be useful here. (These techniques, which are beyond our scope, are discussed in beginning statistics courses.)

Once we know why sales vary, we can usually develop a specific forecast. Sales may be moving directly up as population grows in a specific market segment, for example. So we can just estimate how population is expected to grow and project the impact on sales.

The weakness of the trend extension method is that it assumes past conditions will continue unchanged into the future. In fact, the future isn't always like the past. An agent wholesaler's business may have been on a steady path, but the development of the Internet may have added a totally new factor. The past trend for the agent's sales has changed because now the agent can quickly reach a broader market.

As another example, for years the trend in sales of disposable diapers moved closely with the number of new births. However, as the number of women in the workforce increased and as more women returned to jobs after babies were born, use of disposable diapers increased, and the trend changed. As in these examples, trend extension estimates will be wrong whenever big changes occur. For this reason—although they may extend past behavior for one estimate—most managers look for another way to help them forecast sharp market changes.

Predicting future behavior takes judgment

When we try to predict what will happen in the future instead of just extending the past, we have to use other methods and add more judgment. Some of these methods (to be discussed later) include juries of executive opinion, salespeople's estimates, surveys, panels, and market tests.

Forecasting Company and Product Sales by Extending Past Behavior

Past sales can be extended

At the very least, a marketing manager ought to know what the firm's present markets look like and what it has sold to them in the past. A detailed sales analysis for products and geographic areas helps to project future results.

Just extending past sales into the future may not seem like much of a forecasting method. But it's better than just assuming that next year's total sales will be the same as this year's.

Factor method includes more than time

A simple extension of past sales gives one forecast. But it's usually desirable to tie future sales to something more than the passage of time.

The factor method tries to do this. The **factor method** tries to forecast sales by finding a relation between the company's sales and some other factor (or factors). The basic formula is: Something (past sales, industry sales, etc.) *times* some factor *equals* sales forecast. A **factor** is a variable that shows the relation of some other variable to the item being forecast. For instance, in the preceding example, both the birthrate and the number of working mothers are factors related to sales of disposable diapers.

A bread producer example

The following example, about a bread producer, shows how firms can make forecasts for many geographic market segments using the factor method and available data. This general approach can be useful for any firm—producer, wholesaler, or retailer.

Analysis of past sales relationships showed that the bread manufacturer regularly sold one-tenth of 1 percent (0.001) of the total retail food sales in its various target markets. This is a single factor. By using this single factor, a manager could estimate the producer's sales in a new market for the coming period by multiplying a forecast of expected retail food sales by 0.001.

Sales & Marketing Management magazine makes retail food sales estimates each year. Exhibit B-4 shows the kind of geographically detailed data available.

Let's carry this bread example further—using the data in Exhibit B-4 for the Denver, Colorado, metro area. Denver's food sales were $4,700,116,000 for the previous year. By simply accepting last year's food sales as an estimate of next year's sales and multiplying the food sales estimate for Denver by the 0.001 factor (the firm's usual share of food purchases in such markets), the manager would have an estimate of next year's bread sales in Denver. That is, last year's food sales estimate ($4,700,116,000) times 0.001 equals this year's bread sales estimate of $4,700,116.

Factor method can use several factors

The factor method is not limited to just one factor; several factors can be used together. For example, *Sales & Marketing Management* regularly gives a "buying power index" (BPI) as a measure of the potential in different geographic areas (see Exhibit B-4). This index considers (1) the population in a market, (2) the retail sales in that market, and (3) income in that market. The BPI for the Denver, Colorado, metro area, for example, is 0.9282—that is, Denver accounts for 0.9282 percent of the total U.S. buying power. This means that consumers who live in Denver have higher than average buying power. We know this because Denver accounts for about 0.77 percent of the U.S. population.

Exhibit B-4 Illustrative Page from Sales & Marketing Management's Survey of Buying Power: Metro and County Totals

COLORADO

METRO AREA County City	Total Population (000s)	Population % by Age Group 18-24	25-34	35-49	50+	House-holds (000s)	Total Retail Sales	Food & Beverage Stores	Food Service & Drinking Establishments	General Merchandise	Furniture & Furnish. & Electron. & Appliances	Motor Vehicle & Parts Dealers	Total EBI ($000)	Median Hsld. EBI	EBI A $20,000–$34,999	B $35,000–$49,999	C $50,000 & Over	Buying Power Index
BOULDER-LONGMONT	**303.7**	**13.6**	**15.5**	**25.6**	**22.4**	**119.6**	**5,081,227**	**1,001,555**	**556,377**	**448,453**	**247,484**	**1,220,087**	**7,716,546**	**51,714**	**17.8**	**17.1**	**51.6**	**0.1360**
BOULDER	303.7	13.6	15.5	25.6	22.4	119.6	5,081,227	1,001,555	556,377	448,453	247,484	1,220,087	7,716,546	51,714	17.8	17.1	51.6	0.1360
• Boulder	97.1	26.4	18.8	20.0	20.1	40.7	2,147,663	439,133	267,970	121,357	117,520	447,464	2,480,204	43,427	21.5	15.8	43.1	0.0479
• Longmont	74.6	8.7	14.6	25.3	23.6	28.0	1,126,804	250,483	122,965	125,103	40,068	272,449	1,531,271	47,526	19.4	20.1	46.8	0.0290
COLORADO SPRINGS	**537.3**	**10.7**	**14.4**	**24.7**	**22.6**	**200.1**	**7,883,675**	**819,826**	**647,101**	**984,753**	**465,488**	**2,031,112**	**10,259,019**	**42,082**	**24.7**	**21.0**	**38.7**	**0.1994**
EL PASO	537.3	10.7	14.4	24.7	22.6	200.1	7,883,675	819,826	647,101	984,753	465,488	2,031,112	10,259,019	42,082	24.7	21.0	38.7	0.1994
• Colorado Springs	373.1	10.5	14.8	24.5	23.7	146.5	6,786,693	628,690	549,107	858,048	443,063	1,848,219	7,353,670	41,212	24.9	20.9	37.5	0.1515
DENVER	**2,199.5**	**9.2**	**16.2**	**25.0**	**23.8**	**858.0**	**33,750,880**	**4,700,116**	**3,232,590**	**3,615,646**	**2,518,616**	**9,368,057**	**52,585,220**	**49,109**	**18.8**	**19.0**	**48.9**	**0.9282**
ADAMS	382.9	10.5	16.7	23.3	21.1	134.3	4,558,882	670,017	371,810	416,956	385,657	1,466,781	6,459,840	42,802	22.5	22.4	38.7	0.1253
• Thornton	87.3	9.7	17.5	24.9	18.0	30.6	707,386	152,812	68,679	122,478	40,099	150,769	1,594,293	48,053	19.4	22.4	46.8	0.0270
• Westminster	105.4	9.9	17.1	26.0	20.3	40.1	1,052,771	153,993	114,177	292,678	100,304	88,416	2,368,971	51,512	17.0	21.6	51.9	0.0384
ARAPAHOE	505.4	8.8	15.1	25.7	23.9	197.5	9,846,119	1,160,676	872,314	963,885	595,477	4,049,257	13,314,002	52,887	18.2	18.6	53.1	0.2422
• Aurora	286.8	10.4	17.2	24.0	21.0	109.0	3,889,713	531,245	378,269	566,166	225,882	1,236,463	5,874,943	47,398	20.7	21.5	46.0	0.1076
DENVER	568.5	11.1	19.9	22.1	25.1	243.7	9,287,630	1,235,129	1,270,413	730,810	755,362	1,701,220	13,899,851	42,540	21.6	18.4	41.0	0.2474
• Denver	568.5	11.1	19.9	22.1	25.1	243.7	9,275,551	1,224,988	1,269,426	730,810	755,362	1,700,815	13,899,851	42,540	21.6	18.4	41.0	0.2474
DOUGLAS	200.7	4.9	15.9	29.6	18.0	69.6	2,725,601	517,182	250,871	294,714	289,389	355,394	5,147,699	59,715	10.9	18.2	64.7	0.0851
JEFFERSON	542.0	8.3	13.2	26.8	26.5	212.9	7,332,648	1,117,112	467,182	1,209,281	492,731	1,795,405	13,763,828	54,470	16.4	18.6	55.1	0.2282
• Arvada	104.0	8.0	11.9	25.9	28.1	40.0	1,007,245	224,707	79,597	133,524	60,181	74,965	2,332,241	51,557	18.1	19.3	51.8	0.0376
• Lakewood	148.0	9.9	15.2	24.1	28.8	62.3	2,065,827	297,529	149,068	292,832	124,879	567,796	3,451,207	46,782	20.8	21.1	45.2	0.0599
DENVER-BOULDER-GREELEY																		
CONSOLIDATED AREA	**2,695.8**	**10.0**	**15.9**	**24.9**	**23.6**	**1,044.7**	**40,882,936**	**5,939,206**	**3,929,714**	**4,267,775**	**2,809,610**	**11,246,398**	**63,170,157**	**48,397**	**19.2**	**19.0**	**47.9**	**1.1216**

Using several factors rather than only one uses more information. And in the case of the BPI, it gives a single measure of a market's potential. Rather than falling back on using population only, or income only, or trying to develop a special index, the BPI can be used in the same way that we used the 0.001 factor in the bread example.

Predicting Future Sales Calls for More Judgment and Some Opinions

These past-extending methods use quantitative data—projecting past experience into the future and assuming that the future will be like the past. But this is risky in competitive markets. Usually, it's desirable to add some judgment to other forecasts before making the final forecast yourself.

Jury of executive opinion adds judgment

One of the oldest and simplest methods of forecasting—the **jury of executive opinion**—combines the opinions of experienced executives, perhaps from marketing, production, finance, purchasing, and top management. Each executive estimates market potential and sales for the *coming years*. Then they try to work out a consensus.

The main advantage of the jury approach is that it can be done quickly and easily. On the other hand, the results may not be very good. There may be too much extending of the past. Some of the executives may have little contact with outside market influences. But their estimates could point to major shifts in customer demand or competition.

Estimates from salespeople can help too

Using salespeople's estimates to forecast is like the jury approach. But salespeople are more likely than home office managers to be familiar with customer reactions and what competitors are doing. Their estimates are especially useful in some business markets where the few customers may be well known to the salespeople. But this approach can be useful in any type of market.

However, managers who use estimates from salespeople should be aware of the limitations. For example, new salespeople may not know much about their markets. Even experienced salespeople may not be aware of possible changes in the economic climate or the firm's other environments. And if salespeople think the manager is going to use the estimates to set sales quotas, the estimates may be low!

Surveys, panels, and market tests

Special surveys of final buyers, retailers, or wholesalers can show what's happening in different market segments. Some firms use panels of stores—or final consumers—to keep track of buying behavior and to decide when just extending past behavior isn't enough.

Surveys are sometimes combined with market tests when the company wants to estimate customers' reactions to possible changes in its marketing mix. A market test might show that a product increased its share of the market by 10 percent when its price was dropped 1 cent below competition. But this extra business might be quickly lost if the price were increased 1 cent above competition. Such market experiments help the marketing manager make good estimates of future sales when one or more of the Four Ps is changed.

Accuracy depends on the marketing mix

Forecasting can help a marketing manager estimate the size of possible market opportunities. But the accuracy of any sales forecast depends on whether the firm selects and implements a marketing mix that turns these opportunities into sales and profits.

KEY TERMS

operating statement, 564
gross sales, 566
return, 566
allowance, 567
net sales, 567
cost of sales, 567
gross margin (gross profit), 567
expenses, 567
net profit, 567
purchase discount, 568
stockturn rate, 569
operating ratios, 570
markup, 570
markdown ratio, 571
markdown, 572
return on investment (ROI), 572
balance sheet, 572
return on assets (ROA), 573
market potential, 573
sales forecast, 573
trend extension, 574
factor method, 575
factor, 575
jury of executive opinion, 577

QUESTIONS AND PROBLEMS

1. Distinguish between the following pairs of items that appear on operating statements: (a) gross sales and net sales, and (b) purchases at billed cost and purchases at net cost.

2. How does gross margin differ from gross profit? From net profit?

3. Explain the similarity between markups and gross margin. What connection do markdowns have with the operating statement?

4. Compute the net profit for a company with the following data:

Beginning inventory (cost)	$ 150,000
Purchases at billed cost	330,000
Sales returns and allowances	250,000
Rent	60,000
Salaries	400,000
Heat and light	180,000
Ending inventory (cost)	250,000
Freight cost (inbound)	80,000
Gross sales	1,300,000

5. Construct an operating statement from the following data:

Returns and allowances	$ 150,000
Expenses	20%
Closing inventory at cost	600,000
Markdowns	2%
Inward transportation	30,000
Purchases	1,000,000
Net profit (5%)	300,000

6. Compute net sales and percentage of markdowns for the following data:

Markdowns	$ 40,000
Gross sales	400,000
Returns	32,000
Allowances	48,000

7. (a) What percentage markups on cost are equivalent to the following percentage markups on selling price: 20, 37½, 50, and 66⅔? (b) What percentage markups on selling price are equivalent to the following percentage markups on cost: 33⅓, 20, 40, and 50?

8. What net sales volume is required to obtain a stockturn rate of 20 times a year on an average inventory at cost of $100,000 with a gross margin of 25 percent?

9. Explain how the general manager of a department store might use the markdown ratios computed for her various departments. Is this a fair measure? Of what?

10. Compare and contrast return on investment (ROI) and return on assets (ROA) measures. Which would be best for a retailer with no bank borrowing or other outside sources of funds (that is, the retailer has put up all the money that the business needs)?

11. Explain the difference between a forecast of market potential and a sales forecast.

12. Suggest a plausible explanation for sales fluctuations for (a) computers, (b) ice cream, (c) washing machines, (d) tennis rackets, (e) oats, (f) disposable diapers, and (g) latex for rubber-based paint.

13. Explain the factor method of forecasting. Illustrate your answer.

14. Based on data in Exhibit B-4, discuss the relative market potential of the city of Boulder, Colorado, and the city of Lakewood, Colorado, for (a) prepared cereals, (b) automobiles, and (c) furniture.

Career Planning in Marketing

APPENDIX C

LEARNING OBJECTIVES

One of the hardest decisions facing most college students is the choice of a career. Of course you are the best judge of your own objectives, interests, and abilities. Only you can decide what career you should pursue. However, you owe it to yourself to at least consider the possibility of a career in marketing. This chapter's objectives will help organize your thinking about a career and the marketing plan you could develop to help you achieve your career goals.

When you finish this appendix, you should be able to

1. rest assured there is a job or career for you in marketing.
2. know that marketing jobs can be rewarding, pay well, and offer opportunities for growth.
3. understand how to conduct your own personal analysis for career planning.
4. evaluate the many marketing jobs from which you can choose.
5. prepare your own personal marketing plan.

There's a Place in Marketing for You

LO C.1

We're happy to tell you that many opportunities are available in marketing. There's a place in marketing for everyone, from a service provider in a fast-food restaurant to a vice president of marketing in a large company such as Microsoft or Procter & Gamble. The opportunities range widely, so it will help to be more specific. In the following pages, we'll discuss (1) the typical pay for different marketing jobs, (2) how to set your own objectives and evaluate your interests and abilities, and (3) the kinds of jobs available in marketing.

There Are Many Marketing Jobs, and They Can Pay Well

There are many interesting and challenging jobs for those with marketing training. You may not know it, but more than half of graduating college students take their initial job in a sales, marketing, or customer service position regardless of their stated major. So you'll have a head start because you've been studying marketing and companies are always looking for people who already have skills in place. In terms of upward mobility, more CEOs have come from the sales and marketing side than all other fields combined. The sky is the limit for those who enter the sales and marketing profession prepared for the future!

Further, marketing jobs open to college-level students do pay well. At the time this book went to press, the most recent salary surveys from the National Association of Colleges and Employers found that marketing graduates were being offered starting salaries around $40,000, with a range from $25,000 to more than $60,000. Students with a master's in marketing averaged about $60,000; those with an MBA averaged about $75,000. Starting salaries can vary considerably, depending on your background, experience, and location.

Starting salaries in marketing compare favorably with many other fields. They are lower than those in such fields as computer science and electrical engineering where college graduates are currently in demand. But there is even better opportunity for personal growth, variety, and income in many marketing positions. The *American Almanac of Jobs and Salaries* ranks the median income of marketers number 10 in a list of 125 professions. Marketing also supplies about 50 percent of the people who achieve senior management ranks.

How far and fast your career and income rise above the starting level, however, depends on many factors, including your willingness to work, how well you get along with people, and your individual abilities. But most of all, it depends on *getting results*—individually and through other people. This is where many marketing jobs offer the newcomer great opportunities. It is possible to show initiative, ability, creativity, and judgment in marketing jobs. Some young people move up very rapidly in marketing. Some even end up at the top in large companies or as owners of their own businesses.

Marketing is often the route to the top

Marketing is where the action is! In the final analysis, a firm's success or failure depends on the effectiveness of its marketing program. This doesn't mean the other functional areas aren't important. It merely reflects the fact that a firm won't have much need for accountants, finance people, production managers, and so on if it can't successfully meet customers' needs and sell its products.

Because marketing is so vital to a firm's survival, many companies look for people with training and experience in marketing when filling key executive positions. In general, chief executive officers for the nation's largest corporations are more likely to have backgrounds in marketing and distribution than in other fields such as production, finance, and engineering.

Develop Your Own Personal Marketing Strategy

Now that you know there are many opportunities in marketing, your problem is matching the opportunities to your own personal objectives and strengths. Basically the problem is a marketing problem: developing a marketing strategy to sell a product—yourself—to potential employers. Just as in planning strategies for products, developing your own strategy takes careful thought. Exhibit C-1 shows how you can organize your own strategy planning. This exhibit shows that you should evaluate yourself first—a personal analysis—and then analyze the environment for opportunities. This will help you sharpen your own long- and short-run objectives, which will lead to developing a strategy. Finally, you should start implementing your own personal marketing strategy. These ideas are explained more fully in the next section.

Exhibit C-1 Organizing Your Own Personal Marketing Strategy Planning

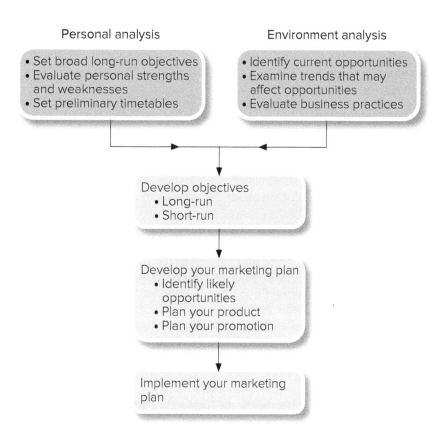

Conduct Your Own Personal Analysis

LO C.3

You are the product you are going to include in your own marketing plan. So first you have to decide what your long-run objectives are—what you want to do, how hard you want to work, and how quickly you want to reach your objectives. Be honest with yourself—or you will eventually face frustration. Evaluate your own personal strengths and weaknesses—and decide what factors may become the key to your success. Finally, as part of your personal analysis, set some preliminary timetables to guide your strategy planning and implementation efforts. Let's spell this out in detail.

Set broad long-run objectives

Your strategy planning may require some trial-and-error decision making. But at the very beginning, you should make some tentative decisions about your own objectives—what you want out of a job and out of life. At the very least, you should decide whether you are just looking for a job or whether you want to build a career. Beyond this, do you want the position to be personally satisfying—or is the financial return enough? And just how much financial return do you need? Some people work only to support themselves (and their families) and their leisure-time activities. These people try to find job opportunities that provide adequate financial returns but aren't too demanding of their time or effort.

Other people look first for satisfaction in their job—and they seek opportunities for career advancement. Financial rewards may be important too, but these are used mainly as measures of success. In the extreme, the career-oriented individual may be willing to sacrifice a lot, including leisure and social activities, to achieve success in a career.

Once you've tentatively decided these matters, then you can get more serious about whether you should seek a job or a career in marketing. If you decide to pursue a career, you should set your broad long-run objectives to achieve it. For example, one long-run objective might be to pursue a career in marketing management (or marketing research). This might require more academic training than you planned, as well as a different kind of training. If your objective is to get a job that pays well, on the other hand, then this calls for a different kind of training and different kinds of job experiences before completing your academic work.

What kind of job is right for you? Evaluate personal strengths and weaknesses

Because of the great variety of marketing jobs, it's hard to generalize regarding the aptitudes you should have to pursue a career in marketing. Different jobs attract people with various interests and abilities. We'll give you some guidelines about what kinds of interests and abilities marketers should have. However, if you're completely lost about your own interests and abilities, see your campus career counselor and take some vocational aptitude and interest tests. These tests will help you compare yourself with people who are now working in various career positions. They will *not* tell you what you should do, but they can help, especially in eliminating possibilities you are less interested in or less able to do well in.

Are you people-oriented or thing-oriented?

One useful approach is to decide whether you are basically "people-oriented" or "thing-oriented." This is a very important decision. A people-oriented person might be very unhappy in an inventory management job, for example, whereas a thing-oriented person might be miserable in a personal selling or retail management job that involves a lot of customer contact.

Marketing has both people-oriented and thing-oriented jobs. People-oriented jobs are primarily in the promotion area—where company representatives must make contact with potential customers. This may be direct personal selling or customer service activities—for example, in technical service or installation and repair. Thing-oriented jobs focus more on creative activities and analyzing data—as in advertising and marketing research—or on organizing and scheduling work—as in operating warehouses, transportation agencies, or the back-end of retailers.

People-oriented jobs tend to pay more, in part because such jobs are more likely to affect sales, the lifeblood of any business. Thing-oriented jobs, on the other hand, are often seen as cost generators rather than sales generators. Taking a big view of the whole company's operations, the thing-oriented jobs are certainly necessary—but without sales, no one is needed to do them.

Thing-oriented jobs are usually done at a company's facilities. Further, especially in lower-level jobs, the amount of work to be done and even the nature of the work may be spelled out quite clearly. The time it takes to design questionnaires and tabulate results, for example, can be estimated with reasonable accuracy. Similarly, running a warehouse, analyzing inventory reports, scheduling outgoing shipments, and so on are more like production operations. It's fairly easy to measure an employee's effectiveness and productivity in a thing-oriented job. At the least, time spent can be used to measure an employee's contribution.

A sales rep, on the other hand, might spend all weekend thinking and planning how to make a half-hour sales presentation on Monday. For what should the sales rep be compensated—the half-hour presentation, all of the planning and thinking that went into it, or the results? Typically, sales reps are rewarded for results—and this helps account for the sometimes extremely high salaries paid to effective order getters. At the same time, some people-oriented jobs can be routinized and are lower paid. For example, salespeople in some retail stores are paid at or near the minimum wage.

Managers needed for both kinds of jobs

Here we have oversimplified deliberately to emphasize the differences among types of jobs. Actually, of course, there are many variations between the two extremes. Some sales reps must do a great deal of analytical work before they make a presentation. Similarly, some marketing researchers must be extremely people-sensitive to get potential customers to reveal their true feelings. But the division is still useful because it focuses on the primary emphasis in different kinds of jobs.

Managers are needed for the people in both kinds of jobs. Managing others requires a blend of both people and analytical skills—but people skills may be the more important of the two. Therefore, people-oriented individuals are often promoted into managerial positions more quickly.

What will differentiate your product?

After deciding whether you're generally people-oriented or thing-oriented, you're ready for the next step—trying to identify your specific strengths (to be built on) and weaknesses (to be avoided or remedied). It is important to be as specific as possible so you can develop a better marketing plan. For example, if you decide you are more people-oriented, are you more skilled in verbal or written communication? Or if you are more thing-oriented, what specific analytical or technical skills do you have? Are you good at working with numbers, using a computer, solving complex problems, or coming to the root of a problem? Other possible strengths include past experience (career-related or otherwise), academic performance, an outgoing personality, enthusiasm, drive, and motivation.

It is important to see that your plan should build on your strengths. An employer will be hiring you to do something—so promote yourself as someone who is able to do something *well*. In other words, find your competitive advantage in your unique strengths—and then communicate these unique things about *you* and what you can do. Give an employer a reason to pick you over other candidates by showing that you'll add superior value to the company.

While trying to identify strengths, you also must realize that you may have some important weaknesses, depending on your objectives. If you are seeking a career that requires technical skills, for example, then you need to get those skills. Or if you are seeking a career that requires independence and self-confidence, then you should try to develop those characteristics in yourself—or change your objectives.

Set some timetables

At this point in your strategy planning process, set some timetables to organize your thinking and the rest of your planning. You need to make some decisions at this point to be sure you see where you're going. You might simply focus on getting your first job, or you might decide to work on two marketing plans: (1) a short-run plan to get your first job and (2) a longer-run plan—perhaps a five-year plan—to show how you're going to accomplish your long-run objectives. People who are basically job-oriented may get away with only a short-run plan, just drifting from one opportunity to another as their own objectives and opportunities change. But those interested in careers need a longer-run plan. Otherwise, they may find themselves pursuing attractive first-job opportunities that satisfy short-run objectives but quickly leave them frustrated when they realize that they can't achieve their long-run objectives without additional training or other experiences that require starting over again on a new career path.

Opportunities in Marketing

LO C.4

Strategy planning is a matching process. For your own strategy planning, this means matching yourself to career opportunities. So let's look at opportunities available in the market environment. (The same approach applies, of course, in the whole business area.) Exhibit C-2 shows some of the possibilities and salary ranges. The exhibit is assembled from a few years ago from salary data at www.salary.com. The salary ranges reflect the 25th percentile to the 75th percentile. You can also find information about other career paths common to marketing majors—including logistics and purchasing/procurement management.

Keep in mind that the salary ranges in Exhibit C-2 are rough estimates. Salaries for a particular job often vary depending on a variety of factors, including company size, industry, and geographic area. People in some firms also get big bonuses that are not counted in their salaries. There are many other sources of salary information. For example, *Advertising Age* publishes an annual survey of salary levels for different marketing and advertising jobs, with breakdowns by company size and other factors. Many trade associations, across a variety of different industries, also publish surveys, and the U.S. government's Bureau of Labor Statistics (www.bls.gov) includes salary data. If you use the search engine at www.google.com and do a search on "salary survey," you will find dozens of such surveys for a number of different industries.

Exhibit C-2 Some Career Paths and Compensation Ranges*

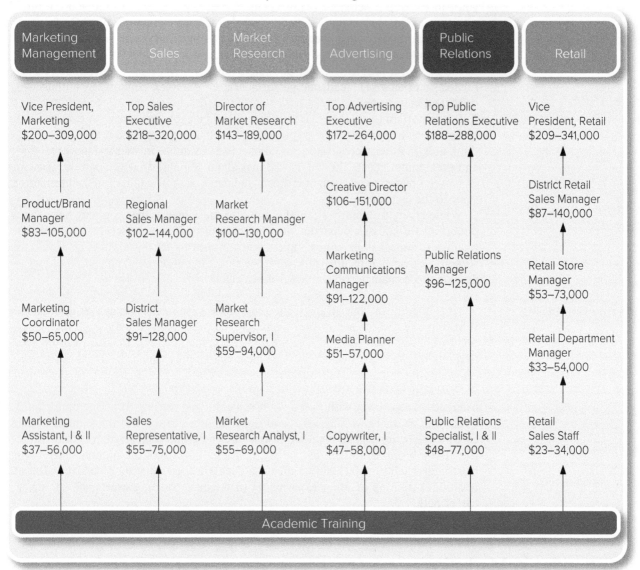

*Compensation is estimated based on data reported on Salary.com as of April 2017, the latest date we could find. So we have added 10% to those numbers to better reflect estimated salaries for each job category as of 2021. Salary for retail vice president was not available and is estimated. Note that these represent averages and estimates that can vary widely by geographic labor market trends, size of company, and economic conditions. For the most current information, talk to an advisor at your career management center.

Identify current opportunities in marketing

Because of the wide range of opportunities in marketing, it's helpful to narrow your possibilities. After deciding on your own objectives, strengths, and weaknesses, think about where in the marketing system you might like to work. Would you like to work for manufacturers, or wholesalers, or retailers? Or does it really matter? Do you want to be involved with consumer products or business products? By analyzing your feelings about these possibilities, you can begin to zero in on the kind of job and the functional area that might interest you most.

One simple way to get a better idea of the kinds of jobs available in marketing is to review the chapters of this text—this time with an eye for job opportunities rather than new concepts. The following paragraphs contain brief descriptions of job areas that marketing graduates are often interested in with references to specific chapters in the text. Some, as noted, offer good starting opportunities, whereas others do not. While reading these paragraphs, keep your own objectives, interests, and strengths in mind.

Marketing manager (Chapter 2)
This is usually not an entry-level job, although some students may move quickly into this role in a smaller company. A marketing manager can be responsible for much of the marketing in an organization. It usually requires experience in a wide range of areas. Marketing managers may also have to coordinate activities with other functional area managers (see Bonus Chapter 2).

Customer or market analyst (Chapters 3 through 5)
Opportunities as consumer analysts and market analysts are commonly found in large companies, marketing research organizations, advertising agencies, and some consulting firms. Investment banking firms also hire entry-level analysts; they want to know what the market for a new business is like before investing. Beginning market analysts start in thing-oriented jobs until their judgment and people-oriented skills are tested. The job may involve collecting or analyzing secondary data or preparation of reports and plans. Because knowledge of statistics, computer software, Internet search techniques, or behavioral sciences is very important, marketing graduates often find themselves competing with majors from statistics, sociology, computer science, and economics. Graduates who have courses in marketing *and* one or more of these areas may have the best opportunities.

Purchasing agent/buyer (Chapter 6)
Entry-level opportunities are commonly found in large companies, and there are often good opportunities in the procurement area. Many companies are looking for bright newcomers who can help them find new and better ways to work with suppliers. To get off on the right track, beginners usually start as trainees or assistant buyers under the supervision of experienced buyers. That's good preparation for a promotion to more responsibility.

Marketing research opportunities (Chapter 7)
There are entry-level opportunities at all levels in the channel (but especially in large firms where more formal marketing research is done in-house), in advertising agencies, and in marketing research firms. Some general management consulting firms also have marketing research groups. Quantitative and behavioral science skills are extremely important in marketing research, so some firms are more interested in business graduates who have studied statistics or psychology as electives. These opportunities have grown significantly in recent years with the explosion of marketing metrics and marketing analytics. But there still are many opportunities in marketing research for marketing graduates, especially if they have some experience in working with computers and statistical software. A recent graduate might begin in a training program—conducting interviews or summarizing open-ended answers from questionnaires and helping prepare electronic slide presentations for clients—before being promoted to a position as an analyst, assistant project manager, account representative, and subsequent management positions.

Packaging specialists (Chapter 8)
Packaging manufacturers tend to hire and train interested people from various backgrounds—there is little formal academic training in packaging. There are many sales opportunities in this field—and with training, interested people can become specialists fairly quickly in this growing area.

Product/brand manager (Chapters 8 and 9)
Many multiproduct firms have brand or product managers handling individual products—in effect, managing each product as a separate business. Some firms hire marketing graduates as assistant brand or product managers, although larger firms typically recruit MBAs for these jobs. Many firms prefer that recent college graduates spend some time in the field doing sales work or working with an ad agency or sales promotion agency before moving into brand or product management positions.

Product planner (Chapter 9)

This is usually not an entry-level position. Instead, people with experience on the technical side of the business or in sales might be moved onto a new-product development team as they demonstrate judgment and analytical skills. However, new employees with winning ideas for new products don't go unnoticed—and they sometimes have the opportunity to grow fast with ideas they spearhead. Having a job that puts you in contact with customers is often a good way to spot new needs.

Distribution channel management (Chapter 10)

This work is typically handled or directed by sales managers and therefore is not an entry-level position. However, many firms form teams of specialists who work closely with their counterparts in other firms in the channel to strengthen coordination and relationships. Such a team often includes new people in sales or purchasing because it gives them exposure to a different part of the firm's activities. It's also not unusual for people to start working in a particular industry and then take a different job at a different level in the channel. For example, a graduate who has trained to be a store manager for a chain of sporting goods stores might go to work for a manufacturers' representative that handles a variety of sports equipment.

Logistics opportunities (Chapter 11)

There are many sales opportunities with physical distribution specialists—but there are also many thing-oriented jobs involving traffic management, warehousing, and materials handling. Here training in accounting, finance, and computer methods could be very useful. These kinds of jobs are available at all levels in channels of distribution.

Retailing opportunities (Chapter 12)

Not long ago, most entry-level marketing positions in retailing involved some kind of sales work. That has changed rapidly in recent years because the number of large retail chains is expanding and they often recruit graduates for their management training programs. Retailing positions tend to offer lower-than-average starting salaries—but they often provide opportunities for very rapid advancement. In a fast-growing chain, results-oriented people can move up very quickly. Most retailers require new employees to have some selling experience before managing others—or buying. A typical marketing graduate can expect to work as an assistant manager or do some sales work and manage one or several departments before advancing to a store management position—or to a staff position that might involve buying, advertising, location analysis, and so on.

Wholesaling opportunities (Chapter 12)

Entry-level jobs with merchant wholesalers typically fall into one of two categories. The first is in the logistics area—working with transportation management, inventory control, distribution customer service, and related activities. The other category usually involves personal selling and customer support. Agent wholesalers typically focus on selling, and entry-level jobs often start out with order-taking responsibilities that grow into order-getting responsibilities. Many wholesalers are moving much of their information to the Internet, so marketing students with skills and knowledge in this arena may find especially interesting opportunities.

Personal selling opportunities (Chapter 14)

Because there are so many different types of sales jobs and so many people are employed in sales, there are many good entry-level opportunities in personal selling. This might be order-getting, order-taking, customer service, or missionary selling. Many sales jobs now rely on sales technology, so some of the most challenging opportunities will go to students who know how to prepare spreadsheets and presentation materials using software programs such as Microsoft Office. Many students are reluctant to get into personal selling—but this field offers benefits that are hard to match in any other field. These include the opportunity to earn high salaries and commissions quickly, a chance to develop your self-confidence and resourcefulness, an opportunity to work with minimal supervision—almost to the point of being your own boss—and a chance to acquire product and customer knowledge that many firms consider necessary for a successful career in product/brand management, sales management, and marketing management. On the other hand, many salespeople prefer to spend their entire careers in selling. They like the freedom and earning potential that go with a sales job over the headaches and sometimes lower salaries of sales management positions.

Customer service opportunities (Chapter 14)

As this book points out, marketing managers are recognizing the growing importance of providing customers with service after the sale. There are a number of different opportunities in customer service. Many firms need qualified customer service representatives who work with customers to fulfill their needs and help ensure customer satisfaction. Customer service reps may interact with customers on the phone, by online chat, or via e-mail. But other service representatives help customers at their businesses with installations or equipment repair. Although some of the entry-level positions that require only high school education are being outsourced to other countries, positions requiring strong communication skills and a good education provide an opportunity for marketing majors. There are also management positions that develop customer service strategies, control costs, and focus on hiring, training, and retaining customer service reps and enhancing customer satisfaction.

Advertising opportunities (Chapters 13, 15, and 16)

Job opportunities in this area are varied and highly competitive. Because the ability to communicate well and a knowledge of the behavioral sciences are important, marketing and advertising graduates often find themselves competing with majors from fields such as English, communications, psychology, and sociology. There are thing-oriented jobs such as copywriting, media buying, art, computer graphics, and so on. Competition for these jobs is very strong—and they go to people with a proven track record. So the entry-level positions are as assistant to a copywriter, media buyer, or art director. There are also people-oriented positions involving work with clients, which are probably of more interest to marketing graduates. This is a small but glamorous and extremely competitive industry where young people can rise very rapidly—but they can also be as easily displaced by new bright young people. Entry-level salaries in advertising are typically low. There are sometimes good opportunities to get started in advertising with a retail chain that prepares its advertising internally. Another way to get more experience with advertising is to take a job with one of the media, perhaps in sales or as a customer consultant. Selling advertising space on a website or cable TV station or newspaper may not seem as glamorous as developing TV ads, but media salespeople help their customers solve promotion problems and get experience dealing with both the business and creative sides of advertising.

Publicity and public relations opportunities (Chapters 13 and 16)

Although the number of entry-level positions in publicity and public relations is expected to continue to grow rapidly, competition for jobs will be stiff. College graduates with degrees in journalism, public relations, communications, and marketing may qualify for these openings. Many are attracted to these high-profile jobs. In addition to the popular press, more firms are placing useful content on the Internet to inform and influence customers. To a large degree, the tools used in these fields rely on emerging technologies. This bodes well for recent college graduates who may be more comfortable using newer technologies than "old hands" in publicity and public relations. International experience and proficiency in a second language can sometimes help job candidates stand out.

Sales promotion opportunities (Chapters 13 and 15)

The number of entry-level positions in the sales promotion area is growing because the number of specialists in this area is growing. For example, specialists might help a company plan a special event for employees, figure out procedures to distribute free samples, or perhaps set up a database to send customers a newsletter. Because clients' needs are often different, creativity and judgment are required. It is usually difficult for an inexperienced person to show evidence of these skills right out of school, so entry-level people often work with a project manager until they learn the ropes. In companies that handle their own sales promotion work, a beginner usually starts by getting some experience in sales or advertising.

Earned, owned, and social media management (Chapters 13, 15, and 16)

The opportunities for managing earned, owned, and social media are growing rapidly. As we saw in Chapters 13, 15, and 16, marketing managers are investing an increasing share of their promotion budgets into these forms of media. Although the media cost is relatively low, there is a need for people to create, update, and manage the content on owned media, including the company's web pages and/or social media platforms. Young people have some advantage here, as they are often consumers of social media and are not intimidated (as some of the "old guard" might be). A social media manager must be comfortable creating content—so he or she should be a good writer and be creative as well as understand the company's customers and products. The manager may be required to write blog posts or write about products on web pages. There are opportunities to analyze the data generated by websites. Sometimes these jobs require someone to facilitate a company's online community or to respond to customer comments and reviews.

Pricing opportunities (Chapters 17 and 18)

Pricing decisions are usually handled by experienced executives. However, in some large companies and consulting firms there are opportunities as pricing analysts for marketing graduates who have quantitative skills. These people work as assistants to higher-level executives and collect and analyze information about competitors' prices and costs as well as the firm's own costs. Thus, being able to work with accounting numbers and computer spreadsheets is often important in these jobs. However, sometimes the route to these jobs is through experience in marketing research or product management.

Credit management opportunities

Specialists in credit have a continuing need for employees who are interested in evaluating customers' credit ratings and ensuring that money gets collected. Both people skills and thing skills can be useful here. Entry-level positions normally involve a training program and then working under the supervision of others until your judgment and abilities are tested.

International marketing opportunities

Many marketing students are intrigued with the adventure and foreign travel promised by careers in international marketing. Some firms hire recent college graduates for positions in international marketing, but more often these positions go to MBA graduates. However, that is changing as more and more firms pursue international markets. It's an advantage in seeking an international marketing job to know a second language and to know about the culture of countries where you would like to work. Your college may have courses or international exchange programs that would help in these areas. Graduates aiming for a career in international marketing usually must spend time mastering the firm's domestic marketing operations before being sent abroad. So a good way to start is to focus on firms that are already involved in international marketing, or that are planning to move in that direction soon. On the other hand, there are many websites with listings of international jobs. For example, you might want to visit www.overseasjobs.com.

Study trends that may affect your opportunities

A strategy planner should always be evaluating the future because it's easier to go along with trends than to buck them. This means you should watch for political, technical, or economic changes that might open or close career opportunities.

If you can spot a trend early, you may be able to prepare yourself to take advantage of it as part of your long-run strategy planning. Other trends might mean you should avoid certain career options. For example, technological changes in computers and communications, including the Internet, are leading to major changes in retailing and advertising as well as in personal selling. Cable television, telephone selling, and direct-mail selling may reduce the need for routine order takers, while increasing the need for higher-level order getters. More targeted and imaginative sales presentations for delivery by mail, e-mail, phone, or Internet websites may be needed. The retailers that prosper will have a better understanding of their target markets. And they may need to be supported by wholesalers and manufacturers that can plan targeted promotions that make economic sense. This will require a better understanding of the production and physical distribution side of business, as well as the financial side. And this means better training in accounting, finance, inventory control, and so on. So plan your personal strategy with such trends in mind.

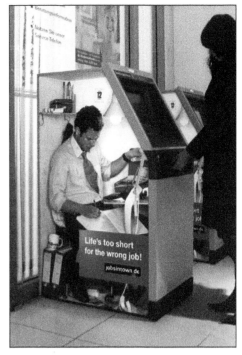

This humorous ad on the side of a vending machine for jobsintown.de says, "Life's too short for the wrong job!" Planning for the right job—like planning a marketing strategy—pays off.
Source: Scholz & Friends Berlin, Jobsintown.de

One good way to get more detailed analysis is to go to the U.S. Bureau of Labor Statistics website at http://stats.bls.gov and do a search for "occupational outlook." The bureau provides detailed comments about the outlook for employment and growth in different types of jobs, industries, and regions.

Evaluate business practices

Finally, you need to know how businesses really operate and the kind of training required for various jobs. We've already seen that there are many opportunities in marketing—but not all jobs are open to everyone, and not all jobs are entry-level jobs. Positions such as marketing manager, brand manager, and sales manager are higher rungs on the marketing career ladder. They become available only when you have a few years of experience and have shown leadership and judgment. Some positions require more education than others. So take a hard look at your long-run objectives—and then see what degree you may need for the kinds of opportunities you might like.

Develop Objectives

LO C.5

Once you've done a personal analysis and environment analysis—identifying your personal interests, your strengths and weaknesses, and the opportunities in the environment—define your short-run and long-run objectives more specifically.

Develop long-run objectives

Your long-run objectives should clearly state what you want to do and what you will do for potential employers. You might be as specific as indicating the exact career area you want to pursue over the next 5 to 10 years. For example, your long-run objective might be to apply a set of marketing research and marketing management tools to the food manufacturing industry, with the objective of becoming director of marketing research in a small food manufacturing company.

Your long-run objectives should be realistic and attainable. They should be objectives you have thought about and for which you think you have the necessary skills (or the capabilities to develop those skills) as well as the motivation to reach.

Develop short-run objectives

To achieve your long-run objective(s), you should develop one or more short-run objectives. These should spell out what is needed to reach your long-run objective(s). For example, you might need to develop a variety of marketing research skills *and* marketing management skills—because both are needed to reach the longer-run objective. Or you might need an entry-level position in marketing research for a large food manufacturer to gain experience and background. An even shorter-run objective might be to take the academic courses that are necessary to get that desired entry-level job. In this example, you would probably need a minimum of an undergraduate degree in marketing, with an emphasis on marketing research. (Note that, given the longer-run objective of managerial responsibility, a business degree would probably be better than a degree in statistics or psychology.)

Developing Your Marketing Plan

Now that you've developed your objectives, move on to developing your own personal marketing plan. This means zeroing in on likely opportunities and developing a specific marketing strategy for these opportunities. Let's talk about that now.

Identify likely opportunities

An important step in strategy planning is identifying potentially attractive opportunities. Depending on where you are in your academic training, this can vary all the way from preliminary exploration to making detailed lists of companies that offer the kinds of jobs that interest you. If you're just getting started, talk to your school's career counselors and placement officers about the kinds of jobs being offered to your school's graduates. Your marketing instructors can help you be realistic about ways you can match your training, abilities, and interests to job opportunities. Also, it helps to read business publications such as *Bloomberg Businessweek, Fast Company, Fortune, The Wall Street Journal,* and *Advertising Age.* If you are interested in opportunities in a particular industry, check at your library or on the Internet to see if there are trade publications or websites that can bring you up to speed on the marketing issues in that area. Your library or college may also have an online service to make it easier to search for articles about specific companies or industries. And many companies have their own websites that can be a very useful source of information.

Don't overlook the business sections of your local newspapers to keep in touch with marketing developments in your area. And take advantage of any opportunity to talk with marketers directly. Ask them what they're doing and what satisfactions they find in their jobs. Also, if your college has a marketing club, join it and participate actively in the club's programs. It will help you meet marketers and students with serious interest in the field. Some may have had interesting job experiences and can provide you with leads on part-time jobs or exciting career opportunities.

If you're far along in your present academic training, list companies that you know something about or are willing to investigate, trying to match your skills and interests with possible opportunities. Narrow your list to a few companies you might like to work for.

If you have trouble narrowing down to specific companies, make a list of your personal interest areas—sports, travel, reading, music, or whatever. Think about the companies that compete in markets related to these interests. Often your own knowledge about these areas and interest in them can give you a competitive advantage in getting a job. This helps you focus on companies that serve needs you think are important or interesting. A related approach is to do a search on the Internet for websites related to your areas of interest. Websites often display ads or links to firms that are involved in that specific interest area. Further, many companies post job openings on their own websites or at websites that specialize in promoting job searches by many companies.

Then do some research on these companies. Find out how they are organized, their product lines, and their overall strategies. Try to get clear job descriptions for the kinds of positions you're seeking. Match these job descriptions against your understanding of these jobs and your objectives. Jobs with similar titles may offer very different opportunities. By researching job positions and companies in depth, you should begin to have a feel for where you would be comfortable as an employee. This will help you narrow your target market of possible employers to perhaps five firms. For example, you may decide that your target market for an entry-level position consists of large corporations with (1) in-depth training programs, (2) a wide product line, and (3) a wide variety of marketing jobs that will enable you to get a range of experiences and responsibilities within the same company.

Plan your product

Just like any strategy planner, you must decide what product features are necessary to appeal to your target market. Identify which credentials are mandatory and which are optional. For example, is your present academic program enough, or will you need more training? Also, identify what technical skills are needed, such as computer programming or accounting. Further, are there any business experiences or extracurricular activities that might help make your product more attractive to employers? This might involve active participation in college organizations or work experience, either on the job or in internships.

Plan your promotion

Once you identify target companies and develop a product you hope will be attractive to them, you have to tell these potential customers about your product. You can write directly to prospective employers, sending a carefully developed résumé that reflects your strategy planning. Or you can visit them in person (with your résumé). Many colleges run well-organized interviewing services. Seek their advice early in your strategy planning effort.

Implementing Your Marketing Plan

When you complete your personal marketing plan, you have to implement it, starting with working to accomplish your short-run objectives. If, as part of your plan, you decide that you need specific outside experience, then arrange to get it. This may mean taking a low-paying job or even volunteering to work in political organizations or volunteer organizations where you can get that kind of experience. If you decide that you need skills you can learn in academic courses, plan to take those courses. Similarly, if you don't have a good understanding of your opportunities, then learn as much as you can about possible jobs by talking to professors, taking advanced courses, and talking to businesspeople. Of course, trends and opportunities can change—so continue to read business publications, talk with professionals in your areas of interest, and be sure that the planning you've done still makes sense.

Strategy planning must adapt to the environment. If the environment changes or your personal objectives change, you have to develop a new plan. This is an ongoing process—and you may never be completely satisfied with your strategy planning. But even trying will make you look much more impressive when you begin your job interviews. Remember, although all employers would like to hire a Superman or a Wonder Woman, they are also impressed with candidates who know what they want to do and are looking for a place where they can fit in and make a contribution. So planning a personal strategy and implementing it almost guarantee you'll do a better job of career planning, and this will help ensure that you reach your own objectives, whatever they are.

Whether or not you decide to pursue a marketing career, the authors wish you the best of luck in your search for a challenging and rewarding career, wherever your interests and abilities may take you.

APPENDIX D: Hillside Veterinary Clinic Marketing Plan[1]

TABLE OF CONTENTS

1. **Executive Summary**
2. **Situation Analysis**
   - 2.1 Company Analysis
     - 2.1.1 Company Objectives and Marketing Objectives
     - 2.1.2 Screening Criteria
     - 2.1.3 Company Resources
     - 2.1.4 Marketing Plans (Marketing Program)
     - 2.1.5 Present Marketing Strategy
     - 2.1.6 Marketing Collaborators—Current and Potential
   - 2.2 Customer Analysis
     - 2.2.1 Product-Market
     - 2.2.2 Demographic Data on the Target Market
     - 2.2.3 Current Customers
       - 2.2.3.1 Database of Current Customers
       - 2.2.3.2 Survey of Current Customers
     - 2.2.4 Customer Needs—Possible Segmenting Dimensions
     - 2.2.5 Identification of Qualifying Dimensions and Determining Dimensions
     - 2.2.6 Identification of Target Market(s) (One or More Specific Segments)
   - 2.3 Competitor Analysis
   - 2.4 Analysis of the Market Context—External Market Environment
     - 2.4.1 Economic Environment
     - 2.4.2 Technological Environment
     - 2.4.3 Political and Legal Environment
     - 2.4.4 Cultural and Social Environment
   - 2.5 S.W.O.T.
     - 2.5.1 HVC's Strengths
     - 2.5.2 HVC's Weaknesses
     - 2.5.3 HVC's Opportunities
     - 2.5.4 HVC's Threats

3. **Marketing Plan Objectives**
4. **Differentiation and Positioning**
5. **Marketing Strategy**
    5.1 Target Market #1—Noncustomers
        5.1.1 Product—Bundled Service Package for Puppies and Kittens
        5.1.2 Promotion—Logo and Slogan
        5.1.3 Promotion—Participation in Community Events
        5.1.4 Promotion—Create a Large Vinyl Sign
        5.1.5 Promotion—*Yellow Pages* Advertising
        5.1.6 Promotion—Google AdWords
        5.1.7 Promotion—Search Engine Optimization
        5.1.8 Promotion—Earned Media Management
        5.1.9 Place—Extended Clinic Hours
        5.1.10 Place—New Facility (Long Term)
        5.1.11 Price—Prices to Be Adjusted for Inflation
    5.2 Target Market #2—Current Customers
        5.2.1 Product—Compassionate and Personalized Care
        5.2.2 Product—Bundled Service Packages for Dental Care and Geriatric Pet Care
        5.2.3 Promotion—In-Store Signage and Point-of-Purchase Materials
        5.2.4 Promotion—Owned Media Management/Website Redesign
        5.2.5 Promotion—Customer Database Enhancement
        5.2.6 Promotion—Brochures to Describe the Clinic and Key Special Services
        5.2.7 Promotion—Swag
        5.2.8 Promotion—Social Media
6. **Implementation and Control**
    6.1 Special Implementation Problems to Overcome
    6.2 Control
    6.3 Budget, Sales Forecasts, and Estimates
        6.3.1 Budget for Total Marketing Plan
        6.3.2 Monthly Expenses for Total Marketing Plan
        6.3.3 Pro Forma Income Statement
    6.4 Timing of Implementation Activities
    6.5 Contingency Plans
7. **Appendices**
    7.1 Appendix A: Survey of Current Customers
    7.2 Appendix B: Competitor Matrix

1. Executive Summary

Hillside Veterinary Clinic (HVC) is a small-animal veterinary clinic operating in Wellington, Colorado, owned and operated by Dr. Kelly Hardy. The clinic has one full-time veterinarian, four full-time vet techs, and an office manager who provide service in a small but fast-growing community in northern Colorado. The clinic has been in operation for six years and has tentative plans (two to three years from now) to move into a new building—where it will add the sale of pet supplies and kennel services to its veterinary services. To fund its new building, HVC must grow sales. The clinic prides itself on offering compassionate and personal care to pets and pet owners.

Current customers appear to be very satisfied with Hillside Veterinary Clinic, and customer retention is high. There are many competitors in nearby Fort Collins—some of whom serve Wellington residents. The rapid growth of the town of Wellington creates many opportunities for acquiring customers who are new to the area.

This plan focuses only on the veterinary services offered to customers and Hillside's growth objectives over the next two years. The specific objectives of this marketing plan are:

- Within two years, the number of customers (currently 1,200) will increase by 15 percent or by 180 customers (to 1,380).
  - Most of this new growth will come from new homeowners in the area. However, additional growth will occur by converting local customers to HVC through promotional efforts.
- The retention rate of current customers will remain at 95 percent (not counting those who leave the area).
  - This is an estimate of the current customer retention rate.
- Within two years, 30 percent of customers will have annual dental examinations and teeth cleanings done on their dogs (up from 5 to 10 percent today).
- Within two years, 30 percent of customers will have well-pet health test screening on their dogs or cats at least every two years (up from 7 percent today).

This results in two related marketing strategies—the first of which targets acquiring new customers. HVC will use a variety of marketing tactics to generate awareness, especially among those moving to Wellington. The second set of tactics targets current customers and focuses on retaining these customers, thus growing its business. In particular, HVC hopes to enhance sales of its prevention services by bundling related services with attractive pricing and promotion tools (including signs and brochures). HVC will also make sure that it retains its current positioning as a vet clinic that is compassionate with pets and their owners.

2. Situation Analysis

This marketing plan begins with an analysis of the market facing Hillside Veterinary Clinic (HVC). It includes an assessment of the company, customers, competitors, and external market environment. This information is then summarized with a S.W.O.T. (Strengths, Weaknesses, Opportunities, and Threats) analysis.

2.1 Company Analysis

Hillside Veterinary Clinic is located in the town of Wellington, Colorado. Wellington, with a population of around 6,800, is located about 10 miles north of Fort Collins (a city of 152,000) in Larimer County. Wellington is also located 65 miles north of Denver, Colorado, and 25 miles south of the Wyoming border. The nearest city north of Wellington is Cheyenne, Wyoming, about 35 miles north.

Hillside Veterinary Clinic has been in operation for six years—opened by Dr. Kelly Hardy a few years after she earned her Doctorate in Veterinary Medicine in 2010. The clinic focuses on small animals—most of its patients being dogs and cats with some other small pets such as rabbits, ferrets, hamsters, and guinea pigs. The clinic does not handle larger animals such as horses, cows, or pigs. HVC now has six full-time employees—including one veterinarian (Dr. Kelly Hardy), four veterinary technicians, and one office manager. More details about the clinic are noted in the following sections.

2.1.1 Company Objectives and Marketing Objectives

In evaluating any new opportunities, Dr. Hardy tries to keep focused on the values and goals she has for her business. Her company has developed a mission statement, a set of general goals, and a more specific set of marketing objectives.

Dr. Hardy aims to keep her business focused with the following mission statement:

It is the mission of the Hillside Veterinary Clinic—doctors and staff—to provide progressive, high-quality veterinary medical and surgical services to our clients and their pets. Our team is committed to improving the quality of life of pets and their owners by strengthening the human-animal bond.

Dr. Hardy describes the general goals of her organization as follows:

- To place its greatest emphasis on compassionate care for pets and their owners, no matter how large HVC becomes. (This will always be HVC's highest priority because it is consistent with Dr. Hardy's personal values.)
- To build a larger, more accommodating facility and to offer more pet-related services to the community—including kennel services and the sale of pet supplies.
- To purchase the latest state-of-the-art anesthesia machines and monitoring equipment so HVC can continue to provide the most humane care for its clients.
- To hire more staff so that individual pet care can be maintained and enhanced.
- To invest in more in-clinic diagnostic equipment.
- To provide greater information to customers about preventive health care such as vaccinations, parasite prevention, regular physical exams and lab screening, dental prophylaxis, as well as weight control/nutrition programs for pets.

When asked about the objectives for her organization, Dr. Hardy indicated the following:

- By July, two years from the date of this plan, HVC will add on to the facility. This add-on will include additional kennel space and a remodel of the basement apartment, allowing a staff member to live on site and provide after-hours care.
- By July, five years from the date of this plan, HVC will increase its current sales volume by 50 percent.
    - Sales growth is needed to justify the facility addition.
    - Most of this new growth will come from new clients planned to coincide with additional services. These services are not the focus of this marketing plan, but they are key to HVC's marketing program.
- By July, two years from the date of this plan, HVC will increase its sales of veterinary services by 30 percent.
- More generally, it is hoped that within two years, in its market area, HVC will continue to be recognized as the *premier veterinary* clinic in the area—known for high-quality and compassionate care. This will primarily be recognized by the American Animal Hospital Association (AAHA), which conducts accreditation inspections every three years.

2.1.2 Screening Criteria

HVC, through past analysis of its own strengths and weaknesses, evaluation of the market, and its own goals and objectives, has developed criteria for the clinic to evaluate opportunities. The screening criteria help Dr. Hardy to evaluate and compare opportunities, and then decide which ones to pursue. These criteria are as follows:

- Investments in new equipment and services should break even within one year.
- Investments in marketing should contribute to the desired annual growth target of 15 percent.
- Marketing investments should earn a return on investment (ROI) of at least 25 percent before taxes.
- Our reputation for compassionate animal care remains our highest priority, and no investments, new personnel, or strategy decisions should jeopardize that reputation.

2.1.3 Company Resources

The company resources include human resources, financial resources, and facility/equipment resources. The current status is briefly outlined below.

Human Resources HVC has a full-time staff of six people, including four veterinary technicians, one office manager (Karen), and one veterinarian. It should be noted that one of the vet techs—Rick—particularly enjoys interacting with people and has volunteered to do more outside activities. Dr. Hardy currently works about 60–70 hours per week, mostly providing veterinary services to customers. A local veterinary acupuncture specialist comes to the office to provide acupuncture services on an as-needed basis.

The clinic has ready access to highly qualified personnel from the graduates and students at the nearby Colorado State University College of Veterinary Medicine.

In addition, students from out of state often move to Fort Collins to work for a couple of years before attending the College of Veterinary Medicine. This allows the applicant to gain state residency and earn money—these people often seek employment as veterinary technicians. These people have proven to be excellent employees.

In emergencies and for vacations, Dr. Hardy brings in a friend and former veterinarian who currently stays home with two young children. This woman has indicated an interest in some regular part-time work. This may be an opportunity to expand capacity—without the cost of adding a full-time vet.

Facility and Equipment Resources HVC's current building is located just off the highway in Wellington, which offers significant exposure. The building is large and recently built. HVC's location on Main Street makes awareness of the clinic very high among Wellington residents.

The clinic includes a variety of state-of-the art equipment—more than the average veterinary clinic. Equipment and facilities include:

- One digital radiograph machine to give in-house x-ray services
- Some in-house laboratory testing equipment
- Gas anesthesia (many veterinary practices use injection only)
- Complete anesthesia monitoring
    - ECG, pulse oximeter, Doppler blood pressure, respiratory monitor
    - Trained technicians
- IV infusion pump systems
- Recovery kennels—complete with blankets for ultimate comfort (many clinics just use newspaper)

Emergencies that cannot be handled in-house are referred to Colorado State University's Teaching Hospital or Animal Emergency Services (another local clinic).

2.1.4 Marketing Plans (Marketing Program)

Hillside Veterinary Clinic has two marketing plans that are currently being implemented or refined. Its two marketing plans are both *product development* strategies—which primarily focus on providing new products for present markets.

One marketing plan focuses on adding kennel services. That marketing plan calls for providing a higher level of service than competitors do. Animal pens will be larger than those of competitors, and dog walking services will be standard fare.

The other marketing plan calls for HVC to offer pet supplies to its customers through an attached retail store.

These additional services have been requested by customers, and the situation analysis conducted for these marketing plans provides support for both plans. Both plans are likely to build additional awareness of HVC and its traditional veterinary services.

2.1.5 Present Marketing Strategy

HVC's current marketing strategy for veterinary services involves the following elements:

Target Market The target market includes all families within a 10-mile radius of Wellington with small animals.

Product HVC currently offers a wide range of veterinary services including the following:

- Preventive Health Care
    - Vaccinations
    - Spay/Neuter (represents about 10–15 percent of the sales)
    - Parasite Control
    - Health Screens through diagnostic laboratory testing
    - Dentistry, including cleaning and oral surgery (noted as its most profitable service, currently utilized by only 5–10 percent of customers)

- Internal Medicine (includes dermatology—allergy and infections, neurology, orthopedics, pediatrics, endocrinology, ophthalmology, gastroenterology, infectious disease, and oncology-related services)
- General Surgery (orthopedics and soft tissue)
- Individualized Care
  - Each pet has its own nurse after surgery
- Acupuncture (offered by a part-time veterinarian who comes in on call)
- In-Depth Client Education (for disease diagnosis and treatment)
- Puppy or Kitten Package (customers pay a set price for all the services their pets need in their first year and receive a 20 percent discount. This helps retain customers who continue to bring them into the HVC clinic.)
- Behavioral Counseling
- Euthanasia
- Grief Support

From observation of the clinic and communication with the staff, the main thing that stands out is the high-quality, personalized, and compassionate care each animal (and owner) receives. HVC takes a personalized approach in dealing with its clients' needs. Dr. Hardy spends extensive time with her clients explaining procedures and giving them options. A survey of current customers (details in the Customer section) suggest that customers view the clinic as a place that offers warm and compassionate care in a friendly and "homey" environment. Further, as is evidenced by the large amount of new clients based on referrals, HVC's customers are happy with the services they receive and recommend the clinic to their family and friends as a result. This appears to be a key distinguishing factor of the clinic.

Dr. Hardy is particularly interested in increasing spay and neutering services. These provide a great introduction to the clinic, and customers who contract for such are usually retained. They have a high lifetime value because the services are given to young pets. She also notes that dental services are highly profitable and underutilized by most pet owners. She feels there is a good opportunity to increase sales of these services.

Place Services are primarily offered through the storefront location on the main street of Wellington. House calls are available upon request for additional charges on an as-available basis.

- Office hours:
  - 8:00 a.m. to 5:30 p.m. Monday, Wednesday, Thursday, and Friday
  - 8:00 a.m. to 7:00 p.m. Tuesday
  - 7:00 a.m. admissions are accepted any day if more convenient for the client
  - House calls are available (including euthanasia)

Promotion The Hillside Veterinary Clinic currently utilizes a number of promotional techniques to acquire and retain customers. These include the following:

- One of the clinic's biggest strategies includes a large, lighted sign on the front of the building that can easily be seen on Main Street in Wellington.
- Over the years, the clinic has given out a variety of promotional items—refrigerator magnets, business cards, leashes, bandannas, and food bowls with the clinic name and phone number.
- The clinic recently participated, for the first time, in the Wellington Parade of Lights, which occurs each December. This appears to be an effective and worthwhile venture. Because Wellington is a small town with small-town attitudes, many of its residents attend this parade.
- Last year, several pet-related companies—a groomer and a local general-goods retailer that sells pet supplies—and HVC participated in a *Pet Fair*. Participation involved vet techs answering questions and handing out promotional items.
- The clinic also has a small (1-inch) in-column advertisement located in the *Yellow Pages* telephone directory.

- The clinic tries to maintain regular contact with its current customer base. It keeps a database that includes owner names, addresses, e-mail addresses, pet names, and information about each pet. Christmas cards are sent to all customers, and in the spring each receives a heartworm reminder letter—most by e-mail. It also sends out a magazine, purchased from another company but customized with HVC's name and phone number, called *Healthy Pet*.
- HVC maintains an active social media presence where it publicizes events and promotions. It also uses social media to help return missing pets to their owners. At times, HVC's Facebook page features popular animal videos to help drive traffic to the page and create more "Likes." Photos and news are shared through Instagram, Twitter, and Pinterest.
- HVC maintains a website that communicates services, hours, and prices to potential customers. There is extensive information about HVC's services. The website also provides biographies of staff members, who highlight their own contributions toward HVC's mission statement and HVC's focus on compassionate care.
- The following limitations are noted:
  - HVC has no pamphlet or web page listing the services it offers with price ranges.
  - HVC does not have optimal search engine optimization (SEO), and its website is not always the first result for a search of "Wellington Vet Clinic."
  - HVC has a few negative reviews on Google and struggles to convince pleased clients to be as vocal as unhappy clients.

Price As shown in the Competitor Analysis section of this plan, prices charged by HVC tend to be about 20 percent higher than those of competitors. The higher prices are not believed to be a threat to the business. Hardy believes the higher prices are justified by the company's use of the highest-quality technology and assistance. If asked, staff members take time and effort to inform their customers as to why their prices are higher than other clinics' prices. The staff do not believe that HVC loses many clients due to its higher prices. In general, the only price shopping the staff observe tends to be for spay/neuter services.

2.1.6 Marketing Collaborators—Current and Potential

HVC works closely with a few outside organizations that help it complete its marketing effort. There are also firms in related businesses that HVC could work with if it wanted to enhance its market offering.

Specialized Veterinary Service Providers In order to expand its available services, HVC refers clients to various specialists when Dr. Hardy does not have the expertise or time to work with a particular patient. These amenities are used for critical care and specialty services such as internal medicine, neurology, oncology, critical care, and complex soft tissue and orthopedic surgeries. Experts include Emergency Animal Service, The Eye Doctor for Animals, Colorado State University Veterinarians, Rocky Mountain Veterinary Specialists, and Peak Veterinary Imaging. Peak Veterinary Imaging is a highly valued source of radiological specialists. This company consists of two boarded veterinary radiologists who have a mobile practice and travel to the clinic to perform ultrasounds, echoes, and ultrasound-guided biopsies.

Diagnostic Service Center HVC uses Heska Diagnostic Laboratory in Fort Collins to test routine samples.

Large-Animal Vet Clinics These clinics are useful to HVC because of their association with animals. The clinics act as indirect competitors because they often fulfill basic small-animal health needs such as vaccinations and spay/neuters. Typically, these services are offered to pet owners who also have large animals—as a convenience to those customers. These clinics often refer customers to the HVC. Close ties between the HVC and large-animal clinics can be beneficial to both parties.

Dog Shows/Dog Trials By being involved with dog trials and dog shows, HVC has acquired new customers and raised clinic awareness. Because these dog owners care about their pets a great deal, they are willing to invest time and money in their pets' health.

Pet Stores Because of its plans to eventually offer pet supplies, HVC has not attempted to work closely with any pet stores. PetSmart and Petco, two superstore chains, have locations in Fort Collins, and there are a number of smaller pet stores in the vicinity. PetSmart offers veterinary services and is unlikely to seek a relationship with a competitor—but Petco does not offer veterinary services and may be open to working with HVC.

Animal Shelters HVC does not currently have any relationship with Larimer County Humane Society or the Dumb Friends League. These organizations provide spay/neuter certificates—but the reimbursement on these is less than half of HVC's usual charges and at this time HVC does not honor them.

2.2 Customer Analysis

Practical limitations point to restricting initial customer analysis to a 20-mile radius from downtown Wellington, Colorado. While there are few vet clinics north of Wellington, the area is also sparsely populated—most residents have large animals if they live north of Wellington. It would also be difficult to target these groups. South of Wellington, there are many veterinary clinics in the Fort Collins area. Most customers value the convenience of a local clinic, so the focus is on Wellington.

2.2.1 Product-Market

The product-market for this marketing plan consists of the following elements:

- **Product type:** the veterinary services outlined in the Company section under "Product."
- **Customer needs:** customers need medical services and advice to care for their pets. They also need these services delivered in a caring and compassionate manner.
- **Customer types:** the final consumers are pet owners and caregivers.
- **Geographic area:** more than 90 percent of HVC's customers currently reside within a 10-mile radius of the city of Wellington.

2.2.2 Demographic Data on the Target Market

Customers are located in the geographic area of Wellington, which is situated 10 miles north of Fort Collins. The town's population is about 6,800. The town is growing rapidly because of its relatively affordable housing as compared to nearby Fort Collins. The population has grown 66 percent in the last five years—similar growth is projected for the next five years. Real estate professionals predict 800 new homes will be built in the next five years—160 new houses per year. Most of these homes are so-called starter homes, often bought by young families. From past observation, the clinic's staff believe that many of these homeowners do not own pets when they move in, but often buy a new pet within the first year or two of home ownership. The median household income in Wellington is $66,411, up by $18,494 in the last 10 years.

Pet ownership is an ongoing trend in the United States, with pet-owning households increasing by more than 1 million each year. Based on a Harris Interactive poll, the following was learned: more than 3 in 5 Americans (62 percent) have pets; among pet owners, 7 in 10 have dogs (69 percent), half have cats (51 percent), 1 in 10 have fish (11 percent), and less than 1 in 10 pet owners have birds (7 percent) or some other type of pet (8 percent).

From experience, the clinic staff believe that Wellington families tend to own dogs at a slightly higher rate and cats at a slightly lower rate than the national numbers.

The large majority of clients come from within a 10-mile radius of the city limits and will be our main focus for current and potential customers. HVC's current client base numbers 1,200. Each customer owns an average of two pets. Dogs make up 70 percent of vet visits, cats about 28 percent, and other small pets (e.g., rabbits, gerbils, hamsters) the final 2 percent. Clients' pets visited the HVC two times on average per year.

2.2.3 Current Customers

To get a better understanding of current customers, HVC's office manager (Karen) and Dr. Hardy were interviewed, and 40 current customers completed a survey.

2.2.3.1 Database of Current Customers According to HVC's office manager, current customers could be put into two broad categories—farmers/ranchers and town dwellers.

About a quarter of HVC's current customers are from the farming and ranching community—living outside the city limits. These customers tend to be more price sensitive and do not see much reason to spend money on preventive health care or dental services.

Town dwellers are the bulk of the business—and the fastest-growing group. As more tract housing is built in Wellington, this will continue to be the growth market for HVC. These customers tend to be less price sensitive than the farmers and ranchers and view their pets as part of their family. They seem to value personalized care and attention from the doctor and vet techs. They are also more open to suggestions for preventive health care and dental services. However, the staff have found that it usually takes them some time to embrace these ideas—it's not until after the second or third time preventive health and dental care are mentioned that customers proactively request these services.

A review of the customer database showed 1,200 active pet owner customers who have made visits to the clinic within the last two years. The majority of its customers own dogs and cats; from those 1,200 customers HVC has provided care for 1,306 dogs and 571 cats.

2.2.3.2 Survey of Current Customers In order to get a better understanding of its current customers, a paper-and-pencil survey was designed and handed out to customers visiting the clinic over a two-week period. The purpose of the research was to learn a bit more about customers, to find out their attitudes toward their pets, to evaluate their receptivity to some potential new services, and to find out about their access to and use of the Internet.

The survey was completed by a convenience sample of 40 HVC customers who were asked to complete the survey while they waited during visits to the clinic in March. Almost all of those asked actually completed the survey. A copy of the survey and a tabulation of its results is shown in Appendix A. Some of the key findings include the following:

- 100 percent of these customers own dogs and 48 percent own cats.
- 93 percent of the customers considered their pets to be a "family member."
- Five factors were all found to be important to customers—but "Expertise of the vet staff," and "Friendly customer service" had the most listings as "very important."
- 88 percent indicated that having a "Location close to my home" was important or very important. And 83 percent lived within 10 miles of the clinic—suggesting a focus on customers in Wellington or a short drive away.
- Among possible new services, customers were most interested in seeing HVC add "Evening hours" (38 percent) and "Boarding" services (33 percent).
- 85 percent of respondents indicated the current hours are sufficient.
- Satisfaction level appears to be quite high, with only 5 percent indicating they had ever been dissatisfied with the clinic—and 90 percent indicating that HVC was the only vet clinic they used.
- 98 percent of customers had access to the Internet and only one of those did not have an e-mail address—so 95 percent of all these customers have an e-mail address.
- Social media usage varied considerably by platform—with 61 percent of current customers on Facebook, 19 percent on Pinterest, 14 percent on Instagram, and 12 percent using Twitter.

2.2.4 Customer Needs—Possible Segmenting Dimensions

Dr. Hardy asked her staff to generate a list of possible customer needs as they related to veterinary services. She understood that the clinic should not try to meet all customer needs, but wanted to know what the staff believed were customer needs.

The following list includes customer needs that were identified—but were not chosen as needs that HVC would try to address:

- Low costs/low prices
  - Although this might appeal to a limited set of customers, the clinic's traditional emphasis on care and taking time with owners and pets makes it difficult to be profitable while also offering low prices. This option also is not consistent with current screening criteria.
- Mobile clinic—visiting pets at their owners' homes
  - Competitors provide this service.
  - Although this would fit with the clinic's goal of compassionate, personal care, it would sacrifice the ability to provide good care to more customers at the clinic.
- Add services for large animals
  - Another vet could be hired to provide these services (Dr. Hardy does not have this specialty), but most current customers do not have these needs, and the current facility does not fit well with providing good service to large animals.
- Convenient location
  - Moving to a new location in the short run is not practical. A new location is being planned for and will be in the same general area so that it remains convenient to most customers.
  - Other options—such as adding a satellite office—were considered too costly.
  - The current location is convenient to the vast majority of current customers and allows some to fit in some shopping on the same trip.
- Treatment of exotic animals
  - Some veterinarians specialize in the treatment of exotic animals. This requires additional training, which Dr. Hardy does not currently have.
  - The local market has very limited call for this need—given the small size of the market, the clinic chose not to address these needs directly but to make referrals to clinics that could provide these services.

HVC feels that it can deliver on the following customer needs—and that these needs are consistent with its current screening criteria:

- Compassionate treatment of animals
  - This is part of the current approach to the market and consistent with Dr. Hardy's personal values.
  - Although this is a customer need, customers learn that HVC delivers on this need only after visiting the clinic. Marketing communications do not emphasize this benefit of HVC's services.
- Compassionate treatment of owners
  - This is part of the current approach to the market and consistent with Dr. Hardy's personal values.
  - Although this is a customer need, customers learn that HVC delivers on this need only after visiting the clinic. Marketing communications do not emphasize this benefit of HVC's services.
- High-quality medical care
  - This also fits with the current approach and facilities. Over the last few years the clinic has invested in state-of-the-art diagnostic and treatment equipment.
  - Although this is a customer need, customers learn that HVC delivers on this need only after visiting the clinic. Marketing communications do not emphasize this benefit of HVC's services.
- Treatment of small animals
  - This is the current focus of the clinic and is consistent with the clinic's expertise and facilities.

2.2.5 Identification of Qualifying Dimensions and Determining Dimensions

HVC believes the following market segments could be combined and approached with a similar marketing strategy:

- Qualifying dimension: owners of cats, dogs, and other small animals
- Qualifying dimension: living within 10 miles of Wellington, Colorado
- Determining dimension: pet owners that value high-quality care that is compassionate and personalized
- Determining dimension: currently a client of HVC or not

HVC believes that different marketing strategies are needed for customers who already utilize HVC's services and pet owners who use another veterinarian, are new to the area, or who obtain a new pet.

2.2.6 Identification of Target Market(s) (One or More Specific Segments)

HVC's marketing strategy may have to be adapted to these two target markets:

1. Pet owners with small animals living within 10 miles of Wellington, Colorado, who *are not currently* customers of HVC.
2. Pet owners with small animals living within 10 miles of Wellington, Colorado, who *are currently* customers of HVC.

2.3 Competitor Analysis

The northern Colorado market can be pretty competitive for veterinary services. Each year more than a hundred new veterinarians graduate from the Colorado State University College of Veterinary Medicine in nearby Fort Collins. Many of these graduates want to stay in the local area—so they open a practice or join a local clinic, often for below-market wages.

As noted in the Customer section, most customers make an initial screening of veterinarians based on location—choosing a vet that is located relatively close to their homes. We identified four veterinary practices that seemed to target the same geography and addressed similar customer needs. Key information about each of the major competitors—and similar information about Hillside Veterinary Clinic—is shown in the competitor matrix in Appendix B. Competitive information was gathered by reviewing each competitor's website and *Yellow Pages* ad, and by contacting each by phone and asking questions about prices and services offered. Each clinic was asked for spay and neuter prices because the services are widely used, relatively comparable from clinic to clinic, and these represent important services that help a clinic acquire new customers.

Comparing HVC to its primary competitors, the matrix in Appendix B points out competitive advantages and competitive disadvantages. These are outlined below.

Competitive disadvantages for HVC:

- When these competitors were compared to the HVC, some disadvantages for HVC are the hours of operation and the price of the neutering and spaying operations. HVC is not open on the weekends, and East Side Animal Hospital and Flatiron Veterinary Clinic are open on the weekends.
- The cost for the spaying and neutering operations was higher for HVC than for the other clinics—although this does not take into account the discounted puppy/kitten package of bundled services. HVC tests blood work before surgery using its in-house lab facility. The other reason the price is higher is because of the more expensive type of anesthesia that is used. Dr. Hardy feels it is the best and safest type to use.

Competitive advantages for HVC include:

- HVC is the only clinic that schedules appointments at 7 a.m.
- HVC utilizes the services of an acupuncture specialist—this is offered by only one other competitor.
- HVC gives individualized care (holds animal when it comes out of surgery).
- HVC offers more services (grief support, dentistry, in-house laboratory testing, and IV infusion pump system).
- HVC's social media presence and the information on its website are superior to all competitors in quantity and quality.

2.4 Analysis of the Market Context—External Market Environment

2.4.1 Economic Environment

Regional and local economies continue to show strength, following a severe decline during the recession of 2008 to 2011. Pet care has been known to be robust to economic conditions—it is considered recession resistant. On the other hand, a poor economy can make it more difficult to sell customers on preventive health care products.

2.4.2 Technological Environment

The technological environment is changing rapidly in a variety of ways that may affect HVC's business:

- Veterinary medicine relies to a much greater extent on technology as new diagnostic equipment becomes available. Lower prices for this equipment make it easier for veterinary clinics to provide more services in-house. This may be a greater advantage for larger clinics, which can utilize the equipment more often, and therefore more easily cost-justify highly advanced/expensive equipment. An example of this advanced technology is the laser technology that declaws animals. It is a faster and easier procedure and is less painful for pets. This will eventually become a common piece of equipment for vets to have.
- The growth of the Internet continues to impact how customers shop and how business can interact with potential and current customers.
- There are a growing number of affordable off-the-shelf software products that allow small service firms to develop customer relationship management databases. Some of these are customized for veterinary clinics.

2.4.3 Political and Legal Environment

State and local laws regulate some of the care that must be given to pets. For example, dogs and cats are required to obtain rabies vaccinations at 4 months and then 2 months later at 6 months of age. Regulations also specify that the owner must obtain a license within 14 days of obtaining the animal.

Currently, there are no limitations on the number of animals a person may own. However, there are specific regulations enforced with regard to the care of animals. One such regulation specifies that a person should only have animals that they can properly care for in healthy conditions, and these animals should not be bothersome to the neighbors. These rules emphasize the importance of the community and animals living together in a safe and healthy environment.

Recently, an animal cruelty law was passed that made any animal abuse a felony. It is now much easier to prosecute pet owners who abuse animals. Therefore, there will be more pressure on pet owners to care for, feed, and treat their animals with respect. It will also be crucial for pet owners to get the necessary vaccines and vet treatments for their pets.

2.4.4 Cultural and Social Environment

HVC reflects many of Wellington's small-town values. HVC customers value that the staff greet them by name when they come into the clinic. The friendliness of the staff keeps the customers happy and feeling valued and welcomed. Many customers walk to the clinic; they enjoy the fact that they don't have to drive to the vet's office. Some clients have been with HVC since the start of the clinic, and have enjoyed going to the same location for years and watching it grow and change.

There is some risk that growth and an eventual move to a larger location will make it more difficult to maintain the small-town values. It may become more difficult for the staff to remember customer names and pet names. As the facility expands, an additional vet may need to be hired. Adding a second vet means some current customers will have to see a veterinarian other than Dr. Hardy, which may jeopardize established relationships. Pets and their owners get attached to their vet, and it can be difficult for people to change.

Pet health care expenditures spiked immediately following the terrorist attacks in 2001. It was suggested that this was a side effect of the "cocooning" that followed these attacks—with more people staying close to home and perhaps bonding more with their pets. According to a recent Harris poll, 91 percent of respondents considered their pets to be members of their families.

A recent story of animal abuse brought animal care to the local headlines. A woman was charged with cruelty to animals when she was found to be running a puppy mill with 60 puppies crammed into 10 cages.

2.5 S.W.O.T.

This situation analysis gives insights about the company, customers, competitors, and external marketing environment in which Hillside Veterinary Clinic operates. In reviewing this information, the data can be summarized in a S.W.O.T. analysis that highlights Hillside's **S**trengths and **W**eaknesses—and the **O**pportunities and **T**hreats in its current market. These are summarized in the following sections.

2.5.1 HVC's Strengths

HVC's strengths include the following:

- Location on Wellington's main street creates high awareness
- Staff members are well informed, highly trained, and have great people and communication skills
- About 70 percent of customers come from referrals
- Good reputation in the local community
- Wide range of services offered
- Strong ethical principles
- Emphasis on quality care and customized attention
- Quality and technologically advanced equipment
- High standards for staff and services
- The clinic's atmosphere—comfortable and homey, not intimidating (for example, pictures of patients posted on wall)
- Excellent "bedside manner"
- Differentiation comes from service level offered, including
  - One nurse assigned to each patient
  - Safer type of anesthesia used
  - Fluid support given while animal recovers
  - Outpatient care has pets' and owners' best interests at heart (for example, they use plastic kennels rather than metal and give blankets instead of newspaper)
- Willingness to increase marketing budget if justified

2.5.2 HVC's Weaknesses

HVC's weaknesses include the following:

- Small facility—not enough exam rooms for amount of business performed
- In the short term, "smallish" facility and veterinarian availability (Dr. Hardy is the only vet) constrain growth options—adding an additional vet alone will not help, because there are limited examination rooms
- Lack of parking availability
- Higher-than-average pricing
- Understaffed—people have to work very hard to offer the level of service they strive to give
- Lack of business knowledge
- Lack of attention focused on marketing/promotional strategies

2.5.3 HVC's Opportunities

Opportunities for HVC include the following:

- Direct competitors in Wellington and surrounding areas are few
- Fast-growing community—particularly in nearby new housing developments
- Many cross-selling opportunities—customers do not seem to buy other preventive health services for their pets

- First mover advantage—they are the first clinic in town
- Competitor entry barriers are high because HVC has established itself as a clinic that has high standards for staff and services
- Some local customers use other vet clinics—there are opportunities to attract new customers from the local market
- Constant advances in technology open doors for HVC to distinguish itself with the latest and greatest equipment
- Staff members are comfortable using web-based marketing strategies to access young home buyers in the area

2.5.4 HVC's Threats

Threats for HVC include the following:

- As Wellington grows, it is increasingly likely to attract more competitors—particularly with Colorado State University graduating so many vets each year
- A Fort Collins-based clinic could set up a satellite branch or a new clinic may open in this area
- The relatively lackluster local economy—rising interest rates threaten the housing boom that has fueled Wellington's growth
- The slowing economy could reduce discretionary income and cause pet owners to become more price sensitive and to reduce spending on preventive health care for their pets
- Potential loss of small-town ambience
- Potential loss of existing clients (if facility moves to a new location)

3. Marketing Plan Objectives

This marketing plan focuses specifically on the primary veterinary services offered by Hillside Veterinary Clinic. After reviewing the situation analysis, HVC believes that two opportunities stand out as the best options to grow the clinic. First, it hopes to take advantage of the population growth in Wellington and attract customers who move to the area. Once customers come to HVC, they tend to be very loyal to the clinic—so it is important that anyone new to the area choose HVC for vet services. To maintain that loyalty, the clinic must make sure that it continues to provide the high-quality customer service, compassionate treatment, and veterinary services that create this loyalty.

Second, HVC believes that current customers do not utilize enough of the clinic's preventive health care services. Dr. Hardy feels most pets would have a better quality of life and live longer if owners took advantage of some of these services. She particularly believes that two services are underutilized: (1) dental services (annual teeth cleaning) for dogs and cats, and (2) health test screenings, which are now used almost exclusively when an animal is sick, but could be used as part of an annual checkup. The clinic will attempt to grow these services.

Taking advantage of these opportunities, HVC sets the following objectives for their marketing plan:

- Within two years, the number of customers (currently 1,200) will increase by 15 percent, or by 180 (to 1,380).
  - Most of this new growth will come from new homeowners in the area. But additional growth will occur by converting local customers to HVC through promotional efforts.
- The retention rate of current customers will remain at 95 percent (not counting those who leave the area).
  - This is an estimate of the current customer retention rate.
- Within two years, 30 percent of customers will have annual dental examinations and teeth cleanings done on their dogs (up from 5 to 10 percent today).
- Within two years, 30 percent of customers (up from 7 percent today) will have had a well-pet health test screening on their dogs or cats—and this screening will be repeated at least every two years.

4. Differentiation and Positioning

An evaluation of Hillside Veterinary Clinic's strengths and weaknesses, its competitors, and its customers provides insights into opportunities for differentiation and positioning. This review suggests that HVC can be differentiated from the competition by its high level of customer service, compassionate health care, and location in Wellington. These points of differentiation lead to the following statement that best describes HVC's desired positioning:

> *For all small-pet owners within a 10-mile radius of Wellington, Colorado, Hillside Veterinary Clinic provides the most compassionate and personalized care of all local veterinary clinics, because its staff members love animals and provide excellent customer service for both pets and pet owners.*

This positioning and differentiation will help drive the marketing strategy decisions that follow.

5. Marketing Strategy

Because two different, but related, target markets are identified for HVC's veterinary services, two strategies are proposed. The first target market includes noncustomers and involves a marketing strategy to attract these customers to try Hillside Veterinary Clinic. The second target market focuses on current customers—and objectives focus on increasing their use of health screening and dental services. HVC also wants to continue to have high retention of its current customers. Some of the elements of the marketing strategy directed at one target market will also serve the other target market—those situations are noted.

5.1 Target Market #1— Noncustomers

The first target market for HVC will be pet owners with small animals living within 10 miles of Wellington, Colorado, who are not currently customers of HVC.

A quick summary of the marketing tactics directed toward this target market is followed by a more detailed explanation of each tactic:

- Product—Bundled Service Package for Puppies and Kittens
- Promotion—Logo and Slogan
- Promotion—Participation in Community Events
- Promotion—Create a Large Vinyl Sign
- Promotion—*Yellow Pages* Advertising
- Promotion—Google AdWords
- Promotion—Search Engine Optimization
- Promotion—Earned Media Management
- Place—Extended Clinic Hours
- Place—New Facility (Long Term)
- Price—Prices to Be Adjusted for Inflation

5.1.1 Product—Bundled Service Package for Puppies and Kittens

HVC currently offers a bundled service package for puppies and kittens. Because this program targets young pets, the lifetime customer value will rise per patient if this program is a success. Success can be defined as retaining those clients for the life of the animal—assuming the person continues to live in the local area. The primary goal for this program will be to acquire new customers for the clinic. Many of the families buying a new home in Wellington follow that purchase by getting a pet—typically a puppy or kitten. The cost savings will provide an incentive for them to come to the clinic and experience firsthand the high-quality care from HVC. This should be the first step in creating a lifelong relationship between the clinic and the pet owner.

This package includes spay or neuter, all vaccinations, a book on puppy or cat care and training, and coupons for grooming services. The package also includes a free one-year-old checkup. Each package will continue to be priced at $129, a 10 percent discount on the services and goods if ordered separately.

5.1.2 Promotion—Logo and Slogan
Hillside Veterinary Clinic does not currently have a logo or slogan. HVC will work with a local design firm to develop a logo that reinforces the positioning themes of compassionate and personalized care offered by its staff. HVC has used the phrase "we treat pets like family" in the past because of its simple, direct message. This slogan captures its positioning and will become its official slogan. A designer will be hired to develop the new logo—and to design other elements of this plan including brochures, refrigerator magnets, bandannas, letterhead, business cards, and signage.

5.1.3 Promotion—Participation in Community Events
Events in the community of Wellington are a way for HVC to enhance its reputation and create more awareness of its business. Residents of Wellington appreciate the small town and tend to want to support local businesses that also support the town.

Previously HVC participated in the *Parade of Lights* and *Pet Fair*. These have been very positive experiences—creating a positive image in the community and introducing new customers to the clinic. The community events program will be expanded and formalized in the coming year, and HVC will participate in the following events:

- Community *Easter Egg Hunt,* held annually on Easter weekend
- *Wellington Town Garage Sale,* held annually in May
- *Family Fun Fest,* held annually on July 4th
- *Wellington Pet Fair,* October
- *Parade of Lights,* held annually during the Christmas season

When involved in these events, HVC must not only interact with the public but also proactively sell itself through various tactics. One of these tactics would be to use banners to generate customer awareness or serve as a reminder. The banners would be displayed somewhere visible to people passing by such as wrapped around the front of a table at the *Wellington Pet Fair*. The banners would encourage people to stop by the conveniently located clinic. Business cards, as well as color brochures, would be available or handed out to people. The clinic's dogs could also be present wearing the HVC bandannas—the dogs can help create a face for HVC, support the "we treat pets as family" slogan, and serve as conversation "ice breakers" with passersby. When it is possible to have some sort of table or booth at events, the clinic could also have brochures and swag (e.g., bandannas and refrigerator magnets).

These activities will help remind current customers (the second target market) and demonstrate HVC's commitment to the community.

5.1.4 Promotion—Create a Large Vinyl Sign
HVC will purchase two large (3-by-5-foot) vinyl signs. The sign will include the name of the clinic, the company's slogan, and its new logo. This size sign can be placed on the front of a table and/or around any special events where HVC participates.

5.1.5 Promotion—*Yellow Pages* Advertising
Some new residents to an area are known to use the *Yellow Pages* to find goods and services in their new community. HVC knows that about 10 percent of current customers learn about the clinic from these ads. While most customers probably search online, HVC feels it needs some minimum level of exposure in the *Yellow Pages* phone directory.

Current customers may also look in the *Yellow Pages* to find HVC's phone number, so the ad will also serve this second target market.

5.1.6 Promotion—Google AdWords
Many customers are expected to use their phones or computers to search online for veterinary services. While search engine optimization (see strategy 5.1.7) will help HVC

to appear high on organic search results, HVC believes it should also have a monthly advertising budget for online advertising for people searching for "Vet" or "Veterinarian" from a 10-mile radius of the clinic.

5.1.7 Promotion—Search Engine Optimization
Most residents of Wellington are more likely to use search engines such as Google or Bing to find veterinary service providers. HVC will work with a web-marketing specialist to ensure its dominance in the search results when locals are looking for veterinary clinics.

5.1.8 Promotion—Earned Media Management
A large number of new customers discover the clinic from an online search. Given that these searches often include a Google reviews page, HVC is going to make a greater effort to monitor and manage online reviews. Consequently, it will check for new reviews on the site monthly. In addition, customers will be asked to provide a review for HVC on either its Facebook page or at Google reviews.

5.1.9 Place—Extended Clinic Hours
HVC operates its current facility at near capacity. There is no way to expand the current facility—there are occupied buildings on either side, and ordinances prevent building out the back. Without the new facility, which is not certain or anticipated for at least two more years, the only way to increase capacity will be to offer longer hours. This change also addresses something that some customers indicated they wanted to see when they filled out the survey.

Therefore, the clinic will increase the hours it is open from the current 49 hours per week to 57.5 hours per week. The new hours will be Monday–Thursday, 8:00 a.m. to 7:00 p.m.; Friday, 8:00 a.m. to 5:30 p.m.; and Saturday, 8:00 a.m. to 12:00 p.m.

New staff members will be needed to implement this element of the plan. It will require additional hours from another veterinarian, as Dr. Hardy will be unable to work all these hours. At least for the first few months, vet techs could have their schedules adjusted, but additional hours will not be needed. A part-time receptionist will also be hired.

- Clinic hours will not be extended until September—to save money and in time for the busier fall season (summer tends to be a slower time of year). Promotion of the expanded clinic hours will occur in the first year with signs inside the clinic.
- Dr. Hardy believes that a veterinarian friend who currently comes in occasionally—to fill in for vacations or in other emergencies—will want to have more stable and regular part-time hours.
- A part-time receptionist will be hired in August.

This element of the marketing strategy will also appeal to current customers, some of whom have indicated a desire for evening and weekend hours.

5.1.10 Place—New Facility (Long Term)
HVC intends to build a new facility in about two years. While this marketing plan anticipates the new facility, it does not directly address those plans. The new facility will attract new customers and help retain customers as well.

5.1.11 Price—Prices to Be Adjusted for Inflation
HVC believes that it is important to raise prices every one to two years—so that prices do not become so low that large increases must be mandated to catch up with inflation. HVC does not want to compete on a price basis, and current prices are higher than the competition. Further, higher prices are necessary for the staffing required to deliver the caring and compassionate service that HVC strives to maintain. Higher prices can also carry higher quality expectations from customers—many of whom are willing to pay

higher prices for the convenience, service, and quality that HVC is known to deliver. Finally, current prices are already about 20 percent above the local competition.

Prices were raised about 5 percent about six months ago—so price increases will not be implemented until the end of the coming year. Based on inflation, price increases are anticipated to be about 5 percent.

This element of the marketing strategy also addresses the current customers target market.

5.2 Target Market #2—Current Customers

The second target market in this marketing plan involves retaining and growing current customers.

- Product—Compassionate and Personalized Care
- Product/Price—Bundled Service Packages for Dental Care and Geriatric Pet Care
- Promotion—In-Store Signage and Point-of-Purchase Materials
- Promotion—Owned Media Management/Website Redesign
- Promotion—Customer Database Enhancement
- Promotion—Brochures to Describe the Clinic and Key Special Services
- Promotion—Swag
- Promotion—Social Media

Some of the elements of the marketing strategy that primarily target customer acquisition will have the additional benefit of helping retain and grow business with current customers. These elements are described in detail in Section 5.1:

- Promotion—Logo and Slogan
- Promotion—Participation in Community Events
- Promotion—Create a Large Vinyl Sign
- Promotion—*Yellow Pages* Advertising
- Promotion—Google AdWords
- Promotion—Search Engine Optimization
- Place—Extended Clinic Hours
- Price—Prices to Be Adjusted for Inflation

5.2.1 Product—Compassionate and Personalized Care

HVC delivers compassionate and personalized care. As HVC grows and hires additional staff, it is important that all employees continue to deliver on the high standards initially set by Dr. Hardy. This will be achieved by carefully hiring employees who share HVC's core values.

5.2.2 Product—Bundled Service Packages for Dental Care and Geriatric Pet Care

Bundled service packages create value for HVC, customers, and their pets. The bundled packages will be offered at price discounts as compared to the purchases of individual services. This creates value for the customer. The package encourages pet owners to focus more on their pets' health, and increases the revenue received by HVC. Best of all, the pets receive better-quality care.

With only limited promotion, HVC has had success with its Puppy and Kitten Bundles.

Similar packages will be introduced for dental services—and later, one targeting older pets with specialized geriatric pet services. The specifics on these future packages are described below:

Dental Service Package Dental services provide significant health benefits for cats and dogs—and provide a good profit margin for HVC. The dental package includes the following:

- Standard medical examination
- Health screen six and CBC (complete blood count)

- Dental radiographs (x-rays)
- Teeth cleaning from Dr. Hardy or the vet tech trained in this area
- A short training session on teeth brushing for the dog/cat owner
- Dental kit by Pet Dentist (a kit that includes toothbrush, toothpaste, and breath mints designed especially for dogs/cats)
- Sample package of Greenies dog/cat treats, which are designed to clean teeth and freshen breath

The total cost of these bundled goods and services is $189. If purchased separately, they would amount to $215. The package provides higher perceived value to customers and helps them get more actively involved in their pets' health care. Once customers get into the habit of this service, there will be less veterinary work required at the time of the annual checkup.

The Dental Service Package will be introduced in the next two months. A promotional brochure and point-of-purchase signs that describe the benefits will be created to support this product. The vet techs and Dr. Hardy will make efforts to mention the service to customers who are in for other reasons.

Geriatric Health Care Package for Dogs and Cats This package is designed for dogs and cats six years of age or older. The package would be targeted at pet owners with aging animals and consists of the following services:

- Standard medical examination
- Health screen six and CBC (complete blood count)
- Urinalysis
- ERD screen (kidney screen)

Customers will receive these bundled services, a $165 value if purchased separately, for $139. This gives customers cost savings, encourages purchasing more services for their pets, and enhances their pets' quality of life.

This product will be introduced about one year from now. Adaptations to the product or promotion will be made based on experiences with the dental services package.

A promotional brochure and point-of-purchase sign describing the benefits and procedures will be created to support this product.

5.2.3 Promotion—In-Store Signage and Point-of-Purchase Materials

Currently, HVC has no signage or point-of-purchase materials. The store's large countertop—where customers check in and check out—is currently bare. Small 8.5-by-5-inch laminated stand-up signs (which work like picture frames) will be placed on the countertops. Four signs will be purchased, one each promoting the three bundled packages (puppy/kitten, dental services, and geriatric pet care), and one promoting the clinic. The signs will be rotated—keeping two on the counter at any one time.

A shelf will be created to hold the clinic's new brochures (see 5.2.6). The shelf will be placed next to the countertop at checkout and will hold four different brochures.

5.2.4 Promotion—Owned Media Management/Website Redesign

A newly revised website will include existing features and some additions for current customers:

- A home page will give basic information about the clinic—phone number, office hours, e-mail address, logo, slogan, and so on.
- The home page would include links to social media as well.
- Separate pages will show photos of staff members and their pets—to help personalize the people and clinic.
- Photos and descriptions of the clinic mascots will be added—Dr. Hardy's two dogs and three cats.

- A customer photo page will be provided—which allows for photos of customers and/or their pets.
- A page will list the main services offered by the clinic and provide links to detailed descriptions of each service. This allows for customer education about the services, demonstration of the high-quality and state-of-the-art equipment the clinic utilizes, and to emphasize the extra efforts the clinic uses to provide compassionate and individualized care for pets and their owners.

This plan results in approximately 25 web pages. The larger site can be expanded to meet customer needs.

The website will also help attract new customers—the other target market in this plan.

5.2.5 Promotion—Customer Database Enhancement

The initial cost of development and investment in a customer relationship management (CRM) program is beyond HVC's available time and financial resources at this time. On the other hand, it is anticipated that such a program will be cost justified in a few years. The company currently has a very simple customer database that tends to include only limited information for each customer. In the next couple of years, HVC must make sure that the data on each customer are complete in anticipation of migrating data to a sophisticated CRM package. Specific data to be collected on each customer:

- Name, address, phone, and e-mail
- Detailed information about each pet
  - Name
  - Breed
  - Vaccination records
  - Prescription/medication records
  - Services purchased for each pet
  - Food—brand/type
  - Other pertinent information

At some point in the future, this information may be included in a yet more sophisticated program that allows for scheduling reminders and mailing of customized promotions. This remains an opportunity that has yet to be funded.

5.2.6 Promotion—Brochures to Describe the Clinic and Key Special Services

The clinic will create three different 8.5-by-11-inch trifold brochures—two black-and-white brochures will promote the puppy/kitten services and dental health services, and a full-color brochure will be used to promote the clinic in general. The brochures will be designed by the same designer hired to create the logo and other promotional materials. The brochure should include contact information, various services, benefits to the pet and the owner, and possibly testimonials.

Brochures will be made available at the front counter of the clinic so customers can easily grab one to take and read. They will also be provided at any type of community/promotional event HVC attends. Brochures can be placed on the tables or other areas where they are clearly accessible for people to take.

Three different brochures will be created:

- Overview of HVC (services and staff)
  - Available in the clinic
  - Handed out at community events
  - Distributed in the new residence packet—this is a packet given to each new resident
- Puppy/Kitten Service Bundle (description and prices)
  - Available in the clinic
  - Mailed to customers upon request
  - Distributed at the local Humane Society and by interested breeders in the area who currently use HVC

- Dental Health Care Service Bundle (description and prices)
  - Available in the clinic
  - Handed out to interested customers following a discussion of the importance of regular dental care for dogs and cats

5.2.7 Promotion—Swag

HVC will use two different types of swag—free promotional materials—as giveaways at the clinic and when appearing at various community events. The items will include:

- Refrigerator magnets. Business card-sized refrigerator magnets with the HVC name, logo, phone number, and website address.
- Bandannas. Bandannas with the HVC name and logo—for the many local dog owners who like to put bandannas around their dogs' necks.

5.2.8 Promotion—Social Media

HVC will utilize slow periods in the workday to generate more online content for the website and social media pages. This initiative will include promotional pricing, community events, public service announcements (e.g., lost pets), national reminders (i.e., World Rabies Day), and fun and lighthearted pictures or videos of pets from the clinic and around the world. Animal-related videos have a strong following on the Internet and would likely move web traffic toward HVC's social media sites.

HVC's online presence currently includes:

- Website—a modern clinic website that is simple and attractive while providing all the necessary information.
- Facebook—this is how HVC keeps in touch with its clientele on a day-to-day basis. Also, through its connected Instagram account, HVC posts humorous but appropriate content and veterinary tips and information.
- Twitter—HVC maintains an active Twitter feed to provide short news clips and to focus spreading the word about lost pets.
- Pinterest—although a less important part of the social media presence, vet tech Rick Woo enjoys photography and takes many excellent photos of the clinic's patients for sharing on Instagram and Pinterest.

Although there was some discussion of pulling back on Pinterest and Twitter (where followers are fewer), the additional cost of these social media tools is relatively modest.

6. Implementation and Control

This section on implementation and control outlines key details related to implementing the marketing plan.

- Special Implementation Problems to Overcome
- Control
- Budget, Sales Forecasts, and Estimates
  - Budget for Total Marketing Plan
  - Monthly Expenses for Total Marketing Plan
  - Pro Forma Income Statement
- Timing of Implementation Activities
- Contingency Plans

6.1 Special Implementation Problems to Overcome

The marketing plan will require the hiring of two additional part-time employees. These employees are necessary to staff extended office hours. One new employee would be a part-time veterinarian who has already been identified. This person would work about 15 hours per week. Annual cost for this additional person would be about $1,250 per month.

It would also be necessary to hire a receptionist to work the extra hours. Currently the office manager also works as a receptionist. The new receptionist would work about 20 hours per week. Incremental cost would be about $750 per month for the new receptionist.

6.2 Control

The objectives for this marketing plan outline specific targets for increased sales of some products, overall sales, and number of additional customers. These goals will be monitored on a monthly basis with a report developed by the office manager (Karen) for Dr. Hardy.

HVC also wants to be sure that it continues to deliver compassionate and individualized care to its customers. This important aspect of the clinic will be monitored through customer feedback. Beginning in January, HVC will give each customer a postcard addressed to HVC following each visit. The postcard will have three questions on the back:

- Would you recommend Hillside Veterinary Clinic to a friend? Yes/No
- On this visit to Hillside Veterinary Clinic, did you receive compassionate and individualized care for your pet? Yes definitely/Yes somewhat/No
- Do you have any comments for Hillside Veterinary Clinic? [create a box for comments]

HVC will monitor this information on a monthly basis to ensure that sales targets and customer service delivery maintain high standards.

Control will also be done with routine monitoring and management of the company's owned media—in particular its website and social media pages. Weekly reports will be prepared to determine which pages get the most hits and the longest visits, and which social media get liked, favorited, and forwarded.

6.3 Budget, Sales Forecasts, and Estimates

6.3.1 Budget for Total Marketing Plan

Marketing Tactic	Cost
Customer satisfaction survey postcards	$ 100
Brochure for puppy/kitten bundle (500)	125
Brochure for all clinic services (1,000)	500
Brochure for dental services bundle (500)	125
Design services for logo and various promotional materials	500
Website maintenance and updating (including purchase of URL and one-year maintenance)	750
Participation in five community events (costs for planning/attendance)	750
Vinyl signs and point-of-purchase signs	175
Costs for running ad and training new personnel	250
Yellow Pages advertisement ($50 per month)	600
Google AdWords ($160 per month)	1,920
Personnel needed for extended clinic hours (4 months)	8,332
Refrigerator magnets (1,000)	224
Bandannas (600)	455
Costs for maintaining and monitoring earned, owned, and social media*	0
Total costs for implementing marketing plan	$14,806

*Lab tech Rick Woo really enjoys social media and photography. He feels that in 3-4 hours per week he will be able to maintain a good social media presence for HVC. Owned media management and earned media management will be handled by the receptionist as part of her regular duties; therefore, there is no incremental cost.

6.3.2 Monthly Expenses for Total Marketing Plan

For cash flow planning purposes, the anticipated expenses are shown on a monthly basis.

Month	Major Expenses*	Cost
January	Customer satisfaction surveys, design services	$ 810
February	Brochures	960
March	Signs, magnets, bandannas	1,064
April	Community event	360
May	Community event	360
June	Website update and maintenance—annual contract due	585
July	Website design, community event	735
August	Ad and training for part-time receptionist	460
September	Personnel for longer hours	2,293
October	Personnel for longer hours, community event	2,443
November	Personnel for longer hours	2,293
December	Personnel for longer hours, community event	2,443
Monthly	Regular posting to website and social media—done on as-needed basis by everyone	0

*Yellow Pages ad and Google AdWords bill monthly at $210.

6.3.3 Pro Forma Income Statement

	Next Year	Two Years	Three Years
Sales revenue	$375,000	$430,000	$485,000
Expenses			
Labor costs	250,000	285,000	320,000
Operating expenses	75,000	77,500	80,000
Marketing expenses	3,174	6,000	7,500
Depreciation	7,300	7,500	8,000
Net income	$ 39,526	$ 54,000	$ 69,500

Notes:
- Sales revenue growth assumes the marketing plan objectives outlined in this plan are achieved.
  - 15 percent growth in total customers over two years will increase revenues by an estimated $60,000.
  - More aggressive promotion of dental services package and geriatric dog care package will result in an additional $30,000 in sales growth over two years.
  - Price increases of 5 percent over that time will yield about $20,000.
- Routine operating expenses include various fixed costs like lease payments and loan payments on equipment and insurance.
- The costs of additional personnel, their hiring expenses and training, are included in operating costs—even though earlier analysis notes these costs are part of implementing the marketing plan.

6.4 Timing of Implementation Activities

The implementation of many of the marketing activities will be done by Karen (office manager), who already orders other supplies. In addition, Rick is the vet tech who most enjoys getting out among people—so he will coordinate and attend all event activities with Karen.

Date	Activity	Responsibility	Vendor
November	Sign and *Yellow Pages* ad	• Karen	*Yellow Pages*
Weekly	Manage social media: daily post on Facebook, four times a week Instagram	• Karen	
Monthly	Review and update website, monitor earned media and owned media	• Dr. Hardy & Karen	Jay's Website Design Service
January	Train receptionist to ensure complete understanding of office culture and procedures	• Dr. Hardy & Karen (receptionist)	
January	Contract for logo and brochure design • Design new logo • Design two brochures • Design refrigerator magnets • Design bandannas • Design signs • Design letterhead • Design business cards	• Dr. Hardy will coordinate with Lisa (our designer)	Specialisa Design Service
February	Print brochures • Puppy/kitten package • Dental services package • Clinic list of services	• Karen	FedEx Office Wellington
March	Order new business cards	• Karen	
March	Order refrigerator magnets	• Karen	magnets.com
March	Order bandannas	• Karen	BandannaMan
March	Order vinyl signs and point-of-purchase signs to promote dental services	• Karen	FasterSigns Fort Collins
March	Develop plan for Easter Egg Hunt (April) and Townwide Garage Sale (May)	• ~2 hours Rick and Karen	
April	Attend Easter Egg Hunt event	• Rick and Karen (maybe Dr. Hardy)	
May	Attend Wellington Town Garage Sale event	• Rick and Karen (maybe Dr. Hardy)	
June	Plan for Family Fun Fest event (July)	• Rick and Karen (maybe Dr. Hardy)	
June	Talk to part-time vet to determine interest; if needed, run ads	• Dr. Hardy	
July	Attend Family Fun Fest event	• Rick and Karen (maybe Dr. Hardy)	
July	Develop and run ad for part-time receptionist; inform staff to spread word of mouth for opening	• Karen	*Coloradoan* newspaper ad
September	Start extended office hours	• New hires will be key to starting • Dr. Hardy and Karen to work later to help transition	
September	Plan for Wellington Pet Fair	• Rick and Karen	
October	Attend Wellington Pet Fair	• Rick, Karen, and Dr. Hardy	
November	Plan for Parade of Lights	• Rick and Karen	
December	Attend Parade of Lights	• Rick, Karen, and Dr. Hardy	
	Extended clinic hours	• $1,250/month (vet) • $833/month (receptionist)	Midyear

6.5 Contingency Plans

If market conditions change, it may be necessary to reevaluate elements of this plan or to develop an alternative plan. Potential environmental changes and the plans for addressing them follow:

- Difficulty hiring a part-time veterinarian. Although we anticipate our current part-time veterinarian will desire more stable and regular hours, this may not happen. We would then need to look on the open job market where it may not be possible to find a suitable vet.
  - If that is the case, the extended clinic hours program may be delayed.
- Slowdown in housing growth. Growth projections are largely based on continuing growth in the local housing market—which will bring new families and pets to the local area. If this slows down, it may
  - Delay the extended clinic hours proposal.
  - Delay the new building, currently planned for about two years out.
- Economic slowdown. Customers are less likely to be interested in health care prevention services—so this aspect of the marketing plan will be toned down.
- Faster growth in the housing market. Although a housing slowdown may come suddenly, the long lead time for new housing makes faster growth something we can more easily foresee. But if growth appears to be picking up:
  - We will speed up plans for the new building.
- New competitor enters the market. The fast growth of the town of Wellington and having CSU's local College of Veterinary Medicine may result in a new competitor coming into the Wellington market.
  - Extended hours will be instituted earlier—to try to attract more customers.
  - Carefully reevaluate the entire plan depending upon the nature of the new competitor—work hard to gather reliable information.
  - Consider more aggressive acquisition activities.

7. Appendices

7.1 Appendix A: Survey of Current Customers

To gather additional information in advance of the development of the marketing plan, a survey of current customers was conducted during this past March. A convenience sample of 40 current customers completed the survey during visits to the clinic. The complete survey and percentage responding to each question are included below. Because some questions allowed respondents to circle more than one answer or due to rounding, percentages may not always total 100 percent.

Hillside Veterinary Clinic Customer Survey

This survey has been designed to give us more information about your needs. Please fill out this survey to the best of your ability.

   Sincerely . . . Dr. Kelly Hardy, Hillside Veterinary Clinic

1. How many pets do you bring to Hillside Veterinary Clinic? (circle one)

   1 **(38%)**   2 **(35%)**   3 **(10%)**   4 **(8%)**   5+ **(15%)**

2. What type of pet(s) do you have? (circle all that apply)

Dog	Cat	Bird	Hamster/Guinea pig	Ferret	Other
100%	**48%**	**0%**	**0%**	**0%**	**5%** (rabbit, chameleon)

3. Please list your pet's type, name, and age.

   Type: Name: Age:
   _____   _____   _____
   _____   _____   _____
   _____   _____   _____

4. Which of the following best describe(s) how you view your pet(s)? (circle all that apply)

Family member	Companion	Work animal	Show animal	Breeding
93%	**15%**	**8%**	**3%**	**3%**

Other ___ **5%** (rescue dog, hunting companion)

5. How many times in the last 12 months have you visited Hillside Veterinary Clinic? (circle one)

1-2	3-4	5-6	7+
38%	**23%**	**23%**	**18%**

6. About how much do you spend each year on your pet(s)? (circle one)

$0-60	$61-150	$151-300	$301+	No Reply
3%	**10%**	**28%**	**53%**	**8%**

7. How important are each of the following to you? (please rank)

	Not at all important	Not very important	Somewhat important	Important	Very important
Location close to my home	1 **(0%)**	2 **(5%)**	3 **(8%)**	4 **(35%)**	5 **(53%)**
Convenient hours	1 **(0%)**	2 **(5%)**	3 **(10%)**	4 **(50%)**	5 **(35%)**
Expertise of vet staff	1 **(5%)**	2 **(5%)**	3 **(0%)**	4 **(13%)**	5 **(78%)**
Quality of equipment	1 **(8%)**	2 **(0%)**	3 **(10%)**	4 **(28%)**	5 **(55%)**
Friendly customer service	1 **(5%)**	2 **(0%)**	3 **(3%)**	4 **(30%)**	5 **(63%)**

8. What services do you use at Hillside Veterinary Clinic? (circle all that apply)

House Calls	5%	Whelping	0%	Dermatology	13%
Vaccinations	80%	Tick/Flea Control	15%	General Surgery	25%
Spay/Neuter	40%	X-Rays	28%	Behavioral Counseling	3%
Parasite Control	35%	Heartworm	53%	Grief Support	5%
Health Screens	68%	Dentistry	10%	Hip Certifications	5%

9. What additional services would you like to see at Hillside Veterinary Clinic? (circle all that apply)

Boarding	33%	Daycare	10%
Weekend hours	20%	Training	3%
Acupuncture	10%	Other _____	5% (ultrasound, emergency hours)
Evening hours	38%		

10. About how many miles do you travel to get to Hillside Veterinary Clinic? (circle one)

0-5 **(70%)**   6-10 **(13%)**   11-15 **(5%)**   16-20 **(10%)**   21-25 **(0%)**   26+ **(3%)**

11. If Hillside Veterinary Clinic were to expand its facility in the general area, how would this affect your decision to do business with the clinic? (circle one)

Would not switch clinics	90%
Would consider switching clinics	5%
Would probably switch clinics	0%
Would definitely switch clinics	0%
No reply	3%

12. Are the hours of the clinic sufficient for you? (circle one)

    Yes **(88%)**   No **(12%)**

    If No, what hours would best serve you? (circle all that apply)

    | Mornings **(3%)** | Days **(3%)** | Evenings **(38%)** | Saturdays **(18%)** | Sundays **(0%)** |

13. Have you ever been dissatisfied with the Hillside Veterinary Clinic? (circle one)

    Yes **(5%)**   No **(95%)**

    If Yes, explain briefly, _____ ("personal—do not want to explain," "poor treatment from an employee who is no longer working here")

14. Is the Hillside Veterinary Clinic the only clinic you take your pet to? (circle one)

    Yes **(90%)**   No **(10%)**

    If No, what other clinic(s) do you use? **CSU Vet Clinic (2), East Side (2)**
    If No, why? **(specialized services at CSU, I like East Side, convenience, hours)**

15. How did you hear about Hillside Veterinary Clinic? (circle all that apply)

Word of mouth (e.g., friend, neighbor, relative)	35%		
Referral from other clinic	5%	Pet Fair	0%
Magazine	0%	Sign	**40%**
Parade of Lights	3%	*Yellow Pages*	**10%**
Other _____		8% newspaper article (2), Heska (1)	

16. Have you recommended the Hillside Veterinary Clinic to anyone? (circle one)

    Yes **(88%)**   No **(12%)**

17. Where do you purchase your pet food and/or supplies? (circle all that apply)

Grocery store	**20%**
Pet specialty store	**68%**
Discount store (e.g., Walmart)	**35%**

18. Do you have access to the Internet? (circle one)

    Yes   No
    98%   3%

    If Yes, do you have an e-mail address?

    Yes **(95%)**   No **(3%)**   No reply **(3%)**

19. Which of the following social media platforms do you currently use? (circle all that apply)

    Facebook **(61%)**   Twitter **(12%)**
    Instagram **(14%)**   Pinterest **(19%)**
    I don't use any social media platforms **(38%)**

APPENDIX D

Hillside Veterinary Clinic Marketing Plan

7.2 Appendix B: Competitor Matrix

	Hillside Veterinary Clinic	East Side Small Animal Hospital	Baker & Reid Veterinary Clinic	Flatirons Veterinary Clinic	Fossil Creek Veterinary Services
Nature of Practice	• Small animals, birds, and exotics	• Small animals only	• Equine and small animals	• Small animals only	• Dogs, cats, and some large animals
Location	• Downtown Wellington	• 10 miles southwest of HVC in Fort Collins	• 1 mile west of HVC	• 8 miles south of HVC in Fort Collins	• Mobile service for Wellington and outlying areas
# of Doctors	• One	• Five	• Two	• One	• One
Extra Services	• House calls • Acupuncture	• Emergency services • Acupuncture	• Ambulatory services • Equine reproduction • Equine embryo transfer	• Grooming • 24-hour emergency care • Specialize in dentistry	• Mobile—it only makes house calls
Gas or Anesthesia	• Anesthesia	• Both—depends on animal weight	• Anesthesia	• Gas	• Gas
Boarding	• No	• No	• No	• Yes	• No
Hours of Operation	• MWThF 8–5:30; Tu 8–7	• M–Th 7:30–8 F 7:30–6 Sat 9–1 Sun 1–3	• Services by appointment only	• M–F 8–5:30 Sat 9–12 noon	• By appointment only
Prices Spay Dog/Cat*	• $200/$150	• $136–210/$78	• $125/$90	• $110/$85	• Unknown
Advertising	• One-inch in-column Y/P ad	• Quarter-page Y/P ad, occasional direct mail, about 3–4/year	• One-inch in-text Y/P ad 2–3/year in Valpak mailer	• One-inch in-text Y/P ad	• Single line in Y/P
Website	• Extensive site	• Extensive site	• Modest site	• Extensive site	• Basic site
Social Media	• Facebook page with 602 Likes and daily posts • Instagram posts images 3–5 times/week • Twitter 3–5 tweets/week • Pinterest growing	• Facebook page with 698 Likes—post 1–2 times/day • Twitter feed with 107 likes • Instagram occasional post • YouTube channel	• Facebook with 240 Likes weekly posts • Big Instagram page with mostly large-animal photos	• Facebook page with 187 Likes and weekly posts • Pinterest page that appears dormant	• No Facebook page • Blog page with weekly posts
Positioning	• Try to emphasize compassionate and caring atmosphere	• Emphasizes "in area since 1961" • Focuses on safety/comfort • Photo of large staff, 5 vets—promotes size	• Personalized business • Emphasizes equine health—small animals as convenience to horse customers	• "I treat your pet like my own." • Emphasizes small-town atmosphere	• Promotes convenience of mobile—we come to you • Difficult to infer

*All respondents indicated that the size of the animal made pricing somewhat variable.

BONUS CHAPTERS

Now available to all adopters of *Essentials of Marketing 17e* are two chapters that previously resided only in our hardcover book, *Basic Marketing*. These are available online and can be bound in a printed version of the text through McGraw-Hill's Create custom publishing. We are calling them "Bonus Chapters" because they are something extra and optional for instructors seeking this extended coverage. You will be able to access these optional chapters online within the Student Resources.

Bonus Chapter 1: Implementing and Controlling Marketing Plans: Metrics and Analysis

This chapter builds on implementation and control, two concepts introduced in Chapter 2. The chapter goes into more depth on these concepts and offers how-to approaches for making implementation and control more effective. The chapter discusses how new information technology tools facilitate these practices and demonstrates how firms use sales analysis, performance analysis, and cost analysis to control marketing strategies and plans.

LEARNING OBJECTIVES

When you finish this chapter, you should be able to

1. understand how information technology speeds up feedback for better implementation and control.
2. know why effective implementation is critical to customer satisfaction and profits.
3. understand how sales analysis can aid marketing strategy planning.
4. understand the differences in sales analysis, performance analysis, and performance analysis using performance indexes.
5. understand the difference between the full-cost approach and the contribution-margin approach.
6. understand how planning and control can be combined to improve the marketing management process.
7. describe measures used to monitor and control marketing strategy planning.
8. understand what a marketing audit is and when and where it should be used.
9. understand the important new terms (shown in red).

Chapter outline

- Good Plans Set the Framework for Implementation and Control
- Speed Up Information for Better Implementation and Control
- Effective Implementation Means That Plans Work as Intended
- Control Provides Feedback to Improve Plans and Implementation
- Sales Analysis Shows What's Happening
- Performance Analysis Looks for Differences
- Performance Indexes Simplify Human Analysis
- A Series of Performance Analyses May Find the Real Problem
- Marketing Cost Analysis—Controlling Costs Too
- Planning and Control Combined

- Marketing Metrics and Analytics Can Guide Marketing Strategy Planning
- The Marketing Audit
- *Conclusion*
- *Key Terms*
- *Questions and Problems*
- *Marketing Planning for Hillside Veterinary Clinic*
- *Suggested Cases*

Bonus Chapter 2: Managing Marketing's Link with Other Functional Areas

This revised chapter covers some of the important ways that marketing links to other functional areas. The emphasis is not on the technical details of these other functional areas, but rather on the most important ways that cross-functional links impact your ability to develop marketing strategies and plans that really work. The chapter includes separate sections that describe how finance, production and operations, accounting, information systems, and human resources interact with marketing to create and implement successful marketing plans.

LEARNING OBJECTIVES

When you finish this chapter, you should be able to

1. understand why turning a marketing plan into a profitable business requires money, information, people, and a way to get or produce goods and services.
2. understand the ways that marketing strategy decisions need to be adjusted in light of available financing.
3. understand how a firm can implement and expand a marketing plan using internally generated cash flow.
4. understand how different aspects of production capacity, flexibility, and the cost and location of production affect marketing strategy planning.
5. describe how marketing managers and accountants work together to analyze the costs and profitability of specific products and customers.
6. understand how information systems enable marketing strategy.
7. know some of the human resource issues that a marketer should consider when planning a strategy and implementing a plan.
8. understand the important new terms (shown in **red**).

Chapter outline

- Marketing in the Broader Context
- The Finance Function: Money to Implement Marketing Plans
- Production Must Be Coordinated with the Marketing Plan
- Accounting Data Can Help in Understanding Costs and Profit
- Enabling Marketing Strategy with Information Technology
- People Put Plans into Action
- *Conclusion*
- *Key Terms*
- *Questions and Problems*
- *Marketing Planning for Hillside Veterinary Clinic*
- *Suggested Cases*

Video Cases

Essentials of Marketing includes two different types of marketing cases: the 5 special video cases in this section and the 34 traditional cases in the next section. All of the cases offer you the opportunity to evaluate marketing concepts at work in a variety of real-world situations. However, the video cases add a multimedia dimension because we have produced a special video to accompany each of the following written cases. The videos are available to professors who adopt *Essentials of Marketing* for use in their courses. (These case-based videos are in addition to the teaching videos we have custom produced and made available to instructors for possible use with other parts of the text.)

The videos bring to life many of the issues considered in each case. However, you can read and analyze the written case descriptions even if there is no time or opportunity to view the video. Either way, you'll find the cases interesting and closely tied to the important concepts you've studied in the text.

The set of questions at the end of each case will get you started thinking about the marketing issues in the case. Further, we provide instructors with a number of suggestions on using the video cases—both for group discussion in class or individual assignments. Thus, as is also true with the traditional cases in the next section, the video cases can be used in many different ways and sequences. You can analyze all of the cases, or only a subset. In fact, the same case can be analyzed several times for different purposes. As your understanding of marketing deepens throughout the course, you'll "see" many more of the marketing issues considered in each case.

1. Potbelly Sandwich Works Grows through "Quirky" Marketing*

Want an inexpensive gourmet sandwich served in a fun and funky place by friendly young people? Lots of people do, including celebrities Will Ferrell, Sandra Bullock, and Keanu Reeves. In fact, Potbelly Sandwich Works was ranked as one of the top fastest-growing chains in the restaurant industry.[1-1]

Chances are you have never eaten at a Potbelly, but you probably will in the near future. Under the leadership of Bryant Keil, chair and CEO, Chicago-based Potbelly Sandwich Works is expanding rapidly (see Table 1). Keil bought the original Potbelly in 1996. Prior to that, it was a very successful neighborhood sandwich shop run by a couple who had originally started the business as an antique store in 1977. They added homemade sandwiches and desserts to bolster the business, and soon the food became more popular than the antiques. As time went on, booths were added along with ovens for toasting the sandwiches and the antiques became "décor" rather than merchandise for sale. A prominently placed antique potbelly stove provided inspiration for the sandwich shop's name.

Although Potbelly has many sandwich shop rivals in the fierce quick-serve segment of the restaurant industry, Potbelly has more than held its own. Billed as "a unique and quirky sandwich joint," it has a unique appeal. Potbelly's core strategy elements include *product, place, promotion,* and *price.*

PRODUCT

According to Keil, "Anyone can sell a sandwich. You need to sell an experience." Industry observers point to several aspects of the Potbelly experience that make it the first choice for young professionals on a quick lunch break. First is the menu, which features made-to-order toasted sandwiches, soups, homemade desserts, malts, shakes, and yogurt smoothies. Toasting is part of what makes Potbelly's sandwiches distinctive. Quality ingredients, including a freshly baked Italian sub roll and freshly sliced meats and cheese, also contribute to superior value.

---

*This case was prepared by Dr. J. B. Wilkinson, professor emeritus, Youngstown State University.

Friendly service and an upbeat atmosphere, live music, antique fixtures, real books for customers to read or borrow, and vintage memorabilia create a homey environment for customers. The idea behind Potbelly is simple: superior value; fun-filled atmosphere; warm, comfy décor; and quick, friendly service.

PLACE

Potbelly stores average 2,200 square feet but can top 4,000 square feet. Most units have indoor seating for more than 50 and outdoor seating during warmer months. Geographic locations are selected carefully. Keil looks for cities that are not saturated with sandwich chains and have an urban/suburban density of core customers—professionals younger than 35 years old. Locations must be convenient for them because Potbelly stores highly rely on repeat business. All units are corporate owned and operated.

PROMOTION

Historically, Potbelly Sandwich Works has not had an ad budget. Promotions are keyed to events like store openings and National Sandwich Day. For example, on National Sandwich Day, Potbelly hosts a "Belly Buster" sandwich-eating contest at Potbelly stores. Prizes are awarded to winners and runners-up. Diners randomly receive free meals. Other event promotions raise money for local charities such as food banks and community-based reading and music appreciation programs.

PRICE

Potbelly sandwiches sell for $3.79. Sheila's Dream Bar, made of oatmeal, caramel, and chocolate, is $1.29, while homemade chocolate chip or sugar cookies go for only 99 cents. Checks average about $6.50. Pricing is an integral part of the value Potbelly offers customers and can be summed up as "Just good food at good prices!"

Considered separately, any one of Potbelly's marketing strategy elements may not seem overly powerful as a competitive weapon, but combined and implemented with zeal, they are a

Table 1 Potbelly Store Openings by Year and City, 1997–2005

Market	1997	1998	1999	2000	2001	2002	2003	2004	2005	Total
Chicago	1	1	1	2	3	8	11	9	11	48*
Washington, DC						8	3	4	3	18
Michigan							4	2	1	7
Minnesota							1	3	2	6
Wisconsin							2		2	4
Indiana								1	1	2
Texas								3	8	11
Ohio									6	6
Total	1	1	1	2	3	16	21	22	34	102

*Includes the original Potbelly opened in 1977 and bought by Keil in 1996.

significant competitive threat to national, regional, and local competitors. Brands such as Subway, Quiznos, Cosi, Panera, Jimmy John's, and Schlotzsky's Deli are wary that Potbelly will become a major national competitor. The Potbelly experience appears to be difficult to duplicate. For example, Quiznos may have similar food quality but not similar atmosphere; Cosi may have the hangout ambience down but not the food! Subway is especially vulnerable because it has neither the food nor the warm and comfy store environment.

Yet Potbelly has a tough road ahead. It requires management to maintain the superior performance of current units while creating new Potbelly units in new markets. Each Potbelly is special in terms of location, décor, and staff. The unit must be tailored to its neighborhood and community. Attentive, enthusiastic workers must be found. Food quality and a fun atmosphere must become an integral part of the store culture. Not an easy task!

1. Identify and describe Potbelly's strategy in terms of product (present or new) and market (present or new).
2. How would you describe Potbelly's positioning strategy?
3. What types of environmental opportunities and threats do you see in Potbelly's external environment? How might they affect Potbelly's current strategy?

2. Suburban Regional Shopping Malls: Can the Magic Be Restored?*

The suburban regional shopping mall is regarded by many as the "crown jewel" of shopping experiences. In a single location, shoppers can visit more than a hundred stores, go to a movie, eat, walk, and lounge for an entire day in a secure, pleasant atmosphere sheltered from undesirable weather and the demands of everyday life. Most Americans at one time or another have escaped for the day to such a mall and felt "uplifted" in spirit by the experience. So pervasive is the suburban regional shopping mall that William Kowinski in the *The Malling of America* (1985) claims that in the United States alone there are more enclosed malls than cities, four-year colleges, or television stations! Indeed, few of us can remember a time when shopping was a trip to "downtown," or the central business district (CBD) of a large city.

Many suburban regional shopping malls are more than one million square feet in size, contain more than a hundred stores, and offer shoppers free parking, restaurants, play facilities, lounge facilities, restrooms, and movie theaters. Some centers even provide amusement rides and other entertainment opportunities. One of the dominant features of these large shopping complexes is the presence of multiple department stores that "anchor" the extreme points of the mall's layout and "pull" shoppers to the mall from surrounding suburban areas. Department stores also encourage shoppers to walk through the mall. In fact, department stores were the driving force behind the original development of suburban regional shopping malls and have played a critical role in their continuing success.

The movement of traditional department stores from CBD locations to the suburbs, complete with large "full-line" departments, contributed greatly to the explosive growth of suburban regional shopping malls during the post–World War II era. At its inception, the suburban regional shopping mall was designed to be a substitute, or even a replacement, for a city's CBD, but without the usual congestion or parking difficulties. This strategy was particularly attractive after the opening of Southdale Center in suburban Minneapolis in 1956 (www.southdale.com), which demonstrated the viability of a regional shopping mall with multiple anchors.

In 2002 there were a total of 46,336 shopping centers in the United States, of which about 1,200 could be considered regional or superregional malls (www.icsc.org). In addition to regional and superregional malls, numerous types of shopping centers have evolved since the 1950s. The International Council of Shopping Centers has defined eight principal shopping center types: neighborhood, community, regional, superregional, fashion/specialty, power, theme/festival, and outlet.

The suburban regional shopping malls and their department store anchors enjoyed great success for almost 50 years and seemed virtually invincible to threats until the final decade of the twentieth century. During that era, several chinks developed in the competitive armor of this type of retail institution, and the problems seem to be getting worse. Shopper activity is declining: the number of tenant vacancies is increasing; and the delinquency rate on mall mortgages is disturbing. Increasingly larger percentages of consumer discretionary income are being spent elsewhere. To make matters worse, many of the older malls need renovating to remain attractive to shoppers. Renovation of an older mall can cost tens of millions of dollars.

Changes in consumers' wants and needs appear to be the major factor that underlies the woes of suburban regional shopping malls. Because the first multi-anchor center opened in 1956, the lifestyles of American families have changed significantly. In 1950, for instance, only 24 percent of wives worked outside the home; today, that percentage exceeds 60 percent. Women between the ages of 18 and 45, the mainstay of mall shoppers, simply do not have the time to shop like they once had. As a result, shopping has become much more purposedriven. Shopping statistics bear this out. Shoppers are visiting suburban regional shopping malls less frequently, visiting fewer stores when they do shop, and also spending less time at the mall when they shop. Mall visitors, however, are more likely to make a purchase when they shop. Yet the typical suburban regional shopping mall was designed for a "shop-all-day" or a "shop-'til-you-drop" philosophy.

Another consumer trend that spells trouble for suburban regional shopping malls is increased shopper price sensitivity. A wider selection of shopping alternatives from which to choose and the desire to make the family income go further (which, in essence, is equal to a pay increase) have proven to be strong forces pushing shoppers to comparison shop between retail establishments—something that most malls are not designed to

---

*This case and the script for the accompanying video were prepared by Dr. J. B. Wilkinson, professor emeritus, Youngstown State University, and Dr. David J. Burns, Xavier University.

facilitate. Despite the large number of stores contained within a regional suburban shopping mall, comparison shopping between stores is not an easy task. Most malls are laid out to cater to a leisure-oriented shopper. Similar stores are located in different wings of the shopping mall to encourage shoppers to walk through the entire center. Shoppers may walk upwards of a quarter mile in their quest to compare products! This is not consistent with the desire for shopping convenience and efficiency on the part of most consumers.

Competition also has played a role in the problems that plague suburban regional shopping malls. High levels of competition characterize most mature industries, and the shopping center industry is no exception. Regional malls have both direct and indirect competitors. Direct competitors are nearby shopping centers with either similar or dissimilar formats. Indirect competitors comprise other types of retail store sites like freestanding or clustered sites and nonstore retailing sites. Nonstore retailing includes online shopping, catalog shopping, home TV shopping, telemarketing, and other forms of direct marketing, all of which have made considerable inroads into retail store sales. Suburban regional shopping malls have been especially vulnerable to both forms of competition.

Many of the more successful retailers (e.g., Kohl's, Home Depot) are located on freestanding sites or in large open-air centers—locations that have greater appeal to time-pressed, purpose-driven shoppers than mall locations. Similarly, many outlet malls, which cater to price-sensitive shoppers, are typically open-air centers to facilitate store access. In addition, some of the newer small shopping centers cater to a focused lifestyle (teen or professional woman) or have an organizing theme (home décor, hobby) that satisfies the specific needs of a market niche by offering a more focused product assortment than what can be found in a suburban regional shopping mall.

Oversupply of retail space has posed considerable problems to all shopping centers. The United States has 20 square feet for every man, woman, and child, compared to 1.4 square feet per person in Great Britain. Sales per square foot of retail space is declining in the United States. In fact, revenue from retail sales is contracting. It grew an average of 2.5 percent in the 1970s; 1.3 percent in the 1980s; and only 0.8 percent in the 1990s, adjusted for inflation. The result has been retail consolidation, store closings, and bankruptcies, leaving shopping centers fighting for a shrinking base of retail tenants.

Finally, department stores, the primary traffic generators for suburban regional shopping malls, are experiencing serious competitive problems. Over the past two decades, department stores have lost half of their market share to discounters and specialty stores. They also have suffered a significant sales revenue decline, causing store closings and consolidation. Given the role department stores have played as traffic generators for shopping malls, the problems of department stores have added to the problems of suburban regional shopping malls. Quite simply, fewer department store shoppers have meant fewer shoppers in the mall. To make matters worse, an empty department store space in a mall gives shoppers less reason to visit that portion of the mall and often leads to the closure of nearby stores. Besides being unproductive, the resulting empty retail space is unsightly, projecting the same image that empty storefronts in the CBDs of cities do—decay and decline.

The predictable outcome of all these changes is that construction of new suburban regional shopping malls has virtually come to a halt. Furthermore, a significant number of existing centers are being "decommissioned"—converted into alternative uses such as office space, learning centers, and telemarketing call centers, or torn down to be replaced by other forms of retail centers. More than 300 malls have been decommissioned since the mid-1990s, a trend that is expected to continue.

The dim outlook for suburban regional shopping malls has stimulated much creative thought about turnaround strategies for those still in operation. One turnaround strategy that has been suggested deals with the way suburban regional shopping malls are traditionally configured and involves changing the way stores in the center are arranged with respect to one another. The traditional layout locates similar stores in different wings or corridors of the center to encourage shoppers to travel through the entire center in their quest to locate and compare products. This type of layout maximizes customer interchange between stores but does not address shopping efficiency. Zonal merchandising represents a different approach to a center's layout. Under zonal merchandising, similar tenants are located in close proximity to one another. This reduces shopping time for shoppers who come to the mall to purchase a specific product. It also creates opportunities for differentiating mall areas in terms of décor, music, amenities, and special events to suit the tastes of shoppers who are most likely to be visiting stores in those areas.

Zonal merchandising has been used most commonly for fast food. Called "food courts," these clusters of fast-food providers have been very successful. Food court tenants have experienced higher levels of sales than under traditional layouts. Food courts also have shown that they are able to draw shoppers from other locations in the mall, similar to the traffic-generating role of a traditional anchor store.

Based on the success of the food court, several attempts to implement zonal merchandising on a wider scale have been made. Beginning with Bridgewater Commons in New Jersey (www.bridgewatercommons.com), several new projects have incorporated zonal merchandising principles, including Rivertown Crossings in Grand Rapids, Michigan (www.rivertowncrossings.com), which has grouped some categories of stores by product line carried, and Park Meadows in Denver, Colorado (www.parkmeadows.com), which has grouped stores by customer lifestyle. The results of these endeavors have been promising, and General Growth (www.generalgrowth.com), the developer of Rivertown Crossings, plans to implement some form of clustering at all of its future projects. Attempts to reconfigure existing centers around zonal merchandising ideas, such as the changes at Glendale Galleria in Glendale, California (www.glendalegalleria.com), seem to be successful as well.

An alternative strategy, which has been proposed for turning around traditional suburban regional shopping malls, is the incorporation of entertainment within the center. The idea behind this strategy is quite simple: Add value to the shopper's visit to a mall and give shoppers additional reasons to shop in the mall rather than at home. Entertainment can run the gamut from simple play areas or a carousel for children to video arcades and virtual golf courses to a full-scale amusement park, such as the Mall of America (www.mallofamerica.com). However, adding entertainment offerings to suburban regional shopping malls does not guarantee success. The entertainment must be something that will attract shoppers and keep their interest for a lengthy period of time—not something that shoppers tire of easily. Also, the effect of the entertainment activities on a center's retailing activities must be considered. Entertainment centers in suburban regional malls often attract people with social goals instead of shopping goals, which does not benefit a center's merchandise-based stores.

Some industry analysts suggest that the key to revitalizing the suburban regional shopping mall is to make the shopping experience itself more exciting. Even at Mall of America, the home of the largest mall-based entertainment facility in the United States, the primary attraction of the center is the entertainment and excitement provided by the shopping experience itself; shoppers find stores and products that they cannot find elsewhere in the region.

Most suburban regional shopping malls are unexciting. They offer shoppers a relatively nondescript homogeneous shopping experience. They look alike, possess the same stores, and sell the same products. What has been forgotten by mall managers is that entertainment, in a mall sense, is not necessarily what activities can be added to the center, but what entertainment is provided by the shopping experience. Shoppers are searching for shopping experiences that are fresh, different, and fun. To provide this experience, suburban regional shopping malls need to attract stores and sell products that are unique, interesting, and ever-changing. The recent addition of the Build-A-Bear Workshop to several suburban regional shopping malls is one such example. The Build-A-Bear Workshop (www.buildabear.com) is a novel retail concept that provides a playful, creative environment. The challenge for mall managers is to find new and exciting retailing concepts like Build-A-Bear Workshop on a continual basis.

The Easton Town Center in Columbus, Ohio (www.eastontowncenter.com), is an example of a suburban regional shopping center that was explicitly designed to provide shoppers with a fun, exciting, entertaining place to shop. Easton Town Center was designed as an open-air center that mimics small-town America over 50 years ago. The center possesses an entertainment-oriented product mix with numerous restaurants, 30 movie theaters, spas, a comedy club, a cabaret, specialty stores, and Nordstrom and Lazarus as anchor department stores. The center has a "town square" and special event areas. It is considered to be one of the most successful retail centers in the region.

The challenge to mall developers and managers is clear. Since the opening of Southdale Center in 1956, changes in competition, retailing, and consumer shopping behavior have resulted in significant threats and opportunities. If suburban regional shopping malls are to enjoy continued success, they must creatively adapt to the new industry and shopping environment. Managers and owners of suburban regional shopping malls must determine the change strategy that is best for them. A number of considerations should guide their thinking—the competition, the needs of shoppers in their area, the opportunities available, and the center's resources. Just as one size does not fit all, the same turnaround strategy will not suit all suburban regional shopping malls.

1. Imagine yourself as the manager of a struggling local suburban regional shopping mall. What do you think the mall should do to improve its performance?
2. What shopping trends do you foresee over the next 10 years? How might these trends affect suburban regional shopping malls?
3. What new retail concepts can you identify? How might you learn about more? What strategies do you suggest for learning about new retail concepts?

3. Strategic Marketing Planning in Big Brothers Big Sisters of America*

For more than a century Big Brothers Big Sisters of America (BBBSA) has helped children reach their potential through professionally supported one-on-one mentoring. With a network of about 400 agencies in all 50 states, BBBSA serves approximately 260,000 children between the ages of 6 and 18. A sister organization, BBBS International, has a global scope with agencies in 12 countries and similar plans to expand and improve international programs.

PROGRAMS
BBBSA has two core programs:

1. **Community-based mentoring** that requires a mentor ("big") to meet with a child ("little") at least one hour each week to engage in community activities.
2. **School-based mentoring** that encourages "bigs" to meet with their "littles" in schools, libraries, and community centers at least once a week to talk and play.

BBBSA "matches" between children ("littles") and mentors ("bigs") are carefully administered and supported by rigorous standards and trained personnel. Professional staff ensure that matches are safe and well suited to the individuals involved. Each potential mentor ("big") is evaluated and trained before a match takes place. With each match, the intent is to provide a satisfying experience for both parties and to help the child develop positive outcomes.

BBBSA attempts to meet the needs of the most vulnerable children in a community. Because the demographics of these vulnerable children vary from community to community, BBBSA has developed different programs that consider these racial, ethnic, and social demographics. Each local affiliate launches a specific program that meets the needs of its community. The targets of these programs include:

1. **African American mentoring.** BBBSA recognizes that African American boys constitute one of the most at-risk populations in the United States. So it makes a concerted effort to recruit male African American "bigs" to serve this population by partnering with the Alpha Phi Alpha fraternity and United Methodist Men.

---
*This case was prepared by Dr. Jane S. Reid, Youngstown State University.

2. **Hispanic mentoring.** Given the huge increases in the Hispanic population, BBBSA also recognizes the many needs of Hispanic children in the United States. It actively recruits Latino "bigs" by building collaborations within Hispanic communities.
3. **Native American mentoring.** With more than 4 million Native Americans and Alaska Natives living in the United States and more than a quarter of them living at or below poverty, BBBSA began a recruitment initiative in 2007 in Native American communities. The BBBSA program is run under the guidance of both formal and informal elders, members of Native American community programs, and American Indian/Alaska Native board members with 30 agencies participating in 16 states.
4. **Amachi program (mentoring children of promise).** BBBSA developed this program to serve children who have a parent in prison. *Amachi* is a Nigerian word of hope—"who knows but what God has brought us through this child." BBBSA focuses on recruiting "bigs" from local congregations.
5. **Military children.** BBBSA established a partnership with the T. Boone Pickens Foundation to serve children with a parent in the military and to provide civic engagement for military personnel returning from deployment. A BBBSA Military Community Advisory Council has been set up for this program.

ORGANIZATION

Like most national nonprofit organizations, BBBSA has two levels of operation. The national office in Philadelphia has an executive staff and national board of directors that provide strategic direction and support to local agencies across the United States. Each local agency has an executive director, board of directors, and program staff who address the specific needs of children within its community. Each agency must operate under an affiliation agreement with BBBSA, which gives it the rights to:

1. Operate an organization using the BBBSA name.
2. Receive a designated service community area which is a specific geographical area in the United States. The local agency is restricted to recruitment and fundraising within that geographical area.
3. Associate with BBBSA long term, unless the affiliation is terminated by the local board or by the national office.
4. Collaborate with BBBSA and other member agencies.

Membership fees are paid quarterly and are based on the total expenditures of the agency. Four deductions are allowed: fees paid to BBBSA, capital purchases, depreciation, and fundraising expenses. The fee is then calculated by taking 3.8 percent of the first $100,000; 2.25 percent of the next $100,000; 1 percent of the next $300,000; and 0.5 percent of the remaining expenditures. The affiliation fee is directly related to the size of the agency, with larger agencies paying more for the national affiliation than smaller agencies. Agencies are also given discounts if they pay their entire affiliation fee up front: 5 percent discount if paid by January 31, 4.5 percent discount if paid by February 28, or 4 percent discount if paid by March 31. Otherwise, affiliation fees are due on March 31, June 30, September 30, and December 31. In order to better manage its own annual budget, the national office encourages early payments from the affiliated agencies.

Local agencies benefit from their national affiliation with BBBSA. First and foremost, the national office provides strategic direction to local agencies. Second, BBBSA sets standards that local agencies must follow in order to use the BBBSA brand, but through networking, local agencies are able to share "best practices." Local agencies also have access to research that the national office conducts. In addition, the national office provides a management information system (MIS) which allows local agencies to process information about volunteers, children, families, and donors. Further, the MIS helps local agencies manage their efforts and measure outcomes. Last, and probably the most important from a marketing standpoint, local agencies benefit from a nationally recognized brand with high credibility, which helps them in recruiting "bigs" and "littles" and attracting donors. This is especially important during economic downturns when funding from foundations, the public sector, and individual donors declines. Agencies with proven programs and high brand recognition and credibility are more likely to survive. Moreover, BBBSA has a competitive advantage compared to other nonprofit organizations that provide programs for youth. It is the only national organization that focuses on one-on-one mentoring.

STRATEGIC PLANNING INITIATIVES

In 2000, BBBSA launched its strategic-planning initiative by hiring Bridgespan Consulting (www.bridgespan.org). The Bridgespan Group helps nonprofit and philanthropic leaders to develop strategic plans and build organizations that inspire and accelerate social change. The use of a professional consulting service to develop a strategic planning process was made possible by the generous support of the Edna McConnell Clark Foundation (www.emcf.org), a foundation that seeks to improve the lives of people in poverty, especially low-income youth in the United States.

The first step was to establish a steering committee to work with the consulting firm. The steering committee consisted of four local agency leaders, one BBBSA board member who also served on a local BBBSA board, and five BBBSA leaders. One of the first activities of the committee was to collect information about the problems, practices, opinions, and attitudes of local agencies. The committee surveyed more than 160 local agencies and received input from leaders at more than 70 agencies, representing all of the different regions, sizes, and types of communities served by BBBSA.

As part of the process, Bridgespan Consulting guided BBBSA through a S.W.O.T. analysis that included examining its mission statement: "To help children reach their potential through professionally supported, one-to-one relationships with measurable impact." As a result of the analysis, the vision statement became: "Successful mentoring relationships for all children who need and want them, contributing to better schools, brighter futures, and stronger communities for all."[3-1] BBBSA also included a promise statement as part of its guiding principles, which commits BBBSA to a culture of diversity and inclusion, partnership and collaboration, continuous learning, people development, and high performance.

In setting its organizational goals, BBBSA built on its past strengths of accountability for outcomes in three areas: (1) the number of children served; (2) how well those children are served; and (3) which children are served, with a focus on children who are most in need and most likely to benefit from BBBSA programs. From there, BBBSA developed four goals with specific outcomes:

1. Quality growth—to serve more than 300,000 children by 2010 (a 25 percent increase over its 2006 level).
2. Positive outcomes for a higher percentage of children served—increasing the average match length from 10 to 13 months and increasing the 6-month retention rate from 80 percent to 85 percent. The overall intent of this goal is to increase the strength of the relationship between "bigs" and "littles." During this planning period, BBBSA also committed to lead the development of new real-time outcome measures for local affiliates.
3. Increased intentionality about which children are served. Although each local agency serves a unique population, the overriding national goals would be to
   a. Increase the percentage of boys served and the percentage of male volunteers (from 38 percent to 41 percent of all volunteers). The focus on males is to counteract the recent increase in the number of female matches and to better serve the large number of boys on the waiting lists.
   b. Increase the percentage of Hispanic youth served from 17 to 22 percent of all matches.
   c. Increase the percentage of African American and Hispanic youth served by same-ethnicity volunteers to at least 57 percent by increasing African American volunteers from 15 to 18 percent of all volunteers and by increasing Hispanic volunteers from 7 to 13 percent of all volunteers.
   d. Establish clear baseline data for measuring the percent of "littles" who are in moderate-to-high need.
4. Strengthen local agencies' capacity for sustainable, quality growth. The specific goals for making that happen include
   a. 10 percent annual total revenue growth to $385 million by 2010.
   b. 10 percent revenue growth by 50 percent of local agencies (up from 35 percent).
   c. 3 months' cash reserves by 65 percent of local agencies (up from 49 percent).
   d. $1 million in revenues by 80 percent of large market agencies (up from 67 percent).
   e. $500,000 in revenues by 40 percent of regional and smaller market agencies (up from 20 percent).
   f. Board and leadership development plans in 60 percent of agencies. In addition, each agency with $500,000 or more in revenue should have four key leadership positions: CEO, VP Program, Chief Development Officer, and VP Partnership/Recruitment.

The goals outlined above are detailed, measurable, and a "stretch" for both the local agencies and the national office. However, BBBSA outlined specific implications and expectations for the national office and agencies.

AGENCIES

- Sustained, quality growth should be at the core of each agency's effort.
- Each agency should work toward maximizing its impact and community support by being intentional about whom it serves.
- Both the Community-Based Mentoring and School-Based Mentoring programs should be strong and robust in terms of quality and growth.
- Resource development (funds and volunteer recruitment) should be a central leadership focus for all agencies.
- Agencies should ensure that their talent and organizational structures are strong enough to support the complex challenges of sustained quality growth.

NATIONAL OFFICE

- Leadership on the definition of success (agreement on "what is achieved" versus "what is done").
- Increased direct support to agencies through financial grants.
- Maintenance of current level of support to agencies with a focus on technology, outcome data, and best practices sharing.
- Enhanced capacity in key central functions including marketing, corporate partnerships, and human resources.
- Emphasis on talent and leadership in the five key roles—board, executive director, vice president of program, chief development officer, and vice president of partnerships.

Research plays an important role within BBBSA. Its MIS is designed to measure the effectiveness of its programs. In a nationwide study, BBBSA found that "littles" were

- 46 percent less likely to begin using illegal drugs.
- 27 percent less likely to begin using alcohol.
- 52 percent less likely to skip school.
- 37 percent less likely to skip a class.

BBBSA has encouraged and supported similar strategic planning efforts in local agencies. Local agencies are encouraged to identify appropriate targets within their communities and to adapt national programs and strategies with measurable objectives. One example of how the national organization supports the marketing efforts of local agencies is public service announcements (PSAs). BBBSA has produced several PSAs that address the specific goals of recruiting male "bigs" and "littles" of both Hispanic and African American ethnicities. However, local agencies must decide which PSAs to use and then get local broadcast stations to run them—perhaps to coincide with major fundraising events like Bowl for Kids' Sake. Local agencies also develop collateral material such as posters, flyers, and brochures.

CURRENT AND FUTURE STRATEGIC INITIATIVES

BBBSA has committed to reviews of its strategic plan on a three-year cycle because it believes the key to an organization's survival is recognizing opportunities and threats and adjusting to them.

BBBSA has already begun its second planning cycle and it faces some specific challenges. The first is to achieve greater consistency in its brand image. Although local agencies are autonomous, BBBSA needs to have a consistent image across all communication efforts: ads, PSAs, events, web pages, and so on.

Another key concern (identified in focus group research) is that BBBSA is now viewed as a "volunteer agency"—not an agency to which people would consider donating money. Consequently, this perception needs to be modified so that, when people think of BBBSA, they think of becoming donors as well as volunteers.

BBBSA recruitment is (and will continue to be) constrained by economic concerns and family demands. These constraints make it hard to attract not only volunteers, but also active and talented people to serve on boards of local agencies or to hire competent staff at modest salaries.

To address some of these concerns, BBBSA has initiated a repositioning campaign (Project: BigChange$) at the national level. It is working with Publicis Groupe on a pro bono basis to redefine mentoring in a more powerful way and to establish itself as a donor-supported, as well as volunteer-supported, organization. To coordinate the campaign, the national office has provided local agencies with specific strategies for communicating to potential donors and volunteers. Some key points are

- Emphasizing BBBSA's commitment to measurable outcomes.
- Connecting donors to its program similar to the connection that volunteers feel.
- Describing services as donor-funded and not "free."
- Ensuring that each staff member understands his/her position is funded by a donor.
- Explaining to each potential "big" that donors make matches possible.
- Using adult "littles" (alumni) to demonstrate successes.
- Focusing on stewardship—the impact that the agency and the donors have on the community.

BBBSA is a strong nonprofit organization. It prides itself on being accountable to its stakeholders by providing meaningful and effective programs to children. Under a new president and CEO, BBBSA seems poised to be a significant influence during the next decade. It's focusing on its strengths and addressing its problems through strategic planning.

1. Compare strategic planning by BBBSA with planning by for-profit organizations. What are the similarities? What are the differences?
2. Using Exhibit 2-9 on the marketing strategy planning process in Chapter 2, identify the components of BBBSA's strategic planning process from the information presented in this case. Who are BBBSA's customers and competitors? How has BBBSA chosen to segment its market? Describe BBBSA's target markets. How does BBBSA differentiate itself from the "competition"? How would you describe the Four Ps of BBBSA's marketing strategy?
3. What are the benefits of a local nonprofit organization affiliating with a national nonprofit organization? Can you identify any disadvantages?
4. Refer to the BBBSA website (www.bbbs.org). What changes have occurred in the mission, vision, goals, and strategies? Has BBBSA successfully met the goals stated in this case? Has the organization successfully overcome the challenges of brand image and recruitment of volunteers?

4. Invacare Says "Yes, You Can!" to Customers Worldwide*

"On this vote the yeas are two hundred and nineteen, the nays are two hundred and twelve. The motion to concur in the Senate amendment is adopted." As he watched the U.S. House of Representatives pass Health Care Reform Act on March 23, 2010, Invacare chair and CEO Malachi "Mal" Mixon contemplated the impact of this landmark legislation on his company.

COMPANY AND INDUSTRY HISTORY

Modern devices that provide mobility where it is needed can trace their history to the Fay Manufacturing Company in Elyria, Ohio, in the 1880s. Originally a bicycle and tricycle manufacturer, the company changed directions when Winslow Fay recognized the need for mobility among wounded veterans and children. He redesigned his tricycles to enable people considered "crippled" at that time to move about. After achieving nationwide distribution of these products, Fay sold out to the Worthington Manufacturing Company in the early 1890s. After a series of mergers and acquisitions, the company became Invacare and, in 1979, an entrepreneurial group led by Mal Mixon and J. B. Richey purchased the company.

Changes in the company paralleled changes in the wheelchair industry. Although wheelchair-like devices were used in ancient times, something akin to modern wheelchairs were not developed until the 16th century. However, because chronically ill or disabled people had short life spans, the need for wheelchairs was not perceived to be important. By the late 19th century, survival rates were higher, but most people needing a wheelchair were considered to be invalids—best kept at home or in an institution. Two world wars and a polio pandemic in the first half of the 20th century greatly increased the number of people with limited mobility, but attitudes about their care had not changed much.

When Franklin Delano Roosevelt, president of the United States from 1933 to 1945, became a victim of polio at the age of 39, perceptions started to change.[4-1] Determined to lead an active life, Roosevelt developed his own wheelchair—a lightweight chair on rollers that could pass through normal-sized doors.

In the 1970s, the Vietnam War produced large numbers of young wounded soldiers who wanted normal civilian lives. At the same time, there was greater awareness of the need to break down socioeconomic boundaries between groups, including the disabled. Heavy, bulky chairs were not suitable for people who wanted to work, participate in sports, dance, and travel with family and friends.

Other changes in the 1970s helped the medical devices industry grow and evolve. Up until that time, Everest & Jennings, then the major supplier of wheelchairs, set the standards for the rest of the industry through its contractual arrangements with the U.S. government. These government contracts determined styles and materials for the products that would be purchased directly—and indirectly influenced what would be acceptable for reimbursement under many insurance programs. This type of market power enabled Everest & Jennings to command a dominant market share without the necessity of adapting to the rapidly changing needs and wants of its target market.

The Medical Devices Act of 1976 changed the dynamics of the industry. The Bureau of Medical Devices was created within

---

*This case was prepared by Dr. Douglas Hausknecht, The University of Akron.

the Food and Drug Administration to oversee production of a vast array of devices from catheters to wheelchairs to MRI machines. This new agency took a fresh look at the needs and wants of individual users, as well as the needs of physicians and insurers. Change did not come immediately, but Invacare began to see new market opportunities.

Under its new leadership of Mixon and Richey, Invacare reexamined its strategy and found itself deficient in many areas. Product quality and delivery were poor; no new products were under development; production costs were high; competition from new producers in Asia was growing; and Invacare had done little to penetrate foreign markets. Moreover, the company had no real strategic plan.

Mixon addressed these issues by first working through the distributor network to improve product design, quality, and delivery times. Distributors were offered better financing support to offset high interest rates, which were a problem at the time. New products, new product lines, and the acquisition of related companies followed. Invacare became a major player in the design, construction, and sale of therapeutic beds and oxygen concentrators, in addition to its traditional product lines. Invacare also entered the "prescription wheelchair" business. Prescription chairs are custom products, designed for individual users. Today, Invacare is known for its customized wheelchair products.

In a related change of strategy, Invacare brought new vitality to the self-powered wheelchair in 1981 with the introduction of a lighter, more capable model. The Rolls IV, as it was known, provided users with a much improved sense of independence and mobility. Wheelchair users began to recognize and communicate their needs and wants to distributors who conveyed this market information to Invacare's sales force. In turn, company engineers added desirable features and attributes, which improved products and increased customer satisfaction.

As a result, Invacare pioneered the idea of "One Stop Shopping" for medical devices. "One Stop Shopping" continues to define the company's product and distribution strategies to this day.

CURRENT STRATEGIES AND CHALLENGES

Today, Invacare is the world's leading manufacturer and distributor for medical equipment used in the home: power and manual wheelchairs, personal mobility, seating and positioning products, home care bed systems, and home oxygen systems. Vertical integration, Statistical Process Control, and total quality management (TQM) have all contributed to the success of Invacare's multiple product lines. In its modern manufacturing facilities in Ohio, Mexico, and elsewhere, the company produces and assembles many of the components for its products—whether standard or customized. Invacare also supplies the components necessary for service and repair of chairs, beds, oxygen concentrators, and other products. Its new-product development process includes representatives from marketing, quality, manufacturing, purchasing, finance, and design.

Because cost has become increasingly important in the medical devices industry, Invacare actively pursues a cost-containment strategy. While the need for specialized products has increased, the availability of insurance or government programs to pay for them has not. Prior to WWII, the cost of medical devices typically was borne by the family or community contributions. As medical insurance developed and evolved into an employee benefit, some of the costs were covered that way. Beginning in 1972, the federal government covered costs for the permanently disabled through Medicare, but because products were intended to be affordable, design took a back seat. Changes in government regulation, most notably the 1990 Americans with Disabilities Act (ADA), encouraged improvements in product design and extensions of existing product lines. Medicare policies were slow to change, however, and even in the 1980s insureds were only reimbursed for equipment necessary to "perform the activities of daily living within the four walls of the home." Customers were not permitted to upgrade from basic models unless they paid the entire cost themselves. Medicare rules at the time did not permit augmenting the basic cost with private funds.

Currently, a greater range of options in government support, insurance, and flexible spending plans—as well as private and foundation money—help to fund purchases of customized equipment. Invacare's domestic revenues are equally dependent on private insurance and Medicare: together contributing about 80 percent of its income. However, cost-containment language in the 2010 Health Care Reform Act may once again limit the availability of funds for high-end equipment.

Invacare sells its products to more than 25,000 home health care and medical equipment providers, distributors, and government locations in the United States, Australia, Canada, Europe, New Zealand, and Asia through its sales force, telesales associates, and various independent agent distributors. Distributors in individual markets work with medical care facilities, doctors, rehabilitation clinics, and retailers to identify potential new customers and current users. Invacare's sales force works with distributors to monitor and meet demand. Awareness of the available range of products is enhanced through a variety of means. Invacare has sponsored individual wheelchair races, a wheelchair racing circuit, and the Wounded Warrior Project. Team Invacare competes in racing, basketball, handcycling, and tennis. Special-purpose chairs encourage participation in sports as demanding as rugby—even to the extent of having specially designed chairs for offense and defense. These sponsorships and participation in special events heighten awareness of customized equipment among potential customers and with the general public. Rather than featuring the attributes of specific pieces of equipment, Invacare's advertisements and promotional materials emphasize people enjoying life with its products.

As Mal Mixon considered Invacare's past strategies, he wondered about the future. How might the new environment for health care affect Invacare and its competitors? Everest & Jennings' sturdy and utilitarian chairs are still being manufactured under the umbrella of GF Health Products. They look the same as they always have. But other competitors have carved out positions in all-terrain accessibility (e.g., Permobil), lightweight portability (e.g., Lifecare Medical), niche uses (e.g., Colours in Motion), and increasingly low price (imports). Dean Kamen, inventor of the Segway, developed the IBOT—a pricey wheelchair that can climb stairs. Would future price pressures and regulatory changes allow all of these competitors to coexist?

Invacare is also a leader in the oxygen concentrator market. The HomeFill combines the efficiency of stationary oxygen concentrators with the convenience of pre-filled bottles to enable a greater range of travel for those who are dependent on supplementary oxygen. Will new regulations favor less costly "standard" products, such as those offered also by DeVilbiss and Respironics? Will pricing pressures allow for further innovation?

The Health Care Reform Act that was passed in 2010 broadly refers to changes in the managed care medical insurance system in the United States. It is expected to affect all Americans in some way and will impact all companies in the medical equipment, services, and insurance industries. In general, it requires all Americans to be covered by health insurance. Insurers are no longer allowed to deny coverage to individuals with preexisting conditions, and young adults will be able to stay on their parents' plans until the age of 26. These changes are expected to be very costly, and some critics predict a shortage of primary care physicians. In addition, implementation of the bill will require cost reductions and savings that are likely to affect the availability of health care services and payments. However, some aspects of the act remain in flux. A sales tax on home medical devices was delayed. A provision that would reduce the ability to sell directly to the customer in favor of leasing arrangements has not been implemented. So the overall impact of the Health Care Reform Bill on Invacare is unknown. Will the promise of greater access to health care and life-improving treatment and devices create a larger domestic market, or will greater restrictions on prices and product design reverse decades of product innovation and improvements?

Markets outside the United States are a concern as well. As of 2010, Invacare produces components or complete products in several countries. It also distributes products in 80 countries worldwide (a range of coverage difficult for a company that prides itself on meeting the needs of its customers), but not all of the company's product lines are available in all markets. In addition, many of its products sold in foreign markets are still based in part on designs from acquired companies. So while quality, reliability, and service are emphasized throughout the global corporation, design has been allowed to fluctuate. This "globalization" of the company's strategy has led to nonstandard components and problems with "cataloging" the product line worldwide.

Further, in the global marketplace, cultural factors account for some differences in local markets. Take China, for example. Basic health care is provided by the government in China, but private health care and individual personal benefits (not available to everyone) are viewed negatively. In recent years, some of these attitudes have changed, but only slowly. Still, income and a willingness to spend on personal items is increasing among the middle-to-upper classes. At the same time, care of the elderly is becoming less reliant on immediate family. In addition, the "one child per household" policy and the preference for male children will continue to strain the ability of earning generations to support their parents. Invacare's competitors in China include the Jiangsu Yuyue Company with product lines very similar to its own. Invacare has some manufacturing and distribution in China now and is contemplating future strategy in light of the culture and competition.

Brazil is another example. It is a large emerging market, but the problems Invacare faces there are somewhat different from the ones in China. Brazil has a history of being a very protectionist economy. High tariffs and duties restrict the ability to import finished goods of many types, and there is a strong preference for locally manufactured products. Nevertheless, the country's abundant natural resources permit global corporations to compete through direct foreign investment in manufacturing and distribution. The downside is that direct investment requires duplication of production facilities and other corporate resources within the country. But Brazil is a major exporter to other South American markets and that could make it an attractive market for Invacare.

Mal Mixon believes in his company and its future. He thinks it has the resources and energy to grow and to serve a customer base that needs and appreciates its products. Invacare's motto of "Yes, you can!" truly conveys the positive corporate attitude that underlies its mission: to remove real and perceived barriers that may restrict the lives of its customers. But Mal wonders how the company can best meet the challenges both within the United States and in foreign markets. How will the company grow and prosper in the next decade?

1. How might the anticipated changes in U.S. health care coverage affect Invacare's current marketing strategies and demand for its products?
2. Conduct a S.W.O.T. analysis for Invacare. What types of opportunities and threats do you see? What types of marketing strategies are suggested by your S.W.O.T. analysis?
3. What have been the "keys to success" for Invacare in the past? Are these likely to change in the future?
4. What do you think will be necessary for Invacare to conduct operations successfully in Brazil? What about China? Which of these two countries is a better "fit" for Invacare's business model?

5. Segway Finds Niche Markets for Its Human Transporter Technology*

Amid heavy media coverage and much speculation, "Ginger" made its debut on ABC's *Good Morning America* on December 3, 2001. It thrilled some and disappointed others, but the technology was breathtaking to all.

Now known as the Segway Human Transporter (HT), Ginger was the brainchild of inventor and entrepreneur Dean Kamen, who is best known for his inventions in the medical field. While in college, Kamen invented the first wearable drug-infusion pump. In the following years, he invented the first portable insulin pump, the first portable dialysis machine, and an array of heart stents. This string of successes established Kamen's reputation and turned DEKA Research and Development Corp., the R&D lab he founded, into a premier company for medical-device design.

The inspiration for Ginger occurred during the development of the IBOT wheelchair at DEKA. Developed for and funded by Johnson & Johnson, the IBOT wheelchair is a gyro-stabilized, microprocessor-controlled wheelchair that gives disabled people the same kind of mobility the rest of us take for granted. Officially called the Independence IBOT mobility system, this six-wheel machine can go up and down curbs, cruise effortlessly

---

*This case was prepared by Dr. J. B. Wilkinson, professor emeritus, Youngstown State University.

through sand or gravel, climb stairs, and rise up on its wheels to lift its occupant to eye level while maintaining balance and maneuverability with such stability that it can't be knocked over. The IBOT wheelchair has been likened to a sophisticated robot.

As Kamen and his team at DEKA were working on the IBOT, it dawned on them that they could build a device using similar technology for pedestrians—one that could go farther, move more quickly, and carry more. The IBOT was also the source of the "Ginger" code name for the Segway HT during its development stage. Watching the IBOT "dance up the stairs," Kamen's team likened it to Fred Astaire—hence the name Ginger for its smaller partner with only two wheels!

Segway's breakthrough technology is based on dynamic stabilization. A self-balancing, electric-powered transport device, the Segway HT contains gyroscopes and sensors that monitor a user's center of gravity and respond to subtle shifts in weight. Lean forward, go forward; lean back, go back; lean back a bit more and you go in reverse; turn by twisting your wrist; arch your back a tad and you slow to a halt. Exactly how the Segway achieves this is difficult to explain, but in every Segway, there are gyroscopes that act like your inner ear, a computer that acts like your brain, motors that act like your muscles, and wheels that act like your feet. You step aboard and it "oscillates" for a few seconds, getting the feel of you, and then it's fully cruiseable at 6 miles per hour in "learning mode," and 12.5 miles per hour "flat-out." It has a range of about 17 to 25 miles per battery charge and can support package weights of 80 pounds. It has no brakes, no engine, no throttle, no gearshift, no steering wheel, and gives off no emissions. It is much cleaner than a car and faster than a bike. It is more pedestrian-friendly than bikes or scooters and safer than a skateboard.

The commercial potential for Segway is enormous. For this reason, Kamen decided to move development and manufacturing of the Segway HT to a new company with the vision to develop highly efficient, zero-emission transportation solutions using dynamic stabilization technology. The new company, Segway LLC (changed to Segway, Inc., in 2005) is headquartered in Bedford, New Hampshire, and construction of a manufacturing and assembly plant was completed in 2001.

Kamen has openly stated that the Segway "will be to the car what the car was to the horse and buggy!" He imagines them everywhere: in parks, on battlefields, on factory floors, zipping around distribution centers, and especially on downtown sidewalks. But market acceptance and penetration has been slow for this engineering marvel.

The Segway HT was initially marketed to major corporations and government agencies. Kamen personally demonstrated the Segway to the postmaster general, who was keen to put letter carriers on Segways, and to the head of the National Park Service, who wanted to do the same with park rangers. Both were among Segway's first customers. The Pentagon's research agency, DARPA, bought Segways to give to robotics labs. The objective was to use Segway bases in the development of robots to do menial tasks (cleaning, picking crops), specialized jobs (nurses aids), and dangerous missions (bomb removal or rescue work in earthquake debris). Amazon.com, GE Plastics, and Delphi Automotive Systems purchased Segways for use in warehouses and manufacturing plants. Police departments, cities, and airports bought Segways for use by foot patrols, security personnel, and meter readers. Customer studies by these early buyers showed double-digit productivity gains and reduced reliance on motorized transport. For postal workers, the Segway has increased their carrying capacity and delivery speed. Test studies with police officers and security personnel have shown faster response times and better sightlines from being higher in the air than pedestrians.

In 2002, Segways went on sale to the public for the first time on Amazon.com at a price of around $5,000. In 2003, the retail store Brookstone became the first retailer to sell the Segway HT. By January 2004, it was estimated by industry observers that about 6,000 Segways had been sold. Given Kamen's stated ambition of replacing automobiles and other forms of personal transport alternatives, initial consumer sales were disappointing.

However, the Segway HT is still in the early stages of its product life cycle. According to product life-cycle theories and related empirical studies on diffusion of innovations, products move through their life cycles in different ways, and the length of any particular stage varies according to many factors. In general, the speed with which a product moves through its introduction and growth stages varies according to product characteristics, market characteristics, competition, and environmental factors. The initial marketing strategy for a product will also affect market acceptance (adoption). A product will move quickly through the introduction and growth stages when it has high relative advantage—cost and benefits—compared to alternative products and is highly compatible with a buyer's current attitudes, lifestyle, and usage situation. The more complex a product is to understand and use, the slower will be its rate of adoption. In addition, if the product can be tried in small amounts, potential buyers perceive less risk associated with initial trial and this will speed adoption. Communication of the new product will affect market acceptance because buyers must first learn about a new product before the buying process can be initiated. Thus, a company's communication strategies and the degree to which favorable word-of-mouth opinion occurs will impact sales. Finally, new products will have higher rates of market acceptance if market conditions, competition, and environmental trends are favorable.

Industrial and government applications for the Segway HT continue to be positive. It's ideally suited for use in large-scale manufacturing plants and warehousing operations, for use by the police and security personnel in certain types of situations, for meter reading, for corporate and university campus transportation, and for package and mail delivery. And it has found a number of new niches in recent years. For example, tour groups are using Segways to move tourists between attractions in cities where tour buses and cars have operational difficulties. For example, Bill and Emily Neuenschwander's tours use Segways equipped with a radio on the handlebars that provide a running commentary to guide customers through tours of the Minneapolis, Minnesota, riverfront and surrounding historic landmarks. Visitors to theme parks, museums, and islands (e.g., Amelia Island Plantation Resort) use them. Some lawyers use them between offices and courtrooms. Some medics use them to reach injured people faster.

However, in the consumer market, the Segway HT is still an unsought product for many people. Industry observers believe that the automobile is a preferred alternative for a number of reasons, including the infrastructure of cities, commuting distances, weather, and the American family lifestyle. Most people agree that the use of a Segway HT on highways and busy streets is dangerous. Also, many commuters live more than 5 miles from work. The need to recharge Segways raises the question of public sources of electricity. Issues related to public safety also have slowed consumer sales. Although the Segway HT is approved in most states for use on sidewalks, restrictions have been placed on speed, helmet use, minimum operating age, and use on streets and highways. At the local level, additional regulations apply. For example, San Francisco banned their use on city sidewalks, and they are outlawed in subways in Washington, DC. Many see the mix of pedestrians, bicycles, skateboards, roller skates, ATVs, scooters, and Segways as a dangerous mix on sidewalks, hiking trails, and other public areas.

On the other hand, the Segway HT has obvious advantages to walking and other types of personal transport in situations where people need to get from one place to another rapidly and efficiently using a clean, quiet, and environment-friendly transport device. The market potential for the product has barely been tapped. Early consumer buyers were mostly "techies" who like to own new, high-tech products, and they have played an important role in communicating the benefits and fun of owning a Segway HT. They have established local clubs across America and have started a number of websites devoted to Segway. In 2003, owners and enthusiasts held the first SegwayFest to celebrate all things Segway. The Fest has been held every year since.

In 2004, Segway launched a repositioning campaign to change the image of the Segway HT from "staid," "high-tech," and "serious" to "fun, smart transportation." The new and more traditional marketing campaign included new customer materials, dealer displays, and cable TV advertising with the tagline "Get Moving." Segway also has been increasing the number of dealerships and distributors worldwide. The new distribution strategy allows potential buyers to see the different models and to try them out before buying. The dealer network increases the visibility of the product and lowers buyers' perceived risk associated with servicing. Publicity, media appearances, and product placement on television shows such as *Frasier, Arrested Development,* and *The Simpsons* continue to provide important market exposure.

The new positioning strategy emphasizes the leisure aspect of the Segway HT. New models aimed at the recreation market have been introduced. For example, the Segway GT is geared toward golf enthusiasts and includes a golf bag carrier. The Segway XT is the company's off-road vehicle, which can perform well in a variety of environments. The Centaur is a four-wheeled ATV-like vehicle that can pick up its wheels to climb over obstacles or simply glide on two wheels at 25 miles per hour.

To further develop the market for Segway technology, the company has started to market Segway Smart Motion Technology through licensing and partnering with third parties to co-develop new products. Under this program, any number of specialized products could emerge.

The future for Segway is unlimited. America's commitment to end its dependence on petroleum and reduce harmful emissions is not likely to change. Segway's technology supports this commitment. The consumer market for Segway has developed slowly, but the future is assured. For this "new-to-the-world" product, the life cycle will be long and classically configured. Its impact on all sectors of the economy will be profound. How we work, play, and get around will radically change in the 21st century. From the company's perspective, this is just dandy! Its ultimate goal is to "be to the car what the car was to the horse and buggy!"

1. *How might Segway, Inc. further develop the market for Segway technology?* Hint: *What types of marketing strategies are associated with sales growth?*
2. What would be some advantages and disadvantages of using a Segway HT to get around on campus?
3. What types of applications and usage situations are there for Segway HTs in your area?
4. What kinds of problems would the use of Segway HTs create in your area? What are some possible solutions for these problems? Explain.

Cases

Guide to the Use of These Cases

Cases can be used in many ways. And the same case can be analyzed several times for different purposes.

"Suggested cases" are listed at the end of each chapter, but these cases can also be used later in the text. The main criterion for the order of these cases is the amount of technical vocabulary—or text principles—that is needed to read the case meaningfully. The first cases are "easiest" in this regard. This is why an early case can easily be used two or three times—with different emphasis. Some early cases might require some consideration of Product and Price, for example, and might be used twice, perhaps regarding product planning and later pricing. In contrast, later cases, which focus more on Price, might be treated more effectively *after* the Price chapters are covered.

In some of the cases, we have disguised certain information—such as names or proprietary financial data—at the request of the people or firms involved in the case. However, such changes do not alter the basic substantive problems you will be analyzing in a case.

1. McDonald's "Seniors" Restaurant

Lisa Aham manages a McDonald's restaurant in a city with many "seniors." She has noticed that some senior citizens have become not just regular patrons, but patrons who come for breakfast and stay on until about 3 p.m. Many of these older customers were attracted initially by a monthly breakfast special for people aged 55 and older. The meal costs $1.99, and refills of coffee are free. Every fourth Monday, between 100 and 150 seniors jam Lisa's McDonald's for the special offer. But now almost as many of them are coming every day—turning the fast-food restaurant into a meeting place. They sit for hours with a cup of coffee, chatting with friends. On most days, as many as 100 will stay from one to four hours.

Lisa's employees have been very friendly to the seniors, calling them by their first names and visiting with them each day. In fact, Lisa's McDonald's is a happy place—with her employees developing close relationships with the seniors. Some employees have even visited customers who have been hospitalized. "You know," Lisa says, "I really get attached to the customers. They're like my family. I really care about these people." They are all "friends," and it is part of McDonald's corporate philosophy (as reflected in its website, www.mcdonalds.com) to be friendly with its customers and to give back to the communities it serves.

These older customers are an orderly group and very friendly to anyone who comes in. Further, they are neater than most customers and carefully clean up their tables before they leave. Nevertheless, Lisa is beginning to wonder if anything should be done about her growing "non-fast-food" clientele. There's currently not a crowding problem during the time when the seniors like to come. But if the size of the senior citizen group continues to grow, crowding could become a problem. Further, Lisa is concerned that her restaurant might come to be known as an "old people's" restaurant—which might discourage some younger customers. And if customers felt the restaurant was crowded, some might feel that they wouldn't get fast service. On the other hand, a place that seems busy might be seen as a "good place to go" and a "friendly place."

Lisa also worries about the image she is projecting. McDonald's is a fast-food restaurant (there are more than 35,000 of them serving more than 65 million people in more than 119 countries every day), and normally customers are expected to eat and run. Will allowing people to stay and visit change the whole concept? In the extreme, Lisa's McDonald's might become more like a European-style restaurant where the customers are never rushed and feel very comfortable about lingering over coffee for an hour or two! Lisa knows that the amount her senior customers spend is similar to the average customer's purchase—but the seniors do use the facilities for a much longer time. However, in actuality, most of the older customers leave McDonald's by 11:30, before the noon crowd comes in.

Lisa is concerned about another possibility. If catering to seniors is OK, then should she do even more with this age group? In particular, she is considering offering bingo games during the slow morning hours—9 a.m. to 11 a.m. Bingo is popular with some seniors, and this could be a new revenue source—beyond the extra food and drink purchases that probably would result. She figures she could charge $5 per person for the two-hour period and run it with two underutilized employees. The prizes would be coupons for purchases at her store (to keep it legal) and would amount to about two-thirds of the bingo receipts (at retail prices). The party room area of her McDonald's would be perfect for this use and could hold up to 150 persons.

Evaluate Lisa Aham's current strategy regarding senior citizens. Does this strategy improve this McDonald's image? What should she do about the senior citizen market—that is, should she encourage, ignore, or discourage her seniors? What should she do about the bingo idea? Explain.

2. Nature's Own Foods, Inc.

It is 2019, and Neal Middleton, newly elected president of Nature's Own Foods, Inc., faces a severe decline in profits. Nature's Own Foods, Inc. is a 127-year-old California-based food processor. Its multiproduct lines are widely accepted under the Nature's Own Foods brand. The company and its subsidiaries prepare, package, and sell canned and frozen foods, including fruits, vegetables, pickles, and condiments. Nature's Own Foods, which operates more than 30 processing plants in the United States, is one of the larger U.S. food processors—with annual sales of about $650 million.

Until 2019, Nature's Own Foods was a subsidiary of a major midwestern food processor, and many of the present managers came from the parent company. Nature's Own Foods' last president recently said:

> The influence of our old parent company is still with us. As long as new products look like they will increase the company's sales volume, they are introduced. Traditionally, there has been little, if any, attention paid to margins. We are well aware that profits will come through good products produced in large volume.

Alex May, a 25-year production manager, agrees with the multiproduct-line policy. As he puts it, "Volume comes from satisfying needs. We will can or freeze any vegetable or fruit we think consumers might want." May also admits that much of the expansion in product lines was encouraged by economics. The typical plants in the industry are not fully used. By adding new products to use this excess capacity, costs are spread over greater volume. So the production department is always looking for new ways to make more effective use of its present facilities.

Nature's Own Foods has a line-forcing policy, requiring any store that wants to carry its brand name to carry most of the 65 items in the Nature's Own Foods line. This policy, coupled with its wide expansion of product lines, has resulted in 88 percent of the firm's sales coming from major supermarket chain stores, such as Safeway, Kroger, and Winn-Dixie.

Smaller stores are generally not willing to accept the Nature's Own Foods policy. May explains, "We know that only large stores can afford to stock all our products. But the large stores are the volume! We give consumers the choice of any Nature's Own Foods product they want, and the result is maximum sales." Many small retailers have complained about Nature's Own Foods' policy, but they have been ignored because they are considered too small in potential sales volume per store to be of any significance.

In late 2018, a stockholders' revolt over low profits (in 2018, profits were only $500,000) resulted in Nature's Own Foods' president and two of its five directors being removed. Neal Middleton (introduced earlier), an accountant from the company's outside auditing firm, was brought in as president. One of the first things he focused on was the variable and low levels of profits in the past several years. A comparison of Nature's Own Foods' results with similar operations of some large competitors supported his concern. In the past 13 years, Nature's Own Foods' closest competitors had an average profit return on shareholders' investment of 5 to 9 percent, whereas Nature's Own Foods averaged only 1.5 percent. Further, Nature's Own Foods' sales volume has not increased much from the 2005 level (after adjusting for inflation)—while operating costs have soared upward. Profits for the firm were about $8 million in 2005. The closest Nature's Own Foods has come since then is about $6 million—in 2009. The outgoing president blamed his failure on an inefficient sales department. He said, "Our sales department has deteriorated. I can't exactly put my finger on it, but the overall quality of salespeople has dropped, and morale is bad. The team just didn't perform." When Middleton e-mailed Shelley Walton, the vice president of sales, with this charge, her reply was:

> It's not our fault. I think the company made a key mistake in the late '80s. It expanded horizontally—by increasing its number of product offerings—while major competitors were expanding vertically, growing their own raw materials and making all of their packing materials. They can control quality and make profits in manufacturing that can be used in promotion. I lost some of my best people from frustration. We just aren't competitive enough to reach the market the way we should with a comparable product and price.

In a lengthy e-mail from Shelley Walton, Middleton learned more about the characteristics of Nature's Own Foods' market. Although all the firms in the food-processing industry advertise heavily, the size of the market for most processed foods hasn't grown much for many years. Further, most consumers are pressed for time and aren't very selective. If they can't find the brand of food they are looking for, they'll pick up another brand rather than go to another store. No company in the industry has much effect on the price at which its products are sold. Chain store buyers are very knowledgeable about prices and special promotions available from all the competing suppliers, and they are quick to play one supplier against another to keep the price low. Basically, they have a price they are willing to pay—and they won't exceed it. Then the chains will charge any price they wish on a given brand sold at retail. That is, a 48-can case of beans might be purchased from any supplier for $23.10, no matter whose product it is. Generally, the shelf price for each is no more than a few pennies different, but chain stores occasionally attract customers by placing a well-known brand on sale.

Besides insisting that processors meet price points, like for the canned beans, some chains require price allowances if special locations or displays are desired. They also carry non-advertised brands and/or their own brands at a lower price—to offer better value to their customers. And most willingly accept producers' cents-off coupons, which are offered by Nature's Own Foods as well as most of the other major producers of full lines.

At this point, Neal Middleton is trying to decide why Nature's Own Foods, Inc. isn't as profitable as it once was. And he is puzzled about why some competitors are putting products on the market with low potential sales volume. (For example, one major competitor recently introduced a line of exotic foreign vegetables with gourmet sauces.) And others have been offering frozen dinners or entrees with vegetables for several years. Apparently, Nature's Own Foods' managers considered trying such products several years ago but decided against it because of the small potential sales volumes and the likely high costs of new-product development and promotion.

Evaluate Nature's Own Foods' present situation. What would you advise Neal Middleton to do to improve Nature's Own Foods' profits? Explain why.

3. NOCO United Soccer Academy

Wesley Diekens came to the United States from the UK in 2009 on a soccer scholarship. Wesley grew up playing soccer on many competitive teams throughout high school and had a brief professional career in England. When St. Albans College recruited him to play soccer, he thought it would open his life to a grand adventure. That adventure changed his life.

While at St. Albans, Diekens met his future wife, Alyce Bilski, who also played soccer there. She graduated a year ahead of him and went to Fort Collins, Colorado, where she played on the semiprofessional Fort Collins Force women's soccer team. When Diekens finished college, he followed Bilski to northern Colorado. Bilski was captain of the Force and worked for the sports marketing company that owned the team.

Diekens got a job at a local meatpacking plant, but soccer was his passion. He made the practice squad for the Colorado Rapids Major League Soccer team, but injuries cut his professional career short. Teaching soccer to kids became a new passion for Diekens. He has a natural talent for coaching. Diekens is charismatic, kids enjoy his easygoing demeanor and British accent, and he really knows soccer and how to teach the game to youngsters.

In 2013, Diekens founded the NOCO United Soccer Academy (NOCO standing for NOrthern COlorado). At first he trained small groups of young players aged 7 to 14. He grouped them by age and skill level and conducted training sessions for small groups of five to seven at a local park. The first kids he attracted came by word-of-mouth as they quickly told friends and teammates about "this British guy who teaches soccer and makes it fun." His small after-school camps quickly grew to include more than 50 kids. Word continued to get around, and by the following summer Diekens conducted 10 different camps—and quit his job at the meatpacking plant. He also trained 11 different NOCO United 3v3 soccer teams that competed in tournaments across the state and nation during the summer. All of his players had bright blue jerseys with the NOCO United name across the front, and the success of these teams made the jerseys a great promotion vehicle. In 2016, four of his teams competed in the national 3v3 soccer tournament, with one winning a national championship.

To keep up with the rapid growth, Diekens brought a few friends over from England to assist with training. Will Bowman moved to the United States to become Diekens' assistant director of coaching. Diekens and Bowman planned to work year-round as trainers and hire a couple of local coaches to help them conduct training sessions. During the summer he added a couple of local college soccer players and a few former teammates from England. The summer season works well for his British mates, because that is the off-season for those still playing professionally. Diekens is confident he can hire and train more coaches if he needs them to handle future growth.

Youth soccer is big in Colorado and across much of the United States. It is the largest participation sport for kids. Fort Collins is a soccer hotbed, and this has helped Diekens' business grow. He now trains about 600 kids per year. But he has even greater ambitions. For example, he would like to build a training facility; the space he currently rents is not always well suited to soccer. However, he figures he would need to double his business to justify the cost of the soccer complex he wants to build. So he is now wondering how to grow his business.

About 90 percent of his current customers live in Fort Collins, which has a population of about 150,000 people. Diekens believes awareness of his program is close to 100 percent among competitive soccer players ages 11 to 14—and is probably at about 40 percent among families with soccer-playing kids ages 6 to 10. Most of his customers are 10 to 13 years old and enroll in two to three NOCO United programs per year. He has also run a few camps in Boulder and Northglenn—both are about 50 miles from Fort Collins. These have been successful but are currently limited.

There are several small cities within 25 miles of Fort Collins. Loveland, a city of about 70,000, borders Fort Collins on the south. Greeley and Longmont, each with about 90,000 to 100,000 people, are about 25 miles away by interstate highway. These areas have very limited soccer training programs except for their competitive teams, and awareness of NOCO United is not very high. Those who have heard of his academy are often not familiar with its philosophy and programs. Diekens is not sure if parents in these communities would be willing to drive their kids to Fort Collins for training. If not, he would have to run his programs there.

Diekens knows that he wants to grow his business but wonders how he can accomplish his goal. He currently sees a few options:

1. His current customer retention rate is pretty high: about 80 percent. However, when the kids reach 14 or 15 years old, other high school sports and activities make them less interested in extra soccer training. One option is to try to increase retention by developing programs targeted at kids older than 14.
2. Another option is to develop a marketing strategy that would encourage his current customers to buy more. He wonders if they have other needs that he might be able to serve.
3. Diekens could try to grow the business by entering new markets and acquiring new customers. His market penetration with kids 6 to 9 years old is still quite modest. He might develop new programs to better meet this group's needs.
4. Another new market option would be to serve more kids from Loveland, Longmont, and Greeley.

Evaluate Diekens' different options for growing NOCO United's customer equity. Develop a set of marketing strategy ideas for each of the options. What marketing research could Diekens perform to better assess his options?

4. Petoskey Tech Support

Claire Kelly is getting desperate about her new business. She's not sure she can make a go of it—and she really wants to stay in her hometown of Petoskey, Michigan, a beautiful summer resort area along the eastern shore of Lake Michigan. The area's permanent population of 10,000 more than triples in the summer months and doubles at times during the winter skiing and snowmobiling season.

Claire spent four years in the Navy after college graduation, returning home in June 2019. When she couldn't find a good job in the Petoskey area, she decided to go into business for herself and set up Petoskey Tech Support. Claire's plan was to work by herself and basically serve as a "for hire" computer consultant and troubleshooter for her customers. She knew that many of the upscale summer residents relied on home computers to keep in touch with business dealings and friends at home, and it seemed that someone was always asking her for computer advice. She was optimistic that she could keep busy with a variety of on-site services—setting up a customer's new computer, repairing hardware problems, installing software or upgrades, creating a wireless network, correcting problems created by viruses, and the like.

Claire thought that her savings would allow her to start the business without borrowing any money. Her estimates of required expenditures were $7,000 for a used SUV; $1,125 for tools, diagnostic equipment, and reference books; $1,700 for a laptop computer, software, and accessories; $350 for an initial supply of fittings and cables; and $500 for insurance and other incidental expenses. This total of $10,675 still left Claire with about $6,500 in savings to cover living expenses while getting started.

Claire chose the technology services business because of her previous work experience. She worked at a computer "help desk" in college and spent her last year in the Navy troubleshooting computer network problems. In addition, from the time Claire was 16 years old until she finished college, she had also worked during the summer for Eric Steele. Eric operates the only successful computer services company in Petoskey. (There was one other local computer store that also did some on-location service work when the customer bought equipment at the store, but that store recently went out of business.)

Eric prides himself on quality work and has been able to build up a good business with repeat customers. Specializing in services to residential, small-business, and professional offices, Eric has built a strong customer franchise. For 20 years, Eric's major source of new business has been satisfied customers who tell friends or coworkers about his quality service. He is highly regarded as a capable person who always treats clients fairly and honestly. For example, seasonal residents often give Eric the keys to their vacation homes so that he can do upgrades or maintenance while they are away for months at a time. Eric's customers are so loyal, in fact, that Fix-A-Bug—a national computer service franchise—found it impossible to compete with him. Even price-cutting was not an effective weapon against Eric.

From having worked with Eric, Claire thought that she knew the computer service business as well as he did; in fact, she had sometimes been able to solve technical problems that left him stumped. Claire was eager to reach her $70,000-per-year sales objective because she thought this would provide her with a comfortable living in Petoskey. While aware of opportunities to do computer consulting for larger businesses, Claire felt that the sales opportunities for her services were limited because many firms had their own computer specialists or even IT departments. As Claire saw it, her only attractive opportunity was direct competition with Eric.

To get started, Claire spent $1,400 to advertise her business in the local newspaper and using Google Ads. With this money she bought two announcement ads and 52 weeks of daily ads in the classified section, listed under "Miscellaneous Residential and Business Services." The ads from Google Ads appeared when locals searched for "computers," "tech support," or "computer repair." She also listed her business under "Computer Services" at Craigslist for Northern Michigan—updating this notice and information once a month. Further, she built a small website with just a basic home page, a page with her picture and experience, and a third page that lists the services she offers. She thought about creating a Facebook page and blog, but she hasn't done that yet. She put magnetic sign boards on her SUV and has waited for business to take off.

At the end of three months, Claire had a few customers, but much of the time she wasn't busy and she was able to gross only about $300 a week. Of course, she had expected much more. Many of the people who did call were Eric's regular customers who had some sort of crisis when he was already busy. While these people agreed that Claire's work was of the same quality as Eric's, they preferred Eric's "quality care" image and they liked the fact that they had an ongoing relationship with him.

On occasion, Claire did get more work than she could handle. This happened during April and May, when seasonal businesses were preparing for summer openings and owners of summer homes and condos were ready to "open the cottage." The same rush occurred in September and October, as many of these places were being closed for the winter; those customers often wanted help backing up computer files or packing up computer equipment so they could take it with them. During these months, Claire was able to gross about $150 to $200 a day.

Toward the end of her discouraging first year in business, Claire Kelly is thinking about quitting. While she hates to think about leaving Petoskey, she can't see any way of making a living there with her independent technology services business. Eric seems to dominate the market, except in the rush seasons and for people who need emergency help. And the resort market is not growing very rapidly, so there is little hope of a big influx of new businesses and homeowners to spur demand.

Evaluate Claire Kelly's strategy planning for her new business. Why isn't she able to reach her objective of $70,000? What should Claire do now? Explain.

5. Resin Dynamics

Paige Chen, a chemist in Resin Dynamics' resins laboratory, is trying to decide how hard to fight for the new product she has developed. Chen's job is to find new, more profitable applications for the company's present resin products—and her current efforts are running into unexpected problems.

During the last four years, Chen has been under heavy pressure from her managers to come up with an idea that will open up new markets for the company's foamed polystyrene.

Two years ago, Chen developed the "foamed-dome concept"—a method of using foamed polystyrene to make dome-shaped roofs and other structures. She described the procedure for making domes as follows: The construction of a foamed dome involves the use of a specially designed machine that bends, places, and bonds pieces of plastic foam together into a predetermined dome shape. In forming a dome, the machine head is mounted on a boom, which swings around a pivot like the hands of a clock, laying and bonding layer upon layer of foam board in a rising spherical form.

According to Chen, polystyrene foam boards have several advantages, such as:

1. Foam board is stiff—but can be formed or bonded to itself by heat alone.
2. Foam board is extremely lightweight and easy to handle. It has good structural rigidity.
3. Foam board has excellent and permanent insulating characteristics. (In fact, the major use for foam board is as an insulator.)
4. Foam board provides an excellent base on which to apply a variety of surface finishes, such as a readily available concrete-based stucco that is durable and inexpensive.

Using her good selling abilities, Chen easily convinced her managers that her idea had potential.

According to a preliminary study by the marketing research department, the following were areas of construction that could be served by the domes:

1. Bulk storage
2. Cold storage
3. Educational construction
4. Covers for industrial tanks
5. Light commercial construction
6. Planetariums
7. Recreational construction (such as a golf course starter house)

The marketing research study focused on uses for existing dome structures. Most of the existing domes are made of cement-based materials. The study showed that large savings would result from using foam boards due to the reduction of construction time.

Because of the new technology involved, the company decided to do its own contracting (at least for the first four to five years). Chen thought this was necessary to make sure that no mistakes were made by inexperienced contractor crews. (For example, if not applied properly, the plastic may burn.)

After building a few domes in the United States to demonstrate the concept, Chen contacted some leading U.S. architects. Reactions were as follows:

"It's very interesting, but we're not sure the fire marshal of Chicago would ever give his OK."

"Your tests show that foamed domes can be protected against fires, but there are no good tests for unconventional building materials as far as I am concerned."

"I like the idea, but foam board does not have the impact resistance of cement."

"We design a lot of recreational facilities, and kids will find a way to poke holes in the foam."

"Building codes in our area are written for wood and cement structures. Maybe we'd be interested if the codes change."

After this unexpected reaction, management didn't know what to do. Chen still thinks they should go ahead with the project. She wants to build several more demonstration projects in the United States and at least three each in Europe and Japan to expose the concept in the global market. She thinks an interactive website and online video will also help spread the idea. She thinks architects outside the United States may be more receptive to really new ideas. Further, she says, it takes time for potential users to "see" and accept new ideas. She is sure that more exposure to more people will speed acceptance. And she is convinced that a few reports of well-constructed domes in leading trade papers and magazines will go a long way toward selling the idea. She is working on getting such reports right now. But her managers aren't sure they want to OK spending more money on "her" project. Her immediate boss is supportive, but the rest of the review board is less sure about more demonstration projects or going ahead at all—either in the United States or in global markets.

Evaluate how Resin Dynamics got into the present situation. What should Paige Chen do? What should Chen's managers do? Explain.

6. Dynamic Steel

Dynamic Steel is one of two major producers of wide-flange beams in the United States. The other producer is USX. A number of small firms also compete, but they tend to compete mainly on price in nearby markets where they can keep transport costs low. Typically, all interested competitors charge the same delivered price, which varies some depending on how far the customer is from either of the two major producers. In other words, local prices are higher in more remote geographic markets.

Wide-flange beams are one of the principal steel products used in construction. They are the modern version of what are commonly known as I-beams. USX rolls a full range of wide flanges from 6 to 36 inches. Dynamic Steel entered the field about 30 years ago, when it converted an existing mill to produce this product. Dynamic Steel's mill is limited to flanges up to 24 inches, however. At the time of the conversion, Dynamic Steel felt that customer usage of sizes over 24 inches was likely to be small. In recent years, however, there has been a definite trend toward the larger and heavier sections.

The beams produced by the various competitors are almost identical—because customers buy according to standard dimensional and physical-property specifications. In the smaller size range, there are a number of competitors. But above 14 inches, only USX and Dynamic Steel compete. Above 24 inches, USX has no competition.

All the steel companies sell these beams through their own sales forces. The customer for these beams is called a structural fabricator. This fabricator typically buys unshaped beams and other steel products from the mills and shapes them according to the specifications of each customer. The fabricator sells to the contractor or owner of the structure being built.

The structural fabricator usually must sell on a competitive-bid basis. The bidding is done on the plans and specifications prepared by an architectural or structural engineering firm and forwarded to the fabricator by the contractor who wants the bid. Although thousands of structural fabricators compete in the United States, relatively few account for the majority of wide-flange tonnage in the various geographic regions. Because the price is the same from all producers, they typically buy beams on the basis of availability (i.e., availability to meet production schedules) and performance (i.e., reliability in meeting the promised delivery schedule).

Several years ago, Dynamic Steel's production schedulers saw that they were going to have an excess of hot-rolled plate capacity in the near future. At the same time, development of a new production technology allowed Dynamic Steel to weld three plates together into a section with the same dimensional and physical properties and almost the same cross section as a rolled wide-flange beam. This development appeared to offer two key advantages to Dynamic Steel: (1) it would enable Dynamic Steel to use some of the excess plate capacity, and (2) the company could offer larger sizes of wide-flange beams. Cost analysts showed that by using a fully depreciated plate mill and the new welding process, it would be possible to produce and sell larger wide-flange beams at competitive prices—that is, at the same price charged by USX.

Dynamic Steel's managers were excited about the possibilities because customers usually appreciate having a second source of supply. Also, the new approach would allow the production of up to a 60-inch flange. With a little imagination, these larger sizes might offer a significant breakthrough for the construction industry.

Dynamic Steel decided to go ahead with the new project. As the production capacity was converted, the salespeople were kept well informed of the progress. They, in turn, promoted this new capability to their customers, emphasizing that soon they would be able to offer a full range of beam products. Dynamic Steel sent several general information letters to a broad mailing list but did not advertise. The market development section of the sales department was very busy explaining the new possibilities of the process to fabricators at engineering trade associations and shows.

When the new production line was finally ready to go, the market reaction was disappointing. No orders came in and none were expected. In general, customers were wary of the new product. The structural fabricators felt they couldn't use it without the approval of their customers, because it would involve deviating from the specified rolled sections. And as long as they could still get the rolled section, why make the extra effort for something unfamiliar, especially with no price advantage. The salespeople were also bothered with a very common question: How can you take plate that you sell for about $460 per ton and make a product that you sell for $470 per ton? This question came up frequently and tended to divert the whole discussion to the cost of production rather than to the way the new product might be used or its value in the construction process.

Evaluate Dynamic Steel's situation. What should Dynamic Steel do?

7. Lake Pukati Lodge

Nestled in the high country of New Zealand's South Island is a getaway adventure playground unashamedly aimed at the world's very wealthy. Presidents, movie stars, and other such globe-trotters are the prime targets of this fledgling tourism business developed by Lake Pukati Lodge. The lodge offers this exclusive niche the opportunity of a secluded holiday in a little-known paradise. Guests, commonly under public scrutiny in their everyday lives, can escape such pressures at a hunting retreat designed specifically with their needs in mind.

A chance meeting between a New Zealand Department of Conservation investigator and the son of the former Indonesian president marked the beginning of this specialty tourist operation. Recognizing that "filthy rich" public figures are constantly surrounded by security and seldom have the luxury of going anywhere incognito, the New Zealander, Peter Slater, suggested that he and his new friend purchase a high-country station and hunting-guide company that was for sale. Slater believed that the facilities, and their secluded and peaceful environment, would make an ideal holiday haven for this elite group. His Indonesian partner concurred.

Slater, who was by now the company's managing director, developed a carefully tailored package of goods and services for the property. Architecturally designed accommodations, including a game trophy room and eight guest rooms, were constructed using high-quality South Island furniture and fittings to create the ambiance necessary to attract and satisfy the demands of their special clientele.

Although New Zealand has an international reputation for being sparsely populated and green, Slater knew that rich travelers frequently complained that local accommodations were below overseas standards. Because the price (NZ$700 a night) was not a significant variable for this target market, sumptuous guest facilities were built. These were designed to be twice the normal size of most hotel rooms, with double-glazed windows that revealed breathtaking views. Ten full-time staff and two seasonal guides were recruited to ensure that visitors received superior customized service, in keeping with the restrained opulence of the lodge.

The 28,000 hectares of original farmland that make up the retreat and back onto the South Island's Mount Cook National Park were converted into a big-game reserve. All merino sheep on the land were sold, and deer, elk, chamois, and wapiti were brought in and released. This was a carefully considered plan. Slater, the former conservationist, believed that financially and environmentally this was the correct decision. Not only do tourists, each staying for one week and taking part in safari shooting, inject as much cash into the business as the station's annual wool clip used to fetch, but the game does less harm to the environment than sheep. Cattle, however, once part of the original station, were left to graze on lower river-flat areas.

For those high-flying customers seeking less bloodthirsty leisure activities, Lake Pukati Lodge developed photographic "safaris" and other product-line extensions. Horse-trekking, golfing on a nearby rural course (with no need for hordes of security forces), helicopter trips around nearby Lake Tekapo, nature walks, and other such activities formed part of the exclusive package.

While still in the early stages of operation, this retreat has already attracted a steady stream of visitors. To date, the manager has relied solely on positive word-of-mouth, publicity, and public relations to draw in new customers. Given the social and business circles in which his potential target market moves, Slater considers these to be the most appropriate forms of marketing communication. The only real concern for Lake Pukati Lodge has been the criticism of at least one New Zealand lobby group that the company is yet another example of local land passing into "foreign" hands, and that New Zealanders are prevented from using the retreat and excluded from its financial returns. However, this unwelcome attention has been fairly short-lived.

Identify the likely characteristics of the market segment being targeted by the company. Why are most target customers likely to be foreigners rather than New Zealanders? Suggest what expectations target customers are likely to have regarding the quality, reliability, and range of services. What are the implications for Lake Pukati Lodge? How difficult is it for Lake Pukati Lodge to undertake marketing research? Elaborate.

8. Carmine's Italian Restaurant

Rita Carmine, the owner and manager of Carmine's Italian Restaurant, is reviewing the slow growth of her restaurant. She's also thinking about the future and wondering if she should change her strategy. In particular, she is wondering if she should join a fast-food or family restaurant franchise chain. Several are located near her, but there are many franchisors without local restaurants. After doing some online research, she has learned that with help from the franchisors, some of these places gross $500,000 to $1 million a year. Of course, she would have to follow someone else's strategy and thereby lose her independence, which she doesn't like to think about. But those sales figures do sound good, and she has also heard that the return to the owner-manager (including salary) can be more than $150,000 per year. She has also considered creating a Facebook page for Carmine's Italian Restaurant but is not sure how that will help. She knows people go to Carmine's web page for directions and to see the menu—but why might she need Facebook or some other social media?

Carmine's Italian Restaurant is a fairly large restaurant—about 3,000 square feet—located in the center of a small shopping center completed early in 2015. Carmine's sells mainly full-course "home-cooked" Italian-style dinners (no bar) at moderate prices. In addition to Carmine's Italian Restaurant, other businesses in the shopping center include a supermarket, a hair salon, a liquor store, a computer repair business, and a vacant space that used to be a hardware store. The hardware store failed when a Home Depot located nearby. Rita has learned that a pizzeria is considering locating there soon. She wonders how that competition will affect her. Ample parking space is available at the shopping center, which is located in a residential section of a growing suburb to the east, along a heavily traveled major traffic route.

Rita graduated from a local high school and a nearby university and has lived in this town with her husband and two children for many years. She has been self-employed in the restaurant business since her graduation from college in 1998. Her most recent venture before opening Carmine's was a large restaurant that she operated successfully with her brother from 2006 to 2012. In 2012, Rita sold her share because of illness. Following her recovery, she was eager for something to do and opened the present restaurant in April 2015. Rita feels her plans for the business and her opening were well thought out. When she was ready to start her new restaurant, she looked at several possible locations before finally deciding on the present one. Rita explained, "I looked everywhere, but here I particularly noticed the heavy traffic. This is the crossroads for three major interstate highways. So obviously the potential is here."

Having decided on the location, Rita signed a 10-year lease with the option to renew for 10 more years, and then eagerly attacked the problem of outfitting the almost empty store space in the newly constructed building. She tiled the floor, put wood paneling on the walls, installed plumbing and electrical fixtures, added an extra washroom, and purchased the necessary restaurant equipment. All this cost $240,000—which came from her own cash savings. She then spent an additional $3,000 for glassware, $4,000 for an initial food stock, and $4,250 to advertise the opening of Carmine's Italian Restaurant in the local newspaper. The paper serves the whole metro area, so the $4,250 bought only three quarter-page ads. These expenditures also came from her personal savings. Next she hired five servers at $550 a week and one chef at $1,100 a week. Then, with a $48,000 cash reserve for the business, she was ready to open. Reflecting her sound business sense, Rita knew she would need a substantial cash reserve to fall back on until the business got on its feet. She expected this to take about one year. She had no expectations of getting rich overnight. (Her husband, a high school teacher, was willing to support the family until the restaurant caught on.)

The restaurant opened in April and by August had a weekly gross revenue of only $4,800. Rita was a little discouraged by this, but she was still able to meet all her operating expenses without investing any new money in the business. She also got a few good customer reviews on Yelp. By September business was still slow, and Rita had to invest an additional $6,000 in the business just to survive.

Business had not improved by November, so Rita stepped up her advertising—hoping this would help. In December, she spent $2,400 of her cash reserve for radio advertising—10 late-evening spots on a news program at a station that aims at middle-income America. Rita also spent $1,600 more during the next several weeks for some Google Local advertising.

By April 2017, the situation had begun to improve, and by June her weekly gross was up to between $6,100 and $6,600. By March 2018, the weekly gross had risen to about $8,400. Rita increased the working hours of her staff from six to seven hours a week and added another cook to handle the increasing number of customers. Rita was more optimistic for the future because she was finally doing a little better than breaking even. Her full-time involvement seemed to be paying off. She had not put any new money into the business since summer 2017 and expected business to continue to rise. She had not yet taken any salary for herself, even though she had built up a small surplus of about $18,000. Instead, she planned to put in a bigger air-conditioning system at a cost of $10,000 and was also planning to use what salary she might have taken for herself to hire two new servers to handle the growing volume of business. And she saw that if business increased much more she would have to add another cook.

Evaluate Rita's past and present marketing strategy. What should she do now? Should she seriously consider joining some franchise chain?

9. Quiet Night Motel

Tristan Knaus is trying to decide whether he should make some minor changes in the way he operates his Quiet Night Motel or if he should join either the Days Inn or Holiday Inn motel chains. Some decision must be made soon because his present operation is losing money. But joining either of the chains will require fairly substantial changes, including new capital investment if he goes with Holiday Inn.

Tristan bought the recently completed 60-room motel two years ago after leaving a successful career as a production manager for a large producer of industrial machinery. He was looking for an interesting opportunity that would be less demanding than the production manager job. The Quiet Night Motel is located at the edge of a very small town near a rapidly expanding resort area and about one-half mile off an interstate highway. It is 10 miles from the tourist area, which has several nationally franchised full-service resort motels suitable for "destination" vacations. There is a Best Western, a Ramada Inn, and a Hilton Inn, as well as many mom-and-pop and limited-service, lower-priced motels—and some quaint bed-and-breakfast facilities—in the tourist area. The interstate highway near the Quiet Night Motel carries a great deal of traffic, because the resort area is between several major metropolitan areas. No development has taken place around the turnoff from the interstate highway. The only promotion for the tourist area along the interstate highway is two large signs near the turnoffs. They show the popular name for the area and that it is only 10 miles to the west. These signs are maintained by the area's tourist bureau. In addition, the state transportation department maintains several small signs showing (by symbols) that near this turnoff one can find gas, food, and lodging. Tristan does not have any signs advertising Quiet Night Motel except the two on his property. He has been relying on people finding his motel as they go toward the resort area.

Initially, Tristan was very pleased with his purchase. He had traveled a lot himself and stayed in many different hotels and motels—so he had some definite ideas about what travelers wanted. He felt that a relatively plain but modern room with a comfortable bed, standard bath facilities, and free cable TV would appeal to most customers. Further, Tristan thought a swimming pool or any other non-revenue-producing additions were not necessary. And he felt a restaurant would be a greater management problem than the benefits it would offer. However, after many customers commented about the lack of convenient breakfast facilities, Tristan started serving a free continental breakfast of coffee, juice, and rolls in a room next to the registration desk.

Day-to-day operations went fairly smoothly in the first two years, in part because Tristan and his wife handled registration and office duties as well as general management. During the first year of operation, occupancy began to stabilize around 55 percent of capacity. But according to industry figures, this was far below the average of 68 percent for his classification—motels without restaurants.

After two years of operation, Tristan was concerned because his occupancy rates continued to be below average. He decided to look for ways to increase both occupancy rate and profitability and still maintain his independence.

Tristan wanted to avoid direct competition with the full-service resort motels. He stressed a price appeal in his signs and brochures and was quite proud of the fact that he had been able to avoid all the "unnecessary expenses" of the full-service resort motels. As a result, Tristan was able to offer lodging at a very modest price—about 40 percent below the full-service hotels and comparable to the lowest-priced resort area motels. The customers who stayed at Quiet Night Motel said they found it quite acceptable. The hotel's online reviews at sites such as TripAdvisor.com and Hotels.com, although not numerous, were generally pretty positive. But he was troubled by what seemed to be a large number of people driving into his parking lot, looking around, and not coming in to register.

Tristan was particularly interested in the results of a recent study by the regional tourist bureau. This study revealed the following information about area vacationers:

1. 68 percent of the visitors to the area are young couples and older couples without children.
2. 40 percent of the visitors plan their vacations and reserve rooms more than 60 days in advance.
3. 66 percent of the visitors stay more than three days in the area and at the same location.
4. 78 percent of the visitors indicated that recreational facilities were important in their choice of accommodations.
5. 13 percent of the visitors had family incomes of less than $27,000 per year.
6. 38 percent of the visitors indicated that it was their first visit to the area.

After much thought, Tristan began to seriously consider affiliating with a national motel chain in hopes of attracting more customers and maybe protecting his motel from the increasing competition. There were constant rumors that more motels were being planned for the area. After some investigating, he focused on two national chain possibilities: Days Inn and Holiday Inn. Neither had affiliates in the area even though they each have about 2,000 units nationwide.

Days Inn of America, Inc. is an Atlanta-based chain of economy lodgings. It has been growing rapidly and is willing to take on new franchisees. A major advantage of Days Inn is that it would not require a major capital investment by Tristan. The firm is targeting people interested in lower-priced motels, in particular, senior citizens, the military, school sports teams, educators, and business travelers. In contrast, Holiday Inn would probably require Tristan to upgrade some of his facilities, including adding a swimming pool. The total new capital investment would be between $300,000 and $500,000, depending on how fancy he got. But then Tristan would be able to charge higher prices, perhaps $75 per day on average rather than the $45 per day he's charging now.

The major advantages of going with either of these national chains would be their central reservation systems and their national names. Both companies offer nationwide, toll-free reservation lines, which produce about 40 percent of all bookings in affiliated motels. Both companies also offer websites (www.daysinn.com and www.holidayinn.com) that help find a specific hotel by destination, rate, amenities, quality rating, and availability.

A major difference between the two national chains is their method of promotion. Days Inn uses little TV advertising and less print advertising than Holiday Inn. Instead, Days Inn emphasizes sales promotions. In one campaign, for example, Blue Bonnet margarine users could exchange proof-of-purchase seals for a free night at a Days Inn. This tie-in led to the Days Inn system selling an additional 10,000 rooms. Further, Days Inn operates a September Days Club for travelers 50 and older who receive such benefits as discount rates and a quarterly travel magazine.

Days Inn also has other membership programs, including its InnCentives loyalty club for frequent business and leisure travelers. Other programs targeted to business travelers include two Corporate Rate programs and its new Days Business Place hotels. Not to be outdone, Holiday Inn has a membership program called Priority Club Worldwide. Both firms charge 8 percent of gross room revenues for belonging to their chain—to cover the costs of the reservation service and national promotion. This amount is payable monthly. In addition, franchise members must agree to maintain their facilities and make repairs and improvements as required. Failure to maintain facilities can result in losing the franchise. Periodic inspections are conducted as part of supervising the whole chain and helping the members operate more effectively.

Evaluate Tristan Knaus' present strategy. What should he do? Explain.

10. Cousin's Ice Center

Mark Cousin, the manager of Cousin's Ice Center, is trying to decide what strategies to use to increase profits.

Cousin's Ice Center is an ice-skating rink with a conventional hockey rink surface (85 feet × 200 feet). It is the only indoor ice rink in a northern U.S. city of about 450,000. The city's recreation department operates some outdoor rinks in the winter, but it doesn't offer regular ice-skating programs because of weather variability.

Mark runs a successful hockey program that is more than breaking even—but this is about all he can expect if he offers only hockey. To try to increase his profits, Mark is trying to expand and improve his public skating program. With such a program, he could have as many as 700 people in a public session at one time, instead of limiting the use of the ice to 12 to 24 hockey players per hour. While the receipts from hockey can be as high as $200 an hour (plus concession sales), the receipts from a two-hour public skating session—charging $5 per person—could yield up to $3,500 for a two-hour period (plus much higher concession sales). The potential revenue from such large public skating sessions could make Cousin's Ice Center a really profitable operation. But, unfortunately, just scheduling public sessions doesn't mean that a large number will come. In fact, only a few prime times seem likely: Friday and Saturday evenings and Saturday and Sunday afternoons.

Mark has included 14 public skating sessions in his ice schedule, but so far they haven't attracted as many people as he had hoped. In total, they generate only a little more revenue than if the times were sold for hockey use. Offsetting this extra revenue are extra costs. More staff people are needed to handle a public skating session—guards, a ticket seller, skate rental, and more concession help. So the net revenue from either use is about the same. He could cancel some of the less attractive public sessions—like the noontime daily sessions, which have very low attendance—and make the average attendance figures look a lot better. But he feels that if he is going to offer public skating he must have a reasonable selection of times. He has noticed the wide variation in the kinds of people attracted to the different skating sessions.

The Saturday and Sunday afternoon public skating sessions have been the most successful, with an average of 200 people attending during the winter season. Typically, this is a "kid-sitting" session. More than half of the patrons are young children who have been dropped off by their parents for several hours, but there are also some family groups.

In general, the kids and the families have a good time—and a fairly loyal group comes every Saturday and/or Sunday during the winter season. In the spring and fall, however, attendance drops by about half, depending on how nice the weather is. (Mark schedules no public sessions in the summer, focusing instead on hockey clinics and figure skating.)

The Friday and Saturday evening public sessions are a big disappointment. The sessions run from 8 until 10, a time when he had hoped to attract teenagers and young adult couples. At $5 per person, plus $1.50 for skate rental, this would be an economical date. In fact, Mark has seen quite a few young couples—and some keep coming back. But he also sees a surprising number of 8- to 14-year-olds who have been dropped off by their parents. The younger kids tend to race around the rink playing tag. This affects the whole atmosphere, making it less appealing for dating couples and older patrons.

Mark has been hoping to develop a teenage and young adult market for a "social activity," adapting the format used by roller-skating rinks. Their public skating sessions feature a variety of couples-only and group games as well as individual skating to dance music. Turning ice-skating sessions into such social activities is not common, however, although industry newsletters suggest that a few ice rink operators have had success with the roller-skating format. Seemingly, the ice-skating sessions are viewed as active recreation, offering exercise or a sports experience.

Mark installed some soft lights to try to change the evening atmosphere. The music was selected to encourage people to skate to the beat and couples to skate together. Some people complained about the "old" music, but it was "danceable," and some skaters really liked it. For a few sessions, Mark even tried to have some couples-only skates. The couples liked it, but this format was strongly resisted by the young boys who felt that they had paid their money and there was no reason why they should be kicked off the ice. Mark also tried to attract more young people, especially couples, by bringing in a local rock radio station disc jockey to broadcast from Cousin's Ice Center—playing music and advertising the Friday and Saturday evening public sessions. Mark's son even set up Facebook and Instagram pages for Cousin's, but only a dozen or so people "Liked" the pages. All of this appeared to have no effect on attendance, which varied from 50 to 100 per two-hour session during the winter.

Mark seriously considered the possibility of limiting the Friday and Saturday evening sessions to people aged 14 and older—to try to change the environment. He knew it would take time to change people's attitudes. But when he counted the customers, he realized this would be risky. More than a quarter of his customers on an average weekend night appeared to be 13 or younger. This meant that he would have to make a serious commitment to building the teen and young adult market. And, so far, his efforts haven't been successful. He has already invested more than $3,000 in lighting changes and more than $9,000 promoting the sessions over the rock music radio station, with very disappointing results. Although the station's sales rep said the station reached teenagers all over town, an on-air offer for a free skating session did not get a single response!

Some days, Mark feels it's hopeless. Maybe he should accept that most public ice-skating sessions are a mixed bag. Or maybe he should just sell the time to hockey groups. Still he keeps hoping that something can be done to improve weekend evening public skating attendance, because the upside potential is so good. And the Saturday and Sunday afternoon sessions are pretty good moneymakers.

Evaluate Cousin's Ice Center's situation. What should Mark Cousin do? Why?

11. Running On

Carla Gomez is the owner of Running On—a retail store that sells shoes and accessories to runners. Carla is trying to decide what she should do with her retail business and how committed she should be to her current target market.

Carla started Running On retail store in 2002 when she was only 24 years old. At that time, she was a nationally ranked runner and felt that the growing interest in jogging offered real potential for a store that provided serious runners with the shoes and advice they needed. The jogging boom quickly turned Running On into a profitable business selling high-end running shoes—and Carla made a very good return on her investment for the first 10 years. From 2002 until 2012, Carla emphasized Nike shoes, which were well accepted and seen as top quality. Nike's aggressive promotion and quality shoes resulted in a positive image that made it possible to get a $5 to $7 per pair premium for Nike shoes. Good volume and good margins resulted in attractive profits for Carla.

Committing so heavily to Nike seemed like a good idea when its marketing and engineering were the best available. In addition to running shoes, Nike had other athletic shoes Carla could sell. So even though they were not her primary focus, Carla did stock other Nike shoes, including walking shoes, shoes for aerobic exercise, basketball shoes, tennis shoes, and cross-trainers. She also added more sportswear to her store and put more emphasis on fashion rather than just function.

Even with this broadened product line, Carla's sales flattened out—and she wasn't sure what to do to get her business back in growth mode. She realized that she was growing older and so were many of her longer-term customers. Many of them were finding that jogging isn't just hard work—it's hard on the body, especially the knees. Some were not running as often—and buying shoes less often. Other of her previously loyal runner-customers were switching to other, less demanding exercise programs. However, when she tried to orient her store and product line more toward these people, she wasn't as effective in serving the needs of serious runners—still an important source of sales for the store.

She was also facing more competition on all fronts. Many consumers who don't really do any serious exercise buy running shoes as their day-to-day casual shoes. As a result, many department stores, discount stores, and regular shoe stores have put more and more emphasis on athletic shoes in their product assortment. Many customers were growing more comfortable buying shoes online. When Carla added other brands and put more emphasis on fashion, she found that she was in direct competition with a number of other stores, which put more pressure on her to lower prices and cut her profit margins. For example, in Carla's area there are a number of local retail chains offering lower-cost and lower-quality versions of similar shoes as well as related fashion apparel. Walmart also expanded its assortment of athletic shoes—and it offers rock-bottom prices. Other chains, such as Foot Locker, have focused their promotion and product lines on specific target markets. Still, all of them (including Carla's Running On, the local chains, Walmart, and Foot Locker) are scrambling to catch up with rival category killers whose selections are immense.

In the spring of 2018 Carla tried an experiment. She took on a line of high-performance athletic shoes that were made to order. The distinctive feature of these shoes was that the sole was molded to precisely fit the customer's foot. A pair of these custom-made shoes cost about $170, so the market was not large. Further, Carla didn't put much promotional emphasis on this line. However, when a customer came in the store with a serious interest in high-performance shoes, Carla's salesclerks would tell them about the custom shoe alternative and show a sample. When a customer was interested, a mold of the customer's bare foot was made at the store, using an innovative material that hardened in just a few minutes without leaving a sticky mess. Carla sent the mold off to the manufacturer by UPS, and about two weeks later the finished shoes arrived. Customers who tried these shoes were delighted with the result. However, the company that offered them ran into financial trouble and went out of business.

Carla recently learned about another company that is offering a very similar custom shoe program. However, that company requires more promotion investment by retailers and in return provides exclusive sales territories. Another requirement is that the store establish a website promoting the shoes and providing more detail on how the order process works. Running On had a pretty basic website, so Carla knew she would have to spend some money to make this happen. In addition, all of a retailer's salesclerks are also required to go through a special two-day training program so that they know how to present the benefits of the shoe and do the best job creating the molds. The training program is free, but Carla would have to pay travel, hotel, and food expenses for her salespeople. So before even getting started, the new program would cost her several thousand dollars.

Carla is uncertain about what to do. Although sales have dropped, she is still making a reasonable profit and has a relatively good base of repeat customers, with the serious runners still more than half of her sales and profits. She thinks that the custom shoe alternative is a way to differentiate her store from the mass-merchandisers and to sharpen her focus on the target market of serious runners. On the other hand, that doesn't really solve the problem that the "runners" market seems to be shrinking. It also doesn't address the question of how best to keep a lot of the aging customers she already serves who seem to be shifting away from an emphasis on running. She also worries that she'll lose the loyalty of her repeat customers if she shifts the store further away from her running niche and more toward fashionable athletic shoes or fashionable casual wear. Yet athletic wear—women's, in particular—has come a long way in recent years. Designers such as Donna Karan, Calvin Klein, Georgio Armani, and Ralph Lauren are part of the fast-growing women's athletic wear business.

So Carla is trying to decide if there is anything else she can do to better promote her current store and product line, or if she should think about changing her strategy in a more dramatic way. Any change from her current focus would involve retraining her current salespeople and perhaps hiring new salespeople. Adding and maintaining a website isn't an insurmountable challenge, but it is not an area where she has either previous experience or skill.

Clearly, a real shift in emphasis would require that Carla make some hard decisions about her target market and her whole marketing mix. She's got some flexibility—it's not like she's a manufacturer of shoes with a big investment in a factory that can't be changed. On the other hand, she's not certain she's ready for a big change, especially a change that would mean starting over again from scratch. She started Running On because she was interested in running and felt she had something special to offer. Now she worries that she's just grasping at straws without a real focus or any obvious competitive advantage. She also knows that she is already much more successful than she ever dreamed when she started her business—and in her heart she wonders if she wasn't just spoiled by growth that came fast and easy at the start.

Evaluate Carla Gomez's present strategy. Evaluate the alternative strategies she is considering. Is her primary problem her emphasis on running shoes, her emphasis on trying to hang on to her current customers, or is it something else? What should she do? Why?

12. DrV.com—Custom Vitamins

Dr. Victoria van der Walt has to decide how to handle a complaint letter from a customer. When she received the letter, she passed it along to Paul Zimbalist, the firm's customer service manager, to get his recommendation. Now van der Walt has a reply from Zimbalist, and she must decide how to respond to the customer and determine if changes are needed in her company's customer service operations.

Dr. van der Walt has a reputation as a health and nutrition guru. Her fame grew after she published two books—both of which were very popular and received a lot of attention in the press. Five years ago, Internet entrepreneur Tania Cox approached her with the idea of creating a website to sell custom vitamins under Dr. Victoria's name.

van der Walt and Cox became partners, and the business enjoyed success in its first four years of operation. Cox handles the website technology, inventory, production, and shipping. Dr. van der Walt is the health expert, creates content provided on the website, and is in charge of marketing and customer service. The complaint letter and reply from Zimbalist follow:

Dear Dr. van der Walt,

I am a longtime fan of your books and like to visit your website for health tips. As a new grandmother, my health is even more important to me. I want to see my grandchildren graduate from high school, go to college, and have kids of their own.

You have made me a bit of a guru as well. I work out every day. People always ask me how I stay so fit and healthy. Having read both your books, I tell them they should exercise regularly and take vitamins and supplements for long-term health. I always recommend your DrV.com website, especially the section on your custom vitamins. That is my favorite part of your website. I really like that you take information about me and my medical history—and then recommend custom vitamins and supplements. I also like how you send me packages that each contain a daily dose.

But after my recent experiences, my loyalty to you and your company is now in jeopardy. Here is my story.

Six months ago, I went to the website to reorder my vitamins and supplements. The home page announced a new and improved health survey and custom health program. So I went through the survey and filled out all the details—it would have been nice if you had saved some of them from my previous survey. At the end of the survey the website offered me a 90-day supply of a custom set of vitamins and supplements selected for my specific needs. The $212 price was about $50 more than my previous 90-day supply, but I trusted your advice so I placed the order.

About two-and-a-half months later, I phoned your 800 number for DrV.com to place a refill order. The person on the phone was very nice and asked if I wanted to set up automatic refills. I said no because I hate those automatic programs. They remind me of those book clubs that automatically send you books you don't want if you do not reply fast enough. About a week later my order arrived—then two days later another identical order arrived. I did not understand this, but I figured I would eventually use them up and I kept everything (and I was billed for both orders—$424 on my credit card). I should mention there was no e-mail explaining this mystery delivery. I was a little annoyed and sent an e-mail to customer service seeking an explanation. I received an automated response, "Thank you for your inquiry; someone will get back to you within 24 hours." No one ever replied, but I forgot about it.

Then two weeks ago I received an e-mail from DrV.com telling me my refill order had been shipped and would arrive in a few days. But I did not place a refill order! I did not even need more vitamins because I was still working off the two 90-day supplies that I received three months ago. So I replied by e-mail that I did not want the order and to cancel it. A reply e-mail (from Sally) told me that I had signed up for automatic refills six months ago. I replied that I certainly had not and that I would not pay for the order that was being sent.

A few days later I received a call from Paul Zimbalist, your director of customer service. He told me that I had originally signed up for automatic refills and that was why I received vitamins. I told him that was impossible, and he told me that unless I checked some box on my original order that this was done automatically "for my convenience." He said there were also several warnings and that I must have missed those. *Basically, I think he told me this was my fault. I did not like that one bit!!!*

Mr. Zimbalist told me the vitamins were on their way, but I could refuse delivery of them. He offered to let me have them for 20 percent off if I simply kept them. Unfortunately, the vitamins were on my doorstep when I arrived home that day. I had to take the vitamins to UPS to get them shipped back to you.

You guys are no longer very good at your business. You might have a great product, but I am now seeing other vitamin companies offering the same products. I have no doubt these other companies offer better customer service. If you want me back as a customer, I would expect a formal apology from Mr. Zimbalist and a free 90-day supply. Otherwise, I figure my business will be welcomed at one of your competitors. I will be sure all my friends know about my experience at DrV.com, and I intend to post a bad review at resellerratings.com, too.

Sincerely,
Maxine Slezak

Next is the reply that Paul Zimbalist sent to Dr. van der Walt concerning Maxine Slezak's letter.

Dear Victoria,

As per your request, I reviewed Mrs. Slezak's order history. Yes, she is a very good customer who spent almost $800 with us last year. And she is a member of our referral program—and we can count at least seven new customers she has directed to us in the last 18 months. But I want to clarify some of this particular situation.

- You might recall that our automatic refill program has been a big success. Since we instituted the program a year ago, our customer retention rate has jumped by 10 percent. There are occasional complaints, but given the large number of customers we serve, the complaints are really just a "drop in the bucket."
- When Mrs. Slezak placed her order six months ago, there were at least two different warnings about the automatic refill program—customers have to check a box at the bottom of the screen to "opt out" of the program. We all agreed that it was better to make them part of the program automatically, but to give them two chances to remove themselves from automatic delivery.
- I do not know if we replied to her e-mail asking for customer service help.
- I did not tell Mrs. Slezak that this was her fault, but I did tell her that when she signed up there were two chances for her to choose not to be part of the automatic refill program.
- Mrs. Slezak did not get an e-mail notifying her of the first refill order because that system was not yet in place. But this has now been fixed, and the e-mail notifying her when we ship shows that this works.
- I offered her 20 percent off, as is our standard policy when we make a mistake. Considering this was her mistake, I thought this was generous.
- If Mrs. Slezak had called and asked, I could have had UPS come out and pick up the package for return to us.

I do not recommend giving her a free 90-day supply. This may simply encourage her to complain again in the future. Besides, this was not our mistake. We may be better off without certain customers—and I think Mrs. Slezak falls into this category.

Feel free to call me if you have any more questions.
Paul

Assess the customer service operations at DrV.com. What should Dr. van der Walt do about Mrs. Slezak? What changes, if any, should van der Walt make to customer service and ordering operations?

13. Paper Products, Inc. (PPI)*

Stasia Acosta, marketing manager for Paper Products, Inc., must decide whether she should permit her largest customer to buy some of PPI's commonly used file folders under the customer's brand rather than PPI's own FILEX brand. She is afraid that if she refuses, this customer—Business Center, Inc.—will go to another file folder producer and PPI will lose this business.

Business Center, Inc. is a major distributor of office supplies and has already managed to put its own brand on more than 45 high-sales-volume office supply products. It distributes these products—as well as the branded products of many manufacturers—through its nationwide distribution network, which includes 150 retail stores. Now Lance Richardson, vice president of marketing for Business Center, is seeking a line of file folders similar in quality to PPI's FILEX brand, which now has more than 60 percent of the market.

This is not the first time that Business Center has asked PPI to produce a file folder line for Business Center. On both previous occasions, Stasia turned down the requests and Business Center continued to buy. In fact, Business Center not only continued to buy the file folders but also the rest of PPI's product lines. And total sales continued to grow as Business Center built new stores. Business Center accounts for about 30 percent

*Adapted from a case by Professor Hardy, University of Western Ontario.

of Stasia Acosta's business. And FILEX brand file folders account for about 35 percent of this volume.

In the past, PPI consistently refused such dealer-branding requests as a matter of corporate policy. This policy was set some years ago because of a desire (1) to avoid excessive dependence on any one customer and (2) to sell its own brands so that its success is dependent on the quality of its products rather than just a low price. The policy developed from a concern that if it started making products under other customers' brands, those customers could shop around for a low price and the business would be very fickle. At the time the policy was set, Stasia realized that it might cost PPI some business. But it felt wise, nevertheless, to be better able to control the firm's future.

PPI has been in business 28 years and now has a sales volume of $40 million. Its primary products are file folders, file markers and labels, and a variety of indexing systems. PPI offers such a wide range of size, color, and type that no competitor can match it in its part of the market. About 40 percent of PPI's file folder business is in specialized lines such as files for oversized blueprint and engineer drawings; see-through files for medical markets; and greaseproof and waterproof files for marine, oil field, and other hazardous environmental markets. PPI's competitors are mostly small paper converters. But excess capacity in the industry is substantial, and these converters are always hungry for orders and willing to cut prices. Further, the raw materials for the FILEX line of file folders are readily available.

PPI's distribution system consists of 10 regional stationery suppliers (40 percent of total sales), Business Center, Inc. (30 percent), and more than 40 local stationers who have wholesale and retail operations (30 percent). The 10 regional stationers each have about six branches, while the local stationers each have one wholesale and three or four retail locations. The regional suppliers sell directly to large corporations and to some retailers. In contrast, Business Center's main volume comes from sales to local businesses and walk-in customers at its 150 retail stores.

Stasia has a real concern about the future of the local stationers' business. Some are seriously discussing the formation of buying groups to obtain volume discounts from vendors and thus compete more effectively with Business Center's 150 retail stores, the large regionals, and the superstore chains, which are spreading rapidly. These chains—for example, Staples and Office Depot—operate stores of 16,000 to 20,000 square feet (i.e., large stores compared to the usual office supply stores) and let customers wheel through high-stacked shelves to supermarket-like checkout counters. These chains stress convenience, wide selection, and much lower prices than the typical office supply retailers. They buy directly from manufacturers, such as PPI, bypassing wholesalers like Business Center. It is likely that the growing pressure from these chains is causing Business Center to renew its proposal to buy a file line with its own name. For example, Staples offers its own dealer brand of files and many other types of products.

None of Stasia's other accounts is nearly as effective in retailing as Business Center, which has developed a good reputation in every major city in the country. Business Center's profits have been the highest in the industry. Further, its brands are almost as well known as those of some key producers—and its expansion plans are aggressive. And now, these plans are being pressured by the fast-growing superstores, which are already knocking out many local stationers.

Stasia is sure that PPI's brands are well entrenched in the market, despite the fact that most available money has been devoted to new-product development rather than promotion of existing brands. But Stasia is concerned that if Business Center brands its own file folders, it will sell them at a discount and may even bring down the whole market price level. Across all the lines of file folders, Stasia is averaging a 35 percent gross margin, but the commonly used file folders sought by Business Center are averaging only a 20 percent gross margin. And cutting this margin further does not look very attractive to Stasia.

Stasia is not sure whether Business Center will continue to sell PPI's FILEX brand of folders along with Business Center's own file folders if Business Center is able to find a source of supply. Business Center's history has been to sell its own brand and a major brand side by side, especially if the major brand offers high quality and has strong brand recognition.

Stasia is having a really hard time deciding what to do about the existing branding policy. PPI has excess capacity and could easily handle the Business Center business. And she fears that if she turns down this business, Business Center will just go elsewhere and its own brand will cut into PPI's existing sales at Business Center stores. Further, what makes Business Center's offer especially attractive is that PPI's variable manufacturing costs would be quite low in relation to any price charged to Business Center—that is, there are substantial economies of scale, so the extra business could be very profitable—if Stasia doesn't consider the possible impact on the FILEX line. This Business Center business will be easy to get, but it will require a major change in policy, which Stasia will have to sell to Ramon Torres, PPI's president. This may not be easy. Ramon is primarily interested in developing new and better products so the company can avoid the "commodity end of the business."

Evaluate PPI's current strategy. What should Stasia Acosta do about Business Center's offer? Explain.

14. Schrock & Oh Design

Kevin Schrock and Katie Oh met while studying architecture in college. They became good friends and after graduating decided to start a modern architecture firm in Los Angeles. Schrock & Oh Design opened in 2014, specializing in contemporary residential and commercial projects.

Starting an architecture business in the midst of a recession was not easy. The business struggled in those first few years. But they loved what they did, scrimped and saved, and took whatever work they could get. The economy gradually improved. They soon built a modest business based mostly on word-of-mouth; Kevin noted that their only marketing investment was business cards and a small five-page website. They did not have the money to invest in much more. They survived and eventually built a bit of a reputation for their work.

Reputation and word-of-mouth were carrying Schrock & Oh Design only so far. They wanted to move their business to the next level. They started by advertising on Yahoo! and Google. Although this led to some initial inquiries, very few panned out into actual work. People seemed most interested in just asking questions, seeing photographs of their work, and learning more; they were just gathering ideas. These customers were early in the buying process. Kevin and Katie spent lots of time talking to customers but not much time designing and billing out. "It feels like we spend so much time educating these customers that in the end, we hardly make any money even if they do work with us on a design project," noted Kevin.

Their next foray was to start blogging. Katie was a pretty good writer and started a blog on the SchrockOhDesign.com website. She blogged about the industry—"Development Fact of the Day—China," which discussed a magazine article she read about architecture in *Metropolis Magazine,* and "Is Prefab the Future," which described issues around the modern prefabricated home. Prefab homes were one of their competitors. Although low in cost, they are not easily customized to a customer's specific needs. Some of her posts generated traffic to the website, but few turned into leads for the design firm. When they looked at clickstream data, they found that most customers read the blog and then left the company's website.

They were stumped about where to go next. "We tried all the marketing we knew. We started with advertising, which did not seem to work. So we went to this new 'owned media' approach but that isn't getting us any business either," claimed Katie. "It doesn't seem like a Facebook page or Twitter feed is our best route either. Our goal is to attract qualified leads. We would like to attract customers who already know they want to build a contemporary home or commercial building—and then we can use our time with them to discuss what our design firm can do."

Evaluate Schrock & Oh Design's strategy so far. What would you recommend to help the architecture firm attract more qualified leads?

15. The Scioto Group

Theresa Campana owns The Scioto Group (named after a local river)—a manufacturers' rep agency. The Scioto Group sells similar products for noncompeting producers in the technology industry. She is deciding whether to add a new product line—serving another producer. She is very concerned because although she wants more lines, she feels that something is wrong with her latest candidate.

Theresa graduated from a large midwestern university 10 years ago with a BS in business. She worked selling cell phones for a year. Then Theresa decided to go into business for herself and formed The Scioto Group. Looking for opportunities, Theresa placed several ads in her local newspaper in Columbus, Ohio, announcing that she was interested in becoming a sales representative in the area. She was quite pleased to receive a number of responses. Eventually, she became the sales representative in the Columbus area for three local computer software producers: Accto Company, which produces accounting-related software; Saleco, Inc., a producer of sales management software; and Invo, Inc., a producer of inventory control software. All of these companies were relatively small and were represented in other areas by other sales representatives like Theresa. The companies often sent her leads when customers from her area expressed interest at a trade show or through the company's website.

Theresa's main job was to call on possible customers. Once she made a sale, she would fax the signed license agreement to the respective producer, who would then UPS the programs directly to the customer or, more often, provide a key code for a website download. The producer would bill the customer, and Theresa would receive a commission varying from 5 to 10 percent of the dollar value of the sale. Theresa was expected to pay her own expenses. And the producers would handle any user questions, either by using 800 numbers for out-of-town calls or by e-mail queries to a technical support group.

Theresa called on anyone in the Columbus area who might use the products she sold. At first, her job was relatively easy, and sales came quickly because she had little competition. Many national companies offer similar products, but at that time they were not well represented in the Columbus area. Most small businesses needed someone to demonstrate what the software could do.

Five years ago, Theresa sold $250,000 worth of Accto software, earning a 10 percent commission; $100,000 worth of Saleco software, also earning a 10 percent commission; and $200,000 worth of Invo software, earning a 7 percent commission. She was encouraged by her progress and looked forward to expanding sales in the future. She was especially optimistic because she had achieved these sales volumes without overtaxing herself. In fact, she felt she was operating at about 60 percent of her capacity and could easily take on new lines. So she began looking for other products she could sell in the Columbus area.

A local software company has recently approached Theresa about selling its newly developed software, which is basically a network security product. It is designed to secretly track all of the keystrokes and mouse clicks of each employee as he or she uses the computer—so that an employer can identify inappropriate uses of its computers or confidential data. Theresa isn't too enthusiastic about this offer because the commission is only 2 percent on potential annual sales of about $150,000—and she also doesn't like the idea of selling a product that might undermine the privacy of employees who are not doing anything wrong.

Now Theresa is faced with another decision. The owner of the MetalCoat Company, also in Columbus, has made what looks like an attractive offer. She called on MetalCoat to see if the firm might be interested in buying her accounting software. The owner didn't want the software, but he was very impressed with Theresa. After two long discussions, he asked if she would like to help MetalCoat solve its current problem. MetalCoat is having trouble with marketing, and the owner would like Theresa to take over the whole marketing effort.

MetalCoat produces solvents used to make coatings for metal products. It sells mainly to industrial customers in the mid-Ohio area and faces many competitors selling essentially the same products and charging the same low prices.

MetalCoat is a small manufacturer. Last year's sales were $500,000. It could handle at least four times this sales volume with ease and is willing to expand to increase sales—its main objective in the short run. MetalCoat's owner is offering Theresa a 12 percent commission on all sales if she will take charge of its pricing, advertising, and sales efforts. Theresa is flattered by the offer, but she is a little worried because it is a different type of product and she would have to learn a lot about it. The job also might require a great deal more traveling than she is doing now. For one thing, she would have to call on new potential customers in mid-Ohio, and she might have to travel up to 200 miles around Columbus to expand the solvent business. Further, she realizes that she is being asked to do more than just sell. But she did take marketing courses in college and thinks the new opportunity might be challenging.

Evaluate Theresa Campana's current strategy and how the proposed solvent line fits in with what she is doing now. What should she do? Why?

16. Hanratty Company

Joe Hanratty, owner of Hanratty Company, feels his business is threatened by a tough new competitor. And now Joe must decide quickly about an offer that may save his business.

Joe has been a sales rep for lumber mills for about 20 years. He started selling in a clothing store but gave it up after two years to work in a lumberyard because the future looked much better in the building materials industry. After drifting from one job to another, Joe finally settled down and worked his way up to manager of a large wholesale building materials distribution warehouse in Richmond, Virginia. Eight years ago, he formed Hanratty Company and went into business for himself, selling carload lots of lumber to lumberyards in southeastern Virginia.

Joe works with five large lumber mills on the West Coast. They notify him when a carload of lumber is available to be shipped, specifying the grade, condition, and number of each size board in the shipment. Joe isn't the only person selling for these mills—but he is the only one in his area. He isn't required to take any particular number of carloads per month—but once he tells a mill he wants a particular shipment, title passes to him and he has to sell it to someone. Joe's main function is to find a buyer, buy the lumber from the mill as it's being shipped, and have the railroad divert the car to the buyer.

Having been in this business for 20 years, Joe knows all of the lumberyard buyers in his area very well and is on good working terms with them. He does most of his business over the telephone or by e-mail from his small office, but he tries to see each of the buyers about once a month. He has been marking up the lumber between 4 and 6 percent—the standard markup, depending on the grades and mix in each carload—and has been able to make a good living for himself and his family. The going prices are widely publicized in trade publications and are listed on the Internet, so the buyers can easily check to be sure Joe's prices are competitive.

In the last few years, a number of Joe's lumberyard customers have gone out of business—and others have lost sales. The main problem is competition from several national home improvement chains that have moved into Joe's market area. These chains buy lumber in large quantities direct from a mill, and their low prices, available inventory, and one-stop shopping are taking some customers away from the traditional lumberyards. Some customers think the quality of the lumber is not quite as good at the big chains, and some contractors stick with the lumberyards out of loyalty or because they get better service, including rush deliveries when they're needed. Then came the mortgage crisis, and the residential housing market really slowed down, though not as bad in Richmond as in other parts of the country. Fortunately for Joe, the commercial market remained pretty strong and he had good relationships there—or Joe's profits would have taken an even bigger hit.

Six months ago, though, things got even worse. An aggressive young salesman set up in the same business, covering about the same area but representing different lumber mills. This new salesman charges about the same prices as Joe but undersells him once or twice a week in order to get the sale. On several occasions he even set up what was basically an online auction to quickly sell excess wood that was not moving fast enough. Many lumber buyers—feeling the price competition from the big chains and realizing that they are dealing with a homogeneous product—seem to be willing to buy from the lowest-cost source. This has hurt Joe financially and personally—because even some of his old friends are willing to buy from the new competitor if the price is lower. The near-term outlook seems dark, because Joe doubts that there is enough business to support two firms like his, especially if the markup gets shaved any closer. Now they seem to be splitting the shrinking business about equally, as the newcomer keeps shaving his markup.

A week ago, Joe was called on by Amy Balderas of Arbor Door and Window Co., a large manufacturer of windows, raised-panel doors, and accessories. Arbor doesn't sell to the big chains and instead distributes its quality line only through independent lumberyards. Amy knows that Joe is well acquainted with the local lumberyards and wants him to become Arbor's exclusive distributor (sales rep) of residential windows and accessories in his area. Amy gave Joe several brochures on Arbor's product lines. She also explained Arbor's new support program, which will help train and support Joe and interested lumberyards on how to sell the higher-markup accessories. Later, in a lengthy e-mail, Amy explained how this program will help Joe and interested lumberyards differentiate themselves in this very competitive market.

Most residential windows of specified grades are basically "commodities" that are sold on the basis of price and availability, although some premium and very low-end windows are also sold. The national home-improvement chains usually stock and sell only the standard sizes. Most independent lumberyards do not stock windows because there are so many possible sizes. Instead, the lumberyards custom-order from the stock sizes each factory offers. Stock sizes are not set by industry standards; they vary from factory to factory, and some offer more sizes. Most factories can deliver these custom orders in two to six weeks, which is usually adequate to satisfy contractors who buy and install them according to architectural plans. This part of the residential window business is well established, and most lumberyards buy from several different window manufacturers—to ensure sources of supply in case of strikes, plant fires, and so on. How the business is split depends on price and the personality and persuasiveness of the sales reps. And given that prices are usually similar, the sales rep–customer relationship can be quite important.

Arbor gives more choice than just about any other supplier. It offers many variations in 1/8-inch increments—to cater to remodelers who must adjust to many situations. Arbor's online ordering system is state-of-the-art. The lumberyard can connect to the website, enter the specs for a window online, and within seconds get a price quote and estimated delivery time.

One reason Amy has approached Joe is because of Joe's many years in the business. But the other reason is that Arbor is aggressively trying to expand—relying on its made-to-order windows, a full line of accessories, and a newly developed factory support system to help differentiate it from the many other window manufacturers. To give Joe a quick big picture of the opportunity she is offering, Amy explained the window market as follows:

1. For commercial construction, the usual building code ventilation requirements are satisfied with mechanical ventilation. So the windows do not have to operate to permit natural ventilation. They are usually made with heavy-grade aluminum framing. Typically, a distributor furnishes and installs the windows. As part of its service, the distributor provides considerable technical support, including engineered drawings and diagrams to the owners, architects, and/or contractors.
2. For residential construction, on the other hand, windows must be operable to provide ventilation. Residential windows are usually made of wood, frequently with light-gauge aluminum or vinyl on the exterior. The national chains get some volume with standard-sized windows, but lumberyards are the most common source of supply for contractors in Joe's area. These lumberyards do not provide any technical support or engineered drawings. A few residential window manufacturers do have their own sales centers in selected geographic areas, which provide a full range of support and engineering services, but none are anywhere near Joe's area.

Arbor feels a big opportunity exists in the commercial building repair and rehabilitation market (sometimes called the *retrofit market*) for a crossover of residential windows to commercial applications—and it has designed some accessories and a factory support program to help lumberyards get this "commercial" business. For applications such as nursing homes and dormitories (which must meet commercial codes), the wood interior of a residential window is desired, but the owners and architects are accustomed to commercial grades and building systems. And in some older facilities, the windows may have to provide supplemental ventilation for a deficient mechanical system. So what is needed is a combination of the residential *operable* window with a heavy-gauge commercial exterior frame that is easy to specify and install. And this is what Arbor is offering with a combination of its basic windows and easily adjustable accessory frames. Two other residential window manufacturers offer a similar solution, but neither has pushed its products aggressively and neither offers technical support to lumberyards or trains sales reps like Joe to do the necessary job. Amy feels this could be a unique opportunity for Joe.

The sales commission on residential windows would be about 5 percent of sales. Arbor would do the billing and collecting. By getting just 20 to 30 percent of his lumberyards' residential window business, Joe could earn about a third of his current income. But the real upside would come in the long term by increasing his residential window share. Joe is confident that the housing market will turn around soon, and when it does he will be well positioned for growth. To do this, he will have to help the lumberyards get a lot more (and more profitable) business by invading the commercial market with residential windows and the bigger-markup accessories needed for this market. Joe will also earn a 20 percent commission on the accessories, adding to his profit potential.

Joe is somewhat excited about the opportunity because the retrofit market is growing. And owners and architects are seeking ways of reducing costs (which Arbor's approach does—over usual commercial approaches). He also likes the idea of developing a new line to offset the slow-growing market for new-construction housing. But he is concerned that a lot of sales effort will be needed to introduce this new idea. He is not afraid of work, but he is concerned about his financial survival.

Joe thinks he has three choices:

1. Take Amy's offer and sell both window and lumber products.
2. Take the offer and drop lumber sales.
3. Stay strictly with lumber and forget the offer.

Amy is expecting an answer within one week, so Joe has to decide soon.

Evaluate Joe Hanratty's current strategy and how the present offer fits in. What should he do now? Why?

17. Wise Water, Inc.

Morton Rinke established his company, Wise Water, Inc., to market a product designed to purify drinking water. The product, branded as the PURITY II Naturalizer Water Unit, is produced by Environmental Control, Inc., a corporation that focuses primarily on water purification and filtering products for industrial markets.

Wise Water is a small but growing business. Morton started the business with an initial capital investment of only $20,000, which came from his savings and loans from several relatives. Morton manages the company himself. He has a secretary and six full-time salespeople. In addition, he employs two college students part-time; they make telephone calls to prospect for customers and set up appointments for a salesperson to demonstrate the unit in the consumer's home. He has built a small website that primarily provides detailed information and allows customers to request a call from a salesperson. By holding spending to a minimum, Morton has kept the firm's monthly operating budget at only $4,500—and most of that goes for rent, his secretary's salary, and other necessities such as computer supplies and telephone bills.

The PURITY II system uses a reverse osmosis purification process. Reverse osmosis is the most effective technology known for improving drinking water. The device is certified by the Environmental Protection Agency to reduce levels of most foreign substances, including fluoride, mercury, rust, sediment, arsenic, lead, phosphate, bacteria, and most insecticides.

Each PURITY II unit consists of a high-quality 1-micron sediment removal cartridge, a carbon filter, a sediment filter, a housing, a faucet, and mounting hardware. The compact system fits under a kitchen or wet bar sink. A Wise Water salesperson can typically install the PURITY II in about a half hour. Installation involves attaching the unit to the cold water supply line, drilling a hole in the sink, and fastening the special faucet. It works equally well with water from a municipal system or well water, and it can purify up to 15 gallons daily, so most people use it for drinking water. Wise Water sells the PURITY II to consumers for $395, which includes installation.

The system has no movable parts or electrical connections, and it has no internal metal parts that will corrode or rust. However, the system does use a set of filters that must be replaced after about two years. Wise Water sells the replacement filters for $80. Taking into consideration the cost of the filters, the system provides drinking water at a cost of approximately $0.05 per gallon for the average family.

There are two major benefits from using the PURITY II system. First, water treated by this system tastes better. Blind taste tests confirm that most consumers can tell the difference between water treated with the PURITY II and ordinary tap water. Consequently, the unit improves the taste of coffee, tea, frozen juices, ice cubes, mixed drinks, soup, and vegetables cooked in water. Perhaps more important, the PURITY II's ability to remove potentially harmful foreign matter makes the product of special interest to the growing number of people who are concerned about health and the safety of the water they consume. For example, there is growing controversy surrounding public fluoridation of drinking water—and many consumers are looking for filters that remove fluoride.

The number of people with health and safety concerns is growing. In spite of increased efforts to protect the environment and water supplies, there are still many problems. Hundreds of new chemical compounds—ranging from insecticides to industrial chemicals to commercial cleaning agents—are put into use each year. Some of the residue from chemicals and toxic waste eventually enters water supply sources. Further, floods and hurricanes have damaged or completely shut down water treatment facilities in some cities. Problems like these have led to rumors of possible epidemics of such dread diseases as cholera and typhoid—and more than one city has recently experienced near-panic buying of bottled water.

Given these problems and the need for pure water, Morton believes that the market potential for the PURITY II system is very large. Residences, both single-family homes and apartments, are one obvious target. The unit is also suitable for use in boats and recreational vehicles; in fact, the PURITY II is standard equipment on several upscale RVs. And it can be used in taverns and restaurants, in institutions such as schools and hospitals, and in commercial and industrial buildings.

There are several competing ways for customers to solve the problem of getting pure water. Some purchase bottled water. Companies such as Ozarka deliver water monthly for an average price of $0.60 per gallon. The best type of bottled water is distilled water; it is absolutely pure because it is produced by the process of evaporation. However, it may be *too pure.* The distilling process removes needed elements such as calcium and phosphate—and there is some evidence that removing these trace elements contributes to heart disease. In fact, some health-action groups recommend that consumers not drink distilled water.

A second way to obtain pure water is to use some system to treat tap water. PURITY II is one such system. Another system uses an ion exchange process that replaces ions of harmful substances such as iron and mercury with ions that are not harmful. Ion exchange is somewhat less expensive than the PURITY II process, but it is not well suited for residential use because bacteria can build up before the water is used. In addition, there are a number of other filtering and softening systems. In general, these are less expensive and less reliable than the PURITY II. For example, water softeners remove minerals but do not remove bacteria or germs.

Morton's first year with his young company has gone quite well. Customers who have purchased the system like it, and there appear to be several ways to expand the business and increase profits. For example, so far he has had little time to make sales calls on potential commercial and institutional users or residential builders. He also sees other possibilities such as expanding his promotion effort or targeting consumers in a broader geographic area.

At present, Wise Water distributes the PURITY II in the 13-county Gulf Coast region of Texas. Because of the Robinson-Patman Act, the manufacturer cannot grant an exclusive distributorship. However, Wise Water is currently the only PURITY II distributor in this region. In addition, Wise Water has the right of first refusal to set up distributorships in other areas of Texas. The manufacturer has indicated that it might even give Wise Water distribution rights in a large section of northern Mexico.

The agreement with the manufacturer allows Wise Water to distribute the product to retailers, including hardware stores and plumbing supply dealers. Morton has not yet pursued this channel, but a PURITY II distributor in Florida reported some limited success selling the system to retailers at a wholesale price of $275. Retailers for this type of product typically expect a markup of about 33 percent of their selling price.

Environmental Control, Inc. ships the PURITY II units directly from its warehouse to the Wise Water office via UPS. The manufacturer's $200 per unit selling price includes the cost of shipping. Wise Water needs to keep only a few units on hand because the manufacturer accepts faxed orders and then ships immediately—so delivery never takes more than a few days. Further, the units are small enough to inventory in the back room of the Wise Water sales office. Several of the easy-to-handle units will fit in the trunk of a salesperson's car.

Morton is thinking about recruiting additional salespeople. Finding capable people has not been a problem so far. However, there has already been some turnover, and one of the current salespeople is complaining that the compensation is not high enough. Morton pays salespeople on a straight commission basis. A salesperson who develops his or her own prospects gets $100 per sale; the commission is $80 per unit on sales leads generated by the company's telemarketing people. For most salespeople, the mix of sales is about half and half. Wise Water pays the students who make the telephone contacts $4 per appointment set up and $10 per unit sold from an appointment. A growing number of leads are coming from the company's website, largely due to search ads placed using Google Ads and on Facebook.

An average Wise Water salesperson easily sells 30 units per month. However, Morton believes that a really effective and well-prepared salesperson can sell much more, perhaps 50 units per month.

Wise Water and its salespeople get good promotion support from Environmental Control, Inc. For example, Environmental Control supplies sales training manuals and sales presentation flip charts. The materials are also well done, in part because Environmental Control's promotion manager previously worked for Electrolux vacuum cleaners, which are sold in a similar way. The company also supplies print copy for magazine and newspaper advertising and tapes of commercials for radio and television. Thus, all Wise Water has to do is buy media space or time. In addition, Environmental Control furnishes each salesperson with a portable demonstration unit, and the company recently gave Wise Water three units to be placed in models of condominium apartments.

Morton has worked long hours to get his company going, but he realizes that he has to find time to think about how his strategy is working and to plan for the future.

Evaluate Morton Rinke's current marketing strategy for Wise Water. How do you think he's doing so far, and what should he do next? Why?

18. West Tarrytown Volunteer Fire Department (WTVFD)

Cheyne Morgan raced out the front door of the Target store where he worked as soon as his beeper sounded. In his pickup truck, he heard the call on his special radio scanner: "Highway 18 Fire Department, there is a grass fire at the old McCullough place. That's a mile down the old dirt road just past the Wilson house." The directions might appear cryptic to someone who had not grown up around West Tarrytown, but it was all Cheyne needed to know. Upon arriving at the fire, Cheyne quickly pulled on his fire-retardant bunker pants and boots. He left his Nomex hood, helmet, and fire pants in the back of his truck—he would not need them for this fire.

Less than 10 minutes from the time the call was placed, Cheyne and 20 other members of the West Tarrytown Volunteer Fire Department (WTVFD) had arrived at the old McCullough place. They were able to put out the fire in less than half an hour, but not before a football-field-size patch of grass was scorched. Their quick response saved the neighbor's barn and kept the fire from spreading to a nearby forest. A third straight year of drought has the crew on high alert.

Cheyne threw his gear in the back of his truck and headed back to finish his shift at Target. He had worked there as a department manager for two years, ever since he graduated from the local state college with a marketing degree. As he drove, Cheyne thought about what WTVFD chief Fran Holland recently asked him to do. Over the last few years, the fire department had more to do but fewer people to do it with. So Chief Holland asked Cheyne to draw up a marketing plan to recruit new volunteers.

Cheyne had already started to gather information for the marketing plan. From an online search he found that WTVFD was one of an estimated 30,000 volunteer fire departments in the United States and that these departments had almost a million volunteers. Cheyne was surprised that more than 75 percent of all U.S. firefighters were volunteers. West Tarrytown, a small city of just over 100,000 and hours from a big city, had only volunteer firefighters.

There are 48 firefighters currently in the WTVFD—down from 55 five years ago. Although there was a surge of interest in the year following the terrorist attacks in 2001, only a few of those volunteers remain. Over time, WTVFD has found that about half of new recruits quit before their three-year anniversary. Those who remain usually stay with the department until they can't keep up with the job's physical demands. Fran Holland has been chief for the last three years. She replaced longtime chief Ken Reeb who retired after being with WTVFD for more than 40 years—the last 10 as chief.

The current volunteers include 44 men and 4 women; more than half of the force is older than 40 years old, and Cheyne is 1 of only 5 volunteers younger than 30. Almost all of them started volunteering while still in their 20s or early 30s. The crew represents all walks of life, and their ranks include a lawyer, a real estate agent, a college professor, a carpenter, a stay-at-home mom, and a few guys from the local factory. Many entered firefighting for the thrill of it or because they hoped the experience might help them land a paid firefighting job in a bigger city. But most of the crew stay with it because they feel good about giving back to their community, view it as a hobby, and enjoy the camaraderie with the other firefighters.

Being a volunteer firefighter is much different than Cheyne thought it would be when he started. Last year he counted 238 hours volunteering for WTVFD, but less than a third of that was actually responding to emergencies. And fewer than half of the emergencies were actual fires. He spent about a quarter of this time in required training and drills. He had to be trained for the many different possible calls, including car accidents, hazardous chemical spills, and terrorist attacks. Another 20 percent of his time was spent in meetings and a similar amount helping with fundraising. Depending on financial needs, WTVFD holds at least four fundraising events a year—some years six or eight. These include annual activities such as a chili cook-off, pancake breakfast, and booth at the county fair—and as-needed events such as pie auctions, turkey shoots, and basketball tournaments.

The biggest requirement to be a volunteer firefighter is the willingness to make the time commitment. Cheyne's time commitment and allocation of hours is typical of all the firefighters at WTVFD. Volunteers have to be able to attend at least 80 percent of the twice-monthly drills—scheduled on the second and fourth Tuesdays of every month. Firefighters also have to live or work near the town of West Tarrytown so they can respond quickly to emergencies. They also have to be at least 18 years old and have a valid driver's license. There is a physical ability test to make sure that firefighters can stand the rigors of the job. Although one doesn't have to be a weightlifter, the job requires volunteers to be in good physical shape.

WTVFD has never really had a marketing strategy or any formal promotion efforts. Most of Cheyne's fellow volunteers heard about WTVFD through word-of-mouth. People are always curious when a volunteer firefighter runs out the door from work or suddenly leaves a party. These occasions give volunteers a chance to tell others about what they do. Sometimes those questions bring someone out to see drills and to apply to become a volunteer. One of Cheyne's high school friends was a volunteer and he encouraged Cheyne to join up while Cheyne was in his junior year of college. But still, awareness of volunteer firefighting in the West Tarrytown area is very low. When Cheyne tells friends about his volunteer work, most are surprised and think the town has full-time paid staff fighting fires.

Cheyne thinks his marketing strategy should focus mostly on promotion. From his studies, he remembers the AIDA model and integrated marketing communications. WTVFD does not really have an advertising budget and right now has only a one-page informational website. He is pretty sure he could get some money to build a more robust website—especially because one of the recent recruits would do the work for free. Cheyne also wonders if his plan should focus only on gaining new recruits or if current volunteers might also be the target of new promotion. Cheyne also considers how and what he should communicate to his target market. He knows he will have limited funding for his efforts. Should he consider starting a Facebook page or blog? Maybe a website would be better—but what kind of information would be on the website? What about advertising or information sessions? Who should be his target market?

What are the promotion objectives Cheyne Morgan should include in his plan? What promotion methods should he use to achieve those objectives?

19. MyOwnWedding.com

Sterling Bassett is happy with her life but disappointed that the idea she had for starting her own business hasn't taken off as expected. Within a few weeks, she either has to renew the contract for her Internet website or decide not to put any more time and money into her idea. She knows that it doesn't make sense to renew the contract if she can't come up with a plan to make her website-based business profitable—and she doesn't like to plan. She's a "doer," not a planner.

Sterling's business, MyOwnWedding.com, started as an idea 18 months ago as she was planning her own wedding. She attended a bridal fair at the convention center in Raleigh, North Carolina, to get ideas for a wedding dress, check out catering companies and florists, and in general learn more about the various services available to newlyweds. While there, she and her fiancé went from one retailer's booth to another to sign up for their wedding gift registries. Almost every major retailer in the city—ranging from the Home Depot warehouse to the Belk department store to the specialty shops that handle imported crystal glassware—offered a gift registry. Some had computers set up to provide access to their online registries. Being listed in all of the registries improved the odds that her wedding gifts would be items she wanted and could use—and it saved time and hassle for gift givers. On the way back from the fair, Sterling and her fiancé discussed the idea that it would be a lot easier to register gift preferences once on a central Internet site than to provide lots of different stores with bits and pieces of information. A list at a website would also make it easier for gift givers, at least those who were computer users.

When Sterling got home, she did an Internet search and found several sites that focused on weddings. The biggest seemed to be www.theknot.com. It had features for couples who were getting married, including a national gift registry. The site featured products from a number of companies, especially large national retail chains; however, there was a search feature to locate people who provide wedding-related services in a local zip code area. Sterling thought that the sites she found looked quite good, but they were not as helpful as a site could be with a more local focus.

The more Sterling and her fiancé discussed the idea of a website offering local wedding-related services, the more it looked like an interesting opportunity. Except for the annual bridal fair, there was no other obvious local place for consumers to get information about planning a wedding and buying wedding-related services. And for local retailers—florists, catering companies, insurance agents, home builders, and many other types of firms—there was no other central place to target promotion to newlyweds. Further, the amount of money spent on weddings and wedding gifts is substantial, and right before and after getting married many young couples make important purchase decisions for everything from life insurance to pots and pans. Spending on the wedding alone can easily exceed the cost of a year of college.

Sterling was no stranger to the Internet. She worked as a website designer for a small firm whose only client was IBM. That IBM was the only client was intentional rather than accidental. A year earlier IBM had decided that it wanted to outsource certain aspects of its website development work and have it handled by an outside contractor. After negotiating a three-year contract to do IBM's work, several IBM employees quit their jobs and started the business. IBM was a good client, and all indications were that IBM could give the firm as much work as it could handle as it hired new people and prospected for additional accounts over the next few years. Sterling especially liked the creative aspects of designing the "look" of a website, and technical specialists handled a lot of the subtle details.

Before joining this new company, Sterling had several marketing-related jobs—but none had been the glamorous ad agency job she dreamed of in college as an advertising major. Her first job as a college graduate was with an ad agency, but she was in a back room operation handling a lot of the arrangements for printing and mailing large-scale direct-mail promotions. In spite of promises that it was a path to other jobs at the agency, the pay was bad, the work was always pressured, and every aspect of what she had to do was boring. After six punishing months, she quit and went looking for something else.

When a number of job applications didn't turn up something quickly, she took a part-time job doing telemarketing calls for a mortgage refinance company. Sterling's boss told her that she was doing a great job reeling in prospects—but she hated disturbing people at night and just didn't like making sales pitches. Fortunately for her, that pain didn't last long. A neighbor in Sterling's apartment complex got Sterling an interview for a receptionist position at an ad agency. That, at least, got her foot in the door. Her job description wasn't very interesting, but in a small agency she had the opportunity to learn a lot about all aspects of the business—ranging from working on client proposals and media plans to creative sessions for new campaigns. In fact, it was from a technician at that agency that she learned to work with the graphics software used to create ad layouts and website pages. When the website design job came open at the new firm, her boss gave her a glowing recommendation, and in two days she was off on her new career.

Although Sterling's jobs had not been high-profile positions, they did give her some experience in sales promotion, personal selling, and advertising. Those skills were complemented by the technical computer skills of her fiancé (now husband), who made a living as a database programmer for a large software consulting firm. Taking everything as a whole, they thought that they could get a wedding-related website up and running and make it profitable.

There were several different facets to the original plan for MyOwnWedding.com. One focused on recruiting local advertisers and "sponsors" who would pay to be listed at the website and be allocated a web page (which Sterling would design) describing their services, giving contact information, and linking to their own websites. Another focused on services for people who were planning to be married. In addition to an online wedding gift registry, sections of the website provided information about typical wedding costs, planning checklists, details about how to get a required marriage license, and other helpful information (including a discussion forum with comments about the strengths and weaknesses of various local suppliers). Sterling also started a blog that helped foster more feedback from customers. A couple could sign up for the service online and could pay the modest $20 "membership" fee for a year by credit card. Friends, family, and invited guests could visit the website at no charge and get information about wedding preferences, local hotels, discounts on local car rentals, and even printable maps to all of the churches and synagogues in the area.

When Sterling told friends about her plan, they all thought it sounded like a great idea. In fact, each time she discussed it, someone came up with another idea for a locally oriented feature to add to the website. Several friends said that they had tried national websites but that the information was often too general. But generating more new ideas was not the problem. The problem was generating revenue. Sterling had already contracted for space from an Internet service provider and created some of the initial content for the website, but she only had four paying sponsors, two of whom happened to be family friends.

Sterling started by creating a colorful flyer describing the website and sent it to most of the firms that had participated in the bridal fair. When no one sent back the reply coupon for more information, Sterling started to make calls (mainly during her lunch hour at her full-time job). Some stores seemed intrigued by the concept, but no one seemed ready to sign up. One reason was that they all seemed surprised at the cost to participate and get ad space at the website—$2,400 a year (about the same as a 1/16-page display ad in the Raleigh *Yellow Pages*). Another problem was that no one wanted to be the first to sign up. As one florist shop owner put it, "If you pull this off and other florists sign up, then come back and I will too."

Getting couples to sign up went slowly too. At first, Sterling paid for four display ads in local Sunday newspapers in the society section, sent information sheets about the website to clergy in the area, and sent carefully crafted press releases announcing the service to almost every publication in the area. She read a bit about search engine optimization and tried to design her site to be more easily found when potential customers were searching. One article that resulted from a press release got some attention, and for a few weeks there was a flurry of e-mail inquiries about her web page. But after that it slowed to a trickle again. More recently, she tried to use Google Ads, which places ads next to Google search results when someone from the greater Raleigh area searches on the keyword *wedding*. She got a few more hits from this and wondered if she should increase the number of keywords—and wondered what the best keywords could be. She thought about creating a Facebook page but wondered how it would help. She knows that many brides are on Pinterest and wasn't sure if she should start her own page there.

Sterling's diagnosis of the problem was simple. Most people thought it was a great idea, but few couples knew where to look on the Internet for such a service. Similarly, potential advertisers—many of them small local businesses—were not accustomed to the idea of paying for Internet advertising. They didn't know if the cost was reasonable or if her site would be effective in generating business.

Sterling's life as a married person was going great, and her job as a web page designer kept her very busy. Her free time outside of work was always in short supply because the young crowd at her office always had some scheme for how to keep entertained. So she wasn't about to quit her job to devote full-time to her business idea. Further, she thought that once it got rolling she would only have to devote 10 hours a week to it to earn an extra $30,000 a year. She didn't have delusions of becoming a "dot-com millionaire." She just wanted a good locally oriented business.

However, it still wasn't clear how to get it rolling. After a year of trying on and off, she had only four paying ad sponsors, and one of them had already notified her that he didn't plan to sign up again because it wasn't clear that the website had generated any direct leads or sales. Further, it looked like anything she could do to attract more "members" would end up being expensive and inefficient.

Her parents offered her $10,000 to keep the idea alive. Sterling believes that proper promotion is the key. Now the question is how to budget this latest investment. She was thinking about doing some search advertising—that can be narrowly targeted to local people searching under "wedding," "wedding planning," and other keywords. She was also thinking about building up her owned media content on the site. While Sterling could create the pages at no cost, she felt she would need help creating content. She wondered about maybe going with some user-generated content. Social media were another possibility—Twitter? Facebook? So many choices, and time was Sterling's scarcest resource (oh yeah, money is scarce too).

Sterling thinks the idea has real potential, and she's willing to do the work. But she's not certain if she can make it pay off. She doesn't want to lose her parents' money.

What is Sterling's strategy? What should she do? If she were to move forward, what strategy would you recommend? Does her financial goal seem realistic? Why?

20. Lake Russell Marine & Camp

Rusty Steuben, owner of Lake Russell Marine & Camp, is worried about his business' future. He has tried various strategies for two years, and he's barely breaking even.

Two years ago, Rusty bought the inventory, supplies, equipment, and business of Lake Russell Marine & Camp, located on the edge of Minneapolis, Minnesota. The business is in an older building along a major highway leading out of town, several miles from any body of water. The previous owner had sales of about $500,000 a year but was just breaking even. For this reason—plus the desire to retire to Arizona—the owner sold to Rusty for roughly the value of the inventory.

Lake Russell Marine & Camp had been selling two well-known brands of small pleasure boats, a leading outboard motor, two brands of snowmobiles and jet skis, and a line of trailer and pickup-truck campers. The total inventory was valued at $250,000—and Rusty used all of his own savings and borrowed some from two friends to buy the inventory and the business. At the same time, he took over the lease on the building—so he was able to begin operations immediately. Rusty had never operated a business of his own before, but he was sure that he would be able to do well. He had worked in a variety of jobs—as a used-car salesman, an auto repairman, and a jack-of-all-trades in the maintenance departments of several local businesses.

Soon after starting his business, Rusty hired his friend, Ginny Wooten. She had worked with Rusty selling cars and had experience as a receptionist and in customer service. Together, they handle all selling and setup work on new sales and do maintenance work as needed. Sometimes the two are extremely busy—at the peaks of each sport season. Then both sales and maintenance keep them going up to 16 hours a day. At these times it's difficult to have both new and repaired equipment available as soon as customers want it. At other times, however, Rusty and Ginny have almost nothing to do.

Rusty usually charges the prices suggested by the various manufacturers, except at the end of a weather season when he is willing to make deals to clear the inventory. He is annoyed that some of his competitors sell mainly on a price basis—offering 10 to 30 percent off a manufacturer's suggested list prices—even at the beginning of a season! Rusty doesn't want to get into that kind of business, however. He hopes to build a loyal following based on friendship and personal service. Further, he doesn't think he really has to cut prices because all of his lines are exclusive to his store. No stores within a five-mile radius carry any of his brands, although nearby retailers offer many brands of similar products. Right now, the Internet does not provide much competition, but he fears future price competition from online boat shows.

To try to build a favorable image for his company, Rusty occasionally places ads in local papers and buys some radio spots. He knows he should probably do some online advertising, but he has no idea how to do it. The basic theme of this advertising is that Lake Russell Marine & Camp is a friendly, service-oriented place to buy the equipment needed for the current season. Sometimes he mentions the brand names he carries, but generally Rusty tries to build an image for concerned, friendly service—both in new sales and repairs—stressing, "We do it right the first time." He chose this approach because, although he has exclusives on the brands he carries, there generally are 10 to 15 different manufacturers' products being sold in the area in each product category—and most of the products are quite similar. Rusty feels that this similarity among competing products almost forces him to try to differentiate himself on the basis of his own store's services.

The first year's operation wasn't profitable. In fact, after paying minimal salaries to Ginny and himself, the business just about broke even. Rusty made no return on his $250,000 investment.

In hopes of improving profitability, Rusty jumped at a chance to add a line of lawn mowers, tractors, and trimmers as he was starting his second year of business. This line was offered by a well-known equipment manufacturer that wanted to expand into the Minneapolis area. The equipment is similar to that offered by other lawn equipment manufacturers. The manufacturer's willingness to do some local advertising and to provide some point-of-purchase displays appealed to Rusty. And he also liked the idea that customers probably would want this equipment sometime earlier than boats and other summer items. So he thought he could handle this business without interfering with his other peak selling seasons.

It's been two years since Rusty bought Lake Russell Marine & Camp—and he's still only breaking even. Sales have increased a little, but costs have gone up too because he had to hire some part-time help. The lawn equipment helped to expand sales—as he had expected—but unfortunately it did not increase profits as he had hoped. Rusty needed part-time helpers to handle this business—in part because the manufacturer's advertising had generated a lot of sales inquiries. Relatively few inquiries resulted in sales, however, because many people seemed to be shopping for deals. So Rusty may have even lost money handling the new line. But he hesitates to give it up because he doesn't want to lose that sales volume, and the manufacturer's sales rep has been most encouraging, assuring Rusty that things will get better and that his company will be glad to continue its promotion support during the coming year.

Rusty is now considering the offer of a mountain bike producer that has not been represented in the area. The bikes have become very popular with students and serious bikers in the last several years. The manufacturer's sales rep says industry sales are still growing (but not as fast as in the past) and probably will grow for many more years. The sales rep has praised Rusty's service orientation and says this could help him sell lots of bikes because many mountain bikers are serious about buying a quality bike and then keeping it serviced. He says Rusty's business approach would be a natural fit with bike customers' needs and attitudes. As a special inducement to get Rusty to take on the line, the sales rep says Rusty will not have to pay for the initial inventory of bikes, accessories, and repair parts for 90 days. And, of course, the company will supply the usual promotion aids and a special advertising allowance of $10,000 to help introduce the line to Minneapolis.

Rusty likes the idea of carrying mountain bikes because he has one himself and knows that they do require some service year-round. But he also knows that the proposed bikes are very similar in price and quality to the ones now being offered by the bike shops in town. These bike shops are service- rather than price-oriented, and Rusty feels that they are doing a good job on service; consequently, he is concerned with how he can be "different."

Evaluate Rusty Steuben's overall strategy(ies) and the mountain bike proposal. What should he do now?

21. GeoTron International (GTI)

GeoTron International (GTI) is a multinational producer of various chemicals and plastics with plants in the United States, England, France, and Germany. Its headquarters are in New Jersey.

Kenneth Shibata is marketing manager of GTI's plastics business. Kenneth is reconsidering his promotion approach. He is evaluating what kind of promotion—and how much—should be directed to car producers and to other major plastics customers worldwide. Currently, Kenneth has one salesperson dedicated to the automobile industry. This man is based in the Detroit area and focuses on GM, Ford, and Chrysler—as well as the various firms that mold plastics to produce parts to supply the car industry. This approach worked well when relatively little plastic was used in each car and the auto producers did all of the designing themselves and then sent out specifications for very price-oriented competitive bidding. But now the whole product planning and buying system is changing—and of course foreign producers with facilities in the United States are much more threatening as competitors.

How the present system works can be illustrated in terms of the team approach Ford used on its project to design the Flex, the full-sized crossover introduced as a 2009 model. For the Flex, representatives from all the various functions—planning, design, engineering, purchasing, marketing, and manufacturing—worked together. In fact, representatives from key suppliers were involved from the outset. The whole team takes final responsibility for a car. Because all of the departments are involved from the start, problems are resolved as the project moves on—before they cause a crisis. Manufacturing, for example, can suggest changes in design that will result in higher productivity or better quality, which is especially important at a time when Ford's initial quality ratings are beating those of Honda and Toyota.

The old approach was different. It involved a five-year process of creating a new vehicle in sequential steps. Under the old system, product planners would come up with a general concept and then expect the design team to give it artistic form. Next, engineering would develop the specifications and pass them on to manufacturing and suppliers. There was little communication between the groups and no overall project responsibility.

In the Flex project, Ford engineers followed the Japanese lead and did some reverse engineering of their own. They dismantled several competitors' cars, piece by piece, looking for ideas they could copy or improve. This helped them learn how the parts were assembled and how they were designed. Eventually, Ford incorporated or modified some of the best features into its design of the Flex. For example, the Flex uses a new design to seal the doors and eliminate wind noise.

In addition to reverse engineering, Ford researchers conducted a series of market studies. This led to positioning the Flex as an "anti-minivan." That positioning resulted in a decision to eliminate the sliding side doors and instead to use traditional hinged doors. That cut costs, but the savings were used for dress-up features, such as 19-inch aluminum wheels and a special new seat design that reduces movement in the seat and gives the car an even smoother drive. The Flex's optional refrigerator/freezer is another example of a feature that did well in concept tests.

Ford also asked assembly-line workers for suggestions before the car was redesigned and then incorporated their ideas into the new car. Most bolts have the same-sized head, for example, so workers don't have to switch from one wrench to another.

Finally, Ford included its best suppliers as part of the planning effort. Instead of turning to a supplier after the car's design was completed, Ford invited them to participate in product planning. For example, Microsoft's Sync system provides the Flex with voice control of the entertainment system.

Most other vehicles are now developed with an approach similar to this. GM, for example, used a very similar team approach to redesign its new Malibu. And major firms in many other industries are using similar approaches. A major outgrowth of this effort has been a trend by these producers to develop closer working relationships with a smaller number of suppliers. To some extent, this is a direct outgrowth of the decision to try to reduce unnecessary costs by using the same components for different vehicles. For example, the powertrain for the Flex is the same as the one used in Ford's Edge.

Many of the suppliers selected for the Flex project had not only the facilities, but also the technical and professional managerial staff who could understand—and become part of—the program management approach. Ford expected these major suppliers to join in its total quality management push and to be able to provide just-in-time delivery systems. Ford dropped suppliers whose primary sales technique was to entertain buyers and then submit bids on standard specifications.

Because many firms have moved to these team-oriented approaches and developed closer working relationships with a subset of their previous suppliers, Kenneth is trying to determine if GTI's present effort is still appropriate. Kenneth's strategy has focused primarily on responding to inquiries and bringing in GTI's technical people as the situation seems to require. Potential customers with technical questions are sometimes referred to other noncompeting customers already using the materials or to a GTI plant—to be sure that all questions are answered. But basically, all producer-customers are treated more or less alike. The sales reps make calls and try to find good business wherever they can.

Each GTI sales rep usually has a geographic area. If an area like Detroit needs more than one rep, each may specialize in one or several similar industries. But GTI uses the same basic approach—call on present users of plastic products and try to find opportunities for getting a share (or bigger share) of existing purchases or new applications. The sales reps are supposed to be primarily order getters rather than technical specialists. Technical help can be brought in when the customer wants it, or sometimes the sales rep simply sets up a conference call between GTI's technical experts, the buyer, and the users at the buyer's facility.

Kenneth sees that some of his major competitors are becoming more aggressive. They are seeking to affect specifications and product design from the start rather than after a product design is completed. This takes a lot more effort and resources, but Kenneth thinks it may get better results. A major problem he sees, however, is that he may have to drastically change the nature of GTI's promotion. Instead of focusing primarily on buyers and responding to questions, it may be necessary to try to contact *all* the multiple buying influences and not only answer their questions but help them understand what questions to raise—to find solutions. Such a process may even require more technically trained sales reps. In fact, it may require that people from GTI's other departments—engineering, design, manufacturing, R&D, and distribution—get actively involved in discussions with their counterparts in customer firms. Further, use of e-mail and a website might make ongoing contacts faster and easier.

While Kenneth doesn't want to miss the boat if changes are needed, he also doesn't want to go off the deep end. After all, many of the firm's customers don't seem to want GTI to do anything very different from what it's been doing. In fact, some say that they're very satisfied with their current supply arrangements and really have no interest in investing in a close relationship with a single supplier. Even with the Flex project, Ford wasn't 100 percent dedicated to the team approach. For example, when Ford's research showed that the target market viewed quiet and more comfortable seats as an especially important factor in purchases, Ford didn't turn to a supplier for help but rather assigned a team of its own design engineers to develop and test them in-house. Now some of what was learned on the Flex project is going to be used in redesigning other models.

Contrast Ford's previous approach of designing and producing cars to Flex's program management approach, especially as it might affect suppliers' promotion efforts. Given that many other major producers have moved in the program management direction, what promotion effort should Kenneth Shibata develop for GeoTron International? Should every producer in every geographic area be treated alike, regardless of size? Explain.

22. Bright Light Innovations: The Starlight Stove*

The top management team of Bright Light Innovations is preparing to meet and review its market situation. The team is a combination of students and faculty from Colorado State University's (CSU) Colleges of Business and Engineering: Dr. Bryan Wilson, Paul Hudnut, Ajay Jha, Sachin Joshi, Katie Lucchesi, Dan Mastbergen, Ryan Palmer, and Chaun Sims. They are excited about the Starlight Stove product they have developed—and passionate about the opportunity that it provides to improve the quality of life for some of the world's poorest people. They know they have a great technology, but they need a marketing plan to bring this product to market.

Every day, more than 2.4 billion people—more than one-third of the world's population—burn solid biomass fuel (wood, charcoal, dung, and coal) for cooking and heating. These fuel sources are usually burned indoors in open pits or traditional cook stoves. About two-thirds of the people using biomass fuel also have no electricity, so the open fires often burn into the night to provide light. These fires create indoor air pollution, which is a leading contributor to respiratory diseases. UN Secretary General Kofi Annan has called for greater energy efficiency and noted that "indoor air pollution has become one of the top 10 causes of mortality and premature death." It is estimated that this source of pollution contributes each year to the deaths of 1 million children younger than the age of 5, and it is a leading cause of miscarriage and women's health problems.

Hoping to address these consumers' needs for safe cooking and electricity, CSU's Engines and Energy Conversion Laboratory developed the Starlight Stove. The Starlight Stove's improved technology requires 50 to 70 percent less biomass fuel than traditional stoves. It also has a thermoelectric generator that converts heat from the stove into electricity that can power a small light bulb or be stored in a rechargeable battery for later use. The technology has been refined, and the team believes it is ready to go to market.

There are other competing enclosed cook stoves, but none produce electricity. Solar panels can provide electricity, but they are expensive—costing $360 each. Micro-hydropower allows households to convert the power from streams and rivers into electricity, but homes must be close to a river, and water flow in many areas of the country is seasonal. These technologies—solar panels and hydropower—are understood by many consumers and are already in use in some areas. The Starlight Stove, on the other hand, offers a new technology, and that may slow its adoption.

The management team decided on Nepal as the initial target market for the Starlight Stove. Several factors made this market particularly attractive. The climate is relatively cold, and only 11 percent of the households have access to electricity, so the heat and electricity production of the stove are particularly beneficial. Eighty-eight percent of the population uses firewood as their main source of energy. In addition, deforestation creates environmental problems in Nepal because it contributes to erosion and flooding. So the social benefits of the Starlight Stove will be particularly appealing to the Nepalese government and aid organizations.

There are approximately 9.2 million households in Nepal, but the gross national income per capita is only about $400, with most adults making between $1 and $3 per day. Nepal is largely rural, with only 17 percent of the population living in urban areas. The country is divided into 75 districts. Each district is further divided into about 60 village development committees (a sort of local government) consisting of about 450 households. The similar characteristics of northern India—immediately south of Nepal—make it a logical follow-up market.

The Starlight Stove offers several benefits to this population. For example, the longer hours with light—thanks to the electricity—and less time required to collect wood or other fuel could allow families to earn money by weaving, farming, or producing other crafts. Family productivity could increase 20 percent or more per day. Or the added hours with light might allow children to gain an education. If the product were manufactured locally, it could provide jobs for the population and help them learn the benefits of technology.

With obvious benefits for such a large number of people, the Bright Light Innovations team could look to donations to subsidize the Starlight Stove for the Nepalese people. But the team has concerns about this traditional form of aid. Financing in the form of grants, government relief, or donations is

---

*This case is based on a business plan written by Ajay Jha, Sachin Joshi, Katie Lucchesi, Dan Mastbergen, Ryan Palmer, and Chaun Sims.

unreliable. If it is not renewed, projects wallow or die. Further, grants often fail to teach disadvantaged people skills and responsibility. So the team wants to create a sustainable venture that provides benefits for all—and has set up Bright Light Innovations as a for-profit business.

The management team has to make a number of marketing decisions. For example, it has to decide how to price the Starlight Stove. It estimates that the stove will cost about $60 to manufacture after setting up a plant in Nepal and expects that microfinancing organizations will provide loans for families. If units are sold for $80, the loan can be financed at 20 percent interest for three years with payments of $0.68 per week (microfinancing institutions typically collect on a weekly, or sometimes daily, basis). The team thinks that it will be easy to find a microfinancing institution to provide these loans. But the team is still unsure about whether this price will provide adequate margins for distributors.

The team also has to decide how to promote the stove to a population where less than half of adults can read. However, the team does have contacts with some business leaders, government officials, and nongovernmental organizations that may be able to provide advice and help.

What should be the marketing strategy of the Bright Light Innovations team for the Starlight Stove? Why?

23. Wire Solutions

Wendy Keating, marketing manager of consumer products for Wire Solutions, is trying to set a price for her most promising new product: a space-saving shoe rack suitable for small homes or apartments.

Wire Solutions—located in Fort Worth, Texas—is a custom producer of industrial wire products. The company has a lot of experience bending wire into many shapes and also can chrome- or gold-plate finished products. The company was started 16 years ago and has slowly built its sales volume to $3.6 million a year. Just one year ago, Keating was appointed marketing manager of the consumer products division. It is her responsibility to develop this division as a producer and marketer of the company's own branded products—as distinguished from custom orders, which the industrial division produces for others.

Keating has been working on a number of different product ideas for almost a year now and has developed several designs for DVD holders, racks for soft drink cans, plate holders, doll stands, collapsible bookends, laptop stands, and other such products. Her most promising product is a shoe rack for crowded homes and apartments. The wire rack attaches to the inside of a closet door and holds eight pairs of shoes.

The shoe rack is very similar to one the industrial division produced for a number of years for another company. That company sold the shoe rack and hundreds of other related items out of its "products for organizing and storing" mail-order catalog. Managers at Wire Solutions were surprised by the high sales volume the catalog company achieved with the rack. In fact, that is what interested Wire Solutions in the consumer market and led to the development of the separate consumer products division.

Keating has sold hundreds of the shoe racks to various local hardware, grocery, and general merchandise stores and wholesalers on a trial basis, but each time she has negotiated a price—and no firm policy has been set. Now she must determine what price to set on the shoe rack, which she plans to push aggressively wherever she can. Actually, she hasn't decided on exactly which channels of distribution to use. But trials in the local area have been encouraging, and as noted earlier, the experience in the industrial division suggests that there is a large market for this type of product. Further, she has noticed that a Walmart store in her local area is selling a similar rack made of plastic. When she talked casually about her product with the store manager, he suggested that she contact the chain's houseware buyers in the home office in Arkansas.

The manufacturing cost of her rack—when made in reasonable quantities—is approximately $2.80 if it is painted black and $3.60 if it is chromed. Similar products have been selling at retail in the $9.95 to $19.95 range. The sales and administrative overhead to be charged to the division will amount to $95,000 a year. This will include Keating's salary and some travel and office expenses. She expects that a number of other products will be developed in the near future. But for the coming year, she hopes the shoe rack will account for about half the consumer products division's sales volume.

Evaluate Wendy Keating's strategy planning so far. What should she do now? What price should she set for the shoe rack?

24. Fresh Harvest

Chelsea Skye-Rice, president of Fresh Harvest, is not sure what she should propose to the board of directors. Her recent strategy change isn't working. And Don Bartley, Fresh Harvest's only sales rep (and also a board member), is so frustrated that he refuses to continue his discouraging sales efforts. Don wants Chelsea to hire a sales force or do *something*.

Fresh Harvest is a long-time processor in the highly seasonal vegetable canning industry. Fresh Harvest packs and sells canned beans, peas, carrots, corn, mixed peas and carrots, and kidney beans. It sells mainly through food brokers to merchant wholesalers, supermarket chains (such as Kroger, Safeway, and Jewel), cooperatives, and other outlets, mostly in the Midwest. Of less importance, by volume, are sales to local institutions, grocery stores, and supermarkets—and sales of dented canned goods at low prices to walk-in customers.

Fresh Harvest is located in Wisconsin's beautiful Devil's Valley. The company has more than $28 million in sales annually (exact sales data are not published by the closely held corporation). Plants are located in strategic places along the valley, with the main offices in the valley as well. The Fresh Harvest brand is used only on canned goods sold in the local market. Most of the goods are sold and shipped under a retailer's label or a broker's/wholesaler's label.

Fresh Harvest is well known for the consistent quality of its product offerings. And it's always willing to offer competitive prices. Strong channel relations were built by Robert Skye, Fresh Harvest's former board chair and chief executive officer. Robert—who still owns controlling interest in the firm—worked the Chicago area as the company's sales rep in its earlier years, before his father stepped down as president and he took over in 1982. Robert was an ambitious and hardworking top manager—the firm prospered under his direction. He became well known within the canned food processing industry for technical/product innovations.

During the off-canning season, Robert traveled widely. In the course of his travels, he arranged several important business deals. His 1994 and 2002 trips resulted in the following two events: (1) inexpensive pineapple was imported from Taiwan and sold by Fresh Harvest, primarily to expand the product line; and (2) a technically advanced continuous process cooker (65 feet high) was imported from England and installed at one of Fresh Harvest's plants. It was the first of its kind in the United States and cut processing time sharply while improving quality.

Robert retired in 2018 and named his daughter, 35-year-old Chelsea Skye-Rice, as his successor. Chelsea is intelligent and hardworking. She has been primarily concerned with the company's financial matters and only recently focused on marketing problems. During her seven years as financial director, the firm received its highest credit rating and was able to borrow working capital ($5 million to meet seasonal can and wage requirements) at the lowest rate ever.

The fact that the firm isn't unionized allows some competitive advantage. However, changes in minimum wage laws have increased costs. These and other rising costs have squeezed profit margins. This led to the recent closing of two plants as they became less efficient to operate. Fresh Harvest expanded capacity of the remaining two plants (especially warehouse facilities) so they could more profitably operate with maximum use of existing processing equipment.

Shortly after Robert's retirement, Chelsea reviewed the company's situation with her managers. She pointed to narrowing profit margins, debts contracted for new plants and equipment, and an increasingly competitive environment. Even considering the temporary labor-saving competitive advantage of the new cooker system, there seemed to be no way to improve the status quo unless the firm could sell direct—as it does in the local market—thereby eliminating the food brokers' 5 percent commission on sales. This was the chosen plan, and Don Bartley was given the new sales job. An inside salesperson was retained to handle incoming orders and do some telemarketing to smaller accounts.

Don, the only full-time outside sales rep for the firm, lives in Devil's Valley. Other top managers do some selling but not much. Because he is Robert's nephew, Don is also a member of the board of directors. He is well qualified in technical matters and has a college degree in food chemistry. Although Don formerly did call on some important customers with the brokers' sales reps, he is not well known in the industry or even by Fresh Harvest's usual customers.

It is now five months later. Don is not doing very well. He has made several selling trips, placed hundreds of telephone calls, and maintained constant e-mail contact with prospective customers—all with discouraging results. He is unwilling to continue sales efforts on his own. There seem to be too many potential customers for one person to reach. And much negotiating and "wining and dining" seems to be needed—certainly more than he can or wants to do.

Don insists that Fresh Harvest hire a sales force to continue its present way of operating. Sales are down, both in comparison to expectations and to the previous year's results. Some regular supermarket chain customers have stopped buying—though basic consumer demand has not changed. Further, buyers for some supermarket chains that might be potential new customers have demanded quantity guarantees much larger than Fresh Harvest can supply. Expanding supply would be difficult in the short run—because the firm typically must contract with growers to ensure supplies of the type and quality it normally offers.

Robert, still the controlling stockholder, has asked for a special meeting of the board in two weeks to discuss the present situation.

Evaluate Fresh Harvest's past and current strategy planning. What should Chelsea Skye-Rice tell Robert Skye? What should Fresh Harvest do now?

25. QXR Tools (QXR)

Pedro Gomez, president and marketing manager of QXR Tools, is deciding what strategy, or strategies, to pursue.

QXR Tools (QXR) is a manufacturer of industrial cutting tools. These tools include such items as lathe blades, drill press bits, and various other cutting edges used in the operation of large metal cutting, boring, or stamping machines. Pedro takes great pride in the fact that his company—whose $5.7 million sales last year is small by industry standards—is recognized as a producer of a top-quality line of cutting tools.

Competition in the cutting-tool industry is intense. QXR competes not only with the original machine manufacturers, but also with many other larger domestic and foreign manufacturers offering cutting tools as one of their many different product lines. Over the years, this has had the effect of standardizing the price, specifications, and, in turn, the quality of the competing products from all manufacturers. It has also led to fairly low prices on standard items.

About a year ago, Pedro was tiring of the financial pressure of competing with larger companies enjoying economies of scale. At the same time, he noted that more and more potential cutting-tool customers were turning to small tool-and-die shops that used computer-controlled equipment to meet specialized needs that could not be met by the mass production firms. Pedro thought perhaps he should consider some basic strategy changes. Although he was unwilling to become strictly a custom producer, he thought that the recent trend toward buying customized cutting edges suggested new markets might be developing—markets too small for the large, multi-product-line companies to serve profitably but large enough to earn a good profit for a flexible company of QXR's size.

Pedro hired a marketing research company, MResearchPro, to study the feasibility of serving these markets. The initial results were encouraging. It was estimated that QXR might increase sales by 65 percent and profits by 90 percent by serving the emerging markets. The research showed that there are many large users of standard cutting tools who buy directly from large cutting-tool manufacturers (domestic or foreign) or wholesalers who represent these manufacturers. This is the bulk of the cutting-tool business (in terms of units sold and sales dollars). But there are also many smaller users all over the United States who buy in small but regular quantities. And some of these needs are becoming more specialized. That is, a special cutting tool may make a machine and/or worker much more productive, perhaps eliminating several steps with time-consuming setups. This is the area that the research company sees as potentially attractive.

The next strategy change made by Pedro was to have the sales manager hire two technically oriented marketing researchers (at a total cost of $85,000 each per year, including travel expenses) to maintain continuous contact with potential cutting-tool customers. The researchers were supposed to identify any present or future needs that might exist in enough cases to make it possible to profitably produce a specialized product. The researchers were not to take orders or sell QXR's products to the potential customers. Pedro felt that only through this policy could these researchers talk to the right people.

The initial feedback from the marketing researchers was most encouraging. Many firms (large and small) had special needs—although it often was necessary to talk to the shop supervisor or individual machine operators to find these needs. Most operators were making do with the tools available. Either they didn't know customizing was possible or doubted that their supervisors would do anything about it if they suggested that a more specialized tool could increase productivity. But these operators were encouraging because they said that it would be easier to persuade supervisors to order specialized tools if the tools were already produced and in stock than if they had to be custom made. So Pedro decided to continually add high-quality products to meet the ever-changing, specialized needs of users of cutting tools and edges.

QXR's potential customers for specialized tools are located all over the United States. The average sale per customer is likely to be less than $500, but the sale will be repeated several times within a year. Because of the widespread market and the small order size, Pedro doesn't think that selling direct—as is done by small custom shops—is practical. At the present time, QXR sells 90 percent of its regular output through a large industrial wholesaler—Superior Mill Supply—which serves the area east of the Mississippi River and carries a very complete line of industrial supplies (to "meet every industrial need"). Superior carries more than 10,000 items. Some sales come from customers who know exactly what they want and just place orders directly by fax or at the firm's website. But most of the selling is by Superior's sales reps, who work from an electronic catalog on a laptop computer. Superior, although very large and well known, is having trouble moving cutting tools. It's losing sales of cutting tools in some cities to newer wholesalers specializing in the cutting-tool industry. The new wholesalers are able to give more technical help, and therefore better service, to potential customers. Superior's president is convinced that the newer, less-experienced concerns will either realize that a substantial profit margin can't be maintained along with their aggressive strategies, or they will eventually go broke trying to overspecialize.

From Pedro's standpoint, the present wholesaler has a good reputation and has served QXR well in the past. Superior has been of great help in holding down Pedro's inventory costs—by increasing the inventory in Superior's 35 branch locations. Although Pedro has received several complaints about the lack of technical assistance given by Superior's sales reps—as well as their lack of knowledge about QXR's new special products—he feels that the present wholesaler is providing the best service it can. All of its sales reps have been told about the new products at a special training session, and new pages have been added to the electronic catalog on their laptops. So regarding the complaints, Pedro says, "It's just the usual things you hear when you're in business."

Pedro thinks there are more urgent problems than a few complaints. Profits are declining, and sales of the new cutting tools are not nearly as high as forecast—even though all research reports indicate that the company's new products meet the intended markets' needs perfectly. The high costs involved in producing small quantities of special products and in adding the marketing research team—together with lower than expected sales—have significantly reduced QXR's profits. Pedro is wondering whether it is wise to continue to try to cater to the needs of many specific target markets when the results are this discouraging. He also is considering increasing advertising expenditures, including some search engine advertising, in the hope that customers will pull the new products through the channel.

Evaluate QXR's situation and Pedro Gomez's present strategy. What should he do now?

26. AAA Custom Castings, Inc.

Mallory Rizocki, marketing manager for AAA Custom Castings, Inc., is trying to figure out how to explain to her boss why a proposed new product line doesn't make sense for the company. Mallory is sure it's wrong for AAA Custom Castings, but she isn't able to explain why.

AAA Custom Castings, Inc. is a producer of malleable iron castings for automobile and aircraft manufacturers and a variety of other users of castings. Last year's sales of castings amounted to more than $70 million.

AAA Custom Castings also produces about 30 percent of all the original equipment bumper jacks installed in new U.S.-made automobiles each year. This is a very price-competitive business, but AAA Custom Castings has been able to obtain a large market share with frequent personal contact between the company's executives and its customers—supported by very close cooperation between the company's engineering department and its customers' buyers. This has been extremely important because the wide variety of models and model changes frequently requires alterations in the specifications of the bumper jacks. All of AAA Custom Castings' bumper jacks are sold directly to automobile manufacturers. No attempt has been made to sell bumper jacks to final consumers through hardware and automotive channels—although they are available through the manufacturers' automobile dealers.

Tim Kingston, AAA Custom Castings' production manager, now wants to begin producing hydraulic garage jacks for sale through auto-parts wholesalers to auto-parts retailers. Tim saw a variety of hydraulic garage jacks at a recent automotive show and knew immediately that his plant could produce these products. This especially interested him because of the possibility of using excess capacity. Further, he says "jacks are jacks," and the company would merely be broadening its product line by introducing hydraulic garage jacks. (*Note:* Hydraulic garage jacks are larger than bumper jacks and are intended for use in or around a garage. They are too big to carry in a car's trunk.)

As Tim became more enthusiastic about the idea, he found that AAA Custom Castings' engineering department already had a patented design that appeared to be at least comparable to the products now offered on the market. Further, Tim says that the company would be able to produce a product that is better made than the competitive products (i.e., smoother castings)—although he agrees that most customers probably wouldn't notice the difference. The production department estimates that the cost of producing a hydraulic garage jack comparable to those currently offered by competitors would be about $48 per unit.

Mallory has just received an e-mail from Jesse Zachary, the company president, explaining the production department's enthusiasm for broadening AAA Custom Castings' present jack line into hydraulic jacks. Jesse seems enthusiastic about the idea too, noting that it would be a way to make fuller use of the company's resources and increase its sales. Jesse's e-mail asks for Mallory's reaction, but Jesse already seems sold on the idea.

Given Jesse's enthusiasm, Mallory isn't sure how to respond. She's trying to develop a good explanation for why she isn't excited about the proposal. The firm's six sales reps are already overworked with their current accounts. And Mallory couldn't possibly promote this new line herself—she's already helping other reps make calls and serving as sales manager. So it would be necessary to hire someone to promote the line. And this sales manager would probably have to recruit manufacturers' agents (who probably will want 10 to 15 percent commission on sales) to sell to automotive wholesalers who would stock the jack and sell to the auto-parts retailers. The wholesalers will probably expect trade discounts of about 20 percent, trade show exhibits, some national advertising, and sales promotion help (catalog sheets, mailers, and point-of-purchase displays). Further, Mallory sees that AAA Custom Castings' billing and collection system will have to be expanded because many more customers will be involved. It will also be necessary to keep track of agent commissions and accounts receivable.

Auto-parts retailers are currently selling similar hydraulic garage jacks for about $99. Mallory has learned that such retailers typically expect a trade discount of about 35 percent off the suggested list price for their auto parts.

All things considered, Mallory feels that the proposed hydraulic jack line is not very closely related to the company's present emphasis. She has already indicated her lack of enthusiasm to Tim, but this made little difference in Tim's thinking. Now it's clear that Mallory will have to convince the president or she will soon be responsible for selling hydraulic jacks.

Contrast AAA Custom Castings, Inc.'s current strategy and the proposed strategy. What should Mallory Rizocki say to Jesse Zachary to persuade him to change his mind? Or should she just plan to sell hydraulic jacks? Explain.

27. Canadian Mills, Ltd.*

Valerie Boudreau, marketing manager of Canadian Mills, Ltd.—a Canadian company—is being urged to approve the creation of a separate marketing plan for Quebec. This would be a major policy change because Canadian Mills' international parent is trying to move toward a global strategy for the whole firm, and Boudreau has been supporting Canada-wide planning.

Boudreau has been the marketing manager of Canadian Mills, Ltd. for the last four years—since she arrived from international headquarters in Minneapolis. Canadian Mills, Ltd., headquartered in Toronto, is a subsidiary of a large U.S.-based consumer packaged food company with worldwide sales of more than $2.8 billion in 2007. Its Canadian sales are more than $450 million, with the Quebec and Ontario markets accounting for 69 percent of the company's Canadian sales.

The company's product line includes such items as cake mixes, puddings, pie fillings, pancakes, prepared foods, and frozen dinners. The company has successfully introduced at least six new products every year for the last five years. Products from Canadian Mills are known for their high quality and enjoy much brand preference throughout Canada, including the Province of Quebec.

The company's sales have risen every year since Boudreau took over as marketing manager. In fact, the company's market share has increased steadily in each of the product categories in which it competes. The Quebec market has closely followed the national trend; however, in the past two years, total sales growth in that market began to lag.

According to Boudreau, a big advantage of Canadian Mills over its competitors is the ability to coordinate all phases of the food business from Toronto. For this reason, Boudreau meets at least once a month with her product managers—to discuss developments in local markets that might affect marketing plans. Although each manager is free to make suggestions and even to suggest major changes, Boudreau has the responsibility of giving final approval for all plans.

One of the product managers, Jackie Provence, expressed great concern at the last monthly meeting about the poor performance of some of the company's products in the Quebec market. Although a broad range of possible reasons—ranging from

---

*This case was adapted from one written by Professor Roberta Tamilia, University of Windsor, Canada.

Table 1 Per Capita Consumption Index, Province of Quebec (Canada = 100)

Cake mixes	107	Soft drinks	126
Pancakes	87	Pie fillings	118
Puddings	114	Frozen dinners	79
Salad dressings	85	Prepared packaged foods	83
Molasses	132	Cookies	123

Note: An index shows the relative consumption as compared to a standard. In this table, the standard is all of Canada. The data show that per capita consumption of cake mixes is 7 percent higher in Quebec and pancake consumption 13 percent lower compared to all of Canada.

inflation and the threat of job losses to politics—were reviewed to try to explain the situation, Provence insisted that it was due to a basic lack of understanding of that market. She felt not enough managerial time and money had been spent on the Quebec market—in part because of the current emphasis on developing all-Canada plans on the way to having one global strategy.

Provence felt the current marketing approach to the Quebec market should be reevaluated because an inappropriate marketing plan may be responsible for the sales slowdown. After all, she said, "80 percent of the market is French-speaking. It's in the best interest of the company to treat that market as being separate and distinct from the rest of Canada."

Provence supported her position by showing that Quebec's per capita consumption of many product categories in which the firm competes is above the national average (see Table 1). Research projects conducted by Canadian Mills also support the "separate and distinct" argument. Over the years, the firm has found many French-English differences in brand attitudes, lifestyles, usage rates, and so on.

Provence argued that the company should develop a unique Quebec marketing plan for some or all of its brands. She specifically suggested that the French-language advertising plan for a particular brand be developed independently of the plan for English-speaking Canada.

Currently, the Toronto agency assigned to the brand just translates its English-language ads for the French market. Boudreau pointed out that the present advertising approach assured Canadian Mills of a uniform brand image across Canada. Provence said she knew what the agency is doing, and that straight translation into Canadian French may not communicate the same brand image. The discussion that followed suggested that a different brand image might be needed in the French market if the company wanted to stop the brand's decline in sales.

The managers also discussed the food distribution system in Quebec. The major supermarket chains have their lowest market share in that province. Independents are strongest there—the mom-and-pop food stores that are fast disappearing outside Quebec remain alive and well in the province. Traditionally, these stores have stocked a higher proportion (than supermarkets) of their shelf space with national brands, an advantage for Canadian Mills.

Finally, various issues related to discount policies, pricing structure, sales promotion, and cooperative advertising were discussed. All of these suggested that things were different in Quebec and that future marketing plans should reflect these differences to a greater extent than they do now. After the meeting, Boudreau stayed in her office to think about the situation. Although she agreed with the basic idea that the Quebec market was in many ways different, she wasn't sure how far the company should go in recognizing this fact. She knew that regional differences in food tastes and brand purchases existed not only in Quebec but in other parts of Canada as well. But people are people, after all, with far more similarities than differences, so a Canadian and, eventually, a global strategy makes some sense too.

Boudreau was afraid that giving special status to one region might conflict with top management's objective of achieving standardization whenever possible—one national strategy for Canada, on the way to one worldwide global strategy. She was also worried about the long-term effect of such a policy change on costs, organizational structure, and brand image. Still, enough product managers had expressed their concern over the years about the Quebec market to make her wonder if she shouldn't modify the current approach. Perhaps they could experiment with a few brands—and just in Quebec. She could cite the language difference as the reason for trying Quebec rather than any of the other provinces. But Boudreau realizes that any change of policy could be seen as the beginning of more change, and what would Minneapolis think? Could she explain it successfully there?

Evaluate Canadian Mills, Ltd.'s present strategy. What should Valerie Boudreau do now? Explain.

28. Kingston Home Health Services (KHHS)

Chelsea Kingston, executive director of Kingston Home Health Services, is trying to clarify her strategies. She's sure some changes are needed, but she's less sure about how *much* change is needed and/or whether it can be handled by her people.

Kingston Home Health Services (KHHS) is a nonprofit organization that has been operating—with varying degrees of success—for 25 years, offering nursing services in clients' homes. Some of its funding comes from the local United Way to provide emergency nursing services for those who can't afford to pay. The balance of the revenues—about 90 percent of the $2.2 million annual budget—comes from charges made directly to the client or to third-party payers, including insurance companies, health maintenance organizations (HMOs), and the federal government for Medicare or Medicaid services.

Chelsea has been executive director of KHHS for two years. She has developed a well-functioning organization able to meet most requests for service that come from local doctors and from the discharge officers at local hospitals. Some business also comes by self-referral—the client finds the KHHS name in the yellow pages of the local phone directory.

The last two years have been a rebuilding time—because the previous director had personnel problems. This led to a weakening of the agency's image with the local referring agencies. Now their image is more positive. But Chelsea is not completely satisfied with the situation. By definition, Kingston Home Health

Services is a nonprofit organization. Yet it still must cover all its costs: payroll, rent payments, phone expenses, and so on, including Chelsea's own salary. She can see that although KHHS is growing slightly and is now breaking even, it doesn't have much of a cash cushion to fall back on if (1) the demand for KHHS nursing services declines; (2) the government changes its rules about paying for KHHS's kind of nursing services, either cutting back what it will pay for or reducing the amount it will pay for specific services; or (3) new competitors enter the market. In fact, the last possibility concerns Chelsea greatly. Some hospitals, squeezed for revenue, are expanding into home health care—especially nursing services—as patients are being released earlier from hospitals because of payment limits set by government guidelines. For-profit organizations (e.g., Kelly Home Care Services) are expanding around the country to provide a complete line of home health care services, including nursing services of the kind offered by KHHS. These for-profit organizations appear to be efficiently run, offering good service at competitive and sometimes even lower prices than some nonprofit organizations. And they seem to be doing this at a profit, which suggests that it would be possible for these for-profit companies to lower their prices if nonprofit organizations try to compete on price.

Chelsea is considering whether she should ask her board of directors to let her offer a complete line of home health care services—that is, move beyond just nursing services into what she calls "care and comfort" services.

Currently, KHHS is primarily concerned with providing professional nursing care in the home. But KHHS nurses are much too expensive for routine home health care activities—helping fix meals, bathing and dressing patients, and other care and comfort activities. The full cost of a nurse to KHHS, including benefits and overhead, is about $65 per hour. But a registered nurse is not needed for care and comfort services. All that is required is someone who is honest, can get along with all kinds of people, and is willing to do this kind of work. Generally, any mature person can be trained fairly quickly to do the job—following the instructions and under the general supervision of a physician, a nurse, or family members. The full cost of aides is $9 to $16 per hour for short visits and as low as $75 per 24 hours for a live-in aide who has room and board supplied by the client.

The demand for all kinds of home health care services seems to be growing. With more dual-career families and more single-parent households, there isn't anyone in the family to take over home health care when the need arises due to emergencies or long-term disabilities. Further, hospitals send patients home earlier than in the past. And with people living longer, there are more single-survivor family situations where there is no one nearby to take care of the needs of these older people. But often family members—or third-party payers such as the government or insurers—are willing to pay for some home health care services. Chelsea now occasionally recommends other agencies or suggests one or another of three women who have been doing care and comfort work on their own, part-time. But with growing demand, Chelsea wonders if KHHS should get into this business, hiring aides as needed.

Chelsea is concerned that a new, full-service home health care organization may come into her market and be a single source for both nursing services *and* less-skilled home care and comfort services. This has happened already in two nearby but somewhat larger cities. Chelsea fears that this might be more appealing than KHHS to the local hospitals and other referrers. In other words, she can see the possibility of losing nursing service business if KHHS does not begin to offer a complete home health care service. This would cause real problems for KHHS—because overhead costs are more or less fixed. A loss in revenue of as little as 10 percent would require some cutbacks—perhaps laying off some nurses or secretaries, giving up part of the office, and so on.

Another reason for expanding beyond nursing services—using paraprofessionals and relatively unskilled personnel—is to offer a better service to present customers *and* make more effective use of the computer systems and organization structure that she has developed over the last two years. Chelsea estimates that the administrative and office capabilities could handle twice as many clients without straining the system. It would be necessary to add some clerical help if the expansion were quite large. But this increase in overhead would be minor compared to the present proportion of total revenue that goes to covering overhead. In other words, additional clients or more work for some current clients could increase revenue and ensure the survival of KHHS, provide a cushion to cover the normal fluctuations in demand, and ensure more job security for the administrative personnel.

Further, Chelsea thinks that if KHHS were successful in expanding its services—and therefore could generate some surplus—it could extend services to those who aren't now able to pay. Chelsea says one of the worst parts of her job is refusing service to clients whose third-party benefits have run out or who can no longer afford to pay for other reasons. She is uncomfortable about having to cut off service, but she must schedule her nurses to provide revenue-producing services if she's going to meet the payroll every two weeks. By expanding to provide more services, she might be able to keep serving more of these nonpaying clients. This possibility excites Chelsea because her nurse's training has instilled a deep desire to serve people in need, whether they can pay or not. This continual pressure to cut off service because people can't pay has been at the root of many disagreements and even arguments between the nurses serving the clients and Chelsea, as executive director and representative of the board of directors.

Chelsea knows that expanding into care and comfort services won't be easy. Some decisions would be needed about relative pay levels for nurses, paraprofessionals, and aides. KHHS would also have to set prices for these different services and tell current customers and referral agencies about the expanded services.

These problems aren't bothering Chelsea too much, however—she thinks she can handle them. She is sure that care and comfort services are in demand and could be supplied at competitive prices.

Her primary concern is whether this is the right thing for Kingston Home Health Services—basically a nursing organization—to do. KHHS's whole history has been oriented to supplying *nurses' services*. Nurses are dedicated professionals who bring high standards to any job they undertake. The question is whether KHHS should offer less-professional services. Inevitably, some of the aides will not be as dedicated as the nurses might like them to be. And this could reflect unfavorably on the nurse image. At a minimum, she would need to set up some sort of training program for the aides. As Chelsea worries about the future of KHHS, and her own future, it seems that there are no easy answers.

Evaluate KHHS's present strategy. What should Chelsea Kingston do? Explain.

29. Kennedy & Gaffney (K&G)

The partners of Kennedy & Gaffney are having a serious discussion about what the firm should do in the near future.

Kennedy & Gaffney (K&G) is a medium-sized regional certified public accounting firm based in Grand Rapids, Michigan, with branch offices in Lansing and Detroit. Kennedy & Gaffney has nine partners and a professional staff of approximately 105 accountants. Gross service billings for the fiscal year ending June 30, 2020, were $6.9 million. See Table 1, which presents financial data for 2020, 2019, and 2018.

K&G's professional services include auditing, tax preparation, bookkeeping, and some general management consulting. Its client base includes municipal governments (cities, villages, and townships), manufacturing companies, professional organizations (attorneys, doctors, and dentists), and various other small businesses. A good share of revenue comes from the firm's municipal practice. Table 1 details K&G's gross revenue by service area and client industry for 2020, 2019, and 2018.

At the monthly partners' meeting held in July 2020, Robert Kennedy, the firm's managing partner (CEO), expressed concern about the future of the firm's municipal practice. Robert's presentation to his partners follows:

> Although our firm is considered to be a leader in municipal auditing in our geographic area, I am concerned that as municipals attempt to cut their operating costs, they will solicit competitive bids from other public accounting firms to perform their annual audits. Three of the four largest accounting firms in the world have local offices in our area. Because they concentrate their practice in the manufacturing industry—which typically has December 31 fiscal year-ends—they have "available" staff during the summer months.
>
> Therefore, they can afford to low-ball competitive bids to keep their staffs busy and benefit from on-the-job training provided by municipal clientele. I am concerned that we may begin to lose clients in our most established and profitable practice area.*

Sherry Gaffney, a senior partner in the firm and the partner in charge of the firm's municipal practice, was the first to respond to Robert's concern.

> Robert, we all recognize the potential threat of being underbid for our municipal work by our large accounting competitors. However, K&G is a leader in municipal auditing in Michigan, and we have much more local experience than our competitors. Furthermore, it is a fact that we offer a superior level of service to our clients—which goes beyond the services normally expected during an audit to include consulting on financial and other operating issues. Many of our less sophisticated clients depend on our nonaudit consulting assistance. Therefore, I believe, we have been successful in differentiating our services from our competitors. In many recent situations, K&G was selected over a field of as many as 10 competitors even though our proposed prices were much higher than those of our competitors.

The partners at the meeting agreed with Sherry's comments. Yet even though K&G had many success stories regarding its ability to retain its municipal clients—despite being underbid—it had lost three large municipal clients during the past year. Sherry was asked to comment on the loss of those clients. She explained that the lost clients are larger municipalities with a lot of in-house financial expertise and therefore less dependent on K&G's consulting assistance. As a result, K&G's service differentiation went largely unnoticed. Sherry explained that

---

*Organizations with December fiscal year-ends require audit work to be performed during the fall and in January and February. Those with June 30 fiscal year-ends require auditing during the summer months.

Table 1 Fiscal Year Ending June 30

	2020	2019	2018
Gross billings	$6,900,000	$6,400,000	$5,800,000
Gross billings by service area:			
Auditing	3,100,000	3,200,000	2,750,000
Tax preparation	1,990,000	1,830,000	1,780,000
Bookkeeping	1,090,000	745,000	660,000
Other	720,000	625,000	610,000
Gross billings by client industry:			
Municipal	3,214,000	3,300,000	2,908,000
Manufacturing	2,089,000	1,880,000	1,706,000
Professional	1,355,000	1,140,000	1,108,000
Other	242,000	80,000	78,000

the larger, more sophisticated municipals regard audits as a necessary evil and usually select the low-cost reputable bidder.

Robert then requested ideas and discussion from the other partners at the meeting. One partner, Rosa Basilio, suggested that K&G should protect itself by diversifying. Specifically, she felt a substantial practice development effort should be directed toward manufacturing. She reasoned that because manufacturing work would occur during K&G's off-season, K&G could afford to price very low to gain new manufacturing clients. This strategy would also help counter (and possibly discourage) low-ball pricing for municipals by the three large accounting firms mentioned earlier.

Another partner, Wade Huntoon, suggested that "if we have consulting skills, we ought to promote them more, instead of hoping that the clients will notice and come to appreciate us. Further, maybe we ought to be more aggressive in calling on smaller potential clients."

Another partner, Stan Walsh, agreed with Wade, but wanted to go further. He suggested that they recognize that there are at least two types of municipal customers and that two (at least) different strategies be implemented, including lower prices for auditing only for larger municipal customers and/or higher prices for smaller customers who are buying consulting too. This caused a big uproar from some who said this would lead to price cutting of professional services, and K&G didn't want to be price cutters: "One price for all is the professional way."

However, another partner, Isabel Ventura, agreed with Stan and suggested they go even further—pricing consulting services separately. In fact, she suggested that the partners consider setting up a separate department for consulting—like the large accounting firms have done: "This can be a very profitable business. But it is a different kind of business and eventually may require different kinds of people and a different organization. For now, however, it may be desirable to appoint a manager for consulting services—with a budget—to be sure it gets proper attention." This suggestion too caused serious disagreement. Partners pointed out that having a separate consulting arm had led to major conflicts, especially in some larger accounting firms. The initial problems were internal. The consultants often brought in more profit than the auditors, but the auditors controlled the partnership, and the successful consultants didn't always feel that they got their share of the rewards. There had also been serious external problems and charges of unethical behavior based on the concern that big accounting firms had a conflict of interest when they did audits on publicly traded companies that they in turn relied on for consulting income. Because of problems in this area, the Securities and Exchange Commission created new guidelines that have changed how the big four accounting firms handle consulting. On the other hand, several partners argued that this was really an opportunity for K&G because the firm handled very few companies listed with the SEC, and the conflict of interest issues didn't even apply with municipal clients.

Robert thanked everyone for their comments and encouraged them to debate these issues in smaller groups and to share ideas by e-mail before coming to a one-day retreat (in two weeks) to continue this discussion and come to some conclusions.

Evaluate K&G's situation. What strategy(ies) should the partners select? Why?

30. Paglozzi's Pizza Pies

Cassidy Newman, manager of the Paglozzi's Pizza Pies store in Flint, Michigan, is trying to develop a plan for the "sick" store she just took over.

Paglozzi's Pizza Pies is an owner-managed pizza take-out and delivery business with three stores located in Ann Arbor, Southfield, and Flint, Michigan. Paglozzi's business comes from telephone, fax, or walk-in orders. Each Paglozzi's store prepares its own pizzas. In addition to pizzas, Paglozzi's also sells and delivers a limited selection of soft drinks.

Paglozzi's Ann Arbor store has been very successful. Much of the store's success may be due to being close to the University of Michigan campus. Most of these students live within five miles of Paglozzi's Ann Arbor store.

The Southfield store has been moderately successful. It serves mostly residential customers in the Southfield area, a largely residential suburb of Detroit. Recently, the store advertised—using direct-mail flyers—to several office buildings within three miles of the store. The flyers described Paglozzi's willingness and ability to cater large orders for office parties, business luncheons, and so on. The promotion was quite successful. With this new program and Paglozzi's solid residential base of customers in Southfield, improved profitability at the Southfield location seems ensured.

Paglozzi's Flint location has had mixed results during the last three years. The Flint store has been obtaining only about half of its orders from residential delivery requests. Cassidy, the Flint store's new manager, believes the problem with residential pizza delivery in Flint is due to the location of residential neighborhoods in the area. Flint has several large industrial plants (mostly auto industry related) located throughout the city. Small, mostly factory-worker neighborhoods are distributed in between the various plant sites. As a result, Paglozzi's store location can serve only two or three of these neighborhoods on one delivery run. Competition is also relevant. Paglozzi's has several aggressive competitors who advertise heavily, distribute cents-off coupons, and offer 2-for-1 deals. This aggressive competition is probably why Paglozzi's residential sales leveled off in the last year or so. And this competitive pressure seems likely to continue as some of this competition comes from aggressive national chains that are fighting for market share and squeezing little firms like Paglozzi's. For now, anyway, Cassidy feels she knows how to meet this competition and hold on to the present residential sales level.

Most of the Flint store's upside potential seems to be in serving the large industrial plants. Many of these plants work two or three shifts, five days a week. During each work shift, workers

Table 1 Practical Capacities and Sales Potential of Current Equipment and Personnel

	11 a.m. Break	8 p.m. Break	2:30 a.m. Break	Daily Totals
Current capacity (pizzas)	48	48	48	144
Average selling price per unit	$ 12.50	$ 12.50	$ 12.50	$ 12.50
Sales potential	$600	$600	$600	$1,800
Variable cost (approximately 40 percent of selling price)*	240	240	240	720
Contribution margin of pizzas	360	360	360	1,080
Beverage sales (2 medium-sized beverages per pizza ordered at 75¢ apiece)†	72	72	72	216
Cost of beverages (30 percent per beverage)	22	22	22	66
Contribution margin of beverages	50	50	50	150
Total contribution of pizza and beverages	$410	$410	$410	$1,230

*The variable cost estimate of 40 percent of sales includes variable costs of delivery to plant locations.
†Amounts shown are not physical capacities (there is almost unlimited physical capacity), but potential sales volume is constrained by number of pizzas that can be sold.

are allowed one half-hour lunch break—which usually occurs at 11 a.m., 8 p.m., or 2:30 a.m., depending on the shift.

Customers can order by phone, fax, e-mail, or at the Paglozzi's website. About 30 minutes before a scheduled lunch break Paglozzi's can expect an order for several (5 to 10) pizzas for a work group. Paglozzi's may receive many orders of this size from the same plant (i.e., from different groups of workers). The plant business is very profitable for several reasons. First, a large number of pizzas can be delivered at the same time to the same location, saving transportation costs.

Second, plant orders usually involve many different toppings (double cheese, pepperoni, mushrooms, hamburger) on each pizza. This results in $11 to $14 revenue per pizza. The delivery drivers also like delivering plant orders because the tips are usually $1 to $2 per pizza.

Despite the profitability of the plant orders, several factors make it difficult to serve the plant market. Paglozzi's store is located 5 to 8 minutes from most of the plant sites, so Paglozzi's staff must prepare the orders within 20 to 25 minutes after the telephone order is received. Often, inadequate staff and/or oven capacity means it is impossible to get all the orders heated at the same time.

Generally, plant workers will wait as long as 10 minutes past the start of their lunch break before ordering from various vending trucks that arrive at the plant sites during lunch breaks. (Currently, no other pizza delivery stores are in good positions to serve the plant locations and have chosen not to compete.) But there have been a few instances when workers refused to pay for pizzas that were only 5 minutes late! Worse yet, if the same work group gets a couple of late orders, they are lost as future customers. Cassidy believes that the inconsistent profitability of the Flint store is partly the result of such lost customers.

In an effort to rebuild the plant delivery business, Cassidy is considering various methods to ensure prompt customer delivery. She thinks that potential demand during lunch breaks is significantly above Paglozzi's present capacity. Cassidy also knows that if she tries to satisfy all phone or fax orders on some peak days, she won't be able to provide prompt service and may lose more plant customers.

Cassidy has outlined three alternatives that may win back some of the plant business for the Flint store. She has developed these alternatives to discuss with Paglozzi's owner. Each alternative is briefly described below:

Alternative 1: Determine practical capacities during peak volume periods using existing equipment and personnel. Accept orders only up to that capacity and politely decline orders beyond. This approach will ensure prompt customer service and high product quality. It will also minimize losses resulting from customers' rejection of late deliveries. Financial analysis of this alternative—shown in Table 1—indicates that a potential daily contribution to profit of $1,230 could result if this alternative is implemented successfully. This would be profit before promotion costs, overhead, and net profit (or loss). *Note:* Any alternative will require several thousand dollars to reinform potential plant customers that Paglozzi's has improved its service and "wants your business."

Alternative 2: Buy additional equipment (one oven and one delivery car) and hire additional staff to handle peak loads. This approach would ensure timely customer delivery and high product quality as well as provide additional capacity to handle unmet demand. Table 2 is a conservative estimate of potential daily demand for plant orders compared to current capacity

Table 2 Capacity and Demand for Plant Customer Market

	Estimated Daily Demand	Current Daily Capacity	Proposed Daily Capacity
Pizza units (1 pizza)	320	144	300

Table 3 Cost of Required Additional Assets

	Cost	Estimated Useful Life	Salvage Value	Annual Depreciation*	Daily Depreciation†
Delivery car (equipped with pizza warmer)	$11,000	5 years	$1,000	$2,000	$5.71
Pizza oven	$20,000	8 years	$2,000	$2,250	$6.43

*Annual depreciation is calculated on a straight-line basis.

†Daily depreciation assumes a 350-day (plant production) year. All variable expenses related to each piece of equipment (e.g., utilities, gas, oil) are included in the variable cost of a pizza.

and proposed increased capacity. Table 3 gives the cost of acquiring the additional equipment and relevant information related to depreciation and fixed costs.

Using this alternative, the following additional pizza delivery and preparation personnel costs would be required:

	Hours Required	Cost per Hour	Total Additional Daily Cost
Delivery personnel	6	$6	$36.00
Preparation personnel	8	$6	48.00
			$84.00

The addition of even more equipment and personnel to handle all unmet demand was not considered in this alternative because the current store is not large enough.

Alternative 3: Add additional equipment and personnel as described in alternative 2, but move to a new location that would reduce delivery lead times to 2 to 5 minutes. This move would probably allow Paglozzi's to handle all unmet demand—because the reduction in delivery time will provide for additional oven time. In fact, Paglozzi's might have excess capacity using this approach.

A suitable store is available near roughly the same number of residential customers (including many of the store's current residential customers). The available store is slightly larger than needed, and the rent is higher. Relevant cost information on the proposed store follows:

Additional rental expense of proposed store over current store	$1,600 per year
Cost of moving to new store (one-time cost)	$16,000

Cassidy presented the three alternatives to Paglozzi's owner, Skipper Kipnis. Skipper was pleased that Cassidy had done her homework. He decided that Cassidy should make the final decision on what to do (in part because she had a profit-sharing agreement with Skipper) and offered the following comments and concerns:

1. Skipper agreed that the plant market was extremely sensitive to delivery timing. Product quality and pricing, although important, were of less importance.
2. He agreed that plant demand estimates were conservative. "In fact, they may be 10 to 30 percent low."
3. Skipper expressed concern that under alternative 2, and especially under alternative 3, much of the store's capacity would go unused more than 80 percent of the time.
4. He was also concerned that Paglozzi's store had a bad reputation with plant customers because the prior store manager was not sensitive to timely plant delivery. So Skipper suggested that Cassidy develop a promotion plan to improve Paglozzi's reputation in the plants and be sure that everyone knows that Paglozzi's has improved its delivery service.

Evaluate Cassidy's possible strategies for the Flint store's plant market. What should Cassidy do? Why? Suggest possible promotion plans for your preferred strategy.

31. Silverglade Homes*

Brian Silverglade, who seven years ago founded Silverglade Homes in Asheville, North Carolina, is excited that he'll complete his first LEED-certified "green" home this month. The LEED (Leadership in Energy and Environmental Design) rating means that the home uses 30 percent less energy and 20 percent less water than a conventional house; it also means that construction waste going into landfills must also be reduced. The house will be the model home to showcase Brian's new development, which includes four more homes that he hopes to complete in the next six months.

Although Brian is excited, he is also nervous. Rising interest rates and an uncertain economy have reduced demand in the local housing market. People who do buy homes are more price sensitive. That's a problem because building a green house

*Erik Hardy did the research for an earlier version of this case.

usually increases construction costs—but customers are not always aware of the benefits that come with the higher price tag. So Brian has to figure out how to find home buyers who are willing to pay a premium for his "green" homes.

Prior to building this home, Brian tried to make environmentally responsible building choices that didn't increase his costs. However, two years ago while at the National Association of Home Builders' convention in Orlando, Florida, Brian visited a booth that described LEED Certification, and it appealed to him. He also met a number of new suppliers who were offering sustainable building materials. When Brian returned from the convention in Orlando, Silverglade Homes salesperson Karen Toller told him that more home buyers were asking about environmental and energy-saving features. Brian thought the time was right to commit to building at least a few homes that met higher environmental standards.

Brian kept his eye out for a good piece of property for his project. Before long he found a three-acre parcel of land about 15 minutes from downtown Asheville. The land had a nice mix of hardwood trees and a small stream, but it lacked the panoramic mountain views expected by high-end home buyers in the area. Nevertheless, Brian thought the property would be ideal for a small neighborhood of moderate-sized green homes. He purchased the land—and his first green project was under way.

Brian worked closely with a local architect, Katie Kelly, who had won several awards from the Green Building Council for her innovative designs. Katie proposed that each home follow a theme based on a classic Appalachian farmhouse design that would blend well with the rural surroundings and fit the concept of clustered development. Clustered development allows a builder to increase the number of home sites allowed if land is set aside for open space. For example, regulations usually required at least one acre for a rural home site. However, in a clustered development, the county would allow Brian to build five homes on 1.5 acres of his land if he dedicated the remaining 1.5 acres as open space controlled by a conservation easement.

Initially, Brian thought the additional two home sites and preservation of open space would be a huge benefit for his project. However, to get the development permit, he had to provide the county with a special land survey and biological inventory of the site. This extra survey work cost Brian $25,000 more than was normal.

Brian is behind schedule with construction because working with new types of materials has slowed him down. In some cases, his workers even had to be trained by factory representatives on the proper installation of materials. In addition, many of the materials in a LEED home—such as low-E windows, blown foam insulation, a high-efficiency furnace and water heater, and Energy Star appliances—have premium prices. As a result, Brian's LEED-certified homes cost about 10 percent more than conventionally built homes of the same size.

However, LEED homes do offer buyers a number of benefits. Toxin-free building materials help combat indoor air pollution—and green homes are less likely to have problems with mold or mildew. Energy and water savings for the homes Brian is building should be $2,000 to $3,000 per year. Plus, buyers can feel good that their homes produce fewer greenhouse gases, reduce dependence on fossil fuels, and send less construction waste to the local landfill.

Brian priced the five homes he is building at $250,000 to $300,000—about 10 percent more than similar non-LEED-certified homes in the area. So far, the homes are getting a few looks, but none have sold and he hasn't had an offer. The feedback that Karen Toller hears is that people like the homes but think that the price seems to be high; they like the general idea of owning a green home and saving money on energy, but they don't focus on the benefits. Perhaps that is because real estate agents in the area have little experience with green building and are used to talking about value in terms of "cost per square foot."

Karen Toller works full-time as a real estate agent—and Silverglade Homes is one of her clients. She receives a 1 percent sales commission for every Silverglade home that is sold. She typically writes the listing that all real estate agents can read on the Multiple Listing Service (MLS) website and also handles the contract to complete a sale. With no movement on Brian's new houses and with prime spring selling season coming up fast, Brian has asked Karen to meet with him to discuss ways to spark more interest in his development. Brian wants a plan that will bring people out to see his new homes and development—and a way to ensure that they are aware of the benefits he is offering.

Brian wonders what needs to be done to tell customers about the benefits of green building. Some other builders are advertising in local media and developing brochures to leave in a rack outside of their homes—but Brian has not needed to do that before. He wonders if he might be able to generate some inexpensive publicity by working with the local newspaper; it has run several feature stories about the environment but none about green building. He also thinks he needs to do more to convince real estate agents about the benefits of green housing. One idea is to ask someone at the Green Building Council or possibly his architect to put on a seminar, but Brian doesn't know if real estate agents will show up. He also wonders if putting signs in his model home to point out the environmental benefits might be helpful.

When Brian started his project, he thought his green homes would sell themselves, but now he wonders if he isn't ahead of his time.

What do you think of Brian Silverglade's marketing strategy so far? What promotion objectives should Brian set for his marketing strategy? What are the advantages and disadvantages of targeting communications at consumers as compared to real estate agents? What would you recommend as a promotion blend?

32. Mallory's Lemonade Stand (A)

Eleven-year-old Mallory is a budding entrepreneur. She is always coming up with new business ideas. In early summer 2019, her latest venture is a lemonade stand. Mallory is focused on making a high level of profits—she is saving for a new bike. Mallory is trying to to develop a marketing strategy to help her achieve this goal.

Last year, Mallory helped a friend with a lemonade stand. This year she wants to run her own business. Her parents expect her to pay for almost everything related to her business. Her dad built her a small lemonade stand—and although he donated his time, he did ask for the $12 in materials needed to build the stand. Mallory had to purchase other supplies as well. She bought a nice one-gallon pitcher for $5, and cups cost her $0.10 each. She figured out that sugar and lemons were costing her $0.30 for each 12-ounce cup she sold. Her parents let her use ice from the freezer—no cost there. Mallory initially planned to charge $0.75 per cup for her lemonade. Oh, and Mallory has a secret weapon. Her grandma had a little spice combination she put into lemonade that people just love—she knows from experience that once customers taste *her* lemonade they will be coming back for more.

Mallory is looking for help with some of her calculations, especially the final determination of an appropriate price. Answer the following questions:

1. Which of Mallory's costs are **fixed costs**? These are the costs that will not change whether Mallory sells few or many cups of lemonade.
2. What are the **variable costs**? These are costs that change directly with Mallory's sales. They go up proportionately to Mallory's increased number of sales.
3. What is Mallory's **fixed-cost contribution per unit** (assumed selling price minus the variable cost per unit)?
4. How many cups of lemonade does Mallory need to sell to reach her **break-even point (BEP)**? Recall that the break-even point (in units) = (total fixed costs)/(fixed cost contribution per unit).
5. After conferring with her sister (a marketing major in college), Mallory considered four marketing strategies in addition to her original plan. Calculate the break-even point (in cups) for each (assume for the moment that each is an independent alternative):
    a) Mallory's family lives on a cul-de-sac and her parents said she has to sell from her driveway. This means her street gets limited traffic. She thinks creating two signs and placing them at the entrance of the cul-de-sac would make more people aware of her stand. The materials to make the signs cost $5. With this change what is her BEP?
    b) Mallory saw some fancier plastic cups that she thought might make her lemonade look nicer. The larger cups would go for $0.15 each, and because they were a little larger, costs for lemons and sugar would rise $0.05 per cup. With this change what is her BEP?
    c) Her sister said she was charging too much for a cup of lemonade, so Mallory considered charging $0.50 per cup instead of the planned $0.75. With this change what is her BEP?
    d) She also wanted to consider a higher price of $1 per cup instead of the planned $0.75. With this change what is her BEP?
6. What other information would be helpful to know before you make your final recommendation?
7. What else could Mallory do to deliver more value to her customers?

More information on the terms in bold can be found in Chapter 18.

33. Mallory's Lemonade Stand (B)

It is March 2020 and Mallory has come a long way since last summer. Her lemonade was a hit—everyone raved about her secret formula (she added some special spices suggested by her grandmother). When Damien Cuban (billionaire investor) happened to drive through the neighborhood, he had a cup of Mallory's lemonade and offered to invest in the company. So Mallory's Lemonade Stand Lemonade will soon be found in a limited number of local Whole Foods Markets. Mallory had to figure out pricing once again. She started with the knowledge that retailers would want to sell her 20-oz. bottles for $2.00 at retail.

Mallory figured her costs as follows:

- $10,000 in fixed costs for various costs including some local advertising to support the Whole Foods launch.
- A co-packer (company that manufactures food products to specifications) will charge $0.50 for each 20-oz. bottle—including bottles, packages, and delivery.
- Mallory assumes she can sell 30,000 bottles.
- Mallory plans to charge Whole Foods $1.00 each for the bottles. Whole Foods will charge $2.00.
- Using average-cost pricing, she is counting on a $5,000 profit.

1. Assuming the numbers shown here, what is Whole Foods' **markup (percent)** on each bottle of Mallory's Lemonade Stand Lemonade?
2. Using Mallory's projected sales of 30,000 bottles, calculate Mallory's (a) **average fixed cost**, (b) **average variable cost**, (c) **margin** in dollars per bottle, and (d) total profit.
3. What is Mallory's profit if sales turn out to be only 15,000 bottles?

When MLS Lemonade failed to meet sales targets, Whole Foods quietly dropped the line. But all was not lost. Damien Cuban was able to help her obtain a presence in a convenience store chain. Cuban suggested that Mallory use **marginal analysis** (which focuses on the changes in total revenue and total cost from selling one more unit to find the most profitable price and quantity). This approach recognized that price and quantity sold change together. Cuban called in a favor from a friend who is an expert in the beverage market. This woman was able to provide a well-informed estimate of how much Mallory could sell during her test at different prices. Fill out the table and suggest an appropriate price. Because the convenience store manager knew it would have a 50 percent markup, the first column is actually the store's purchase price and Mallory's selling price.

Mallory has partially filled out a spreadsheet with the numbers (see below). Finish this spreadsheet based on the following questions.

4. *Assuming the expert's estimates are accurate, what price maximizes revenue? What is that revenue?*
5. *Assuming the expert's estimates are accurate, what price maximizes profit? What is that profit?*
6. *What would help Mallory reduce the price sensitivity some customers may have for her lemonade?*

More information on the terms in bold can be found in Chapter 18.

Price (P)	Quantity	Revenue (P * Q)	Total VC	Total FC	Total Cost	Profit (Rev − Cost)
$3.00	0	$ 0	$ 0	$10,000	$10,000	−$10,000
$1.75	10,000	$17,500	$ 5,000		$15,000	$ 2,500
$1.40	15,000	$21,000			$17,500	$ 3,500
$1.20	20,000	$24,000	$10,000			
$1.00	30,000	$30,000	$15,000			
$0.85	45,000	$38,250				
$0.77	51,000	$39,270				
$0.70	55,000				$37,500	$ 1,000

34. Working Girl Workout

Twenty-four-year-old Tatiana Rogers is finally pursuing her dream to become a veterinarian. After a few years working as a veterinary technician, Tatiana was excited to be accepted into a combination Doctor of Veterinary Medicine and MBA program. The coursework and student assistantship kept her pretty busy, but Tatiana always found time for her hobby as a bodybuilder. She had always been a fitness junkie, and in the last few years Tatiana enjoyed competing in bodybuilding competitions. This usually meant year-round training and attending four to six competitions per year. It was a great stress release.

Tatiana was pretty successful in these competitions. She enjoyed sharing pictures of herself working out and competing on Instagram. Soon friends and Instagram followers started asking, "How do you do it?" and "What's your secret?" They wondered how she found the time to complete a demanding school program, work a part-time job, stay in shape, *and* compete.

Tatiana had developed some tricks to stay in shape over the years. She thought there might be an opportunity to share her knowledge and experience and turn it into a side business. She was familiar with some workout brands that were built through the social media platform Instagram—many were based on the personal brand of an individual. One of these brands had more than a million followers. Tatiana wondered if she could build a business like this. She grabbed the Working.Girl.Workout handle at Instagram and the workinggirlworkout.com domain and started to work.

To get a better idea of the market, Tatiana checked out a couple of other Instagram-based workout coaching businesses. The two leading online fitness plans designed for women were Massy Arias' MA30DAY program and Kathy Drayton's 30daytransformationteam. Arias is an Instagram celebrity whose personal page has 2.5 million followers. She set up MA30DAY as a "business page" on Instagram and invited the followers from her personal page who were interested in fitness to follow her business page. Kathy Drayton is a celebrity fitness trainer. She uses her personal page to solicit followers.

Tatiana created a competitor matrix (shown in Table 1) to compare the two workout coaches. This helped her understand what it might take to start her business—and what she might get in return. She noticed both programs' products include two 30-day plans—each designed for different fitness goals (for example, weight loss or fit and curvy). For a one-time fee, customers receive a detailed meal and workout plan designed to meet specific fitness goals. Both coaches use contests to build their following. Followers enter the contests by posting their progress, workouts, and meals to their Instagram during the contest time frame. In addition, Tatiana noted the frequency and nature of the coaches' posts. The way they posted gave a different feel to the two sites, with Drayton's personal page having a more "friend" feeling; more than half of her posts contained family photos.

In developing her marketing strategy, Tatiana remembered her MBA marketing class. She knew a marketing strategy involved decisions about the target market and a plan for each of the four Ps (Product, Price, Place, and Promotion). She started by thinking about her target market. What kind of customers did Tatiana want to appeal to? Tatiana thought about her vet school; 87 percent of students were female, and the average student was 25 years old. In a room full of future doctors, Tatiana recognized the changing nature of the U.S. workforce; more women are participating in demanding careers, leaving them less time to work out. As a future professional herself, Tatiana decided she would target young, career-driven women in their mid-twenties and thirties. She thought about the unmet needs

Table 1 Competitor Matrix

	MA30DAY	30daytransformationteam
Type of page	Business	Personal
Founder/personal	Massy Arias	Kathy Drayton
Started	October 2015	November 2015
Focus of exercise plan	Performance and wellness	Building a curvy physique
Number of 30-day plans	2	2
Price	$150	$315
Meal plan options	Carnivore, vegan, non-vegan	Carnivore, vegetarian, vegan, pescatarian
Contest prizes	1st Place: $2,000 2nd Place: $1,000 3rd place: 6 months of TRU supplements	$5,000
Instagram posting frequency	7 times per week	16 times per week
Total posts	1,027	5,020
Followers	76,100	1.3 million
Content	42% gym selfies 28% motivational photos 15% customer transformation photos 15% workout videos	53% Kathy lifestyle and family photos 27% customer transformation photos 20% program promotion photos 26% gym selfies

of this target market, remembering her days working at the veterinary hospital during college, before she took an interest in her health. As a new hire, Tatiana adopted a diet of microwaved burritos and instant noodles, a common practice in the fast-paced medical field. Although she enjoyed fitness on her days off, with such a nutritionally poor diet, Tatiana found herself too exhausted to enjoy these activities after work. After reflecting on this experience, she recognized her target market of working professionals had an unmet need; they don't know how to include getting healthy and in shape in their busy schedule.

Next, Tatiana figured out her product. With limited time available outside of the workplace, Tatiana recognized people were interested in learning how to get fit despite a busy schedule. She felt her secret was in her ability to develop extremely efficient workout and nutrition plans, so she decided to formalize these. She created several digital 30-day workout and meal plans tailored toward different fitness goals such as weight loss, full-body toning, lower-body growth, home workouts, and a beginner's gym guide. Each workout was designed to maximize workout effectiveness and efficiency and could be completed in less than an hour. Each exercise in the workout had a link to a tutorial video to ensure customers applied the proper form and did the exercise correctly. Additionally, each plan included quick-prep, healthy, and tasty recipes that could be made and stored for the week. Recognizing the importance of marketing research, Tatiana shared these plans with her friends. They took some before and after photos and gave her feedback for improving her product. Tatiana had a product line.

The next step was thinking about what her business model and pricing should be. One option was a freemium model—just give away all of the products for free. Free might build a big following, and she could then use her popularity to make influencer posts for workout and nutrition brands. She read that Instagram influencers were paid about $1,000 per post per 100,000 followers.

Another strategy would be to charge for her content. This model would increase expenses and generate fixed costs such as website development and maintenance, payment systems, and e-mail marketing systems, among others. Additionally, this model exposed Tatiana to the price sensitivity of her young, working customers who might be facing high levels of educational debt. However, with her competitors pricing their workout and nutrition plans between $150 and $315, selling even one plan per month at the $150 price point could cover her fixed overhead and place her in a position to make a profit. Additionally, once customers started sharing their Working.Girl.Workout program results, her Instagram following could increase to a point where she could become an influencer. Tatiana decided to charge $150 for her 30-day plan.

With target market and pricing model decisions largely made, Tatiana began to think about promotion. How should she let people know about her product? Obviously as an Instagram-based business, this social media platform was going to be central to any plan. As Tatiana began planning her Instagram posting schedule and marketing plan, she wondered if she should also consider other social media. Should she start Facebook, Twitter, or Snapchat accounts? For the launch, Tatiana decided to focus on Instagram and share all posts on a business Facebook page.

Place was pretty straightforward. She knew the online sales model would work—as all of her products were digital. She would have to set up ways to collect payments, but there were models for that. So using only a direct-to-consumer model made sense.

Promotion was trickier. Tatiana wanted to position her product as a "Fitness Boot Camp." How could she build awareness and interest? What would make people part with $150? Would her target market view spending that much money on an unknown product as a risk? As a student, she did not have a lot of money, but she was willing to spend up to $2,000 on promotion. She knew the small budget would require some creativity. She didn't have a lot of extra cash to make high-quality owned media like commercials or sponsored posts. Instagram advertising costs can vary anywhere from $5 to $100 per day. A $5 ad results in about 300 views and two site clicks in 12 hours.

She wondered if there was another way to spread the word and encourage word-of-mouth. She could start by reaching out to different Instagram web pages, like Strongfitnessmag (with 93,000 followers on Instagram) or Womenshealthmag (1.6 million Instagram followers) for a shout-out. An Instagram shout-out would add credibility to her program and would build customer trust. She has previously been featured in *Strong Fitness Magazine*'s "Women to Watch" section, and the magazine also shouted her out on its Instagram. But that was a year ago and before she started Working Girl Workout. Tatiana also wondered about engaging some micro-influencers—perhaps some lifestyle Instagrammers who focused on working women and health. She would have to research that idea.

Tatiana wondered if there was another way to build trust with her target audience. She knew customers were more likely to trust referrals. Was there a way to encourage her followers to invite other people to join? Was there a way to get more user-generated content? How could she generate this earned media? Tatiana thought about how she could encourage her customers to tag Working.Girl.Workout in their transformation photos and workouts.

She also thought about sales promotion. One idea, based on contests run by similar sites, was to announce a 30-day challenge where customers who post about the program on Instagram have an opportunity to win prizes. This challenge would give Tatiana content to post on her own page and would generate word-of-mouth.

How do you recommend Tatiana approach her promotion? What are your specific recommendations? What other ideas do you have for building a following for Working.Girl.Workout? Justify your suggestions based on what you read in the Promotion chapters.

Glossary

2/10, net 30 Allows a 2 percent discount off the face value of the invoice if the invoice is paid within 10 days.

Accessories Short-lived capital items—tools and equipment used in production or office activities.

Accumulating Collecting products from many small producers.

Acquisition cost The expense required to acquire a new customer.

Administered channel systems Various channel members informally agree to cooperate with one another.

Administered prices Consciously set prices aimed at reaching the firm's objectives.

Adoption curve Shows when different groups accept ideas.

Adoption process The steps individuals go through on the way to accepting or rejecting a new idea.

Advertising Any *paid* form of nonpersonal presentation of ideas, goods, or services by an identified sponsor.

Advertising agencies Specialists in planning and handling mass-selling details for advertisers.

Advertising allowances Price reductions to firms in the channel to encourage them to advertise or otherwise promote the firm's products locally.

Advertising managers Managers of their company's mass-selling effort in television, newspapers, magazines, online, and advertising on social media sites.

Advertising media The various means by which a message is communicated to its target market.

Agent wholesalers Wholesalers that do not own (take title to) the products they sell.

AIDA model Consists of four promotion jobs: (1) to get *Attention*, (2) to hold *Interest*, (3) to arouse *Desire*, and (4) to obtain *Action*.

Allowance (accounting term) Occurs when a customer is not satisfied with a purchase for some reason and the seller gives a price reduction on the original invoice (bill), but the customer keeps the goods or services.

Allowances Reductions in price given to final consumers, customers, or channel members for doing something or accepting less of something.

Artificial intelligence (AI) Having machines operate like humans with respect to learning and decision making.

Assorting Putting together a variety of products to give a target market what it wants.

Attitude A person's point of view toward something.

Auction companies Agent wholesalers that provide a place where buyers and sellers can come together and complete a transaction.

Augmented reality (AR) An overlay of a computer-generated image, sound, text, or video onto a user's view of the physical world.

Automatic vending Selling and delivering products through vending machines.

Average cost (per unit) The total cost divided by the related quantity.

Average fixed cost (per unit) The total fixed cost divided by the related quantity.

Average variable cost (per unit) The total variable cost divided by the related quantity.

Average-cost pricing Adding a reasonable markup to the average cost of a product.

B Corporation (B Corp) certification A private certification that a corporation meets a high standard for social and environmental performance.

Baby Boomers People born between 1946 and 1964.

Bait pricing Setting some very low prices to attract customers but trying to sell more expensive models or brands once the customer is in the store.

Balance sheet An accounting statement that shows a company's assets, liabilities, and net worth.

Basic list prices The prices that final customers or users are normally asked to pay for products.

Basic sales tasks Order-getting, order-taking, and supporting.

Battle of the brands The competition between dealer brands and manufacturer brands.

Belief A person's opinion about something.

Benefit corporation A legal corporate structure that allows for goals that may include positive impacts on society, employees, the community, and the environment.

Big data Data sets too large and complex to work with typical database management tools.

Blog A regularly updated website, usually managed by one person or a small group and written in an informal, conversational style.

Brand community A group of customers joined around a particular brand or common set of shared interests.

Brand equity The value of a brand's overall strength in the market.

Brand familiarity How well customers recognize and accept a company's brand.

Brand insistence Customers insist on a firm's branded product and are willing to search for it.

Brand managers Manage specific products, often taking over the jobs formerly handled by an advertising manager. See also *Product managers*.

Brand name A word, letter, or a group of words or letters.

Brand nonrecognition Final customers don't recognize a brand at all—even though intermediaries may use the brand name for identification and inventory control.

Brand preference Target customers usually choose the brand over other brands, perhaps because of habit or favorable past experience.

Brand recognition Customers remember the brand.

Brand rejection Potential customers won't buy a brand unless its image is changed.

Branded apps Sponsored software applications that benefit customers by providing entertainment, solving a problem, and/or saving time.

Branded services Valued services a brand provides that are not directly connected to a core product offering.

Branding The use of a name, term, symbol, or design—or a combination of these—to identify a product.

Break-even analysis An approach to determine whether the firm will be able to break even—that is, cover all its costs—with a particular price.

Break-even point (BEP) The sales quantity where the firm's total cost will just equal its total revenue.

Breakthrough opportunities Opportunities that help innovators develop hard-to-copy marketing strategies that will be very profitable for a long time.

Brokers Agent wholesalers who specialize in bringing buyers and sellers together.

Bulk-breaking Dividing larger quantities into smaller quantities as products get closer to the final market.

Business and organizational customers Any buyers who buy for resale or to produce other goods and services.

Business products Products meant for use in producing other products.

Buying center All the people who participate in or influence a purchase.

Buying function Looking for and evaluating goods and services.

Capital item A long-lasting product that can be used and depreciated for many years.

Case studies Success stories about how a company helped another customer.

Cash discounts Reductions in the price to encourage buyers to pay their bills quickly.

Cash-and-carry wholesalers Like service wholesalers, except that the customer must pay cash.

Catalog wholesalers Sell out of catalogs that may be distributed widely to smaller industrial customers or retailers that might not be called on by other wholesalers.

Channel captain A manager who helps direct the activities of a whole channel and tries to avoid, or solve, channel conflicts.

Channel of distribution Any series of firms or individuals who participate in the flow of products from producer to final user or consumer.

Click-through rate The number of people who click on the ad divided by the number of people the ad is presented to.

Close The salesperson's request for an order.

Clustering techniques Approaches used to try to find similar patterns within sets of data.

Collaborators Firms that provide one or more of the marketing functions other than buying.

Combination export manager A blend of manufacturers' agent and selling agent—handling the entire export function for several producers of similar but noncompeting lines.

Combined target market approach Combining two or more submarkets into one larger target market as a basis for one strategy.

Combiners Firms that try to increase the size of their target markets by combining two or more segments.

Command economy Government officials decide what and how much is to be produced and distributed by whom, when, to whom, and why.

Communication process A source trying to reach a receiver with a message.

Comparative advertising Advertising that makes specific brand comparisons using actual product names.

Competitive advantage A firm has a marketing mix that the target market sees as better than a competitor's mix.

Competitive advertising Advertising that tries to develop selective demand for a specific brand rather than a product category.

Competitive bids Terms of sale offered by different suppliers in response to the buyer's purchase specifications.

Competitive environment The number and types of competitors the marketing manager must face and how they may behave.

Competitive rivals A firm's closest competitors.

Competitor analysis An organized approach for evaluating the strengths and weaknesses of current or potential competitors' marketing strategies.

Complementary product pricing Setting prices on several related products as a group.

Components Processed expense items that become part of a finished product.

Concept testing Getting reactions from customers about how well a new-product idea fits their needs.

Confidence interval The range on either side of an estimate from a sample that is likely to contain the true value for the whole population.

Consultative selling approach A type of sales presentation in which the salesperson develops a good understanding of the individual customer's needs before trying to close the sale.

Consumer panel A group of consumers who provide information on a continuing basis.

Consumer Product Safety Act A 1972 law that set up the Consumer Product Safety Commission to encourage more awareness of safety in product design and better quality control.

Consumer products Products meant for the final consumer.

Consumer surplus The difference to consumers between the value of a purchase and the price they pay.

Containerization Grouping individual items into an economical shipping quantity and sealing them in protective containers for transit to the final destination.

Continuous improvement A commitment to constantly make things better one step at a time.

Continuous innovations New products that don't require customers to learn new behaviors.

Contractual channel systems Various channel members agree by contract to cooperate with one another.

Convenience (food) stores A convenience-oriented variation of the conventional limited-line food stores.

Convenience products Products a consumer needs but isn't willing to spend much time or effort shopping for.

Cooperative advertising Producers sharing in the cost of ads with wholesalers or retailers.

Copy thrust What the words and illustrations of an ad should communicate.

Corporate chain A firm that owns and manages more than one store—and often it's many.

Corporate channel systems Corporate ownership all along the channel.

Corrective advertising Ads to correct deceptive advertising.

Cost of sales Total value (at cost) of the sales during the period.

Cues Products, signs, ads, and other stimuli in the environment.

Cultural and social environment Affects how and why people live and behave as they do.

Culture The whole set of beliefs, attitudes, and ways of doing things of a reasonably homogeneous set of people.

Cumulative quantity discounts Reductions in price for larger purchases over a given period, such as a year.

Customer equity The expected earnings stream (profitability) of a firm's current and prospective customers over some period of time.

Customer journey map The story and graphic diagram of a customer's experience in the buying process from need awareness through the purchase process and post-purchase relationship.

Customer lifetime value (CLV) Total profits a single customer contributes to a firm over the length of the relationship.

Customer relationship management (CRM) An approach where the seller fine-tunes the marketing effort with information from a detailed customer database.

Customer satisfaction The extent to which a firm fulfills a consumer's needs, desires, and expectations.

Customer service A personal communication between a seller and a customer who wants the seller to resolve a problem with a purchase; often the key to building repeat business.

Customer service level How rapidly and dependably a firm can deliver what customers want.

Customer service reps Supporting salespeople who work with customers to resolve problems that arise with a purchase, usually after the purchase has been made.

Customer value The difference between the benefits a customer sees from a market offering and the costs of obtaining those benefits.

Data warehouse A place where databases are stored so that they are available when needed.

Dealer brands Brands created by intermediaries. See also *Private brands*.

Decision support system (DSS) A computer program that makes it easy for marketing managers to get and use information *as they are making decisions*.

Decoding The receiver in the communication process translating the message.

Demand curve A graph of the relationship between price and quantity demanded in a market, assuming all other things stay the same.

Demand-backward pricing Setting an acceptable final consumer price and working backward to what a producer can charge.

Department stores Larger stores that are organized into many separate departments and offer many product lines.

Derived demand Demand for business products derives from the demand for final consumer products.

Determining dimensions The dimensions that actually affect the customer's purchase of a *specific* product or brand in a product-market.

Differentiation The marketing mix is distinct from and better than what's available from a competitor.

Direct competitive advertising Competitive advertising that aims for immediate buying action.

Direct investment A parent firm has a division (or owns a separate subsidiary firm) in a foreign market.

Direct marketing Direct communication between a seller and an individual customer using a promotion method other than face-to-face personal selling.

Discontinuous innovations New products that require that customers adopting the innovation significantly change their behavior.

Discounts Reductions from list price given by a seller to buyers who either give up some marketing function or provide the function themselves.

Discrepancy of assortment The difference between the lines a typical producer makes and the assortment final consumers or users want.

Discrepancy of quantity The difference between the quantity of products it is economical for a producer to make and the quantity final users or consumers normally want.

Discretionary income What is left of disposable income after paying for necessities.

Dissonance Tension caused by uncertainty about the rightness of a decision.

Distribution center A special kind of warehouse designed to speed the flow of goods and avoid unnecessary storing costs.

Diversification Moving into totally different lines of business—perhaps entirely unfamiliar products, markets, or even levels in the production-marketing system.

Door-to-door selling Going directly to the consumer's home.

Drive A strong stimulus that encourages action to reduce a need.

Drop-shippers Wholesalers that own (take title to) the products they sell but do not actually handle, stock, or deliver them.

Dumping Pricing a product sold in a foreign market below the cost of producing it or at a price lower than in its domestic market.

Dynamic behavioral segmentation The use of real-time data to continuously update a customer's placement in a market segment.

Dynamic pricing Pricing products at a particular customer's perceived ability to pay.

Dynamically continuous innovations New products that require minor changes in customer behavior.

Early adopters The second group in the adoption curve to adopt a new product; these people are usually well respected by their peers and often are opinion leaders.

Early majority A group in the adoption curve that avoids risk and waits to consider a new idea until many early adopters try it and like it.

Earned media Promotional messages *not* directly generated by the company or brand, but rather by third parties such as journalists or customers.

Economic buyers People who know all the facts and logically compare choices to get the greatest satisfaction from spending their time and money.

Economic environment Macroeconomic factors, including national income, economic growth, and inflation, that affect patterns of consumer and business spending.

Economic needs Needs concerned with making the best use of a consumer's time and money—as the consumer judges it.

Economic system The way an economy organizes to use scarce resources to produce goods and services and distribute them for consumption by various people and groups in the society.

Economies of scale As a company produces larger numbers of a particular product, the cost of each unit of the product goes down.

Elastic demand If prices are dropped, the quantity demanded will stretch enough to increase total revenue.

Elastic supply The quantity supplied does stretch more if the price is raised.

Electronic data interchange (EDI) An approach that puts information in a standardized format easily shared between different computer systems.

Emergency products Products that are purchased immediately when the need is great.

Empowerment Giving employees the authority to correct a problem without first checking with management.

Empty nesters People whose children are grown and who are now able to spend their money in other ways.

Encoding The source in the communication process deciding what it wants to say and translating it into words or symbols that will have the same meaning to the receiver.

Equilibrium point The quantity and the price sellers are willing to offer are equal to the quantity and price that buyers are willing to accept.

Everyday low pricing Setting a low list price rather than relying on frequent sales, discounts, or allowances.

Exclusive distribution Selling through only one intermediary in a particular geographic area.

Expectation An outcome or event that a person anticipates or looks forward to.

Expense item A product whose total cost is treated as a business expense in the period it's purchased.

Expenses All the remaining costs that are subtracted from the gross margin to get the net profit.

Experimental method A research approach in which researchers compare the responses of two or more groups that are similar except on the characteristic being tested.

Export agents Manufacturers' agents who specialize in export trade.

Export brokers Brokers who specialize in bringing together buyers and sellers from different countries.

Exporting Selling some of what the firm produces to foreign markets.

Extensive problem solving When consumers put much effort into deciding how to satisfy a need.

Facebook An online social networking website that allows registered users to create profiles, upload photos and video, and send messages to friends, family, and colleagues.

Factor A variable that shows the relation of some other variable to the item being forecast.

Factor method An approach to forecast sales by finding a relation between the company's sales and some other factor (or factors).

Fad An idea that is fashionable only to certain groups who are enthusiastic about it—but these groups are so fickle that a fad is even more short-lived than a regular fashion.

Family brand A brand name that is used for several products.

Farm products Products grown by farmers, such as oranges, sugarcane, and cattle.

Fashion Currently accepted or popular style.

Federal Fair Packaging and Labeling Act A 1966 law requiring that consumer goods be clearly labeled in easy-to-understand terms.

Federal Trade Commission (FTC) Federal government agency that polices antimonopoly laws.

Financing Provides the necessary cash and credit to produce, transport, store, promote, sell, and buy products.

Fixed-cost (FC) contribution per unit The selling price per unit minus the variable cost per unit.

Flexible-price policy Offering the same product and quantities to different customers at different prices.

Focus group interview An interview of 6 to 10 people in an informal group setting.

Foreign Corrupt Practices Act A law passed by the U.S. Congress in 1977 that prohibits U.S. firms from paying bribes to foreign officials.

Franchise operation A franchisor develops a good marketing strategy, and the retail franchise holders carry out the strategy in their own units.

Free trade Refers to agreements between countries to not restrict imports and exports.

Freemium A pricing strategy that is a combination of free and premium. It refers to providing a product for no charge, while money is charged for additional features that enhance the product's use.

Full-line pricing Setting prices for a whole line of products.

General Data Protection Regulation (GDPR) A set of laws on data protection and privacy for all individuals within the European Union.

General merchandise wholesalers Service wholesalers that carry a wide variety of nonperishable items such as hardware, electrical supplies, furniture, drugs, cosmetics, and automobile equipment.

General stores Early retailers that carried anything they could sell in reasonable volume.

Generation X (Gen X) People born between 1965 and 1977.

Generation Y (Gen Y) People born between 1978 and 1994. Also called *Millennials*.

Generation Z (Gen Z) People born since 1995.

Generic market A market with *broadly* similar needs and sellers offering various, *often diverse,* ways of satisfying those needs.

Generic products Products that have no brand at all other than identification of their contents and the manufacturer or intermediary.

Greenwashing Making false claims that imply a company or its products do more for the environment.

Gross domestic product (GDP) The total market value of all goods and services provided in a country's economy in a year by both residents and nonresidents of that country.

Gross margin (gross profit) The money left to cover the expenses of selling the products and operating the business.

Gross national income (GNI) A measure that is similar to GDP, but GNI does not include income earned by foreigners who own resources in that nation.

Gross sales The total amount charged to all customers during some time period.

Heterogeneous shopping products Shopping products the customer sees as different and wants to inspect for quality and suitability.

Homogeneous shopping products Shopping products the customer sees as basically the same and wants at the lowest price.

Horizontal channel conflict Conflict that occurs between firms at the same level in a distribution channel.

Hypotheses Educated guesses about the relationships between things or about what will happen in the future.

Ideal market exposure When a product is available widely enough to satisfy target customers' needs but not exceed them.

Implementation Putting marketing plans into operation.

Import agents Manufacturers' agents who specialize in import trade.

Import brokers Brokers who specialize in bringing together buyers and sellers from different countries.

Impulse products Products that are bought quickly as *unplanned* purchases because of a strongly felt need.

Indirect competitive advertising Competitive advertising that points out product advantages—to affect future buying decisions.

Individual brands Separate brand names used for each product.

Individual product A particular product within a product line.

Inelastic demand Although the quantity demanded increases if the price is decreased, the quantity demanded will not stretch enough to avoid a decrease in total revenue.

Inelastic supply The quantity supplied does not stretch much (if at all) if the price is raised.

Influencers Trusted or well-known figures who can sway attitudes or purchase decisions among a particular target market—to promote a brand.

Infographic A visual image such as a chart or diagram used to represent information or data.

Informational privacy Anything that limits others' access to personal data that people consider sensitive or confidential.

Innovation The development and spread of new ideas, goods, and services.

Innovators The first group to adopt new products.

Inside sales force A sales force that meets with customers in a manner that is not face-to-face.

Instagram A free online photo- and video-sharing service geared to mobile phones.

Installations Important capital items such as buildings, land rights, and major equipment.

Institutional advertising Advertising that tries to promote an organization's image, reputation, or ideas rather than a specific product.

Integrated marketing communications The intentional coordination of every communication from a firm to a target customer to convey a consistent and complete message.

Intelligent agent A device that observes an environment and acts to achieve a goal.

Intensive distribution Selling a product through all responsible and suitable wholesalers or retailers who will stock or sell the product.

Intermediary Someone who specializes in trade rather than production.

Introductory price dealing Temporary price cuts to speed new products into a market and get customers to try them.

Inventory The amount of goods being stored.

Job description A written statement of what a salesperson is expected to do.

Joint venture In international marketing, a domestic firm entering into a partnership with a foreign firm.

Jury of executive opinion Forecasting by combining the opinions of experienced executives, perhaps from marketing, production, finance, purchasing, and top management.

Just-in-time delivery (JIT) Reliably getting products there *just* before the customer needs them.

Laggards Prefer to do things the way they have been done in the past and are very suspicious of new ideas. See also *Nonadopters* and *Adoption curve*.

Landing page A customized web page that logically follows from clicking on an organic search result, online advertisement, or other link.

Lanham Act A 1946 law that spells out what kinds of marks (including brand names) can be protected and the exact method of protecting them.

Late majority A group of adopters who are cautious about new ideas. See also *Adoption curve*.

Law of diminishing demand If the price of a product is raised, a smaller quantity will be demanded—and if the price of a product is lowered, a greater quantity will be demanded.

Leader pricing Setting some very low prices—real bargains—to get customers into retail stores.

Learning A change in a person's thought processes caused by prior experience.

Lease An agreement that gives a customer the right to use something for a specified period of time in exchange for regular payments.

Licensed brand A well-known brand that sellers pay a fee to use.

Licensing Selling the right to use some process, trademark, patent, or other right for a fee or royalty.

Lifestyle analysis The analysis of a person's day-to-day pattern of living as expressed in that person's *A*ctivities, *I*nterests, and *O*pinions (AIOs). See also *Psychographics*.

Limited problem solving When a consumer is willing to put *some* effort into deciding the best way to satisfy a need.

Limited-function wholesalers Merchant wholesalers that provide only *some* wholesaling functions.

Limited-line stores Stores that specialize in certain lines of related products rather than a wide assortment. See also *Single-line stores*.

LinkedIn A social networking website for businesspeople who create personal or company profiles.

Logistics The transporting, storing, and handling of goods in ways that match target customers' needs with a firm's marketing mix—both within individual firms and along a channel of distribution. See also *Physical distribution (PD)*.

Low-involvement purchases Purchases that have little importance or relevance for the customer.

Machine learning A type of computer algorithm where a software application becomes more accurate in predicting outcomes without actual programming.

Macro-marketing A social process that directs an economy's flow of goods and services from producers to consumers in a way that effectively matches supply and demand and accomplishes the objectives of society.

Major accounts sales force Salespeople who sell directly to large accounts such as major retail chain stores.

Management contracting The seller provides only management skills—others own the production and distribution facilities.

Manufacturer brands Brands created by producers.

Manufacturers' agents Agent wholesalers that sell similar products for several noncompeting producers for a commission on what is actually sold.

Manufacturers' sales branches Separate warehouses that producers set up away from their factories.

Marginal analysis Evaluating the change in total revenue and total cost from selling one more unit to find the most profitable price and quantity.

Markdown A retail price reduction that is required because customers won't buy some item at the originally marked-up price.

Markdown ratio A tool used by many retailers to measure the efficiency of various departments and their whole business.

Market A group of potential customers with similar needs who are willing to exchange something of value with sellers offering various goods or services—that is, ways of satisfying those needs.

Market development Trying to increase sales by selling present products in new markets.

Market growth A stage of the product life cycle when industry sales grow fast—but industry profits rise and then start falling.

Market information function The collection, analysis, and distribution of all the information needed to plan, carry out, and control marketing activities.

Market introduction A stage of the product life cycle when sales are low as a new idea is first introduced to a market.

Market maturity A stage of the product life cycle when industry sales level off and competition gets tougher.

Market penetration Trying to increase sales of a firm's present products in its present markets—probably through a more aggressive marketing mix.

Market potential What a whole market segment might buy.

Market segment A relatively homogeneous group of customers who will respond to a marketing mix in a similar way.

Market segmentation A two-step process of (1) *naming* broad product-markets and (2) *segmenting* these broad product-markets in order to select target markets and develop suitable marketing mixes.

Market share The portion of total sales in a product category accounted for by a particular brand.

Market-directed economy The individual decisions of the many producers and consumers make the macro-level decisions for the whole economy.

Marketing The performance of activities that seek to accomplish an organization's objectives by anticipating customer or client needs and directing a flow of need-satisfying goods and services from producer to customer or client.

Marketing analytics The practice of measuring, managing, and analyzing marketing performance to maximize its efficiency and effectiveness.

Marketing automation software Software that tracks individual customers' behavior and triggers actions in response to specific customer actions.

Marketing company era A time when, in addition to short-run marketing planning, marketing people develop long-range plans—sometimes five or more years ahead—and the whole-company effort is guided by the marketing concept.

Marketing concept The idea that an organization should aim *all* of its efforts at satisfying its *customers*—at a *profit*.

Marketing dashboard Displays up-to-the-minute marketing information in an easy-to-read format—much like a car's dashboard shows the speedometer and fuel gauge.

Marketing department era A time when all marketing activities are brought under the control of one department to improve short-run policy planning and to try to integrate the firm's activities.

Marketing ethics The moral standards that guide marketing decisions and actions.

Marketing information system (MIS) An organized way of continually gathering, accessing, and analyzing information that marketing managers need to make ongoing decisions.

Marketing management process The process of (1) *planning* marketing activities, (2) directing the *implementation* of the plans, and (3) *controlling* the plans.

Marketing metrics Numeric data that allow marketing managers to evaluate performance, often against a set target or goal.

Marketing mix The controllable variables that the company puts together to satisfy a target group.

Marketing model A statement of relationships among marketing variables.

Marketing orientation Trying to carry out the marketing concept.

Marketing plan A written statement of a marketing strategy *and* the time-related details for carrying out the strategy.

Marketing program Blends all of the firm's marketing plans into one big plan.

Marketing research Procedures to develop and analyze new information to help marketing managers make decisions.

Marketing research process A five-step application of the scientific method that includes (1) defining the problem, (2) analyzing the situation, (3) getting problem-specific data, (4) interpreting the data, and (5) solving the problem.

Marketing strategy Specifies a target market and a related marketing mix.

Markup A dollar amount added to the cost of products to get the selling price.

Markup (percent) The percentage of selling price that is added to the cost to get the selling price.

Markup chain The sequence of markups firms use at different levels in a channel—determining the price structure in the whole channel.

Mass marketing The typical production-oriented approach that vaguely aims at everyone with the same marketing mix.

Mass selling Communicating with large numbers of potential customers at the same time.

Mass-merchandisers Large, self-service stores with many departments that emphasize soft goods (housewares, clothing, and fabrics) and staples (like health and beauty aids) and selling on lower margins to get faster turnover.

Mass-merchandising concept The idea that retailers should offer low prices to get faster turnover and greater sales volume by appealing to larger numbers.

Merchant wholesalers Wholesalers that own (take title to) the products they sell.

Message channel The carrier of the message.

Micro-macro dilemma What is good for some producers and consumers may not be good for society as a whole.

Mission statement A statement that sets out the organization's basic purpose for being.

Missionary salespeople Supporting salespeople who work for producers by calling on intermediaries and their customers.

Mobile payments Payments made at the point of purchase using a mobile device (usually a cell phone or tablet computer).

Modified rebuy The in-between process where some review of the buying situation is done—though not as much as in new-task buying or as little as in straight rebuys.

Monopolistic competition A market situation that develops when a market has (1) different (heterogeneous) products and (2) sellers who feel they do have some competition in this market.

Multichannel distribution When a producer uses several competing channels to reach the same target market—perhaps using several intermediaries in addition to selling directly.

Multichannel shoppers Shoppers who use different channels as they move through the purchase process.

Multiple buying influence Several people share in making a purchase decision—perhaps even top management.

Multiple target market approach Segmenting the market and choosing two or more segments, then treating each as a separate target market needing a different marketing mix.

Nationalism An emphasis on a country's interests before everything else.

Natural products Products that occur in nature—such as timber, iron ore, oil, and coal.

Needs The basic forces that motivate a person to do something.

Negotiated contract buying Agreeing to a contract that allows for changes in the purchase arrangements.

Net An invoice term meaning that payment for the face value of the invoice is due immediately. See also *Cash discounts*.

Net profit What the company earns from its operations during a particular period.

Net sales The actual sales dollars the company receives.

New product A product that is new *in any way* for the company concerned.

New unsought products Products offering really new ideas that potential customers don't know about yet.

New-task buying When an organization has a new need and the buyer wants a great deal of information.

Noise Any distraction that reduces the effectiveness of the communication process.

Nonadopters Prefer to do things the way they have been done in the past and are very suspicious of new ideas. See also *Laggards* and *Adoption curve*.

Noncumulative quantity discounts Reductions in price when a customer purchases a larger quantity on an *individual order*.

Nonprice competition Aggressive action on one or more of the Ps other than Price.

North American Industry Classification System (NAICS) codes Codes used to identify groups of firms in similar lines of business.

Odd-even pricing Setting prices that end in certain numbers.

Oligopoly A special market situation that develops when a market has (1) essentially homogeneous products, (2) relatively few sellers, and (3) fairly inelastic industry demand curves.

Omnichannel A multichannel selling approach where a single retailer provides a seamless customer shopping experience from desktop computer, mobile device, telephone, or brick-and-mortar store.

One-price policy Offering the same price to all customers who purchase products under essentially the same conditions and in the same quantities.

Online retailers Stores that sell exclusively or almost exclusively online.

Operating ratio Ratio of items on the operating statement to net sales.

Operating statement A simple summary of the financial results of a company's operations over a specified period of time.

Operational decisions Short-run decisions to help implement strategies.

Opinion leader A person who influences others.

Order getters Salespeople concerned with establishing relationships with new customers and developing new business.

Order takers Salespeople who sell to regular or established customers, complete most sales transactions, and maintain relationships with their customers.

Order-getting Seeking possible buyers with a well-organized sales presentation designed to sell a good, service, or idea.

Order-taking The routine completion of sales made regularly to target customers.

Outsource When the buying organization chooses to contract with an outside firm to produce goods or services rather than producing them internally.

Owned media Promotional messages generated by a brand (or company or nonprofit organization) communicated through a message channel the brand directly controls.

Packaging Promoting, protecting, and enhancing the product.

Paid media Messages generated by a brand (or company or nonprofit organization) and communicated through a message channel the brand pays to access.

Pass-along When one customer passes information on to one or more other customers.

Patent Grants the inventor the ability to "exclude others from making, using, offering for sale, or selling the invention."

Pay-per-click An advertiser pays media costs only when a customer clicks on the ad that leads to the advertiser's website.

Penetration pricing policy Trying to sell the whole market at one low price.

Perception How we gather and interpret information from the world around us.

Personal data Information that can be used by itself or in combination with other information to identify someone.

Personal needs An individual's need for personal satisfaction unrelated to what others think or do.

Personal selling Direct spoken communication between sellers and potential customers, usually in person but sometimes over the telephone or even via an online conference.

Phony list prices Misleading prices that customers are shown to suggest that the price they are to pay has been discounted from list.

Physical distribution (PD) The transporting, storing, and handling of goods in ways that match target customers' needs with a firm's marketing mix—both within individual firms and along a channel of distribution. See also *Logistics*.

Physical distribution (PD) concept All transporting, storing, and product-handling activities of a business and a whole channel system should be coordinated as one system that seeks to minimize the cost of distribution for a given customer service level.

Physiological needs Biological needs such as the need for food, drink, rest, and sex.

Pinterest A website that allows registered users to share ideas and images they find online with fellow users.

Pioneering advertising Advertising that tries to develop primary demand for a product category rather than demand for a specific brand.

Place Making goods and services available in the right quantities and locations—when customers want them.

Population In marketing research, the total group you are interested in.

Positioning An approach that refers to how customers think about proposed or present brands in a market.

Positioning statement Statement that concisely identifies the firm's desired target market, product type, primary benefit or point of differentiation, and the main reasons a buyer should believe the firm's claims.

Predictive analytics A process used to analyze data to make predictions about unknown future events.

Prepared sales presentation A memorized presentation that is not adapted to each individual customer.

Prestige pricing Setting a rather high price to suggest high quality or high status.

Price The amount of money that is charged for "something" of value.

Price discrimination Injuring competition by selling the same products to different buyers at different prices.

Price fixing Competitors illegally getting together to raise, lower, or stabilize prices.

Price lining Setting a few price levels for a product line and then marking all items at these prices.

Price sensitivity The degree to which customers' purchase decisions are affected by the price.

Primary data Information specifically collected to solve a current problem.

Primary demand Demand for the general product idea, not just the company's own brand.

Private brands Brands created by intermediaries. See also *Dealer brands*.

Private warehouses Storing facilities owned or leased by companies for their own use.

Product The need-satisfying offering of a firm.

Product advertising Advertising that tries to sell a specific product.

Product assortment The set of all product lines and individual products that a firm sells.

Product development Offering new or improved products for present markets.

Product liability The legal obligation of sellers to pay damages to individuals who are injured by defective or unsafe products.

Product life cycle The stages a new-product idea goes through from beginning to end.

Product line A set of individual products that are closely related.

Product line length The number of individual products in a product line.

Product managers Manage specific products, often taking over the jobs formerly handled by an advertising manager. See also *Brand managers*.

Product-bundle pricing Setting one price for a set of products.

Product-market A market with *very* similar needs and sellers offering various *close substitute* ways of satisfying those needs.

Production Actually *making* goods or *performing* services.

Production era A time when a company focuses on production of a few specific products—perhaps because few of these products are available in the market.

Production orientation Making whatever products are easy to produce and *then* trying to sell them.

Professional services Specialized services that support a firm's operations.

Profit maximization objective An objective to get as much profit as possible.

Promotion Communicating information between the seller and potential buyer or others in the channel to influence attitudes and behavior.

Prospecting Following all the leads in the target market to identify potential customers.

Prototype An early sample or model built to test a concept.

Psychographics The analysis of a person's day-to-day pattern of living as expressed in that person's *A*ctivities, *I*nterests, and *O*pinions (AIOs). See also *Lifestyle analysis*.

Psychological pricing Setting prices that have special appeal to target customers.

Public relations Communication with noncustomers—including the press, labor, public interest groups, stockholders, and the government.

Public warehouses Independent storing facilities.

Publicity Any *unpaid* form of nonpersonal presentation of ideas, goods, or services.

Pulling Using promotion to get consumers to ask intermediaries for the product.

Purchase discount A reduction of the original invoice amount for some business reason.

Purchase situation Takes into account the purpose, time available, and location where a purchase is made.

Purchasing managers Buying specialists for their employers.

Purchasing specifications A written (or electronic) description of what the firm wants to buy.

Pure competition A market situation that develops when a market has (1) homogeneous (similar) products, (2) many buyers and sellers who have full knowledge of the market, and (3) ease of entry for buyers and sellers.

Pure subsistence economy Each family unit produces everything it consumes.

Push money (or prize money) allowances Allowances (sometimes called *PMs* or *spiffs*) given to retailers by manufacturers or wholesalers to pass on to the retailers' salesclerks for aggressively selling certain items.

Pushing Using normal promotion effort—personal selling, advertising, and sales promotion—to help sell the whole marketing mix to possible channel members.

Qualifying dimensions The dimensions that are relevant to including a customer type in a product-market.

Qualitative research Seeks in-depth, open-ended responses, not yes or no answers.

Quality A product's ability to satisfy a customer's needs or requirements.

Quantitative research Seeks structured responses that can be summarized in numbers—such as percentages, averages, or other statistics.

Quantity discounts Discounts offered to encourage customers to buy in larger amounts.

Rack jobbers Merchant wholesalers that specialize in hard-to-handle assortments of products that a retailer doesn't want to manage—and they often display the products on their own wire racks.

Raw materials Unprocessed expense items—such as logs, iron ore, and wheat—that are moved to the next production process with little handling.

Rebates Refunds to consumers after a purchase.

Receiver The target of a message in the communication process, usually a potential customer.

Reference group The people to whom an individual looks when forming attitudes about a particular topic.

Reference price The price a consumer expects to pay.

Referral program Offers a current customer an incentive for recommending a new customer to a business.

Regrouping activities Adjusting the quantities or assortments of products handled at each level in a channel of distribution.

Regularly unsought products Products that stay unsought but not unbought forever.

Reinforcement Occurs in the learning process when the consumer's response is followed by satisfaction—that is, reduction in the drive.

Reminder advertising Advertising to keep the product's name before the public.

Requisition A request to buy something.

Research proposal A plan that specifies what marketing research information will be obtained and how.

Response An effort to satisfy a drive.

Response rate The percentage of people contacted in a research sample who complete the questionnaire.

Retailing All of the activities involved in the sale of products to final consumers.

Retargeting Displaying ads to a web user based on sites he or she has previously visited; also called *behavioral retargeting*.

Retention rate The percentage of customers retained as compared to the total number of customers.

Return When a customer sends back purchased products.

Return on assets (ROA) The ratio of net profit (after taxes) to the assets used to make the net profit—multiplied by 100 to get rid of decimals.

Return on investment (ROI) Ratio of net profit (after taxes) to the investment used to make the net profit, multiplied by 100 to get rid of decimals.

Reverse channels Channels used to retrieve products that customers no longer want.

Risk taking Bearing the uncertainties that are part of the marketing process.

Robinson-Patman Act A 1936 law that makes illegal any price discrimination if it injures competition.

Routinized response behavior When consumers regularly select a particular way of satisfying a need when it occurs.

S.W.O.T. analysis Identifies and lists the firm's strengths and weaknesses and its opportunities and threats.

Safety needs Needs concerned with protection and physical well-being.

Sale price A temporary discount from the list price.

Sales decline A stage of the product life cycle when new products replace the old.

Sales era A time when a company emphasizes selling because of increased competition.

Sales forecast An estimate of how much an industry or firm hopes to sell to a market segment.

Sales managers Managers concerned with managing personal selling.

Sales presentation A salesperson's effort to make a sale or address a customer's problem.

Sales promotion Those promotion activities—other than advertising, publicity, and personal selling—that stimulate interest, trial, or purchase by final customers or others in the channel.

Sales promotion managers Managers of their company's sales promotion effort.

Sales quota The specific sales or profit objective a salesperson is expected to achieve.

Sales territory A geographic area that is the responsibility of one salesperson or several working together.

Sales-oriented objective An objective to get some level of unit sales, dollar sales, or share of market—without referring to profit.

Sample A part of the relevant population.

Scientific method A decision-making approach that focuses on being objective and orderly in *testing* ideas before accepting them.

Scrambled merchandising Retailers carrying any product lines they think they can sell profitably.

Search engine optimization (SEO) The process of designing a website so that it ranks high in a search engine's unpaid results.

Seasonal discounts Discounts offered to encourage buyers to buy earlier than present demand requires.

Secondary data Information that has been collected or published already.

Segmenters Aim at one or more homogeneous segments and try to develop a different marketing mix for each segment.

Segmenting An aggregating process that clusters people with similar needs into a market segment.

Selective demand Demand for a company's own brand rather than a product category.

Selective distribution Selling through only those intermediaries who will give the product special attention.

Selective exposure Our eyes and minds seek out and notice only information that interests us.

Selective perception People screen out or modify ideas, messages, and information that conflict with previously learned attitudes and beliefs.

Selective retention People remember only what they want to remember.

Selling agents Agent wholesalers who take over the whole marketing job of producers, not just the selling function.

Selling formula approach A sales presentation that starts with a prepared presentation outline—much like the prepared approach—and leads the customer through some logical steps to a final close.

Selling function Promoting the product.

Senior citizens People older than 65.

Sentiment analysis An automated process of analyzing and categorizing social media to determine the amount of positive, negative, and neutral online comments a brand receives.

Service An intangible offering involving a deed, performance, or effort.

Service mark Those words, symbols, or marks that are legally registered for use by a single company to refer to a service offering.

Service wholesalers Merchant wholesalers that provide all the wholesaling functions.

Shopping products Products that a customer feels are worth the time and effort to compare with competing products.

Simple trade era A time when families traded or sold their surplus output to local distributors.

Single target market approach Segmenting the market and picking one of the homogeneous segments as the firm's target market.

Single-line (or general-line) wholesalers Service wholesalers that carry a narrower line of merchandise than general merchandise wholesalers.

Single-line stores Stores that specialize in certain lines of related products rather than a wide assortment. See also *Limited-line stores*.

Situation analysis An informal study of what information is already available in the problem area.

Skimming price policy Trying to sell the top of the market—the top of the demand curve—at a high price before aiming at more price-sensitive customers.

Snapchat A mobile app and service for sharing photos, videos, and texts with other people.

Social class A group of people who have approximately equal social position as viewed by others in the society.

Social media Websites or software applications that allow users to create and share ideas, information, photos, and videos and interact in a social network.

Social needs Needs concerned with love, friendship, status, and esteem—things that involve a person's interaction with others.

Social responsibility A firm's obligation to improve its positive effects on society and reduce its negative effects.

Sorting Separating products into grades and qualities desired by different target markets.

Source The sender of a message.

Specialty products Consumer products that the customer really wants and makes a special effort to find.

Specialty shop A type of conventional limited-line store—usually small and with a distinct personality.

Specialty wholesalers Service wholesalers that carry a very narrow range of products and offer more information and service than other service wholesalers.

Standardization and grading Sorting products according to size and quality.

Staples Products that are bought often, routinely, and without much thought.

Statistical packages Easy-to-use computer programs that analyze data.

Status quo objectives "Don't-rock-the-*pricing*-boat" objectives.

Stocking allowances Allowances given to wholesalers or retailers to get shelf space for a product—sometimes called *slotting allowances*.

Stockturn rate The number of times the average inventory is sold during a year.

Storing The marketing function of holding goods.

Storing function Holding goods until customers need them.

Straight rebuy A routine repurchase that may have been made many times before.

Strategic (management) planning The managerial process of developing and maintaining a match between an organization's resources and its market opportunities.

Subscription pricing Customers pay on a periodic basis for access to a product.

Substitutes Products that offer the buyer a choice.

Supercenters (hypermarkets) Very large stores that try to carry not only food and drug items, but all goods and services that the consumer purchases routinely.

Supermarkets Large stores specializing in groceries—with self-service and wide assortments.

Supplies Expense items that do not become part of a finished product.

Supply chain The complete set of firms and facilities and logistics activities that are involved in procuring materials, transforming them into intermediate and finished products, and distributing them to customers.

Supply curve The quantity of products that will be supplied at various possible prices.

Supporting salespeople Salespeople who help the order-oriented salespeople but don't try to get orders themselves.

Sustainability The idea that it's important to meet present needs without compromising the ability of future generations to meet their own needs.

Sustainable competitive advantage A marketing mix that customers see as better than a competitor's mix and cannot be quickly or easily copied.

Target market A fairly homogeneous (similar) group of customers to whom a company wishes to appeal.

Target marketing A marketing mix is tailored to fit some specific target customers.

Target return objective A specific level of profit as an objective.

Task method An approach to developing a budget—basing the budget on the job to be done.

Team selling Different sales reps working together on a specific account.

Technical specialists Supporting salespeople who provide technical assistance to order-oriented salespeople.

Technology The application of science to convert an economy's resources to output.

Telemarketing Using the telephone to call on customers or prospects.

Total cost The sum of total fixed and total variable costs.

Total cost approach Evaluating each possible PD system and identifying all of the costs of each alternative.

Total fixed cost The sum of those costs that are fixed in total—no matter how much is produced.

Total quality management (TQM) The philosophy that everyone in the organization is concerned about quality, throughout all of the firm's activities, to better serve customer needs.

Total variable cost The sum of those changing expenses that are closely related to output—such as expenses for parts, wages, packaging materials, outgoing freight, and sales commissions.

Trade (functional) discount A list price reduction given to channel members for the job they are going to do.

Trade promotion Sales promotion aimed at intermediaries.

Trade-in allowance A price reduction given for used products when similar new products are bought.

Trademark Those words, symbols, or marks that are legally registered for use by a single company.

Traditional channel systems A channel in which the various channel members make little or no effort to cooperate with one another.

Transporting The marketing function of moving goods.

Transporting function The movement of goods from one place to another.

Trend extension Extends past experience to predict the future.

Triple bottom line A measure of long-term success that includes an organization's economic, social, and environmental outcomes.

Truck wholesalers Wholesalers that specialize in delivering products that they stock in their own trucks.

Trust The confidence a person has in the promises or actions of another person, brand, or company.

Twitter A social networking microblogging service that allows registered users to send out short (280 characters or less) messages called "tweets."

Unfair trade practice acts Legislation that puts a lower limit on prices, especially at the wholesale and retail levels.

Universal functions of marketing Buying, selling, transporting, storing, standardizing and grading, financing, risk taking, and market information.

Unsought products Products that potential customers don't yet want or know they can buy.

User-generated content Any type of communication created by customers for other customers. User-generated content can take many forms.

Validity The extent to which data measure what they are intended to measure.

Value in use pricing Setting prices that will capture some of what customers will save by substituting the firm's product for the one currently being used.

Value pricing Setting a fair price level for a marketing mix that really gives the target market superior customer value.

Vendor analysis Formal rating of suppliers on all relevant areas of performance.

Vertical channel conflict Conflict that occurs between firms at different levels in the channel of distribution.

Vertical integration Acquiring firms at different levels of channel activity.

Vertical marketing systems Channel systems in which the whole channel focuses on the same target market at the end of the channel.

Wants Needs that are learned during a person's life.

Warranty What the seller promises about its product.

Waste Electrical and Electronic Equipment (WEEE) Directive The European Community's law that requires producers to take back waste electrical and electronic equipment.

Wheel of retailing theory New types of retailers enter the market as low-status, low-margin, low-price operators and then, if successful, evolve into more conventional retailers offering more services with higher operating costs and higher prices.

Wheeler-Lea Act Law that bans unfair or deceptive acts in commerce.

White paper An authoritative report or guide that addresses important issues in an industry and offers solutions.

Wholesalers Firms whose main function is providing *wholesaling activities*.

Wholesaling The *activities* of those persons or establishments that sell to retailers and other merchants, or to industrial, institutional, and commercial users, but who do not sell in large amounts to final consumers.

YouTube A video-sharing website where users upload, view, rate, share, create playlists, and comment on videos.

Notes

CHAPTER 1

1. See www.nike.com; "Protecting the Future of Sport," *Nike News*, February 6, 2019; "Nike Is Ad Age's Marketer of the Year for 2018," *Advertising Age*, December 3, 2018; "Snapchat's Ecommerce Strategy Hit a New High When It Sold Out the New Air Jordans in Minutes," *Adweek*, February 22, 2018; "Watch Out, Nike, the Germans Are Coming," *Fortune*, June 15, 2017; Nike Raises Its Retail Game," *Chain Store Age*, January 5, 2017; "Nike's New Marketing Mojo," *Fortune*, February 13, 2012; "Is Nike's Flyknit the Swoosh of the Future?," *Bloomberg Businessweek*, March 15, 2012; "In China, Nike Sets Out to Alter Sports Mindset," *Wall Street Journal*, October 12, 2011; "Nike's Sustainability Report Takes Top Ceres-ACCA Award," *GreenBiz.com*, May 11, 2011; "How Nike Rules the World," *Fortune*, February 3, 2011; "Adidas' Brand Ambitions," *Portland Business Journal*, December 10, 2010; "Nike Releases Apparel Sustainability Software," *Waste and Recycling News*, December 6, 2010; "Nike, Inc.," *Datamonitor*, December 1, 2010; "Nike, Adidas Readjust Marketing Strategy," *China Daily*, August 30, 2010; "Despite Its Way with Women, Underdog Reebok Fixates on Nike," *AdAge.com*, July 12, 2010; "Letter to Shareholders," July 1, 2010; "Nike Women's Biz Gets Pounded as Toning Footwear Kicks Butt," *AdAge.com*, June 7, 2010; "Nike Looks beyond 'Swoosh' for Growth," *Wall Street Journal*, May 20, 2010; "Nike Launches GreenXchange for Corporate Idea-Sharing," *Fast Company*, February 2, 2010; "Li Ning and Ziba Design Want to Build China's First Truly Global Brand," *Fast Company*, October 1, 2009; "Best in Show: Nike's Scrappy Trash Talk Shoes," *Bloomberg Businessweek*, July 29, 2009; "Cool Runnings," *Time*, October 4, 2007; "Nike's Big Gay-Marketing Coup," *Bloomberg Businessweek*, April 12, 2013; "Nike CEO Mark Parker on His Company's Digital Future: Body-Controlled Music, Color-Coded Heart Rates," *Fast Company*, February 12, 2013; "The World's Most Valuable Brands, 2015," *Forbes*, www.forbes.com/powerful-brands/list/#tab:rank; "How Nike Became King of Endorsements," *CNNMoney*, June 5, 2015; "Nike Is Now Making Most of Its Shoes from Its Own Garbage," *Huffington Post*, May 11, 2016; "Nike Annual Income Statement" *Amigobulls.com*, retrieved February 9, 2017; "Nike's Challenge: Stay Ahead of the Pack," *Wall Street Journal*, September 20, 2015; "Sustainable Innovation Is a Powerful Engine for Growth," *FY14/15 Nike, Inc. Sustainable Business Report*; "From Skinny to Strong: Nike, Adidas Bet on China," *Wall Street Journal* video, December 21, 2015.

2. See www.learnthe4ps.com/category/personal-marketing-plan/ for articles related to developing your own personal marketing plan.

3. Gregory D. Upah and Richard E. Wokutch, "Assessing Social Impacts of New Products: An Attempt to Operationalize the Macromarketing Concept," *Journal of Public Policy & Marketing* 4 (1985).

4. An American Marketing Association committee recently defined marketing as "the activity, set of institutions, and processes for creating, communicating, delivering, and exchanging offerings that have value for customers, clients, partners, and society at large." The definition of marketing can be a source of debate; see "AMA's Definition of Marketing Stirs Debate," *B to B*, February 11, 2008; see also David Glen Mick, "The End(s) of Marketing and the Neglect of Moral Responsibility by the American Marketing Association," *Journal of Public Policy & Marketing*, Fall 2007; Debra Jones Ringold and Barton Weitz, "The American Marketing Association Definition of Marketing: Moving from Lagging to Leading Indicator," *Journal of Public Policy & Marketing*, Fall 2007; George M. Zinkhan and Brian C. Williams, "The New American Marketing Association Definition of Marketing: An Alternative Assessment," *Journal of Public Policy & Marketing*, Fall 2007; Gregory T. Gundlach, "The American Marketing Association's 2004 Definition of Marketing: Perspectives on Its Implications for Scholarship and the Role and Responsibility of Marketing in Society," *Journal of Public Policy & Marketing*, Fall 2007; Jagdish N. Sheth and Can Uslay, "Implications of the Revised Definition of Marketing: From Exchange to Value Creation," *Journal of Public Policy & Marketing*, Fall 2007; William L. Wilkie and Elizabeth S. Moore, "What Does the Definition of Marketing Tell Us about Ourselves?," *Journal of Public Policy & Marketing*, Fall 2007. Marketing can also include political dimensions; see Kelly D. Martin, Brett W. Josephson, Gautham G. Vadakkepatt, and Jean L. Johnson, "Political Management, Research and Development, and Advertising Capital in the Pharmaceutical Industry: A Good Prognosis?," *Journal of Marketing*, May 2018.

5. "Nielsen Breakthrough Innovation Report," *European Edition*, September 2014.

6. "Rustic Wisdom: Unilever to Take Project Shakti Global," *Economic Times*, January 19, 2009; "Strategic Innovation: Hindustan Lever Ltd.," *Fast Company*, December 19, 2007; "Shakti–Changing Lives in Rural India," http://hll.com/-citizen_lever/project_shakti.asp; P. Indu, Komal Chary, and Vivek Gupta, "Reviving Hindustan Lever Limited," ICFAI Center for Management Research, Case Study #306-410-1, 2006; V. Kasturi Rangan and Rhithari Rajan, "Unilever in India: Hindustan Lever's Project Shakti–Marketing FMCG to the Rural Consumer," Harvard Business School, Case #9-505-056; C. K. Prahalad, *The Fortune at the Bottom of the Pyramid* (Philadelphia: Wharton School Publishing, 2005).

7. Robert Bartels and Roger L. Jenkins, "Macromarketing," *The Journal of Marketing*, 1977; Shelby D. Hunt and John J. Burnett, "The Macromarketing/Micromarketing Dichotomy: A Taxonomical Model," *The Journal of Marketing*, 1982; Donald L. Shawver and William G. Nickels, "A Rationalization for Macromarketing Concepts and Definitions," *Journal of Macromarketing* 1, no. 1 (1981); William L. Wilkie and Elizabeth S. Moore, "Macromarketing as a Pillar of Marketing Thought," *Journal of Macromarketing* 26, no. 2 (2006).

8. Based on William McInnes, "A Conceptual Approach to Marketing," *Theory in Marketing*, second series, ed. Reavis Cox, Wroe Alderson, and Stanley J. Shapiro (Homewood, IL: Richard D. Irwin, 1964); Graham Hooley, Tony Cox, John Fahy, David Shipley, et al., "Market Orientation in the Transition Economies of Central Europe: Tests of the Narver and Slater Market Orientation Scales," *Journal of Business Research*, December 2000; Saeed Samiee, "Globalization, Privatization, and Free Market Economy," *Journal of the Academy of Marketing Science*, Summer 2001.

9. Kevin D. Bradford, Gregory T. Gundlach, and William L. Wilkie, "Countermarketing in the Courts: The Case of Marketing Channels and Firearms Diversion," *Journal of Public Policy & Marketing*, Fall 2005; Marian Friestad and Peter Wright, "The Next Generation: Research for Twenty-First-Century Public Policy on Children and Advertising," *Journal of Public Policy & Marketing*, Fall 2005; Kathleen Seiders and Ross D. Petty, "Taming the Obesity Beast: Children, Marketing, and Public Policy Considerations," *Journal of Public Policy & Marketing*, Fall 2007; James M. Carman and Robert G. Harris, "Public Regulation of Marketing Activity, Part III: A Typology of Regulatory Failures and Implications for Marketing and Public Policy," *Journal of Macromarketing*, Spring 1986.

10. For a recent perspective on market orientation, see Johanna Frösén, Jukka Luoma, Matti Jaakkola, Henrikki Tikkanen, et al., "What Counts versus What Can Be Counted: The Complex Interplay of Market Orientation and Marketing Performance Measurement," *Journal of Marketing*, May 2016. For a perspective on the marketing concept, see Christian Homburg, Danijel Jozić, and Christina Kuehnl, "Customer Experience Management: Toward Implementing an Evolving Marketing Concept," *Journal of the Academy of Marketing Science*, May 2017.

11. Mark Jeffery, *Data-Driven Marketing: The 15 Metrics Everyone in Marketing Should Know* (New York: Wiley, 2010); "Marketing Performance Metrics," *Boundless.com*, accessed May 12, 2017.

12. John Elkington, *Cannibals with Forks: The Triple Bottom Line of 21st Century Business* (New York: Wiley, 1997); "Triple Bottom Line," *The Economist*, November 17, 2009; Angus Loten, "Can Firms Aim to Do Good If It Hurts Profit?," *Wall Street Journal*, April 10, 2013.

13. Fernando Angulo-Ruiz, Naveen Donthu, Diego Prior, and Josep Rialp, "The Financial Contribution of Customer-Oriented Marketing Capability," *Journal of the Academy of Marketing Science* 42, no. 4 (2014): 380-99; A. Grinstein, "The Effect of Market Orientation and Its Components on Innovation Consequences: A Meta-Analysis," *Journal of the Academy of Marketing Science* 2 (2008); Ahmet H. Kirca, Satish Jayachandran, and William O. Bearden, "Market Orientation: A Meta-Analytic Review and Assessment of Its Antecedents and Impact on Performance," *Journal of Marketing*, April 2005; Jagdish N. Sheth, Rajendra S. Sisodia, and Arun Sharma, "The Antecedents and Consequences of Customer-Centric Marketing," *Journal of the Academy of Marketing Science*, Winter 2000; Stanley F. Slater and John C. Narver, "Market Orientation and the Learning Organization," *Journal of Marketing*, July 1995; R. W. Ruekert, "Developing a Market Orientation: An Organizational Strategy Perspective," *International Journal of Research in Marketing*, August 1992; Bernard J. Jaworski and Ajay K. Kohli, "Market Orientation: Antecedents and Consequences," *Journal of Marketing*, July 1993; Neil Morgan, "Marketing and Business Performance," *Journal of the Academy of Marketing Science*, January 2012; Son K. Lam, Florian Kraus, and Michael Ahearne, "The Diffusion of Market Orientation throughout the Organization: A Social Learning Theory Perspective," *Journal of Marketing* 74 (September 2012); V. Kumar, Eli Jones, Rajkumar Venkatesan, and Robert P. Leone, "Is Market Orientation a Source of Sustainable Competitive Advantage or Simply the Cost of Competing?," *Journal of Marketing* 75 (January 2010); Christopher P. Blocker, Daniel J. Flint, Matthew B. Myers, and Stanley F. Slater, "Proactive Customer Orientation and Its Role for Creating Customer Value in Global Markets," *Journal of the Academy of Marketing Science*, 2011.

14. Ronald L. Hess Jr., Shankar Ganesan, and Noreen M. Klein, "Service Failure and Recovery: The Impact of Relationship Factors on Customer Satisfaction," *Journal of the Academy of Marketing Science*, Spring 2003; Stanley F. Slater and John C. Narver, "Market Orientation, Customer Value, and Superior Performance," *Business Horizons*, March-April 1994; Sharon E. Beatty, "Keeping Customers," *Journal of Marketing*, April 1994.

15. "The Elements of Value," *Harvard Business Review*, September 2016. For an academic perspective on customer value, see Kumar Ranjan and Stuart Read, "Value Co-Creation: Concept and Measurement," *Journal of the Academy of Marketing Science*, May 2016; V. Kumar and Werner Reinartz, "Creating Enduring Customer Value," *Journal of Marketing*, November 2016; A. F. Payne, K. Storbacka, and P. Frow, "Managing the Co-Creation of Value," *Journal of the Academy of Marketing Science* 36, no. 1 (2008); "Companies Strive Harder to Please Customers," *Wall Street Journal*, July 27, 2009; Xie Chunyan, Richard P. Bagozzi, and Sigurd V. Troye, "Trying to Prosume: Toward a Theory of Consumers as Co-Creators of Value," *Journal of the Academy of Marketing Science*, Spring 2008; Min Soonhong, John T. Mentzer, and Robert T. Ladd, "A Market Orientation in Supply Chain Management," *Journal of the Academy of Marketing Science*, Winter 2007; Wolfgang Ulaga and Andreas Eggert, "Value-Based Differentiation in Business Relationships: Gaining and Sustaining Key Supplier Status," *Journal of Marketing*, January 2006; Roger Baxter, Sheelagh Matear, James C. Anderson, James A. Narus, et al., "Customer Value Propositions in Business Markets," *Harvard Business Review*, March 2006.

16. "We Drive the $30K Chevy Bolt, GM's Tesla-Walloping Electric Car," *Wired*, September 13, 2016.

17. Kyle Stock, "REI's Crunchy Business Model Is Crushing Retail Competitors," *Bloomberg*, March 27, 2015; Mallory Schlossberg, "One Unlikely Sporting Goods Store Is Thriving as Retailers Implode Everywhere," *Business Insider*, June 8, 2016. For an academic perspective of value co-creation, see Avinash Malshe and Scott B. Friend, "Initiating Value Co-Creation: Dealing with Non-Receptive Customers," *Journal of the Academy of Marketing Science*, September 2018.

18. REI.com; Mallory Schlossberg, "One Unlikely Sporting Goods Store Is Thriving as Retailers Implode Everywhere," *Business Insider*, June 8, 2016; Kyle Stock, "REI's Crunchy Business Model Is Crushing Retail Competitors," *Bloomberg*, March 27, 2015.

19. Farnoosh Khodakarami, J. Andrew Petersen, and Rajkumar Venkatesan, "Developing Donor Relationships: The Role of the Breadth of Giving," *Journal of Marketing*, July 2015.

20. "NYC Promotes Free Water," *Brandweek*, July 9, 2007. See also Kathleen Cleeren, Marnik Dekimpe, and Harald van Heerde, "Marketing Research on Product-Harm Crises: A Review, Managerial Implications, and an Agenda for Future Research," *Journal of the Academy of Marketing Science*, September 2017.

21. For a range of discussion on social responsibility, see C. B. Bhattacharya and Sankar Sen, *Leveraging Corporate Social Responsibility: The Stakeholder Route to Maximizing Business and Social Value* (Cambridge, UK: Cambridge University Press, 2011); O. C. Ferrell, John Fraedrich, and Linda Ferrell, *Business Ethics: Ethical Decision Making & Cases* (Cincinnati, OH: South-Western, 2015); "Debate Club: Is Fracking a Good Idea?," *U.S. News & World Report*, www.usnews.com/debate-club/is-fracking-a-good-idea; "Fracking: Gas Drilling's Environmental Threat," *Pro Publica*, www.propublica.org/series/fracking; "Fracking: Pro and Con," *Tufts Now*, December 11, 2012; Daniel Korschun, C. B. Bhattacharya, and Scott D. Swain, "Corporate Social Responsibility, Customer Orientation, and the Job Performance of Frontline Employees," *Journal of Marketing* 78, no. 3 (2014): 20-37; Christian Homburg, Marcel Stierl, and Torsten Bornemann, "Corporate Social Responsibility in Business-to-Business Markets: How Organizational Customers Account for Supplier Corporate Social Responsibility Engagement," *Journal of Marketing* 77, no. 6 (2013): 54-72; Alexander Chernev and Sean Blair, "Doing Well by Doing Good: The Benevolent Halo of Corporate Social Responsibility," *Journal of Consumer Research* 41, no. 6 (2015): 1412-25; Xueming Luo and C. B. Bhattacharya, "Corporate Social Responsibility, Customer Satisfaction, and Market Value," *Journal of Marketing*, October 2006; Alan R. Andreasen, "Social Marketing: Its Definition and Domain," *Journal of Public Policy & Marketing*, Spring 1994; Carlos J. Torelli, Alokparna Basu Monga, and Andrew M. Kaikati, "Doing Poorly by Doing Good: Corporate Social Responsibility and Brand Concepts," *Journal of Consumer Research*, 2012; William Wilkie and Elizabeth Moore, "Expanding Our Understanding of Marketing in Society," *Journal of the Academy of Marketing Science*, January 2012; C. B. Bhattacharya and Sankar Sen, "Doing Better at Doing Good: When, Why, and How Customers Respond to Corporate Social Initiatives," *California Management Review*, Fall 2004.

22. "VW Scion Returns to Lash Out at Top Managers over Scandal," *Bloomberg*, February 8, 2017; "VW Emissions Scandal Ran Deep and Wide, State Alleges," *Wall Street Journal*, July 19, 2016; "VW Papers Shed Light on Emissions Scandal," *BBC News*, January 12, 2017.

23. Kelly Martin, Jean Johnson, and Joseph French, "Institutional Pressures and Marketing Ethics Initiatives: The Focal Role of Organizational Identity," *Journal of the Academy of Marketing Science*, July 2011; Nidhi Agrawal and Adam Duhachek, "Emotional Compatibility and the Effectiveness of Antidrinking Messages: A Defensive Processing Perspective on Shame and Guilt," *Journal of Marketing Research*, April 2010; Andrew V. Abela and Patrick E. Murphy, "Marketing with Integrity: Ethics and the Service-Dominant Logic for Marketing," *Journal of the Academy of Marketing Science*, Spring 2008; Lawrence B. Chonko and Shelby D. Hunt, "Ethics and Marketing Management: A Retrospective and Prospective Commentary," *Journal of Business Research*, December 2000; Jeffrey G. Blodgett, Long-Chuan Lu, Gregory M. Rose, and Scott J. Vitell, "Ethical Sensitivity to Stakeholder Interests: A Cross-Cultural Comparison," *Journal of the Academy of Marketing Science*, Spring 2001; George G. Brenkert, "Ethical Challenges of Social Marketing," *Journal of Public Policy & Marketing*, Spring 2002.

24. O. C. Ferrell, John Fraedrich, and Linda Ferrell, *Business Ethics: Ethical Decision Making & Cases* (Cincinnati, OH: South-Western, 2015); C. B. Bhattacharya and Daniel Korschun, "Stakeholder Marketing: Beyond the Four Ps and the Customer," *Journal of Public Policy & Marketing,* Spring 2008.

25. "Rethinking Marketing: Ajay Kohli," and "Rethinking Marketing: Linda Price," presentations at the 2018 Summer AMA Conference; "Profiting the Planet," *Fortune,* September 1, 2015.

26. Examples from "Change the World List," *Fortune,* September 1, 2015; "Change the World," *Fortune,* September 15, 2017.

CHAPTER 2

1. See www.cirquedusoleil.com; "Soccer Circus: Cirque du Soleil Will Create a Lionel Messi Show," *New York Times,* October 10, 2018; "2017-2020 Corporate Social Responsibility Strategy," Cirque du Soleil Entertainment Group, "The Future of Cirque du Soleil Isn't the Circus," *Bloomberg Businessweek,* November 29, 2017; "Cirque du Soleil Shows Big Brands How to Be Remarkable–And Magical," *Forbes,* February 27, 2012; "Viva Viva Elvis!," *Time,* February 22, 2010; "Cirque's Midlife Crisis," *Canadian Business,* May 10, 2010; "U.S. Casinos' Bet on Macau Pays Off," *USA Today,* January 12, 2010; "Stanley Ho's Last Laugh in Macao," *BusinessWeek,* May 11, 2009; "Big Top to Burlesque: A Circus Sampler," *Wall Street Journal,* January 24, 2008; "Talent Scouts for Cirque du Soleil Walk a Tightrope," *Wall Street Journal,* September 8, 2007; "CMO Strategy: Unleash Emotions for Business Growth," *Advertising Age,* March 12, 2007; "Super Pregame Show," *MiamiHerald.com,* January 10, 2007; "Cirque du Success," *U.S. News and World Report,* November 20, 2006; "The Oliness of Strong Brands," *BusinessWeek,* November 16, 2006; "Crazy for 'Delirium,'" *Billboard,* July 15, 2006; "The Disney of the New Age," *Maclean's,* June 26, 2006; "Cirque du Balancing Act," *Fortune,* June 12, 2006; "How Cirque du Soleil Uses Email to Sell Out Shows at Local Cities," *Marketing Sherpa,* June 1, 2006; W. Chan Kim and Renée Mauborgne, *Blue Ocean Strategy* (Boston: Harvard Business School Press, 2005); "Join the Circus," *Fast Company,* July 1, 2005; "Lord of the Rings," *The Economist,* February 5, 2005; "Best Launch: Zumanity," *Marketing Magazine,* November 11, 2004; "Big Top Television," *Marketing Magazine,* August 9-16, 2004; "Mario D'Amico–Cirque du Soleil," *Reveries,* July 2002; "Michael Jackson's New Vegas Show 'One' Will Double the Fun," *Forbes.com,* February 22, 2013; "Cirque's Dramatic Road Work," *Billboard,* March 16, 2013; "Cirque du Soleil's Next Act: Rebalancing the Business," *Wall Street Journal,* December 1, 2014; "Cirque du Soleil Marketing Chief: Digital Will Help Us Make Customer Experiences More Intimate," *CMO.com,* March 30, 2016; "Cirque du Soleil Is about to Open the World's Most Incredible Theme Park," *Maxim.com,* October 14, 2016.

2. Rajan Varadarajan, "Strategic Marketing and Marketing Strategy: Domain, Definition, Fundamental Issues and Foundational Premises," *Journal of the Academy of Marketing Science,* March 2012; J. Burnett, "Hidden in Plain Sight: How to Find and Execute Your Company's Next Big Growth Strategy," *Journal of Advertising Research* 48, no. 1 (2008); George S. Day, "Is It Real? Can We Win? Is It Worth Doing?," *Harvard Business Review,* December 2007; John G. Singer, "What Strategy Is Not," *Sloan Management Review,* Winter 2008; Rebecca J. Slotegraaf and Peter R. Dickson, "The Paradox of a Marketing Planning Capability," *Journal of the Academy of Marketing Science,* Fall 2004; Kwaku Atuahene-Gima and Janet Y. Murray, "Antecedents and Outcomes of Marketing Strategy Comprehensiveness," *Journal of Marketing,* October 2004; Gail J. McGovern, David Court, John A. Quelch, and Blair Crawford, "Bringing Customers into the Boardroom," *Harvard Business Review,* November 2004; George S. Day, "Marketing's Contribution to the Strategy Dialogue," *Journal of the Academy of Marketing Science,* Fall 1992; P. Rajan Varadarajan and Terry Clark, "Delineating the Scope of Corporate, Business, and Marketing Strategy," *Journal of Business Research,* October-November 1994.

3. See www.herbalessences.com/us/default.jsp; "The Issue: How P&G Brought Back Herbal Essences," *BusinessWeek,* June 17, 2008; "Herbal Essence Snags 2007 ReBrand 100 Global Award," *Marketing Daily,* July 11, 2007; "Nature Boy," *Brandweek,* September 7, 2008; Ram Charan and A. G. Lafley, *The Game-Changer* (New York: Crown Business, 2008).

4. For information on Silverman's new business, see www.preschoolians.com. For more on Toddler University, see "Genesco Names New Executive of Children's Footwear Division," press release, Genesco, July 15, 1994; "Toddler University Ends Up in Westport," *Westport News,* March 8, 1991; "Whiz Kid," *Connecticut,* August 1989; "The Young and the Restless," *Children's Business,* May 1989.

5. Orville C. Walker Jr. and Robert W. Ruekert, "Marketing's Role in the Implementation of Business Strategies: A Critical Review and Conceptual Framework," *Journal of Marketing,* July 1987; Thomas V. Bonoma, "A Model of Marketing Implementation," *1984 AMA Educators' Proceedings,* Chicago: American Marketing Association, 1984.

6. "Marketing Performance Metrics," *Boundless.com,* https://goo.gl/CRjoW5, retrieved February 11, 2017.

7. This coverage of customer lifetime value is relatively simple. It does not account for the time value of money or different customer segments that may have different averages. For the interested student, more details can be found in Elie Ofek, "Customer Profitability and Lifetime Value," Harvard Business School Background Note 503-019, August 2002. For more detailed treatments of customer lifetime value, see V. Kumar, "A Theory of Customer Valuation: Concepts, Metrics, Strategy, and Implementation," *Journal of Marketing,* January 2018; Tracy Meyer, Donald C. Barnes, and Scott B. Friend, "The Role of Delight in Driving Repurchase Intentions," *Journal of Personal Selling & Sales Management,* March 2017; Nita Umashankar, Morgan K. Ward, and Darren W. Dahl, "The Benefit of Becoming Friends: Complaining after Service Failures Leads Customers with Strong Ties to Increase Loyalty," *Journal of Marketing,* November 2017; Sarang Sunder, V. Kumar, and Yi Zhao, "Measuring the Lifetime Value of a Customer in the Consumer Packaged Goods Industry," *Journal of Marketing Research,* December 2016; Sungwook Min, Xubing Zhang, Namwoon Kim, and Rajendra K. Srivastav, "Customer Acquisition and Retention Spending: An Analytical Model and Empirical Investigation in Wireless Telecommunications Markets," *Journal of Marketing Research,* October 2016; Yi-Chun Ou, Peter Verhoef, and Thorsten Wiesel, "The Effects of Customer Equity Drivers on Loyalty across Services Industries and Firms," *Journal of the Academy of Marketing Science,* May 2017; George Watson, Joshua Beck, Conor Henderson, and Robert Palmatier, "Building, Measuring, and Profiting from Customer Loyalty," *Journal of the Academy of Marketing Science,* November 2015; Jeff S. Johnson and Scott B. Friend, "Contingent Cross-Selling and Up-Selling Relationships with Performance and Job Satisfaction: An MOA-Theoretic Examination," *Journal of Personal Selling & Sales Management,* 2015; V. Kumar, Yashoda Bhagwat, and Xi (Alan) Zhang, "Regaining 'Lost' Customers: The Predictive Power of First-Lifetime Behavior, the Reason for Defection, and the Nature of the Win-Back Offer," *Journal of Marketing,* July 2015.

8. www.ikea.com; www.jetblue.com; www.in-n-out.com; Youngme Moon, *Different: Escaping the Competitive Herd* (New York: Crown Business, 2010); "Manland: Ikea's Day-Care for Husbands," *The Week,* September 16, 2011.

9. "Original Superfood–Friendship Dairies," Partners + Napier, 2016, Vimeo, https://vimeo.com/177125940, retrieved February 13, 2017.

10. Ruth N. Bolton, Katherine N. Lemon, and Peter C. Verhoef, "Expanding Business-to-Business Customer Relationships: Modeling the Customer's Upgrade Decision," *Journal of Marketing,* January 2008; Douglas W. Vorhies and Neil A. Morgan, "A Configuration Theory Assessment of Marketing Organization Fit with Business Strategy and Its Relationship with Marketing Performance," *Journal of Marketing,* January 2003.

11. "Uber Targets Trucking with New Trailer-Rental Business," *Wall Street Journal,* October 17, 2018.

12. For more on JLG, see www.jlg.com; "JLG Industries Acquisition Gives a Big Lift to Oshkosh Truck Sales," *Investor's Business Daily*, October 31, 2007; "The Secret of U.S. Exports: Great Products," *Fortune*, January 10, 2000.

13. For more on international marketing, see "Novelty of Mexican Food in India Is a Hit for Yum," *Wall Street Journal*, March 30, 2010; "Yum Brands Bet on India's Young for Growth," *Wall Street Journal*, December 17, 2009; "McDonald's Creates Asian-Inspired Versions of Food as Part of Olympics Blitz," *Wall Street Journal*, July 1, 2008; "In Vietnam, Fast Food Acts Global, Tastes Local," *Wall Street Journal*, March 12, 2008; "Foreign Distribution? Just Ask Mom (& Pop)," *Brandweek*, November 19, 2007; "Scrambling to Bring Crest to the Masses," *BusinessWeek*, June 25, 2007; Naresh K. Malhotra, Francis M. Ulgado, and James Agarwal, "Internationalization and Entry Modes: A Multitheoretical Framework and Research Propositions," *Journal of International Marketing*, 2003; Shaoming Zou and S. Tamer Cavusgil, "The GMS: A Broad Conceptualization of Global Marketing Strategy and Its Effect on Firm Performance," *Journal of Marketing*, October 2002; Susan P. Douglas, "Exploring New Worlds: The Challenge of Global Marketing," *Journal of Marketing*, January 2001.

14. "Change the World," *Fortune*, September 1, 2015.

15. "Change the World," *Fortune*, September 15, 2017.

CHAPTER 3

1. See www.amazon.com; "Products Yanked from Amazon in India to Comply with New E-Commerce Rules," *Wall Street Journal*, February 1, 2019; "Amazon, with Little Fanfare, Emerges as an Advertising Giant," *Wall Street Journal*, November 27, 2018; "Amazon Forecast to Be No. 3 Digital Advertising Player in 2018," *Wall Street Journal*, September 19, 2019; "Amazon Knows What You Buy. And It's Building a Big Ad Business from It," *New York Times*, January 20, 2019; *Amazon.com: The Hidden Empire*, faberNovel, May 2011; "No Holiday Cheer for Amazon.com," *Chicago Tribune*, November 27, 2011; "Amazon Pressured on Sales Tax," *New York Times*, March 13, 2011; "Retailers Struggle in Amazon's Jungle," *Wall Street Journal*, February 22, 2011; "Price Check for iPhone," *Wired*, November 23, 2010; "What's in Amazon's Box? Instant Gratification," *Bloomberg Businessweek*, November 23, 2010; "Wal-Mart Fires Shot in Toy War," *Wall Street Journal*, November 8, 2010; "Study: Amazon.com Is Most Trusted Brand in U.S," *CNET News*, February 22, 2010; "Now, It's Google vs. . . . Amazon.com?," *Investor's Business Daily*, November 19, 2009; "Walmart Making Push Online, Where It's a Pygmy vs. Amazon," *Investor's Business Daily*, October 20, 2009; "Walmart Strafes Amazon in Book War," *Wall Street Journal*, October 16, 2009; "Amazon Packages Green Site," *Investor's Business Daily*, February 25, 2009; "Amazon.com's Third-Party Sales Balloon," *Investor's Business Daily*, April 8, 2009; "When Service Means Survival: How Amazon Aims to Keep You Clicking," *BusinessWeek*, March 2, 2009; "The World's Most Innovative Companies," *Fast Company*, March 2009. For more on Kindle, see "Hachette Enters the E-Book Fray, Saying It, Too, Wants New Pricing," *Wall Street Journal*, February 5, 2010; "Trying to Avert a Digital Horror Story," *BusinessWeek*, January 11, 2010; "Sony E-Reader Opens New Chapters in Kindle Rivalry," *Wall Street Journal*, January 14, 2010; "More Makers Jump into the E-Reader Fray," *Wall Street Journal*, January 8, 2010; "Publishers Aren't Happy with Amazon's Pricing," *USA Today*, December 11, 2009; "iPod Touch, iPhone Join Kindle's Book Club," *USA Today*, March 4, 2009; "Amazon's Kindle 2 Improves the Good, Leaves Out the Bad," *Wall Street Journal*, February 26, 2009; "Kindle's New, but Price Is Old," *USA Today*, February 10, 2009; "As the Kindle Turns Five, Amazon Girds for a New Fight," *Bloomberg Businessweek*, November 26, 2012; "Amazon's (Not So Secret) War on Taxes," *Fortune*, May 23, 2013; "Amazon Bets on Web Groceries, Expands AmazonFresh to L.A.," *TechCrunch*, June 10, 2013; "Once Refined, Same-Day Delivery Will Be Commonplace," *Wired*, May 2013; "Amazon Preps Faster Deliveries by Bike and by Drone," *Wall Street Journal*, December 8, 2014; "Amazon's First Staffed College Campus Store Should Have Retailers Worried," *TechCrunch*, February 3, 2015; "Amazon Makes a Push on College Campuses," *Wall Street Journal*, February 1, 2015; "Can Amazon's Jeff Bezos Save Planet Earth?," *NPR*, January 8, 2014; "Amazon Is Bringing Prime to China, Where It Has Struggled for Market Share," *Recode.net*, October 28, 2016; "Amazon's Living Lab: Reimagining Retail on Seattle Streets," *New York Times*, February 12, 2017; "6-Year-Old Orders $160 Dollhouse, 4 Pounds of Cookies with Amazon's Echo Dot," *CBSNews.com*, January 5, 2017; "The Real Story of How Amazon Built the Echo," *Bloomberg*, April 19, 2016; "The Market Is Underestimating Amazon," *Forbes*, May 27, 2016; "Earnings Preview: Amazon Accounts for 43% of All US Online Sales in 2016," *Internet Retailer*, February 1, 2017; "Amazon's Prime Challenge Is International Growth," *Financial Times*, March 3, 2016; "Amazon Go Is a Real-World, Checkout-Free Shopping Experience. One That Only Amazon Could Pull Off," *Wired*, December 6, 2016; "Amazon.com Inc.'s Vision Statement & Mission Statement (An Analysis)," *Panmore Institute*, February 12, 2017.

2. The American National Red Cross, "Red Cross Mission, Vision, and Fundamental Principles," 2019, www.redcross.org; Lance Leuthesser and Chiranjeev Kohli, "Corporate Identity: The Role of Mission Statements," *Business Horizons*, May–June 1997; Christopher K. Bart, "Sex, Lies, and Mission Statements," *Business Horizons*, November–December 1997. Red Cross mission statement from the "Mission & Principles" of the American Red Cross. Copyright © by the American Red Cross. All rights reserved. Used with permission. For more general discussions of marketing strategy, see Peter F. Drucker, *Management: Tasks, Responsibilities, Practices, and Plans* (New York: Harper & Row, 1973); Sev K. Keil, "The Impact of Business Objectives and the Time Horizon of Performance Evaluation on Pricing Behavior," *International Journal of Research in Marketing*, June 2001; George S. Day and Christine Moorman, *Strategy from the Outside In* (New York: McGraw-Hill, 2010).

3. www.usaa.com; "USAA's Battle Plan," *BusinessWeek*, February 18, 2010; "Customer Service Champs," *BusinessWeek*, March 5, 2007; "USAA Soldiering On in Insurance," *BusinessWeek*, March 5, 2007.

4. Irina Kozlenkova, Stephen Samaha, and Robert Palmatier, "Resource-Based Theory in Marketing," *Journal of the Academy of Marketing Science*, January 2014.

5. Worldwide sales data from "Smartphone OS Sales Market Share," *Kantar Worldpanel*, www.kantarworldpanel.com/global/smartphone-os-market-share/, retrieved February 17, 2018.

6. For more on corporate spying and competitive intelligence, see "Cybercrooks Stalk Small Businesses," *USA Today*, December 31, 2009; "Hounding the Hackers," *BusinessWeek*, September 14, 2009; "How Team of Geeks Cracked Spy Trade," *Wall Street Journal*, September 4, 2009; "Accusations of Snooping in Ink-Cartridge Dispute," *Wall Street Journal*, August 11, 2009; "Today's Spies Find Secrets in Plain Sight," *USA Today*, August 1, 2008; "Snooping on a Shoestring," *Business 2.0*, May 2003. See also Preyas S. Desai, Woochoel Shin, and Richard Staelin, "The Company That You Keep: When to Buy a Competitor's Keyword," *Marketing Science*, August 2014; Adam Rapp, Raj Agnihotri, Thomas Baker, and James Andzulis, "Competitive Intelligence Collection and Use by Sales and Service Representatives: How Managers' Recognition and Autonomy Moderate Individual Performance," *Journal of the Academy of Marketing Science*, May 2015; Zahra A. Shaker, "Unethical Practices in Competitive Analysis: Patterns, Causes and Effects," *Journal of Business Ethics*, January 1994.

7. For a real-world example of similar behavior, see "Huawei Accused of Offering Bonuses to Staff to Steal Secrets," *Financial Times*, January 29, 2019.

8. For more on economic environment, see Marnik G. Dekimpe and Barbara Deleersnyder, "Business Cycle Research in Marketing: A Review and Research Agenda," *Journal of the Academy of Marketing Science*, January 2018; Rajdeep Grewal, "Building Organizational Capabilities for Managing Economic Crisis: The Role of Market Orientation and Strategic Flexibility," *Journal of Marketing*, April 2001.

9. "How Autonomous Cars Will Reshape Our World," *Wall Street Journal,* October 27, 2018; "The Self-Driving Car Generation Gap," *Slate,* June 22, 2016; "Auto Insurers Are Bracing for Upheaval from Self-Driving Cars," *Business Insider,* July 29, 2016.

10. "Five Pioneering Examples of How Brands Are Using Chatbots," *Econsultancy,* August 18, 2016; "In the Age of Intelligent Agents, How Does Marketing Work?," *MarTech Today,* January 11, 2017; "How Artificial Intelligence Creates a Better Customer Service Experience," *Forbes,* November 5, 2016; "10 Examples of Artificial Intelligence You're Using in Daily Life," *Beebom,* September 26, 2016; "10 Examples of AI in Marketing," *Inc.,* August 9, 2016; nest.com; kayak.com.

11. Definition of machine learning from https://searchenterpriseai.techtarget.com/definition/machine-learning-ML; "Artificial Intelligence in Oncology: Current Applications and Future Directions," *Cancer Network,* February 15, 2019; "The Skin Cancer Apps That Can Check Your Suspicious-Looking Moles," *Noted,* June 14, 2018; "Deep Learning Algorithm Does as Well as Dermatologists in Identifying Skin Cancer," *Stanford News,* January 25, 2017.

12. "Japan Lays Groundwork for Boom in Robot Carers," *The Guardian,* February 5, 2018; "Toyota, SoftBank Join Forces to Build Self-Driving Cars That Deliver Meals, Health Care," *Wall Street Journal,* October 4, 2018.

13. "Chrysler's 'Imported from Detroit' Super Bowl Ad Inspires Political Debate," *CBSNews,* February 7, 2011; CA US LLC.; for other examples of nationalism, see "Coming Home: Appliance Maker Drops China to Produce in Texas," *Wall Street Journal,* August 24, 2009; "Buy American Clause Stirs Up Controversy," *USA Today,* February 4, 2009; "Which Is More American?," *USA Today,* March 22, 2007. See also Durairaj Maheswaran and Cathy Yi Chen, "Nation Equity: Incidental Emotions in Country-of-Origin Effects," *Journal of Consumer Research,* October 2006; George Balabanis and Adamantios Diamantopoulos, "Domestic Country Bias, Country-of-Origin Effects, and Consumer Ethnocentrism: A Multidimensional Unfolding Approach," *Journal of the Academy of Marketing Science,* Winter 2004; Zeynep Gurhan-Canli and Durairaj Maheswaran, "Determinants of Country-of-Origin Evaluations," *Journal of Consumer Research,* June 2000.

14. "Free Trade Agreements with Their Pros and Cons," *The Balance,* February 14, 2019.

15. "Ford vs. Firestone: A Corporate Whodunit," *BusinessWeek,* June 10, 2001. For more on auto safety, see "States Go After Texting Drivers: Safety Concerns Fuel New Legislation," *USA Today,* January 25, 2010; "The Number of Small-Car Owners Is Growing Even though More Drivers Die in Small Cars," *USA Today,* August 20, 2007. For more on CPSC, see "More Paper Tiger Than Watchdog?" *BusinessWeek,* September 3, 2007, p. 45; "Product-Safety Cops Handcuffed," *Wall Street Journal,* December 18, 2007. For an overview of the toy, food, and pet product recalls of 2007, see "The New China Price," *Fortune,* November 12, 2007; "Spate of Recalls Boosts Potency of User Reviews," *Advertising Age,* October 29, 2007.

16. For more on Beech-Nut, see "What Led Beech-Nut down the Road to Disgrace," *BusinessWeek,* February 22, 1988. See also Kersi D. Antia, Mark E. Bergen, Shantanu Dutta, and Robert J. Fisher, "How Does Enforcement Deter Gray Market Incidence?," *Journal of Marketing,* January 2006; Debra M. Desrochers, Gregory T. Gundlach, and Albert A. Foer, "Analysis of Antitrust Challenges to Category Captain Arrangements," *Journal of Public Policy & Marketing,* Fall 2003; "Europe: A Different Take on Antitrust," *BusinessWeek,* June 25, 2001; "In Archer-Daniels-Midland Saga, Now the Executives Face Trial," *Wall Street Journal,* July 9, 1998. See also Mary W. Sullivan, "The Role of Marketing in Antitrust," *Journal of Public Policy & Marketing,* Fall 2002; Jeff Langenderfer and Steven W. Kopp, "Which Way to the Revolution? The Consequences of Database Protection as a New Form of Intellectual Property," *Journal of Public Policy & Marketing,* Spring 2003.

17. "Tesla Clashes with Car Dealers," *Wall Street Journal,* June 18, 2013; "Judge Rules Airbnb Illegal in New York City," *CNNMoney,* May 21, 2013.

18. "The New Sustainability Regeneration," *J. Walter Thompson Intelligence,* September 2018; "Consumer-Goods' Brands That Demonstrate Commitment to Sustainability Outperform Those That Don't," *Nielsen Press Room,* October 12, 2015.

19. "The World's Biggest Brands Want You to Refill Your Orange Juice and Deodorant," *Wall Street Journal,* January 24, 2019; "McDonald's Transitions to Sustainably Sourced Coffee," *brandchannel,* October 12, 2016; "Beyond Sustainable: The Growing Demand for Ethical Fashion," *The Robin Report,* April 3, 2018.

20. For more information, see Tamer S. Cavusgil, "Measuring the Potential of Emerging Markets: An Indexing Approach," *Business Horizons,* January–February 1997; C. Riegner, "Wired China: The Power of the World's Largest Internet Population," *Journal of Advertising Research* 48, no. 4 (2008).

21. See also World Bank Open Data for more https://data.worldbank.org/

22. See also Madhubalan Viswanathan, Jose Antonio Rosa, and James Edwin Harris, "Decision Making and Coping of Functionally Illiterate Consumers and Some Implications for Marketing Management," *Journal of Marketing,* January 2005; Natalie Ross Adkins, Julie L. Ozanne, Dawn Iacobucci, and Eric Arnould, "The Low Literate Consumer," *Journal of Consumer Research,* June 2005.

23. A. M. Chircu and V. Mahajan, "Perspective: Revisiting the Digital Divide: An Analysis of Mobile Technology Depth and Service Breadth in the BRIC Countries," *Journal of Product Innovation Management* 26, no. 4 (2009).

24. "In Kerala, Cell Phones Go Fishing," *Liberation,* October 2, 2006; "ITC eChoupal Initiative," Harvard Business School Case #604016, 2003.

25. "mMitra: Connecting More Moms via Mobile," *JnJ.com,* January 31, 2017; "Change the World," *Fortune,* September 15, 2017.

26. See also "2010 America: What the 2010 Census Means for Marketing and Advertising," *Advertising Age,* White Paper, October 12, 2009; "No More Joe Consumer in America," *Advertising Age,* October 12, 2009.

27. "Boomer Consumer Part 3: Buying Habits of Baby Boomers," *Dr. Alexis,* January 1, 2019; "Top Baby-Boomer Shopping Habits Retailers Can't Afford to Ignore," *Chain Store Age,* January 6, 2017; "From Gen X to Baby Boomers, What Every Generation Loves to Buy," *Bloomberg Businessweek,* October 19, 2016; "Target Revamps Groceries for Millennials," *Wall Street Journal,* March 2, 2015; "Millennials—Breaking the Myths," *The Nielsen Company,* 2014.

28. "What You've Been Told about Millennials Is (Mostly) Wrong," *Advertising Age,* August 17, 2015.

29. "Millennials Are Making Frozen Food Hot Again, but Can They Stop Killing Everything Else?," *The Guardian,* May 14, 2018; "Target Revamps Groceries for Millennials," *Wall Street Journal,* March 2, 2015; "Millennials—Breaking the Myths," *The Nielsen Company,* 2014; "From Gen X to Baby Boomers, What Every Generation Loves to Buy," *Bloomberg Businessweek,* October 19, 2016.

30. "MNI Targeted Media Releases Data to Help Marketers Win Gen Z-ers," *MNI Targeted Media,* May 8, 2018; "Engaging Generation Z: Marketing to a New Brand of Consumer," *Adweek,* November 27, 2017; Frank N. Magid Associates, "The First Generation of the Twenty First Century," April 30, 2012; "Generation Z: Rebels with a Cause," *Forbes,* May 28, 2013.

31. "What You've Been Told about Millennials Is (Mostly) Wrong," *Advertising Age,* August 17, 2015.

32. Frank R. Bacon Jr. and Thomas W. Butler Jr., *Planned Innovation,* rev. ed. (Ann Arbor: Institute of Science and Technology, University of Michigan, 1980).

33. Paul F. Anderson, "Marketing, Strategic Planning and the Theory of the Firm," *Journal of Marketing,* Spring 1982; Ronnie Silverblatt and Pradeep Korgaonkar, "Strategic Market Planning in a Turbulent Business Environment," *Journal of Business Research,* August 1987.

34. Keith B. Murray and Edward T. Popper, "Competing under Regulatory Uncertainty: A U.S. Perspective on Advertising in

the Emerging European Market," *Journal of Macromarketing,* Fall 1992.

35. David Tormsen, "9 Ways KFC Is Completely Different in Other Countries," *Mashed,* no date.

36. "Sensitive Export: Seeking New Markets for Tampons, P&G Faces Cultural Barriers," *Wall Street Journal,* December 8, 2000; "Pizza Queen of Japan Turns Web Auctioneer," *Wall Street Journal,* March 6, 2000; Kamran Kashani, "Beware the Pitfalls of Global Marketing," *Harvard Business Review,* September–October 1989.

CHAPTER 4

1. www.lego.com; *LEGO Responsibility Report 2018;* "How LEGO Filled the Toys 'R' Us Void," *CNN.com,* February 27, 2019; "LEGO Returns to Growth as It Builds on U.S. Momentum," *Wall Street Journal,* February 27, 2019; "LEGO Partners with Tencent to Create Games and a Social Network for Children in China," *CNBC.com,* January 15, 2018; "LEGO Group Introduces LEGO Hidden Side, Combining Building with Augmented Reality to Create a New Way to Play," *LEGO.com,* February 14, 2019; "How to Sell Toys in a Culture Where Play Is Viewed Negatively," *Advertising Age,* March 20, 2017; "Innovation Almost Bankrupted LEGO—Until It Rebuilt with a Better Blueprint," *Knowledge@Wharton,* July 18, 2012; *The Brick* (annual magazine), 2010; *Company Profile: An Introduction to the LEGO Group, 2009;* "Toymaker Grows by Listening to Customers," *Advertising Age,* November 9, 2009; "Innovating a Turnaround at LEGO," *Harvard Business Review,* September 2009; "LEGO Proves Itself a Blockbuster Once Again," *The Times,* February 24, 2009; Charlene Li and Josh Bernoff, *Groundswell* (Boston: Harvard Business School Press, 2008); "Innovation at the LEGO Group (A) and (B)" (case series), International Institute for Management Development, 2008; "Please Don't Touch Daddy's LEGO," *Maclean's,* October 6, 2008; "LEGO Introduces WeDo Package for Education," *Macworld,* June 30, 2008; "LEGO Learns a Lesson," *Change Agent,* June 2008; "Geeks in Toyland," *Wired,* February 2006; Majken Schultz and Mary Jo Hatch, "The Cycles of Corporate Branding: The Case of LEGO Company," *California Management Review,* Fall 2003. See also H. J. Schau, A. M. Muniz, and E. J. Arnould, "How Brand Community Practices Create Value," *Journal of Marketing* 73, no. 5 (2009); "LEGO Builds New Billionaires as Toymaker Topples Mattel," *Bloomberg,* March 13, 2013; Annual Report 2012, The LEGO Group; "LEGO Has Become the World's Largest Toy Brand by Constructing a Highly Lucrative TV and Film Model," *Adweek,* April 15, 2013; "For LEGO, Pink Is the New Black," *Bloomberg Businessweek,* February 22, 2013; Morten Berg Jensen, Christoph Hienerth, and Christopher Lettl, "Forecasting the Commercial Attractiveness of User-Generated Designs Using the LEGO User Community," *Journal of Product Innovation Management,* December 31, 2014, 75–93; "How LEGO Became the Apple of Toys," *Fast Company,* February 2015; "LEGO's VP of Marketing on Listening to Customers, Movies, and 360 Marketing," *Marketing Week,* March 23, 2015; "LEGO Bucks Industry Trend with Profit Growth," *Wall Street Journal,* February 25, 2015; "LEGO Builds Stronger Ties to Girls," *Wall Street Journal,* December 29, 2015; "LEGO Targets Pre-Mindstorms Minds with Its Boost Educational Kit," *TechCrunch,* January 3, 2017; "LEGO Education Builds a Winning Marketing Strategy," *Baseline,* July 3, 2015; "LEGO Life Is a Social Network for Kids to Share Their LEGO Creations," *The Verge,* January 31, 2017; "LEGO Bricks Come Alive," *Wall Street Journal,* January 4, 2017; "LEGO Boost Will Let You Program a Walking, Talking Cat," *CNET,* January 4, 2017; "LEGO Education WeDo: Best of CES 2016: The Wondrous and Wacky Year Ahead in Gadgets," *Wall Street Journal,* January 5, 2016.

2. See www.hallmark.com; 2011 Annual Report, Hallmark; "Hallmark Rethinks Approach to Greeting Cards," *Chain Drug Review,* March 16, 2015; "Writing Mother's Day Cards at Hallmark: An Inside Look," *Bloomberg.com,* May 1, 2014.

3. Glen L. Urban and John R. Hauser, "Listening In to Find and Explore New Combinations of Customer Needs," *Journal of Marketing,* April 2004; Terri C. Albert, "Need-Based Segmentation and Customized Communication Strategies in a Complex-Commodity Industry: A Supply Chain Study," *Industrial Marketing Management,* May 2003; Rajendra K. Srivastava, Mark I. Alpert, and Allan D. Shocker, "A Customer-Oriented Approach for Determining Market Structures," *Journal of Marketing,* Spring 1984.

4. "How HP Survives in a Shrinking PC Market," *Wall Street Journal,* May 26, 2018.

5. See www.kaepa.com; "Tapping Into Cheerleading," *Adweek's Marketing Week,* March 2, 1992.

6. Björn Frank, Tadao Enkawa, and Shane Schvaneveldt, "How Do the Success Factors Driving Repurchase Intent Differ between Male and Female Customers?," *Journal of the Academy of Marketing Science* 42, no. 2 (2014): 171–85; Michael Haenlein and Barak Libai, "Targeting Revenue Leaders for a New Product," *Journal of Marketing* 77, no. 3 (2013): 65–80; Thomas Reutterer, Andreas Mild, Martin Natter, and Alfred Taudes, "A Dynamic Segmentation Approach for Targeting and Customizing Direct Marketing Campaigns," *Journal of Interactive Marketing,* Summer 2006; S. Tamer Cavusgil, Tunda Kiyak, and Sengun Yeniyurt, "Complementary Approaches to Preliminary Foreign Market Opportunity Assessment: Country Clustering and Country Ranking," *Industrial Marketing Management,* October 2004; Malaika Brengman, Maggie Geuens, Bert Weijters, Scott M. Smith, et al., "Segmenting Internet Shoppers Based on Their Web-Usage-Related Lifestyle: A Cross-Cultural Validation," *Journal of Business Research,* January 2005; Ruth N. Bolton and Matthew B. Myers, "Price-Based Global Market Segmentation for Services," *Journal of Marketing,* July 2003; Philip A. Dover, "Segmentation and Positioning for Strategic Marketing Decisions," *Journal of the Academy of Marketing Science,* Summer 2000.

7. For more on buyer personas, see "The Beginner's Guide to Segmentation and Buyer Personas," *HubSpot Academy Blog,* March 19, 2014, http://blog.hubspot.com/customers/getting-started-with-segmentation; "Buyer Personas and Segmentation: One Is Not Like the Other," *ClickZ,* March 25, 2015.

8. Example from "Change the World," *Fortune,* September 15, 2017; "Change the World," *Fortune,* September 1, 2016.

9. Stephen D. Ross, "Segmenting Sports Fans Using Brand Associations: A Cluster Analysis," *Sports Marketing Quarterly* 1 (2007); Rajan Sambandam, "Cluster Analysis Gets Complicated," *Marketing Research,* Spring 2003; Girish Punj and David W. Stewart, "Cluster Analysis in Marketing Research: Review and Suggestions for Application," *Journal of Marketing Research,* May 1983.

10. "Forbes Consulting Group Releases 'Holiday Shopping Unwrapped: Shopper Motivations Revealed' Study," press release, December 16, 2015.

11. "How Companies Learn Your Secrets," *New York Times,* February 16, 2012; Charles Duhigg, *The Power of Habit* (New York: Random House, 2012); "Data Mining and Predictive Analytics," *Abbott Analytics,* February 18, 2012; "How Target Knew a High School Girl Was Pregnant before Her Parents Did," *Time,* February 17, 2012.

12. "Netflix's Grand, Daring, Maybe Crazy Plan to Conquer the World," *Wired,* March 27, 2016.

13. For more on CRM, see Timothy Bohling, Douglas Bowman, Steve Lavalle, Vikas Mittal, et al., "CRM Implementation: Effectiveness Issues and Insights," *Journal of Service Research,* May 2006; H. Wilson, M. Clark, and B. Smith, "Justifying CRM Projects in a Business-to-Business Context: The Potential of the Benefits Dependency Network," *Industrial Marketing Management,* June 2007; Katherine N. Lemon, Tiffany Barnett White, and Russell S. Winer, "Dynamic Customer Relationship Management: Incorporating Future Considerations into the Service Retention Decision," *Journal of Marketing,* January 2002; "How Big Data Is Changing the Whole Equation for Business," *Wall Street Journal,* March 8, 2013. For more on targeting, see Hernán A. Bruno, Javier Cebollada, and Pradeep K. Chintagunta,

"Targeting Mr. or Mrs. Smith: Modeling and Leveraging Intrahousehold Heterogeneity in Brand Choice Behavior," *Marketing Science,* July 2018.

14. This example and exhibit were inspired by "6 Smart Behavioral Marketing Examples for E-Commerce," *E-Commerce Revenue Optimization Blog,* October 18, 2017.

15. "Retailers on Quest to Rekindle the Personal Touch of a Bygone Era," *Advertising Age,* February 14, 2011.

16. "What the Airline Knows about the Guy in Seat 14C," *Wall Street Journal,* June 20, 2018.

17. J. Miguel Villas-Boas, "A Dynamic Model of Repositioning," *Marketing Science,* March 2018; Adrian Payne, Pennie Frow, and Andreas Eggert, "The Customer Value Proposition: Evolution, Development, and Application in Marketing," *Journal of the Academy of Marketing Science,* July 2017; "Positioning Brands against Large Competitors to Increase Sales," *Journal of Marketing Research* 51, no. 6 (2014): 647–56; Garrett Sonnier and Andrew Ainslie, "Estimating the Value of Brand-Image Associations: The Role of General and Specific Brand Image," *Journal of Marketing Research,* June 2011; V. Kumar and Denish Shah, "Uncovering Implicit Consumer Needs for Determining Explicit Product Positioning: Growing Prudential Annuities' Variable Annuity Sales," *Marketing Science* 4 (2011); Ji Kyung Park and Deborah Roedder John, "Got to Get You into My Life: Do Brand Personalities Rub Off on Consumers?," *Journal of Consumer Research,* August 2010; Charles Blankson, Stavros P. Kalafatis, Julian Ming-Sung Cheng, and Costas Hadjicharalambous, "Impact of Positioning Strategies on Corporate Performance," *Journal of Advertising Research,* March 2008; Girish Punj and Junyean Moon, "Positioning Options for Achieving Brand Association: A Psychological Categorization Framework," *Journal of Business Research,* April 2002; Kalpesh Kaushik Desai and S. Ratneshwar, "Consumer Perceptions of Product Variants Positioned on Atypical Attributes," *Journal of the Academy of Marketing Science,* Winter 2003; David A. Aaker and J. Gary Shansby, "Positioning Your Product," *Business Horizons,* May–June 1982; Al Ries and Jack Trout, *Positioning: The Battle for Your Mind* (New York: McGraw-Hill, 1981).

18. "At Hyundai, Branding Is Job 2," *Bloomberg Businessweek,* May 20, 2007; "Hyundai Gains in Measure of Brand Loyalty," *Los Angeles Times,* July 18, 2011.

19. Douglas B. Holt, *How Brands Become Icons* (Cambridge, MA: Harvard Business Press, 2004), p. 63 (quote slightly modified).

CHAPTER 5

1. "How Apple's AirPods Became a 'Flex'—and Why It Matters," *NBCNews.com,* February 13, 2019; "Apple's Challenge: Win Over China with Pricey New iPhones," *Wall Street Journal,* September 14, 2018; "Apple's China Problem May Require a New iPhone," *Wall Street Journal,* January 30, 2019; "Everyone Got AirPods for Christmas—and the Memes Are Rolling In," *The Daily Dot,* December 27, 2018; "How Many iPhones Have Been Sold Worldwide?," *Lifewire,* February 8, 2017; "Apple and Google Know What You Want before You Do," *Wall Street Journal,* August 3, 2015; "Apple Pushes to Bolster Market Share in India," *Wall Street Journal,* January 20, 2016; "Apple Music vs Spotify—Which Is Better?," *Whathifi.com,* December 2, 2016; "FastForward 2017: Making Sense of a Data-Driven Industry Culture," *Forbes,* February 28, 2017; "Spotify, Apple Music, Tidal, Amazon Music Unlimited and Google Play Music: Which Music Streaming App Is Right for You?," *CNET,* October 21, 2016; "5 Extremely Private Things Your iPhone Knows about You," *Huffington Post,* March 19, 2015; "How Loyal Are Consumers to Apple's Platform?," *Marketing Profs,* March 12, 2015; "Apple Watch Review: The Smartwatch Finally Makes Sense," *Wall Street Journal,* April 8, 2015; "What Exactly Is an Apple Watch For?," *Wall Street Journal,* February 16, 2015; "Why China Loves Apple Best," *Campaign,* February 26, 2015; "Has Apple Lost Its Cool to Samsung?" *Wall Street Journal,* January 28, 2013; "5 Truths That Explain Our Love-Hate Affair with Apple," *Fast Company,* April 2013; "Can Apple Win Over China?," *Fortune,* October 11, 2012; "How Tim Cook Is Changing Apple," *Fortune,* June 11, 2012; "Apple Macs Land on More Corporate Desks," *Wall Street Journal,* January 18, 2012; Walter Isaacson, *Steve Jobs* (New York: Simon & Schuster).

2. U.S. Census Bureau, "Percentage Distribution of Household Income in the United States in 2013," n.d., Statista, www.statista.com/statistics/203183/percentage-distribution-of-household-income-in-the-us/; retrieved March 1, 2017: U.S. Census Bureau, "Median Household Income in the United States from 1990 to 2015 (in U.S. Dollars)," n.d., Statista, www.statista.com/statistics/200838/median-household-income-in-the-united-states/, retrieved March 1, 2017.

3. For more on economic factors and consumer behavior, see Leonard Lee, Michelle P. Lee, Marco Bertini, Gal Zauberman, et al., "Money, Time, and the Stability of Consumer Preferences," *Journal of Marketing Research,* April 2015; Brett R. Gordon, Avi Goldfarb, and Yang Li, "Does Price Elasticity Vary with Economic Growth? A Cross-Category Analysis," *Journal of Marketing Research* 50, no. 1 (2013): 4–13; Wagner A. Kamakura and Rex Yuxing Du, "How Economic Contractions and Expansions Affect Expenditure Patterns," *Journal of Consumer Research,* 2013, S92–S110; Robin L. Soster, Andrew D. Gershoff, and William O. Bearden, "The Bottom Dollar Effect: The Influence of Spending to Zero on Pain of Payment and Satisfaction," *Journal of Consumer Research* 41, no. 3 (2014): 656–77; Stephanie M. Tully, Hal E. Hershfield, and Tom Meyvis, "Seeking Lasting Enjoyment with Limited Money: Financial Constraints Increase Preference for Material Goods over Experiences," *Journal of Consumer Research* 42, no. 1 (2015): 59–75; Jeffrey S. Larson and Ryan Hamilton, "When Budgeting Backfires: How Self-Imposed Price Restraints Can Increase Spending," *Journal of Marketing Research,* April 2012; R. Y. Du and W. A. Kamakura, "Where Did All That Money Go? Understanding How Consumers Allocate Their Consumption Budget," *Journal of Marketing* 6 (2008); R. Saini and A. Monga, "How I Decide Depends on What I Spend: Use of Heuristics Is Greater for Time Than for Money," *Journal of Consumer Research* 6 (2008); Ravindra Chitturi, Rajagopal Raghunathan, and Vijay Mahajan, "Delight by Design: The Role of Hedonic versus Utilitarian Benefits," *Journal of Marketing,* June 2008.

4. Patti Williams, "Emotions and Consumer Behavior," *Journal of Consumer Research,* February 2014; Kristina D. Frankenberger, "Consumer Psychology for Marketing," *Journal of the Academy of Marketing Science,* Summer 1996; A. H. Maslow, *Motivation and Personality* (New York: Harper & Row, 1970). For academic research on psychological influences, see Jennifer Escalas, Katherine White, Jennifer J. Argo, Jaideep Sengupta, et al., "Self-Identity and Consumer Behavior," *Journal of Consumer Research* 39, no. 5 (2013): xv–xviii; Rosellina Ferraro, Amna Kirmani, and Ted Matherly, "Look at Me! Look at Me! Conspicuous Brand Usage, Self-Brand Connection, and Dilution," *Journal of Marketing Research* 50, no. 4 (2013): 477–88; Christopher K. Hsee, Yang Yang, Xingshan Zheng, and Hanwei Wang, "Lay Rationalism: Individual Differences in Using Reason versus Feelings to Guide Decisions," *Journal of Marketing Research* 52, no. 1 (2015): 134–46; Chris Janiszewski, Andrew Kuo, and Nader T. Tavassoli, "The Influence of Selective Attention and Inattention to Products on Subsequent Choice," *Journal of Consumer Research* 39, no. 6 (2013): 1258–74; Yuping Liu-Thompkins and Leona Tam, "Not All Repeat Customers Are the Same: Designing Effective Cross-Selling Promotion on the Basis of Attitudinal Loyalty and Habit," *Journal of Marketing* 77, no. 5 (2013): 21–36; Patti Williams, "Emotions and Consumer Behavior," *Journal of Consumer Research,* 2014, S268–S69.

5. Bob Gilbreath, *The Next Evolution of Marketing* (New York: McGraw-Hill, 2010).

6. Joel Rubinson, "The Persuadables: How Advertisers Can Use the Principles of Recency and Spend Level to Boost ROAS," White Paper, Viant.

7. See Jonathan Hasford, Blair Kidwell, and David M. Hardesty, "Emotional Ability and Associative Learning: How Experiencing and Reasoning about Emotions Impacts Evaluative Conditioning," *Journal*

of Consumer Research, December 2018; Darron Billeter, Ajay Kalra, and George Loewenstein, "Underpredicting Learning after Initial Experience with a Product," *Journal of Consumer Research,* October 2011; Xavier Drèze and Joseph C. Nunes, "Recurring Goals and Learning: The Impact of Successful Reward Attainment on Purchase Behavior," *Journal of Marketing Research,* April 2011; Marcus Cunha Jr., Chris Janiszewski, and Juliano Laran, "Protection of Prior Learning in Complex Consumer Learning Environments," *Journal of Consumer Research,* April 2008; Les Carlson, Russell N. Laczniak, and Ann Walsh, "Socializing Children about Television: An Intergenerational Study," *Journal of the Academy of Marketing Science,* Summer 2001; Elizabeth S. Moore and Richard J. Lutz, "Children, Advertising, and Product Experiences: A Multimethod Inquiry," *Journal of Consumer Research,* June 2000; Thomas W. Gruen, Talai Osmonbekov, and Andrew J. Czaplewski, "eWOM: The Impact of Customer-to-Customer Online Know-How Exchange on Customer Value and Loyalty," *Journal of Business Research,* April 2006; C. A. de Matos and C. A. V. Rossi, "Word-of-Mouth Communications in Marketing: A Meta-Analytic Review of the Antecedents and Moderators," *Journal of the Academy of Marketing Science* 4 (2008).

8. Charles Duhigg, *The Power of Habit* (New York: Random House, 2012); Dipayan Biswas, Lauren I. Labrecque, Donald R. Lehmann, and Ereni Markos, "Making Choices While Smelling, Tasting, and Listening: The Role of Sensory (Dis)similarity When Sequentially Sampling Products," *Journal of Marketing,* January 2014; Anna S. Mattila and Jochen Wirtz, "Congruency of Scent and Music as a Driver of In-Store Evaluations and Behavior," *Journal of Retailing,* Summer 2001.

9. Jerry Kirkpatrick, *In Defense of Advertising: Arguments from Reason, Ethical Egoism, and Laissez-Faire Capitalism* (Claremont, CA: TLJ Books, 2008).

10. "Sweet Success," *Brandweek,* May 12, 2003.

11. "The End of Car Ownership," *Wall Street Journal,* June 20, 2017; "'Peak Car' and the End of an Industry," *Bloomberg,* August 16, 2018; "My Days as a Zipster: It's Easy—and It Works," *Automotive News,* December 20, 2010; "Renting Cars by the Hour Is Becoming Big Business," *The Economist,* September 4, 2010; "Car Sharing Grows," *Wall Street Journal,* March 24, 2010; "Zipcar Campus Car-Sharing Program Booms," *Design News,* December 2009; "Zipcar," *Advertising Age,* November 16, 2009; "An iPhone Gets Zipcar Drivers on Their Way," *USA Today,* September 30, 2009; "The Best New Idea in Business," *Fortune,* September 14, 2009; "Temporary Plates," *Brandweek,* July 12, 2007; "The Rise of Collaborative Consumption and the Experience Economy," *The Guardian,* January 3, 2014; "Why Don't Young Americans Buy Cars?," *The Atlantic,* March 25, 2012; Rachel Botsman and Roo Rogers, *What's Mine Is Yours: The Rise of Collaborative Consumption* (New York: HarperBusiness, 2010); "What That Car Really Costs to Own," *Consumer Reports,* August 2012; "RelayRides Reinvents the Rental-Car Relationship," *Denver Post,* March 18, 2015; "RelayRides Revs Up Car-Sharing, Fueled by $52M in Financing to Date," *Tnooz,* January 12, 2015; "Why Uber Isn't the Only Future for the Business of Other People's Cars," *Wired,* December 2014; "Would You Rent Your Car to a Total Stranger?," *Bloomberg Businessweek,* October 5, 2016.

12. "Nike Makes Shoes from Trash, Nokia Envisions Remade Cell Phones," *Greenbiz,* February 15, 2008; "Nothing but Green Skies," *Inc.,* November 2007; "UPS: Getting Big Brown Machines to Go Green," *Investor's Business Daily,* November 26, 2007; "Subway's Diet: Less Oil, More Recycling," *Wall Street Journal,* November 21, 2007. For an academic perspective, see Andrew D. Gershoff and Judy K. Frels, "What Makes It Green? The Role of Centrality of Green Attributes in Evaluations of the Greenness of Products," *Journal of Marketing,* January 2015; Michael Giebelhausen, HaeEun Helen Chun, J. Joseph Cronin Jr., and G. Tomas M. Hult, "Adjusting the Warm-Glow Thermostat: How Incentivizing Participation in Voluntary Green Programs Moderates Their Impact on Service Satisfaction," *Journal of Marketing,* July 2016; Mitchell C. Olsen, Rebecca J. Slotegraaf, and Sandeep R. Chandukala, "Green Claims and Message Frames: How Green New Products Change Brand Attitude," *Journal of Marketing,* September 2014; Aaron R. Brough, James E. B. Wilkie, Jingjing Ma, Mathew S. Isaac, et al., "Is Eco-Friendly Unmanly? The Green-Feminine Stereotype and Its Effect on Sustainable Consumption," *Journal of Consumer Research,* December 2016.

13. "Frozen Dinners Make a Comeback," *Wall Street Journal,* September 8, 2018; "Frozen Food Comes In from the Cold," *Bloomberg.com,* October 27, 2016.

14. For more on service expectations, see "Can I Get a Smile with My Burger and Fries?," *Wall Street Journal,* September 23, 2003. See also Chezy Ofir and Itamar Simonson, "The Effect of Stating Expectations on Customer Satisfaction and Shopping Experience," *Journal of Marketing Research,* February 2007.

15. "Samsung Looks to Repair Consumer Trust," *Wall Street Journal,* January 24, 2017. For an academic perspective, see Maferima Touré-Tillery and Ann L. McGill, "Who or What to Believe: Trust and the Differential Persuasiveness of Human and Anthropomorphized Messengers," *Journal of Marketing,* July 2015; Janet Schwartz, Mary Frances Luce, and Dan Ariely, "Are Consumers Too Trusting? The Effects of Relationships with Expert Advisers," *Journal of Marketing Research,* 2011; K. Grayson, D. Johnson, and D. F. R. Chen, "Is Firm Trust Essential in a Trusted Environment? How Trust in the Business Context Influences Customers," *Journal of Marketing Research* 2 (2008); P. Gupta, M. S. Yadav, and R. Varadarajan, "How Task-Facilitative Interactive Tools Foster Buyers' Trust in Online Retailers: A Process View of Trust Development in the Electronic Marketplace," *Journal of Retailing* 2 (2009).

16. "It's Mind Vending . . . in the World of Psychographics," *Time,* October 2003; "Lifestyles Help Shape Innovation," *DSN Retailing Today,* August 4, 2003; "Generation Next," *Advertising Age,* January 15, 2001; "Head Trips," *American Demographics,* October 2000; "Join the Club," *Brandweek,* February 15, 1999; "The Frontier of Psychographics," *American Demographics,* July 1996; W. D. Wells, "Psychographics: A Critical Review," *Journal of Marketing Research,* May 1975.

17. For academic research on various social influences on consumer behavior, see Darren Dahl, "Social Influence and Consumer Behavior," *Journal of Consumer Research,* June 2014; Jiewen Hong and Hannah H. Chang, "'I' Follow My Heart and 'We' Rely on Reasons: The Impact of Self-Construal on Reliance on Feelings versus Reasons in Decision Making," *Journal of Consumer Research,* April 2015; Andre F. Maciel and Melanie Wallendorf, "Taste Engineering: An Extended Consumer Model of Cultural Competence Constitution," *Journal of Consumer Research,* February 2017; Michelle F. Weinberger and Melanie Wallendorf, "Intracommunity Gifting at the Intersection of Contemporary Moral and Market Economies," *Journal of Consumer Research,* June 2014.

18. Based on U.S. Census data and "The American Consumer 2006," *Advertising Age,* January 2, 2006; "The State of Our Unions," *USA Today,* February 26, 2004; "Do Us Part," *American Demographics,* September 2002; "The Ex-Files," *American Demographics,* February 2001. See also James A. Roberts, Chris Manolis, and John F. Tanner Jr., "Family Structure, Materialism, and Compulsive Buying: A Reinquiry and Extension," *Journal of the Academy of Marketing Science,* Summer 2003; Patrick E. Murphy and William A. Staples, "A Modernized Family Life Cycle," *Journal of Consumer Research,* June 1979.

19. "Who's the Boss? Teens Influence Household Spending Worldwide," *eMarketer,* January 13, 2017.

20. Based on U.S. Census data and "The Older Audience Is Looking Better Than Ever," *New York Times,* April 20, 2009; "Why Grown Kids Come Home," *USA Today,* January 11, 2005.

21. For more on kids' influence in purchase decisions, see A. M. Epp and L. L. Price, "Family Identity: A Framework of Identity Interplay in Consumption Practices," *Journal of Consumer Research* 1 (2008); L. A. Flurry and Alvin C. Burns, "Children's Influence in Purchase Decisions: A Social Power Theory Approach," *Journal of Business Research,* May 2005; "Kiddies' *Wired* Wish Lists," *Wall Street Journal,* December 19, 2007; "Inconvenient Youths," *Wall Street Journal,* September 29, 2007; "This Is the Car We Want, Mommy,"

Wall Street Journal, November 9, 2006. See also Sharon E. Beatty and Salil Talpade, "Adolescent Influence in Family Decision Making: A Replication with Extension," *Journal of Consumer Research,* September 1994; Marketingvox, "Rand Youth Poll, Seventeen," Packaged Facts, September 8, 2012.

22. "Consumers Enjoy Lap of Luxury," *Investor's Business Daily,* October 27, 2003; "Downsized Luxury," *Wall Street Journal,* October 14, 2003. See also W. Amaldoss and S. Jain, "Trading Up: A Strategic Analysis of Reference Group Effects," *Marketing Science* 5 (2008); Lan Nguyen Chaplin and Deborah Roedder John, "Growing Up in a Material World: Age Differences in Materialism in Children and Adolescents," *Journal of Consumer Research,* December 2007; Tamara F. Mangleburg, Patricia M. Doney, and Terry Bristol, "Shopping with Friends and Teens' Susceptibility to Peer Influence," *Journal of Retailing,* Summer 2004; Cornelia Pechmann and Susan J. Knight, "An Experimental Investigation of the Joint Effects of Advertising and Peers on Adolescents' Beliefs and Intentions about Cigarette Consumption," *Journal of Consumer Research,* June 2002; David B. Wooten and Americus Reed II, "Playing It Safe: Susceptibility to Normative Influence and Protective Self-Presentation," *Journal of Consumer Research,* December 2004.

23. For more on opinion leaders and word-of-mouth publicity, see Chapter 16. See also "Beware Social Media Snake Oil," *BusinessWeek,* December 3, 2009; "Why Brands Love Mommy Bloggers," *Adweek,* March 30, 2009. See also Dee T. Allsop, Bryce R. Bassett, and James A. Hoskins, "Word-of-Mouth Research: Principles and Applications," *Journal of Advertising Research,* December 2007; Ed Keller, "Unleashing the Power of Word of Mouth: Creating Brand Advocacy to Drive Growth," *Journal of Advertising Research,* December 2007; Girish N. Punj and Robert Moore, "Smart versus Knowledgeable Online Recommendation Agents," *Journal of Interactive Marketing,* Fall 2007; Judith A. Chevalier and Dina Mayzlin, "The Effect of Word of Mouth on Sales: Online Book Reviews," *Journal of Marketing Research,* August 2006; Thomas W. Gruen, Talai Osmonbekov, and Andrew J. Czaplewski, "Customer-to-Customer Exchange: Its MOA Antecedents and Its Impact on Value Creation and Loyalty," *Journal of the Academy of Marketing Science,* Winter 2007. For more research on social influence on consumer behavior, see Maura L. Scott, Martin Mende, and Lisa E. Bolton, "Judging the Book by Its Cover? How Consumers Decode Conspicuous Consumption Cues in Buyer-Seller Relationships," *Journal of Marketing Research* 50, no. 3 (2013): 334–47; Caleb Warren and Margaret C. Campbell, "What Makes Things Cool? How Autonomy Influences Perceived Coolness," *Journal of Consumer Research* 41, no. 2 (2014): 543–63.

24. "Art of Noise," *Brandweek,* November 1, 1999. For academic perspective on culture and consumer behavior, see Ashok K. Lalwani and Jessie J. Wang, "How Do Consumers' Cultural Backgrounds and Values Influence Their Coupon Proneness? A Multimethod Investigation," *Journal of Consumer Research,* February 2019; Reo Song, Sangkil Moon, Haipeng (Allan) Chen, and Mark B. Houston, "When Marketing Strategy Meets Culture: The Role of Culture in Product Evaluations," *Journal of the Academy of Marketing Science,* May 2018; Ela Veresiu and Markus Giesler, "Beyond Acculturation: Multiculturalism and the Institutional Shaping of an Ethnic Consumer Subject," *Journal of Consumer Research,* October 2018; Anna J. Vredeveld and Robin A. Coulter, "Cultural Experiential Goal Pursuit, Cultural Brand Engagement, and Culturally Authentic Experiences: Sojourners in America," *Journal of the Academy of Marketing Science,* March 2019; Fang Wu, Qi Sun, Rajdeep Grewal, and Shanjun Li, "Brand Name Types and Consumer Demand: Evidence from China's Automobile Market," *Journal of Marketing Research,* February 2019.

25. Data in the following paragraphs comes from U.S. Census Bureau, Statista, www.statista.com/statistics/233324/median-household-income-in-the-united-states-by-race-or-ethnic-group/; U.S. Census Bureau, "2010 Census Interactive Population Search," www.census.gov/2010census/popmap/ipmtext.php; "Ranking Latino Populations in the States," Pew Research Center, August 29, 2013; Nielsen, *The Multicultural Edge: Rising Super-Economies*, March 2015. For an academic perspective, see Samantha N. Cross and Mary C. Gilly, "Cultural Competence and Cultural Compensatory Mechanisms in Binational Households," *Journal of Marketing,* May 2014.

26. "How BuzzFeed Is Winning with Asian-Americans," *Advertising Age,* August 5, 2015.

27. "Women in Italy Like to Clean but Shun the Quick and Easy," *Wall Street Journal,* April 25, 2006. See also Tuba Üstuner and Douglas B. Holt, "Toward a Theory of Status Consumption in Less Industrialized Countries," *Journal of Consumer Research,* February 2010; Eric J. Arnould and Craig J. Thompson, "Consumer Culture Theory (CCT): Twenty Years of Research," *Journal of Consumer Research,* March 2005; John L. Graham, "How Culture Works," *Journal of Marketing,* April 1996; Grant McCracken, "Culture and Consumption: A Theoretical Account of the Structure and Movement of the Cultural Meaning of Consumer Goods," *Journal of Consumer Research,* June 1986.

28. For more research on cultural influences on consumer behavior, see Donnel Briley, Robert S. Wyer Jr., and En Li, "A Dynamic View of Cultural Influence: A Review," *Journal of Consumer Psychology* 24, no. 4 (2014): 557–71; Samantha N. Cross and Mary C. Gilly, "Cultural Competence and Cultural Compensatory Mechanisms in Binational Households," *Journal of Marketing* 78, no. 3 (2014): 121–39; Tuba Üstuner and Douglas B. Holt, "Toward a Theory of Status Consumption in Less Industrialized Countries," *Journal of Consumer Research,* 2014, S248–67; "Fuel and Freebies," *Wall Street Journal,* June 10, 2002; Madhu Viswanathan, José Antonio Rosa, and Julie A. Ruth, "Exchanges in Marketing Systems: The Case of Subsistence Consumer–Merchants in Chennai, India," *Journal of Marketing,* May 2010; Kelly D. Martin and Ronald Paul Hill, "Life Satisfaction, Self-Determination, and Consumption Adequacy at the Bottom of the Pyramid," *Journal of Consumer Research,* December 2011.

29. "The Slower You Shop, the More You Spend," *Wall Street Journal,* October 20, 2015. For an academic perspective, see Kelley Gullo, Jonah Berger, Jordan Etkin, and Bryan Bollinger, "Does Time of Day Affect Variety-Seeking?," *Journal of Consumer Research,* June 2019.

30. For research on situational factors influencing consumer behavior, see Timothy J. Gilbride, J. Jeffrey Inman, and Karen Melville Stilley, "The Role of Within-Trip Dynamics in Unplanned versus Planned Purchase Behavior," *Journal of Marketing,* May 2015; Ernest Baskin, Cheryl J. Wakslak, Yaacov Trope, and Nathan Novemsky, "Why Feasibility Matters More to Gift Receivers Than to Givers: A Construal-Level Approach to Gift Giving," *Journal of Consumer Research,* June 2014; Amit Bhattacharjee, Jonathan Z. Berman, and Americus Reed II, "Tip of the Hat, Wag of the Finger: How Moral Decoupling Enables Consumers to Admire and Admonish," *Journal of Consumer Research,* August 2014; Cindy Chan and Cassie Mogilner, "Experiential Gifts Foster Stronger Social Relationships Than Material Gifts," *Journal of Consumer Research,* April 2017; Alain Debenedetti, Harmen Oppewal, and Zeynep Arsel, "Place Attachment in Commercial Settings: A Gift Economy Perspective," *Journal of Consumer Research,* February 2014; Anouk Festjens and Chris Janiszewski, "The Value of Time," *Journal of Consumer Research,* August 2015; Daniel M. Bartels, Nicholas Reinholtz, and Jeffrey R. Parker, "On the Mental Accounting of Restricted-Use Funds: How Gift Cards Change What People Purchase," *Journal of Consumer Research* 42, no. 4 (2015): 596–614; Jordan Etkin, Ioannis Evangelidis, and Jennifer Aaker, "Pressed for Time? Goal Conflict Shapes How Time Is Perceived, Spent, and Valued," *Journal of Marketing Research* 52, no. 3 (2015): 394–406; Leonard Lee, Michelle P. Lee, Marco Bertini, Gal Zauberman, et al., "Money, Time, and the Stability of Consumer Preferences," *Journal of Marketing Research* 52, no. 2 (2015): 184–99; Gabriele Paolacci, Laura M. Straeter, and Ilona E. De Hooge, "Give Me Your Self: Gifts Are Liked More When They Match the Giver's Characteristics," *Journal of Consumer Psychology* 25, no. 3 (2015): 487–94; Morgan K. Ward and Susan M. Broniarczyk, "It's Not Me, It's You: How Gift Giving Creates Giver Identity Threat as a Function of Social Closeness," *Journal of Consumer Research,* 2013, S270–87; Niklas Woermann and Joonas

Rokka, "Timeflow: How Consumption Practices Shape Consumers' Temporal Experiences," *Journal of Consumer Research* 41, no. 6 (2015): 1486–508; Mary Steffel and Robyn A. Le Boeug, "Overindividuation in Gift Giving: Shopping for Multiple Recipients Leads Givers to Choose Unique but Less Preferred Gifts," *Journal of Consumer Research* 40, no. 6 (2014): 1167–80; Lei Su and Leilei Gao, "Strategy Compatibility: The Time versus Money Effect on Product Evaluation Strategies," *Journal of Consumer Psychology* 24, no. 4 (2014): 549–56; Yanping Tu and Dilip Soman, "The Categorization of Time and Its Impact on Task Initiation," *Journal of Consumer Research* 41, no. 3 (2014): 810–22; Michelle F. Weinberger and Melanie Wallendorf, "Intracommunity Gifting at the Intersection of Contemporary Moral and Market Economies," *Journal of Consumer Research,* January 2012; C. Page Moreau, Leff Bonney, and Kelly B. Herd, "It's the Thought (and the Effort) That Counts: How Customizing for Others Differs from Customizing for Oneself," *Journal of Marketing,* September 2011; Markus Giesler, "Consumer Gift Systems," *Journal of Consumer Research,* August 2006; Jennifer Chang Coupland, Dawn Iacobucci, and Eric Arnould, "Invisible Brands: An Ethnography of Households and the Brands in Their Kitchen Pantries," *Journal of Consumer Research,* June 2005; S. Christian Wheeler, Richard E. Petty, and George Y. Bizer, "Self-Schema Matching and Attitude Change: Situational and Dispositional Determinants of Message Elaboration," *Journal of Consumer Research,* March 2005; Russell W. Belk, "Situational Variables and Consumer Behavior," *Journal of Consumer Research* 2 (1975); John F. Sherry Jr., "Gift Giving in Anthropological Perspective," *Journal of Consumer Research,* September 1983.

31. Adapted and updated from James H. Myers and William H. Reynolds, *Consumer Behavior and Marketing Management* (Boston: Houghton Mifflin, 1967). See also Christopher Berry, Scot Burton, and Elizabeth Howlett, "It's Only Natural: The Mediating Impact of Consumers' Attribute Inferences on the Relationships between Product Claims, Perceived Product Healthfulness, and Purchase Intentions," *Journal of the Academy of Marketing Science,* September 2017; Sean T. Hingston and Theodore J. Noseworthy, "Why Consumers Don't See the Benefits of Genetically Modified Foods, and What Marketers Can Do about It," *Journal of Marketing,* September 2018; Rebecca Hamilton, Debora Thompson, Sterling Bone, Lan Nguyen Chaplin, et al., "The Effects of Scarcity on Consumer Decision Journeys," *Journal of the Academy of Marketing Science,* May 2019; Zhongqiang (Tak) Huang, Yitian (Sky) Liang, Charles B. Weinberg, and Gerald J. Gorn, "The Sleepy Consumer and Variety Seeking," *Journal of Marketing Research,* April 2019; Christina Kuehnl, Danijel Jozic, and Christian Homburg, "Effective Customer Journey Design: Consumers' Conception, Measurement, and Consequences," *Journal of the Academy of Marketing Science,* May 2019; Jifeng Mu, Ellen Thomas, Jiayin Qi, and Yong Tan, "Online Group Influence and Digital Product Consumption," *Journal of the Academy of Marketing Science,* September 2018; Christina Schamp, Mark Heitmann, and Robin Katzenstein, "Consideration of Ethical Attributes along the Consumer Decision-Making Journey," *Journal of the Academy of Marketing Science,* March 2019; Caleb Warren, Adam Barsky, and A. Peter McGraw, "Humor, Comedy, and Consumer Behavior," *Journal of Consumer Research,* October 2018; Elizabeth C. Webb and Suzanne B. Shu, "The Effect of Perceived Similarity on Sequential Risk Taking," *Journal of Marketing Research,* December 2018; Sarah C. Whitley, Remi Trudel, and Didem Kurt, "The Influence of Purchase Motivation on Perceived Preference Uniqueness and Assortment Size Choice," *Journal of Consumer Research,* December 2018; Raquel Castaño, Mita Sujan, Manish Kacker, and Harish Sujan, "Managing Consumer Uncertainty in the Adoption of New Products: Temporal Distance and Mental Simulation," *Journal of Marketing Research* 3 (2008); Anna Lund Jepsen, "Factors Affecting Consumer Use of the Internet for Information Search," *Journal of Interactive Marketing,* Summer 2007; Kineta H. Hung and Stella Yiyan Li, "The Influence of eWOM on Virtual Consumer Communities: Social Capital, Consumer Learning, and Behavioral Outcomes," *Journal of Advertising Research,* December 2007; "How and What Do Consumers Maximize?," *Psychology & Marketing,* September 2003; Ronald E. Goldsmith, "A Theory of Shopping," *Journal of the Academy of Marketing Science,* Fall 2000; Judith Lynne Zaichkowsky, "Consumer Behavior: Yesterday, Today, and Tomorrow," *Business Horizons,* May–June 1991; James R. Bettman, *An Information Processing Theory of Consumer Choice* (Reading, MA: Addison-Wesley, 1979); Richard W. Olshavsky and Donald H. Granbois, "Consumer Decision Making: Fact or Fiction?" *Journal of Consumer Research,* September 1979; Craig J. Thompson, "Consumer Risk Perceptions in a Community of Reflexive Doubt," *Journal of Consumer Research,* September 2005; Alice M. Tybout, Brian Sternthal, Prashant Malaviya, Georgios A. Bakamitsos, et al., "Information Accessibility as a Moderator of Judgments: The Role of Content versus Retrieval Ease," *Journal of Consumer Research,* June 2005; Laura L. Pingol and Anthony D. Miyazaki, "Information Source Usage and Purchase Satisfaction: Implications for Product-Focused Print Media," *Journal of Advertising Research,* March 2005; Stijn M. J. Van Osselaer, Joseph W. Alba, and Puneet Manchanda, "Irrelevant Information and Mediated Intertemporal Choice," *Journal of Consumer Psychology* 14, no. 3 (2004); Judi Strebel, Tulin Erdem, and Joffre Swait, "Consumer Search in High Technology Markets: Exploring the Use of Traditional Information Channels," *Journal of Consumer Psychology* 1 (2004); Ram D. Gopal, Bhavik Pathak, Arvind K. Tripathi, and Fang Yin, "From Fatwallet to eBay: An Investigation of Online Deal-Forums and Sales Promotions," *Journal of Retailing,* June 2006; Raj Arora, "Consumer Involvement—What It Offers to Advertising Strategy," *International Journal of Advertising* 4, no. 2 (1985); J. Brock Smith and Julia M. Bristor, "Uncertainty Orientation: Explaining Differences in Purchase Involvement and External Search," *Psychology & Marketing,* November–December 1994.

32. For research on the consumer decision process, see Katherine N. Lemon and Peter C. Verhoef, "Understanding Customer Experience throughout the Customer Journey," *Journal of Marketing,* November 2016; Daniel M. Bartels and Oleg Urminsky, "To Know and to Care: How Awareness and Valuation of the Future Jointly Shape Consumer Spending," *Journal of Consumer Research,* April 2015; Cassie Mogilner, Jennifer Aaker, and Sepandar D. Kamvar, "How Happiness Affects Choice," *Journal of Consumer Research,* June 2014; Jeffrey R. Parker and Donald R. Lehmann, "How and When Grouping Low-Calorie Options Reduces the Benefits of Providing Dish-Specific Calorie Information," *Journal of Consumer Research,* June 2014; Lauren Block, Nilüfer Z. Aydinoğlu, Aradhna Krishna, Koert Van Ittersum, et al., "Food Decision Making," *Journal of Consumer Research* 39, no. 5 (2013): iv–vi; Christopher K. Hsee, Jiao Zhang, Liangyan Wang, and Shirley Zhang, "Magnitude, Time, and Risk Differ Similarly between Joint and Single Evaluations," *Journal of Consumer Research* 40, no. 1 (2013): 172–84; Liu (Cathy) Yang, Olivier Toubia, and Martijn G. De Jong, "A Bounded Rationality Model of Information Search and Choice in Preference Measurement," *Journal of Marketing Research* 52, no. 2 (2015): 166–83; Jeffrey V. Rayport, "Demand-Side Innovation: Where IT Meets Marketing," *InformationWeek,* February 1, 2007; "Web Stores Tap Product Reviews," *Wall Street Journal,* September 11, 2007; David R. Bell, Daniel Corsten, and George Knox, "From Point of Purchase to Path to Purchase: How Preshopping Factors Drive Unplanned Buying," *Journal of Marketing,* January 2011; Wendy Attaya Boland, Merrie Brucks, and Jesper H. Nielsen, "The Attribute Carryover Effect: What the 'Runner-Up' Option Tells Us about Consumer Choice Processes," *Journal of Consumer Research,* October 2012; Uzma Khan, Meng Zhu, and Ajay Kalra, "When Trade-offs Matter: The Effect of Choice Construal on Context Effects," *Journal of Marketing Research,* February 2011.

33. For academic research on post-purchase, see Yany Grégoire, Fateme Ghadami, Sandra Laporte, Sylvain Sénécal, et al., "How Can Firms Stop Customer Revenge? The Effects of Direct and Indirect Revenge on Post-Complaint Responses," *Journal of the Academy of Marketing Science,* November 2018; J. Jeffrey Inman and Marcel Zeelenberg, "Regret in Repeat Purchase versus Switching Decisions: The Attenuating Role of Decision Justifiability," *Journal of Consumer Research,* June 2002; Michael Tsiros and Vikas Mittal, "Regret: A

Model of Its Antecedents and Consequences in Consumer Decision Making," *Journal of Consumer Research,* March 2000; Thomas A. Burnham, Judy K. Frels, and Vijay Mahajan, "Consumer Switching Costs: A Typology, Antecedents, and Consequences," *Journal of the Academy of Marketing Science,* Spring 2003; Ziv Carmon, Klaus Wertenbroch, and Marcel Zeelenberg, "Option Attachment: When Deliberating Makes Choosing Feel Like Losing," *Journal of Consumer Research,* June 2003; Alan D. J. Cooke, Tom Meyvis, and Alan Schwartz, "Avoiding Future Regret in Purchase-Timing Decisions," *Journal of Consumer Research,* March 2001; Anna S. Mattila, "The Impact of Cognitive Inertia on Postconsumption Evaluation Processes," *Journal of the Academy of Marketing Science,* Summer 2003; William Cunnings and Mark Venkatesan, "Cognitive Dissonance and Consumer Behavior: A Review of the Evidence," *Journal of Marketing Research,* August 1976.

34. C. Aaron, E. Edwards, and X. Lanier, "Turning Blogs and User-Generated Content into Search Engine Results." *SES,* May 2009. For an academic perspective, see Andrea Kähr, Bettina Nyffenegger, Harley Krohmer, and Wayne D. Hoyer, "When Hostile Consumers Wreak Havoc on Your Brand: The Phenomenon of Consumer Brand Sabotage," *Journal of Marketing,* May 2016; George Knox and Rutger van Oest, "Customer Complaints and Recovery Effectiveness: A Customer Base Approach," *Journal of Marketing,* September 2014.

35. For more on this topic, see Eric Siegel, *Predictive Analytics: The Power to Predict Who Will Click, Buy, Lie, or Die* (New York: Wiley, 2016); Joseph Turow, *The Aisles Have Eyes: How Retailers Track Your Shopping, Strip Your Privacy, and Define Your Power* (New Haven, CT: Yale University Press, 2017); "What Is Predictive Analytics?," *Predictive Analytics Today,* n.d.

36. Adapted from E. M. Rogers, *Diffusion of Innovation* (New York: Free Press, 2003). See also Marius Claudy, Rosanna Garcia, and Aidan O'Driscoll, "Consumer Resistance to Innovation—A Behavioral Reasoning Perspective," *Journal of the Academy of Marketing Science,* July 2015; Zhenfeng Ma, Zhiyong Yang, and Mehdi Mourali, "Consumer Adoption of New Products: Independent versus Interdependent Self-Perspectives," *Journal of Marketing,* March 2014; Jennifer Labrecque, Wendy Wood, David Neal, and Nick Harrington, "Habit Slips: When Consumers Unintentionally Resist New Products," *Journal of the Academy of Marketing Science,* January 2017; Rex Yuxing Du and Wagner A. Kamakura, "Measuring Contagion in the Diffusion of Consumer Packaged Goods," *Journal of Marketing Research,* February 2011; Donald Lehmann and Mercedes Esteban-Bravo, "When Giving Some Away Makes Sense to Jump-Start the Diffusion Process," *Marketing Letters,* October 2006; Stephen J. Hoch, "Product Experience Is Seductive," *Journal of Consumer Research,* December 2002.

CHAPTER 6

1. See www.deere.com and www.metokote.com; "The AI Farm Experiment," *Axios,* May 3, 2018; "Now Cropping Up: Robo-Farming," *Wall Street Journal,* June 1, 2018; "Deere, Siemens Adapt to Industrial AI," *Wall Street Journal,* January 11, 2019; "America's Farmers Turn to Bank of John Deere," *Wall Street Journal,* July 18, 2017; "The Amazing Ways John Deere Uses AI and Machine Vision to Help Feed 10 Billion People," *Forbes,* March 15, 2019; *2015 Annual Report,* Deere & Company; "Deere Hurt by Weak Farm-Equipment Demand," *Wall Street Journal,* February 20, 2015; "Brazil Raises Ag Equipment Funding, Deere Benefits?," *Barron's Emerging Markets Daily* blog, June 4, 2015; "Deere Should Reap Harvest in Due Time," *Wall Street Journal,* February 19, 2015; "Deere's Big Green Profit Machine," *Bloomberg Businessweek,* July 5, 2012; "Farmers Prepare for the Data Harvest," *Wall Street Journal,* June 14, 2012; "Teaching Drones to Farm," *Wall Street Journal,* September 20, 2011; "American Farmers Venture into New Field: Social Media," *Technorati,* September 7, 2010; "John Deere India Introduces Two New Tractor Models," *Economic Times,* August 13, 2010; "Global Innovation: John Deere's Farm Team," *Fortune,* April 14, 2008; "John Deere's Big-Wheel Rally," *Investor's Business Daily,* September 24, 2007; "Why Deere Is Weeding Out Dealers Even as Farms Boom," *Wall Street Journal,* August 14, 2007; "Deere's Revolution on Wheels," *BusinessWeek,* July 2, 2007; "Deere," *Fortune,* April 30, 2007; "John Deere Cultivates Its Image," *Advertising Age,* July 25, 2005; "MetoKote Develops Customized Systems for Applying High Tech Coatings; More Than Just Painting a Part," *Diesel Progress,* April 2003.

2. Forrester Research, *US Online Retail Forecast, 2011 to 2016;* Forrester Reasearch, *Key Trends in B2B eCommerce for 2013.*

3. "Takata's Future Clipped by Steep Recall Costs," *Wall Street Journal,* May 20, 2015.

4. "New Research: What Motivates Buyers to Receive and Engage with Vendor Email?," *Marketing Sherpa,* February 10, 2010.

5. See www.woolworthsholdings.co.za; "Change the World," *Fortune,* September 1, 2018; "Change the World," *Fortune,* September 15, 2017; "Can Levi's Make Life Better for Garment Workers?," *Fortune,* September 8, 2017; "Levi's Expanding Worker Education Efforts to Improve Quality of Life," *Fortune,* June 26, 2018; "Miners Test Greener Ways to Dig," *Wall Street Journal,* June 16, 2018.

6. "J&J Is Acccused of Kickbacks to Omnicare on Drug Sales," *Wall Street Journal,* January 16, 2010; "J&J Penalty May Total $2.2 Billion," *Wall Street Journal,* July 19, 2012; Richard F. Beltramini, "Exploring the Effectiveness of Business Gifts: Replication and Extension," *Journal of Advertising,* Summer 2000; Jeanette J. Arbuthnot, "Identifying Ethical Problems Confronting Small Retail Buyers during the Merchandise Buying Process," *Journal of Business Ethics,* May 1997; Gail K. McCracken and Thomas J. Callahan, "Is There Such a Thing as a Free Lunch?," *International Journal of Purchasing & Materials Management,* Winter 1996; Robert W. Cooper, Garry L. Frank, and Robert A. Kemp, "The Ethical Environment Facing the Profession of Purchasing and Materials Management," *International Journal of Purchasing & Materials Management,* Spring 1997; I. Fredrick Trawick, John E. Swan, Gail W. McGee, and David R. Rink, "Influence of Buyer Ethics and Salesperson Behavior on Intention to Choose a Supplier," *Journal of the Academy of Marketing Science,* Winter 1991; J. A. Badenhorst, "Unethical Behaviour in Procurement: A Perspective on Causes and Solutions," *Journal of Business Ethics,* September 1994.

7. Scott B. Friend and Jeff S. Johnson, "Familiarity Breeds Contempt: Perceived Service and Sales Complacency in Business-to-Business Relationships," *Journal of Personal Selling and Sales Management,* March 2017; M. D. Steward, F. N. Morgan, L. A. Crosby, and A. Kumar, "Exploring Cross-National Differences in Organizational Buyers' Normative Expectations of Supplier Performance," *Journal of International Marketing* 1 (2010); Kelly Hewett, R. Bruce Money, and Subhash Sharma, "An Exploration of the Moderating Role of Buyer Corporate Culture in Industrial Buyer-Seller Relationships," *Journal of the Academy of Marketing Science,* Summer 2002; Jae-Eun Chung, Brenda Sternquist, and Zhengyi Chen, "Retailer-Buyer Supplier Relationships: The Japanese Difference," *Journal of Retailing,* December 2006; Masaaki Kotabe and Janet Y. Murray, "Global Sourcing Strategy and Sustainable Competitive Advantage," *Industrial Marketing Management,* January 2004.

8. Stephen J. Newell, Bob Wu, Duke Leingpibul, and Yang Jiang, "The Importance of Corporate and Salesperson Expertise and Trust in Building Loyal Business-to-Business Relationships in China," *Journal of Personal Selling & Sales Management,* June 2016; Gregory T. Gundlach and Joseph P. Cannon, "Trust but Verify? The Performance Implications of Verification Strategies in Trusting Relationships," *Journal of the Academy of Marketing Science,* July 2010; E. Fang, R. Palmatier, L. K. Scheer, and N. Li, "Trust at Different Organizational Levels," *Journal of Marketing* 2 (2008); Jeffrey E. Lewin and Naveen Donthu, "The Influence of Purchase Situation on Buying Center Structure and Involvement: A Select Meta-Analysis of Organizational Buying Behavior Research," *Journal of Business Research,* October 2005; Jae H. Pae, Namwoon Kim, Jim K. Han, and Leslie Yip, "Managing Intraorganizational Diffusion of Innovations: Impact of Buying Center Dynamics and Environments," *Industrial Marketing Management,* November

2002; R. Venkatesh, Ajay K. Kohli, and Gerald Zaltman, "Influence Strategies in Buying Centers," *Journal of Marketing,* October 1995; "Major Sales: Who Really Does the Buying?," *Harvard Business Review,* May–June 1982.

9. Patrick J. Robinson, Charles W. Faris, and Yoram Wind, *Industrial Buying Behavior and Creative Marketing* (Boston: Allyn & Bacon, 1967).

10. Jessica J. Hoppner and David A. Griffith, "The Role of Reciprocity in Clarifying the Performance Payoff of Relational Behavior," *Journal of Marketing Research,* October 2011; Jason Shachat and Wei Lijia, "Procuring Commodities: First-Price Sealed-Bid or English Auctions?" *Marketing Science* 2 (2012); S. Khan and B. Schroder, "Use of Rules in Decision-Making in Government Outsourcing," *Industrial Marketing Management* 4 (2009); Neeraj Bharadwaj, "Investigating the Decision Criteria Used in Electronic Components Procurement," *Industrial Marketing Management,* May 2004; Jacques Verville and Alannah Halingten, "A Six-Stage Model of the Buying Process for ERP Software," *Industrial Marketing Management,* October 2003; Carol C. Bienstock, "Understanding Buyer Information Acquisition for the Purchase of Logistics Services," *International Journal of Physical Distribution & Logistics Management* 8 (2002); David Tucker and Laurie Jones, "Leveraging the Power of the Internet for Optimal Supplier Sourcing," *International Journal of Physical Distribution & Logistics Management* 3 (2000); H. L. Brossard, "Information Sources Used by an Organization during a Complex Decision Process: An Exploratory Study," *Industrial Marketing Management,* January 1998; Michele D. Bunn, "Taxonomy of Buying Decision Approaches," *Journal of Marketing,* January 1993; Mark A. Farrell and Bill Schroder, "Influence Strategies in Organizational Buying Decisions," *Industrial Marketing Management,* July 1996; Edward F. Fern and James R. Brown, "The Industrial/Consumer Marketing Dichotomy: A Case of Insufficient Justification," *Journal of Marketing,* Spring 1984.

11. Anindita Chakravarty, Alok Kumar, and Rajdeep Grewal, "Customer Orientation Structure for Internet-Based Business-to-Business Platform Firms," *Journal of Marketing* 78, no. 5 (2014): 1–23; Mahima Hada, Rajdeep Grewal, and Gary L. Lilien, "Supplier-Selected Referrals," *Journal of Marketing* 78, no. 2 (2014): 34–51.

12. For more on these trends, see *The Next Generation of B2B Buyers,* report by Sacunas, Inc., March 2016; *The Changing Face of B2B Marketing,* Think With Google, March 2015; *To Buy or Not to Buy? How Millennials Are Reshaping B2B Marketing,* IBM Institute for Business Value, September 2015. For examples and other reading, see "How B2B Marketers Use Facebook to Reach Customers," *Social Media Examiner,* January 11, 2016; "GE's CMO on Redefining B2B Marketing at the 'Pretty Damn Cool' Brand," *Marketing Week,* January 11, 2017; "3 of the Most Successful Social Media Campaigns for B2B," *New Breed Blog,* June 11, 2014.

13. "The Role of Search in Business to Business Buying Decisions," *Enquiro Search Solutions,* October 27, 2004. See also "10 Great Websites," *B2B,* September 10, 2007; "Usability Problems Plague B-to-B Sites," *BtoBonline.com,* April 23, 2007.

14. See www.spiceworks.com; "Reaching Folks on Their Turf," *New York Times,* March 5, 2009.

15. Ernan Haruvy and Sandy D. Jap, "Differentiated Bidders and Bidding Behavior in Procurement Auctions," *Journal of Marketing Research* 50, no. 2 (2013): 241–58; Daniel J. Flint, Robert B. Woodruff, and Sarah Fisher Gardial, "Exploring the Phenomenon of Customers' Desired Value Change in a Business-to-Business Context," *Journal of Marketing,* October 2002.

16. "The New Golden Rule of Business," *Fortune,* February 21, 1994; Janet L. Hartley and Thomas Y. Choi, "Supplier Development: Customers as a Catalyst of Process Change," *Business Horizons,* July–August 1996.

17. "Ford Seeks Big Savings by Overhauling Supply System," *Wall Street Journal,* September 29, 2005; "Turning Vendors into Partners," *Inc.,* August 2005; "A 'China Price' for Toyota," *BusinessWeek,* February 21, 2005; "Ford to Suppliers: Let's Get Cozier," *BusinessWeek,* September 20, 2004; "How Would You Like Your Ford?," *BusinessWeek,* August 9, 2004; "Walmart's Low-Price Obsession Puts Suppliers through Wringer," *Investor's Business Daily,* January 30, 2004; "Push from Above," *Wall Street Journal,* May 23, 1996; Jan B. Heide, "Plural Governance in Industrial Purchasing," *Journal of Marketing,* October 2003; Shibin Sheng, James R. Brown, Carolyn Y. Nicholson, and Laura Poppo, "Do Exchange Hazards Always Foster Relational Governance? An Empirical Test of the Role of Communication," *International Journal of Research in Marketing,* March 2006; Erin Anderson and Sandy D. Jap, "The Dark Side of Close Relationships," *Sloan Management Review,* Spring 2005.

18. Much of the discussion in this section is based on research reported in Joseph P. Cannon and William D. Perreault Jr., "Buyer–Seller Relationships in Business Markets," *Journal of Marketing Research,* November 1999. See also D. Eric Boyd and P. K. Kannan, "(When) Does Third-Party Recognition for Design Excellence Affect Financial Performance in Business-to-Business Markets?," *Journal of Marketing,* May 2018; Scott B. Friend and Jeff S. Johnson, "Familiarity Breeds Contempt: Perceived Service and Sales Complacency in Business-to-Business Relationships," *Journal of Personal Selling & Sales Management,* March 2017; Aditya Gupta, Alok Kumar, Rajdeep Grewal, and Gary L. Lilien, "Within-Seller and Buyer–Seller Network Structures and Key Account Profitability," *Journal of Marketing,* January 2019; Ashwin Joshi, "OEM Implementation of Supplier-Developed Component Innovations: The Role of Supplier Actions," *Journal of the Academy of Marketing Science,* July 2017; Stephanie M. Mangus and Ayalla Ruvio, "Do Opposites Attract? Assimilation and Differentiation as Relationship-Building Strategies," *Journal of Personal Selling & Sales Management,* March 2019; Bert Paesbrugghe, Arun Sharma, Deva Rangarajan, and Niladri Syam, "Personal Selling and the Purchasing Function: Where Do We Go from Here?," *Journal of Personal Selling & Sales Management,* March 2018; Shibin Sheng, Kevin Zheng Zhou, Julie Juan Li, and Zhaoyang Guo, "Institutions and Opportunism in Buyer-Supplier Exchanges: The Moderated Mediating Effects of Contractual and Relational Governance," *Journal of the Academy of Marketing Science,* November 2018; Kenneth H. Wathne, Jan B. Heide, Erik A. Mooi, and Alok Kumar, "Relationship Governance Dynamics: The Roles of Partner Selection Efforts and Mutual Investments," *Journal of Marketing Research,* October 2018; Stefan Worm, Sundar Bharadwaj, Wolfgang Ulaga, and Werner Reinartz, "When and Why Do Customer Solutions Pay Off in Business Markets?," *Journal of the Academy of Marketing Science,* July 2017; Anthony deLeon and Sharmila Chatterjee, "B2B Relationship Calculus: Quantifying Resource Effects in Service-Dominant Logic," *Journal of the Academy of Marketing Science,* May 2017; Lin Guo, Thomas Gruen, and Chuanyi Tang, "Seeing Relationships through the Lens of Psychological Contracts: The Structure of Consumer Service Relationships," *Journal of the Academy of Marketing Science,* May 2017; Thomas Hollmann, Cheryl Jarvis, and Mary Bitner, "Reaching the Breaking Point: A Dynamic Process Theory of Business-to-Business Customer Defection," *Journal of the Academy of Marketing Science,* March 2015; Girish Mallapragada, Rajdeep Grewal, Raj Mehta, and Ravi Dharwadkar, "Virtual Interorganizational Relationships in Business-to-Business Electronic Markets: Heterogeneity in the Effects of Organizational Interdependence on Relational Outcomes," *Journal of the Academy of Marketing Science,* September 2015; Anindita Chakravarty, Alok Kumar, and Rajdeep Grewal, "Customer Orientation Structure for Internet-Based Business-to-Business Platform Firms," *Journal of Marketing,* September 2014; Inge Geyskens, Katrijn Gielens, and Stefan Wuyts, "United We Stand: The Impact of Buying Groups on Retailer Productivity," *Journal of Marketing,* July 2015; Mahima Hada, Rajdeep Grewal, and Gary L. Lilien, "Supplier-Selected Referrals," *Journal of Marketing,* March 2014; Colleen M. Harmeling, Robert W. Palmatier, Mark B. Houston, Mark J. Arnold, et al., "Transformational Relationship Events," *Journal of Marketing,* September 2015; Emma K. Macdonald, Michael Kleinaltenkamp, and Hugh N. Wilson, "How Business Customers

Judge Solutions: Solution Quality and Value in Use," *Journal of Marketing,* May 2016; Jonathan Z. Zhang, George F. Watson IV, Robert W. Palmatier, and Rajiv P. Dant, "Dynamic Relationship Marketing," *Journal of Marketing,* September 2016; Stephen A. Samaha, Joshua T. Beck, and Robert W. Palmatier, "The Role of Culture in International Relationship Marketing," *Journal of Marketing* 78, no. 5 (2014): 78-98; Anita Luo and V. Kumar, "Recovering Hidden Buyer-Seller Relationship States to Measure the Return on Marketing Investment in Business-to-Business Markets," *Journal of Marketing Research* 50, no. 1 (2013): 143-60; Robert W. Palmatier, Mark B. Houston, Rajiv P. Dant, and Dhruv Grewal, "Relationship Velocity: Toward a Theory of Relationship Dynamics," *Journal of Marketing* 77, no. 1 (2013): 13-30; Qiong Wang, Julie Li, William Ross, and Christopher Craighead, "The Interplay of Drivers and Deterrents of Opportunism in Buyer-Supplier Relationships," *Journal of the Academy of Marketing Science* 41, no. 1 (2013): 111-31; Christopher P. Blocker, Mark B. Houston, and Daniel J. Flint, "Unpacking What a 'Relationship' Means to Commercial Buyers: How the Relationship Metaphor Creates Tension and Obscures Experience," *Journal of Consumer Research,* 38 (October 2012); Shankar Ganesan, Steven P. Brown, Babu John Mariadoss, and Hillbun (Dixon) Ho, "Buffering and Amplifying Effects of Relationship Commitment in Business-to-Business Relationships," *Journal of Marketing Research,* April 2010; Alok Kumar, Jan B. Heide, and Kenneth H. Wathne, "Performance Implications of Mismatched Governance Regimes across External and Internal Relationships," *Journal of Marketing,* March 2011; Kapil R. Tuli, Sundar G. Bharadwaj, and Ajay K. Kohli, "Ties That Bind: The Impact of Multiple Types of Ties with a Customer on Sales Growth and Sales Volatility," *Journal of Marketing Research,* February 2010; Rajdeep Grewal, Anindita Chakravarty, and Amit Saini, "Governance Mechanisms in Business-to-Business Electronic Markets," *Journal of Marketing,* July 2010; R. W. Palmatier, "Interfirm Relational Drivers of Customer Value," *Journal of Marketing* 4 (2008); R. W. Palmatier, L. K. Scheer, K. R. Evans, and T. J. Arnold, "Achieving Relationship Marketing Effectiveness in Business-to-Business Exchanges," *Journal of the Academy of Marketing Science* 2 (2008); R. N. Bolton, K. N. Lemon, and P. C. Verhoef, "Expanding Business-to-Business Customer Relationships: Modeling the Customer's Upgrade Decision," *Journal of Marketing* 1 (2008); Jan B. Heide and Kenneth H. Wathne, "Friends, Businesspeople, and Relationship Roles: A Conceptual Framework and a Research Agenda," *Journal of Marketing,* July 2006; Robert W. Palmatier, Rajiv P. Dant, Dhruv Grewal, and Kenneth R. Evans, "Factors Influencing the Effectiveness of Relationship Marketing: A Meta-Analysis," *Journal of Marketing,* October 2006; Kapil R. Tuli, Ajay K. Kohli, and Sundar G. Bharadwaj, "Rethinking Customer Solutions: From Product Bundles to Relational Processes," *Journal of Marketing,* July 2007; Joseph P. Cannon and Christian Homburg, "Buyer-Supplier Relationships and Customer Firm Costs," *Journal of Marketing,* January 2001; Das Narayandas and V. Kasturi Rangan, "Building and Sustaining Buyer-Seller Relationships in Mature Industrial Markets," *Journal of Marketing,* July 2004.

19. "Purchasing's New Muscle," *Fortune,* February 20, 1995.

20. "Early Warnings in the Supply Chain," *Financial Times,* March 24, 2009; "United They'll Stand," *Wall Street Journal,* March 23, 2009.

21. For more on outsourcing, see "Big Three's Outsourcing Plan: Make Parts Suppliers Do It," *Wall Street Journal,* June 10, 2004; see also M. Ahearne and P. Kothandaraman, "Impact of Outsourcing on Business-to-Business Marketing: An Agenda for Inquiry," *Industrial Marketing Management* 4 (2009); "HP Will Direct More U.S. Customers to U.S. Call Centers," *Investor's Business Daily,* February 28, 2008; "The Future of Outsourcing," *BusinessWeek,* January 30, 2006; "Pulling the Plug," *Fortune,* August 8, 2005, 96B-D; "Outsourcing with a Twist," *Wall Street Journal,* January 18, 2005.

22. "H-P Wields Its Clout to Undercut PC Rivals," *Wall Street Journal,* September 25, 2009. For other examples of big-small partnerships, see "A Fruitful Relationship," *BusinessWeek,* November 20, 2000, EB94-EB96; "Automating an Automaker," *Inc.* Tech 2000, no. 4. See also David Griffith, Jessica J. Hoppner, Hannah S. Lee, and Tobias Schoenherr, "The Influence of the Structure of Interdependence on the Response to Inequity in Buyer-Supplier Relationships," *Journal of Marketing Research,* February 2017.

23. "Thai Floods Jolt PC Supply Chain," *Wall Street Journal,* October 18, 2011.

24. Jeffrey K. Liker and Thomas Y. Choi, "Building Deep Supplier Relationships," *Harvard Business Review,* December 2004; see also G. L. Frazier, E. Maltz, K. D. Antia, and A. Rindfleisch, "Distributor Sharing of Strategic Information with Suppliers," *Journal of Marketing* 4 (2009).

25. "Top 20 Facts about Manufacturing," National Association of Manufacturers.

26. See U.S. Census data including "Industry Statistics for Subsectors and Industries by Employment Size: 2007," from *2007 Economic Census–Manufacturing, Summary Series, General Summary* (Washington, DC: U.S. Government Printing Office, 2010) and available online at www.census.gov. See also "Radical Shifts Take Hold in U.S. Manufacturing," *Wall Street Journal,* February 3, 2010; "Manufacturing Index Hits 5-Year High," *Investor's Business Daily,* February 2, 2010; "Growth Hits 6-Year High," *Wall Street Journal,* January 30, 2010.

27. See www.naics.com.

28. "Change the World," *Fortune,* September 15, 2017.

29. "The Long Road to Walmart," *Wall Street Journal,* September 19, 2005. See also Hannah S. Lee and David A. Griffith, "Social Comparison in Retailer-Supplier Relationships: Referent Discrepancy Effects," *Journal of Marketing,* March 2019; Peter Kaufman, Satish Jayachandran, and Randall L. Rose, "The Role of Relational Embeddedness in Retail Buyers' Selection of New Products," *Journal of Marketing Research,* November 2006.

30. Based on U.S. Census data and "Eyes Wide Open," *Fortune,* September 19, 2005, 316B-D; "How Bikers' Water Backpack Became Soldiers' Essential," *Wall Street Journal,* July 19, 2005; "Targeting the DHS," *Inc.,* May 2005; "Heavy Load Now Lighter with Bleex," *Investor's Business Daily,* December 29, 2004; "Pentagon Sales Give Powerful Jolt to Ultralife's Business," *Investor's Business Daily,* April 6, 2004; "Uncle Sam Is a Tough Customer," *Investor's Business Daily,* October 20, 2003; "Super Soldiers," *BusinessWeek,* July 28, 2003.

31. For academic research on government customers, see Brett W. Josephson, Ju-Yeon Lee, Babu John Mariadoss, and Jean L. Johnson, "Uncle Sam Rising: Performance Implications of Business-to-Government Relationships," *Journal of Marketing,* January 2019.

32. "Not Just Wal-Mart: Dozens of U.S. Companies Face Bribery Suspicions," *Fortune,* April 26, 2012. For a detailed discussion of business ethics in a number of countries, see *Journal of Business Ethics,* October 1997. See also Francesca Sotgiu and Katrijn Gielens, "Suppliers Caught in Supermarket Price Wars: Victims or Victors? Insights from a Dutch Price War," *Journal of Marketing Research,* December 2015; Uchenna Uzo and Jude O. Adigwe, "Cultural Norms and Cultural Agents in Buyer-Seller Negotiation Processes and Outcomes," *Journal of Personal Selling & Sales Management,* June 2016; Stephen A. Samaha, Joshua T. Beck, and Robert W. Palmatier, "The Role of Culture in International Relationship Marketing," *Journal of Marketing,* September 2014; John B. Ford, Michael S. LaTour, and Tony L. Henthorne, "Cognitive Moral Development and Japanese Procurement Executives: Implications for Industrial Marketers," *Industrial Marketing Management,* November 2000; "U.S., Other Nations Step Up Bribery Battle," *Wall Street Journal,* September 12, 2008; "How Can a U.S. Company Go International, Avoid the Economic Disaster of a Thai Baht . . . and Pay No Bribes?," *USA Today,* November 17, 1997; Larry R. Smeltzer and Marianne M. Jennings, "Why an International Code of Business Ethics Would Be Good for Business," *Journal of Business Ethics,* January 1998.

CHAPTER 7

1. See www.dunkinvip.com; telldunkin.com; dunkindonuts.com; *Dunkin' Brands 2015-2016 Corporate Social Responsibility Report*; "Netflix Tops New List of Brands Generating Positive Word of Mouth among Millennials," *YouGov*, September 24, 2018; "Dunkin's CEO Explains the Two Big Reasons for the Name Change," *Delish*, September 28, 2018; "Dunkin' CIO to Lead Corporate Strategy," *Wall Street Journal*, March 25, 2015; "Dunkin' Blends Big Data with Intuition," *Data Informed*, March 31, 2014; "Retailers Tap Software to Pick Best Locations for New Stores," *The Boston Globe*, August 29, 2013; "When It Comes to Ads in Games, These Guys Aren't Playing Around," *Adweek*, June 19, 2012; "Marketing Case Studies: Dunkin' Donuts," *Copernicus Marketing*; "Brand Transformation: Dunkin' Donuts," *Ivy Cohen, Marketing Coach*; "High Five! Dunkin' Donuts Is Number One in Customer Loyalty," press release, Dunkin' Donuts, February 16, 2011; "The Cold Hard Facts: New Dunkin' Donuts Survey Shows Popularity of Iced Coffee in the Winter Is Heating Up," press release, Dunkin' Donuts, January 18, 2011; "Singular Pursuit: Dunkin' Donuts Tests Expansion in Home Market," *Boston.com*, October 9, 2010; "CareerBuilder and Dunkin' Donuts Survey Finds Which Professions Need Coffee the Most," press release, Dunkin' Donuts, September 27, 2010; "Getting a Grip," *Marketing Management*, Spring 2010; "Dunkin' Donuts Tests Facebook," *Social Media Optimization*, March 16, 2009; "Facebook Fans and Dunkin' Donuts," *eMarketer*, February 27, 2009; "It's Not about the Doughnuts," *Fast Company*, December 19, 2007; "Takeout-Breakfast War Reheats," *Wall Street Journal*, September 5, 2007; "Dunkin' Donuts Going Zero Grams Trans Fat," *MSNBC.com*, August 27, 2007; "Dunkin' Donuts Whips Up a Recipe for Expansion," *Wall Street Journal*, May 3, 2007; "Dunkin' Donuts Uses Business Intelligence in War against Starbucks," *InformationWeek*, April 16, 2007; "Brand New Buzz," *Time*, March 9, 2007; "Dunkin' Begins New Push into China," *Wall Street Journal*, January 17, 2007; "Dunkin' Donuts Cheap Chic," *BusinessWeek*, April 12, 2006; "Dunkin' Donuts Tries to Go Upscale, but Not Too Far," *Wall Street Journal*, April 8, 2006; William Rosenberg, *Time to Make the Donuts* (2001); "Dunkin' Donuts Is Taking On Starbucks to Win Over a Growing $2 Billion Industry," *Business Insider*, September 29, 2016; "Digital Dunkin': Non-Tech Firms Crash CES, Looking to Connect," *Wall Street Journal*, January 5, 2017; "Learn How Dunkin' Donuts Runs on Data Analytics," *Big Data Analytics Info*, October 18, 2016.

2. "'It's Been a Rout': Apple's iPhones Fall Flat in World's Largest Untapped Market," *Wall Street Journal*, December 18, 2018.

3. "Questions and Answers," *Marketing Insights*, Fall 2013.

4. For more on big data, see "Big Data: The Next Frontier for Innovation, Competition, and Productivity," *McKinsey Global Institute*, May 2011; "Data Analytics: Crunching the Future," *Bloomberg Businessweek*, September 8, 2011; "Clouds, Big Data, and Smart Assets: Ten Tech-Enabled Business Trends to Watch," *McKinsey Quarterly*, August 2012; "Unlocking Hidden Profits with Big Data Analytics," *Forbes*, September 27, 2012; "Seizing the Potential of 'Big Data,'" *McKinsey Quarterly* 4 (2011); "Are You Ready for the Era of 'Big Data?,'" *McKinsey Quarterly* 4 (2011); "How Intuit Uses Big Data for the Little Guy," *Forbes*, April 26, 2012; "Big Data Requires Complex Analysis, but Don't Be Scared Off," *Advertising Age*, May 8, 2012; Daria Dzyabura and John R. Hauser, "Active Machine Learning for Consideration Heuristics," *Marketing Science* 5 (2011); "Big Data Broadens Its Range," *Wall Street Journal*, March 13, 2013; Pradeep Chintagunta, Dominique M. Hanssens, and John R. Hauser, "Editorial-Marketing Science and Big Data," *Marketing Science*, June 2016.

5. IBM Big Data & Analytics Hub, "The Four V's of Big Data," www.ibmbigdatahub.com/infographic/four-vs-big-data, accessed March 15, 2017.

6. "The Technology That Unmasks Your Hidden Emotions," *Wall Street Journal*, January 28, 2015; "Phone Firms Sell Data on Customers," *Wall Street Journal*, May 21, 2013; "Smile! Marketing Firms Are Mining Your Selfies," *Wall Street Journal*, October 9, 2014.

7. Quote from William Bruce Cameron, *Informal Sociology: A Casual Introduction to Sociological Thinking* (New York: Random House, 1963).

8. "Age of Big Data," *New York Times*, February 12, 2012; "What Is Big Data?," *O'Reilly.com*, January 11, 2012; "Big Bets on Big Data," *Forbes*, June 22, 2012; "Big Data and Its Myths," *Forbes*, June 21, 2012.

9. "Amazon Investigates Employees Leaking Data for Bribes," *Wall Street Journal*, September 16, 2018.

10. For more on this framework, see Gene Gellinger, Durval Castro, and Anthony Mills, "Data, Information, Knowledge, and Wisdom," 2004, www.systems-thinking.org/dikw/dikw.htm; Jennifer Rowley, "The Wisdom Hierarchy: Representations of the DIKW Hierarchy," *Journal of Information Science* 2 (2007).

11. Oded Netzer, Ronen Feldman, Jacob Goldenberg, and Moshe Fresko, "Mine Your Own Business: Market-Structure Surveillance through Text Mining," *Marketing Science* 3 (2012); Martin Reimann, Oliver Schilke, and Jacquelyn S. Thomas, "Customer Relationship Management and Firm Performance: The Mediating Role of Business Strategy," *Journal of the Academy of Marketing Science* 38, no. 3 (2010); Venkatesh Shankar and Russell S. Winer, "When Customer Relationship Management Meets Data Mining," *Journal of Interactive Marketing*, Summer 2006; Amy Miller and Jennifer Cioffi, "Measuring Marketing Effectiveness and Value: The Unisys Marketing Dashboard," *Journal of Advertising Research*, September 2004; "Making Marketing Measure Up," *BusinessWeek*, December 13, 2004; "How Verizon Flies by Wire," *CIO Magazine*, November 1, 2004.

12. Gary L. Lilien, Arvind Rangaswamy, Gerrit H. van Bruggen, and Berend Wierenga, "Bridging the Marketing Theory-Practice Gap with Marketing Engineering," *Journal of Business Research*, February 2002; "Virtual Management," *BusinessWeek*, September 21, 1998; John T. Mentzer and Nimish Gandhi, "Expert Systems in Marketing: Guidelines for Development," *Journal of the Academy of Marketing Science*, Winter 1992; William D. Perreault Jr., "The Shifting Paradigm in Marketing Research," *Journal of the Academy of Marketing Science*, Fall 1992.

13. "Big Data Broadens Its Range," *Wall Street Journal*, March 13, 2013.

14. "Big Data Is Revolutionizing the Music Industry. Here Are the Lessons for Your Business," *Inc.*, June 7, 2018; "Most Innovative Companies 2015, Next Big Sound," *Fast Company*, March 2015; "Moneyball for Music: The Rise of Next Big Sound," *Forbes*, March 3, 2013; "The Shazam Effect," *The Atlantic*, December 2014; "The Scientific Formula for Predicting a Hit," *Gizmag*, December 27, 2011; "Rock Music's Latest New Heroes: Lady Gaga and . . . Big Data?," *CNBC.com*, October 11, 2013; "Predicting What You Want to Hear: Music and Data Get It On," *Billboard*, March 25, 2014; "How Shazam Uses Big Data to Predict Music's Next Big Artists," *The Guardian*, December 10, 2013; "A Trillion Data Points: The Growth of Music Analytics," *Performer*, February 11, 2016; "The Data That Can Predict a Pop Hit," *Wired*, June 2015; "How Big Data Will Change the Music Industry Forever," *LinkedIn*, January 13, 2016.

15. For an excellent resource on data privacy, see Robert W. Palmatier and Kelly D. Martin, *The Intelligent Marketer's Guide to Data Privacy* (Palgrave Macmillan, 2019); see also Kelly Martin and Patrick Murphy, "The Role of Data Privacy in Marketing," *Journal of the Academy of Marketing Science*, March 2017.

16. This entire section draws on Chapter 3 in Robert W. Palmatier and Kelly D. Martin, *The Intelligent Marketer's Guide to Data Privacy* (Palgrave Macmillan, 2019).

17. "You Give Apps Sensitive Personal Information. Then They Tell Facebook," *Wall Street Journal*, February 22, 2019.

18. "You Give Apps Sensitive Personal Information. Then They Tell Facebook," *Wall Street Journal*, February 22, 2019.

19. J. Rubinson, "The New Marketing Research Imperative: It's about Learning," *Journal of Advertising Research* 1 (2009); Larry Selden and Ian C. MacMillan, "Manage Customer-Centric Innovation Systematically," *Harvard Business Review*, April 2006; Bruce H. Clark, "Business Intelligence Using Smart Techniques," *Journal of the Academy of

Marketing Science, Fall 2003; Mark Peyrot, Nancy Childs, Doris Van Doren, and Kathleen Allen, "An Empirically Based Model of Competitor Intelligence Use," *Journal of Business Research,* September 2002; Stanley F. Slater and John C. Narver, "Intelligence Generation and Superior Customer Value," *Journal of the Academy of Marketing Science,* Winter 2000; Satish Jayachandran, Subhash Sharma, Peter Kaufman, and Pushkala Raman, "The Role of Relational Information Processes and Technology Use in Customer Relationship Management," *Journal of Marketing,* October 2005; Peter M. Chisnall, "The Effective Use of Market Research: A Guide for Management to Grow the Business," *International Journal of Market Research* 42, no. 2 (Summer 2000); Seymour Sudman and Edward Blair, *Marketing Research: A Problem Solving Approach* (Burr Ridge, IL: Irwin/McGraw-Hill, 1998). See also Christine Moorman, Rohit Deshpande, and Gerald Zaltman, "Factors Affecting Trust in Market Research Relationships," *Journal of Marketing,* January 1993.

20. See www.kiwicare.com; "Kiwi Goes beyond Shine in Effort to Step Up Sales," *Wall Street Journal,* December 20, 2007.

21. "Internet, Intuition Can Help Businesses Spot Next Big Thing," *York Dispatch,* March 6, 2019.

22. David A. Schweidel and Wendy W. Moe, "Listening In on Social Media: A Joint Model of Sentiment and Venue Format Choice," *Journal of Marketing Research,* August 2014; Seshadri Tirunillai and Gerard J. Tellis, "Mining Marketing Meaning from Online Chatter: Strategic Brand Analysis of Big Data Using Latent Dirichlet Allocation," *Journal of Marketing Research,* August 2014; Aron Culotta and Jennifer Cutler, "Mining Brand Perceptions from Twitter Social Networks," *Marketing Science,* June 2016.

23. "All You Need to Know about Customer Journey Mapping," *Smashing Magazine,* January 15, 2015. For some academic perspectives, see Rebecca Hamilton and Linda L. Price, "Consumer Journeys: Developing Consumer-Based Strategy," *Journal of the Academy of Marketing Science,* March 2019; Anne-Madeleine Kranzbühler, Mirella H. P. Kleijnen, and Peeter W. J. Verlegh, "Outsourcing the Pain, Keeping the Pleasure: Effects of Outsourced Touchpoints in the Customer Journey," *Journal of the Academy of Marketing Science,* March 2019; Thomas P. Novak and Donna L. Hoffman, "Relationship Journeys in the Internet of Things: A New Framework for Understanding Interactions between Consumers and Smart Objects," *Journal of the Academy of Marketing Science,* March 2019.

24. For more on focus groups, see "Mucus to Maxi Pads: Marketing's Dirtiest Jobs," *Advertising Age,* February 16, 2009; "Hypnosis Brings Groups into Focus," *Brandweek,* March 24, 2008; "The Perils of Packaging: Nestlé Aims for Easier Openings," *Wall Street Journal,* November 17, 2005; "Puppet's Got a Brand-New Bag," *Business 2.0,* October 2005; "Focus Groups Should Be Abolished," *Advertising Age,* August 8, 2005; "Web Enhances Market Research," *Advertising Age,* June 18, 2001. See also William J. McDonald, "Focus Group Research Dynamics and Reporting: An Examination of Research Objectives and Moderator Influences," *Journal of the Academy of Marketing Science,* Spring 1993; Thomas Kiely, "*Wired* Focus Groups," *Harvard Business Review,* January–February 1998; see also James R. Stengel, Andrea L. Dixon, and Chris T. Allen, "Listening Begins at Home," *Harvard Business Review,* November 2003; Shalini Bahl and George R. Milne, "Talking to Ourselves: A Dialogical Exploration of Consumption Experiences," *Journal of Consumer Research,* February 2010; "How to Run an Online Focus Group to Discover What Business Prospects Really Want (Tips on Launching an E-Commerce Website)," *Marketing Sherpa,* September 28, 2005; "The New Focus Groups: Online Networks," *Wall Street Journal,* January 14, 2008; "Expand Your Brand Community Online," *Advertising Age,* January 7, 2008; "Design It before You Buy It," *Wall Street Journal,* August 2, 2007; "P&G Plunges into Social Networking," *Wall Street Journal,* January 8, 2007; "It Takes a Web Village," *BusinessWeek,* September 4, 2006; "Shoot the Focus Group," *BusinessWeek Online,* November 14, 2005.

25. www.schoology.com; "Using Customer Feedback to Grow Business," *SurveyMonkey,* accessed March 17, 2017.

26. Pinterest story from "How'd You Get That Job? Questions for a User Experience Researcher at Pinterest," *Wall Street Journal,* January 4, 2019; For more on surveys, see John Hulland, Hans Baumgartner, and Keith Marion Smith, "Marketing Survey Research Best Practices: Evidence and Recommendations from a Review of JAMS Articles," *Journal of the Academy of Marketing Science,* January 2018; "Forget Phone and Mail: Online's the Best Place to Administer Surveys," *Advertising Age,* July 17, 2006; "Electrolux Cleans Up," *BusinessWeek,* February 27, 2006; "Chrysler's Made-Up Customers Get Real Living Space at Agency," *Wall Street Journal,* January 4, 2006; "VW's American Road Trip," *Wall Street Journal,* January 4, 2006; "The Only Question That Matters," *Business 2.0,* September 2005. See also Pierre Chandon, Vicki G. Morwitz, and Werner J. Reinartz, "Do Intentions Really Predict Behavior? Self-Generated Validity Effects in Survey Research," *Journal of Marketing,* April 2005; Stanley E. Griffis, Thomas J. Goldsby, and Martha Cooper, "Web-Based and Mail Surveys: A Comparison of Response, Data, and Cost," *Journal of Business Logistics* 2 (2003); Terry L. Childers and Steven J. Skinner, "Toward a Conceptualization of Mail Survey Response Behavior," *Psychology & Marketing,* March 1996; J. B. E. M. Steenkamp, M. G. De Jong, and H. Baumgartner, "Socially Desirable Response Tendencies in Survey Research," *Journal of Marketing Research* 2 (2010).

27. "Old Ketchup Packet Heads for Trash," *Wall Street Journal,* September 19, 2011.

28. Paco Underhill, *Why We Buy* (New York: Simon & Schuster, 1999); see also N. Diamond, J. F. Sherry, A. M. Muniz, M. A. McGrath, et al., "American Girl and the Brand Gestalt: Closing the Loop on Sociocultural Branding Research," *Journal of Marketing* 3 (2009); Stephen R. Rosenthal and Mark Capper, "Ethnographies in the Front End: Designing for Enhanced Customer Experiences," *Journal of Product Innovation Management,* May 2006; Pierre Berthon, James Mac Hulbert, and Leyland Pitt, "Consuming Technology: Why Marketers Sometimes Get It Wrong," *California Management Review,* Fall 2005. See also "Why Are Tech Gizmos So Hard to Figure Out?," *USA Today,* November 2, 2005; "Poll: Many Like Tech Gizmos but Are Frustrated," *USA Today,* October 31, 2005; "The Fine Art of Usability Testing," *Dev Source,* March 2005.

29. "Seeing Store Shelves through Senior Eyes," *Wall Street Journal,* September 14, 2009; "They Feel Your Pain," *Brandweek,* June 16, 2008; see also "Architects Live in Senior Spaces to Help Elderly," *USA Today,* August 21, 2014.

30. "Zappos Sells 6%–30% More Merchandise When Accompanied by Video Demos," *Business Insider,* December 4, 2009; see also Joel Barajas, Ram Akella, Marius Holtan, and Aaron Flores, "Experimental Designs and Estimation for Online Display Advertising Attribution in Marketplaces," *Marketing Science,* June 2016.

31. "Pen Proving to Be Mighty for Brands, Consumers," *Brandweek,* May 7, 2007; "More Than Squeaking By: WD-40 CEO Garry Ridge Repackages a Core Product," *Wall Street Journal,* May 23, 2006; "WD-40 Is Well-Oiled for Growth," *BusinessWeek,* April 18, 2006; "P&G Provides Product Launchpad, a Buzz Network of Moms," *Advertising Age,* March 20, 2006; "Loosening the Wheels of Innovation," *Quirk's Marketing Research Review,* March 2006.

32. For more detail on data analysis techniques, see Joe Hair, Rolph Anderson, Ron Tatham, and William Black, *Multivariate Data Analysis* (New York: Prentice-Hall, 2005), or other marketing research texts. See also "Ice Cream Shop Gets Scoop on Locales," *Investor's Business Daily,* February 17, 2006; "Is Your Business in the Right Spot?," *Business 2.0,* May 2004; Michael D. Johnson and Elania J. Hudson, "On the Perceived Usefulness of Scaling Techniques in Market Analysis," *Psychology & Marketing,* October 1996; Milton D. Rosenau, "Graphing Statistics and Data: Creating Better Charts," *Journal of Product Innovation Management,* March 1997.

33. "Careful What You Ask For," *American Demographics,* July 1998. See also John G. Keane, "Questionable Statistics," *American Demographics,* June 1985. Detailed treatment of confidence intervals is beyond the scope of this text, but it is covered in most marketing research texts, such as Donald R. Lehmann and Russ Winer, *Analysis for Marketing Planning* (Burr Ridge, IL: Irwin/McGraw-Hill, 2005).

34. "The Inventory of Customer Satisfaction Surveys Is Sick of Them Too," *Bloomberg Businessweek*, May 4, 2016.

35. For a discussion of ethical issues in marketing research, see "What Makes Tesco, Kroger More Than Just Rivals?," *Wall Street Journal*, December 24, 2007; "Intimate Shopping: Should Everyone Know What You Bought Today?," *New York Times*, December 23, 2007; "Ma Bell, the Web's New Gatekeeper?," *BusinessWeek*, November 19, 2007; "Firm Mines Offline Data to Target Online Ads," *Wall Street Journal*, October 17, 2007; "How 'Tactical Research' Muddied Diaper Debate: A Case," *Wall Street Journal*, May 17, 1994. See also Rita Marie Cain, "Supreme Court Expands Federal Power to Regulate the Availability and Use of Data," *Journal of the Academy of Marketing Science*, Fall 2001; Malcolm Kirkup and Marylyn Carrigan, "Video Surveillance Research in Retailing: Ethical Issues," *International Journal of Retail & Distribution Management*, 11 (2000); Eve M. Caudill and Patrick E. Murphy, "Consumer Online Privacy: Legal and Ethical Issues," *Journal of Public Policy & Marketing*, Spring 2000; John R. Sparks and Shelby D. Hunt, "Marketing Researcher Ethical Sensitivity: Conceptualization, Measurement, and Exploratory Investigation," *Journal of Marketing*, April 1998; Naresh K. Malhotra and Gina L. Miller, "An Integrated Model for Ethical Decisions in Marketing Research," *Journal of Business Ethics*, February 1998.

36. Goutam Challagalla, Brian R. Murtha, and Bernard Jaworski, "Marketing Doctrine: A Principles-Based Approach to Guiding Marketing Decision Making in Firms," *Journal of Marketing*, July 2014.

37. www.cleanteamtoilets.com; "Clean Team: In-Home Toilets for Ghana's Urban Poor," IDEO.org Case Study; "Clean Team, a Human-Centred Approach to Sanitation: Initial Trials in Ghana," WSUP, Water & Sanitation for the Urban Poor, Practice Note, November 2011.

CHAPTER 8

1. See www.underarmour.com; "Company Profile: Under Armour," *MarketLine*, November 9, 2018; "Under Armour Wants to Be a 'Louder Brand' as It Looks to Turn 'Each Marketing Dollar Spent into Three'," *Marketing Week*, February 13, 2019; "Under Armour Makes Strides Despite Weak North American Sales," *Women's Wear Daily*, February 13, 2019; "Under Armour, Nike, Apple: The Top 10 YouTube Ads in March," *Marketing Week*, April 10, 2018; "Under Armour's HOVR Smart Running Shoes Are More Than Just a Gimmick," *Engadget*, February 9, 2018; "Under Armour Wins Trademark Battle in China over 'Uncle Martian'," *Wall Street Journal*, August 4, 2017; David Aaker, "Under Armour Wills Itself to Success," *Marketing News*, April 2015, pp. 26–27; "Why Under Armour Is Making a Costly Bet on Connected Fitness," *Fortune*, April 21, 2015; "Ad Age's 2014 Marketer of the Year: Under Armour," *Advertising Age*, December 8, 2014; "Under Armour to Exploit 'Quiet Shift' for Women Wearing Athletic Product Outside the Gym," *Marketing Week*, September 5, 2014; "Under Armour's Founder on Learning to Leverage Celebrity Endorsements," *Harvard Business Review*, May 2012; "Under Armour Scores $1 Billion in Sales through Laser Focus on Athletes," *Women's Wear Daily*, December 1, 2011; "Tough Mudder Brings in $25 Million, Signs Under Armour," *CNBC*, December 5, 2011; "Under Armour Gets Serious," *Fortune*, November 7, 2011; "Under Armour's Best Idea: A Shirt That Measures Heart Rate and G-Force," *The Atlantic*, November 10, 2011; "Under Armour Signs Groundbreaking Deal with Tough Mudder," *Forbes*, December 8, 2011; "Under Armour Expands Fitness Line Overseas," *Business Insider*, March 16, 2011; "Under Armour's Kevin Plank: Creating 'the Biggest, Baddest Brand on the Planet,'" Knowledge@Wharton, January 5, 2011; "Under Armour's Daring Half-Court Shot," *Bloomberg Businessweek*, November 1, 2010; "Under Armour Can't Live Up to Own Hype," *Advertising Age*, November 2, 2009; "Under Armour Reboots," *Fortune*, February 2, 2009; "Under Armour Enters the Running Shoe Race," *CNBC.com*, February 1, 2009; "Under Armour Hopes to Outrun Nike," *Advertising Age*, April 28, 2008; "Do These Clothes Help You Work Out?," *BusinessWeek*, March 24, 2008; "True Confessions of a Super Bowl Ad Virgin," *Advertising Age*, February 4, 2008; "Nike Seeks to Undercut New Under Armour Line," *Brandweek*, January 28, 2008; "Even Rookie Advertisers Feel Big-Game Pressure," *Wall Street Journal*, January 15, 2008; "No Sugar and Spice Here," *Advertising Age*, June 18, 2007; "Under Armour May Be Overstretched," *BusinessWeek*, April 30, 2007; "Under Armour: Thrown for a Loss," *BusinessWeek*, February 1, 2007; "Under Armour," *Apparel Magazine*, February 2007; "Perspiration Inspiration," *BusinessWeek*, June 5, 2006; "Rag Trade Rivalry," *Forbes*, June 5, 2006; "Under Armour, a Brawny Tee House? No Sweat," *BusinessWeek*, May 25, 2006; "Hot and Cool," *Government Executive*, October 15, 2005; "Protect This House," *Fast Company*, August 2005; "Under Armour Shows Feminine Side," *Women's Wear Daily*, February 24, 2005; "Under Armour's Apparel Appeal," *Brandweek*, February 21, 2005; *2009 Annual Report*, Under Armour; *2009 Annual Report*, Army and Air Force Exchange Service; "Kevin Plank," *Brandweek*, April 12, 2004; Under Armour Finds Feminine Side to Go beyond $2 Billion," *Bloomberg Businessweek*, February 15, 2013; "Moving beyond Shrink It & Pink It," *Marketing News*, February 2013; "No More 'Shrinking It and Pinking It' at Under Armour," *Advertising Age*, September 24, 2012; "Under Armour's 'Soaring' Shoe Sales Reveal a Terrible Truth for the Business," *Business Insider*, February 23, 2017; "Runners Will Love These Heart-Monitoring Headphones," *Time.com*, March 2, 2017; "Kohl's and Under Armour: Behind the Biggest Brand Launch in Kohl's History," *Advertising Age*, March 1, 2017; "How Tesla, Under Armour, and Sonos Do Branding," *HBR.org*, October 8, 2015; "Under Armour Share Loss in This Key Demographic Is 'Disconcerting,'" *Investor's Business Daily*, March 2, 2016; "Under Armour CEO: We're Adding 1 Million App Users Every 8 Days," *Investor's Business Daily*, March 14, 2016; "Under Armour Debuts First-Ever 3D-Printed Shoes," *Fortune*, March 8, 2016.

2. For more on customer experiences, see Alexander Bleier, Colleen M. Harmeling, and Robert W. Palmatier, "Creating Effective Online Customer Experiences," *Journal of Marketing*, March 2019; Joseph K. Goodman and Sarah Lim, "When Consumers Prefer to Give Material Gifts Instead of Experiences: The Role of Social Distance," *Journal of Consumer Research*, August 2018; Christian Homburg, Danijel Jozić, and Christina Kuehnl, "Customer Experience Management: Toward Implementing an Evolving Marketing Concept," *Journal of the Academy of Marketing Science*, May 2017; Bryan A. Lukas, Gregory J. Whitwell, and Jan B. Heide, "Why Do Customers Get More Than They Need? How Organizational Culture Shapes Product Capability Decisions," *Journal of Marketing* 77, no. 1 (2013): 1–12; Peter N. Golder, Debanjan Mitra, and Christine Moorman, "What Is Quality? An Integrative Framework of Processes and States," *Journal of Marketing*, July 2012; Marco Bertini, Luc Wathieu, and Sheena S. Iyengar, "The Discriminating Consumer: Product Proliferation and Willingness to Pay for Quality," *Journal of Marketing Research*, February 2012; T. Erdem, M. P. Keane, and B. Sun, "A Dynamic Model of Brand Choice When Price and Advertising Signal Product Quality," *Marketing Science* 6 (2008); Neil A. Morgan and Douglas W. Vorhies, "Product Quality Alignment and Business Unit Performance," *The Journal of Product Innovation Management*, November 2001.

3. This example was inspired by other case studies: Robert Solomon and Kathleen Higgins, *Case Study: Cement for Sale,* Center for the Study of Ethics, Utah Valley State College; C. B. Fleddermann, *Denver Runway Concrete, Engineering Ethics* (Prentice Hall, 1999).

4. "Antibiotics in Your Food: Should You Be Concerned?," *HealthLine*, retrieved March 30, 2019; https://goo.gl/ZUhduY; "From Delivery to Eco-Friendly Packaging and Tons of Recalls: A Look at 2018's Food Trends," *USA Today*, December 26, 2018.

5. "Younger Consumers Drive Shift to Ethical Products," *Financial Times*, December 22, 2017; "5 Trends for 2019," *Trend Watching*.

6. For academic literature on products and product lines, see Paul W. Miniard, Rama K. Jayanti, Cecilia M. O. Alvarez, and Peter R. Dickson, "What Brand Extensions Need to Fully Benefit from Their

Parental Heritage," *Journal of the Academy of Marketing Science,* September 2018; Donald Ngwe, "Why Outlet Stores Exist: Averting Cannibalization in Product Line Extensions," *Marketing Science,* July 2017; Fabio Caldieraro, Ling-Jing Kao, and Marcus Cunha Jr., "Harmful Upward Line Extensions: Can the Launch of Premium Products Result in Competitive Disadvantages?," *Journal of Marketing,* November 2015; Robert Carter and David Curry, "Perceptions versus Performance When Managing Extensions: New Evidence about the Role of Fit between a Parent Brand and an Extension," *Journal of the Academy of Marketing Science* 41, no. 2 (2013): 253–69; Ian Sinapuelas, Hui-Ming Wang, and Jonathan Bohlmann, "The Interplay of Innovation, Brand, and Marketing Mix Variables in Line Extensions," *Journal of the Academy of Marketing Science* 43, no. 5 (2015): 558–73.

7. For more on services, see Arne Albrecht, Gianfranco Walsh, Simon Brach, Dwayne Gremler, and Erica Herpen, "The Influence of Service Employees and Other Customers on Customer Unfriendliness: A Social Norms Perspective," *Journal of the Academy of Marketing Science,* November 2017; Michael Brady and Todd Arnold, "Organizational Service Strategy," *Journal of the Academy of Marketing Science,* November 2017; Ying Ding and Hean Tat Keh, "Consumer Reliance on Intangible versus Tangible Attributes in Service Evaluation: The Role of Construal Level," *Journal of the Academy of Marketing Science,* November 2017; Delphine Dion and Stéphane Borraz, "Managing Status: How Luxury Brands Shape Class Subjectivities in the Service Encounter," *Journal of Marketing,* September 2017; Beibei Dong and K. Sivakumar, "Customer Participation in Services: Domain, Scope, and Boundaries," *Journal of the Academy of Marketing Science,* November 2017; Leighanne Higgins and Kathy Hamilton, "Therapeutic Servicescapes and Market-Mediated Performances of Emotional Suffering," *Journal of Consumer Research,* April 2019; Jens Hogreve, Nicola Bilstein, and Leonhard Mandl, "Unveiling the Recovery Time Zone of Tolerance: When Time Matters in Service Recovery," *Journal of the Academy of Marketing Science,* November 2017; Linda D. Hollebeek, Rajendra K. Srivastava, and Tom Chen, "S-D Logic-Informed Customer Engagement: Integrative Framework, Revised Fundamental Propositions, and Application to CRM," *Journal of the Academy of Marketing Science,* January 2019; Ming-Hui Huang and Roland Rust, "Technology-Driven Service Strategy," *Journal of the Academy of Marketing Science,* November 2017; V. Kumar, Bharath Rajan, Shaphali Gupta, and Ilaria Dalla Pozza, "Customer Engagement in Service," *Journal of the Academy of Marketing Science,* January 2019; Elison Lim, Yih Lee, and Maw-Der Foo, "Frontline Employees' Nonverbal Cues in Service Encounters: A Double-Edged Sword," *Journal of the Academy of Marketing Science,* September 2017; Jochen Wirtz and Valarie Zeithaml, "Cost-Effective Service Excellence," *Journal of the Academy of Marketing Science,* January 2018; Lisa C. Wan and Robert S. Wyer, "The Influence of Incidental Similarity on Observers' Causal Attributions and Reactions to a Service Failure," *Journal of Consumer Research,* April 2019; Heiko Wieland, Nathaniel Hartmann, and Stephen Vargo, "Business Models as Service Strategy," *Journal of the Academy of Marketing Science,* November 2017; Paul P. Maglio and Jim Spohrer, "Fundamentals of Service Science," *Journal of the Academy of Marketing Science,* Spring 2008; P. C. Verhoef, K. N. Lemon, A. Parasuraman, A. Roggeveen et al., "Customer Experience Creation: Determinants, Dynamics and Management Strategies," *Journal of Retailing* 1 (2009); Alexis Allen, Michael Brady, Stacey Robinson, and Clay Voorhees, "One Firm's Loss Is Another's Gain: Capitalizing on Other Firms' Service Failures," *Journal of the Academy of Marketing Science* 43, no. 5 (2015): 648–62; Andrew Gallan, Cheryl Jarvis, Stephen Brown, and Mary Bitner, "Customer Positivity and Participation in Services: An Empirical Test in a Health Care Context," *Journal of the Academy of Marketing Science* 41, no. 3 (2013): 338–56; Nancy J. Sirianni, Mary Jo Bitner, Stephen W. Brown, and Naomi Mandel, "Branded Service Encounters: Strategically Aligning Employee Behavior with the Brand Positioning," *Journal of Marketing* 77, no. 6 (2013): 108–23; Jaihak Chung and Vithala R. Rao, "A General Consumer Preference Model for Experience Products: Application to Internet Recommendation Services," *Journal of Marketing Research,* June 2012. For more on services as supplementary to goods, see Wolfgang Ulaga and Werner J. Reinartz, "Hybrid Offerings: How Manufacturing Firms Combine Goods and Services Successfully," *Journal of Marketing,* November 2011; Stephen L. Vargo and Robert F. Lusch, "Evolving to a New Dominant Logic for Marketing," *Journal of Marketing,* January 2004; Ruth N. Bolton, George S. Day, John Deighton, Das Narayandas, et al., "Invited Commentaries on 'Evolving to a New Dominant Logic for Marketing,'" *Journal of Marketing,* January 2004; James C. Anderson and James A. Narus, "Capturing the Value of Supplementary Services," *Harvard Business Review,* January–February 1995.

8. "What's Pushing These Tech Giants to Make an $8,500 Smart Fridge?," *Bloomberg Businessweek,* November 4, 2016. See also V. Kumar, Ashutosh Dixit, Rajshekar Javalgi, and Mayukh Dass, "Research Framework, Strategies, and Applications of Intelligent Agent Technologies (IATs) in Marketing," *Journal of the Academy of Marketing Science,* January 2016; Michael Giebelhausen, Stacey G. Robinson, Nancy J. Sirianni, and Michael K. Brady, "Touch versus Tech: When Technology Functions as a Barrier or a Benefit to Service Encounters," *Journal of Marketing,* July 2014.

9. "Apple and Google Know What You Want before You Do," *Wall Street Journal,* August 3, 2015.

10. "Save Time with Smart Reply in Gmail," *The Keyword,* May 17, 2017.

11. "Alibaba Made a Smart Screen to Help Blind People Shop and It Costs Next to Nothing," *TechCrunch,* November 12, 2018.

12. "The Risks of Marrying 'Smart' Technology With 'Dumb' Machines," *Wall Street Journal,* March 22, 2019; "The Coming Era of 'On-Demand' Marketing," *McKinsey Quarterly,* April 2013; "The Futurists's Cheat Sheet: Internet of Things," *ReadWrite.com,* August 31, 2012; *Disruptive Technologies: Advances That Will Transform Life, Business, and the Global Economy,* McKinsey & Company, May 2013; "Sensing the Future before It Occurs," *MIT Sloan Management Review,* December 20, 2012; "Welcome to the Programmed World," *Wired,* May 2013; "The Internet of Things and the Future of Manufacturing," *McKinsey.com,* June 2013; "GE Tries to Make Its Machines Cool and Connected," *Bloomberg Businessweek,* December 6, 2012; I. C. Ng and S. Y. Wakenshaw, "The Internet-of-Things: Review and Research Directions," *International Journal of Research in Marketing,* 2016; P. K. Kannan, "Digital Marketing: A Framework, Review and Research Agenda," *International Journal of Research in Marketing,* 2016.

13. For research on logos, see Jonathan Luffarelli, Antonios Stamatogiannakis, and Haiyang Yang, "The Visual Asymmetry Effect: An Interplay of Logo Design and Brand Personality on Brand Equity," *Journal of Marketing Research,* February 2019; Luca Cian, Aradhna Krishna, and Ryan S. Elder, "This Logo Moves Me: Dynamic Imagery from Static Images," *Journal of Marketing Research* 51, no. 2 (2014): 184–97.

14. For more on Listerine PocketPaks, see "The Strip Club," *Business 2.0,* June 2003; "Marketer of the Year: PocketPaks, a Breath of Minty Fresh Air," *Brandweek,* October 14, 2002, pp. M42–M46. For more on brand extensions, see "A Little Less Salt, a Lot More Sales: Campbell's Line Extension," *Advertising Age,* March 10, 2008; "Like Our Sunglasses? Try Our Vodka!," *Wall Street Journal,* November 8, 2007; "P&G Rekindles an Old Flame: New Febreze Candles," *Wall Street Journal,* June 5, 2007. Academic treatment of branding includes S. Brasel and Henrik Hagtvedt, "Living Brands: Consumer Responses to Animated Brand Logos," *Journal of the Academy of Marketing Science,* September 2016; Liwu Hsu, Susan Fournier, and Shuba Srinivasan, "Brand Architecture Strategy and Firm Value: How Leveraging, Separating, and Distancing the Corporate Brand Affects Risk and Returns," *Journal of the Academy of Marketing Science,* March 2016; Burçak Ertimur and Gokcen Coskuner-Balli, "Navigating the Institutional Logics of Markets: Implications for Strategic Brand Management," *Journal of Marketing,* March 2015; Marcus Cunha Jr., Mark R. Forehand, and Justin W. Angle, "Riding Coattails: When Co-Branding Helps versus Hurts Less-Known Brands," *Journal of Consumer*

Research, February 2015; Aaron M. Garvey, Frank Germann, and Lisa E. Bolton, "Performance Brand Placebos: How Brands Improve Performance and Consumers Take the Credit," *Journal of Consumer Research,* April 2016; Christoph Baumeister, Anne Scherer, and Florian Wangenheim, "Branding Access Offers: The Importance of Product Brands, Ownership Status, and Spillover Effects to Parent Brands," *Journal of the Academy of Marketing Science* 43, no. 5 (2015): 574–88; Amber M. Epp, Hope Jensen Schau, and Linda L. Price, "The Role of Brands and Mediating Technologies in Assembling Long-Distance Family Practices," *Journal of Marketing* 78, no. 3 (2014): 81–101; Susan Fournier and Claudio Alvarez, "Relating Badly to Brands," *Journal of Consumer Psychology* 23, no. 2 (2013): 253–64; Satish Jayachandran, Peter Kaufman, V. Kumar, and Kelly Hewett, "Brand Licensing: What Drives Royalty Rates?," *Journal of Marketing* 77, no. 5 (2013): 108–22; George E. Newman and Ravi Dhar, "Authenticity Is Contagious: Brand Essence and the Original Source of Production," *Journal of Marketing Research* 51, no. 3 (2014): 371–86; Linyun W. Yang, Keisha M. Cutright, Tanya L. Chartrand, and Gavan J. Fitzsimons, "Distinctively Different: Exposure to Multiple Brands in Low-Elaboration Settings," *Journal of Consumer Research* 40, no. 5 (2014): 973–92; Sanjay Sood and Kevin Lane Keller, "The Effects of Brand Name Structure on Brand Extension Evaluations and Parent Brand Dilution," *Journal of Marketing Research,* June 2012; Ryan Hamilton and Alexander Chernev, "The Impact of Product Line Extensions and Consumer Goals on the Formation of Price Image," *Journal of Marketing Research,* February 2010; Tom Meyvis, Kelly Goldsmith, and Ravi Dhar, "The Importance of the Context in Brand Extension: How Pictures and Comparisons Shift Consumers' Focus from Fit to Quality," *Journal of Marketing Research,* April 2012; Franziska Vlckner and Henrik Sattler, "Drivers of Brand Extension Success," *Journal of Marketing,* April 2006; Chris Pullig, Carolyn J. Simmons, and Richard G. Netemeyer, "Brand Dilution: When Do New Brands Hurt Existing Brands?," *Journal of Marketing,* April 2006; Thomas J. Madden, Frank Fehle, and Susan Fournier, "Brands Matter: An Empirical Demonstration of the Creation of Shareholder Value through Branding," *Journal of the Academy of Marketing Science,* Spring 2006; Thomas J. Reynolds and Carol B. Phillips, "In Search of True Brand Equity Metrics: All Market Share Ain't Created Equal," *Journal of Advertising Research,* June 2005. For more on brand equity and brand value, see Florian Stahl, Mark Heitmann, Donald R. Lehmann, and Scott A. Neslin, "The Impact of Brand Equity on Customer Acquisition, Retention, and Profit Margin," *Journal of Marketing,* July 2012; Marc Fischer, Franziska Völckner, and Henrik Sattler, "How Important Are Brands? A Cross-Category, Cross-Country Study," *Journal of Marketing Research,* October 2010; Rajeev Batra, Aaron Ahuvia, and Richard P. Bagozzi, "Brand Love," *Journal of Marketing,* March 2010; J. J. Brakus, B. H. Schmitt, and L. Zarantonello, "Brand Experience: What Is It? How Is It Measured? Does It Affect Loyalty?," *Journal of Marketing* 3 (2009); D. E. Boyd and K. D. Bahn, "When Do Large Product Assortments Benefit Consumers? An Information-Processing Perspective," *Journal of Retailing* 3 (2009); R. J. Slotegraaf and K. Pauwels, "The Impact of Brand Equity and Innovation on the Long-Term Effectiveness of Promotions," *Journal of Marketing Research* 3 (2008).

15. "What's New with the Chinese Consumer?," *McKinsey Quarterly,* October 2008; R. Ahluwalia, "How Far Can a Brand Stretch? Understanding the Role of Self-Construal," *Journal of Marketing Research* 3 (2008).

16. "Rocking the Most Hated Brand in America," *Fast Company,* July–August 2011; "6 Strong Alternatives to Ticketmaster for Buying Event Tickets," *MakeUseOf,* June 30, 2015.

17. "Fueling Growth through Word of Mouth," *BCG Perspectives,* December 2013.

18. "He Really Rolls in the Dough," *Investor's Business Daily,* January 26, 2010; "Rising Dough," *Fast Company,* October 2009; "Slicing the Bread but Not the Prices," *Wall Street Journal,* August 18, 2009; "Panera Bakes a Recipe for Success," *USA Today,* July 23, 2009.

19. Jennifer J. Argo, Monica Popa, and Malcolm C. Smith, "The Sound of Brands," *Journal of Marketing* 74, no. 4 (July 2010); "IKEA's Products Make Shoppers Blush in Thailand," *Wall Street Journal,* June 5, 2012; "Lost in Translation," *Business 2.0,* August 2004; "Global Products Require Name-Finders," *Wall Street Journal,* April 11, 1996; Martin S. Roth, "Effects of Global Market Conditions on Brand Image Customization and Brand Performance," *Journal of Advertising,* Winter 1995.

20. "BrandZ Top 100 Most Valuable Global Brands 2018," Millward Brown. For research on brand value, see Ron N. Borkovsky, Avi Goldfarb, Avery M. Haviv, and Sridhar Moorthy, "Measuring and Understanding Brand Value in a Dynamic Model of Brand Management," *Marketing Science,* July 2017; Yanhui Zhao, Roger J. Calantone, and Clay M. Voorhees, "Identity Change vs. Strategy Change: The Effects of Rebranding Announcements on Stock Returns," *Journal of the Academy of Marketing Science,* September 2018.

21. "The In-N-Out Effect: Trademark Mistakes You Need to Avoid," *Business.com,* March 16, 2015; "McCurry Wins Big McAttack in Malaysia," *Wall Street Journal,* September 9, 2009; "Hey, That NutraSweet Looks Like Splenda," *Advertising Age,* May 19, 2008. For research on this topic, see Larisa Ertekin, Alina Sorescu, and Mark B. Houston, "Hands Off My Brand! The Financial Consequences of Protecting Brands through Trademark Infringement Lawsuits," *Journal of Marketing,* September 2018.

22. For more on piracy, see "Episode 125: Duplitecture," *99% Invisible,* July 29, 2014; "Raids Crack Down on Fake Goods," *USA Today,* December 18, 2009; "HP Declares War on Counterfeiters," *BusinessWeek,* June 8, 2009; "EBay Fined over Selling Counterfeits," *Wall Street Journal,* July 1, 2008; "The Economic Effect of Counterfeiting and Piracy," *Executive Summary,* Organization for Economic Co-Operation and Development, 2007; Ashutosh Prasad and Vijay Mahajan, "How Many Pirates Should a Software Firm Tolerate? An Analysis of Piracy Protection on the Diffusion of Software," *International Journal of Research in Marketing,* December 2003. See also K. H. Wilcox, M. Kim, and S. Sen, "Why Do Consumers Buy Counterfeit Luxury Brands?," *Journal of Marketing Research* 2 (2009); Laurence Jacobs, A. Coskun Samli, and Tom Jedlik, "The Nightmare of International Product Piracy: Exploring Defensive Strategies," *Industrial Marketing Management,* August 2001; Janeen E. Olsen and Kent L. Granzin, "Using Channels Constructs to Explain Dealers' Willingness to Help Manufacturers Combat Counterfeiting," *Journal of Business Research,* June 1993.

23. For more on licensing, see "The Final Frontier for Licensors," *Brandweek,* June 9, 2008; "The Power behind J.D. Power," *Investor's Business Daily,* July 17, 2007; "Licensing Life," *Brandweek,* June 11, 2007; "Food Marketers Hope Veggies Look Fun to Kids," *USA Today,* July 15, 2005; "Testing Limits of Licensing: SpongeBob-Motif Holiday Inn," *Wall Street Journal,* October 9, 2003; "Making Tracks beyond Tires," *Brandweek,* September 15, 2003; "Candy Cosmetics: Licensing's Sweet Spot," *DSN Retailing Today,* August 4, 2003; "The Creative License," *Brandweek,* June 9, 2003; "Procter & Gamble Deals License Several Brand Names," *USA Today,* April 18, 2003.

24. For more on branding organic products, see "When Buying Organic Makes Sense—and When It Doesn't," *Wall Street Journal,* January 26, 2006; "The Organic Myth," *BusinessWeek,* October 16, 2006; "Private Food Labels Also Seeking Organic Growth," *Brandweek,* November 7, 2005; "Soap Can Proudly Display Certified 'Organic' Label," *USA Today,* August 25, 2005; "Health-Food Maker Hain Faces Rivals," *Wall Street Journal,* August 13, 2003; "Big Brand Logos Pop Up in Organic Aisle," *Wall Street Journal,* July 29, 2003; "USDA Enters Debate on Organic Label Law," *New York Times,* February 26, 2003; "Food Industry Gags at Proposed Label Rule for Trans Fat," *Wall Street Journal,* December 27, 2002.

25. Katie Kelting, Adam Duhachek, and Kimberly Whitler, "Can Copycat Private Labels Improve the Consumer's Shopping Experience? A Fluency Explanation," *Journal of the Academy of Marketing Science,* July 2017; S. Chan Choi and Anne T. Coughlan, "Private Label Positioning: Quality versus Feature Differentiation from the National Brand," *Journal of Retailing,* June 2006; David E. Sprott and Terence A. Shimp, "Using Product Sampling to Augment the Perceived Quality of Store Brands," *Journal of Retailing,* Winter 2004.

26. "Big-Name Food Brands Lose Battle of the Grocery Aisle," *Wall Street Journal,* April 30, 2017; "The State of Private Label around the World," Nielsen, November 2014.

27. "Store Brands Step Up Their Game, and Prices," *Wall Street Journal,* January 31, 2012; "Why Grocers Are Boosting Private Labels?," *Bloomberg Businessweek,* November 23, 2011; "Consumers Praise Store Brands," *Brandweek,* April 8, 2010; "Safeway Cultivates Its Private Labels as Brands to Be Sold by Other Chains," *Wall Street Journal,* May 7, 2009. See also Anoca Aribarg, Neeraj Arora, Ty Henderson, and Youngju Kim, "Private Label Imitation of a National Brand: Implications for Consumer Choice and Law," *Journal of Marketing Research,* December 2014; Kristopher O. Keller, Marnik G. Dekimpe, and Inge Geyskens, "Let Your Banner Wave? Antecedents and Performance Implications of Retailers' Private-Label Branding Strategies," *Journal of Marketing,* July 2016; Nicole Koschate-Fischer, Johannes Cramer, and Wayne D. Hoyer, "Moderating Effects of the Relationship between Private Label Share and Store Loyalty," *Journal of Marketing,* March 2014; Inge Geyskens, Katrijn Gielens, and Els Gijsbrechts, "Proliferating Private-Label Portfolios: How Introducing Economy and Premium Private Labels Influences Brand Choice," *Journal of Marketing Research,* October 2010; Jack (Xinlei) Chen, Om Narasimhan, George John, and Tirtha Dhar, "An Empirical Investigation of Private Label Supply by National Label Producers," *Marketing Science* 4 (2010); Tülin Erdem and Sue Chang, "A Cross-Category and Cross-Country Analysis of Umbrella Branding for National and Store Brands," *Journal of the Academy of Marketing Science,* January 2012; Jan-Benedict E. M. Steenkamp, Harald J. Van Heerde, and Inge Geyskens, "What Makes Consumers Willing to Pay a Price Premium for National Brands over Private Labels?" *Journal of Marketing Research,* December 2010; Maureen Morrin, Jonathan Lee, and Greg M. Allenby, "Determinants of Trademark Dilution," *Journal of Consumer Research,* August 2006; Bart J. Bronnenberg, Sanjay K. Dhar, and Jean-Pierre Dub, "Consumer Packaged Goods in the United States: National Brands, Local Branding," *Journal of Marketing Research,* February 2007; Donna F. Davis, Susan L. Golicic, and Adam J. Marquardt, "Branding a B2B Service: Does a Brand Differentiate a Logistics Service Provider?," *Industrial Marketing Management,* February 2008.

28. "Coke, Pepsi, A-B Attracted to Metal," *Brandweek,* December 17, 2007; "Can Wine in a Sippy Box Lure Back French Drinkers?," *Wall Street Journal,* August 24, 2007; "As Costs Rise, Whirlpool Makes a Dent in Dings," *Wall Street Journal,* July 30, 2007. See also Xiaoyan Deng and Raji Srinivasan, "When Do Transparent Packages Increase (or Decrease) Food Consumption?," *Journal of Marketing* 77, no. 4 (2013): 104–17; Nailya Ordabayeva and Pierre Chandon, "Predicting and Managing Consumers' Package Size Impressions," *Journal of Marketing* 77, no. 5 (2013): 123–37; Brian Wansink and Koert van Ittersum, "Bottoms Up! The Influence of Elongation on Pouring and Consumption Volume," *Journal of Consumer Research,* December 2003; U. R. Orth and K. Malkewitz, "Holistic Package Design and Consumer Brand Impressions," *Journal of Marketing* 3 (2008).

29. For more on food labeling, see Christopher L. Newman, Scot Burton, J. Craig Andrews, Richard G. Netemeyer et al., "Marketers' Use of Alternative Front-of-Package Nutrition Symbols: An Examination of Effects on Product Evaluations," *Journal of the Academy of Marketing Science,* May 2018; "Honest: New Ingredient in Food Labels," *USA Today,* June 12, 2014; "The Whole Truth about Whole Grain," *Wall Street Journal,* February 16, 2006; "Major Changes Set for Food Labels," *Wall Street Journal,* December 28, 2005; "FDA Says Food Labels Can Tout Tomatoes' Benefits on Cancer," *Wall Street Journal,* November 10, 2005; "Read It and Weep? Big Mac Wrapper to Show Fat, Calories," *Wall Street Journal,* October 26, 2005; "FDA Reexamines 'Serving Sizes,' May Change Misleading Labels," *Wall Street Journal,* November 20, 2003; "A 'Fat-Free' Product That's 100% Fat: How Food Labels Legally Mislead," *Wall Street Journal,* July 15, 2003. See also Avni M. Shah, James R. Bettman, Peter A. Ubel, Punam Anand Keller et al., "Surcharges Plus Unhealthy Labels Reduce Demand for Unhealthy Menu Items," *Journal of Marketing Research* 51, no. 6 (2014): 773–89; Lauren G. Block and Laura A. Peracchio, "The Calcium Quandary: How Consumers Use Nutrition Labels," *Journal of Public Policy & Marketing,* Fall 2006; Siva K. Balasubramanian and Catherine Cole, "Consumers' Search and Use of Nutrition Information: The Challenge and Promise of the Nutrition Labeling and Education Act," *Journal of Marketing,* July 2002; Bruce A. Silvergrade, "The Nutrition Labeling and Education Act—Progress to Date and Challenges for the Future," *Journal of Public Policy & Marketing,* Spring 1996; Sandra J. Burke, Sandra J. Milberg, and Wendy W. Moe, "Displaying Common but Previously Neglected Health Claims on Product Labels: Understanding Competitive Advantages, Deception, and Education," *Journal of Public Policy & Marketing,* Fall 1997; Christine Moorman, "A Quasi Experiment to Assess the Consumer and Informational Determinants of Nutrition Information Processing Activities: The Case of the Nutrition Labeling and Education Act," *Journal of Public Policy & Marketing,* Spring 1996.

30. For more on food claims, see R. Brennan, B. Czarnecka, S. Dahl, L. Eagle, and O. Mourouti, "Regulation of Nutrition and Health Claims in Advertising," *Journal of Advertising Research* 1 (2008); Paula Fitzgerald Bone and Robert J. Corey, "Ethical Dilemmas in Packaging: Beliefs of Packaging Professionals," *Journal of Macromarketing,* Spring 1992.

31. Edward M. Tauber, "Why Do People Shop?," *Journal of Marketing,* October 1972; Christopher H. Lovelock, "Classifying Services to Gain Strategic Marketing Insights," *Journal of Marketing,* Summer 1983; Tom Boyt and Michael Harvey, "Classification of Industrial Services," *Industrial Marketing Management,* July 1997.

32. Dennis W. Rook, "The Buying Impulse," *Journal of Consumer Research,* September 1987; Cathy J. Cobb and Wayne D. Hoyer, "Planned versus Impulse Purchase Behavior," *Journal of Retailing,* Winter 1986.

33. For example, see "Russian Maneuvers Are Making Palladium Ever More Precious," *Wall Street Journal,* March 6, 2000. See also William S. Bishop, John L. Graham, and Michael H. Jones, "Volatility of Derived Demand in Industrial Markets and Its Management Implications," *Journal of Marketing,* Fall 1984.

34. William B. Wagner and Patricia K. Hall, "Equipment Lease Accounting in Industrial Marketing Strategy," *Industrial Marketing Management* 20, no. 4 (1991); Robert S. Eckley, "Caterpillar's Ordeal: Foreign Competition in Capital Goods," *Business Horizons,* March–April 1989; M. Manley, "To Buy or Not to Buy," *Inc.,* November 1987.

35. P. Matthyssens and W. Faes, "OEM Buying Process for New Components: Purchasing and Marketing Implications," *Industrial Marketing Management,* August 1985; Paul A. Herbig and Frederick Palumbo, "Serving the Aftermarket in Japan and the United States," *Industrial Marketing Management,* November 1993.

36. Ruth H. Krieger and Jack R. Meredith, "Emergency and Routine MRO Part Buying," *Industrial Marketing Management,* November 1985; Warren A. French et al., "MRO Parts Service in the Machine Tool Industry," *Industrial Marketing Management,* November 1985. See also "The Web's New Plumbers," *Ecompany,* March 2001.

CHAPTER 9

1. See www.irobot.com; "IRobot Teams with Google to Advance Smart Homes," *Investor's Business Daily,* October 31, 2018; "Roomba Maker iRobot Expands Fleet with Terra Lawn-Mowing Bot," *Investor's Business Daily,* January 30, 2019; "iRobot's Long-Awaited Terra Robot Does the Lawn Mowing for You," *Fortune,* January 30, 2019; "The Makers of Roomba Want to Clean Up More Than Just Dirt," *Fortune,* July 24, 2017; "Robots Lead the Way for Investigations," *Claims Magazines,* June 2015; "iRobot Introduces System to Help Machines Think for Themselves," *Wall Street Journal,* October 9, 2014; "For iRobot the Future Is Getting Closer," *New York Times,* March 3, 2012; "In the Afghan War, a Little Robot Can Be a Soldier's Best Friend," *Wall Street Journal,* June 13, 2012; "Exploring the Oceans: Fleets of Robot Subma-

rines Will Change Oceanography," *The Economist,* June 9, 2012; "Where Humans Fear to Tread," *Wall Street Journal,* April 18, 2011; "Service Robots: The Rise of Machines (Again)," March 3, 2011; "The Future Is Automated," *Canadian Business,* March 14, 2011; "The *Wired* Interview: iRobot CEO Colin Angle," *Wired.com,* October 23, 2010; "TR50 2010: The World's Most Innovative Companies," *Technology Review,* February 10, 2010; "The Robot Revolution May Finally Be Here," *US News & World Report,* April 9, 2008; "2008 CRM Service Awards: Elite–iRobot," *CRM Magazine,* April 2008; "iRobot," *Fast Company,* March 2008; "iRobot Boots Up 2 New Models," *Investor's Business Daily,* September 28, 2007; "iRobot's Military Business on a Roll," *Investor's Business Daily,* August 22, 2007; "Maker Eyes Mainstream with Its Cleanup Robot," *Investor's Business Daily,* August 22, 2007; "Cleaning Up with Customer Evangelists," *BtoB,* August 13, 2007; "Keep Up with the Jetsons: iRobot's New Scooba," *Fortune,* February 20, 2006; "Robotic Orb Can Scrubba, Dub, Dub," *Raleigh News and Observer,* December 25, 2005; "What Can Scooba Do? Wash the Floor for You," *USA Today,* December 8, 2005; "A Robot That Could Hit the Wall," *BusinessWeek,* September 5, 2005; "Death to Cool . . . iRobot," *Inc.,* July 2005; "How the Roomba Was Realized," *BusinessWeek,* October 6, 2003; "Telepresence Robots Invade Hospitals—Doctors Can Be Anywhere, Anytime," *Singularity Hub,* December 4, 2012; "Robots with Your Face Want to Invade Workplaces and Hospitals," *Forbes,* June 10, 2013; "iRobot's RP-Vita Telepresence Robots Start Work at Seven Hospitals," *Singularity Hub,* May 18, 2013; "iRobot Enters 2016 with Strong Tailwinds," *Seeking Alpha,* January 5, 2016; "iRobot's Roomba 980 Dominates a New Competitor," *The Motley Fool,* December 17, 2016; "iRobot Braava Jet 240," *PC Magazine,* September 28, 2016; "iRobot Expands Connected Product Line with Roomba® 960," *PR Newswire,* August 4, 2016; "Smart Little Suckers: Next-Gen Robot Vacuums," *Wall Street Journal,* September 23, 2015.

2. "Nielsen Breakthrough Innovation Report," *The Nielsen Company,* June 2014, U.S. Edition.

3. "Mobility's Second Great Inflection Point," *McKinsey Quarterly,* February 2019; "We'll Always Eat Meat. But More of It Will Be 'Meat,'" *Bloomberg Businessweek,* January 24, 2019.

4. Raquel Castaño, Mita Sujan, Manish Kacker, and Harish Sujan, "Managing Consumer Uncertainty in the Adoption of New Products: Temporal Distance and Mental Simulation," *Journal of Marketing Research* 3 (2008); Rosanna Garcia, Fleura Bardhi, and Colette Friedrich, "Overcoming Consumer Resistance to Innovation," *Sloan Management Review,* Summer 2007; Shih Chuan-Fong and Alladi Venkatesh, "Beyond Adoption: Development and Application of a Use-Diffusion Model," *Journal of Marketing,* January 2004; Youngme Moon, "Break Free from the Product Life Cycle," *Harvard Business Review,* May 2005; Alina B. Sorescu, Rajesh K. Chandy, and Jaideep C. Prabhu, "Sources and Financial Consequences of Radical Innovation: Insights from Pharmaceuticals," *Journal of Marketing,* October 2003; Christopher M. McDermott and Gina Colarelli O'Connor, "Managing Radical Innovation: An Overview of Emergent Strategy Issues,"*Journal of Product Innovation Management,* November 2002; Ronald W. Niedrich and Scott D. Swain, "The Influence of Pioneer Status and Experience Order on Consumer Brand Preference: A Mediated-Effects Model," *Journal of the Academy of Marketing Science,* Fall 2003; Rajesh K. Chandy and Gerard J. Tellis, "The Incumbent's Curse? Incumbency, Size, and Radical Product Innovation," *Journal of Marketing,* July 2000; George Day, "The Product Life Cycle: Analysis and Applications Issues," *Journal of Marketing,* Fall 1981; Igal Ayal, "International Product Life Cycle: A Reassessment and Product Policy Implications," *Journal of Marketing,* Fall 1981.

5. "Soft Drink War Rages in Kenya—But Not over Cola," *Wall Street Journal,* June 4, 2009.

6. See Jorge Alberto Sousa De Vasconcellos, "Key Success Factors in Marketing Mature Products," *Industrial Marketing Management* 20, no. 4 (1991); Paul C. N. Michell, Peter Quinn, and Edward Percival, "Marketing Strategies for Mature Industrial Products," *Industrial Marketing Management* 20, no. 3 (1991); Peter N. Golder and Gerard J. Tellis, "Pioneer Advantage: Marketing Logic or Marketing Legend?," *Journal of Marketing Research,* May 1993.

7. "Milk Industry's Pitch in Asia: Try the Ginger or Rose Flavor," *Wall Street Journal,* August 9, 2005.

8. Heather McGowan, "Everything Is Accelerated (5 Pictures, 500 Words),"*LinkedIn,* March 25, 2015.

9. For another example involving the portable digital music market, see "An iPod Casualty: The Rio Digital-Music Player," *Wall Street Journal,* September 1, 2005; "When Being First Doesn't Make You #1," *Wall Street Journal,* August 12, 2004. See also Sungwook Min, Manohar U. Kalwani, and William T. Robinson, "Market Pioneer and Early Follower Survival Risks: A Contingency Analysis of Really New versus Incrementally New Product-Markets," *Journal of Marketing,* January 2006; W. Boulding and M. Christen, "Disentangling Pioneering Cost Advantages and Disadvantages," *Marketing Science* 4 (2008). See also footnotes 4 and 9 from this chapter.

10. "The Problem with Being a Trendsetter," *Wall Street Journal,* April 29, 2010; "Forget Fleece? Wool Makes a Comeback," *Wall Street Journal,* November 28, 2009; "Cheap and Trendy Gains as Luxury Fades in Japan," *Wall Street Journal,* November 10, 2009; "Life after Lasik: A Clear-Eyed Urge to Wear Glasses, Specs Are Looking Hot on Runway Models," *Wall Street Journal,* April 26, 2008; "Put a Patent on That Pleat," *BusinessWeek,* March 31, 2008; "Runway to Rack: Finding Looks That Will Sell," *Wall Street Journal,* March 6, 2008; "Work Wear: Designers Who Get It," *Wall Street Journal,* February 14, 2008; "How Fashion Makes Its Way from the Runway to the Rack," *Wall Street Journal,* February 8, 2007; "Men Say Bling It On,"*Wall Street Journal,* November 30, 2005. See also Hema Yoganarasimhan, "Identifying the Presence and Cause of Fashion Cycles in Data," *Journal of Marketing Research,* February 2017; Craig J. Thompson and Diana L. Haytko, "Speaking of Fashion: Consumers' Uses of Fashion Discourses and the Appropriation of Countervailing Cultural Meanings," *Journal of Consumer Research,* June 1997.

11. "The Story of Instagram: The Rise of the #1 Photo Sharing App," *Investopedia,* October 26, 2015; "The Startup That Died So Instagram Could Live," *CNN Money,* September 13, 2011.

12. *2010 Annual Report,* RJR Nabisco; "Oreo, Ritz Join Nabisco's Low-Fat Feast," *Advertising Age,* April 4, 1994; "They're Not Crying in Their Crackers at Nabisco," *BusinessWeek,* August 30, 1993; "Nabisco Unleashes a New Batch of Teddies," *Adweek*'s *Marketing Week,* September 24, 1990.

13. "Going Global by Going Green," *Wall Street Journal,* February 26, 2008.

14. "With Bottle-Fillers in Mind, the Water Fountain Evolves," *Wall Street Journal,* March 24, 2013.

15. "Philly Cream Cheese's Spreading Appeal," *Bloomberg Businessweek,* December 8, 2011.

16. There is an extensive academic literature on new-product planning and new-product development. For some insights, see Paola Cillo, David A. Griffith, and Gaia Rubera, "The New Product Portfolio Innovativeness–Stock Returns Relationship: The Role of Large Individual Investors' Culture," *Journal of Marketing,* November 2018; Kelly B. Herd and Ravi Mehta, "Head versus Heart: The Effect of Objective versus Feelings-Based Mental Imagery on New Product Creativity," *Journal of Consumer Research,* June 2019; Kartik Kalaignanam, Tarun Kushwaha, and Tracey A. Swartz, "The Differential Impact of New Product Development 'Make/Buy' Choices on Immediate and Future Product Quality: Insights from the Automobile Industry," *Journal of Marketing,* November 2017; Neeru Paharia and Vanitha Swaminathan, "Who Is Wary of User Design? The Role of Power-Distance Beliefs in Preference for User-Designed Products," *Journal of Marketing,* May 2019; Kyung M. Park, Pradeep K. Chintagunta, and Inho Suk, "Capital Market Returns to New Product Development Success: Informational Effects on Product Market Advertising," *Journal of Marketing Research,* February 2019; Jeffrey R. Parker, Donald R. Lehmann, Kevin Lane Keller, and Martin G. Schleicher, "Building a Multi-Category Brand: When Should Distant Brand Extensions Be Introduced?," *Journal of the*

Academy of Marketing Science, March 2018; Michel van der Borgh and Jeroen Schepers, "Are Conservative Approaches to New Product Selling a Blessing in Disguise?," *Journal of the Academy of Marketing Science,* September 2018; Tim Brexendorf, Barry Bayus, and Kevin Keller, "Understanding the Interplay between Brand and Innovation Management: Findings and Future Research Directions," *Journal of the Academy of Marketing Science,* September 2015; Anna Cui and Fang Wu, "Utilizing Customer Knowledge in Innovation: Antecedents and Impact of Customer Involvement on New Product Performance," *Journal of the Academy of Marketing Science,* July 2016; Richard Gruner, Christian Homburg, and Bryan Lukas, "Firm-Hosted Online Brand Communities and New Product Success," *Journal of the Academy of Marketing Science,* January 2014; Sven Heidenreich, Kristina Wittkowski, Matthias Handrich, and Tomas Falk, "The Dark Side of Customer Co-Creation: Exploring the Consequences of Failed Co-Created Services," *Journal of the Academy of Marketing Science,* May 2015; Ruth Stock and Ines Reiferscheid, "Who Should Be in Power to Encourage Product Program Innovativeness, R&D or Marketing?," *Journal of the Academy of Marketing Science,* May 2014; Tereza Dean, David A. Griffith, and Roger J. Calantone, "New Product Creativity: Understanding Contract Specificity in New Product Introductions," *Journal of Marketing,* March 2016; Eric Fang, Jongkuk Lee, and Zhi Yang, "The Timing of Codevelopment Alliances in New Product Development Processes: Returns for Upstream and Downstream Partners," *Journal of Marketing,* January 2015; Christian Homburg, Martin Schwemmle, and Christina Kuehnl, "New Product Design: Concept, Measurement, and Consequences," *Journal of Marketing,* May 2015; Rupinder P. Jindal, Kumar R. Sarangee, Raj Echambadi, and Sangwon Lee, "Designed to Succeed: Dimensions of Product Design and Their Impact on Market Share," *Journal of Marketing,* July 2016; Woojung Chang and Steven A. Taylor, "The Effectiveness of Customer Participation in New Product Development: A Meta-Analysis," *Journal of Marketing,* January 2016; Ganesh Iyer and David A. Soberman, "Social Responsibility and Product Innovation," *Marketing Science,* October 2016; Michael J. Barone and Robert D. Jewell, "The Innovator's License: A Latitude to Deviate from Category Norms," *Journal of Marketing* 77, no. 1 (2013): 120-34; Qingsheng Wu, Xueming Luo, Rebecca Slotegraaf, and Jaakko Aspara, "Sleeping with Competitors: The Impact of NPD Phases on Stock Market Reactions to Horizontal Collaboration," *Journal of the Academy of Marketing Science* 43 no. 4 (2015): 490-511; Min Zhao, Darren W. Dahl, and Steve Hoeffler, "Optimal Visualization Aids and Temporal Framing for New Products," *Journal of Consumer Research* 41, no. 4 (2014): 1137-51; Thomas Dotzel, Venkatesh Shankar, and Leonard L. Berry, "Service Innovativeness and Firm Value," *Journal of Marketing Research* 50, no. 2 (2013): 259-76; Zhenfeng Ma, Zhiyong Yang, and Mehdi Mourali, "Consumer Adoption of New Products: Independent versus Interdependent Self-Perspectives," *Journal of Marketing* 78, no. 2 (2014): 101-17.

17. C. Christenson, *The Innovator's Dilemma* (Cambridge MA: Harvard Business School Press, 1997); C. Christenson and M. Overdorf, "Meeting the Challenge of Disruptive Change," *Harvard Business Review,* March-April 2000; S. D. Anthony, M. W. Johnson, J. V Sinfield, and E. J. Altman, *Innovator's Guide to Growth—Putting Disruptive Innovation to Work* (Boston: Harvard Business School Press, 2008); "Ten for '10: New Gadgets for Home and Away," *Wall Street Journal,* January 7, 2010; "The World's Most Inventive Companies," *BusinessWeek,* December 23, 2009; "Book of Tens," *Advertising Age,* December 14, 2009. For more on ChotuKool, see "Clarifying Innovation for Success," *Bloomberg Businessweek,* October 6, 2010; "How Can You Enter an Emerging Market—And Improve the Lives of Millions?," *Innosight.com;* "Cool Products for Small Towns," *The Times of Bangalore,* June 1, 2010; "India's New Retailers," *OutlookBusiness,* July 11, 2009.

18. For more on new-product definition, see "Electronic Code of Federal Regulations: Title 16 (Commercial Practices), Part 502 (Introductory Offers)," http://ecfr.gpoaccess.gov/. For more about types of new products, see G. A. Athaide and R. R. Klink, "Managing Seller-Buyer Relationships during New Product Development," *Journal of Product Innovation Management* 5 (2009); K. Aboulnasr, O. Narashimhan, E. Blair, and R. Chandy, "Competitive Response to Radical Product Innovations," *Journal of Marketing* 3 (2008); D. L. Alexander, J. G. Lynch, and Q. Wang, "As Time Goes By: Do Cold Feet Follow Warm Intentions for Really New versus Incrementally New Products?," *Journal of Marketing Research* 3 (2008).

19. Definition of a patent from the United States Patent and Trademark Office, online glossary, www.uspto.gov/main/glossary/#p. For more on Apple versus Samsung, see "Apple v. Samsung Patent Trial Recap: How It All Turned Out (FAQ)," *CNET.com,* May 7, 2014.

20. Michele Boldrin and David K. Levine, "The Case against Patents," *Journal of Economic Perspectives,* Winter 2013, pp. 3-22.

21. "Reposition: Simplifying the Customer's Brandscape," *Brandweek,* October 2, 2000; "Consumers to GM: You Talking to Me?," *BusinessWeek,* June 19, 2000; "How Growth Destroys Differentiation," *Brandweek,* April 24, 2000; "P&G, Seeing Shoppers Were Being Confused, Overhauls Marketing," *Wall Street Journal,* January 15, 1997.

22. "Ignore the Consumer," *Point,* September 2005; "Too Many Choices," *Wall Street Journal,* April 20, 2001; "New Products," *Ad Age International,* April 13, 1998; "The Ghastliest Product Launches," *Fortune,* March 16, 1998; "Flops: Too Many New Products Fail; Here's Why—and How to Do Better," *BusinessWeek,* August 16, 1993; Brian D. Ottum and William L. Moore, "The Role of Market Information in New Product Success/Failure," *Journal of Product Innovation Management,* July 1997; E. Ofek and O. Turut, "To Innovate or Imitate? Entry Strategy and the Role of Market Research," *Journal of Marketing Research* 5 (2008).

23. "Flavor Experiment for KitKat Leaves Nestle with a Bad Taste," *Wall Street Journal,* July 6, 2006; "Makers of Chicken Tonight Find Many Cooks Say, 'Not Tonight,'" *Wall Street Journal,* May 17, 1994; "Failure of Its Oven Lovin' Cookie Dough Shows Pillsbury Pitfalls of New Products," *Wall Street Journal,* June 17, 1993; Sharad Sarin and Gour M. Kapur, "Lessons from New Product Failures: Five Case Studies," *Industrial Marketing Management,* November 1990.

24. For research on idea generation, see B. J. Allen, Deepa Chandrasekaran, and Suman Basuroy, "Design Crowdsourcing: The Impact on New Product Performance of Sourcing Design Solutions from the 'Crowd,'" *Journal of Marketing,* March 2018; Nuno Camacho, Hyoryung Nam, P. K. Kannan, and Stefan Stremersch, "Tournaments to Crowdsource Innovation: The Role of Moderator Feedback and Participation Intensity," *Journal of Marketing,* March 2019.

25. "Agility in Action: How Four Brands Are Using Agile Marketing," *CMO.com,* May 20, 2013.

26. Martin Schreier and Reinhard Prugl, "Extending Lead-User Theory: Antecedents and Consequences of Consumers' Lead Userness," *Journal of Product Innovation Management,* July 2008. See also Rudy K. Moenaert, Filip Caeldries, Annouk Lievens, and Elke Wauters, "Communication Flows in International Product Innovation Teams," *Journal of Product Innovation Management,* September 2000; Vittorio Chiesa, "Global R&D Project Management and Organization: A Taxonomy," *Journal of Product Innovation Management,* September 2000; John J. Cristiano, Jeffrey K. Liker, and Chelsea C. White III, "Customer-Driven Product Development through Quality Function Deployment in the U.S. and Japan," *Journal of Product Innovation Management,* July 2000; Lisa C. Troy, David M. Szymanski, and P. Rajan Varadarajan, "Generating New Product Ideas: An Initial Investigation of the Role of Market Information and Organizational Characteristics," *Journal of the Academy of Marketing Science,* Winter 2001; X. M. Song and Mitzi M. Montoya-Weiss, "Critical Development Activities for Really New versus Incremental Products," *Journal of Product Innovation Management,* March 1998.

27. This box was inspired by a TEDxBozeman talk by Professor Jakki Mohr, "Biomimicry: Business Innovations Inspired by Nature"; "Inspired by Insect Ears, Soundskrit Wants to Make Microphones Magically Directional,"*TechCrunch,* January 10, 2018; "Bugs Will Actually Make Your Next Phone Better," *Wired,* August 16, 2018; "Biomimicry," *Wikipedia,* accessed July 2012; "15 Coolest Cases of Biomimicry,"

Brainz. For more on other forms of idea generation, see "Pipe Up, People! Rounding Up Staff Ideas," *Inc.,* February 2010; "Twitter Serves Up New Ideas from Its Followers," *New York Times,* October 25, 2009; "Where Do the Best Ideas Come From? The Unlikeliest Sources," *Advertising Age,* July 14, 2008; "The Customer Is the Company: Threadless," *Inc.,* June 2008; "Breakthrough Thinking from Inside the Box," *Harvard Business Review,* December 2007; "Let Consumers Control More Than Just Ads," *Advertising Age,* April 9, 2007; "Turn Customer Input into Innovation," *Harvard Business Review,* January 2002; Judy A. Siguaw, Penny M. Simpson, and Cathy A. Enz, "Conceptualizing Innovation Orientation: A Framework for Study and Integration of Innovation Research," *Journal of Product Innovation Management,* November 2006; V. Blazevic and A. Lievens, "Managing Innovation through Customer Coproduced Knowledge in Electronic Services: An Exploratory Study,"*Journal of the Academy of Marketing Science* 1 (2008); "Like Nature but Better," *Bloomberg Businessweek,* April 11, 2016.

28. "AI Is Reinventing the Way We Invent," *MIT Technology Review,* February 15, 2019; "Making New Drugs with a Dose of Artificial Intelligence," *New York Times,* February 5, 2019.

29. P. N. Bloom, G. T. Gundlach, and J. P. Cannon, "Slotting Allowances and Fees: Schools of Thought and the Views of Practicing Managers," *Journal of Marketing,* April 2000; "Shielding the Shield Makers," *Wall Street Journal,* November 26, 2003; "Gun Makers to Push Use of Gun Locks," *Wall Street Journal,* May 9, 2001; Jennifer J. Argo and Kelley J. Main, "Meta-Analyses of the Effectiveness of Warning Labels," *Journal of Public Policy & Marketing,* Fall 2004.

30. For more details on the calculations underlying ROI, see Appendix B.

31. "Test-Marketing a Modern Princess," *Wall Street Journal,* April 9, 2013.

32. Susumu Ogawa and Frank T. Piller, "Reducing the Risks of New Product Development," *Sloan Management Review,* Winter 2006; Joseph M. Bonner, Robert W. Ruekert, and Orville C. Walker Jr., "Upper Management Control of New Product Development Projects and Project Performance," *Journal of Product Innovation Management,* May 2002; Sundar Bharadwaj and Anil Menon, "Making Innovations Happen in Organizations: Individual Creativity Mechanisms, Organizational Creativity Mechanisms or Both?," *Journal of Product Innovation Management* 6 (2000).

33. C. K. Chua, K. F. Leong, and C. S. Lim, *Rapid Prototyping: Principles and Applications* (River Edge, NJ: World Scientific Publishing Company, 2010); Jeanne Liedtka, "Learning to Use Design Thinking Tools for Successful Innovation," *Strategy & Leadership* 5 (2011); T. Brown, "Design Thinking," *Harvard Business Review,* June 2008; T. Brown, *Change by Design: How Design Thinking Transforms Organizations and Inspires Innovation* (HarperBusiness, 2009); see also www.ideo.com.

34. "Torture Testing," *Fortune,* October 2, 2000; "Industry's Amazing New Instant Prototypes," *Fortune,* January 12, 1998.

35. "Oops! Marketers Blunder Their Way through the 'Herb Decade,'" *Advertising Age,* February 13, 1989.

36. For research on this topic, see Mark B. Houston, Ann-Kristin Kupfer, Thorsten Hennig-Thurau, and Martin Spann, "Pre-Release Consumer Buzz," *Journal of the Academy of Marketing Science,* March 2018; M. Talay, M. Akdeniz, and Ahmet Kirca, "When Do the Stock Market Returns to New Product Preannouncements Predict Product Performance? Empirical Evidence from the U.S. Automotive Industry," *Journal of the Academy of Marketing Science,* July 2017; Amalesh Sharma, Alok R. Saboo, and V. Kumar, "Investigating the Influence of Characteristics of the New Product Introduction Process on Firm Value: The Case of the Pharmaceutical Industry," *Journal of Marketing,* September 2018; James E. Burroughs, Darren W. Dahl, C. Page Moreau, Amitava Chattopadhyay, et al., "Facilitating and Rewarding Creativity during New Product Development," *Journal of Marketing,* November 2011; Donna L. Hoffman, Praveen K. Kopalle, and Thomas P. Novak, "The 'Right' Consumers for Better Concepts: Identifying Consumers High in Emergent Nature to Develop New Product Concepts," *Journal of Marketing Research,* October 2010; Kimmy Wa Chan, Chi Kin (Bennett) Yim, and Simon S. K. Lam, "Is Customer Participation in Value Creation a Double-Edged Sword? Evidence from Professional Financial Services across Cultures," *Journal of Marketing,* May 2010; Ashwin W. Joshi, "Salesperson Influence on Product Development: Insights from a Study of Small Manufacturing Organizations," *Journal of Marketing* 74, no. 1 (2010): 94–107; Rebecca J. Slotegraaf and Kwaku Atuahene-Gima, "Product Development Team Stability and New Product Advantage: The Role of Decision-Making Processes," *Journal of Marketing,* January 2011; Rajesh Sethi, Zafar Iqbal, and Anju Sethi, "Developing New-to-the-Firm Products: The Role of Micropolitical Strategies," *Journal of Marketing,* March 2012; Ji Hoon Jhang, Susan Jung Grant, and Margaret C. Campbell, "Get It?? Got It. Good! Enhancing New Product Acceptance by Facilitating Resolution of Extreme Incongruity," *Journal of Marketing Research,* April 2012; L. C. Troy, T. Hirunyawipada, and A. K. Paswan, "Cross-Functional Integration and New Product Success: An Empirical Investigation of the Findings," *Journal of Marketing* 6 (2008); Albert L. Page and Gary R. Schirr, "Growth and Development of a Body of Knowledge: 16 Years of New Product Development Research, 1989–2004," *Journal of Product Innovation Management,* May 2008; Gloria Barczak, Kenneth B. Kahn, and Roberta Moss, "An Exploratory Investigation of NPD Practices in Nonprofit Organizations," *Journal of Product Innovation Management,* November 2006; Kwaku Atuahene-Gima and Janet Y. Murray, "Exploratory and Exploitative Learning in New Product Development: A Social Capital Perspective on New Technology Ventures in China," *Journal of International Marketing* 2 (2007); Khaled Aboulnasr, Om Narasimhan, Edward Blair, and Rajesh Chandy, "Competitive Response to Radical Product Innovations," *Journal of Marketing,* May 2008; Stephen J. Carson, "When to Give Up Control of Outsourced New Product Development," *Journal of Marketing,* January 2007; Subin Im and John P. Workman Jr., "Market Orientation, Creativity, and New Product Performance in High-Technology Firms," *Journal of Marketing,* April 2004; Steven C. Michael and Tracy Pun Palandjian, "Organizational Learning and New Product Introductions," *Journal of Product Innovation Management,* July 2004; John P. Workman Jr., "Marketing's Limited Role in New Product Development in One Computer Systems Firm," *Journal of Marketing Research,* November 1993.

37. *Innovation Leadership Study,* Capgemini Consulting and IESE Business School, 2014.

38. "Care, Feeding, and Building of a Billion-Dollar Brand," *Advertising Age,* February 23, 2004; "Brands at Work," *Brandweek,* April 13, 1998; "Auto Marketing & Brand Management," *Advertising Age,* April 6, 1998; "P&G Redefines the Brand Manager," *Advertising Age,* October 13, 1997. See also Sanjay K. Dhar, Stephen J. Hoch, and Nanda Kumar, "Effective Category Management Depends on the Role of the Category," *Journal of Retailing,* Summer 2001; Don Frey, "Learning the Ropes: My Life as a Product Champion," *Harvard Business Review,* September–October 1991; Stephen K. Markham, "New Products Management," *Journal of Product Innovation Management,* July 1997; Burçak Ertimur and Gokcen Coskuner-Balli, "Navigating the Institutional Logics of Markets: Implications for Strategic Brand Management," *Journal of Marketing* 79, no. 2 (2015): 40–61; Sungtak Hong, Kanishka Misra, and Naufel J. Vilcassim, "The Perils of Category Management: The Effect of Product Assortment on Multicategory Purchase Incidence," *Journal of Marketing,* September 2016.

39. "Soy Sauce Flavored Kit Kats? In Japan, They're #1," *AdAge.com,* March 4, 2010; "Flavor Experiment for KitKat Leaves Nestlé with a Bad Taste," *Wall Street Journal,* July 6, 2006.

40. "These Cows Will Text You When They're in Heat," *Bloomberg Businessweek,* November 3, 2016.

41. "How Artificial Intelligence Is Bringing Us Smarter Medicine," *Fast Company,* July–August 2016.

42. See "Hotels Train Employees to Think Fast," *USA Today,* November 29, 2006; "Takin' Off the Ritz–a Tad," *Wall Street Journal,*

June 23, 2006; "Hotels Take 'Know Your Customer' to New Level," *Wall Street Journal,* February 7, 2006; Roland T. Rust, Anthony J. Zahorik, and Timothy L. Keiningham, *Return on Quality* (Chicago: Probus, 1994); J. J. Cronin and Steven A. Taylor, "SERVPERF versus SERVQUAL: Reconciling Performance-Based and Perceptions-Minus-Expectations Measurement of Service Quality," *Journal of Marketing,* January 1994; Shirley Taylor, "Waiting for Service: The Relationship between Delays and Evaluations of Service," *Journal of Marketing,* April 1994; Mary J. Bitner, Bernard H. Booms, and Lois A. Mohr, "Critical Service Encounters: The Employee's Viewpoint," *Journal of Marketing,* October 1994; Narayan Janakiraman, Robert J. Meyer, and Stephen J. Hoch, "The Psychology of Decisions to Abandon Waits for Service," *Journal of Marketing Research,* November 2011; Michael K. Brady, Clay M. Voorhees, and Michael J. Brusco, "Service Sweethearting: Its Antecedents and Customer Consequences," *Journal of Marketing,* March 2012; Elisabeth C. Brüggen, Bram Foubert, and Dwayne D. Gremler, "Extreme Makeover: Short- and Long-Term Effects of a Remodeled Servicescape," *Journal of Marketing,* September 2011; Roland T. Rust and Chung Tuck Siong, "Marketing Models of Service and Relationships," *Marketing Science,* November–December 2006; Frances X. Frei, "The Four Things a Service Business Must Get Right," *Harvard Business Review,* April 2008; Matthew L. Meuter, Mary Jo Bitner, Amy L. Ostrom, and Stephen W. Brown, "Choosing among Alternative Service Delivery Modes: An Investigation of Customer Trial of Self-Service Technologies," *Journal of Marketing,* April 2005; D. Todd Donavan, Tom J. Brown, and John C. Mowen, "Internal Benefits of Service-Worker Customer Orientation: Job Satisfaction, Commitment, and Organizational Citizenship Behaviors," *Journal of Marketing,* January 2004; Leonard L. Berry, Venkatesh Shankar, Janet Turner Parish, Susan Cadwallader, and Thomas Dotzel, "Creating New Markets through Service Innovation," *Sloan Management Review,* Winter 2006; K. Sivakumar, Mei Li, and Beibei Dong, "Service Quality: The Impact of Frequency, Timing, Proximity, and Sequence of Failures and Delights," *Journal of Marketing,* January 2014.

43. M. H. Huang and R. T. Rust, "Artificial Intelligence in Service," *Journal of Service Research* 21, no. 2 (2018): 155–72.

44. For more on quality management and control, see Roland T. Rust, Anthony J. Zahorik, and Timothy L. Keiningham, "Return on Quality (ROQ): Making Service Quality Financially Accountable," *Journal of Marketing,* April 1995. See also Mark R. Colgate and Peter J. Danaher, "Implementing a Customer Relationship Strategy: The Asymmetric Impact of Poor versus Excellent Execution," *Journal of the Academy of Marketing Science,* Summer 2000; Simon J. Bell and Bulent Menguc, "The Employee-Organization Relationship, Organizational Citizenship Behaviors, and Superior Service Quality," *Journal of Retailing,* Summer 2002; Pratibha A. Dabholkar, C. David Shepherd, and Dayle I. Thorpe, "A Comprehensive Framework for Service Quality: An Investigation of Critical Conceptual and Measurement Issues through a Longitudinal Study," *Journal of Retailing,* Summer 2000; K. Sivakumar, Mei Li, and Beibei Dong, "Service Quality: The Impact of Frequency, Timing, Proximity, and Sequence of Failures and Delights," *Journal of Marketing* 78, no. 1 (2014): 41–58; Gielis Heijden, Jeroen Schepers, Edwin Nijssen, and Andrea Ordanini, "Don't Just Fix It, Make It Better! Using Frontline Service Employees to Improve Recovery Performance," *Journal of the Academy of Marketing Science* 41, no. 5 (2013): 515–30.

CHAPTER 10

1. See www.dell.com; www.lenovo.com; "Lenovo Reclaims the #1 Spot in PC Rankings in Q3 2018, According to IDC," *IDC,* October 10, 2018; "How HP Survives in a Shrinking PC Market," *Wall Street Journal,* May 21, 2018; "Lenovo, the Treasure Hunter of Tech," *Bloomberg Businessweek,* May 8, 2014; "How a Chinese Company Became a Global PC Powerhouse," *Time,* May 4, 2015; "Lenovo's Mobile Chief Steps Down amid Slowdown in Chinese Smartphone Market," *FierceWireless,* June 2, 2015; "Lenovo Fine-Tunes Partner Strategy as Sales Continue to Defy Gravity," *Channel Partners,* May 22, 2014; "Lenovo: A Chinese Dragon in the Global Village," *Ivey Case #W13085,* 2013; "Personal Computer Business—Lenovo Leading the Pack," *Amity Research Centers, Case Study # 313-302-1,* 2013; "Lenovo: Challenger to Leader," *IBS Center for Management Research #314-094-1,* 2014; "China's Competitiveness: Myths, Reality, and Lessons for the United States and Japan, Case Study: Lenovo," *Center for Strategic & International Studies,* January 2013; "Dell's Reinvention Efforts Hold Promise," *Wall Street Journal,* February 21, 2012; "Dell Outlook Cloudy," *Wall Street Journal,* November 16, 2011; "PC Makers Ready iPad Rivals," *Wall Street Journal,* February 18, 2011; "Dell's Faith in Business Sales Pays Off," *Wall Street Journal,* February 16, 2011; "Dell Spurs Sales by Lending to Hard-Hit Small Businesses," *Wall Street Journal,* March 30, 2010; "Discounting Continues to Haunt Dell as Its Turnaround Struggles," *Wall Street Journal,* February 19, 2010; "Direct Seller Dell Now Offers the Full Middleman Route," *Investor's Business Daily,* April 15, 2009; "Dell to Lift Its Data-Center Game," *Wall Street Journal,* March 25, 2009; "New Dell Notebook Fashion-Model Thin, Unfashionable Price," *Investor's Business Daily,* March 18, 2009; "Former Direct-Only Seller Dell Adds Best Buy," *Investor's Business Daily,* December 7, 2007; "Dell's U.S. Sales Fall, Profits Disappear," *USA Today,* November 30, 2007; "Where Dell Sells with Brick and Mortar," *BusinessWeek,* October 8, 2007; "Direct-Selling PC Specialist Dell Hooks Up with Retailer in China," *Investor's Business Daily,* September 25, 2007; "Dell Pushes Reset Button on Its Image," *Wall Street Journal,* July 10, 2007; "Hmm, Hell Can Freeze Over—Dell to Sell PCs at Retail," *Investor's Business Daily,* May 25, 2007; "Where Dell Went Wrong," *BusinessWeek,* February 19, 2007; "Grudge Match in China," *BusinessWeek,* April 2, 2007; "Can Dell Succeed in an Encore?," *Wall Street Journal,* February 5, 2007; "Consumer Demand and Growth in Laptops Leave Dell Behind," *Wall Street Journal,* August 30, 2006; "Dell May Have to Reboot in China," *BusinessWeek,* November 7, 2005; "What You Don't Know about Dell," *BusinessWeek,* November 3, 2003; "Lenovo's U.S. PC, Server Sales Get Channel Boost," *Channel Insider,* September 2, 2016; "Lenovo Takes the Hassle out of Channel Sales," *Channel Insider,* February 9, 2015; "Lenovo Hones Channel Strategy to Grow Market Share," *Channel Insider,* May 11, 2016; "Lenovo, HP, and Dell Lead the Shrinking PC Market," *Fortune,* January 11, 2017.

2. Some academic research on channels of distribution includes Andrew T. Crecelius, Justin M. Lawrence, Ju-Yeon Lee, Son K. Lam, et al., "Effects of Channel Members' Customer-Centric Structures on Supplier Performance," *Journal of the Academy of Marketing Science,* January 2019; Wesley James Johnston, Angelina Nhat Hanh Le, and Julian Ming-Sung Cheng, "A Meta-Analytic Review of Influence Strategies in Marketing Channel Relationships," *Journal of the Academy of Marketing Science,* July 2018.

3. S. Chan Choi, "Expanding to Direct Channel: Market Coverages as Entry Barrier," *Journal of Interactive Marketing,* Winter 2003; David Shipley, Colin Egan, and Scott Edgett, "Meeting Source Selection Criteria: Direct versus Distributor Channels," *Industrial Marketing Management* 20, no. 4 (1991); R. E. Bucklin, S. Siddarth, and J. M. Silva-Risso, "Distribution Intensity and New Car Choice," *Journal of Marketing Research* 4 (2008).

4. For more on Glacéau, see "Are Sodas Losing Their Fizz?," *Investor's Business Daily,* November 19, 2007; "Coke Unit Adds Some Muscle to Water Line," *USA Today,* October 31, 2007; "Coke Plan Riles Glacéau's Old Network," *Wall Street Journal,* September 14, 2007; "Move Over, Coke: How a Small Beverage Maker Managed to Win Shelf Space . . . ," *Wall Street Journal Reports,* January 30, 2006; "Pepsi Picks Water Fight with Surging Glaceau," *Advertising Age,* October 17, 2005.

5. Bob Stone and Ron Jacobs, *Successful Direct Marketing Methods* (New York: McGraw-Hill, 2008).

6. For a classic discussion of the discrepancy concepts, see Wroe Alderson, "Factors Governing the Development of Marketing

Channels," in *Marketing Channels for Manufactured Goods,* ed. Richard M. Clewett (Homewood, IL: Richard D. Irwin, 1954).

7. For some examples of how channels change to adjust discrepancies, see "Selling Literature Like Dog Food Gives Club Buyer Real Bite," *Wall Street Journal,* April 10, 2002. See also Robert Tamilia, Sylvain Senecal, and Gilles Corriveau, "Conventional Channels of Distribution and Electronic Intermediaries: A Functional Analysis," *Journal of Marketing Channels* 3, no. 4 (2002); Robert A. Mittelstaedt and Robert E. Stassen, "Structural Changes in the Phonograph Record Industry and Its Channels of Distribution, 1946-1966," *Journal of Macromarketing,* Spring 1994; Arun Sharma and Luis V. Dominguez, "Channel Evolution: A Framework for Analysis," *Journal of the Academy of Marketing Science,* Winter 1992; A. Ansari, C. F. Mela, and S. A. Neslin, "Customer Channel Migration," *Journal of Marketing Research* 1 (2008).

8. Sridhar Ramaswami and S. Arunachalam, "Divided Attitudinal Loyalty and Customer Value: Role of Dealers in an Indirect Channel," *Journal of the Academy of Marketing Science,* November 2016; Christian Homburg, Josef Vollmayr, and Alexander Hahn, "Firm Value Creation through Major Channel Expansions: Evidence from an Event Study in the United States, Germany, and China," *Journal of Marketing,* May 2014; Arne Nygaard and Robert Dahlstrom, "Role Stress and Effectiveness in Horizontal Alliances," *Journal of Marketing,* April 2002; M. B. Sarkar, Raj Echambadi, S. Tamer Cavusgil, and Preet S. Aulakh, "The Influence of Complementarity, Compatibility, and Relationship Capital on Alliance Performance," *Journal of the Academy of Marketing Science,* Fall 2001; Jakki J. Mohr, Robert J. Fisher, and John R. Nevin, "Collaborative Communication in Interfirm Relationships: Moderating Effects of Integration and Control," *Journal of Marketing,* July 1996; S. Ganesan, M. George, S. Jap, R. W. Palmatier, et al., "Supply Chain Management and Retailer Performance: Emerging Trends, Issues, and Implications for Research and Practice," *Journal of Retailing* 1 (2009).

9. See hulu.com; "The Future of TV," *Advertising Age,* November 30, 2009; "Searching for Life on Hulu," *AdweekMedia,* May 25, 2009; "Hulu," *Fast Company,* March 2009.

10. "As Sephora Adds Products, Rivalry Heats Up at Its Stores," *Wall Street Journal,* January 8, 2017. For other examples, see "P&G Starts Online Subscription Service for Tide Pods," *Wall Street Journal,* July 20, 2016; "Tempur Sealy's Mattress Nightmare Is Real," *Wall Street Journal,* January 30, 2017.

11. "Fender's Move to Sell Instruments Directly to Musicians Upsets Dealers," *LA Times,* January 20, 2015.

12. "Walmart Just Wanted to Sell Pricey Outdoor Gear. Then 'All Hell Broke Loose.'" *Wall Street Journal,* September 7, 2018.

13. Nirmalya Kumar, "Living with Channel Conflict," *CMO,* October 2004. See also Kersi D. Antia, Xu (Vivian) Zheng, and Gary L. Frazier, "Conflict Management and Outcomes in Franchise Relationships: The Role of Regulation," *Journal of Marketing Research* 50, no. 5 (2013): 577-89; Danny T. Wang, Flora F. Gu, and Maggie Chuoyan Dong, "Observer Effects of Punishment in a Distribution Network," *Journal of Marketing Research* 50, no. 5 (2013): 627-43; Gregory M. Rose and Aviv Shoham, "Interorganizational Task and Emotional Conflict with International Channels of Distribution," *Journal of Business Research,* September 2004.

14. "Get Great Results from Salespeople by Finding What Really Moves Them," *Investor's Business Daily,* July 2, 2001.

15. Yasin Alan, Jeffrey P. Dotson, and Mümin Kurtuluş, "On the Competitive and Collaborative Implications of Category Captainship," *Journal of Marketing,* July 2017; Shailendra Gajanan, Suman Basuroy, and Srinath Beldona, "Category Management, Product Assortment, and Consumer Welfare," *Marketing Letters,* July 2007; Kenneth H. Wathne and Jan B. Heide, "Relationship Governance in a Supply Chain Network," *Journal of Marketing,* January 2004; David I. Gilliland, "Designing Channel Incentives to Overcome Reseller Rejection," *Industrial Marketing Management,* February 2004; James R. Brown, Anthony T. Cobb, and Robert F. Lusch, "The Roles Played by Interorganizational Contracts and Justice in Marketing Channel Relationships," *Journal of Business Research,* February 2006; Keysuk Kim and Changho Oh, "On Distributor Commitment in Marketing Channels for Industrial Products: Contrast between the United States and Japan," *Journal of International Marketing* 1 (2002).

16. For research on control in channels of distribution, see Jody Crosno and James Brown, "A Meta-analytic Review of the Effects of Organizational Control in Marketing Exchange Relationships," *Journal of the Academy of Marketing Science* 43, no. 3 (2015): 297-314; David Gilliland and Stephen Kim, "When Do Incentives Work in Channels of Distribution?," *Journal of the Academy of Marketing Science* 42, no. 4 (2014): 361-79.

17. For some examples of vertical integration, see "Moving On Up: Agricultural Firms Are Looking for New Growth Model . . . Vertical Integration," *Wall Street Journal Reports,* October 25, 2004.

18. Flora F. Gu, Namwoon Kim, David K. Tse, and Danny T. Wang, "Managing Distributors' Changing Motivations over the Course of a Joint Sales Program," *Journal of Marketing,* September 2010; David I. Gilliland, Daniel C. Bello, and Gregory T. Gundlach, "Control-Based Channel Governance and Relative Dependence," *Journal of the Academy of Marketing Science,* July 2010; Vishal Kashyap, Kersi D. Antia, and Gary L. Frazier, "Contracts, Extracontractual Incentives, and Ex Post Behavior in Franchise Channel Relationships," *Journal of Marketing Research,* April 2012; Stephen K. Kim, Richard G. McFarland, Soongi Kwon, Sanggi Son, and David A. Griffith, "Understanding Governance Decisions in a Partially Integrated Channel: A Contingent Alignment Framework," *Journal of Marketing Research,* May 2011; Kevin L. Webb, "Managing Channels of Distribution in the Age of Electronic Commerce," *Industrial Marketing Management,* February 2002; Charles A. Ingene and Mark E. Parry, "Is Channel Coordination All It Is Cracked Up to Be?," *Journal of Retailing,* Winter 2000; Kenneth H. Wathne and Jan B. Heide, "Opportunism in Interfirm Relationships: Forms, Outcomes, and Solutions," *Journal of Marketing,* October 2000; Kersi D. Antia and Gary L. Frazier, "The Severity of Contract Enforcement in Interfirm Channel Relationships," *Journal of Marketing,* October 2001; Aric Rindfleisch and Jan B. Heide, "Transaction Cost Analysis: Past, Present, and Future Applications," *Journal of Marketing,* October 1997.

19. "A Talk with the Man Who Got Rayovac All Charged Up," *BusinessWeek,* February 21, 2000.

20. "Antitrust Issues and Marketing Channel Strategy" and "Case 1—Continental T.V., Inc. et al. v. GTE Sylvania, Inc.," in Louis W. Stern and Thomas L. Eovaldi, *Legal Aspects of Marketing Strategy* (Englewood Cliffs, NJ: Prentice Hall, 1984). See also Debra M. Desrochers, Gregory T. Gundlach, and Albert A. Foer, "Analysis of Antitrust Challenges to Category Captain Arrangements," *Journal of Public Policy & Marketing,* Fall 2003. For some examples of exclusive deals, see "Blue Nile: Online Jewelry Seller Looks to Be a Cut above the Competition," *Investor's Business Daily,* December 15, 2005; "When Exclusivity Means Illegality," *Wall Street Journal,* January 6, 2005; "Bringing Chic to Sheets," *Wall Street Journal,* September 8, 2004.

21. For more on multichannel distribution, see Monika Käuferle and Werner Reinartz, "Distributing through Multiple Channels in Industrial Wholesaling: How Many and How Much?," *Journal of the Academy of Marketing Science,* November 2015; Andreas Fürst, Martin Leimbach, and Jana-Kristin Prigge, "Organizational Multichannel Differentiation: An Analysis of Its Impact on Channel Relationships and Company Sales Success," *Journal of Marketing,* January 2017; Steven H. Dahlquist and David A. Griffith, "Multidyadic Industrial Channels: Understanding Component Supplier Profits and Original Equipment Manufacturer Behavior," *Journal of Marketing,* July 2014; Donald J. Lund, and Detelina Marinova, "Managing Revenue across Retail Channels: The Interplay of Service Performance and Direct Marketing," *Journal of Marketing,* September 2014; Tarun Kushwaha and Venkatesh Shankar, "Are Multichannel Customers Really More Valuable? The Moderating Role of Product Category Characteristics," *Journal of Marketing* 77, no. 4 (2013): 67-85; Hongshuang (Alice) Li and P. K.

Kannan, "Attributing Conversions in a Multichannel Online Marketing Environment: An Empirical Model and a Field Experiment," *Journal of Marketing Research* 51, no. 1 (2014): 40–56; Andrea Godfrey, Kathleen Seiders, and Glenn B. Voss, "Enough Is Enough! The Fine Line in Executing Multichannel Relational Communication," *Journal of Marketing,* July 2011; U. Konus, P. C. Verhoef, and S. A. Neslin, "Multichannel Shopper Segments and Their Covariates," *Journal of Retailing* 4 (2008); Rajkumar Venkatesan, V. Kumar, and Nalini Ravishanker, "Multichannel Shopping: Causes and Consequences," *Journal of Marketing,* April 2007; Scott A. Neslin, Dhruv Grewal, Robert Leghorn, Venkatesh Shankar, et al., "Challenges and Opportunities in Multichannel Customer Management," *Journal of Service Research,* May 2006; Sertan Kabadayi, Nermin Eyuboglu, and Gloria P. Thomas, "The Performance Implications of Designing Multiple Channels to Fit with Strategy and Environment," *Journal of Marketing,* October 2007; Arun Sharma and Anuj Mehrotra, "Choosing an Optimal Channel Mix in Multichannel Environments," *Industrial Marketing Management,* January 2007; Bert Rosenbloom, "Multi-Channel Strategy in Business-to-Business Markets: Prospects and Problems," *Industrial Marketing Management,* January 2007; Asim Ansari, Carl F. Mela, and Scott A. Neslin, "Customer Channel Migration," *Journal of Marketing Research,* February 2008; Jule B. Gassenheimer, Gary L. Hunter, and Judy A. Siguaw, "An Evolving Theory of Hybrid Distribution: Taming a Hostile Supply Network," *Industrial Marketing Management,* May 2007; Peter C. Verhoef, Scott A. Neslin, and Björn Vroomen, "Multichannel Customer Management: Understanding the Research-Shopper Phenomenon," *International Journal of Research in Marketing,* June 2007; Jacquelyn S. Thomas and Ursula Y. Sullivan, "Managing Marketing Communications with Multichannel Customers," *Journal of Marketing,* October 2005.

22. For more information on showrooming, webrooming, and multichannel shopping, see Bing Jing, "Showrooming and Webrooming: Information Externalities between Online and Offline Sellers," *Marketing Science,* May 2018; Evert de Haan, P.K. Kannan, Peter C. Verhoef, and Thorsten Wiesel, "Device Switching in Online Purchasing: Examining the Strategic Contingencies," *Journal of Marketing,* September 2018; Dmitri Kuksov and Chenxi Liao, "When Showrooming Increases Retailer Profit," *Journal of Marketing Research,* August 2018; "Customer Desires vs. Retailer Capabilities: Minding the Omni-Channel Gap," *Forrester Consulting,* January 2014; "The New Digital Divide," *Deloitte Digital,* 2014; "Omnichannel Retail," *L2 Intelligence Report,* July 30, 2014; "Seamless Retailing Research Study 2014," *Accenture,* 2014.

23. "12 Examples of Brands with Brilliant Omni-Channel Experiences," *HubSpot Blog,* March 21, 2019.

24. Gregory T. Gundlach and Patrick E. Murphy, "Ethical and Legal Foundations of Relational Marketing Exchanges," *Journal of Marketing,* October 1993; Craig B. Barkacs, "Multilevel Marketing and Antifraud Statutes: Legal Enterprises or Pyramid Schemes?," *Journal of the Academy of Marketing Science,* Spring 1997; Robert A. Robicheaux and James E. Coleman, "The Structure of Marketing Channel Relationships," *Journal of the Academy of Marketing Science,* Winter 1994; Brett A. Boyle and F. Robert Dwyer, "Power, Bureaucracy, Influence and Performance: Their Relationships in Industrial Distribution Channels," *Journal of Business Research,* March 1995.

25. Thanh Tran, Haresh Gurnani, and Ramarao Desiraju, "Optimal Design of Return Policies," *Marketing Science,* July 2018; Elie Ofek, Zsolt Katona, and Miklos Sarvary, "Bricks and Clicks: The Impact of Product Returns on the Strategies of Multichannel Retailers,"*Marketing Science* 1 (2011); Jeffrey D. Shulman, Anne T. Coughlan, and R. Canan Savaskan, "Optimal Reverse Channel Structure for Consumer Product Returns," *Marketing Science* 6 (2010); J. A. Petersen and V. Kumar, "Are Product Returns a Necessary Evil? Antecedents and Consequences,"*Journal of Marketing* 3 (2009); R. Glenn Richey, Haozhe Chen, Stefan E. Genchev, and Patricia J. Daugherty, "Developing Effective Reverse Logistics Programs," *Industrial Marketing Management,* November 2005; Joseph D. Blackburn, V. Daniel, R. Guide Jr., Gilvan C. Souza, et al., "Reverse Supply Chains for Commercial Returns," *California Management Review,* Winter 2004.

26. "Hyper-Green Products Go Cradle to Cradle," *MSNBC.com,* October 12, 2007; "Can One Green Deliver the Other?," *Harvard Business Review,* 2005; "HP Wants Your Old PCs Back," *BusinessWeek.com,* April 10, 2006.

27. "5 Consumer Trends for 2017," *Trendwatching;* "Capacity Capture: The Next Big Idea in Sustainability," Pronto Marketing, February 27, 2017; "xStorage by Nissan Offers Rebuilt LEAF Batteries for Home Power," *Inside EVs,* May 16, 2018.

28. See Rene B. M. de Koster, Marisa P. de Brito, and Masja A. van de Vendel, "Return Handling: An Exploratory Study with Nine Retailer Warehouses," *International Journal of Retail & Distribution Management* 8 (2002); Chad W. Autry, Patricia J. Daugherty, and R. Glenn Richey, "The Challenge of Reverse Logistics in Catalog Retailing," *International Journal of Physical Distribution & Logistics Management* 1 (2001); Dale S. Rogers and Ronald Tibben Lembke, "An Examination of Reverse Logistics Practices," *Journal of Business Logistics* 2 (2001).

29. "Small Business: The Cost of Expanding Overseas," *Wall Street Journal,* February 26, 2014.

30. "Year of the Grease Monkey in China," *Denver Post,* April 26, 2009; "Danone Pulls Out of Disputed China Venture," *Wall Street Journal,* October 1, 2009; "Special Report: Gone Global, How to Get Ahead in China," *Inc.,* May 2008. For academic research on this topic, see Stavroula Spyropoulou, Constantine S. Katsikeas, Dionysis Skarmeas, and Neil A. Morgan, "Strategic Goal Accomplishment in Export Ventures: The Role of Capabilities, Knowledge, and Environment," *Journal of the Academy of Marketing Science,* January 2018; Rajdeep Grewal, Amit Saini, Alok Kumar, F. Robert Dwyer, et al., "Marketing Channel Management by Multinational Corporations in Foreign Markets," *Journal of Marketing,* July 2018; Rajdeep Grewal, Alok Kumar, Girish Mallapragada, and Amit Saini, "Marketing Channels in Foreign Markets: Control Mechanisms and the Moderating Role of Multinational Corporation Headquarters-Subsidiary Relationship," *Journal of Marketing Research* 50, no. 3 (2013): 378–98; Joseph Johnson and Gerard J. Tellis, "Drivers of Success for Market Entry into China and India," *Journal of Marketing,* May 2008; Neil Morgan, Constantine Katsikeas, and Douglas Vorhies, "Export Marketing Strategy Implementation, Export Marketing Capabilities, and Export Venture Performance," *Journal of the Academy of Marketing Science,* March 2012.

CHAPTER 11

1. "The Guy Who Invented the Segway Is Working with Coca-Cola to Bring His Water Distilling Invention to Third-World Countries," *Business Insider,* June 19, 2014; "Drink Dispenser Analytics: Coca-Cola Goes Freestyle, with Help from SAP BI," *Business Analytics,* June 19, 2009; "What Do You Call a Soda Dispenser with Onboard Data Analytics?," *Spotfire, Trends and Outliers,* October 7, 2010; "Africa: Coke's Last Frontier," *Bloomberg Businessweek,* October 28, 2010; "Eliminating Inefficiencies through Automation," *Beverage Industry,* January 2010; "Coca-Cola Expands Hybrid Electric Fleet in Canada," *Beverage Industry,* January 2010; "Coca-Cola Deal Is U.S. Strategy Shift," *Wall Street Journal,* February 26, 2010; "Coke Takes Juice Lead from Pepsi," *Advertising Age,* February 1, 2010; "Masters of Design," *Fast Company,* October 2009; "Coke: Buy 1 Rival, Get Our Brand Free," *Advertising Age,* March 9, 2009; "Sustainability and a Smile," *Advertising Age,* February 25, 2008; "Iran's Cola War," *Fortune,* February 6, 2008; "Coke Tries to Pop Back in Vital Japan Market," *Wall Street Journal,* July 11, 2006; "Coke Gets a Jolt," *Fortune,* May 15, 2006; "Thirsting for More Variety," *Investor's Business Daily,* April 24, 2006; "In the Newest Snack Packs, Less Is More," *USA Today,* April 13, 2006; "Afghan Coke a Taste of Things to Come," *Financial Times,* December 6, 2005; "EU Limits Coke's Sales Tactics in Settlement of Antitrust Case," *Wall Street*

Journal, June 23, 2005; "To Pitch New Soda, Coke Wants the World to Sing—Again," *Wall Street Journal,* June 13, 2005; "How a Global Web of Activists Gives Coke Problems in India," *Wall Street Journal,* June 7, 2005; "The Soda with Buzz," *Forbes,* March 28, 2005; "Into the Fryer: How Coke Officials Beefed Up Results of Marketing Tests," *Wall Street Journal,* August 20, 2003; "Mooove Over, Milkman," *Wall Street Journal,* June 9, 2003; "Fountain Dispenses Problems," *USA Today,* May 28, 2003; "Coca-Cola Targets Small Retailers with Coke on Demand App," *Wall Street Journal,* March 7, 2017; "The Secret behind Coca-Cola's Success in Africa," *CNN.com,* January 21, 2016; "How Coca-Cola Takes a Refreshing Approach on Big Data," *Datafloq,* accessed March 30, 2017; "Coca-Cola's Unique Challenge: Turning 250 Datasets into One," *MIT Sloan Management Review,* May 27, 2015.

2. "U.S. Logistics Costs Drop for First Time in Six Years, Benchmark Report Says," *Supply Chain Management Review,* June 18, 2009; "2009 Annual Report on State of Logistics: Collaboration Time," *Logistics Management,* 2009; "Latin America: Addressing High Logistics Costs and Poor Infrastructure for Merchandise Transportation and Trade Facilitation," working paper, The World Bank, August 2007; "The Cost of Being Landlocked: Logistics Costs and Supply Chain Reliability," Policy Research working paper, The World Bank, June 1, 2007.

3. "Compaq Stumbles as PCs Weather New Blow," *Wall Street Journal,* March 9, 1998; see also Alexander E. Ellinger, Scott B. Keller, and John D. Hansen, "Bridging the Divide between Logistics and Marketing: Facilitating Collaborative Behavior," *Journal of Business Logistics* 2 (2006); Kofi Q. Dadzie, Cristian Chelariu, and Evelyn Winston, "Customer Service in the Internet-Enabled Logistics Supply Chain: Website Design Antecedents and Loyalty Effects," *Journal of Business Logistics,* January 2005; Elliot Rabinovich and Philip T. Evers, "Product Fulfillment in Supply Chains Supporting Internet-Retailing Operations," *Journal of Business Logistics,* January 2003; Daniel Corsten and Thomas Gruen, "Stock-Outs Cause Walkouts," *Harvard Business Review,* May 2004; William D. Perreault Jr. and Frederick A. Russ, "Physical Distribution Service in Industrial Purchase Decisions," *Journal of Marketing,* April 1976; "Walmart Faces the Cost of Cost-Cutting: Empty Shelves," *Bloomberg Businessweek,* March 28, 2013.

4. "H-P Wields Its Clout to Undercut PC Rivals," *Wall Street Journal,* September 24, 2009; "Walmart's H-P Elves," *Wall Street Journal,* December 15, 2005; "A More Profitable Harvest," *Business 2.0,* May 2005; "Costs Too High? Bring in the Logistics Experts," *Fortune,* November 10, 1997. See also G. Tomas M. Hult, Kenneth K. Boyer, and David J. Ketchen Jr., "Quality, Operational Logistics Strategy, and Repurchase Intentions: A Profile Deviation Analysis," *Journal of Business Logistics* 2 (2007); Thierry Sauvage, "The Relationship between Technology and Logistics Third Party Providers," *International Journal of Physical Distribution & Logistics Management* 3 (2003); James R. Stock, "Marketing Myopia Revisited: Lessons for Logistics," *International Journal of Physical Distribution & Logistics Management* 1 (2002).

5. www.amazon.com; www.walmart.com; "How to use Walmart Grocery Pickup and Delivery," *Offers.com,* January 8, 2019. See also Forrest E. Harding, "Logistics Service Provider Quality: Private Measurement, Evaluation, and Improvement," *Journal of Business Logistics,* 1998; Carol C. Bienstock, John T. Mentzer, and Monroe M. Bird, "Measuring Physical Distribution Service Quality," *Journal of the Academy of Marketing Science,* Winter 1997.

6. "More Green from Green Beans," *Business 2.0,* August 2004. See also Mikko Karkkainen, Timo Ala-Risku, and Jan Holmstrom, "Increasing Customer Value and Decreasing Distribution Costs with Merge-In-Transit," *International Journal of Physical Distribution & Logistics Management* 1 (2003); Marc J. Schniederjans and Qing Cao, "An Alternative Analysis of Inventory Costs of JIT and EOQ Purchasing," *International Journal of Physical Distribution & Logistics Management* 2 (2001).

7. For more on JIT, see "Retailers Rely More on Fast Deliveries," *Wall Street Journal,* January 14, 2004; "Uncertain Economy Hinders Highly Precise Supply System," *New York Times,* March 15, 2003; "Port Tie-Up Shows Vulnerability," *USA Today,* October 9, 2002; "Deadline Scramble: A New Hazard for Recovery, Last-Minute Pace of Orders," *Wall Street Journal,* June 25, 2002; "Parts Shortages Hamper Electronics Makers," *Wall Street Journal,* July 7, 2000. See also Richard E. White and John N. Pearson, "JIT, System Integration and Customer Service," *International Journal of Physical Distribution & Logistics Management* 5 (2001).

8. "A New Industrial Revolution," *Wall Street Journal,* April 28, 2008; "Global Scramble for Goods Gives Corporate Buyers a Lift," *Wall Street Journal,* October 2, 2007.

9. See "Clarity Is Missing Link in Supply Chain," *Wall Street Journal,* May 18, 2009; Martin Christopher and John Gattorna, "Supply Chain Cost Management and Value-Based Pricing," *Industrial Marketing Management,* February 2005; G. Tomas M. Hult, "Global Supply Chain Management: An Integration of Scholarly Thoughts," *Industrial Marketing Management,* January 2004; Daniel J. Flint, "Strategic Marketing in Global Supply Chains: Four Challenges," *Industrial Marketing Management,* January 2004; Daniel C. Bello, Ritu Lohtia, and Vinita Sangtani, "An Institutional Analysis of Supply Chain Innovations in Global Marketing Channels," *Industrial Marketing Management,* January 2004; Brian J. Gibson, John T. Mentzer, and Robert L. Cook, "Supply Chain Management: The Pursuit of a Consensus Definition," *Journal of Business Logistics,* 2005; Theodore P. Stank, Beth R. Davis, and Brian S. Fugate, "A Strategic Framework for Supply Chain Oriented Logistics," *Journal of Business Logistics,* 2005; Terry L. Esper, Thomas D. Jensen, Fernanda L. Turnipseed, and Scot Burton, "The Last Mile: An Examination of Effects of Online Retail Delivery Strategies on Consumers," *Journal of Business Logistics,* 2003; Christopher R. Moberg and Thomas W. Speh, "Evaluating the Relationship between Questionable Business Practices and the Strength of Supply Chain Relationships," *Journal of Business Logistics* 2 (2003); Subroto Roy, K. Sivakumar, and Ian F. Wilkinson, "Innovation Generation in Supply Chain Relationships: A Conceptual Model and Research Propositions," *Journal of the Academy of Marketing Science,* Winter 2004; Rakesh Niraj, "Customer Profitability in a Supply Chain," *Journal of Marketing,* July 2001.

10. For an excellent example of EDI, see "To Sell Goods to Wal-Mart, Get on the Net," *Wall Street Journal,* November 21, 2003. See also Cornelia Droge and Richard Germain, "The Relationship of Electronic Data Interchange with Inventory and Financial Performance," *Journal of Business Logistics* 2 (2000).

11. "A Smart Cookie at Pepperidge," *Fortune,* December 22, 1986.

12. "A New Push against Sweatshops," *Time,* October 7, 2005; "Cops of the Global Village," *Fortune,* June 27, 2005; "Stamping Out Sweatshops,"*BusinessWeek,* May 23, 2005; "Nike Opens Its Books on Sweatshop Audits," *BusinessWeek,* April 27, 2000; "Nike Names Names," *BusinessWeek,* April 13, 2005; "Corporate Social Responsibility—Companies in the News: Nike," www.mallenbaker.net/csr/CSRfiles/nike.html.

13. David Grant, Douglas Lambert, James R. Stock, and Lisa M. Ellram, *Fundamentals of Logistics* (Burr Ridge, IL: Irwin/McGraw-Hill, 2005).

14. For more on transportation security, see "Calif. Port Goes Tech in Security," *Investor's Business Daily,* February 11, 2009; "Technology Roots Out Cargo Risks," *Investor's Business Daily,* October 4, 2005; "Keeping Cargo Safe from Terror," *Wall Street Journal,* July 29, 2005; "Protecting America's Ports," *Fortune,* November 10, 2003.

15. For a more detailed comparison of mode characteristics, see Robert Dahlstrom, Kevin M. McNeilly, and Thomas W. Speh, "Buyer-Seller Relationships in the Procurement of Logistical Services," *Journal of the Academy of Marketing Science,* Spring 1996.

16. "Rail Renaissance," *BusinessWeek,* November 3, 2008; "New Rail-Building Era Dawns," *Wall Street Journal,* February 13, 2008; "Next Stop: The 21st Century," *Business 2.0,* September 2003; "Back on Track: Left for Dead, Railroads Revive by Watching Clock," *Wall Street Journal,* July 25, 2003; "Trains: Industry Report," *Investor's Business Daily,* May 7, 2001.

17. "Truckers Question Traffic-Relief Efforts," *USA Today,* December 27, 2007; "Shipping Hubs Spring Up Inland," *USA Today,* December 17, 2007; "Waterways Could Be Key to Freeing Up Freeways," *USA Today,* October 11, 2007.

18. For more on overnight carriers, see "UPS Battles Traffic Jams to Gain Ground in India," *Wall Street Journal,* January 25, 2008.

19. "Chinese Online Retailer JD.com Is Developing Heavy-Duty Delivery Drones," *Wall Street Journal,* May 22, 2017; "Amazon's Drones Exiled to Canada," *Wall Street Journal,* April 5, 2015; "Deutsche Post DHL to Deliver Medicine via Drone," *Wall Street Journal,* September 25, 2014; "Swiss Postal Service Will Start Using Delivery Drones in Pilot Program This Summer," *Slate,* April 23, 2015.

20. "Foldable Shipping Containers Try to Stack Up," *Wall Street Journal,* April 12, 2010; "The Mega Containers Invade," *Wall Street Journal,* January 26, 2009.

21. "Western Grocer Modernizes Passage to Indian Markets," *Wall Street Journal,* November 28, 2007; "Widening Aisles for Indian Shoppers," *BusinessWeek,* April 30, 2007.

22. "DHL Completes More Eco-Friendly Ship Voyage as Industry Comes under Fire," *GreenBiz.com,* February 15, 2008; *State of Green Business 2008,* Greener World Media, 2008. See also *2009 Annual Report,* Du Pont; *2009 Annual Report,* Matlack; *2009 Annual Report,* Shell; *2009 Annual Report,* FedEx; *2009 Annual Report,* UPS.

23. "Data Analytics: Crunching the Future," *Bloomberg Businessweek,* September 8, 2011.

24. "Why Your Ice Cream Will Ride in a Self-Driving Car before You Do," *Wall Street Journal,* January 5, 2019; "GM, DoorDash to Test Autonomous Food Deliveries," *Wall Street Journal,* January 4, 2019; "Self-Driving Truck Convoy Completes Its First Major Journey across Europe," *The Verge,* April 7, 2016; "$30K Retrofit Turns Dumb Semis into Self-Driving Robots," *Wired,* May 17, 2016; "The Driverless Truck Is Coming, and It's Going to Automate Millions of Jobs," *TechCrunch,* April 25, 2016; "Starsky Robotics Puts New Spin on Driverless Trucks," *Fortune,* February 28, 2017; "A Driverless Future Is Coming—But It Won't Start with Self-Driving Cars," *Business Insider,* September 25, 2016; "Driverless Trucks: The Kings of the Autonomous Road," *WSP Parsons Brinckerhoff, Driverless News,* October 14, 2016; "Otto Cofounder Says AI Will Be Widely Available in Trucks within 10 Years," *MIT Technology Review,* March 27, 2017.

25. Charu Chandra and Sameer Kumar, "Taxonomy of Inventory Policies for Supply-Chain Effectiveness," *International Journal of Retail & Distribution Management* 4 (2001); Matthew B. Myers, Patricia J. Daugherty, and Chad W. Autry, "The Effectiveness of Automatic Inventory Replenishment in Supply Chain Operations: Antecedents and Outcomes," *Journal of Retailing,* Winter 2000.

26. "The Supersizing of Warehouses," *New York Times,* February 4, 2004; "New Warehouses Take On a Luxe Look," *Wall Street Journal,* June 18, 2003; Chad W. Autry, Stanley E. Griffis, Thomas J. Goldsby, and L. Michelle Bobbitt, "Warehouse Management-Systems: Resource Commitment, Capabilities, and Organizational Performance," *Journal of Business Logistics,* 2005.

27. "Veggie Tales," *Fortune,* June 8, 2009; "Sysco Hustles to Keep Restaurants Cooking," *BusinessWeek,* May 18, 2009; "Sysco Website Helps Its Customers Build Business," *Nation's Restaurant News,* May 18, 2009; "Sysco Corporation: Company Profile," *Datamonitor,* June 10, 2008.

28. "Frito-Lay," Short Cases from the CSCMP Toolbox, 2005.

29. "Whirlpool Cleans Up Its Delivery Act," *Wall Street Journal,* September 24, 2009; "Walmart's Need for Speed," *Wall Street Journal,* September 26, 2005.

30. "The Role of Logistics in Disaster Management, *NPR All Things Considered,* September 14, 2018; For more on Haiti, see "Sharing in the USA: Social Websites, 'Apps' among Latest Conduits," *USA Today (Special Report),* April 13, 2010; "Aid Frustration: We're Racing against the Clock," *USA Today,* January 18, 2010; "Haitian Rescue Stymied amid Chaos," *Wall Street Journal,* January 15, 2010; "Quake Severely Strains Telecom Services," *USA Today,* January 15, 2010; "Companies Efficiently Pour Out $16M to Help Haiti," *USA Today,* January 15, 2010. See also Ozlem Ergun, Pinar Keskiocak, and Julie Swann, "Humanitarian Relief Logistics," *OR/MS Today,* December 2007. For more on Katrina, see "Good Logistics Offer Better Relief," *Financial Times,* December 16, 2005; "Disasters Demand Supply Chain Software, Research Shows," *eWeek,* October 5, 2005; "The Only Lifeline Was the Walmart," *Fortune,* October 3, 2005; "For FedEx, It Was Time to Deliver," *Fortune,* October 3, 2005; "They Don't Teach This in BSchool," *BusinessWeek,* September 19, 2005, pp. 46–48: "At Walmart, Emergency Plan Has Big Payoff," *Wall Street Journal,* September 12, 2005; "Walmart Praised for Hurricane Katrina Response Efforts," NewsMax.com, September 6, 2005; *Logistics and the Effective Delivery of Humanitarian Relief* (The Fritz Institute, 2005); "Emergency Relief Logistics—A Faster Way across the Global Divide," *Logistics Quarterly,* Summer 2001; Anisya Thomas and Lynn Fritz, "Disaster Relief, Inc.," *Harvard Business Review,* November 2006.

CHAPTER 12

1. See homedepot.com; "Lowe's Posts Weaker Q3 Same Store Sales, Plans Mexico Exit to Focus on U.S.," *TheStreet,* November 20, 2018; "Home Depot Comparable-Store Sales Top Estimates," Reuters, November 13, 2018; "7 Ways AI Is Changing How You Shop, Eat, and Live," *Fortune,* October 22, 2018; "Home Depot Invests in Crowdsourcing Delivery App," *Marketplace.org,* February 26, 2019; "Grocery Shopping 2.0: You'll Soon Be Able to Scan for Food Allergies with Tap of an iPhone," *USA Today,* February 18, 2019; "The Ten Most Influential Cause Marketing Campaigns—Engage for Good," *Engage for Good,* https://engageforgood.com/the-ten-most-influential-cause-marketing-campaigns/, accessed April 12, 2019; "Tiffany's $250 Million Bet on a 78-Year-Old Store," *Wall Street Journal,* September 1, 2018; "Inside Walmart's Journey to Cashierless Retail," *Digiday,* July 9, 2018; "Home Depot Pledges Millions to Train Veterans," *USA Today,* March 8, 2018; "Home Depot Is Succeeding by Nailing Down Its Digital Growth Plan," *TheStreet,* April 12, 2019; "Bernie Marcus & Arthur Blank," *Entrepreneur,* October 10, 2008; "Home Depot's Total Rehab," *Fortune,* September 29, 2008; *Crain's,* June 21, 2014; "Home Depot—Constructing Fulfilment Options for a DIY Digital Retail World," *Diginomica,* March 13, 2017; "Home Depot—Building Interconnected Retail for Digital DIY," *Diginomica,* March 13, 2017; "Home Depot's Greatest Year in Company History," *RIS News,* March 21, 2017; "Home Depot's Resurrection: How One Retailer Made Its Own Home Improvements," *Forbes,* August 21, 2013; "Not All Big Box Stores Are Dead: Wal-Mart, Home Depot Buck Shopping Slump," *Wall Street Journal,* February 22, 2017; "Home Depot Finds Shelter from Digital Storm Battering Retailers," *Wall Street Journal,* October 3, 2016; "Opportunity Meets Preparation: How Home Depot Became 'Amazon-Proof,'" *Forbes,* October 13, 2016; "Home Depot Hammers Away at Online Growth," *Internet Retailer,* January 20, 2016; "Home Depot Is Wildly Successful Probably Because It's Constantly Paranoid about One Thing," *TheStreet,* March 8, 2017; "Here's How Home Depot's E-Commerce Strategy Is Driving Growth," *Forbes,* February 15, 2017; "Will Home Improvement Sales Improve?," *Investopedia,* February 10, 2017; "Seeking E-commerce Shoots among Retail's Brick-and-Mortar Decline," *Investor's Business Daily,* March 10, 2017; "The Inside Story of Home Depot's Retreat from China," *Atlanta Business Chronicle,* October 19, 2012; "Home Depot's Failure in China: Ignoring Women," *Atlanta Business Chronicle,* April 26, 2013; Chris Roush, *Inside Home Depot* (New York: McGraw-Hill, 1999); hdsupply.com; acehardware.com; "How Ace Hardware Profits in a Home Depot World," www.chicagobusiness.com/article/20140621/ISSUE01/306219989/how-ace-hardware-profits-in-a-home-depot-world.

2. U.S. Census data. See also Yuying Shi, Jeremy M. Lim, Barton A. Weitz, and Stephen L. France, "The Impact of Retail Format Diversification on Retailers' Financial Performance," *Journal of the Academy of Marketing Science,* January 2018.

3. "The Top 10 Franchises of 2015," *Entrepreneur,* January 2015; "Burger King Franchisees Can't Have It Their Way," *Inc.,* January 21, 2010; "Do You Have What It Takes?," *Wall Street Journal Reports,* September 19, 2005; "How to Grow a Chain That's Already Everywhere," *Business 2.0,* March 2005; "When Franchisees Lose Part of Their Independence," *New York Times,* February 3, 2005. See also Moeen Naseer Butt, Kersi D. Antia, Brian R. Murtha, and Vishal Kashyap, "Clustering, Knowledge Sharing, and Intrabrand Competition: A Multiyear Analysis of an Evolving Franchise System," *Journal of Marketing,* January 2018; Surinder Tikoo, "Franchiser Influence Strategy Use and Franchisee Experience and Dependence," *Journal of Retailing,* Fall 2002; Patrick J. Kaufmann and Rajiv P. Dant, "The Pricing of Franchise Rights," *Journal of Retailing,* Winter 2001; Madhav Pappu and David Strutton, "Toward an Understanding of Strategic Inter-Organizational Relationships in Franchise Channels," *Journal of Marketing Channels* 1, no. 2 (2001); Rajiv P. Dant and Patrick J. Kaufmann, "Structural and Strategic Dynamics in Franchising," *Journal of Retailing* 2 (2003); A. K. Paswan and C. M. Wittmann, "Knowledge Management and Franchise Systems," *Industrial Marketing Management* 2 (2009).

4. Data on failure rate found at Small Business Development Center, Bradley University; Statistic Brain Research Institute, www.statisticbrain.com/startup-failure-by-industry/; "5 Ways Small Retailers Can Compete (& Win)," *Inc.,* January 8, 2013.

5. For academic research on this topic, see Dhruv Grewal, Carl-Philip Ahlbom, Lauren Beitelspacher, Stephanie M. Noble, et al., "In-Store Mobile Phone Use and Customer Shopping Behavior: Evidence from the Field," *Journal of Marketing,* July 2018; Xiaoling Zhang, Shibo Li, and Raymond R. Burke, "Modeling the Effects of Dynamic Group Influence on Shopper Zone Choice, Purchase Conversion, and Spending," *Journal of the Academy of Marketing Science,* November 2018; Yuchi Zhang, Michael Trusov, Andrew T. Stephen, and Zainab Jamal, "Online Shopping and Social Media: Friends or Foes?," *Journal of Marketing,* November 2017; Preyas S. Desai, Oded Koenigsberg, and Devavrat Purohit, "Forward Buying by Retailers," *Journal of Marketing Research,* February 2010; Kristin Diehl and Cait Poynor, "Great Expectations?! Assortment Size, Expectations, and Satisfaction," *Journal of Marketing Research,* April 2010; Kinshuk Jerath and Z. John Zhang, "Store within a Store," *Journal of Marketing Research,* August 2010; Hai Che, Xinlei (Jack) Chen, and Yuxin Chen, "Investigating Effects of Out-of-Stock on Consumer Stockkeeping Unit Choice," *Journal of Marketing Research,* August 2012; Simon J. Bell, Bülent Mengüç, and Robert E. Widing II, "Salesperson Learning, Organizational Learning, and Retail Store Performance," *Journal of the Academy of Marketing Science,* March 2010; Douglas E. Hughes and Michael Ahearne, "Energizing the Reseller's Sales Force: The Power of Brand Identification," *Journal of Marketing,* July 2010; Velitchka D. Kaltcheva and Barton A. Weitz, "When Should a Retailer Create an Exciting Store Environment?," *Journal of Marketing,* January 2006; Kathleen Seiders, Glenn B. Voss, Dhruv Grewal, and Andrea L. Godfrey, "Do Satisfied Customers Buy More? Examining Moderating Influences in a Retailing Context," *Journal of Marketing,* October 2005; Ruth N. Bolton and Venkatesh Shankar, "An Empirically Derived Taxonomy of Retailer Pricing and Promotion Strategies," *Journal of Retailing,* Winter 2003; Katherine B. Hartman and Rosann L. Spiro, "Recapturing Store Image in Customer-Based Store Equity: A Construct Conceptualization," *Journal of Business Research,* August 2005; Mark J. Arnold, Kristy E. Reynolds, Nicole Ponder, and Jason E. Lueg, "Customer Delight in a Retail Context: Investigating Delightful and Terrible Shopping Experiences," *Journal of Business Research,* August 2005.

6. See, for example, Dhruv Grewal, Carl-Philip Ahlbom, Lauren Beitelspacher, Stephanie M. Noble, et al., "In-Store Mobile Phone Use and Customer Shopping Behavior: Evidence from the Field," *Journal of Marketing,* July 2018; Yan Dong, Kefeng Xu, Tony Haitao Cui, and Yuliang Yao, "Service Failure Recovery and Prevention: Managing Stockouts in Distribution Channels," *Marketing Science,* October 2015; E. Brocato, Julie Baker, and Clay Voorhees, "Creating Consumer Attachment to Retail Service Firms through Sense of Place," *Journal of the Academy of Marketing Science,* March 2015; Tracey S. Dagger and Peter J. Danaher, "Comparing the Effect of Store Remodeling on New and Existing Customers," *Journal of Marketing,* May 2014; Shai Danziger, Liat Hadar, and Vicki G. Morwitz, "Retailer Pricing Strategy and Consumer Choice under Price Uncertainty," *Journal of Consumer Research,* October 2014.

7. "Company Profile: Foot Locker, Inc.," *MarketLine,* March 2016; footlocker.com; "An Investor's Guide to Foot Locker," *Market Realist,* June 21, 2016; "Foot Locker: The Brand That Spells Trouble," *The Guardian,* February 17, 2012.

8. "Zappos.com: Developing a Supply Chain to Deliver WOW!," Stanford Graduate School of Business Case #GS 65, February 13, 2009; "Tireless Employees Get Their Tribute, Even If It's in Felt and Polyester," *New York Times,* March 4, 2010; "Why Everybody Loves Zappos," *Inc.,* May 2009; "Zappos CEO: How to Build a Brand without Spending Big on Ads," *Brandweek,* December 22, 2008; "Zappos Touts More Than Just Shoes," *Adweek,* February 23, 2009. For more on retailer marketing strategies, see E. Brocato, Julie Baker, and Clay Voorhees, "Creating Consumer Attachment to Retail Service Firms through Sense of Place," *Journal of the Academy of Marketing Science* 43, no. 2 (2015): 200–20; Iana A. Castro, Andrea C. Morales, and Stephen M. Nowlis, "The Influence of Disorganized Shelf Displays and Limited Product Quantity on Consumer Purchase," *Journal of Marketing* 77, no. 4 (2013): 118–33; Tracey S. Dagger and Peter J. Danaher, "Comparing the Effect of Store Remodeling on New and Existing Customers," *Journal of Marketing* 78, no. 3 (2014): 62–80; Shai Danziger, Liat Hadar, and Vicki G. Morwitz, "Retailer Pricing Strategy and Consumer Choice under Price Uncertainty," *Journal of Consumer Research* 41, no. 3 (2014): 761–74; Ryan Hamilton and Alexander Chernev, "Low Prices Are Just the Beginning: Price Image in Retail Management," *Journal of Marketing* 77, no. 6 (2013): 1–20; Arjen Van Lin and Els Gijsbrechts, "Shopper Loyalty to Whom? Chain versus Outlet Loyalty in the Context of Store Acquisitions," *Journal of Marketing Research* 51, no. 3 (2014): 352–70; Mark Vroegrijk, Els Gijsbrechts, and Katia Campo, "Close Encounter with the Hard Discounter: A Multiple-Store Shopping Perspective on the Impact of Local Hard-Discounter Entry," *Journal of Marketing Research* 50, no. 5 (2013): 606–26; Morgan K. Ward and Darren W. Dahl, "Should the Devil Sell Prada? Retail Rejection Increases Aspiring Consumers' Desire for the Brand," *Journal of Consumer Research* 41, no. 3 (2014): 590–609.

9. "The Future of Retail in Asia," *Trendwatching,* 2018; "Can Zero-Waste Grocery Stores Make a Difference?," *Wall Street Journal,* March 8, 2019; "Amazon's Rise Forces Laundry Detergents to Sink," *CBS News,* December 27, 2018.

10. "What Price Virtue? At Some Retailers, 'Fair Trade' Carries a Very High Cost," *Wall Street Journal,* June 8, 2004.

11. "Small Toy Shops Play Up the Perks," *Wall Street Journal,* December 3, 2014; "Malls Race to Stay Relevant in Downturn," *Wall Street Journal,* February 26, 2009; Sharon E. Beatty, Morris Mayer, James E. Coleman, Kristy E. Reynolds, et al., "Customer-Sales Associate Retail Relationships," *Journal of Retailing,* Fall 1996.

12. U.S. Census Bureau, *2007 Economic Census—Retail Trade, Subject Series, Establishment and Firm Size,* www.census.gov. For more on department stores, see "Big Retailers Seek Teens (and Parents)," *USA Today,* April 14, 2008; "The Department Store Rises Again," *Business 2.0,* August 2004.

13. Michael Ruhlman, *Grocery: The Buying and Selling of Food in America* (New York: Abrams Press, 2017).

14. See www.census.gov and www.fmi.org/facts_figs/; "Do-It-Yourself Supermarket Checkout," *Wall Street Journal,* April 5, 2007. See also R. A. Briesch, P. K. Chintagunta, and E. J. Fox, "How Does Assortment Affect Grocery Store Choice?," *Journal of Marketing Research* 2 (2009).

15. "Change the World: 99 Cents Only Stores," *Fortune,* September 1, 2018.

16. For more on supercenters, see Kusum L. Ailawadi, Jie Zhang, Aradhna Krishna, and Michael W. Kruger, "When Wal-Mart Enters: How Incumbent Retailers React and How This Affects Their Sales Outcomes," *Journal of Marketing Research,* August 2010; "Retailers Cut Back on Variety, Once the Spice of Marketing," *Wall Street Journal,* June 26, 2009; "Walmart, the Category King," *DSN Retailing Today,* June 9, 2003; "Price War in Aisle 3," *Wall Street Journal,* May 27, 2003.

17. For more on warehouse clubs, see "Club Stores Accepting Coupons," *Wall Street Journal,* August 20, 2009; "Sam's Club Tests the Big-Box Bodega," *Wall Street Journal,* August 10, 2009; "Thinking Outside the Big Box: Costco," *Fast Company,* November 2008; "Costco's Artful Discounts," *BusinessWeek,* October 20, 2008; "Costco's Well-Off Customers Still Shopping," *Investor's Business Daily,* April 8, 2008; "Why Costco Is So Damn Addictive," *Fortune,* October 30, 2006.

18. For more on category killers, see "Newcomers Challenge Office-Supply Stalwarts," *Wall Street Journal,* April 29, 2009; "Big Boxes Aim to Speed Up Shopping," *Wall Street Journal,* June 27, 2007.

19. See also "How to Calculate Inventory Turnover and Why You Should Care," *The Balance,* September 16, 2016.

20. All stockturn (reported as inventory turn) rates found at *CSI-Market* on April 12, 2018. Each measure was from the most recent quarter reported by the retailer, with quarters ending between November 2018 and February 2019.

21. For more on 7-Eleven, see "Know What Customers Want," *Inc.,* August 2005; "Can 7-Eleven Win over Hong Kong Foodies?," *Time,* October 1, 2009; "Convenience Stores Score in Japan," *Wall Street Journal,* August 19, 2008.

22. For more on vending and wireless, see "Vending Machines Get Smart to Accommodate the Cashless," *Bloomberg Businessweek,* August 29, 2013; "Soda? iPod? Vending Machines Diversity," *USA Today,* September 4, 2007; "Self-Serve Movie Rental Kiosks a Surprise Hit with Consumers," *Investor's Business Daily,* May 31, 2007.

23. "In Digital Era, Marketers Still Prefer a Paper Trail," *Wall Street Journal,* October 16, 2009; "Cutting the Stack of Catalogs," *BusinessWeek,* December 31, 2007; "This Isn't the Holiday Catalog You Remember," *Advertising Age,* October 29, 2007.

24. Numbers do not count automobile and fuel sales. "Quarterly Retail E-Commerce Sales, 4th Quarter 2018," *U.S. Census Bureau News,* March 13, 2019; "We're Starting to Shop Online as Often as We Take Out the Trash," *Washington Post,* July 13, 2017.

25. "E-commerce Share of Total Retail Revenue in the United States as of February 2019, by Product Category," *Statista.com,* accessed April 12, 2019.

26. "The Future of Shopping," *Harvard Business Review,* December 2011.

27. Elke Huyghe, Julie Verstraeten, Maggie Geuens, and Anneleen Van Kerckhove, "Clicks as a Healthy Alternative to Bricks: How Online Grocery Shopping Reduces Vice Purchases," *Journal of Marketing Research,* February 2017; Varsha Verma, Dheeraj Sharma, and Jagdish Sheth, "Does Relationship Marketing Matter in Online Retailing? A Meta-Analytic Approach," *Journal of the Academy of Marketing Science,* March 2016; Girish Mallapragada, Sandeep R. Chandukala, and Qing Liu, "Exploring the Effects of 'What' (Product) and 'Where' (Website) Characteristics on Online Shopping Behavior," *Journal of Marketing,* March 2016; Rajiv Lal, José B. Alvarez, and Dan Greenberg, "The Scale of the Ecommerce Threat," *HBS Working Knowledge,* March 2, 2015.

28. "Same-Day Delivery's Second Act," *Inc.,* March 2014; "Using Selfies to Lower Online Clothing Returns," *Bloomberg Businessweek,* September 4, 2014.

29. "12 Examples of Brands with Brilliant Omni-Channel Experiences," *HubSpot Blog,* March 21, 2019.

30. "Wayfair Is Latest Online Seller to Go Bricks and Mortar," *Wall Street Journal,* March 26, 2019; "Inside Amazon's Battle to Break Into the $800 Billion Grocery Market," *Bloomberg Businessweek,* March 20, 2017; "Amazon Heads Deeper into Brick and Mortar with Books," *USA Today,* January 18, 2017; "Warby Parker to Open 25 Stores This Year, Co-CEO Says," *Wall Street Journal,* January 23, 2017.

31. "The Future of Retail in Asia," *Trendwatching,* May 2018.

32. Chip E. Miller, "The Effects of Competition on Retail Structure: An Examination of Intratype, Intertype, and Intercategory Competition," *Journal of Marketing,* October 1999; "Aspirin, Q-Tips and a New You: Drugstores Try to Sell Glamour, Offering Cosmetics Boutiques, Facials, 'Eyebrow Bars,'" *Wall Street Journal,* March 25, 2010; "Root Canal? Try Aisle Five," *BusinessWeek,* October 13, 2008; "Doing Whatever Gets Them in the Door," *BusinessWeek,* June 30, 2008; "New Options for Saving at the Pump," *Wall Street Journal,* August 16, 2005; Ronald Savitt, "The 'Wheel of Retailing' and Retail Product Management," *European Journal of Marketing* 18, no. 6/7 (1984).

33. For this and other examples, see "Tracking Technology Sheds Light on Shopper Habits," *Wall Street Journal,* December 9, 2013; "Personalization: Show Me You Know Me," *Path to Purchase Institute,* July 22, 2013; "Designer Rebecca Minkoff's New Stores Have Touch Screens for an Online Shopping Experience," *Wall Street Journal,* November 11, 2014; "Retailers Wage War against Long Lines," *Wall Street Journal,* May 1, 2013; "The End of the Coffee Line," *Bloomberg Businessweek,* December 1, 2014; "Big Brother Is Watching You Shop," *Bloomberg Businessweek,* December 15, 2011; "Check Out the Future of Shopping," *Wall Street Journal,* May 18, 2011. For more general information on the effects of technology on retail, see "Retailers Reach Out on Cellphones," *Wall Street Journal,* April 21, 2010; "The Hard Sell," *BusinessWeek,* October 26, 2009; "Ring Up E-commerce Gains with a True Multichannel Strategy," *Advertising Age,* March 10, 2008; "Retailing: What's Working Online," *McKinsey Quarterly* 3 (2005); Jill Avery, Thomas J. Steenburgh, John Deighton, and Mary Caravella, "Adding Bricks to Clicks: Predicting the Patterns of Cross-Channel Elasticities over Time," *Journal of Marketing,* May 2012; Xiaoqing Jing and Michael Lewis, "Stockouts in Online Retailing," *Journal of Marketing Research,* April 2011; J. Ganesh, K. E. Reynolds, M. Luckett, and N. Pomirleanu, "Online Shopper Motivations, and E-Store Attributes: An Examination of Online Patronage Behavior and Shopper Typologies," *Journal of Retailing* 1 (2010); P. Gupta, M. S. Yadav, and R. Varadarajan, "How Task-Facilitative Interactive Tools Foster Buyers' Trust in Online Retailers: A Process View of Trust Development in the Electronic Marketplace," *Journal of Retailing* 2 (2009); W. S. Kwon and S. J. Lennon, "Reciprocal Effects between Multichannel Retailers' Offline and Online Brand Images," *Journal of Retailing* 3 (2009); Joel E. Collier and Carol C. Bienstock, "How Do Customers Judge Quality in an E-tailer?," *Sloan Management Review,* Fall 2006; A. E. Schlosser, T. B. White, and S. M. Lloyd, "Converting Website Visitors into Buyers: How Website Investment Increases Consumer Trusting Beliefs and Online Purchase Intentions," *Journal of Marketing,* April 2006; Yakov Bart, Venkatesh Shankar, Fareena Sultan, and Glen L. Urban, "Are the Drivers and Role of Online Trust the Same for All Web Sites and Consumers? A Large-Scale Exploratory Empirical Study," *Journal of Marketing,* October 2005; S. A. Neslin and V. Shankar, "Key Issues in Multichannel Customer Management: Current Knowledge and Future Directions," *Journal of Interactive Marketing* 1 (2009).

34. "How A.I. Is Making Supermarkets Less Exhausting," *Wall Street Journal,* February 28, 2019.

35. Eric T. Bradlow, Manish Gangwar, Praveen Kopalle, and Sudhir Voleti, "The Role of Big Data and Predictive Analytics in Retailing," *Journal of Retailing* 1 (2017): 79–95; Joseph Turow, *The Aisles Have Eyes: How Retailers Track Your Shopping, Strip Your Privacy, and Define Your Power* (New Haven, CT: Yale University Press, 2017); "Incessant Consumer Surveillance Is Leaking into Physical Stores," *The Atlantic,* October 20, 2016; "'Aisles Have Eyes' Warns That Brick-and-Mortar Stores Are Watching You," *NPR Fresh Air,* February 13, 2017; "Amazon Wants to Ship Your Package Before You Buy It," *Wall Street Journal,* January 17, 2014.

36. "McDonald's Bites on Big Data with $300 Million Acquisition," *Wired,* March 25, 2019.

37. "The Stores That Track Your Returns," *Wall Street Journal*, April 4, 2018; "Most Consumers Reject In-Store Mobile Tracking—Or Do They?," *Marketing Land*, July 23, 2014. See also John Paul Fraedrich, "The Ethical Behavior of Retail Managers," *Journal of Business Ethics*, March 1993; Joseph Turow, *The Aisles Have Eyes: How Retailers Track Your Shopping, Strip Your Privacy, and Define Your Power* (New Haven, CT: Yale University Press, 2017).

38. "A Dollar Store's Rich Allure in India," *Wall Street Journal*, January 23, 2007; "Retailers Still Expanding in China: Walmart, Carrefour, Tesco," *Wall Street Journal*, January 22, 2009; "Walmart: Looking Overseas for Growth," *CNNMoney.com*, October 24, 2007; "Walmart: Struggling in Germany," *BusinessWeek*, April 11, 2005; "Japan Isn't Buying the Walmart Idea," *BusinessWeek*, February 28, 2005. See also Vishal Narayan, Vithala R. Rao, and K. Sudhir, "Early Adoption of Modern Grocery Retail in an Emerging Market: Evidence from India," *Marketing Science*, December 2015; Cindy B. Rippé, Suri Weisfeld-Spolter, Alan J. Dubinsky, Aaron D. Arndt, et al., "Selling in an Asymmetric Retail World: Perspectives from India, Russia, and the US on Buyer–Seller Information Differential, Perceived Adaptive Selling, and Purchase Intention," *Journal of Personal Selling & Sales Management*, December 2016.

39. "Understanding Online Shoppers in Europe," *McKinsey Quarterly*, May 2009; "Forrester Forecast: Online Retail Sales Will Grow to $250 Billion by 2014," *TechCrunch*, March 8, 2010; "Forester Forecast: Double-Digit Growth for Online Retail in the US and Western Europe," *Tekrati*, March 9, 2010.

40. See www.friedas.com; "How Lower-Tech Gear Beat Web 'Exchanges' at Their Own Game," *Wall Street Journal*, March 16, 2001; "Family Firms Confront Calamities of Transfer," *USA Today*, August 29, 2000; "Business, Too Close to Home," *Time*, July 17, 2000; "The Kiwi to My Success," *Hemispheres*, July 1999; "Searching for the Next Kiwi: Frieda's Branded Produce," *Brandweek*, May 2, 1994; "Strange Fruits," *Inc.*, November 1989; "The Produce Marketer," *Savvy*, June 1988; "Branded for Life," *Progressive Grocer*, June 1, 2014. See also Sourav Ray, Mark E. Bergen, and George John, "Understanding Value-Added Resellers' Assortments of Multicomponent Systems," *Journal of Marketing*, September 2016.

41. See U.S. Census data, including "Summary Statistics for the U.S.: 2007" from *2007 Economic Census–Wholesale Trade, Geographic Area Series, United States* (Washington, DC: U.S. Government Printing Office, 2010) and available online at www.census.gov.

42. See www.fastenal.com; "The *BusinessWeek* 50: Fastenal," *BusinessWeek*, April 6, 2009; "Fastenal: Need to Fix Up a Building? This Firm's Got Your One-Stop Shop," *Investor's Business Daily*, November 28, 2005; "Fastenal: In This Sluggish Market, You Gotta Have Faith," *Investor's Business Daily*, August 13, 2001.

43. "Revolution in Japanese Retailing," *Fortune*, February 7, 1994; Arieh Goldman, "Evaluating the Performance of the Japanese Distribution System," *Journal of Retailing*, Spring 1992; "Japan Begins to Open the Door to Foreigners, a Little," *Brandweek*, August 2, 1993.

44. See www.rell.com; "Richardson Electronics Ltd.: Maker of Ancient Tech Finds a Way to Prosper," *Investor's Business Daily*, June 27, 2000; *2009 Annual Report*, Richardson Electronics.

45. See www.inmac.com; www.grainger.com.

46. "Henry Schein Give Kids a Smile!" www.henryschein.com/us-en/Corporate/GiveKidsASmile.aspx, accessed April 13, 2019; "Change the World: Henry Schein," *Fortune*, September 1, 2018.

47. For more on manufacturers' agents being squeezed, see "Philips to End Long Relations with North American Reps—Will Build Internal Sales Force Instead," *EBN*, November 10, 2003. See also Daniel C. Bello and Ritu Lohtia, "The Export Channel Design: The Use of Foreign Distributors and Agents," *Journal of the Academy of Marketing Science*, Spring 1995.

CHAPTER 13

1. See www.geico.com; "Company Profiles: GEICO," *MarketLine*, 2019; "The Top Ten Largest Auto Insurance Companies of 2019," *Value Penguin*, website accessed April 15, 2019; "Cost Efficiency Helps Geico Keep Prices Down, Get Big Bang for Its '15-Minute Message,'" *Advertising Age*, September 2, 2013; "Muscling Past Mayhem: GEICO Rides Giant Ad Budget Past Allstate," *Advertising Age*, July 8, 2013; "GEICO, in Numbers," *Guru Focus*, November 25, 2014; "GEICO Advertising Soars to $1 Billion," *Media Post*, June 26, 2012; "Big Blue Makes Big Moves in Big Data," *Forbes*, May 8, 2012; "Customer Service Playing a Bigger Role as a Marketing Tool," *Advertising Age*, November 7, 2011; "GEICO Launches Social Media Platform for Military Personnel," *InsuranceTech*, July 5, 2011; "Consumers: Who's Who?," *Advertising Age*, February 21, 2011; "How the Insurance Industry Got into a $4 Billion Ad Brawl," *Advertising Age*, February 21, 2011; "Umbrella Coverage for Preventing Your Ruin," *New York Times*, March 18, 2008; "Clan of the Caveman," *Fast Company*, December 19, 2007; "Geico," *Advertising Age*, October 15, 2007; "Leapin' Lizards," *Brandweek*, October 8, 2007; "Cavemen (TV Series)," *Wikipedia*, April 21, 2007; "Will the Cavemen's Evolution Boost Geico?," *Fortune*, April 2, 2007; "The Caveman: Evolution of a Character," *Adweek*, March 12, 2007; "Trio of TV Ad Campaigns Succeeds in Wooing Different Target Audience," *Television Week*, March 12, 2007; "How a Gecko Shook Up Insurance Ads," *Wall Street Journal*, January 2, 2007; "When Geckos Aren't Enough," *BusinessWeek*, October 16, 2006; "10 Breakaway Brands," *Fortune*, September 8, 2006; "Loving the Lizard," *Adweek*, October 24, 2005; "Buffett Wants to Know: Do Geico Ads Get Job Done?," *USA Today*, January 24, 2005; "Meet America's 25 Biggest Advertisers," *Advertising Age*, July 8, 2013; "How Big Data Spawned the Geico Gecko," *Advertising Age*, March 18, 2013; "Muscling Past Mayhem: Why Geico's Marketing Works," *Advertising Age*, July 8, 2013; "You Could Spend 217 Minutes or More on GEICO Ads," *Multichannel News*, February 6, 2017; "GEICO, Amtrust, Progressive Auto Insurance Market Share Winners in 2016," *Repairer Driven News*, April 10, 2017; "You'll Actually Want to Watch Geico's New Ad All the Way to the End," *Time.com*, March 4, 2015.

2. Rajeev Batra and Kevin Lane Keller, "Integrating Marketing Communications: New Findings, New Lessons, and New Ideas," *Journal of Marketing*, November 2016; P. J. Kitchen, I. Kim, and D. E. Schultz, "Integrated Marketing Communication: Practice Leads Theory," *Journal of Advertising Research* 4 (2008); Kim Bartel Sheehan and Caitlin Doherty, "Re-Weaving the Web: Integrating Print and Online Communications," *Journal of Interactive Marketing*, Spring 2001.

3. "An Entrepreneur's Story Can Be the Perfect Marketing Tool," *Wall Street Journal*, April 30, 2017; www.slugger.com/our-history; Keith A. Quesenberry and Michael K. Coolsen, "What Makes a Super Bowl Ad Super? Five-Act Dramatic Form Affects Consumer Super Bowl Advertising Ratings," *Journal of Marketing Theory and Practice* 22, no. 4 (2014): 437–54; "How Great Marketers Tell Stories," *Fortune*, April 18, 2014; Seth Godin, *All Marketers Are Liars* (New York: Portfolio Hardcover, 2005); "Ode: How to Tell a Great Story," *Seth's Blog*, April 27, 2006; "The Best Example of Brand Storytelling Ever: The Lego Movie," *The Sales Lion*, accessed April 12, 2017; "List of Films Considered the Worst," *Wikipedia*, accessed April 11, 2017; "How Emotions Influence What We Buy," *Psychology Today*, February 26, 2013; "Science of Storytelling: Why and How to Use It in Your Marketing," *The Guardian*, August 28, 2014; "5 Great B2B Content Marketing Storytellers from 2016," *Skyword*, September 27, 2016.

4. "In a Shift, Marketers Beef Up Ad Spending Inside Stores," *Wall Street Journal*, September 21, 2005.

5. "Lifebuoy Reduces Diarrhea from 36% to 5% in Thesgora," Hindustan Lever Limited press release, March 19, 2014.

6. "Kendall Jenner's Pepsi Ad Sparks Backlash," *CNN.com,* April 5, 2017; "Why Pepsi's Kendall Jenner Ad Bombed in the Age of Twitter," *Fortune,* April 6, 2017. For an academic perspective on this phenomenon, see Barbara B. Stern, "A Revised Communication Model for Advertising: Multiple Dimensions of the Source, the Message, and the Recipient," *Journal of Advertising,* June 1994; George S. Low and Jakki J. Mohr, "Factors Affecting the Use of Information in the Evaluation of Marketing Communications Productivity," *Journal of the Academy of Marketing Science,* Winter 2001; David I. Gilliland and Wesley J. Johnston, "Toward a Model of Business-to-Business Marketing Communications Effects," *Industrial Marketing Management,* January 1997.

7. Yum Brands, Inc.

8. "Don't Let Your Brand Get Lost in Translation," *Brandweek,* February 8, 2010; "Capturing a Piece of the Global Market," *Brandweek,* June 20, 2005; "Global Branding: Same, but Different," *Brandweek,* April 9, 2001; "Hey, #!@*% Amigo, Can You Translate the Word 'Gaffe'?," *Wall Street Journal,* July 8, 1996; "Lost in Translation: How to 'Empower Women' in Chinese," *Wall Street Journal,* September 13, 1994; "13 Slogans That Got Hilarious When They Were Lost in Translation," *Business Insider,* October 17, 2011.

9. "18 of the Best Email Subject Lines You've Ever Read," *HubSpot Marketing blog,* February 13, 2015. See also Navdeep S. Sahni, S. Christian Wheeler, and Pradeep Chintagunta, "Personalization in Email Marketing: The Role of Noninformative Advertising Content," *Marketing Science,* March 2018.

10. "Brands as Publishers: The Changing Rules of SEO," *Search Engine Watch,* July 11, 2012.

11. We cover word-of-mouth in detail in Chapter 16.

12. KPMG International Cooperative; "How an Accounting Firm Convinced Its Employees They Could Change the World," *Harvard Business Review,* October 6, 2015.

13. As an example of cooperation between direct and indirect communication channels, see Berend Wierenga and Han Soethoudt, "Sales Promotions and Channel Coordination," *Journal of the Academy of Marketing Science,* May 2010.

14. G. Martin-Herran, S. P. Sigue, and G. Zaccour, "The Dilemma of Pull- and Push-Price Promotions," *Journal of Retailing* 1 (2010). For more on advertising and sales, see Douglas Hughes, "This Ad's for You: The Indirect Effect of Advertising Perceptions on Salesperson Effort and Performance," *Journal of the Academy of Marketing Science* 41, no. 1 (2013): 1–18.

15. For more on diffusion of innovation, see the classic work of Everett M. Rogers and F. Floyd Shoemaker, *Communication of Innovations: A Cross-Cultural Approach* (New York: Free Press, 1971); see also Geoffrey A. Moore, *Crossing the Chasm: Marketing and Selling High-Tech Products to Mainstream Customers* (New York: HarperBusiness, 2006). See also Hokey Min and William P. Galle, "E-purchasing: Profiles of Adopters and Nonadopters," *Industrial Marketing Management,* April 2003; Yikuan Lee and Gina Colarelli O'Connor, "The Impact of Communication Strategy on Launching New Products: The Moderating Role of Product Innovativeness," *Journal of Product Innovation Management,* January 2003; Subin Im, Barry L. Bayus, and Charlotte H. Mason, "An Empirical Study of Innate Consumer Innovativeness, Personal Characteristics, and New-Product Adoption Behavior," *Journal of the Academy of Marketing Science,* Winter 2003; Robert J. Fisher and Linda L. Price, "An Investigation into the Social Context of Early Adoption Behavior," *Journal of Consumer Research,* December 1992. See also J. Andrew Petersen, Tarun Kushwaha, and V. Kumar, "Marketing Communication Strategies and Consumer Financial Decision Making: The Role of National Culture," *Journal of Marketing* 79, no. 1 (2015): 44–63; Hans Risselada, Peter C. Verhoef, and Tammo H. A. Bijmolt, "Dynamic Effects of Social Influence and Direct Marketing on the Adoption of High-Technology Products," *Journal of Marketing* 78, no. 2 (2014): 52–68.

16. "All-in-One Dinner Kits Gain Popularity," *Food Retailing Today,* May 5, 2003; "Marketers of the Year: This One's a Stove Topper," *Brandweek,* October 14, 2002.

17. Kusum L. Ailawadi, Paul W. Farris, and Mark E. Parry, "Share and Growth Are Not Good Predictors of the Advertising and Promotion/Sales Ratio," *Journal of Marketing,* January 1994.

18. Ashwin Aravindakshan, Kay Peters, and Prasad A Naik, "Spatiotemporal Allocation of Advertising Budgets," *Journal of Marketing Research,* February 2012; Kissan Joseph and Vernon J. Richardson, "Free Cash Flow, Agency Costs, and the Affordability Method of Advertising Budgeting," *Journal of Marketing,* January 2002; Kim P. Corfman and Donald R. Lehmann, "The Prisoner's Dilemma and the Role of Information in Setting Advertising Budgets," *Journal of Advertising,* June 1994; C. L. Hung and Douglas West, "Advertising Budgeting Methods in Canada, the UK and the USA," *International Journal of Advertising* 10, no. 3 (1991); Pierre Filiatrault and Jean-Charles Chebat, "How Service Firms Set Their Marketing Budgets," *Industrial Marketing Management,* February 1990.

CHAPTER 14

1. Based on author interviews with Ferguson salespeople and management. See also www.ferguson.com; *2011 Annual Report and Accounts,* Wolseley; "Customer Focused," *US Business Review,* May 2005; "Customer Loyalty Pays Off," *T1D,* February 2011; "PP-R Piping Shaves Six Figures at Florida Football Stadium," *Contractor,* February 2012; "Wolseley plc," *Datamonitor Report,* February 16, 2012; "Wolseley Faces the Future," *bmj,* December 2011; "Ferguson Revenues Grow 12%," *Down the Pipe,* November 2015; "Design-Build Team Helps Contractors Win More Projects," *Engineering News-Record,* November 21, 2016.

2. For more on selling in international markets, see "Expanding Abroad? Avoid Cultural Gaffes," *Wall Street Journal,* January 19, 2010; "Cultural Training Has Global Appeal," *USA Today,* December 22, 2009; "Where Yellow's a Faux Pas and White Is Death," *Wall Street Journal,* December 6, 2007; "Doing Business Abroad? Simple Faux Pas Can Sink You," *USA Today,* August 24, 2007; "Why the Chinese Hate to Use Voice Mail," *Wall Street Journal,* December 1, 2005; "Mind Your Manners," *Inc.,* September 2005; Paul A. Herbig and Hugh E. Kramer, "Do's and Don'ts of Cross-Cultural Negotiations," *Industrial Marketing Management,* November 1992.

3. Daniel Pink, *To Sell Is Human: The Surprising Truth about Moving Others* (New York: Riverhead Books, 2013).

4. "Shhh!," *Forbes,* November 24, 2003; "Deliver More Value to Earn Loyalty," *Selling,* May 2003; "Hush: Improving NVH through Improved Material," *Automotive Design and Production,* July 2003.

5. For academic research on this topic, see Sascha Alavi, Johannes Habel, Paolo Guenzi, and Jan Wieseke, "The Role of Leadership in Salespeople's Price Negotiation Behavior," *Journal of the Academy of Marketing Science,* July 2018; Raj Agnihotri, Colin B. Gabler, Omar S. Itani, Fernando Jaramillo, et al., "Salesperson Ambidexterity and Customer Satisfaction: Examining the Role of Customer Demandingness, Adaptive Selling, and Role Conflict," *Journal of Personal Selling & Sales Management,* March 2017; Nathaniel N. Hartmann, Heiko Wieland, and Stephen L. Vargo, "Converging on a New Theoretical Foundation for Selling," *Journal of Marketing,* March 2018; Babak Hayati, Yashar Atefi, and Michael Ahearne, "Sales Force Leadership during Strategy Implementation: A Social Network Perspective," *Journal of the Academy of Marketing Science,* July 2018; Stephanie M. Mangus, Dora E. Bock, Eli Jones, and Judith Anne Garretson Folse, "Gratitude in Buyer-Seller Relationships: A Dyadic Investigation," *Journal of Personal Selling & Sales Management,* September 2017; Jeff Johnson and Ravipreet Sohi, "The Curvilinear and Conditional Effects of Product Line Breadth on Salesperson Performance, Role Stress, and Job Satisfaction," *Journal of the Academy of Marketing Science,* January 2014; Lee Allison, Karen E. Flaherty, Jin Ho Jung, and Isaac Washburn, "Salesperson Brand Attachment: A Job Demands-Resources Theory Perspective," *Journal of Personal Selling & Sales Management,* March 2016; Colin B. Gabler, Adam Rapp, and R. Glenn Richey, "The Effect of Environmental Orientation on Salesperson Effort and

Participation: The Moderating Role of Organizational Identification," *Journal of Personal Selling & Sales Management,* February 2014; Bashar S. Gammoh, Michael L. Mallin, and Ellen Bolman Pullins, "Antecedents and Consequences of Salesperson Identification with the Brand and Company," *Journal of Personal Selling & Sales Management* 34, no. 1 (2014); Jeffrey P. Boichuk, Willy Bolander, Zachary R. Hall, Michael Ahearne, et al., "Learned Helplessness among Newly Hired Salespeople and the Influence of Leadership," *Journal of Marketing,* January 2014; William Cron, Artur Baldauf, Thomas Leigh, and Samuel Grossenbacher, "The Strategic Role of the Sales Force: Perceptions of Senior Sales Executives," *Journal of the Academy of Marketing Science* 42, no. 5 (2014): 471-89; V. Kumar, Sarang Sunder, and Robert P. Leone, "Measuring and Managing a Salesperson's Future Value to the Firm," *Journal of Marketing Research* 51, no. 5 (2014): 591-608; C. R. Plouffe, J. Hulland, and T. Wachner, "Customer-Directed Selling Behaviors and Performance: A Comparison of Existing Perspectives," *Journal of the Academy of Marketing Science* 37, no. 4 (2009); Andris A. Zoltners, Prabhakant Sinha, and Sally E. Lorimer, "Match Your Sales Force Structure to Your Business Life Cycle," *Harvard Business Review,* July-August 2006; George R. Franke and Jeong-Eun Park, "Salesperson Adaptive Selling Behavior and Customer Orientation: A Meta-Analysis," *Journal of Marketing Research,* November 2006; Diane Coutu, "Leveraging the Psychology of the Salesperson," *Harvard Business Review,* July-August 2006; David Mayer and Herbert M. Greenberg, "What Makes a Good Salesman," *Harvard Business Review,* July-August 2006; Christian Pfeil, Thorsten Posselt, and Nils Maschke, "Incentives for Sales Agents after the Advent of the Internet," *Marketing Letters,* January 2008; Dawn R. Deeter-Schmelz, Daniel J. Goebel, and Karen Norman, "What Are the Characteristics of an Effective Sales Manager? An Exploratory Study Comparing Salesperson and Sales Manager Perspectives," *Journal of Personal Selling & Sales Management,* Winter 2008; Jerome A. Colletti and Mary S. Fiss, "The Ultimately Accountable Job: Leading Today's Sales Organization," *Harvard Business Review,* July-August 2006; Robert W. Palmatier, Lisa K. Scheer, and Jan-Benedict E. M. Steenkamp, "Customer Loyalty to Whom? Managing the Benefits and Risks of Salesperson-Owned Loyalty," *Journal of Marketing Research,* May 2007; Judy A. Siguaw, Sheryl E. Kimes, and Jule B. Gassenheimer, "B2B Sales Force Productivity: Applications of Revenue Management Strategies to Sales Management," *Industrial Marketing Management,* October 2003; William C. Moncrief, Greg W. Marshall, and Felicia G. Lassk, "A Contemporary Taxonomy of Sales Positions," *Journal of Personal Selling & Sales Management,* Winter 2006; Eli Jones, Steven P. Brown, Andris A. Zoltners, and Barton A. Weitz, "The Changing Environment of Selling and Sales Management," *Journal of Personal Selling & Sales Management,* Spring 2005.

6. For more on frontline sales, see Alex R. Zablah, George R. Franke, Tom J. Brown, and Darrell E. Bartholomew, "How and When Does Customer Orientation Influence Frontline Employee Job Outcomes? A Meta-Analytic Evaluation," *Journal of Marketing,* May 2012; Vishag Badrinarayanan and Debra A. Laverie, "Brand Advocacy and Sales Effort by Retail Salespeople: Antecedents and Influence of Identification with Manufacturers' Brands," *Journal of Personal Selling & Sales Management,* Spring 2011; Susan Cadwallader, Cheryl Burke Jarvis, Mary Jo Bitner, and Amy L. Ostrom, "Frontline Employee Motivation to Participate in Service Innovation Implementation," *Journal of the Academy of Marketing Science,* March 2010.

7. Afdhel Aziz and Bobby Jones, *Good Is the New Cool: Market Like You Give a Damn* (New York: Regan Arts, 2016); Tony Hsieh, *Delivering Happiness: A Path to Profits, Passion, and Purpose* (New York: Grand Central Publishing, 2013).

8. "A Twitterati Calls Out Whirlpool," *Forbes,* September 2, 2009; "The Customer Strikes Back," *Brandweek,* April 26, 2008; "Hello, Houston . . . We Have a Customer Service Problem," *Investor's Business Daily,* March 31, 2008; "Customer Service Champs," *BusinessWeek,* March 3, 2008; "Consumer Vigilantes," *BusinessWeek,* February 21, 2008; "Comcast Takes Its Whacks on Service," *USA Today,* December 3, 2007; "Comcast Must Die," *Advertising Age,* November 19, 2007, p. 1; "One Tough Customer," *Brandweek,* March 19, 2007, pp. 18-24; "Customer Service Champs," *BusinessWeek,* March 5, 2007, pp. 52-64; "Price Points: Good Customer Service Costs Money," *Wall Street Journal,* October 30, 2006, p. R7; "The Customer Vigilante Files," *Church of the Customer Blog,* November 25, 2003.

9. For some recent research on team selling and sales force structure, see Zachary R. Hall, Ryan R. Mullins, Niladri Syam, and Jeffrey P. Boichuk, "Generating and Sharing of Market Intelligence in Sales Teams: An Economic Social Network Perspective," *Journal of Personal Selling & Sales Management,* December 2017; Christine Jaushyuam Lai and Ying Yang, "The Role of Formal Information Sharing in Key Account Team Effectiveness: Does Informal Control Matter and When," *Journal of Personal Selling & Sales Management,* December 2017; Nikolaos G. Panagopoulos, Ryan Mullins, and Panagiotis Avramidis, "Sales Force Downsizing and Firm-Idiosyncratic Risk: The Contingent Role of Investors' Screening and Firm's Signaling Processes," *Journal of Marketing,* November 2018; Park Thaichon, Jiraporn Surachartkumtonkun, Sara Quach, Scott Weaven, et al., "Hybrid Sales Structures in the Age of E-Commerce," *Journal of Personal Selling & Sales Management,* September 2018; Chen Wang, JoAndrea Hoegg, and Darren W. Dahl, "The Impact of a Sales Team's Perceived Entitativity on Customer Satisfaction," *Journal of the Academy of Marketing Science,* March 2018; Bulent Menguc, Seigyoung Auh, and Aypar Uslu, "Customer Knowledge Creation Capability and Performance in Sales Teams," *Journal of the Academy of Marketing Science* 41, no. 1 (2013): 19-39; Christian Schmitz and Shankar Ganesan, "Managing Customer and Organizational Complexity in Sales Organizations," *Journal of Marketing* 78, no. 6 (2014): 59-77; Christian Schmitz, You-Cheong Lee, and Gary L. Lilien, "Cross-Selling Performance in Complex Selling Contexts: An Examination of Supervisory- and Compensation-Based Controls," *Journal of Marketing* 78, no. 3 (2014): 1-19; Christian Schmitz, "Group Influences of Selling Teams on Industrial Salespeople's Cross-Selling Behavior," *Journal of the Academy of Marketing Science* 41, no. 1 (2013): 55-72.

10. "Wal-Mart and P&G: A $10 Billion Marriage under Strain," *Wall Street Journal,* June 14, 2016. For more about major account selling, see Christopher P. Blocker, Joseph P. Cannon, Nikolaos G. Panagopoulos, and Jeffrey K. Sager, "The Role of the Sales Force in Value Creation and Appropriation: New Directions for Research," *Journal of Personal Selling & Sales Management,* Winter 2011; Kevin D. Bradford, Goutam N. Challagalla, Gary K. Hunter, and William C. Moncrief, "Strategic Account Management: Conceptualizing, Integrating, and Extending the Domain from Fluid to Dedicated Accounts," *Journal of Personal Selling & Sales Management,* Winter 2012; Michelle Steward, Beth Walker, Michael Hutt, and Ajith Kumar, "The Coordination Strategies of High-Performing Salespeople: Internal Working Relationships That Drive Success," *Journal of the Academy of Marketing Science,* September 2010; Michael Ahearne, Scott B. MacKenzie, Philip M. Podsakoff, John E. Mathieu, et al., "The Role of Consensus in Sales Team Performance," *Journal of Marketing Research,* June 2010; Michael Ahearne, Son K. Lam, John E. Mathieu, and Willy Bolander, "Why Are Some Salespeople Better at Adapting to Organizational Change?," *Journal of Marketing,* May 2010; George S. Yip and Audrey J. M. Bink, "Managing Global Accounts," *Harvard Business Review,* September 2007; Paolo Guenzi, Catherine Pardo, and Laurent Georges, "Relational Selling Strategy and Key Account Managers' Relational Behaviors: An Exploratory Study," *Industrial Marketing Management,* January 2007; Eli Jones, Andrea L. Dixon, Lawrence B. Chonko, and Joseph P. Cannon, "Key Accounts and Team Selling: A Review, Framework, and Research Agenda," *Journal of Personal Selling & Sales Management,* Spring 2005; Michael G. Harvey, Milorad M. Novicevic, Thomas Hench, and Matthew Myers, "Global Account Management: A Supply-Side Managerial View," *Industrial Marketing Management,* October 2003; Sanjit Sengupta, Robert E. Krapfel, and Michael A. Pusateri, "An Empirical Investigation of Key Account Salesperson Effectiveness," *Journal of Personal Selling & Sales Management,* Fall 2000; Mark A. Moon and Susan F. Gupta, "Examining the Formation of Selling Centers: A Conceptual Framework," *Journal of Personal Selling & Sales Management,* Spring 1997.

11. Brian N. Rutherford, Greg W. Marshall, and JungKun Park, "The Moderating Effects of Gender and Inside versus Outside Sales Role in Multifaceted Job Satisfaction," *Journal of Business Research* 67, no. 9 (2014).

12. "Rose, the Hotel Chatbot," R/GA case study 2017, https://www.rga.com/work/case-studies/rose-the-hotel-chatbot, retrieved April 17, 2019.

13. "Cloud's Fast Rise Even Surprised Salesforce's CEO," *Investor's Business Daily,* January 13, 2010; "Pfizer Adds New Type of Tablet to Sales Calls," *Wall Street Journal,* December 15, 2009; "Videoconferencing Eyes Growth Spurt," *USA Today,* June 23, 2009; "Salesforce Hits Its Stride," *Fortune,* March 2, 2009; "An Early Adopter's New Idea," *Wall Street Journal,* January 22, 2008; Scott M. Widmier, Donald W. Jackson Jr., and Deborah Brown McCabe, "Infusing Technology into Personal Selling," *Journal of Personal Selling & Sales Management,* November 2002.

14. "Matchmaking with Math: How Analytics Beats Intuition to Win Customers," *MIT Management Review,* Winter 2011.

15. For research on these topics, see Michael Giebelhausen, Stacey G. Robinson, Nancy J. Sirianni, and Michael K. Brady, "Touch versus Tech: When Technology Functions as a Barrier or a Benefit to Service Encounters," *Journal of Marketing* 78, no. 4 (2014): 113–24; S. Albers, M. Krafft, and M. Mantrala, "Special Section on Enhancing Sales Force Productivity," *International Journal of Research in Marketing* 1 (2010); L. Ferrell, T. Gonzalez-Padron, and O. C. Ferrell, "An Assessment of the Use of Technology in the Direct Selling Industry," *Journal of Personal Selling & Sales Management* 30, no. 2 (2010); S. Sarin, T. Sego, A. Kohli, and G. Challagalla, "Characteristics That Enhance Training and Effectiveness in Implementing Technological Change in Sales Strategy: A Field-Based Exploratory Study," *Journal of Personal Selling & Sales Management* 30, no. 2 (2010); G. Wright, K. Fletcher, B. Donaldson, and J. H. Lee, "Sales Force Automation Systems: An Analysis of Factors Underpinning the Sophistication of Deployed Systems in the UK Financial Services Industry," *Industrial Marketing Management,* 8 (2008); V. Crittenden, R. Peterson, and G. Albaum, "Technology and Business-to-Consumer Selling: Contemplating Research and Practice," *Journal of Personal Selling & Sales Management* 30, no. 2 (2010); Michael Ahearne and Adam Rapp, "The Role of Technology at the Interface between Salespeople and Consumers," *Journal of Personal Selling & Sales Management* 15, no. 2 (2010); Gary K. Hunter and William D. Perreault, "Making Sales Technology Effective," *Journal of Marketing,* January 2007; Michael Ahearne, Douglas E. Hughes, and Niels Schillewaert, "Why Sales Reps Should Welcome Information Technology: Measuring the Impact of CRM-Based IT on Sales Effectiveness," *International Journal of Research in Marketing,* December 2007; Earl D. Honeycutt Jr., Tanya Thelen, Shawn T. Thelen, and Sharon K. Hodge, "Impediments to Sales Force Automation," *Industrial Marketing Management,* May 2005; Alan J. Bush, Jarvis B. Moore, and Rich Rocco, "Understanding Sales Force Automation Outcomes: A Managerial Perspective," *Industrial Marketing Management,* May 2005; Richard E. Buehrer, Sylvain Senecal, and Ellen Bolman Pullins, "Sales Force Technology Usage—Reasons, Barriers, and Support: An Exploratory Investigation," *Industrial Marketing Management,* May 2005; Leroy Robinson Jr., Greg W. Marshall, and Miriam B. Stamps, "An Empirical Investigation of Technology Acceptance in a Field Sales Force Setting," *Industrial Marketing Management,* May 2005; Devon S. Johnson and Sundar Bharadwaj, "Digitization of Selling Activity and Sales Force Performance: An Empirical Investigation," *Journal of the Academy of Marketing Science,* Winter 2005; Thomas G. Brashear, Danny N. Bellenger, James S. Boles, and Hiram C. Barksdale Jr., "An Exploratory Study of the Relative Effectiveness of Different Types of Sales Force Mentors," *Journal of Personal Selling & Sales Management,* Winter 2006; Gary L. Hunter, "Information Overload: Guidance for Identifying When Information Becomes Detrimental to Sales Force Performance," *Journal of Personal Selling & Sales Management,* Spring 2004; Cheri Speier and Viswanath Venkatesh, "The Hidden Minefields in the Adoption of Sales Force Automation Technologies," *Journal of Marketing,* July 2002.

16. "Selling Salesmanship," *Business 2.0,* December 2002; "The Art of the Sale," *Wall Street Journal,* January 11, 2001.

17. See Yashar Atefi, Michael Ahearne, James G. Maxham III, D. Todd Donavan, et al., "Does Selective Sales Force Training Work?," *Journal of Marketing Research,* October 2018; Jeffrey P. Boichuk, Willy Bolander, Zachary R. Hall, Michael Ahearne, et al., "Learned Helplessness among Newly Hired Salespeople and the Influence of Leadership," *Journal of Marketing* 78, no. 1 (2014): 95–111; Kirby Shannahan, Alan Bush, and Rachelle Shannahan, "Are Your Salespeople Coachable? How Salesperson Coachability, Trait Competitiveness, and Transformational Leadership Enhance Sales Performance," *Journal of the Academy of Marketing Science* 41, no. 1 (2013): 40–54; Thomas L. Powers, Thomas E. DeCarlo, and Gouri Gupte, "An Update on the Status of Sales Management Training," *Journal of Personal Selling & Sales Management,* Fall 2010; Felicia G. Lassk, Thomas N. Ingram, Florian Kraus, and Rita Di Mascio, "The Future of Sales Training: Challenges and Related Research Questions," *Journal of Personal Selling & Sales Management,* Winter 2012; M. Asri Jantan, Earl D. Honeycutt, Shawn T. Thelen, and Ashraf M. Atria, "Managerial Perceptions of Sales Training and Performance," *Industrial Marketing Management,* October 2004; Karen E. Flaherty and James M. Pappas, "Job Selection among Salespeople: A Bounded Rationality Perspective," *Industrial Marketing Management,* May 2004; Anand Krishnamoorthy, Sanjog Misra, and Ashutosh Prasad, "Scheduling Sales Force Training: Theory and Evidence," *International Journal of Research in Marketing,* December 2005; William L. Cron, Greg W. Marshall, Jagdip Singh, Rosann L. Spiro, et al., "Salesperson Selection, Training, and Development: Trends, Implications, and Research Opportunities," *Journal of Personal Selling & Sales Management,* Spring 2005; Brian P. Matthews and Tom Redman, "Recruiting the Wrong Salespeople: Are the Job Ads to Blame?," *Industrial Marketing Management,* October 2001.

18. Maryse Koehl, Juliet F. Poujol, and John F. Tanner, "The Impact of Sales Contests on Customer Listening: An Empirical Study in a Telesales Context," *Journal of Personal Selling & Sales Management,* September 2016; William H. Murphy, Peter A. Dacin, and Neil M. Ford, "Sales Contest Effectiveness: An Examination of Sales Contest Design Preferences of Field Sales Forces," *Journal of the Academy of Marketing Science,* Spring 2004; Steven P. Brown, Kenneth R. Evans, Murali K. Mantrala, and Goutam Challagalla, "Adapting Motivation, Control, and Compensation Research to a New Environment," *Journal of Personal Selling & Sales Management,* Spring 2005; Charles H. Schwepker Jr. and David J. Good, "Marketing Control and Sales Force Customer Orientation," *Journal of Personal Selling & Sales Management,* Summer 2004; Jeong Eun Park and George D. Deitz, "The Effect of Working Relationship Quality on Salesperson Performance and Job Satisfaction: Adaptive Selling Behavior in Korean Automobile Sales Representatives," *Journal of Business Research,* February 2006; J. K. Sager, H. D. Strutton, and D. A. Johnson, "Core Self-Evaluations and Salespeople," *Psychology & Marketing,* February 2006; Joseph O. Rentz, C. David Shepherd, Armen Tashchian, Pratibha A. Dabholkar, et al., "A Measure of Selling Skill: Scale Development and Validation," *Journal of Personal Selling & Sales Management,* Winter 2002.

19. For research on compensation and motivation, see Raghu Bommaraju and Sebastian Hohenberg, "Self-Selected Sales Incentives: Evidence of Their Effectiveness, Persistence, Durability, and Underlying Mechanisms," *Journal of Marketing,* September 2018; Stacey L. Malek, Shikhar Sarin, and Bernard J. Jaworski, "Sales Management Control Systems: Review, Synthesis, and Directions for Future Exploration," *Journal of Personal Selling & Sales Management,* March 2018; Constantine S. Katsikeas, Seigyoung Auh, Stavroula Spyropoulou, and Bulent Menguc, "Unpacking the Relationship between Sales Control and Salesperson Performance: A Regulatory Fit Perspective," *Journal of Marketing,* May 2018; Ashutosh Patil and Niladri Syam, "How Do Specialized Personal Incentives Enhance Sales Performance? The

Benefits of Steady Sales Growth," *Journal of Marketing,* January 2018; Noah Lim and Hua Chen, "When Do Group Incentives for Salespeople Work?," *Journal of Marketing Research,* June 2014; Olivier Rubel and Ashutosh Prasad, "Dynamic Incentives in Sales Force Compensation," *Marketing Science,* August 2016; Erin Gillespie, Stephanie Noble, and Son Lam, "Extrinsic versus Intrinsic Approaches to Managing a Multi-Brand Salesforce: When and How Do They Work?," *Journal of the Academy of Marketing Science,* November 2016; Stephen Kim and Amrit Tiwana, "Chicken or Egg? Sequential Complementarity among Salesforce Control Mechanisms," *Journal of the Academy of Marketing Science,* May 2016; Sunil Kishore, Raghunath Singh Rao, Om Narasimhan, and George John, "Bonuses versus Commissions: A Field Study," *Journal of Marketing Research* 50, no. 3 (2013): 317–33; C. Miao and Kenneth Evans, "The Interactive Effects of Sales Control Systems on Salesperson Performance: A Job Demands-Resources Perspective," *Journal of the Academy of Marketing Science* 41, no. 1 (2013): 73–90; Manfred Krafft, Thomas E. DeCarlo, F. Juliet Poujol, and John F. Tanner, "Compensation and Control Systems: A New Application of Vertical Dyad Linkage Theory," *Journal of Personal Selling & Sales Management,* Winter 2012; Michael Ahearne, Adam Rapp, Douglas E. Hughes, and Rupinder Jindal, "Managing Sales Force Product Perceptions and Control Systems in the Success of New Product Introductions," *Journal of Marketing Research,* August 2010; Jason Garrett and Srinath Gopalakrishna, "Customer Value Impact of Sales Contests," *Journal of the Academy of Marketing Science,* November 2010; Desmond (Ho-Fu) Lo, Mrinal Ghosh, and Francine Lafontaine, "The Incentive and Selection Roles of Sales Force Compensation Contracts," *Journal of Marketing Research,* August 2011; René Y. Darmon and Xavier C. Martin, "A New Conceptual Framework of Salesforce Control Systems," *Journal of Personal Selling & Sales Management,* Summer 2011; Frank Q. Fu, Keith A. Richards, Douglas E. Hughes, and Eli Jones, "Motivating Salespeople to Sell New Products: The Relative Influence of Attitudes, Subjective Norms, and Self-Efficacy," *Journal of Marketing,* November 2010; Sridhar N. Ramaswami and Jagdip Singh, "Antecedents and Consequences of Merit Pay Fairness for Industrial Salespeople," *Journal of Marketing,* October 2003.

20. "New Software's Payoff? Happier Salespeople," *Investor's Business Daily,* May 23, 2000.

21. Sönke Albers, Kalyan Raman, and Nick Lee, "Trends in Optimization Models of Sales Force Management," *Journal of Personal Selling & Sales Management,* December 2015; Jay P. Mulki, Barbara Caemmerer, and Githa S. Heggde, "Leadership Style, Salesperson's Work Effort and Job Performance: The Influence of Power Distance," *Journal of Personal Selling & Sales Management,* 2015; Willy Bolander, Cinthia B. Satornino, Douglas E. Hughes, and Gerald R. Ferris, "Social Networks within Sales Organizations: Their Development and Importance for Salesperson Performance," *Journal of Marketing,* November 2015; Corrine A. Novell, Karen A. Machleit, and Jane Ziegler Sojka, "Are Good Salespeople Born or Made? A New Perspective on an Age-Old Question: Implicit Theories of Selling Ability," *Journal of Personal Selling & Sales Management,* December 2016; Eric G. Harris, John C. Mowen, and Tom J. Brown, "Re-examining Salesperson Goal Orientations: Personal Influencers, Customer Orientation, and Work Satisfaction," *Journal of the Academy of Marketing Science,* Winter 2005; Dominique Rouziès, Erin Anderson, Ajay K. Kohli, Ronald E. Michaels, Barton A. Weitz, and Andris A. Zoltners, "Sales and Marketing Integration: A Proposed Framework," *Journal of Personal Selling & Sales Management,* Spring 2005; Eric Fang, Kenneth R. Evans, and Shaoming Zou, "The Moderating Effect of Goal-Setting Characteristics on the Sales Control Systems Job Performance Relationship," *Journal of Business Research,* September 2005; Kenneth B. Kahn, Richard C. Reizenstein, and Joseph O. Rentz, "Sales-Distribution Interfunctional Climate and Relationship Effectiveness," *Journal of Business Research,* October 2004; Rolph E. Anderson and Wen-yeh Huang, "Empowering Salespeople: Personal, Managerial, and Organizational Perspectives," *Psychology & Marketing,* February 2006; Thomas G. Brashear, James S. Boles, Danny N. Bellenger, and Charles M. Brooks, "An Empirical Test of Trust-Building Processes and Outcomes in Sales Manager–Salesperson Relationships," *Journal of the Academy of Marketing Science,* Spring 2003; Andrea L. Dixon, Rosann L. Spiro, and Lukas P. Forbes, "Attributions and Behavioral Intentions of Inexperienced Salespersons to Failure: An Empirical Investigation," *Journal of the Academy of Marketing Science,* Fall 2003; Douglas N. Behrman and William D. Perreault Jr., "A Role Stress Model of the Performance and Satisfaction of Industrial Salespersons," *Journal of Marketing,* Fall 1984.

22. For academic research on personal selling, see Simon J. Blanchard, Mahima Hada, and Kurt A. Carlson, "Specialist Competitor Referrals: How Salespeople Can Use Competitor Referrals for Nonfocal Products to Increase Focal Product Sales," *Journal of Marketing,* July 2018; Riley Dugan, Bryan Hochstein, Maria Rouziou, and Benjamin Britton, "Gritting Their Teeth to Close the Sale: The Positive Effect of Salesperson Grit on Job Satisfaction and Performance," *Journal of Personal Selling & Sales Management,* March 2019; Bryan Hochstein, Willy Bolander, Ronald Goldsmith, and Christopher R. Plouffe, "Adapting Influence Approaches to Informed Consumers in High-Involvement Purchases: Are Salespeople Really Doomed?," *Journal of the Academy of Marketing Science,* January 2019; Robert Mayberry, James Sanders Boles, and Naveen Donthu, "An Escalation of Commitment Perspective on Allocation-of-Effort Decisions in Professional Selling," *Journal of the Academy of Marketing Science,* September 2018; C. Fred Miao, Kenneth R. Evans, and Pochien Li, "Effects of Top-Performer Rewards on Fellow Salespeople: A Double-Edged Sword," *Journal of Personal Selling & Sales Management,* December 2017; Jessica Ogilvie, Adam Rapp, Daniel G. Bachrach, Ryan Mullins, and Jaron Harvey, "Do Sales and Service Compete? The Impact of Multiple Psychological Climates on Frontline Employee Performance," *Journal of Personal Selling & Sales Management,* March 2017; Nikolaos G. Panagopoulos, Adam A. Rapp, and Jessica L. Ogilvie, "Salesperson Solution Involvement and Sales Performance: The Contingent Role of Supplier Firm and Customer-Supplier Relationship Characteristics," *Journal of Marketing,* July 2017; Essi Pöyry, Petri Parvinen, and Richard G. McFarland, "Generating Leads with Sequential Persuasion: Should Sales Influence Tactics Be Consistent or Complementary?," *Journal of Personal Selling & Sales Management,* June 2017; Sunil Singh, Detelina Marinova, Jagdip Singh, and Kenneth R. Evans, "Customer Query Handling in Sales Interactions," *Journal of the Academy of Marketing Science,* September 2018; Donald P. St. Clair, Gary K. Hunter, Philip A. Cola, and Richard J. Boland, "Systems-Savvy Selling, Interpersonal Identification with Customers, and the Sales Manager's Motivational Paradox: A Constructivist Grounded Theory Approach," *Journal of Personal Selling & Sales Management,* December 2018; Aaron Arndt, Kenneth Evans, Timothy D. Landry, Sarah Mady, and Chatdanai Pongpatipat, "The Impact of Salesperson Credibility-Building Statements on Later Stages of the Sales Encounter," *Journal of Personal Selling & Sales Management* 34, no. 1 (2014); Emily A. Goad and Fernando Jaramillo, "The Good, the Bad and the Effective: A Meta-Analytic Examination of Selling Orientation and Customer Orientation on Sales Performance," *Journal of Personal Selling & Sales Management* 34, no. 4 (2014); Adam Rapp, Daniel G. Bachrach, Nikolaos Panagopoulos, and Jessica Ogilvie, "Salespeople as Knowledge Brokers: A Review and Critique of the Challenger Sales Model," *Journal of Personal Selling & Sales Management* 34, no. 4 (2014); Zachary R. Hall, Michael Ahearne, and Harish Sujan, "The Importance of Starting Right: The Influence of Accurate Intuition on Performance in Salesperson–Customer Interactions," *Journal of Marketing,* May 2015; Ali Faraji-Rad, Bendik M. Samuelsen, and Luk Warlop, "On the Persuasiveness of Similar Others: The Role of Mentalizing and the Feeling of Certainty," *Journal of Consumer Research,* October 2015; Gabriel R. Gonzalez, Danny P. Claro, and Robert W. Palmatier, "Synergistic Effects of Relationship Managers' Social Networks on Sales Performance," *Journal of Marketing* 78, no. 1 (2014): 76–94; Douglas Hughes, Joël Bon, and Adam Rapp, "Gaining and Leveraging Customer-Based Competitive Intelligence: The Pivotal Role of Social Capital and Sales-

person Adaptive Selling Skills," *Journal of the Academy of Marketing Science* 41, no. 1 (2013): 91-110; Ryan R. Mullins, Michael Ahearne, Son K. Lam, Zachary R. Hall, and Jeffrey P. Boichuk, "Know Your Customer: How Salesperson Perceptions of Customer Relationship Quality Form and Influence Account Profitability," *Journal of Marketing* 78, no. 6 (2014): 38-58; "Chief Executives Are Increasingly Chief Salesmen," *Wall Street Journal*, August 6, 1991; Joe F. Alexander, Patrick L. Schul, and Emin Babakus, "Analyzing Interpersonal Communications in Industrial Marketing Negotiations," *Journal of the Academy of Marketing Science*, Spring 1991.

23. "Managing Technology: Selling Software," *Wall Street Journal*, March 19, 2007. For more on lead management, see Gaurav Sabnis, Sharmila C. Chatterjee, Rajdeep Grewal, and Gary L. Lilien, "The Sales Lead Black Hole: On Sales Reps' Follow-up of Marketing Leads," *Journal of Marketing* 77, no. 1 (2013): 52-67.

24. Christian Homburg, Michael Müller, and Martin Klarmann, "When Should the Customer Really Be King? On the Optimum Level of Salesperson Customer Orientation in Sales Encounters," *Journal of Marketing*, March 2011; Gabriel R. Gonzalez, K. Douglas Hoffman, Thomas N. Ingram, and Raymond W. LaForge, "Sales Organization Recovery Management and Relationship Selling: A Conceptual Model and Empirical Test," *Journal of Personal Selling & Sales Management*, Summer 2010; Dennis B. Arnett, Barry A. Macy, and James B. Wilcox, "The Role of Core Selling Teams in Supplier-Buyer Relationships," *Journal of Personal Selling & Sales Management*, Winter 2005; Kirk Smith, Eli Jones, and Edward Blair, "Managing Salesperson Motivation in a Territory Realignment," *Journal of Personal Selling & Sales Management*, Fall 2000; Andris A. Zoltners, "Sales Territory Alignment: An Overlooked Productivity Tool," *Journal of Personal Selling & Sales Management*, Summer 2000; Ken Grant, "The Role of Satisfaction with Territory Design on the Motivation, Attitudes, and Work Outcomes of Salespeople," *Journal of the Academy of Marketing Science*, Spring 2001.

25. "How to Get Your Company Where You Want It," *American Salesman*, December 2003. See also Robert E. Carter, Conor M. Henderson, Inigo Arroniz, and Robert W. Palmatier, "Effect of Salespeople's Acquisition-Retention Trade-off on Performance," *Journal of Personal Selling & Sales Management* 34, no. 2 (2014); Thomas DeCarlo and Son Lam, "Identifying Effective Hunters and Farmers in the Salesforce: A Dispositional-Situational Framework," *Journal of the Academy of Marketing Science*, July 2016.

26. "Novartis' Marketing Doctor," *BusinessWeek*, March 5, 2001.

27. "Consultative Selling—We See Great Examples Everywhere," *Partners in Excellence Blog*, July 24, 2012.

28. For more on sales presentation approaches, see "Three Strategies to Get Customers to Say Yes," *Wall Street Journal*, May 29, 2009; Richard G. McFarland, Goutam N. Challagalla, and Tasadduq A. Shervani, "Influence Tactics for Effective Adaptive Selling," *Journal of Marketing*, October 2006; "The 60-Second Sales Pitch," *Inc.*, October 1994. See also Daniel M. Eveleth and Linda Morris, "Adaptive Selling in a Call Center Environment: A Qualitative Investigation," *Journal of Interactive Marketing*, Winter 2002; Thomas W. Leigh and John O. Summers, "An Initial Evaluation of Industrial Buyers' Impressions of Salespersons' Nonverbal Cues," *Journal of Personal Selling & Sales Management*, Winter 2002; Kalyani Menon and Laurette Dube, "Ensuring Greater Satisfaction by Engineering Salesperson Response to Customer Emotions," *Journal of Retailing*, Fall 2000; Susan K. DelVecchio, James E. Zemanek, Roger P. McIntyre, and Reid P. Claxton, "Buyers' Perceptions of Salesperson Tactical Approaches," *Journal of Personal Selling & Sales Management*, Winter 2002-2003; Alfred M. Pelham, "An Exploratory Model and Initial Test of the Influence of Firm Level Consulting-Oriented Sales Force Programs on Sales Force Performance," *Journal of Personal Selling & Sales Management*, Spring 2002; Thomas E. DeCarlo, "The Effects of Sales Message and Suspicion of Ulterior Motives on Salesperson Evaluation," *Journal of Consumer Psychology*, 15, no. 3 (2005); Amy Sallee and Karen Flaherty, "Enhancing Salesperson Trust: An Examination of Managerial Values, Empowerment, and the Moderating Influence of SBU Strategy," *Journal of Personal Selling & Sales Management*, Fall 2003.

29. John D. Hansen, Donald J. Lund, and Thomas E. DeCarlo, "A Process Model of Buyer Responses to Salesperson Transgressions and Recovery Efforts: The Impact of Salesperson Orientation," *Journal of Personal Selling & Sales Management*, March 2016; Fernando Jaramillo, Belén Bande, and Jose Varela, "Servant Leadership and Ethics: A Dyadic Examination of Supervisor Behaviors and Salesperson Perceptions," *Journal of Personal Selling & Sales Management*, February 2015; J. Hansen and R. Riggle, "Ethical Salesperson Behavior in Sales Relationships," *Journal of Personal Selling & Sales Management* 29, no. 2 (2009); Thomas N. Ingram, Raymond W. LaForge, and Charles H. Schwepker Jr., "Salesperson Ethical Decision Making: The Impact of Sales Leadership and Sales Management Control Strategy," *Journal of Personal Selling & Sales Management*, Fall 2007; Jay Prakash Mulki, Fernando Jaramillo, and William B. Locander, "Effects of Ethical Climate and Supervisory Trust on Salespersons' Job Attitudes and Intentions to Quit," *Journal of Personal Selling & Sales Management*, Winter 2006; Lawrence B. Chonko, Thomas R. Wotruba, and Terry W. Loe, "Direct Selling Ethics at the Top: An Industry Audit and Status Report," *Journal of Personal Selling & Sales Management*, Spring 2002.

CHAPTER 15

1. biz.dominos.com; "The Quest for 10-Minute Pizza Delivery," *Wall Street Journal*, May 28, 2017; "Why Domino's Is Winning the Pizza Wars," *CNNMoney*, March 6, 2018; "Domino's Wants to Open Thousands of Stores," *CNNMoney*, July 19, 2018; "Campaigns for Good: The Ingredients for Cause Marketing Success," *PRWeek*, September 3, 2018; "Tonight's Dinner? In a Cooler-Sized Robot That Knows Where You Live," *Wall Street Journal*, March 11, 2019; "Domino's Atoned for Its Crimes against Pizza and Built a $9 Billion Empire," *Bloomberg Businessweek*, March 15, 2017; "Here's How Domino's Became America's Favorite Pizza Brand," *Business Insider*, March 1, 2017; "Domino's Pizza Franchise Cost & Fees," *Franchise Direct*, retrieved April 17, 2017; *Domino's Pizza 2016 Annual Report*; "Big Data-Driven Decision-Making at Domino's Pizza," *Forbes*, April 6, 2016; "Domino's Claims Victory with New Strategy," *Advertising Age*, May 10, 2010; "Domino's Does Itself a Disservice by Coming Clean about Its Pizza," *Advertising Age*, January 11, 2010; "Opinion: Domino's Delivers," *Adweek*, May 4, 2010; "Papa John's Aims at Rivals with Pan Pizza, Its Biggest New Product in Years," *Advertising Age*, October 10, 2016; "Domino's Pizza: 'We Have a Much Clearer Brand Identity Than Many of Our Fast Food Rivals,'" *Marketing Week*, March 23, 2016; "The Franchise World Finally Gets the Whole 'Big Data' Thing," *Entrepreneur*, January 8, 2016; "Oh Yes We Did. Domino's Pizza," *Pizzaturnaround.com*, retrieved April 17, 2017; "The Many Acts of Domino's Pizza," *QSR Magazine*, August 2010; "It's Simple: Tweet Emoji. Get Pizza Delivered. Win Grand Prix," *Advertising Age*, June 27, 2015; "Domino's Slams Pizza Hut's 'Dirty' Pies in This New Ad," *Business Insider*, October 1, 2012; "This New Car Ad Isn't for a Car—It's for Pizza," *Advertising Age*, February 22, 2016; "Domino's Ends Fast-Pizza Pledge after Big Award to Crash Victim," *New York Times*, December 22, 1993; "What Pizza Experts and Pizza Loyalists Say about Pizza Brands," *Brand Keys*, May 8, 2014.

2. For insights on advertising effectiveness, see "What Do We Know about Advertising Effectiveness?," *Marketing Science Institute*, May 21, 2013; Leonard M. Lodish, Magid Abraham, Stuart Kalmenson, Jeanne Livelsberger, et al., "How T.V. Advertising Works: A Meta-Analysis of 389 Real World Split Cable T.V. Experiments," *Journal of Marketing Research* 32, no. 2 (1995): 125-39; Demetrios Vakratsas and Prasad Naik, "Essentials of Planning Media Schedules," in *Handbook of Advertising*, ed. Gerard J. Tellis and Tim Ambler (Los Angeles: Sage, 2007), pp. 333-48; Peter J. Danaher, André Bonfrer, and Sanjay Dhar, "The Effect of Competitive Advertising on Sales for Packaged Goods," *Journal of Marketing Research* 45, no. 2 (2008): 211-25; Demetrios Vakratsas, Fred M. Feinberg, Frank M. Bass, and Gurumurthy Kalyanaram,

"The Shape of Advertising Response Functions Revisited: A Model of Dynamic Probabilistic Thresholds," *Marketing Science* 23, no. 1 (2004): 109–19; Raj Sethuraman and Gerard J. Tellis, "An Analysis of the Trade-off between Advertising and Price Discounting," *Journal of Marketing Research* 28, no. 2 (1991): 160–74; Demetrios Vakratsas and Tim Ambler, "How Advertising Works: What Do We Really Know?," *Journal of Marketing* 63, no. 1 (1999): 26–43.

3. "Global Advertising Spending from 2010 to 2019 (in Billion U.S. Dollars)," *Statista*; "China Advertising Spend to Grow 7% in 2019," *The Drum*, January 14, 2019; "Digital Ad Spending to Surpass TV Next Year," *eMarketer*, March 8, 2016; "Global Advertising Forecast," *MAGNA*, December 5, 2016.

4. "Global Advertising Forecast," *MAGNA*, December 5, 2016; "Advertising Expenditure in China from 2015 to 2017," *Statista*, retrieved April 17, 2017.

5. For more on advertising to sales ratios, see AdAge Datacenter, "Advertising to Sales Ratios by Industry," adage.com/datacenter/; for some striking data, see "Marketing Spend: Facts and Stats," CMO Council, www.cmocouncil.org/facts_stats.php.

6. Exact data on this industry are elusive, but see U.S. Census Bureau, *Statistical Abstract of the United States 2010* (Washington, DC: U.S. Government Printing Office, 2009), p. 403. For some striking data, see "Marketing Spend: Facts and Stats," CMO Council, www.cmocouncil.org/facts_stats.php.

7. Yan Ruiliang, "Cooperative Advertising, Pricing Strategy and Firm Performance in the E-Marketing Age," *Journal of the Academy of Marketing Science*, July 2012; Steffen Jorgensen, Simon Pierre Sigue, and Georges Zaccour, "Dynamic Cooperative Advertising in a Channel," *Journal of Retailing*, Spring 2000.

8. For more information on the various kinds of advertising, see "So Sue Me: Why Big Brands Are Taking Claims to Court," *Advertising Age*, January 4, 2010; "AT&T Sues Verizon over 'There's a Map for That' Ads," *Engadget.com*, November 3, 2009; "Comparative Disadvantage," *AdweekMedia*, June 15, 2009; "Feature: Cause Marketing," *Advertising Age*, June 13, 2005; "The Selling of Breast Cancer," *Business 2.0*, February 2003. See also Peter J. Danaher, André Bonfrer, and Sanjay Dhar, "The Effect of Competitive Advertising Interference on Sales for Packaged Goods," *Journal of Marketing Research*, May 2008; Debora Viana Thompson and Rebecca W. Hamilton, "The Effects of Information Processing Mode on Consumers' Responses to Comparative Advertising," *Journal of Consumer Research*, March 2006; Robert D. Jewell, H. Rao Unnava, David Glen Mick, and Merrie L. Brucks, "When Competitive Interference Can Be Beneficial," *Journal of Consumer Research*, September 2003; Diana L. Haytko, "Great Advertising Campaigns: Goals and Accomplishments," *Journal of Marketing*, April 1995; Michael J. Barone, Anthony D. Miyazaki, and Kimberly A. Taylor, "The Influence of Cause-Related Marketing on Consumer Choice: Does One Good Turn Deserve Another?," *Journal of the Academy of Marketing Science*, Spring 2000; Minette E. Drumwright, "Company Advertising with a Social Dimension: The Role of Noneconomic Criteria," *Journal of Marketing*, October 1996; Wilfred Amaldoss, and Chuan He, "Product Variety, Informative Advertising, and Price Competition," *Journal of Marketing Research*, February 2010.

9. For more on various media, see Maren Becker, Nico Wiegand, and Werner J. Reinartz, "Does It Pay to Be Real? Understanding Authenticity in TV Advertising," *Journal of Marketing*, January 2019; Tami Kim, Kate Barasz, and Leslie K. John, "Why Am I Seeing This Ad? The Effect of Ad Transparency on Ad Effectiveness," *Journal of Consumer Research*, February 2019; Michaela Draganska, Wesley R. Hartmann, and Gena Stanglein, "Internet versus Television Advertising: A Brand-Building Comparison," *Journal of Marketing Research*, October 2014; Beth L. Fossen and David A. Schweidel, "Television Advertising and Online Word-of-Mouth: An Empirical Investigation of Social TV Activity," *Marketing Science*, February 2017; Jura Liaukonyte, Thales Teixeira, and Kenneth C. Wilbur, "Television Advertising and Online Shopping," *Marketing Science*, June 2015; Yakov Bart, Andrew T. Stephen, and Miklos Sarvary, "Which Products Are Best Suited to Mobile Advertising? A Field Study of Mobile Display Advertising Effects on Consumer Attitudes and Intentions," *Journal of Marketing Research*, June 2014; Michelle Andrews, Xueming Luo, Zheng Fang, and Anindya Ghosh, "Mobile Ad Effectiveness: Hyper-Contextual Targeting with Crowdedness," *Marketing Science*, April 2016; "Comparing the Relative Effectiveness of Advertising Channels: A Case Study of a Multimedia Blitz Campaign," *Journal of Marketing Research* 50 (2013): 517–34; D. W. Baack, R. T. Wilson, and B. D. Till, "Creativity and Memory Effects Recall, Recognition, and an Exploration of Nontraditional Media," *Journal of Advertising* 4 (2008). For more on the video game medium, see V. Cauberghe and P. De Pelsmacker, "Advergames: The Impact of Brand Prominence and Game Repetition on Brand Responses," *Journal of Advertising* 1 (2010); Victoria Mallinckrodt and Dick Mizerski, "The Effects of Playing an Advergame on Young Children's Perceptions, Preferences, and Requests," *Journal of Advertising*, Summer 2007; Mira Lee and Ronald J. Faber, "Effects of Product Placement in On-Line Games on Brand Memory," *Journal of Advertising*, Winter 2007. For more on the outdoor medium, see "Celebrating the Renaissance of Out-of-Home Advertising," *Advertising Age* (special report), June 22, 2009; "Neighbors Hope to Pull Plug on Signs," *USA Today*, September 5, 2007; "In Billboard War, Digital Signs Spark a Truce," *Wall Street Journal*, February 3, 2007. See also Charles R. Taylor, George R. Franke, and Bang Hae-Kyong, "Use and Effectiveness of Billboards," *Journal of Advertising*, Winter 2006; Charles R. Taylor and John C. Taylor, "Regulatory Issues in Outdoor Advertising: A Content Analysis of Billboards," *Journal of Public Policy & Marketing*, Spring 1994. For more on the radio medium, see "The New Radio Revolution: From Satellite to Podcasts . . . ," *BusinessWeek*, March 14, 2005; Daniel M. Haygood, "A Status Report on Podcast Advertising," *Journal of Advertising Research*, December 2007.

10. For research on Super Bowl advertising, see Deepa Chandrasekaran, Raji Srinivasan, and Debika Sihi, "Effects of Offline Ad Content on Online Brand Search: Insights from Super Bowl Advertising," *Journal of the Academy of Marketing Science*, May 2018; Wesley R. Hartmann, and Daniel Klapper, "Super Bowl Ads," *Marketing Science*, January 2018.

11. "TV's Next Wave: Tuning In to You," *Wall Street Journal*, March 6, 2011.

12. "Media Plan of the Year," *Adweek*, July 23, 2012.

13. "Next Year, People Will Spend More Time Online Than They Will Watching TV. That's a First," *Recode*, June 8, 2018. For research more generally on digital and mobile advertising, see Inyoung Chae, Hernán A. Bruno, and Fred M. Feinberg, "Wearout or Weariness? Measuring Potential Negative Consequences of Online Ad Volume and Placement on Website Visits," *Journal of Marketing Research*, February 2019; Peter Pal Zubcsek, Zsolt Katona, and Miklos Sarvary, "Predicting Mobile Advertising Response Using Consumer Colocation Networks," *Journal of Marketing*, July 2017; Courtney Paulson, Lan Luo, and Gareth M. James, "Efficient Large-Scale Internet Media Selection Optimization for Online Display Advertising," *Journal of Marketing Research*, August 2018; Amin Sayedi, Kinshuk Jerath, and Marjan Baghaie, "Exclusive Placement in Online Advertising," *Marketing Science*, November 2018; Amin Sayedi, "Real-Time Bidding in Online Display Advertising," *Marketing Science*, July 2018; Eric M. Schwartz, Eric T. Bradlow, and Peter S. Fader, "Customer Acquisition via Display Advertising Using Multi-Armed Bandit Experiments," *Marketing Science*, July 2017; Data from "The Nine Slides That Matter from Mary Meeker's State of the Internet," *Bloomberg Business*, May 27, 2015. See also Daniel G. Goldstein, Siddharth Suri, R. Preston McAfee, Matthew Ekstrand-Abueg, et al., "The Economic and Cognitive Costs of Annoying Display Advertisements," *Journal of Marketing Research* 51, no. 6 (December 2014): 742–52; Catherine E. Tucker, "Social Networks, Personalized Advertising, and Privacy Controls," *Journal of Marketing Research*, October 2014; Michael Trusov, Liye Ma, and Zainab Jamal, "Crumbs of the Cookie: User Profiling in Customer-Base Analysis and Behavioral Targeting," *Marketing Science*, June 2016; Glen L. Urban, Guilherme (Gui) Liberali, Erin MacDonald, Robert Bordley, et al.,

"Morphing Banner Advertising," *Marketing Science,* February 2014; Cait Lamberton and Andrew T. Stephen, "A Thematic Exploration of Digital, Social Media, and Mobile Marketing: Research Evolution from 2000 to 2015, and an Agenda for Future Inquiry," *Journal of Marketing,* November 2016; Eric (ER) Fang, Xiaoling Li, Minxue Huang, and Robert W. Palmatier, "Direct and Indirect Effects of Buyers and Sellers on Search Advertising Revenues in Business-to-Business Electronic Platforms," *Journal of Marketing Research* 52, no. 3 (2015): 407-22; Ye Hu, Rex Yuxing Du, and Sina Damangir, "Decomposing the Impact of Advertising: Augmenting Sales with Online Search Data,"*Journal of Marketing Research* 51, no. 3 (2014): 300-19; Isaac M. Dinner, Harald J. Van Heerde, and Scott A. Neslin, "Driving Online and Offline Sales: The Cross-Channel Effects of Traditional, Online Display, and Paid Search Advertising," *Journal of Marketing Research* 51, no. 5 (2014): 527-45; Michaela Draganska, Wesley R. Hartmann, and Gena Stanglein, "Internet versus Television Advertising: A Brand-Building Comparison," *Journal of Marketing Research* 51, no. 5 (2014): 578-90; Paul R. Hoban and Randolph E. Bucklin, "Effects of Internet Display Advertising in the Purchase Funnel: Model-Based Insights from a Randomized Field Experiment," *Journal of Marketing Research* 52, no. 3 (2015): 375-93; Anja Lambrecht and Catherine Tucker, "When Does Retargeting Work? Information Specificity in Online Advertising," *Journal of Marketing Research* 50, no. 5 (2013): 561-76; Jan H. Schumann, Florian Von Wangenheim, and Nicole Groene, "Targeted Online Advertising: Using Reciprocity Appeals to Increase Acceptance among Users of Free Web Services," *Journal of Marketing* 78, no. 1 (2014): 59-75; "Video Consumer Mapping Study," The Council for Research Excellence, March 2009; "Google's Latest Bid to Boost Revenue: Interest-Based Ads,"*Investor's Business Daily,* March 12, 2009; "Online Ads: Beyond Counting Clicks," *BusinessWeek,* March 9, 2009. See also Ashish Agarwal, Kartik Hosanagar, and Michael D. Smith, "Location, Location, Location: An Analysis of Profitability of Position in Online Advertising Markets," *Journal of Marketing Research,* December 2011; Avi Goldfarb and Catherine Tucker,"Implications of Online Display Advertising: Targeting and Obtrusiveness,"*Marketing Science* 3 (2011); Tat Y. Chan, Chunhua Wu, and Ying Xie,"Measuring the Lifetime Value of Customers Acquired from Google Search Advertising," *Marketing Science* 5 (2011); Oliver J. Rutz, Michael Trusov, and Randolph E. Bucklin, "Modeling Indirect Effects of Paid Search Advertising: Which Keywords Lead to More Future Visits?" *Marketing Science* 4 (2011); G. M. Fulgoni and M. P. Morn, "Whither the Click? How Online Advertising Works," *Journal of Advertising Research* 2 (2009); B. J. Calder, E. C. Malthouse, and U. Schaedel, "An Experimental Study of the Relationship between Online Engagement and Advertising Effectiveness," *Journal of Interactive Marketing* 4 (2009); J. Q. Chen, D. Liu, and A. B. Whinston, "Auctioning Keywords in Online Search," *Journal of Marketing* 4 (2009); Juran Kim and Sally J. McMillan, "Evaluation of Internet Advertising Research," *Journal of Advertising,* Spring 2008.

14. "Miller's Bakery Doubles Sales during Peak Hours with Enhanced Campaigns," *Think with Google,* July 29, 2013.

15. For more on programmatic advertising, see "10 Things You Need to Know Now about Programmatic Buying," *Advertising Age,* June 1, 2015; "Debunking the Myths of Programmatic Delivery," *Advertising Age,* January 10, 2017.

16. Exhibit 15-6 and related discussion draws on the following sources: Jun Xu, *Managing Digital Enterprise: Ten Essential Topics* (Amsterdam: Atlantis Press, 2014); E. Turban, D. King, J. K. Lee, T. P. Liang, and D. Turban, *Electronic Commerce 2012: A Managerial and Social Networks Perspective* (Upper Saddle River, NJ: Prentice Hall, 2012); J. F. Rayport, "Advertising's New Medium: Human Experience," *Harvard Business Review,* March 2013, pp. 77-84; J. Kirby, "Creative That Cracks the Code," *Harvard Business Review,* March 2013, pp. 86-89; *IAB Internet Advertising Revenue Report,* 2014 Full Year Results, April 2015, an industry survey conducted by PwC and sponsored by the Interactive Advertising Bureau; "Total US Ad Spending to See Largest Increase since 2004," *eMarketer,* July 2, 2014; "Advantages and Disadvantages of Various Advertising Mediums," *PowerHomeBiz.com;* Andrew Stephen, Yakov Bart, and Miklos Sarvary, "Making Mobile Ads That Work," *Harvard Business Review,* December 2013.

17. "Display Advertising Clickthrough Rates—Smart Insights Digital Marketing Advice," *Smart Insights,* March 8, 2017.

18. "Google AdWords Benchmarks for YOUR Industry [DATA]," *WordStream,* March 21, 2017. See also Wilfred Amaldoss, Kinshuk Jerath, and Amin Sayedi, "Keyword Management Costs and 'Broad Match' in Sponsored Search Advertising," *Marketing Science,* April 2016; Hongshuang (Alice)Li, P.K. Kannan, Siva Viswanathan, and Abhishek Pani, "Attribution Strategies and Return on Keyword Investment in Paid Search Advertising," *Marketing Science,* December 2016; Woochoel Shin, "Keyword Search Advertising and Limited Budgets," *Marketing Science,* December 2015; Sridhar Narayanan and Kirthi Kalyanam, "Position Effects in Search Advertising and Their Moderators: A Regression Discontinuity Approach," *Marketing Science,* June 2015.

19. "Amazon's Rise in Ad Searches Dents Google's Dominance," *Wall Street Journal,* April 4, 2019.

20. "Facebook Ad Benchmarks for YOUR Industry [New Data]," *WordStream,* March 28, 2017.

21. "Are You on Your Phone Too Much? The Average Person Spends This Many Hours on It Every Day," *Inc.,* October 30, 2018.

22. "Mobile Advertising Begins to Take Off," *Wall Street Journal,* October 9, 2013; "Top 10 Mobile Advertising Campaigns of Q2," *Mobile Marketer,* June 5, 2012; "Cell Phone Ads That Consumers Love," *Harvard Business School Working Knowledge,* February 28, 2005; Rama Yelkur, Chuck Tomkovick, and Patty Traczyk, "Super Bowl Advertising Effectiveness: Hollywood Finds the Games Golden," *Journal of Advertising Research,* March 2004. For information on apps and shopping, see "Yes, There's an App for That, Too: How Mobile Is Changing Shopping," *Advertising Age,* March 1, 2010; "Brands Get a Boost by Opening Up APIs to Outside Developers," *Advertising Age,* November 30, 2009. See also Yakov Bart, Andrew T. Stephen, and Miklos Sarvary, "Which Products Are Best Suited to Mobile Advertising? A Field Study of Mobile Display Advertising Effects on Consumer Attitudes and Intentions," *Journal of Marketing Research* 51, no. 3 (2014): 270-85.

23. "Using Google's Data to Reach Consumers," *New York Times,* December 22, 2011.

24. "Ads Tied to Web Searches Criticized as Deceptive," *Wall Street Journal,* October 13, 2014.

25. "Your Location Data Is Being Sold—Often without Your Knowledge," *Wall Street Journal,* March 4, 2018; "The Cookies You Can't Crumble," *Bloomberg Businessweek,* August 21, 2014; "Everybody Hates Mobile Advertising," *Business Insider,* July 1, 2014; see also Catherine E. Tucker, "Social Networks, Personalized Advertising, and Privacy Controls," *Journal of Marketing Research* 51, no. 5 (2014): 546-62; Paul M. Connell, Merrie Brucks, and Jesper H. Nielsen, "How Childhood Advertising Exposure Can Create Biased Product Evaluations That Persist into Adulthood," *Journal of Consumer Research* 41, no. 1 (2014): 119-34.

26. "Instagram's Crackdown on Fake Followers Just Might Work," *Wired,* November 20, 2018; "Unilever Demands Influencer Marketing Business Clean Up Its Act," *Wall Street Journal,* June 17, 2018.

27. "Battle Heats Up over Mobile Ad Blocking," *Wall Street Journal,* February 24, 2016.

28. For academic research on this topic, see Nilüfer Z. Aydınoğlu and Luca Cian, "Show Me the Product, Show Me the Model: Effect of Picture Type on Attitudes toward Advertising," *Journal of Consumer Psychology* 24, no. 4 (2014): 506-19; Michael Barone and Robert Jewell, "How Brand Innovativeness Creates Advertising Flexibility," *Journal of the Academy of Marketing Science* 42, no. 3 (2014): 309-21.

29. For more examples of creative campaigns, see "Edgy Advertising in a Tenuous Time," *BusinessWeek,* January 12, 2009; "Hey, No Whopper on the Menu?," *Wall Street Journal,* February 8, 2008; "Did Telling a Whopper Sell the Whopper?," *Advertising Age,* January 14, 2008; "From Admirable to Addlebrained," *USA Today,* December 31,

2007; "Marketers Get More Creative with TV Ads," *USA Today,* November 12, 2007. For some creative and controversial ads, see "Burger Joints Pull Out Oldest Ad Trick in Book: Sex," *USA Today,* December 21, 2009; "Taste Strips Give Ads a New Flavor," *Advertising Age,* June 2, 2008. See also M. R. Brown, R. K. Bhadury, and N. K. L. Pope, "The Impact of Comedic Violence on Viral Advertising Effectiveness," *Journal of Advertising* 39, no. 1 (2010); Josephine L. C. M., Woltman Elpers, Ashesh Mukherjee, and Wayne D. Hoyer, "Humor in Television Advertising: A Moment-to-Moment Analysis," *Journal of Consumer Research,* December 2004; Edward F. McQuarrie and Barbara J. Phillips, "Indirect Persuasion in Advertising," *Journal of Advertising,* Summer 2005; Ryan S. Elder and Aradhna Krishna, "The 'Visual Depiction Effect' in Advertising: Facilitating Embodied Mental Simulation through Product Orientation," *Journal of Consumer Research* 38, no. 6 (2012); Theodore J. Noseworthy, June Cotte, and Seung Hwan (Mark) Lee, "The Effects of Ad Context and Gender on the Identification of Visually Incongruent Products," *Journal of Consumer Research* 38, no. 2 (2011); Jiemiao Chen, Xiaojing Yang, and Robert Smith, "The Effects of Creativity on Advertising Wear-In and Wear-Out," *Journal of the Academy of Marketing Science,* May 2016.

30. "Quiz: Is This Story an Ad?," *Marketplace,* December 3, 2013; "Here's What Else Is Wrong with Native Advertising," *Advertising Age,* September 2, 2014; "Publishers Are Largely Not Following the FTC's Native Ad Guidelines," *Adweek,* April 8, 2016.

31. "Here's What Else Is Wrong with Native Advertising," *Advertising Age,* September 2, 2014; "*BuzzFeed*–The Promise of Native Advertising," Harvard Business School case study 9-0714-512, August 15, 2014; For a send-up of the practice, see *Last Week Tonight with John Oliver: Native Advertising,* YouTube, https://youtu.be/E_F5GxCwizc.

32. "Paying for Viewers Who Pay Attention," *BusinessWeek,* May 18, 2009; "Guess Which Medium Is as Effective as Ever: TV," *Advertising Age,* February 23, 2009; "Web vs. TV: Research Aims to Gauge Ads," *Wall Street Journal,* March 19, 2008.

33. Major media comparisons were estimated based on data from "Major Media CPM Comparison," Peter J. Solomon Company, March 2016; "Just How Much Does Yelp Cost?," *Mockingbird,* May 2, 2014.

34. For more on international differences, see "P&G's Crest Fined Almost $1 Million over Chinese Ad," *Advertising Age,* March 10, 2015; "Beauty Riskier Than Booze on Spanish TV," *Advertising Age,* January 25, 2010; "Catching the Eye of China's Elite," *BusinessWeek,* February 11, 2008; "In Korea, Ads Become Must-See TV," *Wall Street Journal,* December 13, 2007; "For Reality Shows in China, Rules Have Never Been Tighter," *Advertising Age,* October 8, 2007. See also Young Sook Moon and George R. Franke, "Cultural Influences on Agency Practitioners' Ethical Perceptions: A Comparison of Korea and the U.S.," *Journal of Advertising,* Spring 2000.

35. "What Bayer Campaign Means for Pharma Ads," *Advertising Age,* February 16, 2009.

36. "FTC Cracks Down on Weight-Loss Product Marketers," *Wall Street Journal,* January 7, 2014; "Skechers Settles Deceptive Ad Case with FTC for $40M," *Adweek,* May 16, 2012. For more on KFC, see "Hey, Fast Food: We Love You Just the Way You Are," *Brandweek,* November 24, 2003; "FTC Examines Health Claims in KFC's Ads," *Wall Street Journal,* November 19, 2003; "Garfield's Ad Review: KFC Serves Big, Fat Bucket of Nonsense in 'Healthy' Spots," *Advertising Age,* November 3, 2003. See also "The Plug Stops Here," *Next,* November 2, 2009; "Cheerios First in FDA Firing Line. Who's Next?," *Advertising Age,* May 18, 2009; "Health Claims for Cheerios Break Rules, FDA Warns," *Wall Street Journal,* May 13, 2009; "FDA Makes Viagra Feel Blue about Online Ads," *Brandweek,* September 15, 2008.

37. For more on deceptive advertising, see Adam W. Craig, Yuliya Komarova Loureiro, Stacy Wood, and Jennifer M.C. Vendemia, "Suspicious Minds: Exploring Neural Processes during Exposure to Deceptive Advertising," *Journal of Marketing Research,* June 2012; Alison Jing Xu and Robert S. Wyer Jr., "Puffery in Advertisements: The Effects of Media Context, Communication Norms, and Consumer Knowledge," *Journal of Consumer Research* 17, no. 2 (2010). For more on advertising to kids, see Chapter 19;

G. R. Milne, A. Rohm, and S. Bahl, "If It's Legal, Is It Acceptable? Consumer Reactions to Online Covert Marketing," *Journal of Advertising* 38, no. 4 (2009); Kathy R. Fitzpatrick, "The Legal Challenge of Integrated Marketing Communication (IMC)," *Journal of Advertising,* Winter 2005. For more on ethics, see M. E. Drumwright and P. E. Murphy, "The Current State of Advertising Ethics: Industry and Academic Perspectives," *Journal of Advertising* 38, no. 1 (2009).

38. For research on sales promotion, see Michel Ballings, Heath McCullough, and Neeraj Bharadwaj, "Cause Marketing and Customer Profitability," *Journal of the Academy of Marketing Science,* March 2018; Marc Mazodier, Conor M. Henderson, and Joshua T. Beck, "The Long Reach of Sponsorship: How Fan Isolation and Identification Jointly Shape Sponsorship Performance," *Journal of Marketing,* November 2018; Christopher L. Newman, Melissa D. Cinelli, Douglas Vorhies, and Judith Anne Garretson Folse, "Benefitting a Few at the Expense of Many? Exclusive Promotions and Their Impact on Untargeted Customers," *Journal of the Academy of Marketing Science,* January 2019; Karen Page Winterich, Manish Gangwar, and Rajdeep Grewal, "When Celebrities Count: Power Distance Beliefs and Celebrity Endorsements," *Journal of Marketing,* May 2018.

39. www.red.org; www.biggreen.org; www.omaze.com; "Kimbal Musk's Tesla Model 3 Finds a New Home after Raising $2.1 million in Donations,"*Teslarati,* May 10, 2018; "Win an Aston Martin V8 Vantage GT Roadster,"*Redpants,* April 4, 2019; Afdhel Aziz and Bobby Jones, *Good Is the New Cool: Market Like You Give a Damn* (New York: Regan Arts, 2016); "Coca-Cola Teams with (Red) to Help End Mother-to-Child Transmission of HIV," *The Drum,* November 3, 2014; "How Iron Man, James Bond, Star Wars and Game Of Thrones Helped Omaze Raise $100 Million for Good," *Forbes,* June 7, 2018.

40. "BtoB's Best 2009 Marketers & Creative," *BtoB* (Special Issue), Fall 2009; "CEBA Awards," *Brandweek* (Special Section), October 12, 2009; "The Show Goes On: Online Trade Shows," *Wall Street Journal Reports,* April 28, 2003; "The Cyber-Show Must Go On," *Trade Media,* May 7, 2001; "Getting the Most from a Trade Show Booth," *Investor's Business Daily,* April 25, 2000. See also Li Ling-Yee, "Relationship Learning at Trade Shows: Its Antecedents and Consequences," *Industrial Marketing Management,* February 2006; Timothy M. Smith, Srinath Gopalakrishna, and Paul M. Smith, "The Complementary Effect of Trade Shows on Personal Selling," *International Journal of Research in Marketing,* March 2004; Marnik G. Dekimpe, Pierre Francois, Srinath Gopalakrishna, Gary L. Lilien, et al., "Generalizing about Trade Show Effectiveness: A Cross-National Comparison," *Journal of Marketing,* October 1997; Scott Barlass, "How to Get the Most Out of Trade Shows," *Journal of Product Innovation Management,* September 1997; Srinath Gopalakrishna, Gary L. Lilien, Jerome D. Williams, and Ian K. Sequeira, "Do Trade Shows Pay Off?," *Journal of Marketing,* July 1995; "Trade Promotion Rises," *Advertising Age,* April 3, 2000; Taewan Kim and Tridib Mazumdar, "Product Concept Demonstrations in Trade Shows and Firm Value," *Journal of Marketing,* July 2016.

41. "The Goody-Bag Game," *Wall Street Journal,* December 7, 2005; "High Noon in Aisle Five," *Inc.,* January 2004; "P&G Breaks Out of Its Slump," *USA Today,* October 14, 2003; "Ads: Mmm, Junk Mail," *Newsweek,* August 18, 2003; "Road Shows Take Brands to the People," *Wall Street Journal,* May 14, 2003; "Offbeat Marketing Sells," *Investor's Business Daily,* March 27, 2002. For academic research on sales promotion, see Hannes Datta, Bram Foubert, and Harald J. Van Heerde, "The Challenge of Retaining Customers Acquired with Free Trials," *Journal of Marketing Research,* April 2015; Sandra Laporte and Gilles Laurent, "More Prizes Are Not Always More Attractive: Factors Increasing Prospective Sweepstakes Participants' Sensitivity to the Number of Prizes," *Journal of the Academy of Marketing Science,* May 2015; Lena Steinhoff and Robert Palmatier, "Understanding Loyalty Program Effectiveness: Managing Target and Bystander Effects," *Journal of the Academy of Marketing Science,* January 2016; Scott Thompson, Richard Gooner, and Anthony Kim, "Your Mileage May Vary: Managing Untargeted Consumers' Reactions to Promotions," *Journal of the Academy of Marketing Science,* November 2015; Kirk Kristofferson, Brent

Mcferran, Andrea C. Morales, and Darren W. Dahl, "The Dark Side of Scarcity Promotions: How Exposure to Limited-Quantity Promotions Can Induce Aggression," *Journal of Consumer Research,* February 2017; Ravi Pappu and T. Cornwell, "Corporate Sponsorship as an Image Platform: Understanding the Roles of Relationship Fit and Sponsor-Sponsee Similarity," *Journal of the Academy of Marketing Science* 42, no. 5 (2014): 490-510; G. Martin-Herran, S. P. Sigue, and G. Zaccour, "The Dilemma of Pull- and Push-Price Promotions," *Journal of Retailing* 86, no. 1 (2010); K. L. Ailawadi, J. P. Beauchamp, N. Donthu, D. K. Gauri, et al., "Communication and Promotion Decisions in Retailing: A Review and Directions for Future Research," *Journal of Retailing* 85, no. 1 (2009); M. Tsiros and D. M. Hardesty, "Ending a Price Promotion: Retracting It in One Step or Phasing It Out Gradually," *Journal of Marketing* 74, no. 1 (2010); S. P. Sigue, "Consumer and Retailer Promotions: Who Is Better Off?," *Journal of Retailing* 84, no. 4 (2008); R. J. Slotegraaf and K. Pauwels, "The Impact of Brand Equity and Innovation on the Long-Term Effectiveness of Promotions," *Journal of Marketing Research* 45, no. 3 (2008); Page Moreau, Aradhna Krishna, and Bari Harlam, "The Manufacturer-Retailer-Consumer Triad: Differing Perceptions Regarding Price Promotions," *Journal of Retailing,* Winter 2001; Devon DelVecchio, David H. Henard, and Traci H. Freling, "The Effect of Sales Promotion on Post-Promotion Brand Preference: A Meta-Analysis," *Journal of Retailing,* September 2006; Kusum L. Ailawadi, Bari A. Harlam, Jacques César, and David Trounce, "Promotion Profitability for a Retailer: The Role of Promotion, Brand, Category, and Store Characteristics," *Journal of Marketing Research,* November 2006; Kusum L. Ailawadi, Karen Gedenk, Christian Lutzky, and Scott A. Neslin, "Decomposition of the Sales Impact of Promotion-Induced Stockpiling," *Journal of Marketing Research,* August 2007.

42. For more on the Pampers example, see "P&G Promotion Is Too Successful, Angering Buyers," *Wall Street Journal,* April 2, 2002. For another example, see "Shopper Turns Lots of Pudding into Free Miles," *Wall Street Journal,* January 24, 2000; "The Pudding Guy Flies Again (and Again) over Latin America," *Wall Street Journal,* March 16, 2000.

43. George E. Belch and Michael A. Belch, *Advertising and Promotion: An Integrated Marketing Communication Perspective,* 11th ed. (New York: McGraw-Hill, 2018).

44. George E. Belch and Michael E. Belch, *Advertising and Promotion: An Integrated Marketing Communication Perspective* (Burr Ridge, IL: McGraw-Hill, 2012); Priya Raghubir, J. Jeffrey Inman, and Hans Grande, "The Three Faces of Consumer Promotions," *California Management Review,* Summer 2004.

CHAPTER 16

1. Hubspot.com; thechoppingblock.com; "Creating a More Personal Chopping Block Experience," *HubSpot Case Studies;* F. Asis Martinez-Jerez, Thomas Steenburgh, Jill Avery, and Lisa Brem, "HubSpot: Lower Churn through Greater CHI," Harvard Business School Case Study 9-110-052, 2013; Brian Halligan and Dharmesh Shah, *Inbound Marketing* (New York: Wiley, 2010); David Meerman Scott, *The New Rules of Marketing and PR* (New York: Wiley 2013); Thomas Steenburgh, Jill Avery, and Naseem Dahod, "HubSpot: Inbound Marketing and Web 2.0," Harvard Business School Case Study 9-509-049, 2011; "Repeat after Me: The Customer Is Always Human," *Inc.,* November 2013; "HubSpot: Inbound Marketing and Web 2.0 (2011)," Harvard Business School Case Study Flash Forward #6064, November 2011; "HubSpot IPO Taps into Social Media Advertising Trends," *Investor's Business Daily,* October 9, 2014.

2. For academic research on these topics, see Colleen Harmeling, Jordan Moffett, Mark Arnold, and Brad Carlson, "Toward a Theory of Customer Engagement Marketing," *Journal of the Academy of Marketing Science,* May 2017; Sara Hanson, Lan Jiang, and Darren Dahl, "Enhancing Consumer Engagement in an Online Brand Community via User Reputation Signals: A Multi-Method Analysis," *Journal of the Academy of Marketing Science,* March 2019; Lena Steinhoff, Denni Arli, Scott Weaven, and Irina V. Kozlenkova, "Online Relationship Marketing," *Journal of the Academy of Marketing Science,* May 2019; Axel Berger, Tobias Schlager, David E. Sprott, and Andreas Herrmann, "Gamified Interactions: Whether, When, and How Games Facilitate Self-Brand Connections," *Journal of the Academy of Marketing Science,* July 2018; Shiri Melumad, J. Jeffrey Inman, and Michel Tuan Pham, "Selectively Emotional: How Smartphone Use Changes User-Generated Content," *Journal of Marketing Research,* April 2019; Artem Timoshenko and John R. Hauser, "Identifying Customer Needs from User-Generated Content," *Marketing Science,* January 2019; Debora V. Thompson and Prashant Malaviya, "Consumer-Generated Ads: Does Awareness of Advertising Co-Creation Help or Hurt Persuasion?," *Journal of Marketing* 77, no. 3 (2013): 33-47; Dae-Yong Ahn, Jason A. Duan, and Carl F. Mela, "Managing User-Generated Content: A Dynamic Rational Expectations Equilibrium Approach," *Marketing Science,* April 2016; Xin (Shane) Wang, Feng Mai, and Roger H. L. Chiang, "Market Dynamics and User-Generated Content about Tablet Computers," *Marketing Science,* June 2014; Tanya (Ya) Tang, Eric (Er) Fang, and Feng Wang, "Is Neutral Really Neutral? The Effects of Neutral User-Generated Content on Product Sales," *Journal of Marketing,* July 2014.

3. *Global Trust in Advertising and Brand Messages,* September 2015. This was an online survey, so caution should be used in generalizing to non-Internet users. While the general results were consistent across different parts of the world, there were some differences observed. The interested reader is encouraged to consult the original report for more details.

4. David Meerman Scott, *The New Rules of Marketing & PR* (New York: Wiley, 2015); David Meerman Scott, *The New Rules of Sales and Service* (New York: Wiley, 2014); Ann Handley, *Everybody Writes: Your Guide to Creating Ridiculously Good Content* (New York: Wiley, 2014); Brian Halligan and Dharmesh Shah, *Inbound Marketing, Revised and Updated* (New York: Wiley, 2014).

5. Andrew T. Stephen and Jeff Galak, "The Effects of Traditional and Social Earned Media on Sales: A Study of a Microlending Marketplace," *Journal of Marketing Research* 49, no. 5 (2012): 624-39. There is a growing body of academic research on word-of-mouth. See Aliosha Alexandrov, Bryan Lilly, and Emin Babakus, "The Effects of Social- and Self-Motives on the Intentions to Share Positive and Negative Word of Mouth," *Journal of the Academy of Marketing Science* 41, no. 5 (2013): 531-46; Alixandra Barasch and Jonah Berger, "Broadcasting and Narrowcasting: How Audience Size Affects What People Share," *Journal of Marketing Research* 51, no. 3 (2014): 286-99; Jonah Berger and Raghuram Iyengar, "Communication Channels and Word of Mouth: How the Medium Shapes the Message," *Journal of Consumer Research* 40, no. 3 (2013): 567-79; Jonah Berger, "Word of Mouth and Interpersonal Communication: A Review and Directions for Future Research," *Journal of Consumer Psychology* 24, no. 4 (2014): 586-607; Zoey Chen and Jonah Berger, "When, Why, and How Controversy Causes Conversation," *Journal of Consumer Research* 40, no. 3 (2013): 580-93; Zoey Chen and Nicholas H. Lurie, "Temporal Contiguity and Negativity Bias in the Impact of Online Word of Mouth," *Journal of Marketing Research* 50, no. 4 (2013): 463-76; Stephen X. He and Samuel D. Bond, "Why Is the Crowd Divided? Attribution for Dispersion in Online Word of Mouth," *Journal of Consumer Research* 41, no. 6 (2015): 1509-27; Stephen X. He and Samuel D. Bond, "Word-of-Mouth and the Forecasting of Consumption Enjoyment," *Journal of Consumer Psychology* 23, no. 4 (2013): 464-82; Thorsten Hennig-Thurau, Caroline Wiertz, and Fabian Feldhaus, "Does Twitter Matter?," The Impact of Microblogging Word of Mouth on Consumers' Adoption of New Movies," *Journal of the Academy of Marketing Science* 43, no. 3 (2015): 375-94; Nga N. Ho-Dac, Stephen J. Carson, and William L. Moore, "The Effects of Positive and Negative Online Customer Reviews: Do Brand Strength and Category Maturity Matter?," *Journal of Marketing* 77, no. 6 (2013): 37-53; Barak Libai, Eitan Muller, and Renana Peres, "Decomposing the Value of Word-of-Mouth Seeding Programs: Acceleration versus

Expansion," *Journal of Marketing Research* 50, no. 2 (2013): 161–76; Mitchell J. Lovett, Renana Peres, and Ron Shachar, "On Brands and Word of Mouth," *Journal of Marketing Research* 50, no. 4 (2013): 427–44; Ya You, Gautham G. Vadakkepatt, and Amit M. Joshi, "A Meta-Analysis of Electronic Word-of-Mouth Elasticity," *Journal of Marketing* 79, no. 2 (2015): 19–39; Yinlong Zhang, Lawrence Feick, and Vikas Mittal, "How Males and Females Differ in Their Likelihood of Transmitting Negative Word of Mouth," *Journal of Consumer Research* 40, no. 6 (2014): 1097–108. There is also research on reviews: for example, Eric T. Anderson and Duncan I. Simester, "Reviews without a Purchase: Low Ratings, Loyal Customers, and Deception," *Journal of Marketing Research* 51, no. 3 (2014): 249–69.

6. For more on black hat SEO, see "What Is Black Hat SEO?," *About.com*, http://websearch.about.com/od/seononos/a/spamseo.htm; "Search Engine Optimization," *Wikipedia.com*, https://en.wikipedia.org/wiki/Search_engine_optimization. For research on search, see Raluca M. Ursu, "The Power of Rankings: Quantifying the Effect of Rankings on Online Consumer Search and Purchase Decisions," *Marketing Science*, July 2018.

7. For research on virality and word-of-mouth, see Samuel D. Bond, Stephen X. He, and Wen Wen, "Speaking for 'Free': Word of Mouth in Free- and Paid-Product Settings," *Journal of Marketing Research*, April 2019; Irene Consiglio, Matteo De Angelis, and Michele Costabile, "The Effect of Social Density on Word of Mouth," *Journal of Consumer Research* 45, no. 3 (2018): 511–28; Florian Dost, Ulrike Phieler, Michael Haenlein, and Barak Libai, "Seeding as Part of the Marketing Mix: Word-of-Mouth Program Interactions for Fast-Moving Consumer Goods," *Journal of Marketing*, March 2019; David B. Dose, Gianfranco Walsh, Sharon E. Beatty, and Ralf Elsner, "Unintended Reward Costs: The Effectiveness of Customer Referral Reward Programs for Innovative Products and Services," *Journal of the Academy of Marketing Science*, May 2019; Andrew Wilson, Michael Giebelhausen, and Michael Brady, "Negative Word of Mouth Can Be a Positive for Consumers Connected to the Brand," *Journal of the Academy of Marketing Science*, July 2017; Ezgi Akpinar and Jonah Berger, "Valuable Virality," *Journal of Marketing Research*, April 2017; Catherine E. Tucker, "The Reach and Persuasiveness of Viral Video Ads," *Marketing Science*, April 2015; Ya You, Gautham G. Vadakkepatt, and Amit M. Joshi, "A Meta-Analysis of Electronic Word-of-Mouth Elasticity," *Journal of Marketing*, March 2015. For research on word-of-mouth, see Kelly Hewett, William Rand, Roland T. Rust, and Harald J. van Heerde, "Brand Buzz in the Echoverse," *Journal of Marketing*, May 2016.

8. Bob Gilbreath, *The Next Evolution of Marketing* (New York: McGraw-Hill, 2010); "Samsung Mobile Installs 77 Charging Stations at 18 U.S. Colleges and Universities during 2010 Spring Semester to Help Students and Faculty Stay Connected for Free," *Business Wire*, April 17, 2010.

9. Jonah Berger, *Contagious: Why Things Go Viral* (New York: Simon & Schuster, 2013); see also Christian Schulze, Lisa Schöler, and Bernd Skiera, "Not All Fun and Games: Viral Marketing for Utilitarian Products," *Journal of Marketing* 78, no. 1 (2014): 1–19.

10. D. Eric Boyd, P. K. Kannan, and Rebecca J. Slotegraaf, "Branded Apps and Their Impact on Firm Value: A Design Perspective," *Journal of Marketing Research*, February 2019.

11. See UTEC–Potable Water Generator, https://youtu.be/35yeVwigQcc.

12. David Meerman Scott, *The New Rules of Marketing & PR* (New York: Wiley, 2015); David Meerman Scott, *The New Rules of Sales and Service* (New York: Wiley, 2014); Ann Handley, *Everybody Writes: Your Guide to Creating Ridiculously Good Content* (New York: Wiley, 2014); Brian Halligan and Dharmesh Shah, *Inbound Marketing, Revised and Updated* (New York: Wiley, 2014).

13. David Meerman Scott, *The New Rules of Sales and Service* (New York: Wiley, 2014).

14. David Meerman Scott, *The New Rules of Sales and Service* (New York: Wiley, 2014).

15. Jason Lankow and Josh Ritchie, *Infographics: The Power of Visual Storytelling* (New York: Wiley, 2012).

16. G. van Noort and E. A. van Reijmersdal, "Branded Apps: Explaining Effects of Brands' Mobile Phone Applications on Brand Responses," *Journal of Interactive Marketing* 45 (February 2019): 16–26.

17. Albert M. Muniz Jr. and Thomas C. O'Guinn, "Brand Community," *Journal of Consumer Research*, March **2001**; S. Fournier and L. Lee, "Getting Brand Communities Right," *Harvard Business Review*, 2009; Richard Gruner, Christian Homburg, and Bryan Lukas, "Firm-Hosted Online Brand Communities and New Product Success," *Journal of the Academy of Marketing Science* 42, no. 1 (2014): 29–48; Kelly Tian, Pookie Sautter, Derek Fisher, Sarah Fischbach, et al., "Transforming Health Care: Empowering Therapeutic Communities through Technology-Enhanced Narratives," *Journal of Consumer Research* 41, no. 2 (2014): 237–60.

18. Ian Brodie, *Email Persuasion* (Fairfax, VA: Rainmaker Publishing, 2013); Chad White, *Email Marketing Rules*, CreateSpace Independent Publishing Platform, 2014.

19. Andrew T. Ching, Robert Clark, Ignatius Horstmann, and Hyunwoo Lim, "The Effects of Publicity on Demand: The Case of Anti-Cholesterol Drugs," *Marketing Science*, February 2016.

20. "How Southwest Airlines Sold $1.5 Million in Tickets by Posting Four Press Releases Online," *Marketing Sherpa*, October 27, 2004.

21. https://www.surgeryonsunday.org/; "Sunday Offering: Surgery for the Needy," *CNN Heroes*, March 4, 2010; "KY Doctor Is Champion for the Working Poor," *Courier Journal*, September 17, 2015.

22. "How Marketers Use Online Influencers to Boost Branding Efforts," *Advertising Age*, December 21, 2009; "Big Diaper Makers Square Off," *Wall Street Journal*, April 13, 2009; David Meerman Scott, *The New Rules of Marketing & PR* (New York: Wiley, 2015).

23. "How Brands Secretly Buy Their Way into Forbes, Fast Company, and HuffPost Stories," *The Outline*, December 5, 2017; "Can Newsrooms Boost Traffic without Spoiling Their Brand?," *Columbia Journalism Review*, September–October 2014.

24. "The FTC's Endorsement Guides: What People Are Asking," https://www.ftc.gov/tips-advice/business-center/guidance/ftcs-endorsement-guides-what-people-are-asking, accessed April 22, 2019; "Wal-Mart's Jim and Laura: The Real Story," *Bloomberg Businessweek*, October 9, 2006.

25. For more on opinion leaders, word-of-mouth publicity, and buzz marketing, see Chapter 5 and the following: Chunhua Wu, Hai Che, Tat Y. Chan, and Xianghua Lu, "The Economic Value of Online Reviews," *Marketing Science*, October 2015; Bart De Langhe, Philip M. Fernbach, and Donald R. Lichtenstein, "Navigating by the Stars: Investigating the Actual and Perceived Validity of Online User Ratings," *Journal of Consumer Research*, April 2016; Andrew M. Baker, Naveen Donthu, and V. Kumar, "Investigating How Word-of-Mouth Conversations about Brands Influence Purchase and Retransmission Intentions," *Journal of Marketing Research*, April 2016; Alixandra Barasch and Jonah Berger, "Broadcasting and Narrowcasting: How Audience Size Affects What People Share," *Journal of Marketing Research*, June 2014; David Dubois, Andrea Bonezzi, and Matteo De Angelis, "Sharing with Friends versus Strangers: How Interpersonal Closeness Influences Word-of-Mouth Valence," *Journal of Marketing Research*, October 2016; Ana Babić Rosario, Francesca Sotgiu, Kristine De Valck, and Tammo H. A. Bijmolt, "The Effect of Electronic Word of Mouth on Sales: A Meta-Analytic Review of Platform, Product, and Metric Factors," *Journal of Marketing Research*, June 2016; Inyoung Chae, Andrew T. Stephen, Yakov Bart, and Dai Yao, "Spillover Effects in Seeded Word-of-Mouth Marketing Campaigns," *Marketing Science*, February 2017; Zoey Chen and Jonah Berger, "How Content Acquisition Method Affects Word of Mouth," *Journal of Consumer Research*, June 2016; Jonah Berger and Eric M Schwartz, "What Drives Immediate and Ongoing Word of Mouth?," *Journal of Marketing Research*, October 2011; Sarah G. Moore, "Some Things Are Better Left Unsaid: How Word-of-Mouth Influences the Storyteller," *Journal of Consumer Research* 38, no. 6 (2012); Yubo Chen, Qi Wang, and Jinhong Xie, "Online Social Interactions: A Natural Experiment on Word of Mouth versus Observational Learning," *Journal of Marketing Research*,

April 2011; Amar Cheema and Andrew M Kaikati, "The Effect of Need for Uniqueness on Word of Mouth," *Journal of Marketing Research,* June 2010; Bruce I. Norris, Natasha Zhang Foutz, and Ceren Kolsarici, "Dynamic Effectiveness of Advertising and Word of Mouth in Sequential Distribution of New Products," *Journal of Marketing Research,* August 2012; Michael Trusov, V. Bodapati, and Randolph E. Bucklin, "Determining Influential Users in Internet Social Networks," *Journal of Marketing Research,* August 2010.

26. *Retail Customer Dissatisfaction Study–2006*, Verde Group–Baker Retail Initiative at Wharton, 2006. For academic research on reviews, see Judith A. Chevalier, Yaniv Dover, and Dina Mayzlin, "Channels of Impact: User Reviews When Quality Is Dynamic and Managers Respond," *Marketing Science,* September 2018; Daria Dzyabura, Srikanth Jagabathula, and Eitan Muller, "Accounting for Discrepancies between Online and Offline Product Evaluations," *Marketing Science,* January 2019; Lauren Grewal, Andrew T. Stephen, and Nicole Verrochi Coleman, "When Posting about Products on Social Media Backfires: The Negative Effects of Consumer Identity Signaling on Product Interest," *Journal of Marketing Research,* April 2019; Scott Motyka, Dhruv Grewal, Elizabeth Aguirre, Dominik Mahr, et al., "The Emotional Review-Reward Effect: How Do Reviews Increase Impulsivity?," *Journal of the Academy of Marketing Science,* November 2018; Hao Shen and Jaideep Sengupta, "Word of Mouth versus Word of Mouse: Speaking about a Brand Connects You to It More Than Writing Does," *Journal of Consumer Research,* October 2018; Michelle D. Steward, James A. Narus, and Michelle L. Roehm, "An Exploratory Study of Business-to-Business Online Customer Reviews: External Online Professional Communities and Internal Vendor Scorecards," *Journal of the Academy of Marketing Science,* March 2018; Vijay Viswanathan, Sebastian Tillmanns, Manfred Krafft, and Daniel Asselmann, "Understanding the Quality-Quantity Conundrum of Customer Referral Programs: Effects of Contribution Margin, Extraversion, and Opinion Leadership," *Journal of the Academy of Marketing Science,* November 2018; Jared Watson, Anastasiya Pocheptsova Ghosh, and Michael Trusov, "Swayed by the Numbers: The Consequences of Displaying Product Review Attributes," *Journal of Marketing,* November 2018.

27. For more on BzzAgent, see "BzzAgent Seeks to Turn Word of Mouth into a Saleable Medium," *Advertising Age,* February 13, 2006; "Small Firms Turn to Marketing Buzz Agents," *Wall Street Journal,* December 27, 2005.

28. "The Selfish Truth about Word of Mouth (Why Referrals Don't Happen)," *Seth's Blog,* April 2, 2015.

29. For more on referral programs, see P. Schmitt, B. Skiera, and C. Van den Bulte, "Referral Programs and Customer Value," *Journal of Marketing* 75, no. 1 (2011): 46–59; Ina Garnefeld, Andreas Eggert, Sabrina V. Helm, and Stephen S. Tax, "Growing Existing Customers' Revenue Streams through Customer Referral Programs," *Journal of Marketing* 77, no. 4 (2013): 17–32.

30. Clemens F. Köhler, Andrew J. Rohm, Ko de Ruyter, and Martin Wetzels, "Return on Interactivity: The Impact of Online Agents on Newcomer Adjustment," *Journal of Marketing,* March 2011; Feng Zhu and Xiaoquan Zhang, "Impact of Online Consumer Reviews on Sales: The Moderating Role of Product and Consumer Characteristics," *Journal of Marketing* 74, no. 2 (2010); James C. Ward and Amy L. Ostrom, "Complaining to the Masses: The Role of Protest Framing in Customer-Created Complaint Websites," *Journal of Consumer Research,* August 2006; Shahana Sen and Dawn Lerman, "Why Are You Telling Me This? An Examination into Negative Consumer Reviews on the Web," *Journal of Interactive Marketing,* Fall 2007; V. Kumar, J. Andrew Petersen, and Robert P. Leone, "How Valuable Is Word of Mouth?," *Harvard Business Review,* October 2007; J. Goldenberg, S. Han, D. R. Lehmann, and J. W. Hong, "The Role of Hubs in the Adoption Process," *Journal of Marketing* 73, no. 2 (2009); P. Huang, N. H. Lurie, and S. Mitra, "Searching for Experience on the Web: An Empirical Examination of Consumer Behavior for Search and Experience Goods," *Journal of Marketing* 73, no. 2 (2009). For academic research on reviews, see A. M. Weiss, N. H. Lurie, and D. J. Macinnis, "Listening to Strangers: Whose Responses Are Valuable, How Valuable Are They, and Why?," *Journal of Marketing Research* 45, no. 4 (2008); V. Dhar and E. A. Chang, "Does Chatter Matter? The Impact of User-Generated Content on Music Sales," *Journal of Interactive Marketing* 23, no. 4 (2009); A. Finn, L. M. Wang, and T. Frank, "Attribute Perceptions, Customer Satisfaction and Intention to Recommend E-Services," *Journal of Interactive Marketing* 23, no. 3 (2009); E. Keller and B. Fay, "The Role of Advertising in Word of Mouth," *Journal of Advertising Research* 49, no. 2 (2009).

31. "How Amazon Is Turning Opinions into Gold," *BusinessWeek,* October 26, 2009; "What Do You Think?," *Wall Street Journal,* October 12, 2009; "Global Advertising: Consumers Trust Real Friends and Virtual Strangers the Most," *Nielsenwire,* July 7, 2009.

32. This entire section influenced by Jeff Larson and Stuart Draper, *Internet Marketing Essentials*, Stukent, Inc. 2015. Research on social media is emerging; for example, see Christian Hildebrand and Tobias Schlager, "Focusing on Others before You Shop: Exposure to Facebook Promotes Conventional Product Configurations," *Journal of the Academy of Marketing Science,* March 2019; Vamsi K. Kanuri, Yixing Chen, and Shrihari (Hari) Sridhar, "Scheduling Content on Social Media: Theory, Evidence, and Application," *Journal of Marketing,* November 2018; Francisco Villarroel Ordenes, Dhruv Grewal, Stephan Ludwig, Ko de Ruyter, et al., "Cutting through Content Clutter: How Speech and Image Acts Drive Consumer Sharing of Social Media Brand Messages," *Journal of Consumer Research,* February 2019; Grant Packard, Andrew D. Gershoff, and David B. Wooten, "When Boastful Word of Mouth Helps versus Hurts Social Perceptions and Persuasion," *Journal of Consumer Research,* June 2016; Daniel Mochon, Karen Johnson, Janet Schwartz, and Dan Ariely, "What Are Likes Worth? A Facebook Page Field Experiment," *Journal of Marketing Research,* April 2017; Thorsten Hennig-Thurau, Caroline Wiertz, and Fabian Feldhaus, "Does Twitter Matter? The Impact of Microblogging Word of Mouth on Consumers' Adoption of New Movies," *Journal of the Academy of Marketing Science,* May 2015; V. Kumar, JeeWon Choi, and Mallik Greene, "Synergistic Effects of Social Media and Traditional Marketing on Brand Sales: Capturing the Time-Varying Effects," *Journal of the Academy of Marketing Science,* March 2017; Ashish Kumar, Ram Bezawada, Rishika Rishika, Ramkumar Janakiraman, et al., "From Social to Sale: The Effects of Firm-Generated Content in Social Media on Customer Behavior," *Journal of Marketing,* January 2016; Keith Wilcox and Andrew T. Stephen, "Are Close Friends the Enemy? Online Social Networks, Self-Esteem, and Self-Control," *Journal of Consumer Research,* 2014, pp. S63–S76; Manjit S. Yadav and Paul A. Pavlou, "Marketing in Computer-Mediated Environments: Research Synthesis and New Directions," *Journal of Marketing* 78, no. 1 (2014): 20–40; David A. Schweidel and Wendy W. Moe, "Listening In on Social Media: A Joint Model of Sentiment and Venue Format Choice," *Journal of Marketing Research* 51, no. 4 (2014): 387–402.

33. #MOREROLEMODELS, *The Shorty Awards,* winning entry.

34. See www.facebook.com; definition drawn from http://whatis.techtarget.com/definition/Facebook.

35. See www.instagram.com.

36. See www.pinterest.com.

37. *Statista,* as of March 2018.

38. See www.linkedin.com.

39. "15 Ways to Use Snapchat for Your Business," *Quick Sprout,* April 19, 2019; "For Advertisers, Snapchat's Got the Kids," Reuters, February 12, 2018.

40. See www.twitter.com; definition drawn from http://whatis.techtarget.com/definition/Twitter.

41. See www.hootsuite.com.

42. Average cost for keyword ads are roughly based on data from KeywordSpy on April 22, 2017. Facebook costs were estimated.

43. "31 Cart Abandonment Statistics," *Baymard Institute,* http://goo.gl/gyK8S, accessed May 8, 2015.

44. See "8 Tips for Social Business" at https://hootsuite.com/resources/white-paper/8-tips-for-social-business.

CHAPTER 17

1. www.methodhome.com; Actual price data obtained from Target.com on May 4, 2019; "Method Wants You to Stop Buying Plastic Soap Bottles," *Fast Company,* November 9, 2018; "Target's New Brand Caters to Shoppers Looking for 'Clean' Products," *CNBC,* April 22, 2019; "Method Soap's New Brand Campaign Celebrates Sustainable Chicago Factory," *Forbes,* October 30, 2018; "Can You Look Fabulous While Cleaning? According to Method, the Answer Is Yes," *Adweek,* September 14, 2018; *Method Company Booklet,* 2012; "Method Products: Sustainability Innovation as Entrepreneurial Strategy," Darden Business Publishing, Case # UVA-ENT-0159; "Method Sold to European Green-Cleaning Rival Ecover," *Advertising Age,* September 4, 2012; "The Seven Obsessions behind Method," *TriplePundit.com,* October 10, 2011; "Designing Consumer Products for Home: A Case Study," *Padosa,* March 5, 2009; "Home as Corporate Strategy," *Investor Environmental Health Network Case Studies;* "Design Star Method Takes Cue from BK," *Advertising Age,* November 8, 2004; "Rolling through the Legs of Goliath," *Forbes,* February 2006; "Is Innovation Killing the Soap Business?," *Wall Street Journal,* April 3, 2013; "For the Dishwasher's Sake, Go Easy on the Detergent," *New York Times,* March 12, 2010; "Method Products, Inc.," *Hoover's Company Records,* July 24, 2013; "Eco-Friendly Works for Cleaning-Products Firm," *Washington Post,* March 31, 2013; "Mates Find a Method to Clean Up, Adam Lowry and Eric Ryan Have Combined the Environment, Design, and Great Products," *The Sunday Times (London),* November 20, 2011; "A Soap Maker Sought Compabability in a Merger Partner," *New York Times,* January 16, 2013; "How Method Is Disrupting the Monochrome Cleaning Scene through Colour and Design," *The Drum,* August 22, 2016; "P&G under Pressure as Eco-Friendly Products Surge," *Chicago Tribune,* September 25, 2015.

2. "Car Makers Cut Free Maintenance," *Wall Street Journal,* July 7, 2005; "Detroit's Latest Offer: Pay More, Get Less," *Wall Street Journal,* July 24, 2002; "Sticker Shock: Detroit's Hidden Price Hikes," *Wall Street Journal,* April 10, 2002. For another example involving the car rental industry, see "Car-Rental Agencies Talk of Realistic Total Pricing," *New York Times,* February 10, 2004. See also Srabana Dasgupta, S. Siddarth, and Jorge Silva-Risso, "To Lease or to Buy? A Structural Model of a Consumer's Vehicle and Contract Choice Decisions," *Journal of Marketing Research,* August 2007.

3. For research on price setting in channels of distribution, see Pavel Kireyev, Vineet Kumar, and Elie Ofek, "Match Your Own Price? Self-Matching as a Retailer's Multichannel Pricing Strategy," *Marketing Science,* November 2017; Krista J. Li, "Behavior-Based Pricing in Marketing Channels," *Marketing Science,* March 2018.

4. David M. Szymanski, Sundar G. Bharadwaj, and P. Rajan Varadarajan, "An Analysis of the Market Share-Profitability Relationship," *Journal of Marketing,* July 1993.

5. "The Telecom Follies," *Wall Street Journal,* March 26, 2004; F. C. Schweppe, M. C. Caramanis, R. D. Tabors, and R. E. Bohn, *Spot Pricing of Electricity* (Springer Science & Business Media: 2013).

6. "Business' Focus on Maximizing Shareholder Value Has Numerous Costs," *Washington Post,* September 6, 2013; "B Corps Are Businesses Committed to Using Their Profit for Good—These 10 Are Making Some Truly Great Products," *Business Insider,* March 5, 2019; "First-Ever Study of Maryland Benefit Corps Released," *Forbes,* January 25, 2013.

7. "Aluminum Firms Offer Wider Discounts but Price Cuts Stop at Some Distributors," *Wall Street Journal,* November 16, 1984. For another example of administered pricing, see "Why a Grand Plan to Cut CD Prices Went off the Track," *Wall Street Journal,* June 4, 2004.

8. Moritz Fleischmann, Joseph M. Hall, and David F. Pyke, "Smart Pricing," *Sloan Management Review,* Winter 2004; ManMohan S. Sodhi and Navdeep S. Sodhi, "Six Sigma Pricing," *Harvard Business Review,* May 2005; Eric Anderson and Duncan Simester, "Mind Your Pricing Cues," *Harvard Business Review,* September 2003; Kissan Joseph, "On the Optimality of Delegating Pricing Authority to the Sales Force," *Journal of Marketing,* January 2001; Michael V. Marn and Robert L. Rosiello, "Managing Price, Gaining Profit," *Harvard Business Review,* September–October 1992; Peter R. Dickson and Joel E. Urbany, "Retailer Reactions to Competitive Price Changes," *Journal of Retailing,* Spring 1994.

9. "Dynamic Pricing: The Future of Ticket Pricing in Sports," *Forbes,* January 6, 2012; "How Dynamic Pricing Is Changing Sports Ticketing," *CNBC,* July 18, 2012; "Pricing Baseball Tickets Like Airline Seats," *Bloomberg Businessweek,* May 20, 2010. See also Sucharita Chandran and Vicki G. Morwitz, "Effects of Participative Pricing on Consumers' Cognitions and Actions: A Goal Theoretic Perspective," *Journal of Consumer Research,* September 2005; B. T. Ratchford, "Online Pricing: Review and Directions for Research," *Journal of Interactive Marketing* 1 (2009); Fei Weisstein, Kent Monroe, and Monika Kukar-Kinney, "Effects of Price Framing on Consumers' Perceptions of Online Dynamic Pricing Practices," *Journal of the Academy of Marketing Science* 41, no. 5 (2013): 501–14.

10. "Supermarkets Offer Personalized Pricing," *Bloomberg,* November 15, 2013.

11. "Is Your Friend Getting a Cheaper Uber Fare Than You Are?," *The Guardian,* April 13, 2018.

12. For more on pricing in business markets, see Markus Voeth and Uta Herbst, "Supply-Chain Pricing: A New Perspective on Pricing in Industrial Markets," *Industrial Marketing Management,* January 2006; Howard Forman and James M. Hunt, "Managing the Influence of Internal and External Determinants on International Industrial Pricing Strategies," *Industrial Marketing Management,* February 2005; Eric Matson, "Customizing Prices," *Harvard Business Review,* November–December 1995; Michael H. Morris, "Separate Prices as a Marketing Tool," *Industrial Marketing Management,* May 1987; P. Ronald Stephenson, William L. Cron, and Gary L. Frazier, "Delegating Pricing Authority to the Sales Force: The Effects on Sales and Profit Performance," *Journal of Marketing,* Spring 1979; Jan Wieseke, Sascha Alavi, and Johannes Habel, "Willing to Pay More, Eager to Pay Less: The Role of Customer Loyalty in Price Negotiations," *Journal of Marketing,* November 2014.

13. "Hagglers Thrive in New Economy," *USA Today,* March 16, 2009; "Haggling Starts to Go the Way of the Tail Fin," *BusinessWeek,* October 29, 2007; "Online Stores Charge Different Prices Based on Shoppers' Surfing Habits," *USA Today,* June 2, 2005. For more on Amazon offering different prices to different customers, see "Price? For You, $2; For the Rich Guy, $5," *Investor's Business Daily,* September 29, 2000.

14. Benjamin R. Handel and Kanishka Misra, "Robust New Product Pricing," *Marketing Science,* December 2015.

15. For an interesting take on price discounts and cause marketing, see Michelle Andrews, Xueming Luo, Zheng Fang, and Jaakko Aspara, "Cause Marketing Effectiveness and the Moderating Role of Price Discounts," *Journal of Marketing* 78, no. 6 (November 2014): 120–42.

16. For more on slotting fees, see "Kraft Speeds New Product Launch Times," *Advertising Age,* November 18, 2002; "The Hidden Cost of Shelf Space," *BusinessWeek,* April 15, 2002. See also Ramarao Desiraju, "New Product Introductions, Slotting Allowances, and Retailer Discretion," *Journal of Retailing,* Fall 2001; David Balto, "Recent Legal and Regulatory Developments in Slotting Allowances and Category Management," *Journal of Public Policy & Marketing,* Fall 2002; William L. Wilkie, Debra M. Desrochers, and Gregory T. Gundlach, "Marketing Research and Public Policy: The Case of Slotting Fees," *Journal of Public Policy & Marketing,* Fall 2002.

17. "Sale Sale Sale, Today, Everyone Wants a Deal," *USA Today,* April 21, 2010. For more on P&G's everyday low pricing, see "P&G, Others Try New Uses for Coupon-Heavy Media," *Advertising Age,* September 22, 1997; "Move to Drop Coupons Puts Procter & Gamble in Sticky PR Situation," *Wall Street Journal,* April 17, 1997. See also Derick F. Davis and Rajesh Bagchi, "How Evaluations of Multiple Percentage Price Changes Are Influenced by Presentation Mode and Percentage Ordering: The Role of Anchoring and Surprise," *Journal of*

Marketing Research, October 2018; Sujay Dutta, Abhijit Guha, Abhijit Biswas, and Dhruv Grewal, "Can Attempts to Delight Customers with Surprise Gains Boomerang? A Test Using Low-Price Guarantees," *Journal of the Academy of Marketing Science,* May 2019; Necati Ertekin, Jeffrey D. Shulman, and Haipeng (Allan) Chen, "On the Profitability of Stacked Discounts: Identifying Revenue and Cost Effects of Discount Framing," *Marketing Science,* March 2019; Minah H. Jung, Leif D. Nelson, Uri Gneezy, and Ayelet Gneezy, "Signaling Virtue: Charitable Behavior under Consumer Elective Pricing," *Marketing Science,* March 2017; Samir Mamadehussene, "Price-Matching Guarantees as a Direct Signal of Low Prices," *Journal of Marketing Research,* April 2019; Abhijit Biswas, Sandeep Bhowmick, Abhijit Guha, and Dhruv Grewal, "Consumer Evaluations of Sale Prices: Role of the Subtraction Principle," *Journal of Marketing* 77, no. 4 (2013): 49–66; Sascha Alavi, Torsten Bornemann, and Jan Wieseke, "Gambled Price Discounts: A Remedy to the Negative Side Effects of Regular Price Discounts," *Journal of Marketing* 79, no. 2 (2015): 62–78; Michelle Andrews, Xueming Luo, Zheng Fang, and Jaakko Aspara, "Cause Marketing Effectiveness and the Moderating Role of Price Discounts," *Journal of Marketing* 78, no. 6 (2014): 120–42; Aylin Aydinli, Marco Bertini, and Anja Lambrecht, "Price Promotion for Emotional Impact," *Journal of Marketing* 78, no. 4 (2014): 80–96; Keith S. Coulter and Anne L. Roggeveen, "Price Number Relationships and Deal Processing Fluency: The Effects of Approximation Sequences and Number Multiples," *Journal of Marketing Research* 51, no. 1 (2014): 69–82; Leonard Lee and Claire I. Tsai, "How Price Promotions Influence Postpurchase Consumption Experience over Time,"*Journal of Consumer Research* 40, no. 5 (2014): 943–59; Kusum L. Ailawadi, Donald R. Lehmann, and Scott A. Neslin, "Market Response to a Major Policy Change in the Marketing Mix: Learning from Procter & Gamble's Value Pricing Strategy," *Journal of Marketing,* January 2001; Monika Kukar-Kinney and Rockney G. Walters, "Consumer Perceptions of Refund Depth and Competitive Scope in Price-Matching Guarantees: Effects on Store Patronage," *Journal of Retailing,* Fall 2003; P. B. Ellickson and S. Misra, "Supermarket Pricing Strategies," *Marketing Science* 5 (2008); J. W. Henke, S. Yeniyurt, and C. Zhang, "Supplier Price Concessions: A Longitudinal Empirical Study," *Marketing Letters* 1 (2009); Ashok K. Lalwani and Kent B. Monroe, "A Reexamination of Frequency Depth Effects in Consumer Price Judgments," *Journal of Consumer Research,* December 2005; Chris Janiszewski and Marcus Cunha Jr., "The Influence of Price Discount Framing on the Evaluation of a Product Bundle,"*Journal of Consumer Research,* March 2004; V. Kumar, Vibhas Madan, and Srini S. Srinivasan, "Price Discounts or Coupon Promotions: Does It Matter?,"*Journal of Business Research,* September 2004; David Smagalla, "Does Promotional Pricing Grow Future Business?," *Sloan Management Review,* Summer 2004; Priya Raghubir and Joydeep Srivastava, "Effect of Face Value on Product Valuation in Foreign Currencies," *Journal of Consumer Research,* December 2002.

18. For research on coupons and discounts, see Peter J. Danaher, Michael S. Smith, Kulan Ranasinghe, and Tracey S. Danaher, "Where, When, and How Long: Factors That Influence the Redemption of Mobile Phone Coupons," *Journal of Marketing Research,* October 2015; Bernard Caillaud and Romain De Nijs, "Strategic Loyalty Reward in Dynamic Price Discrimination," *Marketing Science,* October 2014. For more on discounts, see Fengyan Cai, Rajesh Bagchi, and Dinesh K. Gauri, "Boomerang Effects of Low Price Discounts: How Low Price Discounts Affect Purchase Propensity," *Journal of Consumer Research,* February 2016; Sascha Alavi, Torsten Bornemann, and Jan Wieseke, "Gambled Price Discounts: A Remedy to the Negative Side Effects of Regular Price Discounts," *Journal of Marketing,* March 2015; Leonard Lee and Claire I. Tsai, "How Price Promotions Influence Postpurchase Consumption Experience over Time," *Journal of Consumer Research,* February 2014; Aylin Aydinli, Marco Bertini, and Anja Lambrecht, "Price Promotion for Emotional Impact," *Journal of Marketing,* July 2014; Xueming Luo, Michelle Andrews, Yiping Song, and Jaakko Aspara, "Group-Buying Deal Popularity," *Journal of Marketing* 78, no. 2 (2014): 20–33.

19. Keith Wilcox, Lauren G. Block, and Eric M. Eisenstein, "Leave Home without It? The Effects of Credit Card Debt and Available Credit on Spending," *Journal of Marketing Research,* November 2011; Promothesh Chatterjee and Randall L. Rose, "Do Payment Mechanisms Change the Way Consumers Perceive Products?," *Journal of Consumer Research* 38, no. 6 (2012).

20. "Selling Solar Panels on the Installment Plan," *Bloomberg Businessweek,* May 30, 2013.

21. "One Photo Shows That China Is Already in a Cashless Future," *Business Insider,* May 29, 2018; "Alipay vs WeChat Pay: Mobile Payment Giants Driving China's Cashless Transformation," *ChoZan,* May 9, 2018.

22. Neil T. Bendle, Paul W. Farris, Phillip E. Pfeifer, and David J. Reibstein, *Marketing Metrics: The Definitive Guide to Measuring Marketing Performance,* 3rd ed. (Upper Saddle River, NJ: Pearson Education, 2016).

23. For more on price-quality, see Ayelet Gneezy, Uri Gneezy, and Dominique Olie Lauga, "A Reference-Dependent Model of the Price-Quality Heuristic," *Journal of Marketing Research* 51, no. 2 (2014): 153–64; Ashok K. Lalwani and Sharon Shavitt, "You Get What You Pay For? Self-Construal Influences Price-Quality Judgments," *Journal of Consumer Research* 40, no. 2 (2013): 255–67. For more on free trials, see Hannes Datta, Bram Foubert, and Harald J. Van Heerde, "The Challenge of Retaining Customers Acquired with Free Trials," *Journal of Marketing Research* 52, no. 2 (2015): 217–34; Mauricio M. Palmeira and Joydeep Srivastava, "Free Offer ≠ Cheap Product: A Selective Accessibility Account on the Valuation of Free Offers," *Journal of Consumer Research* 40, no. 4 (2013): 644–56.

24. This example uses real brands but the data are fictitious. The average price per liter is calculated using the weighted average method, which factors in the relative sales for each brand.

25. "Intel Fine Jolts Tech Sector," *Wall Street Journal,* May 14, 2009. See also "Europe Inc. Takes Aim at Price-Fixers," *BusinessWeek,* November 2, 2009. For an excellent discussion of laws related to pricing, see Louis W. Stern and Thomas L. Eovaldi, *Legal Aspects of Marketing Strategy: Antitrust and Consumer Protection Issues* (Englewood Cliffs, NJ: Prentice-Hall, 1984). See also Joseph P. Guiltinan and Gregory T. Gundlach, "Aggressive and Predatory Pricing: A Framework for Analysis," *Journal of Marketing,* July 1996. For more general research on price fairness, see Liyin Jin, Yanqun He, and Ying Zhang, "How Power States Influence Consumers' Perceptions of Price Unfairness," *Journal of Consumer Research* 40, no. 5 (2014): 818–33; Christina Kan, Donald R. Lichtenstein, Susan Jung Grant, and Chris Janiszewski, "Strengthening the Influence of Advertised Reference Prices through Information Priming," *Journal of Consumer Research* 40, no. 6 (2014): 1078–96; Alexander Rusetski, Jonlee Andrews, and Daniel Smith, "Unjustified Prices: Environmental Drivers of Managers' Propensity to Overprice," *Journal of the Academy of Marketing Science* 42, no. 4 (2014): 452–69.

26. "In Cancer Care, Cost Matters," *New York Times,* October 14, 2012; "How Much Would You Pay for a Year of Life?," *Radiolab,* December 22, 2104; Experts in Chronic Myeloid Leukemia, "The Price of Drugs for Chronic Myeloid Leukemia (CML) Is a Reflection of the Unsustainable Prices of Cancer Drugs: From the Perspective of CML Experts," *Blood,* May 30, 2013, pp. 121–22; "The Cost of Living," *New York Magazine,* October 20, 2013; "Incredible Prices for Cancer Drugs," *New York Times,* November 12, 2012; "Doctors Blast Ethics of $100,000 Cancer Drugs," *CNNMoney,* April 26, 2013; "Pharma Execs Don't Know Why Anyone Is Upset by a $94,500 Miracle Cure," *Bloomberg Businessweek,* June 3, 2015.

27. Patrick J. Kaufmann, N. Craig Smith, and Gwendolyn K. Ortmeyer, "Deception in Retailer High-Low Pricing: A 'Rule of Reason' Approach," *Journal of Retailing,* Summer 1994.

28. Charlayne Hunter-Gault, "ADM: Who's Next?," *MacNeil/Lehrer Newshour (PBS),* October 15, 1996, accessed October 17, 2007.

29. "Century-Old Ban Lifted on Minimum Retail Pricing," *New York Times,* June 29, 2007; Ayelet Israeli, Eric T. Anderson, and Anne

T. Coughlan, "Minimum Advertised Pricing: Patterns of Violation in Competitive Retail Markets," *Marketing Science,* August 2016.

30. Richard L. Pinkerton and Deborah J. Kemp, "The Industrial Buyer and the Robinson-Patman Act," *International Journal of Purchasing & Materials Management,* Winter 1996.

31. "Firms Must Prove Injury from Price Bias to Qualify for Damages, High Court Says," *Wall Street Journal,* May 19, 1981.

32. "Booksellers Say Five Publishers Play Favorites," *Wall Street Journal,* May 27, 1994; Joseph P. Vaccaro and Derek W. F. Coward, "Managerial and Legal Implications of Price Haggling: A Sales Manager's Dilemma," *Journal of Personal Selling & Sales Management,* Summer 1993; John R. Davidson, "FTC, Robinson-Patman and Cooperative Promotion Activities," *Journal of Marketing,* January 1968; L. X. Tarpey Sr., "Buyer Liability under the Robinson-Patman Act: A Current Appraisal," *Journal of Marketing,* January 1972.

CHAPTER 18

1. www.samsung.com/us/video/tvs; "Best Buy Should Be Dead, but It's Thriving in the Age of Amazon," *Bloomberg Businessweek,* July 19, 2018; "Is Now the Time to Buy a 4K TV?," *CNET,* November 12, 2014; "TVs," *Consumer Reports,* June 2015; "Seoul Fines Six LCD Manufacturers in Price-Fixing Case," *Wall Street Journal,* October 31, 2011; "How Faulty Marketing Has Stalled TV Sales," *Bloomberg Businessweek,* January 20, 2011; "Sony TV Unit Seeks Path to Profitability," *Wall Street Journal,* November 26, 2010; "Sony Pledges to Corral Inventory," *Wall Street Journal,* November 2, 2010; "Samsung Edges Out TV Rivals," *Wall Street Journal,* February 17, 2010; "Sony Swings Back to the Black," *Wall Street Journal,* February 5, 2010; "Sony Bravia Internet Video Brings Social Media to Your 3D TV," *Media Mentalism,* January 15, 2010; "Sony Pins Future on a 3-D Revival," *Wall Street Journal,* January 7, 2010; "Sony's 3D TV Plans Become Clearer," *PC World,* November 29, 2009; "U.S. Upstart Takes on TV Giants in Price War," *Wall Street Journal,* April 15, 2008; "Sony Teams with Sharp in Flat-Panel LCD Plant," *Investor's Business Daily,* February 27, 2008; "TV Makers Aren't Turned Off by Slump," *Wall Street Journal,* January 29, 2008; "The Picture Gets Fuzzy for TV Deals," *Wall Street Journal,* December 6, 2007; "How Walmart's TV Prices Crushed Rivals," *BusinessWeek,* April 23, 2007; "Flat Panels, Thin Margins," *BusinessWeek,* February 26, 2007; "Hefty Discounting of Flat-Panel TVs Pinches Retailers," *Wall Street Journal,* December 12, 2006; "As Flat-Panel TV Sales Soar, Unlikely Retailers Step In," *Wall Street Journal,* September 21, 2006; "Cheaper Flat-Panel TVs—from PC Makers," *Wall Street Journal,* December 23, 2004; "Texas Instruments Inside?," *BusinessWeek,* December 6, 2004; "Flat TV Prices Are Falling, but . . . Not Low Enough . . . Blame Retailers' Margins," *Wall Street Journal,* November 3, 2004; "Selling TVs at Full Price," *Wall Street Journal,* May 23, 2012; "Low Prices Hurt the Core of Samsung," *International Herald Tribune,* April 15, 2006; "Outlook Brighter for LCDs," *International Herald Tribune,* April 6, 2007; "To Change Its Image and Attract Customers, Samsung Electronics Is Putting On a Show," *New York Times,* September 20, 2004; "Samsung Speeds Up Corporate Dynamism," *Korea Times,* June 7, 2004; "The Samsung Way," *BusinessWeek,* June 16, 2003; "Samsung Electronics: Innovation and Design Strategy," *Asian Case Research Centre,* 2009; "The Paradox of Samsung's Rise," *Harvard Business Review,* July-August 2011; "Samsung Reveals Details and Pricing on Its 2017 MU Series 4K UHD Televisions," *Techaeris,* April 5, 2017; "Prices Revealed for Samsung QLED 4K TVs, Starting at $2500: Here's Our Analysis," *4K.com,* February 9, 2017; "Samsung QLED TVs Face Tough Battle against LG's OLED," *CNET,* January 3, 2017.

2. For more on margins, see Kusum L. Ailawadi and Bari Harlam, "An Empirical Analysis of the Determinants of Retail Margins: The Role of Store-Brand Share," *Journal of Marketing,* January 2004; Vincent R. Nijs, Kanishka Misra, and Karsten Hansen, "Outsourcing Retail Pricing to a Category Captain: The Role of Information Firewalls," *Marketing Science,* February 2014.

3. "Jack Bogle Is Gone, but He's Still Saving Investors $100 Billion a Year," *Forbes,* January 16, 2019.

4. Douglas G. Brooks, "Cost-Oriented Pricing: A Realistic Solution to a Complicated Problem," *Journal of Marketing,* April 1975; Stephen M. Shugan, "Retail Product-Line Pricing Strategy When Costs and Products Change," *Journal of Retailing,* Spring 2001.

5. Pranav Jindal and Peter Newberry, "To Bargain or Not to Bargain: The Role of Fixed Costs in Price Negotiations," *Journal of Marketing Research,* December 2018.

6. For more on break-even analysis, see G. Dean Kortge, "Inverted Breakeven Analysis for Profitable Marketing Decisions," *Industrial Marketing Management,* October 1984; Thomas L. Powers, "Breakeven Analysis with Semifixed Costs," *Industrial Marketing Management,* February 1987.

7. Approaches for estimating price-quantity relationships are reviewed in Kent B. Monroe, *Pricing: Making Profitable Decisions* (New York: McGraw-Hill, 2003). See also Michael F. Smith and Indrajit Sinha, "The Impact of Price and Extra Product Promotions on Store Preference," *International Journal of Retail & Distribution Management* 28, no. 2 (2000); Michael H. Morris and Mary L. Joyce, "How Marketers Evaluate Price Sensitivity," *Industrial Marketing Management,* May 1988; David E. Griffith and Roland T. Rust, "The Price of Competitiveness in Competitive Pricing," *Journal of the Academy of Marketing Science,* Spring 1997; Robert J. Dolan, "How Do You Know When the Price Is Right?," *Harvard Business Review,* September-October 1995.

8. For more on price sensitivity (also called *price elasticity*), see Koray Cosguner, Tat Y. Chan, and P. B. (Seethu) Seetharaman, "Behavioral Price Discrimination in the Presence of Switching Costs," *Marketing Science,* May 2017; Raoul Kübler, Koen Pauwels, Gökhan Yildirim, and Thomas Fandrich, "App Popularity: Where in the World Are Consumers Most Sensitive to Price and User Ratings?," *Journal of Marketing,* September 2018; Nita Umashankar, Yashoda Bhagwat, and V. Kumar, "Do Loyal Customers Really Pay More for Services?," *Journal of the Academy of Marketing Science,* November 2017; Koen Pauwels and Richard D'Aveni, "The Formation, Evolution and Replacement of Price-Quality Relationships," *Journal of the Academy of Marketing Science,* January 2016; Doreén Pick and Martin Eisend, "Buyers' Perceived Switching Costs and Switching: A Meta-Analytic Assessment of Their Antecedents," *Journal of the Academy of Marketing Science,* March 2014; Ashok K. Lalwani and Lura Forcum, "Does a Dollar Get You a Dollar's Worth of Merchandise? The Impact of Power Distance Belief on Price-Quality Judgments," *Journal of Consumer Research,* August 2016; Xiaomeng Guo and Baojun Jiang, "Signaling through Price and Quality to Consumers with Fairness Concerns," *Journal of Marketing Research,* December 2016; Daniel M. McCarthy, Peter S. Fader, and Bruce G. S. Hardie, "Valuing Subscription-Based Businesses Using Publicly Disclosed Customer Data," *Journal of Marketing,* January 2017.

9. "Seeking Perfect Prices, CEO Tears Up the Rules," *Wall Street Journal,* March 27, 2007.

10. Robert Carter and David Curry, "Transparent Pricing: Theory, Tests, and Implications for Marketing Practice," *Journal of the Academy of Marketing Science,* November 2010; E. T. Anderson and D. I. Simester, "Does Demand Fall When Customers Perceive That Prices Are Unfair? The Case of Premium Pricing for Large Sizes," *Marketing Science* 27, no. 3 (2008); F. Volckner, "The Dual Role of Price: Decomposing Consumers' Reactions to Price," *Journal of the Academy of Marketing Science* 36, no. 3 (2008); J. H. Chu, P. Chintagunta, and J. Cebollada, "A Comparison of Within-Household Price Sensitivity across Online and Offline Channels," *Marketing Science* 27, no. 2 (2008); Lisa E. Bolton, Hean Tat Keh, and Joseph W. Alba, "How Do Price Fairness Perceptions Differ across Culture?," *Journal of Marketing Research,* June 2010; Lisa E. Bolton and Joseph W. Alba, "Price Fairness: Good and Service Differences and the Role of Vendor Costs," *Journal of Consumer Research,* August 2006; Lan Xia, Kent B. Monroe, and Jennifer L. Cox, "The Price Is Unfair! A

Conceptual Framework of Price Fairness Perceptions," *Journal of Marketing,* October 2004; Narayan Janakiraman, Robert J. Meyer, and Andrea C. Morales, "Spillover Effects: How Consumers Respond to Unexpected Changes in Price and Quality," *Journal of Consumer Research,* October 2006; Shuba Srinivasan, Koen Pauwels, and Vincent Nijs, "Demand-Based Pricing versus Past-Price Dependence: A Cost-Benefit Analysis," *Journal of Marketing,* March 2008; Marc Vanhuele, Gilles Laurent, and Xavier Drèze, "Consumers' Immediate Memory for Prices," *Journal of Consumer Research,* August 2006; Manoj Thomas and Geeta Menon, "When Internal Reference Prices and Price Expectations Diverge: The Role of Confidence," *Journal of Marketing Research,* August 2007; Kirk L. Wakefield and J. Jeffrey Inman, "Situational Price Sensitivity: The Role of Consumption Occasion, Social Context and Income," *Journal of Retailing,* Winter 2003. For the ethics exercise, see D. Talukdar, "Cost of Being Poor: Retail Price and Consumer Price Search Differences across Inner-City and Suburban Neighborhoods," *Journal of Consumer Research* 35, no. 3 (2008). For more on price sensitivity, see Shane Frederick, "Overestimating Others' Willingness to Pay," *Journal of Consumer Research* 39, no. 1 (2012): 1–21; Klaus M. Miller, Reto Hofstetter, Harley Krohmer, and Z. John Zhang, "How Should Consumers' Willingness to Pay Be Measured? An Empirical Comparison of State-of-the-Art Approaches," *Journal of Marketing Research,* February 2011; Hong Yuan and Song Han, "The Effects of Consumers' Price Expectations on Sellers' Dynamic Pricing Strategies," *Journal of Marketing Research,* February 2011; Rashmi Adaval and Robert S. Wyer, "Conscious and Nonconscious Comparisons with Price Anchors: Effects on Willingness to Pay for Related and Unrelated Products," *Journal of Marketing Research,* April 2011; Marcus Cunha Jr. and Jeffrey D. Shulman, "Assimilation and Contrast in Price Evaluations," *Journal of Consumer Research* 37, no. 5 (2011); Joanna Phillips Melancon, Stephanie M. Noble, and Charles H. Noble, "Managing Rewards to Enhance Relational Worth," *Journal of the Academy of Marketing Science,* May 2011.

11. "Game of EpiPens," *Bloomberg Businessweek,* September 5, 2016; "Industry Insiders Estimate EpiPen Costs No More Than $30," *NBC News,* September 6, 2016. For research on fairness in pricing, see Jonathan Z. Zhang, Oded Netzer, and Asim Ansari, "Dynamic Targeted Pricing in B2B Relationships," *Marketing Science,* June 2014; Alexander Rusetski, Jonlee Andrews, and Daniel Smith, "Unjustified Prices: Environmental Drivers of Managers' Propensity to Overprice," *Journal of the Academy of Marketing Science,* July 2014; Sundar G. Bharadwaj and Debanjan Mitra, "Satisfaction (Mis)pricing Revisited: Real? Really Big?," *Journal of Marketing,* September 2016; Johannes Habel, Laura Marie Schons, Sascha Alavi, and Jan Wieseke, "Warm Glow or Extra Charge? The Ambivalent Effect of Corporate Social Responsibility Activities on Customers' Perceived Price Fairness," *Journal of Marketing,* January 2016; Christina Kan, Donald R. Lichtenstein, Susan Jung Grant, and Chris Janiszewski, "Strengthening the Influence of Advertised Reference Prices through Information Priming," *Journal of Consumer Research,* April 2014.

12. "Change the World," *Fortune,* September 1, 2016.

13. Thomas T. Nagle and Reed R. Holder, *The Strategy and Tactics of Pricing* (Englewood Cliffs, NJ: Prentice-Hall, 2002); Andreas Hinterhuber, "Towards Value-Based Pricing—An Integrative Framework for Decision Making," *Industrial Marketing Management,* November 2004; Benson P. Shapiro and Barbara P. Jackson, "Industrial Pricing to Meet Customer Needs," *Harvard Business Review,* November–December 1978; "The Race to the $10 Light Bulb," *BusinessWeek,* May 19, 1980. See also Michael H. Morris and Donald A. Fuller, "Pricing an Industrial Service," *Industrial Marketing Management,* May 1989.

14. For more on auctions, see Gerald Häubl and Peter T. L. Popkowski Leszczyc, "Bidding Frenzy: Speed of Competitor Reaction and Willingness to Pay in Auctions," *Journal of Consumer Research,* April 2019; "Copart: Auctioneer's Web Move Puts It in Fast Lane," *Investor's Business Daily,* June 16, 2004; "Renaissance in Cyberspace," *Wall Street Journal,* November 20, 2003; "Sold! To Save the Farm," *Wall Street Journal,* August 29, 2003. See also Larry R. Smeltzer and Amelia S. Carr, "Electronic Reverse Auctions: Promises, Risks and Conditions for Success," *Industrial Marketing Management,* August 2003; Sandy D. Jap, "An Exploratory Study of the Introduction of Online Reverse Auctions," *Journal of Marketing,* July 2003. For more on bid pricing, see Cong Feng, Scott Fay, and K. Sivakumar, "Overbidding in Electronic Auctions: Factors Influencing the Propensity to Overbid and the Magnitude of Overbidding," *Journal of the Academy of Marketing Science,* March 2016; Caroline Ducarroz, Sha Yang, and Eric A. Greenleaf, "Understanding the Impact of In-Process Promotional Messages: An Application to Online Auctions," *Journal of Marketing,* March 2016.

15. "The Price Is Really Right," *BusinessWeek,* March 31, 2003; "New Software Manages Price Cuts," *Investor's Business Daily,* March 31, 2003; "The Power of Optimal Pricing," *Business 2.0,* September 2002; "Priced to Move: Retailers Try to Get Leg Up on Markdowns with New Software," *Wall Street Journal,* August 7, 2001; "The Price Is Right," *Inc.,* July 2001. See also Ashutosh Dixit, Thomas W. Whipple, George M. Zinkhan, and Edward Gailey, "A Taxonomy of Information Technology-Enhanced Pricing Strategies," *Journal of Business Research,* April 2008; Devon DelVecchio, H. Shanker Krishnan, and Daniel C. Smith, "Cents or Percent? The Effects of Promotion Framing on Price Expectations and Choice," *Journal of Marketing,* July 2007; Kelly L. Haws and William O. Bearden, "Dynamic Pricing and Consumer Fairness Perceptions," *Journal of Consumer Research,* October 2006; Tridib Mazumdar, S. P. Raj, and Indrajit Sinha, "Reference Price Research: Review and Propositions," *Journal of Marketing,* October 2005; Michael J. Barone, Kenneth C. Manning, and Paul W. Miniard, "Consumer Response to Retailers' Use of Partially Comparative Pricing," *Journal of Marketing,* July 2004; Chezy Ofir, "Reexamining Latitude of Price Acceptability and Price Thresholds: Predicting Basic Consumer Reaction to Price," *Journal of Consumer Research,* March 2004; Aradhna Krishna, Mary Wagner, Carolyn Yoon, and Rashmi Adaval, "Effects of Extreme-Priced Products on Consumer Reservation Prices," *Journal of Consumer Psychology,* 2006; Sangkil Moon, Gary J. Russell, and Sri Devi Duvvuri, "Profiling the Reference Price Consumer," *Journal of Retailing,* March 2006; John T. Gourville and Youngme Moon, "Managing Price Expectations Through Product Overlap," *Journal of Retailing,* Spring 2004; Marc Vanhuele and Xavier Dreze, "Measuring the Price Knowledge Shoppers Bring to the Store," *Journal of Marketing,* October 2002; Rashmi Adaval and Kent B. Monroe, "Automatic Construction and Use of Contextual Information for Product and Price Evaluations," *Journal of Consumer Research,* March 2002; Joel E. Urbany, Peter R. Dickson, and Alan G. Sawyer, "Insights into Cross- and Within-Store Price Search: Retailer Estimates vs. Consumer Self-Reports," *Journal of Retailing,* Summer 2000; Ronald W. Niedrich, Subhash Sharma, and Douglas H. Wedell, "Reference Price and Price Perceptions: A Comparison of Alternative Models," *Journal of Consumer Research,* December 2001; Erica Mina Okada, "Trade-Ins, Mental Accounting, and Product Replacement Decisions," *Journal of Consumer Research,* March 2001; Aradhna Krishna, Richard Briesch, Donald R. Lehmann, and Hong Yuan, "A Meta-Analysis of the Impact of Price Presentation on Perceived Savings," *Journal of Retailing,* Summer 2002; Valerie A. Taylor and William O. Bearden, "The Effects of Price on Brand Extension Evaluations: The Moderating Role of Extension Similarity," *Journal of the Academy of Marketing Science,* Spring 2002.

16. "The High Cost of Free," *Fast Company,* October 2009; "The Economics of Giving It Away," *Wall Street Journal,* January 31, 2009; "Would You Pay Money to See Your Favorite Site Ad-Free?," *Advertising Age,* December 8, 2008; "Free! Why $0.00 Is the Future of Business," *Wired Magazine,* March 2008; "Free Love," *Trendwatching.com,* March 2008; "Danger of Free," *ReadWriteWeb,* January 16, 2008. See also J. Y. Kim, M. Natter, and M. Spann, "Pay What You Want: A New Participative Pricing Mechanism," *Journal of Marketing* 73, no. 1 (2009); K. Pauwels and A. Weiss, "Moving from Free to Fee: How Online Firms Market to Change Their Business Model Successfully," *Journal of Marketing* 72, no. 3 (2008); Dominik Papies, Felix Eggers, and Nils Wlömert, "Music for Free? How Free Ad-Funded Downloads

Affect Consumer Choice," *Journal of the Academy of Marketing Science,* September 2012; "The Real Reason Everyone Offered You Free Tax Prep This Year," *Wall Street Journal,* April 7, 2017. For academic research on this model, see Sandeep Arora, Frenkel ter Hofstede, and Vijay Mahajan, "The Implications of Offering Free Versions for the Performance of Paid Mobile Apps," *Journal of Marketing,* November 2017; Yuxin Chen, Oded Koenigsberg, and Z. John Zhang, "Pay-as-You-Wish Pricing," *Marketing Science,* September 2017; Steven K. Dallas, and Vicki G. Morwitz, "'There Ain't No Such Thing as a Free Lunch': Consumers' Reactions to Pseudo-Free Offers," *Journal of Marketing Research,* December 2018; Xian Gu, P. K. Kannan, and Liye Ma, "Selling the Premium in Freemium," *Journal of Marketing,* November 2018; Zijun (June) Shi, Kaifu Zhang, and Kannan Srinivasan, "Freemium as an Optimal Strategy for Market Dominant Firms," *Marketing Science,* January 2019; Pavel Kireyev, Vineet Kumar, and Elie Ofek, "Match Your Own Price? Self-Matching as a Retailer's Multichannel Pricing Strategy," *Marketing Science,* November 2017; Krista J. Li, "Behavior-Based Pricing in Marketing Channels," *Marketing Science,* March 2018.

17. For a classic example applied to a high-price item, see "Sale of Mink Coats Strays a Fur Piece from the Expected," *Wall Street Journal,* March 21, 1980.

18. Franziska Völckner and Julian Hofmann, "The Price-Perceived Quality Relationship: A Meta-Analytic Review and Assessment of Its Determinants," *Marketing Letters,* October 2007; K. Douglas Hoffman, L. W. Turley, and Scott W. Kelley, "Pricing Retail Services," *Journal of Business Research,* December 2002; Steven M. Shugan and Ramarao Desiraju, "Retail Product Line Pricing Strategy When Costs and Products Change," *Journal of Retailing,* Spring 2001; Tung-Zong Chang and Albert R. Wildt, "Impact of Product Information on the Use of Price as a Quality Cue," *Psychology & Marketing,* January 1996; B. P. Shapiro, "The Psychology of Pricing," *Harvard Business Review,* July–August 1968; Lutz Hildebrandt, "The Analysis of Price Competition between Corporate Brands," *International Journal of Research in Marketing,* June 2001.

19. Kenneth C. Manning and David E. Sprott, "Price Endings, Left-Digit Effects, and Choice," *Journal of Consumer Research* 36, no. 2 (2009); Keith S. Coulter and Robin A. Coulter, "Distortion of Price Discount Perceptions: The Right Digit Effect," *Journal of Consumer Research,* August 2007; Lee C. Simmons and Robert M. Schindler, "Cultural Superstitions and the Price Endings Used in Chinese Advertising," *Journal of International Marketing* 11, no. 2 (2003); Robert M. Schindler and Thomas M. Kibarian, "Image Communicated by the Use of 99 Endings in Advertised Prices," *Journal of Advertising,* Winter 2001; Robert M. Schindler and Patrick N. Kirby, "Patterns of Rightmost Digits Used in Advertised Prices: Implications for Nine-Ending Effects," *Journal of Consumer Research,* September 1997.

20. "GM Tries a Subscription Plan for Cadillacs—a Netflix for Cars at $1,500 a Month," *Wall Street Journal,* March 19, 2017.

21. "P&G's Global Target: Shelves of Tiny Stores," *Wall Street Journal,* July 16, 2007.

22. For examples, see "For Landlords, Jewelry Stores Are Big Gems," *Wall Street Journal,* May 20, 2005; "Diamond Store in the Rough: Opening of De Beers Boutique . . . ," *Wall Street Journal,* May 20, 2005; "Reaching Out to Younger Crowd Works for Tiffany," *USA Today,* February 26, 2004; "Tiffany & Co. Branches Out under an Alias," *Wall Street Journal,* July 23, 2003; "The Cocoon Cracks Open," *Brandweek,* April 28, 2003; "Goodbye, Mr. Goodbar: Chocolate Gets Snob Appeal," *Wall Street Journal,* February 13, 2003. See also Xing Pan, Brian T. Ratchford, and Venkatesh Shankar, "Can Price Dispersion in Online Markets Be Explained by Differences in E-Tailer Service Quality?," *Journal of the Academy of Marketing Science,* Fall 2002; G. Dean Kortge and Patrick A. Okonkwo, "Perceived Value Approach to Pricing," *Industrial Marketing Management,* May 1993.

23. "Gillette, Bleeding Market Share, Cuts Prices of Razors," *Wall Street Journal,* April 4, 2017; "Review: Epson Kills the Printer Ink Cartridge," *Wall Street Journal,* August 4, 2015.

24. For academic research on bundle pricing, see Jeffrey Meyer, Venkatesh Shankar, and Leonard L. Berry, "Pricing Hybrid Bundles by Understanding the Drivers of Willingness to Pay," *Journal of the Academy of Marketing Science,* May 2018; Mauricio Mittelman, Eduardo B. Andrade, Amitava Chattopadhyay, and C. Miguel Brendl, "The Offer Framing Effect: Choosing Single versus Bundled Offerings Affects Variety Seeking," *Journal of Consumer Research,* December 2014.

25. For more on bundling, see "Google, Sun Agree to Bundle Programs," *Investor's Business Daily,* October 5, 2005; "Adobe Turns Page with New Software Line," *Investor's Business Daily,* November 13, 2003; "The Allure of Bundling," *Wall Street Journal,* October 7, 2003; "Dining Out at 32,000 Feet," *Wall Street Journal,* June 3, 2003; "Food Flights," *USA Today,* January 17, 2003. See also Subramanian Balachander, Bikram Ghosh, and Axel Stock, "Why Bundle Discounts Can Be a Profitable Alternative to Competing on Price Promotions," *Marketing Science* 29, no. 4 (2010); Haipeng (Allan) Chen, Howard Marmorstein, Michael Tsiros, and Akshay R. Rao, "When More Is Less: The Impact of Base Value Neglect on Consumer Preferences for Bonus Packs over Price Discounts," *Journal of Marketing,* July 2012; Uzma Khan and Ravi Dhar, "Price-Framing Effects on the Purchase of Hedonic and Utilitarian Bundles," *Journal of Marketing Research,* December 2010. See also Michael J. Barone and Tirthankar Roy, "Does Exclusivity Always Pay Off? Exclusive Price Promotions and Consumer Response," *Journal of Marketing,* March 2010; Alexander Chernev, "Differentiation and Parity in Assortment Pricing," *Journal of Consumer Research,* August 2006; Bram Foubert and Els Gijsbrechts, "Shopper Response to Bundle Promotions for Packaged Goods," *Journal of Marketing Research,* November 2007; R. Venkatesh and Rabikar Chatterjee, "Bundling, Unbundling, and Pricing of Multiform Products: The Case of Magazine Content," *Journal of Interactive Marketing,* Spring 2006; Andrea Ovans, "Make a Bundle Bundling," *Harvard Business Review,* November–December 1997; Manjit S. Yadav and Kent B. Monroe, "How Buyers Perceive Savings in a Bundle Price: An Examination of a Bundle's Transaction Value," *Journal of Marketing Research,* August 1993; Dorothy Paun, "When to Bundle or Unbundle Products," *Industrial Marketing Management,* February 1993; "Flying Spirit's 'Dollar Store in the Sky' to Profit," *Wall Street Journal,* November 20, 2012.

CHAPTER 19

1. "PepsiCo Focusing on the Healthy Snacks Business," *Forbes,* September 19, 2017; "Change the World," *Fortune,* September 1, 2016; "What to Do When There Are Too Many Product Choices on the Store Shelves," *Consumer Reports,* January 2014; For more on consumer choice, see Ross D. Petty, "Limiting Product Choice: Innovation, Market Evolution, and Antitrust," *Journal of Public Policy & Marketing,* Fall 2002; Terry Clark, "Moving Mountains to Market: Reflections on Restructuring the Russian Economy," *Business Horizons,* March–April 1994. For more on Tesco, see "Grocery Home Delivery in the UK: Big and Getting Bigger. Why Not the U.S.?," *Food Logistics,* April 15, 2014; For more on obesity concerns, see C. L. Ogden, M. D. Carroll, L. R. Curtin, M. M. Lamb, et al., "Prevalence of High Body Mass Index in U.S. Children and Adolescents, 2007–2008," *Journal of the American Medical Association* 303, no. 3 (2010); National Center for Health Statistics, *Health, United States, 2010: With Special Features on Death and Dying* (Hyattsville, MD: U.S. Department of Health and Human Services, 2011); Gina S. Mohr, Donald R. Lichtenstein, and Chris Janiszewski, "The Effect of Marketer-Suggested Serving Size on Consumer Responses: The Unintended Consequences of Consumer Attention to Calorie Information," *Journal of Marketing,* January 2012; Sekar Raju, Priyali Rajagopal, and Timothy J. Gilbride, "Marketing Healthful Eating to Children: The Effectiveness of Incentives, Pledges, and Competitions," *Journal of Marketing,* May 2010; Koert Van Ittersum and Brian Wansink, "Plate Size and Color Suggestibility: The Delboeuf Illusion's Bias on Serving and Eating Behavior," *Journal of Consumer Research* 39, no. 2 (2012); Tirtha Dhar and Kathy Baylis, "Fast-Food Consumption and the Ban on Advertising Targeting Children: The Quebec Experience," *Journal of Marketing Research,*

October 2011; Jennifer J. Argo and Katherine White, "When Do Consumers Eat More? The Role of Appearance Self-Esteem and Food Packaging Cues," *Journal of Marketing,* March 2012; Caglar Irmak, Beth Vallen, and Stefanie Rosen Robinson, "The Impact of Product Name on Dieters' and Nondieters' Food Evaluations and Consumption," *Journal of Consumer Research* 38, no. 2 (2011): S45–S60; Margaret C. Campbell and Gina S. Mohr, "Seeing Is Eating: How and When Activation of a Negative Stereotype Increases Stereotype-Conducive Behavior," *Journal of Consumer Research* 38, no. 3 (2011); "Taxing the Rich—Foods, That Is," *BusinessWeek,* February 23, 2009; "Obesity of China's Kids Stuns Officials," *USA Today,* January 9, 2007; Debra M. Desrochers and Debra J. Holt, "Children's Exposure to Television Advertising: Implications for Childhood Obesity," *Journal of Public Policy & Marketing,* Fall 2007; Elizabeth S. Moore and Victoria J. Rideout, "The Online Marketing of Food to Children: Is It Just Fun and Games?," *Journal of Public Policy & Marketing,* Fall 2007; Kathleen Seiders and Leonard L. Berry, "Should Business Care about Obesity?," *Sloan Management Review,* Winter 2007; Jaideep Sengupta and Rongrong Zhou, "Understanding Impulsive Eaters' Choice Behaviors: The Motivational Influences of Regulatory Focus," *Journal of Marketing Research,* May 2007; Kelly Geyskens, Mario Pandelaere, Siegfried Dewitte, and Luk Warlop, "The Backdoor to Overconsumption: The Effect of Associating 'Low-Fat' Food with Health References," *Journal of Public Policy & Marketing,* Spring 2007; Pierre Chandon and Brian Wansink, "Is Obesity Caused by Calorie Underestimation? A Psychophysical Model of Meal Size Estimation," *Journal of Marketing Research,* February 2007; Melissa Grills Robinson, Paul N. Bloom, and Nicholas H. Lurie, "Combating Obesity in the Courts: Will Lawsuits against McDonald's Work?," *Journal of Public Policy & Marketing,* Fall 2005; Kathleen Seiders and Ross D. Petty, "Obesity and the Role of Food Marketing: A Policy Analysis of Issues and Remedies," *Journal of Public Policy & Marketing,* Fall 2004; Brian Wansink and Mike Huckabee, "De-Marketing Obesity," *California Management Review,* Summer 2005; Kelly D. Martin and Ronald Paul Hill, "Life Satisfaction, Self-Determination, and Consumption Adequacy at the Bottom of the Pyramid," *Journal of Consumer Research* (2013): S78–S91; Nailya Ordabayeva and Pierre Chandon, "Getting Ahead of the Joneses: When Equality Increases Conspicuous Consumption among Bottom-Tier Consumers," *Journal of Consumer Research* (2014): S48–S62. For recent research in marketing on food consumption, see Blair Kidwell, Jonathan Hasford, and David M. Hardesty, "Emotional Ability Training and Mindful Eating," *Journal of Marketing Research* 52, no. 1 (2015): 105–19; Yu Ma, Kusum L. Ailawadi, and Dhruv Grewal, "Soda versus Cereal and Sugar versus Fat: Drivers of Healthful Food Intake and the Impact of Diabetes Diagnosis," *Journal of Marketing* 77, no. 3 (2013): 101–20; Kelly L. Haws and Karen Page Winterich, "When Value Trumps Health in a Supersized World," *Journal of Marketing* 77, no. 3 (2013): 48–64; Barbara Briers and Sandra Laporte, "A Wallet Full of Calories: The Effect of Financial Dissatisfaction on the Desire for Food Energy," *Journal of Marketing Research* 50, no. 6 (2013): 767–81; Jeffrey R. Parker and Donald R. Lehmann, "How and When Grouping Low-Calorie Options Reduces the Benefits of Providing Dish-Specific Calorie Information," *Journal of Consumer Research* 41, no. 1 (2014): 213–35; Michel Tuan Pham, "Using Consumer Psychology to Fight Obesity," *Journal of Consumer Psychology,* July 2014: 411–12; Brian Wansink and Pierre Chandon, "Slim by Design: Redirecting the Accidental Drivers of Mindless Overeating," *Journal of Consumer Psychology* 24, no. 3 (2014): 413–31.

2. The American Customer Satisfaction Index, www.theacsi.org; "Now Are You Satisfied? The 1998 American Customer Satisfaction Index," *Fortune,* February 16, 1998. See also "Would You Recommend Us?," *BusinessWeek,* January 30, 2006; X. M. Luo and C. Homburg, "Satisfaction, Complaint, and the Stock Value Gap," *Journal of Marketing* 72, no. 4 (2008); L. Aksoy, B. Cooil, C. Groening, T. L. Keiningham, et al., "The Long-Term Stock Market Valuation of Customer Satisfaction," *Journal of Marketing* 72, no. 4 (2008); George Knox and Rutger Van Oest, "Customer Complaints and Recovery Effectiveness: A Customer Base Approach," *Journal of Marketing* 78, no. 5 (2014): 42–57; Lopo L. Rego, Neil A. Morgan, and Claes Fornell, "Reexamining the Market Share–Customer Satisfaction Relationship," *Journal of Marketing* 77, no. 5 (2013): 1–20; Alina Sorescu and Sorin M. Sorescu, "Customer Satisfaction and Long-Term Stock Returns," *Journal of Marketing,* September 2016; Rafael Becerril-Arreola, Chen Zhou, Raji Srinivasan, and Daniel Seldin, "Service Satisfaction–Market Share Relationships in Partnered Hybrid Offerings," *Journal of Marketing,* September 2017.

3. Chiara Orsingher, Sara Valentini, and Matteo de Angelis, "A Meta-Analysis of Satisfaction with Complaint Handling in Services," *Journal of the Academy of Marketing Science,* March 2010; Judy Strauss and Donna J. Hill, "Consumer Complaints by E-Mail: An Exploratory Investigation of Corporate Responses and Customer Reactions," *Journal of Interactive Marketing,* Winter 2001; James G. Maxham III and Richard G. Netemeyer, "Modeling Customer Perceptions of Complaint Handling over Time: The Effects of Perceived Justice on Satisfaction and Intent," *Journal of Retailing,* Winter 2002; James G. Maxham III and Richard G. Netemeyer, "A Longitudinal Study of Complaining Customers' Evaluations of Multiple Service Failures and Recovery Efforts," *Journal of Marketing,* October 2002; James G. Maxham III and Richard G. Netemeyer, "Firms Reap What They Sow: The Effects of Shared Values and Perceived Organizational Justice on Customers' Evaluations of Complaint Handling," *Journal of Marketing,* January 2003; Michael Brady, "Improving Your Measurement of Customer Satisfaction: A Guide to Creating, Conducting, Analyzing, and Reporting Customer Satisfaction Measurement Programs," *Journal of the Academy of Marketing Science,* Spring 2000; David M. Szymanski, "Customer Satisfaction: A Meta-Analysis of the Empirical Evidence," *Journal of the Academy of Marketing Science,* Winter 2001; Thorsten Hennig-Thurau and Alexander Klee, "The Impact of Customer Satisfaction and Relationship Quality on Customer Retention: A Critical Reassessment and Model Development," *Psychology & Marketing,* December 1997; Scott W. Hansen, Thomas L. Powers, and John E. Swan, "Modeling Industrial Buyer Complaints: Implications for Satisfying and Saving Customers," *Journal of Marketing Theory & Practice,* Fall 1997.

4. Xueming Luo, Christian Homburg, and Jan Wieseke, "Customer Satisfaction, Analyst Stock Recommendations, and Firm Value," *Journal of Marketing Research,* December 2010; Casse Mogilner, Jennifer Aaker, and Sepandar D. Kamvar, "How Happiness Affects Choice," *Journal of Consumer Research* 39, no. 2 (2012); Forrest V. Morgeson III, Sunil Mithas, Timothy L. Keiningham, and Lerzan Aksoy, "An Investigation of the Cross-National Determinants of Customer Satisfaction," *Journal of the Academy of Marketing Science,* March 2011; Paul-Valentin Ngobo, Jean-François Casta, and Olivier Ramond, "Is Customer Satisfaction a Relevant Metric for Financial Analysts?," *Journal of the Academy of Marketing Science,* May 2012; Claes Fornell, Roland T. Rust, and Marnik G. Dekimpe, "The Effect of Customer Satisfaction on Consumer Spending Growth," *Journal of Marketing Research,* February 2010; Claes Fornell, Sunil Mithas, Forrest V. Morgeson III, and M. S. Krishnan, "Customer Satisfaction and Stock Prices: High Returns, Low Risk," *Journal of Marketing,* January 2006; Neil A. Morgan, Eugene W. Anderson, and Vikas Mittal, "Understanding Firms' Customer Satisfaction Information Usage," *Journal of Marketing,* July 2005; Eugene W. Anderson, Claes Fornell, and Sanal K. Mazvancheryl, "Customer Satisfaction and Shareholder Value," *Journal of Marketing,* October 2004; Michael Tsiros, Vikas Mittal, and William T. Ross Jr., "The Role of Attributions in Customer Satisfaction: A Reexamination," *Journal of Consumer Research,* September 2004. For a classic discussion of the problem and mechanics of measuring the efficiency of marketing, see Reavis Cox, *Distribution in a High-Level Economy* (Englewood Cliffs, NJ: Prentice-Hall, 1965).

5. "Change the World," *Fortune,* September 15, 2017; "Ant Forest Users Plant 55M Trees in 507 Square Kilometers," *Global Times,* February 2, 2018. See also Nicholas Economides and Przemysław Jeziorski, "Mobile Money in Tanzania," *Marketing Science,* November 2017.

6. For more on criticisms of advertising, see David C. Vladeck, "Truth and Consequences: The Perils of Half-Truths and Unsubstantiated

Health Claims for Dietary Supplements," *Journal of Public Policy & Marketing*, Spring 2000; Barbara J. Phillips, "In Defense of Advertising: A Social Perspective," *Journal of Business Ethics*, February 1997; Charles Trappey, "A Meta-Analysis of Consumer Choice and Subliminal Advertising," *Psychology & Marketing*, August 1996; Karl A. Boedecker, Fred W. Morgan, and Linda B. Wright, "The Evolution of First Amendment Protection for Commercial Speech," *Journal of Marketing*, January 1995; Thomas C. O'Guinn and L. J. Shrum, "The Role of Television in the Construction of Consumer Reality," *Journal of Consumer Research*, March 1997. See also Robert B. Archibald, Clyde A. Haulman, and Carlisle E. Moody Jr., "Quality, Price, Advertising, and Published Quality Ratings," *Journal of Consumer Research*, March 1983; M. L. Capella, C. R. Taylor, and C. Webster, "The Effect of Cigarette Advertising Bans on Consumption—A Meta-Analysis," *Journal of Advertising* 37, no. 2 (2008); J. Craig Andrews, Richard G. Netemeyer, Jeremy Kees, and Scot Burton, "How Graphic Visual Health Warnings Affect Young Smokers' Thoughts of Quitting," *Journal of Marketing Research*, April 2014; Marisabel Romero and Dipayan Biswas, "Healthy-Left, Unhealthy-Right: Can Displaying Healthy Items to the Left (versus Right) of Unhealthy Items Nudge Healthier Choices?," *Journal of Consumer Research*, June 2016.

7. Thomas O. Jones and W. E. Sasser, "Why Satisfied Customers Defect," *Harvard Business Review*, November-December 1995; "The Satisfaction Trap," *Harvard Business Review*, March-April 1996.

8. Peter R. Dickson, Paul W. Farris, and Willem J. M. I. Verbeke, "Dynamic Strategic Thinking," *Journal of the Academy of Marketing Science*, Summer 2001; Gloria Barczak, "Analysis for Marketing Planning," *Journal of Product Innovation Management*, September 1997; William A. Sahlman, "How to Write a Great Business Plan," *Harvard Business Review*, July-August 1997; Rita G. McGrath and Ian C. MacMillan, "Discovery-Driven Planning," *Harvard Business Review*, July-August 1995; Jeffrey Elton and Justin Roe, "Bringing Discipline to Project Management," *Harvard Business Review*, March-April 1998; Andrew Campbell and Marcus Alexander, "What's Wrong with Strategy?," *Harvard Business Review*, November-December 1997.

9. "Rethinking Marketing: Ajay Kohli," and "Rethinking Marketing: Linda Price," presentations at the 2018 Summer AMA Conference; "Profiting the Planet," *Fortune*, September 1, 2015.

10. "Green Goods, Red Flags," *Wall Street Journal*, April 24, 2010; "Going Truly Green Might Require Detective Work," *USA Today*, April 22, 2010; "Milestones in Green Consuming," *Wall Street Journal*, April 17, 2010; "False 'Green' Ads Draw Global Scrutiny," *Wall Street Journal*, January 30, 2008; TerraChoice Environmental Marketing, www.terrachoice.com; "The Six Sins of Greenwashing," *Inc.*, November 2007; "Most Americans and Canadians Say 'Green' Labeling Just a Marketing Tactic," *GreenBiz.com*, October 2, 2007; Gregory C. Unruh, "The Biosphere Rules," *Harvard Business Review*, February 2008; Jesse R. Catlin and Yitong Wang, "Recycling Gone Bad: When the Option to Recycle Increases Resource Consumption," *Journal of Consumer Psychology* 23, no. 1 (2013): 122-27; Constantinos Leonidou, Constantine Katsikeas, and Neil Morgan, "'Greening' the Marketing Mix: Do Firms Do It and Does It Pay Off?," *Journal of the Academy of Marketing Science* 41, no. 2 (2013): 151-70; Remi Trudel and Jennifer J. Argo, "The Effect of Product Size and Form Distortion on Consumer Recycling Behavior," *Journal of Consumer Research* 40, no. 4 (2013): 632-43; Katherine White and Bonnie Simpson, "When Do (and Don't) Normative Appeals Influence Sustainable Consumer Behaviors?," *Journal of Marketing* 77, no. 2 (2013): 78-95.

11. For more on privacy, see "Facebook Draws Protests on Privacy Issue," *USA Today*, May 14, 2010; "Facebook Wants to Know More Than Just Who Your Friends Are," *Wall Street Journal*, April 22, 2010; "Friends No More? For Some, Social Networking Has Become Too Much of a Good Thing," *USA Today*, February 10, 2010; "Rogue Marketers Can Mine Your Info on Facebook," *Wired*, January 2010; "Why Facebook Wants Your ID," *BusinessWeek*, December 28, 2009; "Cracking the Code," *USA Today*, April 23, 2009; "Anti-Virus Software Isn't Only Online Security Tool," *USA Today*, April 9, 2008;

Kelly Martin and Patrick Murphy, "The Role of Data Privacy in Marketing," *Journal of the Academy of Marketing Science*, March 2017; Kelly D. Martin, Abhishek Borah, and Robert W. Palmatier, "Data Privacy: Effects on Customer and Firm Performance," *Journal of Marketing*, January 2017.

12. See The Center for Humane Technology, https://humanetech.com/; "Tech Dealers Now Trying to Save the Tech 'Addicts' They've Created," *Scientific American*, February 19, 2018; "Technology Designed for Addiction," *Psychology Today*, January 4, 2018; "*Candy Crush Saga*: The Science behind Our Addiction," *Time*, November 15, 2013; "'Irresistible' by Design: It's No Accident You Can't Stop Looking at the Screen," *NPR Fresh Air*, March 13, 2018; "Digital Addiction: How Technology Keeps Us Hooked," *The Conversation*, June 12, 2018.

13. While the scenarios described are fictional and the brands listed are not necessarily involved in these practices, the activities described are real. See Joseph Turow, *The Daily You* (New Haven, CT: Yale University Press, 2011); Lori Andrews, *I Know Who You Are and I Saw What You Did* (New York: Free Press, 2012); Nicolas Negroponte, *Being Digital* (New York: Knopf, 1995).

14. To learn more, see "The Most Effective Ways to Curb Climate Change Might Surprise You," *CNN*, April 19, 2019.

15. "Why Chick-fil-A and Other Brands Aren't Being Bullied," *Bloomberg Businessweek*, August 1, 2012.

APPENDIX A

1. Strictly speaking, two curves should not be compared for flatness if the graph scales are different, but for our purposes now we will do so to illustrate the idea of "elasticity of demand." Actually, it would be more accurate to compare two curves for one product on the same graph. Then both the shape of the demand curve and its position on the graph would be important.

APPENDIX B

1. Differences occur because of varied markups and nonhomogeneous product assortments. In an assortment of tires, for example, those with low markups might have sold much better than those with high markups. But with Formula 3, all tires would be treated equally.

APPENDIX D

1. Revised June 2015. Hillside Veterinary Clinic Marketing Plan case has been developed as a tool to facilitate student learning and class discussion. The plan is based on a real veterinary clinic, but the names, locations, financial data, and other information used in the plan have been altered to preserve confidentiality. This plan is not intended to serve as a source of primary data or to illustrate effective or ineffective planning. This plan draws on ideas from many former student marketing plans, especially marketing plans created for a real veterinary clinic by Patrick Akers, Kristin Arnal, Betsy Arneil, Sarah Bigum, JennaRae Hall, Ryan Hilgers, Dhania Iman, Heather Jewett, Brad Kaufman, Tim Montano, Robert Mozer, Chantal Pearson, Anna Prendergast, Teresa Rodriguez, Angela Sackett, Bobbi Thorson, and Doan Winkel.

VIDEO CASES

1-1. "Way to Grow: A Tight Real Estate Market Doesn't Deter Top Chains' Expansion Plans," Restaurants and Institutions, July 2005, p. 121.

3-1. Copyright © by Big Brothers Big Sisters of America. All rights reserved. Used with permission.

4-1. Using iron braces on his hips and legs, he also taught himself to walk a short distance by swiveling his torso while supporting himself with a cane. He funded and developed a hydrotherapy center for the treatment of polio patients in Warm Springs, Georgia. After becoming president, he helped found the National Foundation for Infantile Paralysis (now known as the March of Dimes).

Author Index

A

Aaker, David A., 691n17
Aaker, Jennifer, 693n30, 694n32, 731n4
Aaron, C., 695n34
Abela, Andrew V., 686n23
Aboulnasr, Khaled, 705n18, 706n36
Abraham, Magid, 719-720n2
Abrams, J.J., 431
Acosta, Stasia, 647-648
Adaval, Rashmi, 728-729n10, 729n15
Adigwe, Jude O., 697n32
Adkins, Natalie Ross, 689n22
Agarwal, Ashish, 720-721n13
Agarwal, James, 688n13
Agnihotri, Raj, 688n6, 715-716n15
Agrawal, Nidhi, 686n23
Aguirre, Elizabeth, 725n26
Aham, Lisa, 636
Ahearne, Michael, 97n21, 686n133, 712n5, 715-716n5, 716-717n10, 717-718n19, 717n15, 717n17, 718-719n22
Ahlborn, Carl-Philip, 712n5, 712n6
Ahluwalia, R., 702n15
Ahn, Dae-Yong, 723n2
Ahuvia, Aaron, 701-702n14
Ailawadi, Kusum L., 713n16, 715n17, 722-723n41, 726-727n17, 728n2, 730-731n1
Ainslie, Andrew, 691n17
Akdeniz, M., 706n36
Akella, Ram, 699n30
Akers, Patrick, 732n1
Akpinar, Ezgi, 724n7
Aksoy, Lerzan, 731n2, 731n4
Alan, Yasin, 708n15
Ala-Risku, Timo, 710n6
Alavi, Sascha, 715-716n5, 726-727n17, 726n12, 727n18, 729n11
Alba, Joseph W., 694n31, 728-729n10
Albaum, G., 717n15
Albers, Sönke, 717n15, 718n21
Albert, Terri C., 690n3
Albrecht, Arne, 701n7
Alderson, Wroe, 685n8, 707-708n6
Alexander, D. L., 705n18
Alexander, Joe F., 718-719n22
Alexander, Marcus, 732n8
Alexandrov, Aliosha, 723-724n5
Allen, Alexis, 701n7
Allen, B. J., 705n24
Allen, Chris T., 699n24
Allen, Kathleen, 698-699n19
Allenby, Greg M., 703n27
Allison, Lee, 715-716n5
Allsop, Dee T., 693n23
Alpert, Mark I., 690n3
Altman, E. J., 705n17
Alvarez, Cecilia M. O., 700-701n6
Alvarez, Claudio, 701-702n14
Alvarez, José B., 713n274
Amaldoss, Wilfred, 693n22, 720n8, 721n18
Ambler, Tim, 719-720n2
Anderson, Eric T., 723-724n5, 726n8, 727-728n29, 728-729n10
Anderson, Erin, 696n17, 718n21
Anderson, Eugene W., 731n4
Anderson, James C., 686n15, 701n7
Anderson, Paul F., 689n33
Anderson, Rolph E., 699n32, 718n21
Andrade, Eduardo B., 730n24
Andreasen, Alan R., 686n21
Andrew Petersen, J., 725n30
Andrews, J. Craig, 732n6
Andrews, Jonlee, 727n25, 729n11

Andrews, Lori, 732n13
Andrews, Michelle, 720n9, 726-727n17, 726n15, 727n18
Andzulis, James, 688n6
Angle, Justin W., 701-702n14
Angulo-Ruiz, Fernando, 686n13
Annan, Kofi, 658
Ansari, Asim, 708-709n21, 708n7, 729n11
Anthony, S. D., 705n17
Antia, Kersi D., 689n16, 697n24, 708n13, 708n18, 712n3
Aravindakshan, Ashwin, 715n18
Arbuthnot, Jeanette J., 695n36
Archibald, Robert B., 732n6
Argo, Jennifer J., 691n4, 702n19, 706n29, 730-731n1, 732n10
Arias, Massy, 671
Aribarg, Anoca, 703n27
Ariely, Dan, 692n15, 725n32
Arli, Denni, 723n2
Armani, Georgio, 646
Arnal, Kristin, 732n1
Arndt, Aaron D., 714n38, 718-719n22
Arneil, Betsy, 732n1
Arnett, Dennis B., 719n24
Arnold, Mark J., 696-697n18, 712n5, 723n2
Arnold, T. J., 696-697n18
Arnold, Todd, 701n7
Arnould, Eric J., 689n22, 690n1, 693-694n30, 693n27
Arora, Neeraj, 703n27
Arora, Sandeep, 729-730n16
Arroniz, Inigo, 719n25
Arsel, Zeynep, 693-694n30
Arunachalam, S., 708n8
Aspara, Jaakko, 704-705n16, 726-727n17, 726n15, 727n18
Asri Jantan, M., 717n17
Asselman, Daniel, 725n26
Astaire, Fred, 633
Atefi, Yashar, 715-716n5, 717n17
Athaide, G. A., 705n18
Atria, Ashraf M., 717n17
Atuahene-Gima, Kwaku, 687n2, 706n36
Auh, Seigyoung, 716n9, 717-718n19
Aulakh, Preet S., 708n8
Autry, Chad W., 709n28, 711n25, 711n26
Avery, Jill, 713n33, 723n1
Avramadis, Panagiotis, 716n9
Ayal, Igal, 704n4
Aydinli, Aylin, 726-727n17, 727n18
Aydinoglu, Nilüfer Z., 694n32, 721n28
Azalea, Iggy, 183
Aziz, Afdhel, 716n7, 722n39

B

Baack, D. W., 720n9
Babakus, Emin, 718-719n22, 723-724n5
Bachrach, Daniel G., 718-719n22
Bacon, Frank R., Jr., 689n32
Badenhorst, J. A., 695n36
Badrinarayanan, Vishag, 716n6
Bagchi, Rajesh, 726-727n17, 727n18
Baghaie, Marjan, 720n13
Bagozzi, Richard P., 686n15, 701-702n14
Bahl, Shalini, 699n24, 722n37
Bahn, K. D., 701-702n14
Bakamitsos, Georgios A., 694n31
Baker, Andrew M., 724-725n25
Baker, Julie, 712n6, 712n8
Baker, Thomas, 688n6
Balabanis, George, 689n13
Balachander, Subramanian, 730n25

Balasubramanian, Siva K., 703n29
Baldauf, Artur, 715-716n5
Balderas, Amy, 650-651
Ballings, Michel, 722n38
Balto, David, 726n16
Bande, Belén, 719n29
Barajas, Joel, 699n30
Barasch, Alixandra, 723-724n5, 724-725n25
Barasz, Kate, 720n9
Barczak, Gloria, 706n36, 732n8
Bardhi, Fleura, 704n4
Barkacs, Craig B., 709n24
Barksdale, Hiram C., Jr., 717n15
Barlass, Scott, 722n40
Barnes, Donald C., 687n7
Barone, Michael J., 704-705n16, 720n8, 721n28, 729n15, 730n25
Barsky, Adam, 694n311
Bart, Christopher K., 688n2
Bart, Yakov, 713n33, 720n9, 721n16, 721n22, 724-725n25
Bartels, Daniel M., 693-694n30, 694n32
Bartels, Robert, 685n7
Bartholomew, Darrell E., 716n6
Bartley, Don, 659-660
Basilio, Rosa, 666
Baskin, Ernest, 693-694n30
Bass, Frank M., 719-720n2
Bassett, Bryce R., 693n23
Bassett, Sterling, 654-655
Basuroy, Suman, 705n24, 708n15
Batra, Rajeev, 701-702n14, 714n2
Baumeister, Christoph, 701-702n14
Baumgartner, Hans, 699n26
Baxter, Roger, 686n15
Baylis, Kathy, 730-731n1
Bayus, Barry L., 704-705n16, 715n15
Bearden, William O., 686n13, 691n3, 729n15
Beatty, Sharon E., 686n14, 692-693n21, 712n10, 724n7
Beauchamp, J. P., 722-723n41
Becerril-Arreola, Rafael, 731n2
Beck, Joshua T., 687n7, 696-697n18, 697n32, 722n38
Becker, Maren, 720n9
Behrman, Douglas N., 718n21
Beitelspacher, Lauren, 712n5, 712n6
Belch, George E., 723n43, 723n44
Belch, Michael E., 723n43, 723n44
Beldona, Srinath, 708n15
Belk, Russell W., 693-694n30
Bell, David R., 694n32
Bell, Simon J., 707n44, 712n5
Bellenger, Danny N., 717n15, 718n21
Bello, Daniel C., 708n18, 710n9, 714n47
Beltramini, Richard F., 695n36
Bendle, Neil T., 727n22
Bergen, Mark E., 689n16, 714n40
Berger, Axel, 723n2
Berger, Jonah, 693n29, 723-724n5, 724-725n25, 724n7, 724n9
Berman, Jonathan Z., 693-694n30
Bernoff, Josh, 690n1
Berry, Christopher, 694n31
Berry, Leonard L., 704-705n16, 706-707n42, 730-731n1, 730n24
Berthon, Pierre, 699n28
Bertini, Marco, 691n3, 693-694n30, 700n2, 726-727n17, 727n18
Bettman, James R., 694n31, 703n29
Beyoncé, 115, 183
Bezawada, Ram, 725n32
Bezos, Jeff, 59

Bhadury, R. K., 721-722n29
Bhagwat, Yashoda, 687n7, 728n8
Bharadwaj, Neeraj, 696n10, 722n38
Bharadwaj, Sundar G., 696-697n18, 706n32, 717n15, 726n4, 729n11
Bhattacharjee, Amit, 693-694n30
Bhattacharya, C. B., 686n21, 687n24
Bhowmick, Sandeep, 726-727n17
Bieber, Justin, 131
Bienstock, Carol C., 696n10, 710n5, 713n33
Bigum, Sarah, 732n1
Bijmolt, Tammo H. A., 715n152, 724-725n25
Billeter, Darron, 691-692n7
Bilski, Alyce, 637
Bilstein, Nicola, 701n7
Bingtian, Su, 4
Bink, Audrey J. M., 716-717n10
Bird, Monroe M., 710n5
Bishop, William S., 703n33
Biswas, Abhijit, 726-727n17
Biswas, Dipayan, 692n8, 732n6
Bitner, Mary Jo, 696-697n18, 701n7, 706-707n42, 716n6
Bizer, George Y., 693-694n30
Black, William, 699n32
Blackburn, Joseph D., 709n25
Blair, Edward, 705n18, 706n36, 719n24
Blair, Sean, 686n21
Blanchard, Simon J., 718-719n22
Blank, Arthur, 319
Blankson, Charles, 691n17
Blazevic, V., 705-706n27
Bleier, Alexander, 700n2
Block, Lauren G., 694n32, 703n29, 727n18
Blocker, Christopher P., 686n133, 696-697n18, 716-717n10
Blodgett, Jeffrey G., 686n23
Bloom, Paul N., 706n29, 730-731n1
Bobbitt, L. Michelle, 711n26
Bock, Dora E., 715-716n5
Bodapati, Anand V., 724-725n25
Boedecker, Karl A., 732n6
Bohling, Timothy, 690-691n13
Bohlmann, Jonathan, 700-701n6
Bohn, R. E., 726n5
Boichuk, Jeffrey P., 715-716n5, 716n9, 717n17
Boland, Richard J., 718-719n22
Boland, Wendy Attaya, 694n32
Bolander, Willy, 715-716n5, 716-717n10, 717n17, 718-719n22, 718n21
Boldrin, Michele, 705n20
Boles, James Sanders, 717n15, 718-719n22, 718n21
Bollinger, Bryan, 693n29
Bolton, Lisa E., 693n233, 701-702n14, 728-729n10
Bolton, Ruth N., 687n10, 690n6, 696-697n18, 701n7, 712n5
Bommaraju, Raghu, 717-718n19
Bon, Joël, 718-719n22
Bond, Samuel D., 723-724n5, 724n7
Bone, Paula Fitzgerald, 703n30
Bone, Sterling, 694n31
Bonezzi, Andrea, 724-725n25
Bonfrer, André, 719-720n2, 720n8
Bonner, Joseph M., 706n32
Bonney, Leff, 693-694n30
Bonoma, Thomas V., 687n5
Booms, Bernard H., 706-707n42
Borah, Abhishek, 732n11
Bordley, Robert, 720-721n13
Borkovsky, Ron N., 702n20
Bornemann, Torsten, 686n21, 726-727n17, 727n18
Borraz, Stéphane, 701n7
Botsman, Rachel, 692n11
Boudreau, Valerie, 662-663
Boulding, W., 704n9
Bowerman, Bill, 3
Bowman, Douglas, 690-691n13
Bowman, Will, 638

Boyd, D. Eric, 696-697n18, 701-702n14, 724n10
Boyer, Kenneth K., 710n4
Boyle, Brett A., 709n24
Boyt, Tom, 703n31
Brach, Simon, 701n7
Bradford, Kevin D., 685n9, 716-717n10
Bradlow, Eric T., 713n35, 720-721n13
Brady, Michael K., 701n7, 701n8, 706-707n42, 717n15, 724n7, 731n3
Brady, Tom, 324
Brakus, J. J., 701-702n14
Brasel, S. A., 701-702n14
Brashear, Thomas G., 717n15, 718n21
Brem, Lisa, 723n1
Brendl, C. Miguel, 730n24
Brengman, Malaika, 690n6
Brenkert, George G., 686n23
Brennan, R., 703n30
Brexendorf, Tim, 704-705n16
Briers, Barbara, 730-731n1
Briesch, R. A., 712n14
Briesch, Richard, 729n15
Briley, Donnel, 693n28
Bristol, Terry, 693n22
Bristor, Julia M., 694n31
Britton, Benjamin, 718-719n22
Brocato, E., 712n6, 712n8
Brodie, Ian, 724n18
Broniarczyk, Susan M., 693-694n30
Bronnenberg, Bart J., 703n27
Brooks, Charles M., 718n21
Brooks, Douglas G., 728n4
Brossard, H. L., 696n10
Brough, Aaron R., 692n12
Brown, James R., 696n10, 696n17, 708n15, 708n16
Brown, M. R., 721-722n29
Brown, Stephen, 701n7
Brown, Stephen P., 696-697n18
Brown, Stephen W., 701n7, 706-707n42
Brown, Steven P., 715-716n5, 717n18
Brown, T., 706n33
Brown, Tom J., 706-707n42, 716n6, 718n21
Brucks, Merrie L., 694n32, 720n8, 721n25
Brüggen, Elisabeth C., 706-707n42
Bruno, Hernán A., 690-691n13, 720n13
Brusco, Michael J., 706-707n42
Bucklin, Randolph E., 707n3, 720-721n13, 724-725n25
Buehrer, Richard E., 717n15
Bullock, Sandra, 624
Bunn, Michele D., 696n10
Burke, Sandra J., 703n29
Burnett, John J., 685n7, 687n2
Burnham, Thomas A., 694-695n33
Burns, Alvin C., 692-693n21
Burns, David J., 625
Burroughs, James E., 706n36
Burton, Scot, 694n31, 703n29, 710n9, 732n6
Bush, Alan J., 717n15, 717n17
Butler, Thomas W., Jr., 689n32
Butt, Moeen Naseer, 712n3

C

Cadwallader, Susan, 706-707n42, 716n6
Caeldries, Filip, 705n26
Caemmerer, Barbara, 718n21
Cai, Fengyan, 727n18
Caillaud, Bernard, 727n18
Cain, Rita Marie, 700n35
Calantone, Roger J., 702n20, 704-705n16
Calder, B. J., 720-721n13
Caldieraro, Fabio, 700-701n6
Callahan, Thomas J., 695n36
Camacho, Nuno, 705n24
Cameron, William Bruce, 698n7
Campana, Theresa, 649-650
Campbell, Andrew, 732n8

Campbell, Margaret C., 693n233, 706n36, 730-731n1
Campo, Katia, 712n8
Cannon, Joseph P., 695-696n8, 696-697n18, 706n29, 716-717n10
Cao, Qing, 710n6
Capella, M. L., 732n6
Caplan, Frieda, 340-341
Caplan, Jackie, 340-341
Caplan, Karen, 340-341
Capper, Mark, 699n28
Caramanis, M. C., 726n5
Caravella, Mary, 713n33
Carlson, Brad, 723n2
Carlson, Kurt A., 718-719n22
Carlson, Les, 691-692n77
Carman, James M., 685n9
Carmine, Rita, 642
Carmon, Ziv, 694-695n33
Carr, Amelia S., 729n14
Carrigan, Marylyn, 700n35
Carroll, Dave, 14
Carroll, M. D., 730-731n1
Carson, Stephen J., 706n36, 723-724n5
Carter, Robert E., 700-701n6, 719n25, 728-729n10
Casta, Jean-François, 731n2
Castaño, Raquel, 694n31, 704n4
Castro, Durval, 698n10
Castro, Iana A., 712n8
Catlin, Jesse R., 732n10
Cauberghe, V., 720n9
Caudill, Eve M., 700n35
Cavusgil, S. Tamer, 688n13, 689n20, 690n6, 708n8
Cebollada, Javier, 690-691n13, 728-729n10
César, Jacques, 722-723n41
Chae, Inyoung, 720n13, 724-725n25
Chakravarty, Anindita, 696-697n18, 696n11
Challagalla, Goutam N., 700n36, 716-717n10, 717n15, 717n18, 719n28
Chan, Cindy, 693-694n30
Chan, Kimmy Wa, 706n36
Chan, Tat Y., 720-721n13, 724-725n25, 728n8
Chandon, Pierre, 699n26, 703n28, 730-731n1
Chandra, Charu, 711n25
Chandran, Sucharita, 726n9
Chandrasekaran, Deepa, 705n24, 720n10
Chandukala, Sandeep R., 692n12, 713n27
Chandy, Rajesh K., 704n4, 705n18, 706n36
Chang, E. A., 725n30
Chang, Hannah H., 692n17
Chang, Sue, 703n27
Chang, Tung-Zong, 730n18
Chang, Woojung, 704-705n16
Chaplin, Lan Nguyen, 693n22, 694n31
Charan, Ram, 687n3
Chartrand, Tanya L., 701-702n14
Chary, Komal, 685n6
Chatterjee, Promothesh, 727n18
Chatterjee, Rabikar, 730n25
Chatterjee, Sharmila C., 696-697n18, 719n23
Chattopadhyay, Amitava, 706n36, 730n24
Che, Hai, 712n5, 724-725n25
Chebat, Jean-Charles, 715n18
Cheema, Amar, 724-725n25
Chelariu, Cristian, 710n3
Chen, Cathy Yi, 689n13
Chen, D. F. R., 692n15
Chen, Haipeng (Allan), 693n24, 726-727n17, 730n25
Chen, Haozhe, 709n25
Chen, Hua, 717-718n19
Chen, J. Q., 720-721n13
Chen, Jiemiao, 721-722n29
Chen, Paige, 639-640
Chen, Tom, 701n7
Chen, Xinlei (Jack), 703n27, 712n5
Chen, Yixing, 725n32
Chen, Yubo, 724-725n25

Chen, Yuxin, 712n5, 729-730n16
Chen, Zhengyi, 695n37
Chen, Zoey, 723-724n5, 724-725n25
Cheng, Julian Ming-Sung, 691n17, 707n2
Chernev, Alexander, 686n21, 701-702n14, 712n8, 730n25
Chevalier, Judith A., 693n23, 725n26
Chiang, Roger H. L., 723n2
Chiesa, Vittorio, 705n26
Childers, Terry L., 699n26
Childs, Nancy, 698-699n19
Chin, Ronnie, 300
Ching, Andrew T., 724n19
Chintagunta, Pradeep K., 690-691n13, 698n4, 704-705n16, 712n14, 715n9, 728-729n10
Chircu, A. M., 689n23
Chisnall, Peter M., 698-699n19
Chitturi, Ravindra, 691n3
Choi, JeeWon, 725n32
Choi, S. Chan, 702n25, 707n3
Choi, Thomas Y., 696n16, 697n24
Chonko, Lawrence B., 686n23, 716-717n10, 719n29
Christen, M., 704n9
Christenson, C., 705n17
Christiansen, Ole Kirk, 87
Christopher, Martin, 710n9
Chu, J. H., 728-729n10
Chua, C. K., 706n33
Chuan-Fong, Shih, 704n4
Chun, HaeEun Helen, 692n12
Chung, Jae-Eun, 695n37
Chung, Jaihak, 701n7
Chunyan, Xie, 686n15
Cian, Luca, 701n13, 721n28
Cillo, Paola, 704-705n16
Cinelli, Melissa D., 722n38
Cioffi, Jennifer, 698n11
Clark, Bruce H., 698-699n19
Clark, M., 690-691n13
Clark, Robert, 724n19
Clark, Terry, 687n2, 730-731n1
Claro, Danny P., 718-719n22
Claudy, Marius, 695n36
Claxton, Reid P., 719n28
Cleeren, Kathleen, 686n20
Clewett, Richard M., 707-708n6
Cobb, Anthony T., 708n15
Cobb, Cathy J., 703n32
Cola, Philip A., 718-719n22
Cole, Catherine, 703n29
Coleman, James E., 709n24, 712n10
Coleman, Nicole Verrochi, 725n26
Colgate, Mark R., 707n44
Colletti, Jerome A., 715-716n5
Collier, Joel E., 713n33
Connell, Paul M., 721n25
Consiglio, Irene, 724n7
Cooil, B., 731n2
Cook, Robert L., 710n9
Cooke, Alan D. J., 694-695n33
Coolsen, Michael K., 714n3
Cooper, Martha, 699n26
Cooper, Robert W., 695n36
Copeland, Misty, 203
Corey, Robert J., 703n30
Corfman, Kim P., 715n18
Cornwell, T., 722-723n41
Corriveau, Gilles, 708n22
Corsten, Daniel, 694n32, 710n3
Cosguner, Koray, 728n8
Coskuner-Balli, Gokcen, 701-702n14, 706n38
Costabile, Michele, 724n7
Cotte, June, 721-722n29
Coughlan, Anne T., 702n25, 709n25, 727-728n29
Coulter, Keith S., 726-727n17, 730n19
Coulter, Robin A., 693n24, 730n19
Coupland, Jennifer Chang, 693-694n30
Court, David, 687n2

Cousin, Mark, 644-645
Coutu, Diane, 715-716n5
Coward, Derek W. F., 728n32
Cox, Jennifer L., 728-729n10
Cox, Reavis, 685n8, 731n4
Cox, Tania, 646
Cox, Tony, 685n8
Craig, Adam W., 722n37
Craighead, Christopher, 696-697n18
Cramer, Johannes, 703n27
Crawford, Blair, 687n2
Crecelius, Andrew T., 707n2
Cristiano, John J., 705n26
Crittenden, V., 717n15
Cron, William L., 715-716n5, 717n17, 726n12
Cronin, J. Joseph, Jr., 692n12, 706-707n42
Crosby, L. A., 695n37
Crosno, Jody, 708n16
Cross, Samantha N., 693n25, 693n28
Cuban, Damien, 670
Cui, Anna, 704-705n16
Cui, Tony Haitao, 712n6
Culotta, Aron, 699n22
Cummins, Ryan, 431
Cunha, Marcus, Jr., 691-692n7, 700-701n6, 701-702n14, 726-727n17, 728-729n10
Cunnings, William, 694-695n33
Curry, David, 700-701n6, 728-729n10
Curry, Stephen, 203
Curtin, L. R., 730-731n1
Cutler, Jennifer, 699n26
Cutright, Keisha M., 701-702n14
Czaplewski, Andrew J., 691-692n77, 693n23
Czarnecka, B., 703n30

D

Dabholkar, Pratibha A., 707n44, 717n18
Dacin, Peter A., 717n18
Dadzie, Kofi Q., 710n3
Dagger, Tracey S., 712n6, 712n8
Dahl, Darren W., 687n7, 692n17, 704-705n16, 706n36, 712n8, 716n9, 722-723n41, 723n2
Dahl, S., 703n30
Dahlquist, Steven H., 708-709n21
Dahlstrom, Robert, 708n8, 710n15
Dahod, Naseem, 723n1
Dallas, K., 729-730n16
Damangir, Sina, 720-721n13
Danaher, Peter J., 707n44, 712n6, 712n8, 719-720n2, 720n8, 727n18
Danaher, Tracey S., 727n18
Daniel, V., 709n25
Dant, Rajiv P., 696-697n18, 712n3
Danziger, Shai, 712n6, 712n8
Darmon, René Y., 717-718n19
Dasgupta, Srabana, 726n2
Dass, Mayukh, 701n8
Datta, Hannes, 722-723n41, 727n23
Daugherty, Patricia J., 709n25, 709n28, 711n25
D'Aveni, Richard A., 728n8
Davidson, John R., 728n32
Davis, Beth R., 710n9
Davis, Derrick F., 726-727n17
Davis, Donna F., 703n27
Day, George S., 687n2, 688n2, 701n7, 704n4
de Angelis, Matteo, 724-725n25, 724n7, 731n3
de Brito, Marisa P., 709n28
de Haan, Evert, 709n22
De Hooge, Ilona E., 693-694n30
De Jong, Martijn G., 694n32, 699n26
de Koster, Rene B. M., 709n28
De Langhe, Bart, 724-725n25
de Matos, C. A., 691-692n77
De Nijs, Romain, 727n18
De Pelsmacker, P., 720n9
de Ruyter, Ko, 725n30, 725n32
De Valck, Kristine, 724-725n25
De Vasconcellos, Jorge Alberto Sousa, 704n6

Dean, Tereza, 704-705n16
Debenedetti, Alain, 693-694n30
DeCarlo, Thomas E., 717-718n19, 717n17, 719n25, 719n28, 719n29
Deeter-Schmelz, Dawn R., 715-716n5
Deighton, John, 701n7, 713n33
Deitz, George D., 717n18
Dekimpe, Marnik G., 686n20, 688n8, 703n27, 722n40, 731n4
Deleersnyder, Barbara, 688n8
deLeon, Anthony, 696-697n18
Dell, Michael, 265
DelVecchio, Devon, 722-723n41, 729n15
DelVecchio, Susan K., 719n28
Deng, Xiaoyan, 703n28
Desai, Kalpesh Kaushik, 691n17
Desai, Preyas S., 688n6, 712n5
Deshpande, Rohit, 698-699n19
Desiraju, Ramarao, 709n25, 726n16, 730n18
Desrochers, Debra M., 689n16, 708n20, 726n16, 730-731n1
Dewitte, Siegfried, 730-731n1
Dhar, Ravi, 701-702n14, 730n25
Dhar, Sanjay K., 703n27, 706n38, 719-720n2, 720n8
Dhar, Tirtha, 703n27, 730-731n1
Dhar, V., 725n30
Dharwadkar, Ravi, 696-697n18
Di Mascio, Rita, 717n17
Diamantopoulos, Adamantios, 689n13
Diamond, N., 699n28
Dickson, Peter R., 687n2, 700-701n6, 726n8, 729n15, 732n8
Diehl, Kristin, 712n5
Diekens, Wesley, 637-638
Ding, Ying, 701n7
Dinner, Isaac M., 720-721n13
Dion, Delphine, 701n7
Dixit, Ashutosh, 701n8, 729n15
Dixon, Andrea L., 699n24, 716-717n10, 718n21
Doherty, Caitlin, 714n2
Dolan, Robert J., 728n7
Dominguez, Luis V., 708n7
Donaldson, B., 717n15
Donavan, D. Todd, 706-707n42, 717n17
Doney, Patricia M., 693n22
Dong, Beibei, 701n7, 706-707n42, 707n44
Dong, Maggie Chuoyan, 708n13
Dong, Yan, 712n6
Donthu, Naveen, 686n13, 695-696n8, 718-719n22, 722-723n41, 724-725n25
Dose, David B., 724n7
Dost, Florian, 724n7
Dotson, Jeffrey P., 708n15
Dotzel, Thomas, 704-705n16, 706-707n42
Douglas, Susan P., 688n13
Dover, Philip A., 690n6
Dover, Yaniv, 725n26
Draganska, Michaela, 720-721n13, 720n9
Draper, Stuart, 725n32
Drayton, Kathy, 671
Drèze, Xavier, 691-692n7, 728-729n10, 729n15
Drucker, Peter F., 688n2
Drumwright, Minette E., 720n8, 722n37
Du, Rex Yuxing, 691n3, 695n36, 720-721n13
Duan, Jason A., 723n2
Dub, Jean-Pierre, 703n27
Dube, Laurette, 719n28
Dubinsky, Alan J., 714n38
Dubois, David, 724-725n25
Ducarroz, Caroline, 729n14
Dugan, Riley, 718-719n22
Duhachek, Adam, 686n23, 702n25
Duhigg, Charles, 690n11, 692n8
Dutta, Shantanu, 689n16
Dutta, Sujay, 726-727n17
Duvvuri, Sri Devi, 729n15
Dwyer, F. Robert, 709n24, 709n30
Dzyabura, Daria, 725n26

E

Eagle, L., 703n30
Earhart, Amelia, 459
Eastwood, Clint, 71
Echambadi, Raj, 704-705n16, 708n8
Eckley, Robert S., 703n34
Economides, Nicholas, 732n5
Edell, Julie A., 703n29
Edgett, Scott, 707n3
Edwards, E., 695n34
Egan, Colin, 707n3
Eggers, Felix, 729-730n16
Eggert, Andreas, 686n15, 691n17, 725n29
Eisend, Martin, 728n8
Eisenstein, Eric M., 727n18
Ekstrand-Abueg, Matthew, 720-721n13
Elder, Ryan S., 701n13, 721-722n29
Elkington, John, 686n12
Ellickson, P. B., 726-727n17
Ellinger, Alexander E., 710n3
Ellram, Lisa M., 710n13
Elpers, Josephine Woltman, 721-722n29
Elsner, Ralf, 724n7
Elton, Jeffrey, 732n8
Eminem, 71
Enkawa, Takao, 690n6
Enz, Cathy A., 705-706n27
Eovaldi, Thomas L., 708n20, 727n25
Epp, Amber M., 692-693n21, 701-702n14
Erdem, Tülin, 694n31, 700n2, 703n27
Ergun, Ozlem, 711n30
Ertekin, Larisa, 702n21
Ertekin, Necati, 726-727n17
Ertimur, Burçak, 701-702n14, 706n38
Escalas, Jennifer, 691n4
Esper, Terry L., 710n9
Esteban-Bravo, Mercedes, 695n36
Etkin, Jordan, 693-694n30, 693n29
Evangelidis, Ioannis, 693-694n30
Evans, Kenneth R., 696-697n18, 717-718n19, 717n18, 718-719n22, 718n21
Eveleth, Daniel M., 719n28
Evers, Philip T., 710n3
Eyuboglu, Nermin, 708-709n21

F

Faber, Ronald J., 720n9
Fader, Peter S., 720n13, 728n8
Faes, W., 703n35
Fahy, John, 685n8
Falk, Tomas, 704-705n16
Fandrich, Thomas, 728n8
Fang, Eric, 695-696n8, 704-705n16, 718n21, 720-721n13, 723n2
Fang, Zheng, 720n9, 726-727n17, 726n15
Faraji-Rad, Ali, 718-719n22
Faris, Charles W., 696n9
Farrell, Mark A., 696n10
Farris, Paul W., 715n17, 727n22, 732n8
Fay, B., 725n30
Fay, Scott, 729n14
Fay, Winslow, 630
Fehle, Frank, 701-702n14
Feick, Lawrence, 723-724n5
Feinberg, Fred M., 719-720n2, 720n13
Feldhaus, Fabian, 723-724n5, 725n32
Feldman, Ronen, 698n11
Feng, Cong, 729n14
Fern, Edward F., 696n10
Fernbach, Philip M., 724-725n25
Ferraro, Rosellina, 691n4
Ferrell, Linda, 686n21, 687n24, 717n15
Ferrell, O. C., 686n21, 687n24, 717n15
Ferrell, Will, 624
Ferris, Gerald R., 718n21
Festjens, Anouk, 693-694n30
Filiatrault, Pierre, 715n18
Finn, A., 725n30
Fischbach, Sarah, 724n17
Fischer, Marc, 701-702n14
Fisher, Derek, 724n17
Fisher, Robert J., 689n16, 708n8, 715n15
Fiss, Mary S., 715-716n5
Fitzpatrick, Kathy R., 722n37
Fitzsimons, Gavan J., 701-702n14
Flaherty, Karen E., 715-716n5, 717n17, 719n28
Fleddermann, C. B., 700n3
Fleischmann, Moritz, 726n8
Fletcher, K., 717n15
Flint, Daniel J., 686n133, 696-697n18, 696n15, 710n9
Flores, Aaron, 699n30
Flurry, L. A., 692-693n21
Foer, Albert A., 689n16, 708n20
Folse, Judith Anne Garretson, 715-716n5, 722n38
Foo, Maw-Der, 701n7
Forbes, Lukas P., 718n21
Forcum, Lura, 728n8
Ford, John B., 697n32
Ford, Neil M., 717n18
Forehand, Mark R., 701-702n14
Forman, Howard, 726n12
Fornell, Claes, 731n2, 731n4
Fossen, Beth L., 720n9
Foubert, Bram, 706-707n42, 722-723n41, 727n23, 730n25
Fournier, Susan, 701-702n14, 724n17
Foutz, Natasha Zhang, 724-725n25
Fox, E. J., 712n14
Fraedrich, John Paul, 686n21, 687n24, 714n27
France, Stephen L., 711n2
Francois, Pierre, 722n40
Frank, Björn, 690n6
Frank, Garry L., 695n36
Frank, T., 725n30
Franke, George R., 715-716n5, 716n6, 720n9, 722n34
Frankenberger, Kristina D., 691n4
Frazier, Gary L., 697n24, 708n13, 708n18, 726n12
Frederick, Shane, 728-729n10
Frei, Frances X., 706-707n42
Freling, Traci H., 722-723n41
Frels, Judy K., 692n12, 694-695n33
French, Joseph, 686n23
French, Warren A., 703n36
Fresko, Moshe, 698n11
Frey, Don, 706n38
Friedrich, Colette, 704n4
Friend, Scott B., 686n17, 687n7, 695n37, 696-697n18
Friestad, Marian, 685n9
Fritz, Lynn, 711n30
Frösén, Johanna, 685n10
Frow, Pennie, 686n15, 691n17
Fu, Frank Q., 717-718n19
Fugate, Brian S., 710n9
Fulgoni, G. M., 720-721n13
Fuller, Donald A., 729n13
Fürst, Andreas, 708-709n21

G

Gabler, Colin B., 715-716n5
Gaffney, Sherry, 665
Gailey, Edward, 729n15
Gajanan, Shailendra, 708n15
Galak, Jeff, 723-724n5
Gallan, Andrew, 701n7
Galle, William P., 715n15
Gammoh, Bashar S., 715-716n5
Gandhi, Nimish, 698n12
Ganesan, Shankar, 686n14, 696-697n18, 708n8, 716n9
Ganesh, J., 713n33
Gangwar, Manish, 713n35, 722n38
Gao, Leilei, 693-694n30
Garcia, Rosanna, 695n36, 704n4
Gardial, Sarah Fisher, 696n15
Garnefeld, Ina, 725n29
Garrett, Jason, 717-718n19
Garvey, Aaron M., 701-702n14
Gassenheimer, Jule B., 708-709n21, 715-716n5
Gattorna, John, 710n9
Gauri, Dinesh K., 722-723n41, 727n18
Gedenk, Karen, 722-723n41
Gellinger, Gene, 698n10
Genchev, Stefan E., 709n25
George, M., 708n8
Georges, Laurent, 716-717n10
Germann, Frank, 701-702n14
Gershoff, Andrew D., 691n3, 692n12, 725n32
Geuens, Maggie, 690n6, 713n27
Geyskens, Inge, 696-697n18, 703n27
Geyskens, Kelly, 730-731n1
Ghadami, Fateme, 694-695n33
Ghosh, Anastasiya Pocheptsova, 725n26
Ghosh, Anindya, 720n9
Ghosh, Bikram, 730n25
Ghosh, Mrinal, 717-718n19
Gibson, Brian J., 710n9
Giebelhausen, Michael, 692n12, 701n8, 717n15, 724n7
Gielens, Katrijn, 696-697n18, 697n32, 703n27
Giesler, Markus, 693-694n30, 693n24
Gijsbrechts, Els, 703n27, 712n8, 730n25
Gilbreath, Bob, 691n5, 724n8
Gilbride, Timothy J., 693-694n30, 730-731n1
Gillespie, Erin, 717-718n19
Gilliland, David I., 708n15, 708n16, 708n18, 715n6
Gilly, Mary C., 693n25, 693n28
Gneezy, Ayelet, 726-727n17, 727n23
Gneezy, Uri, 726-727n17, 727n23
Goad, Emily A., 718-719n22
Godfrey, Andrea L., 708-709n21, 712n5
Godin, Seth, 714n3
Goebel, Daniel J., 715-716n5
Goldenberg, Jacob, 698n11, 725n30
Golder, Peter N., 700n2, 704n6
Goldfarb, Avi, 691n3, 702n20, 720-721n13
Goldman, Arieh, 714n43
Goldsby, Thomas J., 699n26, 711n26
Goldsmith, Kelly, 701-702n14
Goldsmith, Ronald E., 694n31, 718-719n22
Goldstein, Daniel G., 720-721n13
Golicic, Susan L., 703n27
Gomez, Carla, 645-646
Gomez, Pedro, 660-661
Gonzalez, Gabriel R., 718-719n22, 719n24
Gonzalez-Padron, T., 717n15
Good, David J., 717n18
Goodman, Joseph K., 700n2
Gooner, Richard, 722-723n41
Gopal, Ram D., 694n31
Gopalakrishna, Srinath, 717-718n19, 722n40
Gordon, Brett R., 691n3
Gorn, Gerald J., 694n311
Gourville, John T., 729n15
Graham, John L., 693n27, 703n33
Granbois, Donald H., 694n31
Grande, Hans, 723n44
Grant, David, 710n13
Grant, Ken, 719n24
Grant, Susan Jung, 706n36, 727n25, 729n11
Granzin, Kent L., 702n22
Grayson, K., 692n15
Greenberg, Dan, 713n274
Greenberg, Herbert M., 715-716n5
Greene, Mallik, 725n32
Greenleaf, Eric A., 729n14
Grégoire, Yany, 694-695n33
Gremler, Dwayne D., 701n7, 706-707n42
Grewal, Dhruv, 696-697n18, 708-709n21, 712n5, 712n6, 725n26, 725n32, 726-727n17, 730-731n1
Grewal, Lauren, 725n26
Grewal, Rajdeep, 688n8, 693n24, 696-697n18, 696n11, 709n30, 719n23, 722n38

Griffis, Stanley E., 699n26, 711n26
Griffith, David A., 696n10, 697n22, 697n29, 704-705n16, 708-709n21, 708n18
Griffith, David E., 728n7
Grinstein, A., 686n13
Groene, Nicole, 720-721n13
Groening, C., 731n2
Grossenbacher, Samuel, 715-716n5
Gruen, Thomas W., 691-692n77, 693n23, 696-697n18, 710n3
Gruner, Richard, 704-705n16, 724n17
Gu, Flora F., 708n13, 708n18
Gu, Xian, 729-730n16
Guenzi, Paolo, 715-716n15, 716-717n10
Guha, Abhijit, 726-727n17
Guide, R., Jr., 709n25
Guiltinan, Joseph P., 727n25
Gullo, Kelley, 693n29
Gundlach, Gregory T., 685n4, 685n9, 689n16, 695-696n8, 706n29, 708n18, 708n20, 709n24, 726n16, 727n25
Guo, Lin, 696-697n18
Guo, Xiaomeng, 728n8
Guo, Zhaoyang, 696-697n18
Gupta, Aditya, 696-697n18
Gupta, Pooja, 379, 692n15, 713n33
Gupta, Shaphali, 701n7
Gupta, Susan F., 716-717n10
Gupta, Vivek, 685n6
Gupte, Gouri, 717n17
Gurhan-Canli, Zeynep, 689n13
Gurnani, Haresh, 709n25

H

Habel, Johannes, 715-716n5, 726n12, 729n11
Hada, Mahima, 693-694n30, 696n11, 718-719n22
Hadar, Liat, 712n6, 712n8
Hadjicharalambous, Costas, 691n17
Hae-Kyong, Bang, 720n9
Haenlein, Michael, 690n6, 724n7
Hahn, Alexander, 708n8
Hair, Joe, 699n32
Halingten, Alannah, 696n10
Hall, JennaRae, 732n1
Hall, Joseph M., 726n8
Hall, Patricia K., 703n34
Hall, Zachary R., 715-716n5, 716n9, 717n17
Halligan, Brian, 439, 723n1, 723n4
Hamilton, Kathy, 701n7
Hamilton, Rebecca W., 694n31, 699n23, 720n8
Hamilton, Ryan, 691n3, 701-702n14, 712n8
Han, Jim K., 695-696n8
Han, Song, 725n30, 728-729n10
Handel, Benjamin R., 726n14
Handley, Ann, 723n4, 724n12
Handrich, Matthias, 704-705n16
Hanratty, Joe, 650-651
Hansen, J., 719n29
Hansen, John D., 710n3, 719n29
Hansen, Karsten, 728n2
Hansen, Scott W., 731n3
Hanson, Sara, 723n2
Hanssens, Dominique M., 698n4
Hardesty, David M., 691-692n7, 730-731n1
Hardesty, Dominique M., 722-723n41
Hardie, Bruce G. S., 728n8
Harding, Forrest E., 710n5
Hardy, Kelly, 593
Harlam, Bari A., 722-723n41, 728n2
Harmeling, Colleen M., 696-697n18, 700n2, 723n2
Harrington, Nick, 695n36
Harris, Eric G., 718n21
Harris, James Edwin, 689n22
Harris, Robert G., 685n9
Hartley, Janet L., 696n16
Hartman, Katherine B., 712n5
Hartmann, Nathaniel N., 701n7, 715-716n5
Hartmann, Wesley R., 720-721n13, 720n9, 720n10

Haruvy, Ernan, 696n15
Harvey, Jaron, 718-719n22
Harvey, Michael G., 703n31, 716-717n10
Hasford, Jonathan, 691-692n7, 730-731n1
Hatch, Mary Jo, 690n1
Häubl, Gerald, 729n14
Haulman, Clyde A., 732n6
Hauser, John R., 690n3, 698n4, 723n2
Hausknecht, Douglas, 630
Haviv, Avery M., 702n20
Haws, Kelly L., 729n15, 730-731n1
Hayati, Babak, 715-716n5
Haygood, Daniel M., 720n9
Haytko, Diana L., 704n10, 720n8
He, Chuan, 720n8
He, Stephen X., 723-724n5, 724n7
He, Yanqun, 727n25
Hean Tat Keh, 701n7
Heggde, Githa S., 718n21
Heide, Jan B., 696-697n18, 696n17, 700n2, 708n15, 708n18
Heidenreich, Sven, 704-705n16
Heijden, Gielis, 707n44
Heitmann, Mark, 694n311, 701-702n14
Helm, Sabrina V., 725n29
Henard, David H., 722-723n41
Hench, Thomas, 716-717n10
Henderson, Conor M., 687n7, 719n25, 722n38
Henderson, Ty, 703n27
Henke, J. W., 726-727n17
Hennig-Thurau, Thorsten, 706n36, 723-724n5, 725n32, 731n3
Henthorne, Tony L., 697n32
Herbig, Paul A., 703n35, 715n2
Herbst, Uta, 726n12
Herd, Kelly B., 693-694n30, 704-705n16
Herpen, Erica, 701n7
Herrmann, Andreas, 723n2
Hershfield, Hal E., 691n3
Hess, Ronald L., Jr., 686n14
Hewett, Kelly, 695n37, 701-702n14, 724n7
Hienerth, Christoph, 690n1
Higgins, Kathleen, 700n3
Higgins, Leighanne, 701n7
Hildebrand, Christian, 725n32
Hildebrandt, Lutz, 730n18
Hilgers, Ryan, 732n1
Hill, Donna J., 731n3
Hill, Ronald Paul, 693n28, 730-731n1
Hingston, Sean T., 694n31
Hinterhuber, Andreas, 729n13
Hirunyawipada, T., 706n36
Ho, Hillbun (Dixon), 696-697n18
Hoban, Paul R., 720-721n13
Hoch, Stephen J., 695n36, 706-707n42, 706n38
Hochstein, Bryan, 718-719n22
Ho-Dac, Nga N., 723-724n5
Hodge, Sharon K., 717n15
Hoeffler, Steve, 704-705n16
Hoegg, JoAndrea, 716n9
Hoffman, Donna L., 699n23, 706n36
Hoffman, K. Douglas, 719n24, 730n18
Hofmann, Julian, 730n18
Hofstede, Frenkel ter, 729-730n16
Hofstetter, Reto, 728-729n10
Hogreve, Jens, 701n7
Hohenberg, Sebastian, 717-718n19
Holder, Reed R., 729n13
Hollebeek, Linda D., 701n7
Hollmann, Thomas, 696-697n18
Holt, Debra J., 730-731n1
Holt, Douglas B., 691n19, 693n27, 693n28
Holtan, Marius, 699n30
Homburg, Christian, 685n10, 686n21, 694n311, 696-697n18, 700n2, 704-705n16, 708n8, 719n24, 724n17, 731n2, 731n4
Honeycutt, Earl D., 717n17

Honeycutt, Earl D., Jr., 717n15
Hong, J. W., 725n30
Hong, Jiewen, 692n17
Hong, Sungtak, 706n38
Hooley, Graham, 685n8
Hoppner, Jessica J., 696n10, 697n22
Horstmann, Ignatius, 724n19
Hosanagar, Kartik, 720-721n13
Hoskins, James A., 693n23
Houston, Mark B., 693n24, 696-697n18, 702n21, 706n36
Howlett, Elizabeth, 694n31
Hoyer, Wayne D., 695n34, 703n27, 703n32, 721-722n29
Hsee, Christopher K., 691n4, 694n32
Hsieh, Tony, 716n7
Hsu, Liwu, 701-702n14
Hu, Ye, 720-721n13
Huang, M. H., 707n43
Huang, Ming-Hui, 701n7
Huang, Minxue, 720-721n13
Huang, P., 725n30
Huang, Wen-yeh, 718n21
Huang, Zhongqiang (Tak), 694n31
Huckabee, Mike, 730-731n1
Hudnut, Paul, 658
Hudson, Elania J., 699n32
Hughes, Douglas E., 712n5, 715n14, 717-718n19, 717n15, 718-719n22, 718n21
Hulbert, James Mac, 699n28
Hulland, John, 699n26, 715-716n5
Hult, G. Tomas M., 692n12, 710n4, 710n9
Hung, C. L., 715n18
Hung, Kineta H., 694n31
Hunt, James M., 726n12
Hunt, Shelby D., 685n7, 686n23, 700n35
Hunter, Gary K., 716-717n10, 717n15, 718-719n22
Hunter, Gary L., 708-709n21, 717n15
Hunter-Gault, Charlayne, 727n28
Huntoon, Wade, 666
Hutt, Michael, 716-717n10
Huyghe, Elke, 713n27

I

Iacobucci, Dawn, 689n22, 693-694n30
Im, Subin, 715n15
Iman, Dhania, 732n1
Indu, P., 685n6
Ingene, Charles A., 708n18
Ingram, Thomas N., 717n17, 719n24, 719n29
Inman, J. Jeffrey, 693-694n30, 694-695n33, 723n2, 723n4, 728-729n10
Iqbal, Zafar, 706n36
Irmak, Caglar, 730-731n1
Irwin, Bindi, 459
Irwin, Richard D., 685n8
Isaac, Mathew S., 692n12
Israeli, Ayelet, 727-728n29
Itani, Omar S., 715-716n15
Iyengar, Raghuram, 723-724n5
Iyengar, Sheena S., 700n2
Iyer, Ganesh, 704-705n16

J

Jaakkola, Matti, 685n10
Jackson, Barbara P., 729n13
Jackson, Donald W., Jr., 717n13
Jacobs, Laurence, 702n22
Jacobs, Ron, 707n5
Jagabathula, Srikanth, 725n26
Jain, S., 693n22
Jamal, Zainab, 712n5, 720-721n13
James, Gareth M., 720n13
James, LeBron, 3
Janakiraman, Narayan, 706-707n42, 728-729n10
Janakiraman, Ramkumar, 725n32
Janiszewski, Chris, 691-692n7, 691n4, 693-694n30, 726-727n17, 727n25, 729n11, 730-731n1

Jap, Sandy D., 696n15, 696n17, 708n8, 729n14
Jaramillo, Fernando, 715-716n15, 718-719n22, 719n29
Jarvis, Cheryl Burke, 696-697n18, 701n7, 716n6
Javalgi, Rajshekar, 701n8
Jaworski, Bernard J., 686n13, 700n36, 717-718n19
Jayachandran, Satish, 686n13, 697n29, 698-699n19, 701-702n14
Jayanti, Rama K., 700-701n6
Jedlik, Tom, 702n22
Jeffery, Mark, 686n11
Jenkins, Roger L., 685n7
Jennings, Marianne M., 697n32
Jensen, Morten Berg, 690n1
Jensen, Thomas D., 710n9
Jepsen, Anna Lund, 694n31
Jerath, Kinshuk, 712n5, 720-721n13, 721n18
Jewell, Robert D., 704-705n16, 720n8, 721n28
Jewett, Heather, 732n1
Jeziorski, Przemyslaw, 732n5
Jha, Ajay, 658
Jhang, Ji Hoon, 706n36
Jiang, Baojun, 728n8
Jiang, Lan, 723n2
Jiang, Yang, 695-696n8
Jin, Liyin, 727n25
Jindal, Pranav, 728n5
Jindal, Rupinder P., 704-705n16, 717-718n19
Jing, Bing, 709n22
Jing, Xiaoqing, 713n33
John, Deborah Roedder, 691n17, 693n22
John, George, 703n27, 714n40, 717-718n19
John, Leslie K., 720n9
Johnson, D., 692n15
Johnson, D. A., 717n18
Johnson, Devon S., 717n15
Johnson, Jean L., 685n4, 686n23, 697n31
Johnson, Jeff S., 687n7, 695n37, 696-697n18, 715-716n5
Johnson, Joseph, 709n30
Johnson, Karen, 725n32
Johnson, M. W., 705n17
Johnson, Michael D., 699n32
Johnston, Wesley James, 707n2, 715n6
Jones, Bobby, 716n7, 722n39
Jones, Eli, 686n133, 715-716n5, 716-717n10, 717-718n19, 719n24
Jones, Laurie, 696n10
Jones, Michael H., 703n33
Jones, Thomas O., 732n7
Jordan, Michael, 3
Jorgensen, Steffen, 720n7
Joseph, Kissan, 715n18, 726n8
Josephson, Brett W., 685n4, 697n31
Joshi, Amit M., 723-724n5, 724n7
Joshi, Ashwin W., 696-697n18, 706n36
Joshi, Sachin, 658
Joyce, Mary L., 728n7
Jozic, Danijel, 685n10, 694n311, 700n2
Jung, Jin Ho, 715-716n5
Jung, Minah H., 726-727n17

K
Kabadayi, Sertan, 708-709n21
Kacker, Manish, 694n31, 704n4
Kaepernick, Colin, 3
Kahn, Kenneth B., 706n36, 718n21
Kähr, Andrea, 695n34
Kaikati, Andrew M., 686n21, 724-725n25
Kalafatis, Stavros P., 691n17
Kalaignanam, Kartik, 704-705n16
Kalmenson, Stuart, 719-720n2
Kalra, Ajay, 691-692n7, 694n32
Kaltcheva, Velitchka D., 712n5
Kalwani, Manohar U., 704n9
Kalyanam, Kirthi, 721n18
Kalyanaram, Gurumurthy, 719-720n2
Kamakura, Wagner A., 691n3, 695n36

Kamen, Dean, 631, 632, 633
Kamvar, Sepandar D., 694n32, 731n4
Kan, Christina, 727n25, 729n11
Kannan, P. K., 696-697n18, 701n12, 705n24, 708-709n21, 709n22, 721n18, 724n10, 729-730n16
Kanuri, Vamsi K., 725n32
Kao, Ling-Jing, 700-701n6
Kapur, Gour M., 705n23
Karan, Donna, 646
Karkkainen, Mikko, 710n6
Kashani, Kamran, 690n36
Kashyap, Vishal, 708n18, 712n3
Katona, Zsolt, 709n25, 720n13
Katsikeas, Constantine S., 709n30, 717-718n19, 732n10
Katzenstein, Robin, 694n311
Käuferle, Monika, 708-709n21
Kaufman, Brad, 732n1
Kaufman, Peter, 697n29, 698-699n19, 701-702n14
Kaufmann, Patrick J., 712n3, 727n27
Keane, John G., 699-700n33
Keane, M. P., 700n2
Keating, Wendy, 659
Kees, Jeremy, 703n29, 732n6
Keh, Hean Tat, 728-729n10
Keil, Bryant, 624
Keil, Sev K., 688n2
Keiningham, Timothy L., 706-707n42, 707n44, 731n2, 731n4
Keller, Ed, 693n23, 725n30
Keller, Kevin Lane, 701-702n14, 704-705n16, 714n2
Keller, Kristopher O., 703n27
Keller, Punam Anand, 703n29
Keller, Scott B., 710n3
Kelley, Scott W., 730n18
Kelly, Claire, 638-639
Kelly, Katie, 669
Kelting, Katie, 702n25
Kemp, Deborah J., 728n30
Kemp, Robert A., 695n34
Kennedy, Robert, 665-666
Keskiocak, Pinar, 711n30
Ketchen, David J., Jr., 710n4
Khan, S., 696n10
Khan, Uzma, 694n32, 730n25
Khodakarami, Farnoosh, 686n19
Kibarian, Thomas M., 730n19
Kidwell, Blair, 691-692n7, 730-731n1
Kiely, Thomas, 699n24
Kim, Anthony, 722-723n41
Kim, Chloe, 459
Kim, I., 714n2
Kim, J. Y., 729-730n16
Kim, Juran, 720-721n13
Kim, Keysuk, 708n15
Kim, M., 702n22
Kim, Namwoon, 687n7, 695-696n8, 708n18
Kim, Stephen K., 708n16, 708n18, 717-718n19
Kim, Taewan, 722n40
Kim, Tami, 720n9
Kim, W. Chan, 687n1
Kim, Youngju, 703n27
Kimes, Sheryl E., 715-716n5
King, D., 721n16
Kingston, Chelsea, 663-664
Kingston, Tim, 662
Kipnis, Skipper, 668
Kirby, J., 721n16
Kirby, Patrick N., 730n19
Kirca, Ahmet H., 686n13, 706n36
Kireyev, Paul, 726n3, 729-730n16
Kirkpatrick, Jerry, 692n7
Kirkup, Malcolm, 700n35
Kirmani, Amna, 691n4
Kishore, Sunil, 717-718n19
Kitchen, P. J., 714n2
Kiyak, Tunga, 690n6

Klapper, Daniel, 720n10
Klarmann, Martin, 719n24
Klee, Alexander, 731n3
Kleijinen, Mirella H. P., 699n23
Klein, Calvin, 459
Klein, Noreen M., 686n14
Kleinaltenkamp, Michael, 696-697n18
Klink, R. R., 705n18
Knaus, Tristan, 643
Knight, Phil, 3
Knight, Susan J., 693n22
Knox, George, 694n32, 695n34, 731n2
Koehl, Maryse, 717n18
Koenigsberg, Oded, 712n5, 729-730n16
Köhler, Clemens F., 725n30
Kohli, A., 717n15
Kohli, Ajay K., 686n13, 695-696n8, 696-697n18, 718n21
Kohli, Chiranjeev, 688n2
Kolsarici, Ceren, 724-725n25
Konus, U., 708-709n21
Kopalle, Praveen K., 706n36, 713n35
Kopp, Steven W., 689n16
Korgaonkar, Pradeep, 689n33
Korschun, Daniel, 686n21, 687n24
Kortge, G. Dean, 728n6, 730n22
Koschate-Fischer, Nicole, 703n27
Kotabe, Masaaki, 695n37
Kothandaraman, P., 97n21
Kowinski, William, 625
Kozlenkova, Irina V., 688n4, 723n2
Krafft, Manfred, 717-718n19, 717n15, 725n26
Kramer, Hugh E., 715n2
Kranzbühler, Anne-Madeleine, 699n23
Krapfel, Robert E., 716-717n10
Kraus, Florian, 686n133, 717n17
Krieger, Ruth H., 703n36
Krishna, Aradhna, 694n32, 701n13, 713n16, 721-722n29, 722-723n41, 729n15
Krishnamoorthy, Anand, 717n17
Krishnan, H. Shanker, 729n15
Krishnan, M. S., 731n4
Kristofferson, Kirk, 722-723n41
Krohmer, Harley, 695n34, 728-729n10
Kruger, Michael W., 713n16
Kübler, Raoul, 728n8
Kuehnl, Christina, 685n10, 694n311, 700n2, 704-705n16
Kukar-Kinney, Monika, 726-727n17, 726n9
Kuksov, Dmitri, 709n22
Kumar, Ajith, 695n37, 716-717n10
Kumar, Alok, 696-697n18, 696n11, 709n30
Kumar, Ashish, 725n32
Kumar, Nanda, 706n38
Kumar, Nirmalya, 708n13
Kumar, Sameer, 711n25
Kumar, V., 686n15, 686n133, 687n7, 691n17, 696-697n18, 701-702n14, 701n7, 701n8, 706n36, 708-709n21, 709n25, 715-716n5, 715n15, 724-725n25, 725n30, 725n32, 726-727n17, 728n8
Kumar, Vineet, 726n3, 729-730n16
Kuo, Andrew, 691n4
Kupfer, Ann-Kristen, 706n36
Kurt, Didem, 694n31
Kurtuluş, Mümin, 708n15
Kushwaha, Tarun, 704-705n16, 708-709n21, 715n15
Kwon, Soongi, 708n18
Kwon, W. S., 713n33

L
Labrecque, Jennifer, 695n36
Labrecque, Lauren I., 692n8
Laczniak, Russell N., 691-692n77
Ladd, Robert T., 686n15
Lafley, A. G., 687n3
Lafontaine, Francine, 717-718n19
LaForge, Raymond W., 719n24, 719n29

Lai, Christine Jaushyuam, 716n9
Lal, Rajiv, 713n274
Laliberté, Guy, 33
Lalwani, Ashok K., 693n24, 726-727n17, 727n23, 728n8
Lam, Simon S. K., 706n36
Lam, Son K., 686n13, 707n2, 716-717n10, 717-718n19, 718-719n22, 719n25
Lamb, M. M., 730-731n1
Lambert, Douglas, 710n13
Lamberton, Cait, 720-721n13
Lambrecht, Anja, 720-721n13, 726-727n17, 727n18
Landry, Timothy D., 718-719n22
Langenderfer, Jeff, 689n16
Lanier, X., 695n34
Lankow, Jason, 724n15
Laporte, Sandra, 694-695n33, 722-723n41, 730-731n1
Laran, Juliano, 691-692n77
Larson, Jeff, 725n32
Larson, Jeffrey S., 691n3
Lassk, Felicia G., 715-716n5, 717n17
LaTour, Michael S., 697n32
Lauga, Dominique Olie, 727n23
Lauren, Ralph, 646
Laurent, Gilles, 722-723n41, 728-729n10
Lavalle, Steve, 690-691n13
Laverie, Debra A., 716n6
Lawrence, Justin M., 707n2
Le, Angelina Nhat Hanh, 707n2
Le Boeuf, Robyn A., 693-694n30
Lee, Hannah S., 697n22, 697n29
Lee, J. H., 717n15
Lee, J. K., 721n16
Lee, Jonathan, 703n27
Lee, Jongkuk, 704-705n16
Lee, Ju-Yeon, 697n31, 707n2
Lee, L., 724n17
Lee, Leonard, 691n3, 693-694n30, 726-727n17, 727n18
Lee, Michelle P., 691n3, 693-694n30
Lee, Mira, 720n9
Lee, Nick, 718n21
Lee, Sangwon, 704-705n16
Lee, Seung Hwan (Mark), 721-722n29
Lee, Yih, 701n7
Lee, Yikuan, 715n15
Lee, You-Cheong, 716n9
Leghorn, Robert, 708-709n21
Lehmann, Donald R., 692n8, 694n32, 695n36, 699-700n33, 701n7, 701-702n14, 704-705n16, 715n18, 725n30, 726-727n17, 729n15, 730-731n1
Leigh, Thomas W., 715-716n5, 719n28
Leimbach, Martin, 708-709n21
Leingpibul, Duke, 695-696n8
Lembke, Ronald Tibben, 709n28
Lemon, Katherine N., 687n10, 690-691n13, 694n32, 696-697n18, 701n7
Lennon, S. J., 713n33
Leone, Robert P., 686n133, 715-716n5, 725n30
Leong, K. F., 706n36
Leonidou, Constantinos, 732n10
Lerman, Dawn, 725n30
Leszczyc, Peter T. L. Popkowski, 729n14
Lettl, Christopher, 690n1
Leuthesser, Lance, 688n2
Levine, David K., 705n20
Lewin, Jeffrey E., 695-696n8
Lewis, Michael, 713n33
Li, Charlene, 690n1
Li, En, 693n28
Li, Hongshuang (Alice), 708-709n21, 721n18
Li, Julie, 696-697n18
Li, Julie Juan, 696-697n18
Li, Krista J., 726n3, 729-730n16
Li, Mei, 706-707n42, 707n44
Li, N., 695-696n8

Li, Pochien, 718-719n22
Li, Shanjun, 693n24
Li, Stella Yiyan, 694n31
Li, Xiaoling, 720-721n13
Li, Yang, 691n3
Liang, T. P., 721n16
Liang, Yitian (Sky), 694n31
Liao, Chenxi, 709n22
Liaukonyte, Jura, 720n9
Libai, Barak, 690n6, 723-724n5, 724n7
Liberali, Guilherme (Gui), 720-721n13
Lichtenstein, Donald R., 724-725n25, 727n25, 729n11, 730-731n1
Liedtka, Jeanne, 706n33
Lievens, Annouk, 705-706n27, 705n26
Lijia, Wei, 696n10
Liker, Jeffrey K., 697n24, 705n26
Lilien, Gary L., 696-697n18, 696n11, 698n12, 716n9, 719n23, 722n40
Lilly, Bryan, 723-724n5
Lim, C. S., 706n33
Lim, Elison, 701n7
Lim, Hyunwoo, 724n19
Lim, Jeremy M., 711n2
Lim, Noah, 717-718n19
Lim, Sarah, 700n2
Ling-Yee, Li, 722n40
Liu, D., 720-721n13
Liu, Qing, 713n274
Liu-Thompkins, Yuping, 691n4
Livelsberger, Jeanne, 719-720n2
Lloyd, S. M., 713n33
Lo, Desmond (Ho-Fu), 717-718n19
Locander, William B., 719n29
Lodish, Leonard M., 719-720n2
Loe, Terry W., 719n29
Loewenstein, George, 691-692n7
Lohtia, Ritu, 710n9, 714n47
Lorimer, Sally E., 715-716n5
Loten, Angus, 686n12
Loureiro, Yuliya Komarova, 722n37
Lovelock, Christopher H., 703n31
Lovett, Mitchell J., 723-724n5
Low, George S., 715n6
Lowry, Adam, 471
Lu, Long-Chuan, 686n23
Lu, Xianghua, 724-725n25
Lucchesi, Katie, 658
Luce, Mary Frances, 692n15
Luckett, M., 713n33
Ludwig, Stephan, 725n32
Lueg, Jason E., 712n5
Luffarelli, Jonathan, 701n13
Lukas, Bryan A., 700n2, 704-705n16, 724n17
Lund, Donald J., 708-709n21, 719n29
Luo, Anita, 696-697n18
Luo, Lan, 720n13
Luo, X. M., 731n2
Luo, Xueming, 686n21, 704-705n16, 720n9, 726-727n17, 726n15, 727n18, 731n4
Luoma, Jukka, 685n10
Lurie, Nicholas H., 723-724n5, 725n30, 730-731n1
Lusch, Robert F., 701n7, 708n15
Lutz, Richard J., 691-692n77
Lutzky, Christian, 722-723n41
Lynch, J. G., 705n18

M

Ma, Jingjing, 692n12
Ma, Liye, 720-721n13, 729-730n16
Ma, Yu, 730-731n1
Ma, Zhenfeng, 695n36, 704-705n16
Macdonald, Emma K., 696-697n18
MacDonald, Erin, 720-721n13
Machleit, Karen A., 718n21
Maciel, Andre F., 692n17
Macinnis, D. J., 725n30
MacKenzie, Scott B., 716-717n10

Macklemore & Lewis, 183
MacMillan, Ian C., 698-699n19, 732n8
Macy, Barry A., 719n24
Madan, Vibhas, 726-727n17
Madden, Thomas J., 701-702n14
Mady, Sarah, 718-719n22
Magid, Frank N., 689n30
Maglio, Paul P., 701n7
Mahajan, Vijay, 689n23, 691n3, 694-695n33, 702n22, 729-730n16
Maheswaran, Durairaj, 689n13
Mahr, Dominik, 725n26
Mai, Feng, 723n2
Main, Kelley J., 706n29
Majic Johnson, 431
Malaviya, Prashant, 694n31, 723n2
Malek, Stacey L., 717-718n19
Malhotra, Naresh K., 688n13, 700n35
Malkewitz, K., 703n28
Mallapragada, Girish, 696-697n18, 709n30, 713n27
Mallin, Michael L., 715-716n5
Mallinckrodt, Victoria, 720n9
Malshe, Avinash, 686n17
Malthouse, E. C., 720-721n13
Maltz, E., 697n24
Mamadehussene, Samir, 726-727n17
Manchanda, Puneet, 694n31
Mandel, Naomi, 701n7
Mandl, Leonhard, 701n7
Mangleburg, Tamara F., 693n22
Mangus, Stephanie M., 696-697n18, 715-716n5
Manley, M., 703n34
Manning, Kenneth C., 729n15, 730n19
Manolis, Chris, 692n18
Mantrala, M., 717n15
Mantrala, Murali K., 717n18
Marcus, Bernie, 319
Mariadoss, Babu John, 696-697n18, 697n31
Marinova, Detelina, 708-709n21, 718-719n22
Markham, Stephen K., 706n38
Markos, Ereni, 692n8
Marmorstein, Howard, 730n25
Marn, Michael V., 726n8
Marquardt, Adam J., 703n27
Marshall, Greg W., 715-716n5, 717n11, 717n15, 717n17
Martin, Kelly D., 685n4, 686n23, 693n28, 698n15, 698n16, 730-731n1, 732n11
Martin, Xavier C., 717-718n19
Martinez-Jerez, F. Asis, 723n1
Martin-Herran, G., 715n14, 722-723n41
Maschke, Nils, 715-716n5
Maslow, A. H., 691n4
Maslow, Abraham, 120-121
Mason, Charlotte H., 715n15
Mastbergen, Dan, 658
Matear, Sheelagh, 686n15
Matherly, Ted, 691n4
Mathieu, John E., 716-717n10
Matthews, Brian P., 717n17
Matthyssens, P., 703n35
Mattila, Anna S., 692n8, 694-695n33
Mauborgne, Renée, 687n1
Maxham III, James G., 717n17, 731n3
May, Alex, 636
Mayberry, Robert, 718-719n22
Mayer, David, 715-716n5
Mayer, Morris, 712n10
Mayzlin, Dina, 693n23, 725n26
Mazodier, Marc, 722n38
Mazumdar, Tridib, 722n40, 729n15
Mazvancheryl, Sanal K., 731n4
Mazzella, Frédéric, 125
Mcafee, R. Preston, 720-721n13
McCabe, Deborah Brown, 717n13
McCarthy, Daniel M., 728n8
McCracken, Gail K., 695n36
McCracken, Grant, 693n27

McCullough, Heath, 722n38
McDermott, Christopher M., 704n4
McDonald, William J., 699n24
McFarland, Richard G., 708n18, 718-719n22, 719n28
Mcferran, Brent, 722-723n41
McGee, Gail W., 695n36
McGill, Ann L., 692n15
McGlynn, Larry, 450
McGovern, Gail J., 687n2
McGowan, Heather, 704n8
McGrath, M. A., 699n28
McGrath, Rita G., 732n8
McGraw, A. Peter, 694n31
McInnes, William, 685n8
McIntyre, Roger P., 719n28
McMillan, Sally J., 720-721n13
McNeilly, Kevin M., 710n15
McQuarrie, Edward F., 721-722n29
Mehrotra, Anuj, 708-709n21
Mehta, Raj, 696-697n18
Mehta, Ravi, 704-705n16
Mela, Carl F., 708-709n21, 708n7, 723n2
Melancon, Joanna Phillips, 728-729n10
Melumad, Shiri, 723n2
Mende, Martin, 693n23
Mengüç, Bülent, 707n44, 712n5, 716n9, 717-718n19
Menon, Anil, 706n32
Menon, Geeta, 728-729n10
Menon, Kalyani, 719n28
Mentzer, John T., 686n15, 698n12, 710n5, 710n9
Meredith, Jack R., 703n36
Meuter, Matthew L., 706-707n42
Meyer, Jeffrey, 730n24
Meyer, Robert J., 706-707n42, 728-729n10
Meyer, Tracy, 687n7
Meyvis, Tom, 691n3, 694-695n33, 701-702n14
Miao, C. Fred, 717-718n19, 718-719n22
Michael, Steven C., 706n36
Michaels, Ronald E., 718n21
Michell, Paul C. N., 704n6
Mick, David Glen, 685n4, 720n8
Middleton, Neal, 636-637
Milberg, Sandra J., 703n29
Mild, Andreas, 690n6
Miller, Amy, 698n11
Miller, Chip E., 713n32
Miller, Gina L., 700n35
Miller, Klaus M., 728-729n10
Mills, Anthony, 698n10
Milne, George R., 699n24, 722n37
Min, Hokey, 715n15
Min, Sungwook, 687n7, 704n9
Miniard, Paul W., 700-701n6, 729n15
Misra, Kanishka, 706n38, 726n14, 728n2
Misra, Sanjog, 717n17, 726-727n17
Mithas, Sunil, 731n4
Mitra, Debanjan, 700n2, 729n11
Mitra, S., 725n30
Mittal, Vikas, 690n13, 694-695n33, 723-724n5, 731n4
Mittelman, Mauricio, 730n24
Mittelstaedt, Robert A., 708n7
Mixon, Malachi "Mal," 630-632
Miyazaki, Anthony D., 694n31, 720n8
Mizerski, Dick, 720n9
Moberg, Christopher R., 710n9
Mochon, Daniel, 725n32
Moe, Wendy W., 699n22, 703n29, 725n32
Moenaert, Rudy K., 705n26
Moffett, Jordan, 723n2
Mogilner, Cassie, 693-694n30, 694n32, 731n4
Mohr, Gina S., 730-731n1
Mohr, Jakki J., 705-706n27, 708n8, 715n6
Mohr, Lois A., 706-707n42
Monaghan, James, 407
Monaghan, Tom, 407
Moncrief, William C., 715-716n5, 716-717n10
Money, R. Bruce, 695n37

Monga, A., 691n3
Monga, Alokparna Basu, 686n21
Monroe, Kent B., 726-727n17, 726n9, 728-729n10, 728n7, 729n15, 730n25
Montano, Tim, 732n1
Montoya-Weiss, Mitzi M., 705n26
Moody, Carlisle E., Jr., 732n6
Mooi, Erik A., 696-697n18
Moon, Junyean, 691n17
Moon, Mark A., 716-717n10
Moon, Sangkil, 693n24, 729n15
Moon, Young Sook, 722n34
Moon, Youngme, 687n8, 704n4, 729n15
Moore, Elizabeth S., 685n4, 685n7, 686n21, 691-692n77, 730-731n1
Moore, Geoffrey A., 715n15
Moore, Jarvis B., 717n15
Moore, Robert, 693n23
Moore, Sarah G., 724-725n25
Moore, William L., 705n22, 723-724n5
Moorman, Christine, 688n2, 698-699n19, 700n2, 703n29
Moorthy, Sridhar, 702n20
Morales, Andrea C., 712n8, 722-723n41, 728-729n10
Moreau, C. Page, 693-694n30, 706n36, 722-723n41
Morgan, Cheyne, 653-654
Morgan, F. N., 695n37
Morgan, Fred W., 732n6
Morgan, Neil A., 686n13, 687n10, 700n2, 709n30, 731n2, 731n4, 732n10
Morgeson, Forrest V., III, 731n4
Morn, M. P., 720-721n13
Morrin, Maureen, 703n27
Morris, Linda, 719n24
Morris, Michael H., 726n12, 728n7, 729n13
Morwitz, Vicki G., 699n26, 712n6, 712n8, 726n9, 729-730n16
Moss, Roberta, 706n36
Motyka, Scott, 725n26
Mourali, Mehdi, 695n36, 704-705n16
Mourouti, O., 703n30
Mowen, John C., 706-707n42, 718n21
Mozer, Robert, 732n1
Mu, Jifeng, 694n311
Mukherjee, Ashesh, 721-722n29
Mulki, Jay Prakash, 718n21, 719n29
Muller, Etian, 723-724n5, 725n26
Müller, Michael, 719n24
Mullins, Ryan R., 716n9, 718-719n22
Muniz, A. M., 690n1, 699n28
Muniz, Albert M., Jr., 724n17
Murphy, Patrick E., 686n23, 692n18, 698n15, 700n35, 709n24, 722n37, 732n11
Murphy, William H., 717n18
Murray, Janet Y., 687n2, 695n37, 706n36
Murray, Keith B., 689-690n34
Murtha, Brian R., 700n36, 712n3
Musk, Elon, 431
Musk, Kimball, 431
Myers, James H., 694n31
Myers, Matthew B., 686n133, 690n6, 711n25, 716-717n10

N
Nagle, Thomas T., 729n13
Naik, Prasad A., 715n18, 719-720n2
Nam, Hyoryung, 705n24
Narasimhan, Om, 703n27, 705n18, 706n36, 717-718n19
Narayan, Vishal, 714n38
Narayanan, Sridhar, 721n18
Narayandas, Das, 696-697n18, 701n7
Narus, James A., 686n15, 701n7, 725n26
Narver, John C., 686n13, 686n14, 698-699n19
Natter, Martin, 690n6, 729-730n16
Neal, David, 695n36

Negroponte, Nicolas, 732n13
Nelson, Leif D., 726-727n17
Neslin, Scott A., 701-702n14, 708-709n21, 708n7, 713n33, 720-721n13, 722-723n41, 726-727n17
Netemeyer, Richard G., 701-702n14, 703n29, 731n3, 732n6
Netzer, Oded, 698n11, 729n11
Neuenschwander, Bill, 633
Neuenschwander, Emily, 633
Nevin, John R., 708n8
Newberry, Peter, 728n5
Newell, Stephen J., 695-696n8
Newman, Cassidy, 666-668
Newman, Christopher L., 703n29, 722n38
Newman, George E., 701-702n14
Ng, I. C., 701n12
Ngobo, Paul-Valentin, 731n4
Ngwe, Donald, 700-701n6
Nicholson, Carolyn Y., 696n17
Nickels, William G., 685n7
Niedrich, Ronald W., 704n4, 729n15
Nielsen, Jesper H., 694n32, 721n25
Nijs, Vincent R., 728-729n10, 728n2
Nijssen, Edwin, 707n44
Niraj, Rakesh, 710n9
Noble, Charles H., 728-729n10
Noble, Stephanie M., 712n5, 712n6, 717-718n19, 728-729n10
Normann, Karen, 715-716n5
Norris, Bruce I., 724-725n25
Noseworthy, Theodore J., 694n31, 721-722n29
Novak, Thomas P., 699n23, 706n36
Novell, Corrine A., 718n21
Novemsky, Nathan, 693-694n30
Novicevic, Milorad M., 716-717n10
Nowlis, Stephen M., 712n8
Nunes, Joseph C., 691-692n7
Nyffenegger, Bettina, 695n34
Nygaard, Arne, 708n8

O
O'Brien, Conan, 183
O'Connor, Gina Colarelli, 704n4, 715n15
O'Driscoll, Aidan, 695n36
Ofek, Elie, 687n7, 705n22, 709n25, 726n3, 729-730n16
Ofir, Chezy, 692n14, 729n15
Ogawa, Susumu, 706n32
Ogbogu, Eric, 103, 204
Ogden, C. L., 730-731n1
Ogilvie, Jessica, 718-719n22
O'Guinn, Thomas C., 724n17, 732n6
Oh, Changho, 708n15
Oh, Katie, 648-649
Okada, Erica Mina, 729n15
Okonkwo, Patrick A., 730n22
Olsen, Janeen E., 702n22
Olsen, Mitchell C., 692n12
Olshavsky, Richard W., 694n31
Oppewal, Harmen, 693-694n30
Ordabayeva, Nailya, 703n28, 730-731n1
Ordanini, Andrea, 707n44
Ordenes, Francisco Villarroel, 725n32
Orsingher, Chiara, 731n3
Orth, U. R., 703n28
Ortmeyer, Gwendolyn K., 727n27
Osmonbekov, Talai, 691-692n77, 693n23
Ostrom, Amy L., 706-707n42, 716n6, 725n30
Ottum, Brian D., 705n22
Ou, Yi-Chun, 687n7
Ovans, Andrea, 730n25
Overdorf, M., 705n17
Ozanne, Julie L., 689n22

P
Packard, Grant, 725n32
Pae, Jae H., 695-696n8
Paesbrugghe, Bert, 696-697n18

Page, Albert L., 706n36
Paharia, Neeru, 704-705n16
Palandjian, Tracy Pun, 706n36
Palmatier, Robert W., 687n7, 688n4, 695-696n8, 696-697n18, 697n32, 698n15, 698n16, 700n2, 708n8, 715-716n5, 718-719n22, 719n25, 720-721n13, 722-723n41, 732n11
Palmeira, Mauricio M., 727n23
Palmer, Ryan, 658
Palumbo, Frederick, 703n35
Pan, Xing, 730n22
Panagopoulos, Nikolaos G., 716-717n10, 716n9, 718-719n22
Pandelaere, Mario, 730-731n1
Pani, Abhishek, 721n18
Paolacci, Gabriele, 693-694n30
Papies, Dominik, 729-730n16
Pappas, James M., 717n17
Pappu, Madhav, 712n3
Pappu, Ravi, 722-723n41
Parasuraman, A., 701n7
Pardo, Catherine, 716-717n10
Parish, Janet Turner, 706-707n42
Park, Jeong Eun, 715-716n5, 717n18
Park, Ji Kyung, 691n17
Park, Jung Kun, 717n11
Park, Kyung M., 704-705n16
Parker, Jeffrey R., 693-694n30, 694n32, 704-705n16, 730-731n1
Parry, Mark E., 708n18, 715n17
Parvinen, Petri, 718-719n22
Paswan, A. K., 706n36, 712n3
Pathak, Bhavik, 694n31
Patil, Ashutosh, 717-718n19
Paulson, Courtney, 720n13
Paun, Dorothy, 730n25
Pauwels, Koen, 701-702n14, 722-723n41, 728-729n10, 728n8, 729-730n16
Pavlou, Paul A., 725n32
Payne, A. F., 686n15
Payne, Adrian, 691n17
Pearson, Chantal, 715n18
Pearson, John N., 710n7
Pechmann, Cornelia, 693n22
Pelham, Alfred M., 719n28
Peracchio, Laura A., 703n29
Percival, Edward, 704n6
Peres, Renana, 723-724n5
Perreault, William D., Jr., 696-697n18, 698n12, 710n3, 717n15, 718n21
Peters, Kay, 715n18
Petersen, J. Andrew, 686n19, 709n25, 715n15
Peterson, R., 717n15
Petty, Richard E., 693-694n30
Petty, Ross D., 685n9, 730-731n1
Peyrot, Mark, 698-699n19
Pfeifer, Phillip E., 727n22
Pfeil, Christian, 715-716n5
Pham, Michel Tuan, 723n2, 730-731n1
Phieler, Ulrike, 724n7
Phillips, Barbara J., 721-722n29, 732n6
Phillips, Carol B., 701-702n14
Pick, Doreén, 728n8
Piller, Frank T., 706n32
Pingol, Laura L., 694n31
Pink, Daniel, 715n3
Pinkerton, Richard L., 728n30
Pitt, Leyland, 699n28
Plank, Kevin, 203-204
Plouffe, Christopher R., 715-716n5, 718-719n22
Podsakoff, Philip M., 716-717n10
Pohlson, Matt, 431
Pomirleanu, N., 713n33
Ponder, Nicole, 712n5
Pongpatipat, Chatdanai, 718-719n22
Popa, Monica, 702n19
Pope, N. K. L., 721-722n29

Popper, Edward T., 689-690n34
Poppo, Laura, 696n17
Posselt, Thorsten, 715-716n5
Poujol, Juliet F., 717n18, 717-718n19
Powers, Thomas L., 717n17, 728n6, 731n3
Poynor, Cait, 712n5
Pöyry, Essi, 718-719n22
Pozza, Ilaria Dalla, 701n7
Prabhu, Jaideep C., 704n4
Prahalad, C. K., 685n6
Prasad, Ashutosh, 702n22, 717-718n19, 717n17
Prendergast, Anna, 732n1
Price, Linda L., 692-693n21, 699n23, 701-702n14, 715n15
Prigge, Jana-Kristin, 708-709n21
Prior, Diego, 686n13
Provence, Jackie, 662-663
Prugl, Reinhard, 705n26
Pullig, Chris, 701-702n14
Pullins, Ellen Bolman, 715-716n5, 717n15
Punj, Girish N., 690n9, 691n17, 693n23
Purohit, Devavrat, 712n5
Pusateri, Michael A., 716-717n10
Pyke, David F., 726n8

Q
Qi, Jiayin, 694n311
Quach, Sara, 716n9
Quelch, John A., 687n2
Quesenberry, Keith A., 714n3
Quinn, Peter, 704n6

R
Rabinovich, Elliot, 710n3
Raghubir, Priya, 723n44, 726-727n17
Raghunathan, Rajagopal, 691n3
Raj, S. P., 729n15
Rajagopal, Priyali, 730-731n1
Rajan, Bharath, 701n7
Rajan, Rhithari, 685n6
Raju, Sekar, 730-731n1
Raman, Kalyan, 718n21
Raman, Pushkala, 698-699n19
Ramaswami, Sridhar N., 708n8, 717-718n19
Ramond, Olivier, 731n4
Ranasinghe, Kulan, 727n18
Rand, William, 724n7
Rangan, V. Kasturi, 685n6, 696-697n18
Rangarajan, Deva, 696-697n18
Rangaswamy, Arvind, 698n12
Ranjan, Kumar, 686n15
Rao, Akshay R., 730n25
Rao, Raghunath Singh, 717-718n19
Rao, Vithala R., 701n7, 714n38
Rapp, Adam, 688n6, 715-716n5, 717-718n19, 717n15, 718-719n22
Rashid, Karim, 471
Ratchford, Brian T., 726n9, 730n22
Ratneshwar, S., 691n17
Ravishanker, Nalini, 708-709n21
Ray, Sourav, 714n40
Rayport, J. F., 721n16
Rayport, Jeffrey V., 694n32
Read, Stuart, 686n15
Redman, Tom, 717n17
Reed, Americus, II, 693-694n30, 693n22
Reeves, Keanu, 624
Rego, Lopo L., 731n2
Reibstein, David J., 727n22
Reid, Jane S., 627
Reiferscheid, Ines, 704-705n16
Reimann, Martin, 698n11
Reinartz, Werner J., 686n15, 696-697n18, 699n26, 701n7, 708-709n21, 720n9
Reinholtz, Nicholas, 693-694n30
Reizenstein, Richard C., 718n21
Rentz, Joseph O., 717n18, 718n21

Reutterer, Thomas, 690n6
Reynolds, Kristy E., 712n5, 712n10, 713n33
Reynolds, Thomas J., 701-702n14
Reynolds, William H., 694n31
Rialp, Josep, 686n13
Richards, Keith A., 717-718n19
Richardson, Lance, 647
Richardson, Vernon J., 715n18
Richey, J. B., 630, 631
Richey, R. Glenn, 709n25, 709n28, 715-716n5
Rideout, Victoria J., 730-731n1
Riegner, C., 689n20
Ries, Al, 691n17
Riggle, R., 719n29
Rindfleisch, Aric, 697n24, 708n18
Ringold, Debra Jones, 685n4
Rink, David R., 695n36
Rinke, Morton, 651-653
Rippé, Cindy B., 714n38
Rishika, Rishika, 725n32
Risselada, Hans, 715n152
Ritchie, Josh, 724n15
Rizocki, Mallory, 661-662
Rizzo, Kelly, 287
Roberts, James A., 692n18
Robicheaux, Robert A., 709n24
Robinson, Leroy, Jr., 717n15
Robinson, Melissa Grills, 730-731n1
Robinson, Patrick J., 696n9
Robinson, Stacey G., 701n7, 701n8, 717n15
Robinson, Stefanie Rosen, 730-731n1
Robinson, William T., 704n9
Rocco, Rich, 717n15
Rodriguez, Teresa, 732n1
Roe, Justin, 732n8
Roehm, Michelle L., 725n26
Rogers, Dale S., 709n28
Rogers, Everett M., 695n36, 715n15
Rogers, Roo, 692n11
Rogers, Tatiana, 671-673
Rogeveen, A., 701n7
Roggeveen, Anne L., 726-727n17
Rohm, Andrew J., 722n37, 725n30
Rokka, Joonas, 693-694n30
Romero, Marisabel, 732n6
Ronaldo, Cristiano, 3
Rook, Dennis W., 703n32
Roosevelt, Franklin Delano, 630
Rosa, José Antonio, 689n22, 693n28
Rosario, Ana Babic, 724-725n25
Rose, Gregory M., 686n23, 708n13
Rose, Randall L., 697n29, 727n18
Rosenau, Milton D., 699n32
Rosenberg, Bill, 173
Rosenbloom, Bert, 708-709n21
Rosenthal, Stephen R., 699n28
Rosiello, Robert L., 726n8
Ross, Stephen D., 690n9
Ross, William, 696-697n18
Ross, William T., Jr., 731n4
Rossi, C. A. V., 691-692n77
Roth, Martin S., 702n19
Roush, Chris, 711n1
Rouziès, Dominique, 718n21
Rouziou, Maria, 718-719n22
Rowley, Jennifer, 698n10
Roy, Subroto, 710n9
Roy, Tirthankar, 730n25
Rubel, Olivier, 717-718n19
Rubera, Gaia, 704-705n16
Rubinson, Joel, 691n6, 698-699n19
Ruekert, Robert W., 686n13, 687n5, 706n32
Ruhlman, Michael, 712n13
Ruiliang, Yan, 720n7
Rusetski, Alexander, 727n25, 729n11
Russ, Frederick A., 710n3
Russell, Gary J., 729n15

Rust, Roland T., 701n7, 706-707n42, 707n43, 707n44, 724n7, 728n7, 731n4
Ruth, Julie A., 693n28
Rutherford, Brian W., 717n11
Rutz, Oliver J., 720-721n13
Ruvio, Ayalla, 696-697n18
Ryan, Eric, 471

S

Sabnis, Gaurav, 719n23
Saboo, Alok R., 706n36
Sackett, Angela, 732n1
Sager, J. K., 717n18
Sager, Jeffrey K., 716-717n10
Sahlman, William A., 732n8
Sahni, Navdeep S., 715n9
Saini, Amit, 696-697n18, 709n30
Saini, R., 691n3
Sallee, Amy, 719n28
Samaha, Stephen A., 688n4, 696-697n18, 697n32
Samiee, Saeed, 685n8
Samli, A. Coskun, 702n22
Samuelsen, Bendik M., 718-719n22
Sangtani, Vinita, 710n9
Sarangee, Kumar R., 704-705n16
Sarin, Sharad, 705n23, 717n15
Sarkar, M. B., 708n8
Sarvary, Miklos, 709n25, 720n9, 720n13, 721n16, 721n22
Sasser, W. E., 732n7
Satornino, Cinthia B., 718n21
Sattler, Henrik, 701-702n14
Sautter, Pookie, 724n17
Sauvage, Thierry, 710n4
Savaskan, R. Canan, 709n25
Savitt, Ronald, 713n32
Sawyer, Alan G., 729n15
Sayedi, Amin, 720-721n13, 721n18
Schaedel, U., 720-721n13
Schamp, Christina, 694n311
Schau, H. J., 690n1
Schau, Hope Jensen, 701-702n14
Scheer, Lisa K., 695-696n8, 696-697n18, 715-716n5
Schepers, Jeroen, 704-705n16, 707n44
Scherer, Anne, 701-702n14
Schilke, Oliver, 698n11
Schillewaert, Niels, 717n15
Schindler, Robert M., 730n19
Schirr, Gary R., 706n36
Schlager, Tobias, 723n2, 725n32
Schleicher, Martin G., 704-705n16
Schlesinger, L. A., 701n7
Schlossberg, Mallory, 686n17, 686n18
Schlosser, A. E., 713n33
Schmitt, B. H., 701-702n14
Schmitt, P., 725n29
Schmitz, Christian, 716n9
Schniederjans, Marc J., 710n6
Schoenherr, Tobias, 697n22
Schöler, Lisa, 724n9
Schons, Laura Marie, 729n11
Schreier, Martin, 705n26
Schrock, Kevin, 648-649
Schroder, Bill, 696n10
Schul, Patrick L., 718-719n22
Schultz, D. E., 714n2
Schultz, Majken, 690n1
Schulze, Christian, 724n9
Schumann, Jan H., 720-721n13
Schvaneveldt, Shane, 690n6
Schwartz, Alan, 694-695n33
Schwartz, Eric M., 720-721n13, 724-725n25
Schwartz, Janet, 692n15, 725n32
Schweidel, David A., 699n22, 720n9, 725n32
Schwemmle, Martin, 704-705n16
Schwepker, Charles H., Jr., 717n18, 719n29
Schweppe, F. C., 726n5

Scott, David Meerman, 723n1, 723n4, 724n12, 724n13, 724n14, 724n22
Scott, Maura L., 693n23
Seetharaman, P. B. (Seethu), 728n8
Sego, T., 717n15
Seiders, Kathleen, 685n9, 708-709n21, 712n5, 730-731n1
Selden, Larry, 698-699n19
Seldin, Daniel, 731n2
Sen, Sankar, 686n21, 702n22
Sen, Shahana, 725n30
Senecal, Sylvain, 708n7, 717n15
Sénécal, Sylvain, 694-695n33
Sengupta, Jaideep, 691n4, 725n26, 730-731n1
Sengupta, Sanjit, 716-717n10
Sequeira, Ian K., 722n40
Sethi, Anju, 706n36
Sethi, Rajesh, 706n36
Sethuraman, Raj, 719-720n2
Shachar, Ron, 723-724n5
Shachat, Jason, 696n10
Shah, Avni M., 703n29
Shah, Denish, 691n17
Shah, Dharmesh, 439, 723n1, 723n4
Shaker, Zahra A., 688n6
Shankar, Venkatesh, 698n11, 704-705n16, 706-707n42, 708-709n21, 712n5, 713n33, 730n22, 730n24
Shannahan, Kirby, 717n17
Shannahan, Rachelle, 717n17
Shansby, J. Gary, 691n17
Shapiro, Benson P., 729n13, 730n18
Shapiro, Stanley J., 685n8
Sharma, Amalesh, 706n36
Sharma, Arun, 686n13, 696-697n18, 708-709n21, 708n7
Sharma, Dheeraj, 713n27
Sharma, Subhash, 695n37, 698-699n19, 729n15
Shavitt, Sharon, 727n23
Shawver, Donald L., 685n7
Sheehan, Kim Bartel, 714n2
Sheeran, Ed, 183
Shen, Hao, 725n26
Sheng, Shibin, 696-697n18, 696n17
Shepherd, C. David, 707n44, 717n18
Sherry, J. F., 699n28
Sherry, John F., Jr., 693-694n30
Shervani, Tasadduq A., 719n28
Sheth, Jagdish N., 685n4, 686n13, 713n27
Shi, Yuying, 711n2
Shi, Zijun (June), 729-730n16
Shibata, Kenneth, 657-658
Shimp, Terence A., 702n25
Shin, Woochoel, 688n6
Shipley, David, 685n8, 707n3
Shocker, Allan D., 690n3
Shoemaker, F. Floyd, 715n15
Shoham, Aviv, 708n13
Shrum, L. J., 732n6
Shugan, Steven M., 728n4, 730n18
Shulman, Jeffrey D., 709n25, 726-727n17, 728-729n10
Siddarth, S., 707n3, 726n2
Siegel, Eric, 695n35
Siguaw, Judy A., 705-706n27, 708-709n21, 715-716n5
Sigue, Simon Pierre, 715n14, 720n7, 722-723n41
Sihi, Debika, 720n10
Silva-Risso, J. M., 707n3, 726n2
Silverblatt, Ronnie, 689n33
Silverglade, Brian, 668-669
Silverglade, Bruce A., 703n29
Silverman, Jeff, 41-42, 687n4
Simester, Duncan I., 723-724n5, 726n8, 728-729n10
Simmons, Carolyn J., 701-702n14

Simmons, Lee C., 730n19
Simonson, Itamar, 692n14
Simpson, Bonnie, 732n10
Simpson, Penny M., 705-706n27
Sims, Chaun, 658
Sinapuelas, Ian, 700-701n6
Sinfield, J. V., 705n17
Singer, John G., 687n2
Singh, Jagdip, 717-718n19, 717n17, 718-719n22
Singh, Sunil, 718-719n22
Sinha, Indrajit, 728n7, 729n15
Sinha, Prabhakant, 715-716n5
Siong, Chung Tuck, 706-707n42
Sirianni, Nancy J., 701n7, 701n8, 717n15
Sisodia, Rajendra S., 686n13
Sivakumar, K., 701n7, 706-707n42, 707n44, 710n9, 729n14
Skarmeas, Dionysis, 709n30
Skiera, B., 725n29
Skiera, Bernd, 724n9
Skinner, Steven J., 699n26
Skye, Robert, 660
Skye-Rice, Chelsea, 659-660
Slater, Peter, 641
Slater, Stanley F., 686n13, 686n14, 698-699n19
Slezak, Maxine, 647
Slotegraaf, Rebecca J., 687n2, 692n12, 701-702n14, 704-705n16, 706n36, 722-723n41, 724n10
Smagalla, David, 726-727n17
Smeltzer, Larry R., 697n32, 729n14
Smith, B., 690-691n13
Smith, Daniel C., 727n25, 729n11, 729n15
Smith, J. Brock, 694n31
Smith, Keith Marion, 699n26
Smith, Kirk, 719n24
Smith, Malcolm C., 702n19
Smith, Michael D., 720-721n13
Smith, Michael F., 728n7
Smith, Michael S., 727n18
Smith, N. Craig, 727n27
Smith, Paul M., 722n40
Smith, Robert, 721-722n29
Smith, Scott M., 690n6
Smith, Timothy M., 722n40
Soberman, David A., 704-705n16
Sodhi, ManMohan S., 726n8
Sodhi, Navdeep S., 726n8
Soethoudt, Han, 715n13
Sohi, Ravipreet, 715-716n5
Sojka, Jane Ziegler, 718n21
Solomon, Robert, 700n3
Soman, Dilip, 693-694n30
Son, Sanggi, 708n18
Song, Reo, 693n24
Song, X. M., 705n26
Song, Yiping, 727n18
Sonnier, Garrett, 691n17
Sood, Sanjay, 701-702n14
Soonhong, Min, 686n15
Sorescu, Alina B., 702n21, 704n4, 731n2
Sorescu, Sorin M., 731n2
Soster, Robin L., 691n3
Sotgiu, Francesca, 697n32, 724-725n25
Souza, Gilvan C., 709n25
Spann, Martin, 706n36, 729-730n16
Sparks, John R., 700n35
Speh, Thomas W., 710n9, 710n15
Speier, Cheri, 717n15
Spiro, Rosann L., 712n5, 717n17, 718n21
Spohrer, Jim, 701n7
Sprott, David E., 702n25, 723n2, 730n19
Spyropolou, Stavroula, 709n30, 717-718n19
Sridhar, Shrihari (Hari), 725n32
Srinivasan, Kannan, 729-730n16
Srinivasan, Raji, 703n28, 720n10
Srinivasan, Shuba, 701-702n14, 728-729n10
Srinivasan, Srini S., 726-727n17
Srivastav, Rajendra K., 687n7, 690n3

Srivastava, Joydeep, 726-727n17, 727n23
Srivastava, Rajendra K., 701n7
St. Clair, Donald P., 718-719n22
Staelin, Richard, 688n6
Stahl, Florian, 701-702n14
Stamatogiannakis, Antonios, 701n13
Stamps, Miriam B., 717n15
Stanglein, Gena, 720-721n13, 720n9
Stank, Theodore P., 710n9
Staples, William A., 692n18
Stassen, Robert E., 708n7
Steele, Eric, 638-639
Steenburgh, Thomas J., 713n33, 723n1
Steenkamp, Jan-Benedict E. M., 699n26, 703n27, 715-716n5
Steffel, Mary, 693-694n30
Steinhoff, Lena, 722-723n41, 723n2
Stengel, James R., 699n24
Stephen, Andrew T., 712n5, 720-721n13, 720n9, 721n16, 721n22, 723-724n5, 724-725n25, 725n26, 725n32
Stephenson, P. Ronald, 726n12
Stern, Barbara B., 715n6
Stern, Louis W., 708n20, 727n25
Sternquist, Brenda, 695n37
Sternthal, Brian, 694n31
Steuben, Rusty, 656
Steward, Michelle D., 695n37, 716-717n10, 725n26
Stewart, David W., 690n9
Stierl, Marcel, 686n21
Stilley, Karen Melville, 693-694n30
Stock, Axel, 730n25
Stock, James R., 710n4, 710n13
Stock, Kyle, 686n17, 686n18
Stock, Ruth, 704-705n16
Stone, Bob, 707n5
Storbacka, K., 686n15
Straeter, Laura M., 693-694n30
Strauss, Judy, 731n3
Strebel, Judi, 694n31
Stremersch, Stefan, 705n24
Strutton, David, 712n3
Strutton, H. D., 717n18
Su, Lei, 693-694n30
Sudhir, K., 714n38
Sujan, Harish, 694n31, 704n4, 718-719n22
Sujan, Mita, 694n31, 704n4
Suk, Inho, 704-705n16
Sullivan, Mary W., 689n16
Sullivan, Ursula Y., 708-709n21
Sultan, Fareena, 713n33
Summers, John O., 719n28
Sun, B., 700n2
Sun, Qi, 693n24
Sunder, Sarang, 687n7, 715-716n5
Surachartkumtonkun, Jiraporn, 716n9
Suri, Siddharth, 720-721n13
Swain, Scott D., 686n21, 704n4
Swait, Joffre, 694n31
Swaminathan, Vanitha, 704-705n16
Swan, John E., 695n36, 731n3
Swann, Julie, 711n30
Swartz, Tracey A., 704-705n16
Swift, Taylor, 115, 183
Syam, Niladri, 696-697n18, 716n9, 717-718n19
Szymanski, David M., 705n26, 726n4, 731n3

T

Tabors, R. D., 726n5
Talay, M., 706n36
Talbot, Gemma, 188
Talpade, Salil, 692-693n21
Talukdar, D., 728-729n10
Tam, Leona, 691n4
Tamilia, Robert, 708n7
Tan, Yong, 694n311
Tang, Chuanyi, 696-697n18
Tang, Tanya (Ya), 723n2

Tanner, John F., 717-718n19, 717n18
Tanner, John F., Jr., 692n18
Tarpey Sr., L. X., 728n32
Tashchian, Armen, 717n18
Tatham, Ron, 699n32
Tauber, Edward M., 703n31
Taudes, Alfred, 690n6
Tavassoli, Nader T., 691n4
Tax, Stephen S., 725n29
Taylor, Charles R., 720n9, 732n6
Taylor, John C., 720n9
Taylor, Kimberly A., 720n8
Taylor, Shirley, 706-707n42
Taylor, Steven A., 704-705n16, 706-707n42
Taylor, Valerie A., 729n15
Teixeira, Thales, 720n9
Tellis, Gerard J., 699n22, 704n4, 704n6, 709n30, 719-720n2
Thaichon, Park, 716n9
Thelen, Shawn T., 717n15, 717n17
Thelen, Tanya, 717n15
Thomas, Anisya, 711n30
Thomas, Ellen, 694n311
Thomas, Gloria P., 708-709n21
Thomas, Jacqueline S., 698n11, 708-709n21
Thomas, Manoj, 728-729n10
Thompson, Craig J., 693n27, 694n31, 704n10
Thompson, Debora Viana, 694n31, 720n8, 723n2
Thompson, Scott, 722-723n41
Thorpe, Dayle I., 707n44
Thorson, Bobbi, 732n1
Tian, Kelly, 724n17
Tikkanen, Henrikki, 685n10
Tikoo, Surinder, 712n3
Till, B. D., 720n9
Tillmanns, Sebastian, 725n26
Timoshenko, Artem, 723n2
Tirunillai, Seshadri, 699n22
Tiwana, Amrit, 717-718n19
Toller, Karne, 669
Tomkovick, Chuck, 721n22
Torelli, Carlos J., 686n21
Tormsen, David, 690n35
Toubia, Olivier, 694n32
Touré-Tillery, Maferima, 692n15
Traczyk, Patty, 721n22
Tran, Thanh, 709n25
Trawick, I. Fredrick, 695n36
Tripathi, Arvind K., 694n31
Trope, Yaacov, 693-694n30
Trounce, David, 722-723n41
Trout, Jack, 691n17
Troy, Lisa C., 705n26, 706n36
Troye, Sigurd V., 686n15
Trudel, Remi, 694n31, 732n10
Trusov, Michael, 712n5, 720-721n13, 724-725n25, 725n26
Tsai, Claire I., 726-727n17, 727n18
Tse, David K., 708n18
Tsiros, Michael, 694-695n33, 701n7, 722-723n41, 730n25, 731n4
Tu, Yanping, 693-694n30
Tucker, Catherine E., 720-721n13, 721n25, 724n7
Tucker, David, 696n10
Tuli, Kapil R., 696-697n18
Tully, Stephanie M., 691n3
Turban, D., 721n16
Turban, E., 721n16
Turley, L. W., 730n18
Turnipseed, Fernanda L., 710n9
Turow, Joseph, 695n35, 713n35, 714n27, 732n13
Turut, O., 705n22
Tybout, Alice M., 694n31

U

Ubel, Peter A., 703n29
Ulaga, Wolfgang, 686n15, 696-697n18, 701n7
Ulgado, Francis M., 688n13

Umashankar, Nita, 687n7, 728n8
Underhill, Paco, 699n28
Unnava, H. Rao, 720n8
Unruh, Gregory C., 732n10
Upah, Gregory D., 685n3
Urban, Glen L., 690n3, 713n33, 720-721n13
Urbany, Joel E., 726n8, 729n15
Urminsky, Oleg, 694n32
Ursu, Raluca M., 724n6
Uslay, Can, 685n4
Uslu, Aypar, 716n9
Üstuner, Tuba, 693n27, 693n28
Uzo, Uchenna, 697n32

V

Vaccaro, Joseph P., 728n32
Vadakkepatt, Gautham G., 685n4, 723-724n5, 724n7
Vakratsas, Demetrios, 719-720n2
Valentini, Sara, 731n3
Vallen, Beth, 730-731n1
van Bruggen, Gerrit H., 698n12
van de Vendel, Masja A., 709n28
Van den Bulte, Christophe, 725n29
van der Borgh, Michel, 704-705n16
van der Walt, Victoria, 646-647
Van Doren, Doris, 698-699n19
van Heerde, Harald J., 686n20, 703n27, 720-721n13, 722-723n41, 724n7, 727n23
van Ittersum, Koert, 694n32, 703n28, 730-731n1
Van Kerckhove, Anneleen, 713n27
Van Lin, Arjen, 712n8
van Noort, G., 724n16
van Oest, Rutger, 695n34, 731n2
Van Osselaer, Stijn M. J., 694n31
van Reijmersdal, E. A., 724n16
Vanhuele, Marc, 728-729n10, 729n15
Varadarajan, P. Rajan, 687n2, 705n26, 726n4
Varadarajan, Rajan, 687n2, 692n15, 713n33
Varela, Jose, 719n29
Vargo, Stephen L., 701n7, 715-716n15
Vendemia, Jennifer M. C., 722n37
Venkatesan, Mark, 694-695n33
Venkatesan, Rajkumar, 686n19, 686n133, 708-709n21
Venkatesh, Alladi, 704n4
Venkatesh, R., 695-696n8, 730n25
Venkatesh, Viswanath, 717n15
Ventura, Isabel, 666
Verbeke, Willem J. M. I., 732n8
Veresiu, Ela, 693n24
Verhoef, Peter C., 687n7, 687n10, 694n32, 696-697n18, 701n7, 708-709n21, 709n22, 715n152
Verlegh, Peeter W. J., 699n23
Verma, Varsha, 713n27
Verstraeten, Julie, 713n27
Verville, Jacques, 696n10
Vilcassim, Naufel J., 706n38
Villas-Boas, J. Miguel, 691n17
Viswanathan, Madhubalan, 689n22, 693n28
Viswanathan, Siva, 721n18
Viswanathan, Vijay, 725n26
Vitell, Scott J., 686n23
Vladeck, David C., 732n6
Vlckner, Franziska, 701-702n14
Voeth, Markus, 726n12
Völckner, Franziska, 701-702n14, 728-729n10, 730n18
Voleti, Sudhir, 713n35
Vollmayr, Josef, 708n8
Von Wangenheim, Florian, 720-721n13
Vonn, Lindsey, 203
Voorhees, Clay M., 701n7, 702n20, 706-707n42, 712n6, 712n8
Vorhies, Douglas W., 687n10, 700n2, 709n30, 722n38
Voss, Glenn B., 708-709n21, 712n5

Vredeveld, Anna J., 693n24
Vroegrijk, Mark, 712n8
Vroomen, Björn, 708-709n21

W

Wachner, T., 715-716n5
Wagner, Mary, 729n15
Wagner, William B., 703n34
Wakefield, Kirk M., 728-729n10
Wakenshaw, S. Y., 701n12
Wakslak, Cheryl J., 693-694n30
Walker, Beth, 716-717n10
Walker, Orville C., Jr., 687n5, 706n32
Wallendorf, Melanie, 692n17, 693-694n30
Walsh, Ann, 691-692n77
Walsh, Gianfranco, 701n7, 724n7
Walsh, Stan, 666
Walters, Rockney G., 726-727n17
Walton, Shelley, 637
Wan, Lisa C., 701n7
Wang, Chen, 716n9
Wang, Danny T., 708n13, 708n18
Wang, Feng, 723n2
Wang, Hanwei, 691n4
Wang, Hui-Ming, 700-701n6
Wang, Jessie J., 693n24
Wang, L. M., 725n30
Wang, Liangyan, 694n32
Wang, Q., 705n18
Wang, Qi, 724-725n25
Wang, Qiong, 696-697n18
Wang, Xin (Shane), 723n2
Wang, Yitong, 732n10
Wangenheim, Florian, 701-702n14
Wansink, Brian, 703n28, 730-731n1
Ward, James C., 725n30
Ward, Morgan K., 687n7, 693-694n30, 712n8
Ward, Robert, 303
Warlop, Luk, 718-719n22, 730-731n1
Warren, Caleb, 693n233, 694n311
Washburn, Isaac, 715-716n5
Wathieu, Luc, 700n2
Wathne, Kenneth H., 696-697n18, 708n15, 708n18
Watson, George, 687n7
Watson, George F., IV, 696-697n18
Watson, Jared, 725n30
Wauters, Elke, 705n26
Weaven, Scott, 716n9, 723n2
Webb, Elizabeth C., 694n31
Webb, Kevin L., 708n18
Webster, C., 732n6
Wedell, Douglas H., 729n15
Weijters, Bert, 690n6
Weinberg, Charles B., 694n31
Weinberger, Michelle F., 692n17, 693-694n30
Weisfeld-Spolter, Suri, 714n38
Weiss, A. M., 725n30, 729-730n16
Weisstein, Fei, 726n9
Weitz, Barton A., 685n4, 711n2, 712n5, 715-716n5, 718n21
Wells, W. D., 692n16
Wen, Wen, 724n7
Wertenbroch, Klaus, 694-695n33
West, Douglas, 715n18
West, Kanye, 115
Westbrook, Russell, 324
Wetzels, Martin, 725n30
Wheeler, S. Christian, 693-694n30, 715n9
Whinston, Andrew B., 720-721n13
Whipple, Thomas W., 729n15
White, Chad, 724n18

White, Chelsea C., III, 705n26
White, Katherine, 691n4, 730-731n1, 732n10
White, Richard E., 710n7
White, Tiffany Barnett, 690-691n13, 713n33
Whitler, Kimberly, 702n25
Whitley, Sarah C., 694n31
Whitwell, Gregory J., 700n2
Widing, Robert E., II, 712n5
Widmier, Scott M., 717n13
Wiegand, Nico, 720n9
Wieland, Heiko, 701n7, 715-716n15
Wierenga, Berend, 698n12, 715n13
Wiertz, Caroline, 723-724n5, 725n32
Wieseke, Jan, 715-716n5, 726-727n17, 726n12, 727n18, 729n11, 731n4
Wiesel, Thorsten, 687n7, 709n22
Wilbur, Kenneth C., 720n9
Wilcox, James B., 719n24
Wilcox, K. H., 702n22
Wilcox, Keith, 725n32, 727n18
Wildt, Albert R., 730n18
Wilkie, James E. B., 692n12
Wilkie, William L., 685n4, 685n7, 685n9, 686n21, 726n16
Wilkinson, Ian F., 710n9
Wilkinson, J. B., 624, 625, 632
Williams, Brian C., 685n4
Williams, Jerome D., 722n40
Williams, Patti, 691n4
Williams, Serena, 4
Wilson, Andrew, 724n7
Wilson, Bryan, 658
Wilson, H., 690-691n13
Wilson, Hugh N., 696-697n18
Wilson, R. T., 720n9
Wind, Yoram, 696n9
Winer, Russell S., 698n11, 699-700n33
Winkel, Doan, 732n1
Winston, Evelyn, 710n3
Winter, Russell S., 690-691n13
Winterich, Karen Page, 722n38, 730-731n1
Winters, Rosa, 467
Wirtz, Jochen, 692n8, 701n7
Wittkowski, Kristina, 704-705n16
Wittmann, C. M., 712n3
Wlömert, Nils, 729-730n16
Woermann, Niklas, 693-694n30
Wokutch, Richard E., 685n3
Wood, Stacy, 722n37
Wood, Wendy, 695n36
Woodruff, Robert B., 696n15
Wooten, David B., 693n22, 725n32
Wooten, Ginny, 656
Workman, John P., Jr., 706n36
Worm, Stefan, 696-697n18
Wotruba, Thomas R., 719n29
Wright, G., 717n15
Wright, Linda B., 732n6
Wright, Peter, 685n9
Wu, Bob, 695-696n8
Wu, Chunhua, 720-721n13, 724-725n25
Wu, Fang, 693n24, 704-705n16
Wu, Qingsheng, 704-705n16
Wuyts, Stefan, 696-697n18
Wyer, Robert S., 701n7, 728-729n10
Wyer, Robert S., Jr., 693n28, 722n37

X

Xia, Lan, 728-729n10
Xie, Jinhong, 724-725n25
Xie, Ying, 720-721n13

Xu, Alison Jing, 722n37
Xu, Jun, 721n16
Xu, Kefeng, 712n6

Y

Yadav, Manjit S., 692n15, 713n33, 725n32, 730n25
Yang, Haiyang, 701n13
Yang, Linyun W., 701-702n14
Yang, Liu (Cathy), 694n32
Yang, Sha, 729n14
Yang, Xiaojing, 721-722n29
Yang, Yang, 691n4
Yang, Ying, 716n9
Yang, Zhi, 704-705n16
Yang, Zhiyong, 695n36, 704-705n16
Yao, Dai, 724-725n25
Yao, Yuliang, 712n6
Yelkur, Rama, 721n22
Yeniyurt, Sengun, 690n6, 726-727n17
Yildirim, Gökhan, 728n8
Yim, Chi Kin (Bennett), 706n36
Yin, Fang, 694n31
Yip, George S., 716-717n10
Yip, Leslie, 695-696n8
Yoganarasimhan, Hema, 704n10
Yoon, Carolyn, 729n15
You, Ya, 723-724n5, 724n7
Young, Shelley, 439
Yuan, Hong, 728-729n10, 729n15

Z

Zablah, Alex R., 716n6
Zaccour, Georges, 715n14, 720n7, 722-723n41
Zachary, Jesse, 662
Zahorik, Anthony J., 706-707n42, 707n44
Zaichkowsky, Judith Lynne, 694n31
Zaltman, Gerald, 695-696n8, 698-699n19
Zarantonello, L., 701-702n14
Zauberman, Gal, 691n3, 693-694n30
Zeelenberg, Marcel, 694-695n33
Zeithaml, Valarie, 701n7
Zemanek, James E., 719n28
Zhang, C., 726-727n17
Zhang, Jiao, 694n32
Zhang, Jie, 713n16
Zhang, Jonathan Z., 696-697n18, 729n11
Zhang, Kaifu, 729-730n16
Zhang, Shirley, 694n32
Zhang, Xi (Alan), 687n7
Zhang, Xiaoquan, 725n30
Zhang, Xubing, 687n7
Zhang, Ying, 727n25
Zhang, Yinlong, 723-724n5
Zhang, Yuchi, 712n5
Zhang, Z. John, 712n5, 728-729n10, 729-730n16
Zhao, Min, 704-705n16
Zhao, Yanhui, 702n20
Zhao, Yi, 687n7
Zheng, Xingshan, 691n4
Zheng, Xu (Vivian), 708n13
Zhou, Chen, 731n2
Zhou, Kevin Zheng, 696-697n18
Zhou, Rongrong, 730-731n1
Zhu, Feng, 725n30
Zhu, Meng, 694n32
Zimbalist, Paul, 646-647
Zinkhan, George M., 685n4, 729n15
Zoltners, Andris A., 715-716n5, 718n21, 719n24
Zou, Shaoming, 688n13, 718n21
Zubcsek, Peter Pal, 720n13

Company Index

A
AAA Custom Castings, Inc., 661-662
Accenture, 165
Accto Company, 649
Ace Electronics, 163
Ace Hardware, 319, 322, 504
Acer Computers, 477
Adero, Inc., 72
Adidas, 4, 203, 204
Adobe, 521
Afterpay, 489
Air Jamaica, 205
Airbnb, 80, 125, 244, 456
Airbus, 270
Alcan, 270
Alcoa Aluminum, 164
Alex & Ani, 336
Alibaba, 212
AlliedSignal, 159
Allstate Insurance Company, 198, 209, 351
Allstates Rubber & Tool, 157
Ally Bank, 47, 118
Aloca, 478
Amazon, 327, 421
Amazon.com, 59-60, 68, 83, 218, 242, 286, 308, 322, 325, 335, 521, 532-533, 633
Amelia Island Plantation Resort, 633
American Dental Association (ADA), 360
American Express, 452
American Marketing Association (AMA), 26
American Red Cross, 62, 147, 226
Amway Global, 331
Andersen Windows, 357
Anytime Fitness, 323
A.O. Smith, 149
Apple, 11, 65, 115-116, 122, 162, 216, 218, 285, 447, 463, 482, 532-533
Apple Music, 115, 240
Arbor Door and Window Co., 650
Archer Daniels Midland (ADM) Company, 496
Arm & Hammer, 53
Assurant Solutions, 395
AT&T, 165, 414
Audi, 414, 417
Auntie Anne's, 434
Auntie Em's Cookie Company, 126
Autolite, 273
AutoZone, 182
Avery Dennison, 412-413
Avis, 66

B
Banana Boat, 102
Banco de Alimentos, 289
Bandolino, 324
Bank of America, 287
Bank of the Wichitas, 16
Barclay Prime, 447
Barnum & Bailey's, 33, 34
BASF Corporation, 152
Baskin Robbins, 322
Bath & Body Works, 322
Bathys Hawaii Watch Co., 371
Bayer Healthcare LLC, 429
BBC Breaking News, 463
Becton, Dickinson and Company (BD), 517
Beech-Nut Nutrition Company, 73
Ben & Jerry's, 25, 122, 425, 488
Benjamin Moore, 211
Berkshire Hathaway, 351
Best Buy, 285, 298, 329, 334, 501
Best Western, 643
Betz Laboratories, 159

BHP Billiton, 311
Big Brothers Big Sisters of America (BBBSA), 627-630
Bing, 363, 446
BlaBlaCar, 125
Black & Decker, 54, 475, 504
Blue Apron, 302, 521
Blue Nile, 449
Blue Ribbon Sports, 3
BMW, 161, 254, 414
Boar's Head, 280
BoConcept, 135
Boeing, 270, 312, 391
Boise Cascade, 382
Bolthouse Farms, 537
BookPacks, 287
Borden Company, 497
Boston Scientific, 227
BP, 218
Brawny, 357
Bridgespan Consulting, 628
Bridgewater Commons, 626
Bright Light Innovations, 658-659
British Airways, 250
Brookstone, 633
Bryan Cave Leighton Paisner (BCLP), 362
Budweiser, 359
Bugatti, 283
Build-A-Bear Workshop, 627
Bureau of Labor Statistics (BLS), 583, 589
Bureau of Medical Devices, 630-631
Burger King, 415, 433, 535
Bush's Baked Beans, 224
Business Center, Inc., 647-648
Buzz Generator, 455
BuzzFeed, 133
BzzAgent, 455

C
CafePress, 250
California Closet Company, Inc., 195-196
California Department of Transportation, 158
California Milk Processing Board, 121
California Skateparks, 168
Canadian Mills, Ltd., 662-663
Canon, 92, 229
Canyon, 4
Capterra, Inc., 401
Cargill, 74
Caribou Coffee, 3
CarMax, 482
Carmine's Italian Restaurant, 642
Carnival Cruise Line, 90
Carpet One, 322
Cartier, 521
Caterpillar, 359, 396
Center for Science in the Public Interest (CSPI), 14
Cessna Aircraft Company, 153
Chaco, 3
Chadder's, 218
Channel Master, 302-303
Charles Schwab, 210
Charmin, 121
Charter Components, 163
Chase Bank, 387
Checkers, 523
Chicago South Loop Hotel, 101
Chick-fil-A, 335
China Southern Airlines, 315
Chipotle, 322
Chiquita Brands L.L.C., 167
Chobani, 3, 271
The Chopping Block, 439-440

Chrysler, 71
Circuit City, 501
Cirque du Soleil, 33-34, 453, 532-533
Cisco Systems, 359
Citi, 432
City of San Jose, 168
ClimateMaster, 518
Clorox, 94, 207
CNN, 418, 463, 519
Coca-Cola Company, 38-39, 70, 77, 91, 121, 178, 207-208, 223, 238, 250, 289, 295-296, 338, 477, 532
Coca-Cola Foods, 255
Coffee-Mate, 102
Colgate-Palmolive Company, 96
Colorado Rapids Major League Soccer, 637
Colorado State University (CSU), 658
Colours in Motion, 631
Columbia Forest, 251
Comcast, 388-389
Commerce Bank, 80
Computer Doctors, 166
ConAgra Foods, 121, 373
Consumer Electronics Show, 432
Consumer Product Safety Commission, 24, 73
Consumer Safety Institute (Netherlands), 122
Cooper Car Company, 412
Cooper Industries, 276
Corti-Care, 484
Cosmopolitan Hotel, 393-394
Costco, 278, 329, 501
Cousin's Ice Center, 644-645
Crayola, 373
Crispin, Porter & Bogusky, 407
Crocs, 207
CSX, 308-309
CVS, 503

D
Daimler, 311
Danone, 55
Dassualt Aviation, 412
Dawn, 362
Days Inn of America, Inc., 643-644
DEKA Research and Development Corp., 632-634
Dell Computers, 265, 387, 477
Delphi Automotive Systems, 633
Delta Air Lines, 414
Deluxe Corporation, 71
DeVilbiss, 631
DHL, 315
DHL Worldwide Express, 308
Diamond Multimedia, 115
DiGiorno, 90
Dillard's, 327
Discover Africa, 365
Dockers, 455
Dodge, 128
Dollar Shave Club, 271, 466, 523
DomiNick's Pizza, 407
Domino's Pizza, 407-408
DoorDash, 302
Doyle Research, 193
Dr Pepper, 285, 296
D-Rev, 255
Drift, 393
Dropbox, 139-140, 369, 519
DrV.com Custom Vitamins, 646-647
DSW, 420
Dunkin' Donuts, 173-174, 322, 521
DuPont, 244, 391, 562

Duracell, 283, 488
Dynamic Steel, 640-641

E
E! Entertainment Television, LLC, 417
EA Sports, 137
East African Breweries, 238
Eastman Chemical Company, 155
Eastman Kodak, 92
Easton Town Center, 627
eBags, 250
eBay.com, 322
Ecoworx, 289
Edmunds.com, 122
Edna McConnell Clark Foundation, 628
Edscha, 161
Edward Jones, 129
Einstein, 3
Ela Family Farms, 270
Electra, 4
Electrolux, 279
Elkay Manufacturing Co., 245, 379
Emery Air Freight, 308
EmpordAigua, 161
Enel, 290
Energizer, 283
Environmental Control, Inc., 653
Environmental Protection Agency (EPA), 309, 653
ESPN, 128, 277, 519
Estée Lauder, 277, 397
E*TRADE, 37
Everest & Jennings, 630
Exxon, 51
EyeBuyDirect.com, 333

F
Facebook, 3, 33, 37, 131, 173, 180, 184, 189, 334, 352, 354, 364, 375, 388, 407, 421-422, 439, 442, 448-449, 457-458, 459, 461, 466, 519, 546
Falconbridge Limited, 150
Fastenal, 342
Fay Manufacturing Company, 630
Febreze, 123
Federal Trade Commission (FTC), 520
FedEx, 213, 241, 270, 274, 308, 309, 315, 340
Fender Guitars, 278
Ferguson, 379-380, 389
Fidelity Investments, 357, 426
Firestone, 73
First Bank, 385
Fisher-Price, 434
Fix-A-Bug, 639
FocusVision Research Inc., 193
Foot Locker, 324, 334, 645
Foothills Family Dental, 166
Forbes Consulting Group, 104
Ford Motor Company, 3, 77, 101, 250, 273, 383, 393, 423, 657
Ford Motorcraft, 423
Fred Meyer, 329
Fresh Harvest, 659-660
FreshDirect, 455
Frieda's, Inc., 340-341
Friendship Dairies (FD), 52
Frito-Lay, 180, 255, 314
Frontline, 485

G
GanQiShi, 3
GE, 3
GE Aviation, 270
GE Plastics, 633
GEICO, 351-352, 356, 532-533
General Electric (GE), 3, 81-82, 241, 309, 475
General Growth, 626
General Mills, 64, 220, 271, 455
General Motors (GM), 21, 161, 270, 657

General Services Administration (GSA), 169
GeoTron International (GTI), 657-658
Gerber, 291
GF Health Products, 631
Gibson Guitars, 278
Gilead Science, 495
Gillette, 523
Girl Scouts of America, 147, 219, 269
Giupetto's, 366
Give Kids A Smile, 344
Glacéau, 271
GlassesUSA.com, 494
Glendale Galleria, 626
Glossier, 462
Godrej, 247
Golf Channel, 418
Goodyear, 269
Google, 12, 83, 157, 188, 212, 218, 251, 254, 311, 363, 364, 419, 421, 422, 423, 446, 487, 516, 519
GoPro, 92, 211
Government Employees Insurance Company. *See* GEICO
Grainger, 230
Grease Monkey, 291
Green Building Council, 669
Green Giant, 229
Green Mountain Coffee, 368
Green Works, 94
Grubhub, 302, 527
Gulfstream, 270

H
Hallmark, 89-90, 415
Hampton Hotels, 323
Handy Dan, 319
Hanes, 64
Hanratty Company, 650-651
Hansen's Natural Sodas, 296
Harley-Davidson, 162, 279, 452
Harry's, 271
Harry's Razors, 521, 523
Harvey Nichols, 487
HBO, 277
HD Supply, 320
Heath, 265
Heinz, 194
Hellmann's, 433
Hello Fresh, 302, 354
Henry Schein, 343, 344
Hershey, 187
Hertz, 298
Hertz Corporation, The, 166
Hewlett-Packard Company, 162, 265, 273, 396, 463, 477
HGTV, 418
HighFly Drones, 163
Hillside Veterinary Clinic, 592-619
Hilton Hotels, 165, 215, 291
Hilton Inn, 643
Hindustan Unilever Limited (HUL), 9
Hit Song Science, 183
Hitachi, 415
Hobart Corporation, 386
Holiday Inn, 643
Home Depot, 319-320, 334, 337, 389, 626
Honda, 45, 149, 163, 254, 457, 492
Hostess Snacks, 375
HP, 298
H&R Block, 452
HubSpot, 439-440, 521
Hulu, 60, 277, 502
Hyatt Hotel, 391
Hytorc, 518
Hyundai, 110

I
IBM Corporation, 265-266, 304, 654
IKEA, 51, 207, 210, 217, 305, 329, 338, 473

Impossible Burger, 413
Inmac, 344
Innocent Drinks, 53
In-N-Out Burger, 51, 218
INOBAT, 288
InSinkErator, 2440245
Instagram, 3, 33, 37, 131, 355, 364, 375, 442, 445, 457, 459, 461-462, 671-673
International, 309
International Council of Shopping Centers, 625
Invacare, 630-632
Invo, Inc., 649
iRobot, 235-236, 532-533
IronPlanet, Inc., 346
ithink.com, 198
iVet Professional Formulas, 268

J
Jack's Wife Freda, 456
Jacuzzi, 379
Jax Fish House, 445
JCPenney, 203, 410
J.D. Powers, 190, 530
JD.com, 308, 335
JetBlue, 51
Jiangsu Yuyue Company, 632
Jiffy Lube, 260
JLG, 54
John Deere, 83, 145-146, 162
Johnson & Johnson (J&J), 78, 151, 632
JPMorgan Chase, 291
JustRide, 481

K
Kaepa, 98
Kaleidoscope, 159
Kay's Kloset, 496
Keebler, 219
Kellogg, 314
Kennedy & Gaffney (K&G), 665-666
Keurig, 272
KFC, 83, 361, 429
Kids Foot Locker, 324
Kimberly-Clark, 37, 454
Kingston Home Health Services (KHHS), 663-664
Kioson, 335
Kiwi, 186, 245
Knorr, 364
Kohler, 379
Kohl's, 203, 410, 626
KPMG, 367
Kraft Foods, 37, 221, 314, 455
Kroger, 221, 271, 322, 336, 636
Kyota, 485

L
LaCroix, 462
Lady Foot Locker, 324
Lake Pukati Lodge, 641
Lake Russell Marine & Camp, 656
Lands' End, 298
LECO Japan, 543
Lee, 51
Leegin Creative Leather Products, 496
Legend, 265
Legend Valve, 505
LEGO, 87-88, 459, 532-533
Lego Group, The, 415
Lenovo, 265-266
Levi Strauss, 151
Levi's, 3, 51, 55
LG, 211
LG Electronics, 103, 502
Li Ning, 4
Lifebuoy soap, 359
Lifecare Medical, 631
LimeBike, 477

LinkedIn, 158, 355, 394, 439, 457, 459, 463, 466
Listerine, 124, 214, 360
L.L.Bean, 463, 491
Local Joe's, 46
Logitech, 274, 488
L'Oréal, 516
Lowe's, 319, 329, 340, 389, 462, 485
Lurpak, 425
Lyft, 477

M

Maclaren, 207
Macy's, 107, 327, 334, 410
Mall of America, 626-627
Mallory's Lemonade Stand, 670-671
Malt-O-Meal, 125
Mandarin Oriental Hotel, 51
Manitowoc Company, Inc., 303-304
Marmite, 90
Mars, Incorporated, 167
Marsh LLC, 304
Martin Agency, 351
Mary Kay, 331
Mastercard, 51
Material Sciences Corporation (MSC), 383
Mattel, 459
Mazda, 122
McCaw Cellular, 83
McDonald's, 51, 54, 75, 83, 207, 218, 252, 282, 284, 338, 432, 433, 636
McGlynn, Clinton & Hall Insurance Agency, 450
Meijer, 329
Memorial Sloan-Kettering Cancer Center, 495
Menasha, 310
Mercedes, 516
Mercedez-Benz, 3
MetalCoat Company, 649-650
Method, 532-533
Method Products, 471
MetoKote Corporation, 145-146, 162
Metro AG, 309
Microsoft, 83, 213, 218, 463, 521, 657
Miller's Bakery, 419
Milwaukee Bucks, 358
Mini USA, 426
MOD-PAC, 192
Moen, 379
Mollie Stone's Market, 471
Morton Salt, Inc., 127
Mothers Against Drunk Driving, 415
Motorola, 475
Mount Cook National Park, 641
MResearchPro, 661
Musco Lighting, 168
My Dollarstore, 339
Mylan, 517
MyOwnWedding.com, 654-655

N

Nabisco, 244
Nada, 325
Namaste Solar, 19
NAPA Auto Parts, 322
National Association of Colleges and Employers, 580
National Association of Home Builders, 669
National Park Service, 633
National Semiconductor (NS), 160
Nature's Own Foods, Inc., 636-637
NBCNews.com, 420
NeoMam, 451
Nescafé, 516
Nespresso, 272, 425
Nestlé S.A., 90, 134, 151, 258, 271
Netflix, 60, 66, 105, 139, 277, 334, 413, 502
New Belgium Brewing Company, 150
New York Public Library, 449

New York Times, The, 463, 519
New Zealand Department of Conservation, 641
Next Big Sound (NBS), 183
Nexteer Automotive, 149
Nickelodeon, 418
Nielsen Company, The, 185
Nike, Inc., 3-4, 41, 98, 139, 203, 204, 291, 324, 461, 464, 532-533, 645
99 Cents Only, 328
Nippon Steel, 310
Nissan Motor Company, 119, 290
NOCO United Soccer Academy, 637-638
Nordstrom, 322, 327, 337, 402, 463
North Coast Maid Service, 166
Northwestern Mutual Life, 121
Northwestern University, 183
Novartis, 401-402, 495

O

Office Depot, 334
Office Depot.com, 480
Office Depot/Office Max, 412-413
Omaze, 431
Omnicare Inc., 151
1-800-GOT-JUNK?, 284
Oracle Corporation, 206
Osprey Packs, 441-442
Otto, 311
Overstock.com, 419
Owens-Corning, 485
Owens-Illinois, 545
Ozarka, 652

P

Paglozzi's Pizza Pies, 666-668
Pampers, 129
Panadol Ultra, 226
Panasonic, 210, 501
Pandora, 115, 183
Panera Bread, 122, 216, 533
Papa John's, 407
Paper Products, Inc. (PPI), 647-648
Park Meadows, 626
Parker Hannifin (PH), 506
Parker Jewelry Company, 391
Passion Life, 535
Paul Masson, 488
PayPal, 103
Pedigree, 239
Peloton Technology, 311
PenAgain, 167
Pentagon, 633
Pep Boys, 273
Pepperidge Farm, 305
Pepsi, 214
PepsiCo, Inc., 83, 296, 361, 428
Permobil, 631
Peterbilt, 309
Peterson Manufacturing Company, 279
Petoskey Tech Support, 638-639
P&G, 415
Philadelphia, 3, 245
Philips, 227
Piggly Wiggly, 322
Pillsbury, 219
Pinterest, 194, 364, 457, 459, 462-463
Pizza Hut, 207, 387, 407
Please Don't Tell, 447
Potbelly Sandwich Works, 624-625
PricewaterhouseCoopers, 165
Procter & Gamble (P&G), 37, 44, 67, 70, 98, 132-133, 148, 257, 389-390, 392, 415, 428, 471, 477
ProjectPro, 467
Proto Labs, 64
Prudential Insurance, 165
Public Broadcasting System, 519
Purina, 127

Q

Quiet Night Motel, 643-644
QXR Tools, 660-661

R

RadioShack, 265
Ramada Inn, 643
Rayovac, 283
Red Bull, 214
Reebok, 98, 429
Reebok International, 285
REI (Recreational Equipment, Inc.), 22-23, 421
Remesh, 195
Republic Records, 183
Resin Dynamics, 639-640
Respironics, 631
Richardson Electronics, 343
Rio Centro, 217
Rio Tinto, 311
Rivertown Crossings, 626
Roadmasters Auto and Tire Centers, 269
Robert A. Bothman, Inc., 168
RocketFuel Inc., 427
Rockwell, 229
Rodan + Fields, 274
Royal Caribbean Cruises Ltd., 398
Rubbermaid, 250
Running On, 645-646

S

Safeway, 221, 475, 480-481, 636
Sainsbury's, 325
Saleco, Inc., 649
Samarin, 361
Sam's Club, 135, 329, 334
Samsung, 122, 127, 446, 501-502
San Francisco Giants, 480
SAS Institute Inc., 179
Sauder Woodworking Company, 301
Save the Children Federation, 333
Schneider Electric, 103
Schoology, 193
Schrock & Oh Design, 648-649
Scioto Group, The, 649-650
Sealed Air Corporation, 213
Sears, 520
Seattle's Best Coffee, 65
Securities and Exchange Commission, 666
Segway Inc., 632-634
Sensodyne, 96
Sephora, 277, 452
7-Eleven, 134, 278, 322, 330, 344
Shaw Floors, 289
Shell, 8, 51
Sherwin-Williams Paint Stores, 281-282, 319
Shieldtox NaturGard, 535
Shopify, 333
Showpo, 357
Signal, 409
Silky Smooth, 103
Siltec, 160
Silverglade Homes, 668-669
Simply Lemonade, 123
Skechers, 4, 429
SkinVision, 69
Slumberland, 479
Snapchat, 37, 131, 354, 360, 422, 457, 463
Snapper, 311
Sodebo, 7-8
Solar Sister, 272
Sonos, Inc., 458
Sono-Tek Corporation, 290-291
Sony, 70, 115, 308, 501, 502
Southdale Center, 625, 627
Southwest Airlines Co., 210, 453
Specialized, 4
Spiceworks Community, 158
Splenda, 124

Spotify, 3, 115, 183, 240
Square D, 272
Square Register, 241
St. Albans College, 637
Standard Hotels, 331
Stan's Gourmet Popcorn, 515
Staples, 329
Starbucks, 20-21, 64, 118, 173, 356, 463-464
Starlight Stove (Bright Light Innovations), 658-659
State Farm Insurance, 351, 422
STIHL, 278
Stop-N-Go, 330
Stride, 422
Subaru, 181-182
Subway, 125, 323
Supercuts, 323
Superior Mill Supply, 661
Survey Sampling International, 198
Sysco, 313-314

T
Taco Bell, 65, 90, 207, 375, 412
Takata, 149
Target, 36, 105, 299, 319, 328, 329, 334, 472
Ten Thousand Villages, 13
Tesco, 245, 280, 327, 492
Tesla Motors, 21, 74
Texaco, 65
Texas Instruments, 240
ThoughtLava, 399
Threadless, 250
3M, 247
Ticketmaster, 215
Tide, 268
Tide to Go, 118
Tiffany's, 492
Tillamook County Creamery Association, 374
Timberland, 324
Toddler University (TU), Inc., 41, 42
Tommy Bahama, 485
Tom's of Maine, 96
Tonka Toys, 312
Toro, 311
Tostitos, 298
Toyota, 133, 354
Toyota Motor Corporation, 118, 298
Toys "R" Us, 329
Trek, 4
TripAdvisor, 457

Tropicana, 214
True-Value Company, 382
Tumblr, 461
TurboTax, 519
Turo.com, 125
Twitter, 3, 131, 173, 188, 189, 352, 354-355, 375, 407, 448-449, 457, 458, 461, 463-464, 466
Tyson Foods, Inc., 130

U
Uber, 21, 48, 53, 80, 481
Uber Eats, 302
UL TerraChoice, 14
UltraCare, 370
Under Armour, 4, 139, 203-204
Unilever, 9, 213
United Airlines, 14, 51, 480
United Nations, 77
United Nations' World Food Programme (WFP), 23
United Way, 23
University of Michigan, 529
UnPackt, 325
UPS, 270, 309, 310, 340
U.S. Air Force, 23
U.S. Army, 418
U.S. Bureau of Labor Statistics (BLS), 583, 589
U.S. Census Bureau, 189
U.S. Department of Agriculture (USDA), 73
USAA, 62-63
USPS, 491
USX, 640

V
Van Heusen, 126
Vanguard, 506
Vegpro Kenya, 302
Verizon, 389
Verizon Wireless, 181, 212
Vicks, 386, 423
ViewSonic, 288
Visa, 51
Vizio, 225, 501, 502
Vodafone Group, 387
Volkswagen, 26
Volvo, 311
VRBO, 125
Vudu, 502

W
Walgreens, 49, 195
Walmart, 12, 36, 37, 138, 148, 166, 167, 271, 279, 280, 299, 302, 305, 315, 319, 322, 327, 328, 329, 334, 339, 340, 389-390, 392, 423, 454, 492-493, 505, 527-528, 645
Wal-Mart Stores, Inc., 169
Wanderlust Outdoor, 300
Warby Parker, 62, 271, 333, 335, 362, 494
Wayfair, 335, 462
Weather Channel LLC, 423
Wells Fargo, 132, 165
Wendy's, 492
West Tarrytown Volunteer Fire Department (WTVFD), 653-654
Western Digital, 162
Weyerhaeuser, 229
Whirlpool, 210, 219
Whole Foods Markets, 118, 222, 335, 670
Wicker Central, 323
Winn-Dixie, 482, 636
Wire Solutions, 659
Wise Water, Inc., 651-653
Wonder Works, 323
Woodward, 270
Woolworths, 151
Working Girls Workout, 671-673
World Food Programme (WFP; United Nations), 23
World Health Organization (WHO), 527
World Wildlife Fund (WWF), 23, 124
Worthington Manufacturing Company, 630

X
Xerox, 265, 289

Y
Yahoo!, 423, 446
Yara, 55
Yardbird, 207
Yelp, 211, 443, 457
YETI, 37
YMCA, 23
YouTube, 14, 352, 357, 442, 457, 519

Z
Zappos, 388
Zappos.com, 118, 195, 324, 478
Zara, 3, 313
Zenni Optical, 494
Zipcar, 125

Subject Index

A

A.1. Steak Sauce, 217
A/B testing, 195
Acai berries, 11
Accessory equipment, 228, 229, 268, 674
Accumulating, of products for distribution, 275, 674
Ace laundry detergent, 522
Acer Liquid Leap, 189
Acquisition, 439
Acquisition cost, 46, 674
Acquisition of new customers, 47
Action, in AIDA model, 358, 424, 425–426
"Active play market," 87, 91
Ad blockers, 421
Adaptation
   environmental sensitivity and, 82–83
   for international markets, 257–258
   menu, for foreign markets, 83
   for My Dollarstore's Indian franchises, 339
   organizational customers/vendors and, 162
   of products, 257–258
   relationship-specific, 162
   of retailers moving to international markets, 338–339
Ad-blocking software, 424
Addictive products, 546
Addressable TV technology, 418
Adhesives, 251
Administered channel systems, 282, 674
Administered prices, 478, 674
Adoption
   of continuous innovations, 246
   of discontinuous innovation, 246–248
Adoption curve, 371, 674
Adoption process
   advertising and, 412
   defined, 674
   new concepts and, 139–140
   promotion and, 358, 370–371
   steps of, 139–140
Advertising. *See also* Paid media; Promotion; Sales promotion
   advocacy ads, 415
   aimed at children, 428
   benefits and challenges of, 443
   big data use in, 418, 423
   careers/compensation, 411, 584, 587
   celebrities in, 365–366
   channel coordination and, 412–413
   comparative, 675
   competitive, 471, 675
   complaints about, 27
   cooperative, 675
   corrective, 429, 676
   cost per one thousand impressions (CPM), 428
   deceptive, 429
   defined, 40, 354, 674
   in different media, 40
   digital, 419–424
   direct-response, 265
   effectiveness of, measuring, 426–427
   employment in, 411
   estimated ad spending for 2017 and 2020, 416
   fairness in, 428
   GEICO insurance example, 351–352
   government regulations, 428–429
   implementation, 412
   international dimensions, 409–410
   international growth of, 410
   kinds needed, 413–415
   linking to consumer lifestyles, 127–128
   marketing *versus*, 7
   measuring, 426–427
   media selection, 416–419
   message planning, 424–426
   misleading, 126
   mobile, 419–424
   objectives for, 411–415
   as a one-to-many model of communication, 458
   outdoor, 416, 444
   pretesting, 427
   programmatic advertising, 420
   as share of promotion costs, 354
   spending in, 410
   sponsored links, 157
   testing, 427
   types of, over adoption process stages, 412
   uneven results, 408
   unfair, avoiding, 428–430
   waste of resources?, 534
Advertising Age, 583, 590
Advertising agencies, 410, 429, 674
Advertising allowances, 413, 486, 674
Advertising impressions, 428
Advertising managers, 355, 435, 538, 674
Advertising media, 674
Advertising specialists, 419
Advocacy, customer, 389, 454–457
Advocacy advertising, 415
Affiliation, with national chains, decision dilemma (case study), 643–644
Africa, 3, 77, 194, 251, 295, 331, 447
Aftermarket, 229
Age, group trends (U.S.), 79–80
Agent wholesalers, 342, 345–346, 674
Agents
   export/import, 345
   in international trade, 345
   manufacturers', 345
   purchasing, careers as, 585
   selling, 346
   in wholesaling, 345–346
Aggregating customers, 94–95, 99
AIDA (*A*ttention, *I*nterest, *D*esire, *A*ction) model, 358, 403, 424–426, 674
AIOs (*A*ctivities, *I*nterests, and *O*pinions), 127–128
Air Force, U.S., 23
Air Jordan shoes, 3
Air pollution, 309
Airbags, defective, 149
Airbnb, 244
Airfreight, 298, 307, 308, 341
Airline industry, price variations in, 478
"Alexa," 60, 235–236
Algeria, 76
Allowance (accounting term), 674
Allowances
   advertising, 674
   for cost variations, 506–507
   defined, 486, 674
   off list prices, 486–487
   in operating statement, 567
   push money (prize money), 486, 682
   special promotion, 497
   stocking, 683
   trade-in, 484, 487, 684
   types of, 486–487
Allrecipes.com, 464
Alternatives, evaluating and identifying, 136, 139
Alvaro (beverage), 238
Amazon Echo, 60, 235–236, 407
Amazon Go, 59
Amazon Marketplace, 59
Amazon Prime, 59
Amazon Prime Now, 59
"Amazoned," 319
American Almanac of Jobs and Salaries, 580
American Customer Satisfaction Index, 529, 530
American Marketing Association Statement of Ethics, 26
American parents, 87
Americans with Disabilities Act (ADA; 1990), 631
Analysis
   competitor, 675
   customer, 539
   lifestyle, 678
   marginal, 679
   in marketing research, 187–190
   personal, in career planning, 581–583
   sentiment, 189
   situation, 187–190, 539, 594–605, 683
Analytics. *See also* Predictive analytics
   as aid in distribution, 295
   as aid to artificial intelligence (AI), 211
   for anticipating customer problems, 388–389
   to improve service levels and cut costs, 299
   in managing customer service, 391
   marketing, 679
   for mass-merchandising, 329–330
   personalized selling with, 334
   predictive, 334, 337, 681
   social profiles from, 547
   for transportation and environment, 310
Android, 212, 480
Annual memberships, to warehouse clubs, 329
Anthropology, 194
"Anticipatory shipping" (Amazon), 337
Antidumping laws, 495
Antimonopoly laws, 72, 496
Antitrust laws, 496
Anytime-anywhere communication, 393
AnyWare, 407, 408
Apparel manufacturing, NAICS code for, 164
Appendices section, in marketing plan, 616–619
Apple
   competitors, 116
   products, 115–116, 178
Apple knockoff store, in China, 218
Apple lawsuit, 218
Apple Watch, 407
Approved suppliers list, 168
Apps
   add services to core products, 211
   Amazon Go, 59
   branded, 448, 451, 674
   Burbn, 243
   Cirque du Soleil, 33
   of Coca-Cola, in Vietnam, 295
   Color Capture (Benjamin Moore), 211
   Discover Moscow, 211
   fitness, Under Armour, 203–204
   "free" for smartphones, 519
   GEICO insurance, 351–352
   Girl Scout Cookie Locator, 269
   Hallmark, 89
   Hopper, The, 480
   iPhones, 115
   iTunes, 115
   Knorr, 364
   Macy's, 107
   MapMyRun, 204
   mapping, 337
   for price comparisons with smartphone, 516
   Record Sensor, 204
   as retailer marketing strategy, 334
   Roomba vacuum, 235–236
   Sam's Club Scan & Go, 334
   Shazam, 183
   SitOrSquat, 451

Apps—Cont.
  Snapchat, 360
  Splice (GoPro), 211
  Subway, 211
  target marketing through, 107
  that predict lower prices, 480
  Uber, 48
  United Nations' World Food Programme meal sharing, 23
Aquafina bottled water, 120
Argentina, 76, 77
Ariel laundry detergent, 257
Arithmetic, for marketing (Appendix B), 564–577
Arousing desire (AIDA model), 358, 424, 425
Artificial intelligence (AI), 68–69, 351–352, 394
  advertising and, 420
  Amazon as leader in, 59
  anytime-anywhere communication via, 393
  customer needs anticipated by, 211–212
  for customer service, 68–69
  defined, 68, 674
  driverless trucks and, 311
  for flight prices, 480
  GEICO use of, 351–352
  in health field, 259
  intelligent agents and, 68
  quality improvement and, 259
  social profiles from, 547
Artificial intelligence (AI), 251
Asia, 33, 220, 240, 265, 410
Asian American consumers, 133
Aspiration levels, 530
Assorting, 275, 674
Assortment
  discrepancy of, 10, 11, 275
  in retailing, 326–327
Assurance TripleTred All-Season tires, 269
Athletic shoes, 3–4
ATMs, 393
Attention, in AIDA model, 358, 424
Attention, Interest, Desire, Action (AIDA) model, 403, 424–426
Attentive Parents, as buyers, 41–42
Attitude
  belief versus, 123–124
  defined, 123, 674
  expectations and, 126
  "green," 124
Attractive opportunities, 35, 48–49
Auction companies, 346, 674
Auction sites, buying behavior and, 134–135. See also E-commerce
Auctions, online, 518
Augmented reality (AR), 211–212, 674
Australia, 33, 59, 76, 465
Auto insurance, 351–352
Autoblog, 457
Automakers
  derived demand for steel, 227
  electric cars, 21
  exclusive distribution, 283, 284
  planning example (case study), 657–658
  rebates offered by, 488
  self-driving electric cars and, 407
  "take-back" and recycling programs in Europe, 288
  3-D printing technology and, 254
Automatic computer ordering, 168
Automatic rebuy, 156
Automatic reorder, 156, 168
Automatic vending, 330–331, 674
Autonomous trucks, 311
Average cost (per unit), 508, 674
Average fixed cost (per unit), 508, 674
Average variable cost (per unit), 508, 674
Average-cost pricing, 506–510, 674
Awareness, of new products, 139, 140
Azerbaijan, 218

B
B Corporation (B Corp) certification, 476, 674
B2B e-commerce sites, as wholesalers, 340
B2B (business-to-business) market, 147, 359, 432
Baby Boomers, 79, 674
Baby shoes (strategy planning process), 41–42, 44
"Bag the Box" campaign (Malt-O-Meal), 125
Bait pricing, 520, 674
Balance, as component of operating statement, 565
Balance sheet, 572–573, 674
Banana Boat Baby and Kids Sunblock Lotion, 102
Bangladesh, 76
Banks, researching customer needs, 16
Banner (display) ads, 420, 421
Banquet Homestyle Bakes, 373
Bar codes, 313
Bar soap, 108, 109
Bar-coded return labels, 332
Bartering, 15
Basic list prices, 484, 674
Basic sales tasks, 674
Batteries, 283
Battle of the brands, 220–221, 674
Behavior. See Consumer behavior
Behavioral dimensions, for segmenting consumer markets, 100
Behavioral needs, of organizational buyers, 149–150
Behavioral retargeting, 419
Behavioral segmentation. See Consumer behavior
Behr paints, 319
Beliefs
  attitude versus, 123–124
  defined, 124, 674
  expectations and, 126
  false, 126
Benchmark price, 493
Benefit corporation, 476, 674
Benefits
  consumer needs and, 120
  core, 101
  customer value and, 20
  of research, 196–197
Betty Crocker brand, 219
Bicycle buyer personas, 102
Bicycle market, 365
Bid pricing (case study), 640–641
Bids, 158, 168
Big data
  from credit card purchases, 337
  defined, 177, 674
  Four Vs of, 177–178
  growth of, 177–178
  increased ticket revenue by sports teams via, 480
  in music industry, 183
  personal values and, 547
  personalized selling with, 334, 337
  pricing applications, 480–481
  proactive customer service via, 388–389
  social profiles from, 547
  targeted advertising from, 418
  timely mobile ads from, 423
"Big food," 527
"Big Sleepover, The," 254
Billboards, 416, 418
Biomimicry, 251
BlaBlaCar members, 125
"Black market," 133
Blak energy drink, 532
Blending, of promotion plans, 366–370
Blends
  of goods and services, 208–210
  in promotion, 355–356, 372–374
Bloggers
  ethics and, 454
  as the new press, 454
Blogs
  case study, 648–649
  defined, 674
  example, 450–451
  how customers find them, 448
  position and, 450–451
  promotion objectives of, 448
  Rookie Moms, 454
  uses of, 439–440
Blood drives, 226
Bloomberg Businessweek, 412, 590
Bluetooth-enabled stethoscope, 247
Body cameras, for law enforcement officers, 53
Body Envy, 36
Boeing 787 Dreamliner, 312
Bonobos, 217
Bonobos pants, 375
Bonus chapter information, 621–622
Bottled water, 24, 296
Bounce rate, 466, 467
Boundaries, for product-markets, 92
Bounty paper towels, 44
Braava floor cleaner, 236
Brainstorming, 93
Brand awareness, 216, 217
Brand communities, 448, 451–452, 674
Brand equity, 218, 674
Brand familiarity, 215–218, 674
Brand insistence, 215, 216, 674
Brand loyalty, 434
Brand managers, 257, 585, 674
Brand name, 212, 217–218, 674
Brand nonrecognition, 215, 674
Brand preference, 215, 216, 674
Brand promotion, 214
Brand recognition, 216, 674
Brand rejection, 215, 674
Branded apps, 448, 451, 674
Branded services, 446–448, 447, 674
Branded websites, 443
Branding
  basic considerations in, 213–214
  conditions favoring, 215
  defined, 212, 675
  heterogeneous shopping products and, 225
  of "nuisance" purchases, 229
  packaging and, 221
  protection of, 218
  as a strategy decision, 213–214
  value of, in different countries, 215
Brands
  achieving familiarity for, 215–218
  battle of, 220–221
  benefits of, 213–214
  as company asset, 218
  "generic," 220
  naming of, 217–218
  national, 220, 471
  needs met by, 213–214
  protecting, 218
  rejection costs, 215
  as resources, 64
  as specialty products, 225–227
  store brands, 319
  strong, 373
  types of, 471
Brazil, 76, 121, 134, 289
Break-even analysis, 510–512, 675
Break-even point (BEP), 510–512, 670, 675
Breakthrough opportunity, 33, 48, 59, 65, 675
Bribery, 169
Brick-and-mortar stores, 332, 337
Brokers, 346, 675
Bubble Wrap, 213
Budgeting, for promotion, 374–375
Budget-oriented shoppers, 492
Budgets, 540
Budweiser's "Lost Dog" Super Bowl ad, 359
Build-to-order approach, 265
Bulgaria, 218
Bulk-breaking, 275, 675

Bundling, 523
Burbn app, 243
Bureaucracy, 532
Business and organizational customers, 146–148, 675. *See also* Business markets; Organizational customers
Business buyers, 147
Business customers, 449
Business data classification, 164–165
Business expansion concerns (case study), 663–664
Business markets. *See also* Organizational customers
   buyer-seller relationship, 154, 159–163
   buying decision in, 155–159
   direct distribution channels, 269–270
   major segments in, 100
   manufacturers as customers, 163–165
   promotion in, 369
   service producers as customers, 165–166
Business product classes, 228–230, 268
Business products
   consumer products *versus*, 227–228
   defined, 223, 675
Business strengths dimension, 81–82
Business-to-business (B2B) markets, 107, 147, 432
"Buy American" policy, 71
Buyer personas, 102
Buyers, 153. *See also* Purchasing functions
Buyer's remorse, 138
Buyer-seller relationship, 145–146, 154
Buying, rationality and, 138
Buying behavior, 128–131
Buying center, 153, 675
Buying committee, 167
Buying function, 11, 675
Buying processes, 155–156
Buying specialist, 152
BuzzFeed, 133, 426
Byssal threads, 251
BzzAgent, 455

C

Cable television advertising, 416
Cable television shopping, 331
California Department of Transportation, 158
California Skateparks, 168
Call center, 390
Cameras, 92, 211
Canada
   advertising regulations in, 428
   Amazon in, 59
   BzzAgent in, 455
   Coca-Cola's gas-electric hybrid delivery trucks, 296
   demographic information, 76
   NAFTA and, 71
   pipelines in, 308
   successful distribution in, 298
   Walmart in, 339
Canned approach, in personal selling, 401–402
Capital item, 228, 675
Car dealers, storage costs, 311
Car leasing, 490
Car makers. *See* Automakers
Carbon emissions, 309
Career opportunities, marketing and, 6
Career planning (Appendix C), 579
   awareness of trends, 589
   career paths and compensation ranges, 584
   implementing personal plan, 591
   in marketing, benefits of, 580
   marketing opportunities, 583–589
   personal analysis, 581–583
   personal marketing plan, 590–591
   personal marketing strategy, 579–591
   personal objectives, 589–590
   starting salaries in marketing, 580
Caribbean Islands, 133
Carpooling, 125

Cars
   computers in, 211
   electric, 21
   leasing of, 490
Cars.com, 421
Car-sharing services, 125, 481
Cascadian Farm brand, 220, 221
Case studies, 449, 675
Cash back, 432
Cash discounts, 486, 675
Cash-and-carry wholesalers, 344, 675
Catalog shopping, 331
Catalog wholesalers, 344, 675
Category killers, 329
Celebrities, in advertising, 365–366
Cell phones. *See also* Smartphones
   Apple's innovations and, 115
   mobile advertising on, 419–424
   skimming price policy and, 482–483
   social profiles from, 547
   tracking of customers via, 334
   use in distribution, 295
   use in selected countries, 77–78
Census Bureau, 342
Center for Science and the Public Interest (CSPI), 15
Central America, 11, 133
Centralized buying, 154
Chain discounts, 486
Chain store buyers, 637
Chains
   in retailing, 322
   voluntary, 340
Change
   constancy of, 541–543
   in retailing, 335–338
Channel captains, 279–280, 675
Channel conflict, 276–277
Channel pricing, 504–505
Channel specialists, 267. *See also* Specialists
Channels of distribution
   Coca-Cola in, 295–296
   complexity, 285–291
   conflicts in, 276–277
   defined, 267, 675
   direct *versus* indirect, 269–274
   effects on marketing strategy plan, 542
   ideal market exposure, 283–284
   intermediaries in, 366
   international marketing and, 290–291
   management careers in, 586
   multichannel, 285–290
   producer-led, 280
   pulling and pushing of products through, 366–368
   relationships within, managing, 276–281
   reverse channels, 288–291
   sharing ad costs between, 412–413
   specialists within, 274–276
   vertical marketing systems and, 281–283
   for virtual products, 275–276
   wholesaler-/retailer-led, 280–281
Charmin toilet paper, 44, 121, 451
Chatbots, 394
Chatter, on websites, monitoring, 189
Checkout scanners, 189, 328
Cheer detergent, 44
Cheerios, 221, 481, 527
Cherry Coke, 208
Chevrolet Bolt, 21
Chex Mix Turtle, 64
Child labor, 3
Childhood obesity, 527
Children's shoes, 41–42, 44
China
   advertising climate in, 428
   advertising expenditures in, 410
   advertising growth in, 410

Amazon in, 59
Apple factories in, 11
Apple market strategy in, 115
counterfeiting of products in, 218
demographic characteristics in, 76
direct investment in, 291
dumping by, 495
health care in, 632
in-home shopping, 331
iPhone market share in, 66
joint ventures in, 291
knockoff Apple store, 218
LEGO popularity in, 87
Lenovo computers and direct selling channel, 265–266
low-value on branded items, 215
as lucrative overseas market, 54
message interpretation may fail, 361
mobile payments, 534
mobile payments in, 490
money spent on branded products, 218
nationalism and rejection of outside firms, 71
Nike's marketing strategy in, 3–4
"one child per household" in, 632
outsourcing of production to, 162
population growth in, 77
reverse channels in, 288
social networks in, 464
U.S. outsourcing of production to, 162
Vizio contract manufacturing in, 501
Chiquita bananas, 167
Chocolate Chex Mix, 64
ChotuKool, 247
Christmas layaway program, 138
Circuses, 33–34
Cirque Alfonse, 34
Citizen "journalists," 454
City of San Jose, 168
Class (social), 130
Classes of products, 223–224
Classified ads, 421
Clayton Act (1914), 72
"Clean food" (Panera), 533
Click fraud, 424
Clickstream, 419
Click-through, 370
Click-through rate, 420, 675
Clorox Bleach, 94
Clorox Scented Bleach, 207
Close, in selling, 401, 675
Close competitors, 109
Close substitutes, 65, 66
Cloud data storage services, 139–140
Clump and Seal kitty litter, 53
Clustering techniques, 104, 105, 675
Clutter ads, 416
CNET.com, 225, 235
Coast soap, 108
Coating services, 145
Coca-Cola lawsuit, 222, 223
Co-creation process, 254
Codes of ethics. *See* Ethics
Coffee sales, 173–174
Coffee-Mate, 102
Coke Zero, 207
Cold calls, 397, 439
Colgate, 96
Colgate Total Gum Defense toothpaste, 96
Collaborators, 12, 675
Collusion, 284
Colombia, 275
Columbia Forest products, 251
Combination export manager, 346, 675
Combination plan (compensation), 397
Combined target market approach, 96, 675
Combiners, 96–97, 675
Combining, segmenting *versus*, 109
Command economy, 13, 675

Commercialization, 255
Commission
   for manufacturers' agent, 345
   for reduction of working capital, 398
   for salespeople, 397–398
Committee buying, 167
Commodity products, 229
Common law, 218
Communication, of promotion to target markets, 360–362
Communication process, 360, 363–366, 675
Communication technologies, 542
Company objectives, 61–63. *See also* Objectives
Comparative advantage, 241
Comparative advertising, 414, 415, 675
Compensation, 397–399
Competition
   assessing, in strategy planning, 65–67
   customer perceptions of, 107–110
   for customer value, 21
   demand and supply as aid in understanding, 560–563
   exclusive distribution *versus*, 284
   legal defense of, 72–74
   nonprice, 478
   nonprofit organizations and, 23
   to satisfy consumer needs, 21
Competitive advantage
   of Cirque du Soleil, 34
   defined, 48–49, 675
   from economies of scale, 54
   finding, 63–64
   patents for innovative products and, 247
   software and hardware for, 394–395
   sustainable, 65
Competitive advertising, 413, 675
Competitive bid, 158, 168, 675
Competitive environment, 65–67, 675
Competitive rivals, 66, 675
Competitor analysis, 65–66, 539, 602, 675
Competitor-free environments, 65
Competitors
   close, 109
   illegal, 218
   product ideas from, 250
   rival, 66
Complaints. *See also* Customer service
   about marketing, 27
   from customers, 457
Complementary product pricing, 523, 675
Component parts and materials, 228, 229
Components, 675
Computer programs. *See* Software
Computer sales, 265–266
Computer system breach, 250
Computer virus, 250
Computer-aided segmentation, 111
Computerization, 542
Concentration of industries, 163–165
Concept testing, 253–254, 675
Confidence, consumer, 119
Confidence intervals, 198, 675
Confirmation, in new-product adoption, 139, 140
Conflict, in distribution channels, 276–277
Conflict of interest, 151
Conservation. *See* Sustainability
Conspiracy, 496
Consultative selling approach, 402, 675
Consulting services, 161
Consumer behavior
   Apple loyalty example, 115–116
   consumer decision process and, 135–140
   expanded model of, 136
   influences on, model of, 117
   in international markets, 134
   learning new behaviors, 246–248
   for marketing strategy planning, 117
   online, 419

   organizational customers, 149–150
   post-purchase regret, 138
   psychological influences, 119–128
   purchase situation and, 134–135
   role of needs in, 118–119
   second thoughts, 138
   situational effects, 134–135
   social influences affecting, 128–131
Consumer confidence and spending, 119
Consumer cynicism, 546
Consumer decision process, 116, 135–140
Consumer Electronics Show (Las Vegas), 432
Consumer market segments. *See* Segmentation
Consumer needs
   competing, to satisfy, 21
   interpreting, 8
   as marketing's starting point, 7–8, 17
Consumer panels, 190, 675
Consumer product classes, 223–227
Consumer Product Safety Act (1972), 73, 252, 675
Consumer Product Safety Commission, 24, 252
Consumer products, 223, 227, 675
Consumer protection laws, 27
Consumer Reports magazine, 127
Consumer Safety Institute (Netherlands), 122
Consumer satisfaction. *See* Customer satisfaction
Consumer surplus, 559, 675
Consumer sweepstakes, 410
Consumer-citizens, 536, 549
Consumerism, 544
Consumers, 3
Contagious message, 447
Containerization, 308–309, 675
Contests, 368, 432, 465
Contingency plans, 540, 616
Continuous improvement, 21, 258, 675
Continuous innovations, 246, 675
Continuum of environmental sensitivity, 82–83
Contract manufacturing, 501
Contracts
   between buyers and sellers, 161–162
   plain-language, 544
Contractual channel systems, 282, 675
Control
   analytical tools for, 44
   as marketing management task, 34
   as section in marketing plan, 42–45, 540, 612–616
Convenience
   high value of, 49
   paying extra for, 118
Convenience products, 224, 225, 675
Convenience (food) stores, 275, 278, 330, 335, 675
Conventional retailers, 325–326
Conversion, of markups, 571
Converting (moving customers to desire and action), 439
Cookies (computer), 60, 334, 419, 423
Cooperation, in marketing research, 185–186
Cooperative advertising, 413, 675
Cooperative relationships, 160
Cooperatives, 322–323
Copy thrust, 424, 425, 443, 675
Copyright law, 247
Core benefits, 101
Core products, 211
Corporate chains, 322, 675
Corporate channel systems, 281, 675
Corrective advertising, 429, 676
Cost of sales, 567, 676
Cost per one thousand impressions (CPM), 428
Cost structure, of a firm, 508
Cost-oriented price setting, 502
Costs. *See also* Prices; Profits
   of brand rejection, 215
   budgeting, for promotion, 374–375
   as component of operating statement, 565
   customer value and, 20
   of different transport modes, 307

   in direct/indirect distribution decision, 273–274
   forms of, 20–21, 507–510
   full-line pricing, 522–523
   inventory, 300
   of logistics, 297
   of lost sales, 300
   marginal analysis and, 512–514
   marketing analytics and, 19
   of missed opportunities, 532
   of poor quality, 149, 258
   production (factory), 504–505
   of research, 196–197
   revenues *versus*, 45–48
   storing and handling, 310–313
   total cost approach, 302
   total inventory costs, 311–312
   transportation, 300
   of transporting as percent of selling price, 306
   of transporting function, 305–306
Cottage cheese, "the original superfood," 52
Counterfeit goods, 218
Coupons, 41, 52, 69, 179, 354, 355, 368, 373, 375, 488
Craigslist, 244, 421
Cranes, 303
Crayola crayons, 215, 373
Creative brands, 463
Creative promotion, 203–204
Credit, from intermediaries, 273
Credit cards, 489, 546
Credit insurance, 395
Credit management, career opportunities in, 588
Credit sales, 489
Crest toothpaste products, 236, 257, 428
Criteria, setting, 136
CRM. *See* Customer relationship management (CRM)
Cross-tabulation, 197
Crowdsourcing, 250
CR-V (Honda), 45
Cues, 123, 676
Cultural and social environment, 676
Culture and ethnicity, 117
Culture and social environment, 74–80
Culture(s). *See also* International marketing
   consumer behavior and, 131–134
   corporate, globally oriented, 501
   defined, 132, 676
   different, in international markets, 134
   message interpretation may vary, 361
   social and cultural environment, 74–80
   target marketing and, 131–134
Cumulative quantity discounts, 485, 676
Customer acquisition, 48, 439
Customer advocacy, 389, 454–457
Customer analysis, in marketing plans, 539, 599–602
Customer analyst, career as, 585
Customer behavior. *See* Consumer behavior
Customer complaints. *See* Complaints
Customer data, ideas from, 249–251
Customer databases. *See* Customer relationship management (CRM); Databases
Customer enhancement, 48
Customer equity, 46–47, 676
Customer experience, 446–448, 455–456
Customer journey maps, 191, 676
Customer lifetime value (CVL), 45–48, 676
Customer loyalty, 45–48, 379
Customer loyalty cards. *See* Loyalty cards
Customer needs. *See also* Consumer behavior
   anticipated, by artificial intelligence and intelligent agents, 211–212
   awareness, 139, 140
   basic role in buying behavior, 118–119
   brands and, 213–214
   competing, to satisfy, 22–23
   in generic *versus* product-markets, 90–91

Customer needs—*Cont.*
  as marketing's starting point, 7–8, 17
  psychological influences, 119–128
  as source of new-product ideas, 250
Customer (user) needs, naming markets for, 91
Customer relationship management (CRM)
  defined, 105, 676
  target marketing with, 362
  use in banks, 385
  use in sales, 399, 400
Customer relationships, building, 8, 21–22
Customer retention, 47, 48, 395, 439
Customer reviews, 59, 250
Customer satisfaction
  defined, 5, 676
  as key to market orientation, 17–18, 20–23
  measuring, 529–531
  in the United States, 529
Customer service
  analytics in, 388–389, 391
  career opportunities in, 587
  case study, 646–647
  defined, 39–40, 676
  digital self-service as, 393
  as driver of satisfaction, 62–63
  as form of personal communication, 353
  handling complaints (case study), 646–647
  information technology (IT) for, 392–395
  as logistics goal, 297–300
  personal selling and, 379–380
  proactive, 388–389
  salespeople as helpers, 381–382
  setting levels for logistics decisions, 297–300
  social media examples, 465
Customer service level, 267, 297–298, 300, 676
Customer service representatives (CSRs), 62–63, 385, 386–387, 676
Customer tracking, 212, 334, 408–410, 421, 422–423
Customer trust, 440, 443
Customer type, naming markets for, 91
Customer value. *See also* Customer satisfaction; Value
  defined, 20, 676
  equals benefits minus costs, 21–22
  Four Ps contribute to, 40
  improving, 21–22
  intelligent agents and, 69
  marketing concept and, 20–23
Customers
  product ideas from, 250
  retaining, 47, 48, 439
  as target of marketing mix, 38
  view of products and markets, 552–557
Customers' perceptions, 20, 107–110
Cyberbrokers, 346
Cynicism, 546

D

Damage awards, 67
Dannon yogurt, 226
Danube River, 308
Dasani bottled water, 296
Dashboard, 181
Data. *See also* Big data
  changing, to information, knowledge, and wisdom, 180
  external, 177
  form marketing research studies, 177
  government, 189
  as guide for marketing strategy, 176
  internal, 176–177
  massive growth in, 177–178
  primary, 187, 190–197
  secondary, 187
Data gathering. *See also* Marketing research; Surveys
  methods of, 190–197
  observing as method of, 194
  searching the web, 189–190
  types of, for marketing research, 190–197

Data interpretation
  principles and tools, 197–199
  for quality improvements, 259
  types of, for marketing research, 197–199
Data mining, 407–408
Data sources
  primary, 187, 190–197
  secondary, 187
Data warehouse, 179, 676
Databases
  customer relationship management (CRM) and, 105
  MIS developments and, 176–184
  for pricing, 479
  use in segmentation, 105
Dawn dish detergent products, 98, 362, 471
DDS. *See* Decision support system (DSS)
"Deal of the Week," 367
Dealer brands, 220–221, 471, 647–648, 676
Dealers, 265
"Deal-prone" customers, 434
Deals, 465, 488–489
Deceptive advertising, 429
Deciders, 153
Decision maker, in family, 129–130
Decision making, 139. *See also* Consumer decision process
Decision support system (DSS), 180, 197, 676
Decision-making process. *See also* Problem solving
  in business and organizational buying, 154
  for organizational customers, 155–159
Decisions, in adoption process, 139–140
Decline stage, 240, 373–374, 415
Decoding, 361, 676
Defects, reducing, 258. *See also* Quality
Demand
  derived, 227, 676
  diminishing, law of, 553, 678
  elastic, 554–557
  heterogeneous, 10, 15
  "if-then" thinking and, 512
  impact on average-cost pricing, 509–510
  inelastic, 554–557
  marginal analysis and, 512–514
  pricing methods based on, 514–522
  primary, 413
  selective, 373, 413
  skimming prices and, 482–483
  types of, in promotion, 372–374
Demand and supply
  nature of competition and, 560–563
  size of market and price level and, 559
Demand curve
  defined, 676
  down-sloping, 553
  in economic fundamentals (Appendix A), 553–557
  estimating, 512–514
  kinked, 561–562
  in prestige pricing, 522
  prices along, evaluation of, 09
  in psychological pricing, 520
  skimming price policy and, 482–483
Demand elasticity, 556–557
Demand schedule, 553, 554
Demand-backward pricing, 521–522, 676
Demand-oriented price setting, 502, 514–522
Demographic segmentation. *See* Segmentation
Demographics
  age distribution, 79–80
  of Big Seven social media platforms, 460
  cell phone and Internet use, 77–78
  clustering techniques and, 104
  dimensions and characteristics, for selected countries, 76
  effects on marketing strategy plan, 542
  lifestyle dimensions and, 128
  literacy, 77

population and trends, 78–79
  as segmenting dimension, 100
  social class, 130
  U.S. income distribution, 118–119
Department stores, 287, 327, 330, 676
Derived demand, 227, 676
Desire, as a need, 120
Desire, in AIDA model, 358, 424, 425
Detailers, 386
Determining dimensions (segmentation), 101, 676
Developing countries. *See also* Less-developed nations
  low labor costs in, 307
  transportation costs in, 309
Dial soap, 108–109
Diapers, markets for, 37, 434
Dictatorships, 528
DieHard batteries, 217
Diesel-hybrid 18-wheelers, 309
Diet cola, 242
Differentiation. *See also* Segmentation; Targeting
  achieving, with product design, 203
  adding services to good for, 210
  by blending goods and services, 208–210
  case study, 665–666
  defined, 51, 676
  examples of, 52
  as marketing focus, 107
  as marketing mix focus, 107
  in marketing plans, 539
  meeting competitors' prices for, 493
  in monopolistic competition, 65
  personal, in career planning, 582–583
  price premium and, 493
  product types and, 207
  as section in marketing plan, 606
Digital advertising. *See also* E-commerce; Internet
  Americans' time spent with digital media, 419
  ethical issues, 423–424
  examples, advantages and disadvantages, 421
  tracking customers, 419, 421
  types of, 420–424
"Digital assistant," 212
Digital dashboard, 183
"Digital natives," 80
Digital pets, 242
Digital products, distribution of, 275–276
Digital self-service, 393
Digital tools, 33
Digital video recorders (DVRs), 443
Direct channels, 269–270
Direct competitive advertising, 413–414, 676
Direct distribution, 269–274
Direct home delivery, 270
Direct investment, 291, 676
Direct mail, 362
Direct market environment
  competitors, 65–67
  firm's objectives, 61–63
  firm's resources, 63–64
Direct marketing, 274, 676
Direct response, 420
Direct selling channel, 265
Direct store delivery, 314
Directories, 421
Direct-response advertising, 265
Direct-response promotion, 362
Direct-to-customer sales, 269, 270
Direct-to-job delivery services, 320
Disaggregation, 93
Disaster relief logistics, 314–315
Discontinuous innovation, 246–248
Discontinuous innovations, 676
Discount policies, 484–486
Discounting products, 433
Discounts, 465, 485, 676
Discover Moscow app, 211
Discrepancies of quantity, 10, 11, 275–276

Discrepancy of assortment, 10, 11, 275, 676
Discrepancy of quantity, 275, 676
Discretionary income, 119, 676
Dish detergent, 493
Display (banner) ads, 420, 421
Disposable diapers
　markets for, 37
　sales promotion problem, 434
Dissonance, 138, 676
Distribution
　case study, 647-648
　channel complexity, 285-291
　channel specialists, 274-276
　competitive advantage in, 241
　conflicts in, 276-277
　direct *versus* indirect, 269-274
　exclusive, 283, 284
　intensive, 283
　management of relationships in, 276-281
　multichannel, 285-290
　objectives, 267-269
　relationships within channels, 276-281
　selective, 283, 284, 293
　target marketing decisions, 39
　vertical marketing systems, 281-283
Distribution centers, 281, 313-314, 676
Distribution channel management, 586
Distribution cost, packaging and, 222
Distributors, 295
Diversification
　defined, 54, 676
　McDonald's example, 54
Diversity, 132
Division of labor, 10
Divorce, consumer behavior and, 129
Dollar Menu (Wendy's), 492
"Don't-rock-the-pricing-boat objectives," 478
Door-to-door selling, 331, 676
Doritos Locos Tacos, 412
Dove chocolates, 167
Dove soap, 108
Down-sloping demand curve, 553
Downy fabric softener, 257
Dr Pepper, 219, 285, 296
Driftbot, 393
Drive, 119, 123, 676
Driverless cars, 68
Driverless trucks, 311
Drone aircraft
　as delivery vehicle, 59, 308
　for hobbyist market, 163
Dropbox, 369
Drop-shippers, 344, 676
Drug companies, pricing of, 495
Dumping, 495, 676
Duncan Hines, 67
Duracell batteries, 283, 488
Dynamic behavioral segmentation, 106-107, 676
Dynamic pricing, 479-480, 676
Dynamically continuous innovations, 247, 676

E
Early adopters, 371-372, 676
Early majority, 372, 676
Earned media, 443
　about, 442
　benefits and challenges of, 443, 444
　bloggers are the new press, 454
　career opportunities in, 588
　consumer trust of, 443
　credibility of, 443
　from customer advocacy, 454-457
　customer trust in, 443, 444
　defined, 443, 676
　examples, 442
　falsifying ethics of, 457
　message source, 442

from public relations and the press, 453-454
　risks, 445
Eastern Europe, 16, 194, 477
Eastgate Center, 251
Eastman Tritan copolyester, 155
"Eat 'em like junk food" campaign (carrots), 537
e-books, 439, 449
eCards, 90
Echo (Amazon), 68, 407, 483
E-commerce. *See also* Auction sites; Websites
　advantages and disadvantages, 12
　amazon.com innovations, 59-60
　automatic rebuys via, 156
　defined, 12
　Dell, Inc. model, 265-266
　direct distribution via, 270-271
　distribution advantages, 270-271
　growth of, 12
　international variations, 329
　just-in-time (JIT) delivery systems and, 303
　small service companies and, 167
Economic buyers, 118, 676
Economic environment, 67-68, 676
Economic needs, 118-119, 677
Economic systems
　defined, 13, 677
　marketing's role in, 13-15
　types of, 13-15
Economic-buyer theory, 118
Economics, fundamentals of (Appendix A), 552-563
Economies of scale
　competitive advantage from, 54
　defined, 10, 677
　in promotion, 374
　transporting function and, 305-306
Ecoworx carpet tiles, 288
Editorial coverage, 443, 453
Educational web pages, 448, 449
Egg Beaters, 121
Egypt, 75, 76, 307
80/20 rule, 284
Elastic demand, 554-557, 677
Elastic supply, 558-559, 677
Elasticity, price setting and, 563
Elder care, disposable diapers market and, 37
Electric cars, 21
Electronic data interchange (EDI), 304-305, 312, 314, 677
Electronic scanners, 190
Electronics fairs, 265
E-mail
　attention-grabbing subject line, 362
　benefits to buyers and sellers, 362
　building relationships with, 452-453
　for direct-response promotion, 362
　newsletters, 448, 452-453
　target marketing with, 362
Emergency product, 224, 225, 677
Emojis, 407
Emotion, 447
Emotional benefits, 20
Empowerment, 260, 677
Empty nesters, 129, 677
Encoding, 361, 677
End benefit, 516
End-of-aisle displays, 386
Endorsements, 3, 203, 454
Energizer batteries, 222, 283
England. *See* United Kingdom
Enhancement of customer value, 47
Entertainment-seekers, 90-91
Entrepreneurs, women, in less-developed nations, 272
Environmental issues. *See also* Green choices
　case study, 658-659
　Coca-Cola's commitment to sustainability, 295-296

　in distribution channels, 29
　generation of trash and, 222
　green packaging, 222
　objectives based on, 19
　in product-market screening, 81, 82-83
　in transportation, 309
Environmental outcome, in triple bottom line, 19
Environmental Protection Agency/SmartWay program, 309
Environmental responsibility, 19
Environmental sensitivity, 82-83
EpiPen, 517
Equal (sweetener), 124
Equilibrium, 559
Equilibrium point, 677
Estée Lauder, 277
Estimate, from a sample, 198
Ethical norms, 26
Ethical values, 26
Ethics
　addictive products, 546
　American Marketing Association Statement of Ethics, 26
　bait pricing, 520
　in competitor analysis, 67
　in a consumer-oriented world, 527-528
　of counterfeit products, 218
　of credit card companies, 489
　defined, 679
　digital advertising and, 423-424
　in distribution arrangements, 287-288
　"fair trade" not always fair, 325
　false impressions in promotion, 126
　of falsifying earned media, 457
　of influencing beliefs, 126
　in interpreting and presenting research results, 198-199
　marketing concept and, 27
　in marketing to foreign governments, 169
　misleading ads, 126
　native advertising and, 426
　needs created by marketing, 123
　in new-product planning, 247-248
　in organizational purchasing, 150-151
　overview of, 27
　in packaging, 223
　paying for opinion leaders, 455
　in personal selling, 403
　in physical distribution, 305
　planned obsolescence, 247-248
　price fixing, 496
　price sensitivity and, 517
　product liability, 252
　promotion and, 365-366
　in selecting segmenting dimensions, 103
　sharing user data, 184
　spying on competitors, 67
　in use of marketing research, 198-199
　value issues and, 494
Ethiopia, 76, 77, 78
Ethnic groups
　buying habits, 132
　increase in consumers among, 132
　stereotyping of, 133
Ethnographic research, 194-195
Euro, 72
Europe
　adaptation of marketing strategies for, 339
　dealer brands in, 220
　garbage disposals in, 244-245
　inland waterway transportation in, 308
　Manitowec in, 303
　market development example, 53
　privacy laws in, 409-410
　QVC and Home Shopping Network in, 331
European Community's Waste Electrical and Electronic Equipment (WEEE) Directive, 288
European markets, 71-72

European Union (EU), 71-72, 529
Evaluation
	of new ideas, 252-254
	of new products, 139
Everyday low pricing, 487, 677
Everyday Value Menu (Wendy's), 492
Evolution, of retailers, 335-338
Evolution locomotives (General Electric), 309
Exaggerated claims, 365-366
Exchange, buying and selling and, 11
Exclusive distribution, 283, 284, 677
Executive summary, in marketing plan, 593-594
Expectations, 126, 677
Expense item, 228, 677
Expenses, 567, 569, 677
Experimental method, 195, 677
Export agents, 345, 677
Export brokers, 346, 677
Export rules, 14
Exporting, 290, 677
Exposure of products, 283-284
Extensive problem solving, 137, 677
External market environment, 539. *See also* Market environment
Eyewear marketing, 62, 333, 335, 362, 494

F

Facebook, 33, 36, 459-461, 677
Facebook Messenger, 464
Factor, 575, 677
Factor method, 575, 677
Factory (production cost), 504-505
Fads, 242, 677
"Fail early and fail often," 254
Failure, of products, 248
Fair trade, 325
Fair trade coffee, 24
Fairness, 26. *See also* Ethics
Fairy brand dish detergent, 493
Fake customers, 424
False beliefs, 126
Family brand, 219, 677
Family influences, on consumer behavior, 128-131
Family life cycles, 128
Fanta soft drink, 295
Farm products, 229, 677
Fashion, 242, 677
Fashion accessories (Harley-Davidson), 279-280
Fast Company, 590
Favoritism, 168
Febreze, 123, 132
FedBizOpps.gov website, 169
Federal Aviation Administration (FAA), 308
Federal Communications Commission (FCC), 518
Federal Fair Packaging and Labeling Act (1966), 222, 677
Federal government, marketing to, 168-169
Federal Trade Commission (FTC)
	on bait pricing, 520
	ban on phony prices, 495
	comparative ads approved, 414
	control of unfair advertising, 429
	defined, 677
	"like grade and quality" of similar products, 497
	native advertising review, 426
	new-product definition, 247
	puffery permitted, 430
	search engine requirements of, 423
Federal Trade Commission Act (1914), 496
Feedback, 163, 361, 362
Financial analysis, 47
Financial resources, 64
Financial strength, 64
Financing, 11, 677
Fines, 73
Finland, 75, 76
Firestone tires tragedy, 73
Firm, defined, 8

Fit (Honda), 45
Fitbit, 178, 189
Fitness activity trackers, 178, 189
Fitness apps, 203-204
Fitness technology. *See* Technology
Fixed costs, 670
Fixed-cost (FC) contribution per unit, 511-512, 677
Fixed-response questionnaires, 192
Flexibility, as company strength, 64
Flexible pricing policies, 479-482
Flexible-price policy, 479, 677
Flickr, 464
Flowers and bulbs, 308
Flyknit shoes (Nike), 3
Focus group interview, 192, 677
Focus groups, 173, 192, 235, 253
Followers, 242
Food allergies, 223
Food and Drug Administration (FDA), 252
Food courts, 626
Football clothing, 203
Forbes magazine, 463
Ford Motorcraft, 423
Forecasting. *See also* Predictive analytics
	approaches to, 574
	of company and product sales, 575-577
	of customer needs, 211-212
	of future behavior, 575
	judgment needed in, 575, 577
	levels of, 573-574
	marketing mix and, 577
	in marketing plan, 540
	past behavior and, 574, 575-577
	of target market potential and sales, 573-575
	types of, 573-574
Foreign Corrupt Practices Act (1977), 169, 677
Foresight, 427
For-profit organizations, 7
Fortune magazine, 295, 590
Four Ps of marketing
	decision areas organized by, 538-540
	delivery of value by, 40
	each contributes to the whole, 40
	marketing managers and, 538
	overview of, 38-42
Four Vs of big data, 177-178
Fracking, 24
Frame of reference, 361
France
	advertising regulations, 428
	Amazon in, 59
	Carrefour expansion, 339
	Coca-Cola in, 295
	demographics, 76
	"soda tax," 527
	Sodebo marketing example, 7-8
Franchise, 407
Franchise operations
	case study, 642
	defined, 677
	as distribution model, 284
	retailing through, 322-323
Franchisors, 284
Frankie the Cat robot, 87
Fraud (click fraud), 424
"Free," 519
Free samples, 368, 407
"Free" strategy, 519
Free trade, 72, 677
Free trade agreements, 71-72
Free trials, 517
Freedom from, as a need, 120
Freedom of choice, 529
Freeloaders, 519
Freemium, 677
"Freemium," 519
Freestyle drink machines, 250
Frontline Plus flea protection, 485

Full-line pricing, 522-523, 677
Full-service wholesalers, 344
Functional benefits, 20
Functional (trade) discount, 486

G

Gain laundry detergent, 44, 257
Galaxy Note 7, 127
Gambling, 410
Gamers, 137
Games
	Get the Glass, 121
	Pokémon Go, 211
Garbage disposals, 244-245
Garmin Vivosmart, 189
Gas-electric hybrid delivery trucks, 296
Gatekeepers, 153
GDP (gross domestic product), 77
GDPR (General Data Protection Regulation), 183
General Data Protection Regulation (GDPR), 183, 677
General merchandise wholesalers, 343, 677
General Services Administration (GSA), 169
General stores, 325, 677
General-line (single-line) wholesalers, 343
Generation X (Gen X), 79-80, 677
Generation Y (Gen Y), 80, 677
Generation Z (Gen Z), 80, 419, 677
Generic brands, 220
Generic market, 677
	customer needs and, 93
	defined, 90, 677
	product-market definitions, 92
Generic products, 677
Geographic areas
	industry concentration by, 164
	naming markets for, 91
Geographic dimensions, for segmenting consumer markets, 100
German parents, 87
Germany
	adapting flavors for, 258
	Amazon in, 59
	demographic characteristics in, 76
	drones, for delivery service in, 308
	iPhone market share in, 66
	LEGO popularity in, 87
	Manitowoc in, 303
	market development in, 53
	Metro AG and distribution in India, 309
	population growth in, 77
	steel industry in, 164
	taste preferences in, 258
	Walmart's struggle in, 339
Get the Glass game, 121
Gillette razors, 257
Girl Scout Cookie Locator app, 269
"Giving Is Love" shopper type, 104
Glance medium, 416
Glass bottles, returnable, 295
Gleevec (cancer drug), 495
Global economy. *See* International marketing
Global positioning data, 310
Global warming, 309, 548
Gmail, 254
GNI (gross national income), 77
GNI per capita, 77
Gold Crown (Hallmark) stores, 89
Golden Arch hotels, 54
Golf products, in distribution, 275
Goods. *See also* Goods and services; Products
	adding services to, for differentiation, 210
	products as, 208
Goods and services
	blends of, 208-210
	differences between, 208-210
Goodwill, eroding of, 258
Google, 254

Google+, 240
Google AdWords, 419, 421
Google Analytics software, 466
Google News, 453
Google Now, 212
Google Shopping, 487
GoPro, 211
GoPro Hero, 212
"Got Milk?" campaign, 121
Government
   advertising regulations by, 428-429
   as customer, 147
   dealing with foreign governments, 169
   Environmental Protection Agency/SmartWay program, 309
   Internet of Things (IoT) and, 213
   purchasing functions, 168-169
   rebate regulations, 488
   role of, in command (planned) economy, 13
   role of, in economic systems, 13-15
   transportation regulations, 306
   U.S. support of companies' environmental initiatives, 309
   value issues and, 494
Grading, of raw materials, 229
Grease money, 169
Grease Monkey, 291
Great Britain. *See* United Kingdom
Great Depression (1930s), 351
Great Lakes, 308
Great Recession of 2008, 67, 80
Green attitude, 124
Green choices. *See also* Environmental issues
   Coca-Cola initiatives, 295-296
   Method Products, 471-472
   in packaging, 70, 222
   reverse channels for, 29
Green corporate citizens, 288-291
Green packaging, 70, 222
Green Works cleaners, 94
Greenwashing, 545, 677
Greenworks lawnmower, 337
Greeting cards, 89
Grocery stores, leader pricing and, 516
Grocery trade, 221
Gross domestic product (GDP), 77, 677
Gross margin (gross profit), 567, 677
Gross national income (GNI), 77, 677
Gross profits, 567
Gross sales, 566, 678
Growth
   costs of, 471
   international marketing as cause of, 54-55
   population, 76, 77, 79-80
   as product life cycle stage, 238-239
Growth of business, paths to, 47
Growth options, examining (case study), 637-638
Guatemala, 226

H

Hackers, 250
Hacking, 67
Hair care essentials, 36
Haiti, 76
Hampton Bay fans, 319
Handling, 312-313
Handwashing saves lives campaign (Lifebuoy soap), 359
Happiness, 447
Happy Meals (McDonald's), 433
Hard-to-copy marketing strategies, 48-49
Hardware, 394-395
Harley Owners Group (HOG), 120, 452
Hashtag (#), 464
Hazardous materials, 288
Head & Shoulders shampoo, 257
Headphones, 447

Health care costs, 494
Health care insurance, 370
Health Care Reform Act (2010), 630, 631, 632
Heat-Gear warm weather products, 203
HeatWave packaging concept, 407
Hellmann's mayonnaise, 433
Hello Hydration, 36
"Help a Child Reach 5" campaign (Lifebuoy soap), 359
Herbal Essences, 36
Heterogeneous (different) between market segment, 95
Heterogeneous demand, 10, 15
Heterogeneous shopping goods, 268
Heterogeneous shopping products, 225, 384, 556, 678
Heterogeneous supply, 10, 15
Heterogeneous supply and demand, 10, 15
HFC-free vending machines, 295-296
Hierarchies of needs, 120-121
Hierarchy of objectives, 62-63
"Highway pilot," 311
Hillside Veterinary Clinic Marketing Plan (Appendix D), 592-619
Hindsight, 427
Hispanic market, 132
Hit-or-miss marketing, avoiding, 49
Hobby market, 163
Holland, 308
Home mortgage crisis, 67
Home page, 449
Home robots, 235-236
Home Shopping Network, 331
"Home Try-On Program," 494
Homogeneous shopping goods, 268
Homogeneous shopping products, 224-225, 556, 561, 678
Homogeneous (similar) within market segment, 95
Honda Crosstour, 457
Honey Whole Grain Pretzel, 434
Hootsuite software, 466, 468
Horizontal arrangements, 284
Horizontal channel conflict, 277, 678
Household cleaners, 94, 471-472
Housing market downturn, 67
Houzz, 464
"How to search on Google," 188
HPShopping.com, 393
HubSpot software, 439-440, 468
Huggies disposable diapers, 37, 434, 454
Hulu, 277
Humor, 447, 465
Hydraulic fracturing (fracking), 24
Hypermarkets, 328-329, 683
Hypotheses, 184-185, 678

I

"I Will What I Want" advertising campaign, 203
Iams dog food, 268
IBOT wheelchair, 632-633
Idea evaluation, 252-254
Idea generation, 249-251
Idea screening, 252
Ideal market exposure, 678
Identifying alternatives, 136
Identity thieves, 254
"If it ain't broke, don't fix it," 545
"If-then" thinking, demand and, 512
IKEA's assembly services, 210
Illiteracy, 77, 193, 226
Images, negative, 215
Imitation, 194
Immigration, 132
Immigration populations, 387
Implementation
   of advertising, 412
   defined, 678
   as marketing management task, 34

marketing plan as blueprint for, 538-540
   of marketing plans, 42-45, 540, 612-616
   of personal marketing plan, 591
Import agents, 345, 678
Import brokers, 346, 678
Impressions, in advertising, 428
Impulse buying, 134-135
Impulse products, 224, 225, 678
Inbound Marketing: Get Found Using Google, Social Media, and Blogs (Halligan and Shah), 439
Inbound promotion, 439
Incentives, 397, 398, 517
Income
   discretionary, 119
   needs affected by, 118-119
   spending and, 119
   U.S. distribution, 118-119
India
   Amazon in, 59
   Apple products in, 115
   cell phone use in, 78
   demographic characteristics in, 76
   dollar stores in, 339
   entrepreneurs in, 9
   handwashing saves lives campaign in (Lifebuoy soap), 359
   LG as favored brand in, 103
   refrigeration in rural markets, 246-248
   rural markets, 9
   transportation systems, 309
   Uber in, 48
   U.S. outsourcing of customer service to, 162
Indirect channels, 273-274
Indirect competitive advertising, 413-414, 678
Indirect *versus* direct distribution, 269-274
Individual brands, 220, 678
Individual product, 207, 678
Individual sellers, 228
Indonesia, 339
Industrial buyers, 147
Industrial Revolution, 15
Industry attractiveness, 81-82
Industry codes, 164-165
Industry demand, 227
Industry sales, 238
Industry sales forecast, 573
Inelastic demand, 554-557, 678
Inelastic supply, 558-559, 678
Inflation, interest rates and, 67-68
Influence peddling, 169
Influencers, 153, 369, 422, 678
Infographics, 448, 451, 678
Informal focus groups, 253
Information. *See also* Marketing Information System (MIS)
   as bridge to markets, 175
   from data, 179
   for effective marketing, 174-176
   separation of, 10
   shared, 160-161
Information search, 136, 139, 157
Information sharing, 160-161
Information technology (IT). *See also* Artificial intelligence (AI); Customer relationship management (CRM); E-commerce; Marketing information system (MIS)
   as aid in personal selling, 392-395
   electronic data interchange (EDI), 304-305, 312, 314
   intelligent agents, 68-69, 210-212
informational privacy, 182-183
Informational privacy, 678
Information-gathering methods, 190-197
Informing objective, of promotion, 357
Inland waterways, 308
Innovation. *See also* Breakthrough opportunity
   continuous, 246
   culture of, 256

Innovation—*Cont.*
  defined, 6, 678
  discontinuous, 246-248
  dynamically continuous, 247
  in laundry detergent, 471-472
  in new-product development, 236-237, 246-248
  of Nike, 4
Innovators, 371, 678
Inside sales force, 390, 678
Insider knowledge, 465
InSinkErator, 244-245
Instagram, 33, 460, 461-462, 678
Installations, 228, 229, 678
Installment plans, 489
Instant gratification, 527
Institutional advertising, 413, 415, 678
In-store kiosks, 300
In-store sampling, 374
In-store tracking, 336
Insurance companies, 209, 250, 252, 476, 495, 517
Intangible products, 209
Integrated direct response, 362
Integrated marketing communications, 355-356, 678
Integrated promotion plans, 366-370
"Intel inside," 516
Intelligent agents, 68-69, 210-212, 393-394, 678
Intensive distribution, 283, 678
Intention to buy, 124
Interest
  in adoption process, 139-140
  in AIDA model, 358, 424, 425
Interest rates, inflation and, 67-68
Inter-European commerce, 72
Intermediaries
  in channels of distribution, 276-281
  as customer, 147
  defined, 12, 678
  exclusive and selective, 283
Intermediate buyers, 147
Internal marketing, 367
International marketing. *See also* individual countries
  adapting products for, 257-258
  advertising, sales promotions, and marketing strategies in, 408-410
  advertising regulations, by country, 428-429
  advertising spending in, 410
  agent wholesalers and, 345
  Amazon's sales in, 59
  Apple growth in, 115
  assessing political environments, 71-72
  brand name selection in, 218
  branding, 218
  career opportunities in, 588
  Coca-Cola in, 295-296
  container shipping, 308-309
  counterfeiting of products in, 218
  cultural variations in, 134
  dealing with foreign governments, 169
  of Dell computers, 265-266
  demographics for selected countries, 76
  direct investment in, 291
  dumping and, 495
  economic conditions and, 68
  effects on marketing strategy plan, 542
  entering, ways of, 290-291
  ethics in, 169, 305
  evaluating opportunities in, 82-83
  export/import brokers, 346
  exporting, 290
  finding opportunities in, 91
International markets, U.S. firms marketing to, 169
Internet. *See also* Apps; Blogs; Digital advertising; E-commerce; Websites
  advertising benefits from, 420
  as advertising medium, 419
  career planning resources on, 590-591
  catalog wholesalers adapted to, 344

  direct distribution via, 270-271
  information gathering via, 157
  information sharing via, 160-161
  intermediaries, 12
  marketing opportunities via, 419
  online communities, 158
  online focus groups, 192
  online surveys, 193
  researching new purchases on, 157-158
  use of, by small service companies, 167
  use of, in selected countries, 77-78
Internet of Things (IoT), 213
Internet retailing, 331-335
Internet use, in selected countries, 77-78
  joint ventures in, 291
  learning foreign regulations, 55
  licensing, 291
  logistics costs, 297
  management contracting, 291
  marketing research in, 199
  message interpretation may vary, 361
  opportunities to consider, 54-55
  retailing in, 338-339
  risks in, 55, 82-83
  shipment modes and carriers, 308-309
  shipping systems for, 308-309
  "shopping" in, for ideas, 250
  social networks in, 464
  trademark protection in, 218
  Uber in, 48
Internet-based
  markup by, 503-504
  overview of role, 12
  promotion aimed at, 366-367
  stocking allowances and, 486
  trade promotion aimed at, 432-433
Internet-based data system, of Coca-Cola, 295
Internet-based intermediaries, 12
Internet-connected sensors, 259
Interpreting research results, 197-199
Interstate commerce, 73
Interviews, in marketing research, 190-191
Introduction stage, features of, 238
Introductory price dealing, 483-484, 678
Introductory price discounts, 517
Introductory promotion, 255
Intuition, 179, 183, 185
Inventors, protection of, 247
Inventory
  administered channel systems and, 282
  defined, 310, 678
  exporting and, 290
  just-in-time (JIT) delivery, 161
  local, of Amazon, 59
  stockturn rate (inventory turnover), 329-330
  storing function, 310-313
Inventory carrying costs, 278
Inventory costs, 299
Inventory replenishment system, 167
Inventory turnover, 329-330
Investment, 249, 573
Invoice, 486
iPad, 115, 242, 285
iPhone, 11, 66, 115-116, 216, 242, 487, 546
iPod, 115, 236, 242, 447
Iran, 76
iRobot, 235-236
Israel, 76, 77
Italy, 76, 134, 339

J

Jail sentences, 73
Japan
  adapting flavors for, 83
  advertising regulations, 428
  cable television shopping, 331
  Coca-Cola in, 295
  container shipping by, 308

  demographics, 76
  distribution of Coca-Cola in, 295
  "do it right the first time," 258
  Fair Trade Committee, 428
  food quality and safety, 543
  institutional advertising in, 415
  licensing of U.S. products to, 291
  merchant wholesalers in, 342
  monitoring of food quality and safety in, 543
  population trends, 77
  price fixing allowed in, 496
  reverse engineering in, 657
  Shinkansen commuter train, 251
  taste preferences, 83, 258
  Walmart's struggle in, 339
JDLink (John Deere), 145
Jelly Belly, 217
Jenga ad, 359
Jif peanut butter, 205
Jif Reduced Fat/Simply Jif, 205
Jiffy Lube, 214
Job description, 396, 678
John Deere, 145-146
John Deere Credit, 146
Joint venture, 291, 678
Journey maps, 191
Judgment, 176
Jury of executive opinion, 577, 678
"Just Do It!," 3
Just for U loyalty program (Safeway), 480
Just-in-time (JIT) delivery, 161, 303, 312, 678

K

Kenya, 76, 77, 295
Kerosene, 490
Keywords, for Internet shopping, 421
Kickback payments, 151, 455
Kids Sunblock Lotion and Baby Sprays, 102
Kinked demand curve, 561-562
Kiosks, 300
KitKat candy bars, adaptation of, 258
Kiwano Melons, 341
Kiwi fruit, 341
Kiwi shoe polish, 186, 245
Knorr, 364
Knowledge, 180, 182
Korea, 103, 110
Kraft shredded cheese packaging, 222

L

Labeling, 222-223
Laggards, 372, 678
Lake Cunningham Regional Skate Park, 168
Lancôme, 277
Landing pages, 448, 449, 678
Lanham Act (1946), 218, 430, 678
"Last mile" shipments, for e-commerce, 307-308
#LastSelfie campaign (World Wildlife Fund), 465
Late majority, 372, 678
Latin America, 68, 194, 308, 339, 490
Laundry detergents, 44, 257, 268-269, 390, 471, 522
Lava soap, 108
Law of diminishing demand, 553, 678
Laws
  about "new" products, 247
  Americans with Disabilities Act (ADA; 1990), 631
  antidumping, 495
  antimonopoly, 72, 496
  antitrust, 496
  Consumer Product Safety Act (1972), 73, 252, 675
  consumer protection, 27
  copyrights, 247
  for distribution arrangements, 284
  Federal Fair Packaging and Labeling Act (1966), 222, 677

Laws—*Cont.*
  Federal Trade Commission Act (1914), 496
  Foreign Corrupt Practices Act (1977), 169, 677
  Health Care Reform Act (2010), 630, 631, 632
  Lanham Act (1946), 218, 430, 678
  Medical Devices Act (1976), 630-631
  Nutrition Labeling and Education Act (1990), 223
  packaging, 222-223
  patents, 247
  price discrimination, 496
  product safety, 252
  Robinson-Patman Act (1936), 496, 497, 682
  Sarbanes-Oxley Act (2002), 169
  Sherman Act (1890), 72, 496
  state and local, 73-74
  "take back" laws (recycling), 288
  Wheeler-Lea Act, 495
  Wheeler-Lea Amendment (1938), 684
Layaway, 138
LCD panel (TV) market, 501
Leader pricing, 519, 678
Learning, 122, 678
Learning management system (LMS), 193
Learning process, 122-123
Lease, 678
Leasing, of cars, 490
LEED-certified homes (case study), 668-669
Legal environment. *See also* Laws
  impact on marketing strategies, 72-74
  in medical mobility services, 630-632
  packaging regulations, 222-223
  pricing policies and, 494-497
LEGO products, 87-88, 359, 415, 418, 452
Less-developed nations
  accumulating by, 275
  Coca-Cola in, 295
  cooking and heating fuel (case study), 658-659
  criticisms of infant formula marketing in, 103
  distribution of Coca-Cola in, 295
  ethics and treatment of overseas workers, 305
  "fair trade" not always fair, 325
  "generic" products in, 220
  population growth in, 76, 77
  product counterfeiting in, 218
  transportation costs in, 309
Less-than-carload (LCL) shipments, 307
Lever 2000 soap, 108
Leveraging, 573
Liability, product, 252
Licensed brand, 219-220, 678
Licensing, 291, 678
Lifebuoy soap, 108, 109, 359
Life-changing benefits, 20
Lifestyle analysis, 127-128, 678
Lifetime value of customers, 45-48
"Like grade and quality," 497
Limited availability products, 430
Limited problem solving, 137, 678
Limited-function wholesalers, 342, 344, 678
Limited-line retailers, 326, 327
Limited-line stores, 678
Limited-line wholesalers, 343
Line-forcing policy, 636
LinkedIn, 355, 460, 463, 678
LinkedIn network, 158
List price, 472, 473-474, 484
Listening, to customers, 402
Listerine mouthwash, 360
Listerine PocketPaks, 124, 214
Literacy, 77
Local ordinances, 73-74
Logistics
  career opportunities in, 586
  Coca-Cola and, 295-296
  conceptual approaches, 300-302
  coordinating activities among firms, 302-305
  costs of, 297
  customer service perspectives of, 297-300

  defined, 296, 678
  disaster relief, 314-315
  distribution centers, 313-314
  effects on marketing strategy plan, 542
  electronic data interchange (EDI) and, 304-305
  ethical issues, 305
  storing function, 310-313
  transporting function, 305-310
Looj gutter cleaner, 236
"Lost Dog" ad, 359
Low-cost marketing research, 196
Low-involvement purchases, 138, 678
Loyalty, customer, 45-48
Loyalty cards, 99, 337, 418, 431, 479, 480, 488, 547
Loyalty programs, 216, 408, 479
Lululemon, 217
Lux soap, 108
Luxuries, 119
Lycra, 203

M
Macbook, 115
Machine learning, 69, 251, 679
Macro-marketing. *See also* Marketing
  defined, 7, 9, 528, 679
  effective system of, 10, 530
  market-directed model, 13-14
  role of, 9-10
  who provides functions, 12-13
Madden NFL, 137
Magazines, 416, 418, 443
Mail surveys, 193
Maintenance, repair, and operating (MRO) supplies, 228, 268
Maintenance supplies, 229
Major accounts sales force, 389, 679
Malling of America, The (Kowinski), 625
Malls (case study), 625-627
Management
  of Cirque du Soleil, 33-34
  in marketing, 34-35
  use of operating statements, 566
Management contracting, 291, 679
Manufacturer brands, 220, 679
Manufacturers. *See also* Production
  branding by, 220
  as business customers, 163-165
Manufacturers' agent, 345, 679
Manufacturers' sales branches, 342, 679
MapMyRun app, 204
Mapping app, 337
Marginal analysis, 512-514, 670, 679
Marital status, 128
Markdown, 679
Markdown ratios, 571-572, 679
Market, defined, 89, 679
Market analyst, 585
Market development, defined, 53, 679
Market environment. *See also* International marketing
  changing, Amazon's opportunities in, 59-60
  company resources and opportunities, 63-64
  competitors and the competitive environment, 65-67
  cultural and social environment, 74-80
  economic environment and, 67-68
  evaluating opportunities in international markets, 82-83
  forces shaping, 60-61
  legal environment and, 72-74
  planning grids and evaluation of opportunities, 81-82
  political environment and, 71-72
  screening criteria to narrow down to strategies, 81
  setting whole company objectives, 61-63
  technological environment and, 68-71
Market exposure, 267
Market grid diagrams, 94

Market growth stage, 336, 373, 679
Market information function, 11, 679
Market introduction stage, 238, 372-373, 679
Market maturity stage, 239, 244-245, 373, 433-434, 679
Market penetration, 53, 679
Market potential, 57, 679
Marketing researchers, 584
Market segmentation. *See also* Segmentation; Target market
  defined, 93, 679
  dimensions used to segment markets, 99-103
  two-step process of, 93-96
Market segments, 94, 95, 370, 679. *See also* Market segmentation; Segmentation; Target marketing
Market share, 66, 679
Market share objectives, 477
Market situations, dimensions regarding, 560
Market testing, 255, 577
Market-directed economy, 13-14, 679
Market-directed macro-marketing systems, 298
Marketers, challenges facing, 541-548
Marketing
  advertising *versus,* 7
  arithmetic for (Appendix B), 564-577
  career opportunities in, 6, 583-589
  career planning in (Appendix C), 579-591
  changing role of, 15-16
  criticisms of, 27
  defined, 679
  defining, 6-8
  effective, 10-11
  evaluation of, 528-529
  evolution of, 15-16
  as facilitator of production and consumption, 10
  focus of, 8
  hit-or-miss, avoiding, 49
  importance of, 5-6
  innovation and, 6
  job hunting and, 6
  macro view, 7
  major tasks in, 4-5
  management job in, 34-35
  micro view, 7
  as reflection of values, 535
  role of, in economic system, 13-15
  specialists in, 12
  standard of living and, 6
  starting salaries in, 580
  toward women, Under Armour example, 203
  universal functions of, 11
Marketing analytics, 18, 44, 679. *See also* Analytics
Marketing automation software, 466-468, 679
Marketing company era, 16, 679
Marketing concept
  adoption of, not universal, 19
  basic ideas/elements of, 17-19
  customer value and, 20-23
  defined, 17, 679
  marketing ethics and, 27
  for nonprofit organizations, 23
  production orientation *versus,* 19
  social responsibility and, 24-25
Marketing dashboard, 181, 679. *See also* Decision support system (DSS)
Marketing department era, 16, 679
Marketing ethics. *See* Ethics
Marketing evolution, stages of, 15-16
Marketing exchange, 8
Marketing information function, 11
Marketing information requirements, in marketing plans, 540
Marketing information system (MIS)
  accessing information in, 176
  big data and, 177-178
  collecting external information, 177
  data warehouse and, 179
  decision support systems (DSS) and, 180

Marketing information system (MIS)—Cont.
  defined, 175, 679
  elements of, 177
  imminent changes to, 176-184
  information technology (IT) and, 175
  major functions of, 176-184
  marketing dashboard and, 181
  marketing model and, 181
  scientific method and, 184-185
  secondary data from, 187
Marketing management process
  defined, 34, 679
  process overview, 34-35
  three jobs of, 34-35
Marketing managers
  advertising objectives set by, 411-412
  appeal to customer needs by, 121
  blending of Four Ps by, 538
  career paths and compensation ranges, 584, 585
  coupons, deals, and rebates as aid to, 488-489
  evaluating new opportunities by, 33-34
  pricing objectives set by, 472
  wide-ranging responsibilities, 355
Marketing metrics, 18, 44, 48, 204, 679
Marketing mix
  advertising success and, 426-427
  antimonopoly laws and, 72
  changes in, over product life cycle, 238, 242-243
  customer as target of, 38
  defined, 35, 679
  developing, for target markets, 37-42
  differentiation and, 107
  forecasting and, 577
  Four Ps of, 38-42
  life cycle changes, 238, 242-243
  pivoting and, 242-243
  publicity via, 441
  retailer decisions and, 323-324
  of sellers, 149
  target marketing and, 35, 37-42
Marketing model, 181, 679
Marketing objectives, 61-63. See also Objectives
Marketing orientation, 17, 18, 679
Marketing plan
  as blueprint for implementation, 538-540
  defined, 679
  as guide to implementation and control, 42-45
  Hillside Veterinary Clinic Marketing Plan (Appendix D), 592-619
  implementation of, 42-45
  major functions of, 42-45
  objectives, 605
  personal, in career planning, 590-591
  reasons for decisions, 541
  sections of, 539-540
  separate, for separate regions of the country (case study), 662-663
  timing in, 541
Marketing plan objectives, 539
Marketing program, 44-45, 679
Marketing research
  career opportunities in, 585
  case study, 638-639
  clarifies budgeting estimates, 374
  defined, 175, 679
  effects on marketing strategy plan, 542
  focus on problems, 177
  international, 199
  in new-product development, 254-255
  opportunities in, 584
  options for obtaining, 174-176, 190-197
  primary and secondary data for, 187
  scientific method and, 184-185
  steps in, 176
  testing ads, 427
Marketing research process
  cooperation in, 185-186
  data gathering methods, 190-197

defined, 185, 679
five-step approach to, 185-199
in overseas markets, 199
problem definition in, 185-186
problem solving in, 199
scientific method and, 184-185
situation analysis, 187-190
steps in, 185-186
Marketing strategy. See also Target marketing
  defined, 35, 679
  hard-to-copy, 48-49
  in marketing plans, 539
  Nike example, 3-4
  as section in marketing plan, 606-612
Marketing strategy planning. See also Product life cycles; Segmentation
  Big Brothers Big Sisters of America example, 627-630
  business and organizational customers, understanding, 147-148
  business product classes and, 223-227, 227-228
  case studies, 645-646, 651-653, 658-659
  changes and trends affecting, 542
  Cirque du Soleil example, 33-34
  demographics important to, 74-80
  differentiation and positioning, 107-110
  environmental overview, 60-61
  focus on efforts, 536-538
  Four Ps, 538
  Friendship Dairies example, 52
  highlighting and selecting opportunities, 49-52
  illustrated, 49
  importance of, 45-48
  international opportunities, 54-55
  LEGO example, 87-88
  logic, creativity and, 536-538
Market-oriented strategies, developing, 96
Markets. See also Segmentation; Target markets
  naming product-markets and generic markets, 91-93
  price level, demand and supply and, 559
  as seen by customers and potential customers, 552-557
  as seen by suppliers, 557-559
  understanding, 81-82, 88-91
Markup, 503, 570-571, 679
Markup (percent), 503-504, 679
Markup chain, 504-505, 679
Markup conversions, 571
Marriage, 128
Maslow's hierarchy of needs, 120-121
Mass market, 37
Mass marketers, 37
Mass marketing, 37, 679
Mass markets, 339
Mass production, 10, 210
Mass sellers, 361
Mass selling, 40, 354, 361, 542, 679
Mass-merchandisers, 327-330, 336, 680
Mass-merchandising concept, 327-330, 680
Mastercard, 489
Match.com, 426
Materialism, 534-535
Materials, component, 228, 229
Maturity stage, 239, 244-245
MaxSnax, 255
"Mayhem" ads (Allstate), 209
McDonald's product liability settlement, 252
Measures of people, planet, and profit, 19
Media, defined, 442
Media cost, 428
Media kit, 453
Median age (U.S.), 79
Median family incomes (U.S.), 118-119
Medicaid, 494
Medical Devices Act (1976), 630-631
Medical devices industry, 630-632
Medicare, 494

Mentoring, by Big Brothers Big Sisters of America, 627-630
Merchandisers, 386
Merchant wholesalers, 342-344, 680
Message channel, 361-362, 363, 680
Method Products, 471-472
MetoKote Corporation, 145
Me-too imitators, 107
Me-too products, 430
Mexico, 33, 71, 76, 133, 169, 308
Michael Jackson ONE (Cirque du Soleil), 33
Michael Jackson THE IMMORTAL World Tour (Cirque du Soleil), 33
Micro Commuter (Honda), 254
Microblogging, 463-464
Microchips, 240
Microfinancing, 659
Micro-influencers, 422
Micro-macro dilemma, 24, 680
Micro-marketing
  changes in, 532-533
  costs of, 531-533
  defined, 6-7, 528
  management-oriented, 8
  measures of effectiveness of, 530-531
Middle East, message interpretation may fail, 361
Middle-class society, 130
Milk promotion, 121, 240
Milk sales, 220-221, 240
Millennials, 80, 334, 419, 456, 677
Mini Coopers, 412
Minimum prices, 73, 494-495
Minute Maid Squeeze-Fresh, 255
Mirra pool cleaner, 236
MIS inputs to
  naming product-markets and generic markets, 91-93
  narrowing down process, 50, 88-91
  opportunities to pursue, 52-54
  overview of, 536
  place decisions and, 266-267
  positioning statement for direction, 110
  product classes in, 223-227
  product decisions for, 205
  for product lines, 207
  resources that guide, 63-64
  screening criteria, 50
  sophisticated segmentation techniques, 105
  target markets and mixes and, 37-42
  tools for, 81-82
  value pricing and, 494
Mission statement, 59, 61-62, 680
Missionary salespeople, 385, 386, 680
Mississippi River, 308
Mobile advertising, 419-424, 421. See also Digital advertising; E-commerce; Internet
Mobile payments, 490, 534, 680
Mobile phones. See Cell phones; Smartphones
Model S60 electric car, 21
Modified rebuy, 155-156, 680
"Mommy bloggers," 454
Money-back guarantees, 407
Monopolistic competition, 65, 239, 492, 560, 562, 680
Monopolists, 65
Monopoly markets, 65, 560
Moral standards. See Ethics
Mortgage market crisis, 67
Mortgage scandal, 67
Mothers Against Drunk Driving, 415
Motivation, in sales, 397-399
Motrin, 373
Mountain Dew, 110, 221
Movie 43, 359
Movie tickets, 210
Mozambique, 527
MP3 players, 115

MRO (maintenance, repair, and operating) supplies, 228, 268
Multichannel distribution, 285-290, 680
Multichannel selling approach, 287
Multichannel shoppers, 286, 334, 680
Multicultural markets, 132. *See also* International marketing
Multiple buying influence, 153, 680
Multiple target market approach, 96, 98, 680
Music industry, 183
Mussels, 251
"My Black Is Beautiful" campaign, 133
My Dollarstore, 339

N

NAFTA (North American Free Trade Agreement), 71
NAICS (North American Industry Classification System), 164-165
Naming markets, 91-93
Narrowing down, for selection strategy, 50, 51-52, 88-91, 93
National Basketball Association (NBA), 203
National brands, 220, 471
National Football League (NFL), 203
National Geographic magazine, 461
National income forecast, 573
Nationalism, 71, 680
Native advertising, 426
Natural products, 229, 680
Nature, product ideas from, 251
NBCNews.com, 420
Needs. *See also* Customer needs
  changing, 534
  defined, 680
  types of, 121
  unmet, 536
  urgency of, 134
  wants *versus*, 119
Needs awareness, 139
Need-satisfaction approach, 402
Negative images, 215
Negative publicity, 138
Negative reviews, 443, 445, 454, 456-457
Negotiated contract buying, 162, 680
Nepal, 62, 658-659
Nescafé ad, 516
Nest Thermostat, 69
Net, 486, 680
Net profit, 567, 680
Net sales, 567
Netherlands, 122
New business, strategy planning for (case studies), 638-639, 649-650, 650-651
New product, defined, 246, 680
New unsought products, 225, 226, 284, 680
New York City, 24
New Zealand, 341, 428
Newman's Own, 120
New-product concept, life cycle of, 238
New-product development. *See* Product development
New-product introduction. *See* Introduction stage
New-product planning, 246-248
Newsletters, 452-453
Newspapers, 286, 416, 443
New-task buying, 155, 156-157, 680
Niche, 471
Nigeria, 75, 76, 238, 272, 322
Nike Flyknit, 3
Nine-box planning grid, 81-82
"No interest or payments for one full year," 489
"No matter what, sweat every day" advertising campaign, 203
No Mess Pens (WD-40), 196
"No tx'g while drv'g!," 450
Noise, 360, 361, 364, 680
Nonadopters, 372, 680

Noncumulative quantity discounts, 485, 680
Nonprice competition, 478, 680
Nonprofit organizations
  challenges of, 23
  competition and, 23
  as customer, 147
  fundraising by, 23
  marketing by, 7
  marketing concept applied to, 23
  measure of success, 18
  prices set to increase market share, 477
  pricing objectives, 475-476
  profit measurement for, 23
  "selling" unsought products of, 226
  supporter satisfaction, 530-531
Nonrecognition of brands, 215
Norms, 26
North America, 33, 71, 410
North American Free Trade Agreement (NAFTA), 71
North American Industry Classification System (NAICS), 164-165, 680
Norway, 76
Novida (beverage), 238
"Nuisance" purchases, 229
Nutrition Labeling and Education Act (1990), 223
Nutritional labeling, 544

O

Oakley sunglasses, 283
Obesity, 527
Objectives
  for advertising, 411-415
  case study, 653-654
  as guide to marketing strategy, 61-63
  hierarchy of, 62-63
  informing (promotion), 357
  long- and short-run, 589-590
  in marketing plans, 594-595, 605
  of nonprofit organizations, 23
  of owned media, 448
  personal, in career planning, 589-590
  place decisions based on, 267-269
  of pricing, 475-478
  for promotion, 356-360
  research, setting, 186
  to set firm's course, 61-63
"Obligated Posers" shopper type, 104
Observation, data gathering from, 188
Odd-even pricing, 520, 680
Odyssey (Honda), 45
OEM (original equipment market), 229
Oil tankers, 309
OkCupid, 426
Olay, 44
Oligopoly, 65, 238, 560, 561-562, 680
Omnichannel, 287, 680
Onboarding phase, in promotion blend, 439-440
One Source service (John Deere), 146
One-price policy, 479, 680
Ongoing phase, in promotion blend, 439-440
Online advertising, 419
Online auction, 518. *See also* E-commerce
Online communities, 158, 203, 451
Online eyeglasses, 494
Online focus groups, 192
Online games, 121
Online monitoring, of one's own business, 187-188
Online music streaming, 443
Online retailing, 329, 331-335, 680
Online review, 158
Online search behavior, 181
Online selling. *See* E-commerce; Websites
Online seminars (webinars), 439
Online shopping, 59-60
Online surveys, 193
Online video ads, 444
On-site auction, 134

OnStartups blog, 439
Open community, 452
Open-ended questions, 191
Operating ratios, 570, 680
Operating statement, 564-565, 680
Operating supplies, 228
Operational decisions, 43-44, 680
Operational linkages, 161
Operational market segments, 95
Operationally optimized, 395
Opinion leader, 130-131, 422, 454-455, 680
Opinion leaders, 454-455
Opportunities. *See also* Breakthrough opportunity
  Amazon example, 59-60
  attractive, 48-49
  from business and organizational customers, 146-148
  Cirque du Soleil and, 33-34
  financial resources and, 64
  finding, 91
  four basic types, 52-54
  from innovation, 236-237
  international, 54-55
  in international markets, 82-83
  LEGO example, 87-88
  limited by company resources, 63-64
  in marketing careers, 583-589
  marketing resources and, 64
  in marketing strategy planning, 52-54
  missed, costs of, 532
  planning grids for evaluation, 81-82
  production and, 64
  for progressive wholesalers, 340
  selecting, in strategy planning, 81-82
  S.W.O.T. analysis and, 50
  technology and, 68
  types to pursue, 52-54
  from understanding markets, 88-91
Option evaluation (case study), 637-638
Orange juice packaging, 214
Order getters, 383-384, 680
Order takers, 384-385, 680
Order-getting, 383-384, 397, 680
Order-taking, 384-385, 680
Oregon State University, 251
Oreo's Super Bowl tweet, 458
Organic search, 446
Organizational buying
  buyer-seller relationship, 154, 159-163
  decision-making process, 154, 155-159
  defining the problem, 154
Organizational buying model, 358
Organizational customers
  behavioral needs of, 149-150
  buyer-seller relationship, 154, 159-163
  buying procedures, standardized, 153-154
  buying processes, 155-156
  decision-making process, 154, 155-159
  distinctive traits, 148-154
  economic needs of, 148-149
  effects on marketing strategy plan, 542
  ethical conflicts and, 150-151
  government purchasing, 168-169
  information sources used by, 156-157
  manufacturers as, 163-165
  model of business and organizational buying, 155-159
  multiple buying influence and, 153
  problem definition, 154-155
  purchasing managers and, 152
  retailer and wholesaler purchasing, 167-168
  seller's marketing mix and, 149
  service producers as, 165-166
  small service firms as, 165-166
  types of, 146-148
  understanding, for marketing strategy planning, 148
Organizations, major segments in, 100

Original equipment market (OEM), 229
"Original superfood" (cottage cheese), 52
Osprey Aura 65 AG EX backpack, 441-442
Outback Steakhouse, 217
Outdoor advertising, 416, 444
Outsourcing
  defined, 680
  specific adaptations and, 162
OVO (Cirque du Soleil), 33
Owned media
  about, 442
  benefits and challenges of, 443, 444
  branded websites as, 443
  career opportunities in, 588
  consumer trust of, 443, 444
  creating, that customers can use, 448-453
  credibility of, 443
  defined, 442, 680
  examples, 442
  Facebook as, 457
  of GEICO insurance, 352
  high message control/lower cost, 443
  how customers find it, 448
  message source, 442
  promotion objectives of, 448
  types of, 448

P

Packaging
  benefits to consumers and marketers, 222
  creative, 537
  defined, 680
  ethics in, 223
  generation of trash and, 222
  green, 222
  laws regarding, 222-223
  of Method Products, 471-472
  safety, of Tylenol, 222
  socially responsible, 70
Packaging specialists, 585
Paid media, 419
  about, 442
  benefits and challenges of, 443, 444
  defined, 442, 680
  examples, 442
  high message control/less effective, 443
  less trusted, 443, 444
  message source, 442
  sponsored message as, 457
Pakistan, 76
Pampers disposable diapers, 37, 358-359, 434
Panels, 577
Parents magazine, 418
Partnerships, between buyer and seller, 162
Pass-along, 440, 441, 446, 447, 680
Patent, 247, 680
Patent law, 247
Payment date, 486
Payment methods, for salespeople, 397-399
Payment terms, 486
Pay-per-click, 680
Pay-per-click advertising model, 419
Peer-to-peer network (car-sharing), 125
Penetration pricing policy, 483, 680
People, planet, and profit, measures of, 19
"People against Dirty" campaign (Method Products), 471
"People-oriented" jobs, 582
Pepto-Bismol, 132
Perception
  customer, of competition, 107-110
  defined, 680
  role of, in behavior, 121
Perceptual mapping, 108
Performance clothing market, 203-204
Perishable products, 308
Persil laundry detergent, 390
Personal analysis, in career planning, 581-583

Personal computer sales, 265-266
Personal data, 182, 681
Personal interview surveys, 194
Personal marketing plan, in career planning, 590-591
Personal marketing strategy, in career planning, 580-581
Personal needs, 121, 681
Personal objectives, in career planning, 589-590
Personal selling. *See also* Salespeople
  advantage, 361
  basic sales tasks, 383-387
  to business customers, 369
  career opportunities in, 586-587
  case study, 649-650
  as communication process, 39
  compensating and motivating salespeople, 397-399
  customer service and, 379-380, 387-389
  defined, 39, 353-354, 681
  developing new relationships, 383-384
  effects on marketing strategy plan, 542
  elements of, 39
  ethical issues, 403
  to final consumers, 368
  importance of, 380-383
  information technology (IT) for, 392-395
  as largest operating expense, 381
  order takers and, 384-385
  order-getter task in, 383-384
  order-taking and, 384-385
  organizational structures for, 389-391
  promotion to intermediaries and, 366-367
  prospecting and presenting, 399-403
  role of, 380-383
  selecting and training representatives, 396-397
  steps in, 399
  structure of, 389-391
  supporting sales force, 385-387
  techniques for, 399-403
Personalization, 107. *See also* Customer relationship management (CRM)
Personas, 102
Persuading objective, of promotion, 357
Persuasive promotion, 239
PET plastic bottles, 70
Pet Pride pet products, 221
Phase-out strategy, 245-246
Philadelphia Cream Cheese, 245
Philippine Islands, 83
Phony list prices, 495, 681
Photo sharing services, 464
Physical distribution (PD)
  coordinating activities among firms, 302-305
  customer service, 297-300
  defined, 296, 681
  electronic data interchange (EDI) and, 304-305
  ethical issues, 305
  focus on whole distribution system, 300-302
  invisible to consumers, 298
  sharing arrangements in, 302-303
  trade-offs and customer service levels, 300
  trade-offs of cost, service, and sales, 298-299
  transporting function, 305-310
Physical distribution (PD) concept
  about, 300-302
  defined, 300, 681
  factors affecting service levels, 301
  total cost approach, 302
Physical handling, 312-313
Physiological needs, 120, 121, 681
Pilot (Honda), 492
Pinboards, 462
Pinterest, 460, 462-463, 681
Pioneering advertising, 413, 681
Pioneers, 236, 242
Pipelines, 307, 308, 309

Pirated goods, 218
Pivoting, to new marketing mix, 242-243
Pizza delivery market, 407-408
Place. *See also* Channels of distribution; Distribution; Logistics
  defined, 39, 266, 681
  elements of, 38
  in marketing plans, 540
  marketing strategy planning decisions for, 266-267
  operational decisions and, 43-44
  Potbelly Sandwich Works' core strategy, 624
  potential target market dimensions and, 99
  strategy policies and, 44
  in target marketing/marketing mix, 38, 39
  Toddler University baby shoes example, 41-42, 44
Place decisions, 267-269
Place system, not automatic, 268-269
Plain-language contract, 544
Planned economies, 13
Planned obsolescence, 247-248
Planning. *See also* Marketing plan; Marketing strategy planning
  in command economy, 13
  as marketing management task, 34
  strategic (management), 33
Planning grids, 81-82
Plant-based plastics, 70
PlantBottle packaging, 70
Plasma Concierge Service (Panasonic), 210
Plastic bottles, 70
PMs, 486
Podcasts, 439
Point-of-purchase advertising, 358-359
Pokémon Go, 211
Poland, 303, 364
Political environment, 71-72
Pop music, 183
Population. *See also* Demographics
  African American, 133
  by age group, 79-80
  Asian American, 133
  growth in selected countries, 76, 77
  Hispanic, 132
  of research sample, 197, 681
  for selected countries, 76
  trends in U.S. consumer market, 78-79
Population growth, 75-79
Pop-up ads, 420, 421
Pop-up blocker, 424, 443
Positioning
  defined, 107, 681
  different views of, by different countries, 428-429
  issues, 107-109
  LEGO example, 87-88
  in marketing plans, 539
  repositioning and, 110
  as section in marketing plan, 606
Positioning analysis, 109
Positioning statement, 110, 681
Positioning technique, 108
Positive reinforcement, 123
Post-purchase regret, 138
PowerPoint, 213
Practical value, 447, 448
Predictive analytics, 138-139, 181-182, 183, 334, 337, 388, 394, 480, 681
Premium dog food, 268
Prepared sales presentation, 401, 681
Presenting, by salespeople, 400-403
Press kit, 453
Prestige pricing, 522, 681
Pretesting, of advertising, 427
Price competition, 478
Price cuts, 497
Price discrimination, 681
Price discrimination laws, 496

Price equation, 474–475
Price fixing, 496, 681
Price flexibility policies, 473
Price lining, 521, 681
Price points, 637
Price premium, 473, 493
Price sensitive customers, 514–515
Price sensitivity, 681
Price setting
   average-cost approach, 506–510
   basic approaches to, 502
   break-even analysis and, 510–512
   case study, 659
   costs, types of, 507–510
   decisions about, 40
   defined, 40
   demand-oriented approaches, 514–522
   elasticity and, 563
   for full product lines, 522–523
   with marginal analysis, 512–514
   markup approach, 503–506
   Samsung example, 501–502
   as strategy decision, 502
Price wars, 60, 501
Price-cutting, 481–482
Price-off coupons, 488
Price-oriented bargain hunters, 329
Prices. *See also* Price setting; Pricing policies
   administered, 478
   allowance policies/off list prices, 486–487
   antimonopoly laws and, 72
   in command economy, 14
   customer perceptions of quality and, 494
   defined, 40, 472, 474, 681
   dimensions of, 479
   discount policies, 484–486
   effects on marketing strategy plan, 542
   elements of, 38
   flexibility in, 479–482, 481–482
   impact of cuts on profits, 481–482
   linked to product life cycles, 482–484
   in marketing plans, 540
   minimum retail prices, 496
   objectives and policies, Method Products example, 471–472
   objectives for, 475–478
   operational decisions and, 43–44
   Potbelly Sandwich Works' core strategy, 624–625
   potential target market dimensions and, 99
   relation to value, 20–21
   strategy dimensions, 472–475
   strategy policies and, 44
   in target marketing/marketing mix, 38, 40
   Toddler University baby shoes example, 42, 44
   value on life and, 495
Price-sensitive shoppers, 472
Pricing
   career opportunities in, 588
   case study, 640–641
Pricing objectives, 475–478
Pricing policies
   allowances/off list prices, 486–487
   case studies, 659–660, 663–664
   customer value in, 491–494
   discount policies, 484–486
   flexible, 479–482
   introductory price dealing, 483–484
   legal issues, 494–497
   over the product's life cycle, 482–484
   penetration pricing, 483
   on price levels, 482–484
   rationale for, 479
   skimming pricing, 482–483
Primary data, 187, 188, 190, 681
Primary data gathering, 190
Primary demand, 370, 413, 681
Prime contractors, 169
Print advertising, 416, 418

"Print Lizard" website, 192
Privacy. *See also* Customer tracking
   consumer tracking and, 424
   cookies and, 419, 423
   legal rights to, 183
   mobile advertising and, 424
   online, 424
   online advertising and, 424
   protecting, 546
Private brands (private label), 219, 220, 681
Private research firms, 190
"Private Selection" brand (Kroger), 221
Private warehouses, 312, 681
Prize money (push money) allowances, 486
Pro Xtra Loyalty Program, 320
Problem customers, 387
Problem definition
   in business and organizational buying, 154–155
   in marketing research, 185–186
Problem recognition, 135–136
Problem solving
   in consumer decision process, 135–140
   for consumer needs, 135–140
   continuum, 137
   customer service role in, 387–389
   decision-making process, 155–159
   defining the problem, 154–155, 185–186
   levels of, 137–138
   as marketing research step, 199
   recognizing a need, 135–136
   steps in, 136
Problem-solving continuum, 137
Procurement department, 152
Procurement website, 158
Producer-led channels of distribution, 280
Producers
   as customer, 147
   of goods and services, 147
   minimum retail prices set by, 496
   order getters of, 383–384
   order takers of, 384–385
   personal selling for, 384–385
   of services, in the United States, 165–166
Product
   defined, 38–39
   in target marketing/marketing mix, 38
   Toddler University baby shoes example, 41–42, 44
Product advertising, 413, 681
Product assortment, 207, 681
Product champion, 257
Product classes, 223–227, 267–268
Product dependability. *See* Quality
Product development
   company support of, 256–257
   defined, 53, 681
   idea generation in, 249–251
   market strategies based on, 53
   new ideas from nature, 251
   new-product planning, 246–248
   process steps, 248–255
   product failure and, 248
   Roomba vacuum example, 235–236
Product ideas, 241
Product launches, 465
Product liability, 252, 681
Product life cycles
   defined, 238, 681
   impact on promotion, 372–374
   management, 238–240
   pricing policies and, 482–484
   retailers and, 336
   stages of, 238–240
   strategy planning, for different stages of, 242–246
Product line length, 207, 681
Product lines, 203, 207, 681
Product managers, 257–258, 585, 681
Product planner, 586

Product quality. *See* Quantity
Product safety, 252
Product space, 108
Product type, naming markets for, 91
Product variety, 257
Product web pages, 449–450
Product-bundle pricing, 523, 681
Product-consumer separation, 10–11
Production. *See also* Manufacturers
   antimonopoly laws and, 73
   in command economy, 13
   cost of (case study), 640–641
   defined, 5, 681
   opportunities and, 64
   specialization, 10–11
   Toddler University baby shoes example, 41–42
Production (factory) cost, 504–505
Production era, 16, 681
Production orientation
   defined, 17, 18, 681
   marketing concept *versus*, 19
Product-market. *See also* Product life cycles
   basic features, 91
   boundaries, 92
   defined, 90, 681
   four-part description of, 91
   generic definitions and, 92–93
   naming, 91–93
   screening criteria for selecting, 81
   segmentation dimensions for, 99–103
   segmentation overview, 93–96
   segmentation process steps, 99–103
   shared channel focus on, 276
Product-market commitment, 276
Product-market life cycles, 240
Products. *See also* Artificial intelligence (AI); Augmented reality (AR); Goods; Intelligent agents; Product development; Product life cycles; Services
   basic attributes of, 205–208
   basic differences between, 208–210
   blends of goods and services in, 208–210
   branding, 212–221
   business product classes, 228–230
   business products, consumer products *versus*, 227–228
   consumer product classes, 223–227
   defined, 206, 681
   dying, and phasing out of, 245–246
   effects on marketing strategy plan, 542
   elements of, 38
   failure of, 248
   goods and services, differences between, 208–210
   impact on innovation of, 236–237
   intangible, 209
   Internet of Things (IoT), 213
   management of, over life cycle, 238–240
   in marketing mix, 38–39
   in marketing plans, 539
   maturing, managing, 244–245
   operational decisions and, 43–44
   outside proven product lines (case study), 660–662
   packaging strategies, 222
   in personal marketing plan, 591
   Potbelly Sandwich Works' core strategy, 624
   potential target market dimensions and, 99
   product classes for marketing strategies, 223–227
   product decisions for marketing strategy planning, 204–205
   as seen by customers and potential customers, 552–557
   services as, 208–210
   strategy policies and, 44
   tangible, 209
   in target marketing/marketing mix, 38–39, 40
   Toddler University baby shoes example, 41–42

Products—Cont.
　　value added to, by technology and intelligent agents, 210-212
　　virtual, 276
　　web pages of, drive desire and action, 449-450
　　where produced, 210
Professional contractors, 320
Professional services, 228, 268, 275, 384, 681.
　　See also Services
Profit
　　as balancing point, 99
　　defined, 18, 19
　　as firm objective, 7
　　marketing analytics and, 19
　　of nonprofit organizations, 23
　　in triple bottom line, 19
Profit and loss statement. See Operating statement
Profit maximization objective, 476, 513-514, 681
Profit range, 514
Profitless "success," 477
Profit-making firms, role of marketing in, 7
Profit-oriented objectives, 475-478
Profits. See also Prices
　　average-cost pricing and, 506-510
　　in break-even analysis, 510-512
　　customer equity and, 47
　　declining, in market maturity, 240
　　defined, 512-513
　　evaluating new product ideas for, 252-254
　　flexible-price policies and, 481-482
　　gross, 567
　　impact of price cuts on, 481-482
　　improving (case study), 656
　　increasing (case study), 644-645
　　in market maturity stage, 239
　　markups and, 505
　　measurement for nonprofits, 18
　　net, 567
　　pricing objectives based on, 475-478
　　relation to sales, 239
　　social responsibilities versus, 24-25
　　strategy screening criteria based on, 81
　　as success measure, 18
　　of supermarkets, 327-328
Program management approach (case study), 657-658
Programmatic (advertising) delivery, 420
Progressive wholesalers, 340
Project Shakti, 9
ProjectPro project management software, 467
Promotion. See also Advertising; Sales promotion
　　adoption processes and, 370-371
　　Airbnb, types used by, 244
　　antimonopoly laws and, 72
　　blended plans, 366-370, 372-374
　　blends, 355-356
　　brand, 214
　　budgeting for, 374-375
　　to business customers, 369
　　case studies, 653-654, 654-655, 671-673
　　communicates to target marketing, 352-353
　　communication process in, 360-366
　　creative, 203-204
　　criticisms of, 365-366
　　customer trust and, 443
　　customer-initiated communication, 363-366
　　defined, 39, 40, 352, 681
　　direct-response, 362
　　elements of, 38
　　ethical issues, 365-366
　　by fast food restaurants, 407-408
　　to final consumers, 368
　　focus of, 39
　　GEICO insurance example, 351-352
　　inbound, 439
　　integrated plans, 366-370
　　introductory, 255
　　in market maturity stage, 239

　　market penetration, 53
　　in marketing plans, 540
　　during maturity stage, 239
　　methods, 353-355, 356-360
　　objectives, 356-360
　　objectives and B2B, 357-358
　　operational decisions and, 44
　　overpromises, 126
　　in personal marketing plan, 591
　　planning, integrating, and managing of, 355-356
　　Potbelly Sandwich Works' core strategy, 624
　　potential target market dimensions and, 99
　　product life cycle and, 372-374
　　special promotion, 497
　　strategy policies and, 44
　　in target marketing/marketing mix, 38, 39-40
　　timing and relevance, 364-365
　　Toddler University baby shoes example, 42, 44
　　by trademarks and symbols, 213
　　for unsought products, 226
Proposals, 158, 167, 190
Prospecting, 399-400, 681
"Protect This House" tagline, 203
Protectionist economy, 632
Protective packaging, 222
Prototype, 254, 681
PRWeb, 453
PSSP hierarchy of needs, 120, 121
"Psychic retailers," 337
Psychographic survey, 173
Psychographics, 127-128, 681
Psychological influences, on consumer behavior, 119-128
Psychological needs, 120
Psychological pricing, 520, 681
Psychological variables, on consumer behavior, 117
Public interest groups, 14
Public relations (PR)
　　career opportunities in, 587
　　career paths and compensation ranges in, 584
　　defined, 355, 681
　　press kits and, 453
Public step, to increase pass-along, 447
Public warehouses, 312, 681
Publicity. See also Advertising; Public relations (PR)
　　career opportunities in, 587
　　Cirque du Soleil and, 33
　　defined, 40, 354-355, 440, 681
　　HubSpot example, 439-440
　　integrating into promotion blend, 441-442
　　negative, 138
　　unpaid media (owned or earned), 419
Publishers, 285
Puffery, 430
Pulling, 681
Pulling policy, 368
Purchase discount, 568, 681
Purchase orders, 403
Purchase situations
　　defined, 681
　　effects on consumer behavior, 117, 134-135
Purchases, satisfaction with, 205
Purchasing agent/buyer, 585
Purchasing functions
　　decision-making process, 155-159
　　of governments, 168-169
　　of manufacturers, 163-165
　　organizational decision-making systems, 155-159
　　relationships between buyers and sellers, 159-163
　　of retailers and wholesalers, 167-168
　　of service producers, 165-166
Purchasing managers, 152, 682
Purchasing specifications, 155, 682
Pure competition, 65, 238, 560-561, 682
Pure services, 12
Pure subsistence economy, 8, 682

PureBond plywood, 251
Push money, 474
Push money (prize money) allowances, 486, 682
Pushing, 682
Pushing policy, 366, 367

Q
QR codes, 490
Q-Tips, 213
Qualifying dimensions (segmentation), 101, 682
Qualitative criteria, product-market screening, 81
Qualitative marketing research, 190-191
Qualitative questioning, 190
Qualitative research, 682
Quality
　　building, into services, 259
　　of component parts and materials, 228
　　defined, 206, 682
　　managing, 258-261
　　poor, cost of, 149, 258
　　price and customer perceptions, 494
　　product, and customer needs, 206
　　relative, 206
　　return on, 261
　　of services, 210
　　total quality management (TQM), 258
　　training for, 260
Quality of life, 527, 535
Quantitative criteria, product-market screening, 81
Quantitative marketing research, 192
Quantitative research, 682
Quantity, discrepancies of, 10, 11, 275
Quantity discounts, 471, 485, 682
Questionnaires, 193
Quirky marketing, 624-625
Quotas (sales), 682
QVC cable television shopping network, 331
Qzone (Chinese social media site), 464

R
Rack jobbers, 344, 682
Radio, 458
Radio advertising, 352
Radio frequency identification (RFID) tags, 313
Rail shipments, 307, 309
Rapid prototyping, 254
Rapid response, 312
Ratings, 456
Rationality, in buying, 138
Ratios
　　markdown, 571-572
　　operating, 570
Raw materials, 228, 229, 312, 682
Rayovac batteries, 283
RC Cola, 242
R&D (research and development), 254-255
Real Madrid, 461
Rebates, 488, 682
Rebuys, 156, 167-168
Receiver (of a message), 360, 682
Recessions, 67, 80
Recognition, of brands, 216, 217
Recommendation, of product or brand, 443, 454, 456
Record labels, 183
Record Sensor app, 204
Recycling, 288
Redistribution centers, 313-314
Reference group, 130, 682
Reference price, 518-519, 682
Referral program, 456, 682
Referrals, 209
Refrigeration market, in rural India, 246-248
Refrigerators, technology and, 211
Regaining lost business (case study), 666-668
Regional shopping malls (case study), 625-627
Registration (trademark), 218
Regrouping activities, 267, 275, 682

Regularly unsought products, 225, 226, 682
Regulations. *See also* Government; Laws
   of advertising, 428-430
   impact on marketing strategies, 72-74
   rebates, 488
   for social benefits, 24-25
   for transportation, 306
Reinforcement, 123, 682
Rejection of brands, 215
Relationship, marketing as, 8
Relationships
   among quantity, cost, and price, using cost-oriented pricing, 510
   building, e-mail as tool for, 452-453
   building, value pricing in, 492
   building, with customers, 351-352
   building and maintaining with value, 21-22, 45-48, 64
   buyer-seller, 154, 159-163
   cooperative, 160
   customer lifetime value and, 45-48
   within distribution channels, 276-281
   between firms and suppliers, 159-163
   as goal of marketing, 8
   via social media, 465
Relationship-specific adaptations, 162
Relative quality, 206
Relay robot, 259
Relevant range, 556
Relevant target market, 492
Reliability, 149
RE/MAX, 215
Reminder advertising, 415, 682
Reminding objective, of promotion, 357
Reorders, 167-168
Repeat buying, 324, 485
Repositioning, 110
Representative samples, 134, 173, 192, 197
Requisition, 153-154, 682
Research (marketing). *See* Marketing research; Marketing research process
Research and development (R&D), 254-255
Research firms, 190
Research objectives, setting, 186
Research proposal, 190, 682
Response, 123, 682
Response rate, for survey, 193, 682
Responses, in learning process, 123
Retail list price, 496
Retail shopping, 59
Retailer life cycles, 336
Retailer-led channels of distribution, 280-281
Retailers
   assortment expansion, 326-327
   avoiding price competition, 326
   benefits of, 12
   convenience and, 330-331
   conventional, 326
   corporate chains, 322
   customers' wants as guide for, 167-168
   evolution and changing of, 335-338
   examples/shoe store strategy, 324
   marketing mix issues, 323-324
   mass-merchandising, 327-330
   multichannel shopping problems, 286
   nature of offerings of, 326
   online features added, 334
   order takers of, 385
   as organizational customers, 167-168
   personal selling for, 384
   purchasing functions, 167-168
   service expansion in, 326-327
   size of, 322
   specialization of, 323
   stockturn rate and, 329-330
   strategy planning by, 319-320, 323-325
   types of, 326

Retailing
   career paths and compensation ranges in, 584, 586
   cooperatives and franchisors, 322-323
   crucial to consumers, 322
   defined, 322, 682
   Internet, 331-335
   nature of, 322-323
   stockturn rate and, 329-330
   wheel of retailing theory, 335
Retargeting, 419, 682
Retention
   customer, 47, 48, 395, 439
   selective, 122
Retention rates, 45-46, 682
Retrofitted market, 651
Return, 566, 572, 682
Return on assets (ROA), 573, 682
Return on investment (ROI)
   defined, 682
   example, 253
   as reflection of asset use, 572-573
   as screening criterion, 81, 252-253
Revenue, marketing dynamics and, 18
Reverse channels, 288-291
Reverse channels of distribution, 288-291, 682
Reverse engineering, 250, 657
Reviews, from customers, 14, 456-457
Rhine River, 308
Ridesharing, 125
"Rifle approach" (target marketing), 36
Rio (MP3 player), 115
Risk reduction, with multiple suppliers, 162
Risk taking, 11, 682
Risks
   in buyer-seller relationship, 162
   in earned media, 445
   environmental sensitivity and, 82-83
   in international markets, 55, 82-83
   marketing plan and, 540
   of online shopping, 332-333
   in shared information, 160-161
Risperdal (drug), 151
Rival firms, 66
Robinson-Patman Act (1936), 496, 497, 682
Robots, 70, 235-236
ROI (return on investment). *See* Return on investment (ROI)
Roku streaming player, 222
Rollout, of new products, 255
Romania, 76
Roofing shingles, 216
Rookie Moms (blog), 454
Roomba vacuum cleaner, 235-236
Routinized response behavior, 137, 682
Ruhr Valley, 164
Rules of the game, 14
Russia, 76
Ryobi power tools, 319

S
Sachets, 9
Safeguard soap, 108
Safe-T-Plus, 127
Safety
   laws, 72-74
   of products, 252
   screening for, 252
Safety needs, 102, 121, 682
Salary, 397-399, 580
Sale price, 487, 682
Sales. *See also* Personal selling
   as component of operating statement, 565
   growth factors, 238-239
   meaning of, 566-567
   rebuilding of lost sales (case study), 666-668
   relation to profits, 238-239
Sales analysis, 683
Sales by channel report, 287

Sales contests, 432
Sales decline, 682
Sales decline stage, 240, 373
Sales era, 16, 682
Sales force, size of, 391
Sales forecasts, 57, 540, 682
Sales managers, 355, 398, 682
Sales presentation, 401-402, 682
Sales promotion. *See also* Advertising; Promotion
   activities, 430
   B2B, 432
   career opportunities in, 588
   defined, 40, 354, 430, 682
   for different targets, 430-434
   effects on marketing strategy plan, 542
   international dimension, 409-410
   as learned skill, 435
   management challenges in, 434-435
   mistakes in, examples, 434-435
   nature of, 430
   objectives for, 430-431
   spending in, 433-434
   types of, 430-434
   uneven results, 408
Sales promotion managers, 355, 682
Sales quota, 398, 682
Sales territory, 390-391, 682
Sales-oriented objectives, 477, 682
Salespeople
   adjusting to cultural influences, 380-381
   basic sales tasks, 383-387
   career paths and compensation ranges, 584
   case study, 650-651
   communication with wholesalers and retailers, 366
   compensating and motivating, 397-399
   compensation and motivation, 397-399
   as customer helpers, 381-382
   forecasting by, 577
   main responsibilities of, 380-383
   price adjustments by, 481
   product dilemma and (case study), 660-662
   prospecting and presenting, 399-403
   as representative of whole company, 382
   selecting and training, 396-397
   as strategy planners, 383
   supporting, 385-387
   techniques for, 399-403
   in telephone selling, 390
   types of, 383-387
Samarin brand, 361
Sample (population), 197, 682
Samples (in marketing research), 192
Samples (product), 134, 368
Samsung Galaxy phone, 443
San Francisco "soda tax," 527
Sanuk, 108
Sarbanes-Oxley Act (2002), 169
Satellite radio, 208
Saudi Arabia, 76, 157
Scanner sales data, 426-427
Scanners, 167, 189-190, 328, 479
Scientific method, 184-185, 682
Scorecards, supplier, 163
Scrambled merchandising, 336, 682
Screening
   of new ideas, 252
   ROI as criterion, 252, 253
   in strategy selection, 50, 81
Screening criteria
   in marketing plans, 595
   product-market, 81
Search engine optimization (SEO), 446, 449, 682
Search engines
   advertising on, 421
   ethical issues, 423-424
   information gathering via, 157
   in marketing information systems, 176

Search engines—*Cont.*
  in marketing research, 176
  to research new purchases, 157
  specialized, 453
Seasonal discounts, 485, 682
Second thoughts, about a purchase, 138
Secondary data, 187, 683
Second-movers, 242
Segmentation
  as aggregating process, 94-95
  dimensions of, 99-103
  dynamic behavioral, 106-107
  identifying market segments (case study), 641
  LEGO example, 87-88
  market grid as visual aid, 94
  in planning process, 93-96
  with pricing strategies, 488-489
  purpose of, 50-51
  sales increased by, 97-98
  sophisticated techniques, 105
  in strategy planning, 50-51
  target market defined by, 93-96
Segmenters, 94-95, 97, 683
Segmenting, 683
Segmenting, combining *versus,* 109
Segmenting dimensions
  for business/organizational markets, 100
  relevant, finding, 101
  submarkets, 94-95, 134
"Segments of one" marketing, 107
Segway Human Transporter (HT), 632-634
Selection, of salespeople, 396-397
Selective demand, 373, 413, 683
Selective distribution, 283, 284, 293, 683
Selective exposure, 122, 683
Selective perception, 122, 683
Selective retention, 122, 683
Self-driving car, 407
Self-driving trucks, 311
Selfie, 351
Self-service, 327, 328, 329
Sellers. *See* Buyer-seller relationship; Suppliers
Seller's marketing mix, 149
Selling, marketing *versus,* 7
Selling agents, 346, 683
Selling formula approach, 402-403, 683
Selling function, 11, 683. *See also* Personal selling; Salespeople
Selling price, 504-505
Selling skills, 397
Senior citizens
  defined, 683
  as powerful demographic force, 79
Sensodyne toothpaste, 96
Sentiment analysis, 189, 683
SEO (search engine optimization), 446, 449
Separation, 274-276
Separation, spatial, 10, 11
Separation in time, 10
Separation in values, 10
Separation of information, 10
Separation of ownership, 10
Sephora Collection, 277
Service customers, referrals and, 209
Service firms, 210, 485
Service mark, 213, 683
Service producers, as organizational customers, 165-166
Service products, distribution of, 275-276
Service provider, training of, 260
Service quality. *See* Quality
Service wholesalers, 342, 343, 683
Services. *See also* Goods and services; Products
  accumulating, 275
  adding to goods, for differentiation, 211
  brand rejection costs, 215
  as business product class, 228-230
  defined, 208, 683

as intangible, 209
as organizational buyer, 165-166
products as, 208-210
professional, 228, 230, 268, 275, 384, 653
quality, 261
specialized, 229
Shakti Ammas, 9
Shaktimaan, 9
Shared activities, 161
Shared information, in business markets, 160-161
Sharing arrangement, 302-303
Shazam app, 183
Shelf space, 296, 471-472
Sherman Act (1890), 72, 496
Shieldtox NaturGard, 535
Shinkansen commuter train, 251
Shipping costs, 305-306
Shipping methods, 307-310
Shoe markets
  Under Armour, 203
  case study, 645-646
  criticisms of target marketing of, 103
  store strategies, 324
  Toddler University baby shoes, 41-42, 44
Shoe polish market, 186, 245
Shoe retailers, 41-42, 44
"Shoe rides," 42
Shoe store strategies (retailing), 324
Shopper types, 104
Shopping habits, technology's impact on, 105
Shopping products, 224-225, 683
Shortages, 298
"Shotgun approach" (mass marketing), 36
Showrooming, 286
Silly Bandz, 242
Simple trade era, 15, 683
Simply for Sports gear, 203
Simply Lemonade, 123
Singapore, 76, 77
Single parents, 129
Single people, 129
Single target market approach, 683
Single-line retailers, 326, 329
Single-line stores, 683
Single-line (general-line) wholesalers, 343, 683
Single-target market approach, 96, 98
SitOrSquat app, 451
Situation analysis, 683
Skeleton statement, 566
"Skimming" of the market, 242
Skimming price policy, 683
Skimming pricing, 482-483
Skittles, 189
Slogans, 412
Slotting allowances, 486
"Smart TVs," 502
"Smartclean technology," 472
Smartphones
  as advertising game changer, 422-423
  credit cards accepted on, 241
  Galaxy Note 7, 127
  game apps for, 211
  Google Now and, 212
  Internet of Things (IoT) and, 213
  mobile advertising on, 416, 421
  as new channel, 334
  price comparison examples, 516
  retailer apps, 334
  Roomba vacuum app, 235-236
  Shazam app, 183
  shopping via, 334
  SitOrSquat app, 451
SmartWay program (EPA), 309
Snack food market, 101
Snapchat, 360, 460, 463, 683
Social class, 130, 683
Social connections, 121
Social currency, 447

Social influences, on consumer behavior, 117, 128-131
Social media
  advertising on, 421-422
  age of users, 457
  as amplification of word-of-mouth, 457
  Under Armour marketing toward women and, 203
  career opportunities in, 588
  content types and examples, 465
  defined, 457, 683
  digital advertising, 421-422
  GEICO insurance and, 352
  in-the-moment reactions, 458
  as a many-to-many model of communication, 458
  numbers of Americans using, 457
  objectives, 465
  opinion leaders' use of, 131
  paid, owned, and earned, 457-458
  proactive customer service via, 388-389
  as social influence on consumer behavior, 131
  software to manage, 466-468
  traditional media *versus,* 457-459
  value offered by, 464-465
Social media manager, 355
Social media platforms
  Facebook, 459-461
  Instagram, 460, 461-462
  international, 464
  LinkedIn, 460, 463
  other social networks, 464
  Pinterest, 460, 462-463
  Snapchat, 460, 463
  Twitter, 460, 463-464
  user demographics, 460
  YouTube, 459, 460
Social Mention, 189
Social needs, 120, 121, 683
Social networking, 421
Social outcome, in triple bottom line, 19
Social profiles, 547
Social responsibility. *See also* Ethics; Green choices
  as basic marketing challenge, 24-25
  defined, 24, 683
  ethics and treatment of overseas workers, 305
  profit maximization and, 476
Sofia the First, 254
Soft drink ban dilemma, 38
Soft drink companies, recycling by, 288
Software. *See also* Customer relationship management (CRM)
  ad-blocking, 424
  analytic, for transportation and environment, 310
  apps add services to core products, 211
  bounce rates, 466, 467
  branded apps, 451
  Google Analytics, 466
  Hootsuite, 466, 468
  HubSpot, 439-440, 468
  to manage social media, 466-468
  marketing automation software, 466-468
  to measure results from online media, 466
  online media and, 466-468
  pop-up blockers, 424
  product bundles in, 523
  programmatic advertising and, 420
  ProjectPro, for project management, 467
  for sales support, 392-395
  statistical packages, 197
Solar panels, 658-659
Solar power, 490
"Solutions selling," 384
Sorting, of products for distribution, 275, 683
Source (sender of a message), 360, 683
South America, 11, 133, 339
Southeast Asia, 75, 477
Sovaldi (drug), 495
Spain, 76, 161, 387
Spam messages, 452

Spatial separation, 10, 11
Special promotion allowances, 497
Specialists
  as collaborators, 12
  in distribution channels, 274-276
Specialization
  in distribution channels, 274-276
  in production, 10-11
  of retailers, 323
  of salespeople, 390
Specialized media, for advertising, 419, 443
Specialized search engines, 453
Specialized services, 229
Specialized social networks, 464
Specialized storing facilities, 312-313
Specialty products, 225-227, 268, 683
Specialty shops, 326-327, 330, 683
Specialty wholesalers, 343, 683
Specifications, for organizational customers, 155
Spending, consumer confidence and, 119
Spiffs, 486
Splenda, 124
Splice (GoPro) app, 211
Sponsored links, 157, 351
Sponsored messages, 457-458
Sponsored search ads, 421
Sponsored software applications (branded apps), 451
Sports teams, ticket pricing by, 480
Spying, on competitors, 67
Square register, 241
Standard markup percent, 503-504
Standardization and grading, 11, 683
Staple (products), 224, 225, 683
Star Wars movie-themed LEGO kits, 87
Starlight Stove, 658-659
Starlight Stoves and (case study), 658-659
State laws, 73-74
Statistical Abstract of the United States, 189
Statistical packages, 197, 683
Status quo pricing objectives, 478, 683
Stay-at-home moms, 454
STEPPS (*S*ocial currency, *T*riggers, *E*motion, *P*ublic, *P*ractical value, *S*tories), 447
Stereotyping, 129, 133
Stocking allowances, 486, 683
Stock-keeping unit (SKU), 207
Stockturn rate, 329, 505, 569-570, 683
Stokes Purple Sweet Potatoes, 341
Stoney Ginger Beer, 295
"Stoplight" evaluation method (General Electric), 81-82
Store brands, 319
Stories, 447
Stories, for attention and interest, 359
Storing, 310, 683
Storing function, 11, 310-313, 683. *See also* Inventory
Stove Top, 108
Straight commission, 397
Straight rebuy, 155, 156, 683
Straight salary, 397
Straight sales growth, 477
Straight-line trend projection, 574
Strategic (management) planning, 33, 683
Strategic planning grids, 81-82
Strategy clarification (case study), 663-664
Strategy planning process, 41-42. *See also* Marketing strategy planning
Straw man, 527
Streaming music services, 115, 183
Strengths. *See also* Competitive advantage
  building on (LEGO), 87
  finding, within firms, 63
Strengths, weaknesses, opportunities, threats (S.W.O.T.) analysis, 50, 252, 539, 604-605, 682
Strong brands, 373
Structured questioning, 192
StubHub, 480
Subcultures, 132

Submarkets, 94-95, 134. *See also* Segmentation; Target market
Subscription data services, 189-190
Subscription pricing, 520-521, 683
Subsistence economies, 8
Substantial market segments, 95
Substitutes
  close, 65, 66
  competitor analysis and, 66
  defined, 683
  demand elasticity and, 556-557
  positioning against, 110
  price sensitivity and, 515
  in various competitive environments, 65
Suburban regional shopping malls (case study), 625-627
Suburban shopping centers, 335
SUMA, 315
Summary, on operating statements, 569
Sunkist brand products, 219-220
Sunlight brand dish detergent, 493
Super Bowl, 458
Super Bowl ads, 359, 417
Supercenters (hypermarkets), 328-329, 683
Superiority claims, 414
Supermarkets, 327-328, 330, 336, 683
Supplier performance, 162-163
Supplier scorecards, 163
Suppliers. *See also* Intermediaries; Wholesalers
  approved supplier list, 168
  evaluating and selecting, 158
  government approved, 168
  markets as seen by, 557-559
  monitoring of performance of, by buyer, 162-163
  proposals from, 158
  relationships with, 159-163
Supplies
  defined, 683
  maintenance, repair, and operating (MRO), 229
Supply
  elastic, 558-559
  heterogeneous, 10, 15
  inelastic, 558-559
Supply and demand
  heterogeneous, 10, 15
  nature of competition and, 560-563
  size of market and price level and, 559
Supply chain, 303-304, 683
Supply chain management, 314
Supply curves, 557-559, 683
Supporting salespeople, 385, 683
"Surge" prices, 481
Surroundings, as influence on buying, 134-135
Survey research, 192-194
Surveys
  for forecasting, 577
  limitations of, 194
  mail and online, 193
  personal interview, 194
  psychographic, 173
  as research, 192-194
  self-administered, 192
  telephone, 193
Sustainability, 75. *See also* Environmental issues; Green choices
  Coca-Cola's commitment to, 295-296
  defined, 683
  Method Products and, 471-472
  in packaging, 70
  reverse channels for, 288-291
Sustainable competitive advantage, 65, 683
Sweden, 361
Sweepstakes, 355, 410
Swiffer, 134
Switching costs, 516-517
Switzerland, 54, 76, 90, 428
S.W.O.T. (strengths, weaknesses, opportunities, threats) analysis, 50, 252, 539, 604-605, 682

Symbols, trademarks and, 213
Symptoms, problems *versus,* 186
Syndicated research, 190
"System of play" (LEGO), 87-88
Systems view, of macro-marketing, 9-10

T

"Tailgating" of self-driving trucks, 311
Taiwan, 42
"Take back" laws, 288
Tangible products, 209
Tanzania, 55, 272
Target marketing. *See also* Segmentation
  defined, 35, 684
  elements of, 37-42
  impact of culture on, 131-134
  lifestyle analysis and, 127-128
  promotion communicates to, 352-353
Target markets
  Under Armour example, 203-204
  communication of promotion to, 360-362
  defined, 35, 684
  defined by market segmentation, 93-96
  of Dell computers, 265-266
  developing marketing mix for, 37-42
  dimensions for segmenting, 99-103
  forecasting of potential and sales, 573-575
  of GEICO auto insurance, 351-352
  hotel example (case study), 643-644
  in marketing plans, 539
  market-oriented strategy as, 36-37
  matching advertising media to, 417
  needs of, 41-42
  relevant, 492
  sales force responsibilities and, 399
  stories that resonate with, 359
  U.S. demographic trends, 78-79
Target return objective, 475, 684
Targeted price discounts, 480
Targeting, 89, 107
Tariffs, 632
Task method, of budgeting, 374, 684
Tastee DeLites, 126
TasteeDeeLIES.com, 126
Tax treatment, impact on business product markets, 228
Team selling, 389, 684
Teamwork, importance to marketing orientation, 17
Technical specialists, 386, 684
Technological environment, assessing, in strategy planning, 68-71. *See also* Information technology (IT); Marketing information system (MIS)
Technology. *See also* Apps; Customer tracking; Information technology (IT); Internet; Websites
  addressable TV, 418
  anticipating and planning for the future, 68
  artificial intelligence (AI) and, 68-69
  challenges of, 70
  of Coca-Cola, in developing countries, 295-296
  defined, 68, 684
  digital, Under Armour example, 203-204
  distribution center efficiency via, 313-314
  driverless cars and trucks and, 68, 311
  drones, for delivery service, 59, 163, 308
  electronic data interchange (EDI), 304-305, 312, 314
  Fitbit, 178, 179
  integrated into a product, 210-212
  international cell phone use, 77-78
  Kroger checkout lanes and, 336
  "psychic retailers," 337
  refrigerators and, 211
  in retail evolution, 336-337
  to save rhinos (Cisco Systems), 359
  use by wholesalers, 344
  wearable, 178
  wise use of, 546

Teddy Grahams, 244
TEDx Youth Conference, 465
Teenagers, consumer behavior of, 129, 130
Tek Gear, 203
Telemarketing, 390, 684
Telephone interviews, 193
Telephone surveys, 193
Television advertising, 409–410, 414, 416, 418, 420
Television market, 501–502
Television sets, innovations in, 501
Temporary price cuts, 483
Termite mounds, as energy model, 251
Test marketing, 249, 255
Testimonial, 454
Testing, of advertising, 427
Testing hypotheses, 184–185
Text ad, 443
Text messaging promotions, 390
Thailand, 77, 162, 217, 295
Theme parks, 33, 87
Thermometer marketing, 423
"Thing-oriented" jobs, 582
ThinkPad (IBM), 266
Third-world countries. *See* Less-developed nations
Thought leaders, 448, 449
3-D printing, 64, 254
3M Post-its, 212
Ticket pricing, 480
Tide laundry detergent, 44, 220, 222, 257, 268–269, 390
TikTok, 464
Time, as influence on purchase situation, 134–135
Time pressure, in shopping, 225
Timeliness of promotion, 364–365
Timing, marketing plan and, 540, 541
Tinder, 426
Toll-free telephone lines, 390
Tomagatchi, 242
Tom's Children's toothpaste, 96
Tone soap, 108
Toothbrushes and toothpaste, education about, 409
Toothpaste market, 96
Top management
  product development support, 256
  responsibility for objectives, 61–63
Tostitos, 298
Total cost, 508, 684
Total cost approach, 302, 684
Total cost curve, 513
Total cost of distribution, 308
Total expenditures, 516
Total fixed cost, 508, 684
Total inventory cost, 312
Total quality management (TQM), 258, 684
Total revenue, 555–556, 559
Total revenue curve, 513
Total variable cost, 508, 684
Touchpoints, 259–260
Trade agreements, 71–72
Trade (functional) discount, 486, 684
Trade magazines, 443, 453
Trade promotion, 432–433, 684
Trade shows, 355, 432, 439
Trade-in allowances, 484, 487, 684
Trademark, 204, 212, 213, 218, 684
Trade-offs, in physical distribution, 298–300
Traditional channel systems, 276, 684
Training
  to improve quality, 260
  of salespeople, 379–380, 396–397
Trains, 298
Transportation analytics, 295, 310
Transporting, defined, 305, 684
Transporting costs, 305–306
Transporting function, 11, 305–310, 684
Trash, packaging and, 222
Trend extension, 574, 684
Trend projection, 574

Trial
  in adoption process, 139–140
  of new products, 139
"Tribes" (Dunkin'), 173
Triggers, 447, 468
TripAdvisor, 457
Triple bottom line, 19, 684
Tritan copolyester, 155
Tropicana orange juice, 214
Truck manufacturers, 309
Truck shipments, 307–308, 309
Truck wholesalers, 344, 684
Trucks, 303
Trust, 127, 160–161, 179, 214, 288, 325, 360–361, 443, 449, 684
Truth, need for, 545–546
TRW airbags, 229
TurboTax, 519
Turf battles, 17
Turkey, 76
Tweets, 354–355
Twitter, 36, 460, 463–464, 684
2/10, net 30, 486, 674
Tylenol, 373
Tylenol Feel Better, 120
Tylenol's safety packaging, 222

U

Uber, 481
Uganda, 272
Ultrasonic spray coating technology, 290–291
Under Armour, 203–204
Unethical behavior, in the workforce, 548
Unfair advertising, 428–430
Unfair trade practice acts, 494, 684
Unique selling proposition, 425
"United Breaks Guitars," 14
United Kingdom, 53
  Amazon in, 59
  BzzAgent in, 455
  Coca-Cola in, 295
  demographic characteristics in, 76
  institutional advertising in, 415
  online shopping in, 339
United Nations, 77
United Nations' World Food Programme meal sharing app, 23
United States
  advertising employment in, 411
  advertising expenditures in, 410
  advertising regulations, 428–429
  Americans' time spent with digital media, 419
  amount spent on logistics-related costs, 297
  BzzAgent in, 455
  cable television shopping, 331
  car ownership in, 125
  Coca-Cola's gas-electric hybrid delivery trucks, 296
  comparative advertising allowed, 414
  copyright law, 247
  dealer brand popularity in, 220
  demographic characteristics in, 76
  demographic trends, 78–79
  disposable diapers market in, 37
  income distribution in, 118–119
  in-home shopping, 331
  inflation rates and, 68
  LEGO popularity in, 87–88
  manufacturer-owned branch operations, 342
  market share of Apple iPhone in, 66
  as multicultural market, 132
  NAFTA and, 71
  need to sell overseas, 54–55
  outsourcing by, 162
  pipelines in, 308
  population growth in, 77
  privacy laws in, 409–410
  products in market maturity stage, 239
  sales force in, 381
  service firms in, 165
  successful distribution in, 298
  "take back" laws, 288
  taste preferences in, 258
  trash generated in, 222
Universal functions of marketing, 11, 684
Universal Product Code (UPC) numbers, 313
Unpaid media, 419
Unplanned purchases, 224
Unsought products, 225, 226, 268, 684
Up & Up brand (Target), 472
Urbanization, 76
Urgency of needs, 134, 556
U.S. Air Force, 23
U.S. Army, advertising by, 418
U.S. class system, 130
U.S. Congress, 517
U.S. Supreme Court, on limiting market exposure, 284
USA Today Super Bowl Ad Meter, 359
USA.gov website, 169
User-generated content, 443, 684
Users
  as influence on buying decision, 153
  researching new products, 156–157
"Utilitarian Grinches" shopper type, 104

V

Vacuum cleaner market, 235–236
Vacuum cleaners, robotic, 235–236
Validity, 684
Validity problems, in research, 198
Value
  added to, by technology and intelligent agents, 210–212
  adding, in mature markets, 245
  customer lifetime, 45–48
  defined, 26
  ethical, 26
  importance of, to marketing orientation, 20–23
  on life, prices and, 495
  of marketing, 3–4
  pricing policies and, 491–494
  societal and ethical dimensions, 494
  of transporting function, 305–310
  wholesaler-added, 341–342
Value chain, 265
Value in use pricing, 518, 684
Value pricers, 492–493
Value pricing, 491–494, 684
Vanilla Coke, 208
Variable costs, 507, 670
Variety, as big data dimension, 178
Velcro, 251
Velocity, as big data dimension, 179
Vending machines, 330–331
Vendor analysis, 158, 684. *See also* Suppliers
Vendors, relationships with, 159–163
Venezuela, 55
Veracity, as big data dimension, 179
Veracruz (Hyundai), 110
Verizon Wireless, 212
Vernie robot, 87
Vertical arrangements, 284
Vertical channel conflict, 277, 684
Vertical integration, 282, 684
Vertical marketing systems, 281–283, 684
Vicks Behind Ear Thermometer, 423
Vicks' cold remedy products, 386
Victoria's Secret, 461
Video cases, 623–634
Video data, 337
Video demonstrations, 368
Video entertainment, distribution of, 276
Video sharing services, 459
Video tutorial, 449
Videoconference, 394
Videos, for product demonstration, 195

Vietnam
   Coca-Cola in, 295
   demographics of, 76
   Ford Motor Company's determining dimensions in, 101
Vietnam War, 630
Vimeo, 464
Viral promotion, 447
Virtual assistant, 351
Virtual products, 276
Virus, 250
Visa, 489
Visual images, 451
Vitaminwater, 271
VK (VKontakte), Russian social media site, 464
Volkswagen scandal, 26
Volume, as big data dimension, 178
Voluntary chains, 340
Volunteers, 23

W

Walkman players, 115
Wall Street Journal, The, 286, 590
Wants
   changing, 534
   defined, 684
   needs *versus,* 119
Warehouse club, 329, 344
Warehousing, 312–313
Warranties, 37, 204, 207, 684
Warranty, 207
Waste Electrical and Electronic Equipment (WEEE) Directive, 684
Water transportation, 307, 308
WD-40, 196, 212
Wearable technology, 178
Weather Channel UK, 423
Web advertising, 419, 420, 424
Webcams, tracking and analysis by, 177–178
Webinars, 439

Webrooming, 286
Websites. *See also* E-commerce; Internet; Search engines
   Airbnb, 456
   Bathys Hawaii Watch Co., 371
   Blue Nile, 449
   branded, 443
   Cirque du Soleil, 33–34
   CNET.com, 235
   customer reviews, 14
   customer satisfaction posted on, 14
   direct distribution via, 270–271
   FedBizOpps.gov, 169
   focus groups on, 192
   GEICO insurance, 351–352
   Gibson guitars, 278
   Green Mountain Coffee, 368
   as important communication medium, 466
   Jiffy Lube, 260
   monitoring chatter on, 189
   online surveys, 193
   pay-per-click advertising, 419
   "Print Lizard," 192
   procurement, 158
   researching new purchases on, 157
   small service companies' use of, 167
   of suppliers, 157
   tailored for export customers, 291
   useful content needed, 157
   Wall Street Journal, The, 286
WeChat, 464
Weekly specials, 479
"What's Beautiful" advertising campaign, 203
WhatsApp, 464
Wheel of retailing theory, 335, 684
Wheelchair industry, 630–632
Wheeler-Lea Act, 495
Wheeler-Lea Amendment (1938), 684
White papers, 157, 439, 448, 449, 684
Wholesale clubs, 329
Wholesalers, 684. *See also* Intermediaries

   added value of, 341–342
   benefits of, 12, 64
   customers' wants as guide for, 167–168
   defined, 329–330
   Frieda's Inc., 340–341
   order takers of, 385
   as organizational customers, 167–168
   personal selling for, 384
   progressive, 340
   purchasing functions, 167–168
   sales promotion to, 354
   strategy planning by, 319–320
   types of, 341–344
Wholesaling
   career opportunities in, 586
   decline of, 340
   defined, 329–330, 684
Wilson Blade tennis racquet, 286
Wisdom, 180, 182
Women's sports apparel, 203
Word-of-mouth communication, 33, 216, 244, 354, 443, 455–456, 457
Working capital, 273, 398
World Wide Web. *See* Internet; Websites
World Wildlife Fund (WWF), 124
Wounded Warrior Project, 631

Y

Yaz, 429
Young couples, consumer behavior of, 129
YouTube, 33, 36, 235, 459, 460, 684

Z

Zaltrap (cancer drug), 495
Zappos Couture, 324
Zappos Running, 324
Zest soap, 108
Zimbabwe, 251
Zipcar, 125
Zonal merchandising, 626